CHEMISTRY

FOR

DEGREE STUDENTS

B.Sc. Third Year

All Indian Universities

As per UGC Model Curriculum

CHEMISTRY
FOR
DEGREE STUDENTS

B.Sc. Third Year

For all Indian Universities

As per UGC Model Curriculum

Part – I : Inorganic Chemistry
Part – II : Organic Chemistry
Part – III : Physical Chemistry

Dr. R.L. MADAN
M.Sc., Ph.D.
Former, Head of the Chemistry Department
Government Postgraduate College, Faridabad
and Principal, Government College, Panchkula

S. CHAND PUBLISHING

S Chand And Company Limited
(ISO 9001 Certified Company)

S Chand And Company Limited

(ISO 9001 Certified Company)

Head Office: D-92, Sector–2, Noida – 201301, U.P. (India), Ph. 91-120-4682700

Registered Office: A-27, 2nd Floor, Mohan Co-operative Industrial Estate, New Delhi – 110 044, Phone: 011-49731800

S. CHAND PUBLISHING

www.**schandpublishing.com;** e-mail: **info@schandpublishing.com**

Marketing Offices:

Chennai : Ph: 23632120; chennai@schandpublishing.com
Guwahati : Ph: 2738811, 2735640; guwahati@schandpublishing.com
Hyderabad : Ph: 40186018; hyderabad@schandpublishing.com
Jalandhar : Ph: 4645630; jalandhar@schandpublishing.com
Kolkata : Ph: 23357458, 23353914; kolkata@schandpublishing.com
Lucknow : Ph: 4003633; lucknow@schandpublishing.com
Mumbai : Ph: 25000297; mumbai@schandpublishing.com
Patna : Ph: 2260011; patna@schandpublishing.com

First Edition 2011

Reprints 2012, 2013, 2014 (Twice), 2015 (Twice), 2016 (Twice), 2018, 2020 (Twice)

Low Priced Students' Paperback Edition (LPSPE), 2021

Reprint 2022 (Twice)

ISBN: 978-81-947-7179-1 **Product Code:** H6CDYCHEM6810ENAA21L

PRINTED IN INDIA

By Vikas Publishing House Private Limited, Plot 20/4, Site-IV, Industrial Area Sahibabad, Ghaziabad – 201 010 and Published by S Chand And Company Limited, A-27, 2nd Floor, Mohan Co-operative Industrial Estate, New Delhi – 110 044.

PREFACE

I feel pleasure in presenting this book to the teachers and students of B.Sc. III year of all the Indian Universities. The book has been prepared keeping in view the syllabi prepared by different universities on the basis of Model UGC Curriculum. The present book has evolved out of my experience of more than three decades of classroom teaching at B.Sc. level.

A long span of interaction with students has guided me to give special attention to dark areas which students find difficult to understand. Mathematics background required is simple and the necessary mathematical techniques have been developed within the text. Some salient features of the book are :

- The book has been written in a simple and comprehensible language.
- A large number of illustrations, pictures and interesting examples have been provided to make the reading interesting and understandable.
- The important laws, concepts and principles have been given either in italics or in bold type to help the students to focus on them.
- A large number of solved examples which have appeared in the examination papers of various universities have been given in each chapter, for the benefit of students.
- The questions that have been provided in the Exercise are in tune with the latest pattern of examination.

I believe that the book contains all that is needed to understand the subject in a systematic manner. However, no work is perfect. I would welcome suggestions from the teaching fraternity and students for further improvement of the book.

I express my gratitude to the staff of S Chand And Company Limited, New Delhi for providing all assistance in bringing out the book in the present shape.

Dr. R.L. Madan
Mobile : 9971775666

UGC MODEL SYLLABUS

CHEMISTRY B.Sc. THIRD YEAR

CH-301 Inorganic Chemistry

I Hard and Soft Acids and Bases (HSAB)

Classification of acids and bases as hard and soft. Pearson's HSAB concept, acid-base strength and hardness and softness. Symbiosis, theoretical basis of hardness and softness, electronegativity and hardness and softness.

II Metal-ligand Bonding in Transition Metal Complexes

Limitations of valence bond theory, an elementary idea of crystal-field theory, crystal field splitting in octahedral, tetrahedral and square planar complexes, factors affecting the crystal-field parameters.

III Magnetic Properties of Transition Metal Complexes

Types of magnetic behaviour, methods of determining magnetic susceptibility, spin-only formula, L-S coupling, correlation of μ_s and μ_{eff} values, orbital contribution to magnetic moments, application of magnetic moment data for 3d-metal complexes.

IV Electron Spectra of Transition Metal Complexes

Types of electronic transitions, selection rules for d-d transitions, spectroscopic ground states, spectrochemical series. Orgel-energy level diagram for d^1 and d^9 states, discussion of the electronic spectrum of $[Ti(H_2O)_6]^{3+}$ complex ion.

V Thermodynamic and Kinetic Aspects of Metal Complexes

A brief outline of thermodynamic stability of metal complexes and factors affecting the stability, substitution reactions of square planar complexes.

VI Organometallic Chemistry

Definition, nomenclature and classification of organometallic compounds. Preparation, properties, bonding and applications of alkyls and aryls of Li, Al, Hg, Sn and Ti, a brief account of metal-ethylenic complexes and homogeneous hydrogenation, mononuclear carbonyls and the nature of bonding in metal carbonyls.

VII Bioinorganic Chemistry

Essential and trace elements in biological processes, metalloporphyrin with special reference to haemoglobin and myoglobin. Biological role of alkali and alkaline earth metal ions with special reference to Ca^{2+}, Nitrogen fixation.

VIII Silicones and Phosphazenes

Silicones and phosphazenes as examples of inorganic polymers, nature of bonding in triphosphazenes.

CH-302 Organic Chemistry

I Spectroscopy

Nuclear magnetic resonance (NMR) spectroscopy.

Proton magnetic resonance (^{1}H NMR) spectroscopy, nuclear shielding and deshielding, chemical shift and molecular structure, spin-spin splitting and coupling constants, areas of signals, interpretation of PMR spectra of simple organic molecules such as ethyl bromide, ethanol, acetaldehyde, 1, 1, 2-tribromoethane, ethyl acetate, toluene and acetophenone. Problems pertaining to the structure elucidation of simple organic compounds using UV, IR and PMR spectroscopic techniques.

II Organometallic Compounds

Organomagnesium compounds : the Grignard reagents-formation, structure and chemical reactions.

Organozinc compounds : formation and chemical reaction.

Organolithium compounds : formation and chemical reactions.

III Organosulphur Compounds

Nomenclature, structural features, Methods of formation and chemical reactions of thiols, thioethers, sulphonic acids, sulphonamides and sulphaguanidine.

IV Heterocyclic Compounds

Introduction : Molecular orbital picture and aromatic characteristics of pyrrole, furan, thiophene and pyridine. Methods of synthesis and chemical reactions with particular emphasis on the mechanism of electrophilic substitution. Mechanism of nucleophilic substitution reactions in pyridine derivatives. Comparison of basicity of pyridine, piperidine and pyrrole.

Introduction to condensed five and six-membered heterocycles. Preparation and reactions of indole, quinoline and isoquinoline with special reference to Fisher indole synthesis, Skraup synthesis and Bischler-Napieralski synthesis. Mechanism of electrophilic substitution reactions of indole, quinoline and isoquinoline.

V Organic Synthesis via Enolates

Acidity of α-hydrogens, alkylation of diethyl malonate and ethyl acetoacetate. Synthesis of ethyl acetoacetate : the Claisen condensation. Keto-enol tautomerism of ethyl acetoacetate.

Alkylation of 1. 3-dithianes. Alkylation and acylation of enamines.

VI Carbohydrates

Classification and nomenclature. Monosaccharides, mechanism of osazone formation, interconversion of glucose and fructose, chain lengthening and chain shortening of aldoses. Configuration of monosaccharides. Erythro and threo diastereomers. Conversion of glucose into mannose. Formation of glycosides, ethers and esters. Determination of ring size of monosaccharides. Cyclic structure of D(+)-glucose. Mechanism of mutarotation.

Structures of ribose and deoxyribose.

An introduction to disaccharides (maltose, sucrose and lactose) and polysaccharides (starch and cellulose) without involving structure determination.

VII Amino Acids, Peptides, Proteins and Nucleic Acids

Classification, structure and stereochemistry of amino acids. Acid-base behavior, isoelectric point and electrophoresis. Preparation and reactions of α-amino acids.

Structure and nomenclature of peptides and proteins. Classification of proteins. Peptide structure determination, end group analysis, selective hydrolysis of peptides. Classical peptide synthesis, solid-phase peptide synthesis. Structures of peptides and proteins. Levels of protein structure. Protein denaturation/renaturation.

Nucleic acids : introduction. Constituents of nucleic acids. Ribonucleosides and ribonucleotides. The double helical structure of DNA.

VIII Fats, Oils and Detergents

Natural fats, edible and industrial oils of vegetable origin, common fatty acids, glycerides, hydrogenation of unsaturated oils, Saponification value, iodine value, acid value. Soaps, synthetic detergents, alkyle and aryl sulphonates.

IX Synthetic Polymers

Addition or chain-growth polymerization. Free radical vinyl polymerization, ionic vinyl polymerization, Ziegler-Natta polymerization and vinyl polymers.

Condensation or step growth polymerization. Polyesters, polyamides, phenol formaldehyde resins, urea formaldehyde resins, epoxy resins and polyurethanes.

Natural and synthetic rubbers.

X Synthetic Dyes

Colour and constitution (electronic concept). Classification of dyes. Chemistry and synthesis of Methyl orange, Congo red, Malachite green, Crystal violet, Phenolphthalein, Fluorescein, Alizarin and Indigo.

CH-303 Physical Chemistry

I Elementary Quantum Mechanics

Black-body radiation, Planck's radiation law, photoelectric effect, heat capacity of solids, Bohr's model of hydrogen atom (no derivation) and its defects, Compton effect.

De Broglie hypotesis, the Heisenberg's uncertainty principle, Sinusoidal wave equation, Hamiltonian operator, Schrodinger wave equation and its importance, physical interpretation of the wave function, postulates of quantum mechanics, particle in a one dimensional box.

Schrodinger wave equation for H-atom, separation into three equations (without derivation), quantum numbers and their importance, hydrogen like wave functions, radial wave functions, angular wave functions.

Molecular orbital theory,m basic ideas–criteria for forming M.O. from A.O. construction of M.O's by LCAO-H_2^+ ion, calculation of energy levels from wave functions, physical picture of bonding and antibonding wave functions, concept of σ, σ^*, π, π^* orbitals and their characteristics. Hybrid orbitals–sp, sp^2, sp^3; calculation of coefficients of A.O.'s used in these hybrid orbitals.

Introduction to valence bond model of H_2, comparison of M.O. and V.B. model.

II Spectroscopy

Introduction : electromagnetic radiation, regions of the spectrum, basic features of different spectrometers, statement of the Born-Oppenheimer approximation, degrees of freedom.

Rotational Spectrum

Diatomic molecules. Energy levels of a rigid rotor (semi-classical principles), selection rules, spectral intensity, distribution using population distribution (maxwell-Boltzmann distribution) determination of bond length, qualitative description of non-right rotor, isotope effect.

Vibrational Spectrum

Infrared spectrum : Energy levels of simple harmonic oscillator, selection rules, pure vibrational spectrum, intensity, determination of force constant and qualitative relation of force constant and bond energies, effect of anharmonic motion and isotope on the spectrum, idea of vibrational frequencies of different functional groups.

Raman Spectrum : concept of polarizability, pure rotational and pure vibrational Raman spectra of diatomic molecules, selection rules.

Electronic Spectrum

Concept of potential energy curves for bonding and antibonding molecular orbitals, qualitative description of selection rules and Franck-Condon principle.

Qualitative description of σ, π- and n M.O., their energy levels and the respective transitions.

III Photochemistry

Interaction of radiation with matter, difference between thermal and photochemical processes. Laws of photochemistry : Grothus–Drapper law, Stark–Einstein law, Jablonski diagram depicting various processes occurring in the excited state, qualitative description of fluorescence, phosphorescence, non-radiative processes (internal conversion, intersystem crossing), quantum yield, photosensitized ractions–energy transfer processes (simple examples).

IV Physical Properties and Molecular Structure

Optical activity, polarization–(Clausius–Mossotti equation), orientation of dipoles in an electric field, dipole moment, induced dipole moment, measurement of dipole moment-temperature method and refractivity method, dipole moment and structure of molecules, magnetic properties–paramagnetism, diamagnetism and ferromagnetics.

V Solutions, Dilute Solutions and Colligative Properties

Ideal and non-ideal solutions, methods of expressing concentrations of solutions, activity and activity coefficient.

Dilute solution, colligative properties, Raoult's law, relative lowering of vapour pressure, molecular weight determination. Osmosis, law of osmotic pressure and its measurement, determination of molecular weight from osmotic pressure. Elevation of boiling point and depression of freezing point, Thermodynamic derivation of relation between molecular weight and elevation in boiling point and depression in freezing point. Experimental methods for determining various colligative properties.

Abnormal molar mass, degree of dissociation and association of solutes.

CONTENTS OF B.Sc. I AND B.Sc. II BOOKS

B.Sc. I

PART–I: INORGANIC CHEMISTRY

1. Atomic Structure
2. Periodic Properties
3. Covalent Bond
4. Ionic Bond
5. *s*-Block Elements
6. *p*-Block Elements
7. Chemistry of Noble Gases

PART–II: ORGANIC CHEMISTRY

1. Structure and Bonding
2. Mechanism of Organic Reactions
3. Stereochemistry of Organic Compounds
4. Alkanes and Cycloalkanes
5. Alkenes, Cycloalkenes, Dienes and Alkynes
6. Arenes and Aromaticity
7. Alkyl and Aryl Halides

PART–III: PHYSICAL CHEMISTRY

1. Mathematical Concepts and Computers
2. Gaseous State
3. Liquid State
4. Solid State
5. Colloidal State
6. Chemical Kinetics
7. Catalysis
8. Thermodynamics – I (Thermochemistry and First Law)

B.Sc. II

PART–I: INORGANIC CHEMISTRY

1. Chemistry of Elements of First Transition Series
2. Chemistry of Elements of Second and Third Transition Series
3. Oxidation and Reduction
4. Coordination Compounds
5. Chemistry of Lanthanide Elements
6. Chemistry of Actinides
7. Acids and Bases
8. Non-Aqueous Solvents

PART–II: ORGANIC CHEMISTRY

1. Electromagnetic Spectrum: Absorption Spectra
2. Alcohols
3. Phenols
4. Ethers and Epoxides
5. Aldehydes and Ketones
6. Carboxylic Acids
7. Carboxylic Acid Derivatives
8. Organic Compounds of Nitrogen

PART–III: PHYSICAL CHEMISTRY

1. Thermodynamics-I (Thermochemistry and First Law)
2. Thermodynamics-II (Second and Third Laws)
3. Chemical Equilibrium
4. Phase Rule
5. Liquid-Liquid Mixures
6. Distribution Law
7. Electrochemistry-I
8. Electrochemistry-II

CONTENTS

PART–I: INORGANIC CHEMISTRY

PART–II: ORGANIC CHEMISTRY

PART–III: PHYSICAL CHEMISTRY

INORGANIC CHEMISTRY

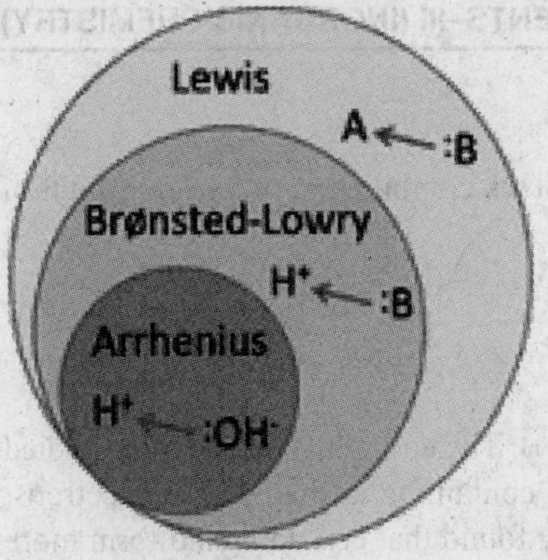

1

Hard and Soft Acids and Bases

1.1. INTRODUCTION

Arrhenius, a Swedish chemist, proposed a theory of acids and bases known as **Arrhenius concept of acids and bases** in 1884. According to **Arrhenius concept** *an acid is a susbstance which gives hydrogen ions (H^+) in aqueous solution and a base is a substance which gives hydroxyl ions (OH^-) in aqueous solution.* The strength of an acid or a base depends upon its capacity of ionisation to give H^+ or OH^- ions respectively.

The Arrhenius concept of acids and bases was extended further by Bronsted and Lowry (1923). According to Bronsted - Lowry concept *an acid is a substance which can donate a proton (H+) and a base is a substance which can accept a proton (H+)*. The acid-base reactions were thus regarded as **proton-transfer reactions**.

G.N. Lewis gave more broad based definitions of acids and bases in 1923 in terms of electron pair donation and acceptance. According to **Lewis concept** *an acid is a substance which can accept a pair of electrons while a base is a substance which can donate a pair of electrons*. In other words, a base is an electron pair donor and an acid is an electron pair acceptor. Thus, any substance which has an unshared pair of electrons can act as a Lewis base while a substance which has an empty orbital that can accommodate a pair of electrons acts as a Lewis acid. There is a donation of a pair of electrons from Lewis base to Lewis acid to form a coordinate bond between the two. Thus, the neutralisation reaction involves the formation of a new coordinate bond between the electron pair donor (Lewis base) and electron pair acceptor (Lewis acid).

Svante Arrhenius

Bronsted-Lowry

G.N. Lewis

1.2. HARD AND SOFT ACIDS AND BASES

According to Lewis concept, an acid-base reaction involves combination of vacant orbital of an acid (A) and a filled unshared orbital on a base (B) as:

$$A + :B \longrightarrow A : B$$

It may also be written as

$$A + :B \longrightarrow A \leftarrow B$$

A strong acid and a strong base form a stable complex A : B. The co-ordination chemists studied the relative reactivities of a large number of ligands (species containing unshared pairs electrons, Lewis bases) with different ions (acting as Lewis acids), They found that certain ligand form more stable complexes with heavier ions such as Ag^+, Hg^{2+}, Pd^{2+}, Pt^{2+} with nearly full *d*-orbital electrons while others prefer to form complexes with ions such as Be^{2+}, Al^{3+}, Ti^{4+} having no *d*-electrons. Based on the preferential bonding they classified the acids and bases into two categories.

Class 1 metal ions include those of *alkali metals, alkaline earth metals like Be^{2+} and lighter transition metals in their higher oxidation states such as Ti^{4+}, Cr^{3+}, Fe^{3+}, Co^{3+}, etc.* as well as hydrogen ion, H^+.

Class 2 metal ions include those of heavier transition metals such as Hg^{2+}, Pd^{2+}, Pt^{2+}, Cd^{2+} and those in lower oxidation states (such as Cu^+, Ag^+, Hg^+, etc.).

Similarly ligands have also been classified as class 1 and class 2 depending upon their preference for metal ions respectively.

Class 1 ligands include those molecules or ions which have the donor atoms N, O, F, Cl, etc. These ligands combine preferably with class 1 metal ions.

Class 2 ligands include those molecules or ions which have the donor atoms P, As, S, Se, I, etc. These ligands combine preferably with class 2 metal ions. Greater the tendency of a ligand to combine with a metal ion, greater will be the stability of the complex. The following table illustrates the two types of ligands.

Those having tendency to complex with class 1 metal ions	Those having tendency to complex with class 2 metal ions
N >> P > As > Sb	N << P > As > Sb
O >> S > Se > Te	O << S < Se ~ Te
F > Cl > Br > I	F < Cl < Br < I

For example, ammonia (NH_3), amines (R_3N), water (H_2O) and fluoride ion (F^-) ligands prefer to coordinate with Ti^{4+}, Fe^{3+}, Co^{3+}, Mg^{2+} metal ions. On the other hand, phosphines (R_3P), thioethers (R_2S) and iodine ion (I^-) ligands prefer to coordinate with Pt^{2+}, Pd^{2+}, Hg^{2+}, Ag^+.

Such a classification has been proved to be very usefull in predicting and explaining stability of coordination compounds.

Pearson in 1963, introduced the terms **hard** and **soft** to describe the members of class 1 and class 2 respectively. For example, **class 1 type** metal ions are called **hard acids** while class 2 type metal ions are called **soft acids**. Similarly, **class 1 type** ligands are named **hard bases** while class 2 type ligands are named **soft bases**.

Classification of Metal Ions (acids)

(*a*) *The metal ions of alkali metals, alkaline earth metals, light transition metals with higher oxidation states (first row transition metal ions such as Ti^{4+}, Cr^{3+}, Fe^{3+}, etc.) and hydrogen, H^+ ion are classified as* **hard acids.**

(*b*) *The metal ions of heavier transition metals (second or third row transition metals) such as Hg^{2+}, Pd^{2+} or the ions of lower oxidation states such as Cu^+, Ag^+ and Hg^+ are classified as* **soft acids.**

Classification of Ligands (Bases)

(*a*) *The ligands which prefer to form stable complexes with hard acids are classified as* **hard bases.** These ligands contain molecules having O, N or F donor atoms. For example, NH_3, H_2O, RNH_2, OH^-, F^-, etc.

(*b*) *The ligands which prefer to form stable complexes with soft acids are classified as* **soft bases.** These ligands contain molecules having P, S, As, Se, etc. as donor atoms. For example, R_3P, $S_2O_3{}^{2-}$, SCN^-, R_3As etc.

Thus phosphine and thioethers (soft bases) have a much greater tendency to co-ordinate with Hg^{2+}, Pd^{2+} or Pt^{2+} (soft acids) while ammonia, water or fluoride ions (hard bases) show a preference for Be^{2+}, Ti^{4+}, Co^{3+} etc.

Some common metal cations are classified into hard and soft acids as shown in Table 1.1.

Table 1.1 Classification of Lewis acids

Hard acids (Type 1)	Soft acids (Type 2)	Border line acids
H^+, Li^+, Na^+, K^+	Cu^+, Ag^+, Au^+, Tl^+, Hg^+	Fe^{2+}, Co^{2+}, Ni^{2+}, Cu^{2+},
Be^{2+}, Mg^{2+}, Ca^{2+}, Sr^{2+}	Pt^{2+}, Cd^{2+}, Pt^{2+}, Hg^{2+}, Pt^{4+}	Zn^{2+}, Sb^{3+}, Bi^{3+}, NO^+
Al^{3+}, Sc^{3+}, Ga^{3+}, In^{3+}, La^{3+}	Ti^{3+}, BH_3, $GaCl_3$, $GaBr_3$,	SO_2, $B(CH_3)_3$, GaH_3
Cr^{3+}, Co^{3+}, Fe^{3+}, As^{3+}, Ce^{3+}	GaI_3, $InCl_3$	
Si^{4+}, Ti^{4+}, Zr^{4+}, Th^{4+}, Pu^{4+}	I^+, Br^+, HO^+, RO^+	
BF_3, BCl_3, $AlCl_3$, AlH_3, $Al(CH_3)_3$	I_2, Br_2, ICN	
Cl^{3+}, Cl^{7+}, I^{5+}, I^{7+}	O, Cl, Br, I, N.	
RCO^+, CO_2, NC^+. HX	M° (Metal atom)	
HX (Hydrogen bonding molectules)	CH_2, carbenes	

Characteristics of Hard Acids:

(*i*) The hard acids are cations with small size.

(*ii*) In case of multiple oxidation states, the hard acids are cations of higher oxidation states (e.g. $\overset{5+}{I}$, $\overset{7+}{I}$, $\overset{3+}{Cl}$, $\overset{7+}{Cl}$).

(*iii*) The hard acids do not possess large number of valence electrons and, therefore, their outer electrons will not be easily distorted or removed. In other words, these ions canot be easily polarized.

To summarise, **hard acids** *are those which bind strongly to a hard base. These are small in size, high in positive charge and will possess no electrons which are easily distorted or removed. In general, alkali metal ions, alkaline earth metal ions, lighter and more highly charged ions belong to this category.*

Characteristics of Soft Acids:

(*i*) The soft acids have cations of large size.

(*ii*) The cations have usually zero or low oxidation states.

(*iii*) The soft acids have large number of valence electrons and therefore, their outer electrons will be easily removed or distorted. In other words, these ions can be easily polarized.

To summarise **soft acids** *are those which bind strongly to highly polarizable or unsaturated bases. The soft acids have acceptor atom of large size, small or zero positive charge and have a* *number of valence electrons that are easily removed. In general, heavier transition metal ions and low valence metal ions belong to this category.*

Having said that, there is no sharp line of demarcation between soft and hard species and a number of border line cases also exist which are termed as neither too hard nor too soft. These are given in Table 1.2. In such cases, the properties like size, oxidation states or polarizability are neither too low nor too high. These properties have intermediate values.

Table 1.2 Classified of Lewis bases as hard and soft bases.

Hard acids	Soft acids	Border line acids
NH_3, RNH_2, N_2H_4	H^-, R^-, C_2H_4, C_6H_6	$C_6H_5NH_2$, C_5H_5N
N_2O, OH^-, O^{2-}, ROH, RO^-, R_2O	CN^-, RNC, CO	N_2, N_3^-, NO_2^-
CH_3COO^-, CO_3^{2-}, NO^{3-},	$S_2O_3^{2-}$, SCN^-, R_3P	SO_3^{2-}, Br^-
PO_4^{3-}, SO_4^{2-}, ClO_4^-, F^-, Cl^-	R_3As, I^-	

Characteristics of Hard Lewis Bases:

1. Hard bases contain donor atoms of high electronegativity.
2. The donor atom of hard bases is of low polarizability.
3. The hard bases are not easily oxidized.
4. Hard bases have full low energy orbitals.
5. All hard bases strongly link to the proton.

Characteristics of Soft Lewis Bases

1. Soft bases have donor atoms of low electronegativity.
2. These have donor atoms which are easily polarized.
3. The soft bases are easily oxidised.
4. They contain empty low energy orbitals.

1.3. PEARSON'S HSAB PRINCIPLE

Pearson in 1963 proposed a simple rule for predicting the stability of compounds formed from acids and bases. This is known as HSAB principle. It states

hard acids prefer to bond to hard bases and soft acids prefer to bond to soft bases.

R.G. Pearson

Applications of HSAB Principle

This principle has a number of applications as discussed below.

1. Stability of complexes

We can explain the relative stability of complexes on the basis of HSAB principle. Consider the following general Lewis acid-base reaction:

$$\underset{\text{Lewis acid}}{A} + \underset{\text{Lewis base}}{:B} \longrightarrow \underset{\text{Complex}}{A:B}$$

The complex AB would be most stable if A and B are either both hard or both soft. On the other hand, the complex would be least stable if A is hard and B is soft or *vice versa.* We fund that AgI_2^-

exists as a stable compound while AgF_2^- does not. The HSAB principle explains this on the basis of different nature of I^- and F^-.

Ag^+ ion is a *soft acid.* Its combines with I^- ion, a *soft base* to forms a stable complex AgI_2^- whereas its combination with F^- ion, a *hard base*, forms an unstable complex, AgF_2^-.

Ag^+	+	$2I^-$	$\longrightarrow$	AgI_2^- (stable complex)
Soft acid		Soft base		Soft-soft
Ag^+	+	$2F^-$	$\longrightarrow$	AgF_2^- (unstable complex)
Soft acid		Hard base		Soft-hard

The reaction between LiI and CsF to give LiF and CsI always proceeds to give LiF and CsI. It is an interesting, example of preferential combination of soft-soft and hard-hard species:

LiI	+	CsF	$\longrightarrow$	LiF	+	CsI
Hard-soft		Soft-hard		Hard-hard		Soft-soft

Similarly, the following reactions go to the right because of the combination of soft-soft and hard-hard species.

LiI	+	CsF	$\longrightarrow$	LiF	+	CsI
Hard-soft		Soft-hard		Hard-hard		Soft-soft
CaS	+	H_2O	$\longrightarrow$	CaO	+	H_2S
Hard-soft		Hard		Hard-hard		Soft
HgF_2	+	BeI_2	$\longrightarrow$	BeF_2	+	HgI_2
Soft-Hard		Hard-soft		Hard-hard		Soft-soft

2. Coordination in complexes of ambidentate ligands

The HSAB principle can predict the formation of various metal ion complexes with ambidentate ligands. An **ambidentate ligand** is a monodentate ligand which can coordinate to metal ion through more than one atoms. SCN^- ion is an ambident ligand as it can coordinates through N or S. It is known to form stable complexes of the type $[M(NCS)_4]^{2-}$ with border line acid cations of Co, Ni, Cu and Zn by coordinating through N. It forms stable complexes of the type $[M(SCN)_4]^{2-}$ with acid cations of Rh, Ir, Pd, Pt and Au by coordinating through S. Their stability can be easily explained according to HSAB priciple. A **soft acid** like Rh, Ir, Pd or Pt prefers to coordinate through softer S atom (base) to form M—SCN bond. On the other hand, a *hard acid* like Co, Ni, Cu or Zn prefers to coordinate through the *harder* N atom (base) to form the M—NCS bond. However, there are many examples of both —SCN and —NCS bonding to the same metal.

3. Classification of acids and bases as soft or hard

The acids or bases can be classfied as hard or soft depending upon their preference for hard or soft reactants. For example, we want to know whether B is a hard or soft base. We shall carry out the following reaction.

$$\underset{}{BH^+} + \underset{\text{Soft acid}}{CH_3Hg^+} \rightleftharpoons CH_3HgB^+ + \underset{\text{Hard acid}}{H^+}$$

If B is a soft base, then the right hand side reaction will be favoured and if it is a hard base, then left hand reaction will be favoured. For example, R_2S is a soft base, so it will combine with soft acid (CH_3Hg^+) and the forward reaction will be favoured.

$$\underset{\text{Soft base}}{R_2SH^+} + \underset{\text{Soft acid}}{CH_3Hg^+} \longrightarrow \underset{\text{Soft acid-soft base}}{CH_3Hg^+\,SR_2} + \underset{\text{Hard acid}}{H^+}$$

On the other hand, NH_3 is a hard base, it will combine with H^+ (hard acid) favouring backward reaction:

$$NH_4^+ + CH_3Hg^+ + \longleftarrow CH_3Hg^+ NH_3 + H^+$$

Soft acid-hard base Hard acid

or

$$CH_3Hg^+NH_3 + H^+ \longrightarrow NH_4^+ + CH_3Hg^+$$

4. Poisoning of metal catalysts

Some metals like Ni, Pt, Pd, Cr Mo, etc (soft acids) act as catalysts. These metal catalysts can be easily poisoned by substances like carbon monoxide, unsaturated hydrocarbons (olefins, arenes, dienes, alkynes), phosphorus and arsenic containing ligands (soft bases). This poisoning occurs due to the soft acid-soft base interactions between soft metal ions **and soft ligands.** These ligands are strongly adsorbed on the surface of the metal and thus block the active sites. However, such soft acid catalysts are not affected by hard bases or ligands containing N, O or F.

5. Occurrence of ores and minerals

We find that certain ores occur as sulphides while others occur as oxides, carbonates or halides. The occurrence of the ores of certain metals can be explained on the basis of HSAB principle. Hard acid metal ions such as Mg^{2+}, Ca^{2+}, Mn^{2+}, Al^{3+}, Fe^{3+}, Cr^{3+}, etc., occur mostly as *their oxides, carbonates or halides* (F^- or Cl^-). This is due to their preferred *hard acid-hard base* interactions because oxides (O^{2-}), carbonates (CO_3^{2-}) and halides (F^- or Cl^-) are *hard bases.* Similarly, soft acid metal ions such as Cu^+, Ag^+, Pb^{2+}, Hg^{2+}, Pd^{2+}, etc. occur as their sulphides because of stable combination of soft acid with soft base (S^{2-} ion). Border line acids like Ni^{2+}, Fe^{2+}, Zn^{2+}, Pb^{2+}, Co^{2+}, etc. have the possibility of occurring both as *sulphides* (soft base) as well as *oxides* or carbonates (hard bases).

1.4. THEORETICAL BASIS OF HARDNESS AND SOFTNESS-HSAB PRINCIPLE

A number of theories have been proposed by different scientists to explain the HSAB principle. A brief description of these is given below.

1. Electrostatic interactions.

A large amount of energy is liberated in the formation of an ionic bond. Hard acids and hard bases form purely ionic compounds. These *hard-hard interactions are purely ionic or electrostatic.* For example, hard acids such as Li^+, Na^+, K^+ and hard bases such as OH^-, F^-, O^{2-} form ionic compounds. The electrostatic energy between a positive ion and a negative ion is inversely proportional to the interatomic distance. Smaller the ions, smaller would be the internuclear distance and greater would be the electrostatic attraction between the ions resulting in the release of energy. Consequently, the resulted complex formed by hard acid and hard base would be highly stable.

2. Polarizing power and polarizability

In the case of soft-soft interactions, we assuming that these are covalent in nature. For example, soft acids such as Ag^+, Hg^+ usually form bond with soft base such as Cl^-. The compound AgCl is much more covalent than the corresponding compounds of alkali metals. We can also explain this in terms of the polarizing power and polarizability of *d*-electrons. Most of the soft acids are transition metals with greater number of *d*-electrons. These *d*-electrons have more polarizing power and the soft bases such as I^-, S^{2-} are easily polarizable. Therefore, bonding between soft acids and soft bases is largely covalent.

3. Electronegativity and hardness and softness

Hardness and softness of acids and bases are related to electronegativity. In general, a species with high electronegativity is hard and the one with low electronegativity is soft. For example, highly electronegative ions Li^+, Na^+, etc. are hard acids while transition metal ions, Cu^+, Ag^+, Au^+ etc., having low electronegativities are soft acids. Similarly, hard and soft bases are defined on this basis. The fact that the trifluoromethyl group (CF_3) is considerably harder than the methyl group (CH_3) and boron trifluoride (BF_3) is harder than borane (BH_3) can be explained on the concept of electronegativity. The concept of electronegativity in explaining hardness/softness of acids and bases is dealt with in detail in the following section.

4. π-bonding contibutions

In soft acid-soft base interactions, π-bonding plays an important role. The soft acids are generally metals in low oxidation states containing loosely held outer *d*-electrons. These can be easily donated to ligands. Many soft bases are π-bond aceptors and many soft-acids are π-bond donors. Soft bases include ligands of phosphorus, arsenic and unsaturated ligands such as CO. The polarizability of soft acids and soft bases further favours π-bonding (back bonding).

1.5. ACID-BASE STRENGTH AND HARDNESS AND SOFTNESS

Strength of an acid or base has nothing to do with hardness or softness of the species.

The hardness or softness simply emphasizes special stability of hard-hard and soft-soft interactions. For example, both OH^- and F^- are hard bases, yet the basicity of the hydroxide ion (OH^-) is very very large (about 10^{13} times) as compared to fluoride ion. Similarly, SO_3^{2-} and Et_3P are soft bases but the latter is much stronger base (about 10^7 times) than the former towards soft acid CH_3Hg^+. It is a limitation of HSAB principle. We cannot estimate inherent strength of an acid or a base on HSAB principle. For example, the following reaction is not expected according to HSAB principle because it does not lead to hard-hard and soft-soft interactions. The stronger softer base, SO_3^{2-} ion can displace the weak hard base, F^- ion from the hard acid, the proton, H^+ :

$$\underset{\text{Soft bases}}{SO_3^{2-}} + \underset{\text{Hard-hard}}{HF} \rightleftharpoons \underset{\text{Hard-soft}}{HSO_3^-} + \underset{\text{Hard base}}{F^-} \qquad K_{eq} = 10^4$$

However, this reaction occurs (favourable $K = 10^4$). Similarly, the *stronger hard* base, OH^- ion can displace weaker *soft base,* SO_3^{2-} from the soft acid, methyl mercury cation as

$$\underset{\text{Hard}}{OH^-} + \underset{\text{Soft-soft}}{CH_3HgSO_3^-} \rightleftharpoons \underset{\text{Soft-hard}}{CH_3HgOH} + \underset{\text{Soft base}}{SO_3^{2-}} \qquad K_{eq} = 10$$

In these cases, the strength factor plays its role. The strengths of the bases ($SO_3^{2-} > F^-$ and $OH^- > SO_3^{2-}$) are sufficient to force these reactions to the forward direction inspite of hard-soft considerations. This happens especially when the inherent strengths of bases (or acids) being compared are significantly different.

Thus, the importance of inherent strengths of acids and bases should not be ignored while considering softness-hardness factor. For example, the following reaction does not occur in the forward direction, although it results into soft-soft combination.

$$\underset{\text{Soft-hard}}{CH_3CdOH} + \underset{\text{Soft}}{SO_3^{3-}} \rightleftharpoons \underset{\text{Soft-soft}}{CH_3CdSO_3^-} + \underset{\text{Hard-base}}{OH^-}$$

This is because OH^- is much stronger base than SO_3^{2-} and therefore cannot be displaced weaker base SO_3^{2-} even though the reaction is favoured by soft-soft interactions.

Thus, *in general, the Lewis acid-Lewis base reactions are governed by two independent principles* **softness and hardness** and **strength of the acid or base.**

Limitations of HSAB principle

HSAB principle has some limitations which are discussed as under.

1. This principle does not provide any quantitative scale of measurement.

2. According to HSAB principle, the reaction,

$$CH_3^+(g) + H_2(g) \longrightarrow CH_4(g) + H^+(g)$$

should occur because of soft-soft ($CH_3^+ — H^-$) interactions. However, this reaction does not take place. It is probably because of greater acidity of the proton (H^+) relative to CH_3^+(g) cation.

3. The hard-soft factors are independent of acidic or basic character of compounds. The factors work independently of each other. However, many examples are known which show interdependence of two concepts. For example, in some cases, hard-hard or soft-soft complexes change to more stable hard-soft system. According to HSAB principle, the following reaction is not favoured.

$$\underset{\text{(Soft base)}}{SO_3^{2-}} + \underset{\text{Hard-acid}}{HF} \longrightarrow \underset{\text{Hard-soft}}{HSO_3^-} + \underset{\text{Hard}}{F^-}$$

However, this reaction occurs in violation of HSAB principle. This is due to the fact that stronger soft base SO_3^{2-} displaces the weak hard base, F^- from the hard acid, H^+.

SYMBIOSIS

The term symbiosis was introduced by Jorgenson in 1868.

Soft ligands have a tendency to combine with a metal ion already having soft ligands and hard ligands have a tendency to combine with a metal ion already having hard ligands. This tendency has been termed as **symbiosis.** For example, F^- ion is a hard ligand and it readily combines with BF_3 because it already has threee hard ligands.

$$BF_3 + F^- \longrightarrow BF_4^-$$

Similarly, H^- ion, a soft ligand, readily combines with BH_3 to form a stable complex BH_4^-.

$$BH_3 + H^- \longrightarrow BH_4^-$$

Due to symbiosis, the symmetrical substituted compounds are preferred than unsymmetrical substituted compounds. Therefore the compounds such as BF_3H^- and BH_3F^-, having mixed substituents (soft acid-hard base or hard acid-soft base) interact spontaneously to yield BH_4^- and BF_4^-, i.e., compounds with same substituents. Thus, the reaction always proceeds to the right.

$$\underset{\text{Hard-soft}}{BF_3H^-} + \underset{\text{Soft-hard}}{BH_3F^-} \longrightarrow \underset{\text{Hard-hard}}{BF_4^-} + \underset{\text{Soft-soft}}{BH_4^-}$$

Fluorinated methane reacts in a similar manner:

$$CH_3F + CHF_3 \longrightarrow CH_4 + CF_4$$

1.6. ELECTONEGATIVITY AND HARDNESS-SOFTNESS

This concept provides the most authentic explanation of the phemonenon of hardness and softness of acids and base. We do not take into consideration, the electronegativity of the element but we consider the electronegativity of the ion participating in a reaction. For example, Li has a low electronegativity, but Li^+ has relatively high tendency to attract the electrons towards itself, and therefore, it has high electronegativity. This is because of small ionic size and extremely high second ionization potential. Consequently, Li^+ is a **hard acid.** On the other hand, transition metal ions in low oxidation states such as Cu^+, Hg^+, Ag^+, Cd^+ *etc.* possess large ionic size and relatively low second ionization energies and therefore possess lower values of electronegativity. Therefore, they are considered as **soft acids.** This way we can consider hard and soft bases.

This relation between hardness-softness and electronegativity helps to explain the greater hardness of trifluoro methyl group (F_3C) as compared to methyl group (H_3C) and greater hardness of BF_3 as compared to BH_3. Similarly, we can explain why F_3P is harder than H_3P which is harder than Me_3P. Ammonia (H_3N) is harder than trimethyl amine (Me_3N).

According to Mulliken-Jaffe definition, electronegativity (χ) is linearly related to the partial charge (δ) on an atom as:

$$\chi = a + b\delta$$

where parameter '*a*' is the neutral atom electronegativity of an atom. The parameter '*b*' measures *the rate of change of electronegativity with charge* and is known as **charge coefficient.**

Large soft and polarizable atoms have low values of b while small, hard and non polarizable atoms possess larger values of b.

In 1983, Parr and Pearson proposed a simple approach to quantitative measure of acid-base reactions. They have introduced the term **absolute hardness** (η) for both neutral and charged species and correlated it with Mulliken-jaffe's definition of electronegativity which they called **absolute electronegativity** (χ).

The **absolute hardness** (η) is defined as one half the difference between the ionization energy and the electron affinity (both in eV).

$$\eta = \frac{I - A}{2} \qquad (1)$$

where I = ionization energy and

A = electron affinity

According to Mulliken, ionization energy, electron affinity and electronegative are related as

$$\chi = \frac{I + A}{2} \qquad (2)$$

According to this concept a hard acid or base is a species which has a large difference between its ionization energy and its electron affinity.

The hardness of base B and acid A may be given as

$$\eta(B) = \eta(\text{species } B^+) = \frac{1}{2}[I(B^+) - A(B^+)] \qquad (3)$$

$$\eta(A) = \eta(\text{species } A) = \frac{1}{2}[I(A) - A(A)] \qquad (4)$$

In this equation, the base B is represented as B^+ in eqn. (3) and acid A as such in eqn. (4). The positive charge on the base and neutral charge on the acid can be understood as follows:

Consider the interaction of acid A and base B to form AB (electrons are donated by : B)

$$A + :B \longrightarrow \overset{\delta-}{A} : \overset{\delta+}{B}$$

In the formation of AB, electrons flow from base B to acid A and therefore, B acquires partial positive charge (δ^+) and A acquires partial negative charge (δ^-). Hardness of B refers to species going (on ionization) from δ^+ to a charge $1 + \delta^+$ while hardness of A refers to a species on ionization from δ^- to a charge of $1 + \delta^-$. If the shift of charge δ is taken as equal to 0.5, then charge on B becomes 1 + 0.5 = 1.50 and that on A becomes 1 + (–0.5) = 0.5.

Therefore, charge on B changes from 0.5 to 1.5 with an average of $\frac{1}{2}$ (0.5 + 1.5) = 1.0 while that on A changes from – 0.5 to + 0.5 with an average of $\frac{1}{2}$(0.5 – 0.5) = 0. Therefore, B is shown as B^+ and A as neutral in equations (3) and (4).

Parr and Pearson argued strongly for the use of the absolute hardness parameter in treating hard-soft acid-base (HSAB) interactions.

Absolute electronegativities (χ) and hardness parameters (η) of certain atoms are given in Table 1.3. Hardness parameters of some Lewis acids and Lewis bases are given in Table 1.4 and 1.5 respectively.

Table 1.3 Hardness parameters (h) for atoms (eV)

Atom	I	A	$\chi = \left(\frac{I+A}{2}\right)$	$\eta = \left(\frac{I-A}{2}\right)$
H	13.59	0.75	7.17	6.42
Li	5.39	0.62	3.00	2.38
B	8.30	0.28	4.29	4.01
C	11.26	1.27	6.27	5.00
N	14.53	−0.07	7.23	7.32
O	13.62	1.46	7.54	6.08
F	17.42	3.40	10.41	7.01
Na	5.14	0.55	2.85	2.30
Al	5.99	0.44	3.21	2.77
Si	8.15	1.39	4.76	3.38
P	10.49	0.75	5.62	4.86
S	10.36	2.08	6.22	4.12
Cl	12.97	3.62	8.30	4.68
K	4.34	0.50	2.42	3.84
Ca	6.11	−0.30	2.90	3.20
Cr	6.77	0.67	3.76	3.05
Fe	7.87	0.16	4.01	3.85
Cu	7.73	1.23	4.48	3.25
Br	11.84	3.36	7.60	4.24
Pd	8.34	0.56	4.45	3.89
Ag	7.58	1.30	4.44	3.14
I	10.45	3.06	6.76	3.70

Table 1.4 Hardness parameters (η_A) for some Lewis acids (eV)

Acid	I_{ACID}	A_{ACID}	χ_A	η_A
Ions				
Al^{3+}	119.99	28.45	74.22	45.77
Li^+	75.64	5.39	40.52	35.12
Mg^{2+}	80.14	15.04	47.59	32.55
Na^+	47.29	5.14	26.21	21.08
Ca^{2+}	50.91	11.87	31.39	19.52
Sr^{2+}	43.60	11.03	27.30	16.30
K^+	31.63	4.34	17.99	13.64

Acid	I_{ACID}	A_{ACID}	χ_A	η_A
Ba^{2+}	35.50	10.00	22.80	12.80
Fe^{3+}	54.80	30.65	42.73	12.08
Zn^{2+}	39.72	17.96	28.84	10.88
Tl^{3+}	50.70	29.80	40.30	10.50
Cd^{2+}	37.48	16.91	27.20	10.29
Cr^{3+}	49.10	30.96	40.00	9.10
Mn^{2+}	33.67	15.64	24.66	9.02
Ni^{2+}	35.17	18.17	26.67	8.50
Cu^{2+}	36.83	20.29	28.50	8.40
Co^{2+}	33.50	17.06	25.28	8.22
Hg^{2+}	34.20	1876	26.50	7.70
TI^{+}	20.40	6.11	13.3	7.2
Ag^{+}	21.49	7.58	14.53	6.96
Pd^{2+}	32.93	19.43	26.18	6.75
Cu^{+}	20.29	7.3	14.01	6.28
Au^{+}	20.50	9.23	14.90	5.60
Br^{+}	21.6	11.8	16.7	4.9
I^{+}	19.1	10.5	14.8	4.3
Molecules				
CO_2	13.8	0.0	6.9	6.9
SO_2	12.3	1.1	6.7	5.6
SO_3	12.7	1.7	7.2	5.5
Cl_2	11.4	2.4	6.9	4.5
I_2	9.3	2.6	6.0	3.4

Table 1.5 Hardness parameters for some bases (eV)

Base	I_{BASE}	A_{BASE}	η_B
Ions			
F^-	17.42	3.40	7.01
OH^-	13.17	1.83	5.67
NH_2^-	11.40	0.74	5.33
CN^-	14.02	3.82	5.10
N_3^-	11.6	1.8	4.9
CH_3^-	9.82	0.08	4.87
Cl^-	13.01	3.62	4.70
NO_2^-	12.9	3.99	4.5
Br^-	11.84	3.36	4.2
SH^-	10.4	2.3	4.1
I^-	10.45	3.06	3.70

Base	I_{BASE}	A_{BASE}	η_B
Molecules			
CO	26.0	14.0	6.0
H_2O	26.6	12.6	7.0
H_2S	21.0	10.5	5.3
NH_3	24.0	10.2	6.9
PH_3	20.0	10.0	5.0

We find from the tables that the absolute hardness values of acids and bases represent their known chemical behaviour. The only exception is hydrogen which may be considered as a special case. The expected increase in softness in going down a column in the periodic e.g. from Mg^{2+} to Ba^{2+} or Ni^{2+} to Pd^{2+} is in accordance with the decrease in η_A values. The expected increase in hardness with increase in oxidation state is explained by η_A values as:

Fe°	Fe^{2+}	Fe^{3+}
3.87	7.3	12.08

There are certain discrepancies also. For example, Tl^+ has η_A = 7.2 and Tl^{3+} has η_A = 10.5 which suggest increase in hardness from Tl^+ to Tl^{3+}. However, Tl^{3+} is known to be chemically softer. However, the camparison of Tl^{3+} with other trivalent ions show the expected soft character.

Hard bases such as F^- and OH^- have large η_B values and soft bases, SH^- and CH_3^- have small values (Table 1.5). The neutral bases are also at the right places.

HSAB Principle and Molecular Orbital Theory

There have been attempts to understand HSAB principle with the help of MO theory. According to Koopman's theorem, the ionization energy gives the energy of the highest occupied molecular orbital (HOMO) while the electron affinity gives the energy of the lowest unoccupied molecular orbital (LUMO) for a given molecule:

$$E(HOMO) = -I$$

$$E(LUMO) = -EA$$

These orbitals are involved in electronegativity and HSAB relationships. The hardness and softness of a species depends upon the average gap between the HOMO and LUMO. The separation of HOMO and LUMO is taken equal to twice the value of η. Hard species have a large HOMO-

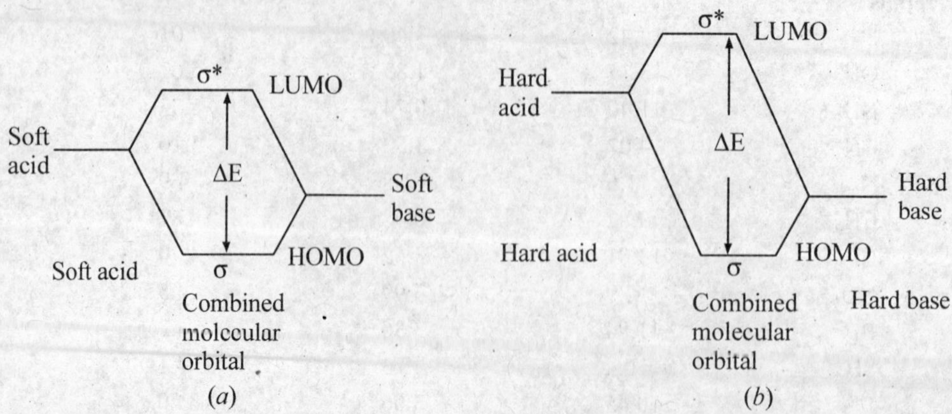

Fig. 1.1. Comparison of orbital energies for (a) Soft-soft and (b) Hard-hard combinations

LUMO gap while soft species have small energy gap. Drago and Wayland have also proposed a quantitative approach of acid-base parameters by taking into account electrostatic and covalent factors. The presence of low lying unoccupied MOs capable of mixing with the ground state orbitals accounts for the polarizability of soft species.

The combination of a soft acid with soft base is analogous to a situation in which the LUMO of the Lewis acid has almost the same energy as the HOMO of the Lewis base. The combination of orbitals of soft acid and soft base gives molecular orbital lower in energy than that of orbital of soft base. This gives a strong covalent bond

In the case of hard acid-hard base combination, there is a wide difference in the energies of bonding and antibonding molecular orbitals or LUMO and HOMO will differ greatly in energy level. This situation is not favourable for covalent bond formation. An ionic combination will result instead.

SOLVED CONCEPTUAL PROBLEMS

Example 1. (*a*) Classify the following as soft or hard acids:

Cu^+, Na^+, Ti^{4+}, Ag^+, Pt^{2+}

(*b*) Classify the following as soft or hard bases:

NH_3, ROH, I^-, CO, O_2^-, C_6H_6

Solution. (*a*) Acids

Cu^+ : Soft acid Na^+ : Hard acid
Ti^{4+} : Hard acid Ag^+ : Soft acid
Pt^{2+} : Soft acid

(*b*) Bases

NH_3 : Hard base ; ROH : Hard base
I^- : Soft base ; CO : Soft base
O_2^- : Hard base ; C_6H_6 : Soft base

Example2. Which of the metals Ag^+, Al^{3+}, Cu^+, Ca^{2+}, Cr^{3+} and Zn^{2+} might be expected to be found in aluminosilicate minerals and which in sulphides?

Solution. Hard acids like Ca^{2+}, Al^{3+} and Cr^{3+} form aluminosilica minerals while soft acids like Cu^+, Ag^+ and Zn^{2+} form sulphides.

Example 3. O^{2-} is a hard base and S^{2-} is a soft base. Which of the following metal ions will occur preferably as oxides or sulphides:

(*i*) Ca^{2+} (*ii*) Ag^+ (*iii*) Al^{3+} (*iv*) Hg^{2+}

Solution. Ca^{2+} and Al^{3+} are hard acids while Ag^+ and Hg^{2+} are soft acids therefore

(*i*) Ca^{2+} occurs as oxides
(*ii*) Ag^+ occurs as sulphides
(*iii*) Al^{3+} occurs as oxides
(*iv*) Hg^{2+} occurs as sulphides.

Example 4. Suggest and justify the way through which the following reactions will go:

(*i*) $$CaS + H_2O \longrightarrow CaO + H_2S$$

(*ii*) $$R_2SBF_3 + R_2O \longrightarrow BF_3OR_2 + R_2S$$

(*iii*) $$CuI_2 + 2CuF \longrightarrow CuF_2 + 2CuI.$$

Solution. (*i*) The reaction will go to the **forward direction.** Hard-soft CaS will be converted to hard-hard CaO.

$$\underset{\text{Hard-soft}}{CaS} + \underset{\text{Hard}}{H_2O} \longrightarrow \underset{\text{Hard-hard}}{CaO} + \underset{\text{Soft}}{H_2S}$$

(*ii*) The reaction will go to the **forward direction.** The hard-soft complex $BF_3.SR_2$ will be converted to a stable hard-hard complex $F_3B.OR_2$.

$$\underset{\text{Hard-soft}}{BF_3.SR_2} + \underset{\text{Hard}}{R_2O} \longrightarrow \underset{\text{Hard-hard}}{BF_3.OR_2} + \underset{\text{Soft}}{R_2S}$$

(*iii*) The reaction will go to **forward direction.** Hard-soft CuI_2 will be converted to soft-soft CuI.

$$\underset{\text{Hard-soft}}{CuI_2} + \underset{\text{Soft-hard}}{2CuF} \longrightarrow \underset{\text{Soft-soft}}{2CuI} + \underset{\text{Hard-hard}}{CuF_2}$$

Example 5. Indicate the feasibility of the following reactions in the forward direction:

(*a*) $\underset{\text{Soft-soft}}{C_2H_5HgSO_3^-} + \underset{\text{Soft base}}{Me_3P} \longrightarrow$

(*b*) $\underset{\text{Soft-hard}}{CH_3CdOH} + \underset{\text{Soft base}}{SO_3^{2-}} \longrightarrow$

The base strength follows the order:

$Me_3P >> SO_3^{2-}$ **and** $OH^- >> SO_3^{2-}$.

Solution. (*a*) This reaction would be feasible because of much stronger Me_3P base compared to SO_3^{2-}. The inherent strength of $Me_3P >> SO_3^{2-}$ will force the reaction in forward direction inspite of formation of hard-soft pair in the product.

(*b*) It is not feasible because OH^- is much stronger base than SO_3^{2-} and hence cannot be displaced by weaker base SO_3^{2-} even though it is favoured by HSAB principle.

Example 6. (*a*) While hard-hard interactions are generally ionic, soft-soft interactions are generally covalent? Why is it so?

(*b*) Classify the following as soft or hard bases:

(*i*) N_2H_4 (*ii*) CN^- (*iii*) CO (*iv*) F^- (*v*) O^{2-} (*vi*) C_6H_6

(*c*) Give two examples each of boarder line between hard and soft acids and bases.

Solution. (*a*) Hard acids combine with hard bases. The compounds formed by their combinations are ionic compounds. For example, hard acids ike Li^+, Na^+, H^+, combine with hard bases like OH^-, F^- or O^{2-} forming ionic compounds. This is because of big energy gap between HOMO and HUMO orbitals. The interactions between these are purely electrostatic or ionic in nature. Soft acid-soft base interactions are covalent in nature. These interactions involve bigger size of the ions.

(*b*) (*i*) N_2H_4 : Hard base (*ii*) CN^- : Soft base (*iii*) CO: Soft base
(*iv*) F^- : Hard base (*v*) O^{2-} : Hard base (*vi*) C_6H_6 : Soft base

(*c*) Zn^{2+}, Bi^{3+} : Border line acids
NO^{2-}, Br^- : Border line bases

Example 7. Classify the following as hard, soft or boarder line acids:

(*i*) Ag^+ (*ii*) Pt^{2+} (*iii*) Te^{4+} (*iv*) Bi^{3+} (*v*) I^{7+}

Solution.

(*i*) Ag^+ : Soft acid (*ii*) Pt^{2+} : Soft acid
(*iii*) Te^{4+} : Hard acid (*iv*) Bi^{3+} : Boarder line acid
(*v*) I^{7+} : Hard acid

Example 8. (*a*) Predict on the basis of HSAB pribciple, whether the following reactions are feasible in the forward direction or not?

(*i*) BF_3H^- + BH_3F^- $\longrightarrow$ BF_4^- + BH_4^-

(*ii*) HgF_2 + BeI_2 $\longrightarrow$ BeF_2 + HgI_2

(*iii*) CH_3Hg^+ + NH_4^+ $\longrightarrow$ $CH_3Hg^+NH_3$ + H^+

(*iv*) CH_3HgF + HSO_3^- $\longrightarrow$ $CH_3HgSO_3^-$ + HF

(*v*) CuI_2 + 2CuF $\longrightarrow$ CuF_2 + 2CuI

(*vi*) LiF + CsI $\longrightarrow$ LiI + CsF

(*b*) O^{2-} is a hard base and S^{2-} is a soft base. Which of the following metal ions will occur preferably as oxides or sulphides: (*i*) Ca^{2+} (*ii*) Ag^+ (*iii*) Al^{3+} (*iv*) Hg^{2+}.

Solution.

(*a*) (*i*) Feasible (*ii*) Feasible (*iii*) Not feasible

(*iv*) Feasible (*v*) Feasible (*vi*) Not feasible.

(*b*) (*i*) Ca^{2+} occurs as oxides (*ii*) Ag^+ occurs as sulphides

(*iii*) Al^{3+} occurs as oxides (*iv*) Hg^{2+} occurs as sulphides

Example 9. Suggest and justify the way through which the following reactions will go

(*i*) CaS + H_2S $\longrightarrow$ CaO + H_2S

(*ii*) R_2SBF_3 + R_2O $\longrightarrow$ BF_3OR_2 + R_2S

(*iii*) CuI_2 + 2CuF $\longrightarrow$ CuF_2 + 2CuI

Solution. (*i*) The reaction will go to the forward direction because hard-soft CaS will be converted to hard-hard product

$$\underset{\text{Hard-soft}}{CaS} + \underset{\text{Hard}}{H_2S} \longrightarrow \underset{\text{Hard-hard}}{CaO} + \underset{\text{Soft}}{H_2S}$$

(*ii*) The reaction will go to the forward direction because the hard-soft complex BF_3SR_2 will be converted to a stable hard-hard product.

$$\underset{\text{Hard-soft}}{BF_3.SR_2} + \underset{\text{Hard}}{R_2O} \longrightarrow \underset{\text{Hard-hard}}{BF_3OR_2} + \underset{\text{Soft}}{R_2S}$$

(*iii*) The reaction will go to forward direction because hard-soft CuI_2 will be converted to soft-soft CuI.

$$\underset{\text{Hard-soft}}{CuI_2} + \underset{\text{Soft-hard}}{2CuF} \longrightarrow \underset{\text{Soft-soft}}{2CuI} + \underset{\text{Hard-hard}}{CuF_2}$$

Example 10. Classify the following into hard, soft and boarder line acids and bases:

Na^+, Pt^{2+}, NH_3, SCN^-, Br^-, Co^{++}

Solution. Na^+ : Hard acid SCN^- : Soft base

Pt^{2+} : Soft acid Br^- : Boarder line base

NH_3 : Hard base Co^{++} : Boarder line acid

Example 11. (*a*) Explain the action of R_2S on C_2H_5HgCl.

(*b*) AgI_2^- complex is stable but AgF_2^- is not. Explain.

Solution. (*a*) R_2S is a soft base. It will interacts with soft acid C_2H_5Hg to form a complex:

$$C_2H_5HgCl + R_2S \longrightarrow [C_2H_5Hg\,(R_2S)]^+\,Cl^-$$

(*b*) Ag^+ is a soft acid. It may combine with soft base I^- ion to give AgI_2^- (soft-soft) whereas it cannot combine with hard base F^- ion. Therefore, AgF_2^- (soft-hard) is not formed as shown in the following equations.

$$\underset{\text{Soft}}{Ag^+} + \underset{\text{Soft base}}{2I^-} \longrightarrow \underset{\text{Stable (soft-soft)}}{AgI_2^-}$$

$$\underset{\text{Soft acid}}{Ag^+} + \underset{\text{Hard base}}{2F^-} \longrightarrow \underset{\text{Unstable (soft-hard)}}{AgF_2^-}$$

Example 12. (*a*) Explain the action of $CdCO_3$ on Na_2S solution.

(*b*) How does HSAB principle explain the validity of the following reactions:

(*i*) $BeF_2 + HgI_2 \longrightarrow HgF_2 + BeI_2$

(*ii*) $BF_3H^+ + BH_3F^- \longrightarrow BF_4^- + BH_4^+$

Solution. (*a*) $CdCO_3$ reacts with Na_2S solution to give the precipitated of CdS. This happens because S^{2-} is a soft base and prefers to combine with soft acid, Cd^{2+}.

$$CdCO_3 + Na_2S \longrightarrow Na_2CO_3 + CdS$$

(*b*) (*i*) $$\underset{\text{Hard-hard}}{BeF_2} + \underset{\text{Soft-soft}}{HgI_2} \longrightarrow \underset{\text{Soft-hard}}{HgF_2} + \underset{\text{Hard-soft}}{BeI_2}$$

This reaction will not proceed because hard-hard and soft-soft combinations will not react to produce soft-hard combinations.

(*ii*) $$\underset{\text{Hard-soft}}{BF_3^+H^-} + \underset{\text{Soft-hard}}{BH_3F^-} \longrightarrow \underset{\text{Hard-hard}}{BF_4^-} + \underset{\text{Soft-soft}}{BF_4^+}$$

This reaction is valid because the compounds such as BF_3H^+ and BH_3F^- having mixed substituents interact to give compounds with same substituents.

Example 13. Calculate absolute electronegativity and absolute hardness for OH^- and Cl^- given that

OH^-	**I = 13.17**	**A = 1.83**
Cl^-	**I = 13.01**	**A = 3.62 (values are in eV)**

Solution. (*i*) OH^-

I 13.17, A = 1.83

Absolute electronegativity, $\chi = \dfrac{13.17 + 1.83}{2} = 7.50$ eV

Absolute hardness $\eta = \dfrac{13.17 - 1.83}{2} = 5.67$ eV

(*ii*) Cl^-

Absolute electronegativity, $\chi = \dfrac{13.01 + 3.62}{2} = 8.31$ eV

Absolute hardness, $\eta = \dfrac{13.01 - 3.62}{2} = 4.70$ eV

Example 14. Which of the following does not proceed to the right?

(*i*) $$\underset{\textbf{Soft-hard}}{HgF_2} + \underset{\textbf{Hard-soft}}{BeI_2} \longrightarrow \underset{\textbf{Hard-hard}}{BeF_2} + \underset{\textbf{Soft-soft}}{HgI_2}$$

(*ii*) $$\underset{\textbf{Hard-hard}}{LiF} + \underset{\textbf{Soft-soft}}{CsI} \longrightarrow \underset{\textbf{Hard-soft}}{LiI} + \underset{\textbf{Soft-hard}}{CsF}$$

Solution. The reaction (*ii*) will not proceed to the right because of unfavoured hard-soft combination.

Example 15. Predict the feasibility of the following reactions:

(*i*) $CH_3HgOH + HSO_3^- \longrightarrow$

(*ii*) $CH_3HgOH + HF \longrightarrow$

(*iii*) $CH_3HgF + HSO_3^- \longrightarrow$

Solution. (*i*) and (*iii*) are favoured because of hard-hard and soft-soft interactions, (*ii*) not feasible because OH^- is much stronger base than F^-.

Example 16. Classify the following bases as hard, soft or border line:

(*i*) N_2H_4 (*ii*) C_2H_4 (*iii*) NO_3^- (*iv*) C_5H_5N
(*v*) F^- (*vi*) Br^- (*vii*) I^- (*viii*) CO
(*ix*) CN^- (*x*) O^{2-}

Solution. Hard bases: N_2H_4, NO_3^-, F^-, O^{2-}
Soft bases: C_2H_4, I^-, CO, CN^-
Border line bases: C_5H_5N, Br^-

Example 17. Which of the following metals will prefer to occur as oxides or sulphides or both types. Give reasons for your answer:

(*i*) Cr^{3+} (*ii*) Ni^{2+} (*iii*) Sn^{2+} (*iv*) Fe^{2+}
(*v*) Cu^+ (*vi*) Al^{3+} (*vii*) Ca^{2+} (*viii*) Hg^{2+}

Solution. Hard acids (metal ions) prefer to occur as oxides (O^{2-}, hard base) while soft acids (metal ions) prefer to occur as sulphides (S^{2-}, soft base). Border line acids form both oxides and sulphides.

Accordingly, Cr^{3+}, Al^{3+} and Ca^{2+} form oxides

Cu^+ and Hg^{2+} form sulphides

Sn^{2+}, Ni^{2+} and Fe^{2+} form oxides as well as sulphides

EXERCISES
(Based on Questions from Different University Papers)

Multiple Choice Questions (Choose the correct option)

1. Out of Sn^{2+}, Ni^{2+} and Fe^{2+}, oxides as well as sulphides are formed by
(*a*) Sn^{2+} and Ni^{2+} (*b*) Ni^{2+} and Fe^2
(*c*) Sn^{2+} and Fe^{2+} (*d*) all the three
2. Which of the following is not a hard base?
(*a*) NH_3 (*b*) H_2O
(*c*) Cl^- (*d*) CN^-
3. Which of the following is not border line acid ?
(*a*) Bi^{3+} (*b*) BMe_3
(*c*) SO_2 (*d*) CO_2
4. CH_3HgOH is
(*a*) Soft-soft (*b*) hard-hard
(*c*) Soft-hard (*d*) hard-soft

5. Which of the following will prefer to occur as sulphide ?

(*a*) Ca^{2+} (*b*) Ni^{2+}

(*c*) Al^{3+} (*d*) Cr^{3+}

6. The term hard and soft acid and base was given by:

(*a*) Bronsted (*b*) Lewis

(*c*) Pearson (*d*) Franklin

7. Which of the following statement is flase ?

(*a*) K^+ is a hard acid.

(*b*) Soft acids are molecules or ions with larger number of valence electrons.

(*c*) Soft bases are easily oxidized.

(*d*) Hard bases have donor atoms with high polarizability.

8. Which of the following reaction will not proceed to the forward direction ?

(*a*) $BF_4^- + BH_4^- \longrightarrow BF_3H^- + BH_3F^-$

(*b*) $BeI_2 + HgF_2 \longrightarrow BeF_2 + HgI_2$

(*c*) $R_2SBF_3 + R_2O \longrightarrow BF_3OR_2 + R_2S$

(*d*) $CaS + H_2O \longrightarrow CaO + H_2S$

9. Which of the following will prefer to occur as oxide ?

(*a*) Ca^{2+} (*b*) Cu^+

(*c*) Cd^{2+} (*d*) Ag^+

10. In which of the following pairs, both the species are not of the same type (hard-hard or soft-soft acid or base) ?

(*a*) NH_3, CO (*b*) H_2O, OH^-

(*c*) ROH, R_2O (*d*) SO_3, CO_2

11. Which of the following is not a hard acid ?

(*a*) Na^+ (*b*) Mg^{2+}

(*c*) Pd^{2+} (*d*) Ti^{4+}

12. Which of the following reactions will not proceed to the right hand side?

(*a*) $LiI + CsF \longrightarrow LiF + CsI$

(*b*) $CaS + H_2O \longrightarrow CaO + H_2S$

(*c*) $CuI_2 + 2CuF \longrightarrow CuF_2 + 2CuI$

(*d*) $BeF_2 + HgI_2 \longrightarrow HgF_2 + BeI_2$

13. Hg^{2+} is classified as

(*a*) Soft acid (*b*) Hard acid

(*c*) Soft base (*d*) Hard base

14. Absolute hardness and absolute electronegativity respectively of OH^- are :

(Given I = 13.17 eV and A = 1.83 eV)

(*a*) 5.67, 7.5 eV (*b*) 7.5, 5.6 eV

(*c*) 0.567, 0.75 eV (*d*) 56.7, 75.0 eV

ANSWERS

1. (*d*)	**2.** (*d*)	**3.** (*d*)	**4.** (*c*)	**5.** (*b*)	**6.** (*c*)
7. (*d*)	**8.** (*a*)	**9.** (*a*)	**10.** (*a*)	**11.** (*c*)	**12.** (*d*)
13. (*a*)	**14.** (*a*)				

Short Answer Questions

1. What are soft acids ? Give one example.

2. Which of the following are not a hard bases ?

$ROH, CN^-, OH^-, S^{2-}, O_2^{2-}$

3. Do you think the followng reaction will proceed on the basis of HSAB principle ?

$LiI + CsF \longrightarrow CsI + LiF$

4. Why is pyridine a border line base while ammonia is a hard base ?

5. Will the following reaction be feasible :

$CH_3 HgSO_3^- + Me_3P \longrightarrow$

$(Me_3P >> SO_3^{2-})$

6. Define soft base and give one example.

7. What is symbiosis.

8. AgI_2^- complex is stable while AgF_2^- is not. Explain.

9. How does HSAB principle govern the occurrence of minerals ?

10. Define absolute hardness.

11. Give two examples of border line acids.

12. Which of the following is odd among the following :

$Li^+, Ga^{3+}, Cd^{2+}, K^+$

13. Which of the following is hard acid

(*i*) Cation of smaller size and high charge

(*ii*) Cation of larger size and small charge ?

14. Draw M.O. diagram showing covalent bonding in soft-soft interactions.

General Questions

1. (*a*) Why is pyridine a border line base while ammonia is a hard base ?

(*b*) What are the limitations of HSAB principle ?

2. (*a*) What are hard and soft acids and bases ? Explain the HSAB principle with suitable examples.

(*b*) Discuss the effect of substituents on hardness and soft ness of an acid.

3. How does HSAB principle explain the validity of the following reactions ?

(*i*) $LiI + CsF \longrightarrow LiF + CsI$

(*ii*) $CuI_2 + 2CuF_2 \longrightarrow CuF_2 + 2CuI.$

4. (*a*) "While hard-hard interactions are generally ionic, soft-soft interactions are generally covalent." Why is it so ?

(*b*) Discuss the contribution of π-bonding in soft-soft interactions.

5. (*a*) Discuss giving examples, the applications of HSAB principle.

(*b*) What are the theoretical justifications of HSAB principle.

6. (*a*) Describe the origin of concept of hard and soft acids and bases.

(*b*) Predict which way the following reactions will proceed:

(*i*) $HI + NaF \longrightarrow HF + NaI$

(*ii*) $CaS + H_2O \longrightarrow CaO + H_2S$

(*iii*) $CuI_2 + 2CuF \longrightarrow CuF_2 + 2CuI.$

Explain your answer.

7. Explain the following.

(*i*) AgI_2^- complex is stable but AgF_2^- is not.

(*ii*) $[Co(NH_3)_5F]^{2+}$ complex is stable while $[Co(NH_5)I]^{2+}$ complex is unstable.

(*iii*) CsF reacts with LiI even though both are ionic.

(*iv*) BF_3 readily combines with F^- to form stable complex BF_4^-.

(*v*) CH_3F and CHF_3 react to form CH_4 and CF_4.

8. (*a*) Suggest and justify the way through which the following reactions will go:

$CaS + H_2O \longrightarrow CaO + H_2S$

$R_2SBF_3 + R_2O \longrightarrow BF_3R_2O + R_2S$

$CuI_2 + 2CuF \longrightarrow CuF_2 + 2CuI.$

(*b*) π-bonding and electronegativity can be used to explain the hardness and softness of acids and bases. Explain.

9. (*a*) What are the characteristics of a soft acid and a soft base ?

(*b*) Classify the following into hard, soft and border line acids and bases:

I^-, K^+, HCl, CO, Ni^{2+}, CO_2, Ag^+, NH_3, Cu^+, I^{7+}

10. (*a*) Explain HSAB principle. Discuss its applications.

(*b*) Explain clearly why hard acids co-ordinate with hard bases and soft acids co-ordinate with soft bases.

11. (*a*) Hard-hard interaction is the major driving force for a reaction to proceed discuss.

12. What is symbiosis ? Give examples. What are its applications ?

13. (*a*) Using HSAB principle, predict whether reactants or products will be formed in each of the following equilibria :

$As_2S_3 + 3HgO \longrightarrow As_2O_3 + 3HgS$

$Ag\,F + LiI \longrightarrow AgI + LiF$

$2CH_3MgF + HgF_2 \longrightarrow (CH_3)_2Hg + 2MgF_2.$

(*b*) Explain the various limitations of HSAB principle.

14. (*a*) State and explain HSAB principle. Give its applications with suitable examples.

(*b*) Using HSAB principle, predict and justify the way the following reactions will go:

(*i*) $CuI_2 + 2CuF \longrightarrow 2CuI + CuF_2$

(*ii*) $CH_3F + CHF_3 \longrightarrow CH_4 + CF_4$

(*iii*) $CH_3HgF + HSO_3^- \longrightarrow CH_3HgHSO_3^- + HF$

15. (*a*) Describe how Mulliken - Jaffe definition of electronegativity is related to hardness of acids and bases, with examples.

(*b*) Give characteristics of hard acids.

(*c*) Write short notes on :

(*i*) Symbiosis.

(*ii*) Pearosn's HSAB principle.

16. (*a*) Define Pearson's HSAB principle. Discuss the validity of folowing reactions on the basis of HSAB principle.

(*i*) $LiI + CsF \longrightarrow LiF + CsI$

(*ii*) $HgF_2 + BeI_2 \longrightarrow HgI_2 + BeF_2$.

(*b*) $[AgI_2]^-$ is stable but $[AgF_2]^-$ is unstable. Why ?

(*c*) Classify the following into hard, soft and border line acids and bases.

I^-, CO, Ni^{2+}, CO_2, Ag^+, NH^{4+}, SO_3^{2-}, BH_3, H_2O, NO_3^-.

17. What is HSAB principle ? Which of the following reactions will proceed in the forward direction and why ?

(*i*) $BF_3H^+ + BH_3F^- \longrightarrow BF_4^- + BH_4^-$

(*ii*) $CaO + H_2S \longrightarrow CaS + H_2O$

18. (*a*) What are hard acids and hard bases ? Give their important characteristics.

(*b*) What is HSAB principle ? What are its uses ?

(*c*) How does HSAB principle explain the validity of the following reactions.

(*i*) $LiI + CsF \longrightarrow LiF + CsI$

(*ii*) $CuI + 2CuF_2 \longrightarrow CuF_2 + 2CuI$

19. (*a*) Define hard and soft acids and bases.

(*b*) How does HSAB principle explain the validity of the following reactions:

(*i*) $LiI + CsF \longrightarrow LiF + CsI$

(*ii*) $CuI_2 + 2\ CuF \longrightarrow CuF_2 + 2\ CuI$

(*c*) Explain how electronegativity can be used to explain hardness and softness of acids and bases.

20. (*a*) Is it correct to say that hard species, both acids and bases, tend to be small, slightly polarizable species and that soft acids and bases tend to be larger and more polarizable ?

(*b*) Which of the following are hard acids and hard bases ?

H^+, Li^+, NH_3, N_2H_4

21. (*a*) How does HSAB principle explain the validity of the folowing reactions ?

(*i*) $BeF_2 + HgI_2 \longrightarrow HgF_2 + BeI_2$

(*ii*) $BF_3H^+ + BH_3F^- \longrightarrow BF_4^- + BH_4^+$

(*b*) Classify the following into hard, soft and boarder line acids and bases :

Na^+, Pt^{2+}, NH_3, SCN^-, Br^-, Co^{2+}

22. Will the following reactions proceed to right hand side ? Justify your answer on the basis of HSAB principle :

(*i*) $Ag^+ + 2F^- \longrightarrow AgF_2^-$

(*ii*) $HgF_2 + BeI_2 \longrightarrow HgI_2 + BeF_2$

(*iii*) $CaS + H_2O \longrightarrow CaO + H_2S$

23. (*a*) How will you determine the relative strength of hard and soft acids and bases ?

(*b*) Why are hard-hard and soft-soft combinations preferred to hard-soft or soft-hard combination ?

(*c*) How does HSAB principle govern the occurrence of minerals ?

24. (*a*) What are hard acids and bases ? Give their important characteristics.

(*b*) How electronegativity can be used to explain hardness and softness of acids and bases ?

(*c*) Define HSAB principle. Discuss the applications of hard soft acid base principle.

25. (*a*) Explain SYMBIOSIS with example.

(*b*) Explain the reaction of

(*i*) $CdCO_3$ and Na_2S solution

(*ii*) R_2S and C_2H_5HgCl

26. (*a*) How does HSAB principle governs the occurrence of minerals and poisoning of metal catalysts ?

(*b*) $[AgI_2]^-$ is stable but $[AgF_2]^-$ is unstable. Why ?

(*c*) Explain on the basis of HSAB principle the action of :

(*i*) $CdCO_3$ on Na_2S

(*ii*) NaOH and $(CH_3)_2Cd\ CO_3$

(*iii*) Action of R_2S on C_2H_5HgCl.

27. (*a*) Comment on the feasibility or non-feasibility of following reaction :

$$BF_4^- + BH_4^- \longrightarrow BF_3H^- + BH_3F^-$$

(*b*) State and explain HSAB principle.

(*c*) Describe the contribution of pi bonding in soft-soft interactions.

28. What is symbiosis ? Discuss theoretical basis of hardness and softness.

2

Metal Ligand Bonding in Transition Metal Complexes

2.1. INTRODUCTION

Werner is known as the Father of Coordination Chemistry. He made the first attempt to explain bonding in coordination compounds, for which he was awarded the Nobel prize in 1913. The features of his theory include the primary valencies and secondary valencies. Later Sidgwick extended the theory by suggesting that the metal ions accept the electron pairs from the ligands in order to achieve the next noble gas configuration. This was followed by the development of valence-bond theory by Linus Pauling. Valence bond theory could explain the stereochemistry and magnetic properties of coordination compounds (Complexes), but it had certain limitations. In this chapter, we plan to deal in detail the Crystal Field Theory. But before that we shall have a brief recap of valence bond theory and its limitations. We shall then explain how crystal field theory addresses some points which are not explained by valence bond theory. But the fact remains that in spite of the limitations of valence bond theory, it is still the simplest theory to explain most of the points.

Linus Pauling

2.2. VALENCE BOND THEORY

Valence bond theory assumes the bonding between metal ion and ligands to be **purely covalent**. The covalent bond in metal complexes is formed *by the overlap of a filled ligand orbital containing a lone pair of electrons with a vacant hybrid orbital of the metal atom or ion,* as shown in Fig. 2.1.

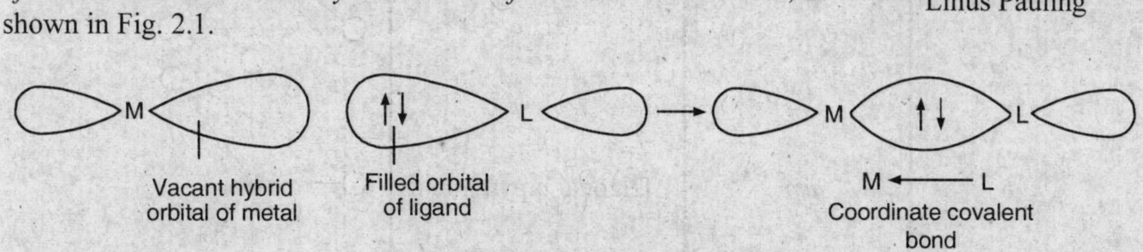

Fig. 2.1. Formation of a metal-ligand bond

This theory gives appropriate explanation for the structures and magnetic properties of coordination compounds. The salient features of this theory are :

1. *The central metal atom or ion in the complex provides a number of empty orbitals for the formation of coordinate bonds with suitable ligands. The number of empty orbitals provided is equal to coordination number of the central metal ion.*

2. *A number of s, p and d atomic orbitals of the metal hybridise to give a set of new orbitals of equivalent energy, called* ***hybrid orbitals.*** *These hybrid orbitals are directed according to definite geometry of the complex such as square planar, tetrahedral, octahedral etc.*

3. *The d-orbitals involved in the hybridisation may be either inner (n – 1) d-orbitals or outer (nd) orbitals. Thus, in case of octahedral hybridisation, the orbitals may involve sp^3d^2 hybridisation or d^2sp^3 hybridisation.*

4. *Each ligand has at least one orbital containing a lone pair of electrons to overlap with the hybrid orbital of the metal.*

5. *The empty hybrid orbitals of metal ion overlap with the filled orbitals of the ligand to form a covalent sigma bond (L $\longrightarrow$ M).*

6. *In addition to a sigma bond, there is also a possibility of π bond.*

Some common types of hybridisation of atomic orbitals and their geometries that are involved are given in Table 2.1.

Table 2.1. Some common types of hybridisations.

Coordination Number	Hybridisation	Shape	Geometry
2	*sp*	Linear	
3	sp^2	Trigonal planar	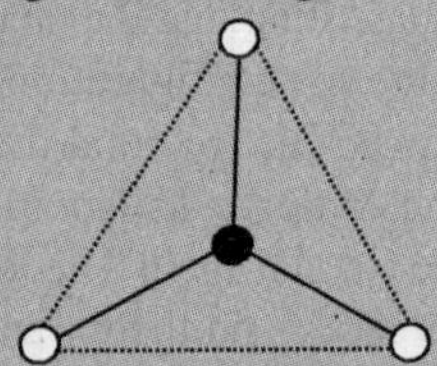
4	sp^3	Tetrahedral	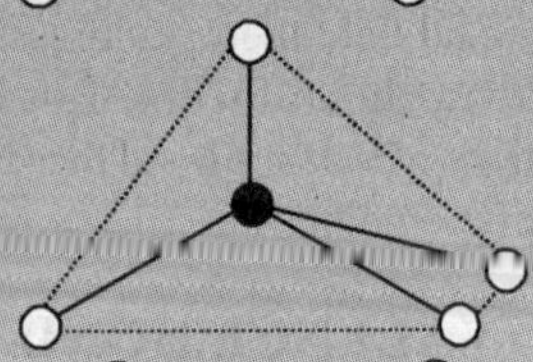
4	dsp^2	Square planar	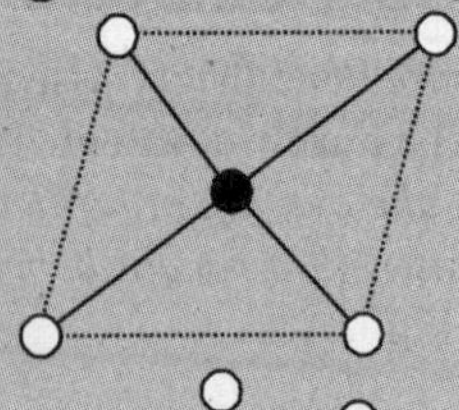
5	dsp^3	Trigonal bipyramidal	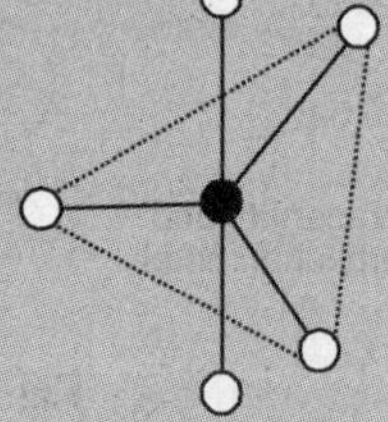
5	sp^3d	Square pyramidal	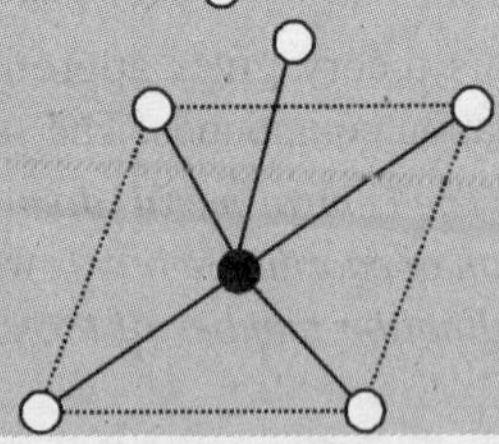

6	d^2sp^3 or sp^3d^2	Octahedral	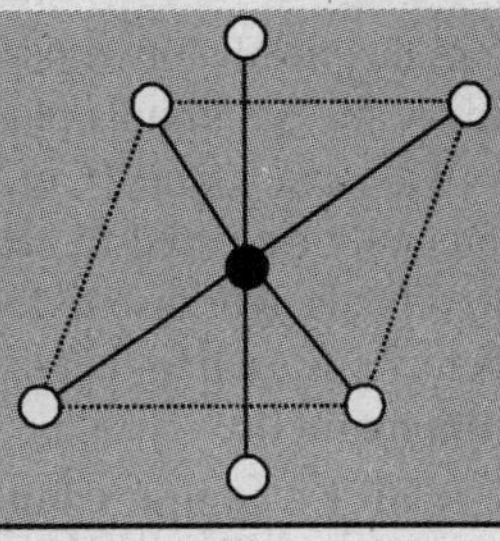

Applications of valence bond theory

We shall illustrate the valence bond theory by taking some examples of complexes.

Complexes of Coordination Number 6

1. $[Cr(NH_3)_6]^{3+}$ complex

Chromium (Z = 24) has the electron configuration $3d^5\ 4s^1$. Chromium in this complex is in +3 oxidation state formed by the loss of one 4s and two of the 3d electrons (Fig. 2.2). The inner *d*-orbitals are already vacant and two vacant 3*d*, one 4*s* and three 4*p* orbitals hybridise to form six d^2sp^3 **hybrid orbitals**. Six pairs of electrons one from each NH_3 molecule, occupy the six hybrid orbitals giving **octahedral geometry**. There are three unpaired electrons in the complex, therefore, we expect it to be paramagnetic.

Configuration of

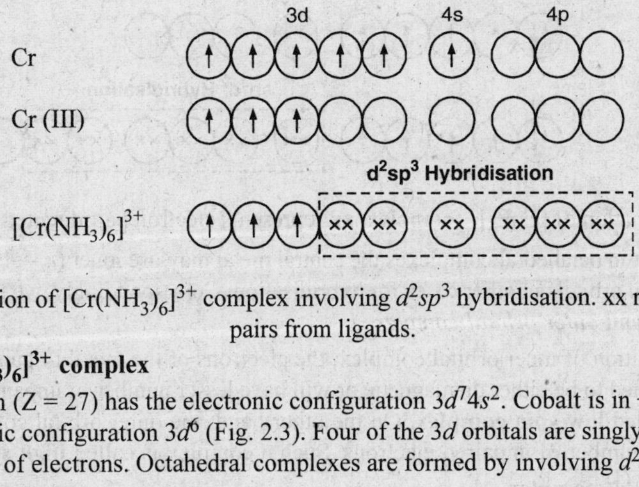

Fig. 2.2. Formation of $[Cr(NH_3)_6]^{3+}$ complex involving d^2sp^3 hybridisation. xx represent electron pairs from ligands.

2. $[Co(NH_3)_6]^{3+}$ complex

Cobalt atom (Z = 27) has the electronic configuration $3d^7 4s^2$. Cobalt is in +3 oxidation state with the electronic configuration $3d^6$ (Fig. 2.3). Four of the 3*d* orbitals are singly filled and one 3*d* orbital has a pair of electrons. Octahedral complexes are formed by involving d^2sp^3 hybridisation.

Configuration of

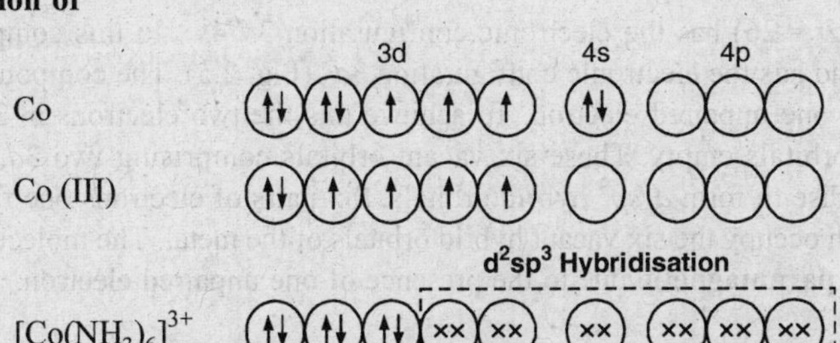

Fig. 2.3. Formation of $[Co(NH_3)_6]^{3+}$ complex ion, xx represents electron pairs from the ligands.

Therefore the metal atom must make available two of its 3*d* orbitals as empty. To achieve this, two electrons in the 3*d* orbitals pair up with two other electrons in 3*d* orbitals making two 3*d* orbitals empty. We expect that $[Co(NH_3)_6]^{3+}$ complex ion will be diamagnetic and this has been experimentally observed to be so. Then, these six vacant orbitals comprising two 3*d*, one 4*s* and three 4*p* orbitals hybridise to form six vacant d^2sp^3 **hybrid orbitals.** Six NH_3 molecules donate one pair of electrons each to these vacant hybrid orbitals. Thus, the complex has **octahedral geometry** and is **diamagnetic.**

3. $[CoF_6]^{3-}$ complex ion

In this complex cobalt is in +3 oxidation state with electronic configuration $3d^6$. This complex has been found to be paramagnetic corresponding to four unpaired dlectrons. This means that it cannot involve d^2sp^3 hybridisation because the complex in that case would be diamagnetic as $[Co(NH_3)_6]^{3+}$.

Huggins (in 1937) solved this puzzle by suggesting that the metal can also use outer *d*-orbitals for hybridisation. Therefore, in the complex ion $[CoF_6]^{3-}$, the 3*d* orbitals are not disturbed *i.e.* they are not paired and the outer 4*d* orbitals are used for hybridisation. The six orbitals comprising one 4*s*, three 4*p* and two 4*d* hybridise forming six sp^3d^2 hybrid orbitals. Six pairs of electrons, one each from F^- ion are donated to the vacant hybrid orbitals forming coordinate bonds as shown in Fig. 2.4.

Configuration of

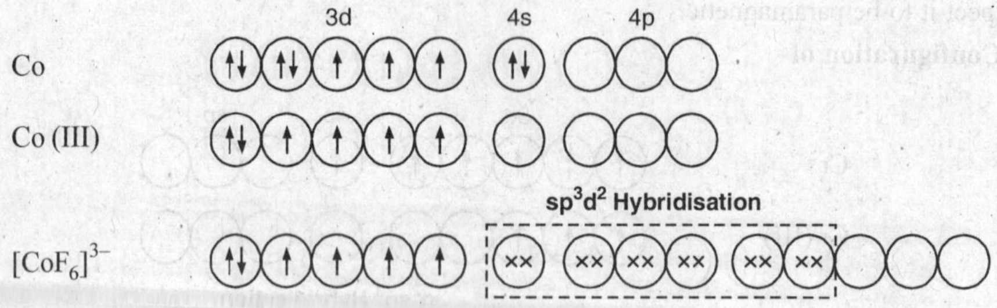

Fig. 2.4. Formation of $[CoF_6]^{3-}$ complex. xx represents electron pair donated by the ligands.

We note that in octahedral complexes the central metal may use inner (*n* —1)d orbitals in some cases and outer *nd* orbitals in other cases for hybridisation. Correspondingly, we use the terms *inner orbital complex* and *outer orbital complex.*

In the formation of inner orbital complex, the electrons of the metal to pair up and hence the complex is expected to be either diamagnetic or will have lesser number of unpaired electrons. Such a complex is called **low spin complex.** On the other hand, the outer orbital complex is expected to have a larger number of unpaired electrons. Such a complex is called **high spin complex.**

4. $[Fe(CN)_6]^{3-}$ complex

Iron atom (Z = 26) has the electronic configuration $3d^64s^2$. In this complex, iron is in +3 oxidation state and has the electronic configuration $3d^5$ (Fig. 2.5). The compound has magnetism corresponding to one unpaired electron. To achieve this the two electrons in 3*d* orbitals pair up leaving two 3*d* orbitals empty. These six vacant orbitals comprising two 3*d*, one 4*s* and three 4*p* orbitals hybridise to form d^2sp^3 hybrid orbitals. Six pairs of electrons one from each CN^- ion are donated which occupy the six vacant hybrid orbitals of the metal. The molecule has **octahedral geometry** and is **paramagnetic** due to the presence of one unpaired electron.

Configuration of

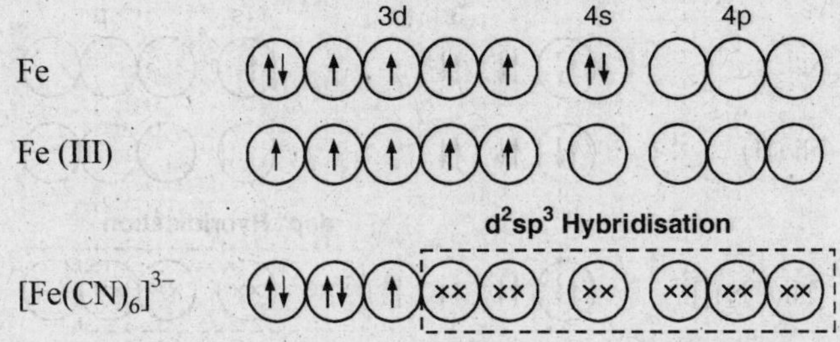

Fig. 2.5. Formation of $[Fe(CN)_6]^{3-}$ complex involving d^2sp^3 hybridisation.

5. $[Fe(H_2O)_6]^{3+}$ ion

In this case also, iron is in +3 oxidation state and has the electronic configuration as $3d^5$ (Fig. 2.6). This complex has been found to be paramagnetic corresponding to five unpaired electrons. We explain it by assuming that the electrons in $3d$-orbitals are not disturbed and the outer $4d$ orbitals are used for hybridisation. As in the case of $[CoF_6]^{3-}$ the six orbitals (one $4s$, three $4p$ and two $4d$) hybridise (sp^3d^2) forming six hybrid orbitals oriented along the corners of an octahedron. Six pairs of electrons, one from each water molecule occupy the six hybrid orbitals. The molecule is **octahedral** and is **paramagnetic,** as explained by Fig. 2.6.

Configuration in

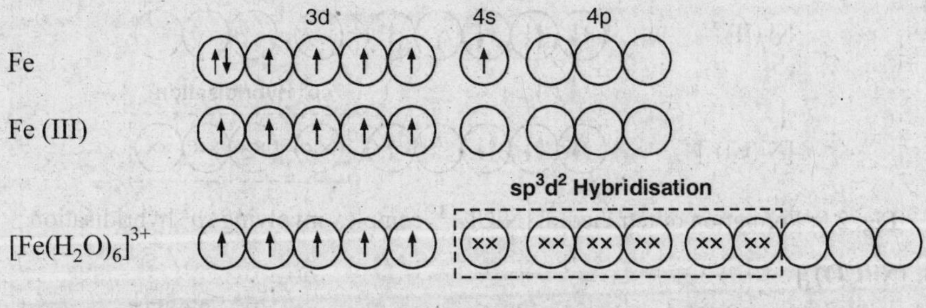

Fig. 2.6. Formation of $[Fe(H_2O)_6]^{3+}$ complex involving sp^3d^2 hybridisation.

Complexes of coordination number 4

There are two ways in which the complexes of coordination number 4 can be formed. If the hybridisation involved is sp^3, we get tetrahedral complexes with the ligands situated at the corners of a tetrahedron. On the other hand, if the hybridisation involved is dsp^2, we get square planar complexes with four ligands occupying four corners of a square. Some examples are discussed below:

1. $[Ni(CN)_4]^{2-}$ ion

The nickel atom has the ground state electronic configuration as $3d^8\ 4s^2$. In this case nickel is in +2 oxidation state and its electronic configuration is $3d^8$ (Fig. 2.7).

The experiments show that the complex $[Ni(CN)_4]^{2-}$ is **diamagnetic** (no unpaired electron) indicating that the hybridisation involved in this case is dsp^2 as represented in Fig. 2.7. Two electrons in $3d$ orbitals pair up to make available one empty $3d$ orbital for dsp^2 hybridisation. Consequently, the structure of the complex would be **square planar**. The empty hybrid orbitals of the metal overlap with the orbitals of cyanide ions, to form metal-ligand coordinate bonds.

Configuration of

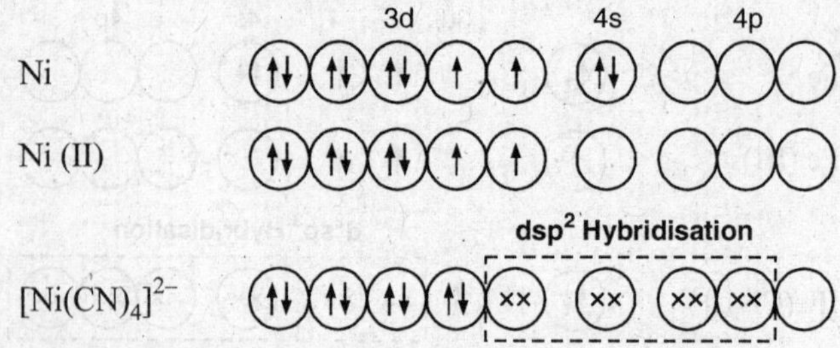

Fig. 2.7. Formation of square planar complex, $[Ni(CN)_4]^{2-}$ involving *dsp*^{2} hybridisation.

2. $[NiCl_4]^{2-}$ ion

The nickel (II) ion has two unpaired electrons, as shown in Fig. 2.8. The magnetic measurements of the complex $[NiCl_4]^{2-}$ show that it is ***paramagnetic*** and *has two unpaired electrons.* Therefore, in this case the 3*d*-orbitals remain undisturbed and *sp*^{3} *hybridisation* occurs resulting in **tetrahedral structure** of the complex. Threre are two unpaired electrons in the complex, therefore it is paramagnetic.

Configuration of

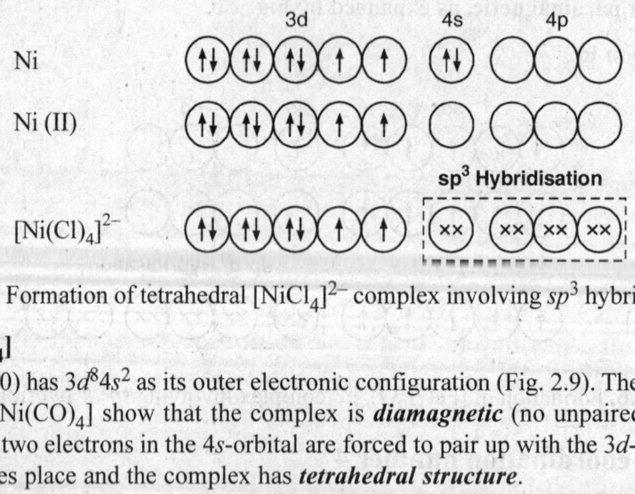

Fig. 2.8. Formation of tetrahedral $[NiCl_4]^{2-}$ complex involving *sp*^{3} hybridisation.

3. $[Ni(CO)_4]$

The nickel (0) has $3d^8 4s^2$ as its outer electronic configuration (Fig. 2.9). The magnetic studies of the complex $[Ni(CO)_4]$ show that the complex is ***diamagnetic*** (no unpaired electrons). This indicates that the two electrons in the 4*s*-orbital are forced to pair up with the 3*d*-orbitals. Thus *sp*^{3} hybridisation takes place and the complex has ***tetrahedral structure***.

Configuration of

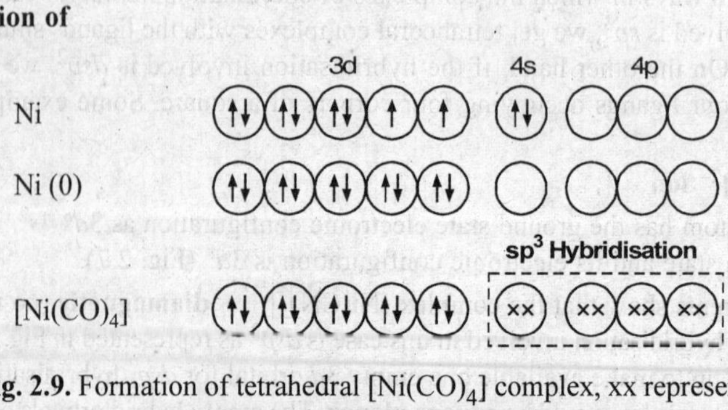

Fig. 2.9. Formation of tetrahedral $[Ni(CO)_4]$ complex, xx represents electron pair donated by ligands.

4. $[Cu(NH_3)_4]^{2+}$ ion

The oxidation state of copper in this complex ion is +2. Therefore, it has $3d^9$ outer configuration (Fig. 2.10). X-ray studies show that this complex ion has a square planar geometry. Therefore, the metal ion must involve dsp^2 hybridisation and one of the $3d$ orbital must be vacated. This can be achieved by promoting one electron from one of the $3d$ orbitals into higher energy vacant $4p$ orbitals. But if the electron occupies higher energy level, it will be easily lost. This means that the complex could be easily oxidised, *i.e.* Cu^{2+} will change into Cu^{3+}. However, this does not happen because Cu^{3+} ions are rare. The spectroscopic studies have also shown that electron is not present in the 4p orbital. To solve the anomaly, it has been suggested that the electrons in $3d$ orbitals remain undisturbed and it involves the use of outer $4d$ orbitals for hybidisation. Thus, this complex is **square planar** involving sp^2d hybrid orbitals.

Configuration of

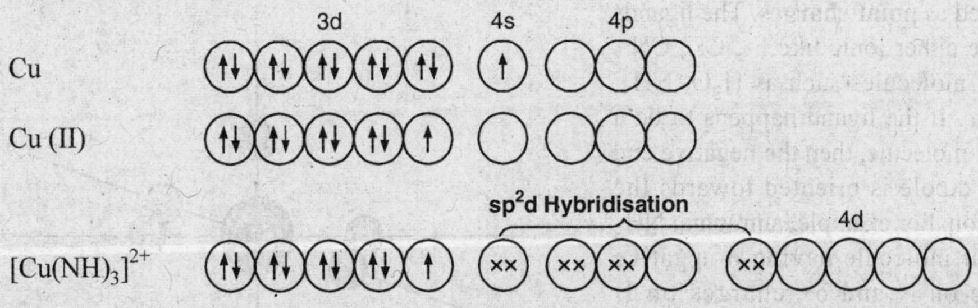

Fig. 2.10. Formation of $[Cu(NH_3)_4]^{2+}$ complex. xx denotes electrons donated by the ligands.

Limitations of valence bond theory

Although valence bond theory has radically improved our understanding of complex compounds still it suffers from many limitations. The main limitations of valence bond theory are:

1. It does not explain the electronic spectra of complexes.
2. It does not predict the relative stabilities of different complexes. Certain complexes are **labile** in which the displacement of ligand takes place rapidly, while some are **inert** in which ligand displacement by another ligand is very slow. Valence bond theory does not explain this.
3. It does not explain the colour of complexes.
4. The valence bond theory does not tell us why water and halide ions commonly form high spin complexes while cyanide ions form low spin complexes.
5. This theory does not explain as to why in some cases inner orbitals and in other cases outer orbitals are used in hybridisation.
6. It is unable to predict the relative stabilities of different ligands.
7. It does not take into account the splitting of *d*-energy levels.

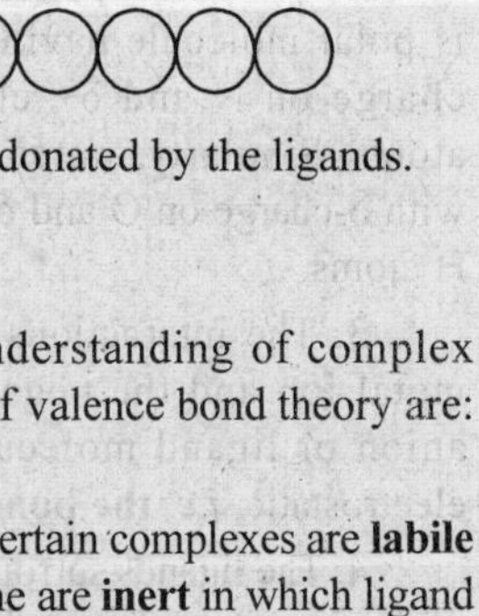

H. Bethe

2.3. CRYSTAL FIELD THEORY (CFT)

The crystal field theory was developed by H. Bethe and V. Bleck. This theory is based on the assumption **that the metal ion and the ligands act as point charges and the interactions between them are purely electrostatic.** In case of negative ligands (anions such as Cl^-, Br^-, CN^-), the interactions with metal ions are ***ion - ion interactions***. If the ligands are neutral molecules (such as

NH_3, H_2O, CO), the interactions with the metal ion are ***ion-dipole interactions***. For example, in the case of complex ion $[CoF_6]^{3-}$ the interactions are between Co^{3+} and F^- ions whereas in $[Co(NH_3)_6]^{3+}$, the interactions are between negatively charged pole of NH_3 molecule and Co^{3+} ion. This is in contrasts the valence bond theory which considers bonding between the metal and ligands as covalent.

The electrons of the metal ion in the environment of ligands within the lattice of a crystal are affected by the non-spherical electric field established by the ligands. The electric field was called the **crystalline field**, but now, it is called as *crystal field*. This is how the theory derived its name.

Basic assumptions (features) of crystal field theory

1. The transition metal ion is surrounded by the ligands with lone pairs of electrons.

2. All types of ligands are regarded as **point charges**. The ligands may be either ionic like F^-, Cl^-, CN^-, *etc.* or molecules such as H_2O, NH_3, CO, *etc.* If the ligand happens to be a neutral molecule, then the negative end of the dipole is oriented towards the metal ion. For example, ammonia, NH_3 is polar molecule having δ-negative charge on N and δ+ charges on H atoms. Similarly, water is also polar with δ-charge on O and δ+ charges on H atoms.

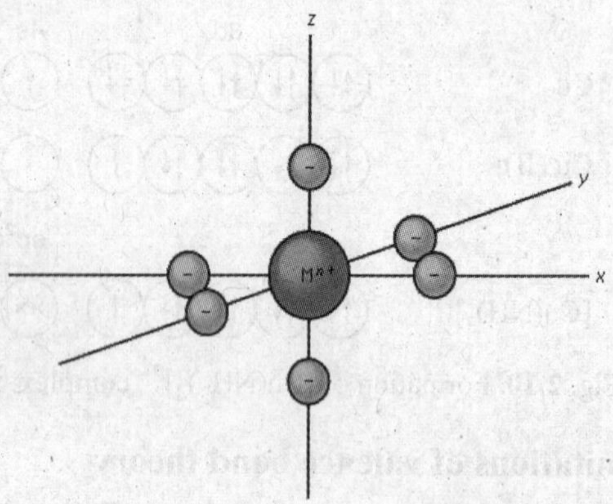

Crystal field

3. The interactions between the metal ion and the negative ends of anion or ligand molecule is purely electrostatic, *i.e.* **the bond between the metal and ligand is considered 100% ionic.**

4. The ligands surrounding the metal ion produce electrical field which influences the energies of the d-orbitals of central metal ion. In a free metal ion, all the five *d*-orbitals have the same energy. These orbitals having the same energies are called ***degenerate orbitals***. This means that an electron can occupy any one of these five *d*-orbitals. However, on the approach of the ligands, the orbital electrons are repelled by the lone pairs of the ligands. The repulsion raises the energy of the *d*-orbitals. Now *d*-orbitals have different orientations and, therefore, these *d*-orbitals will experience different interactions from the ligands. The orbitals lying in the direction of the ligands, will experience greater repulsion and their energies will be raised to greater extent. On the other hand, the orbitals not lying along the approach of the ligands will face smaller interactions and therefore, their energies will be raised to lesser extent.

5. Due to the electrical field of the ligands, the energies of the five *d*-orbitals will split up. This conversion of five degenerate *d*-orbitals of the metal ion into different sets of orbitals having different energies in the presence of electrical field of ligands is called **crystal field splitting**. This concept forms the basis of crystal field theory.

6. We can explain the magnetic properties, spectra and preference for particular geometry in terms of splitting of *d*-orbitals in different crystal fields.

7. The number of ligands (coordination number) and their arrangement (geometry) around the central metal ion will have different effect on the relative energies of the five *d*-orbitals. In other words, the crystal field splitting will be different in different structures with different coordination numbers.

To understand the crystal field theory, it is essential to have a clear picture of the orientation of the five *d*-orbitals in space and the geometrical arrangement of the ligands around the central metal ion. We shall consider the complexes with coordination number 6 and 4, which are very common, to make the theory understandable.

Shapes of *d*-orbitals

There are five *d*-orbitals. These are designated as $d_{xy}, d_{yz}, d_{zx}, d_{x^2-y^2}$ and d_{z^2}. The shapes of these orbitals are given below:

(*i*) The three orbitals d_{xy}, d_{yz} and d_{zx} are similar and each of them consists of four lobes of high electron density lying in *xy*, *yz* and *zx* planes respectively. These lobes lie in between the principal axes. For example, in case of d_{xy} orbital, the four lobes lie in *xy* plane in between the *x* and *y*-axes (Fig. 2.11).

(*ii*) The $d_{x^2-y^2}$ orbital also has four lobes of high electron density along the *x*-axis and *y*-axis. *It may be noted that this orbital is exactly like* d_{xy} *orbital except that it is rotated through* 45° *around the axis.*

(*iii*) The d_{z^2} orbital consists of two lobes along the z-axis with a ring of high electron density in the xy plane.

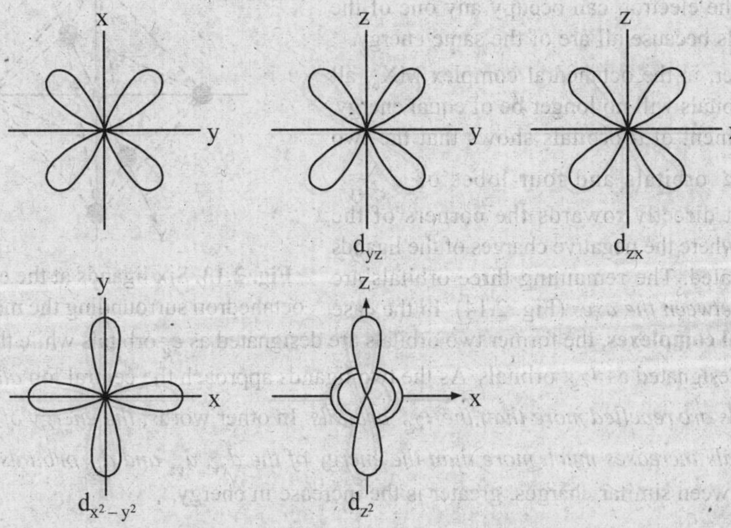

Fig. 2.11. Shapes of five 3*d* orbitals.

There is some thing interesting to be noted about d_{z^2} and $d_{x^2-y^2}$ orbitals. Although d_{z^2} and $d_{z^2-y^2}$ orbitals do not look to be similar yet they are equivalent so far as interactions with the metal ions are concerned. The d_{z^2} orbital may be regarded as a linear combination of two orbitals $d_{z^2-x^2}$ and $d_{z^2-y^2}$, each of which is equivalent to the $d_{x^2-y^2}$ orbital as shown in Fig. 2.12.

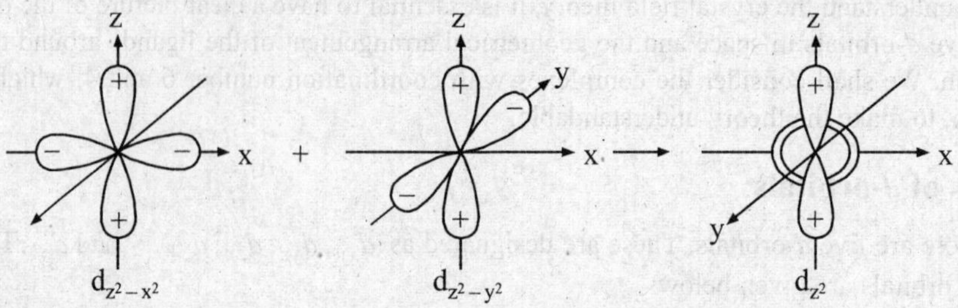

Fig. 2.12. Representation of z^2 orbital as a linear combination of $d_{z^2-x^2}$ and $d_{z^2-y^2}$ orbitals.

However, these two orbitals ($d_{z^2-x^2}$ and $d_{z^2-y^2}$) do not have independent existence and the d_{z^2} orbital can be thought of as having the average properties of the two.

Crystal field Splitting in octahedral complexes (Coordination number 6)

In an octahedral complex the metal ion occupies the centre of the octahedron and the ligands are situated at the six corners as shown in Fig. 2.13. The three axes X, Y and Z point along the corners of the octahedron. Suppose the metal ion, M^{n+}, possesses a single *d*-electron (d^1 configuration, *e.g.* Ti^{3+} in $[TiF_6]^{3-}$). In the free ion, when there are no ligands, the electron can occupy any one of the five *d*-orbitals because all are of the same energy.

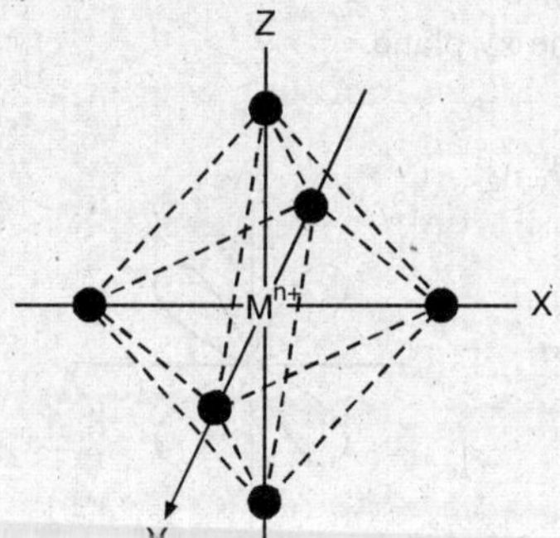

Fig. 2.13. Six ligands at the corners of an octahedron surrounding the metal ion, M^{n+}.

However, in the octahedral complex MX_6, all the five *d*-orbitals will no longer be of equal energy. The arrangement of *d*-orbitals shows that the two lobes of d_{z^2} orbitals and four lobes of $d_{x^2-y^2}$ orbital point directly towards the corners of the octahedron, where the negative charges of the ligands are concentrated. The remaining three orbitals are oriented *in between the axes* (Fig. 2.14). In the case of **octahedral complexes**, the former two orbitals are designated as e_g orbitals while the latter three orbitals are designated as t_{2g} orbitals. As the two ligands approach the central ion *along the axes, the* e_g *orbitals are repelled more than the* t_{2g} *orbitals.* In other words, *the energy of the* d_{z^2} *and* $d_{x^2-y^2}$ *orbitals increases much more than the energy of the* d_{xy}, d_{yz} *and* d_{zx} *orbitals*. Greater the repulsion between similar charges, greater is the increase in energy.

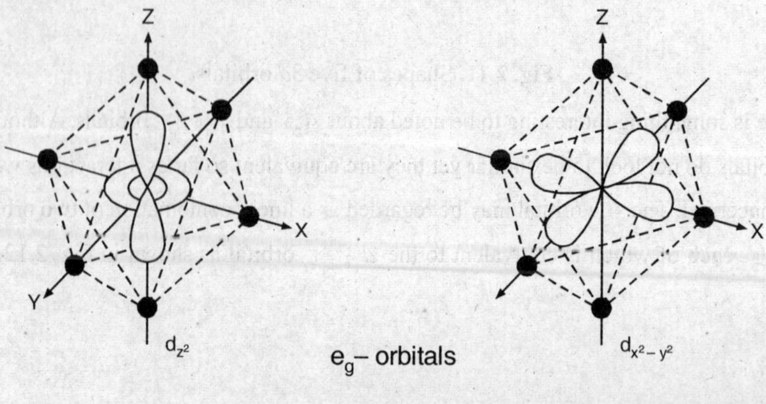

e_g– orbitals

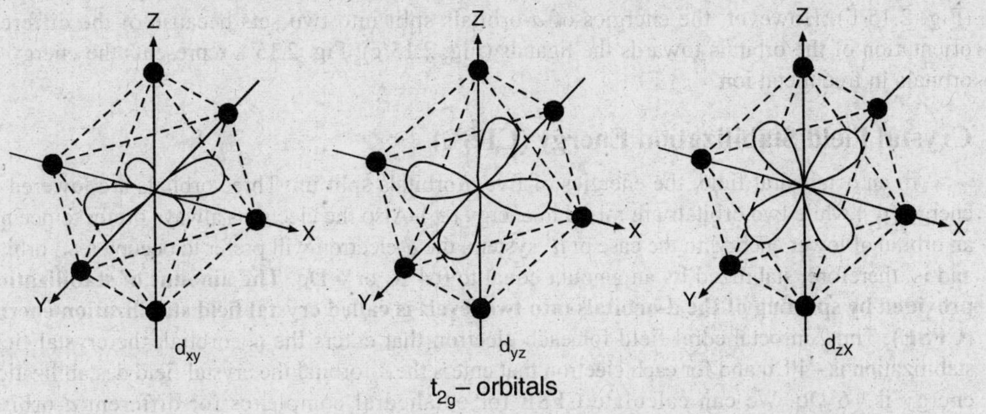

Fig. 2.14. Orientations of different *d*-orbitals in an octahedral field of six ligands.

Thus, in octahedral complexes, the five *d*-orbitals split up into two sets : one set consisting of two orbitals of higher energy (e_g orbitals) and the other set consisting of three orbitals of lower energy (t_{2g} orbitals). This represents crystal field splitting in an **octahedral complex** (Fig. 2.15).

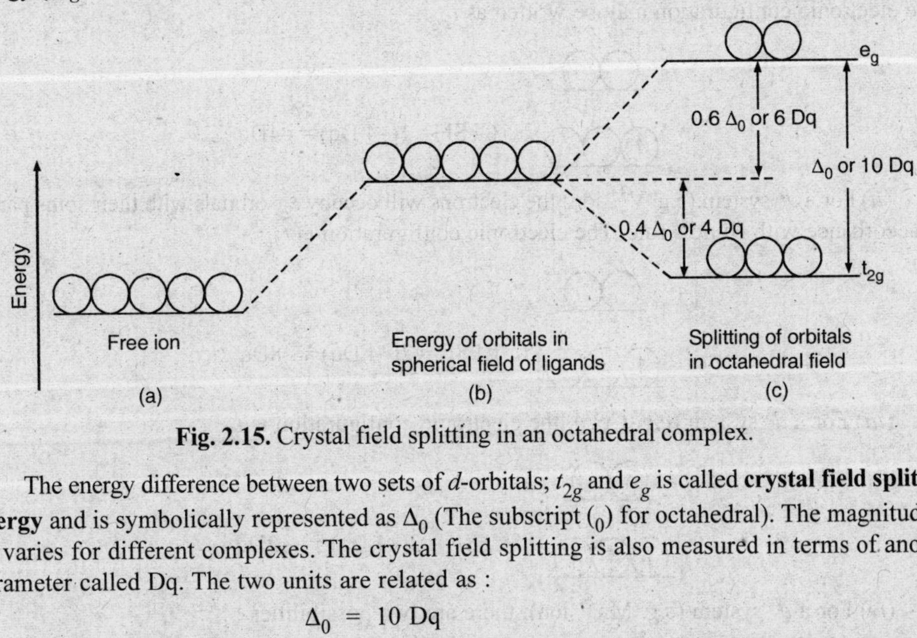

Fig. 2.15. Crystal field splitting in an octahedral complex.

The energy difference between two sets of *d*-orbitals; t_{2g} and e_g is called **crystal field splitting energy** and is symbolically represented as Δ_0 (The subscript ($_0$) for octahedral). The magnitude of Δ_0 varies for different complexes. The crystal field splitting is also measured in terms of another parameter called Dq. The two units are related as :

$$\Delta_0 = 10 \text{ Dq}$$

It may be noted that the crystal field splitting occurs in such a way that the average energy of the *d*-orbitals does not change. This is known as **barycentre rule**, which is similar to *'centre of gravity'* type rule. This requires that the decrease in energy of the set of orbitals that lie at the lower energy level must be balanced by the corresponding increase in other set. Therefore, in terms of total energy 10 Dq, the energy of each of the two e_g orbitals is raised by 6 Dq and the energy of each of the three t_{2g} orbtials is lowered by 4 Dq. In terms of Δ_0, the energy of each of the two e_g orbitals is raised by 3/5 Δ_0 (or 0.6 Δ_0) and the energy of each of the three t_{2g} orbitals is lowered by 2/5 Δ_0 (or 0.4 Δ_0).

The values of 10 Dq or Δ_O are obtained spectroscopically. Initially the energies of all the five *d*-orbitals are raised equally (state II) if the field created by the ligands were spherically symmetrical

(Fig. 2.15 b). However, the energies of *d*-orbitals split into two sets because of the different orientation of the orbitals towards the ligands (Fig. 2.15 c). Fig. 2.15 a represents the energy of orbitals in free metal ion.

Crystal Field Stabilization Energy (CFSE)

In an octahedral field, the energies of five *d*-orbitals split up. Three orbitals are lowered in energy (t_{2g}) while two orbitals are raised in energy (e_g). Also the electrons always prefer to occupy an orbital of lower energy. In the case of d^1 system, the *d*-electron will prefer to occupy a t_{2g} orbital and is, therefore, stabilized by an amount equal to 0.4 Δ_0 or 4 Dq. **The amount of stabilization provided by splitting of the d-orbitals into two levels is called crystal field stabilization energy (CFSE).** Thus, in octahedral field for each electron that enters the t_{2g} orbtial, the crystal field stabilization is –4 Dq and for each electron that enters the e_g orbital the crystal field destabilization energy is +6 Dq. We can calculate CFSE for octahedral complexes for different *d*-orbital configurations.

CFSE for various octahedral complexes

(*i*) For a d^1 system (*e.g.* Ti^{3+} ion) the electron will be present in any one of the three t_{2g} orbitals. The electronic configuration may be written as $t_{2g}{}^1$.

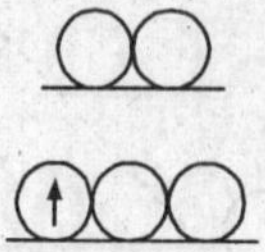

CFSE = 1(–4 Dq) = –4Dq

(*ii*) For a d^2 system (e.g. V^{3+} ion) the electrons will occupy t_{2g} orbitals with their spins parallel in accordance with Hund's rule. The electronic configuration is $t_{2g}{}^2$.

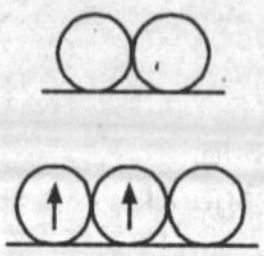

CFSE = 2(–4 Dq) = –8Dq

(*iii*) For a d^3 system (*e.g.* Cr^{3+}), the electronic configuration is $t_{2g}{}^3$.

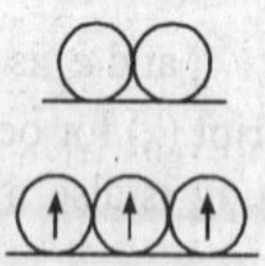

CFSE = 3(–4 Dq) = –12Dq

(*iv*) For a d^4 system (*e.g.* Mn^{3+} ion), there are two possibilities :

(*a*) All the four electrons may occupy t_{2g} orbitals with one electron getting paired with electronic configuration $t_{2g}{}^4$.

(b) Three electrons occupy t_{2g} orbitals and the fourth electron goes to one of the e_g orbitals giving the configuration $t_{2g}{}^3\, e_g{}^1$.

There are two factors that decide the configuration, the CFSE Δ_o and the pairing energy P. While Δ_0 is difference of energy between e_g and t_{2g} orbitals, P is the pairing energy. It is the energy required to pair two electrons together.

The configuration (*a*) is possible if $\Delta_o > P$. These complexes are called **strong field complexes** because Δ_o is large. In this case, the complex has less number of unpaired electrons and is called low **spin complex.** The configuration (*b*) is possible if $\Delta_o < P$. It is called **weak field** and the

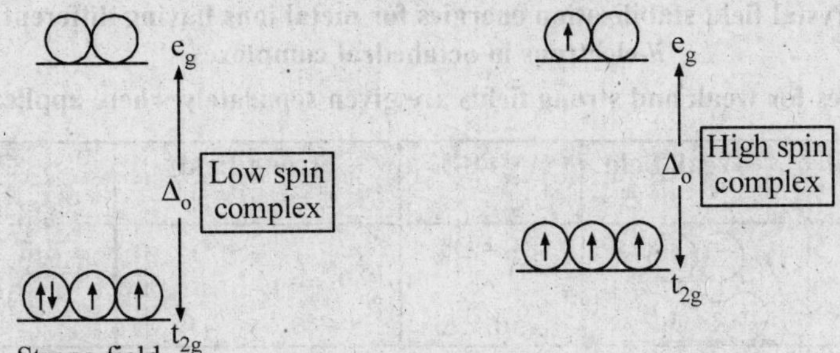

Strong field
$(\Delta_0 > P)$

complexes are called **weak field complexes.** In that case, the maximum number of electrons remain unpaired and the complex is called **high spin complex.** Electrons tend to enter the orbital with smaller energy or involving smaller energy changes. If greater energy is involved in moving to the higher e_g orbital, they prefer to pair within the t_{2g} orbital. On the other hand, if greater energy is involved in pairing of electrons, they prefer to occupy higher energy e_g orbital.

Thus, for a low spin complex having $t_{2g}^{\ 4}$ configuration,

$$\text{CFSE} = 4(-4\text{ Dq}) = -16\text{ Dq}$$

To be precise, we should take into account pairing energy (P) also. Therefore,

$$\text{Net CFSE} = -16\text{ Dq} + \text{P}$$

For a d^4 high spin complex, having the configuration $t_{2g}^{\ 3}\ e_g^{\ 1}$:

$$\text{CFSE} = 3(-4\text{ Dq}) + 1(6\text{ Dq}) = -6\text{ Dq}.$$

(*v*) For a d^5 system (*e.g.* Fe^{3+} ion) again there are two possibilities :

(*a*) All the five electrons may occupy t_{2g} orbitals with two orbitals getting paired. The electronic configuration may be written as $t_{2g}^{\ 5}$. This is possible if $\Delta_0 > P$ and the complex will be called **strong field** or **low spin complex**.

(*b*) The three electrons occupy t_{2g} orbitals and two electrons occupy e_g orbitals. The electronic configuration may be written as $t_{2g}^{\ 3}e_g^{\ 2}$. This will be possible if $\Delta_0 < P$ and the complex is called **weak field** or **high spin complex**.

The CFSE in these two types of complexes will be :

(*a*) **Strong field : $\Delta_o > P$**

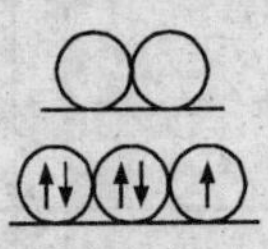

$$\text{CFSE} = 5(-4\text{ Dq}) + 2\text{P}$$
$$= -20\text{ Dq} + 2\text{P}$$

(*b*) **Weak field : $\Delta_o < P$**

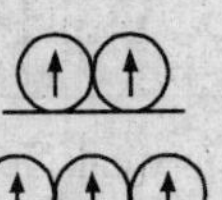

$$\text{CFSE} = 3(-4\text{ Dq}) + 2(6\text{ Dq})$$
$$= 0$$

The CFSE for metal ion for different d^n configurations are given in Table 2.2. It may be noted that d^8, d^9 and d^{10} systems have only one possible configuration.

Table 2.2. Crystal field stabilization energies for metal ions having different numbers of *d*-electrons in octahedral complexes.
(Values for weak and strong fields are given separately where applicable)

Number of electrons	Weak field	CFSE	Strong field	CFSE
d^1	( )( ) (↑)( )( )	– 4 Dq or – 0.4 Δ_0		
d^2	( )( ) (↑)(↑)( )	– 8 Dq or – 0.8 Δ_0		
d^3	( )( ) (↑)(↑)(↑)	– 12 Dq or – 1.2 Δ_0		
d^4	(↑)( ) (↑)(↑)(↑)	– 6 Dq or – 0.6 Δ_0	( )( ) (↑↓)(↑)(↑)	– 16 Dq + P or – 1.6 Δ_0 + P
d^5	(↑)(↑) (↑)(↑)(↑)	0	( )( ) (↑↓)(↑↓)(↑)	– 20 Dq + 2P or – 2.0 Δ_0 + 2P
d^6	(↑)(↑) (↑↓)(↑)(↑)	– 4 Dq or – 0.4 Δ_0	( )( ) (↑↓)(↑↓)(↑↓)	– 24 Dq + 2P or – 2.4 Δ_0 + 2P
d^7	(↑)(↑) (↑↓)(↑↓)(↑)	– 8 Dq or – 0.8 Δ_0	(↑)( ) (↑↓)(↑↓)(↑↓)	– 18 Dq + P or – 1.8 Δ_0 + P
d^8	(↑)(↑) (↑↓)(↑↓)(↑↓)	– 12 Dq or – 1.2 Δ_0	(↑)(↑) (↑↓)(↑↓)(↑↓)	– 12 Dq or – 1.2 Δ_0
d^9	(↑↓)(↑) (↑↓)(↑↓)(↑↓)	– 6 Dq or – 0.6 Δ_0		
d^{10}	(↑↓)(↑↓) (↑↓)(↑↓)(↑↓)	0		

The systems d^1, d^2, d^3, d^9 and d^{10} have same configuration in weak field and strong field and, therefore, have same CFSE values.

Crystal field splitting in tetrahedral complexes (Coordination number 4)

The tetrahedral arrangement of four ligands surrounding a metal ion may be visualized by placing on the four of the eight corners of the cube (Fig. 2.16). The directions X, Y and Z point to the centre of the faces of the cube. From the figure, we find that the orbitals d_{xy}, d_{yz} and d_{zx} point between X, Y and Z-axis (*i.e.* towards the centre of the edes of the cube) whereas the orbitals $t_{x^2-y^2}$ and d_{z^2} point along X, Y and Z axis (*i.e.* to the centres of the faces). Clearly in the tetrahedral field, *none of the d-orbitals point exactly towards the ligands and therefore, smaller splitting of energy will take place compared to that in octahedral field.* The three *d*-orbitals d_{xy}, d_{yz} and d_{zx} point close to the direction in which the ligands are approaching while the two orbitals $d_{x^2-y^2}$ and d_{z^2} lie in between the ligands. Therefore, the energies of the three orbitals will be raised while the energies of the two orbitals will be lowered. Thus, in the presence of tetrahedral field, the splitting of *d*-orbitals will take place as under.

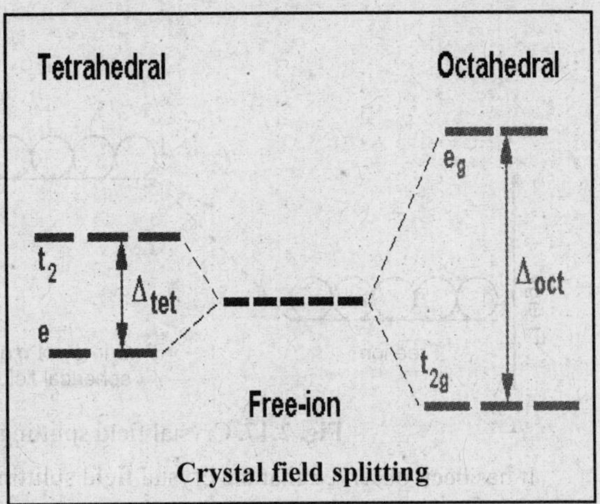

Crystal field splitting

(*i*) The two orbitals, $d_{x^2-y^2}$ and d_{z^2} become stable and their energies are lowered. These are designated as *e* orbitals.

(*ii*) The three orbitals d_{xy}, d_{yz} and d_{zx} become unstable and their energies are raised. These are designated as t_2 orbitals.

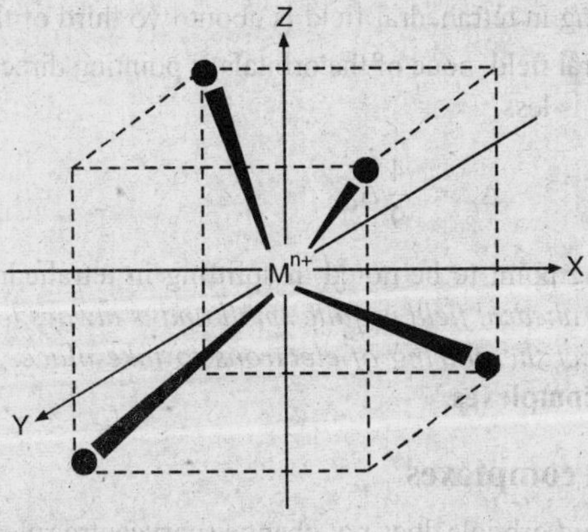

Fig. 2.16. Tetrahedral field of four ligands around the central metal ion.

A point of distinction between splitting of *d*-orbitals in tetrahedral and octahedral fields is that '*g*' is not used. For example, the orbitals $d_{x^2-y^2}$ and d_{z^2} are designated as *e* orbitals whereas the other three orbitals d_{xy}, d_{yz} and d_{zx} are designated as t_2. This is because a tetrahedral geometry has no centre of symmetry. The symbols *g* and *u* have relevance only for fields which have centre of symmetry.

Splitting of *d*-orbitals in tetrahedral field is shown in Fig. 2.17.

The magnitude of crystal field splitting which is the difference between *e* and t_2 orbitals is designated as Δ_t (the subscript *t* indicating tetrahedral complexes). It is also measured in terms of Dq as

$$\Delta_t = 10\ Dq$$

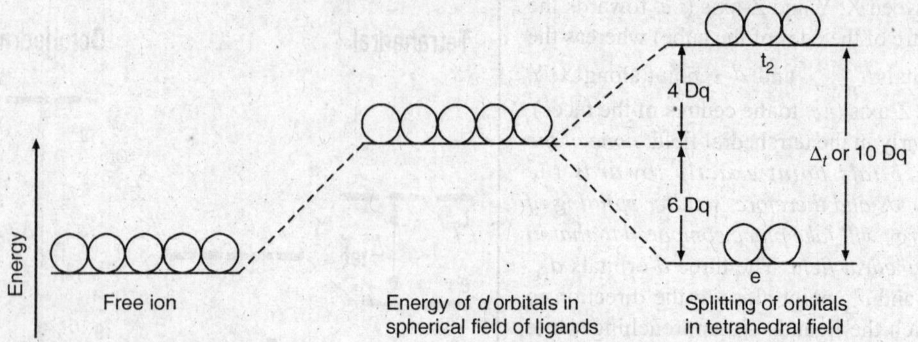

Fig. 2.17. Crystal field splitting in a tetrahedral field.

It has been observed that the crystal field splitting in tetrahedral complexes (Δ_t) is much less than in octahedral complexes (Δ_o). The relation between the two (for the same metal ion and ligands) is

$$\Delta_t = \frac{4}{9}\Delta_o$$

There are two main reasons for the smaller value of Δ_t than Δ_o

(*i*) In tetrahedral complexes there are four ligands while there are six ligands in octahedral complexes. Therefore, lesser ligands produce less crystal field splitting. On this account crystal field splitting in tetrahedral field is about two third of the octahedral field.

(*ii*) Further in tetrahedral field, none of the orbitals is pointing directly towards the ligands and, therefore, splitting is less.

Precisely $$\Delta_t = \frac{4}{9}\Delta_o$$

There is an interesting point to be noted in splitting in tetrahedral field. *The magnitude of crystal field splitting in tetrahedral field is quite small and is always less than the pairing energy. Therefore, we do not expect the pairing of electrons to take place. Thus,* **all the tetrahedral complexes are high spin complexes.**

CFSE for tetrahedral complexes

Average energy of the *d*-orbtials does not change (barycentre rule). Therefore, in terms of 10 Dq the energy of each of the *e* orbital is lowered by 6 Dq and the energy of each of the t_2 orbital is raised by 4 Dq. In terms of Δ_t, the energy of each *e* orbital is lowered by 3/5 Δ_t (or 0.6 Δ_t) while each of the t_2 orbital is raised by 2/5 Δ_t (or 0.4 Δ_t). We calculate the value of Δ_t for different configurations as under.

(*i*) For d^1 system, the electronic configuration is e^1.

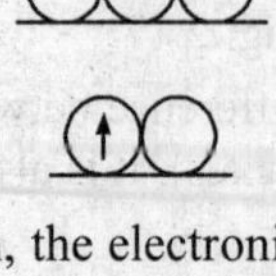

CFSE = 1(– 6 Dq) = – 6Dq

(*ii*) For d^2 system, the electronic configuration is e^2.

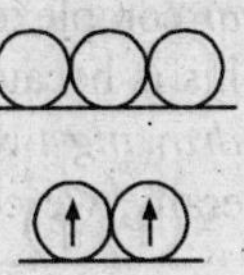

CFSE = 2(– 6 Dq) = – 12Dq

(*iii*) For d^3 system, the electronic configuration is $e^2\ t_2^{\ 1}$. As already discussed, crystal field splitting in tetrahedral field is quite small and is always less than the pairing energy. Therefore, pairing does not occur and most of the complexes are high spin complexes, whether it is a weak field or a strong field.

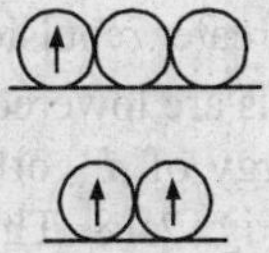

CFSE = 2(– 6 Dq) + 1(4 Dq) = – 8Dq

The CFSE for compelxes with metal ion having different configurations of *d*-orbitals in tetrahedral field are given in Table 2.3.

Table 2.3. Crystal field stabilization energies for tetrahedral complexes with metal ions for different d^n.

d^1	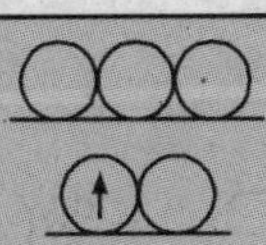	CFSE = – 6Dq	d^2	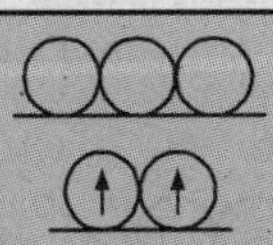	CFSE = – 12Dq
d^3	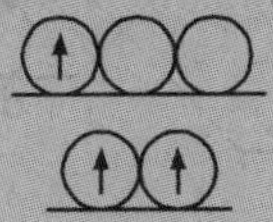	CFSE = – 8Dq	d^4	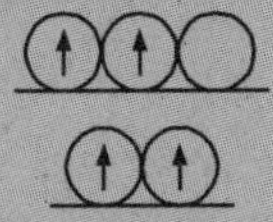	CFSE = – 4Dq
d^5	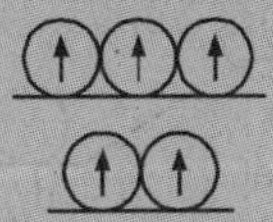	CFSE = 0	d^6	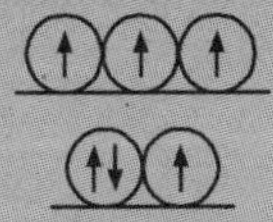	CFSE = – 6Dq
d^7	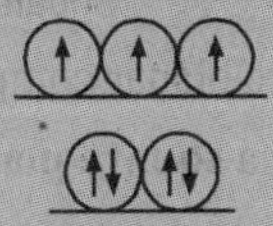	CFSE = – 12Dq	d^8	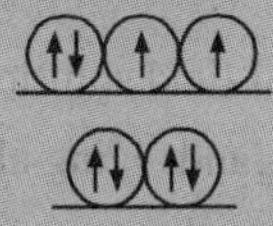	CFSE = – 8Dq
d^9	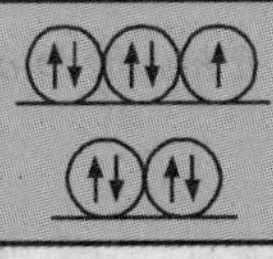	CFSE = – 4Dq	d^{10}	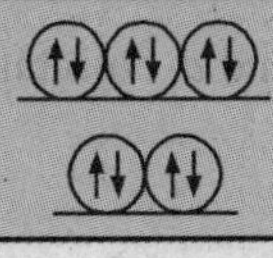	CFSE = 0

Precisely, the pairing energy values should also be taken into consideration. However, to simplify the matter, these have been omitted here.

Crystal field splitting in tetragonal and square planar complexes

Let us first clarify what we mean by tetragonal and square planar complexes. When all the ligands are equidistant from the central metal ion, it is a case of octahedral complex (Fig. 2.13.) However, if some ligands, say two ligands along z-axis are at a different length, then it will become a case of tetragonal complex. In square planar complexes, the four ligands are situated at the corners of a square.

The splitting of *d*-orbitals in *tetragonal* and *square planar* complexes can be visualised from the known splitting of *d*-orbitals in octahedral complexes. This is because tetragonal and square planar geometries can be understood by stretching or *withdrawing two trans ligands from an octahedral complex.* This process is called **elongation**. Generally, we consider the stretching or removal of trans ligands along the Z-axis.

As the ligands lying on the Z-axis are moved away, the ligands in the XY plane are drawn closer to the central ion. Due to increase in metal ligands bond along Z-axis, the repulsions from the ligands to electrons in d_{z^2} orbital decreases and therefore, the energy of $d_z{}^2$ obital is decreased relative to that in octahedral field. At the same time, the metal ligand bond along X and Y are shortened so that the *d*-orbital in XY plane, $d_{x^2-y^2}$ experiences greater repulsion from the ligands and therefore, its energy is raised. Similarly, the d_{xz} and d_{yz} orbitals are lowered in energy because of decrease in repulsion effects along the Z-axis while the energy of d_{xy} orbital is raised. The resulting splitting considering the above forces is shown in Fig. 2.18. This state represents **tetragonally distorted octahedral** structure.

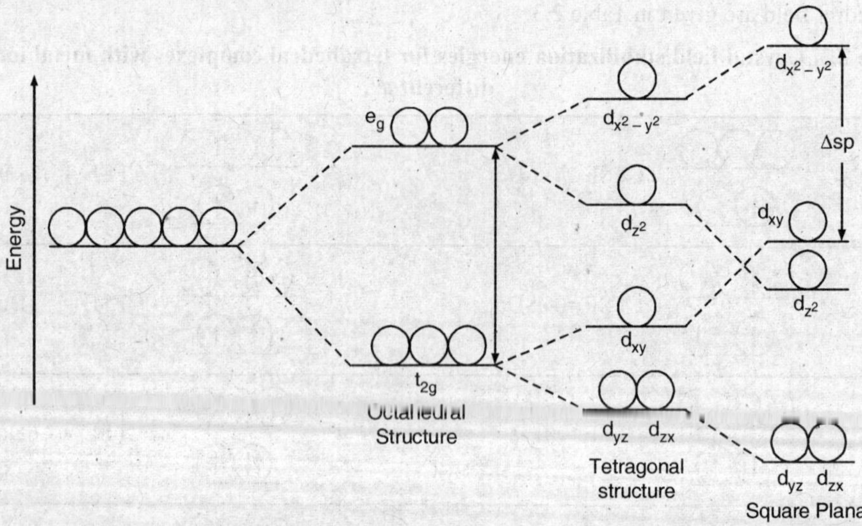

Fig. 2.18. Crystal field splitting in octahedral, tetragonal and square planar complexes.

When the trans ligands lying along Z-axis are completely removed, a **square planar** complex is formed. In the square planar complex, the energies of $d_{x^2-y^2}$ and d_{xy} orbitals rise further and energies of d_{z^2}, d_{xz} and d_{yz} orbitals fall further as shown in Fig. 2.18. *It has been observed that*

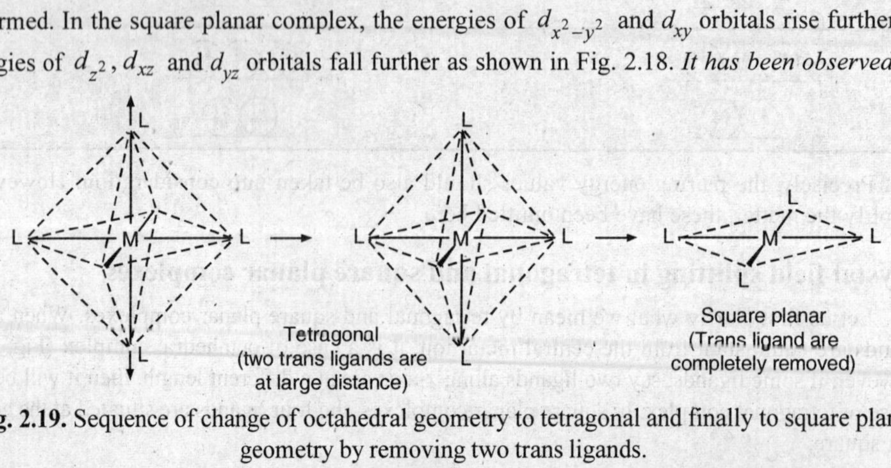

Fig. 2.19. Sequence of change of octahedral geometry to tetragonal and finally to square planar geometry by removing two trans ligands.

in square planar complex, the energy of d_{z^2} orbital falls even below the d_{xy} orbital. The change of an octahedral complex to tetragonally distorted octahedron and finally to square planar arrangement is shown in Fig. 2.19.

Tetragonally distorted structure can be obtained in two ways, by pulling apart trans ligands from the metal ion and by bringing the trans ligands closer to the metal ion Fig. 2.20 (*a*) shows elongation of octahedron. In flattening of octahedron (Fig. 2.20 (*b*), the two trans M—L bonds are shortened while other four M–L bonds become large. The splitting of energies of *d*-orbitals will be reverse of that in the case of elongation.

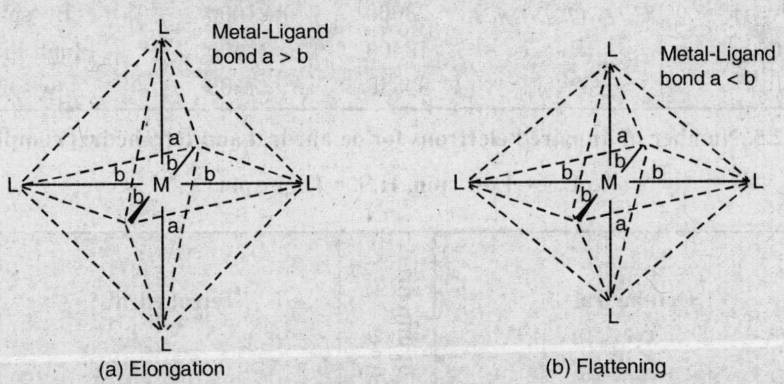

Fig. 2.20. Tetragonally distorted octahedron (*a*) elongation (*b*) flattening.

Crystal field theory and magnetic properties of complexes

Crystal field theory has helped understand the magnetic properties of coordination compounds in a sound manner. The substances which have all paired electrons are called diamagnetic while the substances which contain unpaired electrons are called paramagnetic substances. The number of unpaired electrons, in a given complex of known geometry can be easily predicted if we know whether the complex is low spin or high. Magnitude of Δ_0 and pairing energy P play an important role in this direction.

(*i*) If $P > \Delta_0$, the electrons will not pair up and the complex will be **high spin complex.**

(*ii*) If $\Delta_0 > P$, the electrons will prefer to pair up and the complex will be **low spin complex.**

Thus, the complexes with **weak ligand field** are **high spin complexes** (paramagnetic) and those with **strong ligand field** are **low spin complexes** (diamagnetic or low magnetic character).

This generalisations could not be provided by valence bond theory. Consider the following two complexes of cobalt (III). From experiments, it has been observed that the complex $[Co(NH_3)_6]^{3+}$ is **diamagnetic** while the complex $[CoF_6]^{3-}$ is **paramagnetic.**

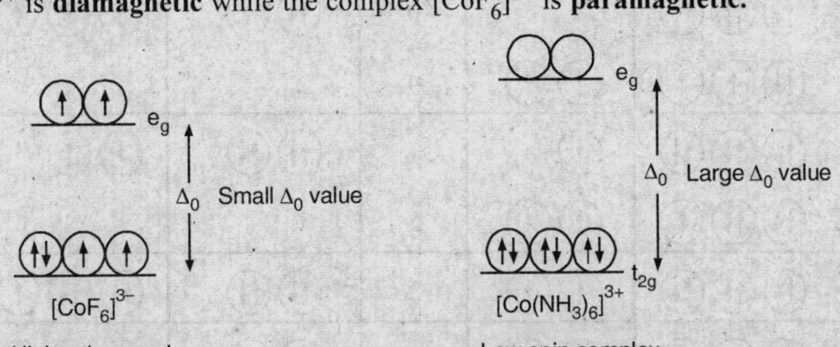

Fig. 2.21. Crystal field splitting and explanation for $[CoF_6]^{3-}$ and $[Co(NH_3)_6]^{3+}$ complexes.

Cobalt in $[Co(NH_3)_6]^{3+}$ and $[CoF_6]^{3-}$ is in +3 oxidation state and it has d^6 configuration. It has been observed that F^- ion is a weak field ligand and therefore Δ_o is small. Therefore, Δ_o will be less than P and the electron will remain unpaired as far as possible. This is obviously because greater energy is required in pairing and smaller energy is required in moving to orbital of higher

Table 2.4. CFSE (Δ_o) and pairing energies (P) for some complexes.

Complex	Configuration	Δ_o (cm^{-1})	P (cm^{-1})	High spin/low spin
$[CoF_6]^{3-}$	d^6	13000	21000	high spin
$[Co(NH_3)_6]^{3+}$	d^6	23000	21000	low spin
$[Fe(H_2O)_6]^{2+}$	d^6	10400	17600	high spin
$[Fe(CN)_6]^{4-}$	d^6	32850	17600	low spin

Table 2.5. Number of unpaired electrons for octahedral and tetrahedral complexes.

L.S = Low spin, H.S. = High spin

Electronic Configuration		Octahedral		No. of unpaired electrons	Tetrahedral		No. of unpaired electrons
		t_{2g}	e_g		e	t_2	
d^1		(↑)		1	(↑)		1
d^2		(↑)(↑)		2			2
d^3		(↑)(↑)(↑)		3	(↑)(↑)	(↑)	3
d^4	L.S.	(↑↓)(↑)(↑)		2	(↑)(↑)	(↑)(↑)	4
	H.S.	(↑)(↑)(↑)	(↑)( )	4			
d^5	L.S.	(↑↓)(↑↓)(↑)		1	(↑)(↑)	(↑)(↑)(↑)	5
	H.S.	(↑)(↑)(↑)	(↑)(↑)	5			
d^6	L.S.	(↑↓)(↑↓)(↑↓)		0	(↑↓)(↑)	(↑)(↑)(↑)	4
	H.S.	(↑↓)(↑)(↑)	(↑)(↑)	4			
d^7	L.S.	(↑↓)(↑↓)(↑↓)	(↑)	1	(↑↓)(↑↓)	(↑)(↑)(↑)	3
	H.S.	(↑↓)(↑↓)(↑)	(↑)(↑)	3			
d^8		(↑↓)(↑↓)(↑↓)	(↑)(↑)	2	(↑↓)(↑↓)	(↑↓)(↑↓)(↑)	2
d^9		(↑↓)(↑↓)(↑↓)	(↑↓)(↑)	1	(↑↓)(↑↓)	(↑↓)(↑↓)(↑)	1
d^{10}		(↑↓)(↑↓)(↑↓)	(↑↓)(↑↓)	0	(↑↓)(↑↓)	(↑↓)(↑↓)(↑↓)	0

energy. The complex will be high spin. As seen from Fig. 2.21 (*a*), the complex is paramagnetic due to the presence of four unpaired electrons. On the other hand, NH_3 is a strong field ligand and therefore, Δ_o will be greater than P. As a result, the electrons *pair up* and this results in a *low spin* complex. This is because smaller energy is required in pairing and greater energy is required in moving to orbital of higher energy. This happens in the case of $[Co(NH_3)_6]^{3+}$, as represented in Fig. 2.21 (*b*). The crystal field splitting energies (Δ_o) and pairing energies for some complexes are given in Table 2.4.

We can estimate the number of unpaired electrons in a given complex if we know whether the complex is a High spin (H.S.) or a Low spin (L.S.) complex. The number of unpaired electrons for tetahedral and tetrahedral complexes for different configurations are given in Table 2.5. for tetrahedral complexes, only high spin complexes have been considered because D_t is always Less than P.

1.4. FACTORS DETERMINING THE MAGNITUDE OF CRYSTAL FIELD SPLITTING

A number of factors influence the magnitude of crystal field splitting. These are discussed as under:

1. Nature of the ligand (spectrochemical series)

Greater the ease with which the ligands can approach the metal ion, greater will be the crystal field splitting caused by it.

(*i*) Small ligands can cause greater crystal field splitting because they can approach the metal ion closely. For example, F^- ion causes greater splitting and gives higher Δ_o value than large Cl^- and Br^- ions.

(*ii*) The ligands containing easily polarizable electron pair will be drawn more closely to the metal ion and cause greater splitting.

(*iii*) The ligands which have greater tendency to form multiple bonds such as CN^- and NO_2^- cause greater crystal field splitting. High crystal field splitting produced by CN^- ligand is attributed to the π bonding in metal. The metal donates electrons from a filled t_{2g} orbital into the vacant orbital on the ligand forming a stable metal-ligand bond.

(*iv*) The following pattern of increasing σ donation is observed :

halide donors < O donors < N donors < C donors

Table 2.6 gives the crystal field splitting (Δ_o) with different ligands attached to Cr (III).

Table 2.6. Crystal field splitting with different ligands attached to Cr(III)

Complex	Ligand	Donor atom	$\Delta_0(cm^{-1})$
$[CrCl_6]^{3-}$	Cl^-	Cl	13640
$[CrF_6]^{3-}$	F^-	F	15200
$[Cr(H_2O)_6]^{3+}$	H_2O	O	17830
$[Cr(NH_3)_6]^{3+}$	NH_3	N	21600
$[Cr (en)_3]^{3+}$	en	N	21900
$[Cr(CN)_6]^{3-}$	CN^-	C	26280

The ligands are arranged in increasing order of crystal field splitting Δ as given below. This regular order is called **spectrochemical series.**

Weak field ligands

$I^- < Br^- < Cl^- < NO_3^- < F^- < OH^- < ox^{2-} < H_2O < EDTA < py \approx NH_3 < en < dipy < o\text{-}phen < NO_2^- < CN^- < CO$

Strong field ligands

Thus CO, CN^- and NO_2^- are strong field ligands whereas I^-, Br^- and Cl^- are weak field ligands.

2. Oxidation state of the metal ion

This is another factor controlling the magnitude of the crystal field splitting. The metal ion with higher oxidation state causes larger crystal field splitting than with lower oxidation state. For example, the crystal field splitting energy Δ_o for $[Co(H_2O)_6]^{3+}$ complex in which the oxidation state of cobalt is +3, is 18,600 cm^{-1}, whereas for the complex $[Co(H_2O)_6]^{2+}$ in which the oxidation state of Co is +2, Δ_o = 9,300 cm^{-1}.

Crystal field splitting energies (Δ_o) for hexaaqua complexes of M^{2+} and M^{3+} ions are given in Table 2.7.

Table 2.7. Crystal field splitting for hexaaqua complexes of M^{2+} and M^{3+} ions

Metal ion	Oxidation state	Δ_o (cm^{-1})	Oxidation state	Δ_o (cm^{-1})
V	II	11800	III	18000
Cr	II	13900	III	17830
Mn	II	7800	III	21000
Fe	II	10400	III	13700
Co	II	9300	III	18600

3. Type of *d*-orbitals (transition series)

The extent of crystal field splitting for similar complexes of a metal in the same oxidation state increases by about 30 to 50% on going from 3*d*-series to 4*d* series. The increase is almost of the same amount (30–50%) on going from 4*d*-series to 5*d*-series.

This is because *4d-orbitals being bigger extend more into space in comparison to 3d-orbitals. As a result, the 4d-orbitals can interact more strongly with the ligands resulting in greater crystal field splitting. On similar considerations 5d orbitals of the metal ion produce greater crystal field splitting than the 4d orbitals.* This is supported by Table 2.8 which gives CFS for the second transition series metals.

Table 2.8. Crystal field stabilisation energies for second transition series metals

Complex ion	Electronic configuration	Δ_o (cm^{-1})
$[Co(NH_3)_6]^{3+}$	$3d^6$	23,000
$[Rh(NH_3)_6]^{3+}$	$4d^6$	34,000
$[Ir(NH_3)_6]^{3+}$	$5d^6$	41,000

4. Geometry of the complex

We have seen earlier that the crystal field splitting energy of tetrahedral complexes (Δ_t) is much lower the value (Δ_o) for octahedral complexes ($\Delta_t \approx 4/9\ \Delta_o$) due to smaller metal-ligand interactions. The value of splitting energy for tetrahedral complexes, in general, is small as compared to the pairing energy P. The tetrahedral complexes are, therefore, mostly *high spin complexes.*

2.5. COLOUR OF TRANSITION METAL COMPLEXES

The transition metals have the property to absorb certain radiations from the visible region of the spectrum and as a result, the transmitted or reflected light shows complementary colour. The visible light consists of radiations of different wavelengths ranging from blue (about 400 nm) to red (about 700 nm). If a substance absorbs blue light, then it appears to us to be red (red is complementary colour of blue light).

In the case of transition metal complexes, the energy difference between t_{2g} and e_g sets of *d*-orbitals is small. When visible light falls on them, the electron gets raised from lower set of orbitals to higher set of orbitals by absorbing light of some wavelength. As a result of absorption of some selected wavelength of visible light corresponding to energy difference between the two energy levels, the transmitted light gives colour to complexes.

The amount of energy absorbed can be calculated from the relation,

$$E = h\nu = \frac{hc}{\lambda}$$

where *c* in the velocity of light and *h* is Planck's constant. We shall explain it by taking the example of the complex $[Ti(H_2O)_6]^{3+}$. This complex is purple.

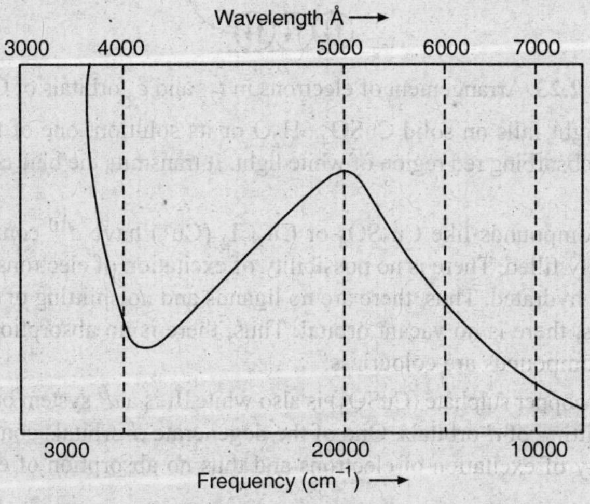

Fig. 2.22. Absorption spectrum of $[Ti(H_2O)_6]^{3+}$.

In this complex, the metal ion has d^1 configuration. In ground state this electron occupies one of the t_{2g} orbitals. The visible spectrum of $[Ti(H_2O)_6]^{3+}$ is shown in Fig. 2.22. We can see from the figure that absorption takes place at 5000 Å or 20,00 cm^{-1}. The energy corresponding to this wavelength can easily be calculated as :

$$E = N_0 h\nu = \frac{N_0 hc}{\lambda}$$

where N_0 = Avogadro number, h = Planck's constant, *c* = Velocity of light

$$E = \frac{6.023 \times 10^{23} \times 6.626 \times 10^{-34} \times 3.0 \times 10^{8}}{5000 \times 10^{-10}} \quad \text{(S.I. units)}$$

$$= 239 \text{ kJ mol}^{-1}$$

This corresponds to Δ_o for $[Ti(H_2O)_6]^{3+}$ complex ion.

The colours corresponding to this wavelength are green and yellow lights which are absorbed from the white light, while the *blue and red poritons are emitted*. The solution of complex $[Ti(H_2O)_6]^{3+}$, therefore, looks purple.

The difference between t_{2g} and e_g orbitals in octahedral complexes depends on the metal ion and the nature of the ligands. Therefore, different complexes absorb different amounts of energies from visible region and transmit different colours.

That is why $[Co(H_2O)_6]^{3+}$ absorbs orange colour and appears blue and $[Co(CN)_6]^{3-}$ absorbs violet colour and appears as yellow.

Colour of copper sulphate

(*i*) Copper sulphate hydrated $CuSO_4.5H_2O$ may be considered as a complex compound with five water molecules as ligands. Copper is present as Cu^{2+} in this compound, with d^9 configuration. In the presence of ligands, the *d*-orbitals are split into two types t_{2g} and e_g. The filling of orbitals is shown in Fig. 2.23.

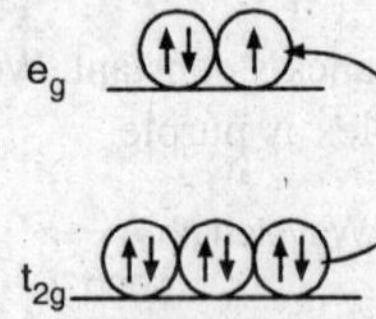

Fig. 2.23. Arrangement of electrons in t_{2g} and e_g orbitals of Cu^{2+}.

When visible light falls on solid $CuSO_4.5H_2O$ or its solution, one of the electrons in t_{2g} is excited to e_g orbital, absorbing red region of white light. It transmits the blue colour and hence looks blue.

(*ii*) Cuprous Compounds like Cu_2SO_4 or Cu_2Cl_2 (Cu^+) have d^{10} configuration. All the *d*-orbitals are completely filled. There is no possibility of excitation of electrons. It may be noted that cuprous salts are not hydrated. Thus, there are no ligands and no splitting of *d*-orbital. Even in the degenerate *d*-orbitals, there is no vacant orbital. Thus, there is no absorption of visible light and therefore, cuprous compounds are colourless.

(*iii*) Anhydrous copper sulphate ($CuSO_4$) is also white. It is a d^9 system but there are no ligands and therefore no splitting of *d*-orbitals. One of the degenerate *d*-orbitals contains 9 electrons. But there is no possibility of excitation of electrons and thus no absorption of energy. Hence $CuSO_4$ anhydrous is white.

In general, we can say that a transition metal complex will be coloured if it possesses incomplete *d*-orbitals. If there are no incomplete *d*-orbitals the complex will be colourless. Thus complexes of Cu^+ (d^{10}), Zn^{2+} (d^{10}), Ag^+ (d^{10}) and Ti^{4+} (d^{10}) are colourless.

Limitations of Crystal Field Theory

The main limitations of crystal field theory are :

(*i*) It does not take into account the partial covalent character of metal-ligand bonds.

(*ii*) It does not explain the relative strengths of ligands. For example, it gives no explanation as to why neutral water appears as a stronger ligand in electrochemical series than negatively charged OH^- ion.

(*iii*) It does not consider the multiple bonding (π bonds) between metal ion and the ligands. Therefore, it does not explain the formation of π bonding in complexes.

(*iv*) It considers only *d*-orbitals of metal ions and does not consider at all the other metal orbitals such as *s*, p_x, p_y, p_z orbitals and the ligand π orbitals.

(*v*) It does not explain the charge transfer bands.

Comparison of Valence Bond Theory and Crystal Field Theory

The main points of comparison between valence bond theory and crystal field theory are given below :

Crystal field theory	Valence bond theory
1. The bond between metal ion and the ligand is taken as purely electrostatic.	1. The bond between metal ion and the ligand is taken as purely covalent.
2. It helps to calculate the stabilization energies of complexes due to crystal field splitting.	2. The concept of crystal field stabilization energy does not exist here.
3. It takes into account the splitting of *d*-orbitals of the metal.	3. It does not take into account the splitting of *d*-orbitals of metal.
4. It explains satisfactorily the colour of complexes.	4. It does not explain the colour of complexes.
5. It does not consider the concept of hybridisation.	5. It considers the concept of hybridisation of the orbitals of the metal ion before complex formation.
6. It explains the effect of distortion in regular geometry of molecules (e.g. tetragonal geometry).	6. It is unable to explain the distortion in regular geometry of molecules.
7. It predicts theoretically the magnetic properties of complexes and predicts the variation in magnetic moments with temperature for complexes where Δ value is close to pairing energy.	7. It does not explain theoretically the number of unpaired electrons. It is unable to predict the effect of temperature on the magnetic properties.

SOLVED CONCEPTUAL PROBLEMS

Example 1. Calculate crystal field stabilization energies for the following :

(*a*) d^6 tetrahedral (*b*) d^4 tetrahedral

(*c*) d^7 weak field octahedral (*d*) d^5 strong field octahedral

Solution. (*a*) d^6 tetrahedral

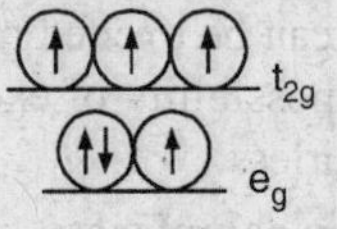

Electronic configuration : $e^3\, t_2^{\,3}$

CFSE = 3(– 6Dq) + 3(4Dq) = – 6 Dq

(*b*) d^4 tetrahedral

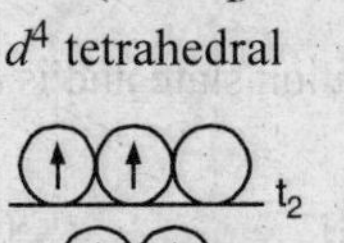

Electronic configuration: $e_g^{\,2}\, t_2^{\,2}$

CFSE = 2(– 6Dq) + 2(4Dq) = – 4 Dq

(*c*) d^7 weak field octahedral

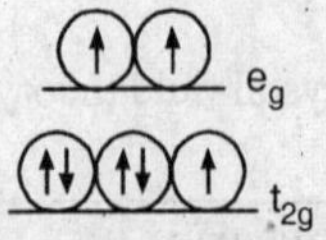

Electronic configuration : $t_{2g}^{5}\, e_g^{2}$

CFSE = 5(– 4 Dq) + 2(6 Dq) = – 8 Dq

(*d*) d^5 system strong field octahedral

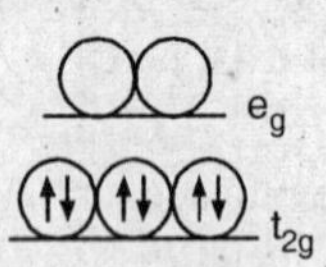

Electronic configuration : t_{2g}^{5}

CFSE = 5 × (– 4Dq) + 2P = – 20 Dq (Ignoring the pairing energy)

Example 2. Why does NH_3 readily form complexes but NH_4^+ does not ? Explain.

Solution. NH_3 contains a lone pair of electrons which coordinate with the metal ion to form the complex compound. However, in NH_4^+ ion, the lone pair is bound to H^+ and therefore, is not available for bonding to metal ion. Therefore, NH_4^+ does not form complexes readily.

Example 3. Calculate CFSE for the following systems :

(*i*) d^4 (high spin octahedral) **(*ii*) d^5 (tetrahedral)**

(*iii*) d^6 (low spin octahedral) **(*iv*) d^7 (high spin octahedral)**

(*v*) d^6 (tetrahedral)

Solution. (*i*) $t_{2g}^{3}\, e_g^{1}$: CFSE = – 6Dq (*ii*) $e^2\, t_2^{3}$: CFSE = 0

(*iii*) t_{2g}^{6} : CFSE = – 24 Dq + 2P (*iv*) $t_{2g}^{5}\, e_g^{2}$: CFSE = – 8 Dq

(*v*) $e^3\, t_2^{3}$: CFSE = – 6 Dq

Example 4. Cu^{2+} ions are coloured and paramagnetic while Zn^{2+} ions are colourless and diamagnetic. Explain.

Solution. The electronic configuration of Cu^{2+} and Zn^{2+} are as under:

Cu^{2+} : $3d^9$, Zn^{2+}: $3d^{10}$

Cu^{2+} has one unpaired electron and therefore it is paramagnetic. On the other hand, Zn^{2+} has no unpaired electrons and therefore it is diamagnetic.

In Cu^{2+}, there is one vacancy in 3d orbitals to which an electron can be excited resulting in *d-d* transition. Therefore, Cu^{2+} ions are coloured. In Zn^{2+}, there is no possibility of excitation of electrons because 3*d* orbitals are completely filled and hence are colourless.

Example 5. Show the *d*-electron configuration in the following complexes:

(*i*) $[Co(NCS)_4]^{2-}$ tetrahedral

(*ii*) $[Fe(CN)_6]^{3-}$ low spin octahedral.

Solution. (*i*) **$[Co(NCS)_4]^{2-}$**. In this complex ion Co is in +2 oxidation state and is d^7 system. Its electronic configuration is :

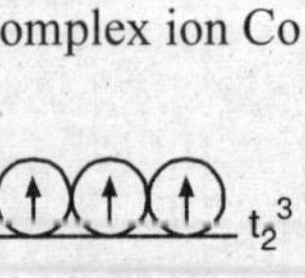

Electronic configuration : $e^4\, t_2^{3}$

(*ii*) **$[Fe(CN)_6]^{3-}$.** In this complex ion, Fe is in +3 oxidation state and is d^6 system. Its electronic configuration is :

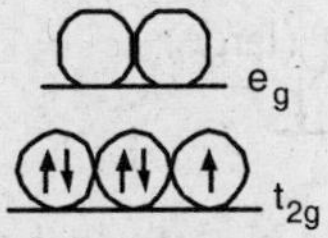

Electronic configuration : $t_{2g}{}^5 e_g{}^0$

Example 6. Which of the following complexes has larger crystal field splitting of *d*-orbitals in each pair ?

(*i*) $[Co(H_2O)_6]^{2+}$ or $[Co(H_2O)_6]^{3+}$ (*ii*) $[Co(NH_3)_6]^{3+}$ or $[Rh(NH_3)_6]^{3+}$

(iii) $[Co(CN)_6]^{3-}$ or $[Co(NH_3)_6]^{3+}$.

Solution. (*i*) $[Co(H_2O)_6]^{3+}$ (higher oxidation state of Co)

(*ii*) $[Rh(NH_3)_6]^{3+}$ (element from 4*d* series)

(*iii*) $[Co(CN)_6]^{3-}$ (strong ligand field of CN^-)

Example 7. Draw the energy level diagram for $[Pd(CN)_4]^{2-}$ square planar diamagnetic complex showing the *d*-electron distribution.

Solution. Palladium in $[Pd(CN)_4]^{2-}$ is in +2 oxidation state. It has d^8 electronic configuration as shown below:

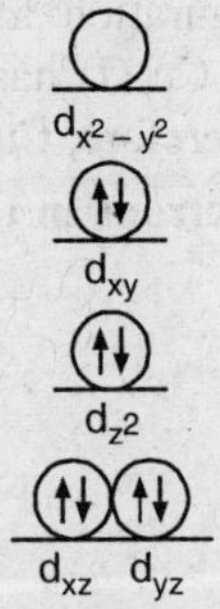

Electronic configuration:
$(d_{xz})^2 (d_{yz})^2 (d_{z^2})^2 (d_{xy})^2$

Example 8. Predict which of the following statements are true or false:

(*a*) Crystal field splitting for tetrahedral field is always less than that for octahedral field.

(*b*) $[CoCl_4]^{2-}$ is tetrahedral and diamagnetic.

(*c*) NO_2^- is weak field ligand as compared to NH_3.

(*d*) Fe^{3+} octahedral weak field has 5 unpaired electrons.

Solution. (*a*) True (*b*) False

(*c*) False (*d*) True

Example 9. Draw crystal field splitting and calculate the number of unpaired electrons in

(*i*) $[Ni(CN)_4]^{2-}$: square planar

(*ii*) $[NiCl_4]^{2-}$: tetrahedral

(*iii*) $[Mn(CN)_6]^{3-}$: octahedral

Solution. (*i*) Ni (II) : $3d^8$

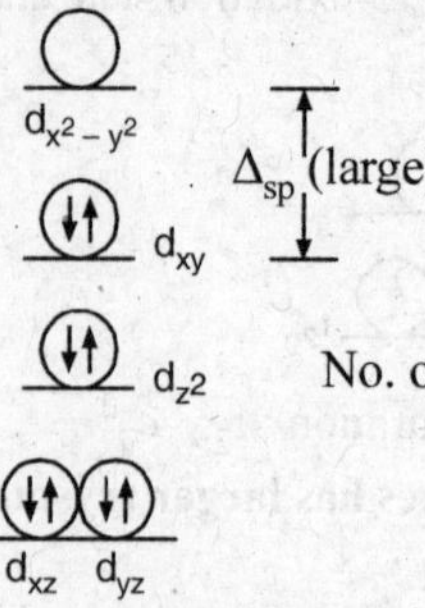

No. of unpaired electrons = 0

(*ii*) Ni (II) : $3d^8$

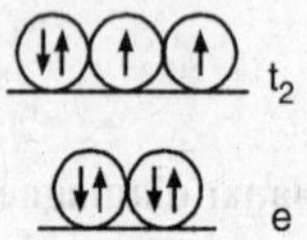

No. of unpaired electrons = 2

(*iii*) $[Mn(CN)_6]^{3-}$

Mn(III) : $3d^4$

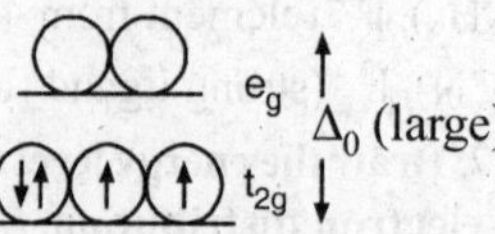

No. of unpaired electrons = 2

Example 10. Cu (I) is diamagnetic whereas Cu (II) is paramagnetic.

Solution. Cu (II) has the outermost electronic configuration $3d^9$ and therefore, it has one unpaired electron. Hence, it is paramagnetic. However, Cu (I) has the outermost electronic configuration as $3d^{10}$. There is no unpaired electron and therefore, Cu (I) is diamagnetic.

Example 11. Calculate the number of unpaired electrons in the following :

(*i*) Cr^{3+} (octahedral strong field)

(*ii*) Co^{3+} (octahedral strong field)

(*iii*) Fe^{3+} (octahedral weak field)

Solution. (*i*) Cr^{3+} : Octahedral strong field

Cr^{3+} (d^3) 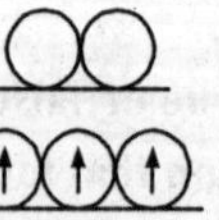 No. of unpaired electrons = 3

(*ii*) Co^{3+} : Octahedral strong field

Co^{3+} (d^6) 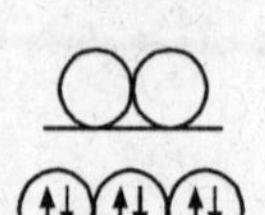 No. of unpaired electrons = 0

(*iii*) Fe^{3+} : Octahedral weak field

Fe^{3+} (d^5) 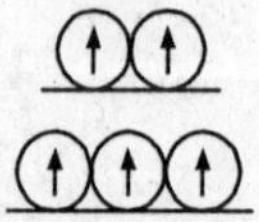 No. of unpaired electrons = 5

Example 12. Zn forms only Zn^{2+} and not Zn^{3+}. Why?

Solution. In the formation of Zn^{2+}, it acquires completely filled ($3d^{10}$) configuration. It is a stable configuration. Therefore, it is difficult to remove an electron from this filled $3d^{10}$ stable configuration and hence Zn^{3+} is not formed.

EXERCISES
(Including Questions from Different University Papers)

Multiple Choice Questions (Select the correct answer)

1. Colour of a complex is satisfactorily explained by
(*a*) Werner's theory (*b*) Valence bond theory
(*c*) Crystal field theory (*d*) ligand field theory

2. The CFSE for a high spin d^4 octahedral complex ion is
(*a*) – 14 Dq (*b*) – 6 Dq
(*c*) – 12 Dq + P (*d*) zero

3. The largest crystal field splitting will be for the ligand (same metal ion)
(*a*) Ox^{2-} (*b*) NO_2^-
(*c*) NH_3 (*d*) CN^-

4. Which of the following systems has maximum number of unpaired electrons ?
(*a*) d^6 (tetrahedral) (*b*) d^9 (octahedral)
(*c*) d^7 (octahedral, high spin) (*d*) d^4 (octahedral, low spin)

5. Which of the following complex ion would have the smallest crystal field splitting ?
(*a*) $[Co(NH_3)_6]^{2+}$ (*b*) $[Rh(NH_3)_6]^{3+}$
(*c*) $[Ir(NH_3)_6]^{3+}$ (*d*) $[Co(NH_3)_6]^{3+}$

6. The number of unpaired electrons in $NiCl_4^{2-}$ (tetrahedral) are
(*a*) two (*b*) zero
(*c*) one (*d*) four

7. Which of the following is a ligand which causes maximum crystal field splitting ?
(*a*) NH_3 (*b*) F^-
(*c*) CO (*d*) Ox^{2-}

8. Which of the following has highest CFSE ?
(*a*) $[TiF_6]^{3-}$ (*b*) $[Mn(H_2O)_6]^{2+}$
(*c*) $[FeCl_4]^{2-}$ (*d*) $[Cr(H_2O)_6]^{3+}$

9. The order of splitting in cubic geometry is same as that in
(*a*) tetrahedral (*b*) square planar
(*c*) octahedral (*d*) tetragonally distorted octahedral

10. The number of unpaired electron in a d^7 tetrahedral configuration is
(*a*) 3 (*b*) 2
(*c*) 1 (*d*) 7

11. Which of the following has no CFSE in octahedral field ?
(*a*) Fe^{3+} (high spin) (*b*) Co^{2+} (low spin)
(*c*) Fe^{3+} (low spin) (*d*) Cr^{3+} (high spin)

12. Choose the correct answer
(*a*) Copper sulphate hydroated solution is colourless
(*b*) Copper sulphate anhydrous in the solid form is white
(*c*) Copper sulphate hydrated in the solid form is white solution
(*d*) Cuprous sulphate solution is blue in colour

ANSWERS

1. (*c*)	**2.** (*b*)	**3.** (*d*)	**4.** (*a*)	**5.** (*a*)	**6.** (*a*)
7. (*c*)	**8.** (*d*)	**9.** (*a*)	**10.** (*a*)	**11.** (*a*)	**12.** (*b*)

Short Answer Questions

1. How will you account for the purple colour of $[Ti(H_2O)_6]^{3+}$?
2. Give one example of an outer orbital complex.
3. Calculate CFSE for d^4 high spin octahedral complex.
4. Draw energy level diagram showing splitting of five *d*-orbitals in octahedral field.
5. Calculate CFSE for d^9 tetrahedral complex.
6. Which complex has larger Δ_0 value $[Co(CN)_6]^{3-}$ or $[Co(NH_3)_6]^{3+}$?
7. Which of the two $[Co(H_2O)_6]^{2+}$ or $[Co(H_2O)_6]^{3+}$ has smaller Δ_0 value ?
8. What is the difference between low-spin and high-spin complexes ?
9. What is meant by spectrochemical series ?
10. What is meant by ligand ? Give five examples of ligand.
11. Define crystal field stabilization energy.

General Questions

1. (*a*) How do the *d*-orbital energy levels split when a transition metal ion is placed in an octahedral crystal field of the ligands ?

(*b*) Explain the effect of nature of ligands and the geometry of the complex on the magnitude of Δ_o.

2. (*a*) Explain how the orientation of d-orbitals in space leads to their splitting in a tetrahedral field of ligands.

(*b*) Define CFSE and calculate its value for d^4 system in octahedral and tetrahedral fields.

(*c*) Explain with examples how the nature of ligands affects the magnitude of Δ.

3. (*a*) What is the difference between inner orbital and outer orbital complexes ? Explain with examples.

(*b*) Calculate the CFSE for the following system :

(*i*) d^4 (high spin octahedral) (*ii*) d^5 (tetrahedral)

(*iii*) d^6 (low spin octahedral)

(*c*) Explain briefly the splitting of *d*-orbitals in the case of a tetrahedral complex.

(*d*) Write short note on ferromagnetism and antiferromagnetism.

4. (*a*) What is crystal field splitting and crystal field stabilisation energy ? Discuss the structures of $[Co(NH_3)_6]^{3+}$ and $[Cu(NH_3)_4]^{2+}$ ions on the basis of crystal field theory.

(*b*) Calculate CFSE for the following :

(*i*) d^4 high spin complex (Octahedral) (*ii*) d^5 strong field Octahedral

(*iii*) d^6 tetrahedral (*iv*) d^9 tetrahedral

5. (*a*) Give the salient features of crystal field theory.

(*b*) Discuss the crystal field splitting of *d*-orbitals in case of octahedral complexes.

(*c*) What is meant by spectrochemical series.

6. (*a*) Explain crystal field splitting in case of square planar examples.

(*b*) Calculate CFSE for the following systems :

(*i*) d^5 (low spin octahedral) (*ii*) d^6 (tetrahedral)

(*iii*) d^4 (high spin octahedral) (*iv*) d^7 (high spin octahedral)

(*c*) How will you account for the purple colour of $[Ti(H_2O)_6]^{3+}$?

7. (*a*) Discuss the factors affecting the magnitude of crystal field splitting (Δ).

(*b*) Define CFSE and calculate its value for the following systems.

(*i*) d^4 – high spin octahedral (*ii*) d^5 – tetrahedral

8. (*a*) What is crystal field splitting and what is crystal field stabilization energy ? Describe the splitting of *d*-orbitals in octahedral and tetrahedral complexes.

(*b*) $CuSO_4$ and Cu_2SO_4 lack colour but $CuSO_4.5H_2O$ is blue. Explain.

9. (*a*) Define CFSE and calculate its value for d^5 system in octahedral and tetrahedral field.

(*b*) Explain crystal field splitting in square planar complexes.

(*c*) How will you account for the purple colour of $[Ti(H_2O)_6]^{3+}$?

10. (*a*) Explain how the orientation of *d*-orbitals in space leads to the splitting in a tetrahedral field of ligands.

(*b*) Calculate CFSE for the following :

(*i*) d^5 strong field Octahedral (*ii*) d^6 tetrahedral

(*iii*) d^7 weak field octahedral (*iv*) d^4 tetrahedral

11. (*a*) Explain Crystal Field Splitting in tetrahedral complexes. Give reasons for smaller value of crystal field splitting in tetrahedral than in octahedral complexes.

(*b*) How does Crystal Field Theory explain the magnetic character of coordinate complexes ? Give the number of unpaired electrons in the following octahedral complexes :

(*i*) Ni (III) low spin (*ii*) Cr (III) low spin (*iii*) Ti (III) low spin.

12. (*a*) Define CFSE and calculate its value for d^4 system in octahedral and tetrahedral fields.

(*b*) How will you account for the purple colour of $[Ti\ (H_2O)_6]^{3+}$?

13. (*a*) Give the salient features of crystal field theory.

(*b*) What is meant by spectrochemical series ? Explain.

(*c*) Explain the crystal field splitting in case of square planar complex.

14. (*a*) How does valence bond theory explain :

(*i*) $[Ni\ (CN)_4]^{2-}$ is diamagnetic and square planar.

(*ii*) $[Ni\ (CO)_4]$ is diamagnetic and tetrahedral.

(*b*) Give the no. of unpaired electrons in octahedral strong field for :

(*i*) Cr^{3+} (*ii*) Co^{3+} (*iii*) Fe^{3+}

(*c*) Why are salts of zinc, cadmium and mercury white ?

15. (*a*) Define crystal field splitting energy and discuss the crystal field splitting of *d*-orbitals in case of octahedral complexes. Also calculate CFSE for d^5 (low spin O_h complex).

(*b*) Give reasons for the smaller value of crystal field splitting in tetrahedral than in octahedral complexes.

(*c*) Discuss the bonding in $[Fe(H_2O)_6]^{3+}$ and $[Fe(CN)_6]^{3-}$ in terms of crystal field theory.

16. (*a*) Explain $FeF_6{}^{3-}$ is colourless whereas $CoF_6{}^{3-}$ is coloured.

(*b*) Write electronic configuration and calculate CFSE for d^4 (high spin octahedral).

(*c*) Calculate magnetic moment for the complex $K_4[Mn(CNS)_6]$ (Spin only value).

17. Define crystal field stabilization energy. Calculate its value for d^5 low and high spin octahedral complexes.

18. (*a*) Explain crystal field splitting of *d*-orbitals in octahedral complexes.

(*b*) Calculate CFSE for the following :

(*i*) d^4 high spin (octahedral) (*ii*) d^5 strong field (octahedral)

(*iii*) d^6 tetrahedral (*iv*) d^9 tetrahedral

(*c*) Explain why the complex ion of transition metals are mostly coloured.

19. (*a*) What is meant by the terms

(*i*) crystal field splitting and

(*ii*) crystal field stabilization energy ? Briefly discuss the crystal field splitting in case of square planar complexes.

(*b*) Describe the bonding in $[FeF_6]^{3-}$ in terms of crystal field theory.

(*c*) Identify the complexes, from amongst the following, which has higher value of crystal stabilisation :

(*i*) $[Co(NH_3)_6]^{3+}$ and $[CoF_6]^{3-}$ (*ii*) $[Fe(CN)_6]^{4-}$ and $[Fe(CN)_6]^{3-}$

20. (*a*) Which of the two, $[Co(NH_3)_6]^{3+}$ or $[Ir(NH_3)_6]^{3+}$ has greater Δ_o value and why ?

21. (*a*) How do the *d*-orbitals energy levels split up of a transition metal ion when placed in octahedral field of ligands ?

(*b*) Write electronic configuration and calculate CFSE for d^4 (high spin, oct) system.

22. (*a*) Give a brief account of limitations of valence bond theory in transition metal complexes.

(*b*) Find out the number of unpaired electrons in strong and weak octahedral complexes of Co^{3+}.

(*c*) Explain why is Ti^{3+} purple in aqueous solution, while Zn^{2+} is colourless ?

(*d*) Calculate the values of CFSE for the following systems :

(*i*) d^2 tetrahedral (*ii*) d^4 octahedral, high spin.

23. (*a*) How are magnitude of crystal field splitting in octahedral and tetrahedral complexes related ?

(*b*) Calculate CFSE for d^7 weak field & strong field (oct.).

(*c*) Discuss splitting of *d*-orbitals of metal atom or ion in octahedral field of ligands.

24. (*a*) Discuss the crystal field splitting in distorted octahedral complexes (Jahn-Teller effect).

(*b*) What is CFSE? Calculate CFSE for the following ions in octahedral field.

(*i*) d^4 strong field (*ii*) d^5 weak field

(*iii*) d^6 weak field (*iv*) d^7 strong field

(*c*) What are inner and outer orbital complexes.

25. (*a*) Show diagrammatically the C.F. splitting of *d*-orbitals in square planar complexes and write the *d*-electronic configuration of the following complexes :

(*i*) Sq. planar $[Pd(CN)_4]^{2-}$ (*ii*) Sq. planar $[Co(NCS)_4]^{2-}$

(*iii*) Sq. planar $[Pt(NH_3)_6]^{4+}$

(*b*) For the ligands CO, NH_3, H_2O and CN^-, wirte their decreasing order of CF splitting with Fe^{3+} ion.

(*c*) Explain on the basis of Valence Bond Theory, why :

(*i*) Ni $(CO)_4$ is diamagnetic and tetrahedral

(*ii*) Tetrahedral complexes are generally high spin.

(*d*) Explain why octahedral complexes with d^4, d^5, d^6 and d^7 system form both types of strong field and weak field complexes.

26. (*a*) Draw diagram showing splitting in square planar complexes.

(*b*) Using simple crystal field theory explain why $[Mn(H_2O)_6]^{2+}$ is colourless.

(*c*) Calculate the crystal field stabilization energy for octahedral complexes of Fe^{3+} in a weak field and also in strong field. Calculate the spin-only magnetic moment for these complexes.

27. (*a*) Discuss the crystal field splitting of *d*-orbitals in case of tetrahedral complexes. Calculate CFSE for d^3 (T*d*) and d^7 (T*d*).

(*b*) What is crystal field theory ? How does this theory account for the fact that $[CoF_6]^{3-}$ is paramagnetic but $[Co(NH_3)_6]^{3+}$ is diamagnetic though both are octahedral ?

(*c*) $CuSO_4.5H_2O$ is coloured compound while $CuSO_4$ (anhydrous) is white. Explain on the basis of CFT.

28. (*a*) Discuss the limitations of valence bond theory.

(*b*) Explain on the basis of CFT that $[Ni(CN)_6]^{2-}$ is diamagnetic but $[NiCl_4]^{2-}$ is paramagnetic.

(*c*) What are low and high spin complexes ?

29. (*a*) The magnetic moment of $[Fe(H_2O)_6]^{3+}$ is 5.92 B.M. and that of $[Fe(CN)_6]^{3-}$ is 1.73 B.M. Explain on the basis of CF theory.

(*b*) Why Ti^{3+} is purple in aqueous solution while Zn^{2+} is colourless.

(*c*) Explain why in tetrahedral complexes the subscript '*g*' has been dropped from the orbital notation.

(*d*) Give the salient features of crystal field theory.

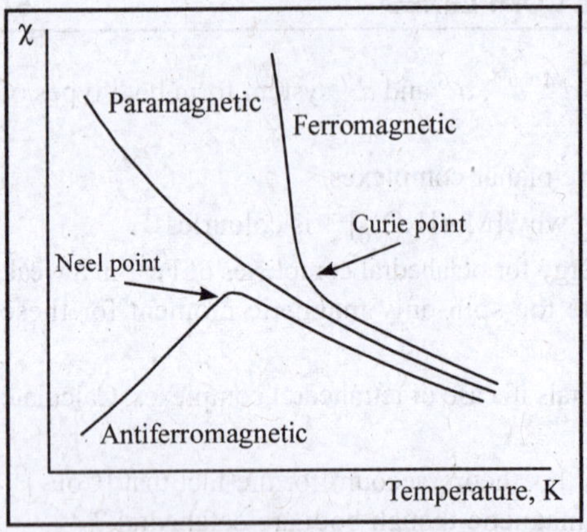

3

Magnetic Properties of Transition Metal Complexes

3.1. INTRODUCTION

By studying the magnetic properties of a substance, we can get an insight into the distribution of electrons in the constituent atoms or ions. We know that the magnetic properties like magnetic moment and magnetic susceptibility arise due to the presence of unpaired electrons. For example, the distinction between high spin and low spin complexes can be made on the basis of their magnetic properties. We observe that most of the transition metals and their salts and complexes have unpaired electrons and show magnetic properties. The unpaired electron spins about its axis and therefore shows magnetic properties. It acts as a tiny magnet The molecules with paired electrons are diamagnetic. This is because the paired electrons have their spins in opposite directions and their magnetic fields cancel each other. The picture is different in the case of molecules having unpaired electrons. The magnetic fields produced by individual electrons in this case are not cancelled, rather they reinforce each other.

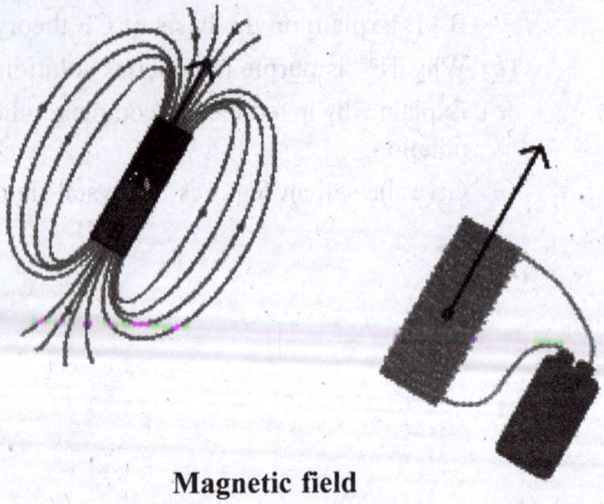

Magnetic field

Broadly there are two types of magnetic behaviour shown by substances.

1. Paramagnetic substances

Substances having unpaired electrons are called paramagnetic substances. Such substances are attracted by the external magnetic field. Greater the number of unpaired electrons, greater the paramagnetic character. Most of the transition metals, their salts and complexes are paramagnetic in nature.

2. Diamagnetic substances

Substances which do not possess unpaired electrons (their electrons are paired) do not show any magnetic moment. Such substances are repelled by external magnetic field. This is because, the magnetic fields produced by paired electrons are neutralised and no net magnetic field is produced by them.

Origin of paramagnetic moments

The magnetic properties of a substance arise from electrons and nucleus. However, the contribution of magnetic behaviour from electrons is 1850 times that of nucleus because mass of proton is about 1850 times that of electron. Therefore, *the contribution of nucleus to magnetic moments is neglected and the magnetic behaviour observed is mainly due to the electrons.*

There are two ways in which the electron creates the magnetic moment.

(*i*) Each electron can be treated as a small sphere of negative charge spinning on its axis. The spinning of charge produces magnetic moment.

(*ii*) An electron travelling in a closed path around a nucleus also produces magnetic moment like an electric current flowing in a circular loop of wire.

The magnetic moment due to spin of the electron on its axis is termed **spin magnetic moment** of the electron and the magnetic moment due to motion of the electron around the nucleus is termed **orbital magnetic moment.** Therefore, observed magnetic moment of an atom or ion is the sum of the spin magnetic moment and orbital magnetic moment.

We shall first discuss the magnetic moment due to the spin of the electron.

The magnetic moment is expressed in terms of units called **Bohr Magneton,** abbreviated as B.M. The magnetic moment for an electron having charge e and mass m is given by the relation.

$$\mu_s = \frac{eh}{4\pi mc}$$

where h is Planck's constant and c is the velocity of light.

The value of μ_s for an electron as obtained by the above expression is 9.274×10^{-21} erg gauss^{-1}. This is taken as one unit of magnetic moment called **Bohr Magneton.** Thus, Bohr magneton is defined as

$$1 \text{ B.M.} = \frac{eh}{4\pi mc} \qquad (1)$$

The magnetic moment, μ_s of a single electron is given by the relation

$$\mu_s = 2\sqrt{s(s+1)} \qquad (2)$$

where s is the spin quantum number $= \frac{1}{2}$.

Substituting the value of s as $\frac{1}{2}$, equation (2) reduces to

$$\mu_s = 2\sqrt{\frac{1}{2}\left(\frac{1}{2}+1\right)} = \sqrt{3} = 1.732 \text{ B.M.}$$

Thus, atom or ion having one unpaired electron (e.g. H, Cu^{2+}) should have spin magnetic moment of 1.732 B.M. from the electron spin alone. For atoms or ions having multiple unpaired electrons, the overall spin moment is given by

$$\mu_s = 2\sqrt{S(S+1)} \qquad (3)$$

where S is the sum of the spin quantum numbers for the individual electrons. For example, Ti^{3+} has only one unpaired electron so that $S = \frac{1}{2}$, Cr^{3+} has four unpaired electrons so that $S = 4 \times \frac{1}{2} = 2$ and so on.

Eq. (3) may also be written as

$$\mu_s = \sqrt{4S(S+1)} \qquad (4)$$

Now we can calculate magnetic moments for different number of unpaired electrons

For one unpaired electron,

$$S = \frac{1}{2}$$

$$\therefore \quad \mu_s = \sqrt{4\times\frac{1}{2}\left(\frac{1}{2}+1\right)} = \sqrt{3} = 1.732 \text{ B.M.}$$

For two unpaired electrons,

$$S = \left(\frac{1}{2}+\frac{1}{2}\right) = 1$$

$$\therefore \quad \mu_s = \sqrt{4\times 1(1+1)} = 2\sqrt{2} = 2 \times 1.414 = 2.828 \text{ B.M.}$$

For three unpaired electrons,

$$S = 3/2$$

$$\therefore \quad \mu_s = \sqrt{4\times\frac{3}{2}\left(\frac{3}{2}+1\right)} = 3.87 \text{ B.M.}$$

There is more conveneint direct method to calculate the magnetic moment using the relation

$$\mu_s = \sqrt{n(n+2)} \text{ B.M.} \qquad (5)$$

where n is the number of unpaired electrons. We can verify the validity of eq. (5) as under :

For one unpaired electron :

$$\mu_s = \sqrt{1(1+2)} = \sqrt{3} = 1.732 \text{ B.M.}$$

For two unpaired electrons :

$$\mu_s = \sqrt{2(2+2)} = 2\sqrt{2} = 2 \times 1.414 = 2.828 \text{ B.M}$$

Table 3.1 gives calculated values of μ_s for the different number of unpaired electrons.

Table 3.1. Spin only magnetic moments for different number of unpaired electrons.

No. of unpaired electrons	μ_s (B.M.)
1	1.73
2	2.84
3	3.87
4	4.90
5	5.92
6	6.93
7	7.94

We observe that there is a fair agreement between the experimental values of magnetic moments of some transition metal ions and the values obtained from the spin only formula. However in some cases, experimental values differ from the spin only values. The reason being that in addition to spin of the electron, **orbital motion of electron** also contribute to the magnetic moment. In these cases, we use the following equation to calculate the magnetic moment.

$$\mu_{S+L} = \sqrt{4S(S+1) + L(L+1)} \quad (6)$$

where S is the resultant spin angular momentum and L is the resultant orbital angular momentum quantum number. A comparison has been made in Table 3.2 between the observed values and values obtained from the spin-only relation for the magnetic moments.

We find from the table that observed values of μ are frequently greater than μ_s and are never more than μ_{S+L}. This means that the contribution to magnetic moments due to orbital motion of electrons is positive although very small. A very small different in the values means that the orbital moments are completely or partially **quenched** (neutralised). This is because the electric field of ligands (atoms, ions or molecules) surrounding the metal ion restrict the orbital motion of the electrons.

Table 3.2. Magnetic moments of some metal ions of transition series

Ion	No. of unpaired electrons	Magnetic moment (B.M.) Spin only value	Observed value
Sc^{2+}	0	0	0
Ti^{3+}	1	1.73	1.7—1.8
Ti^{2+}	2	2.83	2.8—3.1
V^{2+}	3	3.87	3.7—3.9
Cr^{2+}	4	4.90	4.8—4.9
Mn^{2+}	5	5.92	5.7—6.0
Fe^{2+}	4	4.90	5.0—5.6
Co^{2+}	3	3.87	4.3—5.2
Ni^{2+}	2	2.84	2.9—3.4
Cu^{2+}	1	1.73	1.9—2.1
Zn^{2+}	0	0	0

However, the situation is different with second and third row transition elements particularly the lanthanides where the unpaired electrons occupy $4f$ orbitals. Here the orbital motion of the electrons is not quenched. Here, the $4f$ orbitals are shielded by the $5s$ and $5p$ subshells . Thus orbital contribution to the magnetic moment is not quenched. Here both the spin and orbital motion of the electrons contribute to the total magnetic moment.

3.2. MEASUREMENT OF MAGNETIC PROPERTIES

The magnetic moments cannot be measured directly. We first measure the magnetic susceptibility (χ) from which it is possible to measure magnetic moment.

*The **magnetic susceptibility** is a measure of the capacity of a substance to take up magnetisation in an applied magnetic field.* If a substance is placed in a magnetic field of strength H, then magnetic induction or magnetic flux density, B within it, is given by

$$B = H + 4\pi I \quad (1)$$

where I is called the intensity of magnetisation. Intensity of magnetisation (I) represents the extent to which a sample can be magnetised when placed in a magnetic field. It is defined as *the magnetic moment per unit volume of the magnet (magnetic substance)*

$$I = \frac{\text{Magnetic moment}}{\text{Volume}}$$

If a is the area of cross-section of the magnetic substance and $2l$ is the length of the substance then I becomes

$$I = \frac{m \times 2l}{a \times 2l} = \frac{m}{a}$$

Therefore intensity of magnetisation is also defined as the pole strength per unit area of cross-section of the material. The ratio B/H, called **magnetic permeability** of the substance can be given as :

$$\frac{B}{H} = 1 + 4\pi \frac{I}{H} \qquad (2)$$

Dividing both sides by H, we obtain

$$\frac{B}{H} = 1 + 4\pi\,\kappa \qquad (3)$$

κ is called the ***magnetic susceptibility per unit volume***

Magnetic permeability is the ability of a material to permit the passage of magnetic lines of force through it. In other words, magnetic permeability (B/H) gives the ratio of the density of lines of force within the substance to the density of such lines in the same region in the absence of the substance. Therefore, the volume susceptibility of *vacuum* is obviously zero because in vacuum B/H = 1 so that, from eq. (3)

$$1 = 1 + 4\pi\kappa \quad \text{or} \quad \kappa = 0$$

The susceptibility of a diamagnetic substance is negative because lines of force from induced dipole cancel out the lines of force due to applied field and B/H is less than 1.

Paramagnetic substances have the flux is greater within the substance than it would be in vacuum and thus paramagnetic substances have positive susceptibilities.

Susceptibility may be expressed in two forms, specific susceptibility and molar susceptibility given by the following relations

Specific susceptibility , $\chi = \frac{\kappa}{\rho}$

where ρ is the density.

Molar susceptibility, $\chi_M = \frac{\kappa \cdot M}{\rho}$

where M is the molecular mass of the substance.

Measurement of magnetic moment

Two commonly used methods of determination of magnetic moment are described hereunder.

(*i*) Gouy method

(*ii*) Faraday method.

1. Gouy's method

Gouy's balance named after the scientist who devised it is generally used to measure paramagnetism. Finely powdered substance or solution is taken in a pyrex glass tube called *Gouy* *tube* (Fig. 3.1). The substance is weighed first without magnetic field and then in the presence of magnetic field. If paramagnetic, the substance will weigh more in the presence of a magnetic field than in its absence. The increase in weight gives a measure of paramagnetism of the substance. Larger the number of unpaired electrons in a substance, greater is the increase in its weight in a magnetic field. The susceptibility is calculated from the difference in weight of the sample when the magnet is off and when it is on.

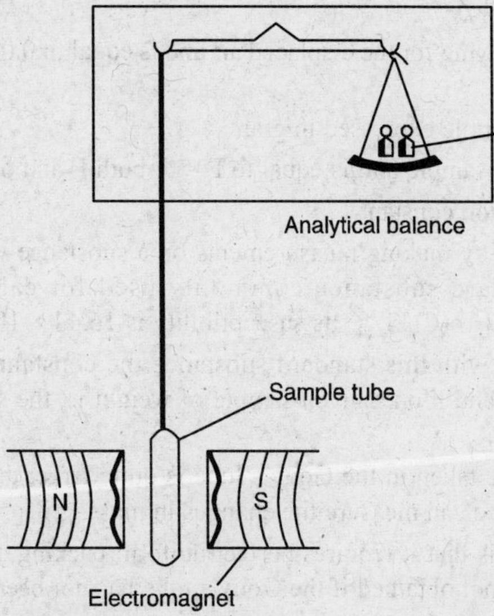

Fig. 3.1. Gouy's method of measuring paramagnetism.

Calculation of magnetic moment

The Gouy procedure offers a convernient method for measuring magnetic susceptibility. The sample in the form of a cylinder is suspended in a non homogeneous magnetic field and the force exerted on the sample is determined by weighing. The force acting on the sample is given by

$$F = \frac{1}{2} A \kappa H^2 \tag{4}$$

where A = is the cross sectional area of the cylinder

H = Intensity in the central homogenous part of the magnetic field

κ = volume susceptibility

The above equation is valid only if the measurements are done in vacuum. However, if the sample is surrounded by air, then the susceptibility of air (κ′) must be subtracted from the measured susceptibility. In that case, we shall use the equation

$$F = \frac{1}{2} AH^2 (\kappa - \kappa') \tag{5}$$

where κ′ = volume susceptibility of air

Now, the Gouy tube itself develops a force which is always present. Therefore, to calculate the actual force acting on the sample, the force acting on the Gouy tube is subtracted from the observed force. This force is negative because of the diamagnetic material of the tube. It is denoted by δ, then eqn. (5) becomes

$$F = \frac{1}{2} AH^2 (\kappa - \kappa') + \delta \qquad (6)$$

For a sample of constant length and cross sectional area, the factor AH^2 is constant. Now if the density of the sample is introduced, the above equation may be rewritten as :

$$10^6 \chi = \frac{\alpha + BF'}{w} \qquad (7)$$

where α = constant allowing for the displaced air and is equal to 0.029 × specimen volume and is expressed in mg.

w = weight of sample expressed in gram.

F' = Force on the sample and is equal to $F - \delta$; both F and δ are given in mg.

β = tube calibration constant.

The apparatus is calibrated by making measurements on a substance whose susceptibility is accurately known. The standard substance commonly used for calibration is mercury tetrathiocynatocobaltate (II), Hg $[Co(CNS)_4]$. Its susceptibility is 16.44×10^{-6} at 20°C.

By making measurements with this standard substance the constant β is first calculated. Substuting the values of α, β, δ and F on a given sample of weight w, the susceptibility χ can be calculated using eq. (4)

Since the amount of sample taken in the Gouy's tube is quite large, the forces are large and therefore, even a chemical balance can measure the changes in mass.

However the disadvantage is that it requires perfect uniform packing of the substance in the Gouy's tube. Correct results are not obtained if the Gouy's tube has not been packed uniformly.

2. Faraday's method

If only a small amount of the substance (in mg) is available, then Faraday's method is perhaps more suitable for magnetic moment measurement. The sample that is suspended in an inhomogeneous field such that $H_0 \left(\frac{\delta H}{\delta x}\right)$ is constant over the entire volume of the sample. The set up of Faraday's apparatus is shown in Fig. 3.2. The sample is suspended between magnet poles that have been carefully shaped so that the value of $H_0 \frac{\delta H}{\delta x}$ remains constant over the region occupied by the sample. A sample (0.1 – 10 mg) of the substance is placed in a magnetic field of constant gradient,

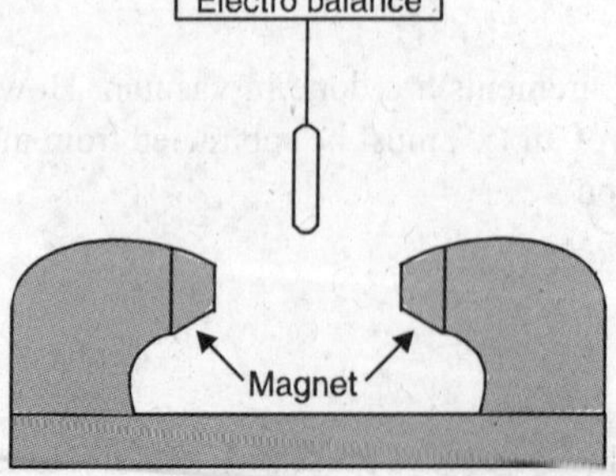

Fig. 3.2. Schematic diagram of Faraday balance for measuring magnetic susceptibility

A modern computerised magnetic measuring system

$\frac{\delta H}{\delta x}$ and the force acting on the sample is measured directly with the help of an electrobalance. If a sample of mass m and specific susceptibility χ is placed in a non-uniform field H, then the sample will experience a force (f) along x as :

$$f = m\chi H_0 \left(\frac{\delta H}{\delta x}\right) \tag{8}$$

The force can be measured by weighing the sample in the absence and presence of the field. The difference between the two weights is equal to f. The matter is simplified by carrying out experiment on a standard of known susceptibility such as $Hg[Co(CNS)_4]$, mercury tetrahiocyanatocobaltate (II) which has susceptibility as $\chi = 16.44 \times 10^{-6}$ at 293 K.

With same magnetic field and gradient for both the standard (s) and unknown (μ), it is not necessary to know the precise values of other quantities. Hence, we have

$$\frac{f_s}{m_s\chi_s} = \frac{f_u}{m_u\chi_u}$$

or

$$\chi_u = \frac{f_u\, m_s\, \chi_s}{f_s \cdot m_u} \tag{9}$$

The molar susceptibility of the sample, χ_m can be calculated using the value χ_u.

This method has the advantage that it can be used with very small quantity of the substance for measurement and the sample need not be homogeneous.

This method gives us specific susceptibility directly. The Gouy's method gives volume susceptibility, which has to be converted to specific susceptibility. The conversion poses a problem because it requires accurate value of density, which is difficult to obtain for solids because the values vary with the packing of the substance.

Diamagnetic correction

The measured magnetic susceptibility will consist of contributions from paramagnetic and diamagnetic susceptibilities. Therefore, the measured susceptibility is corrected by subtracting the diamagnetic contribution from it.

Paramagnetic susceptibility = Measured suscepibility – Diamagnetic susceptibility

3.3. RELATIONSHIP BETWEEN MAGNETIC SUSCEPTIBILITY AND MAGNETIC MOMENT

The magnetic susceptibility obtained after correction is related to magnetic moment, μ (in B.M.) as :

$$\chi_M = \frac{N_0\,\mu^2}{3kT} \tag{1}$$

where k is Boltzmann constant

N_0 is Avogadro number

T is absolute temperature

Rearranging equation (1)

$$\mu^2 = \frac{3k}{N_0} \cdot \chi_M \cdot T$$

or $$\mu = \sqrt{\frac{3k}{N_0}} \cdot \sqrt{\chi_M \cdot T}$$

$$= k\sqrt{\chi_M \cdot T} \qquad \text{(where } k \text{ is constant)}$$

The constant $\sqrt{\frac{3k}{N_0}}$ comes out to be 2.828.

Thus, magnetic moment can be calculated from the magnetic susceptibility using the relation:

$$\mu = 2.828\ \sqrt{\chi_M \cdot T}\ \text{B.M.} \qquad (2)$$

3.4. VARIATION OF MAGNETIC SUSCEPTIBILITY WITH TEMPERATURE

Magnetic susceptibility of a substance is dependent on temperature. Pierre Curie showed that paramagnetic susceptibilities vary inversely with temperature and follow the behaviour as shown by the following equation :

$$\chi_M = \frac{C}{T} \quad \text{or} \quad T = \frac{C}{\chi_M}$$

where T is the absolute temperature and C is a constant characteristic of the substance and is known as Curie constant. The above expression is called **Curie's law.**

Pierre Curie with wife Marie Curie in their laboratory

Thus, if we plot χ_M against absolute temperature, a parabolic curve is obtained as shown in Fig. 3.3(a). However, if we plot reciprocal of χ_M (i.e. $1/\chi_M$) at different temperatures against absolute temperature, a straight line with slope C is obtained (Fig. 3.3(b)). This curve should pass through the origin. However, it is observed that in many cases, the line does not pass through the origin. It cuts the temperature axis either below or above 0°K. This required some modification of Curie equation.

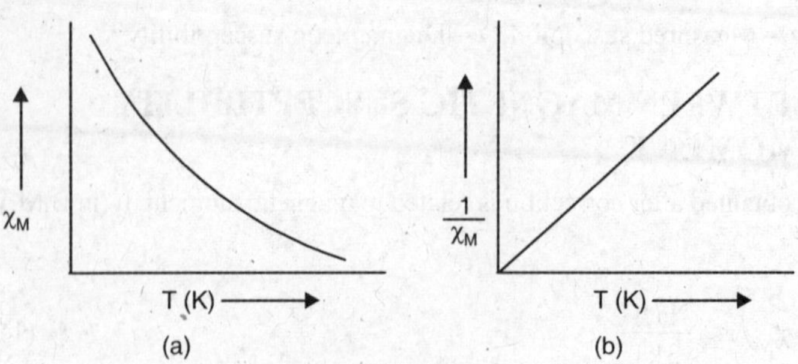

Fig. 3.3. (a) Plot of χ_M vs T(k) (b) Plot of $\frac{1}{\chi_M}$ vs T(k)

It was modified as

$$\chi_M = \frac{C}{T-\theta}$$

where θ is the temperature at which the line cuts the T-axis. This modified equation is called **Curie - Weiss law** and θ is known as **Weiss constant.** The Weiss constant takes into account the intermolecular interactions. Magnetic moments can now be calculated by the equation

$$\mu = 2.828\sqrt{\chi_M (T - \theta)}$$

Ferromagnetism and Anti-ferromagnetism

In addition to paramagnetic and diamagnetic substances, we also come across **ferromagnetic and anti-ferromagnetic substances.** These substances show dependence on both temperature and field strength. Their main features can be understood in terms of Curie or Curie-Weiss law. The temperature dependence of ferromagnetism and anti-ferromagnetism is shown in Fig. 3.4 (a) and 3.4 (b) respectively.

The inspection of curve 3.4(a) shows that there is some discontinuity in the curve at the temperature marked T_c. This temperature is called **Curie temperature.** Above this temperature, the ferromagnetic substance follows the Curie or Curie-Weiss law and therefore, exhibits simple paramagnetism (Refer to Fig. 3.3a). Below this temperature, the magnetic susceptibility behaviour is different and depends upon the field strength.

Inspection of curve 3.4 b shows that for anti-ferromagnetic susbstances, there is a characteristic temperature, T_N called the **Neel temperature.** Above this temperature, the substance follows Curie or Curie-Weiss law (Again refer to Fig. 3.3a) and shows simple paramagnetism. Below Neel temperature the magnetic susceptibility decreases with decrease in temperature.

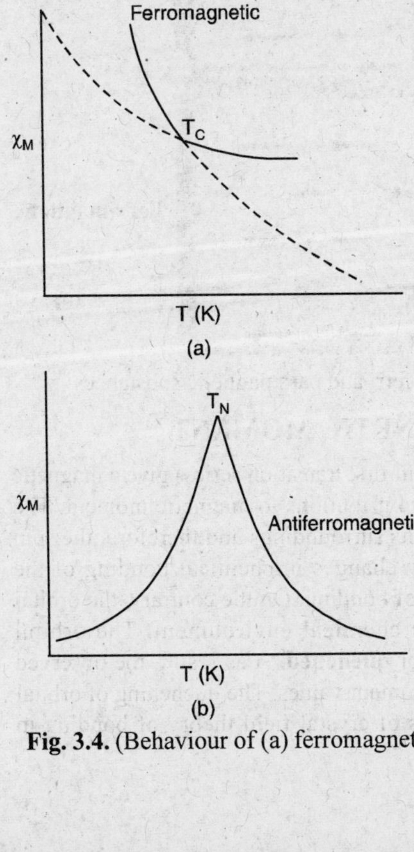

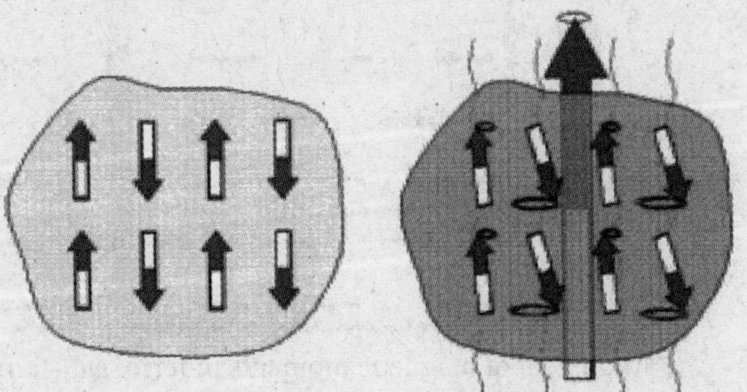

Antiferromagnetic Substances

Fig. 3.4. (Behaviour of (a) ferromagnetic and (b) anti-ferromagnetic substances with temperature.

The abnormal behaviour in ferromagnetic and anti-ferromagnetic substances can be explained in terms of interionic interactions. These interactions have magnitudes comparable to thermal energies at the Curie or Neel temperature and increase in magnitude as the temperature is lowered.

In the case of ***anti-ferromagnetism,*** *the magnetic moments of the ions in the lattice tend to align themselves in a manner so as to cancel one another and give zero magnetic moment.* Above the Neel temperature, thermal energy is large (kT) and prevents the effective alignment of the magnetic moments. MnO, MnO_2, FeO, NiO, etc belong to this category. A pattern of alignment of magnetic moments in anti-ferromagnetic substances is shown in the picture below

In ferromagnetic substances above the Curie temperature, thermal energies are able to randomise the orientations. But, below T_C, alignment becomes dominant and the susceptibility increases rapidly with decreasing temperature. Ni, Co, CrO_2, etc belong to this category.

The arrangement in **paramagnet substances** is quite random as in the picture below

In yet another class of magnetic substances known as **ferrimagnetic substances**, the magnetic moments are aligned in parallel and anti-parallel directions in unequal numbers resulting in net magnetic moment. Magnetic oxide Fe_3O_4 and ferrites belong to this category.

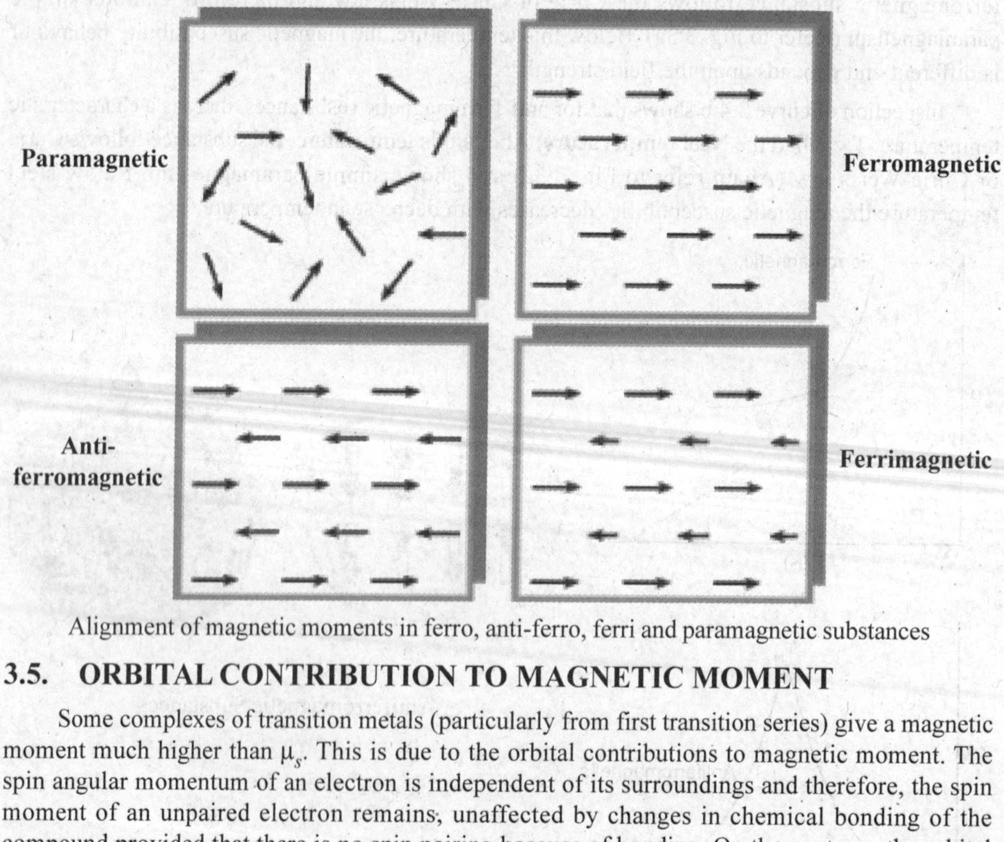

Alignment of magnetic moments in ferro, anti-ferro, ferri and paramagnetic substances

3.5. ORBITAL CONTRIBUTION TO MAGNETIC MOMENT

Some complexes of transition metals (particularly from first transition series) give a magnetic moment much higher than μ_s. This is due to the orbital contributions to magnetic moment. The spin angular momentum of an electron is independent of its surroundings and therefore, the spin moment of an unpaired electron remains, unaffected by changes in chemical bonding of the compound provided that there is no spin pairing because of bonding. On the contrary, the orbital angular momentum of the electron depends upon the chemical environment. The orbital contribution to magnetic moment may get compensated or **quenched.** A as result, the observed magnetic moments are very close to spin-only magnetic moment values. The quenching of orbital angular momentum can be easily explained on the basis of crystal field theory of bonding in transition metal complexes.

The unpaired electrons in a first transition series are present in the $3d$ orbitals. A transition metal ion has five $3d$ orbitals which are degenerate. ***An electron possesses*** as ***angular momentum along a given axis if it is possible to transform its orbital by rotation around this axis into another orbital which is equivalent to it in shape, size and energy.*** This circulation of electron is equivalent to a current flowing and therefore, it produces a magnetic effect.

Thus, **for orbital contribution to magnetic moment, there must be two or more degenerate orbitals which can be inter converted by rotation about a suitable axis and these orbitals must be unequally occupied.**

The orbital angular momentum along the given axis possessed by the electron in such an orbital is equal to the number of times the orbtial gets transformed into the equivalent orbital during a rotation of 90° around that axis. If the orbital degeneracy is lost by chemical bonding or crystal field effects, the orbital contribution to the total magnetic moment is **partially** or **completely quenched.**

Consider a free metal ion in which all the d-orbitals are degenerate. An electron in $d_{x^2-y^2}$ orbital will contribute to orbital angular momentum equal to 2 units of $h/2\pi$ along z-axis because a rotation of $d_{x^2-y^2}$ orbital by 45° around the z-axis takes it to equivalent d_{xy} orbital (Fig. 3.5)

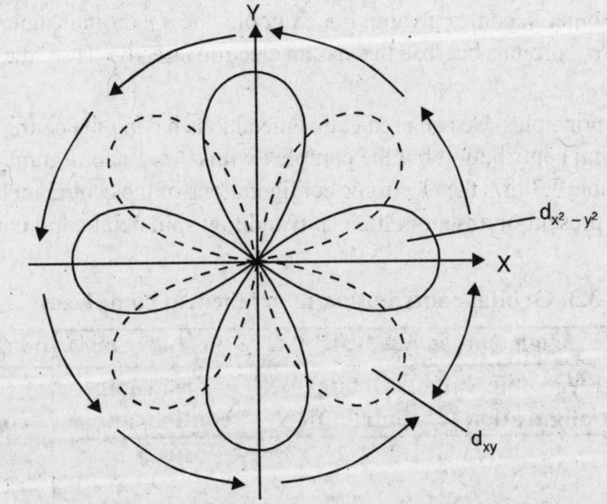

Remember:
$d_{x^2-y^2}$ and d_{xy} are four lobed orbitals.

Fig. 3.5. Circulation of electron density about z-axis (perpendicular to the plane) in d_{xy} and $d_{x^2-y^2}$ orbitals.

This means that a rotation of $d_{x^2-y^2}$ orbital by 90° around z-axis will carry this orbital into d_{xy} orbital two times. Similarly, we can also say that an electron in d_{xy} orbital will contribute to orbital angular momentum of 2 in the units of $\frac{h}{2\pi}$ along z-axis. In a similar manner, an electron in d_{xz} orbital will have an orbital angular momentum equal to 1 unit of $\frac{h}{2\pi}$ along z-axis because d_{xz} orbital gets transformed into an equivalent d_{yz} orbital by rotating the d_{xz} orbital around x-axis by an angle of 90°. The situation is different in d_{z^2}, it cannot be transformed into any other equivalent orbital by rotation around z-axis and therefore, its has **zero orbital angular momentum along** z-axis.

When the metal ion is surrounded octahedrally by six, the degeneracy of *d*-orbitals gets disturbed. The d_{xy} and $d_{x^2-y^2}$ orbitals acquire different energies and therefore, are non equivalent. As a result, an electron in $d_{x^2-y^2}$ orbital cannot be equated with an electron in d_{xy} orbital. In other words, the $d_{x^2-y^2}$ orbital cannot be transformed into d_{xy} orbital and vice versa by rotation of the orbital along z-axis. Thus, the electron in $d_{x^2-y^2}$ orbital will cease to have orbital contribution along the z-axis.

Quenching of Orbital Angular Momentum in Octahedral Complexes

In octahedral field of ligands, the five *d*-orbitals get split into two sets, t_{2g} (d_{xy}, d_{yz}, d_{zx}) and e_g ($d_{x^2-y^2}$, d_{z^2}). Therefore, the equivalence of $d_{x^2-y^2}$ and d_{xy} orbitals disappears and consequently, their orbital contribution gets quenched. However, there will be some contribution of orbital angular momentum along z-direction because d_{xz} and d_{yz} orbitals are still equivalent in energy in octahedral field. Therefore, an electron in d_{xz} or d_{yz} orbital will contribute an orbital angular moment along z-axis in octahedral complexes. Thus, a metal ion with electrons in t_{2g} orbitals would make some contribution along the z-direction to the overall magnetic moment (μ_{eff}) of the complex. The direction of the applied magnetic field is also taken to be z-axis by convention. However, if t_{2g} or e_g orbitals are half filled or fully filled *i.e.* having one or two electrons each, their rotational transformation into one another becomes invalid. For example, the d_{xz} orbital cannot be transformed by rotation into equivalent d_{yz} orbitals because this has an electron already. Thus, the orbital magnetic moment will be zero.

Following the above principles, we can predict that metal ions having the configurations d^1 (t_{2g}^1) and d^2(t_{2g}^2) will have orbital contribution but the complex with d^3(t_{2g}^3) configuration will not have any orbital contribution. Table 3.3 lists the electronic configurations of the central metal atom in which the orbital contribution is present or absent in high spin and low spin octahedral complexes with d^1 to d^2 configurations.

Table 3.3 Orbital contribution in octahedral complexes

No. of electrons	High spin octahedral		Low spin octahedral	
dn	Electronic Configuration	Orbital contribution	Electronic configuration	Orbital contribution
d^1	t_{2g}^1	yes	t_{2g}^1	yes
d^2	t 2g	yes	t_{2g}^2	yes
d^3	t_{2g}^3	no	t_{2g}^3	no
d^4	$t_{2g}^3 e_g^1$	no	t_{2g}^4	yes
d^5	$t_{2g}^3 e_g^2$	no	t_{2g}^5	yes
d^6	$t_{2g}^4 e_g^2$	yes	t_{2g}^6	no
d^7	$t_{2g}^5 e_g^2$	yes	$t_{2g}^6 e_g^1$	no
d^8	$t_{2g}^6 e_g^2$	no	$t_{2g}^6 e_g^2$	no
d^9	$t_{2g}^6 e_g^3$	no	$t_{2g}^6 e_g^3$	no

Orbital contribution in Tetrahedral Complexes

In the case of tetrahedral complexes, the five *d*-orbitals split into two sets t_2 and *e* orbitals. The t_2 orbitals (d_{xy}, d_{yz} and d_z) have higher energies than orbitals ($d_{x^2-y^2}$ and d_z^2). Thus in tetrahedral complexes, the metal ions having the following configurations will not have orbitals contribution :

$d^1(e^1)$, $d^2(e^2)$, $d^5(e^2\ t_2^{\ 3})$, $d^6(e^3\ t_2^{\ 3})$ and $d^7(e^4\ t_2^{\ 3})$

On the otherhand, the tetrahedral complexes having the following configurations will have orbital contribution :

$d^3(e^2\ t_2^{\ 1})$, $d^4(e^2\ t_2^{\ 2})$, $d^8(e^4\ t_2^{\ 4})$ and $d^9(e^4\ t_2^{\ 5})$.

These are given in Table 3.4

Table 3.4. Orbital contribution in tetrahedral complexes

No. of electrons	Electronic Configuration	Orbtial Contribution
d^1	e^1	no
d^2	e^2	no
d^3	$e^2\ t_2^{\ 1}$	yes
d^4	$e^2\ t_2^{\ 2}$	yes
d^5	$e^2\ t_2^{\ 3}$	no
d^6	$e^3\ t_2^{\ 3}$	no
d^7	$e^4\ t_2^{\ 3}$	no
d^8	$e_4\ t_2^{\ 4}$	yes
d^9	$e_4\ t_2^{\ 5}$	yes

The above principle explains why tetrahedal complex of one transition element possesses orbital contribution to the magnetic moment, while another element does not.

For example we find that the magnetic moments of Co(II) tetrahedral complexes (d^7) are nearer to the spin only values as expected. On the other hand, the magnetic moments of Ni(II) complexes of tetrahedral geometry (d^8) are more than the spin only values as expected. In some cases, there may be contribution from **spin-orbital coupling**. In such cases we calculate the magnetic moments by using the following equation :

$$\mu_{eff} = \mu_0\left(1 - \alpha\frac{\lambda}{\Delta}\right)$$

Here μ_0 = Spin only magnetic moment

α = Constant

λ = Spin-orbit coupling constant positive for $n < 5$ and negative

Δ is the difference in energy level between the ground state and higher state. The values of λ and α are given for some metal ions in Table 3.5. It may be noted that the value of λ is positive or negative. This explains as to why the value of the magnetic moment measured is smaller in some cases and greater in other cases than the spin only value–although the number is fairly small in the first category.

Table 3.5. Spin orbital coupling constant (l) and a values of some transition metal ions

Metal ion	No. of electrons	λ	α	Stereochemistry
Ti^{3+}	1	54	2	Tetrahedral
V^{3+}	2	108	4	Tetrahedral
Cr^{3+}	3	91	4	Octahedral
Cr^{2+}	4	57	2	Octahedral
Mn^{3+}	4	88	2	Octahedral
Fe^{3+}	5	88	0	
Mn^{2+}	5	80	0	
Fe^{2+}	6	–102	2	Tetrahedral
Co^{2+}	7	–177	4	Tetrahedral
Ni^{2+}	8	–315	4	Octahedral
Cu^{2+}	9	–829	2	Octahedral

Temperature Independent Paramagnetism (TIP)

The complex ions such as MnO_4^-, CrO_4^{2-}, $[Co(NH_3)_6]^{3+}$, etc. have no unpaired electrons. These systems are not expected to show any orbital magnetic moment and are expected to be diamagnetic. However, they show weak paramagnetism. This happens because of the *coupling of the ground state of the system with the excited states of high energy under the influence of magnetic field.* As a result of coupling, the complexes show some orbital magnetism. As a result such substances are no more diamagnetic. As this paramagnetism does not depend upon the thermal population of levels unlike simple paramagnetism, it is called **temperature independent paramagnetism** or TIP. But it resembles diamagnetism in the sense that it is not due to any magnetic dipole existing in the molecule. But it is induced when the substance is placed in the magnetic field. It also resembles diamagnetism in its temperature independence and it has magnetic susceptibility of the order of diamagnetic substances. For diamagnetic transition metal ions it is very important in determining the position of NMR resonance.

3.6. MAGNETIC BEHAVIOUR OF FIRST ROW TRANSITION METAL COMPOUNDS

An attempt has been made to study the variation of orbital contribution along the first transition series.

The magnetic moments of the compounds of first transition series in common oxidation states are given in Table 3.6. Magnetic moments due to spin only (m_s) as well as magnetic moments due to spin and orbital contributions (m_{S+L}) are also also given.

Table 3.6. Magnetic moments of metal ions of first transition series

Ion	Outer configuration	No. of unpaired electrons	S	L	μ_S	μ_{S+L}	Obversed
Sc^{3+}, Ti^{4+}, V^{5+}	$3d^0$	0	0	0	0	0	
Ti^{3+}, V^{4+}	$3d^1$	1	1/2	2	1.73	3.0	1.7—1.8
Ti^{2+}, V^{3+}	$3d^2$	2	1	3	2.84	4.47	2.76
V^{2+}, Cr^{3+}	$3d^3$	3	3/2	3	3.87	5.20	3.86
Cr^{2+}, Mn^{3+}	$3d^4$	4	2	2	4.90	5.48	4.80
Mn^{2+}, Fe^{3+}	$3d^5$	5	5/2	0	5.92	5.92	5.96
Fe^{2+}	$3d^6$	4	2	2	4.90	5.48	5.1—5.6
Co^{2+}	$3d^7$	3	3/2	3	3.87	5.20	4.4—5.2
Ni^{2+}	$3d^8$	2	1	3	2.84	4.47	2.9—3.4
Cu^{2+}	$3d^9$	1	1/2	2	1.73	3.0	1.7—2.1
Cu^{+}, Zn^{2+}	$3d^{10}$	0	0	0	0	0	0

We observe that the magnetic moments of compounds of first half of transition series are fairly close to their calculated spin only magnetic moments and are quite different from calculated magnetic moments based upon spin and orbital contributions. However, in the later half of the compounds of first transition series, the observed values are slightly higher than the calculated spin only values. This means that *in the first transition series, orbital contribution to the magnetic magnetic moments is negligible.*

Two information can be useful in establishing details of electronic structure and stereochemistry. We discuss below the important features of some complexes of first transition series with different d^n configurations

***d^1* system**

Titanium (III) octahedral complexes having t_{2g}^1 configuration such as $[Ti(H_2O)_6]^{3+}$ are coloured and paramagnetic. These have magnetic moment of 1.8 B.M. at room temperature. This value is higher than the expected spin only value of 1.73 B.M. and is attributed to the **partially quenched orbital contribution.**

***d^2* system**

The metals ions having d^2 system with t_{2g}^2 configuration such as V(III) octahedral complexes are also expected to have some **orbital contribution** to magnetic moment. However, the observed magnetic moment values of V(III) octahedral complexes are in the range of 2.7 to 2.8 B.M. and are even less than the expected spin only value of 2.84 B.M. This observation is not easy to explain and is attributed to the presence of some distortions which split the ground state term $^3T_{2g}$ (refer to electronic spectra of transition metals). (Chapter 4)

***d^3* system**

Chromium (III) octahedral complexes are most extensively studied compounds in this category. The observed magnetic moment values at room temperature lie in the range 3.70 to 3.84 B.M. The magnetic moment arising from spin only value of 3.87 B.M. agrees fairly well with the observed values and is independent of temperature. However, the reduced values of magnetic moments are observed in some bridged complexes such as $[(NH_3)_5\ Cr(\mu\text{–}OH)\ Cr(NH_3)_5]\ X_5$ (where X = Cl or Br) and $[(NH_3)_5\ Cr\text{–}O\text{–}Cr(NH_3)_5]^{4+}$ and are temperature dependent.

d^4 system

Chromium (II) octahedral complexes have d^4 or $t_{2g}^{\,3}\ e_g^{\,1}$ configurations. The complex $[Cr(SO_4)_2]^{2+}$ has magnetic moment of 4.90 B.M. which agrees with the spin-only value. Low spin octahedral complex like $K_4[Cr(CN)_6]$. $3H_2O$ having the configuration $t_{2g}^{\,4}$ shows the magnetic moment in the expected range of 2.74 to 2.84 B.M. (spin only with partial orbital contribution). However, chromium (II) acetate dihydrate, $Cr_2(CH_3COO)_4.2H_2O$ is one of the most stable chromium compounds and is **diamagnetic.** This suggests that all the four unpaired electrons take part in Cr-Cr bonding in the form of quadrupole Cr ≡ Cr bond, thereby leaving no unpaired *d*-electron on Cr-atoms.

d^5 system

Octahedral complexes of Mn(II) and Fe(III) with $t_{2g}^{\,3}\ e_g^{\,2}$ configuration belong to this configuration. These are high spin complexes and show magnetic moments in the expected range of spin only value of 5.92 B.M. However, some oxo bridged species of Fe(II) like those of Cr(III) show values lower than 5.92 B.M. This is explained due to the **antiferromagnetic interactions** between the electron spins of two metal ions transmitted across Fe-O-Fe bridge.

d^6 system

This system includes iron (III) octahedral complexes of high spin having the configuration $t_{2g}^{\,4}$ $e_g^{\,2}$. It has 4 unpaired electrons and has magnetic moment around 5.5 B.M. which agrees well with orbital contribution to spin only value of 4.9 B.M. Some distortions produced values in the range of 5.2 to 5.4 B.M. However, low spin Fe(II) complexes having the configuration $t_{2g}^{\,6}$ are diamagnetic.

Most of the cobalt (II) complexes of $t_{2g}^{\,6}$ configuration show diamagnetism. However, octahedral complexes such as $K_3[CoF_6]$ are high spin having the configuration $t_{2g}^{\,4}\ e_g^{\,2}$ and having the magnetic moment equal to 5.8 B.M. This explained in terms of spin only value having some orbital contribution.

d^7 system

Cobalt (II) having d^7 system forms a number of complexes which are less stable than those of Co(III). These complexes may be tetrahedral or octahedral. Most cobalt (II) complexes are high spin but CN^- produces low spin complexes. A low spin octahedral complex will have the configuration $t_{2g}^{\,6}\ e_g^{\,1}$. The magnetic moments of tetrahedral cobalt (II) having configuration $e^4\ t_2^{\,3}$ lie in the range of 4.4 to 4.8 B.M. at room temperature. The spin only magnetic moment value is 3.87 B.M.

d^8 system

Nickel (II) d^8 octahedral complexes such as $[Ni(H_2O)_6]^{2+}$ with $t_{2g}^{\,6}\ e_g^{\,2}$ configuration have magnetic moment of 3.2 B.M. at room temperature. This is explained on the basis of spin-orbital complexes. Similarly, the $[Ni(acac)_2]$ complex is octahedral and shows a normal magnetic moment of 3.2 B.M. like other Ni(II) complexes. However, it shows normal magnetic moment down to about 80K but below this temperature, the magnetic moment increases from 3.2 B.M. to 4.1 B.M. at 4.3 K. This is attributed to **ferromagnetic coupling.**

Ni(II) tetrahedral complexes having $e^4\ t_2^{\,4}$ configuration have orbital contribution and the observed magnetic moment values in the range 3.2 to 4.1 B.M. can be explained on the basis of orbital contribution to magnetic moment.

The magnetic properties of complexes can be very useful in establishing details of structure and stereochemistry. The Ni(II) complexes provide an example of usefulness of magnetic properties. For example consider the complex $[Ni(py)_4]\ (ClO_4)_2$. This can be prepared under the slightly different conditions and two complexes of this composition have been isolated. One is yellow and diamagnetic while other is blue with a magnetic moment of 2.9 B.M. The blue complex can be assigned an octahedral structure. The yellow complex is most likely the square planar complex in which perchlorate ions are not coordinating. However, five coordinate geometry with a single coordinate ClO_4^- ligand cannot be excluded on the basis of magnetic data.

3.7. MAGNETIC PROPERTIES OF OCTAHEDRAL COMPLEXES BASED ON CRYSTAL FIELD THEORY

The filling of *d*-orbitals in presence of weak ligand and strong ligand is shown in Table 3.7. In a weak ligand field *d*-orbital will be occupied singly before any pairing of electrons takes place. This is because the energy difference between t_{2g} and e_g orbitals will be too small to affect the issue.

In the field of weak ligands (such as F^-, OH^-, Cl^- , Br^- . I^- etc.) the crystal field splitting is small so less energy is required to feed the fourth electron into a e_g orbital than to pair it with one of the t_{2g} orbitals. Hence it will be more favourable energetically to occupy the upper e_g level and have a high spin complex rather than to pair of electrons with one of t_{2g} orbitals. Here Δ_0 (octahedral) is less than pairing energy *P, i.e.,* $\Delta_o < P$ or pairing energy P is more than splitting energy (Δ_o) *i.e.,* $P > \Delta_o$.

In a field of strong ligands such as CN^-, NO_2, en, NH_3 or CNS^-, the crystal field splitting is quite large. Thus Δ_o, the difference between the energies of e_g and t_{2g} orbitals is quite significant, so electron pairing takes place. The electrons would rather pair in lower energy t_{2g} orbitals rather than occupying higher energy e_g orbitals. Splitting energy Δ_o is more than pairing energy *i.e.,* $\Delta_0 > P$ or paring energy is less than splitting energy *i.e.,* $P < \Delta_o$.

There is a significant difference in the arrangement of d^4, d^5, d^6 and d^7 metal ion configurations in a weak and a strong field as shown in Table. 3.7:

Table 3.7. Octahedral complexes

No. of Electrons	*Weak Field (High spin)*					*Strong Field (Low Spin)*				
	t_{2g}			e_g		t_{2g}			e_g	
1	↑					↑				
2	↑	↑				↑	↑			
3	↑	↑	↑			↑	↑	↑		
4	↑	↑	↑	↑		↑↓	↑	↑		
5	↑	↑	↑	↑	↑	↑↓	↑↓	↑		
6	↑↓	↑	↑	↑	↑	↑↓	↑↓	↑↓		
7	↑↓	↑↓	↑	↑	↑	↑↓	↑↓	↑↓	↑	
8	↑↓	↑↓	↑↓	↑	↑	↑↓	↑↓	↑↓	↑	↑
9	↑↓	↑↓	↑↓	↑↓	↑	↑↓	↑↓	↑↓	↑↓	↑

In these arrangements the number of unpaired electrons is greater in a weak field than in a strong field because the strong field forces the electrons to pair. Complexes with weak ligands are high spin complexes and are paramagnetic in nature while those with strong ligand fields are low spin complexes and are diamagnetic in nature, High spin complexes having more unpaired electrons have high paramagnetism than low-spin complexes which have fewer unpaired electrons.

3.8. MAGNETIC PROPERTIES IN TETRAHEDRAL COMPLEXES BASED ON CRYSTAL FIELD THEORY

The filling of *d*-orbitals in certain ions surrounded by tetrahedral ligands in weak and strong fields is shown in Table 3.8.

Table 3.8. Tetrahedral complexes.

No.of Electrons	*Weak Field (High Spin)*					*Strong Field (Low Spin)*				
	e_g			t_{2g}		e_g			t_{2g}	
1	↑					↑				
2	↑	↑				↑	↑			
3	↑	↑	↑			↑↓	↑			
4	↑	↑	↑	↑		↑↓	↑↓			
5	↑	↑	↑	↑	↑	↑↓	↑↓	↑		
6	↑↓	↑	↑	↑	↑	↑↓	↑↓	↑	↑	
7	↑↓	↑↓	↑	↑	↑	↑↓	↑↓	↑	↑	↑
8	↑↓	↑↓	↑↓	↑	↑	↑↓	↑↓	↑↓	↑	↑
9	↑↓	↑↓	↑↓	↑↓	↑	↑↓	↑↓	↑↓	↑↓	↑

It is clear from the table that the configurations of d^3, d^4, d^5 and d^6 ions may form high and low spin complexes. But in actual practice only high spin complexes are formed. This is due to the fact that crystal field splitting Δ_t is quite small and is always less than the pairing energy, *i.e.*, $\Delta_t < P$. Therefore high spin and paramagnetic complexes are formed.

Example. How does crystal field theory explain the diamagnetic behaviour of $[Ni(CN)_4]^{2}$ and paramagnetic behaviour of $[NiC1_4]$.

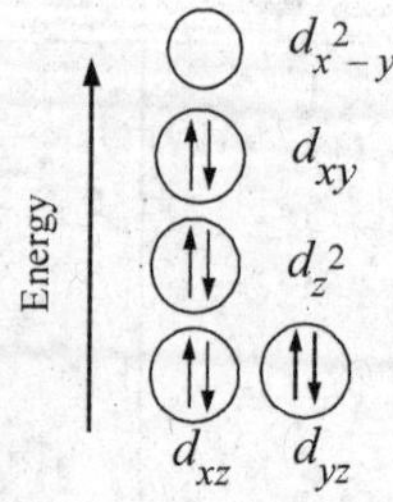

Solution. Cyanide ion is a strong field ligand. It splits the *d* orbitals into three sets of orbitals with varying energies as shown aside:

Eight electrons in Ni^{2+} ions are filled in the orbitals as shown above. Because of large value of splitting energy, electrons don't occupy higher energy $d_{x^2-y^2}$ orbitals. Thus, there are no unpaired electrons and the complex is diamagnetic. Chloride is a weak ligand causing only a small splitting. Electrons occupy different orbitals giving tetrahedral geometry. Complex is paramagnetic due to presence of unpaired electrons.

SOLVED CONCEPTUAL PROBLEMS

Example 1. Cu^{2+} ions are coloured paramagnetic while Zn^{2+} ions are colourless and diamagnetic. Explain.

Solution. The electronic configuration of Cu^{2+} and Zn^{2+} are :

Cu^{2+} : $3d^2$; Zn^{2+} : $3d^{10}$

Cu^{2+} has one unpaired electron and therefore is paramagnetic. On the other hand, Zn^{2+} has all electrons paired and therefore, is diamagnetic.

Example 2. Define magnetic susceptibility. Predict whether the molar magnetic susceptibility is independent or dependent on the applied magnetic field for the following types of substances :

paramagnetic, diamagnetic, ferromagnetic and antiferromagnetic.

Solution. Magnetic susceptibility is a measure of the capacity of a substance to take up magnetisation in applied magnetic field. The magnetic susceptibility per unit volume is represented as κ. The molar susceptibility $\chi_m = \dfrac{\kappa \cdot M}{\rho}$

While ferromagnetic and antiferromagnetic substances are dependent on applied magnetic field, paramagnetic and diamagnetic susbtances are not.

Example 3. Cu (I) is diamagnetic whereas Cu (II) is paramagnetic.

Solution. In Cu (II), the outermost electronic configuration is $3d^9$ and therefore, it has one unpaired electron. Hence, it is paramagnetic. However, in Cu (I), the outermost electronic configuration is $3d^{10}$. There is no unpaired electron and therefore, Cu (I) is diamagnetic.

Example 4. Calculate in Bohr magneton the expected magnetic moment for the following ions (spin magnetic moment) :

(i) **Fe^{3+}** ***(ii)*** **Ni^{2+}** ***(iii)*** **Cu^+**

Solution. Spin magnetic moment is given as :

$$\mu = \sqrt{n(n+2)} \text{ B.M.}$$

where n is number of unpaired electrons.

(*i*) Fe^{3+} : [Ar] $3d^5$ No. of unpaired electrons = 5

$$\mu = \sqrt{5(5+2)} = \sqrt{37} = 5.96 \text{ B.M.}$$

(*ii*) Ni^{2+} : [Ar] $3d^8$ No. of unpaired electrons = 2

$$\mu = \sqrt{2(2+2)} = 2\sqrt{2} = 2.828 \text{ B.M.}$$

(*iii*) Cu^+ : [Ar] $3d^{10}$ No. of unpaired electrons = 0

$$\mu = 0$$

Example 5. Which ion should exhibit a larger magnetic moment : Mn^{2+} or V^{2+} ?

Solution. Mn^{2+} : [Ar] $3d^5$

V^{2+} : [Ar] $3d^3$

Mn^{2+} has five unpaired electrons whereas V^{2+} has three unpaired electrons. Therefore, Mn^{2+} has higher magnetic moment.

Example 6. On the basis of crystal field theory, account for the following : While $[Fe(H_2O)_6]^{3+}$ is throughly paramagnetic, $[Fe(CN)_6]^{3-}$ is less paramagnetic.

Solution. This is because H_2O is a weak ligand and $\Delta_0 < P$ and therefore no spin pairing takes place and the complex obtained is strongly paramagnetic. In the case of CN^- which is a strong ligand $\Delta_0 > P$ and, therefore, electrons pair up leaving only one *d*-electron. This gives a weak paramagnetic complex. This is shown in as under :

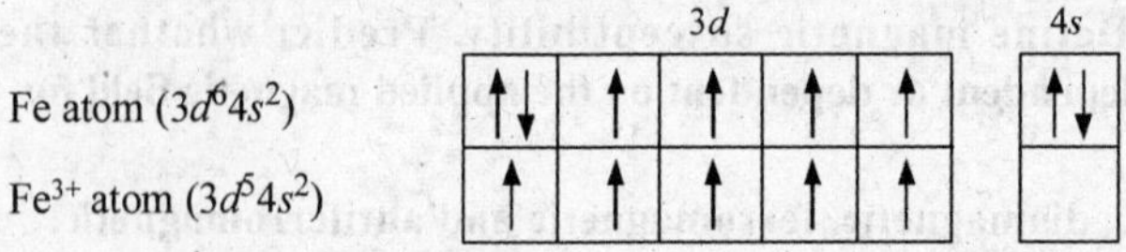

d-electrons in Fe atom and Fe^{3+} ion.

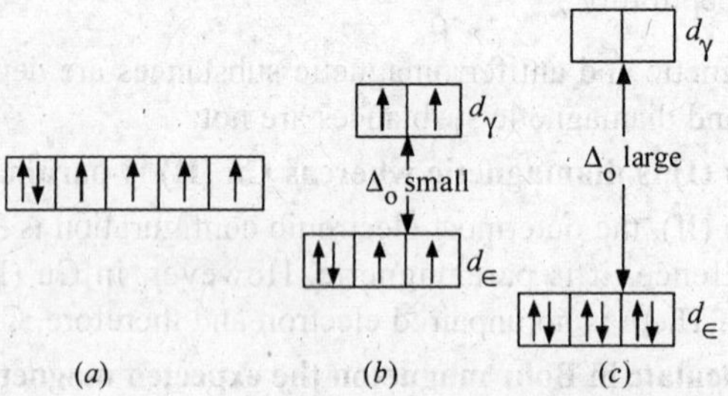

Occupation of *d*-electrons in Fe^{3+} ion under different ligand fields (*a*) Free ion, (*b*), In weak octahedral field, *e.g.*, $[Fe(H_2O)_6]^{3+}$, (*c*) In strong cotahedral field *e.g.*, $[Fc(CN)_6]^{3-}$

Example 7. Predict which of the following configurations are expected to have orbital contribution in tetrahedral field :

(*i*) d^2 (*ii*) d^4 (*iii*) d^8 (*iv*) d^7 (*v*) d^9

Solution. The following configurations are expected to have orbital contribution in tetrahedral field :

(*ii*) d^4 (*iii*) d^8 (*v*) d^9

Example 8. What is magnetic susceptibility ? How does it vary with temperature ?

Solution. Magnetic susceptibility is a measure of the capacity of a substance to take up magnetisation in applied magnetic field. The magnetic field per unit volume is represented as κ. The molar susceptibility is given as :

$$X_m = \frac{\kappa \cdot M}{\rho}$$

where M is the molecular mass and ρ is the density.

Magnetic susceptibility of a substance varies inversely with temperature as :

$$X_m = \frac{1}{T} \quad \text{or} \quad X_m = \frac{C}{T}$$

where T is absolute temperature and C is a constant known as Curie's constant.

The variation of X_m with temperature for paramagnetic substance is shown below :

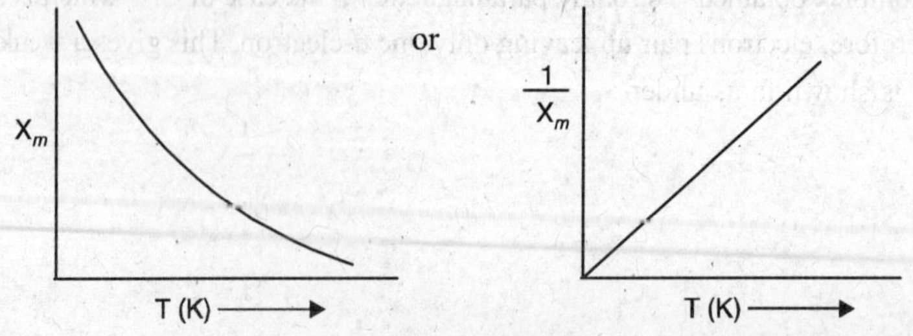

Example 9. Give examples of two species which show temperature independent paramagnetism.

Soution. (*i*) CrO_4^{2-} (*ii*) MnO_4^-

Example 10. Do you expect the complex $[Cr(CH_3COO)_2 \cdot H_2O]_2$ to be diamagnetic or paramagnetic ?

Solution. This complex is diamagnetic due to quadrupole bonding between Cr – Cr.

Example 11. List the main configuration in octahedral field where orbital contributions are

(*i*) quenched (*ii*) not quenched

Solution. Orbital contributions to magnetic moment in octahedral field are quenched in the following configurations :

$(d^3)\ t_{2g}^{\ 3}$, $(d^6)\ t_{2g}^{\ 6}$ due to half filled and completely filled configurations.

Orbital contribution to magnetic moment in octahedral field are quenched :

High spin : $d^3(t_{2g})^3$, $d^4(t_{2g}^{\ 3}\ e_g^{\ 1})$, $d^5(t_{2g}^{\ 3}\ e_g^{\ 2})$, $d^8(t_{2g}^{\ 6}\ e_g^{\ 2})$, $d^9(t_{2g}^{\ 6}\ e_g^{\ 3})$

Low spin : $d^6(t_{2g}^{\ 6})$, $d^7(t_{2g}^{\ 6}\ e_g^{\ 1})$

Orbital contribution to magnetic moment is not quenched.

High spin : $d^1(t_{2g}^{\ 1})$, $d^2(t_{2g}^{\ 2})$, $d^6(t_{2g}4\ e_g^{\ 2})$, $d^7(t_{2g}^{\ 5}\ e_g^{\ 2})$

Low spin : $d^4(t_{2g}^{\ 4})$, $d^5(t_{2g}^{\ 5})$

Example 12. Predict which of the following configurations are expected to have orbital contribution in high spin octahedral field : d^1, d^3, d^6, d^8.

Solution. The following configurations are expected to have orbital contribution in high spin octahedral field : d^1 and d^6

Example 13. On the basis of crystal field theory, account for the following:

While $[CoF_6]^{3-}$ is paramagnetic, $[Co(NH_3)_6]^{3-}$ is diamagnetic.

The $[CoF_6]^{3-}$ ion is an example of high spin (outer orbital) complex while $[Co(CN)_6]^{3-}$ is a low spin (inner orbital) complex. Both have octahedral geometry and the magnetic preperties are explained below on the basis of the field strength of the CN^- ligand.

In a free Co atom ($3d^7\ 4s^2$) and a Co^{3+} ion ($3d^6\ 4s^0$) the electrons are arranged as follows:

	3*d*					4*s*
Co atom	↑↓	↑↓	↑	↑	↑	↑↓
Co^{3+} ion	↑↓	↑	↑	↑	↑	↑

In an octahedral complex, however the five *d*-orbitals split up into t_{2g} and e_g orbitals in the $[CoF_6]^{3-}$ ion. The energy difference Δ_o between the two sets of orbitals is not great enough (as F^- is a weak field ligand) to cause any spin-pairing ($\Delta_o < P$) so that the arrangement of the five *d*-orbitals is as shown in the figure below. There are four unpaired electrons and the complex ion is strongly paramagnetic.

In the $[Co\ (CN)_6]^{3-}$ ions, the stronger field of CN^- ion causes a greater energy difference (Δ_o) between t_{2g} and e_g orbitals. As a result spin pairing occurs because $P < \Delta_o$. The five *d*-orbitals are arranged as show in below :

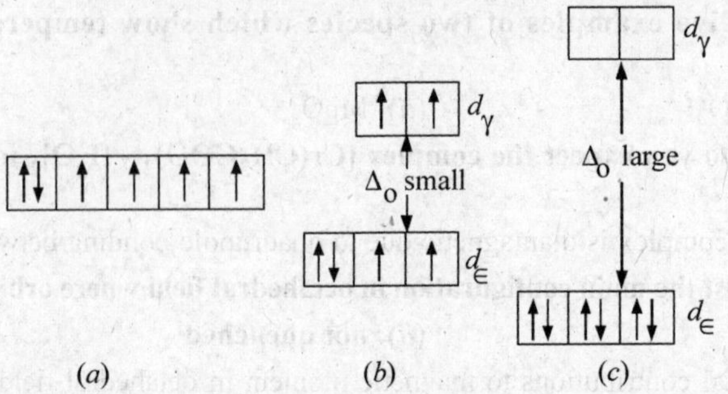

Arrangement of *d* electrons in free Co^{3+} ion and under different ligand fields (*a*) Free ion (*b*), In a weak octahedral field, (*c*) In a strong octahedral field.

EXERCISES
(Including Questions from Different University Papers)

Multiple Choice Questions (Choose the correct answers)

1. Ferromagnetic substance behaves as simple paramagnetics

(*a*) below Curie temperature (*b*) below Neel Temperature
(*c*) above Curie temperature (*d*) above Neel temperature.

2. The number of unpaired electrons in chromous acetals $[Cr(CH_3COO)_2H_2O]_2$ are

(*a*) 0 (*b*) 2 (*c*) 3 (*d*) 4

3. In which of the following configuration, the orbital contribution is quenched in octahedral field ?

(*a*) $t_{2g}^4 e_g^2$ (*b*) $t_{2g}^6 e_g^1$ (*c*) t_{2g}^4 (*d*) $t_{2g}^5 e_g^2$

4. $[Ti(H_2O)_6]$ represents the system

(*a*) d^2 (*b*) d^1 (*c*) d^3 (*d*) d^5

5. The spin only magnetic moment for Co^{2+} ion is

(*a*) 4.90 B.M. (*b*) 3.87 B.M. (*c*) 2.84 B.M. (*d*) 1.73 B.M.

6. Which of the following ion will have maximum μ_s value ?

(*a*) Fe^{2+} (*b*) Cr^{3+} (*c*) Fe^{3+} (*d*) Co^{2+}

7. Ferromagnetic materials show normal paramagnetic behaviour

(*a*) below TC (*b*) below TN (*c*) above TC (*d*) above TN

8. Which of the following is expected to be diamagnetic

(*a*) $CrCl_3$ (*b*) $CuCl_2$ (*c*) $ZnCl_2$ (*d*) $CuSO_4$?

9. Substances having paired electrons are

(*a*) ferromagnetic (*b*) antiferromagnetic
(*c*) diamagnetic (*d*) paramagnetic

10. If *k* is the magnetic susceptibility per unit volume, ρ is the density and M is the molecular weight then molar susceptibility is

(*a*) $\kappa M/\rho$ (*b*) $\kappa\rho/M$ (*c*) $M/\kappa\rho$ (*d*) $\kappa/M\rho$

11. In the first transition series, the divalent compound having maximum magnetic moment is of

(*a*) Mn (*b*) Fe (*c*) Cu (*d*) Cr

12. Which of the following: configuration will not have orbital contribution in tetrahedral geometry ?

(*a*) d^2 (*b*) d^4 (*c*) d^8 (*d*) d^9

ANSWERS

1. (*c*)	**2.** (*a*)	**3.** (*b*)	**4.** (*b*)	**5.** (*b*)	**6.** (*c*)
7. (*c*)	**8.** (*c*)	**9.** (*c*)	**10.** (*a*)	**11.** (*a*)	**12.** (*a*)

Short Answer Questions

1. Give one advantage each of Guoy method and Faraday method.
2. Calculate the magnetic moment (spin only) for the complex $K_4[Mn(NCS)_6]$.
3. Define magnetic susceptibility. Name a method used to measure it.
4. Draw the curves of χ_M vs T(k) and $\frac{1}{\chi_M}$ vs T(K)
5. What is the calculated μ_s value for Fe^{2+} high spin ion ?
6. Define ferromagnetic substance and give one example.
7. State the stereochemistry of Cu^{2+}, Mn^{3+} and V^{3+}
8. How are magnetic permeability and magnetic susceptibility related ?
9. Calculate the spin magnetic moment of V^{3+} ion.
10. What is meant by quenching of orbital angular momentum.

General Questions

1. (*a*) What is the origin of magnetism ? How is magnetic susceptibility measured ?

(*b*) The expected magnetic moment of $K_4[FeF_6]$ is 5 B.M. How will you account for this on the basis of valence bond theory ?

(*c*) Write short note on ferromagnetism and antiferromagnetism.

2. (*a*) Discuss the phenomenon of orbital contribution to magnetic moment. Give electronic configurations in which orbital contribution is quenched in O_h field.

(*b*) Show that $\sqrt{4S(S+1)}$ and $\sqrt{n(n+2)}$ are equivalent expressions.

3. (*a*) What is magnetic susceptibility and how does it vary with temperature ?

(*b*) What do you understand by diamagnetic - paramagnetic equilibrium in solid complexes ?

4. (*a*) What is magnetic susceptibility ? How does it vary with temperature ?

(*b*) Why does Mn(II) ion shows maximum magnetic character amongst the divalent ions of first transition series ?

(*c*) What do you understand by the term diamagnetic correction ?

(*d*) What is temperature independent paramagnetism ?

5. (*a*) Discuss briefly the Gouy's method for measuring magnetic susceptibility.

(*b*) Dicuss the magnetic behaviour of first row transition metal compounds.

(*c*) Discuss the behaviour of ferromagnetism and antiferromagnetism with temperature.

6. (a) Give a brief account of diamagnetism, paramagnetism, ferromagnetism and antiferromagnetism.
   (b) What is meant by temperature independent paramagnetism ? Explain.
7. (a) What is the origin of paramagnetism in transition metal compounds ?
   (b) Calculate in Bohr magneton the expected magnetic moment for the following ions (spin magnetic moment only)
   (i) Fe^{3+} (ii) Mn^{2+} (iii) Ni^{2+} (iv) Cu^{+}
8. (a) How do you explain the ferromagnetic properties of iron, cobalt and nickel ?
   (b) Discuss briefly the Gouy's method for measuring magnetic susceptibility.
9. (a) What is meant by quenching orbital angular momentum and what are its consequences ? Give electronic configuration in which orbital contribution is quenched in octahedral and tetrahedral complexes.
   (b) Calculate μ_s and μ_{S+L} for $3d^2$.
   (**Hint.** S = 1 and L = 3).
10. (a) Define paramagnetism, diamagnetism, ferromagnetism and antiferromagnetism.
    (b) Discuss Gouy's method for measuring magnetic susceptibility.
11. Discuss the phenomenon of orbital contribution to magnetic moment and give brief application of this phenomenon to 3*d*-metal complexes.
12. The magnetic moment of transition elements can be calculated from the relation :

$$\mu(S + L) = \sqrt{4S(S+1) + L(L+1)}$$

Explain the various terms involved.

13. (a) Describe Gouy's method for measuring magnetic susceptibilities.
    (b) What do you understand by the terms paramagnetism and diamagnetism ? Predict the magnetic moment for octahedral complexes of Fe^{2+} with strong field ligands and with weak field ligands.
14. (a) Calculate spin only magnetic moment for Co^{2+} ion.
    (b) What is temperature independent paramagnetism ? Explain.
    (c) Describe briefly quenching of orbital angular momentum in tetrahedral complexes.
    (d) Discuss the variations of magnetic susceptibility with temperature.
15. (a) What do you understand by magnetic susceptibility ?
    How does it vary with temperature ?
    (b) Calculate the magnetic moments (spin only) for the following ions :
    (i) $[Mn(NCS)_6]^{4-}$ (ii) $[Fe(H_2O)_6]^{3+}$ (iii) Ni^{2+}
    Which of the above ions do you expect to be coloured ?
    (c) "Cu^{2+} ions are coloured and paramagnetic while Zn^{2+}-ions are colourless and diamagnetic. Explain.
16. (a) What is meant by temperature independent paramagnetism ?
    (b) Calculate, in Bohr magnetons, the magnetic moment expected from spin only for the ions : V^{4+}, Cr^{+3}, Fe^{3+} and Ni^{3+}.
17. (a) Discuss the variation of magnetic susceptibility with temperature
    (b) What do you understand by the term diamagnetic correction ?
    (c) Why we consider spin only formula for finding the magnetic moment of 3*d*-complexes ?
    (d) What is temperature independent paramagnetism ?

18. (*a*) What is magnetic susceptibility ? How does it vary with temperature ?

(*b*) Calculate the magnetic moment of Cr^{3+} ion by the spin only formula

19. (*a*) Discuss magnetism in

(*b*) Ni complexes

(*c*) Hydroxo, oxo and bridged complexes of Cr(III).

20. (*a*) Discuss origin of magnetism in substances

(*b*) Derive the relationship between magnetic susceptibility and magnetic moment

21. (*i*) Explain antiferromagnetic susbtance

(*ii*) Name one antiferromagnetic substance

22. (*a*) Discuss the relationship between magnetic moment and magnetic susceptibility.

(*b*) The complexes of first transition series are mainly high spin, while those of second and third translation series are of low spin type. Explain.

(*c*) The complex $[Ni(CN)_4]^{2-}$ is diamagnetic but $[NiCl_4]^{2-}$ is paramagnetic

23. (*a*) Calculate magnetic moment (spin only) of $K_4[Mn(CNS)_6]$ complex

(*b*) Discuss variation of magnetic susceptibility with temperature

24. (*a*) What is the origin of paramagnetism ?

(*b*) How magnetic susceptibility varies with temperature ?

(*c*) Calculate magnetic moment for the complex $K_4[Mn(CNS)_6]$ (Spin only value)

25. (*a*) Discuss Gouy's method for measuring magnetic susceptibility. How magnetic susceptibility varies with temperature ?

(*b*) What do you understand by spin and orbital contribution to the magnetic moments ?

(*c*) Calculate the expected values of magnetic moments in B.M. from spin only formula of Ni^{2+}, Zn^{2+}, Cr^{3+} and Fe^{3+} ions.

26. (*a*) The net paramagnetism in a substance is slightly less than the true paramagnetism. How will you explain this behaviour ?

(*b*) Which of the following metal ions are expected to show magnetic moment close to 1.73 B.M.: Cu^+ , Ti^{3+} , Fe^{2+}, Cr^{3+} ?

(*c*) What do you understand by diamagnetic-paramagnetic equilibrium in solid complexes ?

(*d*) Discuss Gouy's method for measuring magnetic susceptibility. How magnetic susceptibility varies with temperature ?

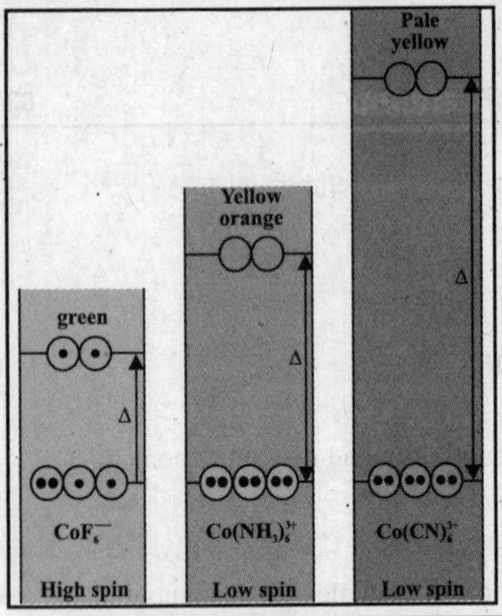

4

Electronic Spectra of Transition Metal Complexes

4.1. INTRODUCTION

Electronic spectra of transition metal complexes help us to gain an insight into the structure and bonding in these compounds. A very exciting characteristic of transition metal complexes is that they exhibit colours of varying intensity throughout the visible range. A blue solution of copper sulphate becomes dark blue on addition of ammonia solution. Perhaps the nature of the ligand has something to do about it. The stronger NH_3 ligands replace the H_2O ligands. Sometimes the same metal displays different colours even with the same ligands in different crystal fields. For example, cobalt (II) forms pink $[Co(NH_3)_6]^{2+}$ in octahedral field but blue $[CoCl_4]^{2-}$ in tetrahedral field. Blood that flows in our body is red because of complexation of heme group with iron. Prussian blue which has been used as a dye since olden times is a coordination compound of iron.

It has been established that colour of complex compounds is because of $d-d$ transitions. The electrons are excited from the lower *d*-orbitals to the higher *d*-orbitals. Energy from the visible range is absorbed in these radiations. Rest of the radiations (energy) are transmitted. It is the transmitted radiations that form the colour of the substance. *It is important to remember that the colour of a substance is not the radiations that are absorbed but the radiations that are transmitted.* We know that visible (white) light consists of different colours like violet, indigo, blue, green, yellow, orange and red. If one or more of the radiations are absorbed by a substance, the resultant of the rest of the colours will be the colour displayed by a substance. If a substance does not absorb any colour, it looks white. If it absorbs all the colours, it looks black.

4.2. BASIS OF ELECTRON–ABSORPTION

In tetrahedral complexes, the electron gets excited from lower set of *d*-orbitals to higher set of *d*-orbitals when visible light is incident on them. As a result of transition some selected wavelength of visible light corresponding to energy difference between t_{2g} and e_g levels is absorbed. The transmitted light gives colour to complexes. This is explained with the help of the following diagram. It may be recalled that degenerate *d*-orbitals split into two sets of *d*-orbitals, t_{2g} and e_g in the presence of ligands in octahedral field.

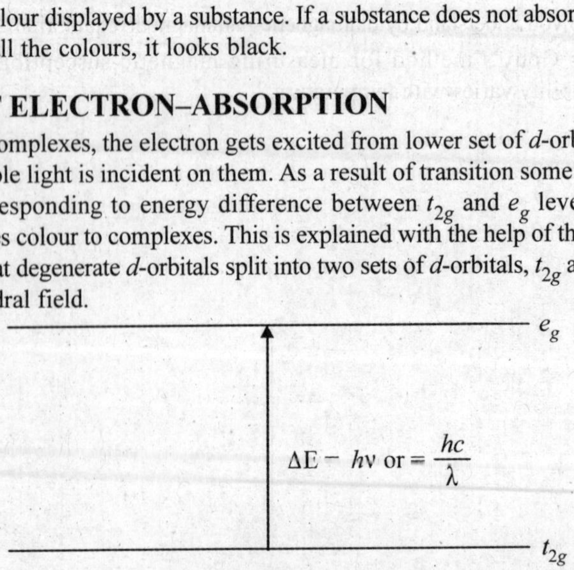

The amount of energy absorbed can be given as :

$$\Delta E = h\nu = \frac{hc}{\lambda}$$

where c is the velocity of light and h is Planck's constant.

The extent of intensity of absorption of light of a particular wavelength (or frequency) by a species in solution can be obtained by **Beer-Lambert** law. If light of intensity, I_0 at a given wavelength passes through a solution, the light transmitted with intensity, I may be measured by a suitable detector. According to Beer–Lambert, law, the light transmitted and incident light are related as :

$$\log \frac{I_0}{I} = A = \in lc$$

where
A = absorbance
c = molar concentration of absorbing species
$\in$ = constant known as **molar extinction coefficient.** It is also called **molar absorptivity** (L mol^{-1} cm^{-1})
l = path length through solution (in cm).

The term $\log \frac{I_0}{I}$ represented by A is known as **absorbance** (or optical density). The molar absorptivity is a characteristic property of the species that is absorbing the light and is highly dependent on wavelength. A plot of molar absorptivity versus wavelength gives a spectrum characteristic of the molecule or ion. This spectrum provides valuable information about the structure and bonding of the molecule or ion.

4.3. TERM SYMBOLS AND LS COUPLING

The complete description of the electrons in an atom can be obtained in terms of **four quantum numbers.** These four quantum numbers are :

(*i*) **Principal quantum number, *n*.** It determines the main energy level or shell in which the electron is present. It can have whole number values as :

$$n = 1, 2, 3, 4 ....$$

(*ii*) **Angular quantum number, *l*.** It is related to the angular momentum of the electron. In multi-electron atoms (an atom having a number of electrons) the energy besides depending upon n, also depends on l. Corresponding to each value of n, there are n possible values of l as :

$$l = 0, 1, 2, 3 .... (n-1)$$

These various values of l (called subshells) are designated as s, p, d, f ....

Value of l	0	1	2	3	4	and so on
Designation	*s*	*p*	*d*	*f*	*g*	and so on

(*iii*) **Magnetic quantum number, *m*.** It refers to the different orientations of the orbitals in space. For a given value of l, m can have $(2l + 1)$ values as :

$$m = -l, -(l-1) .... 0 .... +(l-1), +l$$

Thus if $l = 1$, m will have 3 values viz. $-1, 0, +1$.

If $l = 2$, m will have 5 values viz., $-2, -1, 0, +1, +2$.

(*iv*) **Spin quantum number, *s*.** This describes the spin orientation of the electron. Since the electron can spin only in two ways : clockwise and anticlockwise, *s* can have only two values : $+\frac{1}{2}$ and $-\frac{1}{2}$. The values may also be indicating by arrows pointing up (↑) and down (↓) respectively. It determines the spin angular momentum.

The distribution of electrons in different levels in an atom is known as **electronic configuration.** The filling of orbitals is governed by the three well known rules.

1. Aufbau principle. Electrons enter the orbitals of lowest energy first and subsequent electrons are filled in increasing order of energy (1*s*, 2*s*, 2*p*, 3*s*, 3*p*, 4*s*, 3*d* ....).

2. Paulis exclusion principle. No two electrons in an atom can have the same value for the four quantum numbers.

3. Hund's rule. When several orbitals of the same energy are available (*degenerate orbitals*), the electrons tend to remain unpaired as for as possible *i.e.*, the electrons tend to retain parallel spins as much as possible, on energy consideration.

The electronic configuration of the outershell is generally represented by a box diagram. The boxes represent orbitals and electrons are indicated by arrows. For example, the *s*-orbitals are indicated by a single box, the *p*-orbitals are indicated by three boxes, one each for the p_x, p_y and p_z orbitals.

Similarly the *d*-orbitals are represented by five boxes, one each for d_{xy}, d_{yz}, d_{zx}, $d_{x^2-y^2}$ and d_{z^2} as shown below :

One *s*-orbitials

Three *p*-orbitals

p_x	p_y	p_z

Five *d*-orbitals

d_{xy}	d_{yz}	d_{zx}	$d_{x^2-y^2}$	d_{z^2}

The *f*-orbitals are represented by seven boxes. A spin quantum number of $+\frac{1}{2}$ is indicated by the arrow ↑ and $-\frac{1}{2}$ by the arrow ↓. When only one electron is present in a degenerate energy level such as *s*, *p* or *d* subshell, it can occupy any one of the possible arrangements.

However, for atoms with more than one electron in degenerate orbitals, several different arrangements are possible which may not have same energy. For example, carbon has the electronic configuration as :

$$1s^2\ 2s^2\ 2p^2$$

It has two electrons in 2*p* subshell. There are 15 different arrangements in which the two electrons can be added to a set of degenerate three *p*-orbitals having *l* = + 1, 0, – 1 corresponding to p_x, p_y and p_z

l =	+1	0	–1	+1	0	–1	+1	0	–1	+1	0	–1	+1	0	–1
	↑	↑			↑	↑	↑		↑	↓	↓			↓	↓

l =	+1	0	–1	+1	0	–1	+1	0	–1	+1	0	–1	+1	0	–1
	↓		↓	↑	↓			↑	↓	↑		↓	↓	↑	

l =	+1	0	–1	+1	0	–1	+1	0	–1	+1	0	–1	+1	0	–1
		↓	↑	↑		↓	↑↓				↑↓				↑↓

Some of these arrangements would be of higher energy and some would be degenerate because of different inter electronic-repulsions. The inter-electronic repulsions for the arrangement where two electrons are present in the same orbital are higher than for those arrangements in which they are present in different orbitals. Consequently, the energy for the former arrangements would be larger than for those for the latter arrangements. Thus, unlike in a single electron system, many more energy states exist in multielectron systems. These different arrangements may be grouped into different sets of degenerate energy states. Furthermore, spin angular momentum (s) and orbital angular momentum (l) of the electrons may interact or couple to give a *resultant angular momentum* for the entire multi-electron atom. The resultant angular momentum and the energy of the system is expressed in a **term symbol.** Thus,

term symbols are used to indicate the electronic configuration and the resultant angular momentum of an atom.

The term symbol for a particular atomic state is expressed as :

$$^{2S+1}L_J$$

where S = total spin angular quantum number

L = total orbital angular quantum number

J = total angular momentum quantum number

(2S + 1) is called the spin multiplicity of the state (arrangement)

Just as we have the symbols s, p, d, f .... for l = 0, 1, 2 , 3 ...., the different values of L are designated by S, P, D, F, G, H, I, etc. Thus the state S represents L = 0, P represents L = 1 and so on

L	0	1	2	3	4	5	6	7
State	S	P	D	F	G	H	I	K

It may be noted that we don't assign the state J because this letter is used to represent another quantum number viz. total angular momentum quantum number.

The total angular momentum, J of an atom can be generally determined by two methods. These are :

1. LS or Russell Saunders Coupling

2. JJ Coupling

LS or Russel Saunders Coupling scheme is applicable to systems in which spin orbital interactions are relatively small. In this scheme, the individual orbital angular momenta of the electrons couple (or interact) to give a resultant angular momentum represented by the quantum number, L for the state. The individual electron spin momenta also couple to give a resultant spin momentum described by the quantum number, S. The L and S values together determine the total angular momentum, J which can take quantized positive values ranging from | L – S | to | L + S |. | | indicates the absolute value of | L – S | and no regard is paid to the sign, so that J is always ≥ 0. Thus, the range of J can be written as

$$J = |L + S|, |L + S - 1| .... |L - S|$$

Rules for determining the term symbol according to L – S coupling scheme

The steps involved in determining the term symbol according to L – S coupling scheme are as under :

A. Determining total spin angular momentum (*ss* coupling)

For a single electron we know spin quantum number, $m_s = \frac{1}{2}$. When two or more electrons are present in subshell, then their magnetic fields couple to produce resultant spin quantum numbers, S. The resultant spin quantum number is calculated as :

$$S = (s_1 + s_2), (s_1 + s_2 - 1) \ldots\ldots\ldots |s_1 - s_2|$$

where the modulus sign | | indicates the positive value.

For example, for a subshell having two electrons (p^2 or d^2) :

$$S = \frac{1}{2} + \frac{1}{2}, \frac{1}{2} + \frac{1}{2} - 1$$

or $\quad S = 1, 0$

The **spin multiplicity** is given by the relation 2S + 1 and is written in the upper left hand corner of the term symbol for the state.

Henry Norris Russel

Spin multiplicity of 3 represents **triplet state** and a spin multiplicity of 1 represents a **singlet state.**

Unpaired electrons	S	Spin multiplicity (2S + 1)	Name of state
0	0	1	Singlet
1	$\frac{1}{2}$	2	Doublet
2	$\frac{1}{2} + \frac{1}{2} = 1$	3	Triplet
3	$\frac{1}{2} + \frac{1}{2} + \frac{1}{2} = 1\frac{1}{2}$	4	Quartet
4	$\frac{1}{2} + \frac{1}{2} + \frac{1}{2} + \frac{1}{2} = 2$	5	Quintet

It is clear from above that the spin multiplicity (2S + 1) is always one more than the number of unpaired electrons.

B. Determination of resultant orbital angular momentum (*ll* Coupling)

When two electrons with angular quantum number l_1 and l_2 interact, the resultant angular quantum number, L are given by the relationship :

$$L = (l_1 + l_2), (l_1 + l_2 - 1), (l_1 + l_2 - 2) \ldots\ldots\ldots |l_1 - l_2|$$

where the modulus sign | | indicates the positive value.

For example, for a *p* subshell having two electrons (as in carbon).

$l_1 = 1$ and $l_2 = 1$ so that

L	=	(1 + 1),	(1 + 1 – 1),	....	(1 – 1)
	=	2	1		0
States		D	P		S

Therefore, p^2 electrons (as in carbon) will be represented by three energy states S, P and D corresponding to L = 0, 1 and 2 respectively.

Similarly, for d^2 configuration, the energy states can be obtained as under :

$l_1 = 2$ and $l_2 = 2$

L	=	(2 + 2),	(2 + 2 – 1),	2 + 2 – 2,	(2 + 2 – 3),	2 – 2
	=	4	3	2	1	0
States		G	F	D	P	S

Thus, d^2 configuration is represented by five energy **S, P, D, F** and **G** corresponding to L = 0, 1, 2, 3 and 4 respectively.

C. Spin-orbital or L – S Coupling

The magnetic effects of the resultant angular momentum, L and the resultant spin moment S couple together to give new quantum number J called the **total angular quantum number.** J values can be obtained by the vectorial combination of L and S by the coupling scheme called **Russell Saunders or LS coupling.** The J can acquire all quantized positive values in the range | L – S | to | L + S | as :

$$J = (L + S), (L + S - 1), (L + S - 2) \ldots\ldots\ldots | L - S |$$

The J values are separated by 1. The modulus sign | | indicates the absolute value of | L – S | so that $J \geq 0$.

The J value is written as a subscript in the term symbol.

We shall illustrate these rules by taking the examples of p^2 and d^2 configurations :

p^2–Configuration

Step 1

For the two *p*-electrons $l_1 = 1$ and $l_2 = 1$

L	=	(1 + 1),	(1 + 1 – 1),	(1 + 1 – 2)
	=	2,	1,	0
States		D,	P,	S

It may be noted that the last value of L can also be obtained by $| l_1 - l_2 |$ *i.e.*

$1 - 1 = 0$

The energy states are S, P and D.

Step 2

For two electrons,

$$S = (+\tfrac{1}{2} + \tfrac{1}{2}),\ (+\tfrac{1}{2} + \tfrac{1}{2} - 1),$$
$$= 1,\quad 0$$

It may be noted that the second value of S can also be obtained by $| s_1 - s_2 |$ or $\frac{1}{2} - \frac{1}{2} = 0$

The spin multiplicity (2S + 1) is :

For S = 1, spin multiplicity = 3 (Triplet)

For S = 0, spin multiplicity = 1 (Singlet)

Therefore, we have six states corresponding to states D, P and S and corresponding to spin multiplicity 3 and 1 each which can be written as

^{3}**S**, ^{3}**P**, ^{3}**D**, ^{1}**S**, ^{1}**P and** ^{1}**D.**

Step 3

Calculation of J values;

(*a*) When L = 2, 1 and 0 and S = 1

(*i*) When L = 2, S = 1 *i.e.* ^3D *state*

$$J = (2+1), \quad (2+1-1), \quad (2+1-2)$$
$$= 3, \quad 2, \quad 1$$

We cannot have J = 0 because the range J is from | L + S | to | L – S | *i.e.*

| 2 + 1 | to | 2 – 1 | *i.e.* 3 to 1

∴ Complete spectroscopic term symbols $^{2S+1}L_J$

$^3D_3$, $^3D_2$ and $^3D_1$.

(*ii*) When L = 1 and S = 1 *i.e.* ^3P *state.*

$$J = (1+1), \quad (1+1-1), \quad (1+1-2)$$
$$= 2, \quad 1, \quad 0$$

∴ Complete spectroscopic terms are :

$^3P_2$, $^3P_1$ and $^3P_0$.

(*iii*) When L = 0 and S = 1 *i.e.* ^3S *state*

$$J = 1+0$$
$$= 1$$

We cannot have J = 0 because the range for J is from | L + S | to | L – S |

i.e. | 0 + 1 | to | 0 – 1 | *i.e.* 1 only

Complete spectroscopic term symbol $^{2S+1}L_J$ is $^3S_1$

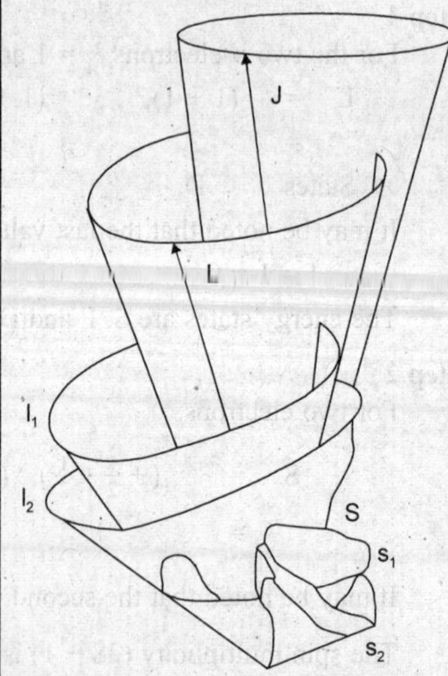

Russel Saunder's Coupling

(*b*) When L = 2, 1 and 0 and S = 0

(*i*) When L = 2 and S = 0 *i.e.* ^1D *states*

$$J = 2+0$$
$$= 2$$

J cannot have values 1 and 0 because the range is from | 2 + 0 | to | 2 – 0 | *i.e.* 2 only

∴ Complete spectroscopic term symbol, $^{2S+1}L_J$ is $^1D_2$

(*ii*) When L = 1 and S = 0 *i.e.* ^1P *states*

$$J = 1+0$$
$$= 1.$$

J cannot have the value 0 because J ranges between | 1 + 0 | and | 1 – 0 | *i.e.* 1 only

∴ Complete spectroscopic term is $^1P_1$

(*iii*) When L = 0 and S = 0 *i.e.* ^1S *state*

$$J = 0$$

∴ Complete spectroscopic term is $^1S_0$

Thus, the total term symbols for p^2 configuration are :

$^3D_3$, $^3D_2$, $^3D_1$, $^3P_2$, $^3P_1$, $^3P_0$, $^3S_1$, $^1D_2$, $^1P_1$ and $^1S_0$.

Each of these value corresponds to an electronic arrangement.

All the spectroscopic terms derived for a p^2 configuration would occur for an excited state of carbon ($1s^2\,2s^2\,2p^1\,3p^1$) in which the two *p* electrons belong to different subshells. However, in the ground state of carbon atom ($1s^2\,2s^2\,2p^2$) the number of states is restricted by the **Pauli exclusion principle** which states that *no two electrons in the same atom can have the same value for all the four quantum numbers.*

In the ground state configuration of carbon, both the *p* electrons have the same value of *n* and *l*, ($n = 2$ and $l = 1$), so they must differ in at least one of the remaining quantum numbers *m* or *s*. Pauli's exclusion principle restricts the number of terms from ^{1}S, ^{1}P, ^{1}D, ^{3}S, ^{3}P, ^{3}D to ^{1}S, ^{3}P and ^{1}D only. Table 4.1 gives all the fifteen allowed values of M_L and M_S for the p^2 configuration (which do not violate Pauli's principle). It can be shown that these 15 arrangements can be represented by only three terms ^{1}S, ^{3}P and ^{1}D by taking suitable values of M_L (2L + 1) and M_S (2S + 1) as illustrated below :

For *p* electrons, the angular quantum number $l = 1$ and magnetic quantum number *m* has values from $+l$ .... 0 .... $-l$, giving in this case values of $m = 1, 0, -1$. The spin quantum number has values of $+\frac{1}{2}$ and $-\frac{1}{2}$ for each value of *m*. The total spin quantum numbers and total orbital quantum numbers can be obtained by adding the appropriate values of *m* and *l* as explained earlier :

M_L has values from + L .... 0 .... – L (a total of 2L + 1 values).

M_S has values from + S .... 0 .... – S (a total of 2S + 1 values).

The L and S quantum numbers which are associated with each electronic configuration or term symbol can be obtained from M_L and M_S quantum numbers as shown in Table 4.1. The values of M_S and M_L can be obtained by adding the appropriate values of m_s and *n*

$$M_S = \Sigma m_s \quad \text{and} \quad M_L = \Sigma m$$

Table 4.1. Allowed values of M_L and M_S for p^2 configuration

S.No.	*m* = + 1	0	– 1	M_S Σm_s	M_L Σm_l	Term symbol
1.	↑↓			0	2	^{1}D
2.			↑↓	0	– 2	^{1}D
3.		↑↓		0	0	
4.	↑		↓	0	0	^{3}P, ^{1}D, ^{1}S
5.	↓		↑	0	0	
6.	↑	↓		0	1	^{3}P, ^{1}D
7.	↓	↑		0	1	
8.		↑	↓	0	– 1	^{3}P, ^{1}D
9.		↓	↑	0	– 1	
10.	↑	↑		1	1	^{3}P
11.	↑		↑	1	0	^{3}P
12.		↑	↑	1	– 1	^{3}P
13.	↓	↓		– 1	1	^{3}P
14.	↓		↓	– 1	0	^{3}P
15.		↓	↓	– 1	– 1	^{3}P

Steps to assign spectroscopic term symbol

(*i*) Write the different values of M_S and M_L for the given configuration of valence shell electrons.

(*ii*) Select the maximum M_S value and then the maximum M_L value associated with it. In the present case, it is $M_S = 1$ and $M_L = 1$ (S.No. 10 in Table). This corresponding to a group of terms where L = 1 and S = 1. Since L = 1, it must be a P state and since S = 1, the multiplicity (2S + 1) = 3 so that it is a ^3P (triplet P) state.

Now if L = 1, M_L may have values + 1, 0 and – 1 and if S = 1, M_S may have the value + 1, 0, and – 1. It gives rise to **nine** combinations of M_L and M_S values as given below

$$M_L = +1 \qquad M_S = +1, 0, -1$$
$$M_L = 0 \qquad M_S = +1, 0, -1$$
$$M_L = -1 \qquad M_S = +1, 0, -1$$

Thus we assign the symbol ^3P to nine of the allowed values in Table 4.1

(*iii*) From the remaining M_S and M_L values, we pick out the next maximum M_S and M_L value. It is $M_S = 0$ and $M_L = 2$. Since L = 2, it must be a D state and S = 0 gives a multiplicity of 2S + 1 = 1, so it is a ^1D (singlet D) state.

With L = 2, M_L may have values of + 2, + 1, 0, – 1 , – 2 and if S = 0, M_S has only one value, *i.e.*, 0.

We get the following five combinations of M_L and M_S which can be assigned ^1D term symbol.

$$M_L = +2 \qquad M_S = 0$$
$$M_L = +1 \qquad M_S = 0$$
$$M_L = +0 \qquad M_S = 0$$
$$M_L = -1 \qquad M_S = 0$$
$$M_L = -2 \qquad M_S = 0$$

So far, we have considered 14 combinations, 9 of ^3P and 5 of ^1D. The remaining configuration corresponds to $M_S = 0$ and $M_L = 0$. This gives a singlet S state ^1S.

Thus we can say that all the 15 electronic arrangements can be represented by three states:

^1S, ^3P and ^1D.

Derivation of Term Symbols for a d^2 configuration

For *d* electrons, the subsidiary quantum number *l* is 2. For two *d*-electrons having $l_1 = 2$ and $l_2 = 2$, the values of L can be obtained as

$$L = (2+2), \quad (2+2-1), \quad (2+2-2), \quad (2+2-3), \quad (2+2-4)$$
$$\quad\ \ 4 \qquad\qquad 3 \qquad\qquad 2 \qquad\qquad 1 \qquad\qquad 0$$

These values of L represent S, P, D, F and G states corresponding to L = 0, 1, 2 , 3 and 4 respectively.

Values of S can be obtained as

$$S = (+\tfrac{1}{2}+\tfrac{1}{2}),\ (\tfrac{1}{2}+\tfrac{1}{2}-1)$$

Therefore, the values of spin multiplicity, (2S + 1) are 3 and 1. So we have ^1S, ^1P, ^1D, ^1F, ^1G and ^3S, ^3P, ^3D, ^3F and ^3G. There are 45 ways in which two *d*-electrons may be arranged without violating the Pauli exclusion principle. These arrangements are given in Table 4.2.

However, these arrangements can be represented by only five states; ^{1}S, ^{3}P, ^{1}D, ^{3}F and ^{1}G. This can be seen in a similar way as for p^2 electrons.

(*i*) Maximum value of M_S = 1 and the maximum value of M_L corresponding to this M_S is 3 (Refer to S.No. 2 in Table 4.2). Since this corresponds to L = 3, it represents F state. This occurs with M_S = + 1, 0, – 1 suggesting a triplet F state *i.e.* ^{3}F.

Now there are 21 **combinations** of M_L and M_S values associated with ^{3}F term for L = 3 as shown below :

$$
\begin{array}{llll}
M_L = +3 & & M_S = +1, 0, -1 \\
M_L = +2 & & M_S = +1, 0, -1 \\
M_L = +1 & & M_S = +1, 0, -1 \\
M_L = 0 & & M_S = +1, 0, -1 \\
M_L = -1 & & M_S = +1, 0, -1 \\
M_L = -2 & & M_S = +1, 0, -1 \\
M_L = -3 & & M_S = +1, 0, -1
\end{array}
$$

(*ii*) The next highest unassigned M_S is 0. The corresponding value of M_L is 4 (Sr. No. 1 in the Table 4.2). This corresponds to G state. Since M_S = 0, S must be 0 and it is a singlet G term; ^{1}G.

Now, if L = 4, M_L can have the values + 4, + 3, + 2, + 1, 0, – 1, – 2, – 3, – 4. If S = 0, M_S = 0 (only one value).

Thus, there are nine configurations associated with this term, ^{1}G.

(*iii*) For M_L value of 2, maximum value of M_S = 0. This M_L = 2 corresponds to D state. Since M_S = 0 for this state, it corresponds to spin multiplicity of 1 suggesting the state ^{1}D.

Table 4.2. Allowed values of M_S and M_L for the d^2 configuration

S.No.	*m* = +2	+1	0	–1	–2	$M_S = \Sigma m_s$	$M_L = \Sigma m_l$	Term symbol
1.	↑↓					0	4	^{1}G
2.	↑	↑				1	3	^{3}F
3.	↑	↓				0	3	^{1}G, ^{3}P
4.	↓	↑				0	3	
5.	↓	↓				– 1	3	^{3}F
6.	↑		↑			1	2	^{3}F
7.	↑		↓			0	2	^{1}G, ^{3}F, ^{1}D
8.	↓		↑			0	2	
9.		↑↓				0	2	
10.	↓		↓			– 1	2	^{3}F
11.	↑			↑		1	1	^{3}F, ^{3}P
12.		↑	↑			1	1	

S.No.	m = +2	+1	0	−1	−2	$M_S = \Sigma m_s$	$M_L = \Sigma m_l$	Term symbol
13.	↓			↓		−1	1	3F, 3P
14.		↓	↓			−1	1	
15.	↑			↓		0	1	^1G, ^3F, ^1D, ^3P
16.	↓			↑		0	1	
17.		↑	↓			0	1	
18.		↓	↑			0	1	
19.	↑				↑	1	0	^3F, ^3P
20.		↑		↑		1	0	
21.	↑				↓	0	0	^1G, ^3F, ^1D, ^3P, ^1S
22.	↓				↑	0	0	
23.		↑		↓		0	0	
24.		↓		↑		0	0	
25.			↑↓			0	0	
26.	↓				↓	−1	0	^3F, ^3P
27.		↓		↓		−1	0	
28.		↑			↓	0	−1	^1G, ^3F, ^1D, ^3P
29.		↓			↑	0	−1	
30.			↑	↓		0	−1	
31.			↓	↑		0	−1	
32.		↑			↑	1	−1	^3F, ^3P
33.			↑	↑		1	−1	
34.		↓			↓	−1	−1	^3F, ^3P
35.			↓	↓		−1	−1	
36.			↑		↑	1	−2	^3F

S.No.	m = +2 +1 0 –1 –2	$M_S = \Sigma m_s$	$M_L = \Sigma m_l$	Term symbol
37.	[ ][ ][↑][ ][↓]	0	– 2	${}^1G, {}^3F, {}^1D$
38.	[ ][ ][↓][ ][↑]	0	– 2	
39.	[ ][ ][ ][↑↓][ ]	0	– 2	
40.	[ ][ ][↓][ ][↓]	– 1	– 2	${}^3F$
41.	[ ][ ][ ][↑][↑]	1	– 3	${}^3F$
42.	[ ][ ][ ][↑][↑]	0	– 3	${}^1G, {}^3F$
43.	[ ][ ][ ][↓][↑]	0	– 3	
44.	[ ][ ][ ][↓][↓]	– 1	– 3	${}^3F$
45.	[ ][ ][ ][ ][↑↓]	0	– 4	${}^1G$

Now, if L = 2, M_L may have values + 2, + 1, 0, – 1 , – 2

and if S = 0, M_S may have (only one value) that is 0.

We can assign this state to five of the combinations (Table 4.2).

(*iv*) We have some combinations, in which maximum M_S = 1 and the maximum M_L value associated with it is 1. Therefore, it corresponds to a P state. Since S = 1, spin multiplicity is 2 × 1 + 1 = 3. So it corresponds to a triplet P state, ${}^3P$.

Now, if L = 1, M_L may have values + 1, 0, – 1 if S = 1, M_S may have the values + 1, 0, – 1.

Thus, there are **nine** combinations of M_L and M_S values :

M_L = + 1 M_S = + 1, 0, – 1

M_L = 0 M_S = + 1, 0, – 1

M_L = – 1 M_S = + 1, 0, – 1

These 9 combinations can thus be assigned the state ${}^3P$.

(*v*) We have considered 21 + 9 + 5 + 9 = 44 combinations so far and only one combination is left. Combination of M_S = 0 and M_L = 0 is left. Here L = 0 and it corresponds to singlet. The term (state) ${}^1S$ can be assigned to it.

Since L = 0 and S = 0, it corresponds to only one term.

Thus, all the allowed values of M_L and M_S for d^2 configuration may be represented by ${}^1S$, ${}^3P$, ${}^1D$, ${}^3F$ and ${}^1G$ states.

Derivation of the Term Symbol for closed shell configuration

If a subsell is completely filled with electrons for example p^6 or d^{10} arrangements, then the values of both M_S and M_L come out to be zero.

	+1 0 –1	M_S	M_L
p^6	[↑↓][↑↓][↑↓]	0	0

	+2 +1 0 –1 –2	M_S	M_L
d^{10}	[↑↓][↑↓][↑↓][↑↓][↑↓]	0	0

Since $M_L = 0$, so L = 0 and it corresponds to 'S' state. Since $M_S = 0$ and so S = 0 and therefore, multiplicity is 2S + 1 = 1.

Thus, **a closed shell of electrons always produces a singlet S state, $^1S_0$.**

4.4. SPECTROSCOPIC GROUND STATES

The different terms in an atom as obtained above can be arranged in order of energy and the ground state term is identified on the basis of Hund's rules. The following rules determine the order of energy for different terms.

1. The terms are arranged depending upon their spin multiplicities, *i.e.* their S-values. **The most stable state has the highest S-value and stability decreases as the value of S decreases.** In other words the most stable state (ground state) has maximum unpaired electrons. This gives the minimum electrostatic repulsion and hence the lowest energy.
2. For a given value of S, **the state with the highest value of L is the most stable state.** This means that if two or more terms have the same value of S (same spin multiplicity), the state with highest value of L will have the lowest energy.
3. For terms having same S and L values,

   the term with smallest J value is most stable if the subshell is less than half filled

   the term with maximum J value is most stable if the subshell is more than half filled.

We shall apply these rules to the terms derived for p^2 and d^2 configurations.

p^2 configuration

The terms for ground state of p^2 configuration are ^3P, ^1D and ^1S. We have to find out the ground state out of these. The most stable state out of these three states is ^3P because its multiplicity is maximum. So, ^3P is the ground state (most stable state).

Out of ^1D and ^1S (both having the same value of S), ^1D is more stable because it has a higher value of L = 2.

No the triplet P state has three terms $^3P_2$, $^3P_1$ and $^3P_0$. According to third rule, $^3P_0$ is the lowest energy term (because $2p^2$ configuration of carbon is less than half filled) and these terms may be arranged in increasing order of energy as

$$^3P_0 < {}^3P_1 < {}^3P_2$$

The different terms of carbon are arranged in the increasing order of energy in Fig. 4.1. This order corresponds to experimentally measured energies shown in brackets.

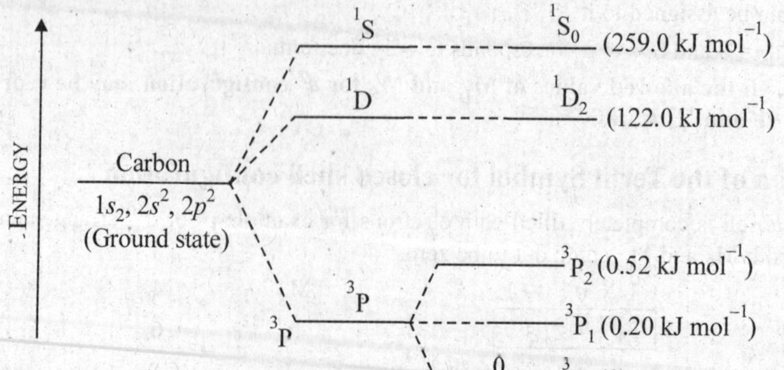

Fig. 4.1. Splitting of terms in ground state of carbon

d^2 configuration

The terms for ground state of d^2 configuration (e.g. of V^{3+} ion) as derived earlier are ^1S, ^3P, ^1D, ^3F and ^1G.

Let us arrange them in increasing order of energy and determine the ground state.

According to rule 1, the stable state corresponds to highest multiplicity. In this case both ^3P and ^3F have common multiplicity 3 (or S = 1). Applying rule 2, ^3F term wil be more stable because it has higher value of L i.e. 3 (^3P state has L = 1). So energies of these two states are arranged as

$$^3F < {}^3P.$$

The remaining three states have same spin multiplicity = 1 (*i.e.* S = 0). These can be arranged according to L values as :

$$^1G < {}^1D < {}^1S.$$

The splitting of terms in ground state of d^2 configuration are shown in Fig. 4.2. It may be noted that the two triplet states ^3F ($^3F_2$, $^3F_3$, $^3F_4$) and ^3P ($^3P_0$, $^3P_1$, $^3P_2$) can further split and arranged according to the rule 3 as shown in Fig. 4.2.

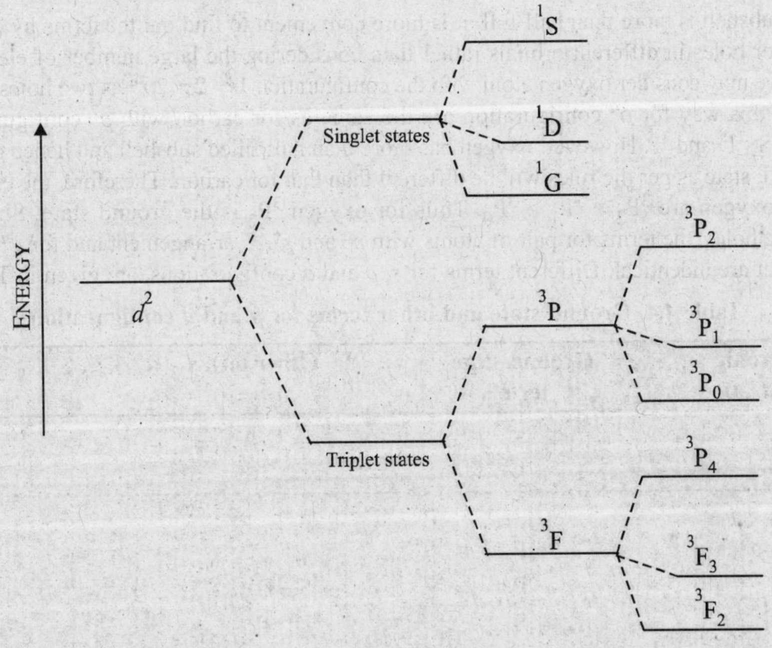

Fig. 4.2. Splitting of terms in d^2 configuration

The ground state terms of d^1 to d^{10} electronic configurations (without splitting) are given in Table 4.3 :

Table 4.3. Ground state terms for d^1 to d^{10} configurations

Configuration	Ground state term	Example
d^1	^2D	Ti^{3+}
d^2	^3F	V^{3+}
d^3	^4F	Cr^{3+}
d^4	^5D	Cr^{2+}
d^5	^6S	Mn^{2+}
d^6	^5D	Fe^{2+}
d^7	^4F	Co^{2+}
d^8	^3D	Ni^{2+}
d^9	^2D	Cu^{2+}
d^{10}	^1S	Cu^+

HOLE FORMULATION (FORMULISM)

If a subshell is more than half full, it is more convenient to find out the terms by considering vacancies or holes in different orbitals rather than considering the large number of electrons. For example, we may consider oxygen atom with the configuration $1s^2\ 2s^2\ 2p^4$ as two holes. The terms derived in this way for p^4 configuration are the same as for carbon with p^2 configuration. The terms are ^1S, ^1D and ^3P. However, oxygen has more than half filled subshell and hence the order of energy of ^3P state as per the rules will be different than that for carbon. Therefore, the energy of ^3P states for oxygen are $^3P_2 > {}^3P_1 > {}^3P_0$. Thus for oxygen $^3P_2$ is the ground state. Similarly, by considering holes, the terms for pair of atoms with p^n and p^{6-n} arrangement and for d^n and d^{10-n} arrangement are indentical. Different terms for *s*, *p* and *d* configurations, are given in Table 4.4.

Table 4.4. Ground state and other terms for *p* and *d* configuration

Electronic configuration	Ground state term	Other terms
s^1	^2S	—
s^2	^1S	—
p^1, p^5	^2P	
p^2, p^4	^3P	$^1S, {}^1D$
p^3	^4S	$^2D, {}^2P$
p^6	^1S	
d^1, d^9	^2D	
d^2, d^8	^3F	$^3P, {}^1G, {}^1D, {}^1S$
d^3, d^7	^4F	$^4P, {}^2H, {}^2G, {}^2F, {}^2D, {}^2P$
d^4, d^6	^5D	$^3H, {}^3G, {}^3F, {}^3D, {}^3P, {}^1I, {}^1G, {}^1F, {}^1D, {}^1S$
d^5	^6S	$^4G, {}^4F, {}^4D\ {}^4P, {}^2I, {}^2H, {}^2G, {}^2D, {}^2P, {}^2P, {}^2S$
d^{10}	^1S	

Calculation of Number of Microstates

The different ways in which the electrons can occupy the orbitals specified by the configuration are called the microstates.

These microstates differ from one another slightly in energy. The number of microstates may be calculated from the number of orbitals and the number of electrons using the formula :

$$\binom{n}{r} = \frac{n!}{r!(n-r)!}$$

where n is twice the number of orbitals and r is the number of electrons. $n!$ and $r!$ are factorial n and factorial r respectively.

For a p^2 configuration, there are three orbitals, so $n = 6$ and $r = 2$.

$$\therefore \text{ No. of microstates, } \binom{n}{r} = \frac{6!}{2!(6-2)!} = \frac{6\times5\times4\times3\times2\times1}{2\times1\times4\times3\times2\times1} = \mathbf{15} \text{ microstates}$$

Similarly, for a p^3 configuration,

$$n = 6, r = 3$$

Using the above relation

$$\therefore \text{ No. of microstates, } \binom{n}{r} = \frac{6!}{3!3!} = \frac{6\times5\times4\times3\times2\times1}{3\times2\times1\times3\times2\times1} = \mathbf{20} \text{ microstates}$$

For a d^2 configuration, there are five orbitals so $n = 10$ and $r = 2$.

$$\therefore \text{ No. of microstate } \binom{10}{2} = \frac{10!}{2!8!} = \frac{10\times9\times8\times7\times6\times5\times4\times3\times2\times1}{2\times1\times8\times7\times6\times5\times4\times3\times2\times1} = \mathbf{45} \text{ microstates}$$

Using the above formula, the number of microstates for all the electronic arrangements p^1 to p^6 and d^1 to d^{10} can be calculated. These are given in Table 4.5.

Table 4.5. Number of microstates for various electron arrangements

Electronic arrangement	No. of micro states	Electronic arrangement	No. of microstates
p^1	6	d^1	10
p^2	15	d^2	45
p^3	20	d^3	120
p^4	15	d^4	210
p^5	6	d^5	252
p^6	1	d^6	210
		d^7	120
		d^8	45
		d^9	10
		d^{10}	1

4.5. ELECTRONIC SPECTRA OF TRANSITION METAL COMPLEXES

The spectra which are obtained when electrons are promoted from one energy level (ground state) to a higher energy level (excited state) are called **electronic transitions.** These are of high energy and are always accompanied by lower energy vibrational and rotational transitions. The vibrational and rotational energy levels are so close in energy that these cannot be resolved into separate absorption bands and they cause considerable broadening of electronic absorption peaks in $d-d$ spectra. The band widths are usually of the order of 1000 to 3000 cm^{-1}. The rules governing these transitions are called selection rules. If the transition of electrons takes place according to a set criteria, it is an **allowed** transition. If not, it is called a **forbidden** transition.

Selection Rules

The main selection rules for electronic spectra are :

1. *Transitions in which there is change in the number of unpaired electrons in going from a lower to a higher energy state are referred* to as spin or multiplicity forbidden. This means that transitions to only those excited states are to considered which have the same spin multiplicity as the ground state. **This is also known as $\Delta S = 0$ rule.**

For example, for a d^2 configuration, we observe transitions from ^3F (ground state) to ^3P (excited state) because both ^3F and ^3P terms have same spin multiplicity 3. The transitions in which $\Delta S \neq 0$(e.g. $\Delta S = \pm 1$) are forbidden or very very weak.

2. *For molecules having a centre of symmetry, transitions which do not involve a change in the subsidiary quantum number ($\Delta l = 0$) are forbidden. These forbidden transitions are* called *Laporte forbidden transitions.*

We can say that the transitions which involve a change in the subsidiary quantum number *i.e.* $\Delta l = \pm 1$ are **Laporte allowed transitions** and have a high absorbance. This rule suggests that transitions from one *d*-level to another *d*-level (*i.e. d–d* transitions) in transition metal complexes are not allowed (Laporte forbidden transitions) because $\Delta l = 0$.

This would suggest that transition metal complexes should not give *d–d* transitions and should not give colours. But, actually it is not so. The transition do take place with the help of slight relaxation in Laporte rule as explained below :

(*i*) If the transition metal complex ion does not have perfect octahedral structure, but is slightly distorted so that the centre of symmetry is destroyed, then mixing of *d* and *p*-orbitals of the metal ion may occur. In such a case, the transitions are no more pure *d-d* transitions but they occur between *d*-levels with varying amounts of *p*-character. The intensity of such transitions is very weak in the range $\in = 20$ to 50. Thus, octahedral complexes in which all the ligands are not same such as $[Co(NH_3)_5Cl]^{2+}$ have irregular octahedral structure and do not possess a centre of symmetry. *d–d* transitions with varying amounts of *p*-character take place giving this compound a colour.

The tetrahedral complexes such as $[MnCl_4]^{2-}$, $[MnBr_4]^{2-}$ *etc.* also do not possess a centre of symmetry because a tetrahedron shape never possesses a centre of symmetry. Thus, intense transitions are observed in these complexes which result in their deep colours. However, mixing of *d* and *p* orbitals does not occur in perfectly octahedral complexes, which have a centre of symmetry such as $[Co(NH_3)_6]^{3+}$.

(*ii*) A complex which has a perfect octahedral structure can also exhibit absorption spectrum because the bonds in the transition metal complexes are not rigid but undergo vibrations that may temporarily change the centre of symmetry. These vibrations continue all the time and at any particular time some ligands may spend an appreciable amount of time out of the centro symmetric equilibrium position. As a result, the molecule may possess distorted octahedral symmetry so that small amount of mixing of *d* and *p*-orbitals occurs and therefore, low intensity ($\in = 5$ to 25) spectra are observed. These transitions are called **vibronically allowed transitions** and the effect is called **vibronic coupling.** For example, in $[Mn(H_2O)_6]^{2+}$ complexes, all transitions are spin multiplicity forbidden and Laporte forbidden. But the complex ion is pale pink in colour. This is explained by vibronic coupling which results in very low intensity transitions.

In short **Laporte allowed** transitions are very **intense** while Laporte forbidden transitions vary from weak intensity, if the complex is non centro symmetric to very weak if it is centro symmetric.

4.6. SPLITTING OF RUSSEL SAUNDERS STATES IN OCTAHEDRAL AND TETRAHEDRAL CRYSTAL FIELDS

To understand the nature of electronic transitions in complexes of transition metals, we need to know how the splitting of electronic energy levels and spectroscopic terms occurs in *s*, *p*, *d* and *f*-orbitals. The following points may be noted.

(*i*) An *s*-orbital is spherically symmetrical and is not affected by an octahedral (or any other) field. Hence no splitting is involved here.

(*ii*) The *p*-orbitals are directional but they have same type of orientation. These are affected by an octahedral field but to equal extent. Therefore, their energy levels remain equal and no splitting occurs.

(*iii*) The five *d*-orbitals are split by an octahedral field into two levels t_{2g} (d_{xy}, d_{yz}, d_{zx}) and e_g ($d_{x^2-y^2}$, d_{z^2}) having different energies. The difference between these two levels is $10D_q$ (or Δ_0). The t_{2g} level is triply degenerate and is $4D_q$ below the bary centre, while e_g level is doubly degenerate and is $6D_q$ above the bary centre. For a d^1 configuration, ground state is a ^2D state and the t_{2g} and e_g levels correspond to T_{2g} and E_g spectroscopic states as shown in Fig. 4.3 (*a*).

(*iv*) There are seven *f*-orbitals and these split by an octahedral field into three levels. For an f^1 arrangement, the ground state is a ^3F state and is split into a triply degenerate T_{1g} state which is $6D_q$ below the vary centre, a triply degenerate T_{2g} level which is $2D_q$ above the bary centre and a single A_{2g} state which is $12D_q$ above the bary centre as shown in Fig. 4.3 (*b*).

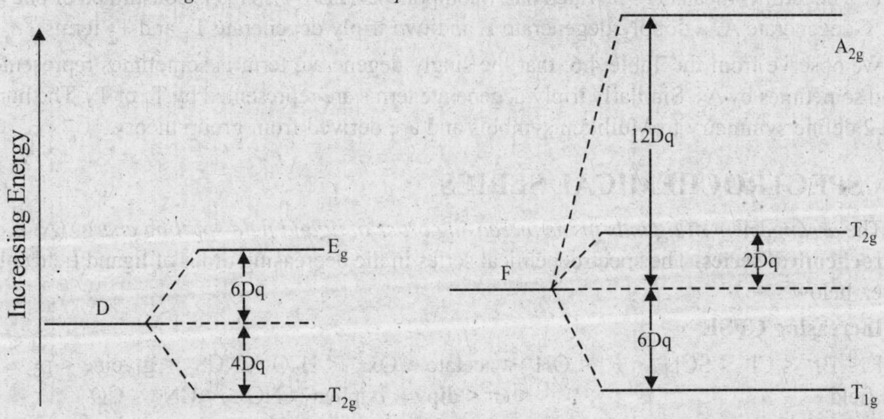

(*a*) (b)

Fig. 4.3. Splitting of spectroscopic terms arsing from (*a*) d^1 electronic arrangement and (*b*) f^1 electronic arrangement

We can summarise that in an octahedral field,

S and P states do not split

D states split into two states, E_g and T_{2g}

F states split into three states, A_{2g}, T_{2g} and T_{1g}

These states split by the external field are called **Mulliken symbols.** These symbols are used in interpreting electronic spectra of transition metal complexes.

Robert S. Mulliken

If may be noted that common Mulliken symbols are used in the octahedral and tetrahedral field these are distinguished by introducing the symmetry symbol '*g*' in octahedral field (Table 4.6)

Table 4.6. Correlation of Spectroscopic terms into Mulliken symbol

Spectroscopic Term	Mulliken Symbol	
	Octahedral field	Tetrahedral field
S	A_{1g}	A_1
P	T_{1g}	T_1
D	$E_g + T_{2g}$	$E + T_2$
F	$A_{2g} + T_{1g} + T_{2g}$	$A_2 + T_1 + T_2$
G	$A_{1g} + E_g + T_{1g} + T_{2g}$	$A_1 + E + T_1 + T_2$

We can obtain Mulliken symbols if we know the spectroscopic terms S, P, D, F, etc. It may be recalled that

(*i*) The term S stands for resultant angular momentum L, when L = 0. The number of components of L is 2L + 1. When L = 0, this is called one component term and is represented by **A**.

(*ii*) The term P means L = 1 which has 3 components (2L + 1). It is triply degenerate and is represented by T.

(*iii*) The term D means L = 2 which has 5 components (2L + 1). This is constituted of doubly degenerate E and a triply degenerate T terms.

(*iv*) The term G means L = 3 which has 7 components (2L + 1). This is constituted of one singly degenerate A, a doubly degenerate E and two triply degenerate T_1 and T_2 terms.

We observe from the Table 4.6, that the singly degenerate term is sometimes represented by A_1 and sometimes by A_2. Similarly triply degenerate terms are represented by T_1 or T_2. The numbers 1 and 2 define symmetry to Mulliken symbols and are derived from group theory.

4.7. SPECTROCHEMICAL SERIES

The *arrangement of ligands in the increasing order of ligand field splitting energy* (Δ) is called **spectrochemical series.** The spectrochemical series in the decreasing order of ligand field splitting is given below :

Increasing CFSE

$I^- < Br^- < Cl^- < SCN^- < F^- < OH^- <$ acetate $< Ox^{2-} < H_2O < NCS^- <$ glycine $<$ py $\simeq NH_3$

Weak field $<$ en $<$ dipy $\approx$ *o*–phen $< NO_2^- < CN^- <$ CO

(small Δ) Strong field (large Δ)

A more compact form of the series involving most common ligands is :

Halides $< C_2O_4^{2-} < H_2O < NH_3 <$ en $< NO_2^- < CN^- <$ CO

Ligands are commonly classified by their donor and acceptor capabilities. The ligands like ammonia are σ donors only, with no orbitals of appropriate symmetry for π-bonding. The ligand field splitting (Δ) depends on the degree of overlap. Ethylenediamine has a stronger effect than ammonia among these ligands generating a larger Δ.

The halide ions ligand field strengths are in the order :

$$F^- > Cl^- > Br^- > I^-$$

The small ligands can cause greater crystal field splitting because they can approach the metal ion closely. For example, F^- ion produces more than large Cl^- ion and Br^- ion.

With ligands having vacant π^* or d-orbitals, there is possibility of π-back bonding with the ligands. Such ligands are called π-acceptors. This interaction to bonding scheme increases Δ. The ligands which can do this effectively are CN^-, CO, etc. With σ-donor strength, ligands can be arranged in terms of their donor atoms as :

halogen < oxygen < nitrogen < carbon

The metal ion also influences although not to the extent of ligands the magnitude of Δ_0 through the overlap and energy match criteria. The following metal ion spectrochemical series is observed.

$Mn^{2+} < Ni^{2+} < Co^{2+} < Fe^{2+} < Fe^{3+} < Cr^{3+} < Co^{3+} < Mn^{4+} < Mo^{3+} < Pd^{4+} < Ir^{3+} < Re^{4+} < Pt^{4+}$

The crystal field splitting is influenced by the oxidation state of transition metal ion. Higher the oxidation state of metal ion greater is the crystal field splitting.

4.8. SPECTRA OF TRANSITION METAL COMPLEXES

1. Spectra of d^1 and d^9 ions

(a) Octahedral complexes of metal ions with d^1 and d^9 configurations: In a free gaseous metal ion, the five d-orbitals are degenerate. We do not expect electronic transition. However, when this ion is surrounded by ligands the electrostatic field of the ligands splits the d-orbital into two groups, t_{2g} and e_g. The simplest example of a d^1 complex is Ti (III) to octahedral complexes such as $[Ti(H_2O)_6]^{3+}$. The splitting of d-orbitals is shown in Fig. 4.4 (a). In the ground state the single electron occupies the lower t_{2g} level and only one transition from $t_{2g} \longrightarrow e_g$ is possible. Therefore spectrum of $[Ti(H_2O)_6]^{3+}$ shows only one band with a peak at 20300 cm^{-1} as shown in Fig. 4.4 (b). Wavelengths corresponding to green and yellow light are absorbed from the white light while the blue and red portions of the light are emitted. Therefore, Ti solution of complex $[Ti(H_2O)_6]^{3+}$ looks purple.

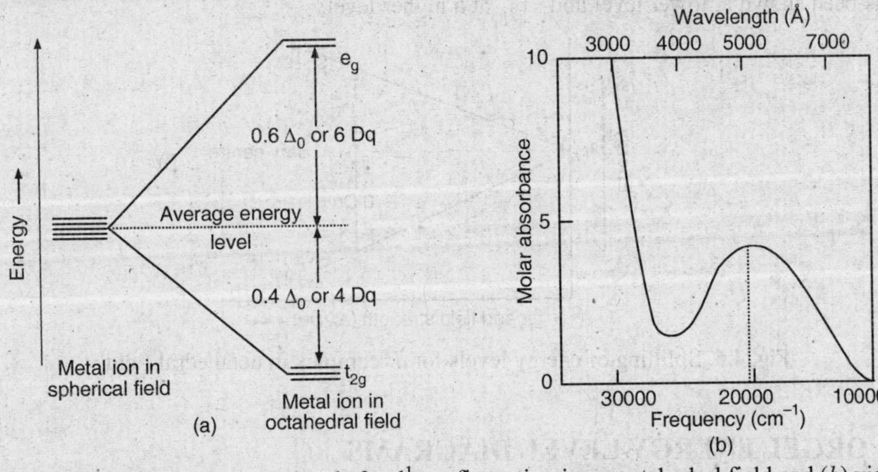

Fig. 4.4 (*a*) Splitting of energy levels for d^1 configuration in an octahedral field and (*b*) visible spectrum of $[Ti(H_2O)_6]^{3+}$ complex ion.

We observe that the intensity of the absorption band is very weak (ϵ = 5–10). This is because it is a forbidden transition. The transitions from one centro symmetric d-orbital to another centro symmetric d-orbital are forbidden. The molar absorbance value of such forbidden transitions are of the order of ϵ = 1 to 10 whereas allowed transitions have ϵ values of about 10,000.

The magnitude of the splitting (Δ_0) depends upon the nature of the ligands. This affects the energy of the transition and the frequency of maximum absorption vary in the spectra of these complexes. For example, $[TiCl_6]^{3-}$ absorbs at 13,000 cm^{-1}, $[TiF_6]^{3-}$ absorbs at 18900 cm^{-1} and $[Ti(H_2O)_6]^{3+}$ absorbs at 20300 cm^{-1}, while $[Ti(CN)_6]^{3-}$ absorbs 22300 cm^{-1}. Consequently these complexes show different colours.

The ground state terms for a free ion with d^1 configuration is ^2D and it is shown on the left (Fig. 4.5). Under the influence of a ligand field, this splits into two states which are described by Mulliken symbols ^2E_g and $^2T_{2g}$. The lower T_{2g} state corresponds to the single *d*-electron occupying one of the T_{2g} orbitals and ^2E_g state corresponds to the electron occupying one of the E_g orbitals. The two states ^2E_g and $^2T_{2g}$ are separated more widely as the strength of the ligand field increases.

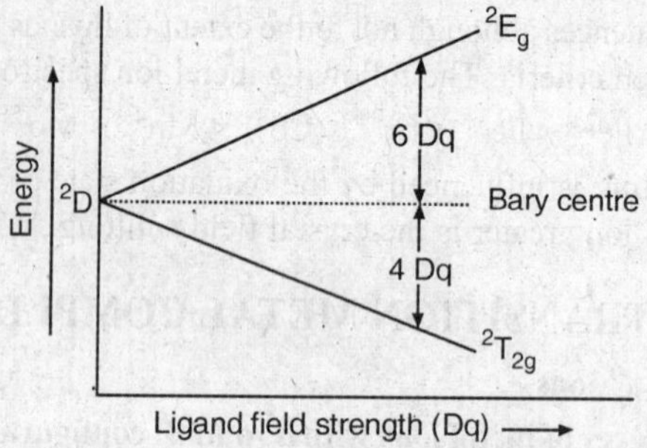

Fig. 4.5 Splitting of *d*-levels for d^1 case in octahedral field

Octahedral complexes of ions with d^9 configuration. Spectra of complexes such as $[Cu(H_2O)_6]^{2+}$ can be explained on the similar lines as the Ti^{3+} octahedral complexes with a d^1 arrangement. In the d^1 complex there is a single electron in the lower T_{2g} level while in the d^9 complex, there is a **single hole in the upper E_g level**. Thus, the transition in the d^1 case corresponds to the promotion of an electron from the T_{2g} to E_g level, whereas the transition in the d^9 case corresponds to the promotion of an electron as the transfer of a hole from E_g to T_{2g}. That is why the energy level diagram for d^9 is the inverse of that for a d^1 configuration (Fig. 4.6). In Fig. 4.6, ^2E_g has been shown at lower level and $^2T_{2g}$ at a higher level.

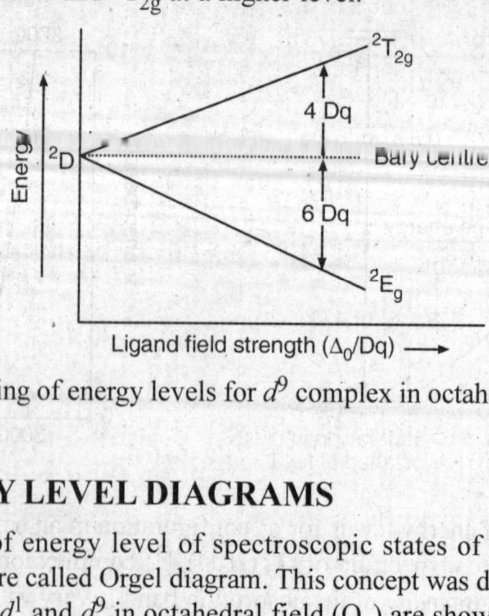

Fig. 4.6. Splitting of energy levels for d^9 complex in octahedral field

4.9. ORGEL ENERGY LEVEL DIAGRAMS

The plots of variation of energy level of spectroscopic states of different symmetry as a function of field strength Dq are called Orgel diagram. This concept was developed of Leslie Orgel. The energy level diagram for d^1 and d^9 in octahedral field (O_h) are shown in Fig. 4.7. This figure is the combination of Fig. 4.5 and 4.6.

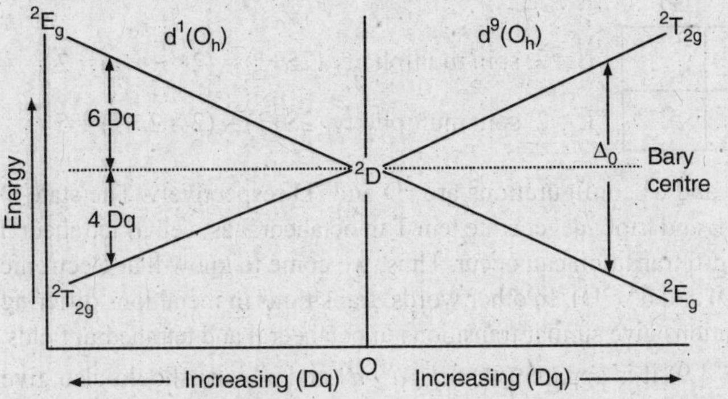

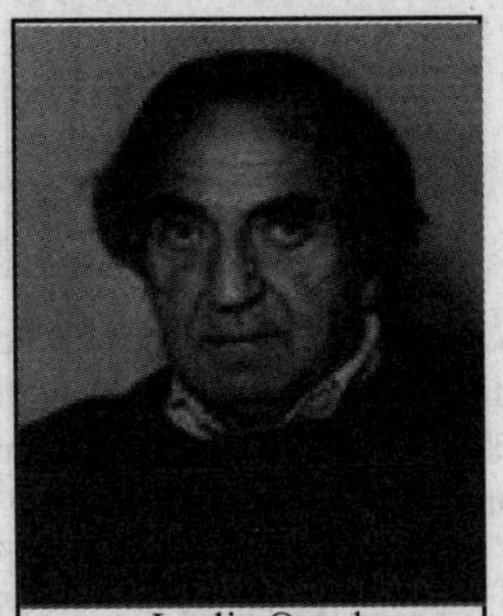
Leslie Orgel

Fig. 4.7 Splitting of energy levels for d^1 and d^9 configurations in an octahedral field (O_h)

Tetrahedral complexes of metal ions with d^1 and d^9 configurations

We have learnt earlier that in tetrahedral field, d-orbitals split into two *e* orbitals of lower energy and three t_2 orbitals of higher energy.

In the case of tetrahedral ligand field, the energy level diagram is inverse of that of octahedral field. The main difference between tetrahedral field and octahedral field is that the splitting is only 4/9 as that of octahedral field ($\Delta_t = 4/9\ \Delta_o$).

Similarly, the tetrahedral complex of d^9 configuration have energy levels inverse of d^1 configuration. So, the energy level diagram for d^9 configurations will also be inverse of that of octahedral field. The splitting of energy levels of d^1 and d^9configurations in tetrahedral field (T_d) are shown in Fig. 4.8. The inverse relation between tetrahedral complexes and octahedral complexes with d^1 configuration and also with d^9 configuration can be understood from Fig. 4.8. Similarly, inverse relation between tetrahedral complexes with d^1 and d^9 configuration can also be understood from this figure.

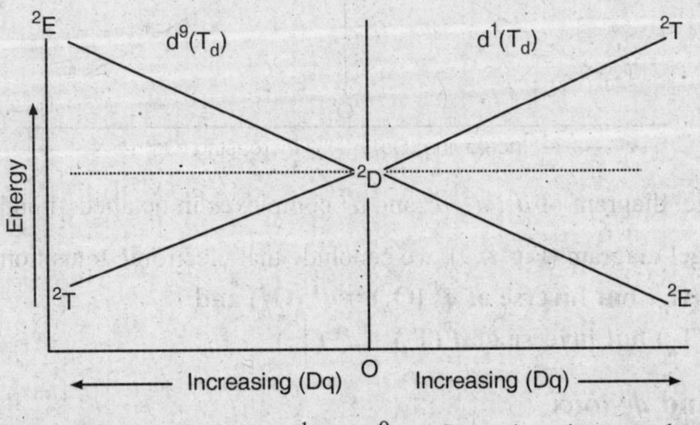

Fig. 4.8. Splitting of energy levels for d^1 and d^9 configurations in a tetrahedral field (T_d)

To summarise, we can say

d^1 (T_d) is inverse of d^9 (T_d) as well as d^1 (O_h)

d^9 (T_d) is inverse of d^1 (T_d) as well as d^9 (O_h)

We can check that the value of L is the same in the case of d^1 and d^6 configurations, but the spin multiplicities are different

	+2	+1	0	−1	−2	
d^1	↑					L = 2, spin multiplicity $(2S+1) = (2\times\frac{1}{2}+1) = 2$
d^6	↑↓	↑	↑	↑	↑	L = 2, spin multiplicity $(2S+1) = (2\times 2+1) = 5$

Thus the term symbols for d^1 and d^2 configurations are ^2D and ^5D respectively. The state D splits into doubly degenerated term E and triply degenerate tem T in octahedral as well as tetrahedral fields. Consequently, only a single d-d transition can occur. Thus, we come to know that electronic transitions will be similar for d^9 (^2D) and d^4 (^5D). In other words, transitions in metal ions differing by five d-electrons in their configurations give similar transitions in octahedral and tetrahedral fields.

From the Orgel diagram (Fig. 4.9) it is apparent that d^1, d^4, d^6 and d^9 ions should also give only one *d-d* absorption band. Splitting of the states as a function of Δ_o for octahedral complexes with d^1 and d^6 electron configurations and tetrahedral complexes with d^4 and d^9 electron configurations are described on the left handside. The spectra of these complexes contain only one band arising from *d-d* transitions and is assigned as $T_{2g} \longrightarrow E_g$.

The right hand side of the Orgel diagram applies to octahedral complexes with d^4 and d^9 electron configurations and tetrahedral complexes with d^1 and d^6 electron configurations. The spectra of these complexes contain only one band arising from single *d-d* transitions and is assigned as $E \longrightarrow T_2$. For tetrahedral complexes we drop the subscripts because a tetrahedron does not have centre of symmetry.

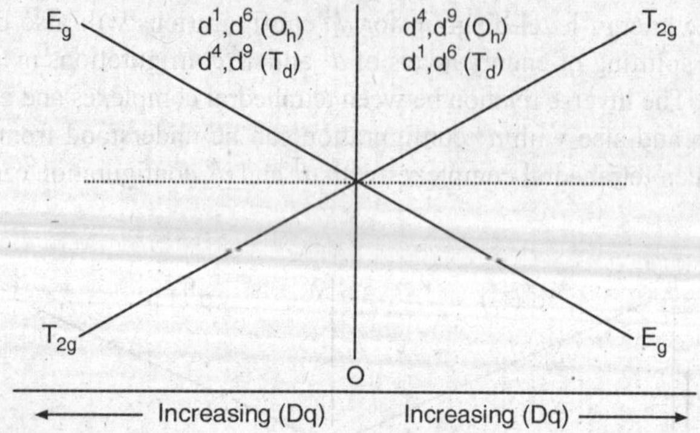

Fig. 4.9. Orgel diagram of d^1, d^4, d^6 and d^9 complexes in octahedral and tetrahedral fields

From the Orgel diagram (Fig. 4.9), we conclude that electronic transitions in:

d^1 (O_h) = d^6(O_h) but inverse of d^9 (O_h) = d^4 (O_h) and
d_1(T_d) = d^6 (T_d) but inverse of d^9(T_d) = d^4 (T_d)

Spectra of d^2 and d^8 ions

(a) d^2 Octahedral field

In the ground state for a d^2 configuration, the two electrons occupy different orbitals. In an octahedral field, the *d*-orbitals are split into three t_{2g} orbitals of lower energy and two e_g orbitals of higher energy. The two *d*-electrons occupy t_{2g} orbitals because of their lower energy. The inter electron repulsions would split the levels giving the spectroscopic terms for d^2 electron configuration as

$$^1S,\ ^3P,\ ^1D,\ ^3F \text{ and } ^1G$$

^{3}F is the ground state with lowest energy and ^{3}P, ^{1}G, ^{1}D and ^{1}S are excited states in accordance with Hund's rule. The transitions from the level ^{3}F are allowed to another triplet state *i.e.* ^{3}P. The transitions from ^{3}F to ^{1}S, ^{1}D and ^{1}G states are spin forbidden and are not observed. *p*-orbitals are not split in an octahedral field and so P states are not split but are transformed into $a^3\ T_{1g}$ state while the *f*-orbitals are split into three levels and so ^{3}F state splits into $^3A_{2g} + {}^3T_{1g} + {}^3T_{2g}$. The energy level diagram for these is shown in Fig. 4.10 (a).

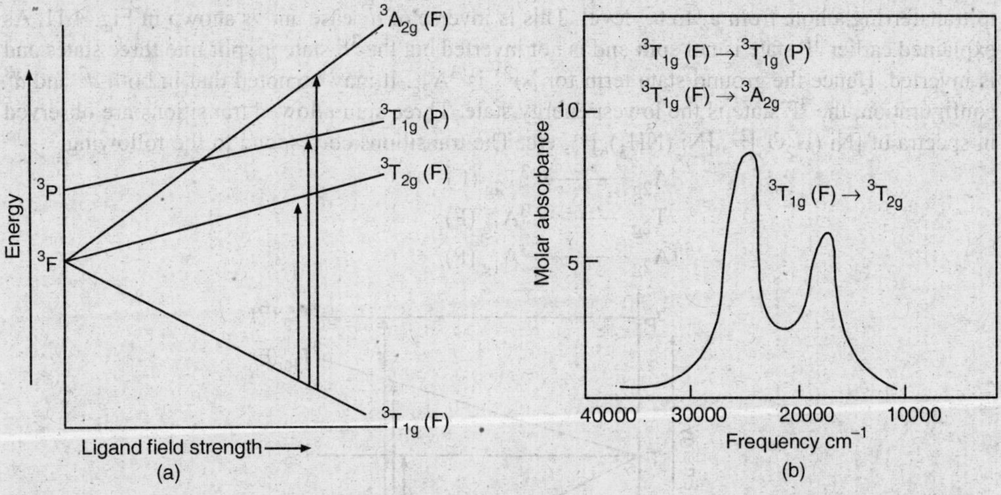

Fig. 4.10 (a) Transitions for V^{3+} (d^2) ion and (*b*) Absorption spectrum of a d^2 complex $[V(H_2O)_6]^{3+}$.

We expect three peaks corresponding to three transitions as shown in Fig. 4.10 (a). The three transitions are from the ground state $^3T_{1g}$ (F) to $^3T_{2g}$, $^3T_{1g}$ (P) and $^3A_{2g}$ respectively. The spectrum of $[V(H_2O)_6]^{3+}$ is shown in Fig. 4.10 (b) which shows only two peaks. The peak at ~17,000 cm^{-1} is assigned to $^3T_{1g}$ (F) $\longrightarrow$ $^3T_{2g}$ (F) and the peak at ~ 24,000cm^{-1} is due to both $^3T_{1g}$ (F) $\longrightarrow$ $^3T_{1g}$(P) and $^3T_{1g}$(F) $\longrightarrow$ $^3A_{2g}$ transitions.

Thus, for $[V(H_2O)_6]^{2+}$, the bands are assigned as:

$$^3T_{1g}\ (F) \longrightarrow {}^3T_{2g}\ (F) \qquad 17{,}000\ cm^{-1}$$

$$^3T_{1g}\ (F) \longrightarrow {}^3A_{2g}\ (F) \qquad 24{,}000\ cm^{-1}$$

The energies corresponding to these two transitions lie very close to each other and therefore, these two transitions are not resolved into two separate peaks.

However, with strong field ligands like NH_3, we get three bonds which may be assigned as:

$$^3T_{1g}\ (F) \longrightarrow {}^3T_{2g}\ (F) \qquad 17{,}200\ cm^{-1}$$

$$^3T_{1g}\ (F) \longrightarrow {}^3T_{1g}\ (P) \qquad 25{,}600\ cm^{-1}$$

$$^3T_{1g}\ (F) \longrightarrow {}^3A_{2g}\ (P) \qquad 36{,}000\ cm^{-1}$$

(b) d^2 Tetrahedral field

A few tetrahedral complexes also exhibit two bands in the range of 9000 cm^{-1} corresponding to $^3A_2$ (F) $\longrightarrow$ $^3T_1$(F) and 15000 cm^{-1} corresponding to $^3A_2$ (F) $\longrightarrow$ $^3T_1$ (P) transitions.

$$^3A_2\ (F) \longrightarrow {}^3T_1\ (F) \qquad 9{,}000\ cm^{-1}$$

$$^3A_2\ (F) \longrightarrow {}^3T_1\ (P) \qquad 15{,}000\ cm^{-1}$$

The transition $^3A_2$ (F) $\longrightarrow$ $^3T_2$ (F) is of very low energy and does not fall in the visible region and hence is not observed in the electronic spectrum of d^2 (T_d).

(c) d^8 Octahedral

The complexes of metal with d^8 configuration can be treated similar to d^2 octahedral complexes. There are two holes in the e_g level and therefore, promoting one electron is equivalent to transferring a hole from e_g to t_{2g} level. This is inverse of d^2 case and is shown in Fig. 4.11. As explained earlier ^3P state is not split and is not inverted but the ^3F state is split into three states and is inverted. Hence the ground state term for Ni^{2+} is $^3A_{2g}$. It may be noted that in both d^2 and d^8 configuration, the ^3F state is the lowest energy state. Three spin allowed transitions are observed in spectra of $[Ni (H_2O_6]^{2+}$, $[Ni (NH_3)_6]^{2+}$, etc. The transitions correspond to the following.

$$^3A_{2g} \longrightarrow {}^3T_{2g}\ (F)$$
$$^3T_{2g} \longrightarrow {}^3A_{1g}\ (F)$$
$$^3TA_{2g} \longrightarrow {}^3A_{1g}\ (P)$$

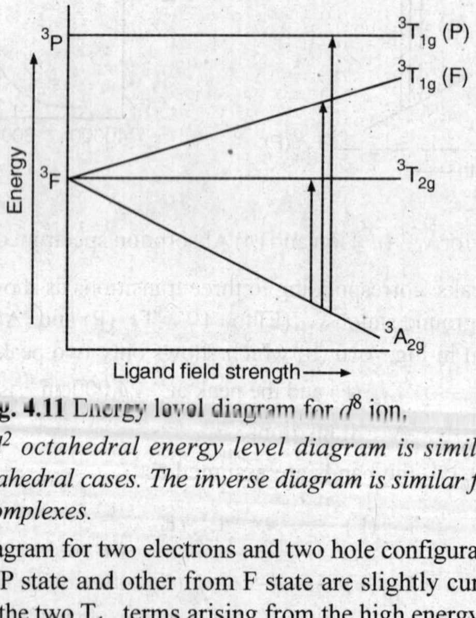

Fig. 4.11 Energy level diagram for d^8 ion.

It may be noted that d^2 octahedral energy level diagram is similar to the high spin d^7 octahedral and d^3 and d^8 tetrahedral cases. The inverse diagram is similar for d^3 and d^8 octahedral and d^2 and d^7 tetrahedral complexes.

Fig. 4.12 gives Orgel diagram for two electrons and two hole configuration 4.12. In this figure, the two T_{1g} states, one from P state and other from F state are slightly curved lines. This may be attributed to mixing between the two T_{1g} terms arising from the high energy P term and low energy F term because of same symmetry possessed by them.

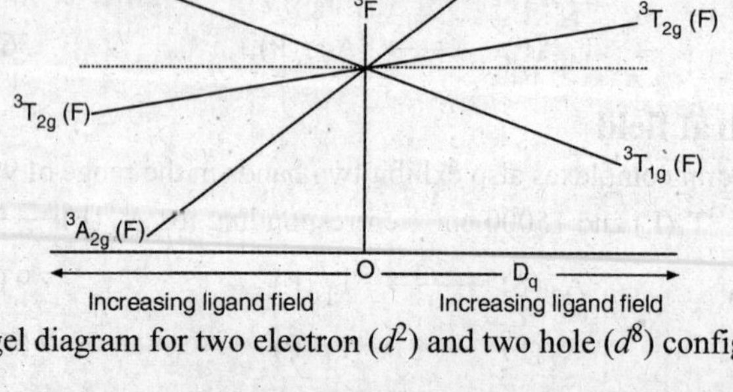

Fig. 4.12. Orgel diagram for two electron (d^2) and two hole (d^8) configuration

ELECTRONIC SPECTRA OF SOME WELL-KNOWN COMPLEXES

1. Cr (III) octahedral complexes : d^3 (O_h)

In the octahedral complex such as $Cr(H_2O)_6]^{3+}$ Cr (III) (d^3) has electron configuration : t_{2g}^3. The d^3 arrangement gives rise to two states ^4F and ^4P. In an octahedral field ^4F state splits into $^4A_{2g}$ (F), $^4T_{2g}$ (F) and $^4T_{1g}$ (F) states while ^4P state does not split out transforms into a $^4T_{1g}$ (P).

Increasing order of energies of different states is:

$^4T_{2g}$ (F) < $^4T_{2g}$ (F) < $^4T_{1g}$ (F) < $^4T_{1g}$ (P)

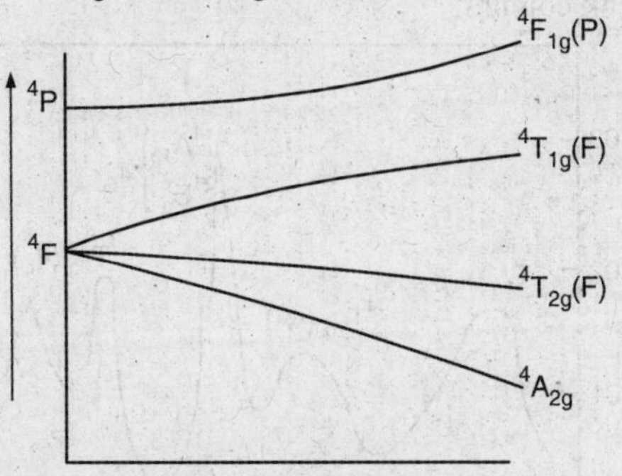

Fig. 4.13 Energy level diagram for a d^3 ion.

The Orgel energy level diagram is shown in Fig. 4.13. The spectra of Cr (III) complexes would be expected to show three absorption peaks corresponding to the following transitions:

For $[Cr(H_2O)_6]^{3+}$

$^4A_{2g}$ (F) $\longrightarrow$ $^4T_{2g}$ (F) 17,400 cm^{-1}

$^4A_{2g}$ (F) $\longrightarrow$ $^4T_{1g}$ (F) 24,500 cm^{-1}

$^4A_{2g}$ (F) $\longrightarrow$ $^4T_{1g}$ (P) 38,600 cm^{-1}

Most of the Cr (III) complexes show two well-defined absorption peaks in the visible region while some complexes show the third band though it is often overlapped by a very intense charge transfer band. Transitions between ground term and higher states and corresponding bands (cm^{-1}) for some Cr (III) complexes are given below in Table 4.7.

Table 4.7 Electronic transitions (in cm^{-1}) of some Cr^{3+} (O_h) complexes.

Transition	Complex				
	$[CrF_6]^{3-}$ Green	$[Cr(H_2O)_6]^{3+}$ Violet	$[Cr(ox)_3]^{3-}$ Red-violet	$[Cr(en)_3]^{3+}$ Yellow	$[Cr(CN)_6]^{3-}$ Yellow
$^4A_{2g} \longrightarrow {^4T_{2g}}$ (F)	14,900	17,400	17,500	21,900	26,700
$^4A_{2g} \longrightarrow {^4T_{1g}}$ (F)	22,700	24,700	23,000	28,500	32,200
$^4A_{2g} \longrightarrow {^4T_{1g}}$ (P)	34,400	38,600	—	—	—

In d^3 tetrahedral complexes, the following transitions ocur:

(*i*) $^4T_1$ (F) $\longrightarrow$ $^4T_2$ (F) (*ii*) $^4T_1$ (F) $\longrightarrow$ $^4A_2$ (F) (*iii*) $^4T_1$ (F) $\longrightarrow$ $^4T_1$ (P)

2. Manganese (II) octahedral complex ions: d^5 (O_h)

In high spin octahedral complexes such as $[MnF_6]^{4-}$, $[Mn(H_2O)_6]^{2+}$, Mn(II) has t_{2g}^3, e_g^2 electronic configuration. **The ground state term is ^{6}S** and there are five unpaired electrons with parallel spins. Any electronic transition within *d* level must involve a reversal of spins and therefore these *d—d transitions are spin forbidden transitions* and are very weak. That is why Mn(II) compounds have pale pink colour.

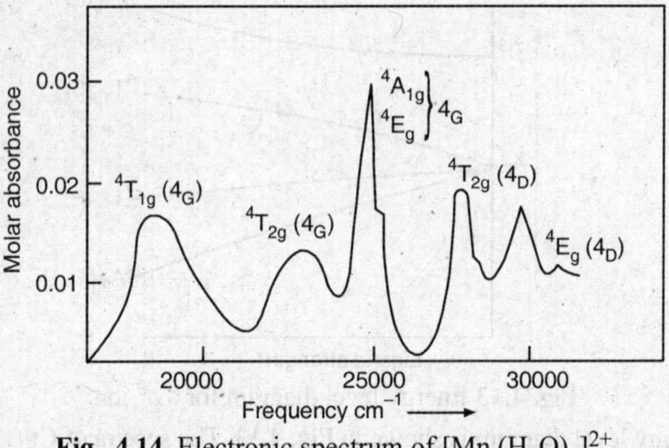

Fig. 4.14. Electronic spectrum of $[Mn(H_2O)_6]^{2+}$

The spectrum of $[Mn(H_2O)_6]^{2+}$ is shown in Fig. 4.14. It has the following features:

(*i*) The molar absorption coefficients, ∈ is about 0.02 – 0.03 *l* mol^{-1} cm^{-1} as compared with 5–10 *l* mol^{-1} cm^{-1} for spin allowed transitions. Thus the bands are very weak.

(*ii*) A large **number of bands** is obtained.

(*iii*) Some of the bands are **sharp** while others are **broad**.

The spectrum can be explained as follows:

(*a*) The ground state term of Mn(II) is ^{6}S. This state does not split but transforms to the A_{1g} state in a weak octahedral field (drawn along the horizontal axis). This is the only sextuplet state and any alteration which may occur on electron excitation results in pairing of two or four spins, thus resulting in doublet or quartet states. Therefore, all excited states will have different spin

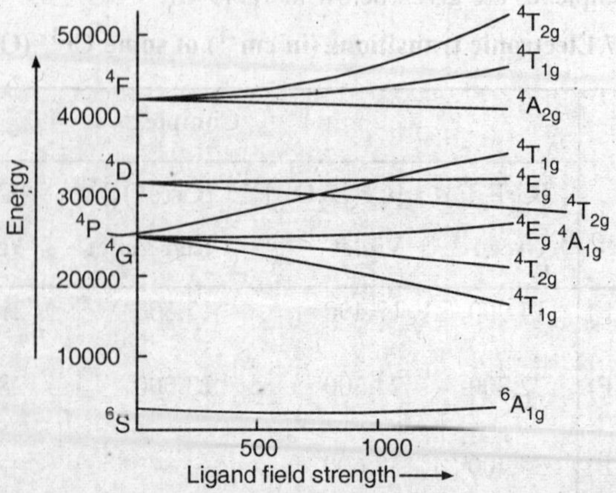

Fig. 4.15. Orgel energy level diagram for Mn^{2+} (d^5) octahedral

multiplicity compared to the ground state and therefore, transition to these excited states are **spin multiplicity forbidden**. However, due to weak spin orbital interactions, these types of transitions are not totally forbidden but give rise to weak absorption bands.

(*b*) For the ground state term ^6S, **there are 11 excited states** (^4G, ^4F, ^4D, ^4P, ^2I, ^2H, ^2G, ^2F, ^2D, ^2P, ^2S. Out of these 11 excited states only four quarters, ^4G, ^4F, ^4D and ^4F involve the reversal of only one spin. Rest of the seven states are doublets and are *doubly spin forbidden* and are unlikely to be observed. In octahedral field, these four states split into ten states and therefore, upto ten weak absorption bands may be observed as shown in Fig. 4.15.

(*c*) Calculations shows that width of the peaks for *d-d* transitions vary directly as the slope of the higher energy state relative to the slope of the ground state in the Orgel diagram. Fig. 4.15 shows that the ground state ^6S does not split but transforms to $^6A_{1g}$ state which is drawn along the horizontal line. The ^4E_g(G), $^4A_{1g}$ (G), ^4E_g(D) and $^4A_{2g}$(F) terms are also horizontal lines. So, their energies are independent of the crystal fields. Since the slope of the ground state terms $^6A_{1g}$ is zero and the slopes of ^4E_g(G), $^4A_{1g}$(G), ^4E_g(D) and $^4A_{2g}$ (F) terms are also zero, transitions from the ground state to these four states should give sharp peaks. Similarly, the peaks due to transitions to $^4T_{1g}$(G) and $^4T_{2g}$ (G) will give broad peaks. These bands are listed as under.

$$^6A_{1g} \longrightarrow {}^4T_{1g} \qquad 18900 \text{ cm}^{-1}$$

$$^6A_{1g} \longrightarrow {}^4T_{2g} \text{ (G)} \qquad 23100 \text{ cm}^{-1}$$

$$\left.\begin{matrix} ^6A_{1g} \longrightarrow {}^4E_g \\ ^6A_{1g} \longrightarrow {}^4A_{1g} \end{matrix}\right\} \qquad 24970 \text{ and } 25300 \text{ cm}^{-1}$$

$$^6A_{1g} \longrightarrow {}^4T_{2g} \text{ (D)} \qquad 28000 \text{ cm}^{-1}$$

$$^6A_{1g} \longrightarrow {}^4E_g \text{ (D)} \qquad 29700 \text{ cm}^{-1}$$

For tetrahedral complexes of d^5 electronic configuration, the same diagram is applicable.

3. Cobalt (III) octahedral complex ions : d^7 (O_h)

In octahedral field of weak ligands, Co(II) has the electronic configuration as : $t_{2g}^{\ 5}\, e_g^{\ 2}$. Thus it has three unpaired electrons. The ground state term is ^4F which is transformed to $^4T_{1g}$ (F), $^4T_{2g}$ and $^4A_{2g}$. The other quartet states is ^4F which does not split but transforms into $^4T_{1g}$(P) state is an octahedral field. The energy level diagram is shown in Fig. 4.16.

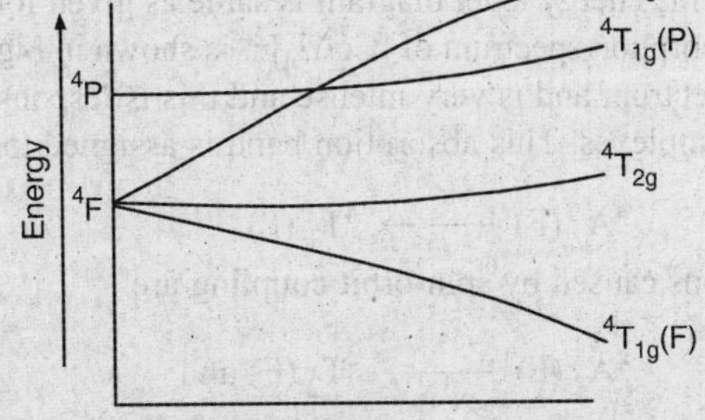

Fig. 4.16 Energy level diagram for Co (II) octahedral complexes.

The figure shows that an octahedrally coordinated Co (II) ion should exhibit the following three spin allowed *d-d* transitions:

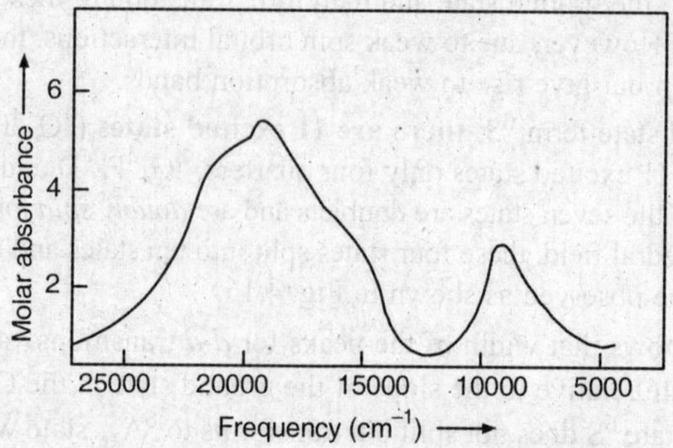

Fig. 4.17 The absorption spectrum of octahedral complex of $[Co(H_2O)_6]^{2+}$

$^4T_{1g}(F) \longrightarrow {}^4T_{2g}(F)$	8700 cm^{-1}	Near IR
$^4T_{1g}(F) \longrightarrow {}^4A_{2g}(F)$	16,000 cm^{-1}	
$^4T_{1g}(F) \longrightarrow {}^4T_{1g}(P)$	19,400 cm^{-1}	

The absorption spectrum of $[Co(H_2O)_6]^{2+}$ (Fig 4.17) is weak and occurs in the blue part of the visible region and therefore accounts for the pale **pink colour** of the $[Co(H_2O)_6]^{2+}$ ion.

The absorption band at 8700 cm^{-1} which falls in the near infra red region is assigned to $^4T_{1g}(F) \longrightarrow {}^4T_{2g}(F)$. A multiple absorption band appears at about 20,000 cm^{-1}. It has three peaks at 16,000, 19400 and 21600 cm^{-1}. The main band at 19,400 cm^{-1} is assigned to $^4T_{1g}(F) \longrightarrow {}^4T_{1g}(P)$ and the shoulder at 16000 cm^{-1} to $^4T_{1g}(F) \longrightarrow {}^4T_{2g}(F)$. It is a two electron process and is therefore very weak as compared to the $^4T_{1g}(F) \longrightarrow {}^4T_{1g}(P)$ transition because two electron transitions lie very close to $^4T_{1g}(F) \longrightarrow {}^4A_{2g}$ transition. Therefore, the transitions are not observed as separate peak.

These transitions are close together because of cross over point between $^4A_{2g}$ and $^4T_{1g}(P)$. The extra shoulder is due to spin orbit coupling in $^4T_{1g}$ (P) state.

Co(II) Tetrahedral complex ions

Co^{2+} in the tetrahedral complex, $[CoCl_4]^{2-}$ has d^7 configuration. In this complex the electron configuration is $e^4t_2^3$. This energy level diagram is same as given for Cr (III) octahedral complex ions (Fig. 4.16). The absorption spectrum of $[CoCl_4]^{2-}$ is shown in Fig. 4.18. The absorption occurs in the red part of the spectrum and is very intense and this is responsible for the deep blue colour of Co(II) tetrahedral complexes. This absorption band is assigned to the transition:

$$^4A_2(F) \longrightarrow {}^4T_1(P)$$

The other transitions caused by spin-orbit coupling are

$$^4A_2(F) \longrightarrow {}^4T_2(F) \text{ and}$$

$$^4A_2(F) \longrightarrow {}^4T_1(F)$$

However, the only band which appears in visible region is at 15,000 cm^{-1} corresponding to $^4A_2(F) \longrightarrow {}^4T_1(P)$ occurring in the red part of the spectrum. The other two transitions (*i*) $^4A_2 \longrightarrow {}^4T_2$ fall in the infrared region (33,00 cm^{-1}) and (*ii*) $^4A_2 \longrightarrow {}^4T_1$ (F) appears in the near infrared region (58,00 cm^{-1}). The various transitions can be listed as under:

$^4A_2$ (F) $\longrightarrow$ $^4T_2$ (F) 33,00 cm^{-1} (infrared region)

$^4A_2$ (F) $\longrightarrow$ $^4T_1$ (F) 58,000 cm^{-1} (near infrared)

$^4A_2$ (F) $\longrightarrow$ $^4T_1$ (P) 15,000 cm^{-1} (visible region)

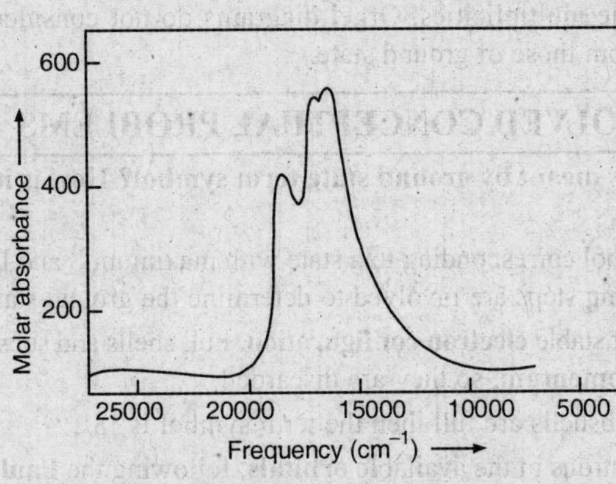

Fig. 4.18 Absorption spectrum of $[CoCl_4]^{2-}$ ion.

4. Nickel (II) octahedral complex ions: $d^8(O_h)$

The spectra of $[Ni(H_2O)_6]^{2+}$ can be understood by considering the energy level diagram (Fig. 4.19). Three spin allowed transitions are expected and these are assigned to the following transitions:

$^3A_{2g}$ (F) $\longrightarrow$ $^3T_{2g}$ (F) 8500 cm^{-1}

$^3A_{2g}$ (F) $\longrightarrow$ $^3T_{1g}$ (F) 15,400 cm^{-1}

$^3A_{2g}$ (F) $\longrightarrow$ $^3T_{1g}$ (P) 26,000 cm^{-1}

Similarly, these band in $[Ni(NH_3)_6]^{2+}$ complex ocur at 10800, 17500 and 28200 cm^{-1} respectively. The bands are quite weak as expected.

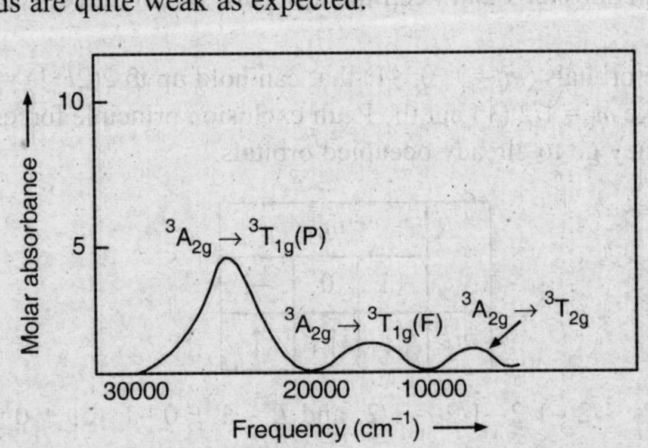

Fig. 4.19. Absorption spectrum of $[Ni(H_2O)_6]^{2+}$ ion.

Limitations of Orgel Energy Level Diagrams

1. The Orgel energy level diagrams consider only the energy levels obtained by splitting the energy terms when the metal ion is subjected to weak octahedral field or tetrahedral field. In other *words, Orgel energy level diagrams consider only weak field (or high spin) cases.*

2. The Orgel energy level diagrams consider spin allowed transitions in which the ground and excited states are of same multiplicities. Orgel diagrams do not consider the excited states of multiplicities different from those of ground state.

SOLVED CONCEPTUAL PROBLEMS

Example 1. What is meant by ground state term symbol? How is it calculated? Explain with an example.

Solution. Term symbol corresponding to a state with maximum S and L is called ground state term symbol. The following steps are involved to determine the ground state term symbol.

1. Start with the most stable **electron configuration.** Full shells and subshells do not contribute to the overall **angular momentum**, so they are discarded.

- If all shells and subshells are full then the term symbol is $^1S_0$.

2. Distribute the electrons in the available **orbitals**, following the **Pauli exclusion principle**. First, we fill the orbitals with highest m_l value with one electron each, and assign a maximal m_s to them (i.e. + 1/2). Once all orbitals in a subshell have one electron, add a second one (following the same order), assigning $m_s = -1/2$ to them.

3. The overall S is calculated by adding the m_s values for each electron. That is the same as multiplying 1/2 times the number of **unpaired** electrons. The overall L is calculated by adding the m_l values for each electron (so if there are two electrons in the same orbital, then we add twice that orbital's m_l).

4. Calculate J as:

- if less than half of the subshell is occupied, take the minimum value $J = |L-S|$;
- if more than half filled, take the maximum value $J = L + S$;
- if the subshell is half filled then L wil be 0, so J = S.

As an example, in the case of fluorine, the electronic configuration is : $1s^2 2s^2 2p^5$.

1. Discard the full subshells and keep the $2p^5$ part. So we have five electrons to place in subshell p (l =1).

2. There are three orbitals (m_l = 1, 0, –1) that can hold up to $2(2l+1) = 6$ electrons. The first three electrons can make $m_s = 1/2$ (↑) but the Pauli exclusion principle forces the next two to have $m_s = -1/2$(↓) because they go to already occupied orbitals.

	m_l		
	+1	0	–1
m_s:	↑↓	↑↓	↑

3. $S = 1/2 + 1/2 + 1/2 - 1/2 - 1/2 = 1/2$; and $L = 1 + 0 - 1 + 1 + 0 = 1$, which is "P" in spectroscopic notation.

4. As fluorine 2p subshell is more than half filled, $J = L + S = 3/2$. Its ground state term symbol is then $^{2S+1}L_J = {}^2P_{3/2}$.

Example 2. The $[Ti(H_2O)_6]^{3+}$ complex absorbs at 5000Å. Calculate Δ_0 for this absorption.

Solution. $E = N_0 h\nu = \frac{N_0 hc}{\lambda}$

For $[Ti(H_2O)_6]^{3+}$

λ = 5000Å = 5000×10^{-10} m, $c = 3.0 \times 10^8$ ms^{-1}, $h = 6.626 \times 10^{-34}$ Js

$N_0 = 6.023 \times 10^{23}$. Substituting the values in the equation above

$$\therefore E = \frac{6.023\times10^{23}\times6.626\times10^{-34}\times3.0\times10^{8}}{5000\times10^{-10}} = 239 \text{ kJ mol}^{-1}$$

$\therefore \Delta_0$ for the complex = 239 kJ mol^{-1}.

Example 3. Derive the term symbols for d^{10} configuration.

Solution. In a d^{10} configuration

+2	+1	0	−1	−2
↑↓	↑↓	↑↓	↑↓	↑↓

$M_L = 0$ so L = 0 and it corresponds to 'S' state.

$M_s = 0$ and so S = 0 and therefore, multiplicity is 2S + 1 = 1

$\therefore$ Term symbol : ^1S

It may be noted that for a closed configuration, the term symbol is always ^1S.

Example 4. The tetrahedral complexes of Mn (II) are more intensely coloured than the octahedral complexes of Mn (II). Explain.

Solution. Manganese (II) has d^5 electronic configuration. Tetrahedral Mn (II) complexes are yellow green. The *d–d* transitions in tetrahedral Mn (II) complexes are quite intense than octahedral complexes. These *d–d* transitions are Laporte allowed because tetrahedral molecules do not possess centre of symmetry. The transitions in Mn (II) d^5 are also multiplicity allowed due to mixing of *d*- and *p*-orbitals in tetrahedral geometry. Hence we observe intense coloured Mn (II) tetrahedral complexes.

On the otherhand, in $[Mn(H_2O)_6]^{2+}$ octahedral complexes, the transitions are both Laporte forbidden (due to centre of symmetry) and multiplicity forbidden because ground term $^6A_{1g}$ is the only term with multiplicity 6 and all other terms are of lower multiplicity. Therefore, the bands are very weak.

Example 5. Identify the ground state terms from each set of terms:

(*i*) ^1S, ^3P, ^1D, ^3F, ^1G

(*ii*) ^1S, ^3P, ^1D

(*iii*) ^1S, ^2P, ^2D, ^3F, ^1G

Solution. (*i*) ^3F because although it has same multiplicity as ^3P but it has higher value of L (=3) and hence more stable.

(*ii*) ^3P because it has highest multiplicity.

(*iii*) ^3F because it has highest multiplicity.

Example 6. Calculate the number of microstates for:

(*i*) d^1 (*ii*) p^3 (*iii*) d^4 configurations

Solution. (*i*) d^1

Here $n = 10$, $r = 1$

Applying the formula to calculate number of microstates

$$\text{No. of microstates } \binom{n}{r} = \frac{n!}{r!(n-r)!} = \frac{10!}{1!(10-1)!} = \frac{10!}{1!9!}$$

$$= \frac{10\times9\times8\times7\times6\times5\times4\times3\times2\times1}{1\times9\times8\times7\times6\times5\times4\times3\times2\times1} = \mathbf{10}$$

(ii) p^3

$n = 6, \quad r = 3$

$$\text{No. of microstates} = \frac{6!}{3!(6-3)!} = \frac{6!}{3!3!} = \frac{6\times5\times4\times3\times2\times1}{3\times2\times1\times3\times2\times1} = \mathbf{20}$$

(iii) d^4

$n = 10, \quad r = 4$

$$\text{No. of microstates} = \frac{10!}{4!(10-4)!} = \frac{10!}{4!6!} = \frac{10\times9\times8\times7\times6\times5\times4\times3\times2\times1}{4\times3\times2\times1\times6\times5\times4\times3\times2\times1} = \mathbf{210}$$

Example 7. Determine the term symbol for ground state of nitrogen.

Solution. The electronic configuration of nitrogen, N is $1s^2\ 2s^2\ 2p^3$. We consider only $2p$ subshell because the inner orbitals are filled.

l	+1	0	−1
$2p$	↑	↑	↑

$M_L = +1 + 0 - 1 = 0$ It corresponds to state S.

$$M_S = +\frac{1}{2}+\frac{1}{2}+\frac{1}{2} = \frac{3}{2}$$

$$\therefore \text{Spin multiplicity} \quad 2S+1 = \frac{2\times3}{2}+1 = 4$$

Since $M_L = 0$, J has only one value equal to M_S i.e. $J = 3/2$

∴ Term symbol for N is $^4S_{3/2}$.

Example 8. Electronic transitions of *d–d* type in octahedral complexes should be forbidden by the Laporte selection rule. Why are moderately strong spectra observed in certain cases?

Solution. Electronic transitions of *d–d* type are Laporte forbidden only under the conditions of centro symmetric octahedral complexes. However, in substituted octahedral complexes such as $[Co(NH_3)_5Cl]^{2+}$, some mixing of *d* and *p*-orbitals may occur. Therefore the transitions in these complexes are no longer pure *d-d* in nature, rather mixed *d* and *p* type which are moderate in intensity. Hence, non centrosymmetric octahedral complexes give strong *d-d* transitions and have moderately strong spectra.

Example 9. Determine the term symbols for

***(i)* s^1p^1 *(ii)* d^1s^1 *(iii)* d^7s^1**

Solution. *(i)* s^1p^1

	+1	0	−1
p	↑		

	0
s	↑

$M_L = +1 + 0 = 1$ $\quad\quad \therefore$ State = P

$M_S = \frac{1}{2} + \frac{1}{2} = 1$ $\quad\quad \therefore S = 1$

Spin multiplicity, $= 2S + 1$

$= 2 \times 1 + 1 = 3$

$\therefore$ Term symbol : $^3\mathbf{P}$

(*ii*) d^1s^1

+2	+1	0	−1	−2		0
↑						↑

$M_L = 2 + 0 = 2$ $\quad\quad \therefore$ State = D

$M_S = \frac{1}{2} + \frac{1}{2} = 1$ $\quad\quad S = 1$

Spin multiplicity, $= 2S + 1 = 2 \times 1 + 1 = 3$

$\therefore$ Term symbol : $^3\mathbf{D}$

(*iii*) d^7s^1

+2	+1	0	−1	−2		0
↑↓	↑↓	↑	↑	↑		↑

$M_L = 4 + 2 + 0 - 1 - 2 + 0 = 3$ $\quad \therefore$ State = F

$M_S = 4 \times \frac{1}{2} = 2$ $\quad\quad \therefore S = 2$

Spin multiplicity, $2S + 1 = 2 \times 2 + 1 = 5$

$\therefore$ Term symbol : $^5\mathbf{F}$

Example 10. What terms arise from p^1d^1 configuration?

Solution.

	+1	0	−1		+2	+1	0	−1	−2
p	↑				↑				

For a p^1d^1 configuration,

$M_L = +1 + 2 = 3$ $\quad\quad \therefore$ Term = F

The electrons may be paired, S = 0 or parallel, S = 1

Spin multiplicity $2S + 1 = 2 \times 0 + 1 = 1$ or $2 \times 1 + 1 = 3$

Terms $^1\mathbf{F}$ and $^3\mathbf{F}$ are possible corresponding to the two spin multiplicities

Example 11. Calculate the number of microstate for
(*i*) p^4 and (*ii*) d^4 configurations.

Solution. Microstates can be calculated using the formula :

No. of microstates : $$\binom{n}{r} = \frac{n!}{r!(n-r)!}$$

n = twice the no. of orbitals and *r* = the number of electrons.

(*i*) for p^4

$$n = 6, r = 4, \quad \text{Substituting the values}$$

$$\text{No. of microstates} = \frac{6!}{4!(6-4)!}$$

$$= \frac{6!}{4!2!} = \frac{6\times5\times4\times3\times2\times1}{4\times3\times2\times1\times2\times1}$$

$$= \textbf{15 microstates}$$

(*ii*) for d^4 configuration,

$$n = 10, r = 4, \quad \text{Substituting the values, we have}$$

$$\text{No. of microstates} = \frac{10!}{4!(10-4)!} = \frac{10!}{4!6!}$$

$$= \frac{10\times9\times8\times7\times6\times5\times4\times3\times2\times1}{4\times3\times2\times1\times6\times5\times4\times3\times2\times1}$$

$$= \textbf{210 microstates}$$

Example 12. What is vibronic coupling? Give one example of this phenomena.

Solution. The *d-d* transition in octahedral complexes are Laporte forbidden on symmetry consideration. However an octahedral (O_h) complex may absorb light in the following manner.

The bonds in the transition metal complexes are not rigid but may undergo vibrations that may temporarily change the symmetry. Some of the molecules may slightly be distorted from octahedral symmetry at a given time. For example, octahedral complexes vibrate in a way in which the centre of symmetry is temporarily lost. This phenomenon is called vibronic coupling. As a result, *d-d* transitions having molar absorptivities in the range of approximately 10 to 50 L mol^{-1} cm^{-1} are commonly observed. These are also responsible for the bright colour of many of these complexes. For example, this phenomenon is observed in $[Mn(H_2O)_6]^{2+}$. In this case, the transitions are both multiplicity and Laporte forbidden. The complex is coloured. Had there been no vibronic coupling it would have been colourless. These transitions are vibronically allowed and the effect is called vibronic coupling.

Example 13. Calculate the term symbols for ground state of
(*i*) Cr ($3d^5$ $4s^1$) and (*ii*) Ni ($3d^8$ $4s^2$).

Solution. (*i*) Cr : $3d^5$ $4s^1$

	+2	+1	0	−1	−2			0
3*d*	↑	↑	↑	↑	↑		4*s*	↑

$$M_L = +2 + 1 + 0 - 1 - 2 + 0 = 0 \qquad \therefore \text{State} = \text{F}$$

$$M_S = +\frac{1}{2}\times 6 = 3, \qquad S = 3$$

$$\text{Multiplicity, } 2S + 1 = 2\times3+1 = 7$$

J = 3 (because L = 0)

∴ Term symbol for ground state = $^7S_3$

(*ii*) Ni : $3d^8\ 4s^2$

	+2	+1	0	−1	−2		0
3d	↑↓	↑↓	↑↓	↑	↑		↑↓

$M_L = +4+2+0-1-2+0 = 3$ State = F

$M_S = +\frac{1}{2} \times 2 = 1$ S = 1

Multiplicity, = 2 × 1 + 1 = 3

J = 3 + 1 = 4

∴ Term symbol for ground state = $^3\mathbf{F}_4$

Example 14. Identify the ground state terms from each set of following giving reasons.

(*a*) $^3F_2, ^3F_3, ^3F_4$ **(*b*)** $^1S, ^2P, ^2D, ^3F, ^1G$ **(*c*)** $^1S, ^3P, ^1D, ^3F, ^1G$

Solution. (*a*) $^3F_2$ is the ground state term. L and S values are same for all the three cases. The ground term is determined by J. For less than half filled (d^2 in the present case), the minimum value of J gives the ground state.

(*b*) ^3F is the ground state term because it has highest multiplicity

(*c*) 3F is the ground state term because it has higher value of L (= 3) but same multiplicity as ^3P (L = 2).

Example 15. Derive states for d^2 configuration.

Solution. For d^2 configuration :

$$l_1 = l_2 = 2$$

L = (2 + 2), (2 + 2 −1), (2 + 2 −2) (2 + 2 −3), (2 + 2 −4)
= 4 3 2 1 0

∴ States = G, F, D, P, S

Since there are two electrons so

S = 1, 0

or 2S + 1 = 3, 1

Hence term states for d^2 configurations are:

$^3G, ^3F, ^3D, ^3P, ^3S, ^1G, ^1F, ^1D, ^1P$, and ^1S.

Example 16. Determine the ground terms for low spin and high spin d^6 octahedral configurations.

Solution. High spin :

+2	+1	0	−1	−2
↑↓	↑	↑	↑	↑

$M_L = +2+2+1+0-1-2 = 2$ ∴ State = D

$M_S = 4+\frac{1}{2} = 2$

Spin multiplicity = 2 × 2 + 1 = 5

∴ Term symbol = $^5\mathbf{D}$

Low spin :

↑↓	↑↓	↑↓		

$M_L = +2+2+1+1+0+0 = 6$, $\therefore$ State $=$ I

$M_S = 0$ $\therefore$ Multiplicity $= 2 \times 0 + 1 = 1$

$\therefore$ Term $= {}^1I$

Example 17. Determine the possible values of J for the term obtained from a d^2 configuration.

Solution. For d^2 configuration:

$$l_1 = 2, l_2 = 2$$

$$L = (2+2), (2+2-1), (2+2-2), (2+2-3), (2+2-4)$$

$$= 4 \quad 3 \quad 2 \quad 1 \quad 0$$

$\therefore$ States $=$ G, F, D, P, S

Since there are two electrons so, $S = 1\left(\frac{1}{2}+\frac{1}{2}\right), 0\left(\frac{1}{2}-\frac{1}{2}\right)$

or Spin multiplicity S $= (2S + 1) = 3, 1$

Possible values of J are:

$L = 4, S = 0, J = 4$ ${}^1G$

$L = 3, S = 1, J, 4, 3$ ${}^3F$

$L = 2, S = 0, J = 2$ ${}^1D$

$L = 1, S = 1, J = 2, 1, 0$ ${}^3P$

$L = 0, S = 0, J = 0$ ${}^1S$

Example 18. Explain the terms:

Term, Level

Solution. The combination of an S value and an L value is called a **term** and has a statistical weight (i.e number of possible microstates) of (2S + 1) (2L + 1).

A combination of S, L and J is called a **level**. A given level has a statistical weight of (2J + 1), which is the number of possible microstates associated with this level in the corresponding term. For example, for S = 1, L = 2, there are $(2 \times 1 + 1)(2 \times 2 + 1) = 15$ different microstates corresponding to the ${}^3D$ term, of which $(2 \times 3 + 1) = 7$ belong to the ${}^3D_3$ (J = 3) level. The sum of (2J + 1) for all levels in the same term equals (2S + 1) (2L + 1). In this case, J can be 1, 2 or 3 so 3 + 5 + 7 = 15.

Example 19. Give the ground state term symbols for H($1s^1$) and He ($1s^2$)

Solution. H ($1s^1$)

Ground state term symbol is ${}^2S_{1/2}$

H ($1s^2$)

Ground state term symbol is ${}^1S_0$

Example 20. Why do tetrahedral complexes of an element give much more intense *d-d* spectra than its octahedral complexes?

Solution. Tetrahedral complexes do not have centre of symmetry and therefore, *d-d* transitions in tetrahedral complexes are Laporte allowed. The Laporte allowed transitions are very intense. Therefore, such transitions give intense bands in tetrahedral complexes. On the other hand, octahedral complexes show only very weak transitions due to possibility of some mixing.

Example 21. Give the term symbols for

(a) He ($1s^2\ 2s^1$) in excited state

(b) B ($1s^2 2s^2 2p^1$)

Solution. (a) He ($1s^2\ 2s^1$)

Term symbols $^1S_0$, $^3S_1$

(b) B ($1s^2\ 2s^2\ 2p^1$)

Term symbols $^2P_{1/2}$, $^2P_{3/2}$

Example 22. Give the number of microstates in N ($1s^2 2s^2 2p^3$). Give the distribution in different term symbols in N.

Solution. No. of microstates = $\dfrac{6!}{3!3!} = 20$

Distribution

$^2D = 10$

$^2P = 6$

$^4S = 4$

Total 20

EXERCISES
(Including Questions from Different University Papers)

Multiple Choice Questions (Choose the correct option)

1. Ground state term symbol for He ($1s^2$) is

(*a*) $^1S_{1/2}$ (*b*) $^1S_0$

(*c*) $^0S_{1/2}$ (*d*) none of these

2. Ground state term for d^2 configuration is:

(*a*) ^3F (*b*) ^3P (*c*) ^1G (*d*) ^1S

3. The colour of the complex ion, $[Ti(H_2O)_6]^{3+}$ is due to:

(*a*) presence of water molecules

(*b*) intermolecular vibrations

(*c*) excitation of electron from t_{2g} to e_g energy level

(*d*) excitation of electron from $3d$ to $4s$ energy level

4. The number of microstates for a p^2 configuration is

(*a*) 10 (*b*) 20 (*c*) 18 (*d*) 15

5. For L = 0 and S = 1, full spectroscopic term is :

(*a*) $^3D_1$ (*b*) $^1S_1$ (*c*) $^3S_1$ (*d*) $^1P_3$

6. Which of the following electronic arrangement has maximum number of microstates?

(*a*) d^5 (*b*) d^3 (*c*) d^6 (*d*) d^9

7. Ground state term of d^5 configuration is :

(*a*) ^6S (*b*) ^4F (*c*) ^2D (*d*) ^3P

8. For Laporte forbidden transitions

(*a*) $\Delta l = 0$ (*b*) $\Delta S = 0$ (*c*) $\Delta l = -1$ (*d*) $\Delta l = \pm 1$

9. Mulliken symbol for spectroscopic term P in octahedral field is

(*a*) A_{1g} (*b*) T_{1g} (*c*) T_{2g} (*d*) E_g

10. The ground state for $2p^3$ is

(*a*) $^4S_3$ (*b*) $^3F_4$ (*c*) $^4S_{3/2}$ (*d*) $^2P_1$

11. The ground state term for p^6 is same for

(*a*) d^{10} (*b*) d^6 (*c*) p^3 (*d*) d^5

12. The lowest energy term for d^2 ion is

(*a*) ^3F (*b*) ^1S (*c*) ^3P (*d*) ^4P

13. Which of the following corresponds to absorption peak of maximum wave number in $[Cr(H_2O)_6]^{3+}$ complex ion?

(*a*) $^4A_{2g} \longrightarrow {}^4T_{2g}$ (F) (*b*) $^4A_{2g} \longrightarrow {}^4T_{1g}$ (P)

(*c*) $^4A_{2g} \longrightarrow {}^4T_{1g}$ (F) (*d*) $^4T_{2g}$(F) $\longrightarrow {}^4T_{1g}$ (F)

14. Solutions of $[CoCl_4]^{2-}$ are deep blue because

(*a*) it is tetrahedral anion

(*b*) it has d^7 configuration

(*c*) it has electronic transition in the blue part of the spectrum

(*d*) of transition $^4A_2$ (F) $\longrightarrow {}^4T_1$(P)

ANSWERS

1. (*b*)	**2.** (*a*)	**3.** (*c*)	**4.** (*d*)	**5.** (*c*)	**6.** (*a*)
7. (*a*)	**8.** (*a*)	**9.** (*b*)	**10.** (*c*)	**11.** (*a*)	**12.** (*a*)
13. (*b*)	**14.** (*d*)				

Short Answer Questions

1. State and explain Laporte selection rule.
2. Calculate the term symbol for d^{10} arrangement of electrons.
3. Derive the term symbol for an atom with s^1 configuration.
4. Give a relation to calculate the number of microstates.
5. Write the Mulliken symbol for spectroscopic terms P and D in octahedral field.
6. Arrange the following in increasing order of energy:
   $^1S_0$, $^1D_2$, $3P_2$, $^3P_0$, $^3P_1$
7. Which out of $^3P_2$, $^3P_1$ and $^3P_0$ has lowest energy?
8. Differentiate between LS coupling and jj coupling.
9. The terms of d^2 configuration are ^1S, ^3P, ^1D, ^3F and ^1G. Which of these has the highest energy?
10. Draw Orgel diagram for a d^9 configuration in an octahedral field.

General Questions

1. (*a*) Why do tetrahedral complexes of an element give much more intense *d-d* spectra than its octahedral complexes?

(b) Discuss the electronic spectra of $[Co(H_2O)_6]^{2+}$ and $[CoCl_4]^{2-}$ complexes.

2. (*a*) Find the ground state term for each of the following configuration:

(*i*) d^3 (*ii*) d^5 (*iii*) d^9

(*b*) What are spin multiplicity forbidden and Laporte forbidden transitions?

3. (*a*) What are selection rules for electronic spectra?

(*b*) Discuss the spectral features of Mn(II) in octahedral complexes of weak and strong ligand fields.

(c) Write briefly about L – S coupling.

4. (*a*) The correlation diagram for a d^2 configuration as in V^{3+} (aq) ion does not give simple Δ_0 value unlike the crystal field splitting diagram. Which transition from the ground state $^3T_{1g}$ would give Δ_0 value?

(*b*) Define spectroscopic term. Find the terms for atoms with the following configurations: (*i*) s^1 (*ii*) s^1p^1 (*iii*) p^1

(*c*) Discuss the absorption spectra for Co (II) octahedral and Co (II) tetrahedral complexes.

(*d*) Discuss the nature of electronic transitions in octahedral complexes of metal ions with d^6 (high spin) configuration.

5. (*a*) Calculate the ground state terms with spin multiplicity for the following octahedral ions: V^{3+}, Ni^{2+}, Cu^{2+}.

(*b*) What is L-S coupling? How does it help to calculate the resultant angular momentum (J) of an atom having two electrons in a *d*- subshell.

(*c*) Draw combined Orgel diagram for d^1, d^4, d^6 and d^9 complexes.

6. (*a*) Discuss the Orgel diagram and absorption spectra for a d^8 ion.

(*b*) According to Laporte selection rule, electronic transitions of the *d-d* type displaced in spectra of octahedral complexes should be forbidden. Why are moderately strong spectra actually observed?

(*c*) The absorption spectrum of Ti^{3+} (*aq*) (d^1) is attributed to a single $t_{2g} \longrightarrow e_g$ transition. Explain the position, intensity and broad nature of this band.

7. (*a*) Determine the ground state term symbol of anion with d^1 configuration.

(*b*) What is meant by spectrochemical series?

(*c*) Draw combined Orgel diagram for d^1, d^4, d^6 and d^9 complexes.

(*d*) Write briefly about L – S coupling scheme of angular momenta.

8. (*a*) Identify the ground state term from each set of the following terms:

(*i*) ^1S, ^3P, ^1D

(*ii*) ^1S, ^2P, ^2D, ^3F, ^1G

(*iii*) ^1S, ^3P, ^1D, ^3F, ^1G

(*b*) For $[V(H_2O)_6]^{3+}$ two absorption bands are observed at 17,800 and 25,700 cm^{-1}. Assign these bands.

(*c*) Explain as to why Co(II) octahedral complexes are light pink while tetrahedral species are intense blue in colour?

9. (*a*) Draw the energy level diagram for d^2 configuration in tetrahedral and octahedral fields showing three possible transitions.

(*b*) How does Hund's rule help to arrange the different spectroscopic terms in order of their increasing energies? How does it help to find the term in ground state?

(*c*) Name the largest energy transition for high spin octahedral complexes for V^{3+}, Ni^{2+}, Cr^{3+}, and Cu^{2+}.

10. (*a*) Discuss Russel-Saunders states for d^2 configurations.

(*b*) Give energy level diagram and electronic spectra of Cr (III) octahedral complexes.

11. (*a*) Discuss the special features of electronic spectra of $[Ni(H_2O)_6]^{2+}$ ion.
(*b*) Draw combined Orgel diagram for d^1, d^4, d^6 and d^9 complexes.

12. (*a*) Write a short note on L-S coupling.
(*b*) Discuss special features of electronic spectra of Cr (III) octahedral and Mn (II) octahedral complex ions.

13. (*a*) Find out the ground state terms with spin multiplicity for Cu^{2+} and V^{3+} ions.
(*b*) Name the factors which govern the magnitude of Δ.
(*c*) Draw an Orgel diagram for d^3 in O_h complex.

14. (*a*) Show how one gets the following terms for d^2 configuration:
^1G, ^3F, ^1D, ^3P and ^1S. Which of these belong to ground state?
(*b*) Will the electronic spectra of $[Co(H_2O)_6]^{2+}$ and $[CoCl_4]^{2-}$ be same? If not explain the differences that one encounters.
(*c*) Electronic transition of the *d-d* type displayed in the spectra of octahedral transition metal complexes should be forbidden by the Laporte selection rule. Why are moderately strong spectra actually observed? Is this the reason that spectrum of $[Co(H_2O)_6]^{2+}$ is less easy to interpret?

15. (*a*) Derive Russel-Saunders states for d^2 configuration. Arrange these in accordance with their increasing energies diagrammatically.
(*b*) What are Orgel diagrams? Draw and discuss the Orgel energy level diagram for $[Cu(H_2O)_6]^{2+}$ ion.
(*c*) Discuss in details the electronic spectrum of $[Ti(H_2O)_6]^{3+}$ complex ion.

16. (*a*) What is L-S coupling scheme to derive the energy states of d^2-system?
(*b*) Discuss special features of electronic spectra of Co(II) octahedral and Co(II) tetrahedral complex ions.

17. (*a*) Arrange the different spectroscopic terms of titanium with the help of Hund's rules.
(*b*) Discuss the special features of the electronic spectrum of $[Co(H_2O)_6]^{2+}$ ion.
(*c*) State and explain the Laporte selection rule.

18. (*a*) Why do tetrahedral complexes of an element give much more intense *d-d* specrta than its octahedral complexes?
(*b*) What are the two important limitations of Orgel energy level diagrams?

19. (*a*) What do you understand by a term symbol? Derive the term symbols for p^2-configuration and calculate the ground state term using Hund's rule or L-S coupling scheme.
(*b*) Draw combined Orgel–Energy level diagram for d^1 and d^9 octahedral complexes.

20. (*a*) Define L–S coupling.
(*b*) What is spectrochemical series.
(*c*) Why is $[Ti(H_2O)_6]^{3+}$ ion violet?
(*d*) Describe electronic spectra of $[Mn(H_2O)_6]^{2+}$.

21. (*a*) Derive the term symbols for d^2 configuration.
(*b*) $[Ti(H_2O)_6]^{3+}$ is purple in colour. Explain.
(*c*) Describe the vibronic coupling with example.

22. (*a*) What are Orgel diagrams? What information is conveyed by these diagrams?
(*b*) Give a detailed account of the selection rules of electronic spectra.

23. (*a*) Why do tetrahedral complexes of *d*-block elements give much more intense *d-d* spectra than its O_h complexes?

(*b*) Derive the ground state term symbol for an atom with the help of L-S coupling for d^2 and d^6 configuration?

24. (*a*) Derive the term symbols for p^6 and d^{10} systems.

(*b*) Draw Orgel diagram for d^1 and d^9 systems.

25. (*a*) Discuss the nature of electronic transitions in d^6 octahedral.

(*b*) Tetrahedral complexes of Mn(II) are more intensely coloured than the octahedral complexes of Mn(II) explain.

(*c*) Draw Orgel diagram for d^1 complexes.

26. (*a*) Draw Orgel diagram for d^1 and d^9 system.

(*b*) State any two selection rules for electronic spectrum of transition metal complexes. Discuss exception if any.

(*c*) Identify ground term assuming L–S coupling for ions.

(*i*) $2p^2$ (*ii*) $3d^8$

27. (*a*) Find out the ground state term for the following octahedral ions: Cu^{+2}, V^{+3} and Ni^{+2}. Write their multiplicity also.

(*b*) Draw a combined Orgel diagram for d^1, d^4, d^6 and d^9 complexes in octahedral and tetrahedral complexes.

(*c*) The absorption Ti^{3+} (*aq*), d^1 is attributed to single (*i*) $t_{2g} \rightarrow e_g$ transition. Explain the position, intensity and broad nature of the band.

28. (*a*) The complex $[Ti(H_2O)_6]^{3+}$ shows a single absorption band at 20,300 cm^{-1}, Calculate Δ (Crystal field splitting energy).

(*b*) What will be the term symbol for ground state for configuration d^3.

(*c*) Draw Orgel diagram for d^2 configuration in octahedral complexes.

29. (*a*) Calculate the term symbol for ground state of Cr.

(*b*) For $[V(H_2O)_6]^{3+}$ two absorption bands are observed at 17780 and 25700 cm^{-1}. Assign these bands.

(*c*) State and explain Hunds rule for assigning ground state spectroscopic term with suitable examples.

30. (*a*) Find the Russel-Saunders terms for p^2 configuration. What will be the ground state? Draw their splitting diagram in terms of energy.

(*b*) Why do tetrahedral complexes give much more intense *d-d* transition than octahedral complexes?

(*c*) Find the ground state term for $[Ti(H_2O)_6]^{3+}$ ion. Comment upon the colour of this complex ion.

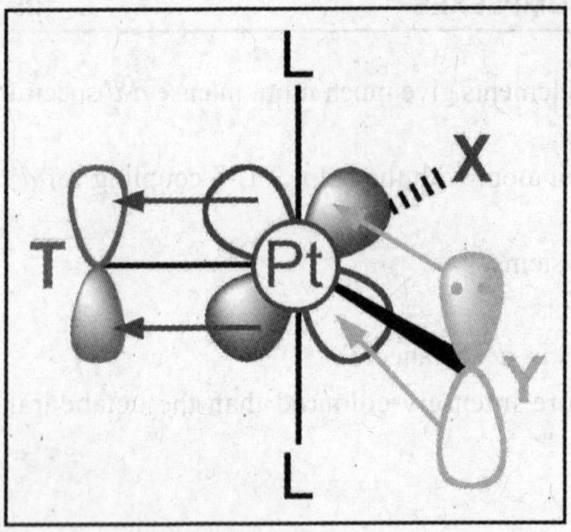

5

Thermodynamic and Kinetic Aspects of Metal Complexes

5.1. INTRODUCTION

The interaction between a hydrated metal ion and a ligand to form a metal complex can be regarded as Lewis acid-base reaction. When we talk of the stability of a complex, we mean the tendency of the metal ion and the ligand to form the complex. We have also to take into consideration the tendency of the complex formed to dissociate into the metal ion and the ligand. Thus, the stability of the complex will be given by the relative tendencies of the two processes.

(*a*) Association of the metal ion and the ligand to form the complex

(*b*) Dissociation of the complex to give back metal ion and the ligand.

If a compound is stable, we can say that it can be stored almost unchanged for a fairly long period. We resort to thermodynamics to asses the stability of a complex. However, one of the limitations of thermodynamics is: It tell us whether a reaction will take place or not, but it does not tell the rate of the reaction. For example, we have a reaction which is thermodynamically possible but is very very slow. Then, for practical purposes, we shall say that the reaction does not take place. So, a reaction may be thermodynamically feasible but kinetically not feasible. Similarly, a complex may be thermodynamically stable but kinetically unstable. We shall read about it in detail in the following sections.

5.2. THERMODYNAMIC STABILITY OF METAL COMPLEXES

The terms stable or unstable are thermodynamic terms which refer to the tendency of species to exist or not.

We can define the thermodynamic stability of a complex as **the measure of the extent of formation of a complex at equilibrium.** If the interaction between the metal ion and ligand is strong, the complex formed would be thermodynamically more stable. Therefore, the thermodynamic stability may be expressed in terms of equilibrium constant for the formation of the complex. For the reaction :

$$Cu^{2+} + 4NH_3 \rightleftharpoons [Cu(NH_3)_4]^{2+}$$

Tetraamminecopper (II) ion

the equilibrium constant, K_s is given by the equation

$$K_s = \frac{[Cu(NH_3)_4]^{2+}}{[Cu^{2+}][NH_3]^4}$$

Larger the numerical value of K_s, more thermodynamically stable will be the complex. Similar expressions can be written for other complexes also and we can compare the stability constants for various complexes formed under similar conditions in terms of K_s. The stability constant, K_s for the above reaction is 4.5×10^{21}.

The stability constant for the similar cyanide complex of Cu^{2+}:

$$Cu^{2+} + 4CN^- \rightleftharpoons [Cu(CN)_4]^{2-}$$

Tetracyanocopper (II) ion

$$K_s = \frac{[Cu(CN)_4]^2}{[Cu^{2+}][CN^-]^4}$$

is 2.0×10^{27} which is higher than the corresponding value for Cu^{2+}—NH_3 complex. *This shows that CN^- is a stronger ligand than NH_3 molecule.*

We have the following data for the complexes of Ag with NH_3 and CN^- as ligands

$$Ag^+ + 2NH_3 \rightleftharpoons [Ag(NH_3)_2]^+ \qquad K_s = 1.6 \times 10^7$$

$$Ag^+ + 2CN^- \rightleftharpoons [Ag(CN)_2]^- \qquad K_s = 5.4 \times 10^{18}$$

Compare the stability constants of two nickel complexes with ligands NH_3 and en (ethylenediamine)

$$Ni^{2+} + 6NH_3 \rightleftharpoons [Ni(NH_3)_6]^{2+} \qquad K_s = 6.1 \times 10^8$$

$$Ni^{2+} + 3en \rightleftharpoons [Ni(en)_3]^{2+} \qquad K_s = 4.6 \times 10^{18}$$

The K_s values indicate that bidentate ligand ethylenediamine forms a much more stable complex than ammonia.

Stability Constants

Consider the formation of the complex, ML_n by the combination of the metal (M) with the ligand L. *n* is the coordination number.

$$M + nL \rightleftharpoons ML_n$$

The reaction proceeds in the following steps :

$$M + L \rightleftharpoons ML$$

The equilibrium constant for this step of the reaction is given by

$$K_1 = \frac{[ML]}{[M][L]}$$

Add another molecule of ligand L to ML

$$ML + L \rightleftharpoons ML_2$$

Equilibrium constant for this step is given by

$$K_2 = \frac{[ML_2]}{[M][L]}$$

Similarly, ML_n is formed by a series of the following reactions :

$$ML_2 + L \rightleftharpoons ML_3 \text{ and } K_3 = \frac{[ML_3]}{[ML_2][L]}$$

$$\vdots$$

$$ML_{n-1} + L \rightleftharpoons ML_n \text{ and } K_n = \frac{[ML_n]}{[ML_{n-1}][L]}$$

There will be *n* such equilibria, where *n* represents the coordination number of the metal ion for the ligand L. The equilibrium constants K_1, K_2, K_3 .... K_n are known as **stepwise stability constants.** It may be realised that the stepwise stability constants will go on decreasing, because the ligands which are already coordinated to the metal ion will repal incoming ligands towards the metal ion because of coulombic repulsions. Therefore,

$$K_1 > K_2 > K_3 ..... > K_n$$

This is illustrated by noting down the values of K or log K for different steps for the formation of complexes of Cu^{2+} and Cd^{2+} with NH_3.

Cation	Ligand	log K_1	log K_2	log K_3	log K_4
Cu^{2+}	NH_3	4.3	3.6	3.1	2.3
Cd^{2+}	NH_3	2.6	2.1	1.4	0.9

We could express **overall stability constant** or commulative stability constant in another way. The equilibrium reaction :

$$M + nL \rightleftharpoons ML_n$$

may be expressed as :

$$M + L \rightleftharpoons ML \qquad \beta_1 = \frac{[ML]}{[M][L]}$$

$$M + 2L \rightleftharpoons ML_2 \qquad \beta_2 = \frac{[ML_2]}{[M][L]^2}$$

$$M + 3L \rightleftharpoons ML_3 \qquad \beta_3 = \frac{[ML_3]}{[M][L]^3}$$

$$\vdots$$

$$M + nL \rightleftharpoons ML_n \qquad \beta_n = \frac{[ML_n]}{[M][L]^n}$$

We call β_n as ***n*th overall stability constant** or **commulative stability constant.** K's and β's describe the same chemical system. They must be related to each other. The relationship between them can be understood as below :

Write the expression for β_3 again

$$\beta_3 = \frac{ML_3}{[M][L]^3}$$

$$\beta_3 = \frac{[ML_3]}{[M][L]^3}\cdot\frac{[ML][ML_2]}{[M L][ML_2]}$$ Multiplying the numerator and denominator by [ML] [ML_2] and rearranging, we have

$$\beta_3 = \frac{[ML]}{[M][L]}\cdot\frac{[ML_2]}{[M L][L]}\cdot\frac{[ML_3]}{[ML_2][L]}$$

or $\beta_3 = K_1.\ K_2.\ K_3$

In general, for the nth overall stability constant, we have

$$\beta_n = \frac{[ML_n]}{[M][L]^n} = \frac{[ML]}{[M][L]}\cdot\frac{[ML_2]}{[ML][L]}\cdots\frac{[ML_n]}{[ML_{n-1}][L]}$$

$$= K_1.\ K_2.\ K_3\ ....\ K_n$$

Thus, $\beta_n = \prod_{i=1}^{n} K_i$ where $\prod_{i=1}^{n}$ stands for cummulative product

There is same number of overall formation constants as stepwise formation constants related as :

$$\beta_1 = K_1$$
$$\beta_2 = K_1.K_2$$
$$\beta_3 = K_1.K_2.K_3$$

In general $\beta_n = K_1.K_2.K_3 \ ....\ K_n$

Taking logarithm of both sides

$$\log \beta_n = \log K_1 + \log K_2 + \log K_3 \ .... + \log K_n$$

Thus, log β may be used as a measure of stability of the complex. For example, for the complex $[Cu(NH_3)_4]^{2+}$, involving four steps,

$$\log \beta_4 = \log K_1 + \log K_2 + \log K_3 + \log K_4$$
$$= 4.3 + 3.6 + 3.1 + 2.3 = 13.3$$

And for $[Cd(NH_3)_4]^{2+}$, $\log \beta_4 = 2.6 + 2.1 + 1.4 + 0.9 = 7.0$

It is again emphasised that β represents the measure of stability of the complex formed. Greater the value of β, greater is the stability of the complex formed.

Thus, the log β values of above two complexes suggest that $[Cu(NH_3)_4]^{2+}$ complex is more stable than $[Cd(NH_3)_4]^{2+}$ complex. That is why in the second group of qualitative analysis of basic radicals, Cu^{2+} does not form precipitate as CuS when H_2S is passed through the solution containing $[Cu(NH_3)_4]^{2+}$ complex ions while Cd^{2+} forms precipitate as CdS because of instability of $[Cd(NH_3)_4]^{2+}$complex ion.

A fair value of 8 for log β points to a stable complex.

5.3. KINETIC AND THERMODYNAMIC STABILITY

There are two kinds of stabilities as applied to complexes, the thermodynamic stability and the kinetic stability. In reality, the two could be different in some cases.

Kinetic stability is explained in terms of **lability** or **inertness. The ability of a complex to permit quick exchange (or substitution) of its one or more ligands in its coordination sphere by other ligands** is called the **lability of the complex.** The reverse of this called **inertness.**

Complexes for which such substitution reactions are rapid are called **labile;** whereas **complexes for which such substitution reactions proceed slowly or do not proceed at all are** called **inert.**

The distinction between these types of stability can be understood by taking the example of $[Co(NH_3)_6]^{3+}$ ion.

The complex $[Co(NH_3)_6]^{3+}$ ion, remains stable for months in acidic medium which indicates its **kinetic inertness.** However, this complex ion is thermodynamically unstable as it has very large equilibrium constant ($K \sim 10^{25}$) for the reaction :

$$[Co(NH_3)_6]^{3+} + 6H_3O^+ \longrightarrow [Co(H_2O)_6]^{3+} + 6NH_4^+ \quad K \simeq 10^{25}$$

This indicates that $[Co(NH_3)_6]^{3+}$ is **thermodynamically unstable** in acid medium but it is **kinetically inert.**

Now let us consider the following reaction :

$$Ni^{2+} + 4CN^- \longrightarrow [Ni(CN)_4]^{2-} \quad \beta = 10^{22}$$

The overall formation constant for the reaction is 10^{22} :

This indicates that this complex is thermodynamically very stable. However, the rate of exchange of CN^- ligands with $^{14}CN^-$ is very fast and therefore, kinetically, the complex is labile.

$$[Ni\ (CN)_4]^{2-} + {}^{14}CN^- \longrightarrow [Ni(CN)_3\ {}^{14}CN]^{2-} + CN^-$$

Thus, the complex $[Ni\ (CN_4]^{2-}$ ion is **thermodynamically stable** but **kinetically it is labile.** We can therefore say that inertness or lability of the complexes has nothing to do with the stability as obtained from thermodynamics.

5.4. FACTORS AFFECTING THE STABILITY OF COMPLEXES

The stability of a complex depends upon the following factors :

1. *Nature of the central metal ion*
2. *Nature of the ligand*

These are separately discussed as under

1. Nature of the central metal ion

***(a)* Charge on the central metal ion.** Metal ions having high charge density form stable complexes. Charge density means ratio of charge to the radius of the ion. Thus, smaller the size and larger the charge of a metal ion, more stable are the complexes. This is because a smaller, more highly charged ions allows closer and faster approach of the ligands and the greater force of attraction results into stable complex. *In general, greater the charge on the central metal ion, greater is the stability of the complex.* For example, the complexes of Fe^{3+} ion are more stable than those of Fe^{2+} ion.

$$Fe^{3+} + 6CN^- \rightleftharpoons [Fe(CN)_6]^{3-} \qquad \log \beta = 31, \text{ very stable}$$

$$Fe^{2+} + 6CN^- \rightleftharpoons [Fe(CN)_6]^{4-} \qquad \log \beta = 8.3, \text{ less stable}$$

***(b)* Size of metal ion.** *As the size of metal ion decreases, the stability of complex increases.* If we consider the bivalent metal ions, than the stability of their complexes increases with decrease in the ionic radius of the central metal ion as :

Ion	Mn^{2+}	Fe^{2+}	Co^{2+}	Ni^{2+}	Cu^{2+}	Zn^{2+}
Ionic radius (pm)	91	83	82	78	69	74

Therefore, the order of stability is :

$$Mn^{2+} < Fe^{2+} < Co^{2+} < Ni^{2+} < Cu^{2+} > Zn^{2+}$$

This is again because the charge density increases from left to right.

This order is called *natural order of stability of Iving William's order.*

(c) Electronegativity or charge distribution of metal ion. This is another factor determining stability of a complex. We can classify the metal ions into two types.

Class 'a' metals: These are electropositive metals and include the alkali metals, alkaline earth metals, most of the non-transition metals and those transition metals having only a few *d*-valence electrons (such as Sc, Ti, V). Such metals have relatively few electrons beyond an inert gas core.

Calss 'b' metals. Less electropositive metals and heavy metals such as Rh, Pd, Ag, Ir, Pt, Au, Hg, Pb, etc., form this category. These have relatively large number of *d* electrons beyond the inert gas core.

Class 'a' metals, which attract electrons weakly, form most stable complexes with those ligands having slightly electronegative atoms such as nitrogen, oxygen and fluorine.

Class 'b' metals form most stable complexes with π acceptor ligands containing P, S, As, Br, and I.

Certain metal ions exert a polarizing effect on the ligands causing an increase in bond stability. A metal ion in the complex may get polarized by the electrical field imposed by the ligands. In such cases, covalent bonding between metal ions becomes important. For example in complexes of electronegative metals such as those of copper group (Cu, Ag, Au) and zinc group (Zn, Cd, Hg), Pt, Pd, Rh, etc., an electrostatic approach to stability is ineffective. For example, silver forms insoluble halide salts AgX and stable halide complex ions AgX_2^- in which order of stability is : $I^- > Br^- > Cl^- > F^-$. This is clear from log β values :

Complex	AgF	AgCl	AgBr	AgI
log β	0.3	3.2	4.5	8.0

2. Nature of Ligand

(a) Size and charge of the ligands. In general small ligands with higher negative charge form stable complexes. Thus, F^- forms more stable complexes with Fe^{3+} than Cl^-, Br^- or I^-. Thus, a small fluoride, F^- ion forms more stable Fe^{3+} complex than the large Cl^- ion. This is due to easy approach of the ligand towards metal ion. This is supported by log β values.

$(FeF)^{2+}$ $\quad$ $\log \beta = 6$ $\quad$ $(FeCl)^{2+}$ $\quad$ $\log \beta = 1.3$

Large perchlorate ion, ClO_4^- has relatively small tendency to form metal complexes than Cl^-.

Similarly, a small dinegative anion O^{2-} forms more stable complexes than does the large S^{2-} ion.

(b) Basic strength. A metal-ligand bond involves the donation of electron pairs from the ligands to the empty orbitals of the central metal ion. Greater the ease of donation of electron pairs by the ligand, greater is the stability of the complexes formed. We can generalise that strong bases perform strongly as ligands, too.

For example, CN^- and NH_3 molecules, which are strong bases are also good ligands and form many stable complexes. NH_3 should be a better ligand than H_2O. Such type of behaviour is observed for class 'a' metals, that is, alkali metals, alkaline earth metals, lanthanides and actinides. However, it is not true for complexes of metals of type b metals (Pt, Au, Pb, Hg, Pd, etc) where other factors like back bonding (π acceptors) also play an important role.

(c) Presence of ring structure. For multidentate ligands, the stability of a complex is also controlled by formation of ring structure, ring size, ring strain, etc.

Chelate effect. We observe that complexes formed by chelating ligands such as ethylene diamine (en), EDTA, etc are more stable than those formed by monodentate ligands such as H_2O or NH_3. *This enhanced stability of complexes containing chelating ligands* is called **chelate effect.**

For example, the nickel complex with chelating ligand, en is more stable than corresponding complex with NH_3 ligand. This is supported by their log β values

$$[Ni(H_2O)_6]^{2+} + 6NH_3 \longrightarrow [Ni(NH_3)_6]^{2+} + 6H_2O \quad \log \beta = 8.61$$

$$[Ni(H_2O)_6]^{2+} + 3en \longrightarrow [Ni(en)_3]^{2+} + 6H_2O \quad \log \beta = 18.28$$

The chelate effect can be explained on the basis of favourable entropy change during chelation. During formation of metal complexes, water molecules in the solvated shell of a metal ion in aqueous solution are replaced by ligands. Several molecules of the solvent in the solvated cation are replaced by one or more number of chelating ligand molecules. This results in entropy increase. Consider the above example of Ni^{2+} complexes of NH_3 and en. In the complexation of Ni^{2+} ion with NH_3, there are seven molecules on both sides of the equation. That is, there is no change in entropy, but in complexation with en, there are four molecules on LHS and seven molecules on RHS of the equation of the reaction. That is, there has been increase in entropy, which is always preferred. Thus, metal ions form more stable complexes with multidentate ligands.

Number of chelate rings. It has also been observed that larger the number of chelate rings in a complex, greater is the stability of a complex. This is illustrated with the help of the following data :

Complex	No. of chelate rings	log β
$[Cu(NH_3)_4]^{2+}$	0	12.6
$[Cu(en)_2]^{2+}$	2	20.0
$[Cu(trien)]^{2+}$	3	20.5

where trien is triethylene tetraamine : H_2 N CH_2 CH_2 NH CH_2 CH_2 NH CH_2CH_2 NH_2

Three chelate rings formed by trien (triethylene tetraammine) are shown below:

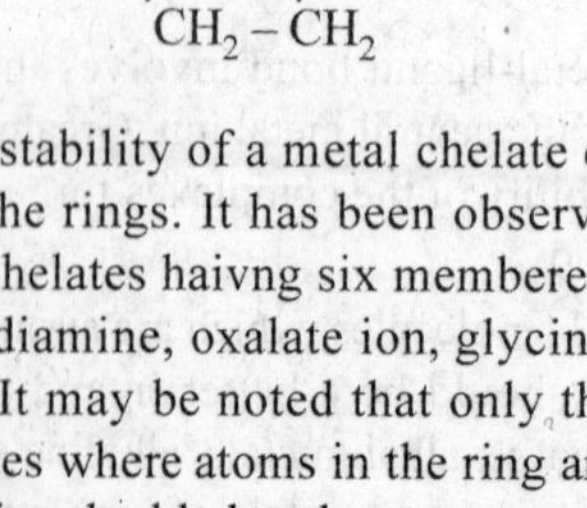

Size of the chelate ring. The stability of a metal chelate depends not only upon the no. of rings but also upon the size of the rings. It has been observed that metal chelates having 5-membered ring are most stable. Chelates haivng six membered rings are slightly less stable. Chelating ligands such as ethylenediamine, oxalate ion, glycine and dimethylglyoxime form quite stable 5-membered chelates. It may be noted that only those 5-membered chelates are more stable than 6-membered chelates where atoms in the ring are joined only by single bonds. However, 6-membered chelates having double bond or unsaturation in the ring are more stable than corresponding 5-membered chelates. For example, acetylacetonate complexes containing 6-membered rings stable complexes, because of the presence of double bonds in the 6-membered ring.

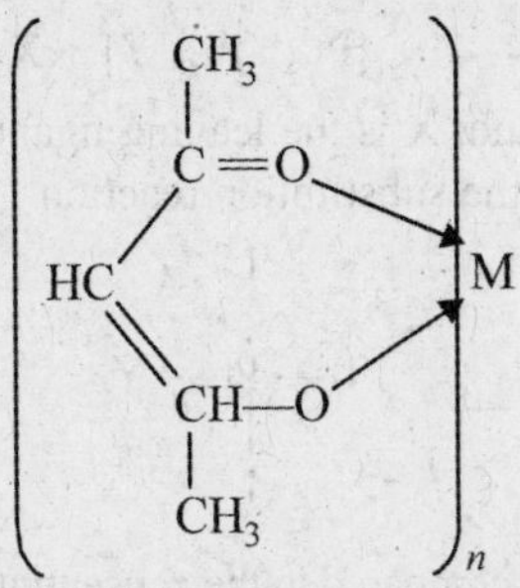

where M = Ti, Cr, Co, etc

Rings of other sizes are known but these are unstable and show little chelate effect for steric reasons.

***(d)* Steric effects.** Substituted ligands containing large bulky groups form less stable complexes than do unsubstituted smaller ligands. This is due to the fact that the bulky groups present near a donor atom cause mutual repulsion (steric effect) among the ligands making metal-ligand bonds to be weak. This reduces the stability of the complex. For example, $H_2N\ CH_2CH_2NH_2$ forms more stable complexes than its substituted derivative, $(CH_3)_2N\ CH_2CH_2N(CH_3)_2$. Similarly, Ni^{2+} forms less stable complex with 2-methyl-8-hydroxy quinoline than with less sterically crowded ligand 8-hydroxy quinoline.

5.5. SUBSTITUTION REACTIONS IN SQUARE PLANAR COMPLEXES

Metal ions with d^8 electronic configuration form four coordinate square planar complexes, with strong field ligands. Among these square planar complexes, the kinetic studies of complexes of Pt (II) have been extensively made because they are stable, relatively easy to synthesise and undergo ligand exchange reactions at conveniently measurable rates.

The substitution reactions bring about the replacement or exchange of one ligand by another ligand. There are three major ways by which substitution reactions of square planar complexes take place. These are, *oxidative addition, electrophilic substitution* and *nucleophilic substitution.*

***(i)* Oxidative addition followed by reductive elimination**

$$PtMeCl(PMe_2Ph)_2 \xrightarrow[\text{Oxidative addition}]{Cl_2} PtMeCl_3(PMe_2Ph)_2 \xrightarrow[\text{elimination}]{\text{Reductive}} PtCl_2(PMe_2Ph)_2 + MeCl$$

In this reaction the oxidation state of Pt (II) changes to Pt (IV) and coordination number changes from 4 to 6 in the intermediate. This intermediate then undergoes reductive elimination of methyl chloride to form again a four coordinated Pt (II) complex. Here the Me group has been substituted by Cl.

***(ii)* Electrophilic substitution can replace a ligand by another ligand**

$$PtMeCl(PMe_2Ph)_2 + HgCl_2 \longrightarrow PtCl_2(PMe_2Ph)_2 + MeHgCl$$

This reaction involves the electrophilic attack of Hg (II) on the platinum–carbon bond.

***(iii)* Square planar complexes may undergo nucleophilic substitution**

$$PtMeCl\ (PMe_2Ph)_2 + N_3^- \longrightarrow PtMe\ (N_3)\ (PMe_2\ Ph)_2 + Cl^-$$

In this case nucleophile N_3^- (azide ion) replaces Cl^- ion.

5.6. RATE LAW FOR NUCLEOPHILIC SUBSTITUTION IN SQUARE PLANAR COMPLEXES

We shall now proceed to derive rate law for the above process.

Consider a general nucleophilic substitution reaction in a square planar complex of platinum, represented by $[PtL_2TX]$ undergoing substitution of ligand X by Y as :

$$[Pt(T)\,L_2\,X] + Y \longrightarrow [Pt\,(T)\,L_2\,Y] + X$$

in which Y is the entering nucleophilic ligand, X is the leaving ligand and T is the ligand trans to X. Configuration at the metal is retained in the substitution reaction

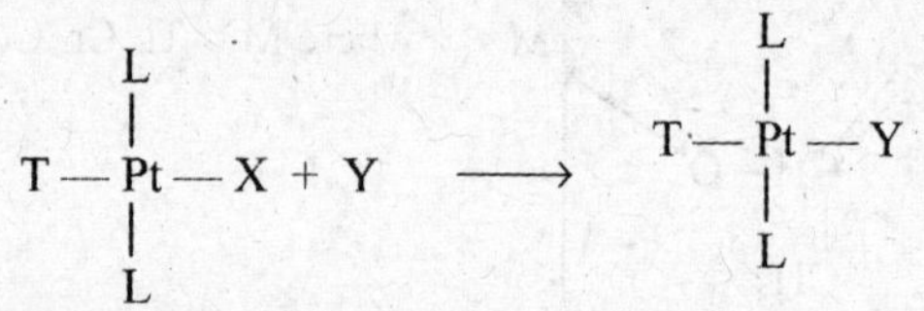

The concentration of Y was kept large as compared to the concentration of the complex so that the concentration of Y *i.e.* [Y] may be assumed to be constant. If the reverse reaction is believed to be insignificant, the observed *pseudo first* order rate law for square planar substitution may be written as :

$$\text{rate} = -\frac{d[PtL_2TX]}{dt} = k_1\,[PtL_2\,TX] + k_2\,[PtL_2\,TX]\,[Y] \quad ...(i)$$

where $[PtL_2\,TX]$ = concentration of the complex and

$[Y]$ = concentration of the incoming ligand

Eq. (*i*) may be rearranged to give :

$$\text{rate} = [k_1 + k_2\,Y]\,[PtL_2\,TX]$$

or $$= k_{obs}\,[PtL_2\,TX] \quad ...(ii)$$

where $$k_{obs} = k_1 + k_2\,[Y] \quad ...(iii)$$

From eq. (*iii*), we can determine both k_1 and k_2 by carrying out the reaction at various concentrations of Y. This gives a plot of k_{obs} versus [Y], which is a straight line and we get k_1 as intercept and k_2 as slope. The plots of rate constants (k_{obs}) as a function of nucleophile concentration [Y] for the reaction of trans $[Pt\,(py)_2\,Cl_2]$ with various nucleophiles in methanol at 30°C are given in Fig. 5.1.

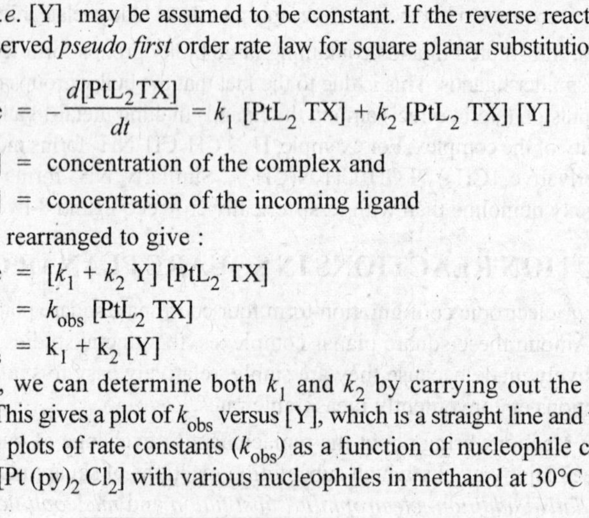

Fig. 5.1. Rate constants (k_{obs}) as a function of nucleophile concentration ([Y]) for reaction of *trans*–$[Pt(Py)_2Cl_2]$ with various nucleophiles in methanol at 30°C

Substitution reactions in inorganic chemistry may be divided into four classes depending upon the rate determining step during bond making and bond breaking. These are :

1. Associative reactions (A)

In this type of reaction, the rate determining step is the formation of intermediate with an increased coordination number. This means that Pt–Y bond is first formed and then Pt–X begins to break.

2. Dissociative reaction (D)

In this type of reaction, the rate determining step is the formation of the intermediate in which one ligand is lost. This means that Pt–X bond is fully broken and then the Pt–Y bond begins to form.

3. Interchange associative (I_a)

In this type of reaction, the Pt–X bond begins to break before Pt–Y bond is fully formed. However, bond *making is more significant than bond breaking.*

4. Interchange dissociative (I_d)

In this type of reaction, the Pt–Y bond begins to form before the Pt–X bond is fully broken. However, *bond breaking is more significant than bond making.*

It has been observed that the values of both k_1 and k_2 were non zero which indicate that PtL_2–TX complex reacts in two different pathways.

The k_2 term indicates the direct nucleophilic displacement of X by Y. The reaction is first order with respect to both complex and Y. This indicates **associative pathway (A),** which is similar to SN^2 (Bimolecular nucleophilic substitution) reaction. The term k_1 represents that the rate of reaction is first order with respect to complex and is independent of Y. This indicates a **dissociative pathway (D).** However, there are strong evidences which support the view that this pathway is also associative due to the nature of the solvent. The solvent (S), in general, also behaves as a nucleophile and competes with Y for the complex [PtL_2 TX] to form [PtL_2 TS] (also called solvento complex).

Thus, the two term rate law may be expressed as

$$\text{rate} = k' [PtL_2\,TX]\,[S] + k_2\,[PtL_2\,TX]\,[Y] \qquad ...(iv)$$

However, since solvent is present in large excess, its concentration, [S] remains practically constant so that

$$k'\,[S] = k_1$$

Therefore, the eqn. (iv) simplifies to eq. (i)

5.7. TRANS EFFECT

In the detailed study of coordination compounds of Pt (II), some interesting observations were made. It has been observed that in square planar complexes of Pt (II), the ligand *trans* to the leaving group has a strong effect on the rate of substitution in comparison to ligand *cis* to the leaving group. In 1926, Chernyaev introduced the concept of **trans effect** in platinum complexes. He studied that in reactions of square planar Pt (II) complexes, the ligands *trans* to chloride ion are more easily replaced than those *trans* to ligands such as NH_3.

The ability of an attached group to direct the substitution into a position trans (*i.e. opposite*) *to itself* is called **trans effect.** Such a group has marked **effect on the rate of reaction.**

This can be illustrated by considering the preparation of two geometric isomers of [$PtCl_2(NH_3)_2$] complex. These can be prepared by one of the following processes.

(*i*) replacement of Cl^- in $[PtCl_4]^{2-}$ by NH_3

(*ii*) replacement of NH_3 in $[Pt(NH_3)_4]^{2+}$ by Cl^-

We obtain different isomers in the two sequences.

First sequence

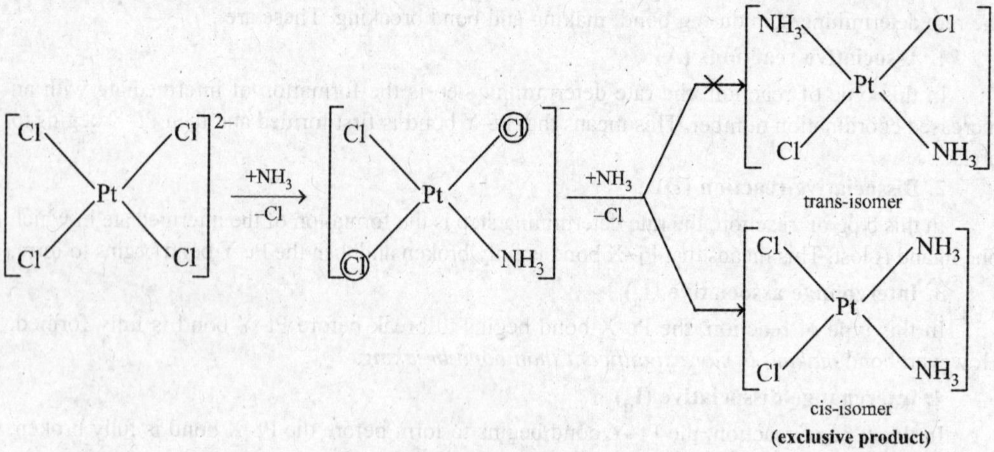

(*i*) In the first step, Cl^- is replaced by NH_3 to form $[PtCl_3(NH_3)]^-$. This is a simple displacement reactions and since all the four groups present are identical only one compound is formed.

(*ii*) In the second step, two products can be formed. However, only one product *cis*-$[PtCl_2(NH_3)_2]$ is actually obtained in which the substitution of a ligand occurs **trans to Cl^- ion.** The ligand trans to Cl^- has been circled.

Second sequence

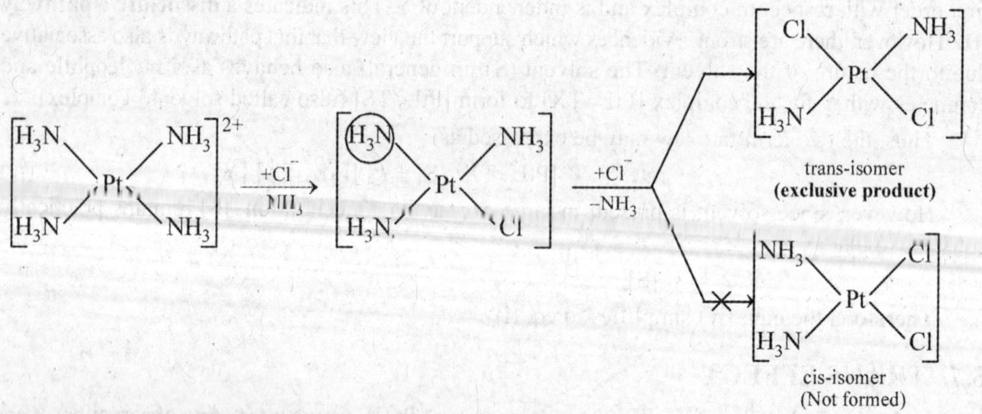

(*i*) In this case, the first step involves the replacement of any one NH_3 molecule by Cl^- because all four groups are identical.

(*ii*) In second step, two products are possible but only one product is actually obtained in which substitution of ligand occurs trans to a Cl^- ion (circled).

So, we observe from these two examples that *the main isomer obtained is the one which is formed by replacement of ligand trans to Cl^- ion.* This is known as **trans effect.**

Trans effect may also be defined as **the labilization of ligands trans to certain other ligands which can then be regarded as trans directing ligands.** By carrying out a large number of reactions, using different ligands and substituting reagents, it is possible to compare the trans directing capabilities of a variety of ligands. Theoretically, if we consider A and B as two ligands labilizing a trans Cl^- with same entering group, NH_3, we get the following products depending upon the capabilities of A or B.

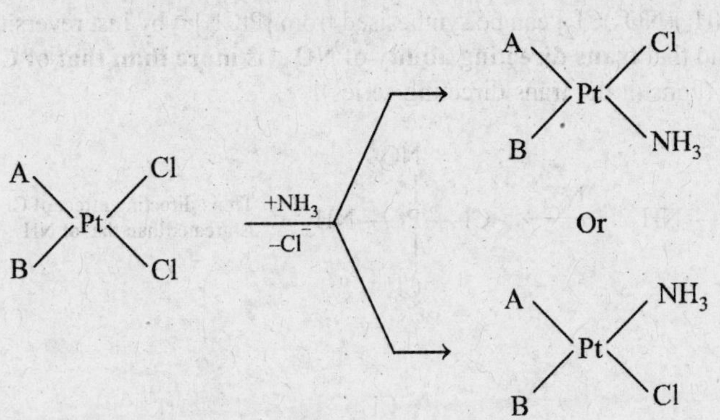

A has greater trans-directing effect than B

B has greater trans-directing effect than A

The different ligands can be arranged in the following decreasing order of **trans-directing effect :**

CN^-, CO, NO, C_2H_4 > PR_3, H^- > CH_3^-, $C_6H_5^-$, SC $(NH_2)_2$, SR_2 > SO_3H^- > NO_2^-, I^-, SCN^- > Br^- > Cl^- > py > RNH_2, NH_3 > OH^- > H_2O

This is also known as **trans directing series.** It is clear from the series that **CN^-, CO and NO are powerful trans directing ligands whereas OH^- and water are very poor trans directing ligands.**

It has been observed that trans-effect can be very large, rates may differ by the order of 10^6 between complexes with very strong trans-effect and with weak trans-effect ligands.

Utility of trans-effect

The trans-effect has been used in synthesising certain specific (cis or trans) complexes. For example, cis- and trans [Pt $(NH_3)_2$ Cl_2] complexes are prepared keeping in view the larger trans-effect of Cl^- than ammonia.

Cis-product

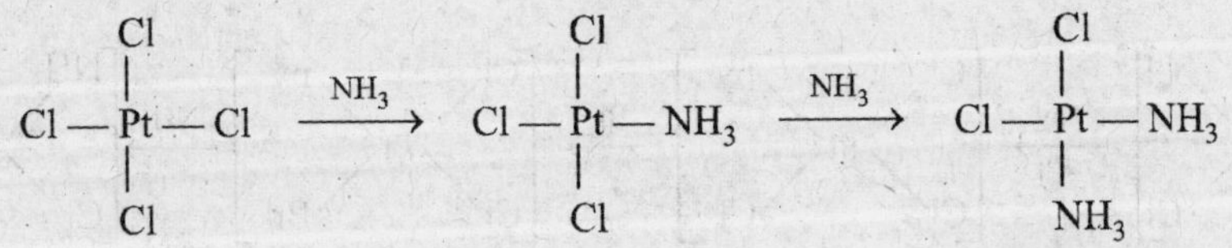

cis-diamminedichloro platinum (II)

Trans-product

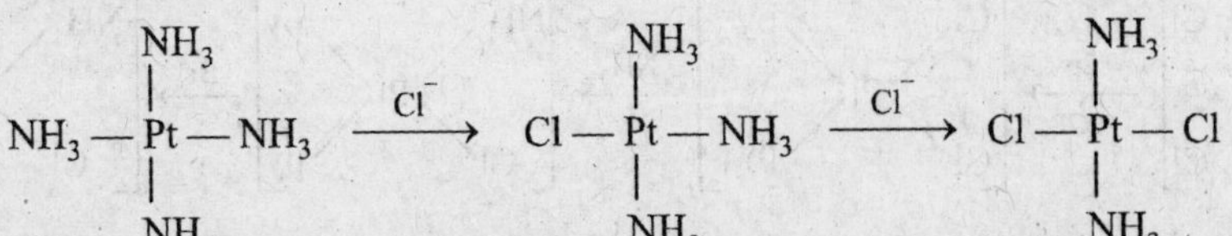

trans-diamminedichloro platinum (II)

In the first reaction, the second NH_3 enters a cis-position because the trans-directing influence of Cl^- is greater than that of NH_3. In second reaction, Cl^- ion replaces the most labile NH_3 which is opposite to the chloro group, forming trans–[Pt $(NH_3)_2$ Cl_2].

Similarly, cis-and trans-$[Pt(NH_3)(NO_2)Cl_2]$ can be synthesised from $[PtCl_4]^{2-}$ by just reversing the order of groups keeping in mind that **trans directing ability of NO_2^- is more than that of Cl^- which is more than that of NH_3** (consult the trans directing series).

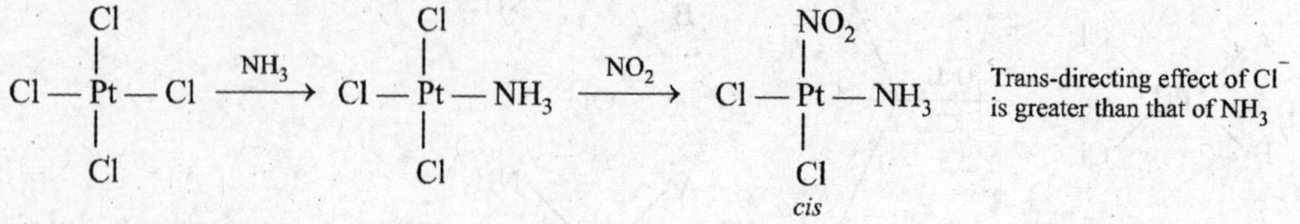

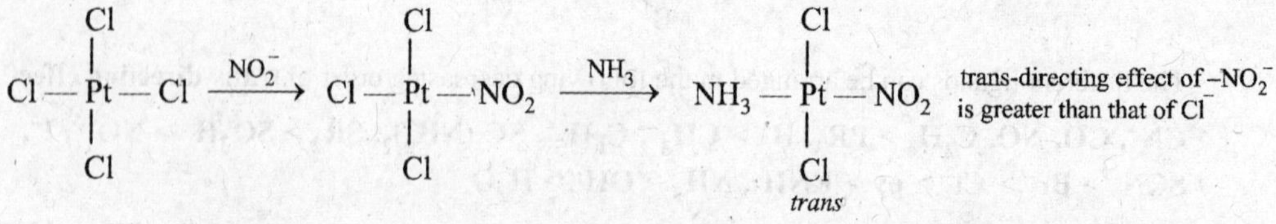

Similarly, three isomers of $[Pt(py)(NH_3)BrCl]$ I, II and III were prepared from $[PtCl_4]^{2-}$ by the following reactions on the basis of greater trans effect of the ligands Br^- and Cl^- compared to py and NH_3 and the fact that the Pt–N bond strength is greater than the Pt–Cl bond strength.

First reaction

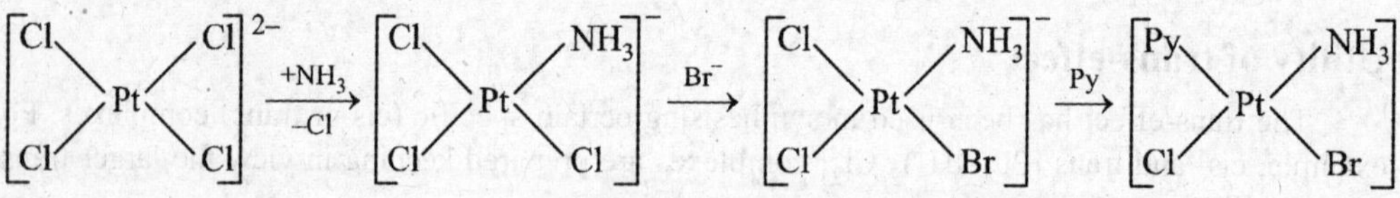

I

Second reaction

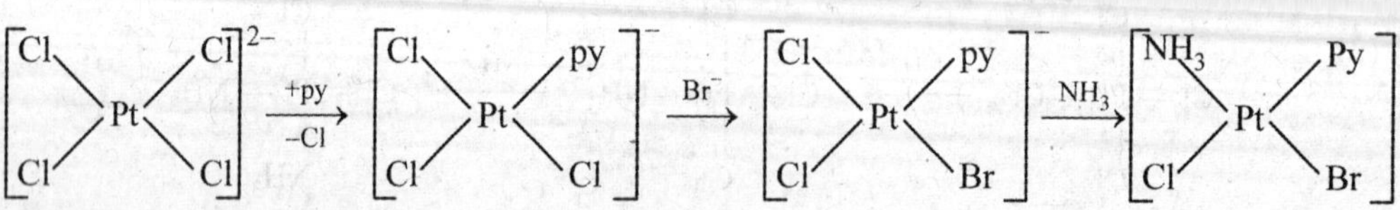

II

Third reaction

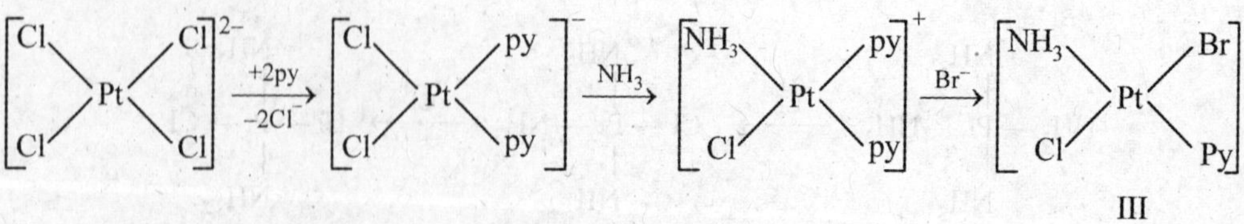

III

It may be noted that the second step in third reaction seems to contradict the trans effect. The trans effect predicts that any ligand *trans* to Cl^- should be replaced rather than any ligand trans to py. However, Cl^- ion trans to py is replaced because the bond strength of Pt–Cl is weaker than that of Pt– N bond.

5.8. MECHANISM OF NUCLEOPHILIC SUBSTITUTION IN SQUARE PLANAR COMPLEXES

As discussed in the previous section the square planar substitution reaction involves a bimolecular displacement reaction. It is reasonable to assume that the substitution reaction proceeds by an expansion of coordination number to include the entering ligand. The metal is exposed for attack either above or below the plane forming coordination number 5. Moreover, these low spin d^8 systems have vacant p_z orbital of relatively low energy which can easily accommodate the pair of electrons donated by the entering ligand.

The ligand Y attacks the metal M from Z-direction (by utilizing the empty p_z orbital of the metal) to form a five coordinated square pyramidal intermediate (structures B). Thus square planar intermediate rearranges itself into a stable trigonal bipyramidal intermediate. This trigonal bipyramidal arrangement has three ligands (Y, T and X) in equatorial plane and two ligands (L) (which were trans to each other in the original complex), in the axial positions as shown in structure C. As X departs from trigonal plane, the T–M–Y angle will open up and the geometry will pass through a square pyramidal intermediate (structure D) into square planar product ML_2 TY (structure E) with release of X.

Square pyramidal intermediate is formed twice in the reaction, once with the entering group Y and second time with the leaving group X.

The mechanism is shown below :

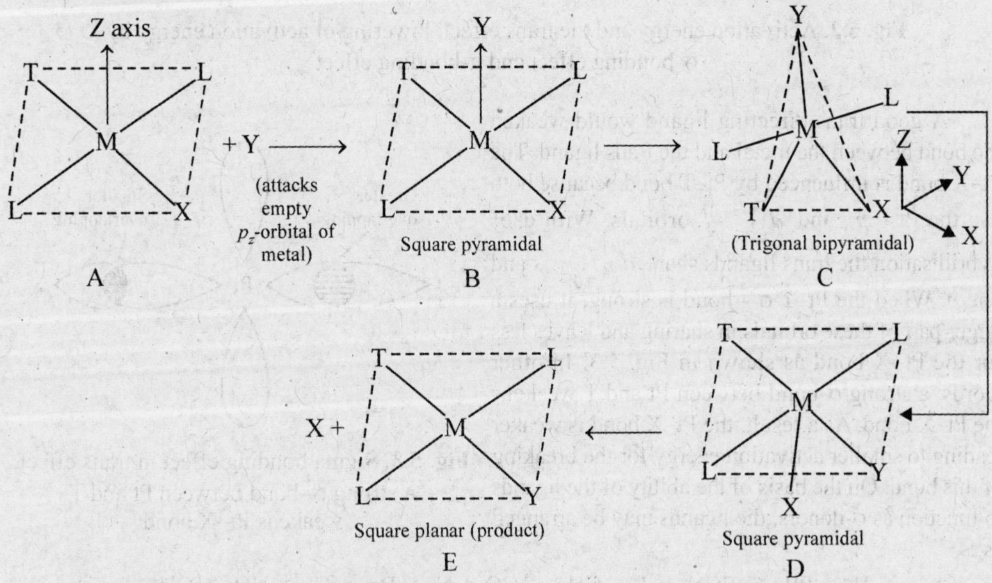

The studies indicate a very strong evidence for 5–coordinated intermediate.

5.9. THEORIES OF TRANS EFFECT

As trans effect represents a kinetic phenomenon, it must be affecting the **activation energy of a reaction,** which is the *difference between the energies of the reactant ground state and the first transition state.* The stability of the ground state before substitution and of the transition state can affect activation energy required for the substitution reaction. The energy relationships are shown in Fig. 5.2.

The activation energy for any reaction can be lowered in either of the two ways (Fig. 5.2)

(*i*) by raising the energy of the ground state of reactants i.e. by weakening the metal ligand bond trans to itself (shown as b)

(*ii*) by lowering the energy of the transition state (shown as c). This can be done by stabilizing the transition state.

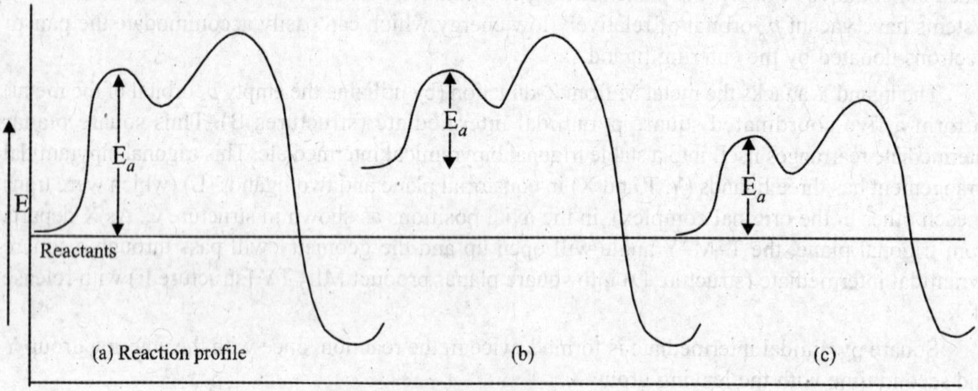

Fig. 5.2. Activation energy and the trans effect, lowering of activation energy by σ-bonding effect and π-bonding effect

A good trans directing ligand would weaken the bond between the metal and the trans ligand. The Pt–X bond is influenced by Pt–T bond because both use the Pt – p_x and $d_{x^2-y^2}$ orbitals. With dsp^2 hybridisation the trans ligands share $d_{x^2-y^2}$, *s* and one *p*. When the Pt–T σ - bond is strong, it uses a larger part of these orbitals in sharing and leaves less for the Pt–X bond as shown in Fig. 5.3. In other words, a strong σ-bond between Pt and T weakens the Pt–X bond. As a result, the Pt–X bond is weaker leading to smaller activation energy for the breaking of this bond. On the basis of the ability of the ligands to function as σ-donors, the ligands may be arranged as :

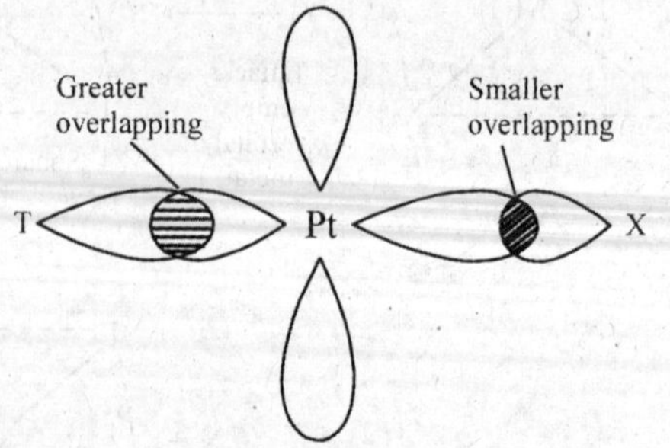

Fig. 5.3. Sigma bonding effect in trans effect. A strong σ–bond between Pt and T weakens Pt–X bond

$$H^- > PR_3 > SCN^- > I^- \sim CH_3^- \sim CO, CN^- > Br^- > Cl^- > NH_3 > OH^-$$

This order goes almost parallel to trans effect series with some difference.

There are two theories of trans-effect of which one is related to ground state and the other to activated state.

1. Polarization Theory

This theory suggested by Grinberg is primarily concerned with the effects on the ground state in terms of weakening of Pt–X bond trans to Pt–T bond. Consider two bonds T–Pt and Pt–X trans to each other in a square planar complex. Suppose the ligand T is more polarizable than the ligand

X. This will bring about a change in the distribution of charge on ligand T and metal. The primary charge on metal ion polarizes the electron cloud on T and induces a dipole in T. The dipole in T in turn induces a dipole in the metal as shown in Fig. 5.4.

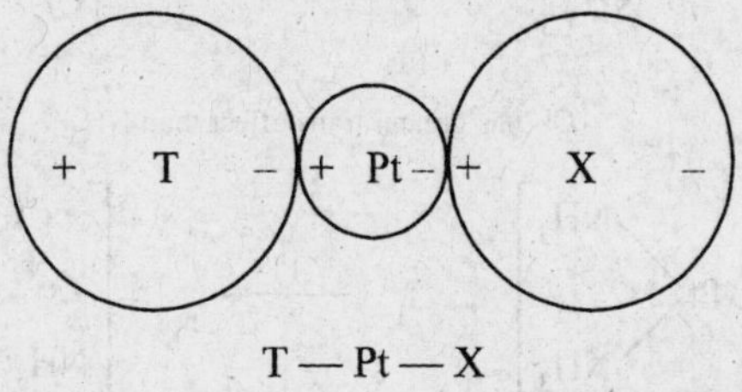

Fig. 5.4. Distribution of charge in induced dipoles in T–Pt–X bond

The orientation of this dipole on the metal ion is such that it repels the negative charge in the ligand X. As a result, the attraction of X for the metal is reduced and the Pt–X bond is weakened.

Thus, according to this theory the polarization of the ligand should be directly related to its trans effect. However, this is valid for those ligands which do not form *pi*-bonds with the metal ions.

2. Pi (π) bonding theory

This theory gives a logical explanation of the trans-effect of those attached ligands which are π-acceptors like CO, PR_3, C_2H_4, etc. According to this theory, when the T ligand forms a strong π-acceptor bond with the metal (*e.g.* Pt) charge is removed from Pt and the attachment of another ligand to form a 5-coordinate species becomes easier.

Thus, the removal of charge from Pt(II) by π bonding of T will increase the chances of addition of Y and will favour a more rapid reaction.

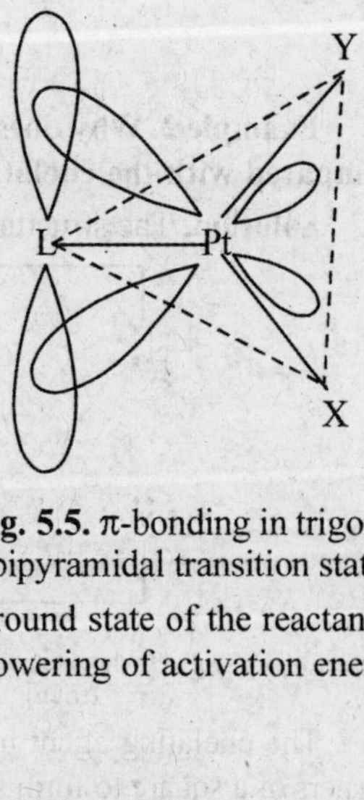

Fig. 5.5. π-bonding in trigonal bipyramidal transition state

In addition to charge effect, the $d_{x^2-y^2}$ orbital, which is involved in σ-bonding in square planar geometry and both the d_{xz} and d_{yz} orbitals can contribute to π-bonding in trigonal bipyramidal transition state (Fig. 5.5). As a result, the effect on the ground state of the reactant is small, but the energy of transition state gets lowered. This results in lowering of activation energy and hence enhances the lability of trans ligand.

The order of π-acceptor ability of the ligands is :

$$C_2H_4 \sim CO > CN^- > NO_2^- > SCN^- > I^- > Br^- > Cl^- > NH_3 > OH^-$$

It can be summarised that the ligands with strong trans effects, affect the ground state as well as the transition state. The polarization effect weakens the bonds in the ground state while π-bonding tendency stabilizes the transition state and both these factors contribute towards the trans-effect.

The ligands highest in the trans-effect series are strong π-acceptors followed by strong σ-donors. Ligands at the low end of the series have neither strong π-acceptor or σ-donor abilities.

SOLVED CONCEPTUAL PROBLEMS

Example 1. Predict the products of the following reactions :

(*i*) $[PtCl_3\ NH_3]^- + I^- \longrightarrow$ **(*ii*) $[Pt\ Cl\ (NH_3)_3]^+ + NO_2^- \longrightarrow$**

(*iii*) $[Pt\ I_4]^{2-} + Cl^- \longrightarrow (a) \xrightarrow{Cl^-} (b)$

Solution. (*i*)

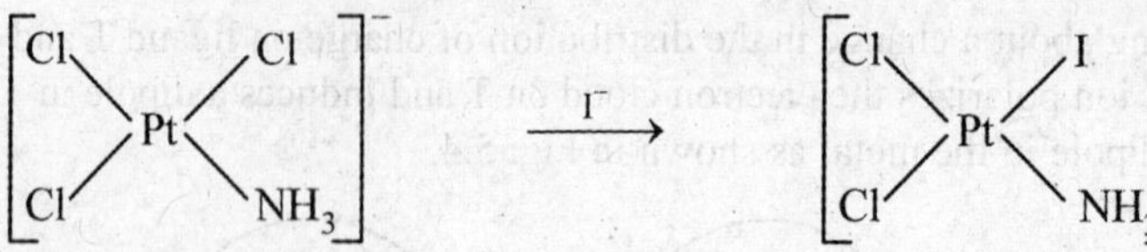

cis-product

Cl^- has greater trans effect than NH_3

(*ii*)

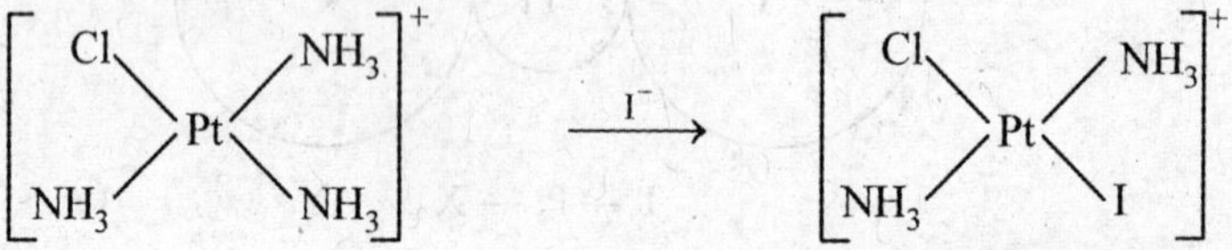

trans-product

Cl^- has greater trans effect than NH_3

(*iii*)

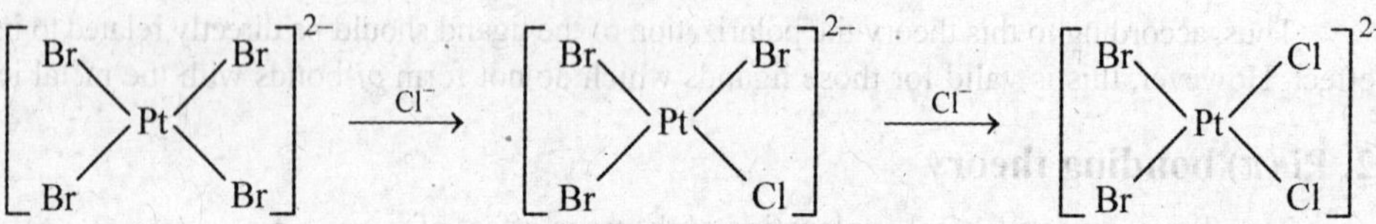

cis-product

Br^- has greater trans effect than Cl^-

Example 2. Why does triethylene tetraammine (trien) form a square planar complex easily compared with the chelating agent 2, 2′, 2″-triamino triethylamine (tren) ?

Solution. The structures of the two chelating agents are written below :

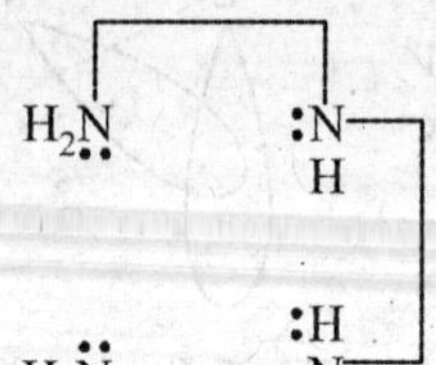

Triethylene tetraammine (trien)

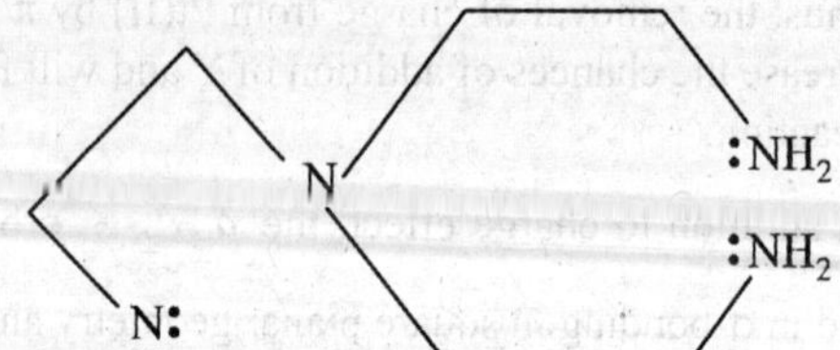

2 , 2′, 2″-triaminotriethylamine (tren)

The chelating agent trien can coordinate to the metal through its four nitrogen atoms at the corners of a square to form square planar complex. In the second compound, the four nitrogen atoms are not situated geometrically to coordinate simultaneously with the metal.

Example 3. Draw stereochemistry of substitution on the following reactions :

(*i*) $[PtCl_4]^{2-} \xrightarrow{+NO_2^-}$ **(*a*)** ; **(*a*)** $\xrightarrow{+NH_3}$ **(*b*)**

(*ii*) $[PtCl_4]^{2-} \xrightarrow{+NH_3}$ **(*c*)** ; **(*c*)** $\xrightarrow{+NO_2^-}$ **(*d*)**

Solution.

(*i*)

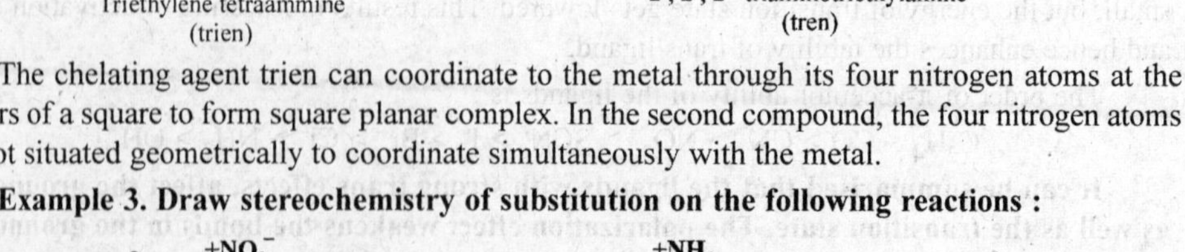

(*a*) (*b*)

(*ii*)

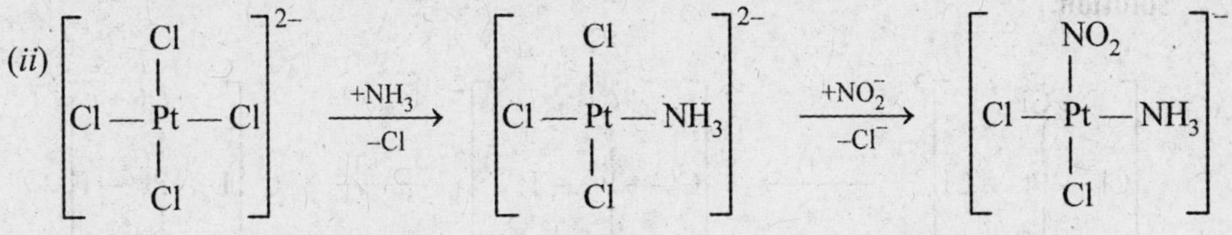

(*c*) (*d*)

Example 4. What is log β ? How is it related to stability of complexes ?

Solution. log β is a measure of stability of the complexes. In general, a complex is regarded a stable complex if its log β is more than 8.

Example 5. What is chelate effect ?

Solution. The complexes formed by chelating ligands (polydentate ligands) such as ethylene diamine (en), EDTA etc. are more stable than those formed by monodentate ligands. This increased stability of complexes containing chelating ligands is called **chelate effect.** For example Ni (II) forms more stable complex with chelating ligand, *i.e.* $[Ni(en)_3]^{2+}$ than with nonchelating ligand such as (NH_3) *i.e.* $[Ni(NH_3)_6]^{2+}$.

Example 6. What is general relation between overall stability constant and stepwise stability constants ?

Solution. The overall stability constant β is related to stepwise stability constants K as

$$\beta = \prod_{i=1}^{n} K_i$$

$$i = K_1 . K_2 . K_3 \text{ ....} K_i$$

Example 7. The log β value for $[FeF]^{2+}$ is 6 and that for $[FeCl]^{2+}$ is 1.3. What do these values indicate about these species ?

Solution. $[FeF]^{2+}$ is more stable compared to $[FeCl]^{2+}$. A small F^- ion forms more stable Fe^{3+} complex than larger Cl^- ion.

Example 8. Name the product for the reaction :

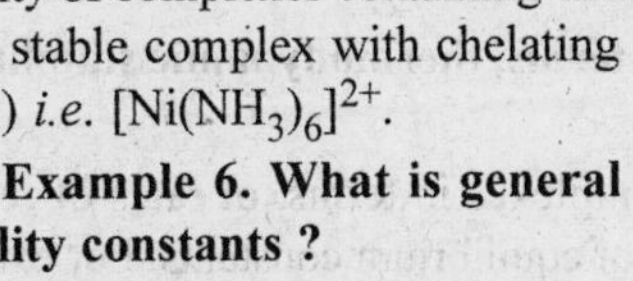

Assume that Y has greater trans effect than X

Solution.

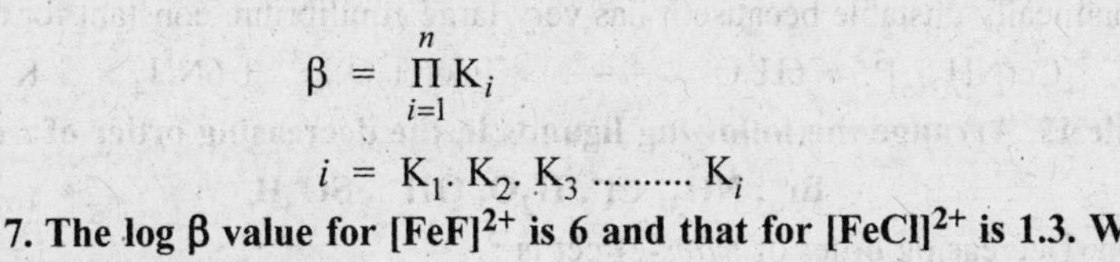

Example 9. Predict the product of the reaction :

$$[Pt\,Cl_4]^{2-} \xrightarrow{I^-} ? \xrightarrow{I^-} ?$$

Solution.

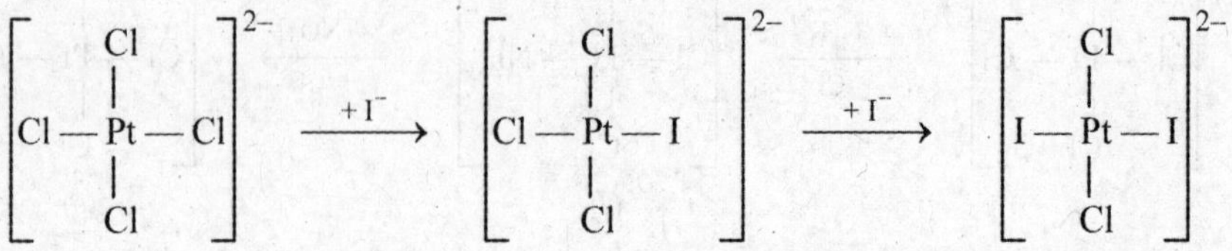

trans-
(I^- has greater trans effect than Cl^-)

Example 10. Name two powerful trans directing ligands and two very poor trans directing ligands.

Solution. (*i*) Powerful trans directing ligands : CO, NO

(*ii*) Poor trans directing ligands : OH^- and H_2O

Example 11. What is the basic difference between the terms, thermodynamic stability and kinetic stability ? Give one example.

Solution. Kinetic stability represents the stability of complexes in terms, of rates of reaction while thermodynamic stability expresses the stability in terms of equilibrium constants. For example, $[Co(NH_3)_6]^{3+}$ ion, presists for months in acidic medium indicating its kinetic inertness. However, it is thermodynamically unstable because it has very large equilibrium constant for the reaction :

$$[Co(NH_3)_6]^{3+} + 6H_2O \longrightarrow [Co(H_2O)_6]^{3+} + 6NH_4^+ \qquad K \sim 10^{25}$$

Example 12. Arrange the following ligands in the decreasing order of *trans*-effect :

Br^-, NH_3, Cl^-, H_2O, OH^- , SO_3H^-

Solution. Decreasing order of *trans*-effect is :

$$SO_3H^- > Br^- > Cl^- > NH_3 > OH^- > H_2O$$

EXERCISES
(Including Questions from Different University Papers)

Multiple Choice Questions (choose the correct answer)

1. Stability of a complex does not depend upon

(*a*) nature of the central metal ion (*b*) number of chelate rings

(*c*) steric effects (*d*) mass of the complex

2. The overall stability constant β_3 is related to stepwise stability constants K_1, K_2 and K_3 as

(*a*) $\beta_3 = K_1 \cdot K_2 \cdot K_3$ (*b*) $\beta_3 = K_1 + K_2 + K_3$

(*c*) $\beta_3 = \dfrac{K_1 + K_2}{K_3}$ (*d*) $\beta_3 = K_3 - K_2 - K_1$

3. Reaction of $[Pt\,Cl_4]^{2-}$ with NH_3 followed by reaction with NO_2^- gives

(*a*) *trans*–$[PtCl_2\,(NO_2)_2]^{2-}$ (*b*) *cis*–$[PtCl_2\,(NO_2)\,(NH_3)]^-$

(*c*) *cis*–$[PtCl_2\,(NO_2)_2]^{2-}$ (*d*) *trans*–$[PtCl_2\,(NO_2)\,(NH_3)]^-$

4. Maximum value of log β is that of

(*a*) AgI (*b*) AgCl (*c*) AgBr (*d*) AgF

5. Which of the following statement is false ?

(*a*) $[Cu(en)_2]^{2+}$ is more stable than $[Cu\,(NH_3)_4]^{2+}$

(*b*) $[FeF]^{2+}$ is more stable than $[FeCl]^{2+}$

(*c*) $[Fe(CN)_6]^{4-}$ is less stable in comparison to $[Fe(CN)_6]^{3-}$

(*d*) $[Cu(NH_3)_4]^{2+}$ is less stable than $[Cd(NH_3)_4]^{2+}$

6. Which of the following statement is wrong about square planar complexes ?

(*a*) Most of the square planar complexes have d^8 system.

(*b*) Most of the square planar substitution reactions follow associative mechanism.

(*c*) Good trans directing must have strong tendency to form strong σ-bonds and weak π-bonds.

(*d*) NH_3 has smaller trans-directing effect than Cl^-.

7. The weakest trans-directing ligand among the following : I^-, Cl^-, NH_3, OH^- is

(*a*) OH^- (*b*) I^- (*c*) NH_3 (*d*) Cl^-

8. Which of the following elements has been extremely studied for square planar substitution reactions

(*a*) Pd(II) (*b*) Ni(II) (*c*) Pt(II) (*d*) Au(III)

9. Which of the following has strongest trans-effect ?

(*a*) py (*b*) SCN^- (*c*) Br^- (*d*) Cl^-

10. In general a metal complex is regarded as stable if its log β value is

(*a*) zero (*b*) less than 8 (*c*) more than 8 (*d*) 14

ANSWERS

1. (*d*) **2.** (*a*) **3.** (*b*) **4.** (*a*) **5.** (*d*) **6.** (*c*) **7.** (*a*) **8.** (*c*)
9. (*b*) **10.** (*c*)

Short Answer Questions

1. What are labile complexes ?
2. What is chelate effect ?
3. Name the type of mechanism commonly observed in square planar substitution reactions.
4. What is thermodynamic stability of a complex ?
5. Define stepwise stability constant and overall stability constant.
6. Name two metal ions which generally form square planar complexes.
7. Which of two is more stable : $[Fe(CN)_6]^{3-}$ or $[Fe(CN)_6]^{4-}$?
8. Define trans effect.
9. How is overall stability constant β_n is related to its stepwise stability constants K_1, K_2, K_3, ... K_n.
10. What do you mean by inert complexes ?
11. What do you understand by the stability of a chelate ring ?

General Questions

1. What do you mean by labile complexes and inert complexes ? Explain.
2. (*a*) How does the presence of ring structure affect the stability of complexes ?

   (*b*) Write short notes on :

   (*i*) Labile and inert complexes

   (*ii*) Stepwise and overall stability constants.

3. What is trans effect ? Discuss the theories for trans effect.
4. (*a*) How does the central metal ion effect the stability of a complex ?
   (*b*) Define trans-effect and ligands with trans-directing effect. Explain these with an example.
   (*c*) Explain the term associative and interchange associative in the reaction :
   $$PtL_2\,TX + Y \longrightarrow PtL_2\,TY + X$$
5. (*a*) How does the following factors affect the stability of complexes :
   (*i*) Nature of central metal ion and
   (*ii*) Nature of ligand.
   (*b*) What do you understand and by the stability of a chelate ring ?
6. What is meant by stability of a complex? Derive relationship between stepwise and commulative stability constants.
7. How does the nature of ligand affect the stability of complexes ?
8. Explain the following :
   (*a*) $[Fe(CN)_6]^{3-}$ is more stable than $[Fe(CN)_6]^{4-}$
   (*b*) $[FeF]^{2+}$ is more stable than $[FeCl]^{2+}$
   (*c*) $[Ni(en)_3]^{2+}$ (aq) is more stable than $[Ni(NH_3)_6]^{2+}$ (aq).
9. What are labile and inert complexes ? Do they convey same meaning as unstable and stable complexes.
10. Define trans effect. Illustrate trans effect using reactions of $PtCl_4^{\,2-}$ and Pt $(NH_3)_4^{\,2+}$.
11. (*a*) Plan a sequence of reactions starting with $[PtCl_4]^{2-}$ which will result in platinum (II) complexes, with four different ligands py, NH_3, NO_2^- and $CH_3\,NH_2$ with two different sets of trans ligands. (CH_3NH_2 is similar to NH_3 in trans effect).
    (*b*) Which theory of trans effect explains larger trans effect of CO compared to pyridine ?
12. (*a*) Explain the electrophilic and nucleophilic substitution in square planar complexes.
    (*b*) Explain how the stability of complexes is increased by chelation.
13. Explain the difference between thermodynamic stability and kinetic stability of the complexes.
14. Name the factors affecting the stability of metal complexes. Discuss any one.
15. $[Ni(CN)_4]^{2-}$ is thermodynamically stable but kinetically labile. What does this statement mean ? Explain.
16. (*a*) Derive relationship between stepwise and overall stability constants.
    (*b*) How does nature of the central metal ion affect the stability of the complex ?
17. What is meant by the terms : inert and labile complexes ? Show that inertness of a complex is different from its thermodynamic stability.
18. (*a*) What are the possible pathways in which a ligand may replace another ligand in square planar complexes ?
    (*b*) Discuss the role of basic solvents in preventing dissociative pathways.
19. What is trans effect ? Which theory of trans effect explains satisfactorily the following order of trans effect : $F^- < Cl^- < Br^- < I^-$
    (*b*) Show stereochemistry of substitution of following reactions;
    (*i*) $[Pt\,Cl_4]^{2-} \xrightarrow{PR_3} \xrightarrow{PR_3}$ (*ii*) $[PtCl_4]^{2-} \xrightarrow{NO_2^-} \xrightarrow{NH_3}$

(*iii*) $[Pt(PR_3)_4]^{2+} \xrightarrow{Cl^-} \xrightarrow{Cl^-}$ (*iv*) trans–$[PtCl_2 (NH_3)_2] \xrightarrow{py} \xrightarrow{py}$

(*v*) *cis* $[Pt (py)_2 (NH_3)_2] \xrightarrow{Cl^-} \xrightarrow{Cl^-}$

20. (*a*) How will you synthesise the three isomers of $[Pt(NH_3) (py) BrI]$ starting from $Pt Cl_4^{2-}$.

(*b*) Discuss mechanism of substitution reactions in square planar complexes.

21. (*a*) Discuss mechanism of nucleophilic substitution reactions in square planar complexes.

(*b*) How do you explain the acceleration of substitution reactions in square planar complexes with ligands like CO, C_2H_4 and CN^- ?

(*c*) Illustrate trans effect reactions of $PtCl_4^{2-}$ and $Pt(NH_3)_4^{2+}$

22. (*a*) Elaborate "Substitution in square planar complexes proceeds with retention of configuration".

(*b*) How does nature of the ligand affect the stability of the complexes ?

23. (*a*) Name two powerful trans directing ligands.

(*b*) What are inert and labile complexes.

(*c*) How do you explain the acceleration of substitution in square planar complexes with ligands such as CN^-, CO etc ?

24. (*a*) What is the basic difference between the terms thermodynamic stability and kinetic stability ? Give one example.

(*b*) $NiCl_4^{2-}$ is thermodynamically stable but kinetically labile. Discuss.

(*c*) Account for the fact that chelate effect is essentially an entropy effect.

(*d*) Name two powerful trans directing ligands.

(*e*) Although Co^{2+} (*aq*) forms tetrahedral tetrachloro complex on treatment with concentrated hydrochloride acid, Ni^{2+} (*aq*) does not. Explain.

25. (*a*) $NiCl_4^{2-}$ is thermodynamically stable but kinetically labile. Discuss.

(*b*) What are the factors which affect the stability of complexes ? Discuss.

(*c*) Define trans effect and ligands with trans direction effect. Why should trans effect be controlled kinetically ?

26. (*a*) What do you mean by labile and inert complexes ?

(*b*) $[Ni (CN)_4]^{2-}$ is thermodynamically stable but kinetically labile ?

(*c*) Why $[Fe (CN)_6]^{3-}$ is more stable than $[Fe (CN)_6]^{4-}$?

(*d*) How chelation increases the stability of a complex?

(*e*) What are stapwise and overall stability constants?

27. Discuss steric and chelate effect to describe the stability of complexes.

28. (*a*) Define stepwise stability constant and overall stability constant giving one example each.

(*b*) What is trans effect? Explain the theories for it.

(*c*) How do the nature of central metal ion as well as ligand affect the stability of the complexes ?

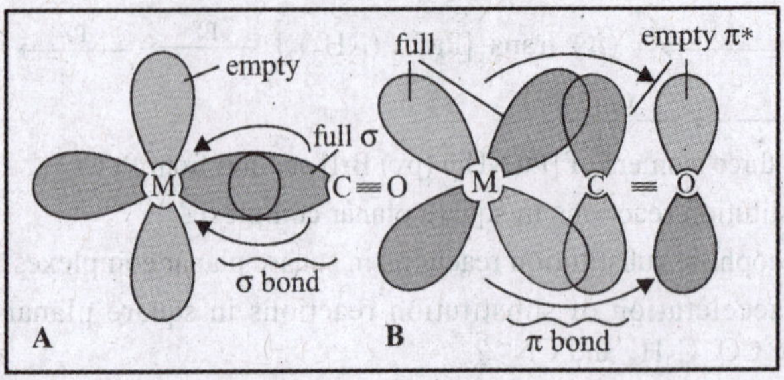

6 Organometallic Chemistry

6.1. INTRODUCTION

Organometallic chemistry is the branch of chemistry that acts as a bridge between organic chemistry and inorganic chemistry. Its importance lies in the production of new reagents and catalysts which are used in the preparation of useful compounds. For example, with the use of Grignard reagents, which are organometallic compounds of magnesium and alkyl lithiums, we can prepare almost any class of organic compounds. Structural determination of ferrocene, an organo metallic compound of iron has added to our knowledge regarding the formation of chemical bond. These are the compounds that contain a direct linkage between a metal atom and one or more carbon atoms. It is a rapidly growing and promising field of chemistry and we can expect to produce substances having versatile properties using organometallic compounds.

6.2. DEFINITION OF ORGANOMETALLIC COMPOUNDS

Organometallic compounds are the compounds which contain one or more metal-carbon linkage. For example, tetraethyl lead, ziese salt, diethyl zinc, ferrocene, etc.

$Pb(CH_2CH_3)_4$
Tetraethyl lead

$Zn(CH_2CH_3)_2$
Diethyl zinc

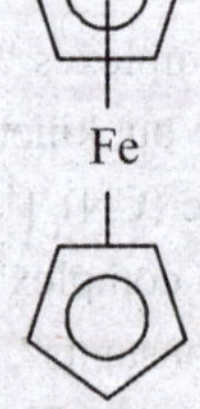

Ferrocene

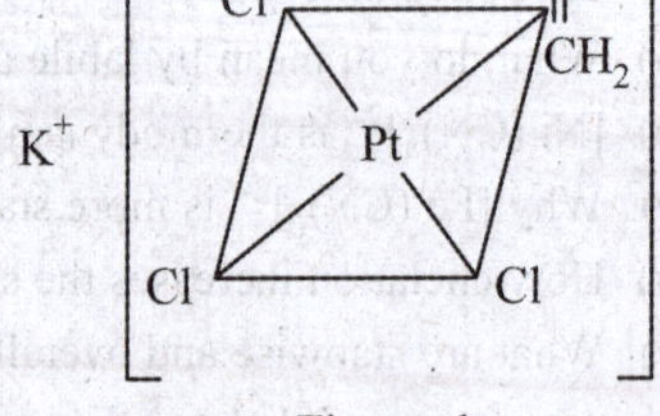

Ziese salt

Compounds like aluminium triethyl, sodium acetate, etc. do not qualify to be organometallic compounds because, the metal is not directly linked to carbon, but to oxygen in the above stated compounds.

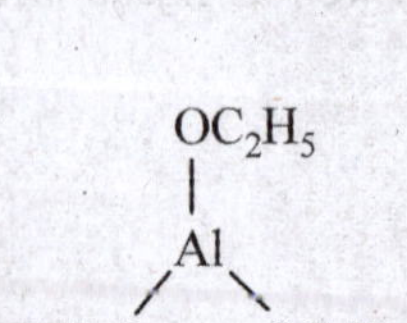

Aluminium triethoxide

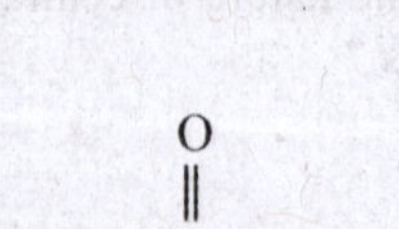

Sodium acetate

Also metal cyanides (M—CN) and metal carbides like CaC_2 and Al_4C_3 are not included in organometallic compounds because they show properties of compounds studied in inorganic chemistry, although these metal cyanides and carbides contain a metal-carbon bond.

6.3. DIFFERENT TYPES OF ORGANOMETALLIC COMPOUNDS

Theorganometallic compounds may be classified into the following types depending upon the nature of metal-carbon bond.

1. Ionic compound. Such compounds are formed when the negative charge on the hydrocarbon anion is delocalised over carbon atoms in the aromatic or unsaturated ring. $K^+ C_5H_5^-$ is a common example of this type where delocalisation of negative charge over the five carbon atoms of cyclopentadienyl ring gives rise to a stable complex. Some other examples of such compounds are: butyl sodium, $Na^+ C_4H_9^-$ phenyl sodium, $Na^+ C_6H_5^-$ trityl sodium, $Na^+ (C_6H_5)_3 C^-$. These compounds exhibit properties expected from ionic compounds. They are insoluble in non-polar solvents. The reactivity of ionic organometallic compound depends upon the stability of the anion. Compounds containing unstable anions are found to be extremely reactive.

2. Covalent compounds. Such compounds have the metal bonded to an organic part by normal two-electron sigma covalent bonds. Such compounds give properties expected from covalent compounds. They are soluble in organic solvents and insoluble in water. Electronegativity difference between the metal atom and carbon determines the polarity of the carbon-metal bond.

Common examples of this type of compounds are:

$$Pb(C_2H_5)_2, (CH_3)_2Cd, (C_6H_5)_2Zn, (CH_3)_3Al$$

3. Electron deficient compounds. Formation of this type of compounds cannot be explained on the basis of formation of usual two electrons–two centre bonds with carbon and metal atom. Compounds of Li, Be, Mg and Al with bridging alkyl groups appear in this category of compounds. Dimeric trialkyl aluminium (Al_2R_6), polymeric dimethyl beryllium $(BeMe_2)_n$ and diethyl magnesium $(MgEt_2)_n$ are some examples of this type of compounds. These compounds possess high charge to mass ratio and thus have strongly polarizing cation. As a result, polar covalent bond results.

Such polar molecules tend to associate and give polymeric structures. Trialkyl aluminium exists as dimeric in which the alkyl group is present as bridge (Fig. 6.1).

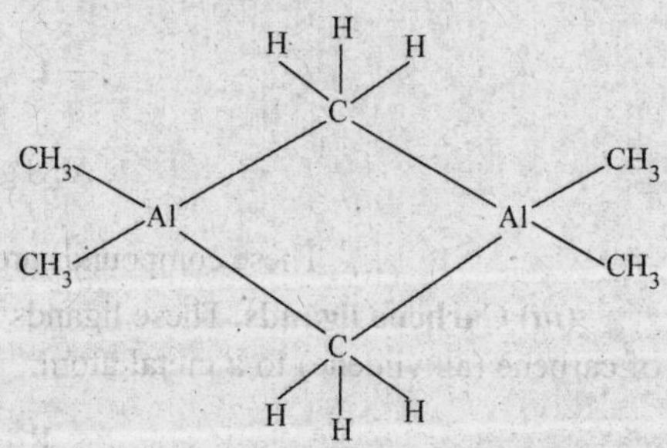

Fig. 6.1. Structure of dimeric Al_2Me_6.

4. Transition metal organometallic compounds. These organometallic compounds have transition metals bonded to unsaturated organic compounds in which the transition metal forms bonds to more than one carbon atoms of the same organic compound. The interaction occurs between the p-orbitals of the organic ligands with the suitable *d* or *p* orbitals of the metal atom.

The common examples of ligands forming such type of organometallic compounds are (*i*) alkene (2-electron donors), (*ii*) butadiene (4-electron donors), (*iii*) cyclopentadiene (5-electron donors) and (*iv*) benzene ring (6-electron donors). The common structures of dibenzene chromium $(C_6H_6)_2Cr$ and ferrocene $(C_5H_5)_2Fe$ are shown in Fig. 6.2. Their structures are known as sandwich structures in which metal atom lies in between two rings.

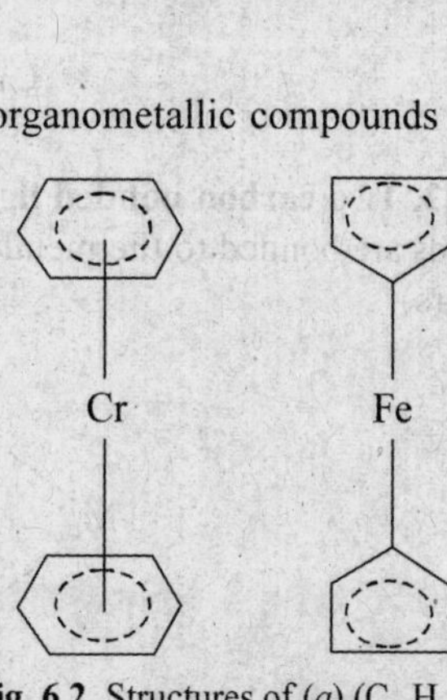

Fig. 6.2. Structures of (*a*) $(C_6H_6)_2Cr$ and (*b*) $(C_5H_5)_2Fe$.

6.4. CLASSIFICATION OF THE LIGANDS

An atom, ion or molecule which is capable of donating a pair of electrons to the metal atom is called a *ligand.* The number of atoms of the ligand which are within bonding distance of the metal atom is called its *hapticity.* It is written by using the symbol η (eta).

The ligands are classified on the basis of number of carbon atoms involved in bonding to the metal atom, *i.e.*, hapticity. The hapticity ranges from 1 to 8.

1. One carbon bonded ligands. These are the molecules in which the carbon atom of the ligand is bonded directly to the metal atom. Such ligands are also called **monohapto ligands** (one electron donors), η^1. For example, $-CH_3$ group attaches by a single M – C bond. These ligands are of the following different types:

***(i)* Hydrocarbon ligands.** These include alkyl ($-CH_3$), aryl ($-C_6H_5$), ($-CR=CR_2$), σ-cyclopentadienyl ($-C_5H_5$), alkynyl ($-C\equiv CR$) group.

For example, cyclopentadiene, C_5H_5, combines with alkali metals to form $(\sigma - C_5H_5)M$ compound as shown below:

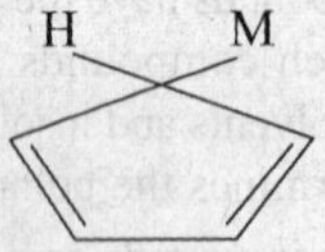

(One carbon bonded cyclopentadienyl compound).

***(ii)* Acyl ligands.** These ligands include acyl group which involves direct bonding of the acyl group (–COR) to a metal atom as shown below:

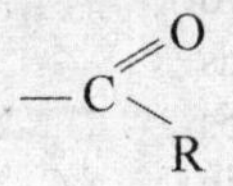

Acyl group

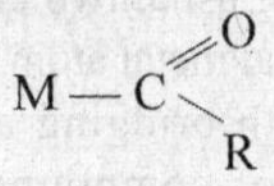

Organometallic compound

These compounds are formed by transition metals only.

***(iii)* Carbene ligands.** These ligands involve the direct metal carbon-bond of the carbon atom of carbene (alkylidene) to a metal atom.

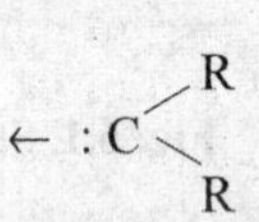

Carbene ligand

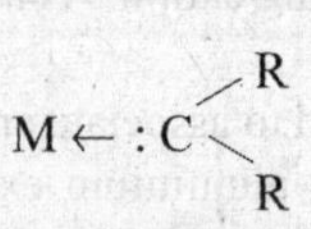

Organometallic compound

2. Two carbon bonded ligands. These are the molecules in which two carbon atoms of the ligands are bonded to the metal atom. These are called **dihapto (η^2) ligands** (or two-electron donors).

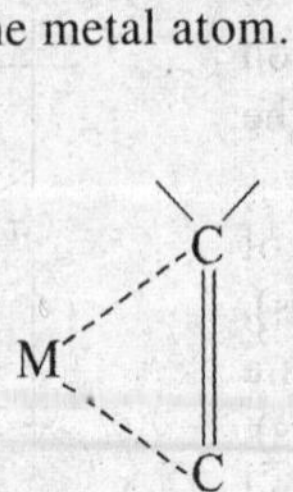

or

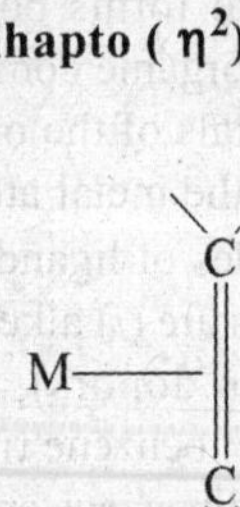

Examples of such ligands are alkenes or alkynes in which both the carbon atoms at each end of the multiple bond are involved in forming the metal-carbon bonds. The bonding in these ligands is indicated either by two dotted lines to the participating carbon atom or a single solid line from the centre of the participating carbon atoms of the ligand to the metal atom.

3. Three carbon bonded ligands. *These are the molecules in which three carbon atoms of the ligands are bonded to the metal atom.* These are called **trihapto ligands** (η^3). Examples of such ligands are allyl group (C_3H_5) known as π -allyl ligands as shown below. These may be distinguished from σ-allyl complexes in which the allyl group is bonded through one carbon atom only (monohapto).

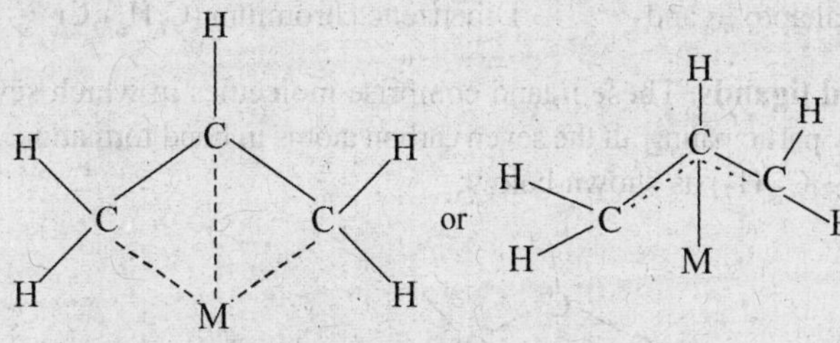

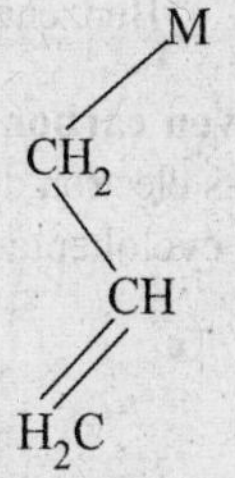

π - Allyl complex σ- Allyl complex

4. Four carbon bonded ligands. These ligands contain four carbon atoms which may be bonded to the metal atom. Therefore, these are called **tetrahapto** (4-electron donors) or **η^4 ligands**. These include acyclic ligands such as 1,3-butadiene and cyclic ligands such as cyclobutadiene.

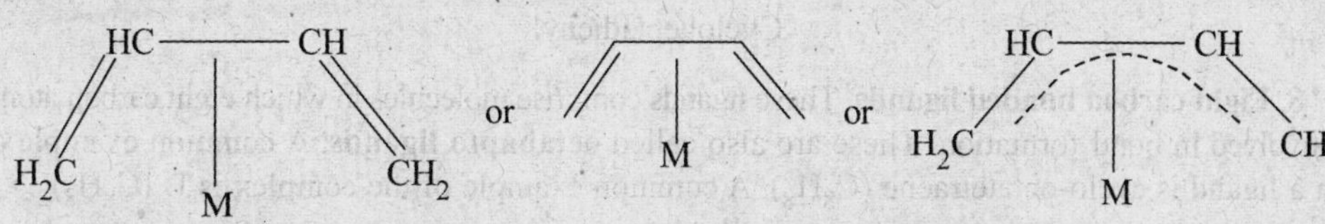

5. Five carbon bonded ligands. These ligands comprise molecules in which five carbon atoms of the ligands are bonded to the metal atom. These are called **pentahapto (η^5) ligands**. The common examples are acyclic dienyls and cyclic dienyls (*e.g.*, cyclopentadienyl).

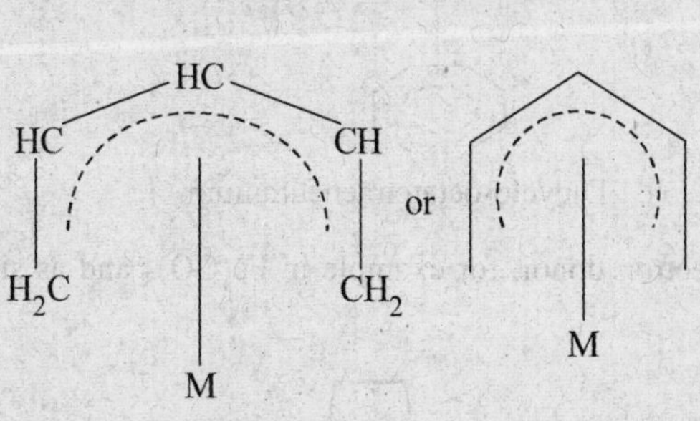

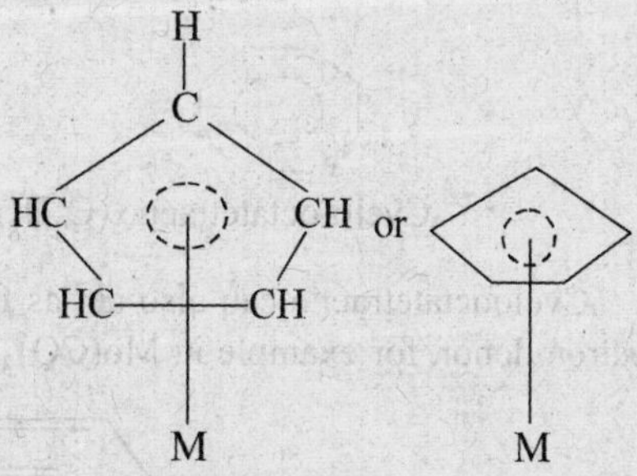

Pentadienyl Cyclopentadienyl

6. Six carbon bonded ligands. The molecules in which six carbon atoms of the ligand are bonded to the metal atom. The common example of such a ligand is benzene. The common example of an organo-metallic compound is dibenzene chromium $(C_6 H_6)_2 Cr$. This is also a *sandwich compound* in which Cr atom is present in between two benzene rings.

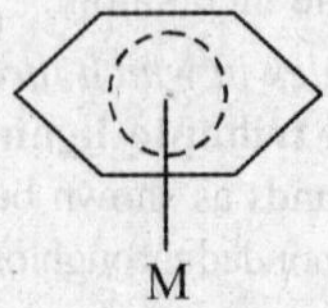

Benzene as hexahapto ligand

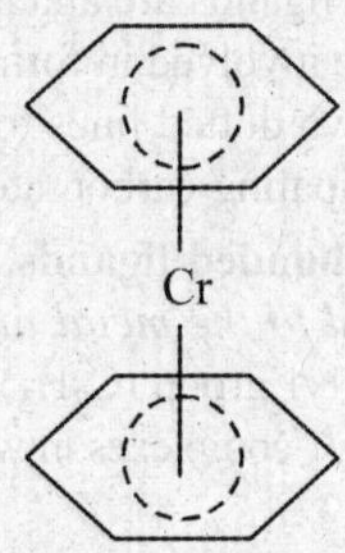

Dibenzene chromium $(C_6H_6)_2Cr$

7. Seven carbon bonded ligands. These ligand comprise molecules in which seven carbon atoms act as electron donors by participating all the seven carbon atoms in bond formation. A common example is cycloheptadienyl ($-$ C_7 H_7) as shown below:

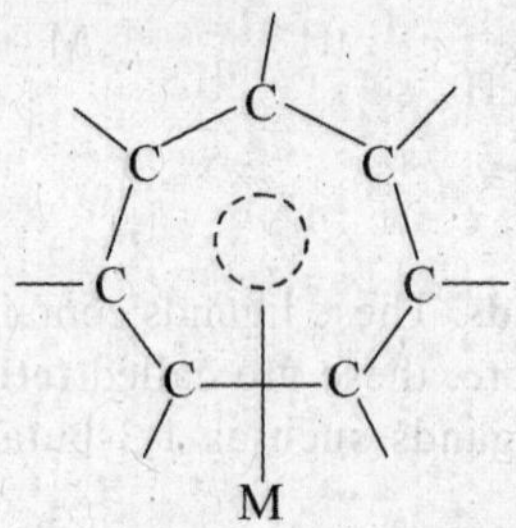

Cycloheptadienyl

8. Eight carbon bonded ligands. These ligands comprise molecules in which eight carbon atoms are involved in bond formation. These are also called **octahapto ligands**. A common example of such a ligand is cyclo-octatetraene (C_8H_8). A common example of the complex is U $(C_8H_8)_2$.

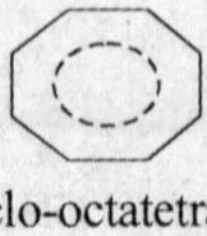

Cyclo-octatetraene (C_8H_8)

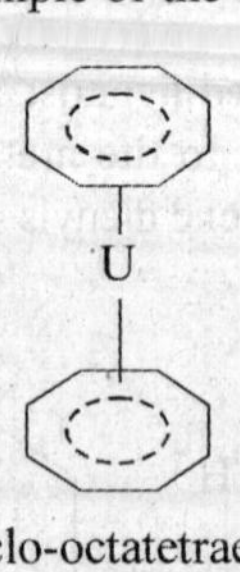

Dicyclo-octatetraeneuranium

Cyclooctatetraene can also act as four electron donor, for example in $Fe(CO)_3$ and as six-electron donor, for example in $Mo(CO)_3$

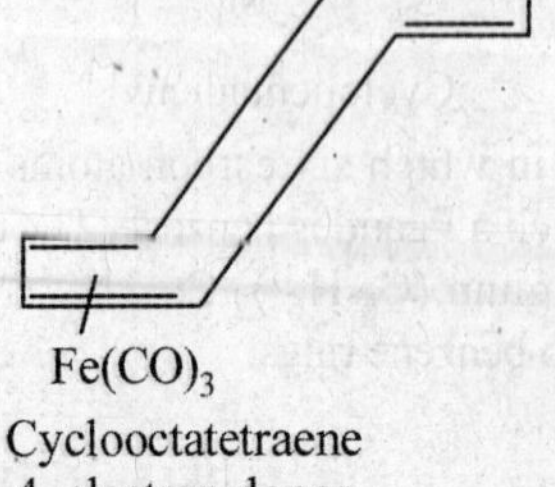

$Fe(CO)_3$

Cyclooctatetraene

4–electron donor

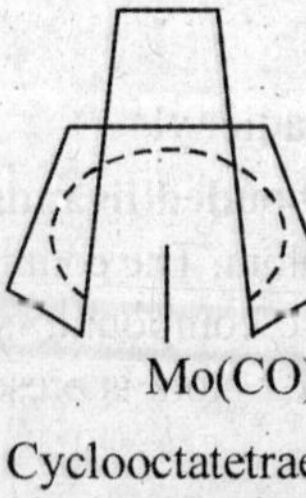

$Mo(CO)_3$

Cyclooctatetraene

(6–electron donor)

6.5. EFFECTIVE ATOMIC NUMBER RULE

Sidgwick suggested that metal ion accepts electron pairs from the ligands until it achieves the next noble gas configuration. This is called **effective atomic number rule.** *The total number of electrons possessed by the central metal ion and the electrons gained by it from ligands* is called the **effective atomic number** (EAN). Thus, according to effective atomic number rule, the effective atomic number in a complex should be equal to 36 (electrons in Kr), 54 (electrons in Xe) and 86 (electrons in radon).

The rule may be illustrated by taking the following examples:

(*i*) **Hexaammine Cobalt (III) ion, $[Co(NH_3)_6]^{3+}$**

Atomic number of Co = 27

Electrons in Co(III) = 24

Six ammonia molecules give two electrons each (6 × 2) = 12

Total number of electrons in the ion (EAN) = 24 + 12 = 36

Thus, the complex obeys EAN rule.

(*ii*) **Tetracarbonyl nickel (0), $Ni(CO)_4$**

Atomic number of Ni = 28

Electrons in Ni(0) in $Ni(CO)_4$ = 28

Four CO groups give two electrons each (4 × 2) = 8

EAN of $[Ni(CO)_4]$ = 36

Thus, the complex $[Ni(CO)_4]$ obeys EAN rule.

(*ii*) **Pentacarbonyl iron (0), $Fe(CO)_5$**

Atomic number of Fe = 26

Electrons in Fe(0) = 26

Five CO groups give two electrons each (5 × 2) = 10

EAN = 36

Table 6.1 gives the list of some other complexes obeying EAN rule.

Table 6.1 Complexes obeying effective atomic number rule.

Complex	Central metal ion	At. No.	No. of electrons in metal ion ligands	Electrons gained from	EAN
$[Fe(CN)_6]^{4-}$	Fe^{2+}	26	24	12	36 (Kr)
$[Fe(CO)_5]$	Fe	26	26	10	36 (Kr)
$[Cr(CO)_6]$	Cr	24	24	12	36 (Kr)
$[Ni(CO)_4]$	Ni	28	28	8	36 (Kr)
$[Pt(NH_3)_6]^{4+}$	Pt^{4+}	78	74	12	86 (Rn)
$[PtCl_6]^{2-}$	Pt^{4+}	78	74	12	86 (Rn)
$[Pd(NH_3)_6]^{4+}$	Pd^{4+}	46	42	12	54 (Xe)

Table 6.2 gives the list of metal carbonyls obeying EAN rule.

Table 6.2 Metal carbonyls obeying EAN rule.

Complex	Metal ion	Electrons in Metal ion	Electrons donated from ligands	EAN
$Cr(CO)_6$	Cr	24	12	36
$Fe(CO)_5$	Fe	26	10	36
$Ni(CO)_4$	Ni	28	8	36
$Mo(CO)_6$	Mo	42	12	54
$Ru(CO)_5$	Ru	44	10	54
$Os(CO)_6$	Os	76	10	86
$W(CO)_6$	W	74	12	86

But we observe that EAN rule is not obeyed in many stable complexes.

Table 6.3 gives the list of such complexes.

Table 6.3 Complexes not obeying effective atomic number rule.

Complex	Central metal ion	At. No.	No. of electrons in metal ion ligands	Electrons gained from	EAN
$[Ni(CN)_4]^{2-}$	Ni^{2+}	28	28–2 = 26	8	34
$[Ni(NH3)_6]^{2+}$	Ni^{2+}	28	28–2 = 26	12	38
$[Pt(NH_3)_4]^{2+}$	Pt^{2+}	78	78–2 = 76	8	84
$[Fe(CN)_6]^{3-}$	Fe^{3+}	26	26–3 = 23	12	35
$[Cr(NH_3)_6]^{3+}$	Cr^{3+}	24	24–3 = 21	12	33
$[Ag(NH_3)_2]^+$	Ag^+	47	47–1 = 46	4	50
$[PdCl_4]^{2-}$	Pd^{2+}	46	46–2 = 44	8	52

EAN rule can explain the stability of metal carbonyls but it cannot be the *sole criteria to form stable carbonyls.* Metals with odd number of electrons cannot obey EAN rule because total number of electrons will always be odd, no matter how many carbonyls are added. Thus, hexacarbonyl vanadium V $(CO)_6$ is a stable complex though it does not obey EAN rule as:

Atomic No. of V	= 23
Electrons in V(0)	= 23
6 CO groups give two electrons each (6 ×2)	= 12
EAN	= 35

Therefore, the complex V $(CO)_6$ does not obey EAN rule.

Nevil Sidgwick

In order to satisfy EAN rule, there are some **options** for the metal ions. These are:

(*i*) They may form carbonylate anion such as $[M(CO)_n]^-$ by gaining an electron from a reducing agent or they may form cationic species, $[M\ (CO_n)]^+$ by losing an electron. For example, the anions $[V(CO)_6]^-$, $[Mn(CO)_5]^-$, $[Co(CO)_4]^-$ obey EAN rule.

$[V(CO)_6]^-$

Atomic number of V = 23

Electrons in V (–1)	= 24
Six CO groups give two electrons each (6 × 2) = 12	
EAN	= 36

Thus, $V(CO)_6^-$ obeys EAN rule.

(*ii*) They may form single covalent bond with an atom or group having a single unpaired electron, *e.g.*, hydrogen ($H^\bullet$) or chlorine ($Cl^\bullet$) [$HM(CO)_n$] or [$M(CO)_n Cl$]. For example,

$Mn(CO)_5Cl$

Electrons in Mn(0)	= 25
5 CO groups give two electrons each (5× 2)	= 10
$Cl^\bullet$	= 1
EAN	= 36

∴ EAN for $Mn(CO)_5Cl$ = 36

Similarly, $HCo(CO)_4$

Electrons in Co(0)	= 27
4 CO groups give two electrons each (4 × 2)	= 8
$H^\bullet$	= 1
EAN	= 36

(*iii*) If no other species is available with which the metal carbonyl containing odd number of electrons can interact, it can dimerize resulting in pairing of electrons. This will lead to the formation of metal-metal bond. While counting the EAN, the electron pair shared between two metal atoms forming metal-metal bond is counted on both the metals.

For example, manganese forms a stable dinuclear carbonyl, $Mn_2(CO)_{10}$ having metal-metal bond, $(CO)_5Mn-Mn(CO)_5$. Let us calculate EAN for this complex.

$Mn_2(CO)_{10}$

Electrons in 2 Mn (2 × 25)	= 50
10 CO groups give two electrons each (10 × 2)	= 20
One Mn–Mn bond	= 2
EAN for $Mn_2(CO)_{10}$	= 72
∴ EAN per Mn atom	= 36 (Kr)

EAN rule for other Organometallic Compounds

EAN rule can also be applied to other organometallic compounds. For this purpose, number of electrons donated by various ligands are taken into consideration.

Let us illustrate the calculation of EAN for some complexes.

(*a*) Alkyl groups as 1 electron donors

$CH_3Re(CO)_5$

Electrons in Re (Rhenium)	= 75
5(CO) give two electrons each (5 × 2)	= 10
CH_3	= 1
EAN	= 86

(*b*) Olefins and alkynes acting as 2-electron donors

$[Mn(CO)_5C_2H_4]^+$

Electrons in Mn^+	= 24
5 (CO) give two electrons each	= 10
Contribution from C_2H_4	= 2
EAN	= 36

(*c*) π-allyls acting as 3-electron donors

$Mn(\pi-C_3H_5)(CO)_4$

Electrons in Mn	= 25
Contribution of $\pi-C_3H_5$	= 3
4 (CO) contribute	= 8
EAN	= 36

$Co(\pi-C_3H_5)(CO)_3$

Electrons in Co	= 27
Contribution of $\pi-C_3H_5$	= 3
3(CO) donate	= 6
EAN	= 36

(*d*) π-cyclo pentadienes as 5-electron donors

$Co(\pi-C_5H_5)(CO)_2$

Electrons in Co	= 27
Contribution of 2 $(\pi-C_5H_5)$	= 5
2 (CO) donate	= 4
EAN	= 36

$Fe(\pi C_5H_5)_2$

Electrons in Fe	= 26
Contribution of 2 $(\pi-C_5H_5)$	= 10
EAN	= 36

Thus, it has been observed that most of the organometallic compounds obey EAN. There are some exceptions also in which EAN rule is not obeyed. For example,

$Co(\pi-C_5H_5)_2$

Electrons in Co	= 27
Contribution of 2 $(\pi-C_5H_5)$	= 10
EAN	= 37

$Cr(C_6H_6)(CO)_4$

Electrons in Cr	= 24
Contribution of C_5H_5	= 6
Contribution of 4 (CO)	= 8
EAN	= 38

6.6. NOMENCLATURE OF ORGANOMETALLIC COMPOUNDS

The rules for the nomenclature of coordination compounds have been explained in the previous class. Most of these rules apply to organometallic compounds. However, some of the rules are modified keeping in view the complex nature of the ligands.

Nomenclature of different types of organometallics is discussed as under.

1. Nomenclature of simple compounds. The simple alkyl or aryl organometallic compounds of metals are named by writing the name of the metal after the name of organic group. For example,

CH_3Li	Methyl lithium
$(C_2H_5)_2Zn$	Diethyl zinc
C_2H_5MgBr	Ethyl magnesium bromide
C_6H_5MgC1	Phenyl magnesium chloride

2. Nomenclature of carbonyls. (*i*) The compounds containing CO as ligands are called *metal* carbonyls. In case the metal has zero oxidation state, it may not be mentioned.

For example,

$Ni(CO)_4$	Tetracarbonylnickel
$Mn_2(CO)_{10}$	Decacarbonyldimanganese
$Fe_2(CO)_9$	Nonacarbonyldiiron

$Co_2(CO)_8$	Octacarbonyldicobalt
$[V(CO)_6]^-$	Hexacarbonylvanadate ion

(*ii*) If the ligands act as bridges between two metal atoms, the Greek letter mu (μ) is written before their names. The prefix μ is repeated before the name of each kind of bridging ligand. For example,

$[(CO)_3 Co (CO)_2 Co (CO)_3]$	Di-μ-carbonyl bis (tricarbonyl cobalt)
$[(CO)_3 Fe(CO)_3 Fe (CO)_3]$	Tri-μ–carbonyl bis (tricarbonyl iron)

(*iii*) When the metal carbonyls contain metal-metal bonds, these may be classified as *symmetrical* or *unsymmetrical*. The symmetrical metal carbonyls are named by the use of multiple prefixes (bis, tris, etc.). In case of unsymmetrical metal carbonyls, one central metal atom and its ligands are treated as a ligand on the other central metal atom. For example,

$[(CO)_4 Co - Co (CO)_4]$ (symmetrical)	Bis (tetracarbonylcobalt)
$[(CO)_4 Co - Re (CO)_5]$ (unsymmetrical)	Pentacarbonyl (tetracarbonyl cobalto) rhenium

3. Nomenclature of σ and π bonded ligands. We use the notations σ and π to distinguish between one carbon bonded ligand and multiple carbon bonded ligands. For example, cyclopentadiene (C_5H_5) is referred to as σ – C_5H_5 (*e.g.*, Li C_5H_5) when it behaves as one carbon bonded ligand. However, when it behaves as five carbon bonded ligand, it is named as π-C_5H_5 (*e.g.* π (C_5H_5)$_2$ Fe). Similarly, alkyl group is referred as σ-ally1 or π–ally1 depending upon whether it behaves as one electron donor or three-electron donor.

For unsaturated molecules or groups, the prefix η (called as eta or hapto) is used. For example, a one carbon bonded ligand is specified by monohapto (or η^1), two carbon bonded ligand as dihapto (or η^2), three carbon bonded ligand as trihapto (or η^3), a four carbon bonded ligand as tetrahapto (or η^4) and so on. According to latest IUPAC convention, the η notation is recommended.

For example,

$Fe (C_5H_5)_2$	Bis (η^5-cyclopentadienyl) iron
$Cr (C_6 H_6)_2$	Bis (η^6-benzene) chromium
$Co (CO)_3(\pi- C_3H_5)$	(η^3-ally1) tricarbonyl cobalt
$(C_6H_6) Cr (CO)_3$	(η^6-benzene) tricarbonyl chromium
$Fe_2 (CO)_4 (C_5 H_5)_2$	Bis (η^5-cyclopentadienyl) tetracarbonyldiiron
$Fe (CO)_2 (\sigma - C_5H_5)(\pi - C_5H_5)$	Dicarbonyl (η^1-cyclopentadienyl) (η^5-cyclopentadienyl) iron
$Fe (CO)_3 (C_4 H_6)$	(η^4-butadiene) tricarbonyl iron
$Mn (CO)_5 (- CH_2 - CH = CH_2)$	(η^3-ally1) tricarbonyl iron

6.7. BONDING IN ORGANOMETALLIC COMPOUNDS

The type of bonding in an organometallic compound determines its properties. The different types of bonding that occur in an organometallic compound are 1. Ionic bonding. 2. σ-covalent bonding. 3. Multiple covalent bonding.

These are separately discussed as under:

1. Ionic bonding. This type of bonding is observed in organometallic compounds of electropositive metals like sodium, potassium, rubidium, cesium, magnesium etc. In such compounds, the hydrocarbon contains carbon carrying negative charge which is strongly attracted by the positively charged metal ion by electrostatic forces of attraction. In some cases, the negative charge is delocalised over the ring containing different number of carbon atoms. For example, in Mg^{2+} $(C_5H_5^-)_2$, the negative charge of cyclopentadienide anion, $C_5H_5^-$, is delocalised over the ring of five carbon atoms.

2. σ-covalent bonding. This type of bonding occurs in organometallic compounds of representative elements. It is generally observed in metal-alkyl bonds. For example, $(CH_3)_4$ Sn, C_4H_9 Li, $(CH_3)_6$ Al_2, etc. The polarity of covalent bond between metal and alkyl group depends upon the difference in electronegativity between metal atom and carbon atom.

Transition metals can also form metal-carbon σ bonds with organic groups although relatively unstable. They can easily decompose even at room temperature. For example, the alkyl derivatives of transition metals such as Ti $(CH_3)_4$, Nb $CH_3)_5$, W $(CH_3)_2$ are well known but they are unstable and explode even below room temperature. The stability of the π-bonded covalent compounds increases if some additional π-bonding ligands are also present. For example, the presence of ligands such as PR_3, CO, amines, etc. increases the stability of these compounds. For example, tetramethyl titanium undergoes spontaneous decomposition violently at −23° C but the addition of dipyridine or diamine increases its stability and it does not decompose up to room temperature.

6.7.1. Factors responsible for kinetic instability of transition metal sigma bonded organometallic compounds

The following factors contribute to kinetic instability of organometallic compounds containing metal-alkyl bond:

***(i)* Formation of reactive species due to cleavage.** The cleavage of metal-carbon bond in transition metal alkyls give reactive alkyl radicals or ions. The reactive species formed react with each other or solvent to give chain reactions. This behaviour is different from that of the other inorganic metal-ligand bonds such as M– OH_2, M –NR_2, M – X (X = halogen), etc. where a cleavage leads to relatively less reactive species like H_2O, R_2N^- or X^- respectively.

***(ii)* Cleavage products form strong bonds.** In case the cleavage of metal-carbon bond occurs, they combine among themselves to form strong bonds such as C—C, C—O, C—N, C—X (X = halogen) etc. depending upon the nature of organometallic compound. During the formation of new bonds, the cleavage reaction is highly exothermic and is thermodynamically favoured. For example, metal-alkyl bond undergoes cleavage as

$$\mathrm{M—R} \longrightarrow \underbrace{\dot{\mathrm{M}} + \dot{\mathrm{R}}}_{\text{Reactive Intermediates}}$$

The formation of alkyl radical leads to subsequent reactions.

***(iii)* Hydrogen transfer from alkyl to metal.** The higher alkyl organometallic compounds particularly those which carry β-hydrogen are unstable. β-hydride elimination reaction provides an opportunity for the decomposition of metal alkyls into a metal hydride and olefin. (β-elimination reaction may be represented as

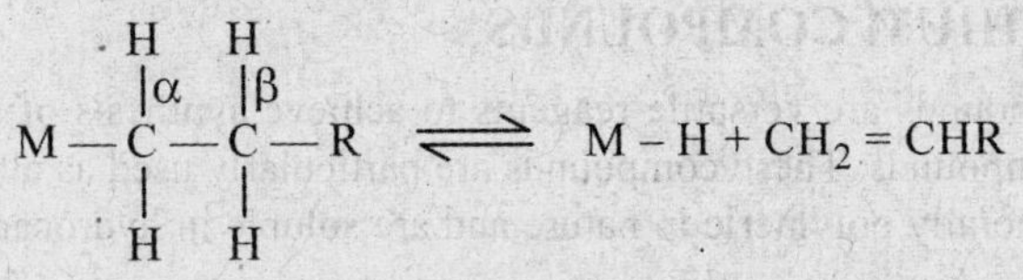

6.7.2. Factors that can increase stability of metal-alkyl compounds

The following measures can increase stability of metal-alkyl compounds.

***(i)* Chacking β-elimination.** In higher alkyl organometallic compounds, the hydrogen present on carbon atom at β-position to the metal is labile (removable). It brings about decomposition of the compound into metal hydride and an alkene. This is called **β-elimination.**

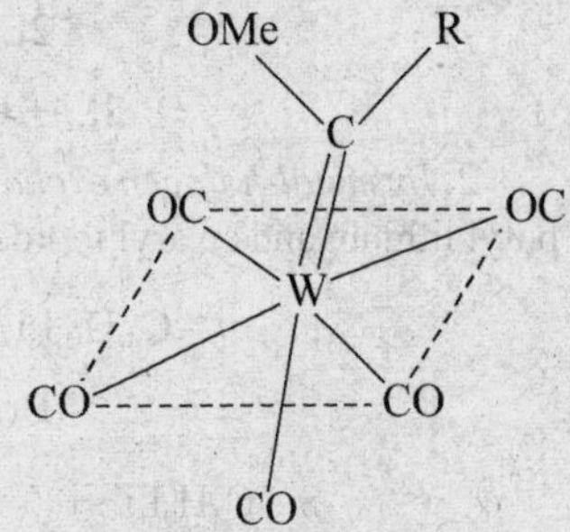

Structure of $(CO)_5WCR\ (OMe)$.

The β-elimination is responsible for the instability of alkyl organometallic compounds. Therefore the blockage of β-elimination reaction can bring about stability in metal alkyls. Alkyls which do not contain replaceable hydrogen atom on β-carbon are therefore stable. For example,

$M - CH_2 - CH_3$, $M - CH_2 - Ar$, $M - CH_2 - NR_2$ etc.

***(ii)* Chelation.** The stability of a metal-carbon sigma bond can be enhanced by chelation. Donation through a terminal group can bring about chelation.

***(iii)* Increase in coordination number.** The stability of metal-alkyl compound can be raised by delocalising the negative charge on carbon $\left(\begin{matrix}\delta+ & & \delta- \\ M & — & C\end{matrix}\right)$. Thus aromatic compounds having metal-carbon linkage are relatively more stable. For example $M—C_6H_5$ is more stable than $M—CH_3$.

6.8. METAL-CARBON MULTIPLE BONDING

If multiple bonding occurs between metal and carbon, the bond strength is increased. For example, carbenes, $= CR_2$ and carbynes, $\equiv CR$ form double and triple bonds with the metal atom.

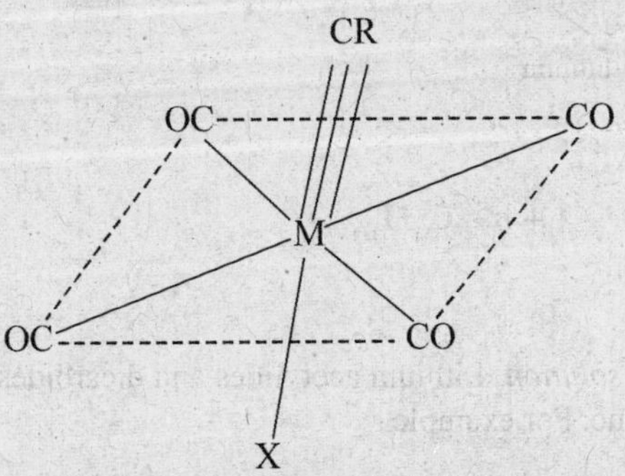

Structure of $(CO)_4\ M\ (=CR)X$

It has been observed that in carbene compounds, the metal-carbon bond is somewhat shorter than normal metal-alkyl σ bond but larger than metal-carbon double bond. This suggests that the bond between metal-carbon in carbenes is in between single and double bond.

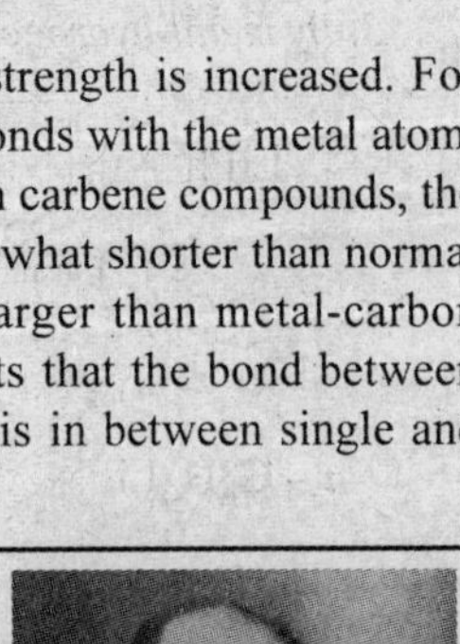

Emil Fischer

The first carbene complex $(CO)_5$ W[: C(R) OMe)] was synthesised by Emil. Fisher in 1964. It has been found to have octahedral arrangement with one of the positions occupied by the carbene, = CR (OMe).

Metal carbynes have triple bonds. For example, $(CO)_4\ M\ (\equiv CR)X$ where M = Cr, Mo or W, R = Me, Et or Ph, X = CI, Br, I). The crystal structure of these types of complexes shows that ≡ CR group occupies position *trans* to halo ligand in an octahedral geometry.

6.9. ORGANOLITHIUM COMPOUNDS

Organolithium compounds are versatile reagents to achieve synthesis of a wide variety of inorganic and organic compounds. These compounds are particularly used as alkylating reagents. These compounds are generally polymeric in nature and are soluble in hydrocarbon solvents.

Preparation

Organolithium compounds can be prepared as follows:

1. *By direct reaction of Li with alkyl or arylchloride in benzene or petroleum.*

$$2Li + RCl \xrightarrow{\text{Solvent}} LiR + LiCl \qquad (R = \text{alkyl or aryl})$$

$$2Li + C_2H_5Cl \xrightarrow{\text{Solvent}} C_2H_5Li + LiCl$$

2. *By metal-halogen exchange in ether.* Lithium aryls are best prepared by this method using *n*-butyl lithium and an aryl iodide.

$$n-C_4H_9Li + ArI \xrightarrow{\text{Ether}} LiAr + n\text{-}C_4H_9I \ (Ar = \text{aryl})$$

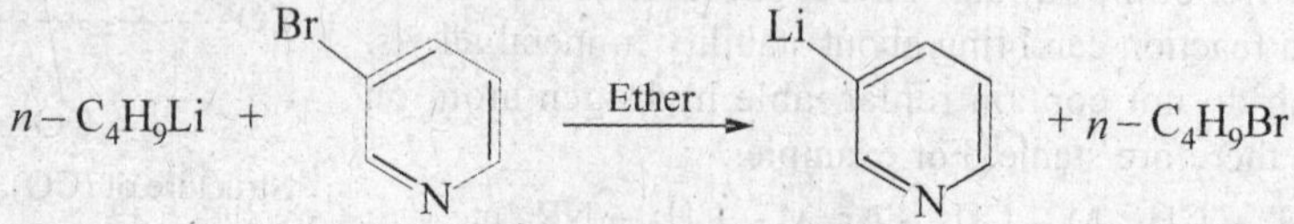

3. Vinyl-allyl and other unsaturated derivatives can be obtained by transmetallation.

$$4LiPh + Sn(CH = CH_2)_4 \xrightarrow{\text{Ether}} 4LiCH = CH_2 + SnPh_4$$

4. Methyllithium is also prepared by exchange through the interaction of n–C_4H_9Li and CH_3I in hexane at low temperatures where it precipitates as insoluble white crystals.

$$n-C_4H_9Li + CH_3I \xrightarrow{\text{hexane}} CH_3Li + n-C_4H_9I$$

5. *By metal-hydrogen exchange (metallation) in ether.*

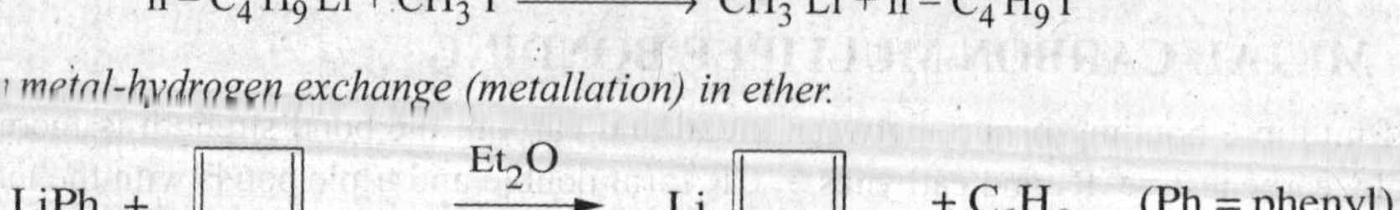

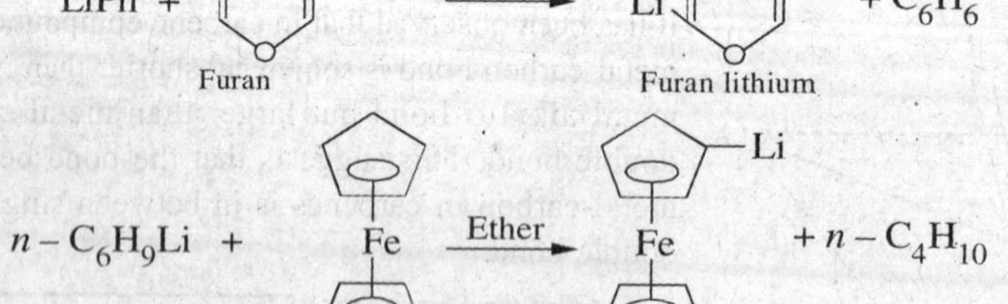

6. *By deprotonation of alkynes in liquid ammonia solution.* Lithium acetylides and dicarbides are obtained by this method using Li metal and acetylene. For example,

$$Li + HC \equiv CH \xrightarrow{\text{Liq. } NH_3} LiC \equiv CH + 1/2\, H_2$$

$$2Li + HC \equiv CH \xrightarrow{\text{Liq. } NH_3} Li_2C_2 + H_2$$

7. *By metal-metal exchange in petroleum or benzene.* Reaction between an excess of *Li* and an organomercury compound is a useful alternative when isolation of the product is required, rather than its direct use in further synthetic work. For example,

$$2Li + HgR_2 \xrightarrow[\text{benzene}]{\text{Petrol or}} 2LiR + Hg$$

Properties. 1. Organolithium compounds tend to be thermally unstable and most of them decompose to LiH and an alkene on standing at room temperature or above.

2. They are generally liquids or low melting solids and are found in molecular association.

3. Organolithium compounds are typical covalent substances, soluble in hydrocarbons or other non-polar liquids.

4. They react rapidly with oxygen and are usually spontaneously flammable in air, with liquid water and with water vapours.

5. LiMe and LiBu are usually synthetic reagents and have been increasingly used in organic syntheses as polymerization catalysts, alkylating agents and precursors to metalated organic reagents.

Some of the reactions with different compounds are given below:

(*a*) Reactions with alkyl iodides give compounds having C – C bonds. For example,

$$\text{LiR} + \text{R}'\text{I} \longrightarrow \text{LiI} + \text{RR}'$$

(*b*) Reactions with proton donor give the corresponding hydrocarbons. For example,

$$\text{LiR} + \text{H}^+ \longrightarrow \text{Li}^+ + \text{RH}$$

(*c*) Reactions with metal carbonyls give aldehydes and ketones. For example,

$$\text{LiR} + [\text{Ni(CO)}_4] \longrightarrow \text{Li}^+\left[\text{R} - \overset{\overset{\displaystyle\text{O}}{\|}}{\text{C}} - \text{Ni(CO)}_3\right]^-$$

The intermediate unstable acyl nickel carbonyl complex can be attacked by electrophiles such as H^+ or R′ Br to give aldehydes or ketones by solvent induced reductive elimination.

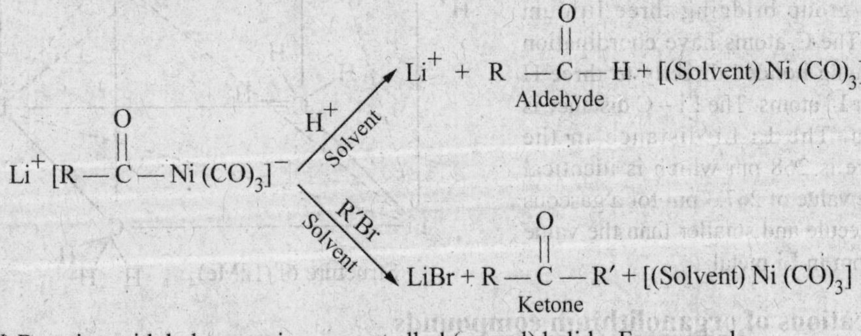

(*d*) Reaction with halogens gives parent alkyl (or aryl) halide. For example,

$$\text{LiR} + \text{X}_2 \longrightarrow \text{LiX} + \text{RX}$$

(*e*) Reaction with CO gives symmetrical ketones. For example.

$$2\text{LiR} + 3\text{CO} \longrightarrow 2\text{LiCO} + \text{R}_2\text{CO}$$

(*f*) Reaction with ethers gives alkenes; the organometallic reacts here as a very strong base (proton acceptor). For example,

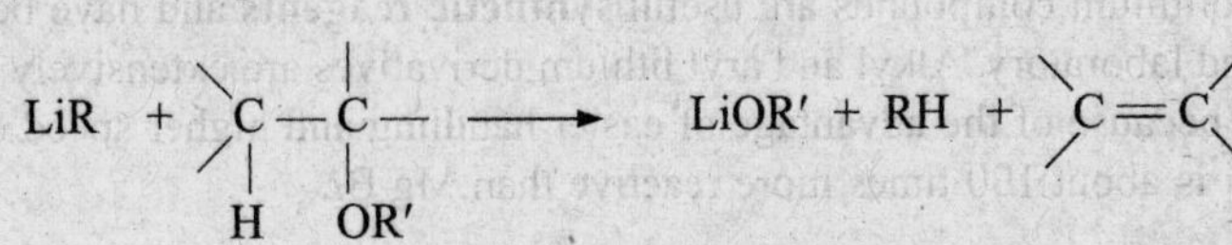

(*g*) Metal-halogen exchange reactions give other organometallic compounds. For example,

$$3LiR + BCl_3 \longrightarrow 3LiCl + BR_3$$

$$4LiR + SnCl_4 \longrightarrow 4LiCl + SnR_4$$

$$3LiR + P(OEt)_3 \longrightarrow 3LiOEt + PAr_3$$

(*h*) Reaction of lithium aryls (behaving as typical carbanion) in non-polar solvents with CO_2 gives carboxylic acid and with aryl ketones gives tertiary carbinols. For example,

$$LiAr + CO_2 \longrightarrow ArCO_2Li \xrightarrow{H_2O} LiOH + ArCO_2H$$

$$LiAr + Ar'_2LO \longrightarrow [Ar'_2C(Ar)OLi] \xrightarrow{H_2O} LiOH + Ar'_2C(Ar)OH$$

(*i*) Reactions with N, N′- disubstituted amides give aldehydes and ketones. For example,

$$LiR + HCONMe'_2 \longrightarrow LiNMe_2 + RCHO$$

$$LiR + R'CONMe_2 \longrightarrow LiNMe_2 + R'COR$$

Structure. Lithium alkyls are polymeric and generally exist as tetrameric units such as $(LiCH_3)_4$. In lithium methyl there is a tetrahedral set of four lithium atoms with a methyl group placed symmetrically above each Li_3 face. The structure clearly involves a methyl group bridging three lithium atoms. The C-atoms have coordination no. 7 being bonded directly to three H and four Li atoms. The Li – C distance is 231 pm. The Li-Li distance in the structure is 268 pm which is identical with the value of 267.3 pm for a gaseous Li_2 molecule and smaller than the value of 304 pm in Li metal.

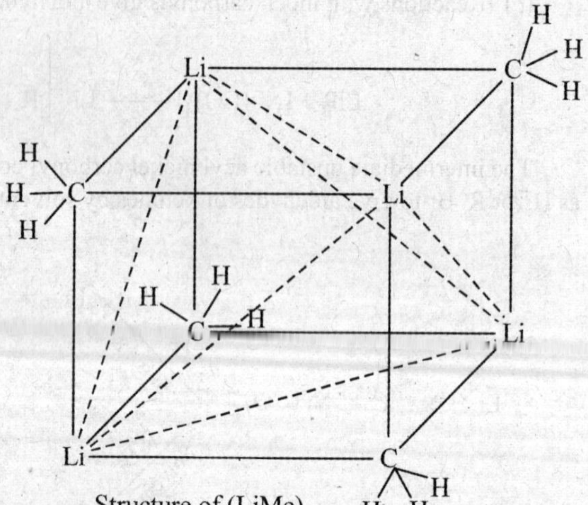

Structure of $(LiMe)_4$

Applications of organolithium compounds

The important applications of organolithium compounds are given below:

1. Organolithium compounds are used as **catalysts** in polymerisation reactions. For example, *n*-BuLi catalyses polymerization of butadienes. If butadiene is polymerized in the presence of catalysts such as Co or Ni some metal impurities are always left in the reaction mixture but with organolithium compounds there is no such problem. These factors provide advantages in certain applications such as impact resistance modifications of plastics.

2. Organolithium compounds are also used as **precursors** to metalated organic reagents.

3. The organolithium compounds are useful **synthetic reagents** and have been increasingly used in industry and laboratory. Alkyl and aryl lithium derivatives are extensively used in place of Grignard reagents because of the advantage of easier handling and higher speed of the reactions. For example, PhLi is about 100 times more reactive than Mg Br.

4. The allyllithium compounds combine with butadiene and tend to copolymerize with monomers such styrene to give **block copolymers** by forming the stable living polymers.

6.10. ORGANO ALUMINIUM COMPOUNDS

Preparation. Organoaluminium compounds may be prepared by the following methods:

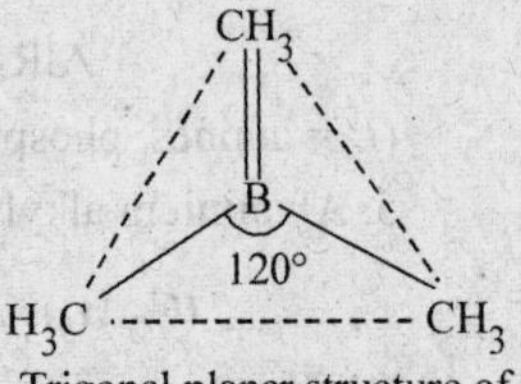

Trigonal planar structure of BMe_3.

1. *By the reaction of Grignard reagents with $AlCl_3$.*

$$RMgCl + AlCl_3 \longrightarrow RAlCl_2, R_2AlCl, R_3Al$$

one, two or three alkyl groups could combine with one Al.

2. *By the reaction of aluminium with an organomercury compound.*

$$2Al + 3HgMe_2 \xrightarrow{363\ K} Al_2Me_6 + 3Hg$$

Al_2Ph_6 can be prepared similarly using $HgPh_2$ in boiling toluene or by the reaction of LiPh on Al_2Cl_6.

$$6LiPh + Al_2Cl_6 \longrightarrow Al_2Ph_6 + 6LiCl$$

3. *On the industrial scale Al* is alkylated by means of *RX* giving sesquichloride $R_3Al_2Cl_3$

$$2Al + 3RCl \xrightarrow[AlCl_3 \text{ or } AlR_3]{\text{Trace of } I_2} R_3Al_2Cl_3$$

4. *Higher trialkyls* are readily prepared on an industrial scale by the alkene route in which H_2 adds to *Al* in the presence of performed AlR_3 acting as catalyst (Ziegler catalyst) to give a dialkyl-aluminium hydride which then readily adds to the alkene.

$$2Al + 3H_2 + 2Al_2Et_6 \xrightarrow{425\ K} (6\ Et_2AlH) \xrightarrow[345\ K]{6\ CH_2CH_2} 3Al_2Et_6$$

Properties. 1. Aluminium trialkyls and triaryls are highly reactive volatile liquids or low melting solids, which burn in air and react violently with water.

2. They react with protonic reagents to liberate alkanes.

$$Al_2R_6 + 6HX \longrightarrow 6RH + 2AlX_3$$

3. Organoaluminium compounds give alkene insertion reactions known as growth reactions to synthesize unbranched long chain primary alcohols and alkenes. Here alkenes insert into the Al – C bond of monomeric AlR_3 (acting as catalyst) at ~425 K and 100 atom to give long chain derivatives whose composition can be closely controlled by the temperature, pressure and contact time.

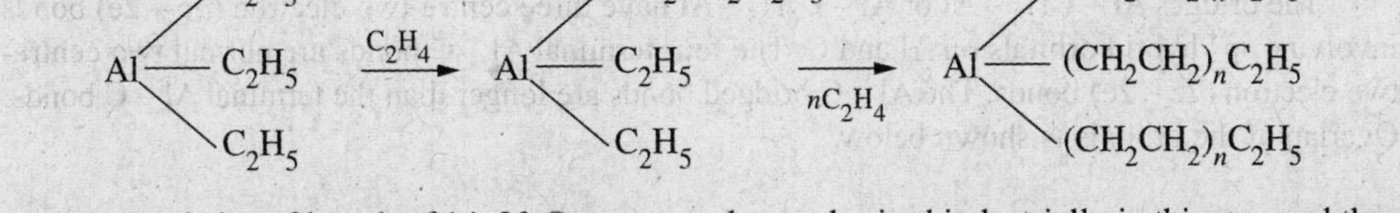

Long chains of length of 14–20 C atoms can be synthesised industrially in this way and then converted to unbranched aliphatic alcohols after oxidizing with air and then hydrolysing with water for use in the synthesis of biodegradable detergents.

$$Al(CH_2CH_2R)_3 \xrightarrow[\text{(ii) } H_3O^+]{\text{(i) } O_2} 3RCH_2CH_2OH$$

4. Al_2R_6 (or AlR_3) react readily with ligands to form adducts and behave as Lewis acids. For example,

$$AlR_3 + L \text{ (ligand)} \longrightarrow Li\,Al\,R_3 \text{ (adduct)}.$$

(*L* = amines, phosphines, ethers and thioethers)

5. Aluminium alkyls also combine with lithium alkyls to give mixed alkyls.

$$(C_2H_5)_3Al + LiC_2H_5 \xrightarrow{\text{In benzene}} Li\,Al\,(C_2H_5)_4$$

6. They react with halides or alkoxides of elements like B, Ca, Si, Sn etc which are less electropositive than Al to form other organometallics.

$$MX_n \longrightarrow MR_n + \frac{n}{3}AlX_3$$

7. Organoaluminium compounds bring about low pressure polymerization of ethene and propene in presence of organometallic mixed catalyst. A Ziegler-Natta catalyst is a mixture of $TiCl_4$ and Al_2Et_6 in heptane giving a brown suspension which rapidly absorbs and polymerises ethene even at room temperature and atmospheric pressure. Polyethene produced in this way has superior qualities.

6.10.1. STRUCTURE OF ORGANOALUMINIUM COMPOUNDS

In the solid and gas phase $AlMe_3$ is dimeric and has the methyl bridged structure. Al_2Ph_6 exhibits the similar dimeric structure. In both cases the bridging Al – C distance is 10% longer than the terminal Al – C distance. Solutions of Al_2Me_6 show only one proton *nmr* signal at room temperature due to the rapid interchange of bridging and terminal Me groups. In Al_2Me_6 the terminal Al – C distances are 195 pm and the bridging distances 212 pm, the bridging Al – C – Al angle is 76°. In general AlR_3 molecules are bridged dimers in the solid state except when R is very bulky such as tertiary butyl. The crystal structure of Al_2Ph_6 shows phenyl as bridging groups and the terminal and bridging Al– C distances being 196 pm and 218 pm respectively. The bridging Al – C – Al angle is measured to be 77°.

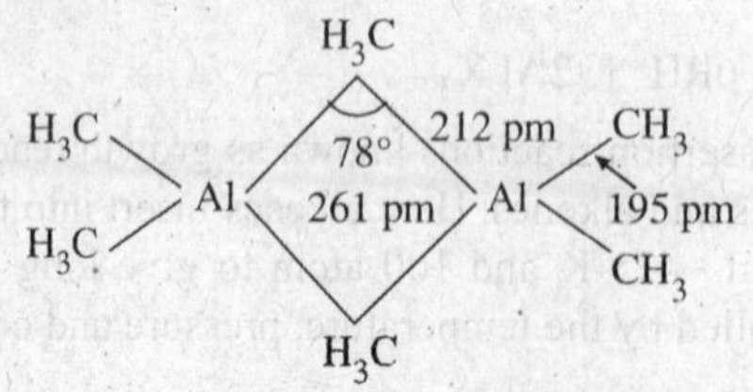

Structure of $Al_2(CH_3)_6$.

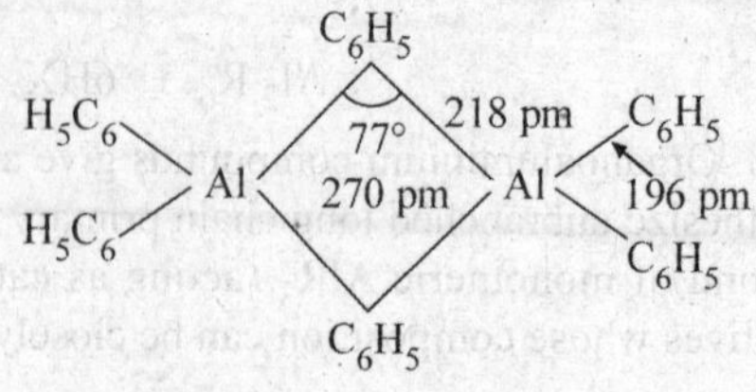

Structure of $Al_2(C_6H_5)_6$

The bridges Al – CH_3 – Al or Al – C_6H_5 – Al have three centre-two electron (3c – 2e) bonds involving sp^3 hybrid orbitals on Al and C. The four terminal Al – C bonds are normal two centre-two electron (2c – 2e) bonds. The Al – C bridged bonds are longer than the terminal Al – C bonds. Overlap of the orbitals is shown below.

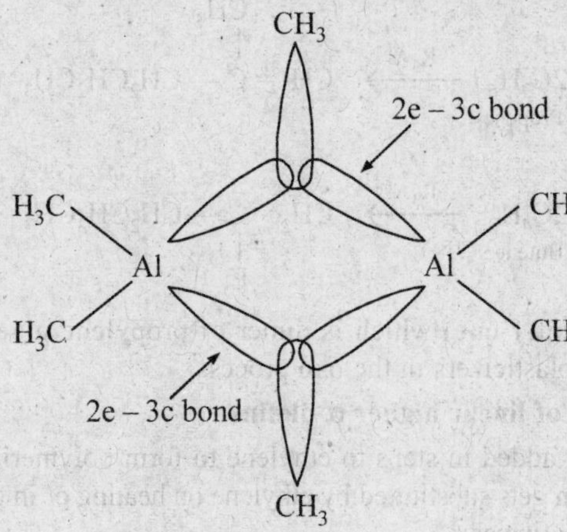

Bonding in Al $(CH_3)_3$

6.10.2. Applications of organoaluminium compounds

Some important reactions of organoluminium compounds are given below.

1. For the preparation of organometallic compounds. Organoaluminium compounds being inexpensive and reactive are used to prepare other organo metallic compounds. For example,

$$4(CH_3)_3\,Al + 3SnCl_4 + 4NaCl \longrightarrow \underset{\text{Tetramethyl tin}}{3\,(CH_3)_4\,Sn} + 4NaAlC_{14}$$

2. In the production of high linear α-alcohols

The high linear trialkyl aluminium formed as an intermediates in the production of high linear α-olefins are oxidised with air and on subsequent hydrolysis give alcohols.

$$[Al(C_2H_4)_n\,Et]_3 + \frac{3}{2}O_2 \longrightarrow Al[O(C_2H_4)_n\,Et]_3 \xrightarrow{H_2O} Al(OH)_3 + 3Et(C_2H_4)_nOH$$

For example, butanol is obtained by air oxidation of tributyl aluminium in toluene at 40°C followed by hydrolysis with 10% aqueous H_2SO_4.

$$Bu_3Al \xrightarrow[\text{toluene}]{40°C,\ O_2(air)} (BuO)_3\,Al \xrightarrow{10\%H_2SO_4} 3\,BuOH$$

3. As high polymer catalysts.

Organoaluminium compounds are used as a component of the Ziegler Natta catalyst for polymerisation of olefins and dienes. The Ziegler Natta catalysts are the main group organometallic compounds of Al and transition metal compounds. Some common examples of such catalysts for obtaining different products are

High density polythenes:	$TiCl_4 - Al\ Et_2Cl$, $TiCl_4 - Al\ Et_3$
Polypropylenes :	$TiCl_3 - Al\ Et_2Cl$
Butadiene rubbers :	$CoCl_2 - Al\ Et_2Cl$
Ethylene – propylene rubbers :	$VOCl_3 - Al_2Et_3Cl_3$

4. In Dimerization of olefins

Alkylaluminium reacts with ethylene as a single catalyst to produce molecular alkyl compounds. However, with propene and butene, dimerization reactions occur as :

$$\underset{\text{Propene}}{2C_3H_6} \xrightarrow{R_3Al} CH_2=\underset{\displaystyle CH_3}{\overset{|}{C}}-CH_2CH_2CH_3$$

$$\underset{\text{Butene}}{2C_4H_8} \xrightarrow{R_3Al} CH_2=\underset{\displaystyle Et}{\overset{|}{C}}-CH_2CH_2CH_3$$

Hexene (2-methyl pent-1-ene) which is dimer of propylene gives heptanol which is the important raw material of plasticizers in the oxo process.

5. In the production of linear higher α-olefins

Trialkyllaluminium is added in steps to ethylene to form polymeric trialkyaluminium. This polymeric trialkylaluminium gets substituted by ethylene on heating or in the presence of nickel salt to produce high linear α-olefins as:

$$AlEt_3 + 3n\,C_2H_4 \xrightarrow[100\text{ atm}]{120°C} Al\,[(C_2H_4)_n\,Et]_3 \xrightarrow[3C_2H_4]{\text{Heat}} 3C_2H_5\,(C_2H_4)_{n-1}\,CH=CH_2 + AlEt_3$$

6.11. ORGANOTIN COMPOUNDS

Preparation

Organotin compounds may be prepared by the following methods:

1. Reaction with organoaluminium compounds. Organotin compounds may be prepared by reaction of tin halides with organoaluminium compounds.

$$4R_3Al + 3\,SnX_4 \longrightarrow 3R_4Sn + 4AlX_3$$

The reaction forms AlX_3 which easily forms the complex $RnSnX_{4-n}$. The formation of this complex lowers the yield of R_4Sn. Therefore, the reaction is carried out in the presence of stronger complexing agents for further alkylation to proceed easily in the presence of AlX_3.

2. Wurtz reaction. Alkyl tin compounds may be prepared by treating tin halides with alkyl halides in the presence of sodium. This reaction is known as Wurtz reaction.

$$Sn\,X_4 + 4\,RX + 8\,Na \longrightarrow R_4Sn + 8\,NaX$$

3. By Grignard reactions. The organotin compounds are prepared by Grignard reactions using ethyl ether or ethyl ether-hydrocarbon mixture as solvent.

$$4\,R\,MgX + SnX_4 \longrightarrow R_4Sn + 4\,MgX_2$$

$$3\,R_4Sn + SnCl_4 \longrightarrow 4R_3SnCl$$

$$R_4Sn + SnCl_4 \longrightarrow 2R_2SnCl_2$$

$$R_4Sn + 3SnCl_4 \longrightarrow 4\,R\,SnCl_3$$

Compounds containing secondary or tertiary aliphatic groups and long chain alkyl phenyl or vinyl groups give poor yield because of steric hindrance.

4. Direct reaction. Alkyltin can be prepared by direct reaction of alkyl halide and tin.

$$2RX + Sn \longrightarrow R_2SnX_2$$

e.g. $$2C_2H_5X + Sn \longrightarrow (C_2H_5)_2SnX_2$$

This reactivity of alkyl halides follows the order

$$RI > RBr > RCl$$

For given halide, the reactivity follows the order

$$MeX > EtX > PrX$$

Properties

Tetraalkyl and tetraaryl compounds of tin $R_n\, Sn\, X_{4-n}$ ($n = 1$ to 4) are colourless, monomeric liquids or solids. They resist hydrolysis and oxidation under normal conditions. However, when ignited, they burn to give SnO_2, CO_2 and water.

Major reactions of organotin compounds are given below:

1. Triorganotin radical reactions. The $R_3Sn - H$, $R_3Sn - CH_3$ and $R_3Sn - SnR_3$ bonds are weak having bond strengths 70 and 65 and 63 k cal mol^{-1}. In the presence of light, these compounds form radicals easily.

$$R_3Sn - SnR_3 \xrightarrow{h\nu} \underset{\text{Radicals}}{R_3Sn^\bullet + R_3Sn^\bullet}$$

$$t\text{–}BuO^\bullet + R_3\text{–}SnSnR_3 \longrightarrow BuOSnR_3 + R_3Sn^\bullet$$

These radicals react with halides to form new radical as

$$R_3Sn^\bullet + R'X \longrightarrow R_3SnX + R'^\bullet$$

Tetraalkyl and tetraaryl compounds of tin i.e. $RnSnX_{4-n}$ are colourless, monomexic liquids or solids. They resist hydrolysis and oxidation under normal conditions. However when ignited they burn to give SnO_2, CO_2 and H_2O.

2. Reactions of organostannylenes. Divalent organotin compounds, R_2Sn (called organostannylenes) containing alkyl or aryl are labile. However, other compounds containing different groups are stable. These provide synthetic routes to obtain higher polystannes as:

$$SnCl_2 + 2[LiCH\,(SiMe_3)_2] \xrightarrow[0°C]{Et_2O} [Sn\{CH\,(SiMe_3)_2\}_2]$$

The compound was obtained as red crystals (m.p. 136°). It is monomeric in benzene.

3. Hydrostannation. The unsaturated compounds such as olefins and acetylenes react with tin hydride. The reaction is known as **hydrostannation reaction.** The compound formed has functional group substituted by organotins in the reaction and tin atom is exclusively attached to the terminal carbon.

$$CH_2 = CR'COOMe + Me_3SnH \longrightarrow Me_3SnCH_2CHR'COOMe$$

4. Hydrostannolysis. The bond between alkyl (or aryl) group and halogen or sulphur is cleaved by an organotin hydride. This is called **hydrostannolysis reaction** and is used for dehalogenation reaction of a halide.

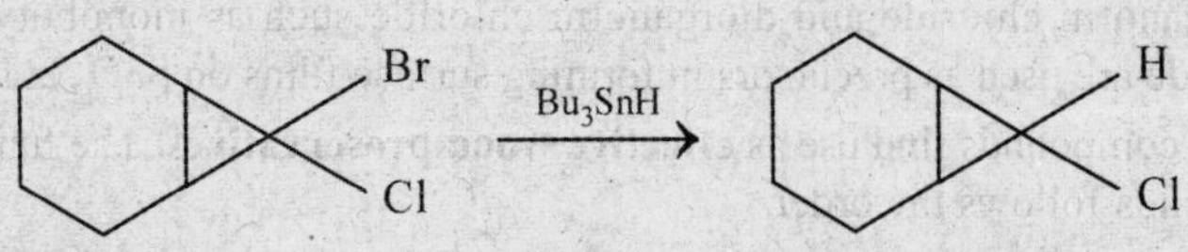

Organostannylenes are reactive compounds and undergo insertion and cyclization reactions as:

$$Me_3Sn\ Sn\ Me_3 + R_2Sn \longrightarrow R_2Sn\,(SnMe_3)_2 \quad \text{(insertion)}$$

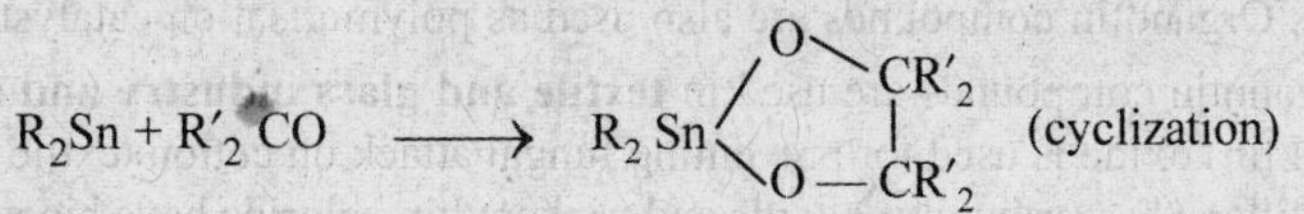

Bonding and structure

The divalent Sn (II) and tetravalent Sn (IV) organotin compounds with coordination numbers of 2, 3, 4, 5, 6 and 7 are known. The dimethyl tin chloride, Me_2SnCl_2 has a tendency to polymerize through tin-halogen bond. The bond parameters in associated Me_2SnCl_2 are given below.

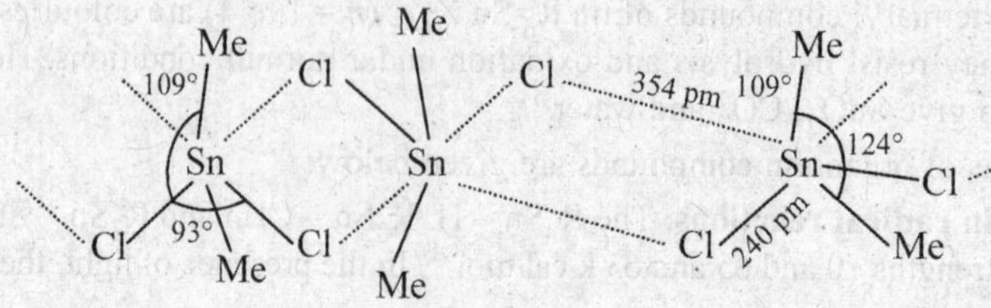

Bond parameters in Me_2SnCl_2

We observe extensive polymerisation in R_3SnX. The associated structure of Me_3 SnF is given below.

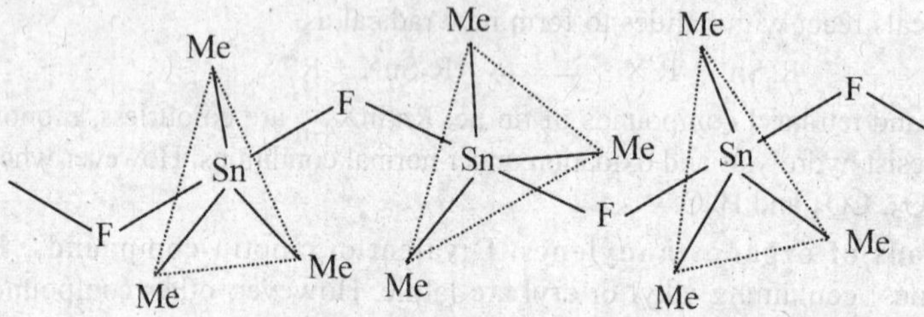

Associated structure of Me_3SnF

Applications

1. The organotin compounds have **bacteriocidal, fungicidal, acaricidal and insecticidal** properties and have strong affinity of tin atoms to a donor ligand atoms such as S, O and N. It is an important indispensable element in human body.

Organotin compounds which are used in **agriculture as fungicides, antifeedants** and **acaricides** are listed separately as under.

Fungicides	:	Ph_3SnX (X = OH, OAc)
Antifeedants	:	Ph_3SnX (X = OH, OAc)
Acaricides	:	$(C_6H_{11})_3$ Sn X(X = OH)

2. mono organotin chloride and diorganotin chloride such as monobutyltin chloride and dimethyltin chloride are used as precursors in forming surface films on SnO_2 at high temperatures.

3. Organotin compounds find use as **effective wood preservatives.** The fungicidal activity of organotin compounds follows the order.

$$R_3SnX > R_4\ Sn \approx R_2SnX_2 > RSnC_3$$

4. Organotin compounds are used as **homogeneous catalysts** for polymerization of urethenes, polymerization of silicones and esterifications because of strong affinity to oxygen of organo compounds. Organotin compounds are also used as polymerisation catalysts for olefins.

5. Organotin compounds are used in **textile and glass industry and as** water repellant etc. Bis (tributyl tin) oxide is used for preventing fungal attack on cotton textile and in cellulose based house hold fillers. Bis (tributyltin) oxide and triphenyltin chloride have biocidal effect against cloth moth.

6. The organotin compounds have also been used successfully as **anthelmintics, disinfectants** and **antitumour drugs** because they have poisonous characteristics in cells. R_2Sn compounds have

also been used as anticancer drugs because they block the growth of tumour cells and are also used for curing hyperbilirubinemia.

Tributyltin compounds are used against gram positive bacteria.

7. Organotin compounds are used as **heat and light stabilizers** for polyvinyl chloride (PVC) plastics. The common examples to this effect are $(R_2SnOCOCH = CHCOO)_n$ (R = Bu, Oct) and $Bu_2Sn(OCOCH = CHCOO\ Oct)_2$, $Bu_2Sn\ (OCOC_{11}\ H_{23})_2$ etc.

6.12. ORGANOTITANIUM COMPOUNDS

Preparation

1. Tetrakisneopentyltitanium is prepared by treating neopentyllithium with Ti Cl_4 in ether solution at room temperature.

$$TiCl_4 + 4LiCH_2-\underset{CH_3}{\overset{CH_3}{\underset{|}{\overset{|}{C}}}}-CH_3 \xrightarrow[\text{ether}]{\text{room temp.}} \left[CH_3-\underset{CH_3}{\overset{CH_3}{\underset{|}{\overset{|}{C}}}}-CH_2\right]_4 Ti$$

2. Cyclopentadienyl compounds such as $[Ti(\eta^5 - C_5H_5)_2X_2)]$ X = Cl, Br, I are more stable. Dicyclopentadienyl titanium halides may be prepared as:

$$TiCl_4 + 2(C_5H_5)M \longrightarrow (\eta^5 - C_5H_5)_2\ TiX_2$$

M = MgCl, Mg Br, Na or Li; X = Cl, Br

The same compound has also been prepared by the reaction of cyclopentadiene with $TiCl_4$ in the presence of diethylamine.

$$TiCl_4 + 4C_5H_6 \xrightarrow[\text{NHEt}_2\text{, dioxane}]{60-80°\ C} (C_5H_5)_2\ TiCl_2$$

3. Organotitanium compounds can be prepared by the reaction of titanium halides with organometallic compounds such as organoalkali metal compounds. For example, tetrachloro titanium is mixed with methyl magnesium chloride at −78°C when a dark green solid is precipitated. The supernatant solution is removed and a solution of ether or hexane as solvent is added to the residue and concentrated at − 48°C when yellow crystals of tetramethyltitanium are formed.

$$TiX_m + RM \longrightarrow RnTiX_{4-n}$$

M = Li, Na, MgX; X = Cl, Br, R = alkyl

$$Ti + 4\ Me\ MgCl \xrightarrow[\text{in ether}]{-78°C} Me_4Ti$$

4. When $TiCl_4$ is added to $Al(C_2H_5)_3$ in Leptone, a brown solid known as Ziegler Natta catalyst is obtained. This is a wonderful catalyst and is used for polymerisation of ethylene to straight chain polyethene.

5. Tetraphenyl titanium is obtained by mixing phenyl-lithium and titanium tetrachloride at −70°:

$$TiCl_4 + 4PhLi \xrightarrow[\text{ether}]{-70°C} Ph_4Ti$$

Tetraphenyl titanium decomposes with rising temperature and at room temperature it decomposes to diphenyl-titanium and diphenyl.

$$Ph_4Ti \xrightarrow{\text{room temp.}} Ph_2Ti + Ph - Ph$$

PROPERTIES

Organotitanium compounds undergo following types of reactions:

1. **Substitution reactions.** Cyclopentadienyltitanium dichloride gives substitution reactions as:

$$(C_5H_5)_2\,TiCl_2 + 2LiNMe_2 \longrightarrow (C_5H_5)_2\,Ti\,(NMe_2)_2$$

2. Tetravalent titanium compounds get reduced to trivalent titanium compounds with metals such as sodium, magnesium or zinc. The reaction with zinc takes place as under.

$$(C_5H_5)_2\,TiCl_2 + \frac{1}{2}Zn \longrightarrow (C_5H_5)_2\,TiCl + \frac{1}{2}ZnCl_2$$

Trivalent titanium compounds, $(C_5H_5)_3$ TiCl and other lower valence organotitanium compounds are easily oxidised and therefore, they have strong reducing power. Therefore, they are used as organosynthetic reagents.

Some common examples of trivalent, divalent and zero valent organotin compounds are

Ti (III) : $(C_5H_5)_2$ TiCl, (C_5H_5) $TiCl_2$

Ti (II) : (C_5H_5) (C_6H_5) Ti $(OEt_2)_2$

Ti (O) : $(\eta^6 - C_6H_6)_2$ Ti

3. **Carbo metalation.** This reaction deals with olefin polymerisation with titanium compounds. Titanium reacts with organoaluminium compounds when titanium atom forms a bridged structure (Ti — R — Al) with an alkyl group of organoaluminium compounds and then Ti—R adds to a carbon-carbon double bond as shown below.

4. Insertion reactions. Carbon monoxide reacts with organotin compound when the former can be readily inserted into Ti—R bond as

$$X(C_5H_5)_2\,TiR + CO \longrightarrow (C_5H_5)_2\,TiCOR(X)$$

R = Me, Et, Ch_2Ph, X = Cl, I

Structure and Bonding

Cyclopentadienyl compounds are more stable and common. The compound $Ti(C_5H_5)_4$ is prepared from $TiCl_4$ and NaC_5H_5 and the structure of the green black titanium compound is given below. It is formulated as Ti $(\eta^1 - C_5H_5)_2$ $(\eta^5 - C_5H_5)_2]$

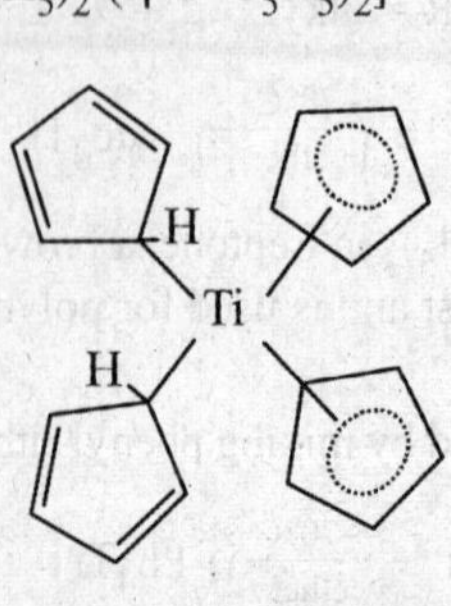

Here, two cyclopentadienyl rings act as monohapto (η^1) and other two rings acts as pentahapto (η^5). The structures of carbonyl compound $[Ti(\eta^5 - C_5H_5)_2\,(CO)_2]$ and dimeric $Ti(C_5H_5)_2$ have the following structures.

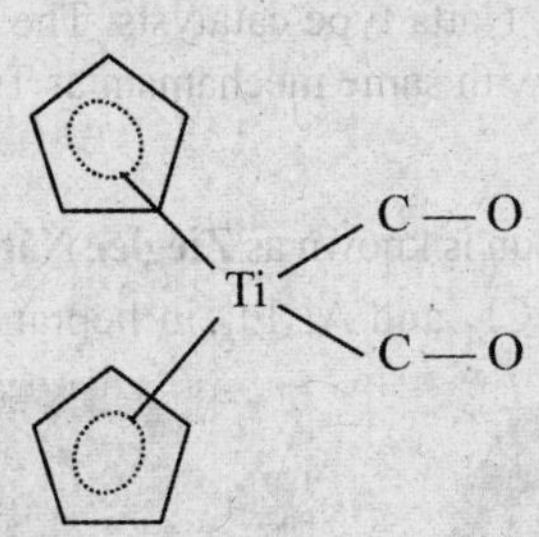

Structure of $[Ti(\eta^5 - (C_6H_5)_2 (CO)_2]$

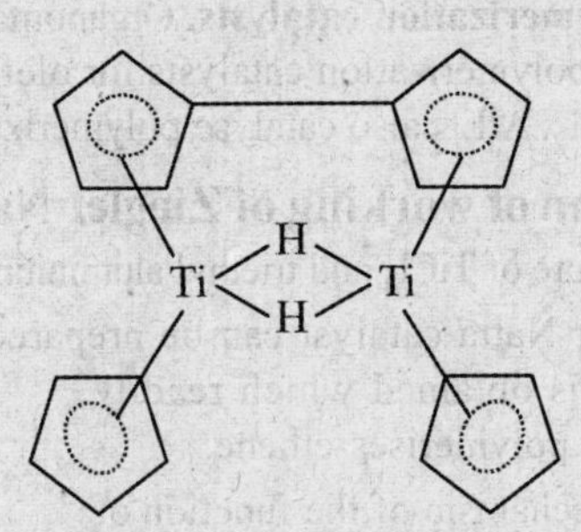

Structure of $Ti(C_5H_5)_2$ dimer

Applications

1. **Thin film forming materials.** The organometallic titanium compounds are used as high purity single metallic materials for forming thin films. This has advantage over the usual techniques of vacuum vaporization, wet electroplating and electrodes plating which are difficult to form a homogeneous pinholeless metal thin film. Organotitanium compounds are used as raw materials for metal carbides. Titanium carbide formed is useful as abrasion resistant coating material such as lining of fusion reaction vessels, cutting tools and bearings. etc. TiC coating is carried out by heating $TiCl_4$, H_2 and CH_4 and this process is called CVD process (Chemical Vapour Deposition process). The thin forming process is believed to occur as:

$$Ti(OCH_2CH_3)_4 + 4LiCH_2C(CH_3)_3 \longrightarrow Ti[CH_2C(CH_3)_3]_4$$

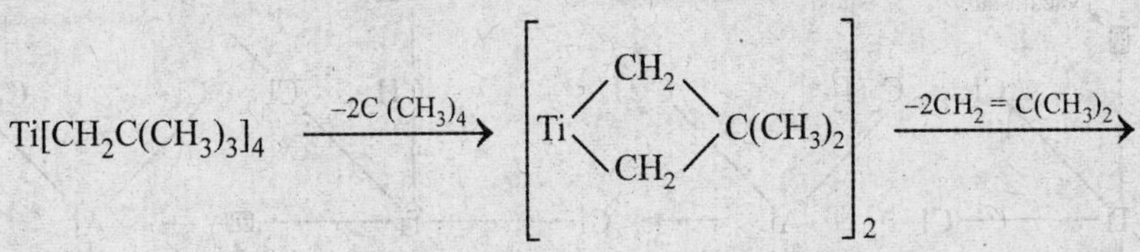

$$—CH_2—Ti—CH_2— \longrightarrow TiC$$

In this reaction, elimination of a neopentyl group is accompanied by the elimination of hydrogen at the γ-position to form titanium cyclobutane. The elimination of isobutylene gives — CH_2—Ti—CH_2—) which finally decomposes to TiC.

2. **Anticancer drugs.** Organotitanium compounds have also been found to act as anticancer drugs. Dicyclopentadienyl titanium dichloride has been used for the treatment of lymphoid leukemia. However, their anticancer activity is lower than those of *cis* platin and organotin compounds.

3. **Surface treatment.** Some titanium organic compounds especially titanium alkoxides are easily synthesised by the reaction of $TiCl_4$ with alcohols in the presence of ammonia or amines.

$$TiCl_4 + 4ROH \xrightarrow{NH_3} \underset{\text{Titanium alkoxide}}{Ti(OR)_4} + 4NH_4Cl$$

These alkoxides absorb moisture from air and get hydrolysed to form —Ti—O—Ti—O which gives transparant film. Therefore, Ti(O—i $Pr)_4$ is used for surface treatment for the prevention of damage and scratching of glass. $Ti(OBu)_4$ is used as heat resistant paint and can be used upto a temperature of 600°C.

4. **Polymerization catalysts.** Organotitanium compounds formed *in situ* from $TiCl_4$ and AlR_3 are used as polymerization catalysts for olefins as Ziegler – Natta type catalysts. The mixtures of $(C_5H_5)_2\ TiCl_2 . AlEt_2$ also catalyse polymerization reactions with same mechanism as $TiCl_3 . AlEt_2$.

Mechanism of working of Ziegler Natta catalyst

A mixture of $TiCl_4$ and triethyl aluminium in a hydrocarbon is known as **Ziegler Natta Catalyst.**

Ziegler Natta catalyst can be prepared by mixing $TiCl_4$ and Al_2Et_6 in heptane. A brown suspension is obtained which readily absorbs and polymerises ethene.

Karl Zeigler G. Natta

The mechanism of the function of the catalyst involves the following steps:

(*i*) $TiCl_4$ and trialkyl aluminium react to give fibrous form of $TiCl_3$. $TiCl_3$ consists of a hexagonal close packed array of Cl^- ions, with Ti^{3+} ions occupying two-thirds of the octahedral holes in each alternate pairs of layers.

(*ii*) The triethyl aluminium fils one of the ethyl groups in the vacant site of Ti^{3+} ion. Therefore, the catalyst has octahedrally coordinated titanium atom at a surface coordinated to four chlorine atoms with another site becoming vacant as shown.

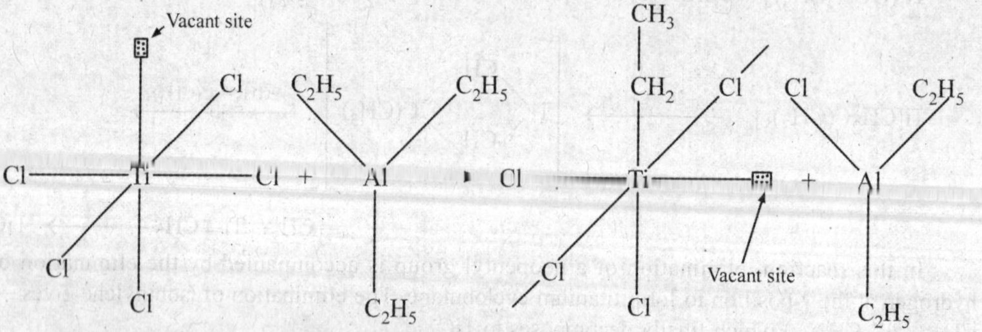

(*iii*) The double bond in alkene molecule attaches itself to the vacant site on a Ti atom on the surface of the catalyst finally forming a four centred transition state. This transition state enables the ethyl group to transfer to coordinated ethylene. As a result, the carbon chain is extended (from two to four in case of ethyl and ethene). As a result of migration, a coordinated site gets vacated.

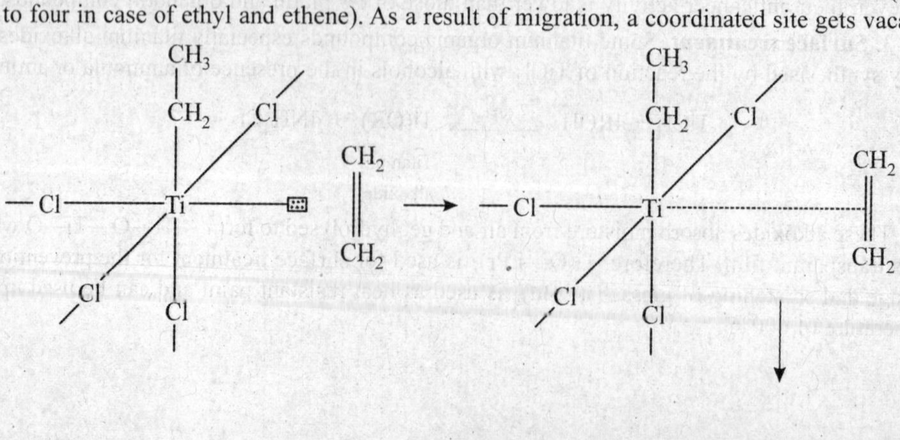

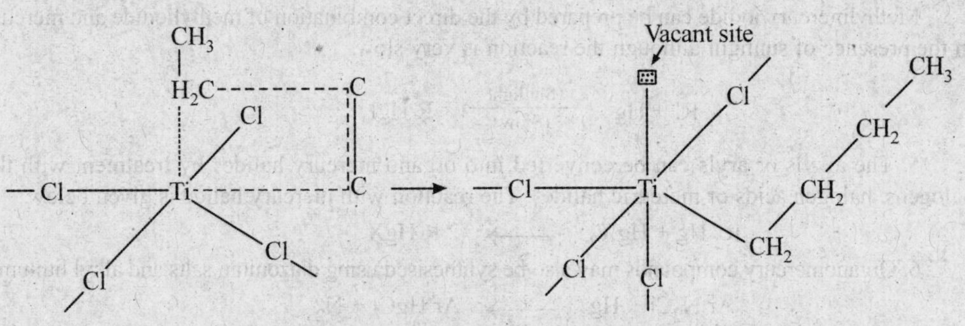

Mechanism of Zeigler Natta catalyst

Another molecule of alkene is then bound to vacant site and the process is repeated again and again as shown figure.

An advantage of the catalyst is that the polymer produced from asymmetric alkenes is **stereoregular**. Such a stereoregular polymer is stronger and has higher melting point than non-regular polymer.

The Zieglar Natta catalyst can be used for polymerisationof alkenes. It can be carried out under mild conditions at room temperature in contrast to the original methods which require high temperature and high pressure. These polymers have high melting points and are crystaline.

6.13. ORGANOMERCURY COMPOUNDS

The first organomercury compound was prepared by E. Frankland by the reaction between methyl iodide and mercury under sunlight radiations.

A large number of organomercury compounds of the types R_2Hg and monomeric R HgX have been prepared since then. These are stable at room temperature and have low reactivity so that these are not affected by air and water.

Preparation

1. Alkyl or aryl mercury compounds can be prepared from Grignard reagents.

$$RMgX + HgCl_2 \longrightarrow RHgCl + MgXCl$$
$$RHgCl + RMgX \longrightarrow HgR_2 + MgXCl$$

2. Disubstituted alkyls or aryls may be prepared by removing the mercury salt by reduction or by conversion to very stable compounds as:

$$2C_6H_5HgI + 2Na \longrightarrow (C_6H_5)_2\ Hg + Hg + 2NaI$$
$$(C_6H_5\ Hg)_2S \xrightarrow{\text{Heat}} (C_6H_5)_2\ Hg + HgS$$
$$2C_6H_5HgOAc + 2NaOH + Na_2SnO_2 \longrightarrow (C_6H_5)_2Hg + Hg + 2NaOAc + Na_2SnO_3 + H_2O$$
$$2C_2H_5\ Hg\ I + 2KCN \longrightarrow (C_6H_5)_2\ Hg + K_2HgI_2\ (CN)_2$$

3. These can be obtained by direct replacement of hydrogen by mercury.

$$RH + HgX_2 \longrightarrow RHgX + HX$$
$$Ph\ H + Hg(OAc)_2 \longrightarrow Ph\ Hg\ OAc + HOAc$$

4. Organomercury compounds are prepared by the action of sodium amalgam on alkyl halide:

$$2RX + 2Hg \xrightarrow{Na/Hg} HgR_2 + HgX_2$$

Methylmercury iodide can be prepared by the direct combination of methyliodide and mercury in the presence of sunlight although the reaction is very slow.

$$RI + Hg \xrightarrow[\text{(Slow)}]{\text{Sunlight}} R\,HgI$$

5. The alkyls or aryls can be converted into organo mercury halides by treatment with the halogens, halogen acids or mercuric halides. The reaction with mercury halide is given below.

$$R_2\,Hg + HgX_2 \longrightarrow 2\,R\,HgX$$

6. Organomercury compounds may also be synthesised using diazonium salts and alkyl lithiums.

$$\underset{\text{Diazonium salt}}{Ar\,N_2Cl + Hg} \longrightarrow Ar\,HgCl + N_2$$

$$HgCl_2 \xrightarrow{RLi} RHgCl \xrightarrow{RLi} R_2Hg$$

Properties of organomercury compounds

1. All of the alkyls except dimethyl mercury decompose on standing giving metallic mercury and hydrocarbon even at room temperature. However, these are known to be sensitive to light.

2. RHgX are crystalline solids while Hg R_2 are toxic liquids. Because of similar electronegativities of Hg and C, the Hg—C bond is essentially covalent.

Mercury alkyls are volatile liquids at room temperature and have a considerable amount of thermal stability.

3. Mercury alkyls or aryls react with halides of more electronegative elements forming new organometallic or organo metalloid compounds.

$$(C_6H_5)_2\,Hg + SiCl_4 \longrightarrow C_6H_5HgCl + C_6H_5SiCl_3$$

$$2(C_6H_5)_2\,Hg + SbCl_3 \longrightarrow (C_6H_5)_3\,SbCl_2 + C_6H_5\,HgCl + Hg$$

$$2(C_6H_5)_2\,Hg + SnCl_4 \longrightarrow 2C_6H_5\,HgCl + (C_6H_5)_2\,SnCl_2$$

$$(C_6H_5)\,Hg + 2AsCl_3 \longrightarrow HgCl_2 + 2C_6H_5\,AsCl_2$$

$$(C_6H_5)_2\,Hg + 2BCl_3 \longrightarrow HgCl_2 + 2C_6H_5\,BCl_2$$

$$3(C_6H_5)_2\,Hg + AsCl_3 \longrightarrow 3C_6H_5\,HgCl + (C_6H_5)_3\,As$$

4. Organo mercury compounds are almost inert towards air and moisture which means that mercury-carbon bond is not really cleaved by water or bases or dilute acids.

Structure and Bonding

R HgX and HgR_2 compounds contain linear R—Hg—X or R—Hg—R units due to *sp* hybridisation of mercury. In some cases, polymerization takes place to achieve this linearity. For example, o-phenylene mercury could be written as

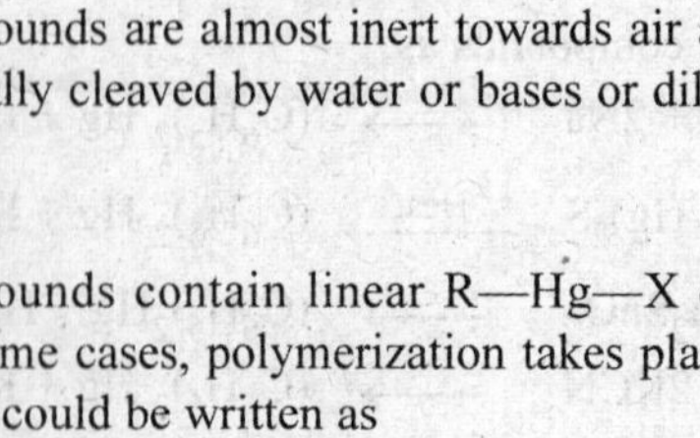

However, trimerisation takes place to maintain the linearity of C—Hg—C bond, as shown below

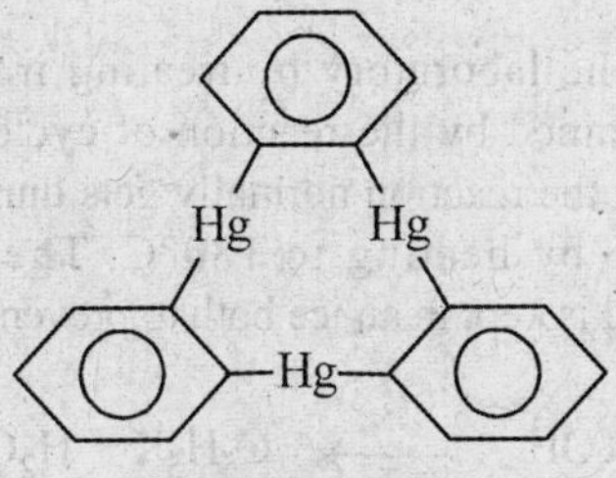

o-Phenylene trimer

Applications of organomercury compounds

1. Organomercury compounds such as dimethyl mercury $Hg(Me)_2$ or MeHgX are used for treating seeds against fungal attack

2. Merthoiolate (sodium ethyl mercury thiosalicylate), metaphor (3–hydroxy mercuric–4–nitro–o–cresol anhydride) and *mercurochrome* (merbromin 2, 7–dibromo–4 – hydroxy mercuric fluoresein) are used externally as antiseptics.

3. Phenyl mercury acetate has been used against apple scab, a fungus diseases of the leaves and fruit and also to kill crabgrass in lawns.

4. Some of the mercurials are used as the most powerful diuretics.

5. Mercury compounds are toxic to humans and animals. These can cause giddiness, lung damage and even brain damage. These have been tested against human's natural enemies.

6.14. FERROCENE*

The compound $Fe(\eta^5 - C_5H_5)_2$ *bis* (cyclopentadienyl) iron popularly known as ferrocene was discovered in 1951. It has a sandwich structure in which the metal lies between two planar cylopentadienyl rings. This compound has contributed a lot towards the understanding of organometallic chemistry.

Model of Ferrocene

Methods of preparation of ferrocene

Ferrocene is prepared by the following methods:

1. **By treating iron (I) chloride with Grignard compound.** Ferrocene can be prepared by treating iron (II) chloride with Grignard compound of cyclopentadiene.

$$FeCl_2 + 2(C_5H_5)\,MgBr \longrightarrow Fe\,(C_5H_5)_2 + MgCl_2 + MgBr_2$$

2. **By reaction of iron halide with cyclopentadiene in the presence of strong base.** Ferrocene is prepared by treating iron (II) chloride with cyclopentadiene in the presence of strong base like diethylamine. The base helps to remove HCl produced during the reaction:

$$2C_5H_6 + FeCl_3 + 2(C_2H_5)_2\,NH \longrightarrow Fe\,(C_5H_5)_2 + 2(C_2H_5)_2\,NH_2Cl$$

3. **By reaction of iron halide with sodium cyclopentadienide.** Cyclopentadiene is first converted into sodium cyclopentadienide by treating it with finely dispersed sodium in tetrahydrofuran (THF).

$$C_5H_6 + Na \xrightarrow{THF} C_5H_5\,Na + \frac{1}{2}\,H_2$$

* Not in U.G.C syllabus

Laboratory method

Ferrocene is prepared in the laboratory by treating iron (II) chloride with potassium cyclopentadienide, which is obtained by the reaction of cyclopentadiene with caustic potash. Cyclopentadiene which is used in the reaction normally gets dimersised at room temperature. It is converted to monomeric state by heating to 180°C. The cracked product (monomeric cyclopentadiene) which is obtained is kept in an ice both to prevent dimerisation again. The reactions may be represented as

$$\underset{\text{Cyclopenta diene}}{C_5H_6} + KOH \longrightarrow C_5H_5K + H_2O$$

$$2\,KC_5H_5 + FeCl_2 \longrightarrow \underset{\text{Ferrocene}}{(C_5H_5)_2\,Fe} + 2KCl$$

Properties of ferrocene

Physical properties. 1. It is an orange crystalline solid.

2. It is soluble in organic solvents like benzene but insoluble in water.

3. Its melting point is 174°C.

Chemical Properties

1. Acetylation. Ferrocene gives acetylation reactions with acetyl chloride in the presence of anhydrous $AlCl_3$. Acetic anhydride in the presence of phosphoric acid can also be used.

This reaction occurs more readily with ferrocene than with benzene. Since the reaction involves the electrophilic attack by the CH_3CO^+ group, therefore, it indicates that there is greater availability of negative charge on the π-cyclopentadienyl rings than on benzene ring.

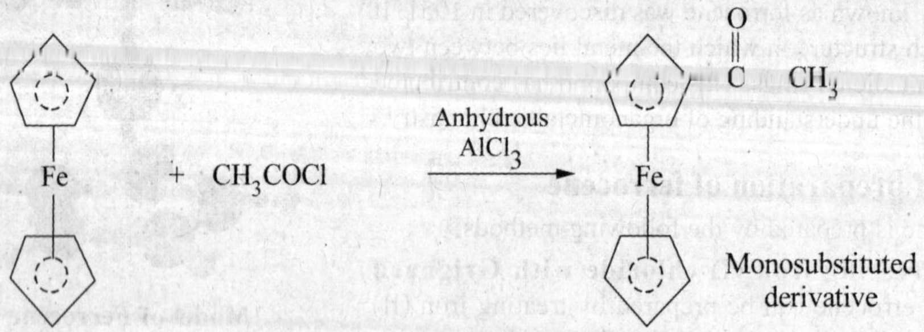

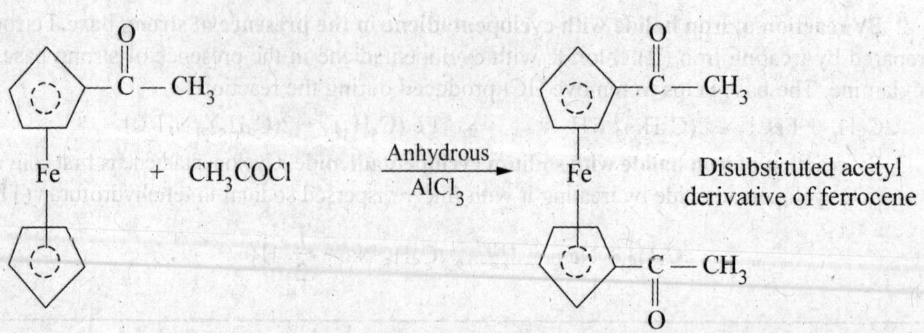

2. Friedel Crafts alkylation. Ferrocene reacts with ethylene in the presence of anhydrous $AlCl_3$ undergoing alkylation of the ring.

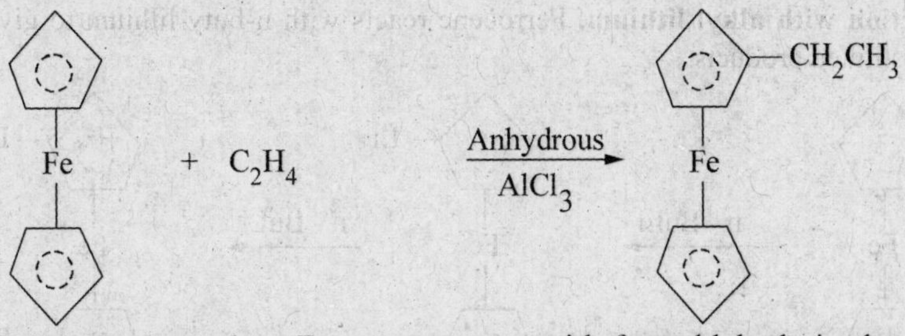

3. Mannich condensation. Ferrocene reacts with formaldehyde in the presence of dimethylamine to give mono-substituted ferrocene. This reaction is called **Mannich condensation reaction.**

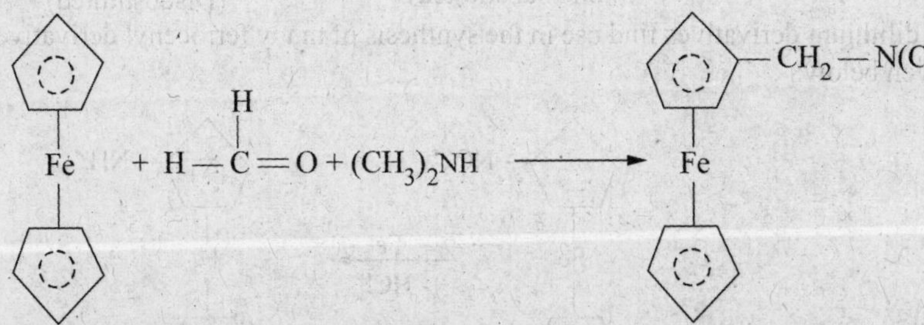

4. Sulphonation. Ferrocene reacts with chlorosulphuric acid or sulphuric acid in presence of acetic anhydride to form mono-substituted and hetero-substituted derivatives.

$$(C_5H_5)_2Fe + HOSO_3H \xrightarrow{(CH_3CO)_2O} C_5H_5 - Fe - C_5H_4.SO_3H + H_2O$$

5. Reaction with mercuric acetate. Ferrocene reacts with mercuric acetate (HgOAc) to give mono and disubstituted derivatives.

The substituted derivatives of mercuric acetate react with iodide to give iodine derivatives. This provides an indirect method for carrying out iodination of ferrocene.

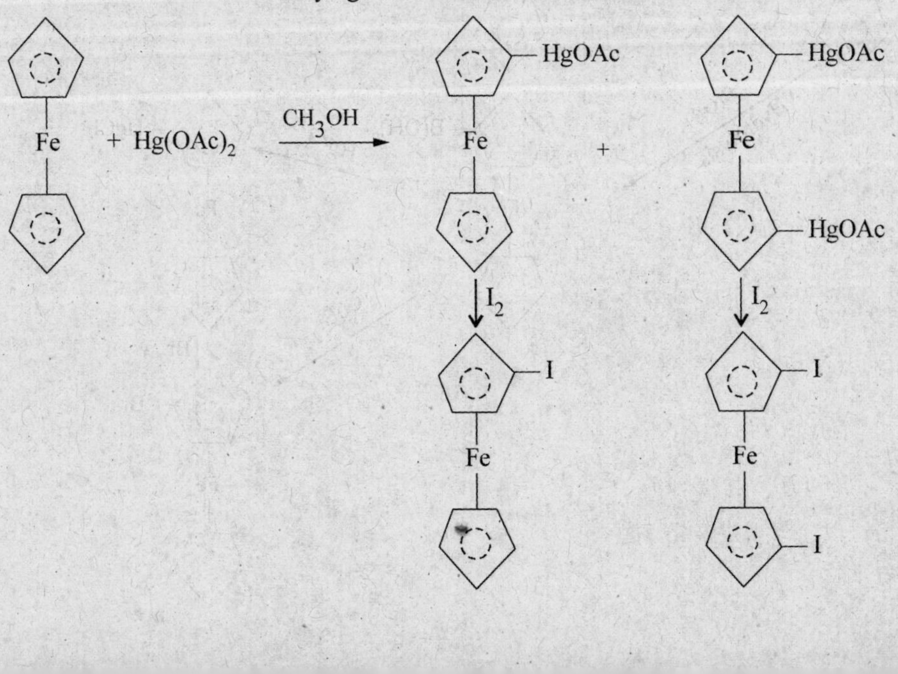

6. Nitration Ferrocene does not undergo direct nitration and halogenation reactions because it gets oxidised to ferricinium ion, Fe $(C_2H_5)_2^+$.

7. Reaction with alkyl lithium. Ferrocene reacts with n-buty lithium to give mono and dilithium substituted products.

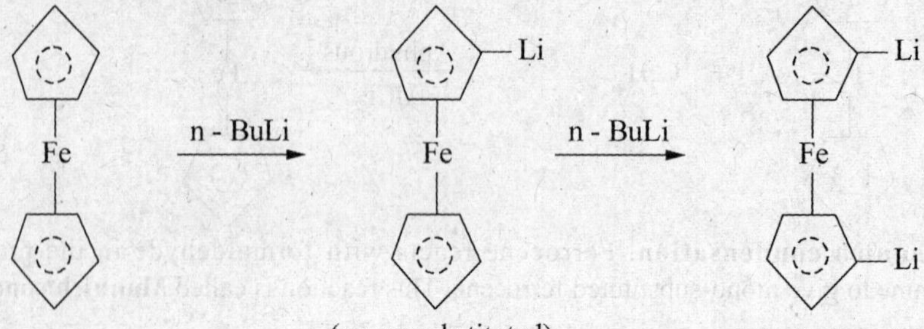

(mono substituted) (Disubstituted)

The mono and dilithium derivatives find use in the synthesis of many ferrocenyl derivatives. Some of these are given below:

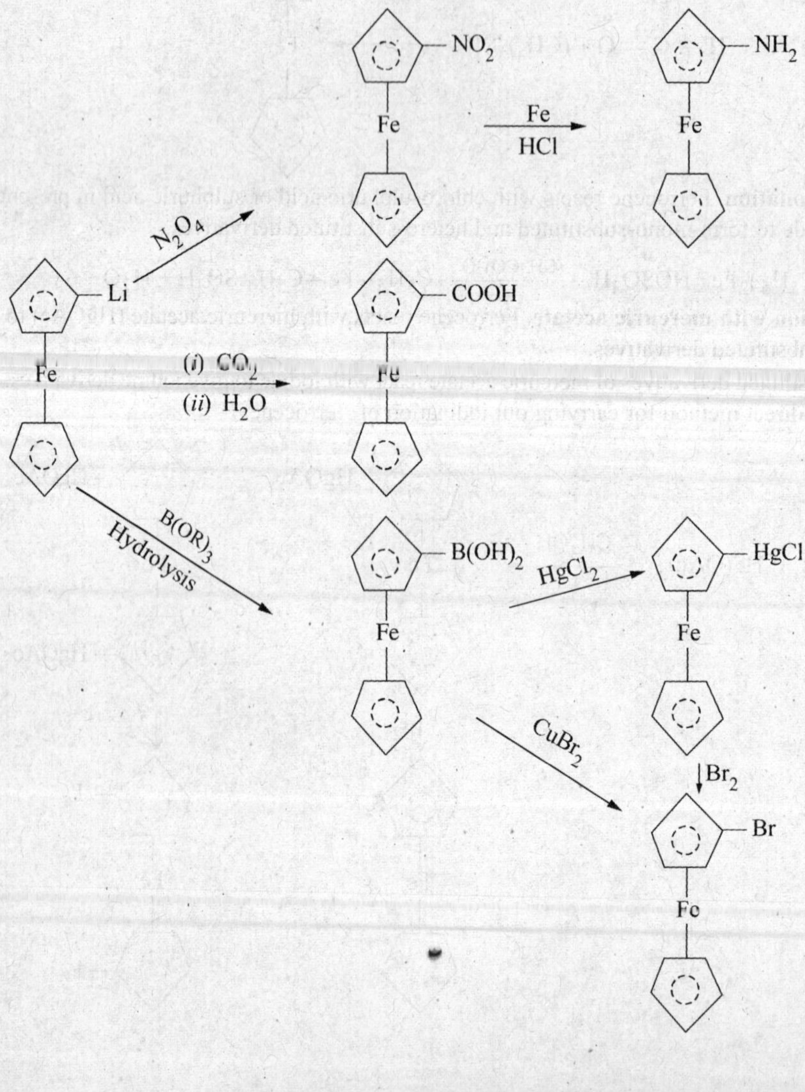

STRUCTURE OF FERROCENE

Ferrocene has a structure in which the iron atom is sandwiched between two C_5H_5 organic rings. The planes of the rings are parallel so that all the carbon atoms are at the same distance from the iron atom.

X-ray diffraction studies reveal that ferrocene has two types of arrangements of the cyclopentadiene rings known as **staggered** or **eclipsed**. It has been observed that in gas phase the structure of ferrocene is eclipsed while at low temperatures the structure of ferrocene is staggered. *The staggered configuration in the solid phase exists because of the crystal packing forces so that carbon-carbon and hydrogen-hydrogen repulsions between the two rings are minimum.* The energy difference between staggered and eclipsed ferrocene configurations is only 4 kJ mol^{-1}.

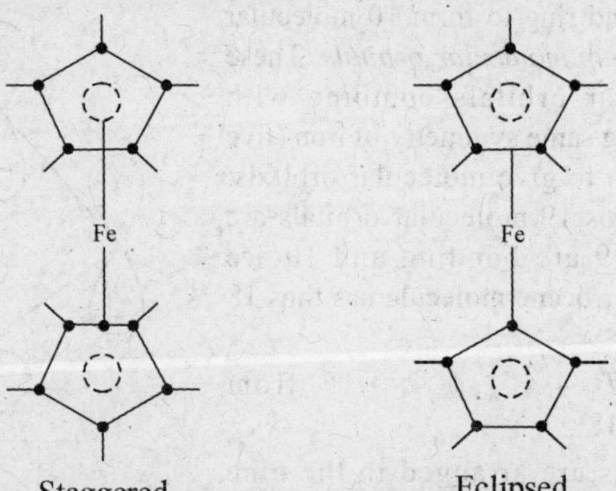

The bonding in ferrocene can be explained on the basis of valence bond theory and molecular orbital theory as discussed below:

Valence bond theory. Valence bond theory does not provide satisfactory description of ferrocene molecule, but it explains the diamagnetic character of a stable 18 electron outer configuration of the molecule. The outer electronic configuration of Fe (II) is $3d^6$ having six electrons in the $3d$ orbitals [Fig. 6.1 (*a*)]. To make available two *d*-orbitals for accommodating electron pairs from the ligands, the six electrons get paired up in three orbitals [Fig. 6.1 (*b*)]. The six vacant orbitals (two $3d$, one $4s$ and three $4p$) get hybridised forming six d^2sp^3 hybrid orbitals which accept 12 electrons; 6 electrons from each C_5H_5 ring. Thus, the molecule is diamagnetic as observed because there is no unpaired electron.

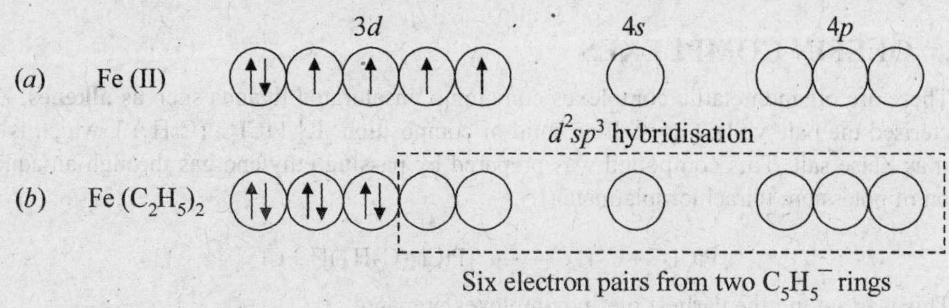

Fig. 6.1. Valence bond description for ferrocene.

Molecular orbitals theory. The molecular orbital theory better explains the bonding in ferrocene.

The basic bond in ferrocene is a single bond between the π-orbitals of C_5H_5 ring and the iron atom. The bond is formed by the overlap of d_{xz} and d_{yz} orbitals on Fe (the z-axis is taken as the molecular axis) with the delocalized $p\pi$ aromatic orbital from each pentadienyl ring.

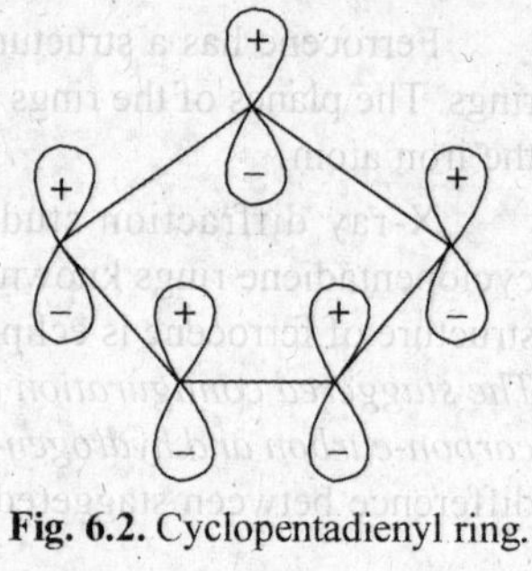

Fig. 6.2. Cyclopentadienyl ring.

The carbon atom involves sp^2 hybridisation in pentadienyl ring and forms three σ bonds (one with H atom and two with neighbouring C atoms(Fig. 6.2). As a result, one $p\ \pi$ orbital is left on each carbon atom containing one electron each. These five atomic orbitals now combine with one another to give five molecular orbitals. These five MOs of σ ring can combine with similar MOs of the second ring to form 10 molecular orbitals called *ligand group molecular orbitals.* These ligand group molecular orbitals combine with appropriate orbitals having same symmetry of iron (five $3d$, one $4s$ and three $4p$) to give molecular orbitals. During their combinations 19 molecular orbitals are formed, out of which 9 are bonding and 10 are antibonding MOs. The ferrocene molecule has thus 18 electrons as:

8 from iron ($3d^6\ 4s^2$) + 2 × 5 from cyclopentadiene rings = 18

These 18 electrons are arranged in the nine bonding and non-bonding molecular orbitals. As a result the 10 anti-bonding MOs remain vacant.

In the ferrocene molecule, the most significant bonds are the bonds of π symmetry formed between Fe atom and the two C_5H_5 rings. These are obtained by the overlap of d_{xz} and d_{yz} atomic orbitals of Fe with appropriate ligand group molecular orbitals of cyclopentadiene rings. For example, the formation of a π-bond between d_{xz} orbital of Fe and the appropriate ligand group orbital is shown in Fig. 6.3. The interaction of the rings with d_{yz} orbitals is similar.

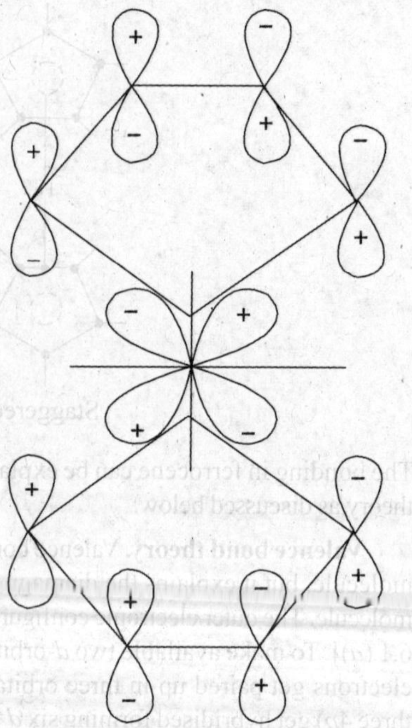

Fig. 6.3. The interaction between the ring orbitals and the d_{xz} orbital of the ion.

6.15. OLEFIN COMPLEXES

These are organometallic complexes containing unsaturated ligands such as **alkenes.** Zeise characterised the pale yellow crystalline solid of composition K[$PtCl_3$. (C_2H_4)], which is now known as Zeise salt. This compound was prepared by passing ethylene gas through an aqueous solution of potassium tetrachloroplatinate (II).

$$PtCl_4^{2-} + C_2H_4 \longrightarrow [PtCl_3(C_2H_4)]^- + Cl^-$$

This was among the earliest olefin complexes prepared.

It was later discovered that certain metal halides or ions formed complexes when treated with a variety of olefins.

Structure and bonding in metal olefin complexes

The structure of the anion of the Zeise salt was properly established only in 1975. The significant features of the structure are:

(*i*) The three Cl atoms and the middle point of the ethylene double bond form a square plane (Fig. 6.4).

(*ii*) The platinum atom is present in the centre of the square and C = C double bond of the ethylene molecule is perpendicular to the plane containing Pt and three Cl atoms.

(*iii*) The C_2H_4 group is significantly distorted from planarity.

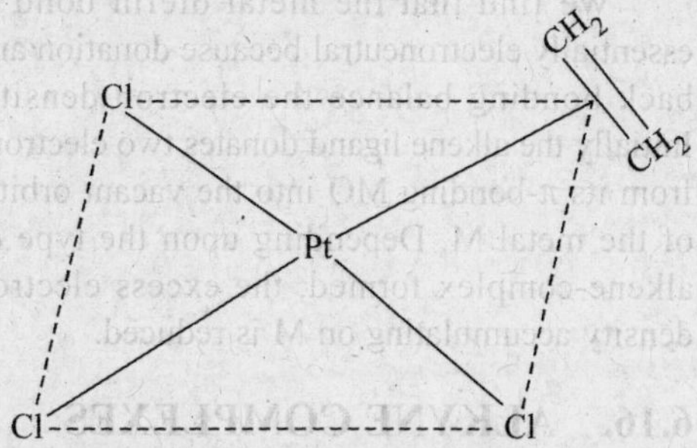

Fig. 6.4. Structure of $[PtCl_3(C_2H_4)]^-$ complex

The metal atom has filled (shown shaded) and empty (shown unshaded) orbitals. The bonding occurs in two ways:

(*i*) *The filled π–molecular orbitals of the ethylene molecule are donated to the suitably direct vacant orbitals of the metal to form alkene* → metal sigma bond. This is shown in Fig. 6.5 (*a*).

(*ii*) *A back bond is formed by the overlap of the filled metal hybrid orbitals with the vacant π MO of the olefin.* This leads to a π-bond. The back bonding will strengthen the Pt-olefin bond but it should weaken the C — C bond in alkene.

This type of bonding is shown in Fig. 6.5 (*b*).

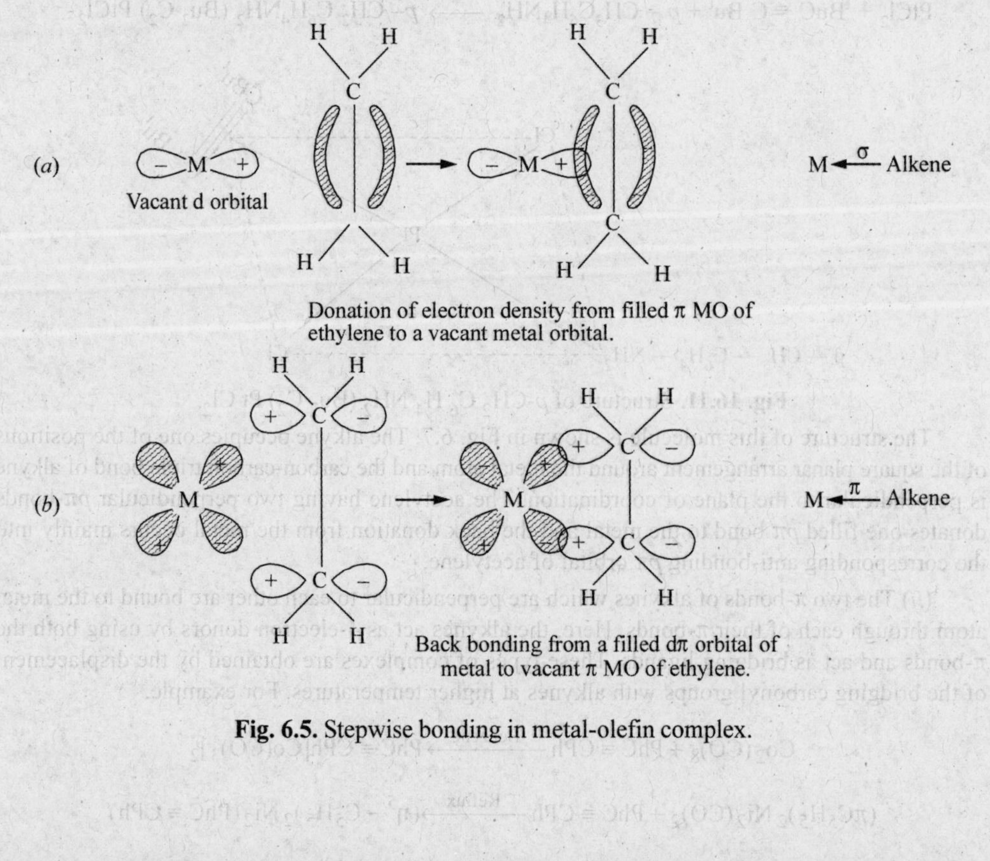

Fig. 6.5. Stepwise bonding in metal-olefin complex.

The complete bonding picture is shown in Fig. 6.6.

We find that the metal olefin bond is essentially electroneutral because donation and back bonding balance the electron density. Initially the alkene ligand donates two electrons from its π-bonding MO into the vacant orbital of the metal M. Depending upon the type of alkene-complex formed, the excess electron density accumulating on M is reduced.

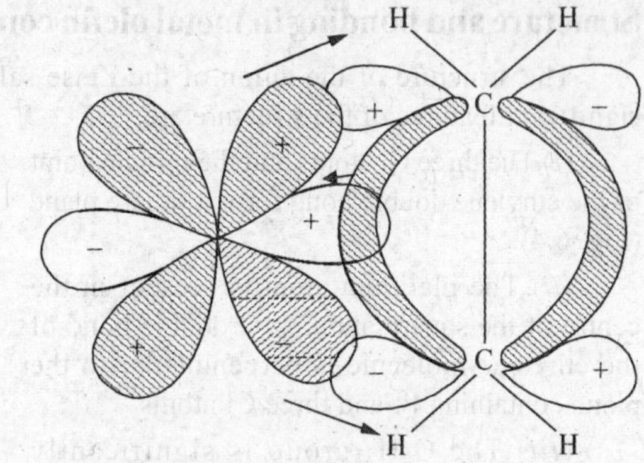

Fig. 6.6. Bonding in metal olefin complex.

6.16. ALKYNE COMPLEXES

Organometallic compounds containing metal atom linked to alkynes are called alkyne complexes. It has been observed that alkynes form π–complexes with a variety of transition metals. In alkynes, there are two π bonds at right angles to each other and a metal atom can be bound to each. Some of the important types of metal-alkyne complexes are considered below:

(*i*) The alkyne may coordinate to only one metal atom and function like alkenes. In these molecules, alkyne molecule occupies one of the positions like that of ethylene in Zeise salt. The carbon-carbon triple bond of the alkyne is perpendicular to the plane of coordination. These compounds may be prepared as shown below

$$PtCl_2 + {}^tBuC \equiv C\,Bu^t + p - CH_3C_6H_4NH_2 \longrightarrow p\text{–}\ CH_2\ C_6H_4NH_2\ (Bu_2\ C_2)\ PtCl_2$$

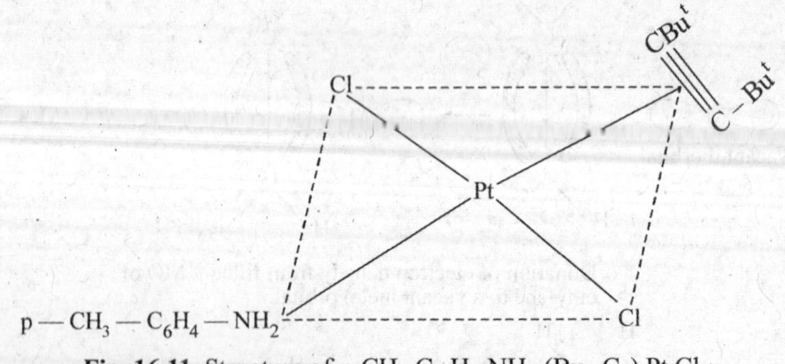

Fig. 16.11. Structure of p-$CH_2\ C_6\ H_4\ NH_2\ (Bu_2\ C_2)\ Pt\ Cl_2$.

The structure of this molecule is shown in Fig. 6.7. The alkyne occupies one of the positions of the square planar arrangement around the metal atom and the carbon-carbon triple bond of alkyne is perpendicular to the plane of coordination. The acetylene having two perpendicular *p*π-bonds donates one filled *p*π bond to the metal and the back donation from the metal occurs mainly into the corresponding anti-bonding *p*π orbital of acetylene.

(*ii*) The two π-bonds of alkynes which are perpendicular to each other are bound to the metal atom through each of their π-bonds. Here, the alkynes act as 4-electron donors by using both the π-bonds and act as bridging ligands. These types of complexes are obtained by the displacement of the bridging carbonyl groups with alkynes at higher temperatures. For example,

$$Co_2(CO)_8 + PhC \equiv CPh \xrightarrow{\text{Benzene}} PhC \equiv CPh[Co(CO)_3]_2$$

$$(\pi C_5H_5)_2\,Ni_2(CO)_2 + PhC \equiv CPh \xrightarrow{\text{Reflux}} (\eta^5 - C_5H_5)_2\,Ni_2\,(PhC \equiv CPh)$$

Fig. 6.8 gives the structures of these complexes. In these structures, the two acetylenic carbons are kept between the two cobalt atoms so that two π-bonds of acetylene, which are perpendicular to each other, interact with a cobalt atom.

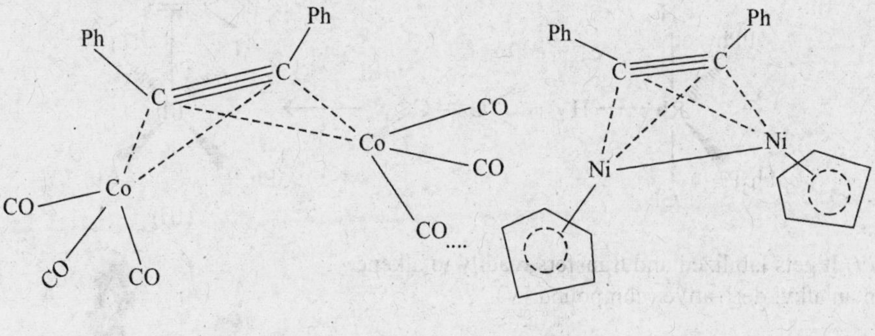

Fig. 6.8. Metal-alkyne complexes.

6.17. HOMOGENEOUS HYDROGENATION

By homogeneous catalyst we mean a catalyst which exists in the same phase as the reacting substances. However, catalysts miscible in the reaction medium come under the same category. Such catalysts have the advantage that they operate at moderate temperatures and pressures. In recent times, a number of catalysts have been developed for the hydrogenation of alkenes, alkynes and other unsatured aromatic and heterocyclic compounds. These are discussed as under.

Geoffrey Wilkinson

1. Wilkinson's catalyst (cis-dihydrido catalyst)

Wilkinson's catalyst, $RhCl\ (PPh_3)_3$ was the first effective homogenous catalyst used for hydrogenation of alkenes at near atmospheric temperatures or pressures.

$$>C=C< \; + \; H_2 \xrightarrow{\text{Catalyst}} -\underset{H}{\overset{|}{\underset{|}{C}}}-\underset{H}{\overset{|}{\underset{|}{C}}}-$$

It is red violet compound having square planar geometry.

The mechanism of the reaction may be understood as follows:

(*i*) In the presence of hydrogen, $RhCl\ (PPh_3)_3$ solution forms octahedral dihydride compound. Because of strong trans effect of H, it readily dissociates a molecule of Ph_3P at room temperature to give a five coordinate rhodium (III) species (compound II).

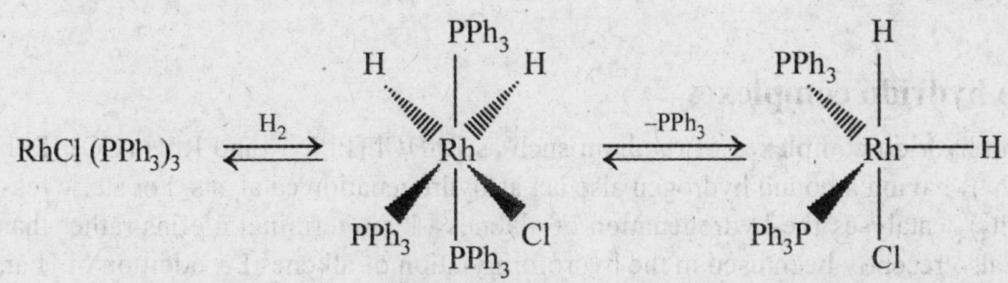

Octahedral dihydride (I)

Five co-ordinated rhodium (III) (II)

(*ii*) The species (II) combines with alkene molecule forming compound which has *cis*-phosphines and ahydride *trans* to PPh_3 compound (III). This is a favoured product on steric considerations.

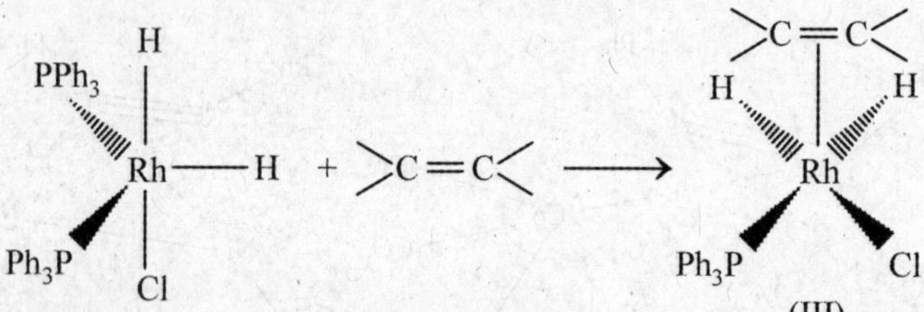

(III)

(*iii*) It gets labilized and transfers readily to alkene to form an alkyl derivative (compound IV).

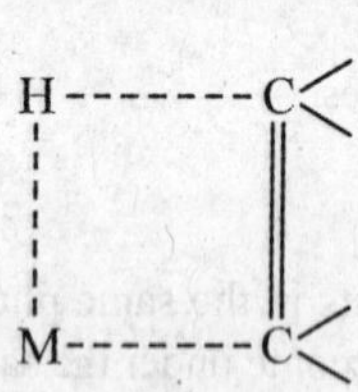

(IV)

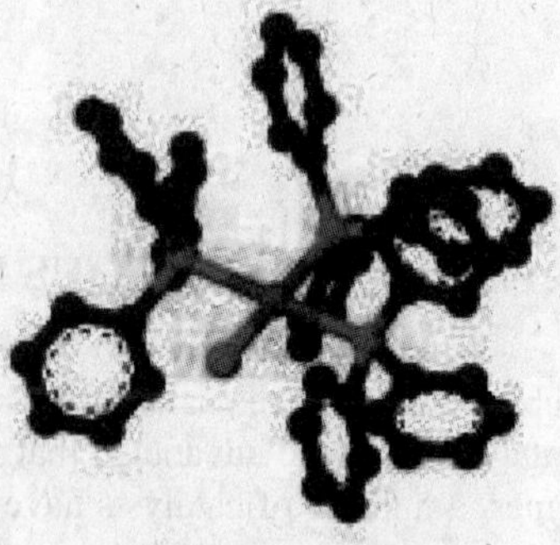

Model of Wilkinson catalyst

A second transfer via a three centre transition state i.e. reductive elimination gives alkane as shown below.

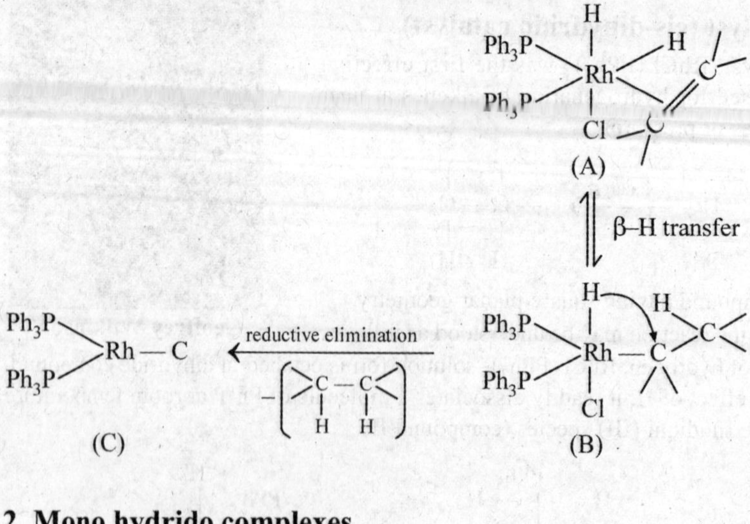

2. Mono hydrido complexes

Mono hydrido complexes of rhodium such as RhHCl $(PPh_3)_3$ and RhH (CO) (PPh_3) and Ir HCO $(PPh_3)_3$ having a bound hydrogen also act as hydrogenation catalysts. For steric reasons RhH (CO) $(PPh_3)_3$ catalyses the hydrogenation of alkenes –1 *i.e.* terminal olefins rather than alkenes –2. It has also recently been used in the hydroformylation of alkenes *i.e.* addition of H and the the formyl group (CHO). The process is known as *oxo process*.

The complete cycle for catalystic hydrogenation and isomerization of alkenes by RhHCO $(PPh_3)_3$ at 25°C and 1 atm pressure is shown below.

The complex undergoes easier dissociation of PPh_3 ligand to form four coordinated intermediate compound (II). The compound (II) then coordinates to an alkene (R—CH = CH_2) to form compound (III). The coordinated 1–alkene in (III) can react with a metal —H bond either by Markownikoff addition or to form a secondary brranched alkyl derivative (IVb) or by anti-Markownikoff addition to form a primary straight chain alkyl derivative (IVa). It may be noted that an alkene–2 can give only a branched chain alkyl.

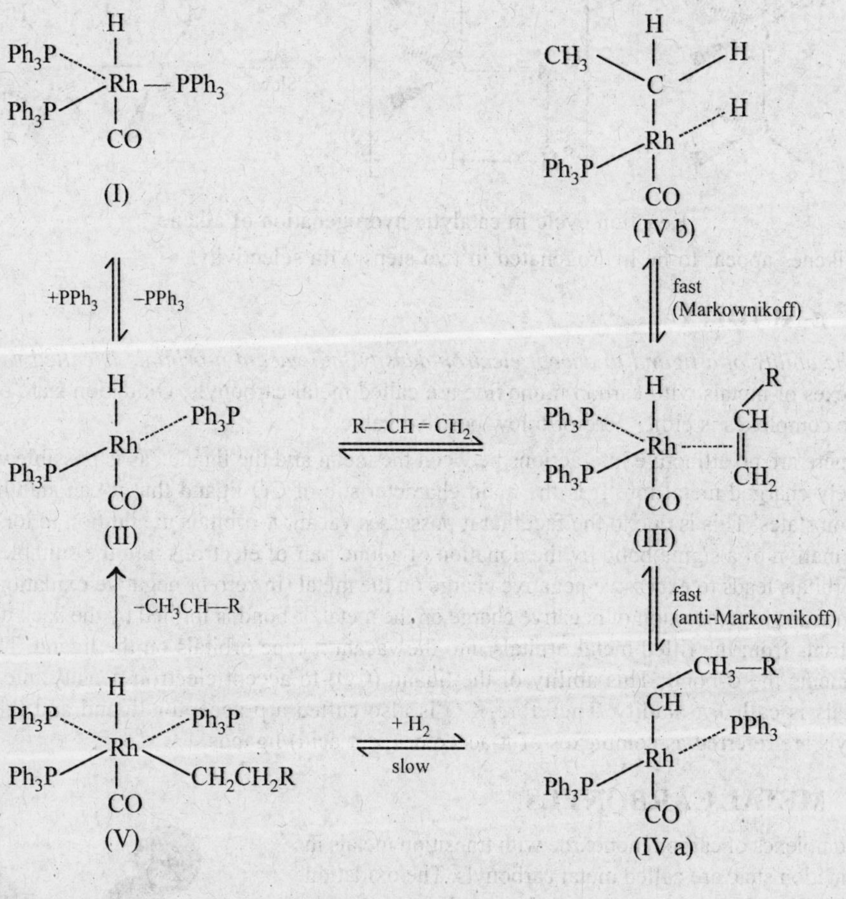

3. Hydrogenation by *f*-block organometallic complexes.

The lanthanide and actinides form some organometallic complexes such as $(Cp_2LuH)_2$, $(Cp_2 Th H_2)_2$ (where Cp = pentamethylcyclopentadiene group, C_5Me_5) which are very active for hydrogenation of alkenes and alkynes. These are about 100 times more reactive. The process proceeds by dissociation to monomer.

The monomer then interacts with alkene undergoing insertion to form an alkyl which undergoes hydrogenolysis. The reaction cycle is shown as under.

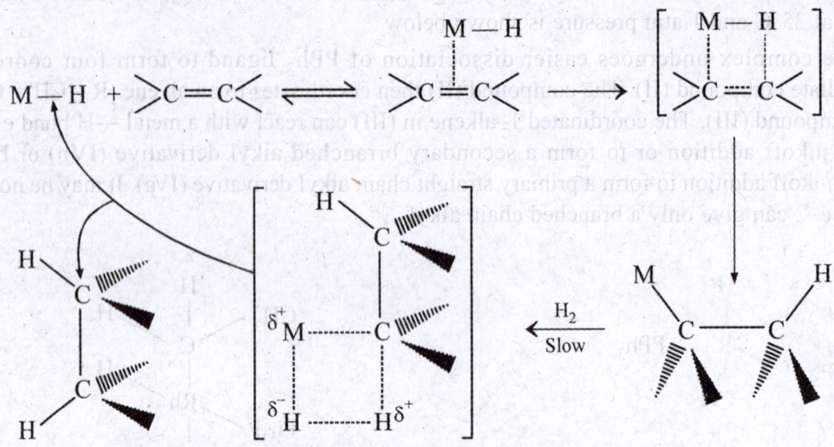

Reaction cycle in catalytic hydrogenation of alkene

Alkenes appear to be hydrogenated in two steps with selectivity.

6.18. π-ACIDITY

The ability of a ligand to accept electron density into vacant p-orbitals is called π-acidity. Complexes of metals with carbon monoxide are called metal carbonyls. Oxidation state of metal in these complexes is either zero or a low positive value.

There are no attractive interactions between the metal and the ligands as is possible with the positively charged metal ion. It is the main characteristic of CO ligand that it can stabilize low oxidation states. This is due to the fact that it possesses vacant π-orbitals in addition to lone pairs. The formation of a sigma bond by the donation of a lone pair of electrons into the suitable vacant metal orbitals leads to excessive negative charge on the metal (in zero or negative oxidation state). To counter the accumulation of negative charge on the metal, π-bond is formed by the back donation of electrons from the filled metal orbitals into the vacant π-type orbitals on the ligand. This also supplements the σ-bond. This ability of the ligand (CO) to accept electron density into vacant π-orbitals is called π-acidity. Therefore, CO is also called a p-acceptor ligand and the metal carbonyls are referred as complexes of π-acceptor (or π acid) ligands.

6.19. METAL CARBONYLS

Complexes of carbon monoxide with transition metals in low oxidation state are called metal carbonyls. The oxidation state of the metals in such complexes is zero, low positive or negative. It is interesting to note that although carbon monoxide is a weak Lewis base, yet it forms a strong bond to the metal. Carbonyl complexes can be mononuclear, binuclear or polynuclear.

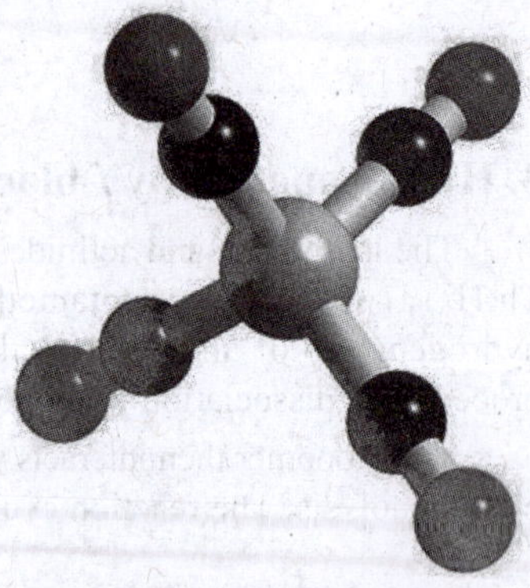

Model of a transition metal carbonyl

Methods of preparation of metal carbonyls

Some of the methods of preparation of metal carbonyls are described as under:

1. Direct synthesis. The metal carbonyls can be prepared by the direct reaction of carbon monoxide with the finely divided metal. For example, nickel carbonyl is prepared by the direct reaction of finely divided nickel metal with CO at 25°C under one atmospheric pressure.

$$Ni + 4CO \xrightarrow{25°C} Ni(CO)_4$$

Some other metal carbonyls require high temperatures and pressures.

For example,

$$Fe + 5CO \xrightarrow[100\ atm]{200°C} Fe(CO)_5$$

$$Mo + 6CO \xrightarrow[250\ atm]{200°C} Mo(CO)_6$$

2. Reduction. The metal salts are reduced in the presence of CO. For example,

$$CrCl_3 + Al + 6CO \xrightarrow[C_6H_6]{AlCl_3} Cr(CO)_6 + AlCl_3$$

$$2CoCO_3 + 8CO + 2H_2 \xrightarrow[250\ atm]{150°C} Co_2(CO)_8 + 2CO_2 + 2H_2O$$

$$WCl_6 + 3Zn + 6CO \xrightarrow[120°C]{AlCl_3} W(CO)_6 + 3ZnCl_2$$

$$OsO_4 + 9CO \xrightarrow{250°C,\ 350\ atm} Os(CO)_5 + 4\ CO_2$$

3. Photolysis or thermolysis. Higher carbonyls can be synthesised by photolysis or thermolysis of lower metal carbonyls.

$$2Fe(CO)_5 \xrightarrow{hv} Fe_2(CO)_9 + CO$$

$$3Os(CO)_5 \xrightarrow{hv} Os_3(CO)_{12} + 3CO$$

Chemical properties of metal carbonyls

1. Displacement reactions. Metal carbonyls undergo displacement reactions in which one or more CO ligands may be displaced in contact with electron donors (Lewis bases) such as pyridine, phosphines (R_3P), isocyanides (RNC), etc. One molecule of CO is replaced by one electron pair donor. A six electron donor (like benzene) could displace three CO molecules. Some of the common displacement reactions are:

$$Fe\ (CO)_5 + Ph_3\ P \longrightarrow Fe(CO)_4\ PPh_3 + CO$$

$$Fe\ (CO)_5 + 2Ph_3\ P \longrightarrow Fe\ (CO)_3\ (PPh_3\ )_2 + 2CO$$

$$Cr(CO)_6 + C_6H_6 \xrightarrow{-3CO} (C_6H_6)\ Cr\ (CO)_3$$

2. Formation of carbonyl anionic complexes. The metal carbonyls can be converted into anions known as carbonylate ions by the following methods:

(*a*) Metal carbonyls react with bases to give carbonylate ion

$$Fe\ (CO)_5 + OH^- \longrightarrow [Fe\ (CO)_4\ ]^{2-} + H^+ + CO_2$$

$$Fe_2\ (CO)_9 \xrightarrow[Et_4\ NI]{KOH,\ MeOH} (Et_4N)_2\ [Fe_2\ (CO)_8]$$

(*b*) Reduction of metal carbonyls with reducing agents produce carbonylate anions. The reduction is carried out by alkali metal amalgams, hydride reagents and Na/K alloy in basic solvents including liquid ammonia.

$$Co_2(CO)_8 \xrightarrow[THF]{Na/Hg} Na[Co(CO)_4]$$

$$Cr(CO)_6 \xrightarrow[THF]{Na/Hg} Na_2[Cr(CO)_5]$$

$$Mn_2(CO)_8 \xrightarrow[THF]{KH} 2K[Mn(CO)_5]$$

(*c*) Many substituted carbonyl anions can be obtained by displacement of carbon monoxide from a metal carbonyl.

$$Me_4\,NI + Mo\,(CO)_6 \longrightarrow Me_4N\,[Mo\,(CO)_5\,I] + CO$$

$$Mo\,(CO)_6 + NaB_3H_8 \longrightarrow Na[Mo(CO)_5\,B_3H_8] + CO$$

(*iv*) Metal carbonylate anions can be prepared by the reaction with CH_3 or C_6H_5 radicals.

$$(CO)_5\,Cr\,(CO) + CH_3^- \longrightarrow \left[(CO)_5\,Cr - C{\overset{O}{\lt}}_{CH_3}\right]^- \xrightarrow[-N_2]{H^+,\,CH_2N_2} (CO_5)\,Cr = C{\overset{OCH_3}{\lt}}_{CH_3}$$

3. Formation of cationic carbonyl complexes. These may be obtained in the formation of complex halides by the reaction of carbon monoxide and a Lewis acid such as $AlCl_3$ or BF_3. For example,

$$Mn\,(CO)_5\,Cl + CO + AlCl_3 \longrightarrow [Mn\,(CO)_6^+\,[AlCl_4]^-$$

$$Re\,(CO)_5\,Cl + AlCl_3 + CO \longrightarrow [Re\,(CO)_6\,][AlCl_4]^-$$

6.20. BONDING IN METAL CARBONYLS

Bonding in linear metal carbonyls (M – C – O)

Carbon monoxide has a triple bond with lone pair of electrons on both C and oxygen atom as

$$:C \equiv O:$$

(*i*) There is a dative overlap of the filled orbital of carbon (of CO) and suitable empty orbital of the metal forming a dative sigma bond (M ← CO). This is shown in Fig. 6.9 (*a*)

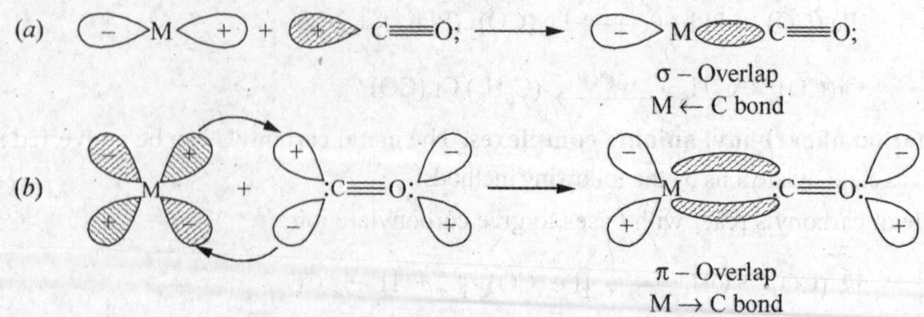

Fig. 6.9. Bonding in metal carbonyls.

(*ii*) There is a π-overlap involving donation of electrons from filled metal *d*-orbitals into vacant antibonding π molecular orbitals. This results into the formation of M → CO bond. This is also called *back donation or back bonding*. Fig. 6.9 (*b*)

The bonding in metal carbonyls is shown in Fig. 6.9. In these figures, the shaded orbitals represent filled orbitals.

The formation of σ dative bond tends to increase the electron density on the metal atom. At the same time, the formation of π bond from metal to carbon tends to decrease the electron density on metal. Then the formation of π bond increases the strength of M—CO bond. This accounts for the fact that CO is a very weak Lewis base towards non-transition metal halides like BX_3, AlX_3, etc., but forms very strong complexes with transition metals.

Evidence in support of bonding

1. The formation of back bonding from metal to CO molecule results in decrease in electron density on metal. This is supported by the dipole moment studies. It has been observed that the dipole moment of M—C bond is only very low, about 0.5 D.

2. The back bonding from metal to CO is expected to increase M—C bond strength with corresponding weakening of C ≡ O. This is due to the fact that electrons from back bonding fill the antibonding MOs of CO. As a result, the bonding ability of CO will decrease. Therefore, as the M—C bond becomes stronger, the C ≡ O bond becomes weaker. Therefore, the multiple bonding should be evidenced by the shorter M—C bond as compared to M—C single bonds and longer C—O bonds as compared to normal C ≡ O triple bonds. It has been observed that the bond length in CO molecule is 1.128 Å and in metal carbonyls, it is of the order of 1.15 Å.

Vibrational spectra of metal carbonyls

Infrared spectroscopy has given valuable support to study bonding in metal carbonyls. These studies provide information regarding bond orders of M—C and C ≡ O bonds. The decrease in C—O bond order or force constant is estimated by studying the CO stretching frequency in infrared spectroscopy.

The infrared spectra is characterised by frequency of vibration, which is related to force constant (k) as

$$\nu = \frac{1}{2\pi}\sqrt{\frac{k}{\mu}}$$

where μ is reduced mass of bonded atoms with mass m_1 and m_2 and is given as

$$\mu = \frac{m_1 m_2}{m_1 + m_2}$$

The force constant is a measure of bond strength. Larger the force constant, stronger is the bond and higher is the vibrational frequency. For example, a triple bond has higher force constant and therefore higher frequency than a double bond (C ≡ C : 2200 cm^{-1} and C = C : 1650 cm^{-1}). In infrared spectrum the CO frequencies are generally very strong and, therefore, can be used to study the force constant or bond strength of CO bond in different bonding situations.

Free CO has an infrared stretching frequency of 2143 cm^{-1}. In the case of terminal carbonyl groups in neutral molecules, the CO frequency has been found to be 2125-1850 cm^{-1}. This suggests that the bond strength of CO has decreased or M—C bond strength has increased. This is only possible if there is back bonding from filled metal *d*-orbitals to the antibonding orbitals of CO.

Structure of mononuclear, binuclear and polynuclear metal carbonyls

Mononuclear metal carbonyls. The mononuclear metal carbonyls have all terminal CO groups. The shapes are those expected from their formulae. Hexacarbonyls like $Cr(CO)_6$ have regular octahedral structures. Tetracarbonyls like $Ni(CO)_4$ have tetrahedral structure. The pentacarbonyls like $Fe(CO)_5$, $Os(CO)_5$ have trigonal bipyramidal shape. The shapes of some mononuclear carbonyls are given below.

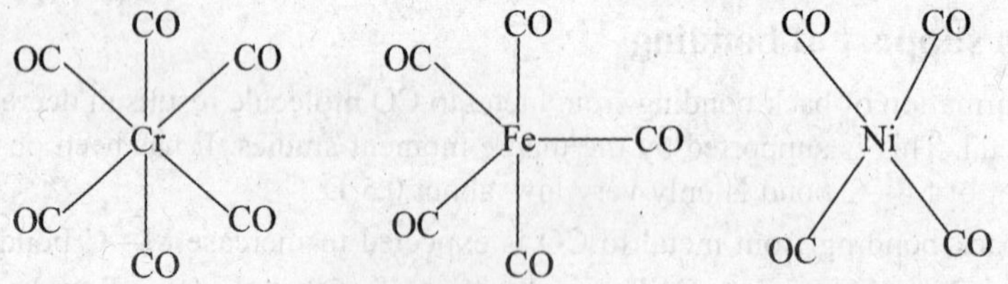

Structures of some mononuclear carbonyls

Binuclear metal carbonyls

The binuclear metal carbonyls have different types of structures. They show metal-metal bonding and bridging CO groups in their molecules. Their structures are described below:

(*i*) The binuclear metal carbonyls, $M_2(CO)_{10}$ where M = Mn, Ti or Re have metal-metal bonding in their structures. Each metal has octahedral arrangement of CO groups. The structure of $Mn_2(CO)_{10}$ is shown below.

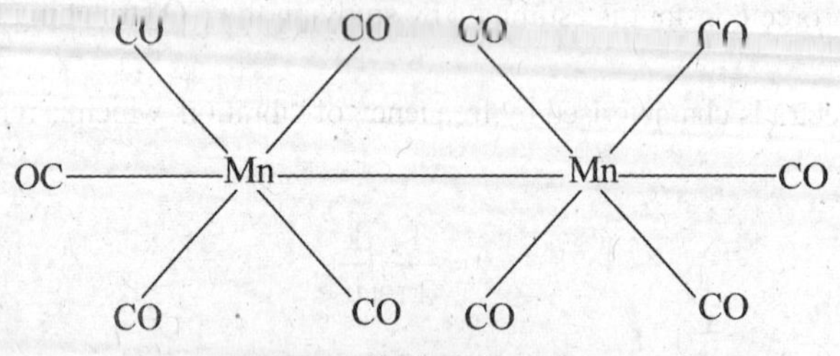

Structure of $Mn_2(CO)_{10}$

(*ii*) The binuclear carbonyls such as $Fe_2(CO)_9$, $Os_2(CO)_9$ and $Co_2(CO)_8$ have bridging carbonyl groups in addition to metal-metal bonds. $Fe_2(CO)_9$ has three bridging CO groups placed symmetrically around the Fe—Fe bond. Each Fe atom is also bonded to three terminal CO groups.

$Os_2(CO)_9$ has only one bridging carbonyl group and each Os atom has four terminal CO groups.

We find that in addition to terminal CO groups, there are some bridging carbonyl groups in dinuclear complexes. The bridging CO groups are symmetrical and have equal M—C distances.

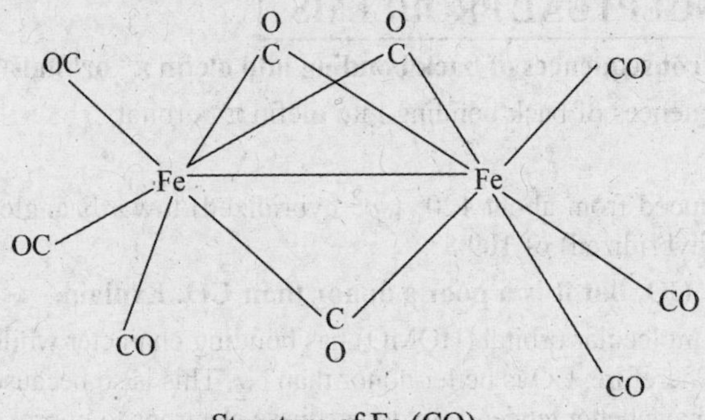

Structure of Fe $(CO)_4$

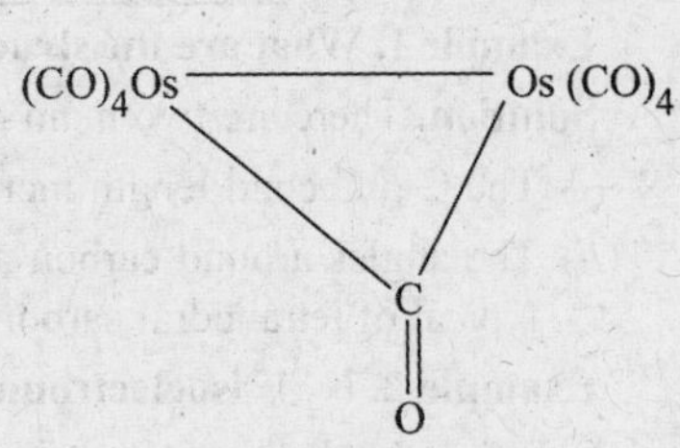

Structure of Os $(CO)_9$

The bonding in bridging carbonyl groups may be regarded as 2-electron 3-centred overlap. It may be noted that the bridge occurs over metal-metal bond. The carbon monoxide bridges in the absence of metal-metal bonds are unstable. Thus, the metal-metal bond is essential for the stability of the bridges. The stability to CO bridges depends upon the size of the metal atom. If the metal atoms are larger in size, the bridged structures become unstable relative to unbridged structures. Therefore, the relative stability of non-bridged structures increases as the size of the metal atom increases. For example, the smaller Fe atom in $Fe_2(CO)_9$ has 3 bridging CO groups whereas bigger osmium in $Os_2(CO)_9$ has one bridging CO group.

Polynuclear metal carbonyls

The structures of polynuclear metal carbonyls are more complex. They contain metal-metal bonds and terminal and bridging carbonyl groups. The structures of some polynuclear carbonyl are discussed below:

Among the trinuculear carbonyls known at present, crystalline $Os_3(CO)_{12}$ and $Ru_3(CO)_{12}$ have the structures as shown below. In this structure, the metal atoms form an equilateral triangle with no bridging CO groups. All the twelve CO groups are terminal. The structure of trinuclear carbonyls, $Fe_3(CO)_{12}$, is different from that of $Os_3(CO)_{12}$ and $Ru_3(CO)_{12}$. It has two bridging CO groups as shown below. The three Fe atoms form isosceles triangle with three Fe—Fe bonds.

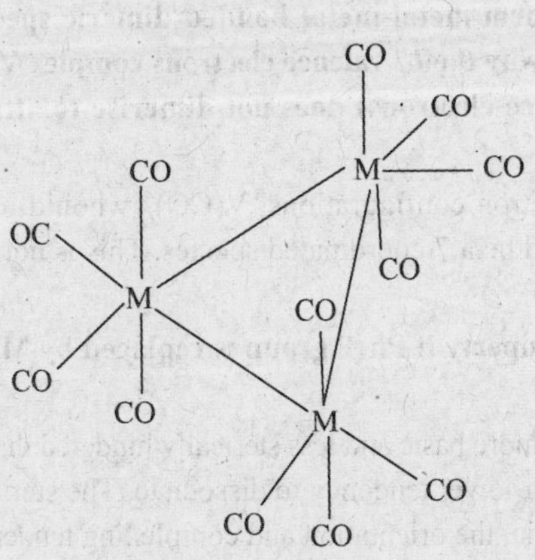

Structure of $M_3 (CO)_{12}$ (M = Os or Ru)

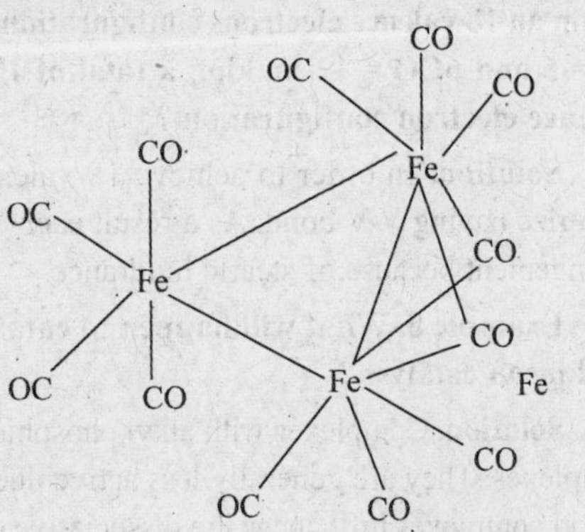

Structure of $Fe_3 (CO)_{12}$

SOLVED CONCEPTUAL PROBLEMS

Example 1. What are the structural consequences of back bonding into olefin π^* orbitals ?

Solution. There are two main consequences of back bonding into olefin π^* orbital:

(*i*) The C = C bond length increases.

(*ii*) The angles around carbon are reduced from about 120° (sp^2 hybridized) towards angles typical of tetrahedral carbon (sp^3 hybridized) of 109.5°.

Example 2. N_2 is isoelectronic with CO, but it is a poor s donor than CO. Explain.

Solution. In N_2 the highest occupied molecular orbital (HOMO) has bonding character while in CO the HOMO is weakly antibonding. Therefore, CO is better donor than N_2. This is so because species having HOMO of antibonding type have better tendency to donate these electrons to improve its bond order.

Example 3. Give two examples of π-acid ligands similar to CO.

Solution. (*i*) Isocyanide (RNC) (*ii*) Nitric oxide (NO)

Example 4. Give one example each of 3 electron donor and 4 electron donor ligand.

Solution. 3 electron donor : π-allyl

4 electron donor : 1, 3-butadiene

Example 5. Give evidence to show that the lone pair on carbon in CO ligand resides in antibonding molecular orbital ?

Solution. When CO is ionized to form CO^+ *i.e.* one of the lone pair of electrons residing in HOMO is removed, the C—O stretching frequency (ν_{CO}) increases from 2143 cm^{-1} to 2184 cm^{-1}. This shows that bond order has increased. Therefore, the electron pair is donated from anti-bonding MO.

Example 6. Which transition elements show preference towards metal-metal bond formation among metal carbonyls ?

Solution Heavy transition elements of second and third series form stable metal-metal bonds than first row transition metals. The large size of the heavier metals prefer formation of metal-metal bond rather than CO bridges.

Example 7. The carbonyls of Mn and Co form metal-metal bonded dimeric species to attain an 18 valence electrons configurations then why the 17 valence electrons complex $V(CO)_6$ (V = 5 and 6CO = 12 making a total of 17 valence electrons) does not dimerise to attain 18 valence electron configuration ?

Solution. In order to achieve 18 valence electron configurations, $V(CO)_6$ whould have to dimerize having V-V bond. As a result each V would be a 7 coordinated species. This is not stable arrangement because of stearic hindrance.

Example 8. What will happen to catalytic property if Ph_3P group is replaced by Me_3P in Wilkinson catalyst ?

Solution. Complexes with alkyl phosphines are more basic and less sterically hindered than aryl complexes. They are generally less active due to their lower tendency to dissociate. The steric bulk in aryl complexes influences the dissociative equilibria, the orientation and complexing tendency of the unsaturated alkene and consequently the stability of the intermediate alkyl derivative.

Example 9. Give the IUPAC names of $Fe(CO)_5$ and $[(CO)_3Co(CO)_2 Co(CO)_3]$

Solution. $Fe(CO)_5$: Pentacaronyl iron

$[(CO)_3 Co(CO)_2Co(CO)_3]$: Di–μ–carbonylbis(tricarbonyl cobalt)

Example 10. Write the IUPAC names of the following :

(a) $[Mn(CO)_5(C_2H_4)]^+$

(b) $[(CO)_5-Mn-Mn(CO)_5]$

(c) $[(CO)_3Co(CO)_2Co(CO)_3]$

(d) $Fe(C_5H_5)_2$

(e) $Fe_2(CO)_4(C_5H_5)_2$

Solution. IUPAC names of the compounds are given as follows:

(a) Pentacarbonyl (ethylene) manganese (+1)

(b) Bis (pentacarbonylmanganese)

(c) Di–μ–carbonylbis(tricarbonylcobalt)

(d) Bis (cyclopentadienyl) iron (II)

(e) Bis (η^5-cyclo pentadienyl) tetracarbonyldi iron

Example 11. Which of the following do and which do not obey EAN rule ?

(i) $Cr(C_6H_6)(CO)_4$

(ii) $CH_3Re(CO)_5$

(iii) $Cr(CO)_6$

(iv) $Co_2(CO)_6(RC\equiv CR)$

(v) $Mo(CO)_3(\pi\text{–}C_5H_5)$

Solution. (i) $24(Cr)+6(C_6H_6)+8(4CO)=38$: Does not obey EAN rule

(ii) $75(Re)+10(5CO)+1(CH_3)=86$ Obeys EAN rule

(iii) $24(Cr)+12(6CO)=36$ Obeys EAN rule

(iv) $54(2Co)+12(6CO)+4(RC\equiv CR)+2(Co-Co)=72$ or 36 per Co Obeys EAN rule

(v) $42(Mo)+6(3CO)+5(\pi\text{–}C_5H_5)=53$ Does not obey EAN rule

Example 12. Both $Mn(CO)_5$ and $V(CO)_6$ do not obey the EAN rule (35 electrons). $Mn(CO)_5$ achieves EAN of 36 by forming metal-metal bond $Mn_2(CO)_{10}$. However $V(CO)_6$ does not demerize to give $V_2(CO)_{12}$. Explain.

Solution. $V(CO)_6$ (EAN = 35) does not form V–V bond to give $V_2(CO)_6$ because then dimer vanadium will acquire coordination number of 7. However, this is not stable because small sized vanadium may face steric hindrance and may not be able to accommodate seven ligands around it. However, each Mn in $Mn_2(CO)_{10}$ acquires coordination number of 6 on forming $Mn_2(CO)_{10}$.

Example 13. Complete the following reactions:

(i) $Co_2(CO)_8 + PhC\equiv CPh \longrightarrow$

(ii) $Fe(CO)_5 + Me_3SiNC \longrightarrow$

(iii) $Mo(CO)_6 + NaBH_4 \longrightarrow$

(iv) $Co_2(CO)_8 \xrightarrow[THF]{Na/Hg}$

(v) $Mo(CO)_6$ + Pyridine

(vi) $KC_5H_5 + FeCl_2 \longrightarrow$

(vii) $K_2PtCl_4 + C_2H_4 \longrightarrow$

Solution. (*i*) $Co_2(CO)_8 + PhC \equiv CPh \longrightarrow Co_2(CO)_6\ (PhC \equiv CPh)$

(*ii*) $Fe(CO)_5 + Me_3SiNC \xrightarrow{-CO} (Me_3SiNC)\ Fe(CO)_4$

(*iii*) $2Mo(CO)_6 + NaBH_4 \longrightarrow Na_2[Mo_2(CO)_{10}]$

(*iv*) $Co_2(CO)_8 \xrightarrow[THF]{Na/Hg} 2Na{+}\ [Co(CO)_4]^-$

(*v*) $Mo(CO)_6 + C_5H_5N \xrightarrow{-CO} C_5H_5NMo\ (CO)_5 \xrightarrow[-CO]{C_5H_5N} (C_5H_5N)_2\ Mo(CO)_4 \xrightarrow{C_5H_5N(-CO)} (C_5H_5N)_3\ Mo(CO)_3$

(*vi*) $2KC_5H_5 + FeCl_2 \longrightarrow (C_5H_5)_2\ Fe + 2KCl$

(*vii*) $K_2PtCl_4 + C_2H_4 \longrightarrow K^+\ [PtCl_3.\ C_2H_4]^-$

Example 14. How will you prepare Wilkinson catalyst ?

Solution. Wilkinson catalyst is RhCl $(PPh_3)_3$. It can be obtained by treating $RhCl_3.H_2O$ with Ph_3P

$$\overset{III}{RhCl_3}.\ H_2O + \overset{III}{Ph_3P} \longrightarrow \overset{I}{RhCl}\ (PPh_3)_3 + \overset{V}{Ph_3PO}$$

EXERCISES
(Including Questions from Different University Papers)

Multiple Choice Questions (Choose the correct option)

1. Out of $Cr(CO)_6$ (A) and $[Ag(NH_3)_2]^+$ (B)

(*a*) (A) obeys EAN rule and (B) does not

(*b*) (B) obeys EAN rule and (A) does not

(*c*) (A) and (B) both obey EAN rule

(*d*) Neither (A) nor (B) obeys EAN rule

2. Which of the following organometalic compound does not obey EAN rule ?

(*a*) $Co(CO)_3\ (\pi\text{-}C_3H_5)$ (*b*) $(C_2H_5)\ Cr(CO)_3\ (\pi\text{-}C_5H_5)$

(*c*) $[Mn(CO)_5\ (C_2H_4)]^+$ (*d*) $Co(C_6H_6)_2$.

3. An example of olefin complex is

(*a*) Ferrocene (*b*) Zeise salt

(*c*) Bis (η^6-benzene) chromium (*d*) $(CO)_6\ Co_2\ (PhC \equiv CPh)$.

4. Which of the following has lowest CO stretching frequency ?

(*a*) $Ni(CO)_4$ (*b*) $[V(CO)_6]^+$

(*c*) $[Co(CO)_4]^-$ (*d*) $[Fe(CO)_4]_2^-$.

5. Which of the following will have highest CO stretching frequency ?

(*a*) $Cr(CO)_6$ (*b*) $Mn(CO)_6^{\ +}$

(*c*) $V(CO)_6$ (*d*) $Fe(CO)_4^{\ 2-}$.

6. Which of the folowing is Wilkinson catalyst ?

(*a*) $(\eta^5 C_5H_5)_2 Ni_2 (PhC \equiv CPh)$ (*b*) $RuHCl (PPh_3)_3$

(*c*) $RhCl (PPh_3)_3$ (*d*) $IrCl (PPh_3)_3$

7. Which of the following is an example of four carbon bonded ligand ?

(*a*) Butadiene (*b*) Cyclopentadienyl

(*c*) Cyclooctatetraene (*d*) C_3H_5.

8. Which of the following has three bridging CO groups ?

(*a*) $Co_4(CO)_{12}$ (*b*) $Fe_3(CO)_{12}$

(*c*) $Co_2(CO)_8$ (*d*) $Fe_2(CO)_9$.

9. Which of the following mononuclear carbonyls does not obey EAN rule ?

(*a*) $Fe(CO)_5$ (*b*) $Ni(CO)_4$

(*c*) $Cr(CO)_6$ (*d*) $V(CO)_6$.

10. Which of the following does not have bridging carbonyl ?

(*a*) $Fe_3(CO)_{12}$ (*b*) $Ru_3(CO)_{12}$

(*c*) $Fe_2(CO)_9$ (*d*) $Co_4(CO)_{12}$.

11. The number of bridging carbonyl groups in $Fe_3(CO)_{12}$ is

(*a*) zero (*b*) one

(*c*) two (*d*) three.

12. Which of the folowing statements is not correct for alkyne complex $PtCl_2(Bu^t{}_2C_2)$?

(*a*) The alkyne molecule behaves as a 2 electron donor.

(*b*) The molecule $^tBu\ C \equiv C\ Bu^t$ is no longer a straight line.

(*c*) The carbon - carbon bond is slightly longer than a triple bond.

(*d*) The molecule is monomer.

13. Which of the following metal carbonyl can be prepared by the direct interaction of finely divided metal with CO ?

(*a*) $Cr(CO)_6$ (*b*) $Mn_2(CO)_{10}$

(*c*) $Fe(CO)_5$ (*d*) $Co_2(CO)_8$.

14. $Fe_2(CO)_9$ has

(*a*) one bridgine carbonyl (*b*) two bridging carbonyls

(*c*) three bridging carbonyls (*d*) none of the above

ANSWERS

1. (*a*)	**2.** (*d*)	**3.** (*b*)	**4.** (*d*)	**5.** (*b*)	**6.** (*c*)
7. (*a*)	**8.** (*a*)	**9.** (*d*)	**10.** (*b*)	**11.** (*c*)	**12.** (*d*)
13. (*c*)	**14.** (*c*)				

Short Answer Questions

1. What are π-bonded organometallic compounds ?

2. Give one example of a homogenous hydrogenation catalyst ?

3. Give one example each of a metal carbonyl which

(*i*) obeys EAN rule (*ii*) does not obey EAN rule.

4. What are sandwiched compounds ? Give one example.
5. Give one method to prepare $Fe_2(CO)_9$.
6. How many bridging CO are present in $Os_3(CO)_{12}$, $Fe_2(CO)_9$?
7. Define EAN rule and give one example.
8. How are Zeiglar – Natta catalysts obtained ? What is their utility ?
9. Give the formula of Wilkinson catalyst.
10. Which of the following has lowest CO group frequency ?

    Terminal, doubly briged, triply bridged
11. Why metal-metal bond distance in $Mn_2(CO)_{10}$ is longer than that in $Fe_2(CO)_9$ although both are binuclear ?
12. Out of Pb $(C_2H_5)_4$ and $(C_4H_9)_3$ SnCl which one is a mixed organometallic compound and why ?
13. How do you define haptacity ?

General Questions

1. Carbon monoxide is considered to be a special type of ligand in organometallic compounds. What is so special about carbon monoxide in these compounds ?
2. Name the different categories of organometallic compounds. Discuss with one example each.
3. What are organometallic compounds ? Give structures of organometallic compounds formed by 5, 6 and 8 electron donor unsaturated molecules.
4. Give IUPAC names of the following:

    (*i*) (C_6H_6) $Cr(CO)_3$ (*ii*) $Fe(C_5H_5)_2$

    (*iii*) $(\sigma\text{–}C_5H_5)_2$ $Ti(\pi\text{-}C_5H_5)_2$ (*iv*) ReH $(\pi\text{-}C_5H_5)$
5. What is EAN rule ? Give two examples each of organometallic compounds in which EAN rule is (*i*) obeyed and (*ii*) not obeyed.
6. What are metal olefin complexes ? Give one method for the preparation of an important metal olefin complex. Discuss the bonding in these complexes.
7. What are organotin compounds ? Discuss their preparations structures and applications.
8. (*a*) What is a ligand ? How are the ligands classified ? Give one example in each case.

    (*b*) Give a brief account of homogeneous hydrogenation catalysts.
9. (*a*) Discuss bonding in metal carbonyls. How does infrared spectroscopy help in explaining bonding in metal carbonyls ?

    (*b*) Can infrared spectroscopy distinguish between the terminal and bridging CO groups in metal carbonyls ? Explain.
10. What are metal-alkyne complexes ? Discuss the different types of bonding in these complexes with or example of each.
11. Define organometallic compound. Give structures of complexes in which cycloctatetraene behave as 4-electron donor, 6-electron donor and 8-electron donor.
12. Calculate the EAN value for each of the following species :

    (*a*) $Cr(C_5H_5)(CO)_3$ (*b*) $Fe(\pi\text{– }C_5H_5)(CO)_3$

    (*c*) $(\sigma\text{– }C_3H_5)$ $Mo(CO)_3$ $\pi(C_5H_5)$ (*d*) (Me_3Si) $W(CO)_3$ $(\pi\text{– }C_5H_5)$

    (*e*) $Co(\pi\text{– }C_5H_5)(CH_3)_2$

13. What are organometallic compounds. Give IUPAC names of the following :

(*a*) $Ni(\pi-C_5H_5)_2$ (*b*) $(CO)_5Mn-Mn(CO)_5$

(*c*) $Ru(CO)_3(C_4H_6)$ (*d*) $[PtCl_3(C_2H_4)]^-$

(*e*) $Co(CO)_3(\pi-C_3H_5)$

14. Write formulae for the following organometallic compounds

(*a*) (η^3 - allyl) tricarbonyl cobalt

(*b*) (η^1 - allyl) tricarbonyl (η^5 - cyclopentadienyl) molybdenum

(*c*) Bis (η^5 - cyclopentadienyl) iron

(*d*) Tricarbonyl (η^5 - cyclopentadienyl) manganese

(*e*) (η^4 - butadiene) tricarbonyl iron

15. (*a*) Give one method for the preparation of each of the following :

(*i*) $Ni(CO)_4$ (*ii*) $Fe_2(CO)_9$ (*iii*) $Co_2(CO)_8$ (*iv*) $Fe(CO)_5$

(*b*) Draw structures of the above metal carbonyls.

16. Discuss the structures of the following compounds :

(*a*) $Mn_2(CO)_{10}$ (*b*) $(\eta^5-C_5H_5)Mn(CO)_2(Ph_2C_2)$

(*c*) $K[PtCl_3(C_2H_4)]$ (*d*) $Fe_3(CO)_{12}$

(*e*) $Co_2(CO)_8$ solid

17. (*a*) What are metal carbonyls ? Discuss a method of preparation and structure of one (*i*) mononuclear (*ii*) dinuclear and (*iii*) trinuclear carbonyl.

(*b*) Discuss the bonding in metal carbonyls and evidence in support of such bonding.

18. (*a*) What are metal carbonyls ? Give structures of two dinuclear metal carbonyls having different structures.

(*b*) How does infrared spectroscopy help in explaining the bonding in metal carbonyls ?

19. (*a*) What is Zeise salt ? Draw its structure and discuss the salient features of this structure.

(*b*) Give an example where an alkyne behaves as a bidentate divalent ligand.

20. (*a*) How is dicobalt decarbonyl prepared ? Discuss its structure.

(*b*) How does infrared spectroscopy help in explaining the type of metal–CO group bonding in the complex ?

21. Complete the chemical equations :

$$FeO + 2C_5H_6 \xrightarrow{523K}$$

$$Fe + 2C_5H_6 \xrightarrow{573K}$$

22. (*a*) Discuss the nature of bonding in metal olefin complexes.

(*b*) Discuss the structures of

(*i*) $Os_2(CO)_9$ (*ii*) $Ni(CO)_4$

23. (*a*) Discuss the nature of bonding in metal-olefin complexes.

(*b*) Explain the structures of

(*i*) $Mn_2(CO)_{10}$ (*ii*) $Fe(CO)_5$ (*iii*) $Ni(CO)_4$ (*iv*) $Fe_3(CO)_{12}$.

24. (*a*) N_2 is isoelectronic with CO, it is a poor σ donor than CO. Explain.

(*b*) Draw the structures of the following :

(*i*) $Co_2(CO)_6 \cdot (Ph_2C_2)$ and (*ii*) $Fe_3(CO)_{12}$.

25. (*a*) Discuss the nature of bonding in metal carbonyls. Give experimental supports in its favour.

(*b*) Cyclopentadiene ring in ferrocene show aromatic character but cyclopentadiene itself has no such aromatic character.

(*c*) Why can lower oxidation states exist in metal complexes of CO, NO, etc.

26. In case of transition metal carbonyls state:

(*i*) Whether the metals are in their lower oxidation states or not?

(*ii*) EAN rule is obeyed or not?

(*b*) Define π acidity and give one example.

(*c*) Describe the structures of $V(CO)_6$, Ni $(CO)_4$, $Mn_2(CO)_{10}$, $Fe_2(CO)_9$, $Os_2(CO)_9$, and $Co(CO)_8$

27. (*a*) Draw the structure of Fe_2 $(CO)_9$.

(*b*) Complete the following equations giving only the major products :

(*i*) $Fe(C_5H_5)_2 + HCHO + HN(CH_3)_2 \longrightarrow$

(*ii*) $Co_2(CO)_8 \xrightarrow[THF]{Na/Hg}$

(*iii*) $K_2PtCl_4 + C_2H_2 \longrightarrow$

(*iv*) $Mo(CO)_6$ + Pyridine $\longrightarrow$

(*v*) $KC_5H_5 + FeCl_2 \longrightarrow$

(*vi*) $(C_5H_5)_2$ Fe + Li $C_4H_9 \longrightarrow$

28. What do you understand by β-elimination in transition metal alkyls ? How can it be avoided ? Explain giving examples.

29. (*a*) What is EAN rule ? Show whether the compounds $Cr(CO)_6$, $Ni(PF_3)_4$ and $Fe(CO)_4$ (PPh_3) obey the EAN rule.

(*b*) Complete the following reactions :

(*i*) $Co_2(CO)_8 \xrightarrow[THF]{Na/Hg}$

(*ii*) $(C_5H_5)_2$ Fe + $H_2SO_4 \xrightarrow{(CH_3CO)_2O}$

(*iii*) $(C_5H_5)_2$ Fe + $LiC_4H_9 \longrightarrow$

30. (*a*) Give two methods of preparation of transition metal carbonyls.

(*b*) What are metal olefin complexes ? Discuss the structure of $[PtCl_3(C_2H_4)]^-$ ion.

31. (*a*) Discuss the bonding in metal-carbonyls. How does infrared spectroscopy help in explaining bonding in metal carbonyls ?

(*b*) Write I.U.P.A.C. names

(*i*) [Mn $(CO)_5$ $(C_2H_4)]^+$ (*ii*) $Fe_2(CO)_9$

(*iii*) [K $SbCl_5$ (C_6H_5)] (*iv*) [V $(CO_6)^-$

32. (*a*) Accidental discovery of ferrocene was a single major event in organometallic chemistry. Discuss with the help of M.O. theory why is it so stable.

(*b*) What is β elimination in metal alkyls ? How can it be avoided ? Explain giving examples.

(*c*) Give the methods of preparation of $Ni(CO_4)$ and $W(CO)_4$

(*d*) Complete the following reactions :

(*i*) $Co_2(CO)_8 + PhC \equiv CPh \longrightarrow$

(*ii*) $Fe(CO)_5 + Me_3SiNC \longrightarrow$

33. (*a*) What do you understand by β-elimination in transition metal alkyls. How can it be avoided ? Explain giving examples.

(*b*) Discuss the structures of the folowing :

(*i*) $Ir(CO)_{12}$ (*ii*) $Os(CO)_9$

(*c*) Write a short note on EAN rule as applied to organometallic compounds.

(*d*) Name the following organometalic compounds according to IUPAC system :

(*i*) $Co_2(CO)_8$ (*ii*) $C_6H_6Cr(CO)_3$

34. (*a*) What are metal olefin complexes ? Discuss the structure of $[PtCl_3(C_2H_4)]^-$ ion.

(*b*) Give the structure of $Fe_2(CO)_9$

(*c*) What is the meaning of η^5 in the formula $Fe(\eta^5 - C_5H_5)_2$?

35. (*a*) Write short notes on the following.

(*i*) The effective atomic number rule.

(*ii*) Back bonding in metal carbonyls.

36. (*a*) Give the structure of $Fe(CO)_9$.

(*b*) Give the structure of $Os_3(CO)_{12}$

(*c*) Give the mode of bonding in metal conjugated diene complexes.

(*d*) Discuss the bonding in π metal olefin complexes. Give examples.

(*e*) Give classification of ligands on the basis of electrons involved.

37. What are metal olefin complexes ? Give one method of preparation of an important metal-olefin complex. Discuss bonding in these complexes.

38. (*a*) Define an organometallic compound. Give the types of organometallic compounds.

(*b*) Explain the mode of bonding in complexes of metals with conjugated dienes.

(*c*) Define π-acidity and give one example.

39. (*a*) Draw the structures of the following :

(*i*) $Fe_2(CO)_9$ (*ii*) $Os_2(CO)_9$ (*iii*) $Mn_2(CO)_{10}$

(*b*) Write IUPAC names of the following compounds :

(*i*) $C_6H_6Cr(CO)_3$ (*ii*) $(C_5H_5)_2$ Fe (*iii*) $Fe_2(CO)_9$

40. (*a*) What is EAN rule ? Give two examples each of organometallic compounds in which EAN rule is

(*i*) obeyed (*ii*) not obeyed

(*b*) Can infrared spectroscopy distinguish between the terminal and bridging CO groups in the metal carbonyls ? Explain.

(*c*) Complete the following equations giving only the major products :

(*i*) $KC_5H_5 + FeCl_2 \longrightarrow$ .... (*ii*) $Mo(CO)_6$ + pyridine $\longrightarrow$ ....

41. (*a*) Based on the IUPAC rules, name the following organometallic compounds :

(*i*) $Fe\ (\eta^3 - C_5H_5)_2$

(*ii*) $Mn\ (\eta^3 - C_3H_5)\ (CO)_4$

(*b*) Identify organometallic compounds amongst the following :

$Na\ OO\ CCH_3$; $Mn\ (C_5H_5)_2$;
$CH_3Sn\ H_2Cl$; $(H_5C_6)_3\ Si–Si\ (C_6H_5)_3$

(*c*) What is the use of IR spectroscopy in the elucidation of structure of meal carbonyls ?

42. (*a*) Discuss briefly bonding in metal-alkyne complexes.

(*b*) Give IUPAC name of the following compounds :

(*i*) $Ni\ (\pi–C_5H_5)_2$
(*ii*) $(CO)_5 – Mn–Mn\ (CO)_5$
(*iii*) $Ru\ (CO)_3\ (C_4H_6)$

(*c*) Calculate EAN vlaue for the following :

(*i*) $Cr\ (C_5H_5)\ (CO)_3$
(*ii*) $Fe\ (\pi–C_5H_5)\ (CO)_3$

43. (*a*) Discuss the structure of $(LiCH_3)_4$.

(*b*) Give important industrial applications of organotin compounds.

(*c*) Briefly discuss the metal-carbon sigma bonding.

44. (*a*) Complete the reactions :

(*i*) $Co_2(CO)_8 \longrightarrow$ Na / Hg, THF
(*ii*) $K_2PtCl_4 + C_2H_2$
(*iii*) $Fe(CO)_5 + NaOH$

(*b*) Why do organolithium compounds prefer to oligomerize than exists as single molecule ?

(*c*) Discuss structure of $Fe_2(CO)_9$.

(*d*) Write names of following organometallic compounds

$[(CO)_3Fe(CO_3)Fe(CO)_3]$; $[Fe(\pi\text{-}C_5H_5)_2]$; $[Cr(C_6H_6)_2]$

45. (*a*) Discuss the nature of bonding in metal-olefin complexes.

(*b*) "Both N_2 and CO are isoelectronic, yet N_2 is a poorer σ donor than CO". Explain.

(*c*) What are metal carbonyls ? Discuss the structure of $V\ (CO)_6$ and $Fe\ (CO)_5$.

46. (*a*) Discuss bonding in metal carbonyl complexes.

(*b*) Complete the reactions :

(*i*) $Co_2\ (CO)_8 \xrightarrow[THF]{Na / Hg}$
(*ii*) $K_2PtCl_4 + C_2H_2 \longrightarrow$

(*c*) Discuss the structure of $Fe_2(CO)_9$.

(*d*) Give mechanism of homogeneous hydrogenation of alkenes with Wilkinson catalyst.

(*e*) Give structure of Zeise salt.

47. (*a*) Discuss preparation, properties, bonding and applications of organoaluminium compounds.

(*b*) What do you mean by homogeneous hydrogenation ? Name three homogeneous catalysts used for homogeneous hydrogenation of alkenes.

48. (*a*) Discuss important applications of organoaluminium compounds.

(*b*) Give two important methods of preparation of metal carbonyls.

(*c*) Briefly discuss the structure of binuclear metal carbonyls.

49. (*a*) Describe bonding in metal carbonyls. How does IR spectroscopy help in explaining bonding in metal carbonyls ?

(*b*) Give methods of preparation of the transition metal carbonyls.

(*c*) Give one method of preparation of metal alkene complex. Also explain bonding in this complex.

50. (*a*) In which of the following complexes EAN rule is objected ?

(*i*) $[Mn(CO)_5C_2H_4]^+$ (*ii*) $[Mo(CO_6)]$

(*iii*) $[Cu(NH_3)_4]$ (*iv*) $[Co(CO)_4]$

(*b*) Give two examples which would distinguish between organometallic and non-organometallic compounds.

(*c*) Discuss nature of bonding in metal carbyne complexes.

(*d*) Write formulae against the names of the following organometallic compounds :

(*i*) Hexacarbonyl niobate (–1)

(*ii*) Tetracarbonyl bis (ethylene) manganese (+1).

51. (*a*) Discuss the bonding in metal-carbonyl complexes. How IR spectra helps in explaining bonding in metal-carbonyls ?

(*b*) How homogeneous hydrogenation of C_2H_4 is carried out by using Wilkinson's catalyst ?

(*c*) Define metal carbonyls. Draw the structures of :

(*i*) $Ni(CO)_4$ (*ii*) $Fe(CO)_5$

(*iii*) $Cr(CO)_6$ (*iv*) $Mn_2(CO)_{10}$

52. (*a*) Classify bonding in organometallic compounds. Give examples of each type.

(*b*) What do you understand by β-elimination in transition metal alkyls ? How can it be avoided ?

(*c*) Why do organolithium compounds prefer to doligomerize rather than exist as single molecule ?

53. (*a*) Discuss the synthesis and bonding of Zeise salt.

(*b*) Write the nomenclature of the following compounds :

(*i*) $PtCl_3\ (\eta^2 - C_2H_4)^-$

(*ii*) $Mn\ (\eta^3 - C_3H_5)\ (CO)_4$

(*iii*) $Rh\ (CO)\ Cl\ (PPh_3)_2$

(*c*) Define haptacity and discuss the classification of organometallic compounds on the basis of haptacity.

54. (*a*) Write the IUPAC names of the following :

(*i*) $Zn\ (C_2H_5)_2$

(*ii*) $CH_3–SnH_2Cl$

(*iii*) $Fe\ (\eta^5 - C_5H_5)$

(*iv*) $[Mn\ (\eta^3 - C_6H_5)\ (CO)_4]$

(*b*) How sepectroscopical evidences support bonding in metal carbonyls ? Explain.

55. (*a*) Discuss the nature of bonding in metal carbonyl.

(*b*) Discuss transmetallation for the synthesis of organometallic compounds.

56. (*a*) Define EAN rule. Which of the following species obey effective atomic number rule and why ?

(*i*) $Co(CO)_4$ (*ii*) $H_3C\,Mn(CO)_5$

(*iii*) $Fe(CO)_3(C_4H_6)$ (*iv*) $Fe(\pi\text{-}C_5H_5)_2$

(*b*) What do you understand by beta-elimination in transition metal-alkyls ? How can it be avoided ? Give examples.

(*c*) Discuss and draw the structure of Zeise's salt ?

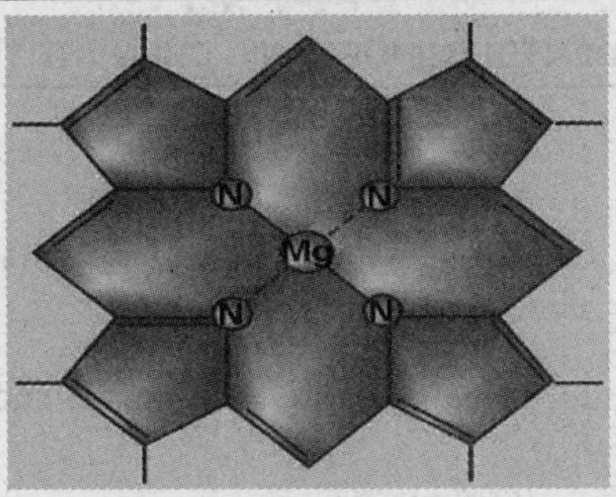

7

Bioinorganic Chemistry

7.1. INTRODUCTION

Bioinorganic chemistry is the branch of chemistry which correlates inorganic materials with biological processes. It is a fast developing branch because of the importance of inorganic substances in biological systems. This branch highlights the role of metal ions in biology. As we understand biological systems more and more, we come to know the importance of the metal ions. This has helped to synthesise new inorganic compounds which resemble those found in biological systems naturally. These ions are present in specific parts of the biomolecule, and are responsible for initiating or inhibiting reactions in biological systems.

7.2. ESSENTIAL AND TRACE ELEMENTS

An essential element is that element which is required for the maintenance of life. Its absence results in death or a severe malfunction of the organism. Of the 112 known elements, 30 elements are believed to be essential for life processes in plants and animals. Nineteen of thirty elements are trace elements (required in milligrams or micrograms), out of which twelve are transition elements. Biological activity of about one half of the essential elements including Fe, Zn, Cu, Mn, Mo, Co, Ni and Se has been properly explained by bioinorganic chemistry while the action of V, Cr, Cd, Pb, Sn, Li, F, Si, As, and B is still to be understood completely.

30 essential elements have been classified as under

1. *Bulk elements*	H, C, N, O, P, S
2. *Macrominerals*	Na, K, Mg, Ca, Cl
3. *Trace elements*	Fe, Zn, Cu
4. *Ultratrace elements*	
Non-metals	F, I, Se, Si, As, B
Metals	Mn, Mo, Co, Cr, Cd, Sn, Pb, Li

Criteria for an essential element

1. An element is considered essential when a deficient intake produces an impairment of function and when restoration of level of that elements cures the impaired function.

2. A specific biochemical function is associated with a particular element. It is quite difficult to define their concentration limits in the system. This varies from system to system. The same element may prove to the inadequate at a particular concentration in one biological system while it may be over dose in another system.

All the essential and trace elements are not required by every plant or animal. Sodium is vital for humans and higher animals but it is not essential for bacteria. Some higher plants require Al, B and V, which is not so much required for humans. Definitely, some essential elements prove to be toxic, even fatal, when taken in higher concentration.

7.3. ESSENTIAL BULK ELEMENTS

Role of Na^+ and K^+ ions in biological systems.

Sodium

Sodium is the major component of extracellular fluid. It mainly exists as chloride and bicarbonate.

1. Its role in biological system is to regulate acid-base equilibrium.
2. It also helps in maintaining osmotic pressure of the body fluid, thereby protecting the body against fluid loss.
3. It has a role to play in preservation of normal irritability of muscle and permeability of the cell.
4. Many enzyme reactions are controlled by Na^+.
5. The injectable medicines are dissolved in sodium chloride before they are injected into human body.

Potassium

Potassium is present as cation in intracellular fluid as well as extracellular fluid.

1. It influences acid base equilibrium like sodium in extracellular fluid.
2. It controls osmotic pressure and water retention.
3. It is important for metabolic functions like protein biosynthesis by ribosomes.

Chemically, sodium and potassium may be similar (alkali metals, Group 1) but there is different biological response for the two ions. While Na^+ ions are pumped out of the cytoplasm, K^+ ions are pumped in. This ion transport is called sodium-pump, which involves active take up of K^+ ions and expulsion of Na^+ ions. The difference in concentration of the two ions inside and outside the cell membrane produces an electrical potential, which is crucial for the functioning of nerve and muscle cells.

It may be noted that higher intake (we require about 1 g of Na^+ per day) of Na^+ ions is a problem, but there is no such risk from higher intake of K^+ ions. Rather potassium deficiency is more common. We can make up the K^+ ions deficiency by consuming more of banana and coffee.

A number of enzymes such as glycolytic enzyme pyruvate kinase require K^+ for maximum activity.

Na^+ and K^+ are present in the red blood cells. The ratio of these ions in mammals such as human beings, rabbits, rats and horses is 1:7. In cats and dogs, this ratio is 15:1.

To establish this ratio also called *concentration gradient* in the cell, biologists have suggested different mechanisms involving sodium pump and potassium pump.

1. The concentration gradient controls the development and functioning of the nerve cells.
2. In restful state, potential of the nerve cell is linked to K^+ concentration across the membranes.
3. During activation of the nerve cells, a chemical called *acetylcholine* is discharged. This discharge is transmitted through the length of the nerve cell by electric impulse.

Role of Mg^{2+} and Ca^{2+} in biological systems

1. Mg^{2+} ions are present in chlorophyll which is the green colouring matter present in plants. Chlorophyll absorbs light from the sun and carries out the process of photosynthesis in plants.

2. Magnesium ions are present in the enzyme *phosphatase* which act upon organic phosphates to hydrolyse them into free phosphates.

3. Magnesium ions are present in the enzyme *aminopeptidase* which hydrolyses polypeptides at the free amino acid end of the chain to form lower peptides and sometimes free amino acids.

4. Calcium ions are present as phosphates in the bones of human beings and animals. These ions play an important role in muscle contraction. The malnutrition in children is due to the deficiency of Ca^{2+}.

5. About 99% of the body calcium is in the skeleton where it is maintained as deposits of calcium phosphates in soft and fibrous matrix. A very small quantity of calcium not present in skeletal structure, is in the body fluids where it is partly ionized. This small amount of ionized calcium in the body fluid is of great importance in blood coagulation, in maintaining the normal excitability of the heart, muscles and nerves and in the differential aspects of membrane permeability.

6. The major inorganic constituent of bone is comprised of a crystalline from of calcium phosphate resembling the mineral *hydroxyapatite*. Additionally, the bone contains a substantial amount of non-crystalline, amorphous calcium phosphate. It appears that the amorphous material is predominant is early life but later on the crystalline form takes over in adult life.

Important minerals of calcium used in biological structures are given in the following table.

Organism	Location	Mineral	Formula
Birds	Egg shells	Calcite	$CaCO_3$
Molluscs	Shell	Argonite	$CaCO_3$
Vertebrates Mammals	Bone	Hydroxyapatite	$Ca_5(PO_4)_3$ (OH)

Some noteworthy points regarding Ca^{2+} and Mg^{2+} ions are :

1. Calcium ions are needed to bring about **blood clotting** and contract of muscles such as **heart beat**.
2. Deficiency of Ca^{2+} ions causes **tetany**, while excess of it causes **Calcification**.
3. Both Mg^{2+} and Ca^{2+} ions are excreted via the large intestine instead of kidney. Calcium deficiency might occur due to its precipitation as calcium oxalate, after reaction with soluble oxalates present in the body.

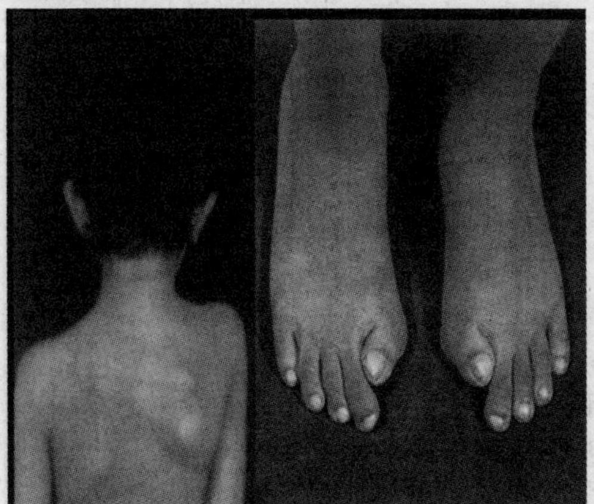

Consequences of Calcification

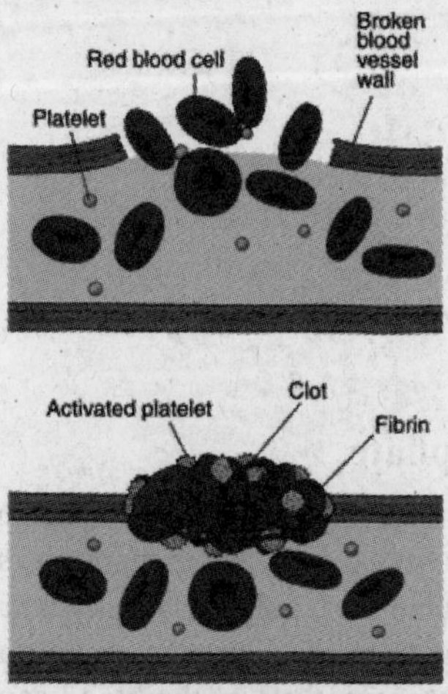

Blood Clothing

Major calcium movements in the body

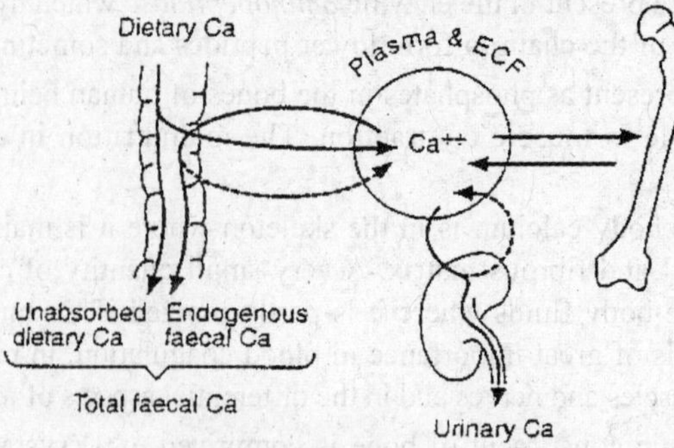

Biological role of calcium

Magnesium, which is essential to all organisms is relatively present in greater concentration in red cells than in plasma. It is a constituent of chlorophyll which plays an important role in photosynthesis.

It has been observed that transphosphorylation reactions involving ATP (adenosine triphosphate) proceed smoothly in the presence of Mg^{2+} ions. ATP is an energy-rich molecule which drives most of the cell reactions. Mg plays a vital role in these reactions.

Free energy of conversion of ATP to ADP (adenosine diphosphate) through hydrolysis is $-31.0\ kJ\ mol^{-1}$

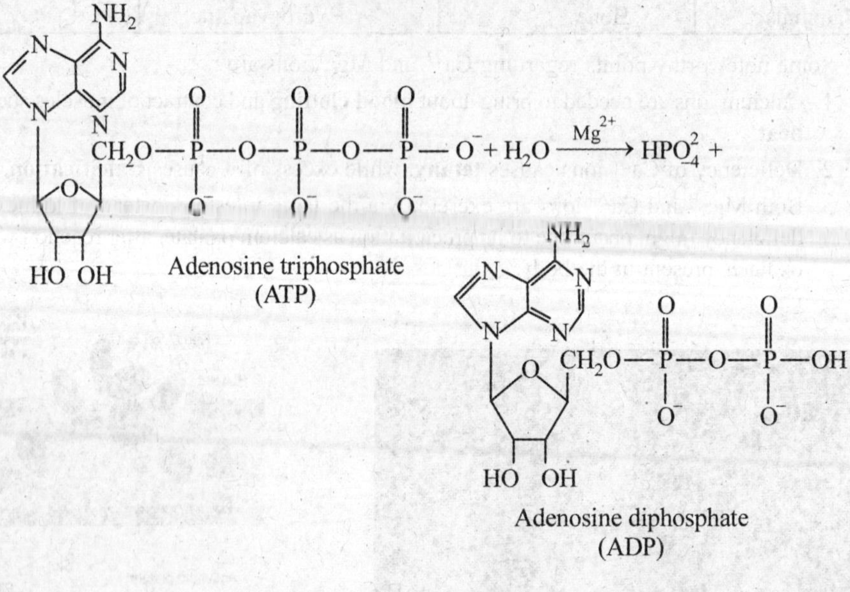

Phosphate

Inorganic phosphorous compounds are involved in the incorporation and release of phosphate groups from organic compounds and skeletal materials (bones). The important phosphate compounds are nucleotides, polynucleotides and phospholipids. Phosphate plays its role in the metabolism of carbohydrates, fats, proteins and vitamins. The presence of phosphorous in bones and its function in regulation of pH of the blood cannot be overlooked. Relative amounts of dietary phosphorus, and calcium in addition to vitamin D, control the formation of normal bones.

7.4. ESSENTIAL TRACE ELEMENTS

Iron

It is an important trace metal required for the growth and survival of cells and organism. Of the total iron in the human body about 70% is present in haemoglobin and about 3% in myoglobin. Most of the remaining iron is stored as **ferritin**. An adult with about 70 kg body weight carries 4.3 g of iron.

Ferritin is major iron storage protein in mammals but it is also found in plant chloroplasts and fungi. It is distributed mainly in the spleen, liver and bone marrow. The iron content of ferritin varies from zero to about 23%. Apoferritin protein may be prepared from ferritin by reduction of ferric ions to ferrous ions at pH = 4.5 followed by removal of Fe^{2+} ions by dialysis. The apoferritin protein can store iron and then transport it to an appropriate site to aid in the biosynthesis of other molecules involving this particular metal. Recommended dietary intake of iron are 10 mg/day for male adult and 18 mg /day for female adult.

Reaction pathway of Fe^{3+} from food stuffs to haemoglobin and to ferritin involves the following processes:

(i) The Fe^{3+} of dietary material is reduced to Fe^{2+} in the gastrointestinal tract.

(ii) After absorption into the cells of the intestinal mucosa, Fe^{2+} is incorporated into ferritin as Fe^{3+}.

(iii) The Fe^{2+} in mucosa is also converted to Fe^{3+} plasma (bound by iron-binding globulin named transferrin).

(iv) Plasma Fe^{3+} is in equilibrium with iron in the liver, spleen and bone marrow.

The changes that occur are depicted as under.

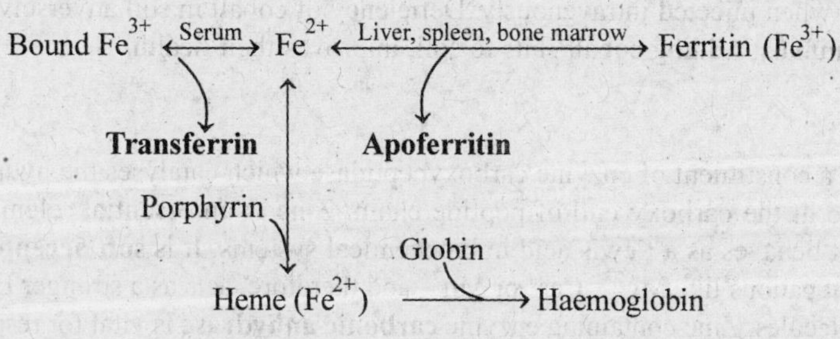

Iron salts in the form of tablets and capsules are given to patients suffering from anaemia.

Copper

Copper is essential to all organisms and is a constituent of redox enzymes and hemocyanin. Copper in hemocyanin is **oxygen carrier** and supplies oxygen to certain aquatic creatures. Copper containing *enzymes play an important role in the pigmentation of skin and functioning of brain.* The deficiency of copper in human system develops an anemia characterized not only by marked decrease in the total iron and heme content in blood and tissues but also by an increased amount of free porphyrin in the erythrocytes.

The excess of copper in human body causes a disease known as ***Wilson disease***. It upsets the copper metabolism of human system in such a way that it absorbs excessive copper from intestinal track and gets deposited in excess in the liver, kidney or brain. Patients suffering from this disease are treated with controlled doses of EDTA which forms a water soluble complex with Cu^{2+} which get excreted through urine. It is a constituent of the enzyme phosphatase which converts organic phosphatase into free phosphates.

About 2 mg of copper per day is the recommended dietary intake for an adult.

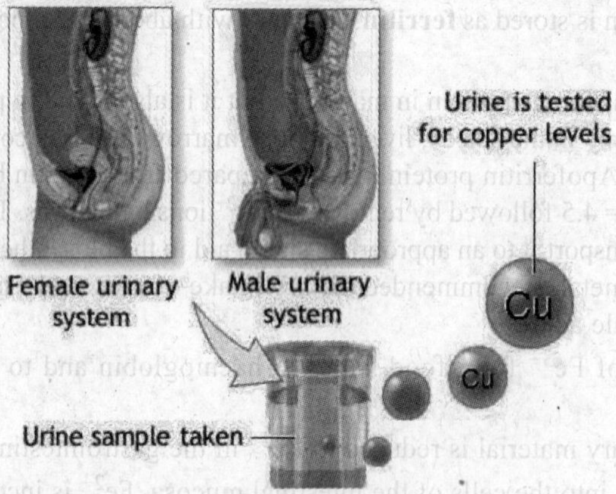

Cobalt

Cobalt is also essential trace element which is essential for many organisms including mammals. It **activates a number of enzymes**. It is constituent of vitamin B_{12} which is required for the formation of haemoglobin in *vivo*. However, it is highly toxic to plants and moderately toxic to mammals when injected intravenously. Deficiency of cobalt in soil adversely affects the health of grazing animals. Adding cobalt salts to soil, improves their health.

Zinc

Zinc is a constituent of enzyme carboxypeptidase which catalyses the hydrolysis of terminal peptide bond at the carboxy end of peptide chain. Zinc is an essential element for almost all organisms. It behaves as a Lewis acid in biochemical systems. It is **soft acceptor** as compared to other divalent cations like Mg^{2+}, Ca^{2+} or Mn^{2+} and therefore, acts as a stronger Lewis acid towards many biomolecules. Zinc containing enzyme **carbonic anhydrase** is vital for respiration in animals because it catalyses the normally slow carbonic acid-carbon dioxide reaction equilibrium.

$$CO_2\,(aq) + H_2O \underset{\text{anhydrase}}{\overset{\text{Carbonic}}{\rightleftharpoons}} H_2CO_3(aq)$$

The enzyme has one atom of zinc per molecule and has a molecular weight of 30,000. It has a common four coordinated environment around Zn^{2+} in which three of the ligands are the immadazole nitrogens of three histidines and fourth is water molecule or hydroxide ion (as shown below). The carbon dioxide hydration and dehydration of carbonic acid equilibrium is pH dependent.

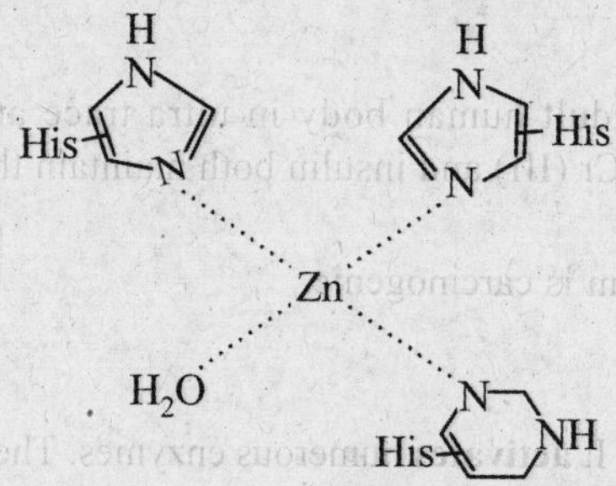

Carboxypeptidase is also zinc containing enzyme having some 300 amino acid residues including three methionyl residues. Their functions are the catalytic hydrolysis of peptide bonds. Its activity is directed specially towards carboxyl terminal peptide bonds.

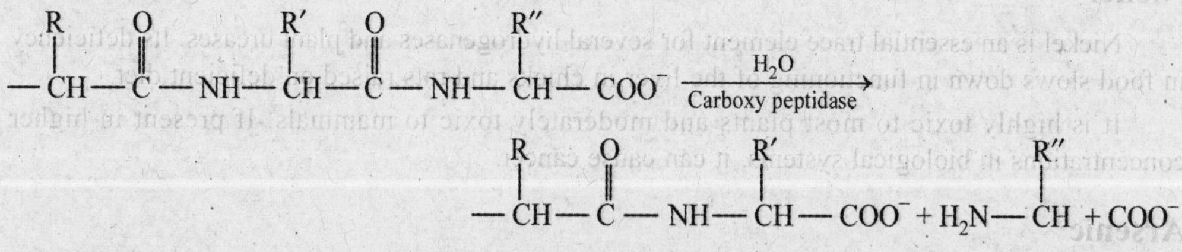

This selective enzyme hydrolyses those polypeptides in which the terminal amino acid segment has an aromatic or a branch chain aliphatic substituent, R″.

The deficiency of this enzyme may cause loss of appetite. It is toxic only in very large amounts. Zinc is essential for normal growth, reproduction and life expectancy of animals and has a major role to play in tissue repair and wound healing.

Selenium

Selenium is essential trace element for mammals and some higher plants. It is a component of glutathione peroxidase. It is an essential constituent of some enzymes and protects biological systems against free radical oxidants and stresses. Livestocks grown on selenium deficient pastures suffer from white muscle disease. However, if grazing in a soil having higher selenium concentration, they suffer from **central nervous system toxinosis**.

The minimum dietary intake of selenium is 70 mg. Sea foods, meats and onion family are the sources of selenium. Selenium deficiency in humans results in degenerative condition of the heart tissue, known as **Keshan disease**.

Molybdenum

It is an essential trace element for all organisms except green algae. It is a constituent of several enzymes used by nature for nitrogen fixation and nitrate reduction.

It is moderately toxic. Molybdenum excesses in biological systems may cause gout like syndrome.

Chromium

This element is present in adult human body in ultra trace amounts. It is involved in glucose metabolism and diabetes. Cr (III) and insulin both maintain the correct level of glucose in the blood.

In +6 oxidation state chromium is carcinogenic.

Manganese

It is essential to all organisms. It **activates** numerous enzymes. The Mn (II) enzyme produced in the liver converts nitrogenous waste products into urea which is carried by the blood to the kidneys which excrete it out into the urine.

The deficiency of manganese in soil inhibits plant growth and leads to infertility in mammals which consume such plants.

Nickel

Nickel is an essential trace element for several hydrogenases and plant ureases. Its deficiency in food slows down in functioning of the liver in chicks and rats raised on deficient diet.

It is highly toxic to most plants and moderately toxic to mammals. If present in higher concentrations in biological systems, it can cause cancer.

Arsenic

It is an essential ultra trace element for biological systems including humans. Its deficiency in chicks results in depressed growth.

It is moderately toxic to plants and highly toxic to mammals when present in more than ultra trace amounts.

Cadmium

Cadmium has practically no significant use in living organisms. It is toxic to all organisms. Its cummulative poison in mammals causes renal failure, hypertension, anemia and disorders of bone marrow.

Mercury

Mercury is a toxic substance. The mercury vapour is absorbed in the lungs, dissolves in the blood and then carried to the brain. This causes irreversible damage to central nervous system.

Phosphorus

It is an essential element for life. It is an important part of energy rich molecules, ATP. It also is an important component of the bones as apatite $Ca_5(PO_4)_3$ in the skeleton of vertebrates. It is also strength ensuring material of the teeth. Partial formation of fluorapatite, $Ca_5(PO_4)_3F$ strengthens the structure and makes it less soluble in the acid formed from fermenting organic material in teeth, and thus protects tooth decay.

Chloride

Chloride ion has a vital role in ion balance. It is essential for higher plants and mammals as NaCl electrolyte. In HCl, it is essential in digestic juices. The deficiency of chloride in infants can cause impaired growth.

Iodine

It is an essential constituent of the thyroid hormone *e.g.* thyroxine which is important in metabolism and growth regulation.

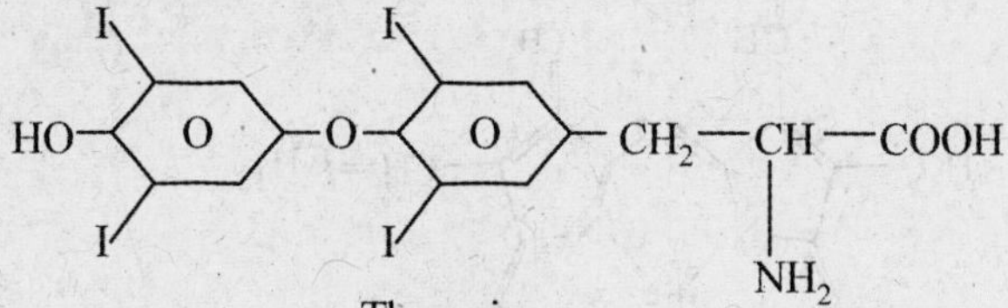

Thyroxine

Nearly 75% of iodine present in our body is located in thyroid gland. Iodine deficiency causes abnormal functioning of thyroid gland and causes goitre.

7.5. METALLOPORPHYRINS

Porphyrins are one of the most important groups of bioinorganic compounds in which a metal ion is surrounded by the four nitrogens of porphin ring. Porphines are made of four pyrrole rings linked together through methene bridges. Therefore, porphines have macrocylic pyrrole system with conjugated double bonds as shown in Fig. 7.1 These porphines act as tetradentate ligands with four nitrogen donor sites. Two of these are tertiary nitrogen donor positions which can form coordinate bonds by donating a pair of electrons each to the metal ion. The other two are secondary nitrogen donor positions, each of which lose a proton in forming a coordinate bond with a metal ion. Thus, a porphin ring acts as a tetradentate dinegative ligand (or dianion). Dipositive cations such as Mg^{2+}, Fe^{2+}, or Ni^{2+} are capable of forming neutral complexes with porphine as shown Fig. 7.2.

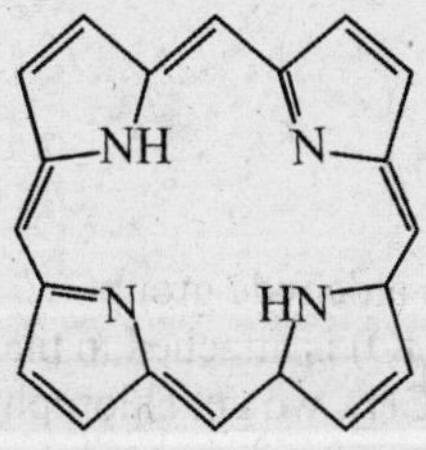

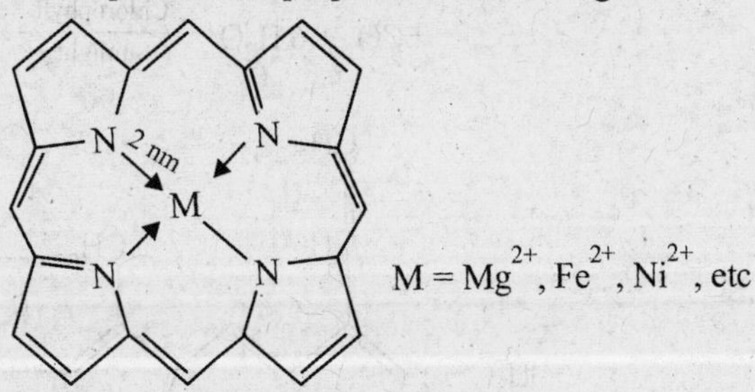

Fig. 7.1 Porphin ligand **Fig. 7.2** Metal complex with porphin ligand

Four pyrrole rings of porphin carrying substituents other than hydrogen are called porphyrins. The complexes in which a metal ion is held in the porphyrin ring system are called **metalloporphyrins**. Such complexes play a vital role in biological systems.

With delocolisation of electrons in the pyrrole rings we obtain stable porphyrin system. The best value of the size of the central hole has been estimated to be of radius 0.2 nm. The size of the central **hole** in the centre of the porphyrin ring is ideal for accommodating metals of first transition series or lighter alkaline earth metals. The metal ion of appropriate size are surrounded by four nitrogens of porphine ring in a square planar geometry and the axial sites are available for other ligands. However, if the size of the metals ion is smaller than required (or ideal) size, the ring becomes ruffled to allow closer approach of the nitrogen atoms to the metal. On the other hand, if the size of the metal ion is lager than the size of the hole, the metal ion cannot fit into the hole and stays out of the plane of four nitrogen atoms. In such a case the metal sites above the ring which becomes domed and acquires square pyramidal configuration.

The structures of two important metalloporphyrins heme and chlorophyll are being considered here: Heme Fig. 7.3 contains iron (II) which is present in hemoglobin and myoglobin.

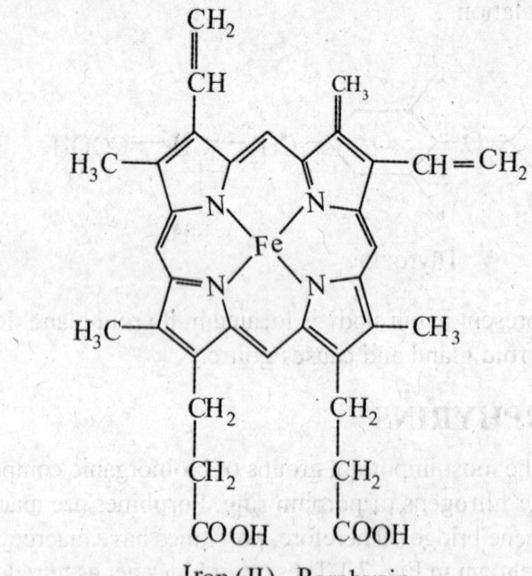

Fig. 7.3. Iron protoporphyrin in heme

Chlorophyll is magnesium complex of porphyrin which plays important role in **photosynthesis**. In this metalloporphyrin, in addition to substituents, a double bond in one of the

$$6CO_2 + 6\,H_2O \xrightarrow[\text{Sunlight}]{\text{Chlorophyll}} \underset{\text{Glucose}}{C_6H_{12}O_6} + 6O_2$$

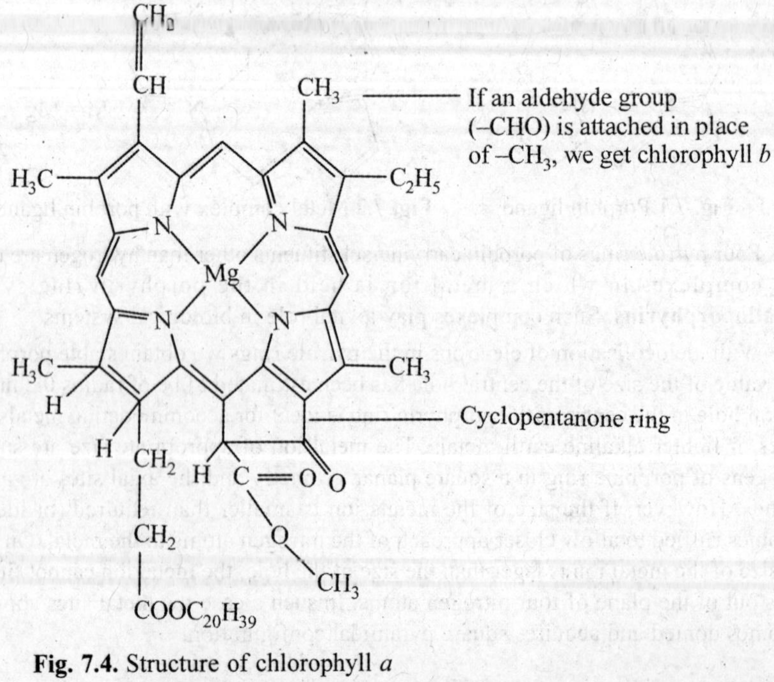

Fig. 7.4. Structure of chlorophyll *a*

pyrrole rings is reduced to form a magnesium dihydrido porphyrin complex as shown in Fig. 7.4. A cyclopentanone ring is also fixed to a pyrrole ring. Photosynthesis is an important redox reaction occurring in nature to convert water and carbon dioxide into carbohydrates and oxygen in the presence of sunlight.

7.6. IRON PORPHYRINS : HAEMOGLOBIN AND MYOGLOBIN

All organisms require oxygen in order to survive. It is formed during photosynthesis involving biologically important redox reactions. Different proteins have different tendencies to bind and transfer oxygen. These proteins are known as **oxygen carriers.** The most important two oxygen carrier proteins are haemoglobin (Hb) and myoglobin (Mb). These are iron porphyrin complexes which are oxygen transfer and oxygen storage agents in the blood and muscle tissues respectively. Myoglobin is a small intracellular protein present in vertibrate muscles while hemoglobin is a large intracellular protein responsible for the red color of the red blood cells. Functions of these two proteins are given as under:

(*i*) Haemoglobin picks up oxygen in the lungs and delivers it to the rest of the body.

(*ii*) Myoglobin accepts oxygen from the haemoglobin in the muscles and stores it until needed for energetic processes.

(*iii*) Deoxygenated hemoglobin uses some of its amino groups to fix up CO_2 and then transports CO_2 back to the lungs.

Structures of myoglobin and haemoglobin

Myoglobin

Myoglobin consists of one polypeptide chain (globin) with one heme group (Fig. 7.3). The peptide chain consists of 150-160 amino acid residues folded about the single heme group as shown below in Fig. 7.5.

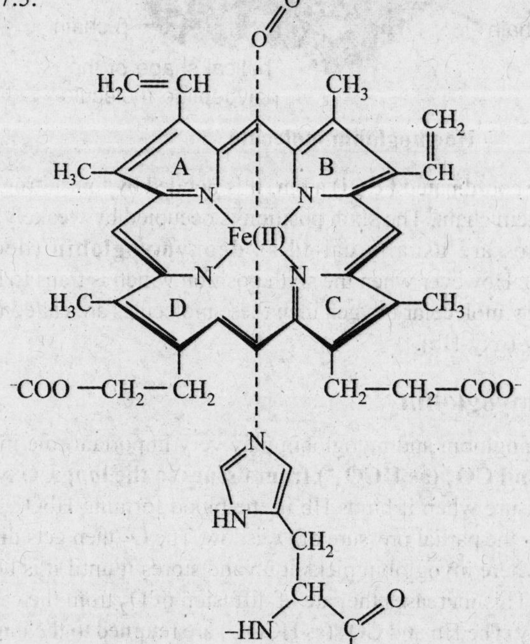

Fig. 7.5. The structure of myoglobin containing heme group and polypeptide chain.

The heterocyclic ring system of heme is porphyrin derivative containing four pyrrole groups joined by methene bridges. The Fe (II) atom present at the centre of the heme is bonded by four porphyrin nitrogen atoms and one nitrogen atom from imidazole side chain of histidine residue which is a part of long protein chain of amino acid residues. This polypeptide chain plays an important role in biological fixation of O_2.

Haemoglobin

Haemoglobin is a large protein with a molecular weight of about 60000. It consists of four sub units each of which contains one heme group (Refer to Fig. 7.3) associated with protein globin. Therefore, there are four heme groups bonded to four protein chains. One heme group with its protein chain is called sub unit. The four sub units are similar but not identical, two sub units form alpha (α) chains of 141 amino acids and two form β chains of 146 amino acids. It may be noted that amino acid sequences of neither α nor β sub units of haemoglobin match the sequences in myoglobin.

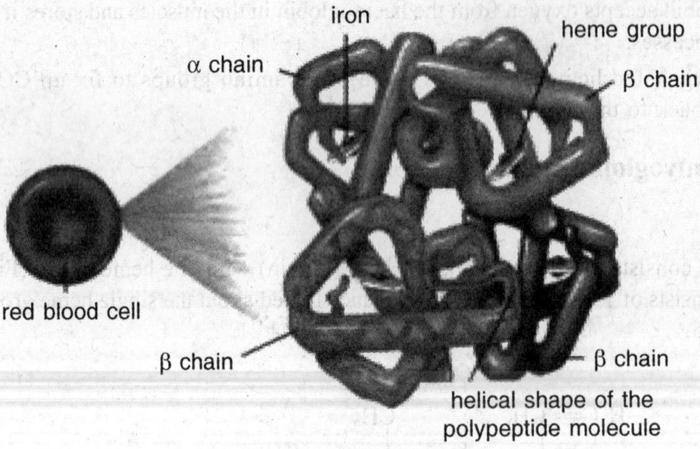

Haemoglobin molecule

Both Hb and Mb have five coordinated Fe (II) atom. It is bonded by four nitrogen atoms from pyrrole rings and fifth from protein chain. The sixth position is occupied by weakely bonded water. Mb and Hb in such molecules are usually called as **deoxymyoglobin** (deoxy-Mb) and **deoxyhaemoglobin** (deoxy-Hb). However when the sixth position which is trans to histidine chain (shown in Fig. 7.4) is occupied by molecular oxygen then these molecules are called **oxymyoglobin** (oxy-Mb) and **oxyhaemoglobin** (oxy-Hb).

Role of hemoglobin and myoglobin

As mentioned earlier haemoglobin and myoglobin play very important role in **transporting oxygen from lungs to tissues and CO_2 (as HCO_3^-) from tissues to the lungs**. Oxygen is inhaled into the lungs at very high pressure when it binds Hb in the blood forming HbO_2. The oxygen is then transported to tissues where the partial pressure of O_2 is low. The O_2 then gets dissociated from Hb and diffuses to the tissues where myoglobin picks it up and stores it until it is needed. Mb has greater affinity for O_2 than Hb. This increases the rate of diffusion of O_2 from the capillaries to the tissues by increasing its solubility. The Hb and CO_2 (as HCO_3^-) are returned to the lungs from where CO_2 is exhaled.

The complete process may be depicted as under:

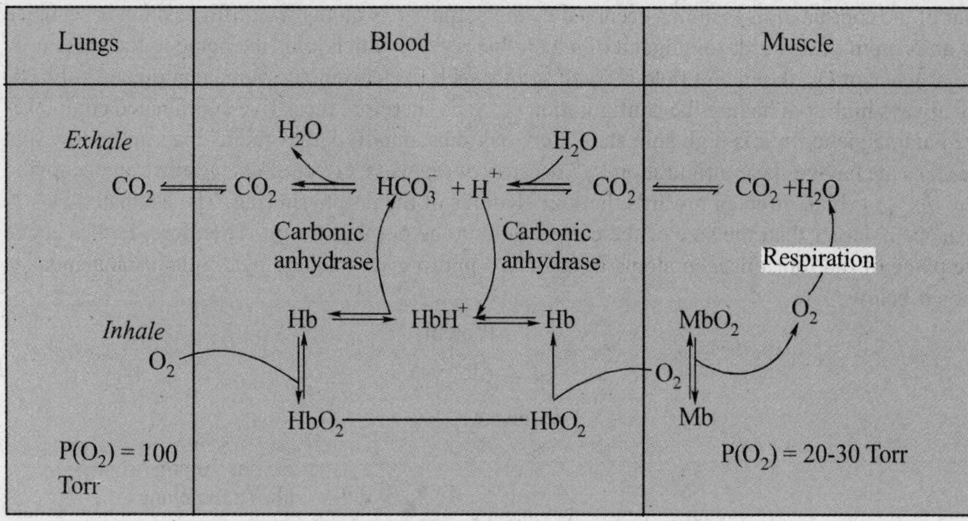

The heme group consists of Fe^{2+} ions enclosed in a porphyrin ligand. Fe^{2+} ion has six coordination sites. The porphyrin ligand takes up only four of the six sites, leaving two free binding sites on opposite sides of the metal ion. If a free heme is present in aqueous solution, the two vacant sites may be occupied by water molecules. A naked heme group binds to oxygen molecule in an irreversible manner when Fe^{2+} is oxidised to Fe^{3+}. This oxidation of iron destroys its oxygen binding capacity. The following reaction takes place

$$\text{Heme (Fe}^{\text{II}}\text{)} \xrightarrow{O_2,\ H_2O} \text{Hematin (Fe}^{\text{III}}\text{)}$$

Hematin which is a dimer has the structure

Hematin
This dimer can no longer act as oxygen carrier

However, it is clear from Fig. 7.5 that porphyrin Fe is protected due to the presence of hydrophobic protein chain around the Fe (II) and blocks the approach of larger molecules to the neighbourhood of Fe (II) and hence prevent oxidation of Fe (II) in haemoglobin to Fe (III). The hydrophobic groups also prevent the solvation of ions. Fe (II) of myoglobin and hemoglobin can be oxidised under certain controlled conditions to Fe (III) forming metmyoglobin of Mb and **methemoglobin** of Hb. These Fe (III) proteins are responsible for brown colour of old meat and dried blood.

Deoxymyoglobin abbreviated as (Deoxy-Mb) is a five coordinate high spin Fe (II) complex with four of the coordination positions occupied by the porphyrin N atoms. The fifth position is occupied by an N atom of an imidazole ligand of a histidine residue which joins the heme to the protein. In the absence of O_2, the ligand field is weak so that such five coordinate heme complexes of Fe (II) are always high spin having the configuration $t_{2g}^4 e_g^2$. Therefore these five coordinated complexes are paramagnetic. In this high spin state, Fe^{2+} has substantially larger radius than in the low spin Fe (II) state having the configuration t_{2g}^6 because of repulsive effect of one e_g electron occupying the $d_{x^2-y^2}$ orbital directed towards the four N atoms of the porphyrin ring. The estimated Fe – N distance is larger than the size of the central hole in the porphyrin ring. Therefore, Fe lies above the plane of the four nitrogen atoms by about 70 pm to give a square pyramidal arrangement as shown below

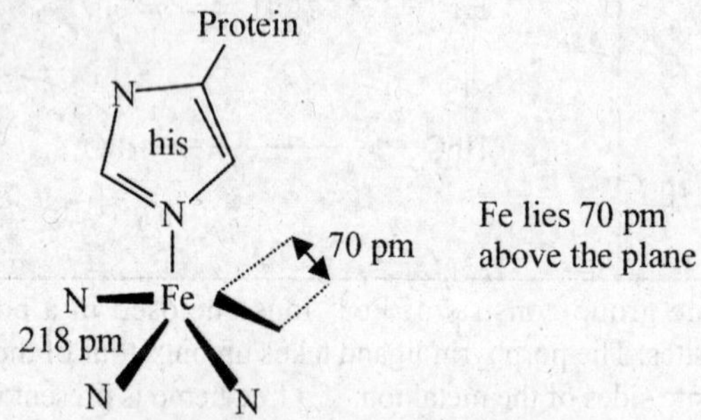

Square pyramidal structure with Fe lying about 70 pm above the plane of the ring

When oxygen bonds to the sixth position the iron becomes **coplanar**. In such a situation oxygen gets coordinated to Fe (II). The ligand field becomes strong resulting in spin pairing giving a low spin t_{2g}^6 complex. The resulting complex is **diamagnetic** and Fe – N bond distance is approximately same as porphyrin hole. Therefore, the iron becomes coplanar. The release of strain energy of the square pyramidal complex compensates the loss of spin pairing energy in going from 5 coordinated deoxy Hb or Mb to six-coordinated oxy-Hb or oxy Mb. Therefore, the coordination of O_2 will result in dropping of Fe in the plane of the heme group. The shrinkage of Fe^{2+} and falling into the plane of porphyrin ligands is depicted in Fig. 7.6.

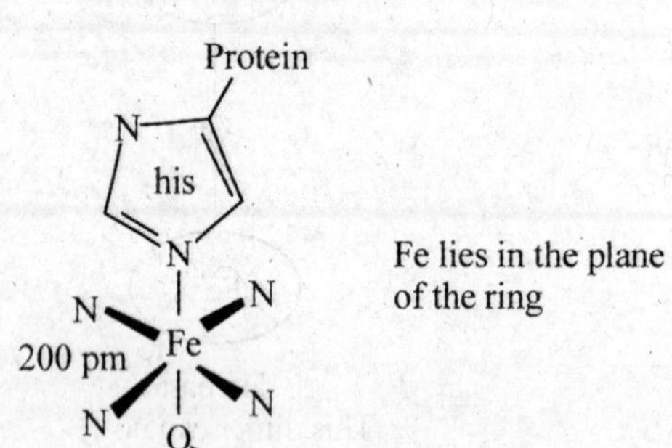

Oxy hemoglobin low spin octahedral complex. Fe lies in the plane of porphyrin ring.

The net result of this interaction in hemoglobin is to increase the affinity of O_2 of the second heme site and so on. In other words, as one iron binds an oxygen molecule, the molecular shape changes to make the binding of additional oxygen molecules easier. This phenomenon is called **cooperativity**.

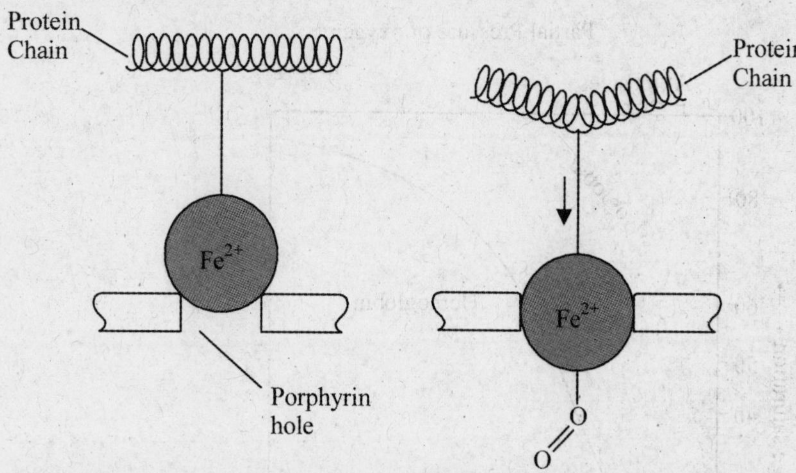

Fig. 7.6 The shrinkage in size of Fe^{2+} and dropping into the plane of porphyrin ligands.

Cooperativity enables Hb to bind and release O_2 more effectively.

Myoglobin and Haemoglobin Functions and Cooperativity

The main function of hemoglobin is to bind O_2 at high partial pressure of O_2 in the lungs and then carry it through the blood to the tissues where myoglobin picks up O_2 from Hb. In the lungs where the pressure O_2 is high and much O_2 is bound, the affinity of heme for O_2 becomes very high. It efficiently loads up with as much O_2 as possible. But when the hemoglobin reaches the cells where the pressure of O_2 is low, O_2 begins to dissociate from the complex. The myoglobin picks up all the O_2. Since myoglobin has only one heme group, it does not have any cooperative binding, so it does not lose its affinity for O_2. This shows the Mb has a higher affinity for O_2 than Hb at low partial pressure of O_2 in the muscle.

Carbon dioxide transport and Bohr effect

A graph between percentage saturation and partial pressure of O_2 is shown in Fig. 7.7.

The graph shows that Hb and Mb have almost similar binding affinity for O_2 at high O_2 pressure but Hb is much poor O_2 binder at lower pressures of O_2 in muscles. As a result, Hb passes its O_2 onto Mb as required. Moreover, the need for O_2 is greatest in tissues which have already consumed oxygen and simultaneously produced CO_2. The CO_2 lowers the pH ($2H_2O + CO_2 \rightleftharpoons HCO_3^- + H_3O^+$) and the increased acidity favours the release of O_2 from oxyhaemoglobin to Mb. The oxygen affinity of haemoglobin varies with the pH of the medium. This pH **sensitivity effect** is called **Bohr effect**. In other words the variation of oxygen affinity with the pH of the medium is called Bohr effect. It can be explained in terms of the effect of pH on the interaction between the heme group and the ionizable groups of the protein.

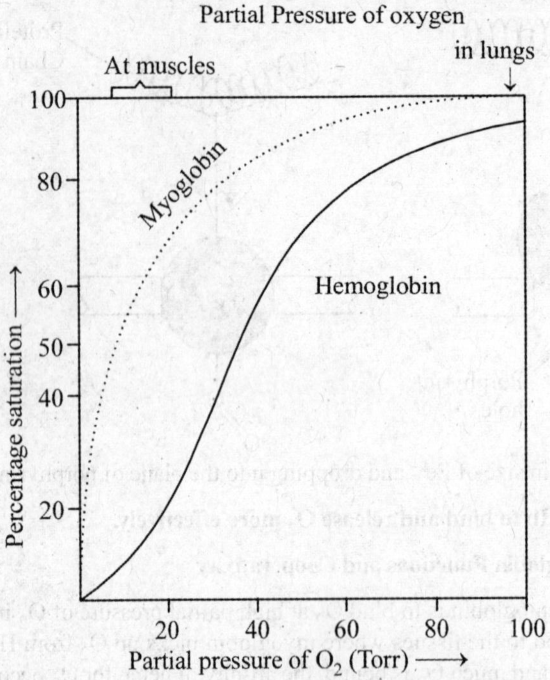

Fig. 7.7 O_2 binding curves for Mb and Hb with partial pressure of O_2.

SOLVED EXAMPLES

Example 1. Explain the mechanism of photosynthesis.

Solution. The process in which carbon dioxide and moisture in the atmosphere combine in the presence of chlorophyll and sunlight to produce carbohydrates is called *photosynthesis*. The reaction for the process can be represented as under:

$$6CO_2 + 6H_2O \xrightarrow[\text{Sunlight}]{\text{Chlorophyll}} C_6H_{12}O_6 + 6O_2 \qquad .....(i)$$

Tracer studies have proved that all the oxygen evolved in the above reaction comes from water and not from carbon dioxide. Hence the following mechanism for photosynthesis is proposed:

First Step. Chlorophyll absorbs solar energy which in used in forming energetic ATP molecules and H_2O molecules are oxidised to O_2.

$$12H_2O \longrightarrow 12H_2 + 6O_2 \qquad .....(ii)$$

Second step. Energetic ATP molecules formed above reduce CO_2 into glucose

$$6CO_2 + 12H_2 \longrightarrow C_6\,H_{12}\,O_6 + 6H_2O \qquad .....(iii)$$

On adding equations (*ii*) and (*iii*), we get equation (*i*). Magnesium makes the entire molecule rigid such that energy is not lost thermally and it also increases the rate at which short lived singlet excited state is transformed into triplet state. Triplet state has a longer life and thus can transfer its excitation energy into the redox chain producing carbohydrates.

Example 2. Give a brief account of the structure of haemoglobin.

Solution. About 65% of iron in the human body (roughly 4g) is present as haemoglobin. Haemoglobin is regarded as red pigment in the red blood cells (RBC's). Haemoglobin binds O_2 at lungs and then arteries carry blood to different parts of the body such as muscles where O_2 is required.

100 ml of the blood of a normal male contains approximately 15 g of haemoglobin. Haemoglobin contains about 0.35% of iron. Thus 100 ml of the blood contains roughly 50 mg Fe. Total iron content in blood volume of 5000 ml in the human body is thus 2.5 g approximately.

Haemoglobin is a conjugated protein having a molecular weight of 64500. It contains four identical units arranged roughly tetrahedrally. Each unit contains one heme group whose structure is shown below.

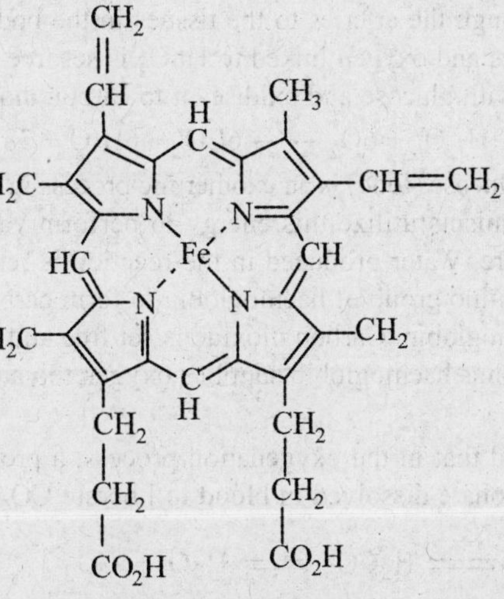

Structure of heme group

Each haemoglobin molecule has four *heme* groups which are bound to globin (a protein) on its surface. Schematic representation of structure of haemoglobin is shown below.

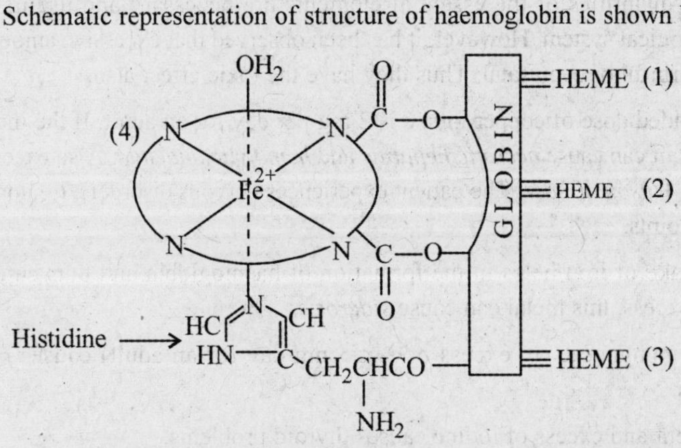

Schematic representation of the structure of hemoglobin.

Fe (II) is in high spin state ($3d^6$ or $t^4_{2g}\ e^2_g$). Thus, iron has four unpaired electrons in haemoglobin molecule.

Thus, haemoglobin is an octahedral complex of iron (II). Iron occupies the central position and the four corners of the square base are occupied by four N-atom of histidine while the other axial position is occupied by H_2O molecule.

Example 3. Discuss the role of haemoglobin in living systems.

Solution. When we breathe in air, oxygen of the air combines with haemoglobin (Hb) present in blood to form oxyhaemoglobin (HbO_2)

$$\text{Haemoglobin} + O_2 \rightleftharpoons \text{Oxyhaemoglobin} + H_2O$$

$$Hb + O_2 \rightleftharpoons HbO_2 + H_2O$$

This process is known as oxygenation of haemoglobin. This takes place in the lungs. Iron is in +3 oxidation state. H_2O molecule present at one of the axial positions in haemoglobin is reversibly replaced by O_2 in this process.

The blood runs through the arteries to the tissues in the body. In the course of this process, oxygen pressure decreases and oxygen linked to HbO_2 is set free. Free oxygen diffuses into body cells where it combines with glucose and oxidises it to carbon dioxide and water.

$$C_6H_{12}O_6 + 6O_2 \longrightarrow 6CO_2 + 6H_2O + 38\ \text{ATP (Energy)}$$

Thus, oxidation of glucose (food) is an exothermic process in which energy in the form of ATP is produced. Living organisms utilize this energy to perform various metabolic process and to maintain body temperature. Water produced in the reaction is retained in the body while carbon dioxide combines with amino group of haemoglobin to form carbamino haemoglobin which later decomposes to give haemoglobin. Carbon dioxide is set free and exhaled out. Haemoglobin goes back to lungs for reuse. Thus haemoglobin supplies oxygen to various parts of the body or its acts as **oxygen carrier.**

It may be mentioned that in the oxygenation process, a proton (H^+) is also produced. This proton reacts with bicarbonate dissolved in blood to liberate CO_2.

$$HCO_3^- + H^+ \rightleftharpoons H_2CO_3 \rightleftharpoons H_2O + CO_2 \uparrow$$

Example 4. Write a note on harmful effect of excess intake of metals on human body.

Solution. Small quantities of the essential elements are necessary and useful for normal functioning of the biological system. However, it has been observed that excessive amounts of these metals might prove harmful or even fatal. Thus they have the toxic effect also.

1. The recommended dose of copper intake is 2 mg per day for an adult. If the intake rises to about 100 times of this, it can cause *necrotic hepatitis* and *hemolytic anaemia.* Also excessive intake of copper metal causes *Wilson disease.* The patient experiences nervous disorders, tremors, difficulty in breathing and stiff joints.

2. Controlled intake of iron helps in the formation of haemoglobin and thus cures anaemia. However, if taken in excess, this metal can cause *siderosis.*

3. Too much of manganese (in excess of 2.5–5 mg/day for an adult) causes *psychic* and *neurological* disorders.

4. Both insufficient and excess of iodine causes thyroid problems.

Example 5. What is the role of metal chelates in living system?

Solution. Most of the essential elements in the human body do not exist in free state. They form complex compounds (called chelates) with other groups. These chelates perform the following functions.

1. They help in the storage and transfer of the metal ions and serve as agents for transmission of energy in the animal and plant metabolism.

2. They catalyse metabolic processes involving (*a*) acid-base reactions in which the metal is the Lewis acid that combines with a substrate to accelerate a reaction (*b*) reduction-oxidation reaction with the change in the valency of the metal in the chelate.

Example 6. Discuss the structure of chlorophyll and explain the mechanism of photosynthesis.

Solution. Chlorophyll is a green coloured pigment present in green plants and algae. The basic unit in the molecule of chlorophyll is porphyrin ring. Chlorophyll is a square planar complex with Mg^{2+} at the centre. It is a naturally occurring chelate complex of Mg.

It has been noticed that Mg-atom in chlorophyll lies slightly above the plane of ring. Two water molecules can be added axially to the molecule of chlorophyll. These coordinated water molecules provide an opportunity to associate with other chlorophyll molecules through hydrogen bonding.

Two types of chlorophyll are found in higher plants, chlorophyll *a* and chlorophyll *b*, which differ in a substituent on ring II, while algae contain four chlorophyll (*a*, *b*, *c*, *d*). Chlorophyll *a* and *b* function most effectively in the blue region and the red region of the visible light at wavelengths of 400 – 500 and 600 – 700 nm but weakly between 500 – 600 nm. Most abundant of chlorophylls is chlorophyll *a*.

Four nitrogen atoms are situated at the four corners of a square.

All types of chlorophyll contain porphyrin ring in which a double bond in one of the pyrrole rings is reduced and a fused cyclopentanone ring is also present.

Example 7. Explain the mechanism of $Na^+ - K^+$ pump

Solution. The basis of the mechanism is that the $Na^+ - K^+$ AT Pase is phosphorylated by ATP in the presence of Na^+ and Mg^{2+} ions. The site for phosphorylation is the side chain of a specific **asparate** residue, E.

The phosphorylated intermediate (E–P) is then hydrolysed. It gets dephosphorylated in the presence of K^+ and gives the original phosphoprotein.

$$E + ATP \xrightleftharpoons{Na^+,\ Mg^{2+}} \underset{\text{Intermediate}}{E - P + ADP} \qquad \text{Phosphorylation}$$

$$E - P + H_2O \xrightleftharpoons{K^+} E + HPO_4^{2-} \qquad \text{Dephosphorylation}$$

It has been found that Na^+ and Mg^{2+} are necessary for the formation of the intermediate and K^+ are necessary for dephosphorylation of the intermediate. During this reaction $3Na^+$ and $2K^+$ are transported per ATP hydrolysed. Therefore, the pump generates an electric current across the plasma. In other words, $Na^+ - K^+$ ATPase pump is **electrogenic**. Each operation of the cell, pumps out larger number of Na^+ ions from the cell than the number of K^+ ions that it pumps into the cell. As a result, the interior of the cell acquires an excessive negative charge while the exterior of the cell acquires an excessive positive charge. As a result electrical potential gradient across the cell membrane is developed which is responsible for the **transmission of nerve signals** in the animals.

ATP is not hydrolysed unless Na^+ and K^+ are transported. In other words, the system is coupled. The energy stored in ATP is only dissipated if the pump works. The $Na^+ - K^+$ pump is also reversible so that it synthesizes ATP.

Example 8. Discuss biological role of calcuim ions

Solution. Ca^{2+} ions plays an important role in biological processes and in skeletal formation. With phosphorus in the mineral hydroxyapatite $Ca_5(PO_4)_3OH$, it is the major constituent of bones, teeth and shells. Ca^{2+} plays many biochemical roles such as

(*i*) It acts as a messenger for hormonal action

(*ii*) It acts as a trigger for muscular contraction

(*iii*) It acts as initiation of blood clotting

(*iv*) It also plays a role in the stabilization of protein structure

(*v*) It helps in the maintenance of rhythm of heart

Role of Ca^{2+} in transport

Ca^{2+} plays an important role in muscles. It triggers a signal that stimulates muscles to contract. In the normal state, the concentration in the intracellular fluids is very low because nearly all the Ca^{2+} ions in muscle is pumped into a complex network of vehicles known as **sarcoplasmic reticulum** (SR). Their concentration in intracellular fluids is about 10,000 times less than their concentration in the extra cellular fluids. The maintenance of low calcium in concentration in the intra cellular fluids is done by biochemical process known as **calcium pump.**

The mechanism of Ca^{2+} pump is as under

$$E + ATP \xrightleftharpoons{Ca^{2+},\ Mg^{2+}} \underset{\text{Intermediate}}{ADP + E - P}$$

$$E - P + H_2O \xrightleftharpoons{Mg^{2+}} E + HPO_4^{2-}$$

During the cycle, two Ca^{2+} are transported for each ATP hydrolysed.

7.7. NITROGEN FIXATION

The conversion of atmospheric nitrogen into useful nitrogenous compounds by natural or artificial methods is called fixation of nitrogen. Nitrogen present in these nitrogenous compounds is called *fixed* or *combined* nitrogen.

Natural methods of the fixation of nitrogen

1. *By lightning discharges.* The nitrogen and oxygen present in air combine together to form nitric oxide under the influence of lightning discharges. Nitric oxide gets oxidised by excess of oxygen present in the atmosphere to form nitrogen peroxide which dissolves in water to form nitric acid. This acid is washed down by rain into the soil, where it reacts with limestone and alkalis of the soil to form nitrates and is stored there as plant food.

$$N_2 + O_2 \xrightarrow{\text{Lightning}} 2NO \ (\textit{Nitric oxide})$$

$$4NO + 3O_2 \longrightarrow 2N_2O_5 \quad (\textit{Nitrogen pentoxide})$$

$$N_2O_5 + H_2O \longrightarrow 2HNO_3$$

$$2HNO_3 + CaCO_3 \longrightarrow Ca(NO_3)_2 + CO_2 + H_2O$$

2. *By symbiotic bacteria.* The atmospheric nitrogen is being constantly transferred to the soil through certain bacteria called *symbiotic bacteria (rhizobium)*. They grow in small nodules (Fig. 7.8) in the roots of plants belonging to the family of *leguminacea* (pea, gram etc.) and directly assimilate atmospheric nitrogen and convert it into products useful for plant growth.

Nitrogen and ammonium salts present in the soil are converted by *nitrosifying bacteria* into *nitrites* and by *nitrifying bacteria* into *nitrates.* The final products nitrates serve as a plant food. This process of oxidation is known as **nitrification** and it causes the combined nitrogen to be available to the plants.

The species of *rhizobium* is specific for each species of plant. Cyanobacteria (blue-green algae) perform nitrogen fixation in the ocean and, to some extent, in fresh water. All these microorganisms produce ammonia through the activity of nitrogenase.

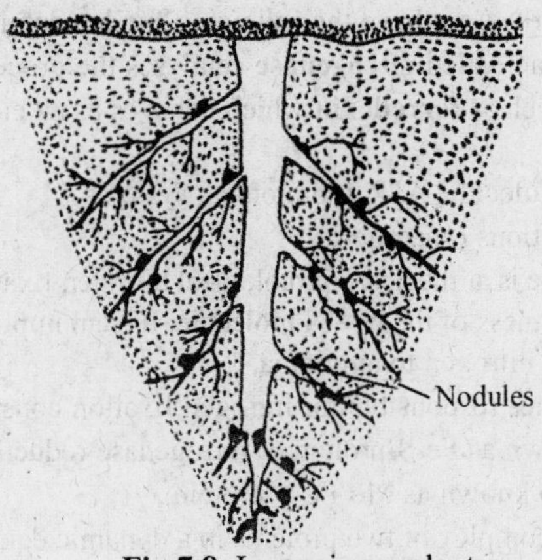

Fig 7.8. Leguminous plants.

Most free nitrogen bacteria perform nitrogen fixation anaerobically whereas cyanobacteria do it aerobically.

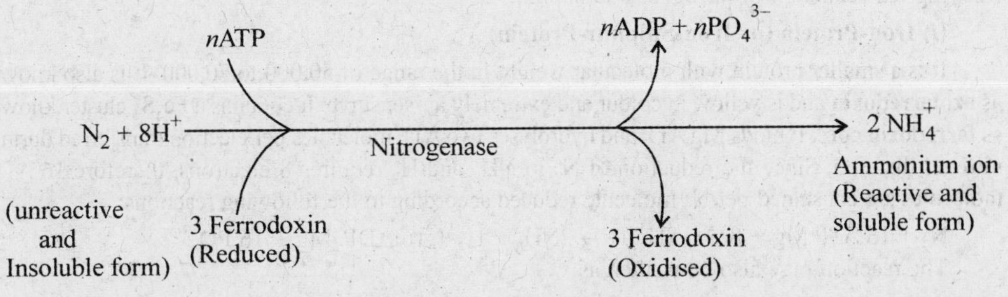

The reaction involves three reduced molecules of ferredoxin which donate six electrons. The source of this reducing power is NADPH. Twelve to eighteen molecules of ATP supply the necessary energy for the production of ammonia at a rate which is regulated by the inhibitory effect of the product on the association of the two components of the nitrogenase.

Denitrification. Nitrogenous compounds present in the soil are acted upon by a bacteria known as **denitrifying bacteria** which convert these compounds into free nitrogen which goes back to the atmosphere. Because of these cycle of changes which are constantly going on in nature the total average amount of the gas in the atmosphere remains unchanged.

Biological nitrogen fixation involves the reduction of nitrogen into ammonia by a biological process with the help of enzymes or microorganisms. The main reaction occurring in the process is

$$N_2 + 6H^+ + 6e^- \longrightarrow 2NH_3$$

The reaction for biological nitrogen fixation may also be presented as:

$$N_2 + 16\ ATP + 8H^+ + 8e^- \longrightarrow 2NH_3 + 16\ ADP + 16\ PO_4^{3-} + H_2$$

The ammonia may be further converted into nitrate or nitrite or directly used in the synthesis of amino acids or other essential compounds. This reaction occurs at atmospheric conditions of temperature and pressure in Rhizobia, bacteria in nodules on the roots of leguminous plants such as peas and beans. In contrast, industrial synthesis of ammonia require high temperatures and pressures and catalysts.

All nitrogen fixation processes have the following fundamental features:

(*i*) The bacterial enzyme called **nitrogenase** catalyses the process.

(*ii*) Strong reductant such as **ferredoxin** which behaves as an electron carrier with a very low redox potential.

(*iii*) The energy rich molecule, ATP in biological systems

(*iv*) Oxygen free conditions (anaerobic).

The enzyme nitrogenase is at the heart of biological nitrogen fixation. It has solved the problem of overcoming the great inertness of the N ≡ N molecule. It is an important enzyme which catalyses the conversion of molecular nitrogen to ammonia.

The nitrogenase enzymes responsible for nitrogen fixation consists of two metalloproteins:

(*i*) Fe-protein also known as Fe-S-protein or nitrogenase reductase

(*ii*) Mo-Fe-protein also known as Mo-Fe-S-protein.

Thus, nitrogenase is a complex of two proteins in a dynamic equilibrium with its components:

$$\text{Mo-Fe-protein} + 2\text{Fe-Protein} \rightleftharpoons \text{Nitrogenase complex.}$$

Thus, *nitrogenases are the complexes of iron-sulphur protein (Fe-S-Protein) and molybdenum-iron-sulphur protein whose activity couples ATP hydrolysis to electron transfer from ferredoxins for carrying out reduction of nitrogen to ammonia.*

(*i*) Iron-Protein (or Iron-Sulphur-Protein)

It is a smaller protein with molecular weight in the range of 50,000 to 70,000. It is also known as **azoferredoxin** and is yellow in colour and extremely air sensitive. It contains a Fe_4S_4 cluster known as **ferredoxin core**. It binds Mg ATP and hydrolyses two ATP molecules per electron transferred during nitrogen fixation. Since the reduction of N_2 to NH_3 and H_2 requires 8 electrons, therefore 16 ATP molecules are consumed per N_2 molecule reduced according to the following reactions:

$$N_2 + 16\text{ ATP Mg} + 8e^- + 8H^+ \longrightarrow 2NH_3 + H_2 + 16\text{ ADP. Mg} + 16\ PO_4^{3-}$$

The reaction may also be written as:

$$N_2 + 8H + 8e \xrightarrow[]{16\text{ ATP} \;\searrow\;\nearrow\; 16\text{ ADP} + 16\ PO_4^{3-}} 2NH_3 + H_2$$

(*ii*) Fe-Mo-Protein (for Mo-Fe-S-Protein)

Fe-Mo-Cofactor. It contains both Mo and Fe and therefore is known as Mo Fe protein or "molybdo ferrodoxin". It is a brown, air sensitive protein and has molecular weight in the range 220,000 to 240,000.

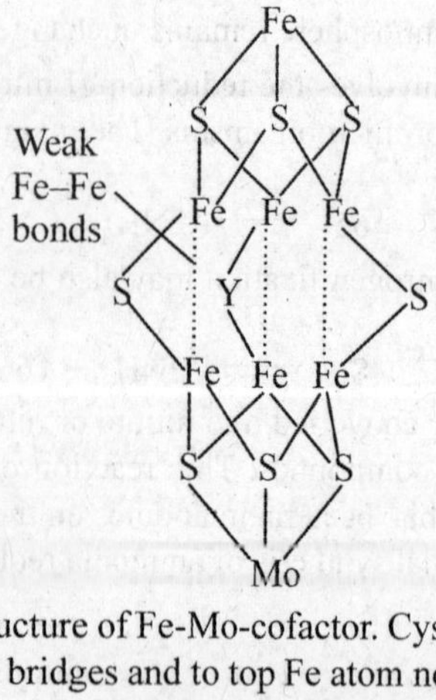

Fig. 7.9. Structure of Fe-Mo-cofactor. Cystine linkages to thiol bridges and to top Fe atom not shown.

This contains molybdenum, iron and sulphur. It consists of two MFe_3S_3(M=Mo or Fe) clusters joined by two bridging sulphides and a third ligand designated as Y which is a less ordered sulphur species. The iron atom in centre interact each other (shown dotted) and have Fe–Fe bond distance of oabout 2.5 Å suggesting Fe-Fe bond. This is shown in Fig. 7.9

On the basis of theoretical calculations and the knowledge of the catalytic properties of nitrogenase, the following model for binding of the substrate N_2 to the Fe-Mo cofactor has been suggested.

The Fe-Mo cofactor contains three weak Fe-Fe bonds (shown by dotted linkages). These are destabilized because of the distortion from ideal tetrahedral geometry. Therefore, it may be assumed that N_2 could bind in the centre of Fe-Mo cofactor (Fig. 7.10) thereby replacing the weak Fe-Fe bonds with multiple Fe-N bonds having approximately sp^3- geometry. As a result of Fe-N interactions, the sp^2-hybridized $N \equiv N$ bond gets weakened, thereby lowering the high activation barrier for N_2 reduction.

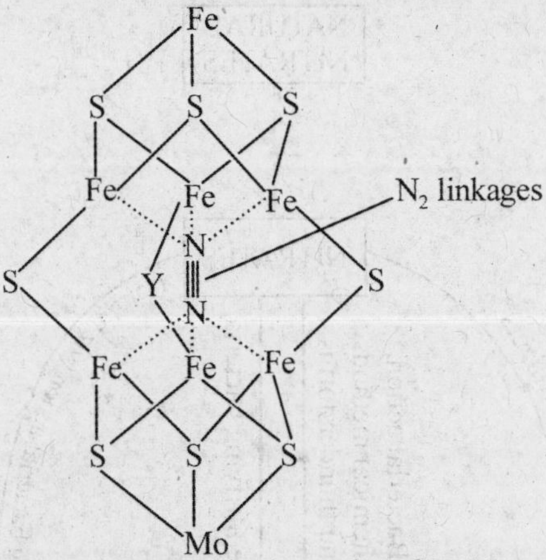

Fig. 7.10 Binding mode of N_2 in the cavity of Fe-Mo-cofactor.

Artificial methods used for the fixation of nitrogen

1. *Fixation of nitrogen as HNO_3.* Under the influence of high tension electric arc where the temperature is high, nitrogen of the air combines with oxygen to form nitric oxide. It combines with more of oxygen to form nitrogen peroxide. This may be absorbed in water in presence of excess of air to give nitric acid which may be used for the manufacture of nitrogenous fertilisers.

$$N_2 + O_2 \rightleftharpoons 2NO \qquad \textit{(Nitric oxide)}$$

$$2NO + O_2 \longrightarrow 2NO_2 \qquad \textit{(Nitrogen peroxide)}$$

$$\underset{\text{Nitrogen peroxide}}{4\,NO_2} + 2H_2O + \underset{\text{From air}}{O_2} \longrightarrow \underset{\text{Nitric acid}}{4HNO_3}$$

2. *Fixation of nitrogen as NH_3 and ammonium salts.* A mixture of nitrogen (manufactured by liquefaction of air) and hydrogen in the ratio 1 : 3 is compressed to a pressure of 200 – 500 atmospheres and is passed over a catalyst (finely divided iron + molybdenum) heated to about 550°C. This forms the *Haber's process* for the manufacture of ammonia which then can be converted into ammonium salts by treatment, with suitable acids.

$$N_2 + 3H_2 \xrightleftharpoons[\text{high press}]{\text{catalyst}} 2NH_3$$

3. *Fixation of nitrogen as calcium cyanamide, $CaCN_2$.* Nitrogen gas obtained by the evaporation of liquid air is passed over calcium carbide heated to 800 – 1000°C. A mixture of calcium cyanamide and carbon is obtained which is extensively used as a fertilizer under the name of *nitrolim.*

$$\underset{\text{Calcium carbide}}{CaC_2} + N_2 \longrightarrow \underset{\text{Calcium cyanamide}}{[CaCN_2 + C]} \; (\textit{Nitrolim})$$

4. *Fixation of nitrogen as nitrides.* Nitrogen combines with magnesium and aluminium at high temperature to give nitrides which are employed as a source of ammonia. These nitrides are decomposed by H_2O and NH_3 is evolved.

$$3Mg + N_2 \longrightarrow \underset{\text{Magnesium nitride}}{Mg_3N_2}$$

$$2Al + N_2 \longrightarrow \underset{\text{Aluminium nitride}}{2AlN}$$

$$AlN + 3H_2O \longrightarrow Al(OH)_3 + NH_3$$

A labelled diagram of nitrogen cycle is given below :

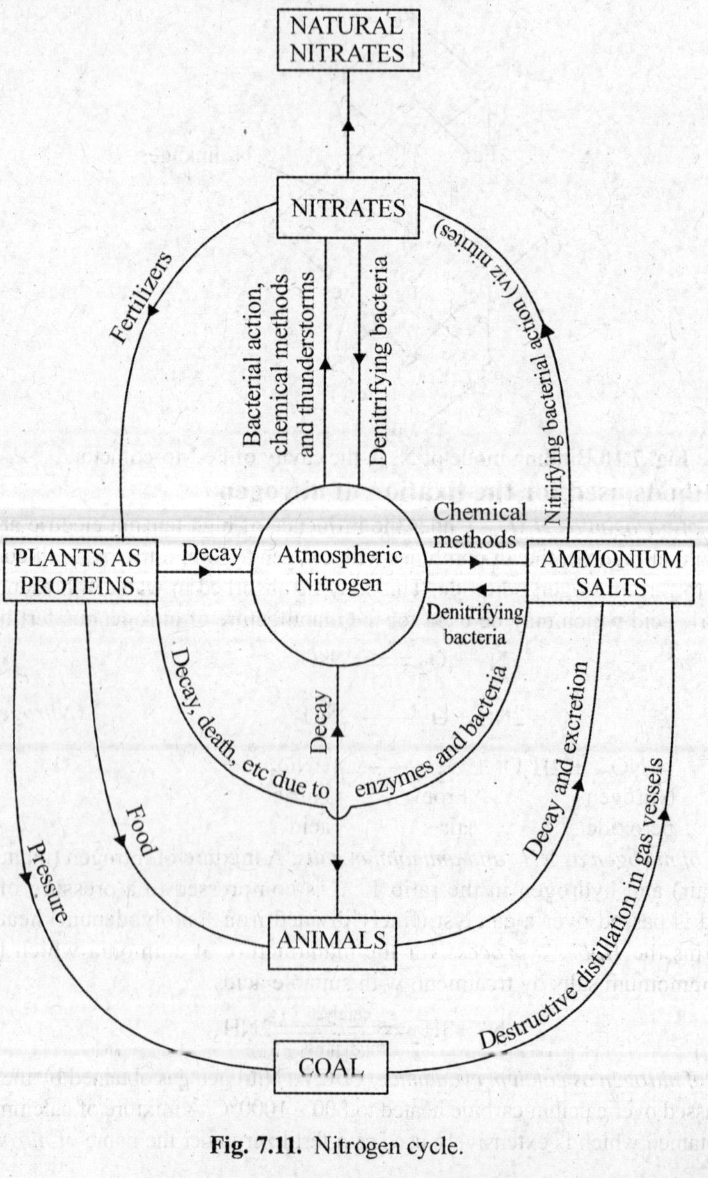

Fig. 7.11. Nitrogen cycle.

Certain bacteria return the reduced form of nitrogen back to environment by oxidising them. NH_4^+ are oxidised to NO_3^- by **nitrifying bacteria**. The nitrates are then reduced to N_2 by certain microorganism called denitrifying bacteria and this process is called **denitrification**.

MISCELLANEOUS SOLVED CONCEPTUAL PROBLEMS

Example 1. Name the two metalloproteins which comprises nitrogenase.

Solution. (i) Fe-protein or Fe-S-protein.

(ii) Mo-Fe-protein or Mo-Fe-S-protein.

Example 2. Compare roles of Ca^{2+} and Zn^{2+} at the active sites of enzymes. In what ways is Ca^{2+} advantageous over alkali metal ions?

Solution. Zn^{2+} is a soft acceptor and therefore acts as a strong Lewis acid towards many biomolecules especially with those having N and S donor sites. These Zn^{2+} complexes are very labile and facile in geometry. On the other hand, Ca^{2+} is a harder ion and therefore, is more selective towards bonding to hard donors such as oxygen rather than N or S. Zinc containing enzymes serve important function of catalysing the hydration-dehydration equilibrium of CO_2.Ca^{2+} is an intermediary between the nerve impulse and muscle contraction.

Example 3. Iron (II) salts undergo easy oxidation in air but it is not so in haemoglobin and myoglobin. Explain.

Solution. Iron (II) salts undergo oxidation readily. However, the structure of deoxy-Mb or deoxy-Hb is such that Fe (II) site is surrounded by protein chain. This blocks the approach of larger molecules to the neighbourhood of Fe(II) and therefore, it is not easily oxidised. This chain also prevents solvation. Therefore, the Fe (II) complex can survive in environment long enough to bind and release O_2.

Example 4. Write the main reaction for biological nitrogen fixation and name the enzyme involved.

Solution. $N_2 + 16\ ATP + 8e^- + 8H^+ \longrightarrow 2NH_3 + 16\ ADP + 16\ PO_4^{3-} + H_2$.

The enzyme nitrogenase is involved in the reaction.

Example 5. In what ways in Ca^{2+} advantageous over alkali metal ions?

Solution. Ca^{2+} has higher charge to radius ratio and therefore has a higher affinity for most hard ligands than alkali metal ions which are also hard acceptors. Therefore, Ca^{2+} should be advantageous when especially high reaction rates are needed alongwith high affinity for oxygen ligands.

Example 6. The reduction of N_2 to NH_3 is thermodynamically favourable (ΔG = –ve). In biological systems, still ATP molecule is required for energy requirement. Comment.

Solution. N≡N bond has very high bond dissociation energy (945 kJ mol^{-1}). This N≡N bond must be broken during nitrogen fixation. Therefore, a very large energy input is required to overcome the large activation energy barrier even though $\Delta G°$ is negative. In biological systems, this energy barrier is overcome by the energy provided by ATP.

Example 7. Give names of three essential trace elements and three essential ultra trace elements.

Solution. Trace elements : Iron, copper, zinc

Ultra trace elements : fluorine, iodine, selenium

Example 8. Oxygen molecule behaves as a π-acceptor ligand during interaction with Fe(II) in haemoglobin or myoglobin. Can other well known π-acceptors like CO, NO or CN^- etc. bind themselves with Mb or Hb?

Solution. Yes any π-acceptor can bind to haemoglobin but to a limited extent. For example CO is most toxic because of its greater π-acceptor character. However, the toxic effect of CO is limited by the steric effects of protein chain which prevents appropriate geometry to the bound CO.

Example 9. What is the oxidation state of iron in haemoglobin and myoglobin?

Solution. The iron in haemoglobin and myoglobin is + 2 oxidation state.

Example 10. Why nitrogenase enzyme perform better than the Haber process in the fixing of nitrogen?

Solution. Nitrogenase enzyme fixes nitrogen to produce NH_4^+ salts from N_2 and water in the presence of O_2 at the temperature of soil and atmospheric pressure. On the other hand, in the Haber process, the fixing of N_2 to NH_3 is carried out in the presence of catalyst at a very high temperature and high pressure. Thus, nitrogenase enzymes perform better than the Haber process.

Example 11. Sodium pump is electrogenic in nature. Comment.

Solution. Sodium pump acts as a transport system to maintain $Na^+ - K^+$ concentration gradient. For each ATP molecule hydrolysed, $3Na^+$ are transported out of the cell and $2K^+$ ions are transported into the cell according to the reaction:

$ATP^{4-} + H_2O + 3Na^+ \text{ (inside)} + 2K^+ \text{ (outside)} \longrightarrow 2K^+ \text{ (inside)} + 3Na^+ \text{ (outside)} + ADP^{3-} + H_2PO_4^-$. The net result is the movement of one positive charge outward per cycle and hence sodium pump is electrogenic in nature.

EXERCISES
(Including Questions from Different University Papers)

Multiple choice questions (Choose the correct option)

1. What is oxidation state of iron in haemoglobin and myoglobin respectively

(*a*) 3, 2 (*b*) 2, 2

(*c*) 2, 3 (*d*) 3, 3

2. O_2 is bound to heme in a

(*a*) bent way (*b*) linear arrangement

(*c*) tetrahedral arrangement (*d*) bridged way

3. Heme is a porphyrin complex of

(*a*) Fe (II) (*b*) Fe (III)

(*c*) Mg (II) (*d*) Zn (II)

4. During biological nitrogen fixation, nitrifying bacteria convert

(*a*) NO_3^- to NH_4^+ (*b*) N_2 to NH_4^+

(*c*) NH_4^+ to NO_3^- (*d*) NO_3^- to N_2

5. During nitrogen fixation, the number of ATP molecules which are consumed per N_2 reduced to NH_3 are

(*a*) 8 (*b*) 4

(*c*) 16 (*d*) 10

6. Oxymyoglobin contains

(*a*) O_2 at trans position to histidine chain.

(*b*) O_2 in the hole of porphyrin.

(*c*) O_2 bonded by coordinate bond to Mg (II)

(*d*) does not contain O_2

7. Acording to Bohr effect

(*a*) affinity of Hb for O_2 increases with decreasing pH

(*b*) affinity of Hb for O_2 decreases with decreasing pH

(*c*) affinity of Hb for Mb changes with pH

(*d*) affinity of Hb for CO_2 does not change with pH

8. The metal present in chlorophyll is

(*a*) Mg (II) (*b*) Ca (II)

(*c*) Zn (II) (*d*) Fe(II)

9. Which symbiotic bacteria is capable of fixing N_2?

(*a*) Clostridium pasteurianum (*b*) Rhizobia

(*c*) Azobacter (*d*) Nitrogenase

10. Irreversible oxidation of Hb occurs

(*a*) in the presence of excess oxygen

(*b*) in the absence of protective polypeptide chain

(*c*) in the absence of Mb in its neighbourhood

(*d*) in the absence of CO_2

11. The energy rich molecule in biological systems is

(*a*) ferredoxin (*b*) nitrogenase

(*c*) ATP (*d*) porphyrin

ANSWERS

1. (*b*)	**2.** (*a*)	**3.** (*a*)	**4.** (*c*)	**5.** (*c*)	**6.** (*a*)
7. (*b*)	**8.** (*a*)	**9.** (*b*)	**10.** (*b*)	**11.** (*c*)	

Short Answer Questions

1. What is the oxidation state of iron in haemoglobin and myoglobin?
2. What are porphines?
3. What does ATP stand for?
4. Name two widely distributed oxygen carrier proteins.
5. Define nitrogen fixation.
6. Name the two metalloproteins which comprise nitrogenase.
7. Under what conditions does Hb gets oxidised?

8. What abnormality is caused due to iodine deficiency?

9. What is cooperativity?

10. Name three essential trace elements.

11. What is the name of enzyme that hydrolyses ATP in $Na^+ - K^+$ pump?

General Questions

1. Discuss the structures of myoglobin and haemoglobin. Discuss in detail the roles played by these bioinorganic compounds in biological systems.

2. (*a*) Oxygen acts as π-acceptor ligand in its interaction with heme. What happens when CO interacts with heme in place of O_2.

(*b*) Discuss and draw Mb binding curve at different partial pressures of O_2.

3. (*a*) Draw and discuss Hb-O_2 bonding curve at different partial pressures of O_2. How is it different from that of Mb-O_2 curve?

(*b*) Discus the biochemical roles of Ca^{2+}.

(*c*) Explain the role of Mg^{2+} in biological systems and in ATP.

4. (*a*) Discuss the biological roles of Na and K ions.

(*b*) Discuss the biological important of Ca^{2+}. How is it different from that of Mg^{2+}?

5. Name the essential nonmetals which constitute the living systems. Discuss the role of any three of these.

6. (*a*) Explain the terms cooperativity effect and Bohr effect. What explanation is offered for the cooperativity effect in haemoglobin?

(*b*) Discuss situations when high spin and low spin Fe (II) are observed in the Hb complexes.

7. (*a*) Discuss Hb–O_2 binding curves at

(*i*) different partial pressures of O_2

(*ii*) different pH.

8. Define Hb, Mb and their deoxy derivatives. Discuss the roles of Hb and Mb in transporting O_2.

9. (*a*) Discus the mechanism of intake of oxygen by haemoglobin and myoglobin.

(*b*) What is $Na^+ - K^+$ pump? Explain.

(*c*) What is Bohr effect? Explain.

10. (*a*) Sodium pump is electrogenic in nature. Explain.

(*b*) Compare roles of Zn^{2+} and Ca^{2+} in the active sites of enzymes. In what way is Ca^{2+} advantageous over alkali metal ions.

(*c*) Write a short note on Ca^{2+}– pump.

11. (*a*) Discuss reasons as to why Fe (II) in Mb does not oxidise.

(*b*) Draw a cyclic process showing roles of Hb and Mb as oxygen and CO_2 transporters.

(*c*) What happens when Fe-porphyrin complex without polypeptide chain comes in contact with O_2?

12. (*a*) Fe (II) salts undergo easy oxidation in air but Hb and Mb cannot. Explain.

(*b*) What is cooperativity in haemoglobin? How is it conveyed?

(*c*) Why is high spin Fe(II) deoxy-Mb bigger than low spin Fe (II) oxy-Mb complex?

13. What is N_2 fixation? Discuss briefly

(*i*) Biological N_2 fixation

(*ii*) Abiological N_2 fixation

14. (*a*) Discuss and compare the Hb–O_2 and Mb–O_2 bonding curves at different partial pressures of O_2.

(*b*) Discuss possible binding and reduction of N_2 in the cavity of Fe-Mo cofactor. Draw the structure of N_2 complex with cofactor.

15. (*a*) Give the biological importance of any one of the following transition elements Cr, Mn, Mo, Fe

(*b*) Predict which way the following equilibrium will lie:

$Hb + Hb\,(O_2)_4 \rightleftharpoons 3Hb\,(O_2)_2$. Explain.

(*c*) Myoglobin contains Iron (II) in high spin state. Is it true?

16. (*a*) Describe the possible mechanism of conversion of N_2 into NH_3 using a Mo or W dinitrogen complex

(*b*) State fundamental requirement for biological N_2 fixation.

17. (*a*) Draw and discuss the structures of:

(*i*) P–cluster pair

(*ii*) Fe–Mo cofactor

(*b*) How N_2 is converted into NH_3 using Mo or W dinitrogen complex?

(*c*) What is porphyrin? Draw structure of Heme.

(*d*) How are met-myoglobin and methemoglobin formed?

18. (*a*) Explain cooperativity in haemoglobin. Discuss its mechanism.

(*b*) What is the biological role of Zn^{2+} ions?

19. (*a*) What is meant by nitrogen fixation? What are the main fundamental requirements of biological N_2 fixation?

(*b*) What kind of bacteria are known to fix N_2 biologically?

20. (*a*) What is the difference between nitrogen fixation and nitrogen assimilation?

(*b*) Discuss biological N_2 fixation.

21. Write short notes on:

(*i*) Ultratrace essential elements

(*ii*) $Na^+ - K^+$ pump

22. (*a*) What do you understand by Fe-Mo protein? What type of metal centre clusters does it contain?

(*b*) Give essentialities of Na^+ and K^+ in living systems.

23. (*a*) Write a short note on nitrogen fixation.

(*b*) What is porphyrin? Draw the structure of heme.

(*c*) How are met-myoglobin and methemoglobin formed.

24. (*a*) Give a simple mechanism of N_2 reduction in nitrogenase reaction.

(*b*) Give orientation of $Na^+ - K^+$ pump with respect to position of alkali metal ion, ATP; cardiotonic steroids and vanadates.

(*c*) Give the biological roles of Ca^{2+}.

25. Write a brief note an abiological N_2 fixation.

26. (*a*) Draw and discuss the structures of Fe-Mo-cofactor

(*b*) Discuss any one case of Ti and V as source of ammonia in vitro.

27. (*a*) Discuss the important features of $Na^+ - K^+$ pump.

(*b*) How is iron stored in the body?

(*c*) Discuss the role of iodine in the human systems.

28. What is nitrogenase? What role does it play in nitrogen fixation?

29. Write a brief account of

(*i*) $Na^+ - K^+$ Pump

(*ii*) Ca^{2+} Pump

(*iii*) Essential trace elements.

30. (*a*) What is porphyrin? Draw the structure of heme.

(*b*) Discuss biological role of Na and K ions.

(*c*) Fe (II) salts undergo easy oxidation in air but it is not so in Mb and Hb. Explain.

(*d*) What is Fe-Me-cofactor? Give its structure.

31. (*a*) List the biological roles of Ca^{2+}.

(*b*) Name two metalloproteins which comprise nitrogenase.

(*c*) Describe situation when high spin and low spin Fe(II) complexes are observed in haemoglobin complexes.

32. (*a*) Discuss possible binding and reduction of N_2 in the cavity of Fe–Mo cofactor. Draw N_2 complex with cofactor.

(*b*) Discuss the physiological functions of Ca^{2+} ions.

(*c*) What are trace elements? Why are they so called?

(*d*) Fe(II) salts are easily oxidised in air but it is not so in Mb or Hb. Explain why?

33. (*a*) Oxygen acts as π-acceptor ligand in its interaction with heme. What happens when CO interact with heme in place of oxygen?

(*b*) What is porphyrin? Draw the structures of Heme.

(*c*) Discuss the important features of $Na^+–K^+$ pump.

34. (*a*) What are metalloporphyrins? Discuss the structure and role played by haemoglobin and myoglobin as O_2 carriers.

(*b*) Discuss the biological importance of Ca^{2+}. How is it different from that of Mg^{2+}?

(*c*) What do you mean by Nitrogen fixation? Discuss the role of nitrogenase reductase in biological N_2 fixation.

35. (*a*) Name the important essential trace elements which are associated with biological processes.

(*b*) Discuss the role of Mg^{2+} ion in biological systems and in ATP.

(*c*) What is the difference between nitrogen-fixation and nitrogen-assimilation? What kind of bacteria are known to fix N_2 biologically?

36. (*a*) What is the function of sodium and potassium? Why is it important to have correct percentage of these ions in human body?

(*b*) What are the functions of haemoglobin and myoglobin? What are similarities and differences in their structures.

37. (*a*) Write a short note on nitrogen fixation.

(*b*) Explain biological role of alkali and alkaline earth metal ions.

(*c*) What are the function of myoglobin?

(*d*) Name one metalloprotin which comprises nitrogenase.

38. (*a*) Draw a cyclic process showing roles of Hb and Mb as oxygen and CO_2 transporters.

(*b*) What do you understand by cooperativity in Hb?

(*c*) What is the biological role of Zn^{2+}?

(*d*) Discuss reasons why Fe (II) in Mb does not oxidise.

39. (*a*) Name at least two oxygen carriers and give their importance in biological systems.

(*b*) What happens when Fe-porphyrin complex, without polypeptide chain, comes in contact with O_2?

40. (*a*) Define porphyrin.

(*b*) Explain biological role of alkali and alkaline earth metal ions.

(*c*) What are the functions of myoglobin?

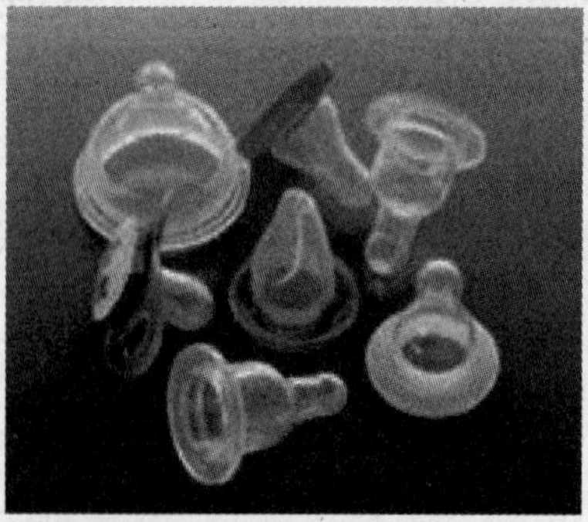

8

Silicones And Phosphazenes

8.1. INTRODUCTION

Polymers touch and influence almost every aspect of modern life from electronics materials to medicines to a wide range of fibres and structural materials. In the organic polymers, the backbone consists of carbon atoms linked together or separated by atoms such as O or N. Inorganic based polymers provide a combination of useful properties. This could be due to the reason that bonds between inorganic elements are longer and stronger and resist free radical cleavage, compared to those formed by carbon atoms. With different valencies of inorganic elements than carbon, there is a possibility of attachment of different group to the atoms of inorganic elements. This is definite to alter the properties of the polymer, their stability at high temperature and solubility in different solvents. Realising the potential of inorganic polymers in different fields, fast developments are taking place in this field. Elements from groups 13, 14, 15 and 16 form the core of this type of polymers. In the coming times, we expect there polymers to provide materials of great utility and act as powerful tools for technology. In this chapter, we shall restrict our treatise to the study of silicones and phosphazenes.

8.2. GENERAL PROPERTIES OF INORGANIC POLYMERS

Some properties of inorganic polymers are given below :

1. With a few exceptions, inorganic polymers do not burn. They soften or melt at high temperatures.

2. Most of the inorganic polymers are built up of highly polar substances. Most of these polymers, however, react with the solvents. There are thus only a few inorganic polymers which actually dissolve in solvents properly.

3. Inorganic polymers having cross-linked structures with a high density of covalent bonds are generally stiffer and harder than the organic polymers.

4. The chain segments between cross links in polymers having cross-linked structures are usually short. Consequently, these structures are not flexible enough to permit interaction of solvent molecules.

5. Inorganic polymers are generally much less ductile than the organic polymers. Thus, while organic polymers such as polyethylene can extend by about 20 per cent or more before breaking, inorganic polymers break even when stretched by about 10 per cent.

6. Inorganic polymers, in general, are stronger, harder and more brittle than the organic polymers.

7. Inorganic polymers can usually be obtained in pure crystalline as well as in pure amorphous forms. Organic polymers, on the other hand, have partial crystalline and partial amorphous structure.

8.3. GLASS TRANSITION TEMPERATURE. T_g

The glass transition temperature is defined as the temperature at which the internal energy of polymer molecules increases to such an extent that the chain segments of the polymer molecules leave their lattice positions. At this temperature, the chain segments start moving apart from one another even on applying a very small strain to the polymer.

The glass transition temperature depends upon the chain length and the degree of cross linking. It also depends upon the barrier restricting internal rotation round the chain links. Below the glass transition temperature, the chain segments of the polymer are fixed on the lattice sites. The polymer in this condition is hard, brittle and hence breakable like glass. Above the glass transition temperature, the segments of the polymer begin to diffuse like the molecules of a liquid. At temperatures sufficiently higher than T_g, the polymer is present in the molten state. Consider the case of a polyphosphate polymer.

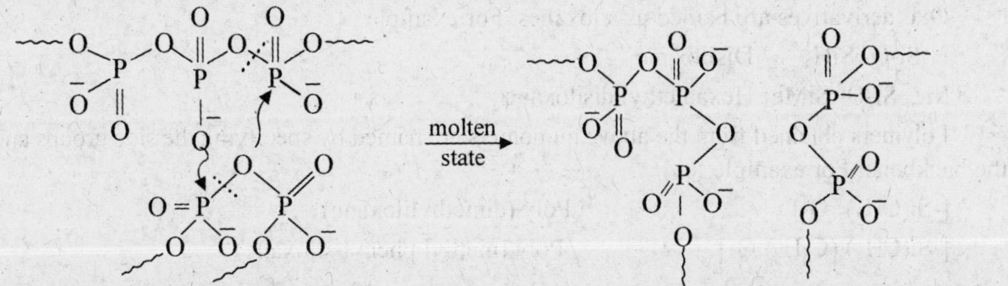

The above exchange reaction shows breaking of P – O bonds at some places and the formation of new P – O bonds at other places giving sufficient mobility to the chain segments of the polymer in the molten state.

The cross-linked inorganic polymers show a wide range in their glass transition temperatures. In this temperature range, they change from rigid solids to leathery solids, from leathery solids to rubbery solids and finally to highly viscous solids.

The glass transition temperature of a linear polymer is fairly sharp. This is because the movement of chain segments from one site to another does not involve the exchange of bonds as happens in the case of cross-linked inorganic polymers.

8.4. SILICONES

These compounds are polymeric organosilicone derivatives containing **Si-O-Si linkages**. These are also called **polysiloxanes.** We can represent them by the general formula $(R_2\ SiO)_n$:

These may be linear, cyclic or cross linked polymers. These polymers have very high thermal stability and are therefore also called **high temperature polymers.** Because of the high thermal stability of Si-O-Si chains, chemists have preapared a large variety of silicone polymers, which are very useful in high temperature processes. Therefore, these find uses in high temperature applications such as heat transfer agents and high performance elastomers. Such qualities are not expected of organic polymers, which cannot withstand high temperatures.

Nomenclature

We follow the IUPAC system, the basic unit, SiH_4 is called **silane** like methane, CH_4. A molecule containing two silicon atoms *i.e.* $H_3Si–SiH_3$ is called **disilane,** $H_3Si–Si(H_2)–SiH_3$ is called **trisilane** and so on. The alkyl, aryl or halo substituted silanes are named by prefixing silane by the specific group. For example :

$Cl_2\ SiH_2$	: Dichlorosilane	$Cl_2\ Si(CH_3)_3$	:	Dichlorodimethylsilane
$Cl_3\ SiCH_3$	: Trichloromethylsilane			

Silanes containing hydroxy groups by adding the suffix, *-ol*, *-diol*, - *triol*, etc. depending upon the number of hydroxyl groups. For example :

$H_3\ SiOH$	: Silanol	$H_2\ Si(OH)_2$	:	Silandiol
$H\ Si(OH)_3$	: Silantriol	Me_3SiOH	:	Trimethylsilanol

Oxo derivatives are named as siloxanes. For example :

H_3Si-O- SiH_3 Disiloxane

Me_3 Si-O- $SiMe_3$ Hexamethyl disiloxane

Polymers obtained from the above monomers are named by specifying the side groups and then the backbone. For example :

[-Si(CH_3)$_2$ O-]	:	Poly (dimethyliloxane)
[-Si(CH_3) (C_6H_5) O-]	:	Poly (methyl phenyl siloxane)

8.5. PREPARATION OF SILICONES

1. The first step in the preparation of silicone polymers is the preparation of substituted silicon halides like R_2SiCl_2 as shown below :

The silicone is converted directly to tetrachlorosilane by the reaction :

$$Si + 2Cl \longrightarrow \underset{\text{Tetrachlorosilane}}{SiCl_4}$$

The organosilane can then by obtained by the Grignard reaction as :

$$SiCl_4 + 2R\ MgX \longrightarrow R_2\ SiCl_2 + 2MgClX$$

A variety of substituted chlorosilanes are produced by this method.

2. However, this process is complicated and it has been replaced by a direct process called **Muller Rochow process.** In this process, the methyl chloride or alkyl chloride treated with elemental silicon at 280-300°C and 200-400 kPa pressure in the presence of copper catalyst.

In this reaction R $SiCl_3$ and R_3 SiCl may also be produced which are removed by distillation.

Compounds of the formula $R_2\ SiCl_2$ are of utmost important because they provide starting materials for the preparation of a wide variety of substances having both organic and inorganic character. The nature of the product formed depends on the conditions such as temperature, pressure, catalyst and silicon purity. **The basic catalyst and high temperatures in general yield higher molecular weight polymers that are linear. On the other hand acidic catalysts tend to form cyclic small molecules or low molecular weight polymers.**

3. Hydrolysis of $R_2\ SiCl_2$. The hydrolysis of R_2SiCl_2 such as dimethyl dichlorosilane gives dimethylsilanol as

$$(CH_3)_2\ SiCl_2 + 2H_2O \longrightarrow \underset{\text{Dimethylsilanol}}{(CH_3)_2\ Si(OH)_2} + 2HCl$$

The polymerisation of dimethylsilanol through the elimination of a molecule of water from two hydroxyl groups of the adjacent dimethyl silanol takes place as under :

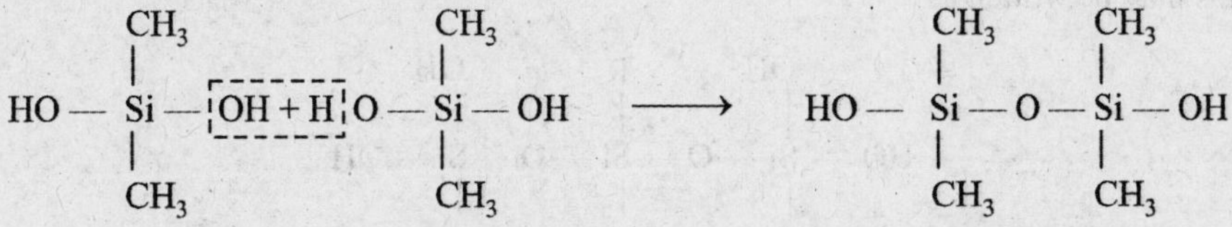

Linear dimer

Polymerisation reaction continues and the length of the chain increases as active OH groups are left at the end of the chain.

4. Formation of linear silicones.

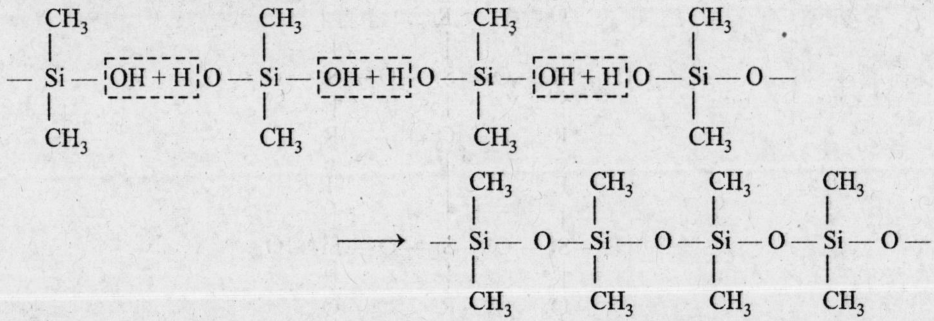

Linear silicone

5. Formation of cross linked silicones.

Alkyl or aryl trichlorosilane gives cross linked silicones on hydrolysis. The alkyl or aryl trichlorosilane are obtained as a fraction in the preparation of dichlorosilanes or can be prepared by any of the following methods.

$$CH_3Cl + Si + 2HCl \xrightarrow{Cu,\ 300°C} CH_3SiCl_3 + H_2$$

$$SiCl_4 + \underset{\text{Phenyl magnesium chloride}}{C_6H_5MgCl} \longrightarrow C_6H_5SiCl_3 + MgCl_2$$

The hydrolysis of alkyl trichlorosilane takes place as under :

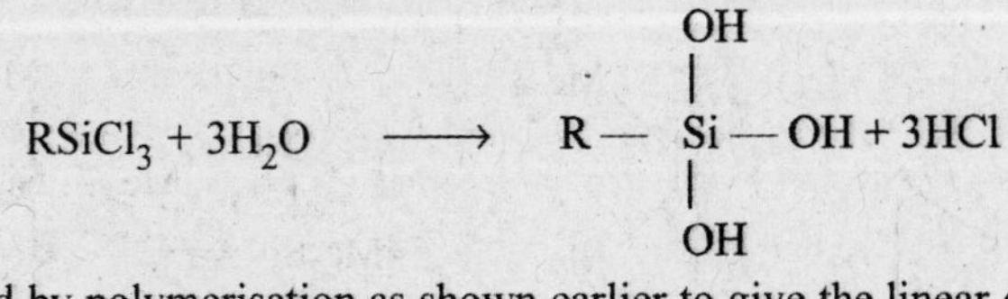

This is followed by polymerisation as shown earlier to give the linear compound first.

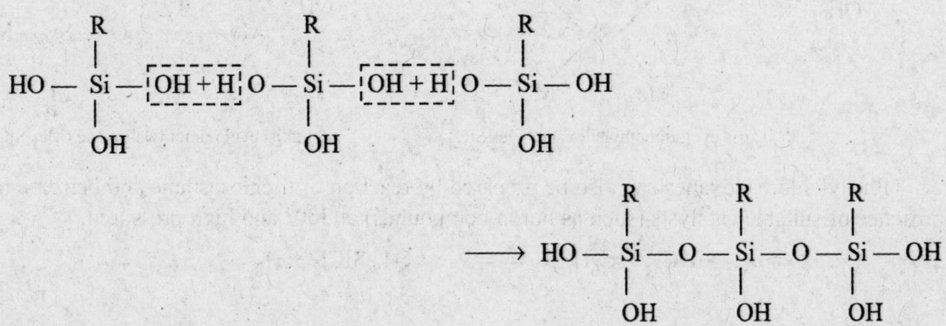

This may be written as

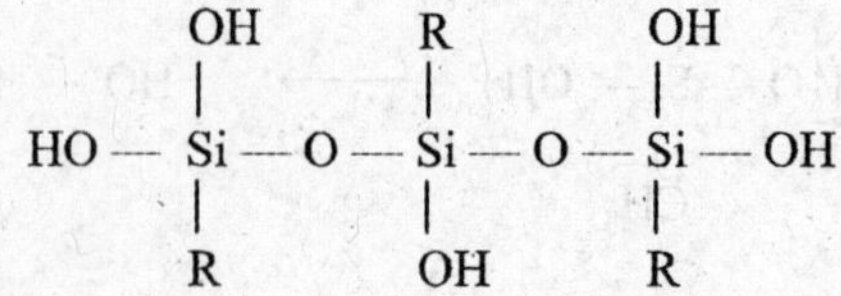

Presence of OH group at the end of the Si chain and below and above gives rise to polymerisation to give a cross linked

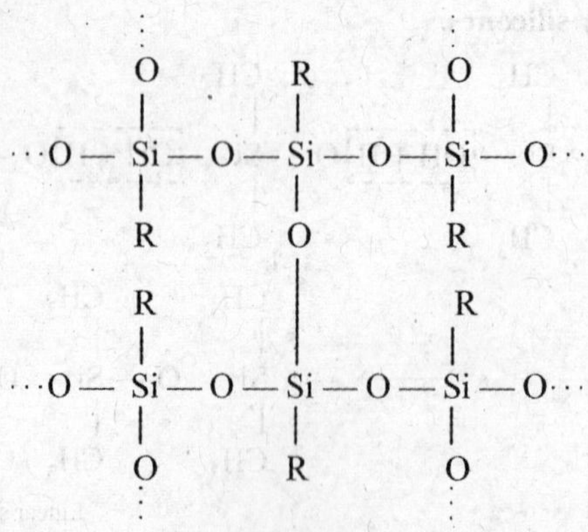

Cross linked polymer

The above structure can be extended in two dimensions to any length and in any manner. Silicon polymers of any type, straight chain, cross linked or even mixed can be obtained by mixing the right amounts of trialkyl chlorosilane and dialkyl dichlorosilanes and subsequent hydrolysis. *The trimethyl silanol obtained by hydrolysis of trimethylchlorosilane contains only one OH group and hence it can be used to block the chain to get controlled chain polymers.*

Therefore, **R_3SiCl is regarded as chain stopping unit.** Thus, ratio of R_3SiCl and R_2SiCl_2 in the starting mixture, will determine the average chain of the polymer.

We can also prepare cyclic polymers with 3, 4, 5 or 6 silicon rings by

For example

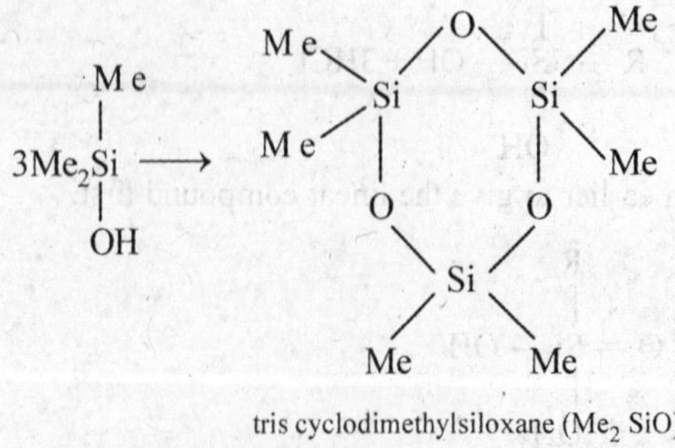

tris cyclodimethylsiloxane $(Me_2\ SiO)_3$

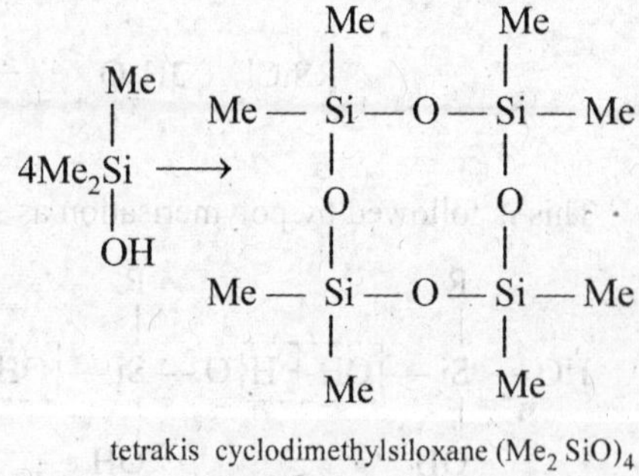

tetrakis cyclodimethylsiloxane $(Me_2\ SiO)_4$

Phenylchlorosiloxanes can also be prepared by reaction of trichlorosilane and benzene in the presence of suitable catalysts (such as boron compounds) at 300° and high pressure.

$$SiHCl_3 + C_6H_6 \longrightarrow C_6H_5\ SiCl_3 + H_2$$

Factors affecting the nature of silicone polymers

The following factors control the properties of silicone polymers.

1. The length of the chain
2. The distribution of organic groups if there are more than one type of R group.
3. The extent of cross-linking and
4. The nature of alkyl or aryl groups,
5. The type and proportion of the structural units.

Generally, short chain silicones are oily liquids, medium chain silicones are viscous liquids like greases and viscous oils while those with very long chains or longer cross-linking are rubber like polymers.

8.6. EQUILIBRATION AND RING OPENING POLYMERISATION (ROP) OF CYCLOSILOXANES

We observe that if cyclic or linear polysiloxanes are treated with a base or acid catalyst, an equilibrium mixture between cyclic and linear species is obtained. *The process of simultaneous cleavage and reformation of siloxane bonds is called equilibration or equilibrium of cyclopolysiloxanes.*

The ring opening of cyclic polysiloxane by heating them alone or in the presence of a catalyst (acid or base) is a general method for preparing straight chain silicones. Two types of mechanisms forming Si-O-Si bonds can take place simultaneously. The active centre can add to another Si-O-Si bond (addition polymerisation mechanism) or by hydrolysis/condensation reactions. All these reactions involve thermodynamic equilibrium between the reactants and the products as shown below :

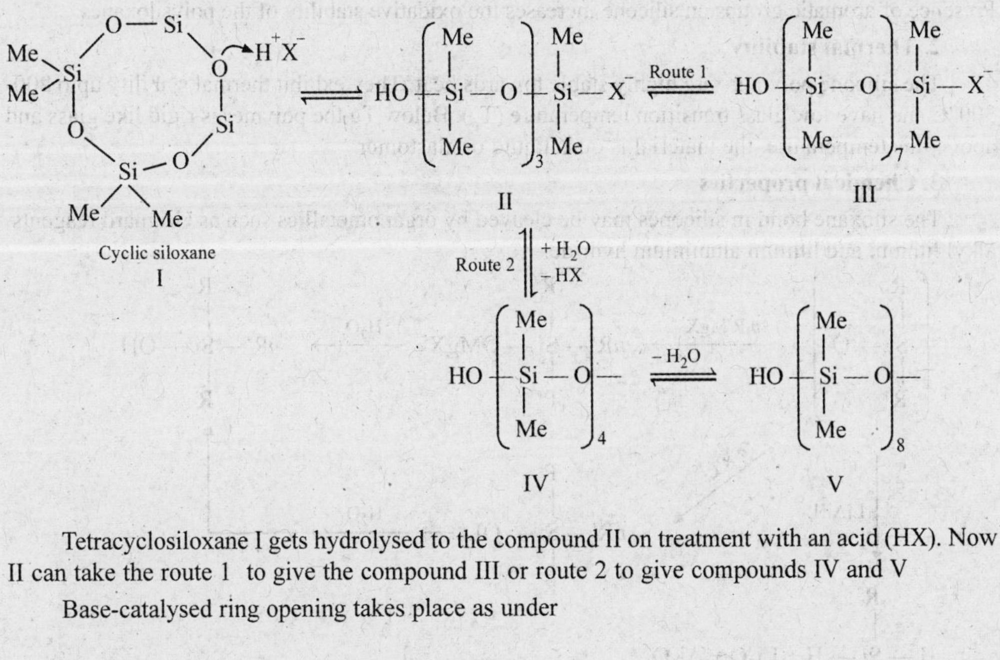

Tetracyclosiloxane I gets hydrolysed to the compound II on treatment with an acid (HX). Now II can take the route 1 to give the compound III or route 2 to give compounds IV and V

Base-catalysed ring opening takes place as under

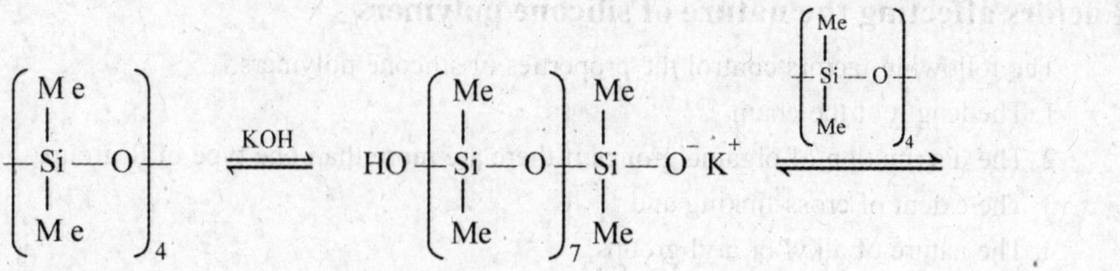

Tetra cyclicsiloxane

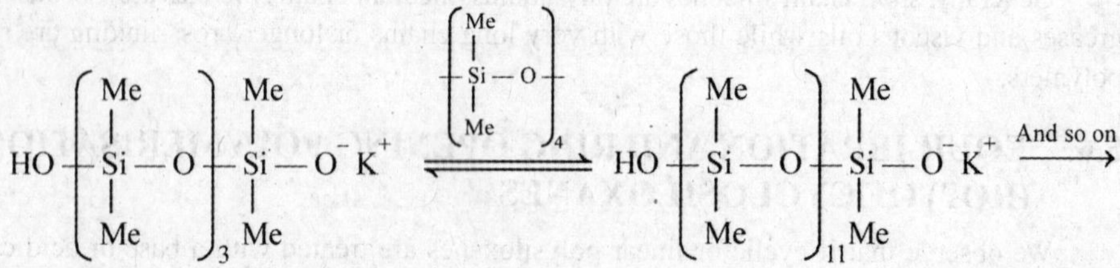

8.7. PROPERTIES OF SILICONES

The important properties of silicones are given a under :

1. Chemical stability

The silicones are stable towards chemical reagents. They are not affected by common chemicals such as weak acids, alkalis, salts solutions or water at room temperature. These are also quite stable to attack by oxygen. However, some oxidation may occur at higher temperatures. At higher temperature, fluid polysiloxanes get cross-linked through oxygen atoms to form a rubber like gel in a few hours. Presence of aromatic groups on silicone increases the oxidative stability of the polysiloxanes.

2. Thermal stability

The silicone polymers are highly stable towards heat. They exhibit thermal stability upto 200-300°C and have low glass transition temperature (T_g). Below Tg the polymer is rigid like glass and above this temperature, the material is quasiliquid or elastomer.

3. Chemical properties

The siloxane bond in silicones may be cleaved by organometallics such as Grignard reagents, alkyl lithium and lithium aluminium hydride.

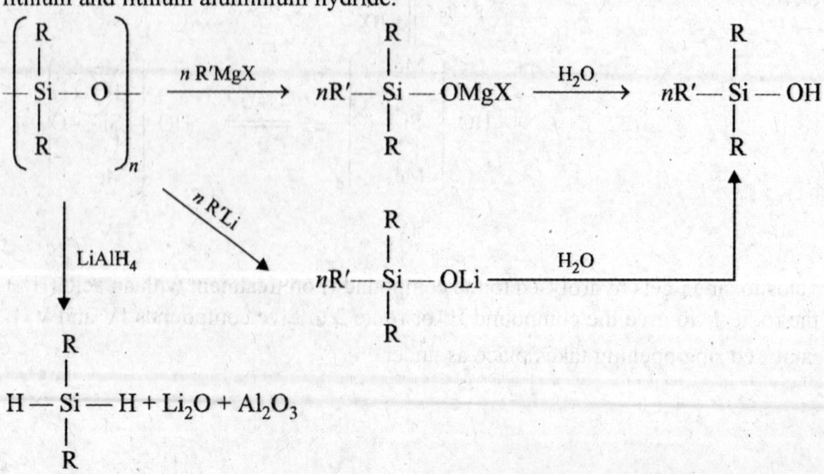

4. They are used for low temperature lubrication. This is in contrast to hydrocarbon oils which become viscous on cooling and therefore, cannot be used.

5. They are good insulators and provide electrical insulators for electrical motors and other appliances because they can withstand high temperatures.

6. They are **water repellants.** Their water repellant because a silicon chain is surrounded by organic side groups.

7. They are **resistant to oxidation** *i.e.* have **oxidation stability.** However, when heated in air to 350 to 400°C, silicones are rapidly oxidized and this leads to cross linking.

8. Low molecular weight silicon polymers are soluble in organic solvents like ether, carbon tetrachloride, benzene, *etc.*

Use of silicone polymers

The important uses of silicone polymers are given below

1. They are used for water-proofing and in electrical condensers.
2. With low toxicity, they are used in medicinal and cosmetic implants.
3. They are used as lubricants at both high and low temperatures.
4. Silicone rubbers are very useful because they retain their elasticity at lower temperatures as compared to other rubbers.
5. Silicone polymers are used as greases, varnishes.

8.8. SILICON FLUIDS OR OILS

The silicon fluids are usually linear polysiloxanes with an average of 50 to 200 units. They are prepared by the hydrolysis of a mixture of $(CH_3)_2$ $SiCl_2$ and $(CH_3)_3$ SiCl. Commercially, these are prepared by treating a mixture of tetrakis cyclodimethyl siloxane [cyclo (Me_2 SiO_4] and hexamethyl disiloxane $(Me)_3$ SiO Si $(Me)_3$ with a small quantity of 100% H_2SO_4.

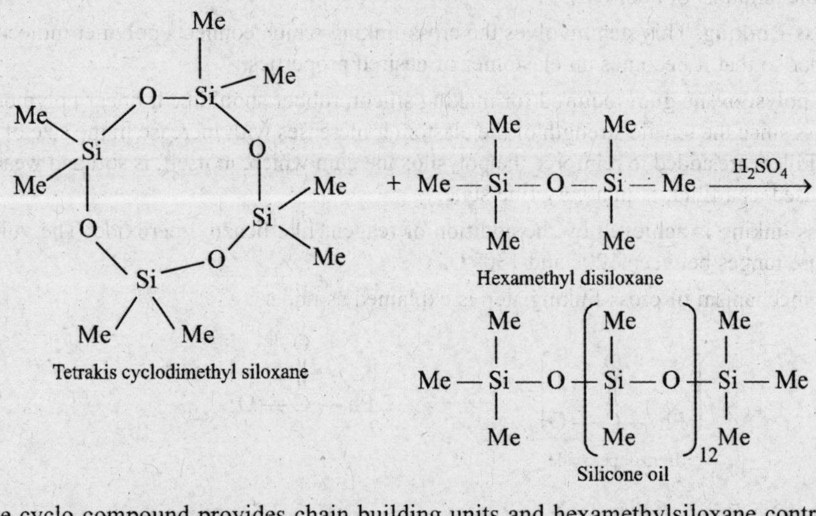

Tetrakis cyclodimethyl siloxane

Hexamethyl disiloxane

Silicone oil

The cyclo compound provides chain building units and hexamethylsiloxane controls chain formation. The average chain length or molecular weight of the polymer is determined by the ratio of these chain building groups and chain ending groups.

The viscosity of the silicone fluids is about 30,000 times that of water and changes slightly with temperature. The boiling point and viscosity of silicon oils increase with chain length. These are inert and non toxic materials and have low surface tension.

Uses of silicon fluids or oils

Some important uses of silicone fluids are given below

1. Silicon oils are used as dielectric insulating materials as capacitor and transformer fluids hydraulic oils, compressible fluids for liquid springs and lubricants.

2. The fluids are used as water repellants for treating masonry and buildings, glassware and fabrics.

3. Methyl silicones can be used as light duty lubricating oils but are not suitable for heavy duty applications such as gear boxes because the oil film breaks down under high pressure.

4. Silicone oils can be used as antifoaming agents and hydrophobizing agents in sewage disposal, textile dyeing, beer making (fermentation) etc.

5. Silicone oils are also used as heat transfer media in heating belts and as components in car polishes, lipstick, shampoo and other cosmetic formulations.

8.9. SILICONE ELASTOMERS (RUBBERS)

Silicone elastomers have very high molecular weights of the order of 10^5–10^7. They are usually produced by the equilibration of cyclic siloxanes followed by hydrolysis of dimethyldichlorosilane with KOH. Great care is taken to exclude chain blocking and cross linking groups. The filler is added to obtain the required reinforcement and affects physical properties such as tensile strength, gear strength and hardness. The commonly used filler is finely divided silica. This is chemically inert and is unaffected by heat.

Preparation of silicon rubbers involves two steps, *viz.*, compounding and cross-linking.

Compounding. It consists of milling of polysiloxane gum, a filler and a cross-linking agent together with certain additives. A typical formulation comprises 100 parts of benzene-soluble polysiloxane gum, 20–50 parts of SiO_2 as filler, 6 parts of benzoyl peroxide as vulcanising agent and suitable amounts of additives.

Cross-Linking. This step involves the cross-linking which connects polymer molecules with one another so that it becomes an elastomer of desired properties.

The polysiloxane gum required for making silicon rubber should be a linear polymer of high molar mass since the tensile strength of the elastomer increases with increase in the size of the gum polymer. Fillers are added to reinforce the polysiloxane gum which, in itself, is soft and weak. Silica, which is used as a filler, is in a very fine state.

Cross-linking is achieved by the addition of reagent like benzoyl peroxide. The vulcanising temperature ranges between 120° and 130° C.

The mechanism of cross-linking step is explained as under :

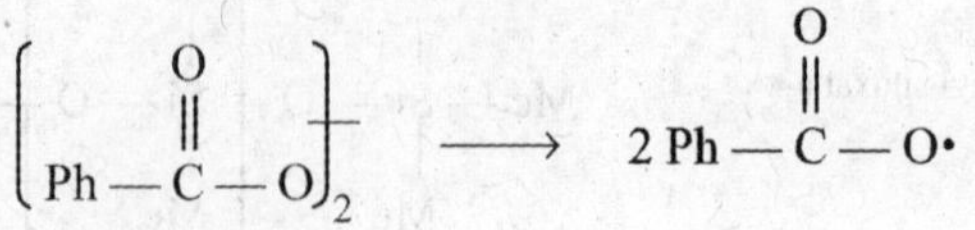

Benzol peroxide

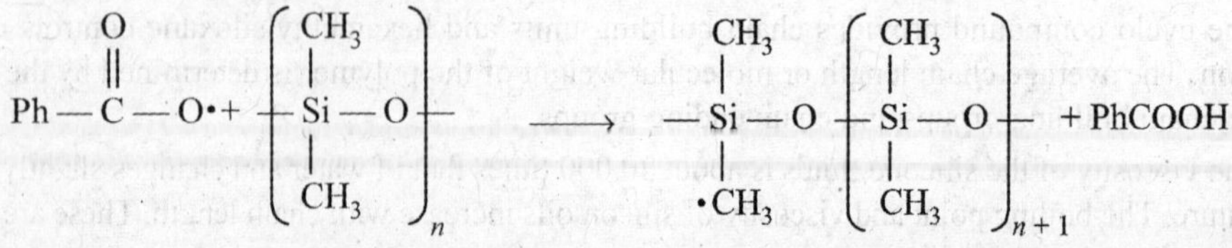

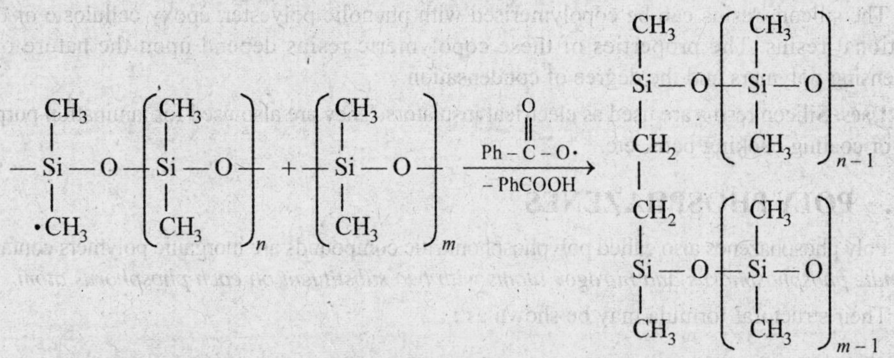

Uses. Silicon rubbers are extremely useful materials on account of their high thermal stability, low temperature performance, chemical stability, water repelling power and flexibility.

8.10. SILICON RESINS

Silicon resins possess ring structures and have a much higher cross-link density than silicon elastomers. In a stretched elastomer, the molecular chains uncoil through rotation about the -Si–O– bonds. In a resin, such rotation is highly restricted and the response to stress occurs largely through bending and stretching of bonds. The properties of resins are dependent upon R/Si ratio where R is the alkyl group. As the R/Si ratio is lowered from 2.0 to 1.0, the polymer becomes progressively less fluid, less soluble and less fusible. Most of the resins have R/Si ratio between 1.00 and 1.6 and their properties are intermediate between those of fluid silicones and those expected of a highly cross-linked structure with R/Si ratio equal to 1.00.

Preparation. For the preparation of silicon resins, a mixture of chlorosilones such as CH_3SiCl_3, $(CH_3)_2SiCl_2$, $C_6H_5SiCl_3$ $(C_6H_5)_2SiCl_2$ are first hydrolysed. During hydrolysis about 90 per cent of silanol groups, Si — OH, condense to form siloxane linkages. In this process of random condensation, some ring closure invariably occurs because silanol groups condense more slowly and with increasing difficulty.

After hydrolysis the product is shaken with an organic solvent and the organic layer is separated. This organic layer is then washed thoroughly to remove the Cl^- ions in order to check the reversibility of the hydrolysis.

The silanol groups (*i.e.*, Si – OH) left uncondensed in the finished resin impart typical properties to the resin. These groups contribute towards viscosity of the product which is an important property that determines the usefulness of resin. For instance, the viscosity of resin used for laminating purposes should be relatively low so that it may impregnate the glass fibres better. The silanol groups also determine the shelf life of the resin. C_2H_5, C_3H_7, etc. groups make the resin softer, increase its solubility in organic solvents and increase its water-repellence. The phenyl groups increase resistance towards heat and so on.

Copolymeric resins of desired properties can be obtained by condensing two polymers, as shown below :

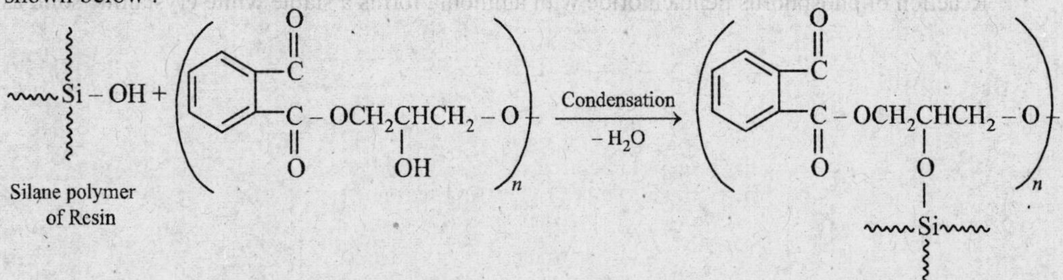

The silicon resins can be copolymerised with phenolic polyester, epoxy cellulosic or other, functional resins. The properties of these copolymeric resins depend upon the nature of the condensing polymers and the degree of condensation.

Uses. Silicon resins are used as electrical insulators. They are also used for lamination purposes and for coating cooking pans, etc.

8.11. POLY PHOSPHAZENES

Poly phosphazenes also called polyphosphonitritic compounds are inorganic polymers containing *alternate phosphosphorus and nitrogen atoms with two substituent on each phosphorus atom.*

Their structural formula may be shown as :

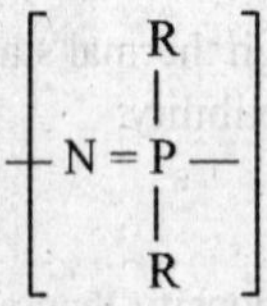

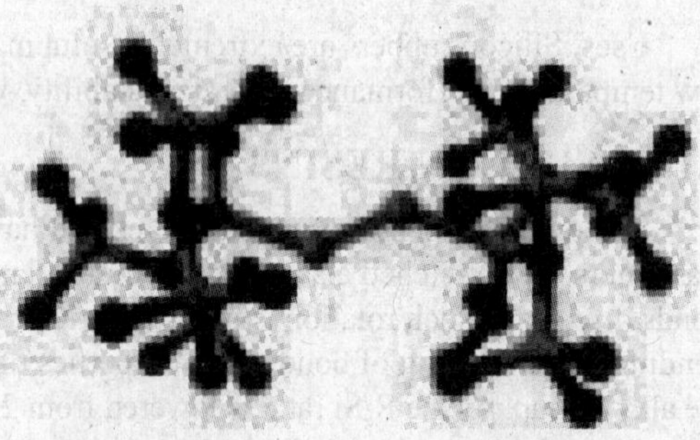

Model of phosphazene

Where R are the side groups attached to each phosphorous. It may also be written as $(NPR_2)_n$. The side group R, may be some alkyl, phenyl or halogen group. Each macromolecule may contain about 15 ,000 or more repeatig units linked end to end giving it the molecular weight in the range of 2 million to 10 million.

Common phosphazenes exist as cyclic (trimers or tetamers) and linear polymers as :

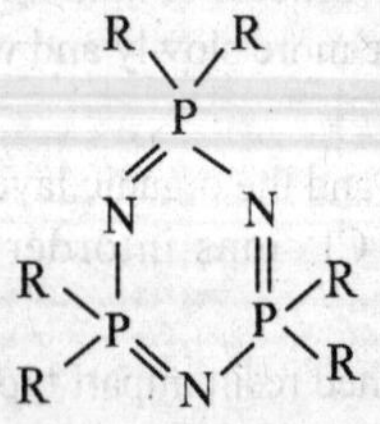

Cyclic trimer $(NPR_2)_3$

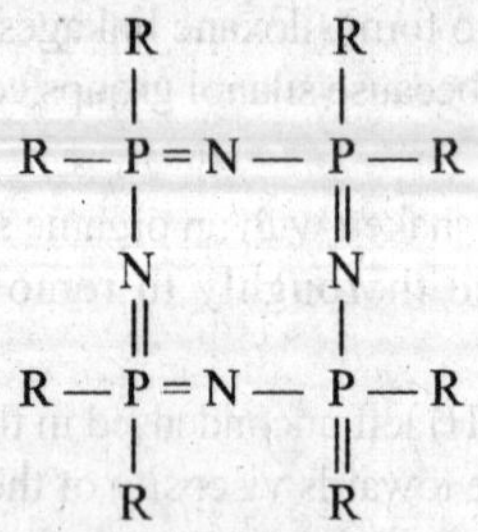

Cyclic tetramer $(NPR_2)_4$

Linear polymer $(NPR_2)_{3n}$

Phosphonitrilic halides have R = F, Cl, Br and among these the best known compounds are chlorides having the general formula $(NPCl_2)_n$ where *n* ranges from 3 to 7 and onwards. Out of these $(NPCl_2)_3$ and $(NPCl_2)_4$ have been studied extensively.

Reaction of phosphorus pentachloride with ammonia forms a stable white crystalline solid as

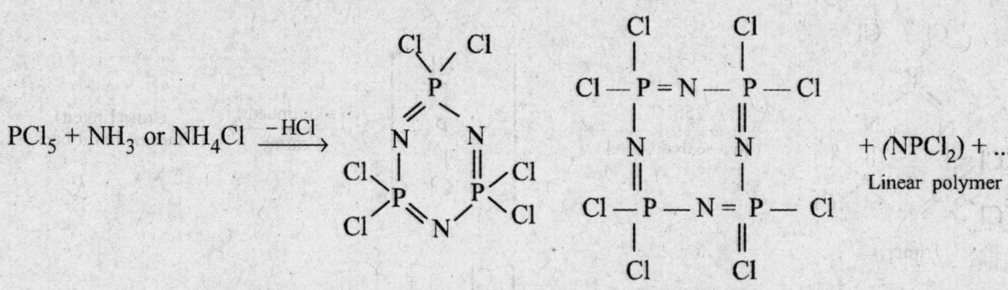

Cyclic trimer Cyclic tetramer

On heating chlorophosphazenes are transformed into an elastomeric material which is known as ' ***inorganic rubber***'. This was a remarkable achievement which laid the foundation of inorganic polymers.

8.12. PREPARATION OF PHOSPHAZENES

We can classify phosphazines mainly into three types. Accordingly, we shall study the methods of preparation of such compounds under three headings.

Methods for the preparation of polyhalophosphazene

1. Phosphonitrilic chlorides can be obtained by heating a mixture of phosphorus pentachloride with a small excess of ammonium chloride in chlorobenzene or tetrachloroethane solvent at 120 – 140°C.

$$nPCl_5 + nNH_4Cl \xrightarrow[120-140^\circ C]{\text{Solvent}} (PNCl_2)_n + 4n\ HCl$$

In this method, a mixture of cyclic phosphonitrilic chlorides $(PNCl_2)_n$ where $n = 3, 4, 5, 6....$ and short linear chain phosphonitrilic chlorides is obtained. The main products of the reaction are the cyclic trimer [triphosphonitrilic chloride, $(PNCl_2)_3$] and cyclic tetramer (tetraphosphonitrilic chloride, $(PNCl_2)_4$] and some linear polymers.

2. By heating $(P_3N_5)_n$ in chlorine at 700°C.

$$(P_3N_5)_n + nCl_2 \longrightarrow (PNCl_2)_n + \text{other products.}$$

3. A chlorofluoropolymer $(PNFCl)_4$ is prepared by heating the trimer $(PNCl_2)_3$ with PbF_2.

$$(PNCl_2)_3 + PbF_2 \longrightarrow (PNFCl)_4 + PbCl_2 + \text{other products}$$

4. Trimer phosphonitrilic chloride can be formed by treating S_4N_4 with $SOCl_2$ in PCl_3

$$S_4N_4 + 6SOCl_2 + 12PCl_3 \longrightarrow 4(PNCl_2)_3 + 10S + 3O_2 + 12Cl_2$$

Methods for the preparation of other polyphosphazenes

(*i*) The trimer $(NPCl_2)_3$ which is the starting material to obtain different polymers is prepared on an industrial scale by the reaction of PCl_5 with NH_4Cl in an organic solvent such as chlorobenzene or tetrachloroethane.

(*i*) The purified compound is heated in the molten state in a sealed vessel in the absence of moisture for several days at temperatures between 200-250°C to bring about ring opening and polymerisation. This results in the ring opening of the cyclic compound. The resulting poly (dichlorophosphazene) is dissolved in a suitable solvent and produces a number of other interesting polymers via nucleophilic substitution reactions. This is explained as under :

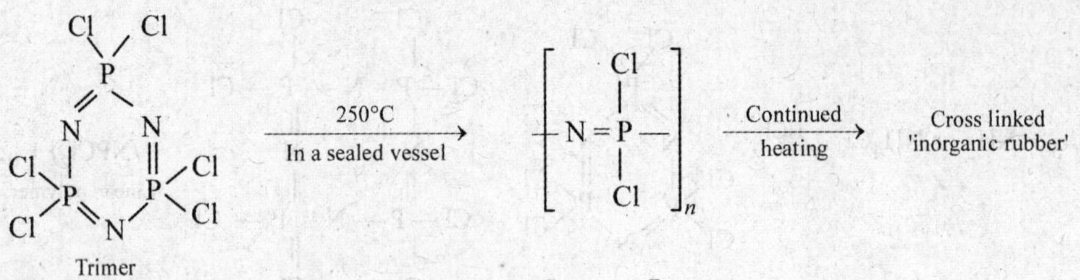

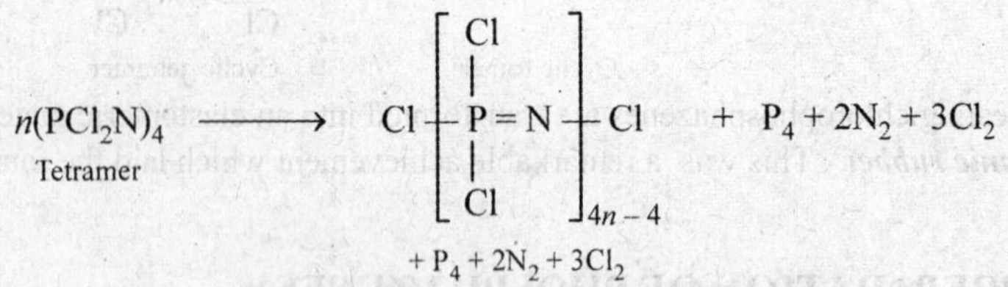

Chlorine atoms in poly dichlorophosphazene can be substituted by alkyl, phenyl, alkoxy, amine or substituted amine groups by treatment with the appropriate reagents.

The product polymers are stable and possess a broad range of useful properties.

This is illustrated as under by takig RONa, RNH_2 R_2NH and MCl (say CH_3 Li), to give new polymers.

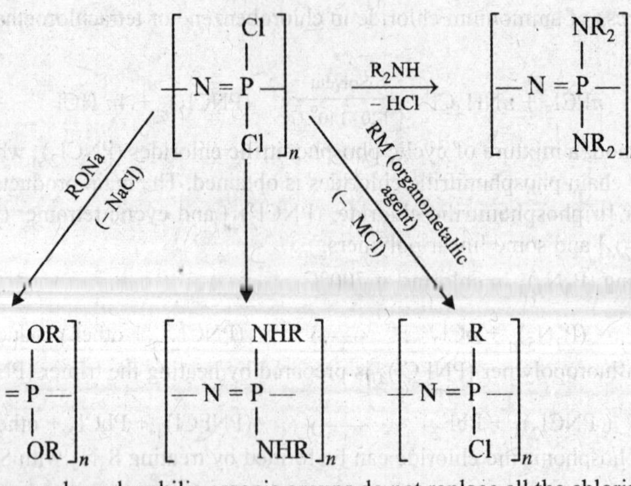

Certain bulky or weak nucleophilic organic groups do not replace all the chlorine atoms at mild reaction conditions due to steric hindrance. In certain cases, after the introduction of a bulky substituent, the second (smaller) group is possible with controlled conditions. Stearic hindrance slows down the reaction rate after a bulky side group has been introduced. For example, when $(NPCl_2)_n$ is reacted with diethylamine, only one chlorine atom per phosphorus is replaced. The remaining chlorine can then be substituted by treatment with less hindered nucleophile such as methylamine or shortchain alkoxides or alkyl or aryl organometallic reagents. The general scheme is given as under :

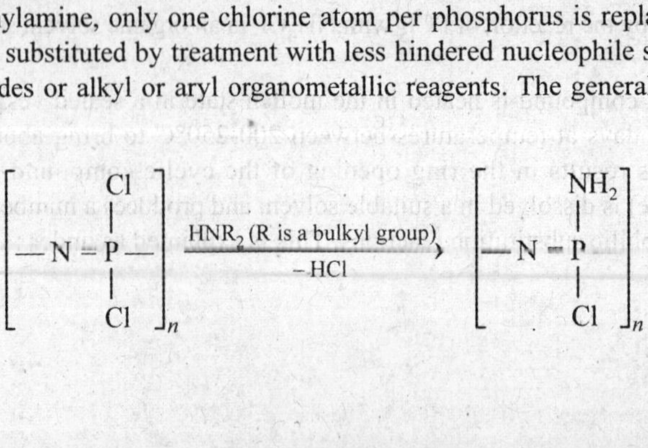

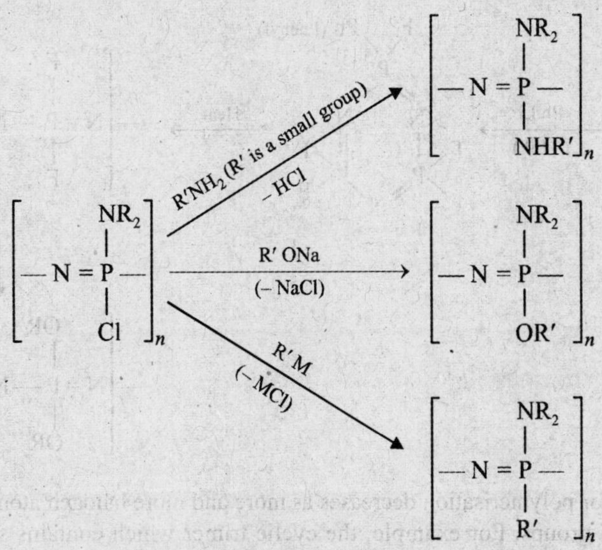

Organometallic nucleophiles such as Grignard reagent or organolithium reagents react in different way and give more complicated reactions. For example, $(NPCl_2)_n$ reacts with RMgX and results in replacement of Cl atom by R group and is accompanied by cleavage of P–N bond bonds in the skeleton. The same happens with alkyl lithium.

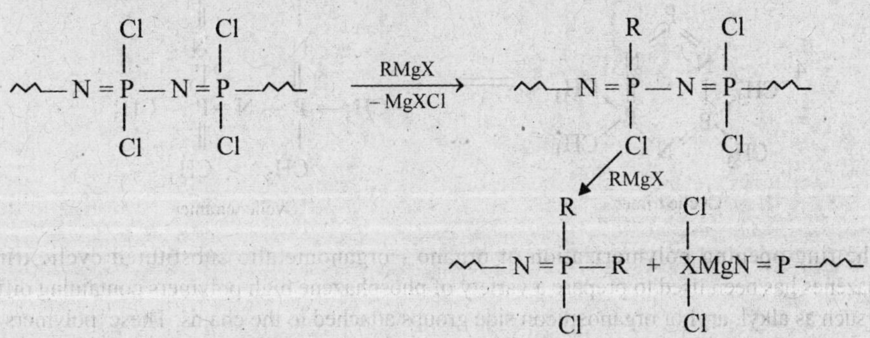

Methods of preparation from susbtituted cyclic phosphazenes

The organic susbtituents are introduced at the cyclic trimer phosphazene followed by ring opening polymerization of the substituted cyclic trimer to a high polymer. However, ring opening tendency decreases as more and more halogen atoms in the trimer are replaced by organic groups.

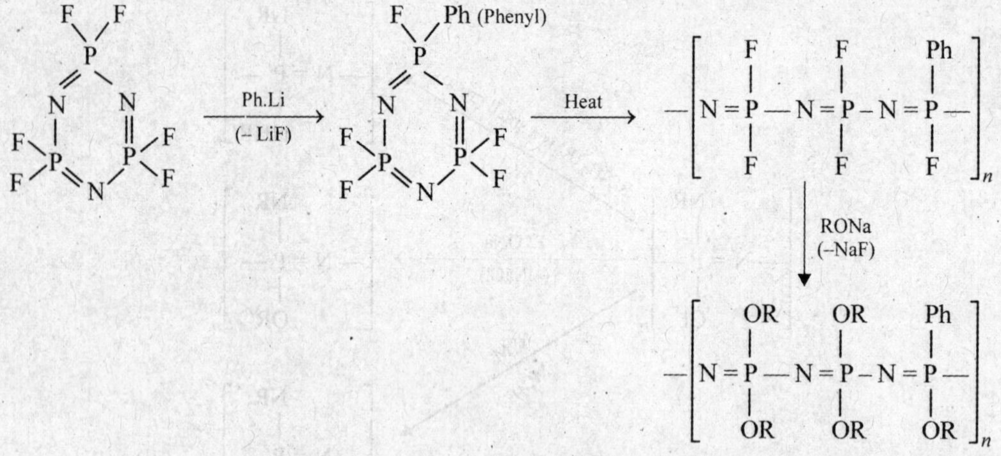

The tendency for polymerisation decreases as more and more halogen atoms in the trimer are replaced by organic groups. For example, the cyclic trimer which contains six methyl groups, $(NPMe_2)_3$ undergoes ring-ring equilibration to the eight membered cyclic tetramer on heating but it does not polymerise to a high polymer. This is the limitation of this method.

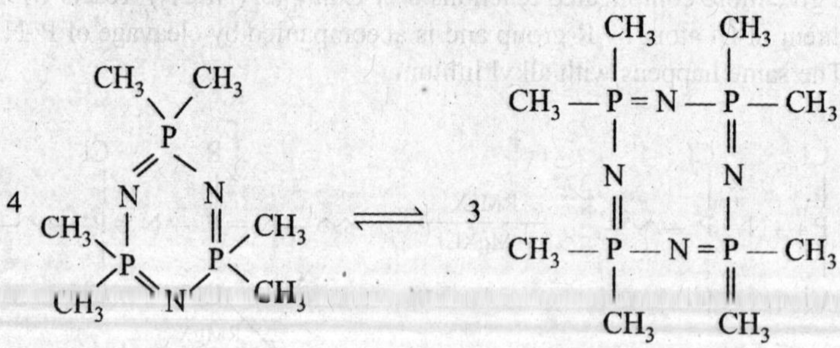

Cyclic trimer   Cyclic tetramer

The ring opening polymerization of organo - organometallo substituted cyclic trimeric phosphazenes has been used to prepare a variety of phosphazene high polymers containing different groups such as alkyl, aryl or organosilicon side groups attached to the chains. These polymers show interesting properties and applications.

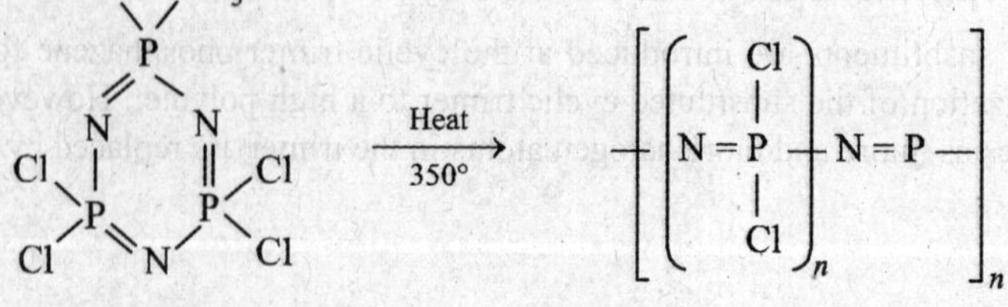

8.13. PROPERTIES AND STRUCTURES

Cyclic phosphonitrilic halides, $(NPCl_2)_3$ and $(NPCl_2)_4$ have been studied in detail. Their properties are discussed below :

Physical properties

1. $(NPCl_2)_3$ has melting point 114°C and boiling point 256°C at 1 atm. pressure. It is readily soluble in ether, benzene and carbon tetrachloride.

2. On the other hand $(NPCl_2)_4$ has melting point 123.5°C and boiling point 328.5°C. It has lower solubility in ether, benzene and carbon tetrachloride than $(NPCl_2)_3$.

3. When the solutions are allowed to stand, the polymers get cross linked and slowly gel formation takes place.

Chemical properties

1. On heating to 150-300°C both $(NPCl_2)_3$ and $(NPCl_2)_4$ polymerise to an elastic product with high molecular weight of the order of 20,000. On further heating to 350°C, the product gets depolymerised.

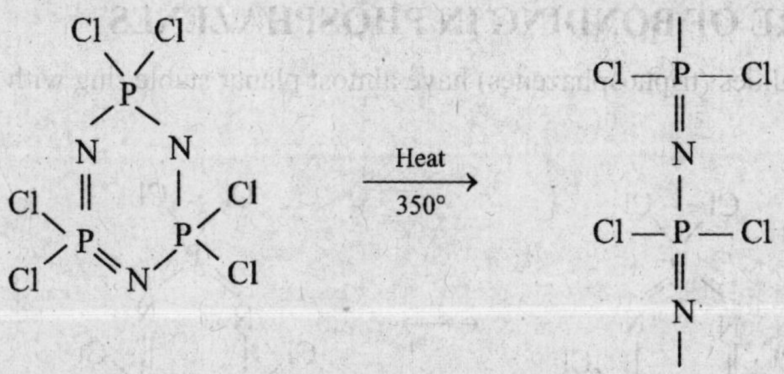

2. Substitution reaction

The chlorine atom in phosphonitrilic chlorides is reactive and can be replaced by other monovalent groups like F, Br, OH, OR, SH, SR, SCN, NH_2, NR_2 *etc.*

$$(NP\ Cl_2)_3 + 6NaF \longrightarrow [NPF_2]_3 + 6NaCl$$

$$(NP\ Cl_2)_3 + 6NaOR \longrightarrow [NP(OR)_2]_3 + 6NaCl$$

$$(NP\ Cl_2)_3 + 6CH_3MgI \longrightarrow [NP(CH_3)_2]_3 + 3MgCl_2 + 3MgI_2$$

$$(NP\ Cl_2)_3 + 6NaSCN \longrightarrow [NP(SCN)_2]_3 + 6NaCl$$

3. Hydrolysis

The cyclic trimer can be hydrolysed to trimetaphosphamic acid which undergoes isomeric change to trimetaimido phosphoric acid.

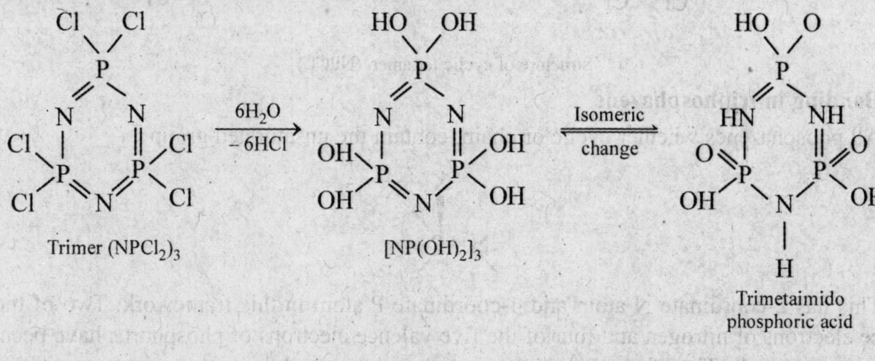

Hydrolysis of tetramer takes place as under

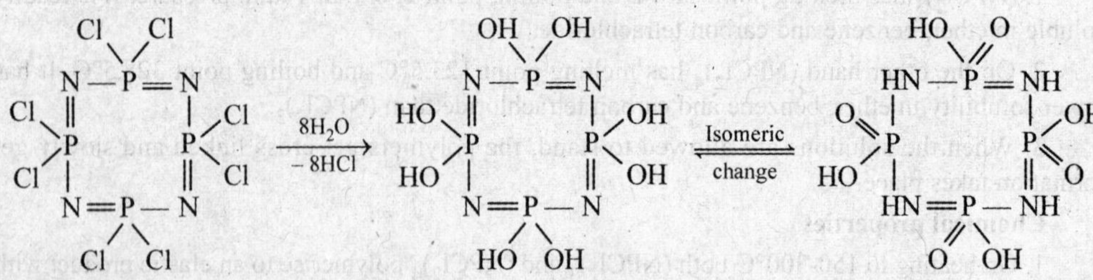

8.14. NATURE OF BONDING IN PHOSPHAZENES

The trimer halides (triphosphazenes) have almost planar stable ring with resonance structures as shown below :

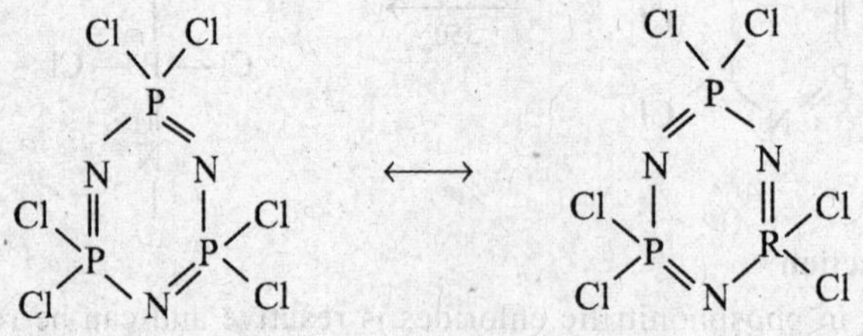

Trimeric phosphazene, $(NPCl_2)_3$

The tetramer $(PNCl_2)_4$ has a puckered ring structure as in the case of cyclohexane with alternate P and N atoms and with two Cl atoms on each P atom. Chair and boat conformations of $(PNCl_2)_4$ have been isolated.

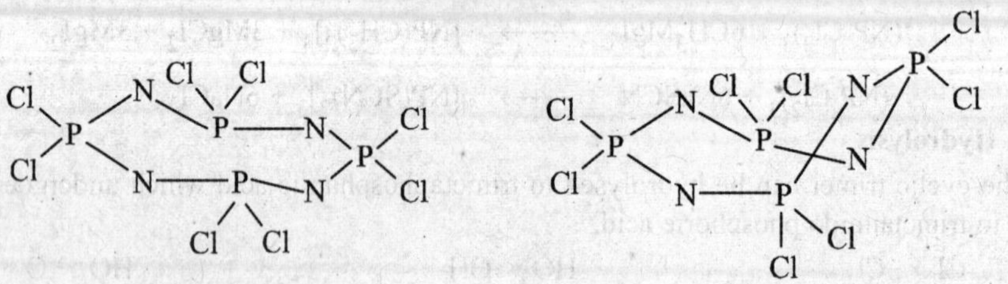

Structure of cyclic tetramer, $(NPCl_2)_4$

Bonding in triphosphazene

All phosphazenes whether cyclic or chain, contain the unsaturated group

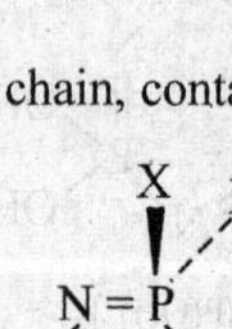

This has 2-coordinate N atom and 4-coordinate P atom in this framework. Two of the five valence electrons of nitrogen and four of the five valence electrons of phosphorus have been used for forming σ-bonds. Therefore, one electron on phosphorus and three electrons from nitrogen are left on these atoms.

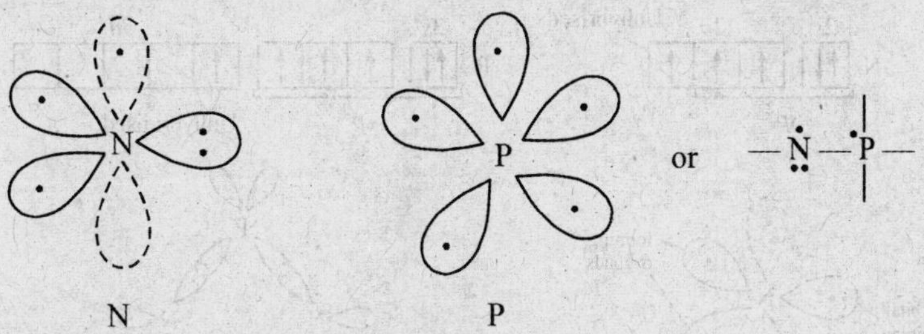

It is assumd that two of the remaining electrons on N atoms occupy a non-bonding lone pair orbital while the third electron from nitrogen interacts with the one electron from P atom to form a π-bond as shown below. The bond parameters are also given

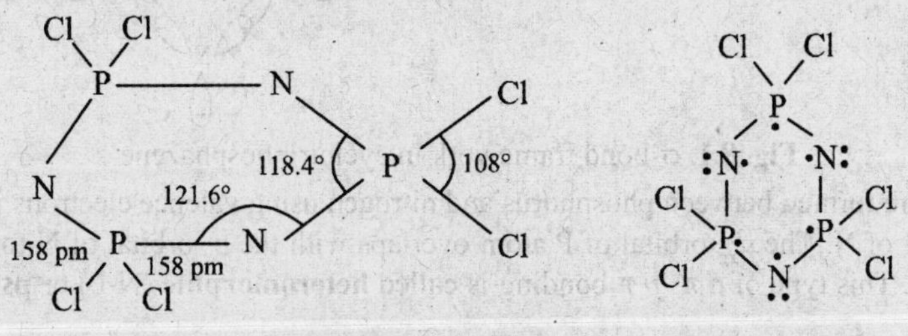

Simple structure of $(NPCl_2)_3$

We obtain the following structural evidence from X-rays studies of phosphonitrilic chlorodes:

(*i*) Phosphazene rings and chains are very stable.

(*ii*) The skeletal P–N bond lengths are equal in the ring.

(*iii*) The P–N bond distances are 158 pm. These are shorter than expected for a covalent single bonds.

(*iv*) Skeletal bond angles at phosphorus in cyclotriphosphazene is approximately equal to 120°; PNP and NPN bond angles in the ring are 121.6° and 118.4° respectively.

(*v*) The skeletal nitrogen atoms in cyclophosphazene are weakly basic and can be protonated or form coordination complexes especially when there are electron releasing groups on P.

(*vi*) Spectral effects such as bathochromic U.V. shift which are normally associated with organic π-electron system due to increased delocalisation are not observed for cyclotriphosphazene.

(*vii*) Cyclotriphosphazene skeleton cannot be easily reduced electrochemically. This behaviour is different from that shown by organic aromatic compounds.

We can therefore say the phosphorus-nitrogen bonds differ significantly from aromatic σ–π system in which there is extensive delocalisation through *p* π -*p* π bonding. To account for this, π-bonding in cyclo triphosphazene was considered as *p* π -*d* π bonding involving *d*-orbitals of phosphorus and *p*-orbitals of nitrogen.

d π -*p* π bonding model for cyclotriphosphazene

The cyclic triphosphazene say $(NPCl_2)_3$ consists of planar six membered ring. The σ-bond framework is made up of overlapping orbitals from phosphorus and nitrogen. The bond angles suggest **sp^2 hybridisation of nitrogen and sp^3 hybridisation of phosphorus.** Two of the sp^2 hybrid orbitals of N containing one electron each are used for σ-bonding and the third contains a lone pair of electrons. This leaves one electron unhybridized in p_z orbital. Similarly, four of the sp^3 hybrid orbitals of P containing one electron each are used for σ-bonding (two with N atoms and two with halogen atoms) leaving one unhybridised orbital containing one electron (Fig. 8.1)

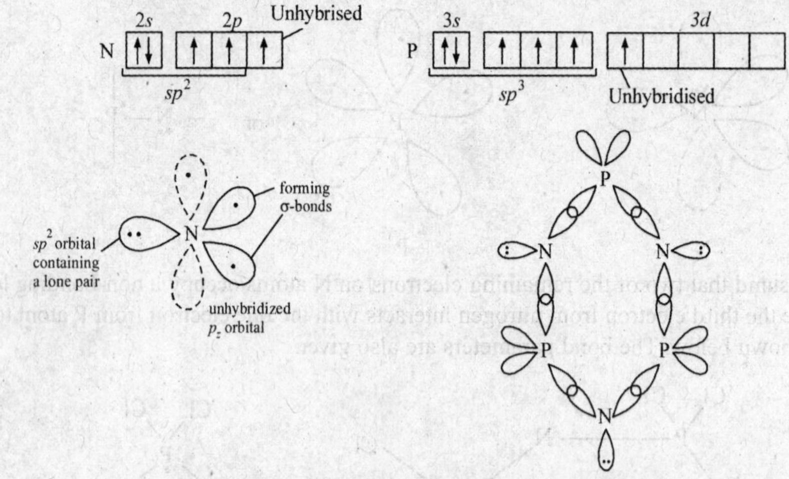

Fig. 8.1. σ-bond framework in cyclotriphosphazene

π-bonds are formed between phosphorus and nitrogen using valence electrons in 3*d* orbital of P and p_z orbital of N. The d_{xz} orbital of P atom overlaps with the p_z orbital of N to form $d\pi$ -$p\pi$ bond (Fig. 8.2). This type of $d\pi$ -$p\pi$ bonding is called **heteromorphic** (N-P) or **pseudoaromatic π bonding.**

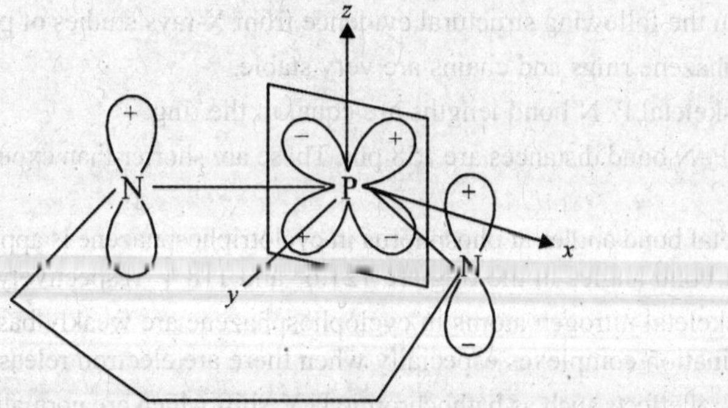

Fig. 8.2. $d\pi$ -$p\pi$ (or π) bonding between d_{xz} orbital of P with p_z orbital of N.

We have a lone pair of electrons on sp^2 hybridised N atom and a single electron on the vacant d_{xy} or $d_{x^2-y^2}$ orbital of P. They can overlap to give additional set of π bonds in the plane of the ring by the donation of the lone pair of electrons on nitrogen atom into the vacant d_{xy} and $d_{x^2-y^2}$ orbitals of phosphorus (Fig. 8.3). This results in a **coordinate bonding** in *xy* plane and is called **π′-bonding.**

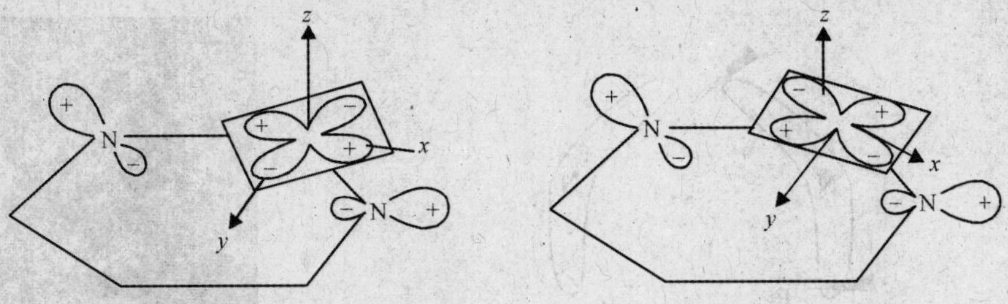

π-interaction in xy plane of sp^2(N) and d_{xy} (P)

π'-interaction in xy plane of sp^2(N) and $d_{x^2-y^2}$ (P)

Fig. 8.3. π'-bonding by donation of lone pair of electrons of N into vacant $d_{x^2-y^2}$ and d_{xy} orbitals of P.

To sum up the bonding in cyclotriphosphazene involves two types of π bonding : **out of plane heteromorphic π bonding and in plane coordinate π'-bonding.**

Dewar and co-workers model

Dewar and coworkers have proposed additional interpretation, accordingly to which d_{yz} orbital which is perpendicular to the d_{xz} orbital can also overlap with the p_z orbital of nitrogen. Thus, both these orbitals should contribute equally to the π-bond as shown in Fig. 8.4. These are **heteromorphic** (N–P) $p\pi$-$d\pi$ bonding with d_{xz} shown in Fig. 8.2 or **homomorphic** (N–P) $p\pi$-$d\pi$ bonding through d_{yz}. Both heteromorphic and homomorphic bonding have been shown in Fig. 8.4 for the sake of comparison.

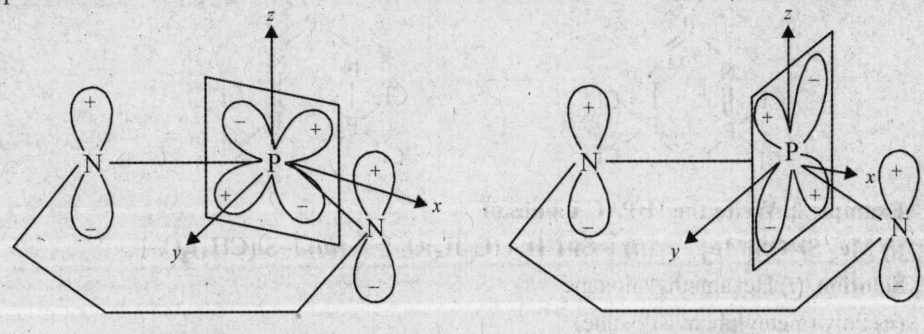

Heteromorphic π-interaction of p_z (N) and d_{xy} (P)

Homomorphic π-interaction of p_z (N) and d_{yz} (P)

Fig. 8.4. π-bonding by overlap of d_{xz} and d_{yz} orbital of P with p_z orbital of N.

This results into localised three-centre bond around each N atom which is known as **island** of **π-character.** This island of π-character is interrupted at each P atom. Hence there will be nodes at each P atom. (Fig. 8.5). **Dewar's model explain as to why cyclotriphosphazenes do not show broad delocalization effects like benzene which has simple $p\,\pi$ -$p\,\pi$ bonds.**

Uses of Phosphazene

1. They are used as catalysts in manufacture of silicones.

2. Thin films of poly (aminophosphazene) are used to cover severe burns and serious wounds because they prevent the loss of body fluids and keep germs out.

3. The phosphonitrilic halides are used as rigid plastics, expanded foam and fibres because they are water proof, fire proof and are unaffected by oil, petrol and other solvents.

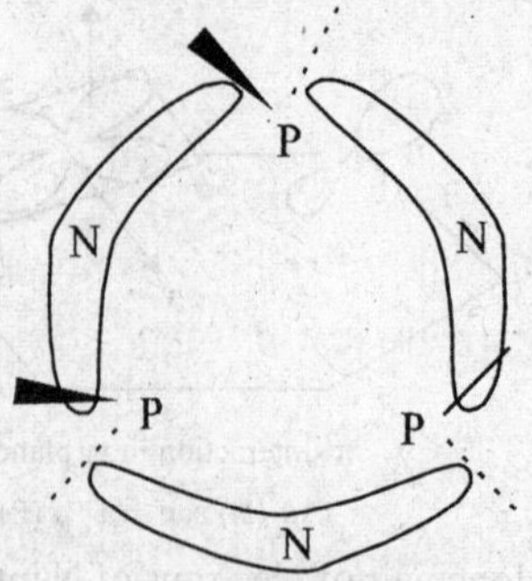

Fig. 8.5. Three centre island of π-bonding in $(NPCl_2)_3$

James Dewar

4. They are also used as flexible plastics which are useful for fuel hoses and gaskets because they retain their elasticity at low temperatures.

SOLVED CONCEPTUAL PROBLEMS

Example 1. Show that the structure of cyclotriphosphazene $(PNCl_2)_3$ is stabilized by resonance.

Solution. Due to the following resonating structures, the compound is stabilised

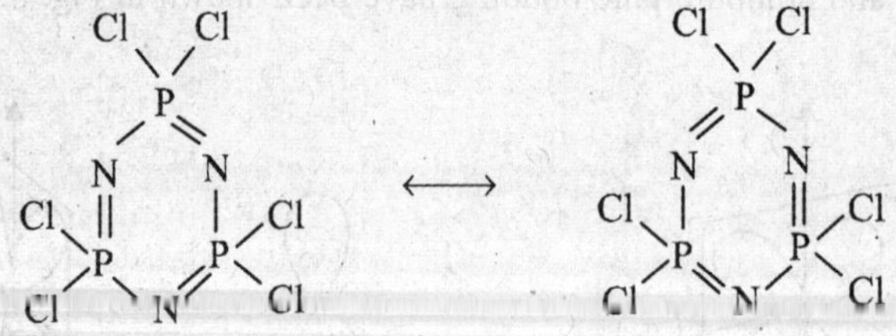

Example 2. Write the IUPAC names of

(*i*) Me_3 Si–OSiMe$_3$ (*ii*) [–Si(CH_3) (C_6H_5)O–] (*iii*) [–Si$(CH_3)_2$O–]

Solution. (*i*) Hexamethylsiloxane

(*ii*) Poly (methylphenylsiloxane)

(*iii*) Poly (dimethylsiloxane)

Example 3. Draw resonance hybrids of straight chain polyphosphazenes $(NPR_2)_n$.

Solution. Since all the N-P bonds in the chain are equal, the alternating single and double bonds may be represented as resonance hybrids of the following structures.

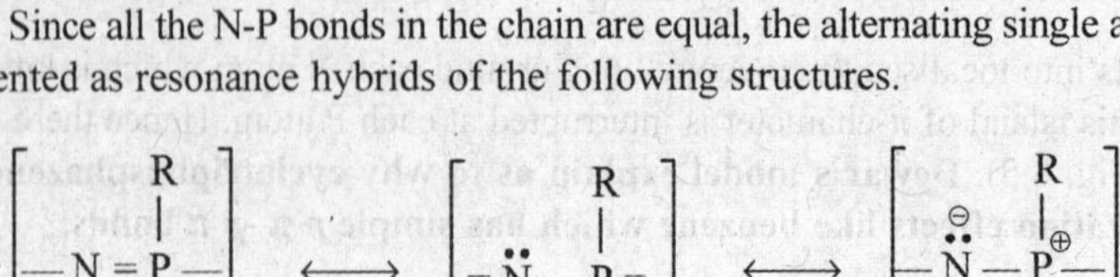

Example 4. Give a method of preparation of $(PNF_2)_3$.

Solution. $(PNF_2)_3$ can be obtained by the reaction between $(PNCl_2)_3$ and NaF.

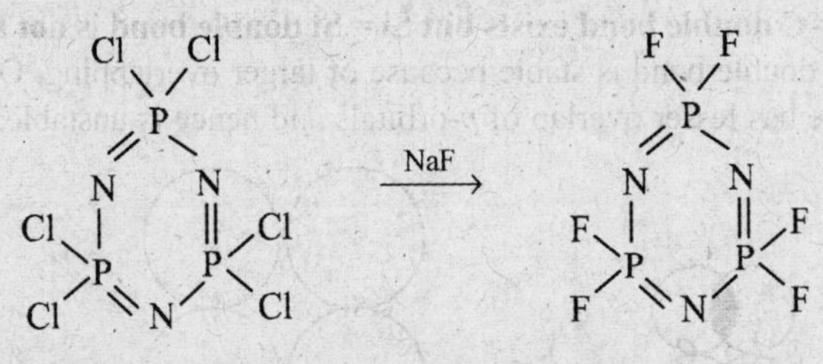

Example 5. Why do polyphosphazenes chain prefer a cis-trans conformation to a trans-trans conformation?

Solution. The conformation of a polyphosphazene is mainly determined by interactions between the substituents present on the neighbouring P atoms. The repulsion is minimum in the *cis-trans* planar conformation as compared to *trans-trans* planar conformation as shown below :

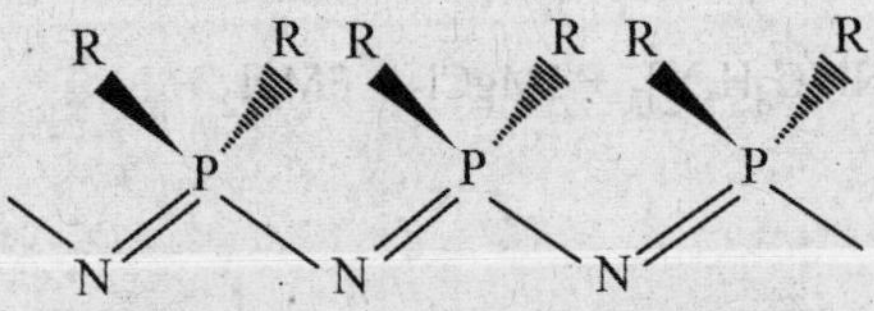

trans-trans conformation

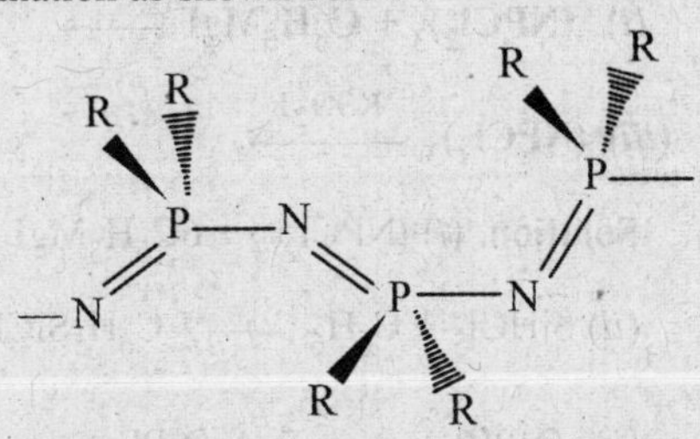

cis-trans conformation less repulsion

Example 6. What is island p-bond ?

Solution. Island π-bond is formed as a result of equal participation of p_z orbital of nitrogen and d_{xz} and d_{yz} orbitals of P. This is also known as 3-centre island bond.

Example 7. Explain island model of bonding taking the example of $(NPCl_2)_3$?

Solution. According to this model proposed by Dewar & Co-workers, d_{xz} and d_{yz} orbitals of P are hybridized to form two orbitals which are directed towards the adjacent N atoms. This allows the formation of three centred bond about each N. This results in delocalization over 3 atoms, PNP in the ring with nodes at each P atom as.

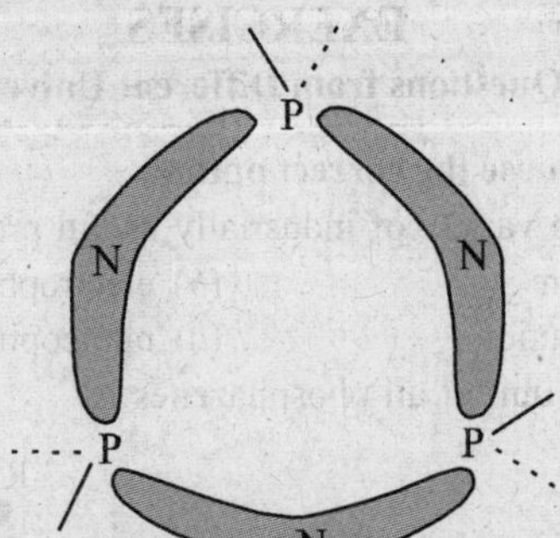

Example 8. Differentiate between the delocalization of π-system in cyclic triphosphazene and π-system in benzene.

Solution. In cyclic triphosphazene, there is island of π-character covering P–N–P atoms, *i.e.*, 3-centre bond with node at each P atom. Therefore, the p-chracter does not spread over whole ring. However, in bezene the π-bond is spreads over whole ring.

Example 9. Why C = C double bond exists but Si = Si double bond is not known. Explain.

Solution. The C = C double bond is stable because of larger overlapping. On the other hand Si atom being larger in size has lesser overlap of *p*-orbitals and hence is unstable.

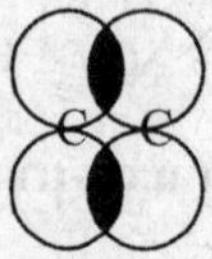

Shorter C=C bond

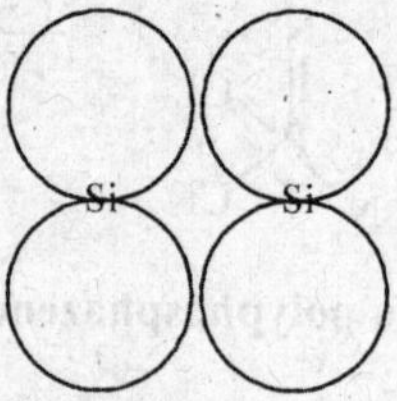

Longer Si=Si bond

Example 10. Complete the following reactions :

(*i*) $(NPCl_2)_3 + C_6H_5MgI \longrightarrow$ (*ii*) $SiHCl_3 + C_6H_6 \longrightarrow$

(*iii*) $(NPCl_2)_3 \xrightarrow{KSO_2F}$ (*iv*) $PCl_5 + NH_4Cl \longrightarrow$

Solution. (*i*) $(NPCl_2)_3 + 6C_6H_5MgI \longrightarrow [NP(C_6H_5)_2]_3 + 3MgCl_2 + 3MgI_2$

(*ii*) $SiHCl_3 + C_6H_6 \longrightarrow C_6H_5SiCl_3 + H_2$

(*iii*) $(NPCl_2)_3 \xrightarrow{KSO_2F} (NPF_2)_3$

(*iv*) $PCl_5 + NH_4Cl \longrightarrow \underset{\text{cyclic}}{(NPCl_2)_3} + \underset{\text{cyclic}}{(NPCl_2)_4} + \underset{\text{linear}}{(NPCl_2)_n}$

Example 11. How is methoxy substituted cyclophosphazene synthesized ?

Solution. Alkoxide ions react with cyclochlorophosphazene to give the desired product.

$$(PNCl_2)_3 + \underset{\text{Sodium methoxide}}{6\,NaOCH_3} \longrightarrow \underset{\text{Methoxy substituted cyclophosphazene}}{[PN(CH_3O)_2]_3} + 6\,NaCl$$

EXERCISES
(Including Questions from Different University Papers)

Multiple Choice Questions (Choose the correct option)

1. Polyphosphazene give a variety of indusrially useful products with the help of
   (*a*) nucleophilic addition (*b*) electrophilic addition
   (*c*) electrophilic substitution (*d*) nucleophilic substitution
2. The common structural unit of all phosphazenes is

(*a*)
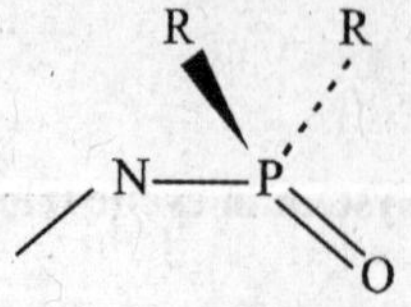

(*b*)
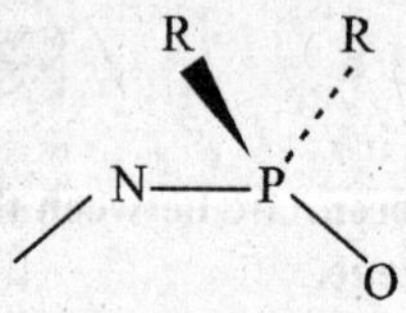

(*c*)
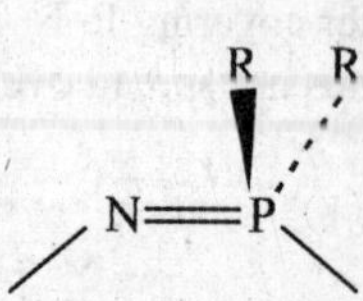

(*d*)
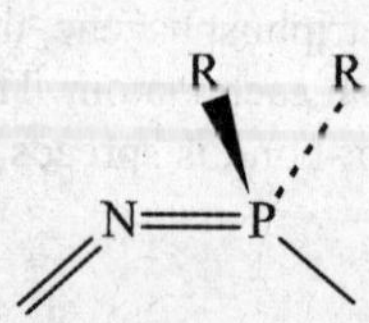

3. Most of the cyclotriphosphazenes have
   (*a*) planar structure (*b*) puckered ring structure
   (*c*) chain form structure (*d*) tetrahedral structure
4. Silicones have the structural unit

(*a*)  (*b*) 

(*c*)  (*d*) 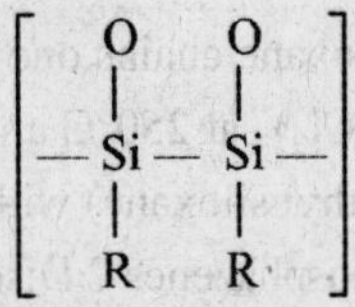

5. The process in which chloroalkanes react with elemental silicon at 200–300°C in the presence of copper to yield chlorosilanes is known as
   (*a*) Dewar process (*b*) Rochow process (*c*) Craig process (*d*) Mitchell process
6. Muller-Rochaw process involves the reaction of
   (*a*) Si and CH_3Cl (*b*) PCl_5 and NH_4Cl
   (*c*) Me_3SiOH and CH_3Cl (*d*) Me_2SiX_2 and Me_2SiO
7. Phosphazene can be prepared by reacting PCl_5 with another compound in a chlorohydro carbon solvent under mild conditions. The other compound is
   (*a*) NH_4OH (*b*) NH_2NH_2 (*c*) NH_4Cl (*d*) $PhNH_2$
8. The ring opening of $(NPCl_2)_3$ occurs by heating in a sealed tube at
   (*a*) 1250° C (*b*) 100° C (*c*) 250° C (*d*) 2500° C

ANSWERS

1. (*d*)	2. (*c*)	3. (*a*)	4. (*b*)	5. (*b*)	6. (*a*)	7. (*c*)	8. (*c*)

Short Answer Questions

1. What is meant by glass transition temperature ?
2. Draw the general repeating unit in silicones.
3. Complete the reaction :
   Cyclic $(NPCl_2)_3 \xrightarrow{250^\circ C}$
4. What is the structure of cyclic $(NPF_2)_3$?
5. Which groups of elements in the periodic table have been explored for the formation of inorganic polymers ?
6. Draw the general repeating unit in phosphazenes.
7. Draw resonance hybrid structures of cyclic $(NPCl_2)_3$.
8. Give one important use of silicone rubbers.
9. Give one important use of silicone oils.

10. What is meant by ring opening polymerisation ?

11. Give the IUPAC names of Cl_2SiH_2 H $Si(OH)_3$ and H_3 Si-O-SiH_3.

General Questions

1. How is cyclic $(NPCl_3)_2$ prepared ? Give a account of its nucleophilic substitution reactions.

2. Name four main structural units of silicones and designate them.

3. What are homomorphic and heteromorphic π-systems ? Explain.

4. Why do polyphosphazene chains prefer cis-trans conformations to a trans-trans conformation? Give three important uses of polyphosphazenes.

5. Give equations to indicate the following reactions :

(*i*) Two poly siloxane chains one with Si-H and other with Si-H = CH_2 group.

(*ii*) Heating $(NPCl_2)_3$ at 250°C and undergoing nucleophilic substitution reaction.

(*iii*) Cyclo (dimethyl siloxane) with Me_2 $SiCl_2$.

6. What are polyphosphazenes ? Discuss the nature of bonding in cyclic triphosphazenes.

7. Discuss Muller Rochow process for the synthesis of silicones.

8. What are polymeric backbones in silicones and phosphazenes ?

9. What are silicones ? Discuss important properties of silicones.

10. Polyphosphazenes are isoelectronic with polysiloxanes. Explain and comment on consequences.

11. Silicones and phosphazenes are isoelectronic. Discuss consequences.

12. (*i*) "Polyphosphazenes are isoelectronic to Polysiloxanes". Comment.

(*ii*) Discuss general features of $d\pi$-$p\pi$ model for bonding in $(NPCl_2)_3$.

13. (*a*) Discuss the island model of bonding in cyclic $(NPCl_2)_3$.

(*b*) Discuss briefly the phosphazenes polymers.

(*c*) Give important applications of silicones.

14. Write a brief account of

(*i*) Silicone rubbers

(*ii*) Silicone resins.

15. (*a*) What is cross-linking ? Explain important consequences of cross-linking in mac romolecules.

(*b*) What are homomorphic and heteromorphic π-systems.

(*c*) Draw polymeric back bones of silicones and phosphazenes.

16. Explain :

(*a*) Why do polyphosphazene chains prefer a cis-trans conformation to a trans- trans conformation ?

(*b*) How do the π-system in cyclic $(NPCl_2)_3$ differ from π-system in C_6H_6 ?

21. (*a*) Draw polymeric backbones of silicones and phosphazenes.

(*b*) Draw resonance hybrids of $(HPCl_2)_3$.

(*c*) Explain important consequences of cross-linking in macromolecules.

(*d*) Draw polymeric backbone of silicones.

22. (*a*) What are silicones ? How are cross linked silicones prepared ?

(*b*) Give islands model of bonding in cyclic $(NPCl_2)_3$.

23. Give a brief account of inorganic polymers with special reference to polyphosphazenes.

24. (*a*) Silicones and phosphazenes are isoelectronic. What are its consequences ?

    (*b*) Name three major classes of silicones elastomers.

25. (*a*) What are silicones ? Discuss their polymerisation.

    (*b*) Give four important properties of silicones.

26. (*a*) Draw polymeric backbones of silicones and phosphazenes.

    (*b*) Why does the p-system in cyclic $(NPCl_2)_3$ differ from p-system in C_6H_6 ?

    (*c*) What are silicones oils, silicones rubbers and silicones resins ?

    (*d*) What is IUPAC name of $[–Si\ (CH_3)_2\ O–]$?

ORGANIC CHEMISTRY

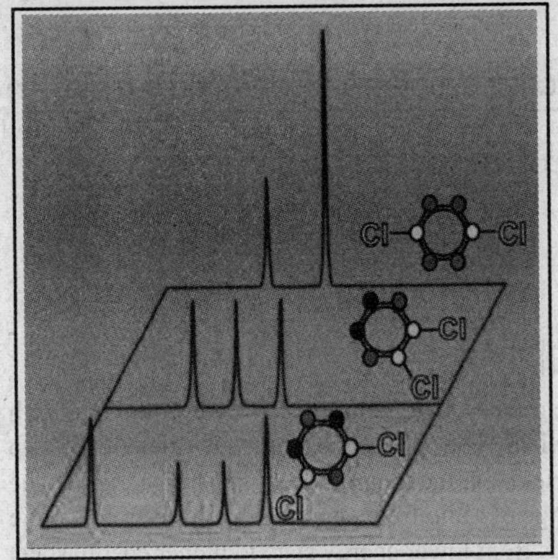

1

Spectroscopy

NUCLEAR MAGNETIC RESONANCE (NMR)

1.1. INTRODUCTION

Like electrons, the protons and neutrons also spin. If the particles in a nucleus do not have their spins paired, there is a net spin. A charged particle like proton, if it is spinning, will produce a magnetic field and magnetic moment along the axis of spin. Such a proton (or nucleus) acts like a tiny magnet. Thus a nucleus spinning in the anticlockwise direction will be associated with a magnetic field with moment acting upwards. If the nucleus is spinning clockwise, the magnetic moment will act downwards. Nuclei of atoms having odd numbered masses such as ^{1}H, ^{17}F, ^{31}P or those having even numbered masses but odd atomic numbers such as $^{2}_{1}H$, $^{10}_{5}B$, $^{14}_{7}N$ are magnetic in nature as their spins are not paired. Nuclei of atoms with even masses and even atomic numbers, such as, $^{12}_{6}C$, $^{16}_{8}B$, $^{32}_{16}S$ have no magnetic properties as there is no resultant spin.

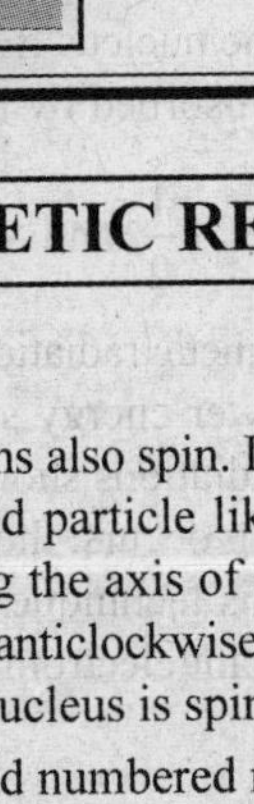

Nuclear magnetic resonance spectroscopy is now carried out under proton magnetic spectroscopy (PMR or ^{1}H–NMR) and carbon magnetic resonance spectroscopy (^{13}C–NMR). However, we shall restrict our treatise to PMR spectroscopy only.

1.2. PRINCIPLE OF NUCLEAR MAGNETIC RESONANCE SPECTROSCOPY

The magnetic properties of the nuclei in most of the organic compounds is the basis of **nuclear magnetic resonance (NMR) spectroscopy.**

If a nucleus like proton is placed in an external magnetic field (H_0), the magnetic moment of proton (H^+) will be oriented either with or against the external field as shown in the figure below:

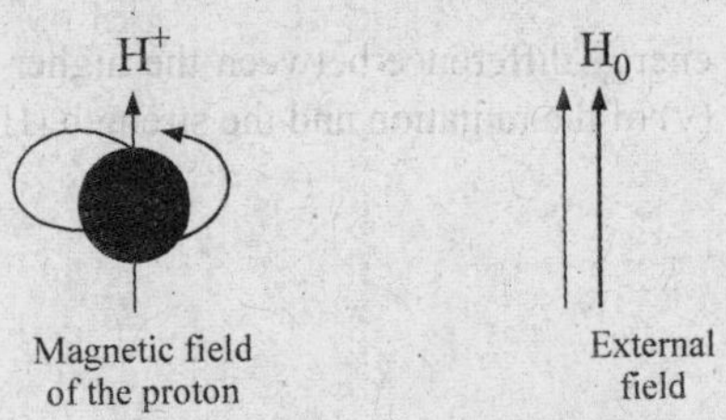

Orientation with the applied (external) field

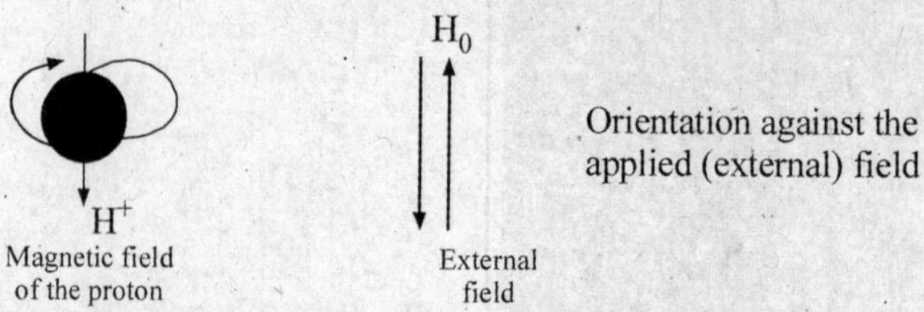

Out of the two orientations the one along the applied field is more stable (or associated with smaller energy) than the one against the applied field. The difference in the energies of the two orientations is designated as ΔE. Its magnitude varies with the magnitude of the applied field. The difference is, however, small and lies within the range of radio frequency region of electromagnetic spectrum. Thus if we desire to flip (shift) the nucleus from lower energy state to higher energy state, an amount of energy ΔE will have to the absorbed by the nucleus.

Like the electron, only two spin states $+\frac{1}{2}$ and $-\frac{1}{2}$ are allowed for a proton.

If the neclei are exposed to electromagnetic radiations of proper frequency, absorption of energy takes place and the nuclei (protons) in lower energy spin state flip (shift) to higher energy state. However, the energy associated with the radiations should match the energy difference between the two nuclear spin states. When such a fliping occurs, the nuclei are said to be in **resonance** with the electromagnetic radiations. That is why this technique of spectroscopy is called nuclear magnetic resonance. As already mentioned, the relevant electromagnetic radiations lie in radio frequency (rf) region.

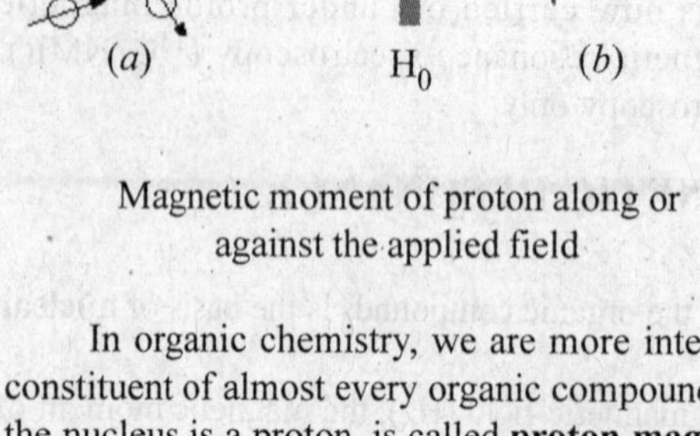

Magnetic moment of proton along or against the applied field

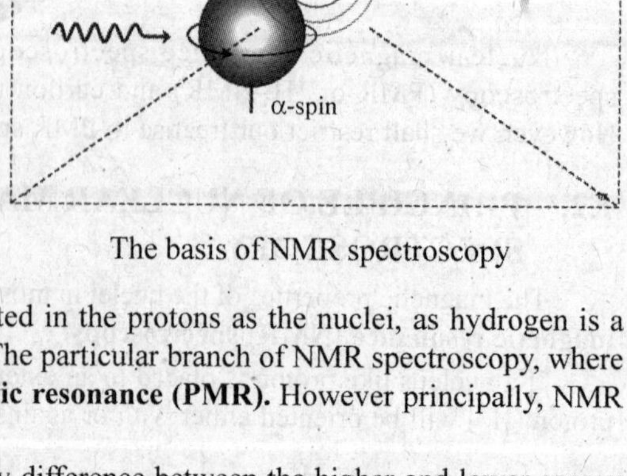

The basis of NMR spectroscopy

In organic chemistry, we are more interested in the protons as the nuclei, as hydrogen is a constituent of almost every organic compound. The particular branch of NMR spectroscopy, where the nucleus is a proton, is called **proton magnetic resonance (PMR).** However principally, NMR and PMR spectroscopies are the same.

The relationship between the ΔE *i.e.* energy difference between the higher and lower energy states of the proton (or nucleus), frequency (ν) of the radiation and the strength (H_0) of the magnetic field is given by the following equations:

$$\Delta E = h\nu$$

and $$h\nu = \frac{\gamma . h . H_0}{2\pi}$$

or $$\nu = \frac{\gamma . H_0}{2\pi} \quad \dots (i)$$

where γ = nuclear constant known as gyromagnetic ratio

h = Plaks's constant

For a proton γ = 26750

From eq. (*i*), it is obvious that higher the value of H_0 (applied magnetic field), higher will be the frequency of radiation required to flip the proton from lower to higher energy state.

A PMR spectrum can be recorded by placing the substance containing hydrogen nuclei in a magnetic field of constant strength and passing radiations of changing frequency through the substance and noting the frequency at which the absorption of energy corresponding to flipping of proton, from lower to higher energy state, takes place. However for ease of operation, the frequency of radiation is kept constant (at 40, 60 or 100 megacycles/sec) and the strength of the magnetic field is changed. We obtain a graph of absorption of energy versus magnetic field strength shown in Fig. 1.2. A signal is obtained which signifies the absorption of energy at a particular field strength.

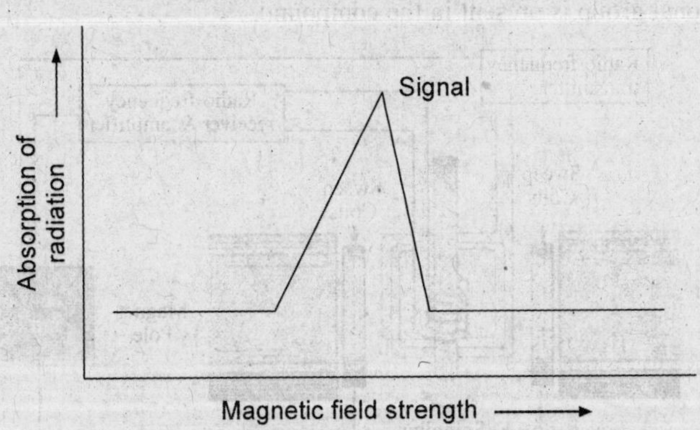

Fig. 1.2. A typical PMR signal

1.3. DIAGRAM OF A PMR SPECTROPHOTOMETER

The main constituents of a PMR spectrophotometer are shown in Fig. 1.3

Absorption of radiofrequency energy (resonance) occurs at a particular field strength when a compound containing protons is placed below the magnetic pole gap and irradiated with radiofrequency radiations. The sample is subjected to rapid rotation to ensure uniform exposure to the radiations. The solution of the compound, if solid, is prepared in a solvent like deuteriochloroform ($CDCl_3$), deuterioacetone (CD_3COCD_3) or deuterium oxide (D_2O). If liquid, the sampleis taken as such.

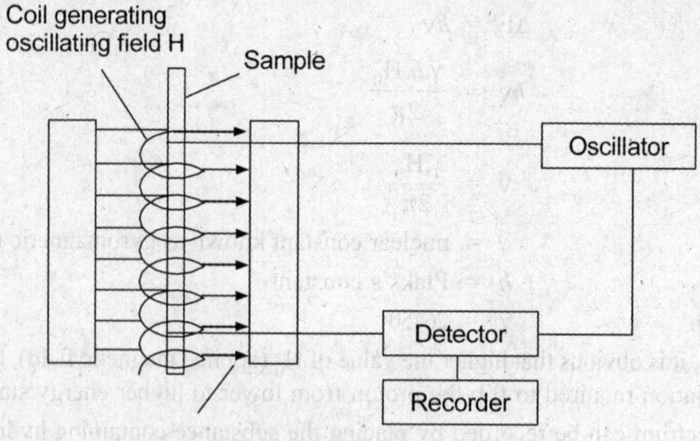

Fig. 1.3. Components of PMR spectrophotometer

1.4. APPLICATIONS OF NMR SPECTROSCOPY

Various applications of NMR spectroscopic technique are given as under:

1. Identification of functional groups. Every functional group gives a characteristic signal in the NMR spectrum. By studying the chemical shift of compound, it becomes possible to establish what kind of functional group is present in the compound.

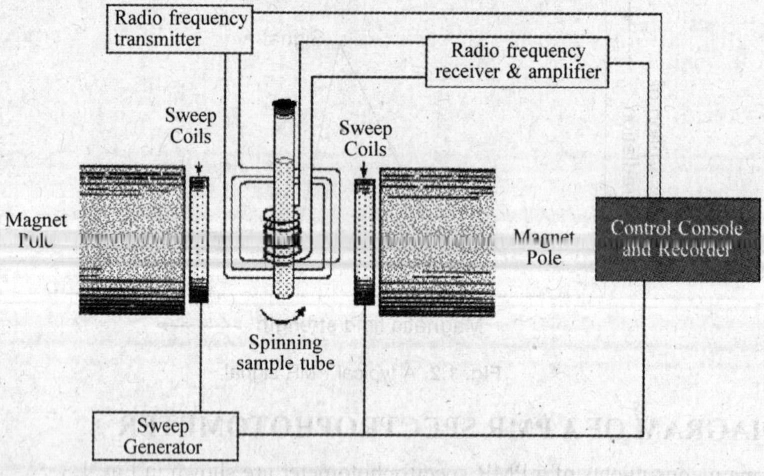

Working of NMR spectrophotometer

2. Structure of an unknown compound. It is possible to elucidate the structure of an unknown compound from the NMR studies. This is because protons under different environments give different chemical shifts. By observing doublets, triplets and multiplets, it is possible to place hydrogens at appropriate place in the formula and hence to establish the structure.

3. Comparison of two compounds. NMR spectrum is like fingerprint of a compound. Two compound showing same NMR spectrum must be structurally identical.

1.5. EQUIVALENT AND NON-EQUIVALENT PROTONS

The absorption by a proton depends upon its local environments such as electron density at the proton and presence of other protons in the neighbourhood. The magnetic field strength experienced by a proton is actually different from the magnetic field applied. The environments modify the applied magnetic field. For the same applied field, different protons having different environments will experience or receive different magnetic fields. Therefore to experience the same magnetic field, different protons have to be subjected to different applied fields. In other words for the same frequencies, protons under different environments absorb at the same effective magnetic field but at different applied fields.

Equivalent protons. Protons in a molecule having the same environments absorb at the same magnetic field strength, such protons are called equivalent protons.

Non-equivalent protons. Protons which have different environments absorb at different magnetic fields, such protons are called non-equivalent protons.

All equivalent protons give rise to one signal in the NMR spectrum. From the number of signals, we can tell how many different types of protons are there in the molecule.

Magnetically equivalent protons are also chemically equivalent and *vice-versa*. We can judge whether the two (or more) protons are chemically equivalent or not, by the isomer number method. Let us take the case of ethyl chloride $CH_3 - CH_2 - Cl$. We have three methyl proton and two methylene protons in the molecule. The methyl protons are evidently different from the methylene protons. This is because, if we substitute a methyl proton and a methylene proton separately by another group, we obtain two products as follows:

$$\underset{\text{I}}{CH_2\,ZCH_2\,Cl} \qquad\qquad \underset{\text{II}}{CH_3CHZCL}$$

Compounds I and II are clearly different compounds therefore, we say that methyl protons are different from methylene protons or in other words, the two types of protons are non-equivalent.

Now let us see whether the three methyl protons amongst themselves are equivalent or not. If we substitute any of the three hydrogens by a group Z, we obtain the same compound, $CH_2\,ZCH_2$ Cl. Therefore, three methyl protons are chemically equivalent. They will provide just one signal in the spectrum because they will absorb at the same field strength.

Again let us see whether the two methylene protons are equivalent or not. If we replace, two hydrogens separately by a group Z, we obtain structures III and IV which are enantiomers of each other: Such protons are also considered chemically equivalent.

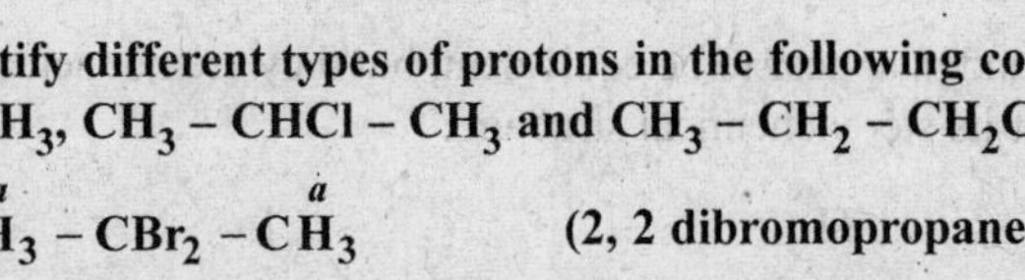

The spectrum does not distinguish between such enantiomeric protons. We get one signal corresponding to these two protons. In all, ethyl chloride will give rise to two signals in NMR spectrum.

Example 1. Identify different types of protons in the following compounds.
$CH_3 - CBr_2 - CH_3$, $CH_3 - CHCl - CH_3$ and $CH_3 - CH_2 - CH_2Cl$

Solution. (*i*) $\overset{a}{CH_3} - CBr_2 - \overset{a}{CH_3}$ **(2, 2 dibromopropane)**

Six methyl protons on the two extremes are equivalent. This has been indicated by putting a letter "*a*" on the methyl protons. All the protons being equivalent, only one signal will be obtained.

(*ii*) $\overset{a}{CH_3} - \overset{b}{CH}Cl - \overset{a}{CH_3}$ **(Isopropyl chloride)**

Three protons on the extreme left are equivalent. Three protons on the extreme right are again equivalent and equivalent to extreme left protons. The middle proton is of different type. Thus there are two types of protons. This has been indicated by putting small letters *a and b* on them.

(*iii*) $\overset{a}{CH_3} - \overset{b}{CH_2} - \overset{c}{CH_2} - Cl$ **(*n* - propyl chloride)**

There are three types of protons in the above molecule as indicated by the letters *a*, *b*, *c*.

Example 2. Identify different protons in the following compounds.

(*i*) 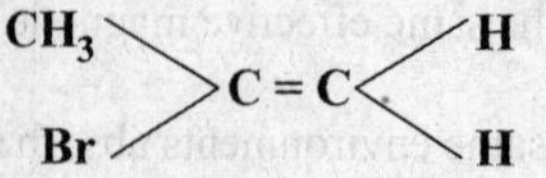 (*ii*) 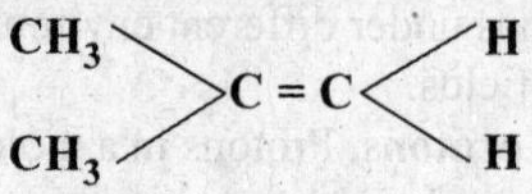

(*iii*) 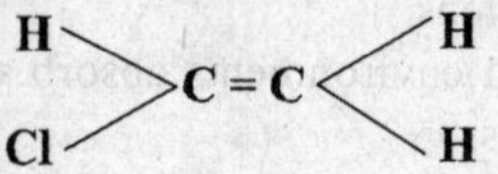 (*iv*) **1, 2 dichloropropene**

Solution. (*i*) 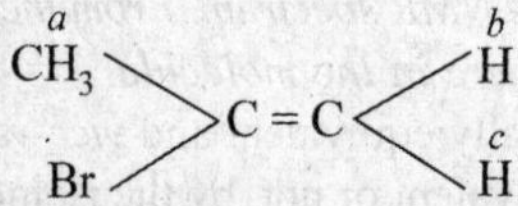 2-bromopropene

A close look at the molecule reveals that there are three different types of protons, indicated by the letters *a, b, c*. We may follow the rules explained earlier to dicide whether any two protons in a molecule are equivalent or not. Replace the two protons separately by another group Z. If the two new products obtained are the same or enantiomers of each other, the protons are equivalent, otherwise not.

(*ii*) 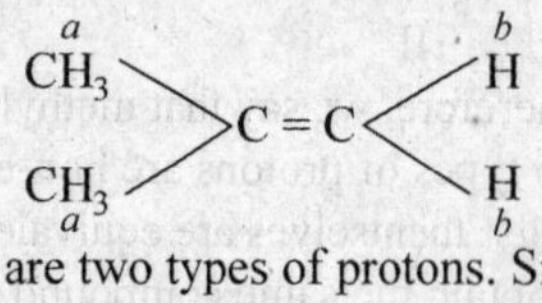

There are two types of protons. Six methyl protons on the L.H.S. are equivalent. Two protons on R.H.S. are again equivalent.

(*iii*) 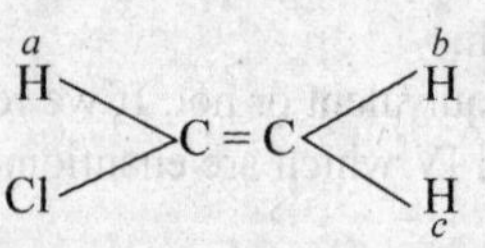

No two proton in the above molecule are equivalent. Thus there are three types of protons in the above molecule indicated by *a, b* and *c*. Correspondingly, three signal are obtained in NMR spectrum.

(*iv*) 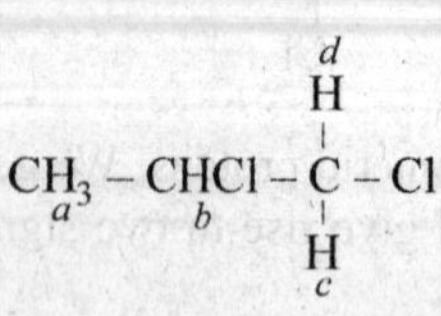 (1, 2, dichloropropane)

Protons marked *c* and *d*, appear to be equivalent, but they are not actually so. It becomes clear when we have a look, at its stereochemical structure.

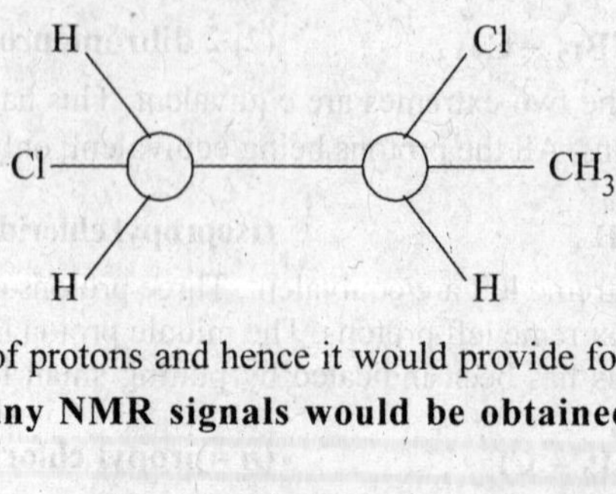

Thus it has four types of protons and hence it would provide four signals.

Example 3. How many NMR signals would be obtained in the case of following compounds?

(*i*) $CH_3OCH_2CH_3$ (*ii*) CH_2ClCH_2Cl (*iii*) CH_3OCH_3

(*iv*) Cis and trans 1, 2 dibromocyclopropane (*v*) Cyclopentane

(*vi*) Cyclohexane **(*vii*) CH_3CH_2CHO** **(*viii*) $C_6H_5-CH_3$**

(*ix*) $(CH_3)_2CH-CH_2-CH_3$ **(*x*) $CH_3-CH=CH_2$**

(*xi*) $CH_3-CH_2-CHCl-CH_3$

Solution. (*i*) $\overset{a}{CH_3}-O\overset{b}{CH_2}-\overset{c}{CH_3}$

There are three kinds of protons as indicated above, hence it would give three signals.

(*ii*) $Cl-\overset{a}{CH_2}-\overset{a}{CH_2}-Cl$

All the four protons are equivalent. Hence only one signal will be obtained.

(*iii*) $\overset{a}{CH_3}-O-\overset{a}{CH_3}$

All the six protons are magnetically and chemically equivalent. Hence one signal will be obtained.

(*iv*)

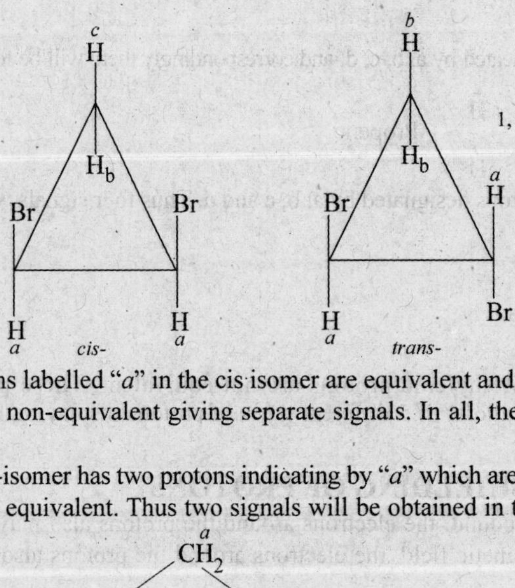

cis and *trans*
1, 2 dibromocyclo-propane

Protons labelled "*a*" in the cis isomer are equivalent and will give one signal. Protons labelled b and c are non-equivalent giving separate signals. In all, there will be three signals from the cis-isomer.

Trans-isomer has two protons indicating by "*a*" which are equivalent. Similarly protons labelled '*b*' are also equivalent. Thus two signals will be obtained in this case.

(*v*)

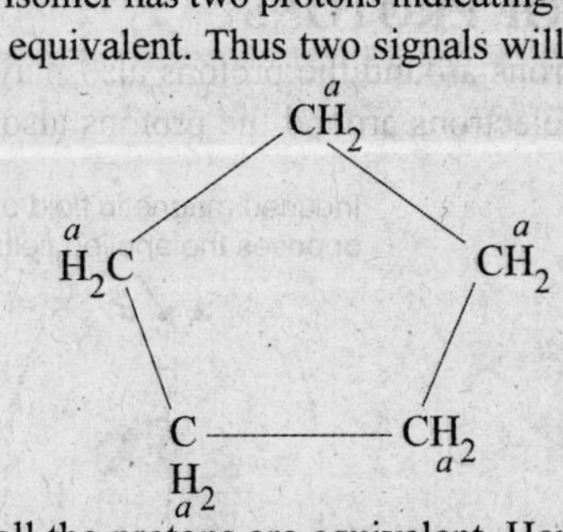

Cyclopentane

Here all the protons are equivalent. Hence only one signal will be obtained.

(*vi*)

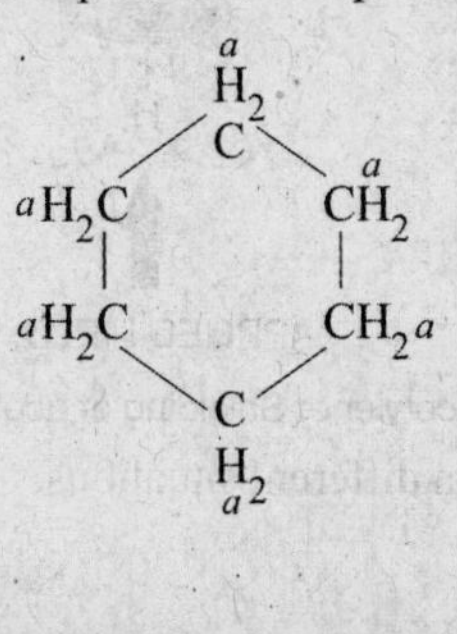

Cyclohexane

Here, again, all the protons are equivalent giving rise to only one PMR signal.

(*vii*) 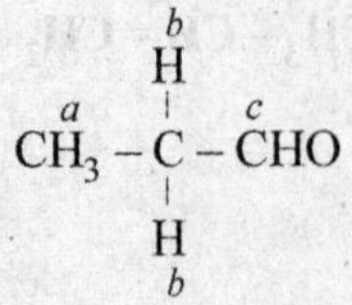 Propionaldehyde

There are three kinds of protons giving rise to three signals.

(*viii*) 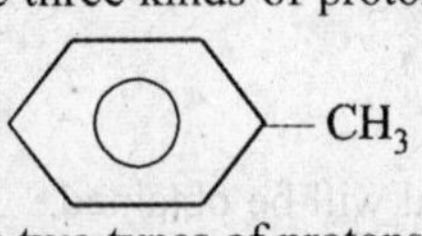 Toluene

There are two types of protons, benzene ring protons and the methyl protons. Two signals will be observed.

(*ix*) 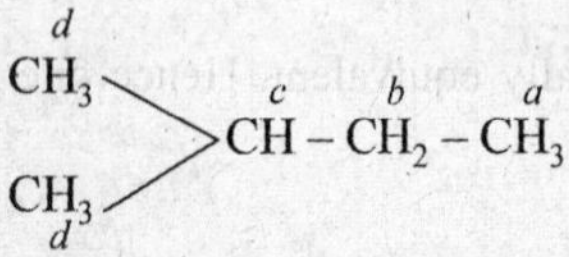 Isopentane

There are four types of protons indicated by a, b, c, d, and correspondingly there will be four signals.

(*x*) 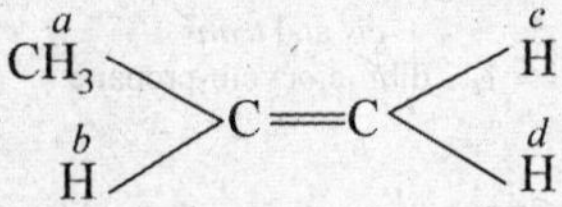 Propene

There are again four types of protons designated by a, b, c and d. Thus four signals will be seen.

(*xi*) 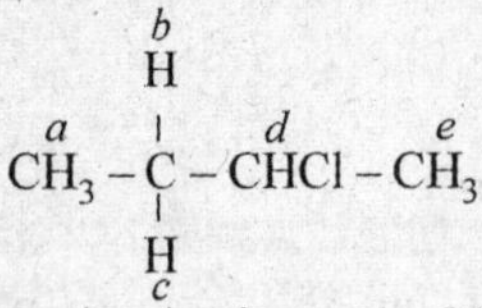

It offers an interesting case. Methylenic protons are not equivalent. These protons are diastereomeric protons. Thus we have a total of 5 different types of protons displaying five signals in PMR spectrum.

1.6. SHIELDING AND UNSHIELDING OF PROTONS

In the NMR spectrum of a compound, the electrons around the protons also play their role. When a compound is placed in a magnetic field, the electrons around the protons also generate a

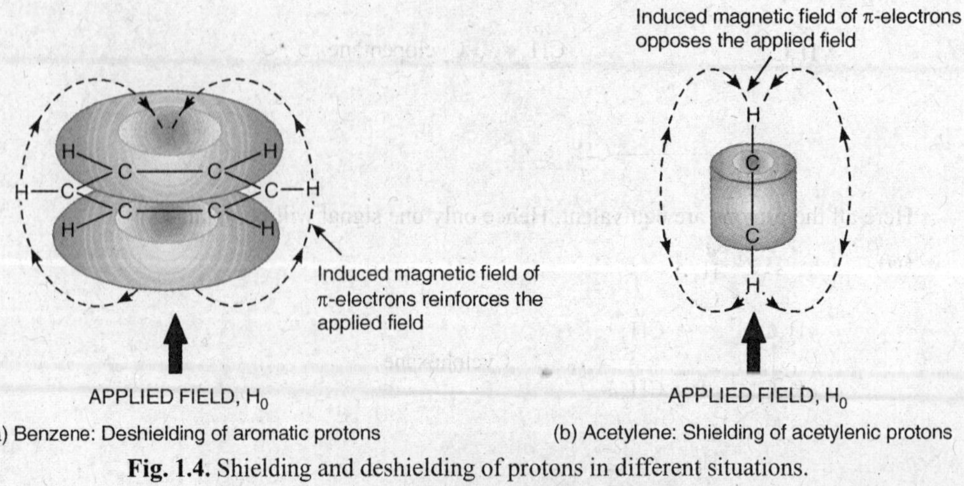

(a) Benzene: Deshielding of aromatic protons (b) Acetylene: Shielding of acetylenic protons

Fig. 1.4. Shielding and deshielding of protons in different situations.

magnetic field called *induced magnetic field.* The induced magnetic field may reinforce (support) or oppose the applied field. Thus two cases arise.

(*i*) If the induced field opposes the applied field, the effective field strength experienced by the protons decreases. Thus a greater applied field is required for the excitation of protons to higher level. It is expressed by saying that the proton absorbs **upfield**. The proton is said to be **shielded** by electrons in this case.

(*ii*) If the induced field reinforces or supports the applied field, an enhanced field strength will be experienced by the proton. Proton is said to be **deshielded** in this case. A deshielded proton absorbs **downfield** as a smaller field strength will be sufficient for absorption of energy to give a signal in NMR spectra. Shielding and deshielding of protons is a result of factors, like the inductive and electromeric effects of groups and hydrogen bonding.

In a case when π electrons are involved, whether the induced field opposes or reinforces the applied field depends upon the manner in which π-electrons circulate under the influence of applied field. For example, when a benzene derivative (Fig. 1.4a) is placed in a magnetic field, the delocalised π-electrons of benzene ring circulate in such a way that induced magnetic field reinforces the applied field at the aromatic protons. As a result, aromatic protons experience a greater magnetic field strength or they get deshielded.

However when a molecule of acetylene (Fig. 1.4b) is placed in a magnetic field the π-electrons circulate around the axis of the molecule in such a way that the induced field works against the applied field at the protons. Consequently, the protons experience weaker magnetic field strength and get shielded.

As a result of shielding and deshielding of protons, there is a **shift** in the position of the signal.

1.7. CHEMICAL SHIFT

The shift in the positions of PMR signal, compared with a standard substance, as a result of shielding and deshielding by electrons, is referred to as **chemical shift.**

Factors influencing chemical shift.

(*i*) Nature of groups. Atoms and groups which are electron withdrawing, deshield the protons. The effective magnetic field experienced by the protons, in such a case, is more than normal. Hence the protons will absorb downfield. Atoms like halogens, oxygen and nitrogen are electron-withdrawing. If such atoms are linked with protons, they deshield the protons resulting in downfield absorption. For example the proton in CH_3OCH_3 shows a downfield abosorption compared to that in CH_3-CH_3. An electron releasing group like methyl increases the electron density around the proton and thus shields it.

Greater the electronegativity of the atom, greater will be the extent of deshielding and hence lower will be the absorption energy. Proton in CH_3-F are deshielded to a greater extent than in CH_3-Cl.

Consider the case of $\overset{b}{CH_3}\overset{a}{CH_2}Cl$. It has two types of protons. The protons labelled *a* are more deshielded than protons labelled *b* because of different distance of Cl from the two types of protons.

Electron withdrawing groups reduce the electron density around the proton or deshield the proton, thereby resulting in downfield absorption. Electron releasing groups increase the electron density around protons and result in upfield absorption.

(*ii*) Hydrogen bonding. If the proton is linked to some electronegative atom in the form of a hydrogen bond, it is deshielded. As such, the absorption will occur downfield.

(*iii*) Space effect. Space effects cause shielding and deshielding on the proton due to induced magnetic fields in other parts of the molecule which operate through space. For example, when a magnetic field is applied to a molecule containing π–electrons, these electrons begin to circulate at right angles to the direction of the applied field thereby producing induced magnetic field. The effect of this field on the nearby proton depends upon the **orientation of the proton** with respect to the π–bond producing the induced field. For example, space effects explain strong deshielding of aldehydic and aromatic protons. On the other hand, the circulation of electrons of π–system has a shielding effect on acetylenic protons as mentioned earlier.

The effect of various atoms, bonds or groups is summarised as under:

(*a*) Protons on a primary carbon atom are shielded to the maximum followed by secondary and tertiary protons.

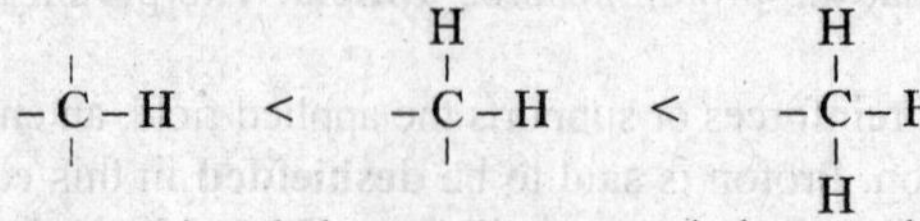

(*b*) Electronegative atoms such as halogens, oxygen and nitrogen deshield the protons resulting in donwfield absorption.

(*c*) π electrons of the carbonyl group, alkene and benzene ring have the deshielding effect on protons. The order of deshielding is

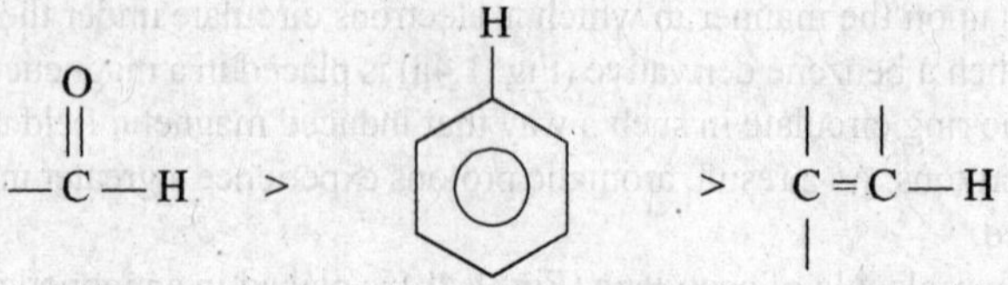

(*d*) Carbonyl group has the effect of deshielding protons on the neighbouring carbon.

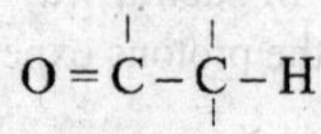

(*e*) Hydrogen bonding deshields the proton.

(*f*) Strongly electropositive atoms such as silicon increase the electron density on the proton and thus shield the proton.

1.8. TETRAMETHYLSILANE (TMS) $(CH_3)_4$ Si AS A STANDARD SUBSTANCE FOR RECORDING CHEMICAL SHIFT

In order to measure or record chemical shifts, we need some standard substance. The choice falls on TMS for the reasons given below.

(*i*) It has 12 equivalent protons and thus gives a sharp signal.

(*ii*) Electronegativity of silicon (1.8) is lower than that of carbon. Consequently the shielding of equivalent protons in TMS is more than that of almost all organic compounds. Therefore the signals of almost all organic compounds appear in the downfield direction with respect to TMS.

(*iii*) TMS is chemically inert and possesses a low b.p. of 300 K. It can be easily evaporated after the experiment to recover the compound.

Units for expressing chemical shifts

Chemical shifts are expressed in frequency units (cycles per second). Frequency and magnetic field are related to each other by the following equation.

$$\nu = \frac{\gamma H_0}{2\pi}$$

Chemical shift expressed in cps (cycles per second) is directly proportional to the strength of applied field which in turn is equivalent to radiofrequency used. Since different PMR spectrophotometer use different radiofrequencies of 40, 60 or 100 magacycle/sec (Mcps) it is more appropriate to use units in which the radiofrequency of the instrument has been cancelled out. This is done by dividing the shift in cps by the radiofrequency employed and multiplying by a factor of 10^6 so that shifts are obtain in **parts per million (ppm).**

If a signal is obtained at 120 cps downfield with reference to TMS using 60 Mcps (or 60×10^6 cps), then the chemical shifts is given by

$$\text{Chemical shift }(\delta) = \frac{120\times10^6}{60\times10^6} = 2 \text{ ppm}.$$

There are two scales for expressing the chemical shift. These are the δ (delta) scale and τ (tau) scale. Both of them are in **parts per million.**

$$\delta = \frac{\Delta\nu\,(\text{in cps})\times10^6}{\text{Radio frequency used in cps}}$$

and

$$\tau = 10 - \delta$$

Taking the position of TMS signal as zero ppm, most chemical shifts have values between 0–10. A small value of δ indicates a small downfield shift whereas a large δ value indicates a large downfield shift.

On the τ scale, the positions of the TMS signal is taken as 10.0 ppm and most compounds have chemical shift values between 10 – 0. This is explained with the help of Fig. 1.5.

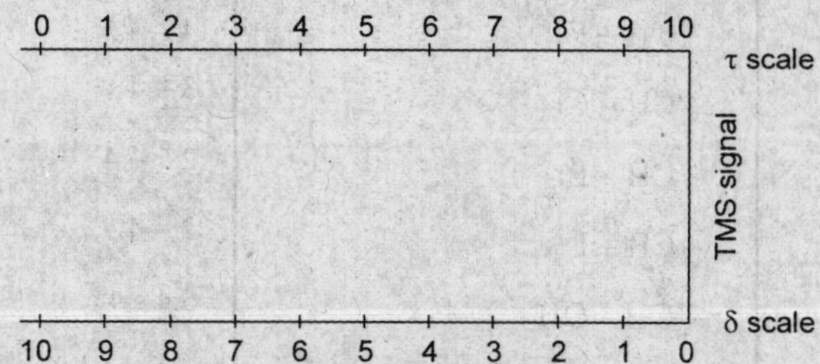

Fig. 1.5. δ and τ scales

1.9. CHEMICAL SHIFTS OF DIFFERENT TYPES OF PROTONS AND POSITIONS OF PMR SIGNALS

Chemical shift of a proton of a proton is determined by its electronic environments. In a given molecule, equivalent protons have the same chemical shift while non-equivalent protons have different chemical shifts. Moreover, a proton with a particular environment will show the same chemical shift regardless of the molecule in which it is present. We can therefore tell from the position of signals in a PMR spectrum of a sample what the electronic environment of each kind of proton is in the sample. The values of some typical proton chemical shifts relative to TMS are listed in Table 1.1.

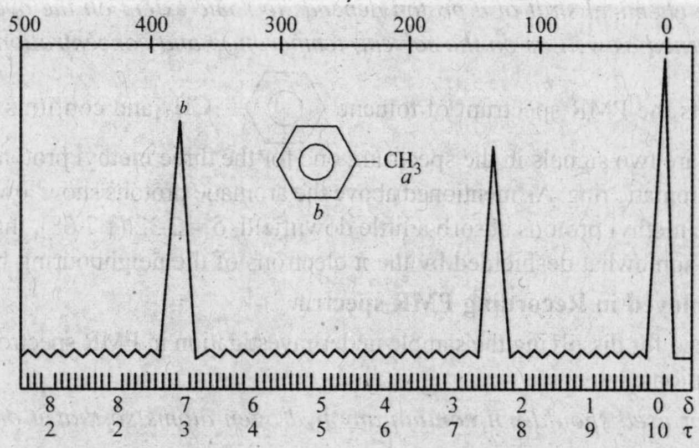

Fig. 1.6. PMR spectrum of toluene,

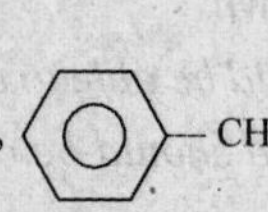

Table 1.1. Typical proton chemical shifts

Types of proton		*Chemical sift ppm*	
		δ	τ
Primary	$R-CH_3$	0.9	9.1
Secondary	R_3CH_2	1.3	8.7
Tertiary	R_3CH	1.5	8.5
Vinylic	$-\overset{\vert}{C}=\overset{\vert}{C}-H$	4.6 – 5.9	4.1 – 5.4
Acetylene	$-C\equiv C-H$	2 – 3	7 – 8
Aromatic	$Ar-H$	6 – 8.5	1.5 – 4
Benzylic	$Ar-\overset{\vert}{\underset{\vert}{C}}-H$	2.2 – 3	7 – 7.8
Allylic	$-\overset{\vert}{C}=\overset{\vert}{C}-\overset{\vert}{\underset{\vert}{C}}-H$	1.7	8.3
Alkyl fluoride	$-\overset{\vert}{C}H-F$	4 – 4.5	5.5 – 6
Alkyl chloride	$-\overset{\vert}{C}H-Cl$	3 – 4	6 – 7
Alkyl bromide	$-\overset{\vert}{C}H-Br$	2.5 – 4	6 – 7.5
Alkyl iodide	$-\overset{\vert}{C}H-I$	2 – 4	6 – 8
Alcohol	$-\overset{\vert}{C}H-OH$	3.4 - 4	6 – 6.6
Ethers	$-\overset{\vert}{C}H-O-R$	1.3 – 4	6 – 6.7
Carbonyl compounds	$-\overset{\vert}{C}H-\overset{O}{\overset{\Vert}{C}}-$	2 – 2.7	7.3 – 8
Acids	$-\overset{\vert}{C}H-COOH$	2 – 2.6	7.4 – 8
Hydroxylic	$R-O-H$	1 – 5.5	4.5 – 9
Phenolic	$Ar-O-H$	4 – 12	(–2) – 6
Aldehydic	$R-\overset{O}{\overset{\Vert}{C}}-H$	9 – 10	0 – 1
Carboxylic	$R-COOH$	10.5 – 12	(–2) – (–0.5)

However, the value given for each type of proton may show a slight variation in different PMR spectra because *the chemical shift of a proton depends to some extent on the overall structure of the molecule concerned as well as on the solvent, temperature and concentration employed.*

Fig. 1.6 depicts the PMR spectrum of toluene 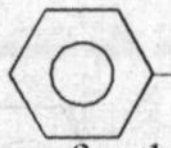 — CH_3 and confirms the observations made above. There are two signals in the spectrum; one for the three methyl protons and one for the five protons of the aromatic ring. As mentioned above the aromatic protons show low field absorption, δ 7.17 (τ 2.83). The methyl protons absorb a little downfield, δ = 2.32 (τ 7.68), than ordinary alkyl protons as they are somewhat deshielded by the π electrons of the neighbouring benzene ring.

Solvents Employed in Recording PMR spectra

The solvent used for dissolving the sample under investigation in PMR spectroscopy must fulfil the following conditions.

(*i*) *The solvent used should not contain any hydrogen atoms so that it does not give any absorption signal of its own.*

(*ii*) *The solvent should be able to dissolve the sample under investigation to reasonable extent.*

(*iii*) *The solvent used should be chemically inert.*

The commonly employed solvents in PMR spectroscopy are given below :

(*i*) Carbon tetrachloride (CCl_4) (*ii*) Carbon disulphide (CS_2)

(*iii*) Deutero chloroform ($CDCl_3$) (*iv*) Hexachloro acetone (CCl_3COCCl_3)

1.10. CHEMICAL SHIFTS OF VARIOUS TYPES OF PROTONS

Chemical shifts for various types of protons are given in Table 1.1.

1.11. PMR SPECTRUM OF BENZYL ALCOHOL

PMR spectrum of benzyl alcohol is given in Fig. 1.7

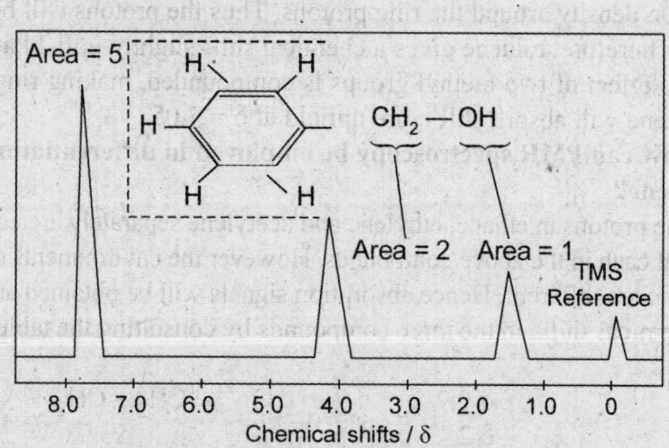

Fig. 1.7. PMR spectrum of benzyl alcohol

The no. of signals in a spectrum tells us how many different types of protons are present and the position of the signals (chemical shift) tells us about the nature of different protons (or their environments).

In the spectrum of benzyl alcohol shown above, we observe four signals which are due to the protons as detailed below:

(*i*) The small but sharp signal at $\delta = 0$ is the reference signal of TMS.

(*ii*) Signal at $\delta = 7.3$ is due to five equivalent ring protons.

(*iii*) The signal at $\delta = 4.6$ is due to two chemically equivalent methylene protons.

(*iv*) The peak at $\delta = 2.4$ is due to hydroxyl proton.

1.12. SIGNIFICANCE OF PEAK AREA

Peaks obtained in the PMR spectrum make different areas with the base line. It is observed that area under a PMR signal is directly proportional to the number of equivalent protons giving rise to that signal. By comparing the areas subtended by different signals, we can calculate the relative proportion of different types of protons. For example, in the case of peaks of benzyl alcohol. The areas under the peaks are in the ratio of 1 : 2 : 5 (See Fig. 1.7) indicating that the three types of protons are in the ratio 1 : 2 : 5. This is actually so. There are one hydroxyl proton, two methylene protons and five ring protons.

The peak areas of different signals are measured by an automatic electronic integrator. Also heights of peaks are proportional to areas.

Example 1. How can you explain the difference in chemical shifts of aromatic protons in the following compounds? Benzene $\delta = 7.37$, Toluene $\delta = 7.17$, *p*-xylene $\delta = 7.05$.

Solution.

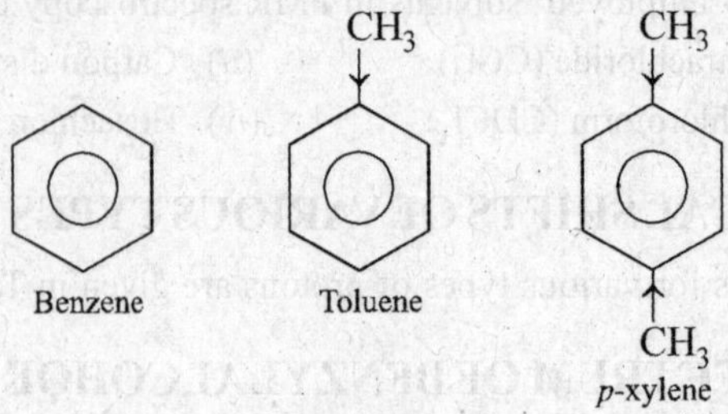

Methyl group is an electron releasing group. Hence in toluene, the effect of methyl group will be to increase electron density around the ring protons. Thus the protons will be shielded and they will absorb upfield. Therefore, toluene gives a chemical shift slightly upfield at $\delta = 7.17$.

In *p*-xylene, the effect of two methyl groups is compounded, making ring protons still more shielded. Thus *P*-xylene will absorb still more upfield at $\delta = 7.05$.

Example 2. How can PMR spectroscopy be employed in differentiating between ethane, ethylene and acetylene?

Solution. All the protons in ethane, ethylene and acetylene separately are equivalent. Hence we expect only one signal each in the above compounds. However the environments of protons in ethane, ethylene and acetylene are different. Hence absorption signals will be obtained at different positions. We can know the chemical shifts in the three compounds by consulting the table of chemical shifts. These are as follows:

$CH_3 - CH_3$ δ 0.9 $CH_2 - CH_2$ δ 5.3
$CH \equiv CH$ δ 2

Position of the signal can help us differentiate one compound from the rest.

Example 3. How many signals do you expect to have in the NMR spectrum of $CH_3 - CH_2 - Cl$? What would be the approximate absorption positions of the signals?

Solution. $CH_3 - CH_2 - Cl$ (Ethyl chloride) has two types of protons, three equivalent methyl protons and two equivalent methylene protons. Hence we expect two signals. Methylene protons are attached to carbon which also has an electron withdrawing chlorine attached to it. Thus the electron withdrawing or shielding effect of chlorine will be more on methylene protons than on the methyl protons. So methylene protons will absorb downfield compared to methyl protons.

The signals are obtained as detailed below:

Methyl protons δ 1– 1.5
Methylene protons δ 3.5

1.13. SPLITTING OF PMR SIGNALS

We assume that one type of protons give rise to one signal or peak in PMR spectrum. In actual practice it is not so. Consider, for example, the following compounds.

$BrCH_2 - CHBr_2$	$CH_3 - CHBr_2$	$CH_3 - CH_2 Br$
1, 1, 2 Tribromoethane	1, 1 dibromoethane	Ethyl bromide

(*a*) An examination of the above compounds reveals that each one of them contains two types of protons. We expect to observe, therefore, two signals in each case but this is actually not the case.

(*b*) We observe that 1, 1, 2 tribromoethane gives five peaks (two peaks split into five), 1, 1 dibromo ethane gives six peaks and ethyl bromide gives seven peaks. This phenomenon of splitting of a peak into several peaks is called **splitting of signals.**

(*c*) Splitting of signals takes place when there are non-equivalent protons on adjacent carbon atoms. Thus in $CH_3 - CH_3$, all the six protons are equivalent, hence no splitting is observed.

(*d*) Splitting of signals is explained on the basis of **spin-spin coupling** of the absorbing and neighbouring protons. Let us consider a case of two adjacent carbon atoms carrying non-equivalent protons, one having two secondary protons and the other a tertiary protons.

$$-\overset{t}{C}H-\overset{s}{C}H_2-$$

Letters *t* and *s* stand for tertiary and secondary respectively. Consider the absorption by one of the secondary protons. Magnetic field produced by neighbouring proton (*i.e.* tertiary proton) may have two possible orientations with respect to applied magnetic field.

(*i*) If the magnetic field produced by the proton is oriented along the applied field, the secondary protons experience some increased magnetic field strength. Consequently, the secondary proton will absorb at lower applied field (downfield) than it would have done in the absence of *t*-protons.

(*ii*) If the magnetic field produced by the proton is oriented against the applied field, the secondary protons experience some decreased magnetic field strength. Consequently, the secondary proton will absorb at higher applied magnetic field (upfield) than it would have done in the absence of *t*-protons.

(*e*) Considering that there is an equal probability of the *t*-proton magnetic field being aligned along or against the applied magnetic field, half of the secondary protons will absorb downfield and the other half will absorb upfield. Thus one peak will be split into two peaks **(doublet)** with equal peak intensities as shown below:

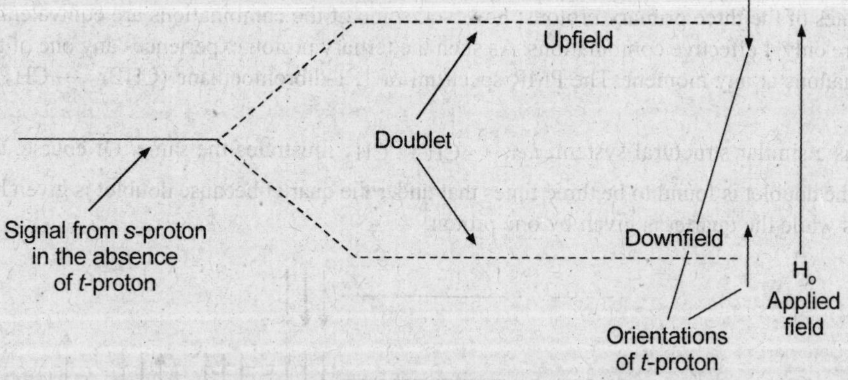

Fig. 1.8. Splitting by one adjacent proton gives to 1 : 1 doublet

(*f*) Now consider the absorption by *t*-protons. In this case, the two neighbouring s-protons may have three probable combinations of spin alignments as given below:

Case (*i*): The two protons are aligned with the applied magnetic field.

Case (*ii*): The two protons are aligned against the applied magnetic field.

Case (*iii*): One proton is aligned along the field and the other proton is aligned against the field.

In the first case, the tertiary proton will absorb downfield.

In the second case, the tertiary proton will absorb upfield.

In the third case, the position of the signal will not change.

As a result of above modification, the peak due to tertiary proton will be split into three peaks **(triplet)**. The relative intensities are in the ratio 1 : 2 : 1. The central peak is of double intensity because it belongs to the third case (as given above) in which one proton is along and the other against the field. If we designate the secondary proton as no. 1 & 2, then proton 1 is along and proton 2 is against the field or proton 2 is along and proton 1 is against the field. The middle peak corresponds to case (*iii*) which takes into account both these possibilities. Hence the middle peak is of double intensity. This is shown in Fig. 1.9.

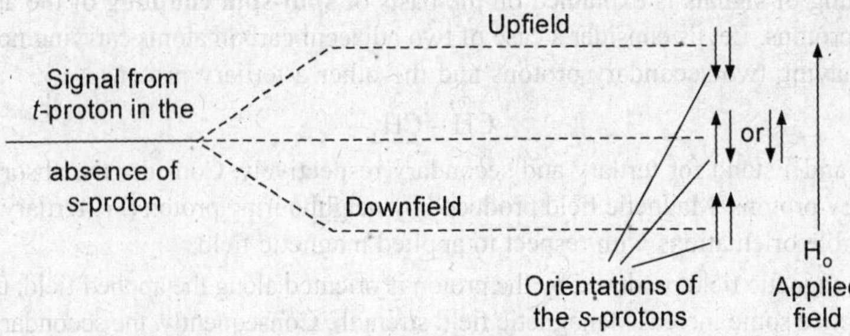

Fig. 1.9. Splitting by two adjancent protons give 1 : 2 : 1 triplet

Now consider the system of two adjacent carbon atoms which carry three primary protons and one tertiary proton as shown below :

$$—\overset{|}{C}H^{t} — CH_3^{p}$$

Following the same argument, we can say that the PMR spectrum of this system would contain a **doublet** (due to three primary protons) with 1 : 1 intensities and a **quartet** (due to tertiary proton) with 1 : 3 : 3 : 1 intensities. It may be seen from Fig. 1.10 that there are 8 possible combinations of spin states of the three primary protons; however, some of the combinations are equivalent so that there are only 4 effective combinations. As such the tertiary proton experiences any one of the four combinations at any moment. The PMR spectrum of 1, 1-dibromoethane ($CHBr_2$ — CH_3) which contains a similar structural system, *i.e.*, $—\overset{|}{C}H — CH_3$ illustrates the same. Of course, the area under the doublet is found to be three times that under the quartet because doublet is given by three protons while the quartet is given by one proton.

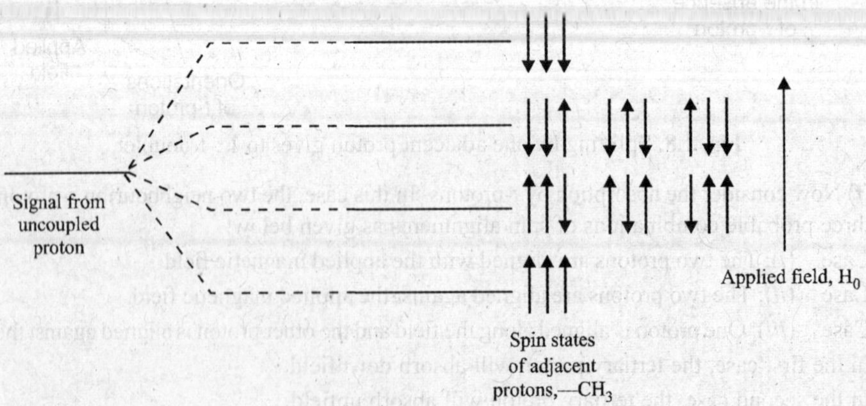

Fig. 1.10. Splitting by three adjacent protons; spin-spin coupling gives rise to a 1 : 3 : 3 : 1 quartet.

1.14. RULES GOVERNING SPLITTING OF SIGNALS

The following rules govern the splitting of signals.

(*i*) Only the neighbouring or vicinal protons cause the splitting of a signal provided these vicinal protons are non-equivalent to the absorbing protons. Thus we don't expect splitting in the following compounds.

(*a*) CH_2Cl-CH_2Cl Adjacent carbon has equivalent protons

(*b*) $CH_3-\underset{CH_3}{\overset{CH_3}{\underset{|}{\overset{|}{C}}}}-CH_2Cl$ There is no proton on the central carbon which is neighbouring to four carbons.

(*ii*) Total area under the doublet would be twice the total area under the triplet because the doublet is due to absorption by two protons and triplet is due to absorption by one proton. (Fig. 1.11)

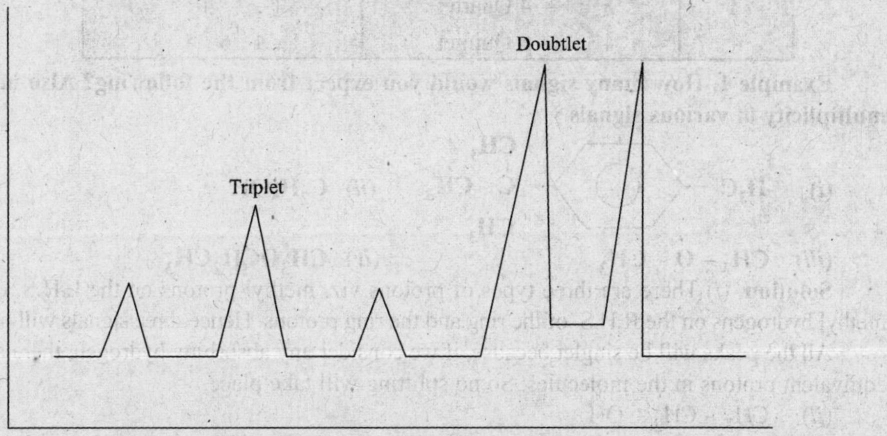

PMR spectrum of 1, 1, 2, tribromomethane

Fig. 1.11. PMR spectrum of 1, 1, 2, tribromomethane

(*iii*) **(*n* + 1) rule.** The no. of peaks obtained after splitting is one more than the no. of non-equivalent protons on the adjacent carbon atoms. This is illustrated below for different types of protons.

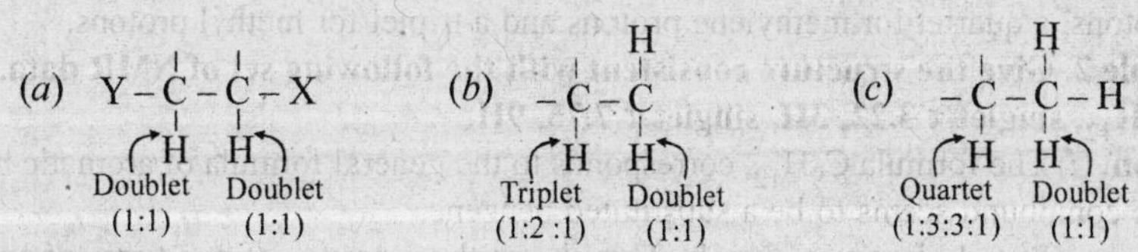

Compound (*a*)

There is one proton each on the two carbon atoms. These two protons are non-equivalent because they are in different environment. The signal of proton on the L.H.S. is split only by one proton on the R.H.S. ($n = 1$). Therefore the signal will be split to (1 + 1) *i.e.* 2 peaks (duplet). Similarly, the signal of proton on R.H.S. will be split by one proton on the L.H.S. Therefore the signal will split to (1 + 1) *i.e.* 2 peaks. We shall get two duplets.

Compound (*b*)

For L.H.S. proton, $n = 2$. Therefore number of peaks = 2 + 1 = 3

Thus L.H.S. proton will show a triplet

For R.H.S. protons $n = 1$. Therefore the number of peaks = 1 + 1 = 2

Thus R.H.S. protons will show a duplet

Compound (*c*)

For L.H.S. proton, $n = 3$. Therefore the number of peaks = 3 + 1 = 4

Thus L.H.S. proton will show a quartet

For R.H.S. proton, $n = 1$. Therefore the number of peaks = 1 + 1 = 2

Thus, R.H.S. proton will show a duplet.

The following Table gives the number of peaks and relative ratio of the intensities for different values of n

n	*No. of peaks*	*Ratio of intensities*
1	1 + 1 = 2 Doublet	1 : 1
2	2 + 1 = 3 Triplet	1 : 2 : 1
3	3 + 1 = 4 Quartet	1 : 4 : 4 : 1
4	4 + 1 = 5 Quintet	1 : 4 : 6 : 4 : 1

Example 1. How many signals would you expect from the following? Also indicate the multiplicity of various signals.

(*i*) 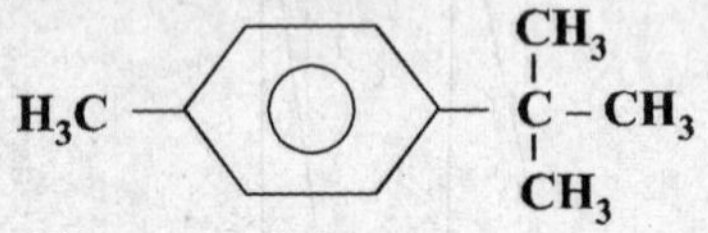 **(*ii*)** C_2H_5OH

(*iii*) $CH_3 - O - CH_3$ **(*iv*)** $CH_3OCH_2CH_3$

Solution. (*i*) There are three types of protons viz. methyl protons on the L.H.S. of the ring, methyl hydrogens on the R.H.S. of the ring and the ring protons. Hence three signals will be obtained.

All the peaks will be singlet because, if we consider any absorbing hydrogen, there are no non-equivalent protons in the molecules. So no splitting will take place.

(*ii*) $CH_3 - CH_2 - OH$

There are three types of protons, hence three signals. It will be a triplet for methyl protons, a quartet for methylene protons and a singlet for OH.

(*iii*) $CH_3 - O - CH_3$

There are only one type of protons, hence one signal will be obtained and it will be a singlet.

(*iv*) $CH_3 - O - CH_2 - CH_3$

There are three types of protons, hence three signals will be obtained. It will be a singlet for methoxy protons, a quartet for methylene protons and a triplet for methyl protons.

Example 2. Give the structure consistent with the following set of NMR data. Molecular formula C_9H_{12}, singlet τ 3.22, 3H, singlet τ 7.75, 9H.

Solution. (*i*) The formula C_9H_{12}, corresponds to the general formula of aromatic hydrocarbon C_nH_{2n-6}. The compound seems to be a substituted benzene.

(*ii*) The compound gives two signals. This shows the presence of two types of protons.

The above conditions are satisfied by 1, 3, 5, trimethyl benzene.

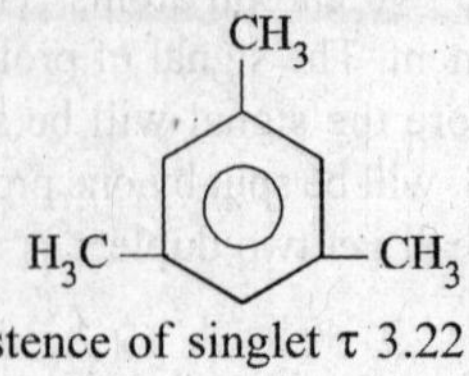

This structure explains the existence of singlet τ 3.22 due to three ring protons and singlet τ 7.75 due to nine methyl protons.

Example 3. How many NMR signals would you expect from the following? Toluene, 1, 1-dichloroethane and allyl alcohol.

Solution. (*i*) 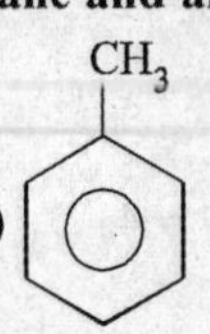 Toluene

It contains two types of protons, the ring protons and methyl protons. Hence two signals will be observed.

(*ii*) $\overset{b}{CHCl_2}-\overset{a}{CH_3}$ 1, 1 dichloroethane

There are two kinds of protons indicated by *a* & *b*. Hence two signals will be obtained.

(*iii*) 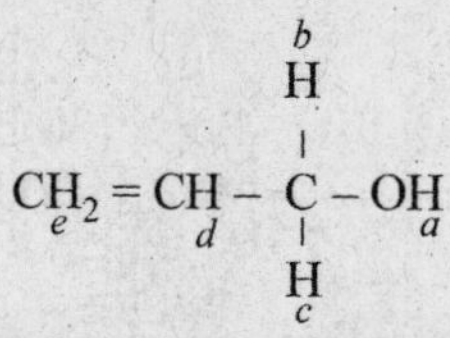

There are five kinds of protons designated as a, b, c, d, e. So five signals will be obtained.

Example 4. Of the following molecules, which will exhibit spin-spin coupling? What will be the multiplicity of each kind of proton in such molecules?

(*i*) $BrCH_2CH_2Br$ (*ii*) CH_3-CHBr_2 (*iii*) $ClCH_2CH_2Br$

(*iv*) 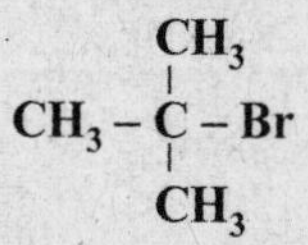 **(*v*)** 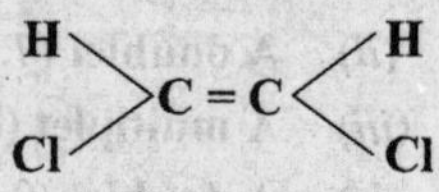

(*vi*) 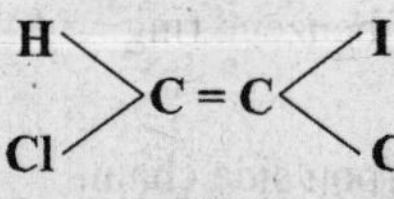 **(*vii*)** 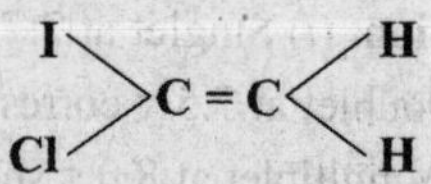

Solution. (*i*) All the protons are of the same type, hence no spin-spin coupling takes place.

(*ii*) The proton of $-CHBr_2$ is split into quartet and the protons of methyl are split into a doublet.

(*iii*) Protons of each methylene group are split into triplets.

(*iv*) No spin-spin coupling takes place because there is no hydrogen on the bromine-bearing carbon. Consequently, there are no non-equivalent protons on the neighbouring carbon with reference to any methyl group.

(*v*) No spin-spin coupling takes place because both the protons in the molecule are equivalent.

(*vi*) Each proton is split into doublet.

(*vii*) As the two protons on one carbon are not equivalent, each is split into doublet.

Example 5. For the compound

$$CH_3-CH_2-\overset{\overset{O}{\|}}{C}-O-\overset{\overset{O}{\|}}{C}-CH_2CH_3$$

predict the number of signals, their positions, relative intensities and splitting.

Solution. (*i*) As there are two types of protons, the methyl and methylene protons, two signals are expected.

(*ii*) Signals for methyl protons are obtained at δ 1–1.5 and for methyl protons at δ 2–2.5 (consult table 1.1.).

(*iii*) Methyl protons are split into triplet and methylene protons into quartet. So the overall spectrum can be expressed as; Triplet δ 1–1.5 (6H), quartet δ 2–2.5 (4H).

Example 6. A compound having molecular formula $C_{10}H_{14}$ gives the following PMR data.

(*a*) Singlet τ 9.12 (δ 0.88), 9H

(*b*) Singlet τ 2.72 (δ 7.28), 5H

Assign a structure to the compound on the basis of above data.

Solution. (*i*) The formula corresponds to the aromatic compounds general formula C_nH_{2n-6}. Hence it appears to be a substituted benzene.

(*ii*) There are nine protons of one kind and 5 protons of another kind. It suggests that there is mono-substitution in the benzene ring.

(*iii*) The two kinds of protons are not spin-spin coupled as we are getting the singlets.

(*iv*) The singlet τ 2.72 (δ 7.28) can be due to C_6H_5 *i.e.* phenyl group, confirming monosubstitution.

(*v*) The remaining part of the molecule *i.e.* C_4H_9 can have nine equivalent protons only if it is in the form of a *t*-butyl group.

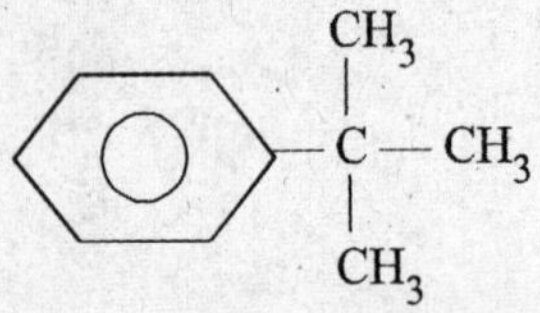

The singlet at τ 9.12 (δ 0.8) is due to the methyl protons.

Example 7. On the basis of the following data, assign the structure to the compound having molecular formula $C_{10}H_{14}$.

NMR **(*i*) A singlet (2.7 τ, 5H)**

(*ii*) A doublet (7.5 τ, 2H)

(*iii*) A multiplet (8.0 τ, 1H)

(*iv*) A doublet (9.0 τ, 6H)

Solution. (*i*) Singlet at 2.7 τ corresponds to hydrogen atoms of benzene ring.

(*ii*) Doublet at 7.5 τ corresponds to – CH_2.

(*iii*) A multiplet at 8.0 τ shows that there is branching in the carbon side chain.

(*iv*) A doublet at 9.0 τ corresponds to hydrogens of terminal methyl groups.

On the basis of above, the compound has the following structure:

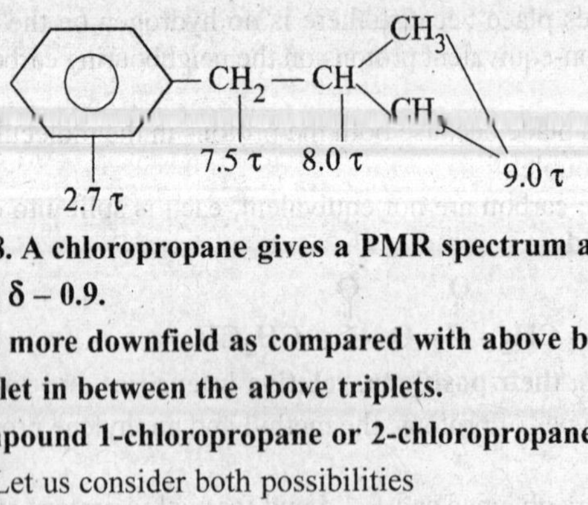

(Isobutyl benzene)

Example 8. A chloropropane gives a PMR spectrum as follows:

(*a*) Triplet δ – 0.9.

(*b*) Triplet more downfield as compared with above but an intensity of 2/3 of above.

(*c*) Multiplet in between the above triplets.

Is the compound 1-chloropropane or 2-chloropropane?

Solution. Let us consider both possibilities

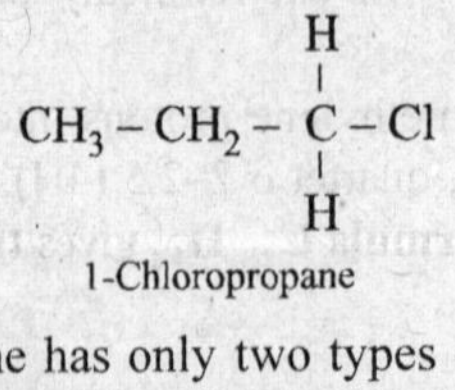

1-Chloropropane

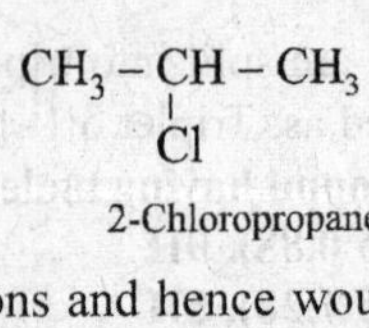

2-Chloropropane

(*i*) 2-chloropropane has only two types of protons and hence would give only two signals whereas we are getting three signals. Hence the possibility of the compound being 2-chloropropane is ruled out and it appears to be 1-chloropropane.

(*ii*) Now 1-chloropropane contains three types of protons and hence three signals are expected.

(*iii*) Triplet δ ~ 0.9 is due to methyl protons (consult table). Triplet is obtained because of methylene protons.

(*iv*) A triplet more downfield as compared to the methyl protons is because of the protons of > C – Cl. This is due to deshielding effect of chlorine. The triplet, which is again due to methylene protons will appears at δ 3–4.

(*v*) A signal due to methylene protons will be obtained at about δ 1.3 (consult table). It will however be a sixlet (multiplet) due to 5 protons on the two sides.

The data given conforms to the structures of 1 chloropropane.

Example 9. Suggest a structure on the basis of PMR data of the following compounds.

(*i*)	**C_3H_7Br**	**δ 1.7 (*d*, 6H),**	**δ 4.3 (septet, 1H)**
(*ii*)	**$C_4H_{10}O$**	**δ 1.28 (*s*, 9H),**	**δ 1.35 (*s*, 1H)**
(*iii*)	**$C_5H_9Cl_3$**	**δ 1.0 (*d*, 6H),**	**1.5 (multiplet, 1H)**
		δ 3.3 (*s*, 2H)	

Solution. (*i*) The structure is $CH_3 - CHBr - CH_3$

There are two types of protons, hence two signals. Six protons of one type and one proton of another type as given in the data conforms to this structure. The middle proton will give a septet because of six surrounding protons and doublet will be obtained in respect of methyl protons due to the middle proton.

(*ii*) $C_4H_{10}O$ is the molecular formula of butyl alcohol. A no. of butyl alcohols are possible with this formula as given below

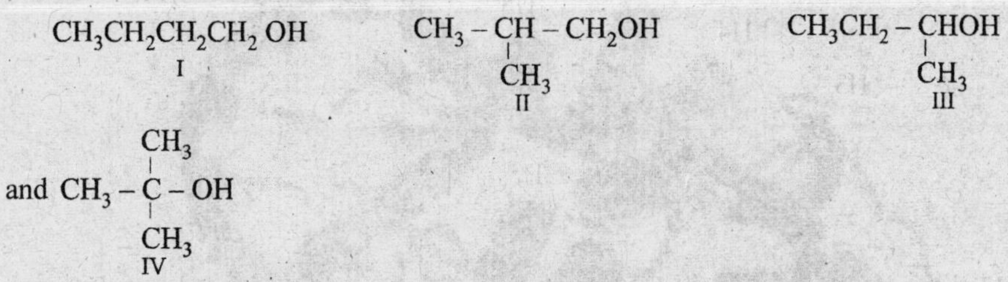

I, II, III and IV

The data given above conforms to structure IV.

Structure IV has only two types of protons and gives rise to two signals. Structures I, II & III would have given more than two signals and are therefore, discarded.

There are nine protons of one type (all methyl protons) and one proton of one type. This is supported by the data. Singlet δ 1.28 is because of methyl protons and singlet δ 1.35 is due to the hydroxy proton.

(*iii*) Out of the various structures possible, the structure which is in conformity with the data is

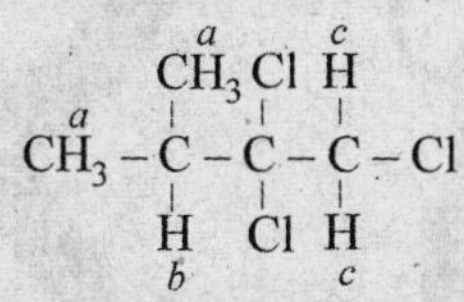

There are three types of protons and hence three signals are obtained. Protons designated "*a*"will give a doublet δ 1.0, protons designated "*b*" will give a multiple at δ 1.5 and protons designated "c" will give singlet. All these properties are satisfied with the structure given above.

Example 10. A compound having a molecular formula C_4H_9Br gave the following PMR data:

(*a*) doublet τ 8.96 (δ 1.04), 6H

(*b*) multiplet τ 8.05 (δ 1.95), 1H

(*c*) doublet τ 6.67 (δ 3.33), 2H

What structure can the compound have?

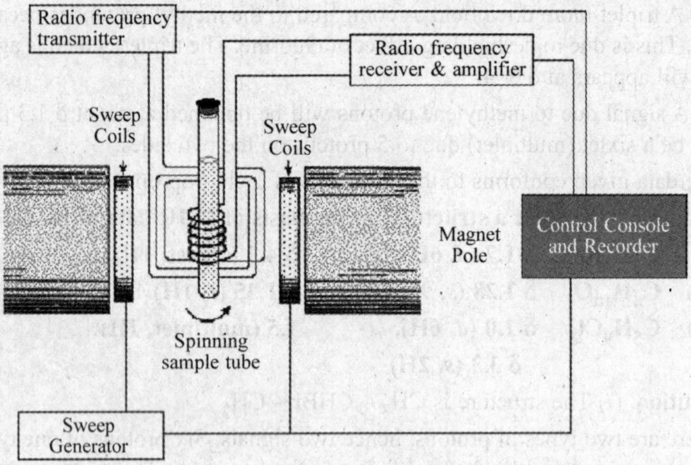

Working of NMR spectrometer

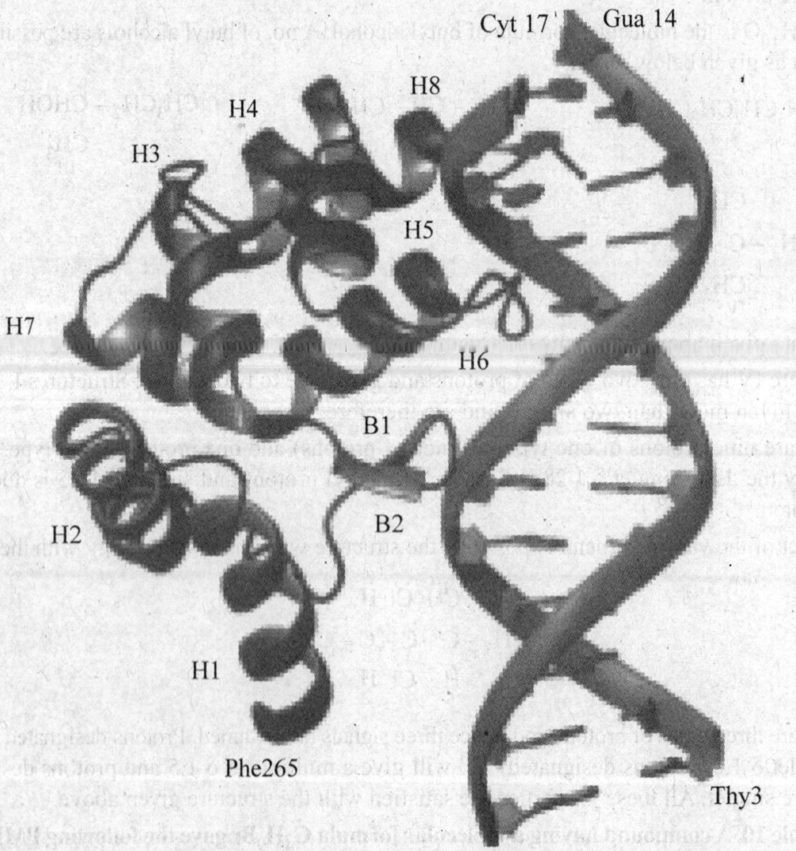

Structure of biomolecules is determined with the help of NMR spectroscopy

Solution. There are four structures possible with this molecular formula

$CH_3 - CH_2 - CH_2 - CH_2Br$
1-Bromobutane (I)

$CH_3 - CH_2 - CH_2Br - CH_3$
2-Bromobutane (II)

$$\begin{array}{c} CH_3 \\ | \\ CH_3 - CH - CH_2\,Br \end{array}$$
1-Bromo-2-methylpropane (III)

$$\begin{array}{c} CH_3 \\ | \\ CH_3 - C - Br \\ | \\ CH_3 \end{array}$$
2-Bromo-2 methylpropane (IV)

(*a*) Structure I and II have four types of protons each. Structure IV has only one type of protons. The data shows that the compound contains three types of protons. Thus structures I, II & IV are ruled out. Structure III satisfies this condition and is the likely structure.

(*b*) Structure III has six protons of one type, one protons of second type and two protons of third type which tallies with the data.

(*c*) Six methyl protons give a doublet due to tertiary proton. The tertiary proton gives a multiplet due to six methyl protons and protons in $-CH_2Br$.

Finally protons in $-CH_2Br$ give a doublet due to *t*-proton.

Thus III is the structure which is consistent with the data.

Example 11. Give a structure consistent with the following set of PMR data. Molecular formula of the compound is $C_9H_{11}Br$.

(*i*) Multiplet τ 7.85 (2H) **(*ii*) Triplet τ 7.25 (2H)**

(*iii*) Triplet τ 6.62, (2H) **(*iii*) Singlet τ 2.78, (5H)**

Solution. The data reveals that

(*a*) There are four types of protons.

(*b*) There is a benzene ring in the compound as shown by singlet τ 2.78 (5H).

(*c*) The side chain in the benzene ring consists of $-C_3H_6Br$ and the six protons in the side chain form three pairs of different types of protons. The following structure satisfies the data completely.

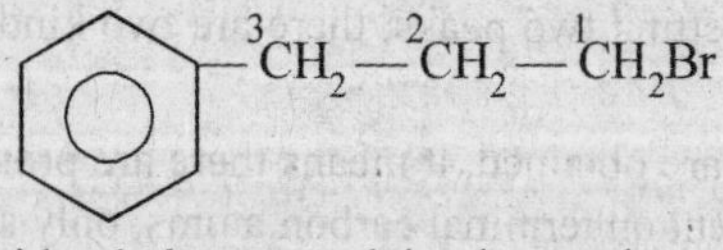

The signals given by the side-chain are explained as under:

(*a*) The two protons on C_3 give a signal τ 7.25. It splits into three signals due to two protons on C_2.

(*b*) The two protons on C_2 give a signal τ 7.85 and it splits into 5 peaks (multiplet) due to 4 neighbouring protons on C_3 and C_1. Slight downfield shift is due to the unshielding effect of bromine.

(*c*) The two protons on C give a downfield signal τ 6.62 due to deshielding effect of bromine. The signal is split into a triplet due to the neighbouring protons on C_2. Thus the structure assigned above is consistent with the given data.

Example 12. A compound having molecular formula $C_9H_{11}Br$ furnished the following set of PMR data:

1. Singlet δ 7.25, 5H **2. Doublet δ 7.25, 2H**

3. Multiplet δ 3.40, 1H **4. Doublet δ 1.45, 3H**

Assign a structure to this compound showing your reasoning.

Solution. (*a*) The data reveals that there are four types of protons.

(*b*) There is a benzene ring in the compound as shown by δ 7.25 (5H) (*c*) The side chain in the benzene ring consists of

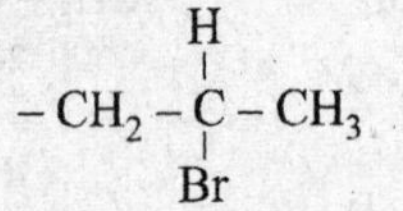

and the six protons in the side chain form three different sets of protons. The following structure for the compound satisfies the data completely.

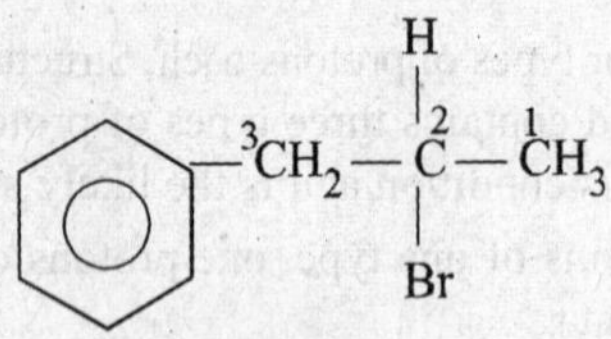

The signals given by the side-chain are explained as under:

(*a*) Three protons on C-1 give the signal δ 1.45. It splits into doublet due to the presence of one proton on C-2.

(*b*) One proton on C-2 gives the signal δ 3.40. It splits into multiplet due to the presence of five protons on neighbouring carbons.

(*c*) Two protons on C-3 give the signal δ 2.75. It splits into doublet due to the presence of one proton on neighbouring C-2.

Example 13. An organic compound with molecular formula $C_3H_3Cl_5$ gave the following NMR data

(*i*) A triplet 4.49 τ (4.52 δ) 1H

(*ii*) A doublet 3.93 τ (6.07 δ) 2H

Assign a structural formula to the compound consistent with the above data

Solution. The molecular formula $C_3H_3Cl_5$ suggests that it is a saturated molecules with no multiple bonds. Since we are getting two peaks, there are two kinds of protons. Hence protons are attached to two carbon atoms.

As a triplet and a doublet are obtained, it means there are protons on the neighbouring carbon atoms. If the protons were present on terminal carbon atoms, only singlet would be obtained. In the light of above, the structure is

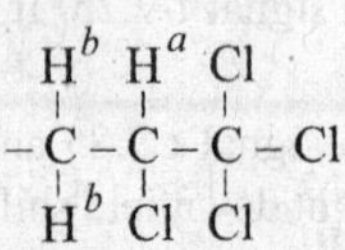

Example 14. A compound with molecular formula $C_{10}H_{12}O$ shows a strong absorption at 1705 cm^{-1} in its IR spectrum and NMR spectrum of the compound shows the following peaks:

δ 7.22 (Singlet 5H) δ 3.59 (Singlet 2H) δ 2.77 (Quartet 2H) δ 0.97 (Triplet, 3H)

Giving reasons assign a structure to the compound.

Solution. (*a*) IR absorption at 1705 cm^{-1} shows the presence of a carbonyl group (–CHO or > C = O)

(*b*) The molecular formula suggests the presence of a benzene ring.

(*c*) A peak at δ 7.22 singlet, 5H suggests that it is a monosubstituted aromatic compound i.e. there is only one chain in the ring, thereby retaining five equivalent protons in the ring.

(*d*) The possibility of an aldehyde is ruled out because in that case, there will be two protons each of three different types. This is not in accordance with the data.

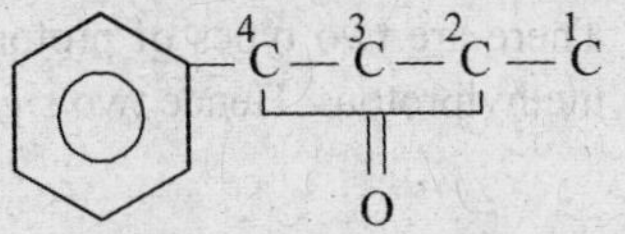

(*e*) The only possibility left is a ketone, with the ketonic position at 2 or 3 or 4

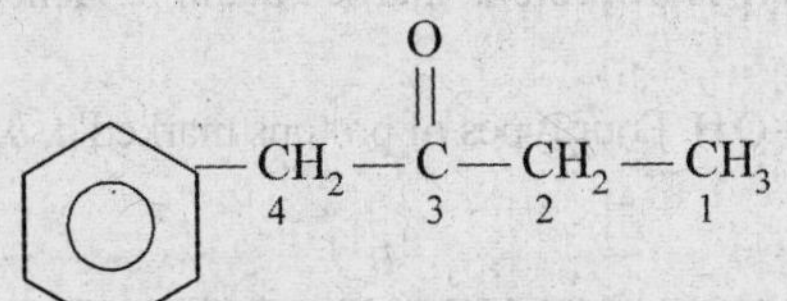

(*f*) A ketonic group at C-3 satisfies the observation.

(*g*) Three protons at C-1 will give a peak, it will be a triplet due to two protons at C-2. Two protons at C-2 will give a quartet due to three protons on C-1. Two protons at C-4 will give a singlet because there is no proton on the neighbouring carbon atoms.

Example 15. A compound C_3H_6O contains a carbonyl group. How could NMR establish whether this compound is an aldehyde or ketone?

Solution. The aldehyde with the above formula will have the structure

$$\overset{c}{CH_3}-\overset{b}{CH_2}-\overset{a}{CHO}$$

It will give three peaks corresponding to three types of protons as follows :

triplet, 1H pentet, 2H triplet, 3H

The ketone with the above formula will have the structure

$$\overset{a}{CH_3}-\overset{\overset{O}{\|}}{C}-\overset{a}{CH_3}$$

Here all the six protons are of one type and will give one peak which will be singlet.

This is how the identity of the compound can be established.

Example 16. Propose structural formulae for the compounds with the following molecular formulae which give only one NMR signal:

***(i)* C_5H_{12} *(ii)* C_2H_6O *(iii)* $C_2H_4Br_2$ *(iv)* C_8H_{18}**

Solution. Since only one NMR signal is obtained, the compounds must possess only one kind of protons. The structural formulae of the compounds accordingly are:

(*i*) $CH_3-\underset{CH_3}{\overset{CH_3}{\overset{|}{\underset{|}{C}}}}-CH_3$

(*ii*) CH_3-O-CH_3
Methoxymethane

(*iii*) $Br-CH_2-CH_2Br$
1, 2-Dibromoethane

(*iv*) $CH_3-\underset{CH_3}{\overset{CH_3}{\overset{|}{\underset{|}{C}}}}-\underset{CH_3}{\overset{CH_3}{\overset{|}{\underset{|}{C}}}}-CH_3$
2, 2, 3, 3-Tetramethylbutane

Example 17. How many NMR signals would you expect from the following?

(*i*) Toluene (*ii*) 1, 1-Dichloroethane (*iii*) Allyl alcohol

Solution. (*i*) 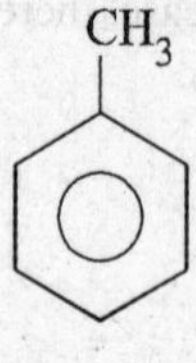 There are two types of protons the ring protons and the methylprotons. Hence two signals will be obtained.

(*ii*) $\overset{b}{CH_3}-\overset{a}{CHCl_2}$ Two types of protons marked a and b. Hence two signals

(*iii*) $\overset{d}{CH_3}-\overset{c}{CH}=\overset{b}{CH_2}-\overset{a}{OH}$ Four types of protons marked *a*, *b*, *c* & *d*. Hence four signals will be obtained.

1.15. COUPLING CONSTANT WHAT ARE THE FACTORS THAT GOVERN COUPLING CONSTANT

The distance between adjacent peaks is called *coupling constant* and denoted by J. Coupling constant gives a measure of the splitting effect. The values of J are independent of applied field strength and depend upon the molecular structure only. The unit of J is Hertz (Hz). Mutually coupled protons show the same value of J.

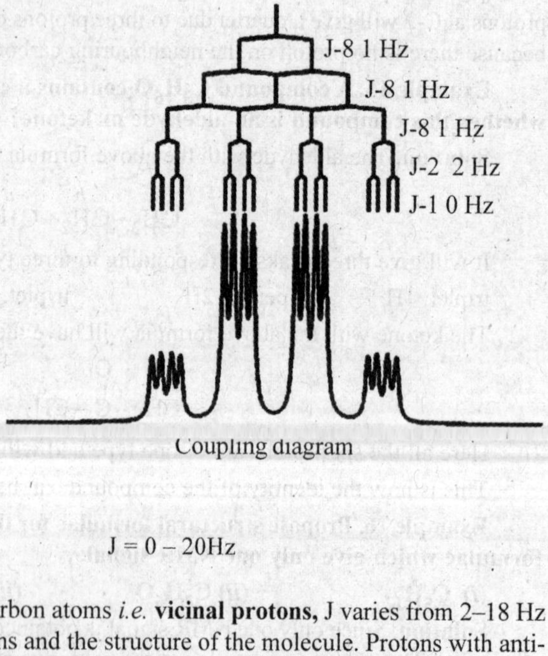

Coupling diagram

It is observed that J depends upon the number and type of bonds in between and the spatial relations between the protons. Following examples will illustrate.

(*i*) For the protons attached to same carbon *i.e.* **geminal protons**, J varies from 0–20 Hz depending upon the dihedral angle and the structure of the molecule.

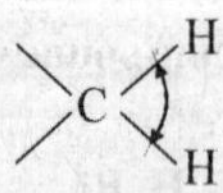

J = 0 – 20Hz

(*ii*) For protons attached to adjacent carbon atoms *i.e.* **vicinal protons,** J varies from 2–18 Hz depending upon the spatial position of protons and the structure of the molecule. Protons with anti-conformation have the value of J between 5 – 12 Hz and those with gauche conformation have J = 2 – 4 Hz.

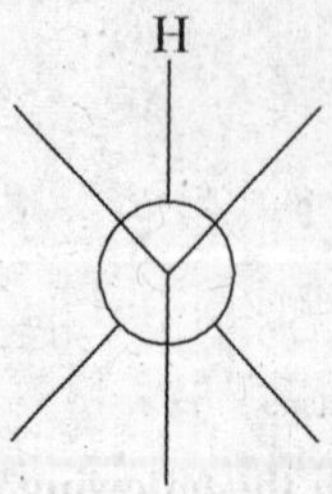

J = 5 – 12 Hz

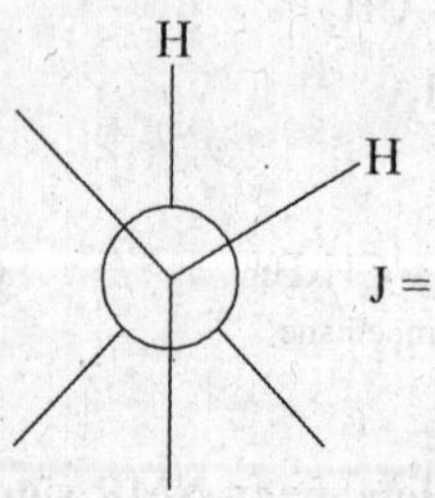

J = 2 – 4 Hz

In case of vinylic protons, where there is a restricted rotation because of the double bond, the *cis* protons have J = 6 – 14 Hz whereas *trans* protons have J = 11 – 18 Hz.

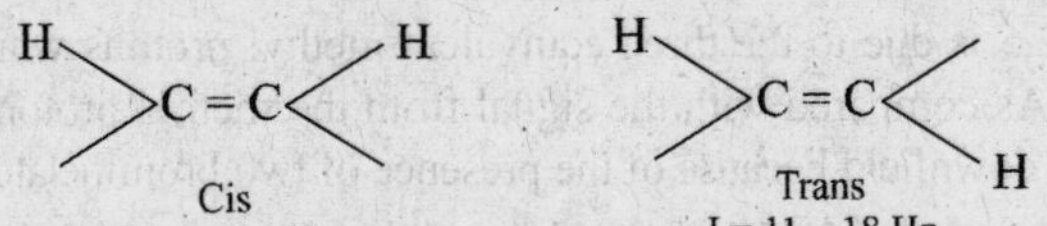

1.16. NMR SPECTRA OF CERTAIN MOLECULES

(1) Ethyl bromide CH_3CH_2Br.

The NMR spectra of ethyl bromide is represented in Fig. 1.12 :

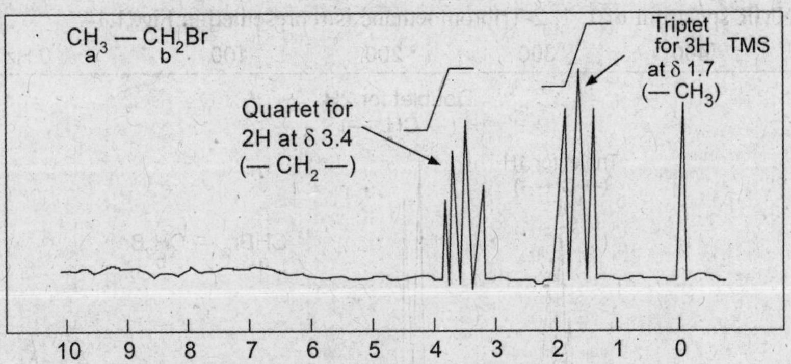

Fig. 1.12. NMR spectrum of ethyl bromide

The following peaks can be identified in the spectrum

(*a*) Triplet, δ 1.7, 3H (*b*) Quartet, δ 3.4, 2H

The triplet at δ 1.7 is given by the three methyl protons which are magnetically equivalent and are coupled with the two methylene protons to give an upfield triplet.

The quartet at δ 3.4 is from the two equivalent methylene protons which are coupled with the three methyl protons to produce a downfield quartet as a result of deshielding influence of bromine.

The relative areas under the respective signals are in the ratio of the number of protons involved *i.e.* 3 : 2.

(2) 1, 1 Dibromoethane, $CH_3–CHBr_2$.

NMR spectrum of 1, 1 Dibromoethane is represented in Fig. 1.13

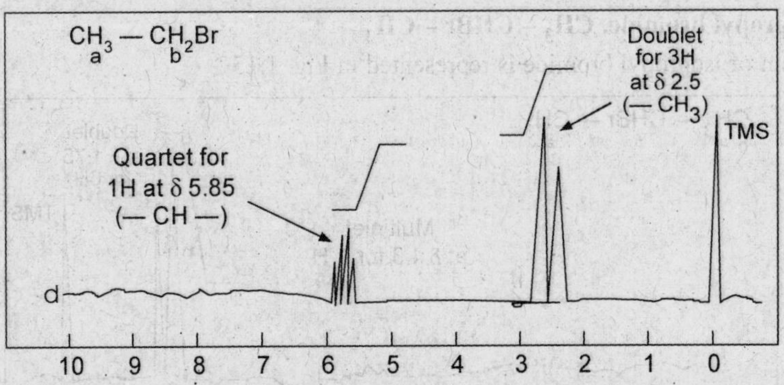

Fig. 1.13. NMR spectrum $CH_3 – CHBr_2$

The following peaks can be identified in the spectrum.

(*a*) Doublet, δ 2.5, 3H (*b*) Quartet, δ 5.85, 1H

The doublet at δ 2.5 is due to the three equivalent methyl protons which are coupled with the methine (CH) protons. As compared with the signal from the methyl protons of $CH_3 - CH_2$ Br, the signal here is somewhat downfield because of the presence of two bromine atoms on adjacent carbons.

The downfield quartet at δ 5.85 is from the methine proton which is coupled with the three methyl protons. Its downfield position is due to the attachment of two bromine atoms to the carbon carrying the protons.

The areas under the peaks are in the ratio of 3 : 1 *i.e.* in the ratio of no. of methyl and methine protons.

(3) 1, 1, 2-Tribromoethane $CH_2Br - CHBr_2$.

The NMR spectrum of 1, 1, 2-Tribromoethane is represented in Fig. 1.14

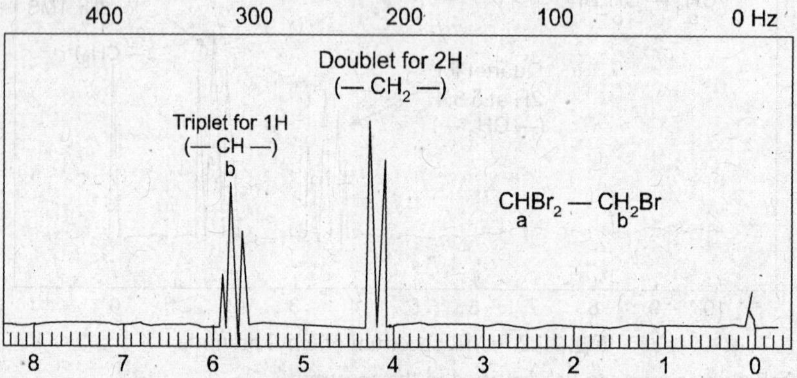

Fig. 1.14. PMR spectrum of 1, 1, 2-Tribromoethane

The following peaks can be identified in the spectrum:

(*a*) Doublet, δ 4.2, 2H (*b*) Triplet, δ 5.7, 1H

The doublet at δ 4.2 is due to two equivalent protons of CH_2 Br– group which is split into a doublet by the neighbouring proton of $-CHBr_2$ group. The downfield position of this doublet can be attributed to the deshielding influence of bromine.

The triplet at δ 5.7 is produced by the proton of $-CHBr_2$ group. It is split into a triplet by the two protons of CH_2 Br groups. Relative to the doublet of CH_2Br-, the triplet of $-CHBr_2$ is more downfield since there are two bromines attached to the proton bearing carbon in this case.

The areas under the two peaks are in the ratio 2 : 1.

(4) Isopropyl bromide, $CH_3 - CHBr - CH_3$

Spectrum of isopropyl bromide is represented in Fig. 1.15

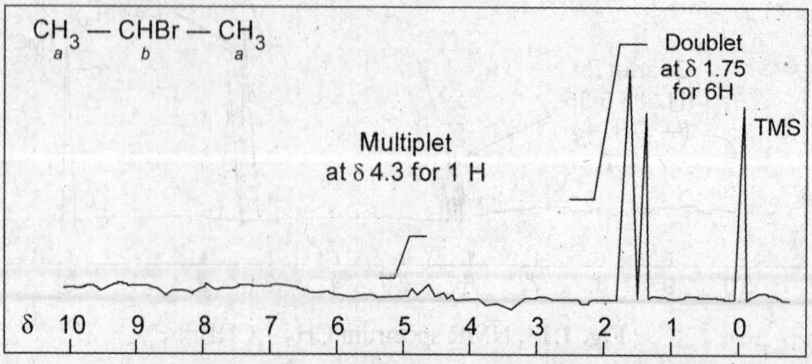

Fig. 1.15. NMR spectrum of CH_3 — CHBr — CH_3

Following peaks can be identified in the spectrum:

(*a*) Doublet, δ 1.75, 6H (*b*) Multiplet, δ 4.3, 1H

It can be seen that the six protons on the two methyl groups are all equivalent and different from the proton of –CHBr– group. These six protons give rise to an upfield signal at δ 1.75 which is split into a doublet due to coupling with the lone proton –CHBr– group.

The proton of –CHBr– gives a downfield signal at 4.3 which is split into a multiplet by the six protons, on adjacent carbons.

The two peak areas are in the ratio 6 : 1.

(5) Ethanol, $CH_3–CH_2OH$

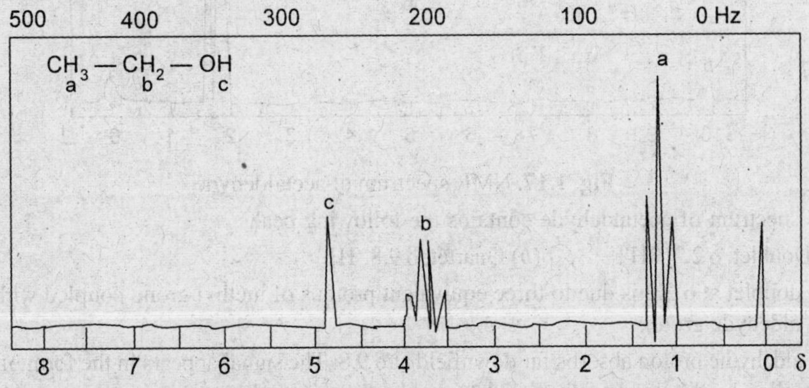

Fig. 1.16. PMR spectrum of an ordinary sample of ethanol

Fig. 1.16 shows the NMR spectrum of ordinary ethanol.

The spectrum of an ordinary sample of ethanol may be described as follows:

(*a*) Triplet, δ 1.2, 3H (*b*) Quartet, δ 3.63, 2H (*c*) Singlet, δ 4.8, 1H

Ethanol contains three kinds of protons and consequently exhibits three signals as shown above.

The upfield triplet of δ 1.2 is due to three equivalent methyl protons. Evidently its splitting into a triplet is due to the two neighbouring protons on the methylene group.

The quartet at δ 3.63 is from the two methylene protons. Its multiplicity of 4 is due to coupling with the three methyl protons. Coupling with hydroxyl proton does not occur.

Finally the downfield singlet at δ 4.8 is due to hydroxyl proton which does not show any coupling with the adjacent methylene protons.

The relative areas under the peaks of methyl, methylene and hydroxyl protons are in the ratio 3 : 2 : 1.

Absence of coupling between methylene and hydroxyl protons is explained in terms of rapid exchange.

$$\underset{\text{Molecule 1}}{CH_3CH_2-\overset{*}{O}H} + \underset{\text{Molecule 2}}{CH_3CH_2-OH} \qquad \underset{\text{Molecule 1}}{CH_3CH_2-OH} + \underset{\text{Molecule 2}}{CH_3CH_2-\overset{*}{O}H}$$

The hydroxyl protons undergoes exchange with another molecule in which the alignment of methylene protons is different from that in the first molecule and so on. In other words the hydroxyl proton does not stay in the same environment long enough for its coupling with the methylene protons to be recorded.

NMR spectrum of pure ethanol shows the signal from hydroxyl proton split into triplet and signal from methylene protons split into octet. This is because coupling between hydroxyl and methylene protons can be recorded now as the exchange process slows down.

(6) Acetaldehyde, CH_3–CHO.

NMR spectrum of acetaldehyde is reproduced in Fig. 1.17

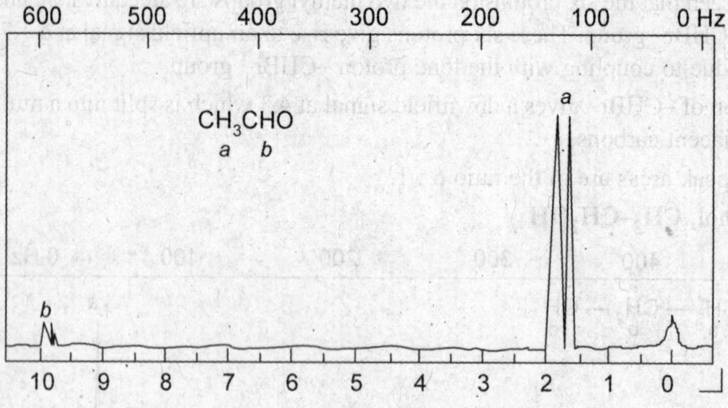

Fig. 1.17. NMR spectrum of acetaldehyde

The spectrum of acetaldehyde contains the following peaks:

(*a*) Doublet, δ 2.2, 3H (*b*) Quartet, δ 9.8, 1H

The doublet at δ 2.2 is due to three equivalent protons of methyl group coupled with a single proton of aldehyde group.

The aldehydic proton absorbs far downfield at δ 9.8. The signal appears in the form of a quartet due to coupling by the three protons of the neighbouring methyl groups.

Peaks areas are in the ratio 3 : 1.

(7) Benzene. NMR spectrum of benzene is reproduced in Fig. 1.18

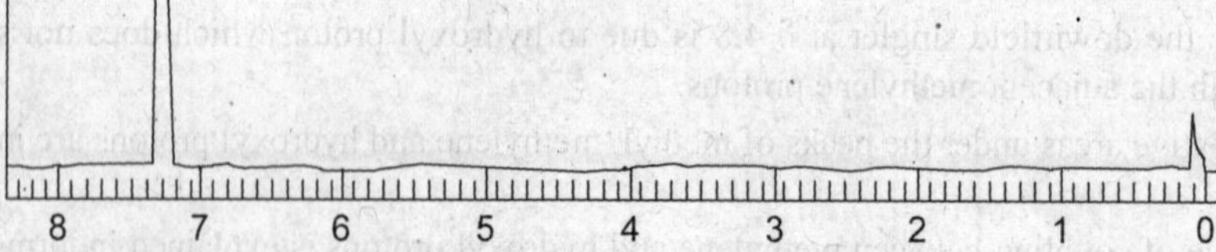

Fig. 1.18. PMR spectrum of benzene

The NMR spectrum of benzene exhibits only a sharp singlet at δ 7.37 from the six equivalent protons.

Such a downfield signal can be explained in terms of π electron system of benzene which is considerably polarisable. Therefore, the application of a magnetic field induces a flow of these electrons around the ring. This ring current in turn generates an induced magnetic field which reinforces the applied field. As a result the benzene protons get deshielded and the absorption moves downfield.

(8) Toluene. 

NMR spectrum of toluene is reproduced in Fig. 1.19

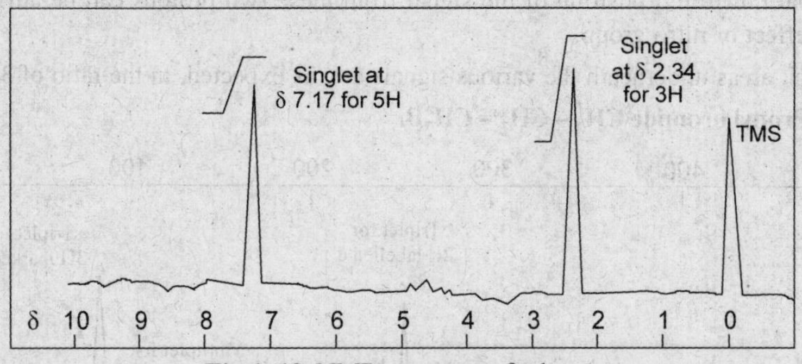

Fig.1. 19. NMR spectrum of toluene

The following peaks are observed in the NMR spectrum of toluene:

(*a*) Singlet, δ 2.34, 3H (*b*) Singlet, δ 7.17, 5H

Toluene has eight protons, five of which are aromatic and remaining three from methyl group.

The signal for three protons of methyl group which is joined to an aromatic ring appears as a singlet at δ 2.34.

In practice all the five aromatic protons are equivalent because they are hardly affected by methyl group. Hence these protons do not couple with each other and give rise to only one signal in the form of a singlet at δ 7.17. It may be seen that relative areas under the two signals in the spectrum are in the ratio 3 : 5.

(9) *p*-Nitrotoluene, O_2N– 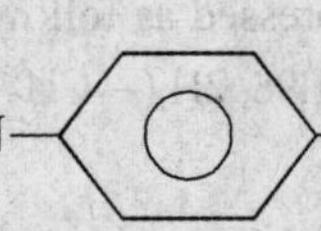 **–CH_3**

NMR spectrum of the compound is reproduced in Fig. 1.20

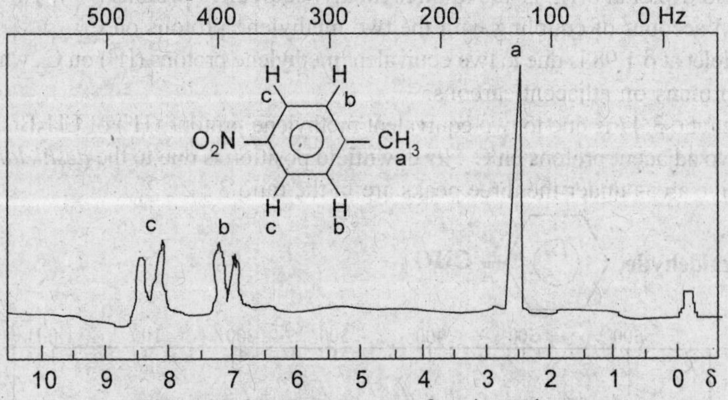

Fig. 1.20. NMR spectrum of *p*-nitro toluene

The spectrum of *p*-nitrotoluene has the following peaks:

(*a*) Singlet, δ 2.4, 3H (*b*) Doublet, δ 8.4, 2H (*c*) Doublet, δ 1.2, 2H

Of the seven protons in p-nitrotoluene, the three protons of methyl group give rise to a singlet at δ 2.4.

The four aromatic protons are of two kinds. Two protons which are *ortho* to methyl group are equivalent to each other while the two protons *ortho* to nitro group are equivalent to each other. The two protons *ortho* to methyl group give rise to an absorption signal at δ 7.4 which appears in the form of a doublet due to coupling with the neighbouring proton *ortho* to nitro group.

Similarly the two protons ortho to nitro group produce a signal at δ 8.2 in the form of a doublet. The somewhat downfield position of the signal from these two protons can be attributed to the deshielding effect of nitro group.

The peak areas underneath the various signals are, as expected, in the ratio of 3 : 2 : 2.

(10) *n*-Propyl bromide $CH_3 - CH_2 - CH_2Br$

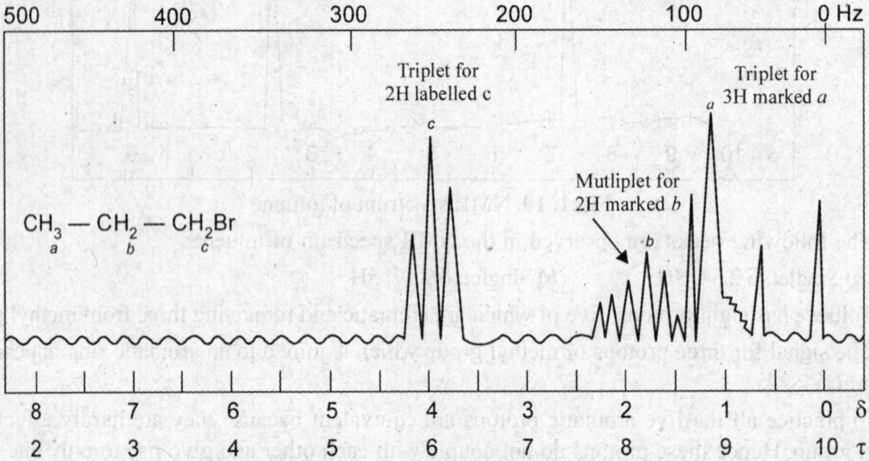

Fig. 1.21. PMR spectrum of $CH_3—CH_2—CH_2Br$

The spectrum of *n*-propyl bromide may be expressed as follows :

(*a*) **Triplet,** δ 1 .2, 3H (CH_3) (*b*) **Multiplet,** δ 1.98, 2H (—CH_2—)

(*c*) **Triplet,** δ 3.47, 2H (CH_2Br)

This can be explained as under :

The upfield triplet at δ 1.2 is due to three methyl equivalent protons (H^a). Their signal is split into a triplet on account of coupling with the two methylene protons on C_2.

The multiplet at δ 1.98 is due to two equivalent methylene protons (H^b) on C_2 which are coupled with the five protons on adjacent carbons.

The triplet at δ 3.47 is due to two equivalent methylene protons (H^c) of CH_2Br. It is split into a triplet by the two adjacent protons on C_2. Its downfield position is due to the *deshielding by bromine.*

The relative areas under the three peaks are in the ratio 3 : 2 : 2.

(11) Benzaldehyde, 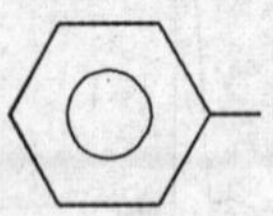

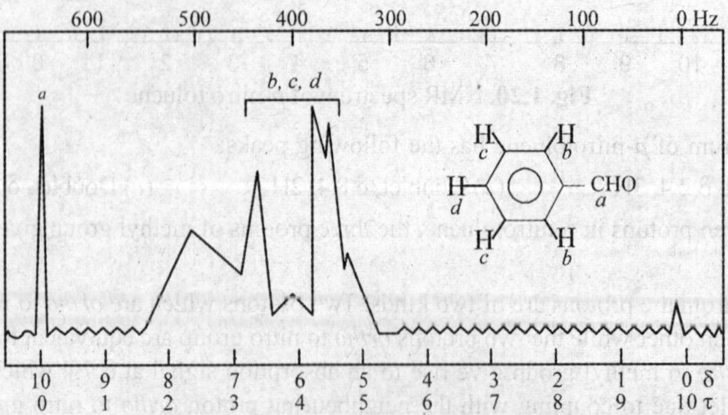

Fig. 1.22. PMR spectrum of benzaldehyde

Fig. 1.22 gives the PMR spectrum of benzaldehyde

The PMR spectrum of benzaldehyde is as follows :

(*a*) **Singlet,** δ 9.85, 1H (Aldehyde proton)

(*b*) **Complex multiple,** δ 7.3–8.2 (Five aromatic protons)

Explanation. The aldehydic proton gives a far downfield singlet at δ 9.85. This is because there is no proton on neighbouring carbon ($n = 0$).

The five aromatic protons give a complex multiplet at δ 7.3–8.2. The five aromatic protons are not equivalent to each other; the two ortho protons are of one kind, the two meta protons of another kind and the para proton of a different kind because of a strongly electron attracting carbonyl group attached to the benzene ring. Its deshielding effect is experienced by ortho, meta and para protons to different extent and therefore they show different chemical shifts. Due to long range coupling, the signal appears as a complicated multiplet which cannot be interpreted in a simple way. The signals obtained can be represented as under :

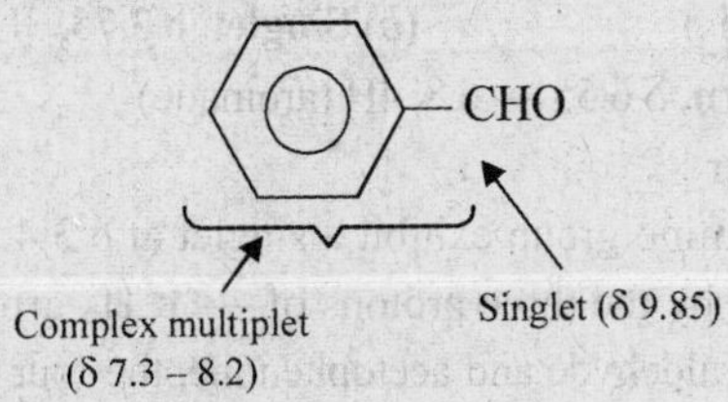

(12) Acetophenone, 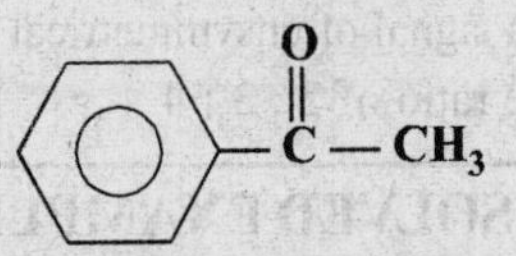

The PMR spectrum of acetophenone exhibits the following peaks (Fig. 1.24)

(*a*) **Singlet,** δ 2.45, 3H (CH_3) (*b*) **Complex multiplet,** δ 7.6–8.2 (Five aromatic protons).

This can be explained as under :

There are eight protons in acetophenone. Of these the three protons of methyl group are equivalent to each other and give rise to singlet at δ 2.45.

Like benzaldehyde the five protons of aromatic ring are of three kind. They give rise to a complex multiplet.

The signals obtained from different protons are represented as under :

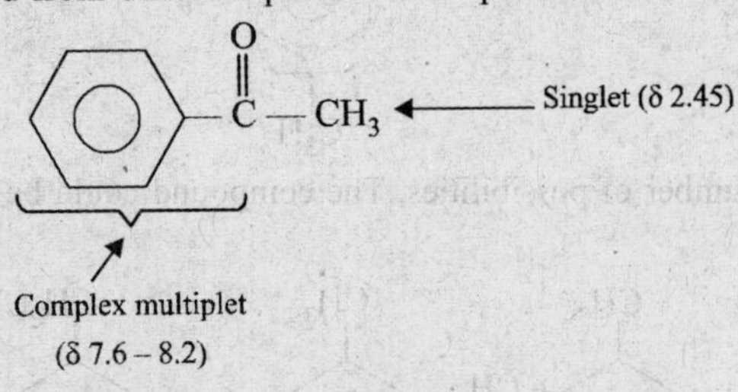

(13) *p*-Anisidine, 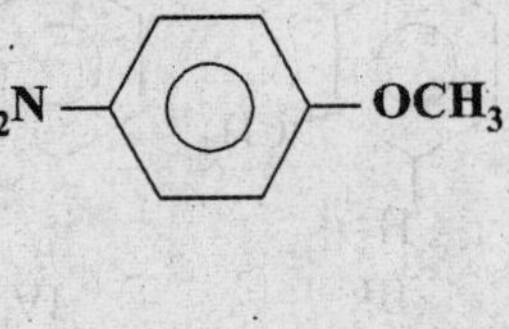

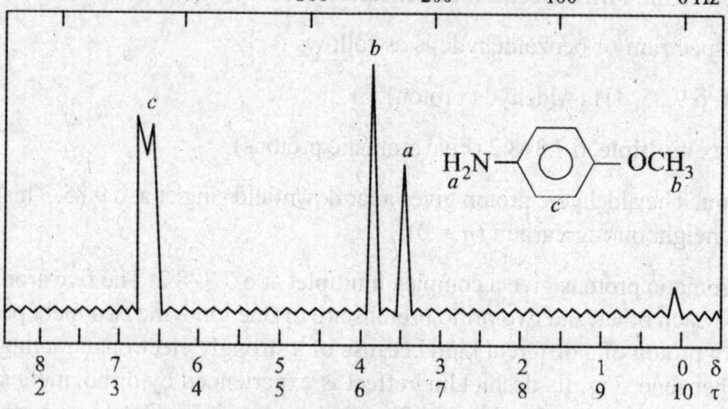

Fig. 1.23. PMR spectrum of *p*-anisidine

The PMR spectrum of *p*-anisidine shows the following peaks (Fig. 1.24)

(*a*) **Singlet,** δ 3.4, 2H (NH_2) (*b*) **Singlet,** δ 3.75, 3H (OCH_3)

(*c*) **Unsymmetrical pattern,** δ 6.55 — 6.8, 4H (aromatic)

This is explained as under :

The two protons of the amino group exhibit a singlet at δ 3.4.

Singlet at δ 3.75 is given by the three protons of $—OCH_3$ group.

Unlike in the case of benzaldehyde and acetophenone, the four protons of the benzene ring are not much different from one another. They give a signal of unsymmetrical pattern at δ 6.55 — 6.8.

The peak areas under the signals are in the ratio of 2 : 3 : 4.

MISCELLANEOUS SOLVED EXAMPLES

Example 1. An organic compound with molecular formula C_9H_{12} gives on strong oxidation with $KMnO_4$ a compound with molecular formula $C_7H_6O_2$ which gives an effervescence with sodium hydrogencarbonate. The compound gives three pmr signals. Deduce the structure of the compound.

Solution. The compound $C_7H_6O_2$ gives effervescence with $NaHCO_3$. It is likely to be a carboxylic acid. One of the compounds with this formula is benzoic acid C_6H_5COOH.

The compound C_9H_{12} is a hydrocarbon. It must contain a benzene ring because on oxidation the compound gives benzoic acid. The likely structure for the compound is

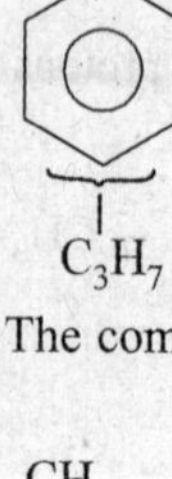

Again there are a number of possibilities. The compound could be

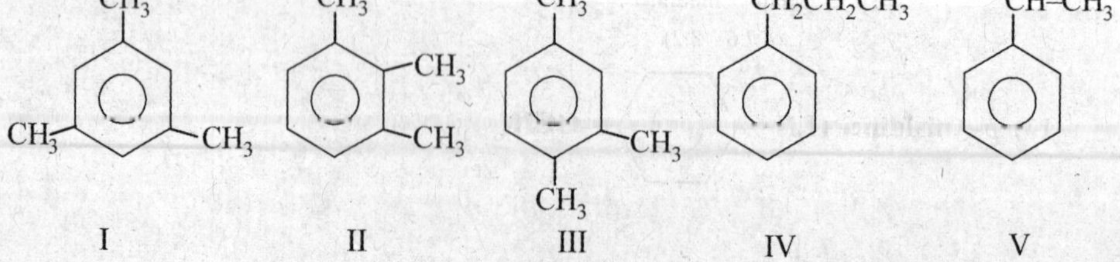

I II III IV V

Compounds I, II, III on oxidation would tricarboxylic acid, which is not the case. Compounds IV and V are the probable compounds. We shall choose out of the two on the basis of number of signals.

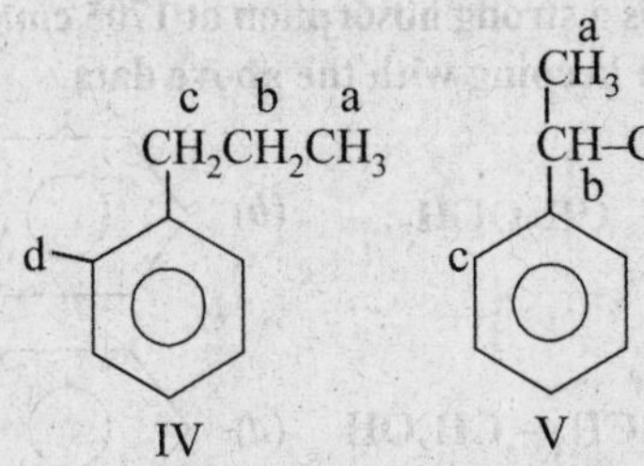

IV V

Compound IV gives four types of signals corresponding to a, b, c and d protons. Compound V gives three signals corresponding to a, b, c protons. Therefore the organic compound C_9H_{12} is V.

Example 2. A compound has the molecular formula $C_7H_6O_2$. It gives an infrared absorption bond at 1771 cm^{-1}. On treatment with $LiAlH_4$, it gets converted into compound B which shows characteristic infrared absorption bands at 3330 cm^{-1} and 1050 cm^{-1}. Assign structures to compound A and B.

Solution. $LiAlH_4$ is a reagent which reduces aldehydes, ketones and carboxy acids into alcohols. The formula $C_7 H_6 O_2$ rules out the possibility of an aldehyde or ketone which contain one oxygen. The compound appears to be benzoic acid with molecular formula $C_7H_6O_2$.

Benzoic acid

An absorption band at 1771 cm^{-1} support this assumption. On treatment with $LiAH_4$, benzoic acid is reduced to benzyl alcohol.

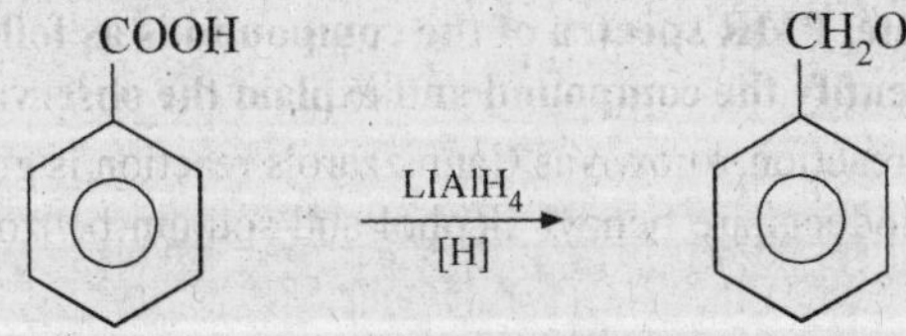

Absorption bands at 3330 cm^{-1} is due to aromatic ring and at 1050 cm^{-1} due to alcoholic group.

Example 3. Give a structure consistent with the following data:

Molecular formula – $C_2H_3Br_2$

a – Doublet at δ 4.15, 2H

b – Triplet at δ 5.77, 1H

What is the name of the compound?

Solution. The NMR data shows that two hydrogens in the compound are equivalent and one hydrogen is of different type.

A probable structure is

$$\overset{a}{CH_2}Br - \overset{b}{CH}Br_2$$

There are two types of hydrogens indicated by *a* and *b*. Doublet at δ 4.15 is due to methylene protons (marked *a*) coupled with methine proton (*b*).

Proton marked *b* coupled with methylene protons gives a downfield triplet.

Methine hydrogen gives the peak more downfield because of the presence of two bromines on the carbon carrying this hydrogen.

Example 4. PMR spectrum of a compound shows the following peaks δ 7.22 (*s*, 5H), δ 3.59 (*s*, 2H), δ 2.77 (*q*, 2H), δ 0.97 (*t*, 3H).

In the IR spectrum there is a strong absorption at 1705 cm^{-1}. Giving reasons, find out which of the following structures is in keeping with the above data.

(*a*)  (*b*) 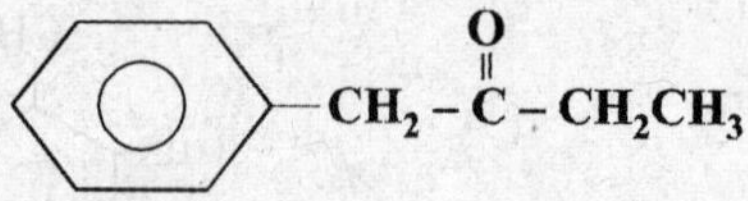

(*c*) 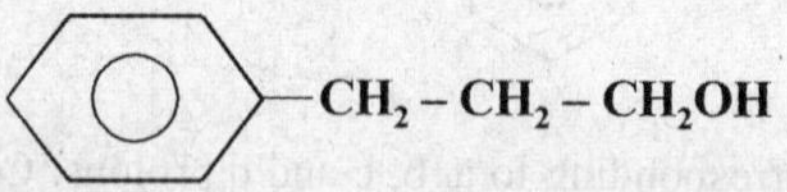 (*d*) 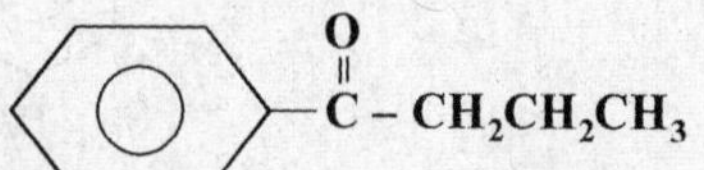

Solution. (*i*) IR absorption band is obtained at 1705 cm^{-1}. This is characteristic of a carbonyl group. Hence structures (*a*) and (*c*) are ruled out as they don't contain the carbonyl group.

The probable structure is out of (*b*) or (*d*)

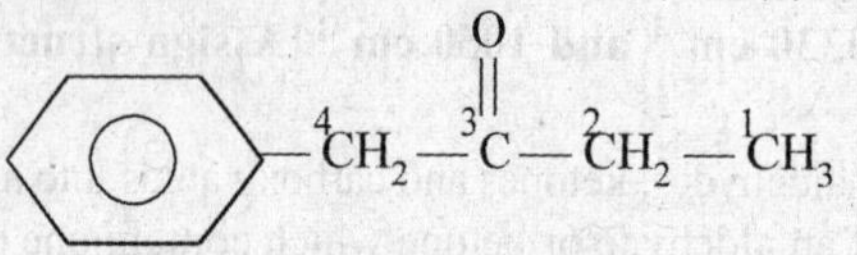

(*b*)

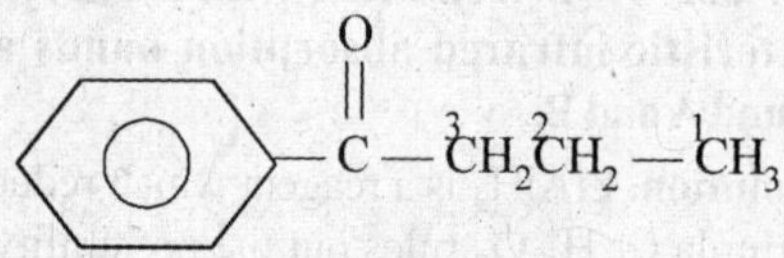

(*d*)

(*ii*) The PMR spectrum shows a peak δ 3.59 (*s*, 2H). This is a singlet in respect of two hydrogens. We don't expect to observe it in structure (*d*) because protons on carbon 1 will give a quartet, protons on carbon 2 will give a sixlet and protons on carbon 3 will give a triplet. Hence structure (*d*) is ruled out.

(*iii*) The only structure left is (*b*). Protons on carbon 2 will give a singlet. There will be no splitting of peak as there is no hydrogen on the neighbouring carbons. Hence structure (*b*) is confirmed.

Example 5. An organic compound C_7H_6O on treatment with concentrated NaOH gives benzyl alcohol and sodium benzoate. PMR spectra of the compound is as follows : singlet δ 9.85, 1 H, multiplet δ 7.3 – 8.2, 5H. Identify the compound and explain the observations.

Solution. This type of reaction, known as Cannizzaro's reaction is given by aldehydes with no α-hydrogen atom. As the products are benzyl alcohol and sodium benzoate, the given compound is likely to be benzaldehyde.

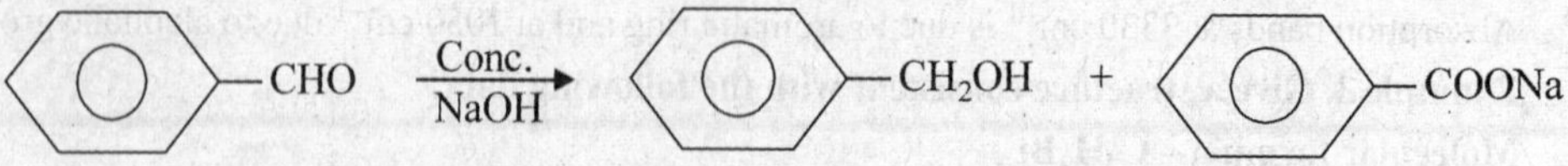

Benzyl alcohol Sod. benzoate

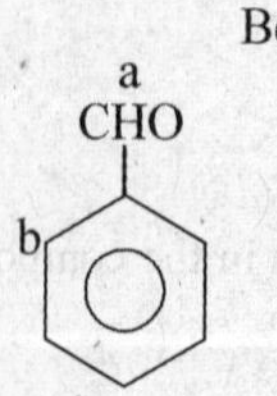

Aldehydic hydrogen gives a singlet because no hydrogen is present on the neighbouring carbon atom. It is a downfield singlet at δ 9.85. Five ring protons are not exactly equivalent because of presence of electron attracting carbonyl group attached to the benzene ring. Its deshielding effect is experienced by ortho, meta and para protons to different extent and they show different chemical shifts. Thus the multiplet δ 7.3–8.2 5H is explained.

Example 6. Both ethyl acetate and methyl propionate have the molecular formula $C_4H_8O_2$. How do they differ in their PMR spectra ?

Solution. $CH_3—\overset{O}{\overset{\|}{C}}—OCH_2CH_3$ (ethyl acetate) shows a slightly downfield singlet at δ ~ 2.0 due to methyl group of acetate part, methylene quartet at δ ~ 2.3 and methyl triplet at δ 1.25 due to ethyl group attached to oxygen.

$CH_3—CH_2—\overset{O}{\overset{\|}{C}}—OCH_3$ (methyl propionate) shows a methylene quartet at δ ~ 4.2 and a methyl triplet at δ ~ 1.2 due to $CH_3—CH_2—$ group of propionate part; downfield methyl singlet at δ ~ 3.7 due to methyl group attached to oxygen.

Example 7. A carbonyl compound containing carbon, hydrogen and oxygen and having a molecular mass of 72 gives a PMR spectrum which shows a triplet, a singlet and a quartet (at increasing values of δ) in the ratio 3 : 3 : 2. What is the structure of the compound ?

Solution. There are three structures possible with molecular mass 72. These are $CH_3COCH_2CH_3$, $CH_3CH_2CH_2CHO$, $(CH_3)_2CH—CHO$.

The structure which gives the signals as given in the example is $CH_3COCH_2CH_3$.

Example 8. A compound having the percentage composition C = 70.6%, H = 13.7% and O = 15.7% exhibits the following PMR spectrum:

Multiplet δ 3.56 (2H); Doublet δ 1.05 (12H)

Deterine its molecular formula and assign a suitable structure to it.

Solution. The molecular formula with the above percentage composition comes out to be $C_6H_{14}O$. The structure which gives the above signals is

$$(CH_3)_2CH—O—CH(CH_3)_2$$

Example 9. PMR spectrum of a compound shows the following peaks : δ 7.22 (*s*, 5H); δ 3.59 (*s*, 2H); δ 2.77 (*q*, 2H); δ 0.97 (*t*, 3H).

Giving reasons find out which of the following structures is in keeping with the above data :

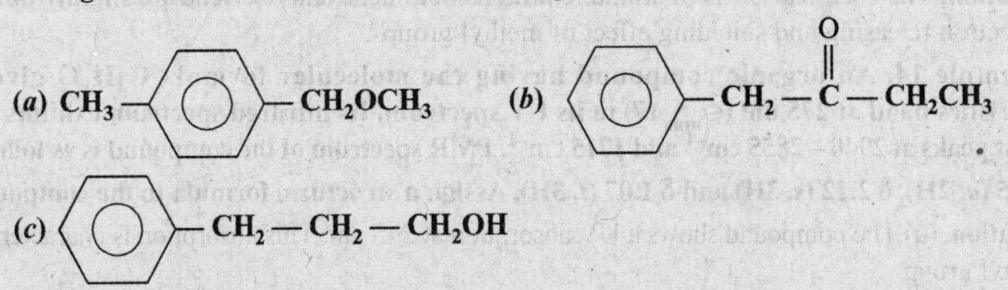

(*a*) $CH_3—C_6H_4—CH_2OCH_3$ (*b*) $C_6H_5—CH_2—\overset{O}{\overset{\|}{C}}—CH_2CH_3$ (*c*) $C_6H_5—CH_2—CH_2—CH_2OH$

Solution. The three compounds are isomers having the same molecular formula. On consulting the table of chemical shift, we find that the data points to structure (*b*)

Example 10. A carbonyl compound having the molecular formula $C_7H_{14}O$ gives the following PMR spectrum. Identify the compound.

δ 1.01 (singlet, 9H) ; δ 2.32 (signlet, 2H) ; δ 2.11 (singlet 3H)

Solution. The structure with the formula $C_7H_{14}O$ is

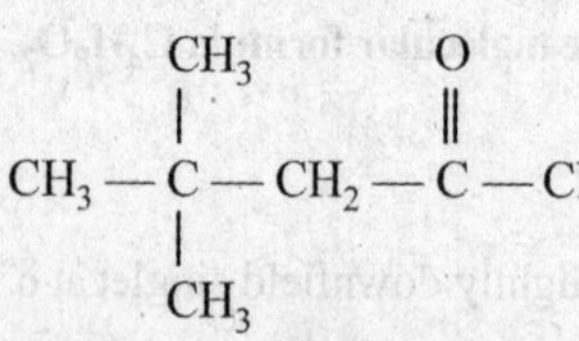

4, 4-Dimethylpentane-2-one

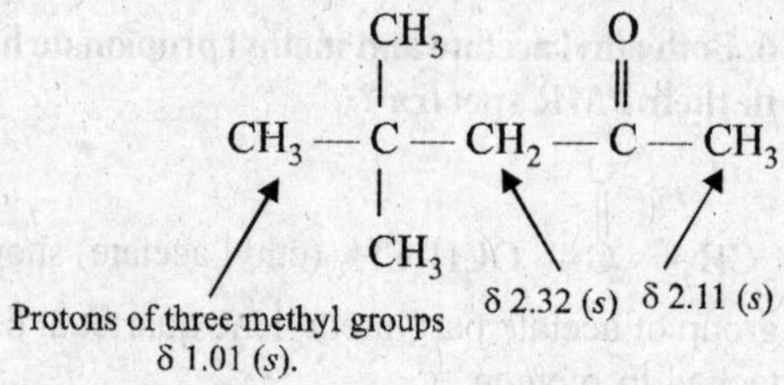

Example 11. Predict the number of signals and the splitting pattern of each signal in the PMR spectrum of $CH_3—\overset{O}{\overset{\|}{C}}—\underset{CH_3}{\underset{|}{CH}}—CH_3$.

Solution. The given molecule has three sets of equivalent protons labelled as *a*, *b* and *c*. Its PMR spectrum shows a singlet, a septet and a doublet as explained below.

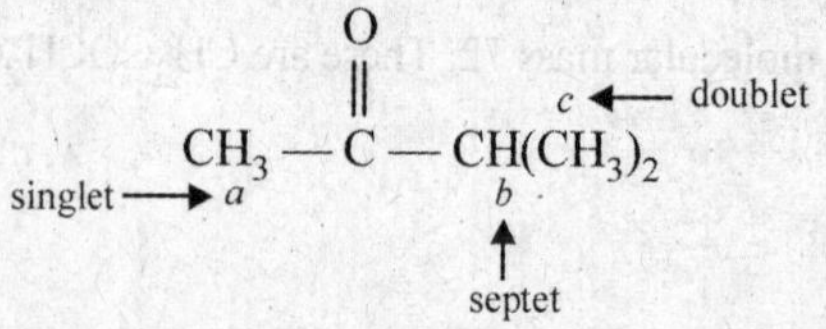

Example 12. Chloroethane contains only two kinds of protons but it shows as many as seven peaks in its ^{1}HNMR spectrum. Justify this observation.

Solution. Chloroethane has only two kinds of protons labelled as *a* and *b* *i.e.* three methyl protons and two methylene protons.

$$CH_3 — CH_2Cl$$

split into triplet ⟶ *a* *b* ⟶ split into quartet

But the three methyl protons would be split into a triplet due to coupling by the two methylene protons. Similarly the two methylene protons would be split into a quartet by the three methyl protons. As a result, chloroethane shows seven peaks in its spectrum.

Example 13. How can you explain the following difference in the chemical shifts of aromatic protons in the following compounds ?

Benzene δ 7.37, Toluene δ 7.17, *p*-Xylene 7.05.

Solution. The chemical shifts of aromatic protons in toluene and *p*-xylene are slightly upfield due to electron releasing and shielding effect of methyl groups.

Example 14. An organic compound having the molecular formula C_4H_8O gives a characteristics band at 275 nm ($\in_{max}$ 17) in its UV spectrum. Its infrared spectrum exhibits two important peaks at 2940 – 2855 cm^{-1} and 1715 cm^{-1}. PMR spectrum of the compound is as follows:

δ 2.5 (*q*, 2H), δ 2.12 (*s*, 3H) and δ 1.07 (*t*, 3H). Assign a structural formula to the compound.

Solution. (*a*) The compound shows a UV absorption at 275 nm. This absorption is characteristic of carbonyl group.

(*b*) IR absorption peaks at 2940 – 2855 cm^{-1} and 1715 cm^{-1} are due to C–H stretching of methyl groups and a saturated ketonic group respectively.

(*c*) With the formula C_4H_8O, there are two possibilities.

$CH_3CH_2CH_2CHO$ (I) $CH_3COCH_2CH_3$ (II)

(*d*) Formula I is ruled out as it contains four types of protons and would give four signals which is not in agreement with the data.

(*e*) The structure left is $\overset{4}{C}H_3\,\overset{3}{C}O\overset{2}{C}H_2\,\overset{1}{C}H_3$ There are two protons of one type, three of second and three of third type and will give rise to three signals. C_2 protons will give a quartet δ 2.5, C_4 protons will give singlet, δ 2.12 and C_1 positions will give a triplet. Observations made with structure II agree with the data.

Example 15. IR spectrum of the compound shows that it is aromatic in nature. PMR spectrum exhibits three signals at δ 1.2 (*d*, 6H), δ 2.8 (*m*, 1H) and δ 7.2 (*s*, 5H). Work out a structure for the compound.

Solution. The molecule contains a benzene ring. The mass of the side-chain/s can be calculated by substracting 77 (mass of phenyl ring) from 120. It comes to 43. The side chain thus could be $-C_3H_7$. The possible structures are:

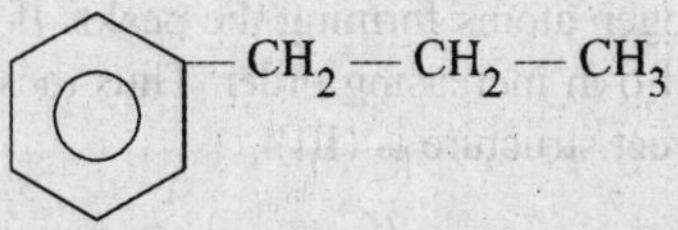

I

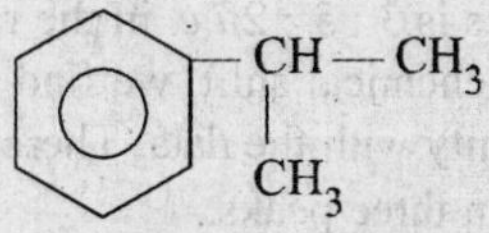

II

Also there is a possibility of the compound being disubstituted benzene. The possible structures are

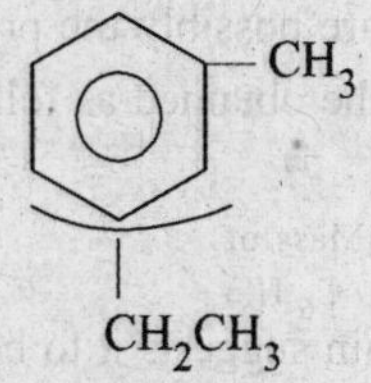

o, m or *para*
(III)

Structures (I) and (III) are ruled out because, each one of them contains four types of protons and thus would give four signals which is inconsistent with the data.

The only structure left is (II)

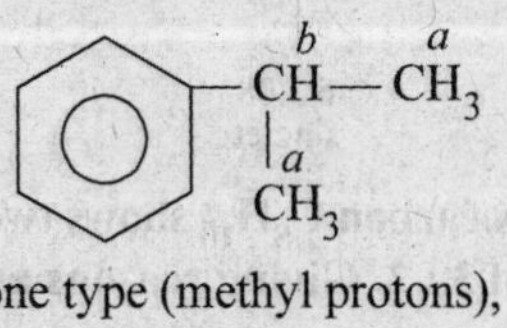

This structure has 6 protons of one type (methyl protons), 1 proton designated *b* of second type and 5 ring protons of third type.

Six methyl protons give a doublet δ 1.2 due to splitting by protons *b*, one proton designated *b* gives a multiplet δ 2.8 due to splitting by methyl protons and 5 ring protons give a singlet δ 7.2.

All the observations made with structure (II) agree with the data. Hence this is the correct structure.

Example 16. A compound contains C, H and O. Its IR spectrum shows a strong band at 1715 cm^{-1}. PMR spectrum shows a triplet, a singlet and a quartet (at increasing values of δ) in the ratio 3 : 3 : 2. What is the structure of the compound with molecular mass of 72?

Solution. The IR spectrum shows an absorption peak at 1715 cm^{-1}. This is characteristic of a carbonyl compound. It could be an aldehyde or ketone. Possible structures with molecular mass 72 are as follows:

$CH_3CH_2CH_2\,CHO$ (I) $\quad$ $(CH_3)_2\,CHCHO$ (II) $\quad$ $CH_3COCH_2CH_3$ (III)

Structure I contains four types of protons. It would therefore give four PMR signals which is inconsistent with the data. Hence structure I is ruled out.

We would expect a doublet, a multiplet and again a doublet from structure (II). Hence this structure is also ruled out.

Structure III is the only structure left

$$\overset{1}{C}H_3\,\overset{2}{C}O\overset{3}{C}H_2\,\overset{4}{C}H_3$$

Three protons on carbon 4 will give a triplet due to splitting by protons of carbon 3. Singlet will be given by three protons on carbon 1. There will be no splitting here. Again two hydrogens on carbon 3 will give quartet due to interaction with the protons of carbon 4. Thus the ratio of various peak intensities is 3 : 3 : 2 *i.e.* in the ratio of hydrogen atoms forming the peaks. If we consult the table of PMR chemical shift, we find the values of δ in increasing order. Thus these observations are in conformity with the data. Therefore the correct structure is (III).

We obtain three peaks.

Example 17. A hydrocarbon with molecular mass 102 exhibits NMR signals δ 7.4 (5H, singlet) and δ 3.08 (1H, singlet). What is the probable structure of the hydrocarbon?

Solution. (*i*) NMR signals show that there are two types of protons, 5 protons of one type and 1 proton of another type. Five protons are possibly the protons of monosubstituted benzene ring.

(*ii*) The size of the side-chain can be obtained as follows :

102	–	77	=	25
(Mol mass of compound)		(Mass of $-C_6H_5$)		Mol. mass of side-chain

A value of 25 for mass of side-chain suggests it to be $-C\equiv CH$

This is verified from NMR spectrum which tells that the second singlet is due to one hydrogen only. Thus, the structure of the hydrocarbon is

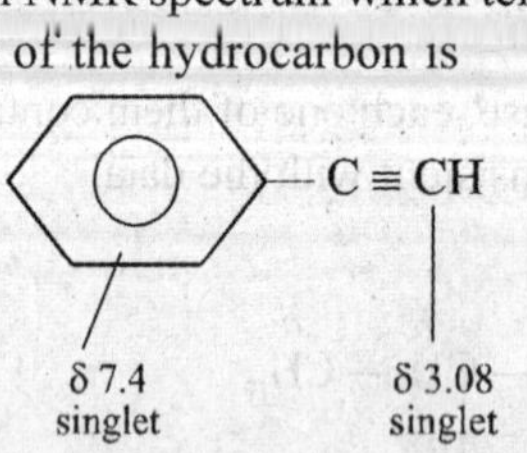

(Phenyl acetylene)

Example 18. An aromatic hydrocarbon C_8H_{10} shows two singlets in its PMR at δ 2.3 and 7.05 with proton counting in the ratio of 3 : 2. Giving reasons assign structure to it.

Solution. (*a*) There are two possible hydrocarbons with the formula C_8H_{10} as given below:

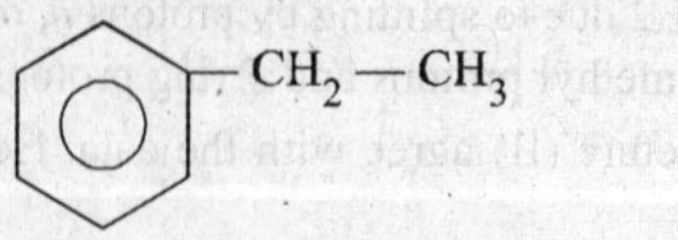

I

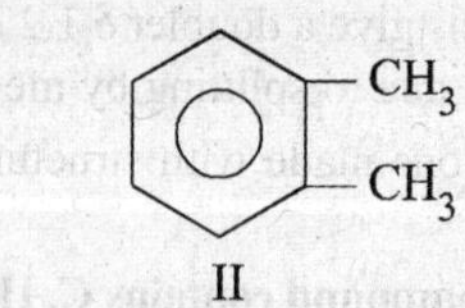

II

Structure I would give three peaks as there are three kinds of protons in it. Structure II would give two signals as there are two kinds of protons. It may be noted that six protons of two methyl groups are equivalent. As the compound under examination gives two peaks, it can be only II.

(*b*) The peak at δ 2.3 corresponds to methyl protons and that at δ 7.05 corresponds to ring hydrogens.

(*c*) There are six methyl protons and four ring protons. Thus, they are in the ratio 6 : 4 or 3 : 2. This ratio agrees with the given value. Thus the structure of the hydrocarbon is

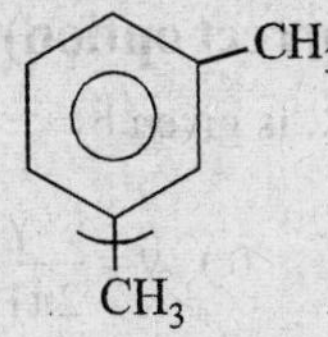

o, *m* or *p*-Xylene

Example 19. A compound was believed to be either diphenyl ether or diphenyl methane. Could PMR spectrum be used to distinguish between these two compounds How?

Solution. Diphenyl ether does not show spin-spin splitting while diphenyl methane shows.

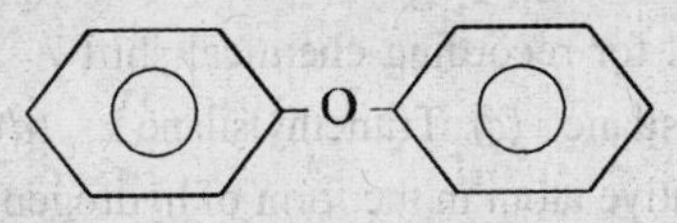

diphenyl ether

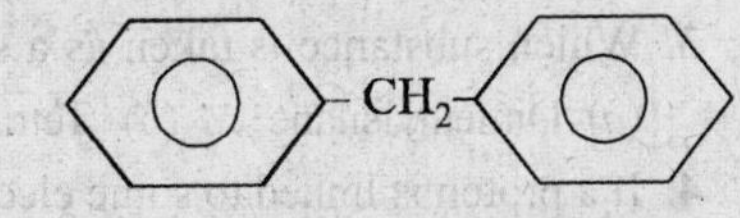

diphenyl methane

Example 20. How will you distinguish between the three dibromobenzens by their NMR spectra.

Solution. *o*-Dibromobenzene will show *two* peaks : *m*-dibromobenzene will show three peaks: *p*-dibromobenzene will show only *one* peak.

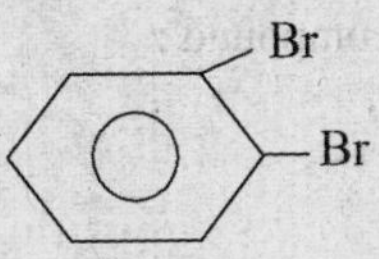

o-Dibromobenzene
(two peaks)
1 : 1

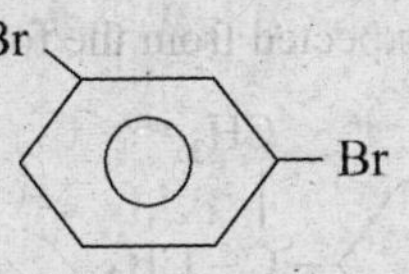

m -Dibromobenzene
(three peaks)
1 : 2 : 1

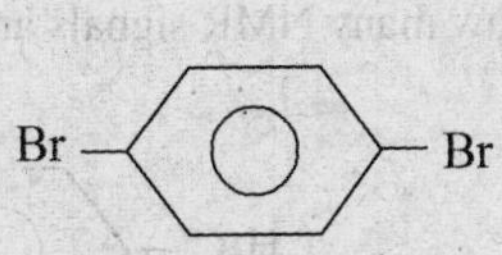

p -Dibromobenzene
(one peaks)

Example 21. Suggest a structure consistent with the following NMR data : Molecular formula = C_9H_{12}

(*a*) Singlet at δ 6.78, 3H **(*b*) singlet at δ 2.25, 9H**

Solution. The compound is

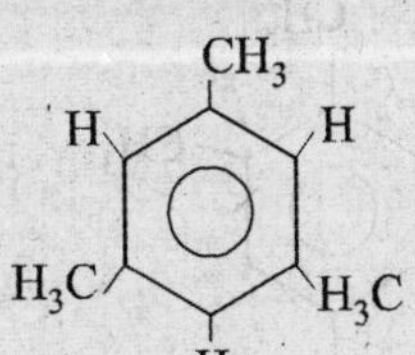

Mesitylene

Example 21. How would you distinguish between the following pairs of compounds by NMR spectroscopy.

(*a*) $CH_3-\overset{O}{\overset{\|}{C}}-CH_3$ and $CH_3CH_2-\overset{O}{\overset{\|}{C}}-H$

(*b*) $CH_3-\overset{O}{\overset{\|}{C}}-CH_3$ and $CH_3-\overset{O}{\overset{\|}{C}}-OCH_3$

Solution. (*a*) CH_3COCH_3 will give only *one* signal while CH_3CH_2CHO will give *three* signals.

(*b*) CH_3COCH_3 will give only *one* signal while CH_3COOCH_3 will give *two* signals.

EXERCISES
(Including Questions from Different University Papers)

Multiple Choice Questions (Choose the correct option)

1. The relation between ν, H_0 and γ in N.M.R. is given by

(*a*) $\nu = \frac{\gamma . H_0}{2\pi}$ (*b*) $\nu = \frac{H_0}{2\pi\gamma}$ (*c*) $\nu = \frac{\gamma}{2\pi H_0}$ (*d*) $\nu = \frac{1}{2\pi\gamma H_0}$

2. How many NMR signals would be given by the compound : $(CH_3)_2CHCH_2CH_3$

(*a*) 3 (*b*) 4 (*c*) 5 (*d*) 2

3. Which substance is taken as a standard for recording chemical shift ?

(*a*) Dimethylsilane (*b*) Tetramethylsilane (*c*) Trimethylsilane (*d*) Methylsilane

4. If a proton is linked to some electronegative atom in the form of hydrogen bond, absorption will occur

(*a*) unchanged (*b*) upfield (*c*) downfield (*d*) cannot say

5. If the chemical shift on the δ scale is 4.4, that on τ scale would be

(*a*) – 4.4 (*b*) $\frac{1}{4.4}$ (*c*) 5.6 (*d*) – 5.6

6. How many NMR signals are expected from the following compound :

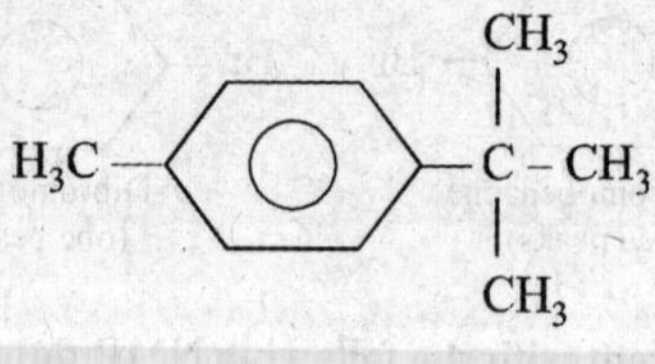

(*a*) 5 (*b*) 2 (*c*) 4 (*d*) 3

7. A compound with molecular formula C_9H_{12} gave the following NMR signals singlet τ 3.22, 3H, singlet τ 7.75 9H. The structure of the compound is

(*a*) 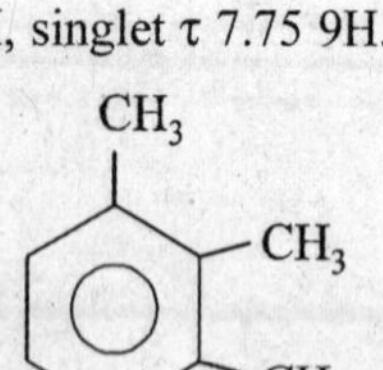

(*b*) 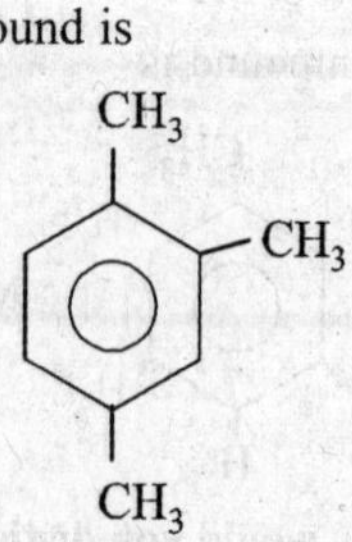

(*c*) 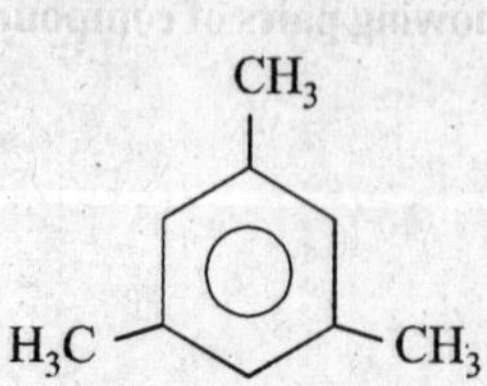

(*d*) 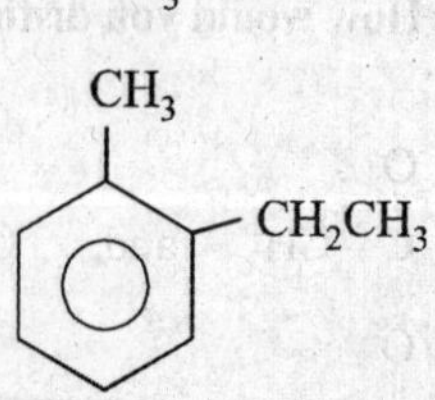

8. Which of the following statements regarding NMR data of *o*-, *m*- and *p*-dibrombenzene is not correct ?

(*a*) *o*-dibromobenzene shows two peaks (*b*) *m*-dibromobenzene shows three peaks

(*c*) *p*-dibromobenzene shows only one peak (*d*) all the above statements are incorrect

ANSWERS

1. (a)	2. (b)	3. (b)	4. (c)	5. (c)	6. (d)	7. (c)	8. (d)

Short Answer Questions

1. Will there be any spin-spin splitting in the spectrum of the molecules ?
   (i) $ClCH_2—CH_2Cl$ (ii) $CH_3—CCl_2—CH_2Cl$
2. Name an important internal reference standard used in NMR spectroscopy. Give its advantages.
3. Will the two protons on C_1 in 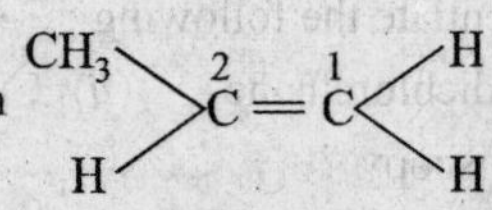 cause mutual splitting of signals ? If so, why ?
4. In PMR spectrum of ethyl bromide which proton will absorb at higher value of δ ?
5. In a given organic compound two kinds of protons exhibit signals at 50 and 200 Hz using a 60 MHz PMR spectrometer. What would be their equivalent positions using 90 MHz spectrometer ? Also convert the positions of the signals at 50 into δ and τ units.
6. Will there by any spin-spin coupling in the spectrum of the molecules :
   (i) $BrCH_2—CH_2Br$ (ii) $CH_3CCl_2CH_2Cl$
7. Predict the structural formulae for the compounds with the following molecular formula showing only one PMR signal.
   (i) C_8H_{18} (ii) C_3H_6O (iii) C_5H_{12} (iv) $C_2H_4Cl_2$
8. What is chemical shift ? What are the scales of measurement with TMS and how are they related ?
9. Why is TMS chosen as a reference compound in PMR studies ?
10. How many signals will you expect from each of the following compounds. Label all sets of equivalent protons :
    (i) $CH_3—O—CH_2—CH_3$ (ii) $CH_3—CH_2—CH_2—CH_3$
11. What property of certain atomic nuclei is involved in NMR phenomenon ?
12. Will the two protons on C_1 in

    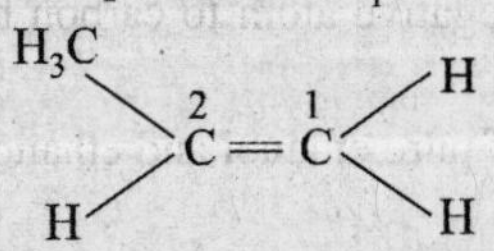

    cause mutual splitting of signals ? If so why ?
13. Indicate "True" or "False" for the following statements :
    (i) Deshielding shifts the absorption downfield in PMR spectrum.
    (ii) I.R. spectrum of a compound is the result of $\pi \rightarrow \pi^*$ transitions.
    (iii) The compound $(CH_3)_2O$ will have 2 signals in NMR spectrum.
14. Explain the PMR spectrum of benzaldehyde.
15. An organic compound with the molecular formula $C_3H_3Cl_5$ gave the following PMR data :
    (i) A triplet 5.48 τ (4.52 δ), 1H (ii) A doublet 3.93 τ (6.07 δ), 2H
    Assign a structural formula to the compound consistent with the above data.
16. Explain the PMR spectrum of ethyl bromide.
17. Discribe principle of NMR spectroscopy.
18. Describe the PMR spectrum of an ordinary sample of ethanol.
19. Explain the PMR spectrum of acetaldehyde.
20. Tetramethylsilane is chosen as a reference compound in PMR studies. Why ?

21. (*a*) Explain the terms :

(*i*) Spin-spin coupling (*ii*) Chemical shift.

22. A compound having the molecular formula $C_{10}H_{14}$ gave the following PMR data :

(*i*) Singlet τ 9.12 (δ 0.88) 9H (*ii*) Singlet τ 2.72 (δ 7.28) 5H

Assign structure to the compound on the basis of the above data.

23. What is chemical shift ? What are the scales of measurement with TMS and how are they related ?

24. Indicate how will you differentiate the following :

(*i*) Chlorobenzene and 1, 2-dichlorethane (*ii*) $CH_2Cl—CH_2Cl$ and $CH_3—CH_2Cl$

on the basis of PMR spectroscopy ?

25. How can PMR spectroscopy be employed in differentiating ethene, ethane and acetylene?

26. A compound having the molecular formula $C_9H_{11}Br$ showed the following set of PMR data:

(*i*) Multiplet (δ 2.25), 2H (*ii*) Triplet (δ 2.72), 2H

(*iii*) Triplet (δ 3.38), 2H (*iv*) Singlet (δ 7.22), 5H

27. What do you understand by the following terms in relation to NMR spectroscopy ?

(*i*) Magnetic anisotropy (*ii*) Spin-spin splitting

28. Name an important internal reference standard in PMR spectroscopy. Give its advantages.

29. Write with one example the shielding and deshielding effects involved in PMR spectroscopy.

30. A compound with molecular formula C_3H_6O can have the following possible structures :

(*i*) CH_3COCH_3 (*ii*) CH_3CH_2CHO (*iii*) $CH_2=CH—CH_2OH$

Show how with the help of PMR spectroscopy you can decide about its structure.

31. Using PMR spectroscopy, how will you differentiate between the following pairs :

(*i*) 1-Bromopropane and 2-Bromopropane (*ii*) Propane and Cyclopropane.

32. Give a structure consistent with the following data :

Molecular formula $C_2H_3Br_3$ (*i*) Double at δ 4.13, 2H

(*ii*) Triplet at d 5.77, 1H

33. Explain that attachment to the strongly electro negative atom to carbon bearing a proton causes a downfield shift in PMR spectra.

34. Use appropriate values and sketch the ^{1}HNMR of pure ethanol and ethanol having a trace of mineral acid.

35. Propose structural formulae for the following compounds which give only one PMR signal :

(*i*) C_5H_{12} (*ii*) C_8H_{18} (*iii*) C_2H_6O

36. How PMR spectroscopy helps to predict whether the given compound is aromatic or not?

37. How can you explain the following difference in the chemical shifts of aromatic protons in the following compounds :

Benzene δ 7.37; Toluene δ 7.17; *p*-Xylene δ 7.05

38. Assign a structure to the compound with molecular formula C_4H_9Br, PMR spectrum of which shows δ values at 1.04 (*d*, 6H) 1.95 (*m*, 1H) and 3.33 (*d*, 2H).

39. A compound having the molecular formula $C_9H_{11}Br$ showed the following set of NMR data :

(*i*) Multiplet, δ 2.25, 2H; (*ii*) Triplet, δ 2.72, 2H;

(*iii*) Triplet, δ 3.38 2H; (*iv*) Singlet, δ 7.22, 5H.

40. Distinguish the following pairs on the basis of NMR data :

(*i*) CH_3OCH_3 and CH_3CH_2OH (*ii*) $CH_3COOC_2H_5$ and $C_2H_5COOCH_3$

41. A compound was believed to be either diphenyl ether or diphenyl methane. Can PMR spectrum be used to distinguish between these two compounds ? How ?

42. How many proton signals would be expected in the PMR spectra of each of the following compounds :

(i) $CH_3 — O — CH_2CH_3$ (ii) 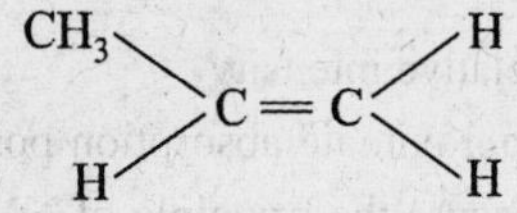

(iii) $CH_3 — \underset{\displaystyle Cl}{\underset{|}{CH}} — CH_2CH_3$

43. What is shielding and deshielding of protons ? Give examples.

44. A compound having molecular formula $C_9H_{11}Br$ showed the following sets of PMR data :

(i) Multiplet, δ 2.26 (2H) (ii) Triplet, δ 2.72 (2H)
(iii) Triplet, δ 3.38 (2H) (iv) Singlet δ 7.23 (5H)

45. Why does benzene display downfield proton resonance signals as compared to cyclohexane?

46. Write a brief account of equivalence and non-equivalence of protons.

General Questions

1. (a) Using PMR spectrum how will you differentiate between the following pairs :
   (i) n-propyl bromide and isopropyl bromide
   (ii) Propane and cyclopropane

   (b) Pridict the number of signals in the PMR spectrum of each of the following :
   (i) 1, 1-Dimethylclopropane (ii) *cis*-1, 2-Dimethylcyclopropane
   (iii) *trans*-1, 2-Dimethylcyclopropane (iv) 1, 2-Dichloropropane.

2. Explain the following terms in relation to PMR spectroscopy :
   (i) Chemical shift (ii) Coupling constants
   (iii) Equivalent and non-equivalent protons (iv) Positions of signals.

3. Using PMR spectroscopy, how will you distinguish between the following pairs :
   (i) CH_3CH_2CHO and CH_3COCH_3
   (ii) 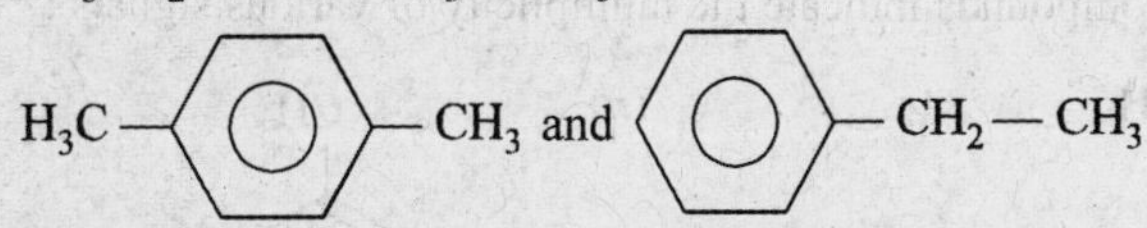

4. Give the structural formula of each of the following compounds (disregarding enantiomers), and label all sets of equivalent protons. How many signals would be shown by each in NMR spectra ?
   (i) the two isomers of $C_2H_4Cl_2$ (ii) *p*-Xylene.

5. Write notes on the following :
   (i) Equivalent and non-equivalent protons
   (ii) Spin-spin coupling and coupling constant.

6. What is chemical shift ? What are scales of measurement with TMS and how are they related ?

7. (a) Explain the following terms in relation to PMR spectroscopy :
   (i) Equivalent and non-equivalent protons (ii) Peak area

(*b*) While interpreting PMR spectrum, what information is given by the following :
(*i*) Position of signals (*ii*) Splitting of signals.

8. How many signals do you expect in PMR spectrum of $CH_3CH_2CH_2Cl$. Predict for each signal :
(*i*) Relative intensity (*ii*) Multiplicity and
(*iii*) Approximate absorption position

9. (*a*) Discuss the principle of NMR spectroscopy.
(*b*) Discuss the PMR spectra of the following molecules :
(*i*) Ethyl bromide (*ii*) Benzaldehyde.

10. (*a*) Using PMR spectroscopy, how will you differentiate between the following pairs :
(*i*) 1-Bromopropane and 2-Bromopropane
(*ii*) Propene and cyclopropane
(*b*) Of the following molecules, which will exhibit spin-spin coupling ? What will be the multiplicity of each kind of protons in such molecules ?
(*i*) $CH_3—CHBr_2$ (*ii*) Neopentyl bromide
(*iii*) 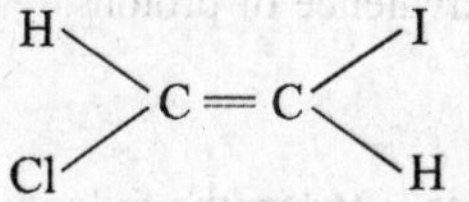

(*iv*) $ClCH_2—CH_2Br$

11. From the following set of PMR data give a structure consistent with molecular formula $C_{10}H_{13}Cl$:
(*i*) Singlet, δ 1.57, 6H (*ii*) Singlet, d 3.07, 2H
(*iii*) Singlet, δ 7.27, 5H

12. Explain the following terms in relation to PMR spectroscopy :
(*i*) Equivalent and non-equivalent protons.
(*ii*) Shielding and deshielding
(*iii*) Coupling Constant.

13. Explain the following terms :
(*a*) (*i*) Coupling constant (*ii*) Spin-spin coupling
(*b*) In the following compounds indicate the multiplicity of various signals :
(*i*) $CH_3—CH_2—OH$ (*ii*)
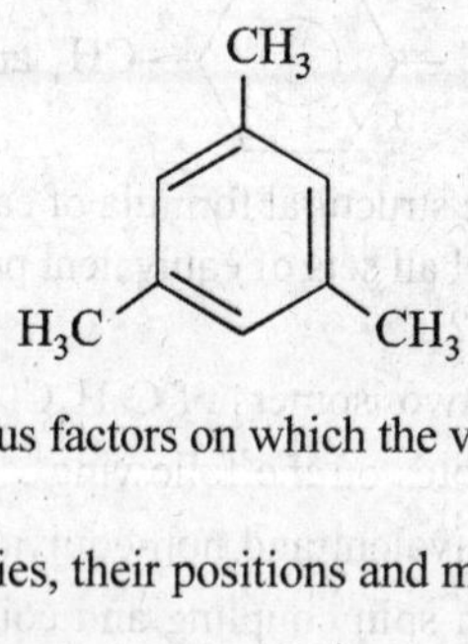

14. What is meant by chemical shift ? Discuss the various factors on which the value of chemical shift depends.

15. Predict the number of signals, their relative intensities, their positions and multiplicity in the PMR spectra of the following compounds :
(*i*) Isopropyl bromide (*ii*) Ethyl bromide.

16. Explain why different protons of the molecule $CH_3CH_2CH_2Cl$ do not give PMR signals at the same position. How many PMR signals do you expect from it ? Predict their relative intensities and multiplicities.

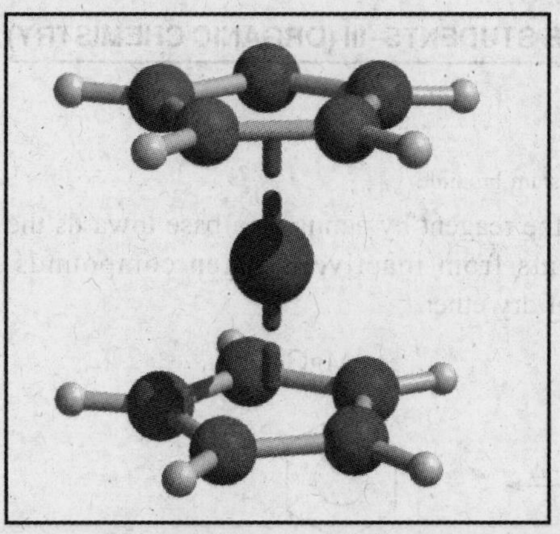

2

Organometallic Compounds

2.1. INTRODUCTION

Organometallic compounds are those organic compounds in which there is a bond between carbon and metal. There is a wide variation in the nature of this bond in different organometallic compounds. Main factor that is responsible for difference in the nature of carbon-metal bond is the nature of metal itself. Highly electropositive metals like sodium and potassium tend to make this bond ionic, with the metal carrying positive charge and carbon carrying negative charge. Organometallic compounds of magnesium and lithium have this bond with partial ionic character.

We come across organometallic compounds with metal constituents such as sodium, potassium, lithium, magnesium, lead and zinc.

Some examples of these compounds are:

CH_3MgBr	C_6H_5MgI	C_2H_5MgCl
Methyl magnesium bromide	Phenyl magnesium iodide	Ethyl magnesium chloride
CH_3Li	C_2H_5Li	$(CH_3)_2Zn$
Methyl lithium	Ethyl lithium	Dimethyl zinc
$(C_2H_5)_2Cd$	$(C_2H_5)_4Pb$	
Diethyl cadmium	Tetraethyl lead	

The following sequence indicates the variation in ionic character in carbon-metal bond.

$$C-K > C-Na > C-Li > C-Mg > C-Zn > C-Cd$$

ORGANOMAGNESIUM COMPOUNDS (GRIGNARD REAGENTS)

Organic compounds having the general formula RMgX are called Grignard reagents. Here R stands for some alkyl or aryl group and X denotes some halogen. These compounds were discovered by Victor Grignard who was awarded Nobel prize in 1912 for this discovery.

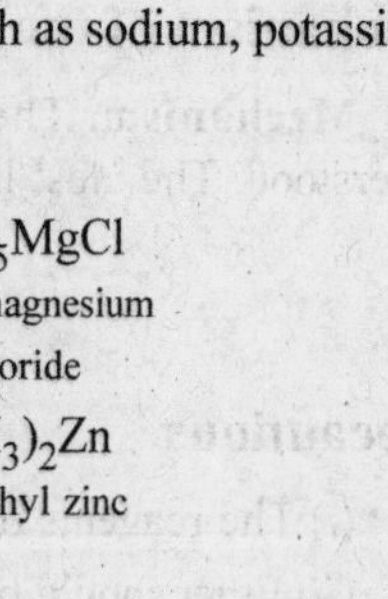

Victor Grignard

2.1. PREPARATION OF GRIGNARD REAGENTS

Grignard reagents are prepared by reacting alkyl or aryl halides and Mg metal in dry ether.

$$R-X + Mg \xrightarrow[\text{ether}]{\text{Dry}} RMgX$$

$$Ar-X + Mg \longrightarrow ArMgX$$

$$\underset{\text{Ethyl bromide}}{C_2H_5-Br + Mg} \xrightarrow{\text{Dry ether}} \underset{\text{Ethyl magnesium bromide}}{C_2H_5MgBr}$$

Ether is used as a solvent because it can solvate the reagent by acting as a base towards the acidic magnesium. For obtaining Grignard reagents from inactive halogen compounds, tetrahydrofuran (THF) is used as the solvent in place of dry ether.

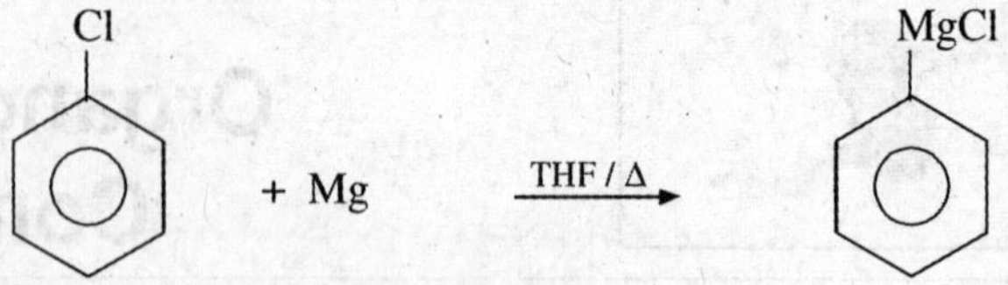

The magnesium metal disappears and Grignard reagents is formed with evolution of heat.

For a given alkyl group, the order of reactivity of halogen is

$$I > Br > Cl$$

Similarly, for a given halogen atom, the order of reactivity of alkyl group is:

$$CH_3 > C_2H_5 > C_3H_7$$

Grignard reagents are seldom isolated in solid state as they ignite spontaneously in air. They are usually used in solution.

Mechanism of the reaction

Mechanism. The mechanism by which Grignard reagents are formed is still not fully understood. The most likely mechanism appears to be free radical mechanism as shown below:

$$R-X + Mg \longrightarrow R\cdot + \cdot MgX$$

$$R\cdot + \cdot MgX \longrightarrow RMgX$$

Precautions

(*i*) The reagents used should be pure and dry as moisture tends to stop the reactions.

(*ii*) Ether should be washed with water to free it from alcohol and dried over anhydrous $CaCl_2$ for 2-3 days to remove alcohol and moisture. It should be finally distilled over sodium and P_4O_{10} to remove last traces of water.

(*iii*) Magnesium turnings should be treated with ether to remove grease and then dilute HCl to remove oxide film. They should be finally dried in an oven at 383-393 K.

(*iv*) Alkyl halide should be dried by distilling over P_4O_{10}

Structure of Grignard reagent

During the preparation of Grignard reagents, ether is used as a solvent. Accordingly it has been assumed that ether molecules are present as ether of crystallisation. Thus, two most probable structures have been proposed for the Grignard reagent.

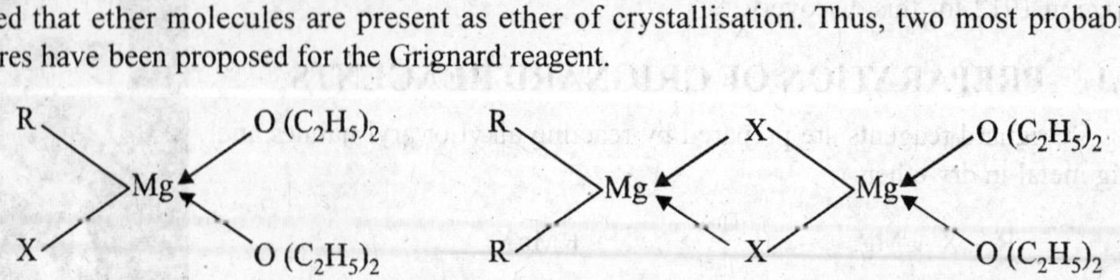

(I) (II)

It has also been proposed that the following equilibria exist for Grignard reagent in ether.

$$R_2Mg \cdot MgX_2 \rightleftharpoons R_2Mg + MgX_2 \rightleftharpoons 2\,RMgX$$

However, the general formula of Grignard reagents is represented as RMgX for the sake of convenience.

2.2. PROPERTIES OF GRIGNARD REAGENTS

Physical Properties

Grignard reagents are colourless non volatile liquids. They are not isolated in free state due to their explosive nature. They are used *in situ* i.e. in the solution form in the ether solvent.

Chemical Properties

The reactions of Grignard reagents can be studied under two headings.

(*i*) Double decomposition with compounds containing an active hydrogen atom.

An active hydrogen atom is one which is joined to oxygen, nitrogen or sulphur. When such compounds are treated with Grignard reagent, the alkyl group is converted into alkane

$$H-O-\boxed{H+R}-Mg-X \longrightarrow R-H+Mg\begin{cases}X\\OH\end{cases}$$

$$R-O-\boxed{H+R}-Mg-X \longrightarrow R-H+Mg\begin{cases}X\\OR\end{cases}$$

$$NH_2-\boxed{H+R}-Mg-X \longrightarrow R-H+Mg\begin{cases}X\\NH_2\end{cases}$$

$$HCl+RMg-X \longrightarrow R-H+Mg\begin{cases}X\\Cl\end{cases}$$

Carbon-magnesium bond in RMgX has considerable ionic character. Because of this, the carbon atom bonded to the metal is a strong base or we can say that Grignard reagents are strong bases. They react with those compounds which have a hydrogen more acidic than the hydrogen of the hydrocarbon from which the Grignard reagent is derived. The reactions of Grignard reagents with above mentioned compounds are acid-base reactions and lead to the formation of conjugate acid and conjugate base:

$$\underset{\text{(Stronger base)}}{\overset{\delta-\;\;\delta+}{R:Mg}}\,X+\underset{\text{(Stronger acid)}}{H-\ddot{O}-R} \longrightarrow \underset{\text{(Weaker acid)}}{R-H}+\underset{\text{(Weaker base)}}{R-\ddot{O}:^-}+Mg^{+2}+X^-$$

$$\underset{\text{(Stronger base)}}{\overset{\delta-\;\;\delta+}{R:Mg}}\,X+\underset{\text{(Stronger acid)}}{H-\ddot{O}-H} \longrightarrow \underset{\text{(Weaker acid)}}{R-H}+\underset{\text{(Weaker base)}}{H-\ddot{O}:^-}+Mg^{+2}+X^-$$

(*ii*) Addition to compounds containing multiple bonds.

In such cases, the addition of R Mg X, takes place in such a way that alkyl group (nucleophile) goes to the atom having the lower electronegativity. For example,

$$\overset{\delta-}{R} : \overset{\delta+}{Mg} X + \begin{matrix} R \\ R \end{matrix}\!\!>C=\ddot{O}: \longrightarrow R-\underset{R}{\overset{R}{\underset{|}{\overset{|}{C}}}}-\ddot{O}:^{-}\overset{+2}{Mg}\overset{-}{X}$$

When water or dilute acid is added to the reaction mixture, an acid-base reaction takes place to produce an alcohol.

$$R-\underset{R}{\overset{R}{\underset{|}{\overset{|}{C}}}}-\ddot{O}:^{-}\overset{+2}{Mg}\overset{-}{X} + H-\ddot{O}-H \longrightarrow R-\underset{R}{\overset{R}{\underset{|}{\overset{|}{C}}}}-\ddot{O}-H + Mg X_2 + H_2O$$

2.3. SYNTHETIC APPLICATIONS OF GRIGNARD REAGENTS

We can prepare almost every type of organic compound from them by selecting a suitable reagent.

(1) **Alkanes.** Grignard reagents are decomposed by water, alcohol, ammonia to form alkanes. Thus:

$$\underset{\text{Methylmag. Iodide}}{CH_3\,Mg-I} + \underset{\text{Water}}{H-OH} \longrightarrow \underset{\text{Methane}}{CH_4} + \underset{\text{Magnesium hydroxy iodide}}{Mg\langle^{I}_{OH}}$$

$$\underset{\text{Methylmag. Iodide}}{C_2H_5-Mg-I} + \underset{\text{Ethanol}}{HOC_2H_5} \longrightarrow \underset{\text{Ethane}}{C_2H_6} + \underset{\text{Magnesium ethoxy iodide}}{Mg\langle^{I}_{OC_2H_5}}$$

$$CH_3-Mg-Br + \underset{\text{Ammonia}}{H-NH_2} \longrightarrow CH_4 + \underset{\text{Magnesium amino bromide}}{Mg\langle^{Br}_{NH_2}}$$

(2) **Alkenes.** These are obtained by reaction with unsaturated halogen derivatives. Thus:

$$CH_3\,Mg\,I + ICH_2-CH=CH_2 \longrightarrow \underset{\text{Butene-1}}{CH_3-CH_2-CH=CH_2} + MgI_2$$

(3) **Alkynes.** Higher alkynes can be obtained by treating lower alkynes with a Grignard reagent and then alkyl halide. Thus:

$$\underset{\text{Propyne}}{CH_3-C\equiv CH} + CH_3\,MgI \longrightarrow CH_3-C\equiv C-Mg\,I + CH_4$$

$$CH_3-C\equiv C-Mg\,I \xrightarrow{CH_3-CH_2-I} \underset{\text{2-Pentyne}}{CH_3-C\equiv C-CH_2-CH_3} + Mg\,I_2$$

(4) **Ethers.** Grignard reagent reacts with halogenated ether to form higher ethers. For example,

$$CH_3CH_2MgI + ClCH_2OCH_3 \longrightarrow \underset{\text{Methoxy propane}}{CH_3CH_2CH_2OCH_3} + MgICl$$

A suitable combination of the Grignard reagent and halogenated ether can produce the desired ether.

(5) **Organolead, organocadmium and organosilicon compounds.** Such compounds can be obtained by the combination of Grignard reagent with metallic salts as illustrated below :

$$4C_2H_5MgI + 2PbCl_2 \longrightarrow \underset{\text{Tetraethyl lead}}{(C_2H_5)_4Pb} + Pb + 4\,MgICl$$

$$2C_2H_5MgBr + CdCl_2 \longrightarrow \underset{\text{Diethylcadmium}}{(C_2H_5)_2Cd} + 2\,MgBrCl$$

$$4CH_3MgBr + SiCl_4 \longrightarrow \underset{\text{Tetramethylsilicon}}{(CH_3)_4Si} + 4MgClBr$$

(6) **Alcohols.** (*a*) *Primary alcohols*. Primary alcohols can be obtained by one of the following methods:

(*i*) *By treating formaldehyde with Grignard reagent followed by decomposition with dilute acid.*

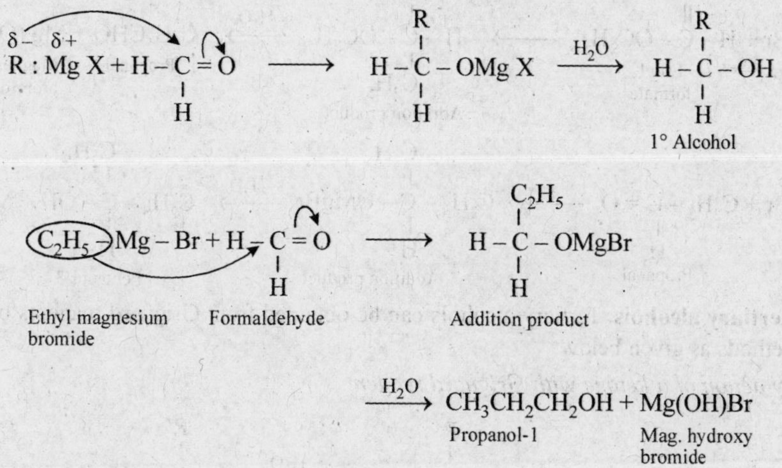

(*ii*) *By treating ethylene oxide with Grignard reagent and subjecting the product to hydrolysis.*

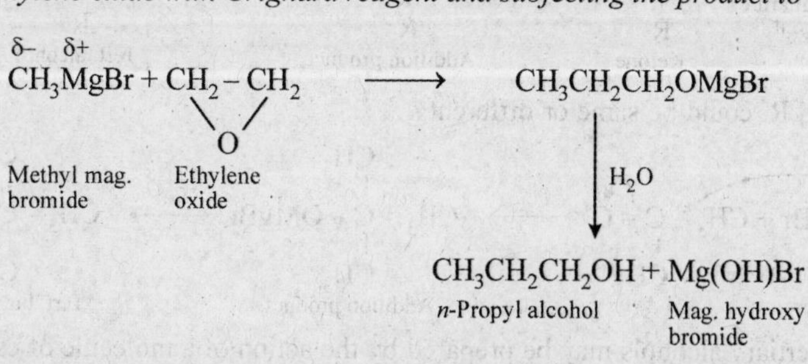

(*iii*) *By the reaction between Grignard reaction and oxygen followed by hydrolysis*

$$CH_3MgBr + \tfrac{1}{2}O_2 \longrightarrow CH_3OMgBr \xrightarrow{H_2O} \underset{\text{Methanol}}{CH_3OH} + Mg(OH)Br$$

(*b*) **Secondary alcohols.** Secondary alcohols can be obtained by one of the following methods.

(*i*) *By the action of an aldehyde other than formaldehyde followed by hydrolysis.*

$$\text{R Mg X} + \text{R}-\underset{\text{H}}{\underset{|}{\text{C}}}=\text{O} \longrightarrow \underset{\text{Addition product}}{\text{R}-\overset{\text{R}}{\overset{|}{\underset{\text{H}}{\underset{|}{\text{C}}}}}-\text{OMgX}} \xrightarrow{H_2O} \underset{\text{Sec. alcohol}}{\text{R}_2\text{CHOH}} + \text{Mg (OH) X}$$

$$\underset{\text{Methyl mag. bromide}}{CH_3MgBr} + \underset{\text{Acetaldehyde}}{CH_3-\underset{\text{H}}{\underset{|}{\text{C}}}=\text{O}} \longrightarrow \underset{\text{Addition product}}{CH_3-\overset{CH_3}{\overset{|}{\underset{\text{H}}{\underset{|}{\text{C}}}}}-\text{OMgBr}} \xrightarrow{H_2O} \underset{\text{Isopropyl alcohol}}{CH_3-\overset{CH_3}{\overset{|}{\underset{\text{H}}{\underset{|}{\text{C}}}}}-\text{OH}} + \text{Mg (OH) Br}$$

(*ii*) *By the action of ethyl formate on two molecules of Grignard reagent.* Ethyl formate with one molecule of Grignard reagent gives a molecule of aldehyde which then reacts further with the second molecule of Grignard reagent to produce secondary alcohol.

$$\underset{\text{Methyl mag. bromide}}{C_2H_5MgBr} + \underset{\text{Ethyl formate}}{\text{H}-\overset{\text{O}}{\overset{||}{\text{C}}}-OC_2H_5} \longrightarrow \underset{\text{Addition product}}{\text{H}-\overset{\text{OMgBr}}{\overset{|}{\underset{C_2H_5}{\underset{|}{\text{C}}}}}-OC_2H_5} \xrightarrow{H_2O} \underset{\text{Propanal}}{C_2H_5CHO} + \underset{\text{Magnesium ethoxy bromide}}{\text{Mg }(OC_2O_5)\text{ Br}}$$

$$\underset{\text{Ethyl mag. bromide}}{C_2H_5MgBr} + \underset{\text{Propanal}}{C_2H_5-\underset{\text{H}}{\underset{|}{\text{C}}}=\text{O}} \longrightarrow \underset{\text{Addition product}}{C_2H_5-\overset{C_2H_5}{\overset{|}{\underset{\text{H}}{\underset{|}{\text{C}}}}}-\text{OMgBr}} \xrightarrow{H_2O} \underset{\text{Pentanol-3}}{C_2H_5-\overset{C_2H_5}{\overset{|}{\underset{\text{H}}{\underset{|}{\text{C}}}}}-\text{OH}} + \text{Mg(OH)Br}$$

(*c*) **Tertiary alcohols.** Tertiary alcohols can be obtained from Grignard reagents by either of the two methods as given below:

(*i*) *By action of a ketone with Grignard reagent*

$$\underset{\text{Grignard reagent}}{\text{R}'\text{MgX}} + \underset{\text{Ketone}}{\text{R}-\underset{\text{R}}{\underset{|}{\text{C}}}=\text{O}} \longrightarrow \underset{\text{Addition product}}{\text{R}-\overset{\text{R}'}{\overset{|}{\underset{\text{R}}{\underset{|}{\text{C}}}}}-\text{OMgX}} \xrightarrow{H_2O} \underset{\text{Tert. alcohol}}{\text{R}-\overset{\text{R}'}{\overset{|}{\underset{\text{R}}{\underset{|}{\text{C}}}}}-\text{OH}} + \text{Mg (OH) X}$$

R and R′ could be same or different

$$(CH_3)\text{MgBr} + CH_3-\underset{CH_3}{\underset{|}{\text{C}}}=\text{O} \longrightarrow \underset{\text{Addition product}}{CH_3-\overset{CH_3}{\overset{|}{\underset{CH_3}{\underset{|}{\text{C}}}}}-\text{OMgBr}} \xrightarrow{H_2O} \underset{\text{Tert. butyl alcohol}}{CH_3-\overset{CH_3}{\overset{|}{\underset{CH_3}{\underset{|}{\text{C}}}}}-\text{OH}} + \text{Mg (OH) Br}$$

(*ii*) Tertiary alcohols may be prepared by the action of a molecule of ester (other than formic ester) with two molecules of Grignard reagent. First one molecule each of the ester and Grignard reagent react to form a ketone, which subsequently reacts with another molecule of Grignard reagent to produce a tertiary alcohol.

$$\underset{\text{Methyl mag. bromide}}{CH_3MgBr} + \underset{\text{Ethyl acetate}}{CH_3-\overset{\text{O}}{\overset{||}{\text{C}}}-OC_2H_5} \longrightarrow \underset{\text{Addition product}}{CH_3-\overset{\text{OMgBr}}{\overset{|}{\underset{CH_3}{\underset{|}{\text{C}}}}}-OC_2H_5} \xrightarrow{H_2O} \underset{\text{Acetone}}{CH_3-\overset{\text{O}}{\overset{||}{\underset{CH_3}{\underset{|}{\text{C}}}}}} + \underset{\text{Magnesium ethoxy bromide}}{\text{Mg }(OC_2H_5)\text{ Br}}$$

$$C_2H_5MgBr + \underset{\text{Acetone}}{CH_3-\underset{CH_3}{\overset{|}{C}}=O} \longrightarrow \underset{\text{Addition product}}{CH_3-\overset{CH_3}{\underset{CH_3}{C}}-OMgBr} \xrightarrow[-Mg(OH)Br]{H_2O} \underset{\text{Tert. butyl alcohol}}{CH_3-\overset{CH_3}{\underset{CH_3}{C}}-OH}$$

(7) Thioalcohols. Grignard reagent reacts with sulphur to form addition compound which on hydrolysis in the presence of an acid produces thioalcohol.

$$C_2H_5MgBr + S \longrightarrow C_2H_5SMgBr \xrightarrow{H_2O/H^+} \underset{\text{Ethanethiol}}{C_2H_5SH} + Mg(OH)Br$$

(8) Aldehydes. (*i*) *By the action of ethyl formate with Grignard reagent (in equimolar amount)*

$$\underset{\text{Gringard reagent}}{R\,MgX} + \underset{\text{Ethyl formate}}{H-\overset{O}{\overset{\|}{C}}-OC_2H_5} \longrightarrow \underset{\text{Addition product}}{H-\overset{OMgBr}{\underset{R}{C}}-OC_2H_5} \xrightarrow{H_2O} \underset{\text{Aldehyde}}{H-\overset{O}{\underset{R}{C}}} + \underset{\text{Mag. ethyoxy bromide}}{Mg(OC_2H_5)Br}$$

(*ii*) *A better yield of aldehyde is obtained if ethyl ortho formate is used instead of ethyl formate*

$$R-MgX + \underset{\text{Ethyl ortho formate}}{H-C(OC_2H_5)_3} \longrightarrow RCH(OC_2H_5)_2 + Mg\begin{matrix}X\\OC_2H_5\end{matrix} \xrightarrow{H_2O} RCHO + 2C_2H_5OH$$

(9) Ketones. Ketones may be obtained by either of the two methods given below:

(*i*) *By the action of acid chloride on Grignard reagent*

$$CH_3MgCl + CH_3-\underset{Cl}{\underset{|}{C}}=O \longrightarrow \underset{\text{Addition product (unstable)}}{CH_3-\overset{CH_3}{\underset{Cl}{C}}-OMgCl} \xrightarrow{\text{Changes to}} \underset{\text{Acetone}}{CH_3-\overset{CH_3}{\overset{|}{C}}=O} + MgCl_2$$

(*ii*) *By the action of alkyl cyanide on Grignard reagent.*

$$CH_3MgBr + CH_3-C\equiv N \longrightarrow \underset{\text{Addition product}}{CH_3-\underset{CH_3}{\underset{|}{C}}=NMgBr} \xrightarrow{H_2O} \underset{\text{Acetone}}{CH_3-\underset{CH_3}{\underset{|}{C}}=O} + \underset{\text{Magnesium amino bromide}}{Mg\begin{matrix}Br\\NH_2\end{matrix}}$$

(10) Carboxylic acids. By the action of CO_2 on Grignard reagent followed by hydrolysis.

$$RMg\,X + O=C=O \longrightarrow \underset{\text{[Addition product]}}{O=\overset{R}{\overset{|}{C}}-OMgX} \xrightarrow{H_2O} \underset{\text{Acid}}{RCOOH} + Mg(OH)\,X$$

(11) Primary amines. Primary amines may be prepared by the action of chloramine on Grignard reagent.

$$R\,MgX + \underset{\text{Chloramine}}{Cl-NH_2} \longrightarrow \underset{\text{Primary amine}}{R-NH_2} + Mg\begin{matrix}X\\Cl\end{matrix}$$

(12) Esters. On treatment with chloroformic ester, Grignard reagents produce esters.

$$Cl-\overset{O}{\overset{\|}{C}}-OC_2H_5 + CH_3MgBr \longrightarrow Cl-\underset{CH_3}{\underset{|}{\overset{OMgBr}{\overset{|}{C}}}}-OC_2H_5 \xrightarrow{-Mg(Br)Cl} \underset{\text{Ethyl acetate}}{CH_3\overset{O}{\overset{\|}{C}}-OC_2H_5}$$

Limitations of Grignard Reagents

Grignard reagents suffer from some limitations as given below :

1. Grignard reagents are highly sensitive and somewhat explosive and therefore difficult to handle
2. Alkyl or aryl halides used in the preparation of Grignard reagent should be free from substituent groups such as $-NH_2$, –COOH, –OH, –CHO, –CN etc, since these groups themselves react with Grignard reagent
3. Nitro compounds cannot be used as solvent or otherwise in reactions with Grignard reagent since they oxidise the Grignard reagents
4. Steric hindrance takes place when the Grignard reagent contains bulky groups or when there is branching near the functional group in the other compound. Thus the reaction does not take place.

Example1. How can you obtain the following from Grignard reagent?

***(i)* Dithionic acid** ***(ii)* Tert. butyl alcohol**

Solution. (*i*) Dithionic acid may be prepared by the action of carbon disulphide on a Grignard reagent, followed by hydrolysis.

$$R-MgX + \overset{S}{\overset{\|}{C}}=S \longrightarrow \underset{\text{Addition product}}{R-\overset{S}{\overset{\|}{C}}-SMgX} \xrightarrow{H_2O} \underset{\text{Dithionic acid}}{R-\overset{S}{\overset{\|}{C}}-SH} + Mg\begin{matrix}OH\\X\end{matrix}$$

$$CH_3MgBr + \overset{S}{\overset{\|}{C}}=S \longrightarrow CH_3-\overset{S}{\overset{\|}{C}}-SMgX \xrightarrow{H_2O} \underset{\text{Dithionic acid}}{CH_3-\overset{S}{\overset{\|}{C}}-SH} + Mg(OH)Br$$

(*ii*) Tert. butyl alcohol

See the previous section under tertiary alcohol.

Example 2. What happens when methyl magnesium bromide is reacted with (*i*) CO_2 (*ii*) $HCOOC_2H_5$ (*iii*) CH_3CHO.

Solution. (*i*)

$$CH_3MgBr + \overset{O}{\overset{\|}{C}}=O \longrightarrow CH_3-\overset{O}{\overset{\|}{C}}-OMgBr \xrightarrow{H_2O} \underset{\text{Acetic acid}}{CH_3-\overset{O}{\overset{\|}{C}}-OH} + Mg\begin{matrix}OH\\Br\end{matrix}$$

(*ii*)

$$CH_3MgBr + H-\overset{O}{\overset{\|}{C}}-OC_2H_5 \longrightarrow H-\underset{CH_3}{\underset{|}{\overset{OMgBr}{\overset{|}{C}}}}-OC_2H_5 \longrightarrow \underset{\text{Acetaldehyde}}{CH_3CHO} + \underset{\text{Mag. ethoxy bromide}}{Mg(OC_2H_5)Br}$$

(*iii*)

$$CH_3MgBr + CH_3-\underset{H}{\underset{|}{C}}=O \longrightarrow CH_3-\underset{H}{\underset{|}{\overset{CH_3}{\overset{|}{C}}}}-OMgBr \xrightarrow{H_2O} \underset{\text{Isopropyl alcohol}}{CH_3-\underset{H}{\underset{|}{\overset{CH_3}{\overset{|}{C}}}}-OH} + Mg(OH)Br$$

Example 3. Selecting a suitable organometallic compound, how will you prepare isobutyric acid ?

Solution. Isobutyric acid will be obtained by treating isopropyl magnesium bromide with CO_2, followed by hydrolysis.

$$\begin{matrix}CH_3\\CH_3\end{matrix}\!\!>CHMgBr + \overset{\overset{\displaystyle O}{\|}}{C}=O \longrightarrow \begin{matrix}CH_3\\CH_3\end{matrix}\!\!>CH-\overset{\overset{\displaystyle O}{\|}}{C}-OMgBr \xrightarrow{H_2O} \underset{\text{Isobutyric acid}}{\begin{matrix}CH_3\\CH_3\end{matrix}\!\!>CH-\overset{\overset{\displaystyle O}{\|}}{C}-OH} + Mg(OH)Br$$

Example 4. Write equations to show the reaction of CH_3MgBr with each of the following:

(*i*) $SiCl_4$ (*ii*) C_2H_5OH (*iii*) $\underset{\displaystyle \searrow O \swarrow}{CH_2-CH_2}$ (*iv*) $CH_3COOC_2H_5$

Solution. (*i*) $4CH_3MgBr + SiCl_4 \longrightarrow \underset{\text{Tetra methyl silane}}{(CH_3)_4Si} + \underset{\text{Magnesium chloro bromide}}{4Mg(Cl)Br}$

(*ii*) $$CH_3MgBr + C_2H_5OH \longrightarrow \underset{\text{Methane}}{CH_4} + Mg\begin{matrix}\diagup OC_2H_5\\ \diagdown Br\end{matrix}$$

(*iii*)

$$CH_3MgBr + \underset{\text{Ethylene oxide}}{\underset{\displaystyle \searrow O \swarrow}{CH_2-CH_2}} \longrightarrow CH_3CH_2CH_2OMgBr \xrightarrow{H_2O} \underset{\text{Propanol-1}}{CH_3CH_2CH_2OH} + Mg(OH)Br$$

(*iv*)

$$CH_3MgBr + CH_3COOC_2H_5 \longrightarrow CH_3\overset{\overset{\displaystyle OMgBr}{|}}{\underset{\underset{\displaystyle CH_3}{|}}{C}}-OC_2H_5 \xrightarrow[\text{to}]{\text{changes}} \underset{\text{Acetone}}{CH_3-\overset{\overset{\displaystyle O}{\|}}{C}-CH_3} + Mg(OC_2H_5)Br$$

Example 5. Using suitable RMg X prepare the following:

(*i*) $CH_3CH_2CHOHCH_3$ (*ii*) $CH_3CH_2CH_2COOH$

Solution. (*i*) $\underset{\text{Butanol-2}}{CH_3CH_2CHOHCH_3}$

Secondary alcohol can be obtained by treating Grignard reagent with an aldehyde other than formaldehyde.

$$CH_3MgBr + CH_3CH_2CHO \longrightarrow CH_3CH_2-\overset{\overset{\displaystyle H}{|}}{\underset{\underset{\displaystyle CH_3}{|}}{C}}-OMgBr \xrightarrow{H_2O} \underset{\text{Butanol-2}}{CH_3CH_2-\overset{\overset{\displaystyle H}{|}}{\underset{\underset{\displaystyle CH_3}{|}}{C}}-OH} + Mg(OH)Br$$

(*ii*) $CH_3CH_2CH_2COOH$ (Butyric acid)

$$\underset{\text{Propyl mag. bromide}}{CH_3CH_2CH_2MgBr} + \overset{\overset{\displaystyle O}{\|}}{C}=O \longrightarrow CH_3CH_2CH_2-\overset{\overset{\displaystyle O}{\|}}{C}-OMgBr$$

$$\xrightarrow{H_2O} \underset{\text{Butyric acid}}{CH_3CH_2CH_2-\overset{\overset{\displaystyle O}{\|}}{C}-OH} + Mg(OH)Br$$

Example 6. How can you obtain thioalcohols and sulphanilic acid using Grignard reagents?

Solution. (*i*) Preparation of thioalcohols.

$$\text{R MgX} + \text{S} \longrightarrow \text{RSMgX} \xrightarrow{H_2O} \underset{\text{Thio alcohol}}{\text{RSH}} + \text{Mg(OH) X}$$

$$CH_3MgBr + S \longrightarrow CH_3SMgBr \xrightarrow{H_2O} \underset{\text{Methyl thioalcohol}}{CH_3SH} + Mg(OH)Br$$

(*ii*) Preparation of sulphanilic acid

$$CH_3MgBr + \overset{O}{\overset{\|}{S}} = O \longrightarrow CH_3 - \overset{O}{\overset{\|}{S}} - OMgBr \xrightarrow{H_2O} \underset{\text{Sulphanilic acid}}{CH_3 - \overset{O}{\overset{\|}{S}} - OH} + Mg\begin{matrix} Br \\ OH \end{matrix}$$

Example 7. Starting from ethyl magnesium iodide, how will you obtain each of the following:

(*i*) 2-Pentyne **(*ii*) Propanoic acid**

(*iii*) 3-Pentanone **(*iv*) 2-Methyl-2-butanol**

(*v*) Ethane **(*vi*) 1-Pentene**

Solution. (*i*) **2-Pentyne**

$$\underset{\text{Propyne}}{CH_3 - C \equiv CH} + CH_3 - CH_2 - MgI \longrightarrow \underset{\text{Propynyl magnesium iodide}}{CH_3 - C \equiv C - MgI} + CH_3 - CH_3$$

$$CH_3 - C \equiv C - MgI + \underset{\text{Ethyl iodide}}{I - CH_2 - CH_3} \longrightarrow \underset{\text{2-Pentyne}}{CH_3 - C \equiv C - CH_2 - CH_3} + MgI_2$$

(*ii*) Propanoic acid

$$O = C = O + \underset{\text{Ethyl magnesium iodide}}{CH_3 - CH_2 - MgI} \longrightarrow \left[CH_3 - CH_2 - \overset{O}{\overset{\|}{C}} - OMgI \right] \xrightarrow[H_2O, H^+]{-Mg(OH)I} \underset{\text{Propanoic acid}}{CH_3 - CH_2 - \overset{O}{\overset{\|}{C}} - OH}$$

(*iii*) 3-Pentanone

$$\underset{\text{Propanenitrile}}{CH_3 - CH_2 - C \equiv N} + CH_3 - CH_2 - MgI \longrightarrow \left[CH_3 - CH_2 - \underset{\underset{CH_3}{|}}{\underset{CH_2}{\underset{|}{C}}} = NMgI \right]$$

Addition product

$$\downarrow -Mg(OH)I \quad HOH, H^+$$

$$\underset{\text{3-Pentanone}}{CH_3 - CH_2 - \underset{\underset{CH_3}{|}}{\underset{CH_2}{\underset{|}{C}}} = O} \xleftarrow[-NH_3]{HOH, H^+} \left[CH_3 - CH_2 - \underset{\underset{CH_3}{|}}{\underset{CH_2}{\underset{|}{C}}} = NH \right]$$

(iv) 2-Methyl-2-butanol

$$\underset{\text{Acetone}}{CH_3-\overset{\overset{CH_3}{|}}{C}=O} + CH_3-CH_2-MgI \longrightarrow \underset{\text{Addition product}}{CH_3-\overset{\overset{CH_3}{|}}{\underset{\underset{\underset{CH_3}{|}}{\underset{CH_2}{|}}}{C}}-OMgI} \xrightarrow[\text{HOH, }H^+]{-\,Mg(OH)} \underset{\text{2-Methyl-2-butanol}}{CH_3-\overset{\overset{CH_3}{|}}{\underset{\underset{\underset{CH_3}{|}}{\underset{CH_2}{|}}}{C}}-OH}$$

(v) Ethane

$$HOH + CH_3-CH_2-MgI \longrightarrow \underset{\text{Ethane}}{CH_3-CH_3} + Mg(OH)I$$

(vi) 1-Pentene

$$CH_2=CH-CH_2-I + CH_3-CH_2-MgI \longrightarrow \underset{\text{1-Pentene}}{CH_2=CH-CH_2-CH_2-CH_3} + MgI_2$$

Example 8. Explain, giving equations, what happens when:

(i) Phenyl magnesium bromide is treated with ethylene oxide and the product formed is hydrolysed.

(ii) Ethyl magnesium bromide is treated with ethyl formate and the product is hydrolysed.

(iii) Ethyl magnesium iodide is treated with chloromethyl ether.

(iv) Ethyl magnesium bromide is treated with cadmium chloride.

(v) Methyl magnesium iodide is treated with ethyl amine.

(vi) Methyl magnesium iodide is treated with acetyl chloride and the product formed is hydrolysed.

Solution.

(i)

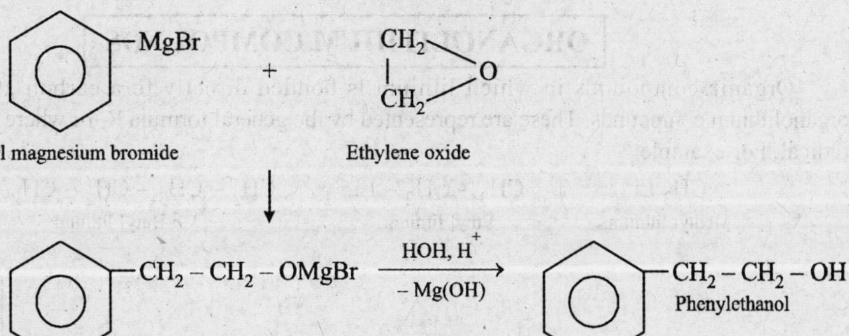

(ii)

$$\underset{\text{Ethyl formate}}{H-\overset{\overset{O}{\|}}{C}-OC_2H_5} + CH_3-CH_2-MgBr \longrightarrow \left[H-\overset{\overset{OMg\,Br}{|}}{\underset{\underset{\underset{CH_3}{|}}{\underset{CH_2}{|}}}{C}}-OC_2H_5\right] \xrightarrow{-\,Mg(OC_2H_5)\,Br}$$

$$\underset{\text{3 - Pentanol}}{CH_3-CH_2-\overset{\overset{OH}{|}}{CH}-CH_2-CH_3} \xleftarrow[\text{HOH, }H^+]{-\,Mg(OH)} H-\overset{\overset{OMgBr}{|}}{\underset{\underset{\underset{CH_3}{|}}{\underset{CH_2}{|}}}{C}}-CH_2-CH_3 \xleftarrow{CH_3\,CH_2\,MgBr} \underset{\text{Propanal}}{H-\overset{\overset{O}{\|}}{C}-CH_2-CH_3}$$

(*iii*) $CH_3 - CH_2 - MgI + Cl - CH_2 - O - CH_3 \longrightarrow CH_3 - CH_2 - CH_2 - O - CH_3 + Mg(I)Cl$
Ethyl magnesium iodide; Chloromethyl ether; 1-Methoxypropane (Methyl n-propyl ether)

(*iv*) $2C_2H_5MgBr + CdCl_2 \longrightarrow (C_2H_5)_2Cd + 2Mg(Br)Cl$
Ethyl magnesium bromide; Diethyl Cadmium

(*v*) $C_2H_5NH_2 + CH_3MgI \longrightarrow CH_4 + Mg(NHC_2H_5)I$
Ethylamine

(*vi*)

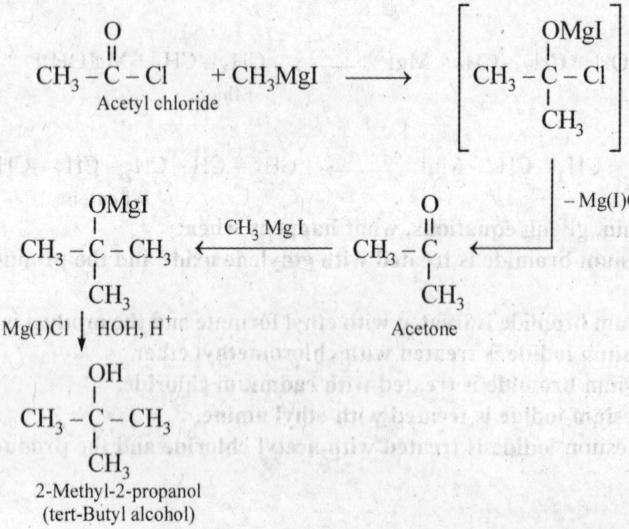

ORGANOLITHIUM COMPOUNDS

Organic compounds in which lithium is bonded directly to a carbon atom are called organolithium compounds. These are represented by the general formula R–Li where R is the organic radical. For example

$CH_3\ Li$	$CH_3 - CH_2 - Li$	$CH_3 - CH_3 - CH_2 - CH_2 - Li$
Methyl lithium	Ethyl lithium	*n*-Butyl lithium

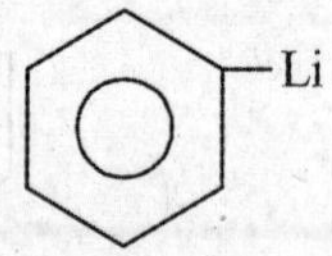

Phenyl lithium

Due to greater ionic character of C–Li bond, organolithium compounds are more reactive than Grignard reagent and thus undergo a few additional useful reactions.

2.4. PREPARATION OF ORGANOLITHIUM COMPOUNDS

These compounds are generally prepared by the following methods:

(1) By reaction between lithium and an alkyl halide. Alkyl lithium can be prepared by treating a suitable alkyl halide with lithium in the presence of dry ether or benzene at low temperature and under an inert atmosphere of nitrogen or argon. For example,

$$R - Br + 2Li \xrightarrow[263\ K]{\text{Dry ether, } N_2} R - Li + LiBr$$

Alkyl lithium

$$CH_3CH_2CH_2CH_2Cl + 2Li \xrightarrow[263\ K]{Benzene,\ N_2} \underset{n\text{-Butyllithium}}{CH_3CH_2CH_2CH_2Li} + LiCl$$

(2) By exchange method. Aryl lithium are usually prepared by treating an aryl halide with an alkyl lithium. For example,

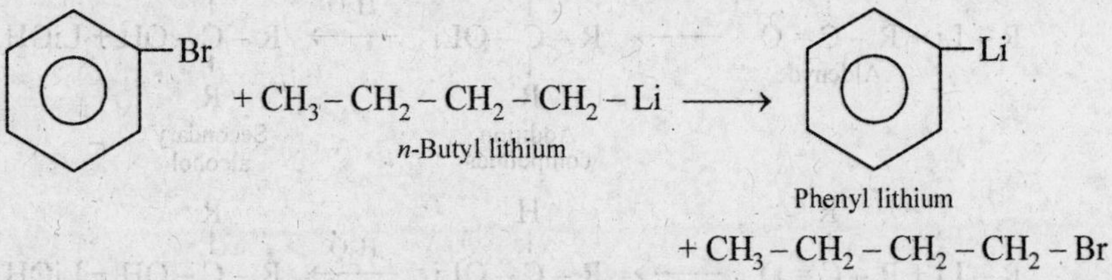

(3) By metal exchange method. Alkyl lithiums can be readily prepared by the action of lithium on organometallic compounds of a less reactive metal such as mercury. For example,

$$\underset{\text{Diethylmercury}}{(CH_3CH_2)_2Hg} + 2Li \longrightarrow \underset{\text{Diethyllithium}}{2CH_3CH_2Li} + Hg$$

Physical properties

1. Organolithium compounds exist as dimers or even higher aggregates of molecules $(RLi)_n$ where n is a fairly large number. This is due to strong polarization of C–Li bond and small size of lithium atom.
2. Organolithium compounds are colourless liquids or solids with low melting points.
3. They are highly soluble in hydrocarbon and ether solvents.

2.5. SYNTHETIC APPLICATIONS (CHEMICAL PROPERTIES) OF ORGANOLITHIUM COMPOUNDS

Organolithium compounds are more reactive than Grignard reagents as the former have greater ionic character than the latter.

Important reactions of organolithium compounds are given below.

1. Reactions with compounds containing active hydrogen.

$$R - Li + H_2O \longrightarrow R - H + LiOH$$

$$\underset{n\text{-Butyl lithium}}{CH_3CH_2CH_2CH_2Li} + H_2O \longrightarrow \underset{n\text{-Butane}}{C_4H_{10}} + LiOH$$

2. Formation of alcohols.

$$R - Li + \underset{\text{Formaldehyde}}{H - \overset{\displaystyle H}{\overset{|}{C}} = O} \longrightarrow \underset{\text{Addition compounds}}{H - \underset{\displaystyle R}{\underset{|}{\overset{\displaystyle H}{\overset{|}{C}}}} - OLi} \xrightarrow{H_2O} \underset{\text{Primary alcohol}}{H - \underset{\displaystyle R}{\underset{|}{\overset{\displaystyle H}{\overset{|}{C}}}} - OH} + LiOH$$

$$\underset{n\text{-Butyl lithium}}{C_4H_9Li} + H - \overset{\displaystyle H}{\overset{|}{C}} = O \longrightarrow H - \underset{\displaystyle C_4H_9}{\underset{|}{\overset{\displaystyle H}{\overset{|}{C}}}} - OLi \xrightarrow[- LiOH]{H_2O} \underset{\text{Pentanol-1 (Primary alcohol)}}{H - \underset{\displaystyle C_4H_9}{\underset{|}{\overset{\displaystyle H}{\overset{|}{C}}}} - OH} + LiOH$$

$$\underset{\text{Phenyl lithium}}{C_6H_5Li} + H-\overset{\displaystyle H}{\overset{|}{C}}=O \longrightarrow H-\overset{\displaystyle H}{\overset{|}{\underset{\displaystyle C_6H_5}{\underset{|}{C}}}}-OLi \xrightarrow[-\,LiOH]{H_2O} \underset{\text{Benzyl alcohol}}{C_6H_5CH_2OH}$$

$$R-Li+\underset{\text{Aldehyde}}{R-\overset{\displaystyle H}{\overset{|}{C}}=O} \longrightarrow \underset{\text{Addition compounds}}{R-\overset{\displaystyle H}{\overset{|}{\underset{\displaystyle R}{\underset{|}{C}}}}-OLi} \xrightarrow{H_2O} \underset{\text{Secondary alcohol}}{R-\overset{\displaystyle H}{\overset{|}{\underset{\displaystyle R}{\underset{|}{C}}}}-OH} + LiOH$$

$$R-Li+\underset{\text{Ketone}}{R-\overset{\displaystyle R}{\overset{|}{C}}=O} \longrightarrow R-\overset{\displaystyle H}{\overset{|}{\underset{\displaystyle R}{\underset{|}{C}}}}-OLi \xrightarrow{H_2O} \underset{\text{Tertiary alcohol}}{R-\overset{\displaystyle R}{\overset{|}{\underset{\displaystyle R}{\underset{|}{C}}}}-OH} + LiOH$$

There is smaller steric hindrance involved in the reactions with organo lithium compounds. While Grignard reagents do not react with ketones containing bulky groups, organolithium compounds react to form tertiary alcohols.

3. Formation of carboxylic acid.

$$R-Li+O=C=O \longrightarrow O=\overset{\displaystyle R}{\overset{|}{C}}-OLi \xrightarrow{H_2O} \underset{\text{Carboxy acid}}{O=\overset{\displaystyle R}{\overset{|}{C}}-OH} + LiOH$$

If excess of alkyl lithium is used, another molecule of R–Li may add to the addition product leading to the formation of ketone.

$$R-Li+R-\overset{\displaystyle O}{\overset{||}{C}}-OLi \longrightarrow R-\overset{\displaystyle OLi}{\overset{|}{\underset{\displaystyle R}{\underset{|}{C}}}}-OLi \xrightarrow{2H_2O} \underset{\text{Ketone}}{R-\underset{\displaystyle R}{\underset{|}{C}}=O} + 2LiOH$$

4. Reaction with epoxides.

$$R-Li+\underset{\diagdown\ O\ \diagup}{CH_2-CH_2} \longrightarrow R-CH_2-CH_2OLi \xrightarrow{H_2O} \underset{\text{Primary alcohol}}{RCH_2CH_2OH} + LiOH$$

5. Reaction with alkenes.

$$R-Li+CH_2=CH_2 \longrightarrow R-CH_2-CH_2\,Li \xrightarrow{CH_2=CH_2} R-CH_2-CH_2-CH_2-CH_2-Li$$

and so on.

This reaction can be used for preparing long-chain alkanes.

6. Formation of aldehydes.

Organolithium compounds react with dimethyl formamide to form aldehydes.

$$C_6H_5Li+H-\overset{\displaystyle O}{\overset{||}{C}}-N\begin{matrix} CH_3 \\ CH_3 \end{matrix} \longrightarrow C_6H_5-\overset{\displaystyle OLi}{\overset{|}{\underset{\displaystyle H}{\underset{|}{C}}}}-\underset{\displaystyle CH_3}{\underset{|}{N}}-CH_3$$

$$\xrightarrow{H_2O} \underset{\text{Benzaldehyde}}{C_6H_5-\overset{\displaystyle O}{\overset{||}{C}}-H} + LiOH + \underset{\text{Dimethyl amine}}{(CH_3)_2NH}$$

Aldehydes are also formed by the reaction of alkyl lithium with hydrogen cyanide or ethyl orthoformate.

$$H-C\equiv N + \overset{\delta-}{CH_3}-\overset{\delta+}{Li} \longrightarrow H-\underset{\substack{|\\ CH_3}}{C}=NLi \xrightarrow{H_3\overset{+}{O}} \underset{\text{Acetaldehyde}}{CH_3CHO} + NH_3 + LiOH$$

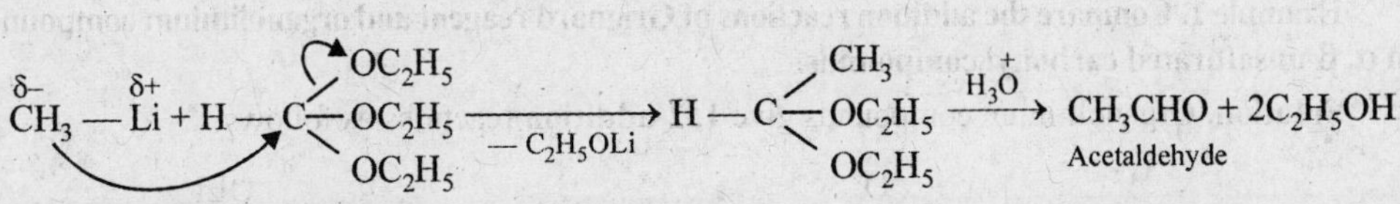

7. Formation of ketones.

$$\underset{\substack{\text{Phenyl}\\ \text{lithium}}}{C_6H_5-Li} + \underset{\text{Methyl cyanide}}{CH_3-C\equiv N} \longrightarrow CH_3-\overset{\substack{C_6H_5\\ |}}{C}=NLi \xrightarrow{H_2O} \underset{\substack{\text{Methyl phenyl}\\ \text{Ketone}}}{CH_3-\overset{\substack{C_6H_5\\ |}}{C}=O} + \underset{\substack{\text{Lithium}\\ \text{amide}}}{LiNH_2}$$

Ketones are also by reaction with orthoesters other than orthoformic ester

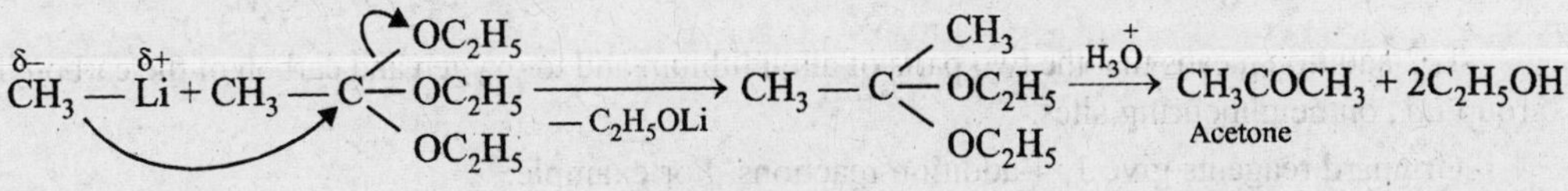

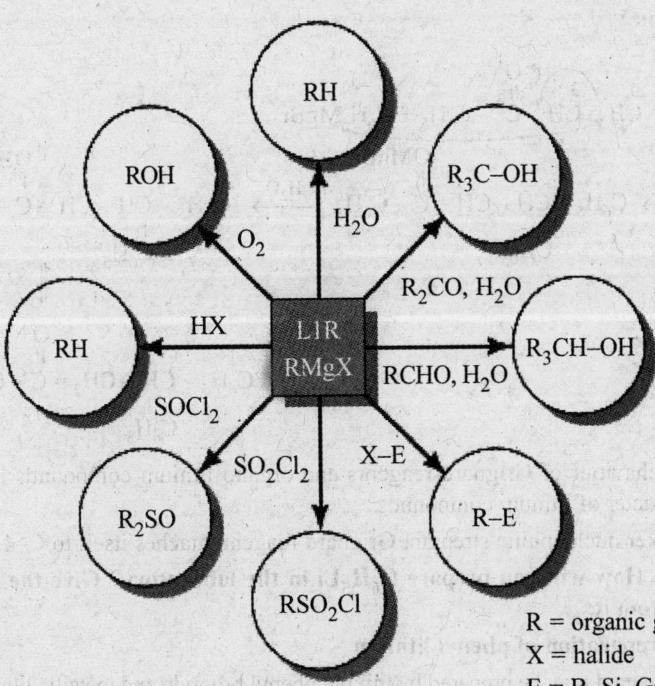

R = organic group
X = halide
E = B, Si, Ge, Sn, Pb, As, Sb

Synthetic applications of Grignard reagents and alkyl lithiums

8. Reaction with pyridine.

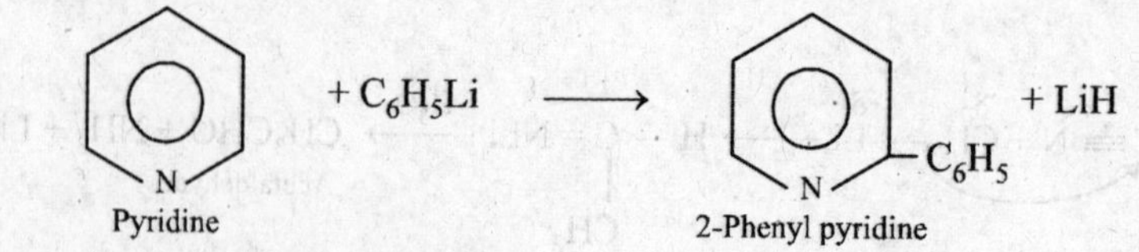

Example 1. Compare the addition reactions of Grignard reagent and organolithium compounds on α, β-unsaturated carbonyl compounds.

Solution. Organolithium compounds give **1, 2-addition** reactions as follows:

$$\underset{\text{Benzal acetophenone}}{C_6H_5-CH=CH-\overset{\overset{1}{O}}{\overset{\|}{^{2}C}}-C_6H_5} \xrightarrow{C_6H_5Li} \underset{\text{Addition product}}{C_6H_5-CH=CH-\underset{C_6H_5}{\overset{OLi}{C}}-C_6H_5}$$

$$\xrightarrow{H_2O} C_6H_5-CH=CH-\underset{C_6H_5}{\overset{OH}{C}}-C_6H_5$$

1, 2-addition means that the two parts of alkyl lithium add to oxygen and carbon of the carbonyl group *i.e.*, on neighbouring sites.

Grignard reagents give 1, 4-addition reactions. For example:

1. 4-addition

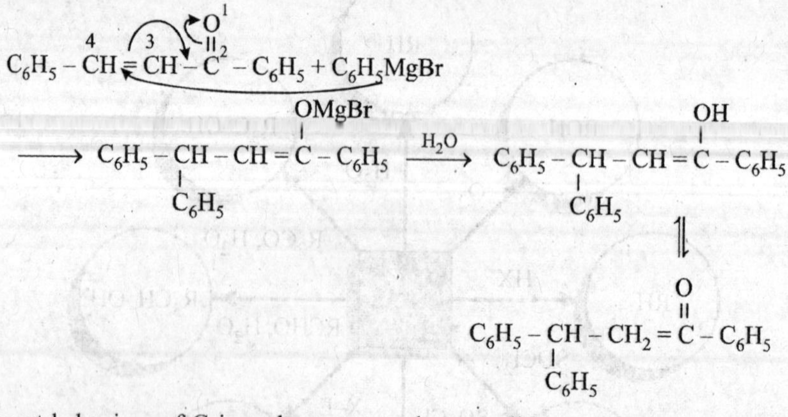

Different behaviour of Grignard reagents and organo lithium compounds is due to greater nucleophilic character of lithium compounds.

Due to weaker nucleophilic strength, Grignard reagent attaches itself to C–4 instead of C–2.

Example 2. How will you prepare C_6H_5Li in the laboratory? Give the synthesis of *m*-phenyl anisole from it.

Solution. Preparation of phenyl lithium

(*i*) This compound may be prepared by mixing phenyl bromide and metallic lithium in the molar ratio of 1 : 2 respectively in dry ether as solvent at –10°C in an inert atmosphere of N_2.

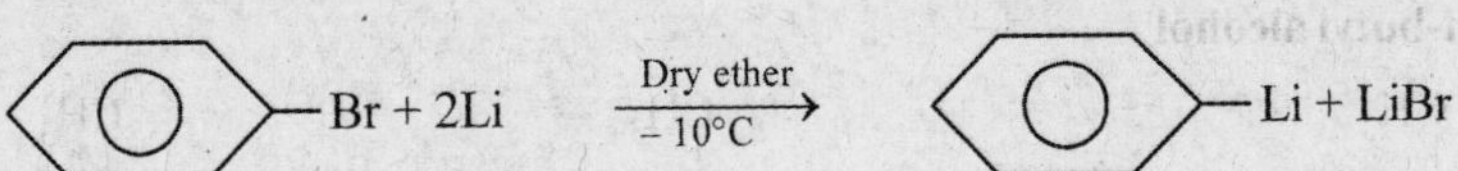

(*ii*) A more convenient and efficient method would be to treat bromobenzene with *n*-butyl lithium.

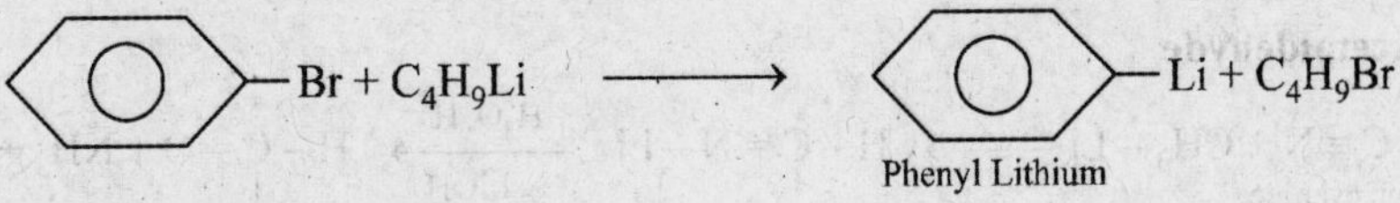

Synthesis of *m*-phenyl anisole

This reaction involves benzyne mechanism

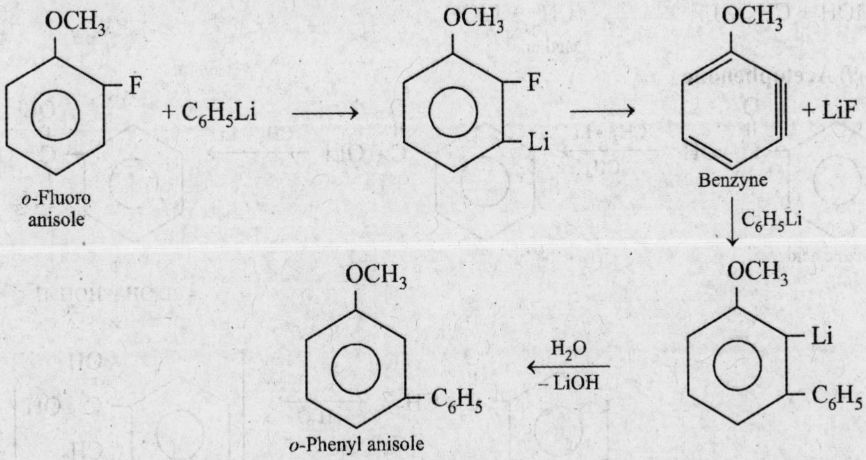

Example 3. With the help of methyllithium how can you prepare the following compounds:

(*i*) Ethanol	**(*ii*) Iso-propyl alcohol**
(*iii*) tert-Butyl alcohol	**(*iv*) Acetaldehyde**
(*v*) Methane	**(*vi*) Acetophenone**

Solution. (*i*) Ethanol

$$\underset{\text{Formaldehyde}}{H-\overset{\overset{H}{|}}{C}=O} + CH_3-Li \longrightarrow H-\underset{\underset{CH_3}{|}}{\overset{\overset{H}{|}}{C}}-OLi \xrightarrow[-\,LiOH]{HOH,\ H^+} \underset{\text{Ethanol}}{CH_3-CH_2-OH}$$

(*ii*) Iso-propyl alcohol

$$\underset{\text{Acetaldehyde}}{CH_3-\overset{\overset{H}{|}}{C}=O} + CH_3-Li \longrightarrow CH_3-\underset{\underset{CH_3}{|}}{\overset{\overset{H}{|}}{C}}-OLi \xrightarrow[-\,LiOH]{HOH,\ H^+} \underset{\text{Iso-propyl alcohol}}{CH_3-\underset{\underset{CH_3}{|}}{\overset{\overset{H}{|}}{C}}-OH}$$

(*iii*) Tert-butyl alcohol

$$\underset{\text{Acetone}}{CH_3-\overset{\overset{CH_3}{|}}{C}=O} + CH_3-Li \longrightarrow CH_3-\underset{\underset{CH_3}{|}}{\overset{\overset{CH_3}{|}}{C}}-OLi \xrightarrow[-\,LiOH]{H_2O,\,H^+} \underset{\text{Tert. Butyl alcohol}}{CH_3-\underset{\underset{CH_3}{|}}{\overset{\overset{CH_3}{|}}{C}}-OH}$$

(*iv*) Acetaldehyde

$$\underset{\text{Hydrogen cyanide}}{H-C\equiv N} + CH_3-Li \longrightarrow H-\underset{\underset{CH_3}{|}}{C}=N-Li \xrightarrow[-\,LiOH]{H_2O,\,H^+} \underset{\text{Acetaldehyde}}{H-\underset{\underset{CH_3}{|}}{C}=O} + NH_3 + LiOH$$

(*v*) Methane

$$HOH + CH_3-Li \longrightarrow \underset{\text{Methane}}{CH_4} + LiOH$$

(*vi*) Acetophenone

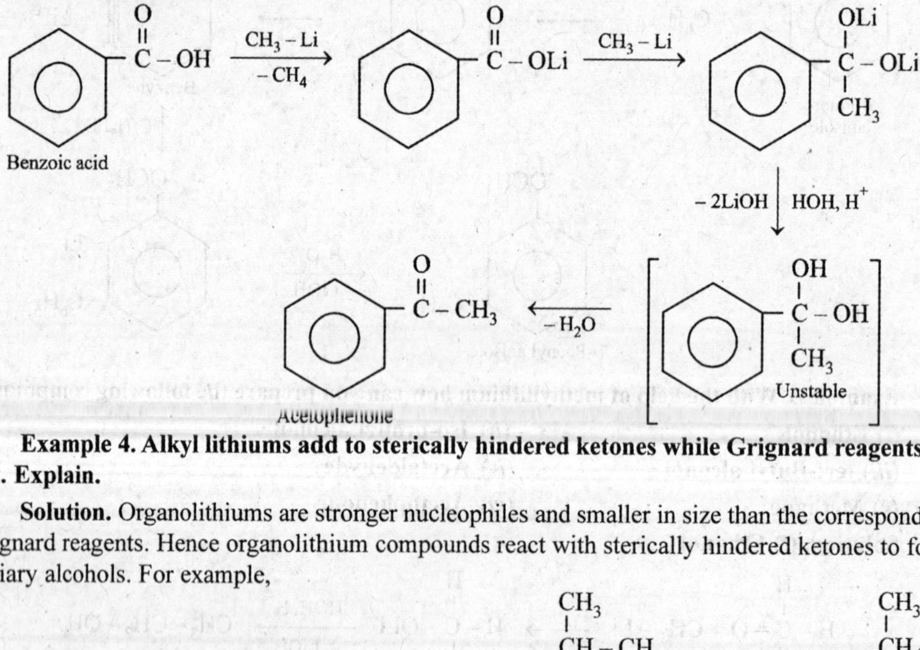

Example 4. Alkyl lithiums add to sterically hindered ketones while Grignard reagents do not. Explain.

Solution. Organolithiums are stronger nucleophiles and smaller in size than the corresponding Grignard reagents. Hence organolithium compounds react with sterically hindered ketones to form tertiary alcohols. For example,

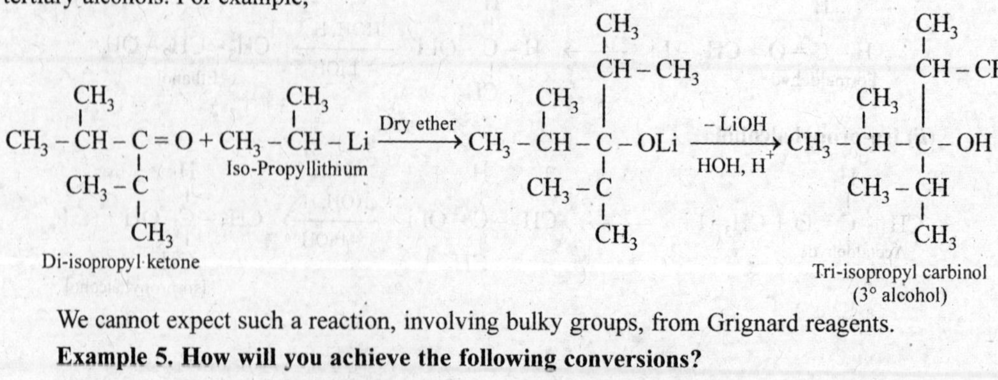

We cannot expect such a reaction, involving bulky groups, from Grignard reagents.

Example 5. How will you achieve the following conversions?

(*i*) Ethyl magnesium bromide into ethanethiol (Ethyl mercaptan).

(*ii*) Methyl magnesium bromide into methyl amine.

(*iii*) Methyl lithium into ethane nitrile.

(iv) Ethyllithium into butanone.

(v) Ethyl magnesium iodide into ethanesulphinic acid.

Solution. (i) Ethanethiol

$$CH_3-CH_2-Mg\,Br+S\longrightarrow CH_3-CH_2-S-MgBr\xrightarrow[-\,Mg\,(OH)\,Br]{HOH,\,H^+}\underset{\text{Ethanethi}}{CH_3-CH_2-SH}$$

(ii) Methylamine

$$CH_3-Mg\,Br+\underset{\text{Chloramine}}{Cl-NH_2}\longrightarrow\underset{\text{Methylamine}}{CH_3-NH_2}+Mg\,(Br)\,Cl$$

(iii) Ethanenitrile

$$CH_3-Li+\underset{\text{Cyanogen chloride}}{Cl\,CN}\longrightarrow LiCl+\underset{\text{Ethanenitrile}}{CH_3\,CN}$$

(iv) Butanone

$$\underset{\substack{\text{Ethanenitrile}\\\text{(Acetonitrile)}}}{CH_3-C\equiv N}+\underset{\text{Ethyl lithium}}{CH_3-CH_2-Li}\longrightarrow CH_3-\underset{\substack{|\\CH_2\\|\\CH_3}}{C}=N-Li\xrightarrow{HOH,\,H^+}\underset{\text{Butanone}}{CH_3-\overset{\substack{O\\\|}}{C}-CH_2-CH_3}+NH_3+LiOH$$

(v) Ethanesulphinic acid

$$\overset{\substack{O\\\|}}{S}=O+CH_3-CH_2-MgI\longrightarrow CH_3-CH_2-\overset{\substack{O\\\|}}{S}-OMgI\xrightarrow[-\,Mg\,(OH)I]{HOH,\,H^+}\underset{\text{Ethanesulphinic acid}}{CH_3-CH_2-\overset{\substack{O\\\|}}{S}-OH}$$

Example 6. How will you prepare the following starting from phenyllithium

(i) 2-Phenyl pyridine (ii) Benzoic acid

Solution. Preparation of 2-Phenylpyridine

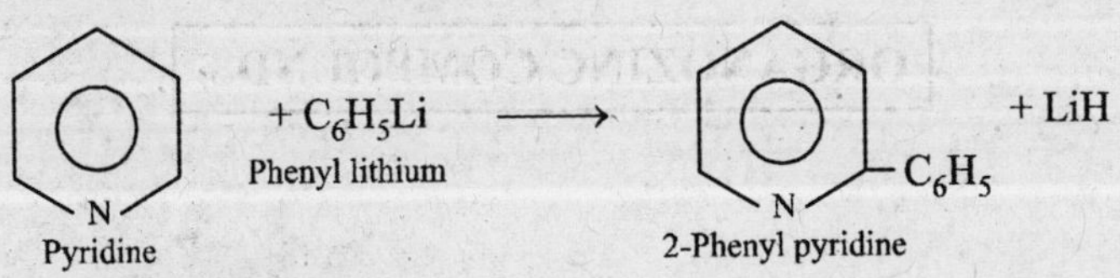

Preparation of benzoic acid

$$\underset{\substack{\text{Phenyl}\\\text{Lithium}}}{C_6H_5Li}+O=C=O\longrightarrow O=\overset{\substack{C\\|}}{C}-OLi\xrightarrow{H_2O}\underset{\text{Benzoic acid}}{C_6H_5COOH}+LiOH$$

Example 7. Starting from a suitable Grignard reagent, how will you prepare the following:

(i) Ethanol (ii) 1-Butene (iii) Ethane (iv) Ethane sulphinic acid (*Kurukshetra, 1996*)

Solution. (i) Ethanol

$$CH_3MgBr+\underset{\substack{|\\H}}{\overset{\substack{H\\|}}{C}}=O\longrightarrow CH_3-\underset{\substack{|\\H}}{\overset{\substack{C_6H_5\\|}}{C}}-O\,MgBr\xrightarrow{H_2O}\underset{\text{Ethanol}}{CH_3CH_2OH}+Mg(OH)Br$$

(*ii*) 1-Butene

$$\underset{\text{Allyl chloride}}{CH_2 = CH - CH_2Cl} + CH_3MgCl \longrightarrow \underset{\text{1-Butene}}{CH_2 = CH - CH_2 - CH_3} + MgCl_2$$

(*iii*) Ethane

$$CH_3CH_2Mg\,Br + H_2O \longrightarrow CH_3CH_3 + Mg\,(OH)\,Br$$

(*iv*) Ethane sulphinic acid

$$\overset{O}{\overset{\|}{S}} = O + CH_3CH_2 - MgBr \longrightarrow CH_3 - CH_2 - \overset{O}{\overset{\|}{S}} - OMg\,Br \xrightarrow{H_2O} \underset{\text{Ethanesulphinic acid}}{CH_3CH_2 - \overset{O}{\overset{\|}{S}} - OH}$$

Example 8. How will you prepare the following compounds using appropriate organometallic compounds:

(*i*) Ethyl methyl ketone (*ii*) Acetic acid (*iii*) Ethyl propionate

Solution. (*i*) Ethyl methyl ketone

$$CH_3CH_2MgBr + CH_3\,C \equiv N \longrightarrow CH_3 - \overset{C_2H_5}{\overset{|}{C}} = N\,Mg\,Br \xrightarrow[2H_2O]{-\,Mg\,(OH)\,Br} \underset{\text{Ethyl methyl ketone}}{CH_3 - \overset{C_2H_5}{\overset{|}{C}} = O} + NH_3$$

(*ii*) Acetic acid

$$CH_3\,Mg\,Br + \overset{O}{\overset{\|}{C}} = O \longrightarrow CH_3 - \overset{O}{\overset{\|}{C}} - OMg\,Br \xrightarrow[2H_2O]{-\,Mg\,(OH)\,Br} \underset{\text{Acetic acid}}{CH_3 - \overset{O}{\overset{\|}{C}} - OH}$$

(*iii*) Ethyl propionate

$$Cl - \overset{O}{\overset{\|}{C}} - OC_2H_5 + CH_3CH_2MgBr \longrightarrow Cl - \underset{CH_2CH_3}{\underset{|}{\overset{OMgBr}{\overset{|}{C}}}} - OC_2H_5 \xrightarrow{-\,Mg(Br)Cl} \underset{\text{Ethyl propionate}}{CH_3\,CH_2\,\overset{O}{\overset{\|}{C}} - OC_2H_5}$$

ORGANOZINC COMPOUNDS

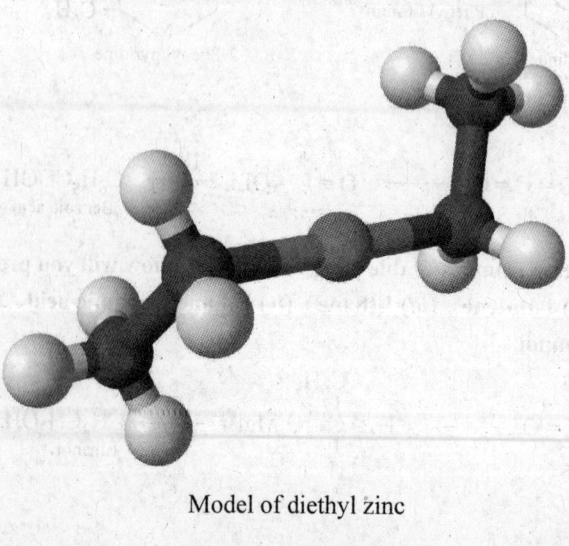

Model of diethyl zinc

2.6. PREPARATION

Dialkyl zinc Compounds were prepared by Frankland in the year 1849 in an attempt to prepare the ethyl radical by removing iodine from ethyl iodide by means of zinc. However, it ended with the formation of dialkyl zinc.

$$RI + Zn \longrightarrow \underset{\text{Alkyl Zinc iodide}}{R - Zn - I}$$

$$2RZnI \longrightarrow \underset{\text{Dialkyl Zinc}}{R_2Zn} + ZnI_2$$

The yield of dialkyl zinc may be increased by carrying out distallation in vacuum

Physical Properties

1. Alkylzinc compounds are volatile liquids, spontaneously inflammable in air.
2. They burn the skin and possess an unpleasant smell

2.7. CHEMICAL PROPERTIES (SYNTHETIC APPLICATIONS)

1. Preparation of hydrocarbons containing a quaternary carbon atom. Neopentane may be prepared by the action of dimethylzinc on *Tert.*-butyl Chloride.

$$\underset{\textit{Tert.}\text{-butyl chloride}}{(CH_3)_3CCl} + \underset{\text{Dimethyl zin}}{(CH_3)_2Zn} \longrightarrow \underset{\text{Neopentane}}{(CH_3)_4C} + \underset{\text{Methylzinc chloride}}{CH_3ZnCl}$$

2. Preparation of Ketones

Alkyl zinc compounds react with acyl chlorides to form ketones

$$\underset{\text{Acetyl chloride}}{CH_3COCl} + \underset{\text{Dimethyl zinc}}{(CH_3)_2Zn} \longrightarrow \underset{\text{Acetone}}{CH_3COCH_3} + CH_3ZnCl$$

$$CH_3COCl + \underset{\text{Diethyl zinc}}{(C_2H_5)_2Zn} \longrightarrow \underset{\text{Methylethyl ketone}}{CH_3COC_2H_5} + C_2H_5ZnCl$$

3. Preparation of long-chain fatty acids (or their esters)

First a heto-ester is prepared by reaction between as alkyl zinc chloride and the acid chloride-ester derivative. The keto-ester is then reduced by Clemmenson reduction

$$\underset{\text{Alkyl zinc chloride}}{CH_3(CH_2)_x\ ZnCl} + \underset{\text{acid chloride-ester derivative}}{ClCO(CH_2)_y\ COOC_2H_5} \longrightarrow$$

$$CH_3(CH_2)_x\ CO(CH_2)_y\ COOC_2H_5 \xrightarrow[HCl]{Zn/Hg} CH_3(CH_2)_x\ CH_2(CH_2)_y\ COOC_2H_5$$

$$\xrightarrow[H^+]{\text{Hydrolysis}} CH_3(CH_2)_{x+y+1}\ COOH + C_2H_5\ OH$$

4. Reformatsky reaction. In this reaction an aldehyde or a ketone is treated with an α-bromoester and metallic zinc in the presence of dry ether or benzene. An organo zinc compound is first formed which adds to the carbonyl group to give an addition product. The addition product thus formed is decomposed with dilute mineral acids to yield β-hydroxyesters. For example,

$$\underset{\text{Acetaldehyde}}{(CH_3)(H)C=O} + \underset{\text{Ethyl } \alpha\text{-bromoacetate}}{BrCH_2COOC_2H_5} + Zn \xrightarrow{\text{Ether}} \underset{\text{Addition product}}{(CH_3)(H)C(OZnBr)(CH_2COOC_2H_5)}$$

$$\xrightarrow{H_3O^+} \underset{\text{Ethyl } \beta\text{-hydroxyl butyrate}}{CH_3CH(OH)CH_2COOC_2H_5} + Zn(OH)Br$$

5. Preparation of a hydrocarbon using water. Water reacts with organozinc compounds to form alkanes. For example,

$$2\,(C_2H_5)_2Zn + 2H_2O \longrightarrow 2\,C_2H_6 + C_2H_5-Zn\underset{OH}{\overset{OH}{\rightleftarrows}}Zn-C_2H_5$$

Example 1. What happens when methyl magnesium bromide is reacted with

(*i*) Lead Chloride (*ii*) Propyne (*iii*) Ethyl formate (*iv*) Allylbromide.

Solution. (*i*) $4CH_3MgBr + 2PbCl_2 \longrightarrow (CH_3)_4\,Pb + Pb + 4Mg\,(Br)Cl.$

(*ii*) See the relevant section in the text.

(*iii*) See relevant section in the text

(*iv*) $CH_3MgBr + BrCH_2-CH=CH_2 \longrightarrow \underset{\text{Butene-1}}{CH_3-CH_2-CH=CH_2} + MgBr_2$

Example 2. Using suitable R-Mg-X, prepare the following compounds:

(*i*) $C_6H_5CH_2-OH$ (*ii*) $CH_3-CO-CH_2-CH_2-CH_3$

(*iii*) $CH_3-C(CH_3)(OH)-CH_2-CH_3$ (*iv*) $(CH_3)_3-C-COOH$

Solution.

(*i*) $$\underset{\text{Phenyl magnesium bromide}}{C_6H_5-MgBr} + H-\overset{H}{C}=O \longrightarrow H-\overset{H}{\underset{C_6H_5}{C}}-OMgBr \xrightarrow{HOH} C_6H_5CH_2OH + Mg\begin{matrix}Br\\OH\end{matrix}$$

(*ii*) $$\underset{n\text{-Propyl chloride}}{CH_3CH_2CH_2Cl} + CH_3\,\underset{Cl}{C}=O \longrightarrow CH_3CH_2CH_2-\overset{CH_3}{\underset{Cl}{C}}-OMgCl \longrightarrow C_3H_7-\overset{CH_3}{C}=O + MgCl_2$$

(*iii*) $$CH_3CH_2MgBr + (CH_3)_2C=O \longrightarrow CH_3-(CH_3)_2C-OMgBr \xrightarrow{H_2O} CH_3-\overset{CH_3}{\underset{OH}{C}}-C_2H_5 + Mg\,(OH)\,Br$$

(*iv*) $$(CH_3)_3CMgBr + O=C=O \longrightarrow O=\underset{C(CH_3)_3}{C}-OMgBr \xrightarrow{HOH} (CH_3)_3C\,COOH$$

Example 3. Using organometallic compound, how will you obtain the following?

(*i*) Dimethyl mercury **(*ii*) Ethyl mercaptan**

(*iii*) 2–Butyne **(*iv*) Acetone**

(*v*) Methyl *n*-propyl ether **(*vi*) Acetaldehyde**

(*vii*) Diethylamine **(*viii*) Trimethyl carbinol**

Solution. (*i*) $2\ C_2H_5MgI + \underset{\text{Mercuric chloride}}{HgCl_2} \longrightarrow \underset{\text{Mercury dimethyl}}{(C_2H_5)_2Hg} + 2Mg(I)\,Cl$

(*ii*) $CH_3-CH_2-MgBr + S \longrightarrow CH_3-CH_2-S-MgBr \xrightarrow[-\,Mg(OH)Br]{HOH,\ H^+} CH_3-CH_2-SH$

(*iii*) $CH\equiv CH + 2CH_3MgI \longrightarrow IMgC\equiv CMgI \xrightarrow{+\,2CH_3I} CH_3C\equiv C-CH_3 + 2MgI_2$

(*iv*) $CH_3MgBr + \underset{\text{Ethyl acetate}}{CH_3-\overset{O}{\overset{\|}{C}}-OC_2H_5} \longrightarrow CH_3-\overset{OMgBr}{\overset{|}{\underset{CH_3}{\underset{|}{C}}}}-OC_2H_5 \xrightarrow{HOH} \underset{\text{acetone}}{CH_3-\overset{O}{\overset{\|}{\underset{CH_3}{\underset{|}{C}}}}} + Mg(OC_2H_5)Br$

(*v*) $CH_3-CH_2-MgI + \underset{\text{Chloromethyl ether}}{Cl-CH_2-O-CH_3} \longrightarrow \underset{\text{Methyl n-propyl ether}}{CH_3-CH_2-CH_2-O-CH_3} + MgI(Cl)$

(*vi*) $CH_3-MgBr + \underset{\text{Ethyl formate}}{H-\overset{O}{\overset{\|}{C}}-OC_2H_5} \longrightarrow H-\overset{OMgBr}{\overset{|}{\underset{CH_3}{\underset{|}{C}}}}-OC_2H_5 \longrightarrow \underset{\text{acetaldehyde}}{CH_3CHO} + Mg(OC_2H_5)Br$

(*vii*) Trimethyl carbinol is the same compound as t-butyl alcohol. See under synthetic applications.

Example 4. Suggest as many different routes as possible for the Grignard synthesis of the following alcohol :

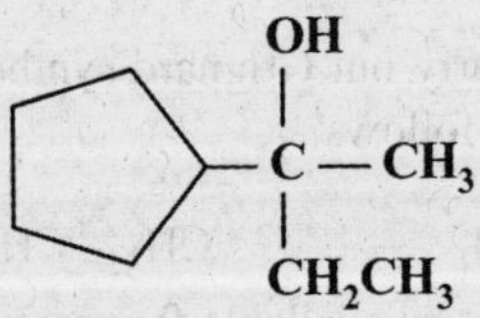

Solution. The given compound is a tertiary alcohol. To prepare it by Grignard synthesis would require the use of a Grignard reagent and a ketone. There can be three different combinations for its synthesis.

(*i*)

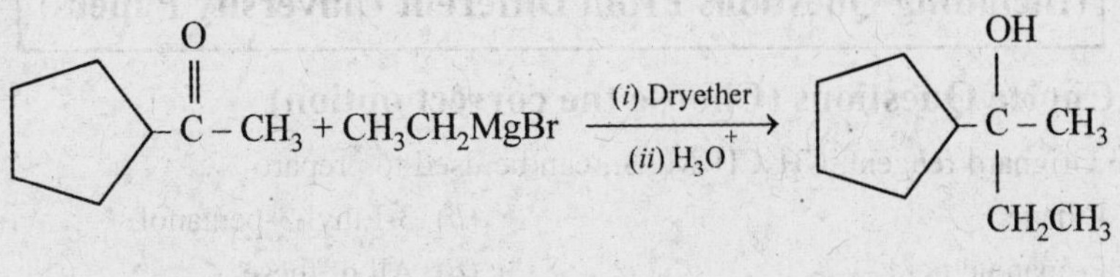

(*ii*)

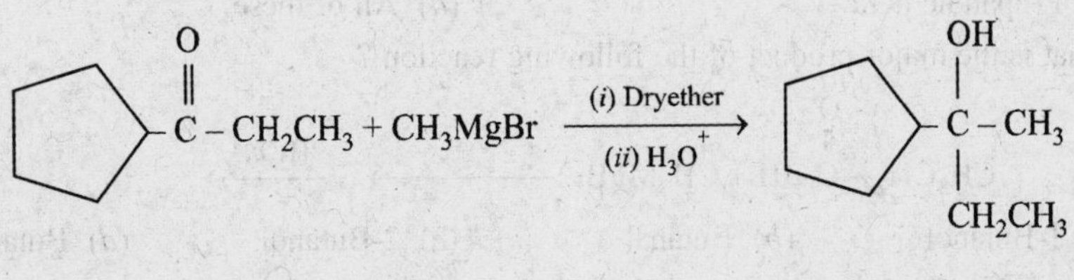

(*iii*) 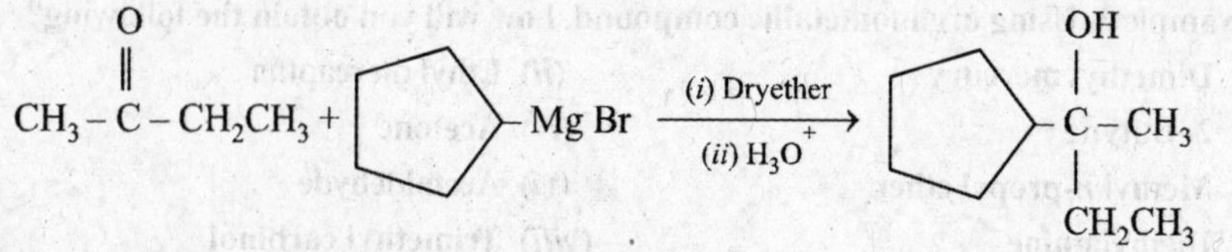

Example 5. Write the structure of Grignard reagent formed by the reaction of *p*-bromofluorobenzene with magnesium in diethyl ether.

Solution. Bromine reacts at a much faster rate than fluorine with magnesium. Therefore carbon-bromine bond is converted into carbon-magnesium bond while fluorine remains attached to benzene ring. Hence, the structure of the Grignard reagent will be.

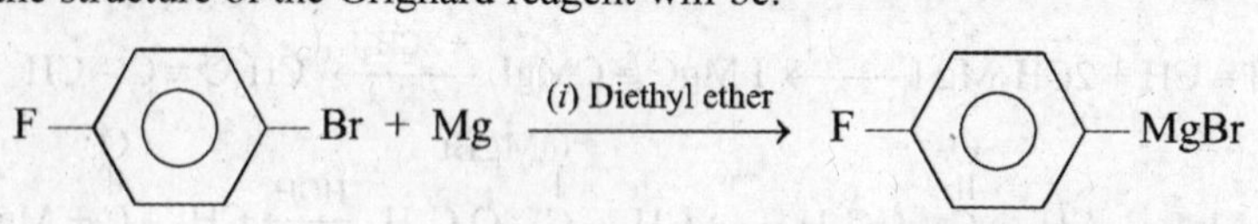

p-Fluorophenyl magnesium bromide

Example 6. Why are organolithium compounds more reactive than Grignard reagents ?

Solution. The reactivity of organometallic compounds depends upon the ionic character of the carbon-metal bond. Greater the ionic character of this bond, more reactive is the organometallic compound. As C – Li bond has a greater ionic character than C – Mg bond, organolithium compounds are more reactive than Grignard reagents.

Example 7. Why is it not possible to carry out the Grignard synthesis as given below :

$$CH_3-CH_2-MgBr \xrightarrow{CH_3-\overset{\overset{\displaystyle O}{\|}}{C}-CH_3,\ H_3\overset{+}{O}} CH_3-CH_2-\underset{\underset{\displaystyle OH}{|}}{\overset{\overset{\displaystyle CH_3}{|}}{C}}-CH_3$$

Solution. It is not possible to carry out Grignard synthesis as outlined above because water would react with Grignard reagent as follows :

$$CH_3-CH_2-MgBr + H.OH \longrightarrow CH_3-CH_3 + Mg(OH)Br.$$

Thus Grignard reagent would not be available for reaction with acetone.

EXERCISES
(Including Questions From Different University Papers)

Multiple Choice Questions (Choose the correct option)

1. The Grignard reagent, CH_3CH_2MgBr, can be used to prepare

(*a*) Ethane (*b*) 3-Ethyl-3-pentanol

(*c*) Propanoic acid (*d*) All of these

2. What is the major product of the following reaction ?

$$CH_3CH_2-\overset{\overset{\displaystyle O}{\|}}{C}-H + CH_3MgBr \longrightarrow \xrightarrow{H_2O/H^+}$$

(*a*) 1-Butanol (*b*) Butanal (*c*) 2-Butanol (*d*) Butanone

3. Ketones react with Grignard reagents to form an addition product which on hydrolysis gives a
(*a*) Primary alcohol (*b*) Tertiary alcohol (*c*) Secondary alcohol (*d*) Ketal

4. Grignard reagents do not show any reaction with
(*a*) Alkoxyalkanes (*b*) Alkanones (*c*) Alkyl alkanoates (*d*) Acyl halides

5. Which is the best reagent to accomplish the following conversion ?

$$CH_3CH_2Br \xrightarrow{(?)} CH_3CH_3$$

(*a*) Conc. H_2SO_4 (*b*) Na (*c*) Conc. HCl (*d*) Mg, then H_2O

6. *n*-Propylmagnesium bromide on treatment with carbon dioxide and further hydrolysis gives :
(*a*) Acetic acid (*b*) Propanoic acid (*c*) Butanoic acid (*d*) Formic acid

7. Ethylmagnesium iodide reacts with formaldehyde to give a product which on acid-hydrolysis forms :
(*a*) an aldehyde (*b*) a primary alcohol
(*c*) a ketone (*d*) a secondary alcohol

8. Which of the following compounds will react with methylmagnesium iodide followed by acid hydrolysis to give ethyl alcohol ?
(*a*) Ethylene (*b*) Acetaldehyde (*c*) Formaldehyde (*d*) Acetone

9. Which of the following compounds will react with methylmagnesium bromide to give *tert*-butyl alcohol ?
(*a*) Acetyl chloride (*b*) Acetone (*c*) Isopropyl alcohol (*d*) Acetaldehyde

10. Which of the following gives a tertiary alcohol when treated with Grignard reagents ?
(*a*) $H-\overset{\overset{\displaystyle O}{\|}}{C}-H$ (*b*) $CH_3-\overset{\overset{\displaystyle O}{\|}}{C}-H$ (*c*) $CH_3-\overset{\overset{\displaystyle O}{\|}}{C}-CH_3$ (*d*) None of these

11. Phenylmagnesium bromide reacts with acetaldehyde to form an addition product which undergoes acid-hydrolysis to give
(*a*) Diphenylcarbinol (*b*) Benzyl alcohol
(*c*) Methylphenylcarbinol (*d*) Benzoic acid

12. Reaction between Grignard reagent and oxygen followed by hydrolysis gives Mg(OH)Br and
(*a*) methanol (*b*) methanal (*c*) methane (*d*) methanoic acid

ANSWERS

1. (*d*)	2. (*c*)	3. (*b*)	4. (*a*)	5. (*d*)	6. (*c*)
7. (*b*)	8. (*c*)	9. (*b*)	10. (*c*)	11. (*c*)	12. (*a*)

Short Answer Questionss

1. How can you prepare the following from methyl magnesium bromide ?
(*i*) 2-Methyl-2-propanol (*ii*) 2-Propanol

2. Starting from a Grignard reagent, how will you prepare each of the following :
(*i*) 2-Butanone (*ii*) Ethanoic acid

3. What happens when
(*i*) *n*-Butyl lithium is treated with bromobenzene
(*ii*) Ethyl magnesium bromide is treated with cadmium chloride.

4. Why are organo lithium compounds more polar, more stable, least reactive among other organometallic compounds of heavier alkali metals ?

5. Arrange Grignard reagents, organolithium compounds and organozinc compound in decreasing order of reactivity

6. How does ethyl acetate react with methyl magnesium bromide ?

7. Name two organolithium compounds. Why are they more reactive than Grignard reagents?

8. Explain what happens when *n*-butyl lithium is treated with bromobenzene.

9. Using suitable RMgX, prepare the following compounds :

(*i*) $CH_3 — CH_2 — C(H)(OH) — CH_3$ (*ii*) $(CH_3)_3C — OH$

(*iii*) $CH_3 — CH_2 — CH_2 — COOH$.

10. Write equations to show the reaction of methyl magnesium bromide with each of the following :

(*i*) $ZnCl_2$ (*ii*) C_2H_5OH

(*iii*) 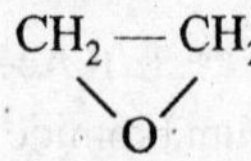 (*iv*) $CH_3COOC_2H_5$.

11. How are the following compounds obtained starting from suitable Grignard reagents :

(*i*) Methyl *n*-propyl ether (*ii*) Acetaldehyde

(*iii*) Acetone (*iv*) Diethyl amine.

12. Using suitable RMgX, prepare the following compounds :

(*i*) CH_3COCH_3 (*ii*) CH_3CH_2COOH

(*iii*) $CH_3CH_2OCH_2CH_3$ (*iv*) CH_3CH_2SH.

13. What happen when :

(*i*) CH_3MgI is heated with SO_2 (*ii*) CH_3Li is treated with acetone.

14. Write the preparation of the following from methyl magnesium bromide :

(*i*) Dithioacetic acid (*ii*) Tert. butyl alcohol

(*iii*) Ethyl acetate.

15. How will you prepare the following starting from ethyl magnesium bromide ?

(*i*) Ethyl methyl ketone (*ii*) Ethyl propionate

(*iii*) Tetraethyl lead.

16. Using suitable organolithium compound, how will you prepare :

(*i*) *n*-Butyl alcohol (*ii*) Diethyl ketone

(*iii*) α, β-usaturated alcohol.

17. Name two organolithium compounds. Why these are more reactive than Grignards reagents ? Complete the reaction.

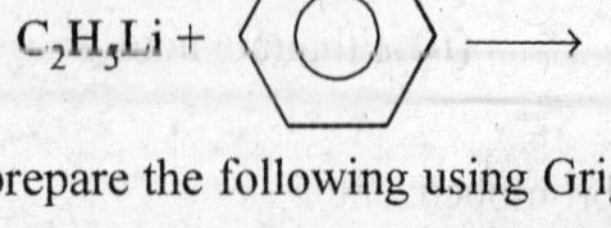

18. How will you prepare the following using Grignard reagents ?

(*i*) 3-Pentanone (*ii*) 2-Propanol

General Questions

1. Name two organolithium compounds. Why are they more reactive than Grignard reagents ? How can you prepare a primary alcohol, a carboxylic acid and a ketone from these compounds ?
2. Starting from ethyl magnesium bromide, how can you obtain each of the following :
   (*i*) Ethane (*ii*) Propanoic acid (*iii*) 1-Pentene
   (*iv*) 2-Pentyne (*v*) 2-Methyl-2-butanol (*vi*) 2-Butanol
   (*vii*) 3-Pentanone (*viii*) Propanol.
3. Complete the following reactions :
   (*i*) $C_2H_5Li + CH_2 - CH_2$ (epoxide ring, O bridging the two CH_2) $\xrightarrow{THF}$ (*ii*) $CH_2 = CH_2 + C_2H_5Li \xrightarrow{THF}$
   (*iii*) $CH_3Li + CO_2 \xrightarrow{H_2O/H^+}$ (*iv*) $C_2H_5Li + HCHO \longrightarrow$
4. Explain, giving equations, what happens when :
   (*i*) *n*-Butyl lithium is treated with carbon dioxide.
   (*ii*) *n*-Butyl lithium is treated with excess of ethylene.
   (*iii*) A Grignard reagent is treated with a carbonyl compound having an α-hydrogen.
   (*iv*) Ethyl magnesium bromide is treated with cadmium chloride.
   (*v*) *n*-Butyl lithium is treated with bromobenzene.
5. With the help of methyl magnesium iodide how can you prepare the following alcohols :
   (*i*) Methanol (*ii*) Ethanol (*iii*) 1-Propanol
   (*iv*) 2-Propanol (*v*) 1-Butanol (*vi*) 2-Butanol
   (*vii*) 2-Methyl-1-butanol (*viii*) 2-Methyl-2-butanol.
6. Starting with CH_3MgBr, how will you prepare :
   (*i*) $CH_3 - CH = CH_2$ (*ii*) CH_3CH_2OH (*iii*) CH_3COCH_3
   (*iv*) $CH_3 - NH_2$ (*v*) $CH_3OCH_2CH_2CH_3$ (*vi*) $CH_3COOC_2H_5$
7. What are organo lithium compounds ? How are they prepared ? In what way do they differ from Grignard reagents ?
8. Compare the reactivity of Grignard reagent and organolithium compounds with α, β-unsaturated carbonyl compounds giving reasons.
9. How will you convert the following using suitable organo-metallic compound?
   (*i*) Propanol-1 into butano-1 (*ii*) Ethylene oxide into propanol-1
   (*iii*) Ethyl acetate into acetone.
10. (*a*) How does methyl magnesium iodide react with :
    (*i*) Ethyl acetate (*ii*) Acetyl chloride (*iii*) CO_2.
    (*b*) Give one method of preparation of dimethyl zinc compound. How does it react with acetyl chloride ?
11. Using suitable Grignard reagents, synthesize the following compounds :
    (*i*) $CH_3 - \underset{\displaystyle OH}{\underset{|}{CH}} - CH_3$ (*ii*) $CH_3 - CH_2 - CH_2 - COOH$
    (*iii*) $CH_3 - \overset{\displaystyle O}{\overset{||}{C}} - CH_2 - CH_3$
12. Discuss the preparation and properties of ethyl magnesium chloride.
13. Starting from C_2H_5MgBr, how will you prepare the following :
    (*i*) 2-Methyl-2-butanol (*ii*) Propan-1-ol (*iii*) Propanoic acid.
14. Discuss formation and reactions of organolithium compounds.

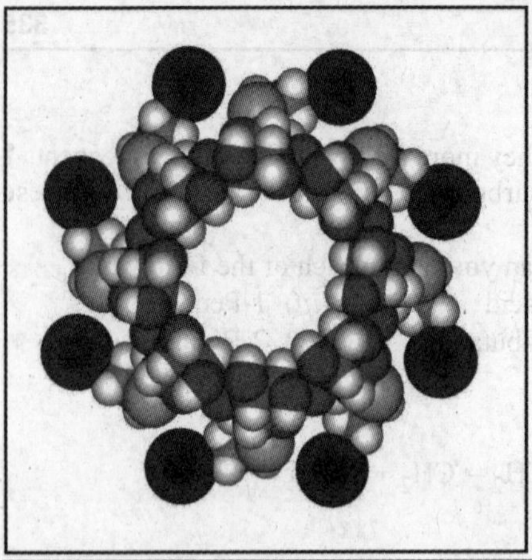

3

Organosulphur Compounds

3.1. INTRODUCTION

Organanosulphur compounds are analogous to organic compounds containing oxygen with the diffrence that oxygen has been replaced by sulphur. These compounds give reactions similar to those given by oxygen contaning compounds. Thiols and thioethers are principal examples of organosulphur compunds which are analogous to alcohols and ethers respectively.

Oxygen-containing organic compounds		**Sulphur-containing organic compounds**	
R – O – H	(Alcohols)	R – S – H	(Thiols)
– O – H	Alcohols group	– S – H	Thiol group
R – O – R	(Ethers)	R – S – R	(Thioethers)
– O – .	Ethers group	– S –	Thioether group

There also exist many organosulphur compounds containing sulphur–oxygen bonds with double bond charcter, for example, sulphoxides such as dimethylsulphoxide and sulphonic acids

$(CH_3)_2$ SO

Dimethylsulphoxide

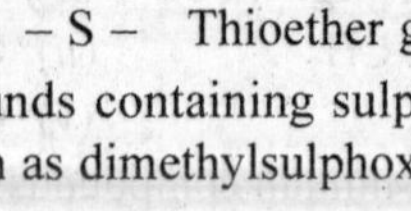

Benzene sulphonic acid

The formation of these compounds is explained in terms of pπ – dπ bonding

THIOLS

Thiols are sulphur analogs of alcohols in which the oxygen has been replaced by sulphur atom.

Thus, the functional group in thiol is —SH. It is also called *mercapto* group. Like H_2S, thiols are weakly acidic They react with mercuric ions to form insoluble salts. Therefore, they were given the name mercaptans (meaning mercury catching). They may be conceived to be derived from hydrogen sulphide as follows

$$H-S-H \xrightarrow[+R]{-H} R-S-H$$

Thiol

3.2. NOMENCLATURE

Thiols are named as alkyl mercaptans according to the common system. The IUPAC names are obtained by adding the suffix *thiol* to the name of the corresponding alkane.

Compound	Common Name	IUPAC Name
CH_3SH	Methyl mercaptan	Methanethiol
CH_3CH_2SH	Ethyl mercaptan	Ethanethiol
$CH_3CH_2CH_2SH$	*n*-Propyl mercaptan	1-Propanethiol

When the compound contains other groups of higher priority, the – SH group is denoted by the prefix **meriapto.** Also – OH group is assigned higher priority than – SH thus compounds containing – SH group along with groups are named as under.

$HSCH_2\,CH_2\,CH_2\,OH$
3– mercaptoethanol

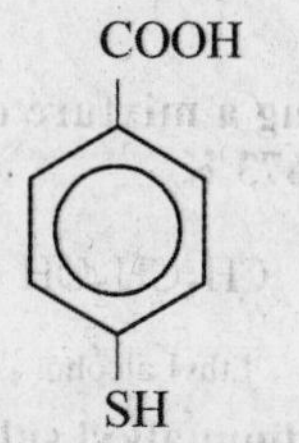

4– mercaptobenzoic acid

3.3. STRUCTURE OF THIOALCOHOLS (THIOLS)

Thiols have a structure similar to alcohols. The polarity of S – H bond is lower as compared to that of O – H bond is alcohols. This is due to lower electronegativity of S compared to O. Thiolcohols are bent molecules with < C S H being 100°

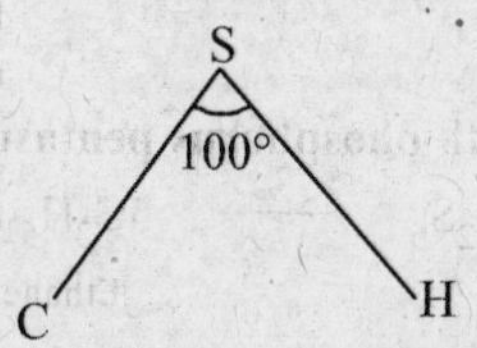

3.4. METHODS OF PREPARATION OF THIOLS

Thiols are obtained by the following methods:

1. By heating alkyl halides with potassium hydrosulphide (KSH) solution.

$$\underset{\text{Ethyl iodide}}{CH_3CH_2I} + KSH \xrightarrow{\text{Heat}} \underset{\text{Ethanethiol}}{CH_3CH_2SH} + KI$$

The mechanism of the reaction is the same as the formation of alcohols from alkyl halides i.e. SN^2.

2. By the reaction of Grignard reagent with sulphur followed by hydrolysis using an acid.

$$\underset{\text{Ethyl magnesium bromide}}{CH_3CH_2MgBr} + S \longrightarrow CH_3CH_2SMgBr \xrightarrow[H^+]{H_2O} \underset{\text{Ethanethiol}}{CH_3CH_2SH} + Mg(OH)Br$$

3. By the addition of hydrogen sulphide to alkene in the presence of sulphuric acid as catalyst

$$CH_3-CH=CH_2 + H-SH \xrightarrow{H_2SO_4} \underset{\text{2–Propanethiol}}{CH_3-\overset{\overset{\large SH}{|}}{C}H-CH_3}$$

The addition takes place according to Markovnikov rule.

4. By heating alkyl halides with thiourea followed by alkaline hydrolysis.

$$\underset{\text{Ethyl iodide}}{CH_3CH_2Br} + \ddot{S}=\underset{NH_2}{\underset{|}{C}}-NH_2 \longrightarrow \underset{\text{Ethyl isothiourea salt}}{\left[CH_3CH_2-S^{+}=\overset{NH_2}{\overset{|}{C}}-NH_2\right]Br^-}$$

$$\xrightarrow{NaOH} \underset{\text{Ethanethiol}}{CH_3CH_2SH} + O=\underset{\underset{\text{urea}}{NH_2}}{\underset{|}{C}}-NH_2 + NaBr$$

5. By passing a mixture of alcohol vapours and hydrogen sulphide over thoria (ThO_2) as a catalyst at 673 K

$$\underset{\text{Ethyl alcohol}}{CH_3CH_2OH} + H_2S \xrightarrow[673K]{ThO_2} \underset{\text{Ethanethiol}}{CH_3CH_2SH} + H_2O$$

6. From sodium alkyl sulphates or p-toluenesulphonic esters

$$\underset{\text{Sodium ethyl sulphate}}{C_2H_5OSO_3Na} + NaSH \longrightarrow \underset{\text{Ethane thiol}}{C_2H_5SH} + Na_2SO_4$$

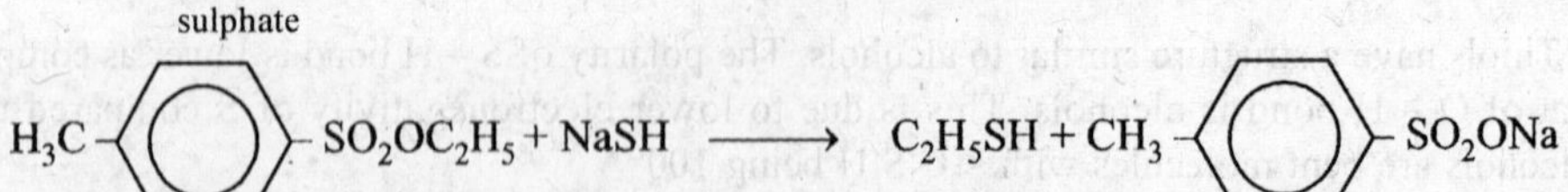

Ethyl p-toluenesulphonate — Ethane thiol — Sodium p-toluenesulphonate

7. By heating an alcohol with phosphorus pentasulphide

$$5\,C_2H_5OH + P_2S_5 \longrightarrow \underset{\text{Ethanethiol}}{5C_2H_5SH} + P_2O_5$$

8. By hydrolysis of thioesters

When treated with a dilute acid or alkali, theoesters yield thiols

$$\underset{\text{Ethylthioacetale}}{CH_3-\overset{O}{\overset{||}{C}}-SC_2H_5} + H_2O \xrightarrow{\overset{+}{H}\text{ or }\overset{-}{OH}} CH_3-COOH + \underset{\text{Ethanethiol}}{C_2H_5SH}$$

3.5. PHYSICAL PROPERTIES OF THIOLS

Important physical properties of thiols are gives below:

1. Thiols possess a strong odour. The smell coming out of the leaking LPG cylinders is because of methanethiol or ethanethiol that has been added in small quantities (less than 1%) to butane gas in the cylinders. This is done for quick detection of leakage in the cylinders.

2. Boiling points of thiols are much lower than those of corresponding alcohols.

Compound	Boiling point (°C)
CH_3SH	6
CH_3OH	56
CH_3CH_2SH	35
CH_3CH_2OH	78
$CH_3CH_2CH_2SH$	68
$CH_3CH_2CH_2OH$	98

This is explained in terms of absence of hydrogen bondings in thiols. Normal alcohols have higher boiling points due to hydrogen bonding.

3. Thiols are practically insoluble in water due to absence of hydrogen bonding with water.

Example 1. Explain why *n*-propyl alcohol is more soluble in water than 1-propanethiol.

Solution. *n*-propyl alcohol is more soluble compared to 1-propane thiol because of stronger hydrogen bonds with water. There is greater polarisation in the O–H bond compared to S–H bond. This is because of greater electronegativity difference between oxygen and hydrogen than between sulphur and hydrogen. Hence the alcohol molecules form stronger hydrogen bonds with water molecules and therefore are more soluble in water.

Example 2. Explain why the boiling point of 1-butanol (bp 117°C) is higher than that of 1-butanethiol (bp 98.5°C).

Solution. Alcohol molecules are associated because of hydrogen bonding

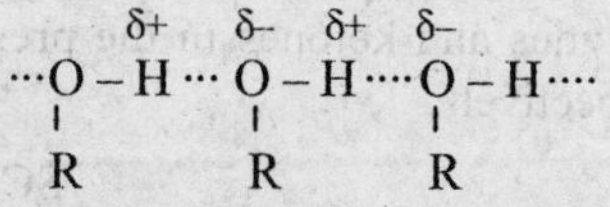

The extent of hydrogen bonding depends upon the separation of charges which will in turn depend upon the electronegativity difference between the two elements.

Hydrogen bonding is weaker in case of thiols because of smaller separation of charges. And this is because of smaller difference in electronegativities of S and H than between those of O and H

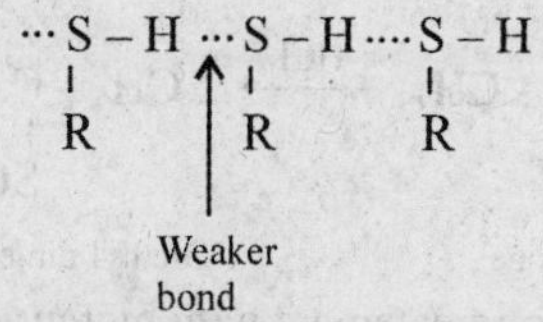

As the hydrogen bonding is stronger is alcohols, they associate to a greater extent resulting in higher b.p. for alcohols.

3.6. CHEMICAL PROPERTIES OF THIOLS

1. Acidic nature

Thiols although weakly acidic show greate acidic character than alcohols, which is rather unexpected because S has smaller electronegativity than O. This could be due to the following reasons:

(*i*) The S – H bond is weaker than O – H bond. Thus, there is greater probability of the release of H^+ from thiols than alcohols

(*ii*) The anion RS^- formed after the release of the proton is more stable than RO^- formed from alcohols, because the former can accommodate the negative change more easily than RO^- due to larger size of sulphur

There is some similarity between the chemical properties of alcohols and thiols.

2. Reaction with active metals like Na, K and Ca

Thiols react with active metals liberating hydrogen gas

$$2CH_3CH_2SH + Na \longrightarrow 2CH_3\,CH_2\,\overset{-}{S}\overset{+}{Na} + H_2 \uparrow$$

3. Reaction with alkalis

Thiols are more acidic than alcohols. Reaction takes place with alkalis forming salts

$$CH_3CH_2SH + NaOH \longrightarrow CH_3CH_2\overset{-}{S}\overset{+}{Na} + H_2O$$

4. Reaction with metallic salts and oxides

Thiols react with metallic salts and oxides to form salts

$$2CH_3CH_2SH + HgO \longrightarrow \underset{\text{Mercurydiethyl mercaptide}}{(CH_3CH_2S)_2Hg} + H_2O$$

$$2CH_3CH_2SH + (CH_3COO)_2Pb \longrightarrow \underset{\text{Lead diethyl mercaptide}}{(CH_3CH_2S)_2Pb} + 2CH_3COOH$$

5. Reaction with aldehydes and ketones

Thiols react with aldehydes and ketones in the presence of hydrochloric acid to form mercaptals and mercaptols respectively.

$$2C_2H_5SH + \underset{\text{Acetaldehyde}}{CH_3CHO} \xrightarrow{HCl} \underset{\text{Diethyl methyl mercaptal}}{CH_3-\overset{\overset{SC_2H_5}{|}}{\underset{\underset{SC_2H_5}{|}}{C}}-H} + H_2O$$

$$2C_2H_5SH + \underset{\text{Acetone}}{CH_3-\overset{\overset{O}{\|}}{C}-CH_3} \xrightarrow{HCl} \underset{\text{Diethyl dimethyl mercaptol}}{CH_3-\overset{\overset{SC_2H_5}{|}}{\underset{\underset{SC_2H_5}{|}}{C}}-CH_3} + H_2O$$

Mercaptals and mercaptols on treatement with mercuric chloride in the presence of cadmium carbonate regenerate the original aldehyde or ketone For example.

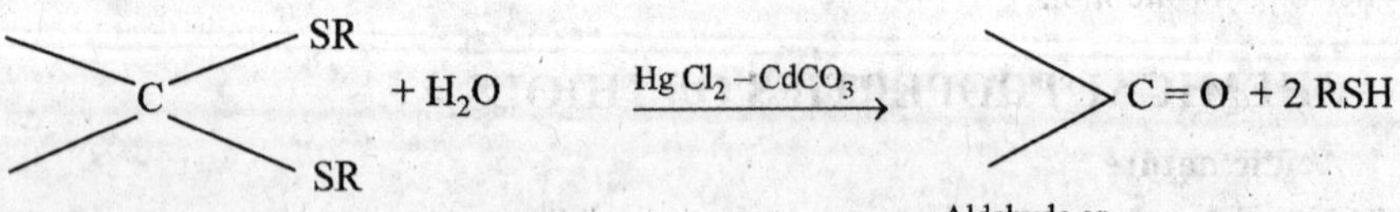

Mercaptal or mercaptol

Aldehyde or ketone

Diethyl dimethyl mercaptol on oxidation with an oxidising agent like potassium permanganate gives *sulphonal* which is an important hypnotic

$$CH_3-\overset{\overset{SC_2H_5}{|}}{\underset{\underset{SC_2H_5}{|}}{C}}-CH_3 + 4[O] \xrightarrow{KMnO_4} \underset{\text{Sulphonal}}{CH_3-\overset{\overset{SO_2C_2H_5}{|}}{\underset{\underset{SO_2C_2H_5}{|}}{C}}-CH_3}$$

6. Reaction with acids and acid chlorides

Thiols react with acids and acid chlorides to form thioesters.

$$CH_3CO\boxed{OH + H}SC_2H_5 \longrightarrow \underset{\text{Ethyl thioacetate}}{CH_3COSC_2H_5} + H_2O$$

$$CH_3CO\boxed{Cl + H}SC_2H_5 \longrightarrow \underset{\text{Ethyl thioacetate}}{CH_3COSC_2H_5} + HCl$$

7. Oxidation

(*a*) Thiols get oxidised easily even with mild oxidising agents like halogens and hydrogen peroxide.

$$2C_2H_5SH + I_2 \longrightarrow \underset{\text{Diethyl disulphide}}{C_2H_5 - S - S - C_2H_5} + I_2$$

$$2C_2H_5SH + H_2O_2 \longrightarrow \underset{\text{Diethyl disulphide}}{C_2H_5 - S - S - C_2H_5} + 2H_2O$$

(*b*) Sulphonic acids are formed when oxidation is brought about by a strong oxidising agent like potassium permanganate or conc. nitric acid.

$$C_2H_5SH + 3[O] \xrightarrow{KMnO_4} \underset{\text{Ethanesulphonic acid}}{C_2H_5SO_3H}$$

8. Reaction with alkyl halides

On heating with alkyl halides, sodium salts of thiols from thioethers.

$$C_2H_5SNa + BrC_2H_5 \longrightarrow \underset{\text{Diethyl thioether}}{C_2H_5SC_2H_5} + HBr$$

9. Desulphurisation

On heating with Raney nickel, thiols undergo desulphurisation and give hydrocarbons (Raney nickel is an active form of nickel).

$$C_2H_5SH \xrightarrow[\text{Heat}]{\text{Ni}} C_2H_6 + NiS$$

Example 1. How will you distinguish between ethanethiol and ethyl alcohol?

Solution. The following two tests can be performe to distinguish between the two compounds.

1. Sodium hydroxide test

Ethanathiol (ethyl mercaptan) dissolves in sodium hydroxide forming a mercaptide

$$C_2H_5SH \xrightarrow{NaOH} \underset{\substack{\text{Sodium ethyl}\\ \text{mercaptide}\\ \text{(clear solution)}}}{C_2H_5\overset{-}{S}\overset{+}{Na}} + H_2O$$

Ethyl alcohol does not react with sodium hydroxide solution and hence does not dissolve in it.

2. Mercuric chloride test

Ethanethiol reacts with mercuric chloride to form a precipitate of mercury diethyl mercaptide.

$$\underset{\text{Ethanethiol}}{2C_2H_5SH} + HgCl_2 \longrightarrow \underset{\substack{\text{Mercury diethyl}\\ \text{mercaptide}}}{(C_2H_5S)_2Hg\downarrow} + 2HCl$$

Ethyl alcohol does not give this reaction.

Thus, by performing these two tests, the two compounds can be distinguished.

Example 2. How will you prepare 1-propanethiol from propene?

Solution. Method 1

$$\underset{\text{Propene}}{CH_3CH=CH_2} \xrightarrow[\text{Peroxide}]{\text{HBr}} CH_3CH_2CH_2Br \xrightarrow[\text{Sod. hydrogen sulphide}]{\text{NaSH}} \underset{\text{1-Propanethiol}}{CH_3CH_2CH_2SH}$$

Method 2

$$CH_3CH=CH_2 \xrightarrow[\text{Peroxide}]{\text{HBr}} CH_3CH_2CH_2Br \xrightarrow[\text{Dry ether}]{\text{Mg}} \underset{\text{Propyl magnesium bromide}}{CH_3CH_2CH_2MgBr}$$

$$\xrightarrow{\text{S}} CH_3CH_2CH_2SMgBr \xrightarrow[\text{H}]{H_2O} \underset{\text{1-Propanethiol}}{CH_3CH_2CH_2SH} + Mg(OH)Br$$

Example 3. How will you synthesis sulphonal from propene?

Solution. The following steps are involved in the synthesis :

$$\underset{\text{Propene}}{CH_3CH=CH_2} \xrightarrow[\text{H}]{H_2O} CH_3\overset{\overset{OH}{|}}{C}H-CH_3 \xrightarrow[K_2Cr_2O_7/H^+]{[O]} CH_3-\overset{\overset{O}{||}}{C}-CH_3$$

$$\xrightarrow[\text{HCl}]{2C_2H_5SH} CH_3-\underset{\underset{SC_2H_5}{|}}{\overset{\overset{SC_2H_5}{|}}{C}}-CH_3 \xrightarrow[KMnO_4]{[O]} \underset{\text{sulphonal}}{CH_3-\underset{\underset{SO_2C_2H_5}{|}}{\overset{\overset{SO_2C_2H_5}{|}}{C}}-CH_3}$$

THIOETHERS

Thioethers are sulphur analogues of ethers. They have general formula R – S – R′ where R and R′ are alkyl groups. These alkyl groups may be same or different. Thioethers may be considered to be derived from H_2S as follows :

$$H-S-H \xrightarrow[+2R]{-2H} R-S-R$$

Functional group in thioethers is – S –.

3.7. NOMENCLATURE

Names of some compounds according to common and IUPAC systems are given in the table below :

Compound	Common name	IUPAC name
CH_3-S-CH_3	Dimethyl sulphide	Methylthiomethane
$CH_3-S-CH_2CH_3$	Ethylmethyl sulphide	Methylthioethane
$CH_3CH_2-S-CH_2CH_3$	Diethyl sulphide	Ethylthioethane
$CH_3-S-CH_2CH_2CH_3$	Methyl-*n*-propyl sulphide	1-Methylthiopropane
$CH_3-\overset{\overset{SCH_3}{\vert}}{C}H-CH_2CH_3$	Methyl-sec.butyl sulphide	2-Methylthiobutane

Smaller alkyl group linked to sulphur atom is regarded as the substituent.

3.8. STRUCTURE OF THIOETHERS

Thioethers have tetrahedral structure with two positions occupied by lone pair of electons as shown below

The angle RSR depends on the size of the alkyl groups attached to S for example, the C – S – C angle in dimethyl thioether is 105°

3.9. METHODS OF PREPARATION

The following methods are generally employed for the preparation of thioethers

1. Reaction of alkyl halides with sodium or potassium merceptide

$$C_2H_5SNa + CH_3Br \longrightarrow \underset{\text{Ethylmethyl thioether}}{C_2H_5 - S - CH_3} + NaBr$$

The reaction takes place by SN^2 mechanism as shown below

$$R\bar{S} + R' - X \longrightarrow R - S - R' + \bar{X}$$

Primary or secondary alkayl halides and aryl halides give this reaction which is similar to Williamson synthesis of ethers. Tertiary alkyl halides are not used because they have a tendency to undergo elimination rather than substituation reaction.

2. From ethers

Ethers on heating with P_2S_5 form thioethers

$$\underset{\text{Diethyl ether}}{5C_2H_5 - O - C_2H_5} + P_2S_5 \longrightarrow \underset{\text{Diethyl thioether}}{5\,C_2H_5 - S - C_2H_5} + P_2O_5$$

3. From alkenes

Addition of thiols to alkenes in the prense of peroxides (anit-Markownikoff's addition) gives thioethers

$$R - CH = CH_2 + R'SH \xrightarrow{\text{peroxide}} R - CH_2 - CH_2 - S - R'$$

In the absence of peroxides, the reactions does not take place.

4. Reaction of alkyl halides or potassium alkyl sulphate with sodium or potassium sulphide

$$2\,C_2H_5\,9 + K_2S \longrightarrow \underset{\text{Diethyl thioether}}{C_2H_5 - S - C_2H_5} + 2KI$$

$$2C_2H_5\,SO_4\,K + K_2S \longrightarrow \underset{\text{Diethyl thioether}}{C_2H_5 - S - C_2H_5} + 2\,K_2\,SO_4$$

5. From thiols

When vapours of a thiol are passed over a mixture of Al_2O_3 and ZnS heated to 575 K, thioethers are obtained.

$$2\,C_2H_5SH \xrightarrow[575\,K]{Al_2O_3,\,ZnS} \underset{\text{Diethyl thioether}}{C_2H_5-S-C_2H_5} + H_2S$$

3.10. PHYSICAL PROPERTIES OF THIOETHERS

1. Thioethers are colourless oily liquids.
2. They have unpleasant smell.
3. Thioethers have higher boiling points compared to those of corresponding ethers.

Compound	Boiling point °C
CH_3SCH_3	37
CH_3OCH_3	24
$C_2H_5SC_2H_5$	92
$C_2H_5OC_2H_5$	35

4. Thioethers are insoluble in water but soluble in ether and alcohol.
5. They show a stretching band at 600–800 cm^{-1}

3.11. CHEMICAL PROPERTIES OF THIOETHERS.

Chemical properties of thioethers are given below :

1. Reaction with halogens

Thioethers react with halogens to give dihalides.

$$\underset{\text{Diethyl sulphide}}{C_2H_5-S-C_2H_5} + Br_2 \longrightarrow \underset{\text{Diethyl sulphide dibromide}}{C_2H_5-\overset{\overset{Br}{|}}{\underset{\underset{Br}{|}}{S}}-C_2H_5}$$

2. Reaction with alkyl halides

Thioethers react with alkyl halides to form sulphonium salts

$$\underset{\text{Diethyl sulphide}}{C_2H_5-S-C_2H_5} + C_2H_5I \longrightarrow \underset{\text{Triethyl sulphonium iodide}}{C_2H_5-\overset{\overset{C_2H_5}{|}}{S^+}-C_2H_5I^-}$$

Triethyl sulphonium iodide reacts with moist silver oxide (AgOH) to form strongly basic triethylsulphonium hydroxide.

$$(C_2H_5)_3\,S^+\,I^- + AgOH \longrightarrow \underset{\text{Triethyl sulphonium hydroxide}}{(C_2H_5)_3\,\overset{+}{S}\overset{-}{O}H} + AgI$$

3. Hydrolysis

Thioethers get hydrolysed with aqueous NaOH to form H_2S and alcohols

$$\underset{\text{Diethyl sulphide}}{C_2H_5-S-C_2H_5} + 2H_2O \xrightarrow{NaOH} \underset{\text{Ethyl alcohol}}{2C_2H_5OH} + H_2S\uparrow$$

4. Oxidation

(*a*) Mild oxidising agents like dilute nitric acid or hydrogen peroxide convert thioethers into sulphoxides at room temperature. Dimethyl sulphoxide is an excellent solvent for polar or non-polar organic compounds. It must be handled carefully as it penetrates the skin readily.

$$\underset{\text{Dimethyl sulphide}}{CH_3-S-CH_3} + H_2O_2 \longrightarrow \underset{\substack{\text{Dimethyl sulphoxide}\\ \text{(DMSO)}}}{CH_3-\overset{\overset{\displaystyle O}{\|}}{S}-CH_3} + H_2O$$

(*b*) Strong oxidising agent potassium permanganate converts thioethers into sulphones. The same reaction is obtained with H_2O_2 at 100°C.

$$\underset{\text{Dimethyl sulphide}}{CH_3-S-CH_3} + H_2O_2 \xrightarrow{100^\circ C} \underset{\text{Dimethyl sulphone}}{CH_3-\overset{\overset{\displaystyle O}{\|}}{\underset{\underset{\displaystyle O}{\|}}{S}}-CH_3}$$

Liquid thioethers can be identified by converting them into sulphones.

5. Reaction with metallic salts

Thioethers react with metallic salts like $HgCl_2$ and $SnCl_4$ to form crystallinic salts. For example

$$(C_2H_5)_2S + HgCl_2 \longrightarrow (C_2H_5)_2\,S^+ - HgCl^-_2$$

$$(C_2H_5)_2\,S + SnCl_4 \longrightarrow (C_2H_5)_2\,S^+ - SnCl^-_4$$

Example. Explain why dimethyl sulphoxide (bp 189°C) has a much higher boiling point than dimethyl sulphide (bp 37°C).

Solution. The bond between S and O in dimethyl sulphoxide is polar because of electronegativity difference.

$$CH_3-\overset{\overset{\displaystyle O}{\|}}{S}-CH_3 \longleftrightarrow CH_3-\overset{\overset{\displaystyle \bar{O}}{|}}{\overset{+}{S}}-CH_3$$

This makes the molecule to associate resulting in increase in its boiling point

$$\ldots\ldots\underset{\underset{\displaystyle CH_3}{|}}{\overset{\overset{\displaystyle CH_3}{|}}{\overset{+}{S}}}-\bar{O}\ldots\ldots\underset{\underset{\displaystyle CH_3}{|}}{\overset{\overset{\displaystyle CH_3}{|}}{\overset{+}{S}}}-\bar{O}\ldots\ldots\underset{\underset{\displaystyle CH_3}{|}}{\overset{\overset{\displaystyle CH_3}{|}}{\overset{+}{S}}}-\bar{O}\ldots\ldots$$

No such polarisation of bond and association between molecules takes place in dimethyl sulphide. Hence, dimethyl sulphoxide has a much higher boiling than dimethyl sulphide.

MUSTARD GAS

Mustard gas is a deadly poisonous gas. It was used as a poison gas in world war (1914-1919). It has the structure :

$$\begin{matrix} CH_2Cl & & CH_2Cl \\ | & & | \\ CH_2 & -S- & CH_2 \end{matrix}$$

3.12. PREPARATION OF MUSTARD GAS

Mustard gas may be prepared by the following methods :

1. $CH_2 = CH_2 \xrightarrow[\text{Addition}]{HOCl} CH_2OH - CH_2Cl$

$$2CH_2OH - CH_2Cl \xrightarrow[-2\,NaCl]{Na_2S} \begin{matrix} CH_2OH & & CH_2OH \\ | & & | \\ CH_2 & -S- & CH_2 \end{matrix} \xrightarrow[-2\,H_2O]{2\,HCl} \underset{\text{Mustard gas}}{\begin{matrix} CH_2Cl & & CH_2Cl \\ | & & | \\ CH_2 & -S- & CH_2 \end{matrix}}$$

2. From ethylene (Addition of Sulphur monochloride)

$$CH_2 = CH_2 + \underset{\text{Sulphur monochloride}}{\begin{matrix} Cl & & Cl \\ | & & | \\ S & - & S \end{matrix}} + CH_2 = CH_2 \longrightarrow \underset{\text{Mustard gas}}{\begin{matrix} CH_2Cl & & CH_2Cl \\ | & & | \\ CH_2 & -S- & CH_2 \end{matrix}} + S$$

3.13. PROPERTIES OF MUSTARD GAS

1. It is an oily liquid with b.p. 215°C (Mustard gas is a misnomer)
2. It is insoluble in water, but soluble in ethyl alcohol and ether.
3. It produces painful blisters on the skin and damages lungs.
4. It has a prolonged action and causes death to the exposed persons after four day

SULPHONIC ACIDS

Sulphonic acids are organic compounds having the general formula R–SO_2OH or R–SO_3H where R is an alkyl or aryl group. It is mainly the aromatic compounds which are prominent and useful.

Sulphonic acids may be regarded as derivatives of sulphuric acid $HOSO_2OH$ in which – OH group is replaced by an aryl group.

Aryl alkyl sulphonic acids

Sulphonic acid has the following structure.

$$\begin{matrix} & O & \\ & | & \\ R - & S & - OH \\ & | & \\ & O & \end{matrix}$$

Studies indicate that there is some double bond character in S–O bonds.

3.14. NOMENCLATURE OF SULPHONIC ACIDS

Sulphonic acids are named by adding the words sulphonic acid to the parent aliphatic or aromatic compound. For example,

Compound	*Name*
CH_3SO_3H	Methanesulphonic acid
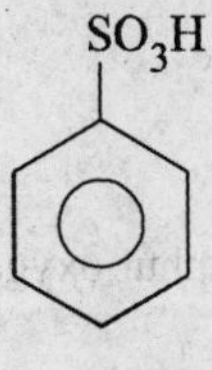	Benzenesulphonic acid
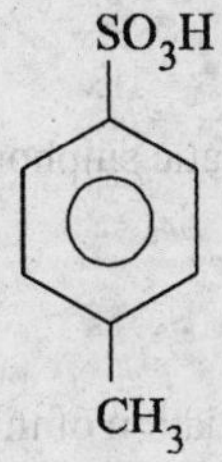	*p*-Toluenesulphonic acid
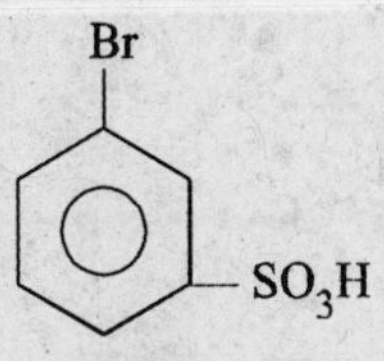	*m*-Bromobenzenesulphonic acid
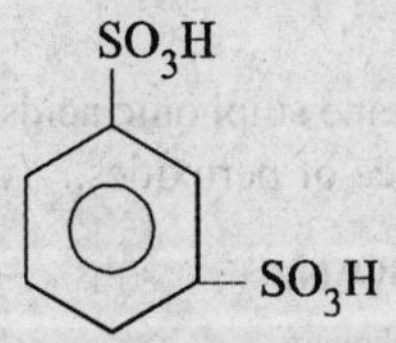	*m*-Benzenedisulphonic acid

3.15. STRUCTURE OF SULPHONIC ACIDS

The hydroxy oxygen is linked to sulphur and hydrogen respectively by single bondes, i.e. S – O – H. However, there is some difference of opinion regarding the representation of the bonds between remaining two oxygens and sulphur. There are two main types of representations in use as given below:

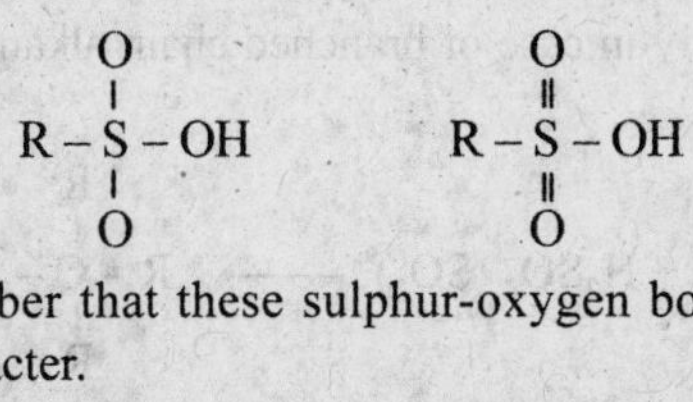

It is important to remember that these sulphur-oxygen bonds are highly polar and have considerable double bond character.

The bonding of sulphur in sulphonic acids can be explained as follows. The sulphur atom in sulphonic acids is in an **excited** and **sp^3 hybridised state**. These hybrid orbitals of sulphur form σ bonds with each of the three oxygens and one carbon (of alkyl or aryl group). At the same time d orbitals of sulphur undergoe dπ -pπ overlapping with the properly oriented *p* orbitals of the two oxygen atoms as illustrated below:

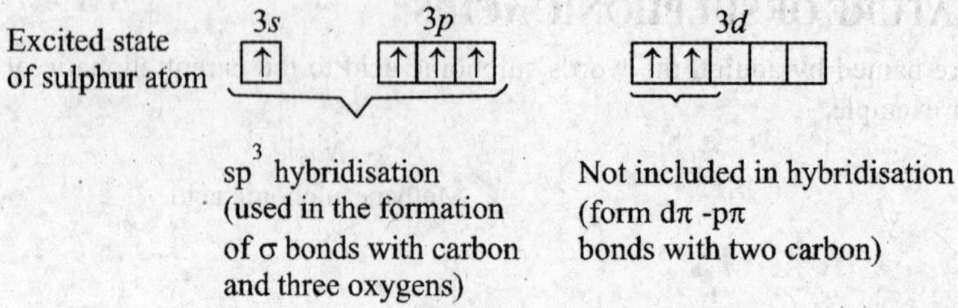

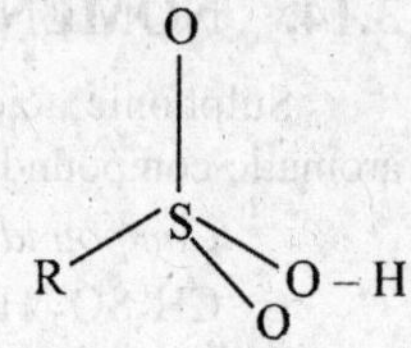

Formation of sulphonic acids

The measurment of bond lengths and bond energies shows that the two sulphur oxygen bonds have bonds order of slightly less than 2.

3.16. METHODS OF FORMATION

The methods generally employed for the preparation of alliphatic and aromatic sulphonic acids are given as under:

Aliphatic sulphonic acids

(1) Oxidation of thiols. Alkane sulphonic acids can be prepared by the oxidation of thiols with strong oxidising agents such as nitric acid and potassium permangenate.

$$\underset{}{RSH} \xrightarrow{HNO_3} \underset{\text{Alkane sulphonic acid}}{RSO_3H}$$

For example:

$$\underset{\text{Ethane thiol}}{C_2H_5SH} \xrightarrow{HNO_3} \underset{\text{Ethane sulphonic acid}}{C_2H_5SO_3H}$$

(2) Additon of sodium bisulphite to alkenes. Soldium salts of alkane sulphonic acids are also obtained by the addition of sodium bisulphate to alkenes in the presence of peroxides.

$$\underset{\text{Propene}}{CH_3-CH=CH_2} + NaHSO_3 \xrightarrow{\text{Peroxide}} \underset{\text{Sodium propane sulphonate}}{CH_3-CH_2-CH_2SO_3Na}$$

(3) Reaction between sodium sulphate and alkyl halide (Strecker reaction). When a mixture of sodium sulphite and alkyl halide is heated, sodium salts of alkane sulphonic acids are obatined.

$$C_2H_5Cl + Na_2SO_3 \longrightarrow \underset{\text{Sod.salt of ethane sulphonic acid}}{C_2H_5SO_3\,Na} + NaCl$$

(4) Sulphonation of alkanes. Direct sulphonation of alkanes may be carried out by reaction with fuming sulphuric acid particularly in case of branched-chain alkanes in which tertiary cabon atom are available

$$R-\overset{\overset{R}{|}}{\underset{\underset{R}{|}}{C}}-H + H_2SO_4\,(SO_3) \longrightarrow R-\overset{\overset{R}{|}}{\underset{\underset{R}{|}}{C}}-SO_3H + H_2SO_4$$

Aromatic sulphonic acids

Aromatic sulphonic acids are prepared by sulphonation. Various sulphonating agents are used depending upon the substance to be sulphonated and the number of sulphonic acid groups to be introduced. Different sulphonating agents are as follows.

1. Concentrated H_2SO_4.
2. Chlorosulphonic acid ($ClSO_3H$).
3. Fuming H_2SO_4 (Oleum) containing varying amount of SO_3 dissolved in conc. H_2SO_4.

Also different temperatures are employed for different substances.

(*a*) Benzene sulphonic acid is prepared by the sulphonation of benzene using oleum as the sulphonating agent at 303–323 K.

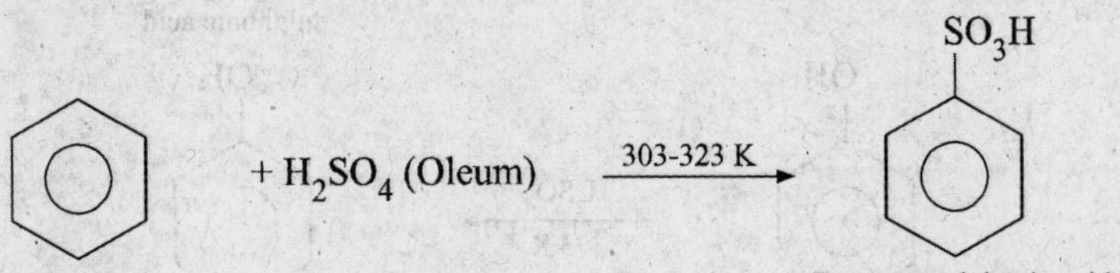

Benzene — Benzene sulphonic acid

(*b*) If the benzene ring contains an activating group like phenolic or alkyl group, conc. H_2SO_4 is used as the sulphonating agent at a temperature of 303 K.

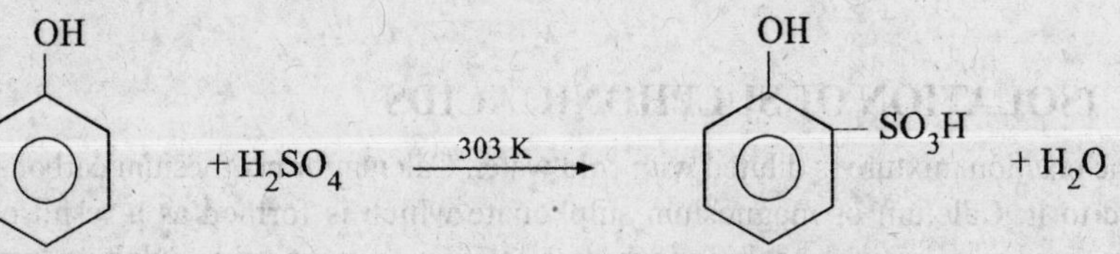

Phenol — *o*-hydroxy benzene sulphonic acid

Sulphonic group takes the *o*- and *p*-position with respect to the activating group already present.

(*c*) If the benzene ring contains a deactivating group like $-NO_2$ or $-COOH$ or $-SO_3H$ itself, the sulphonating agent used is oleum at 473–513 K. It requires heating for several hours. Sulphonic group enters the meta position with respect to the group present.

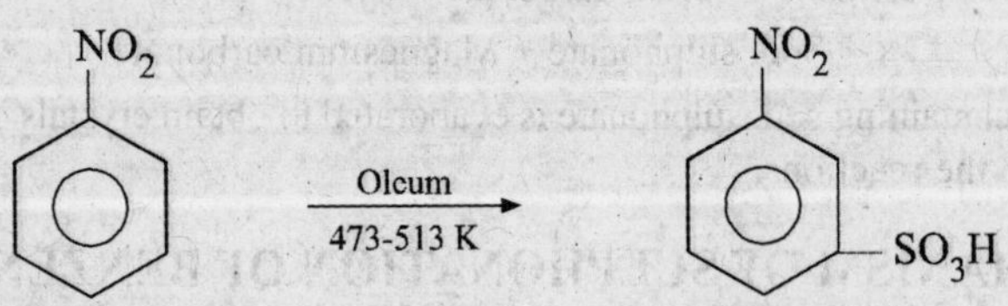

Nitrobenzene — *m*-nitrobenzene sulphonic acid

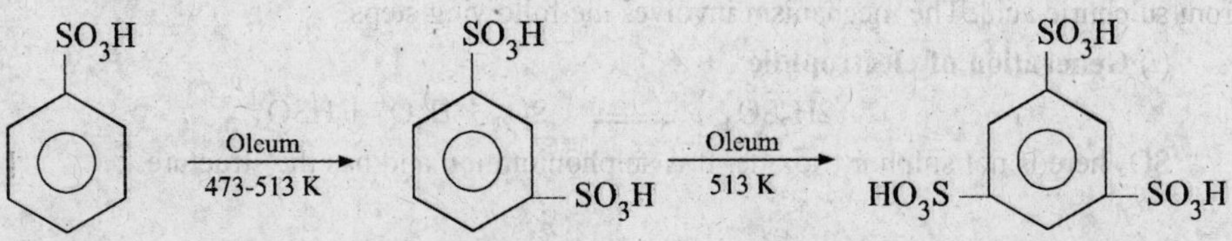

Benzene sulphonic acid — Benzene disulphonic acid — Benzene trisulphonic acid

(*d*) There is a marked influence of temperature on the relative proportion of isomers obtained. In the case of an activating group already present in the benzene ring, we obtain mixture of *o-p* sulphonic acids. It is experienced that a lower temperature favours the formation of ortho sulphonic acid, whereas at higher temperature, para isomer is obtained predominantly.

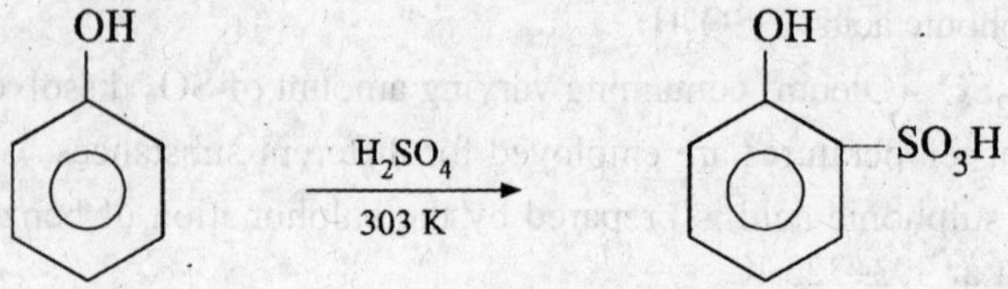

Phenol o-phenol sulphonic acid

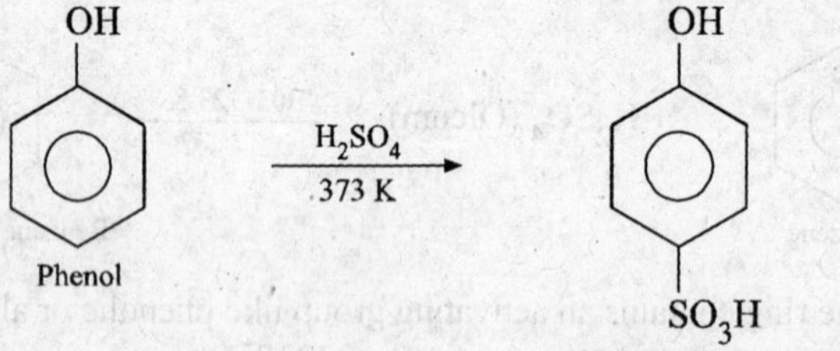

Phenol *p*-phenol sulphonic acid

3.17. ISOLATION OF SULPHONIC ACIDS

The reaction mixture is diluted with cold water. Calcium or magnesium carbonate or hydroxide is added to it. Calcium or magnesium sulphonate which is formed as a result of reaction with sulphonic acid remains in the solution whereas calcium or magnesium sulphate formed by reaction with unused sulphuric acid is thrown out and removed by filtration. The filtrate is then treated with calculated quantity of sodium carbonate when calcium or magnesium carbonate is precipitated and filtered out.

Calcium sulphonate + Sodium carbonate

$\longrightarrow$ Sod. sulphonate + Calcium carbonate

Magnesium sulphonate + Sodium carbonate

$\longrightarrow$ Sod. sulphonate + Magnesium carbonate

The filtrate containing sod. sulphonate is evaporated to obtain crystals of sod. sulphonate which is used as such in the reactions.

3.18. MECHANISM OF SULPHONATION OF BENZENE

Sulphonation of benzene is an electrophilic reaction in which hydrogen of the benzene ring is substituted by sulphonic acid group. The electrophile here is SO_3 sulphonium ion which is produced from sulphuric acid. The mechanism involves the following steps.

(*i*) **Generation of electrophile**

$$2H_2SO_4 \rightleftharpoons SO_3 + H_3O^+ + HSO_4^-$$

SO_3 here is not sulphur trioxide. It is sulphonium ion and has the structure.

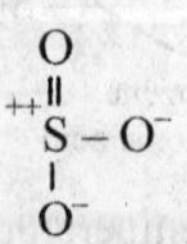

Although there is no overall charge on SO_3, but still it acts as an electrophile i.e. electron-seeking because the positive charges are concentrated at one point whereas negative charges are scattered. Thus it acts as an electrophile.

(*ii*) **Attachment of electrophile to the benzene nucleus**

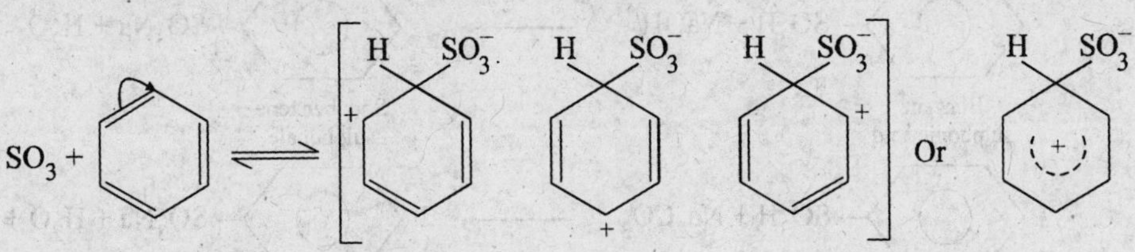

(*iii*) **Removal of proton to form sulphonic acid anion.**

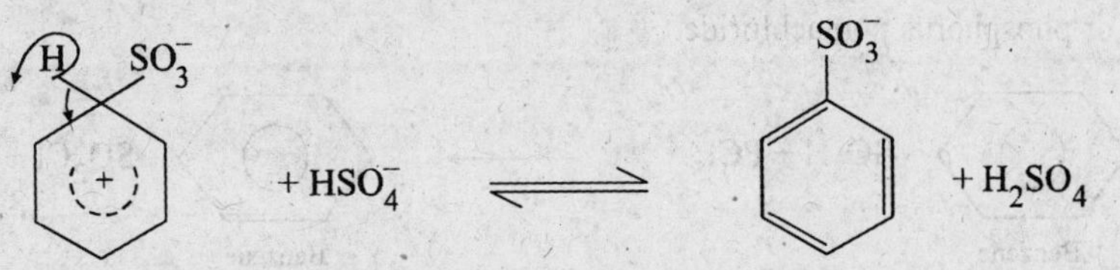

(*iv*) **Reaction between sulphonic acid anion and hydronium ion**

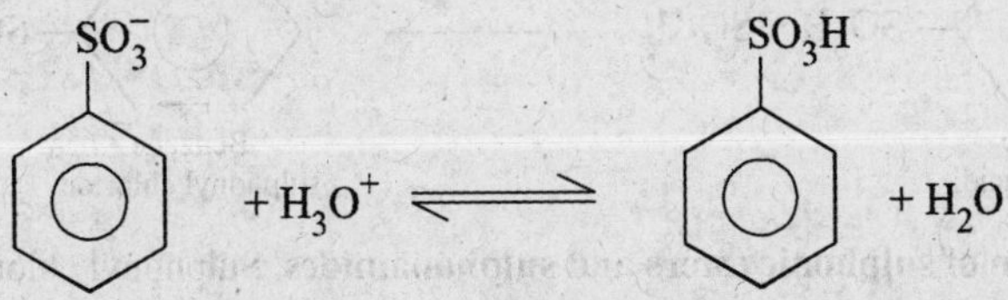

Why are sulphonic acids stronger than carboxylic acids?

Sulphonic acids are strong acids and are completely ionised in water to form sulphonate ion and hydronium ion.

$$C_6H_5SO_3H + H_2O \longrightarrow C_6H_5SO_3^- + H_3O^+$$

The acidity of sulphonic acids is due to the ease of release of protons. This becomes possible due to presence of two strongly electronegative oxygens and formation of a stable sulphonate ion in which the negative charge is dispersed over three oxygen atoms.

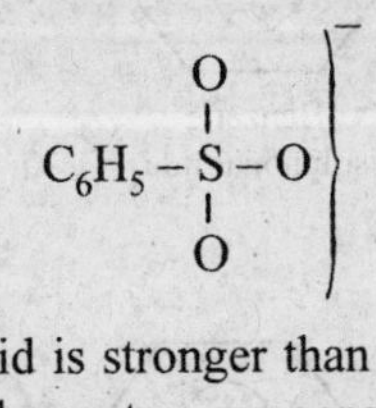

Sulponate ion stabilized by dispersal of negative charge over three oxygens.

Sulphonic acid is stronger than a carboxy acid because in the carboxylate ion, the negative charge is dispersed over two oxygens only

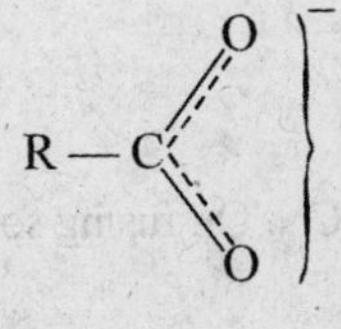

3.19. CHEMICAL PROPERTIES OF BENZENE SULPHONIC ACID

Chemical properties of benzene sulphonic acid are given below:

1. Salt formation. As sulphonic acids are strongly acidic, they form salts with hydroxides, carbonates and bicarbonates

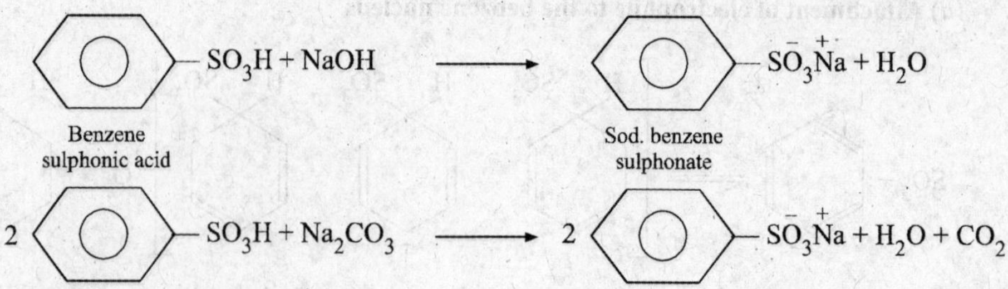

2. Formation of sulphonyl chloride. Sulphonic acid forms sulphonyl chloride with thionyl chloride or phosphorus pentachloride.

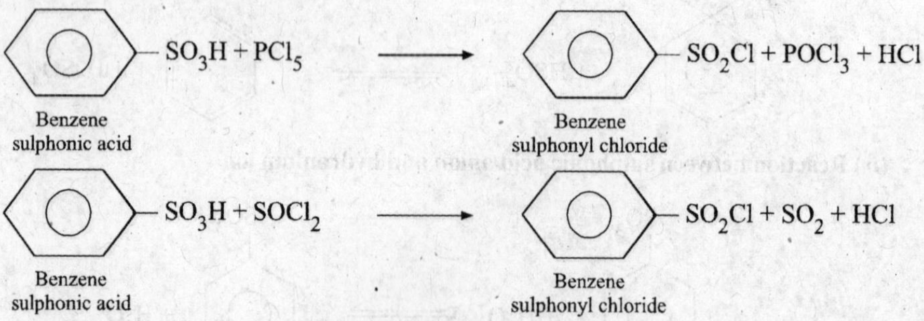

3. Formation of sulphonic esters and sulphonamides. Sulphonyl chloride reacts with alcohols and ammonia to give sulphonic esters and sulphonamide respectively.

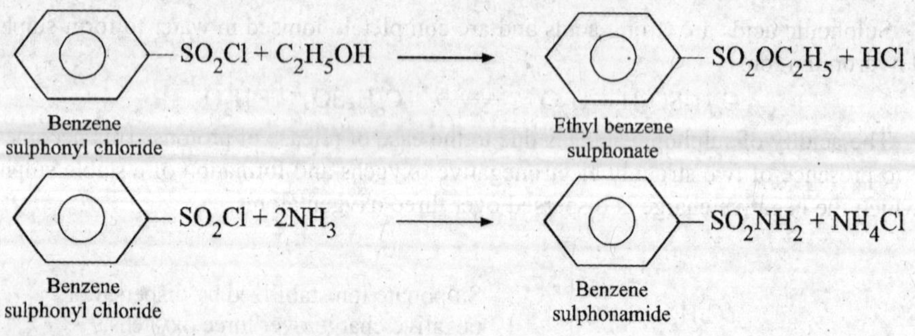

4. Replacement of sulphonic group by phenolic group

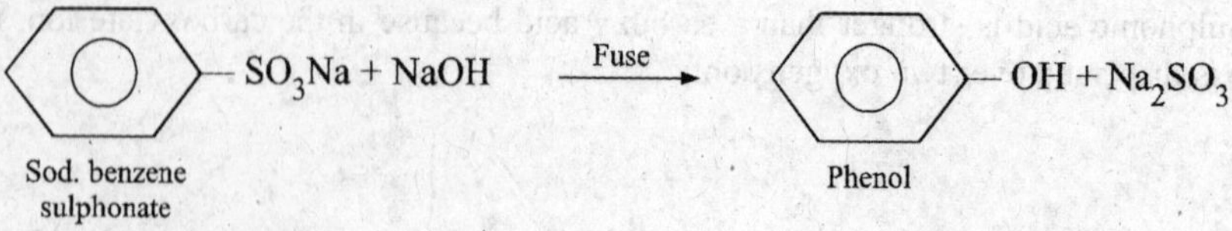

5. Replacement of – SO_3H by – CN. On fusing sod. benzene sulphonate with sodium cyanide, benzonitrile is obtained.

6. Replacement of – SO_3H by – H. When sulphonic acids are heated with mineral acids at 425 K, sulphonic acid group is replaced by hydrogen to give aromatic hydrocarbon. This reaction is known as **desulphonation.**

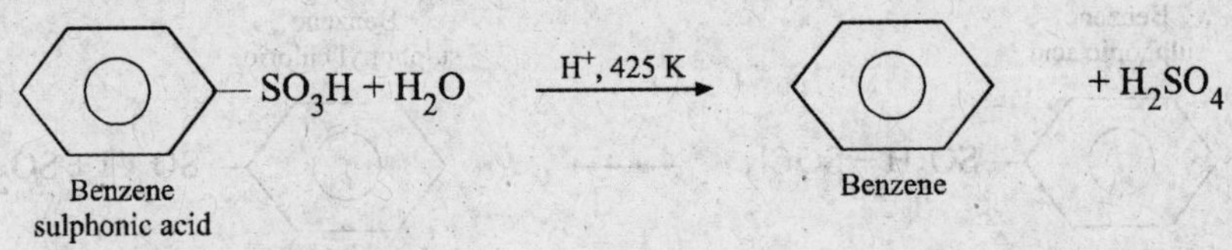

7. Replacement of – SO_3H by – NH_2. The fusion of sodium salt of sulphonic acid with sodamide leads to the formation of amines.

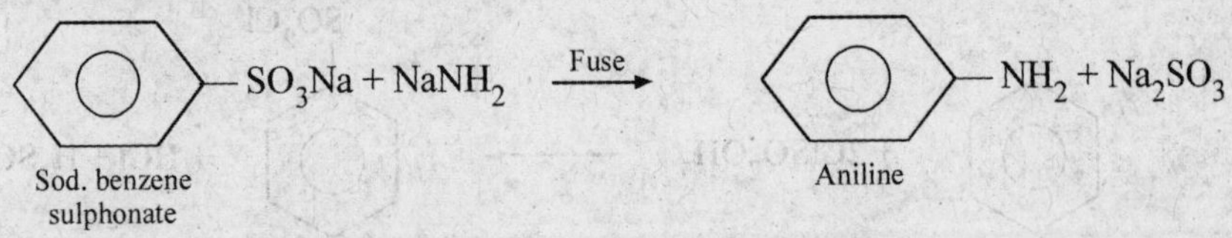

8. Electrophilic substitution reactions. Aromatic sulphonic acids undergo electrophilic substitution reactions like halogenation, nitration and sulphonation. The new group enters the meta position with respect to the group already present. It requires tough conditions for the reactions as sulphonic group is a deactivating group.

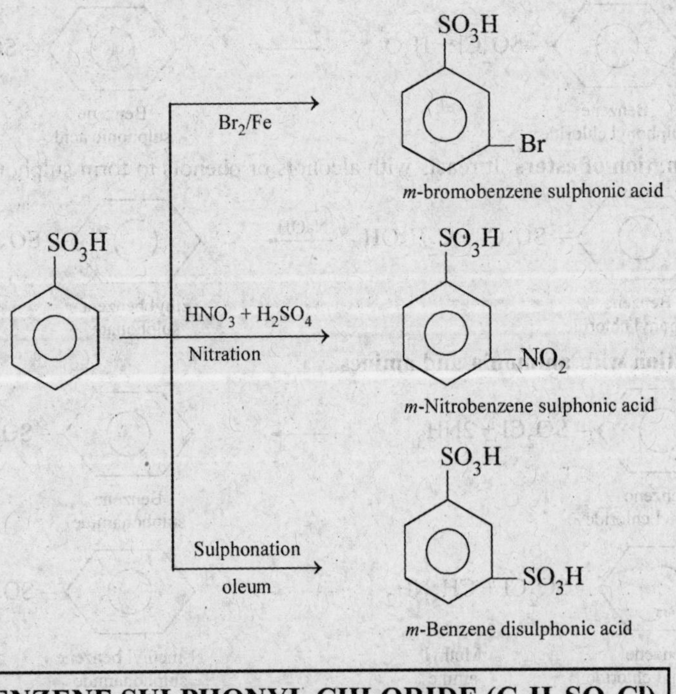

BENZENE SULPHONYL CHLORIDE ($C_6H_5SO_2Cl$)

3.20. METHODS OF PREPARATION

Benzene sulphonyl chloride can be prepared by the following methods.

1. By the action of phosphorus pentachloride or thionyl chloride on benzene sulphonic acid

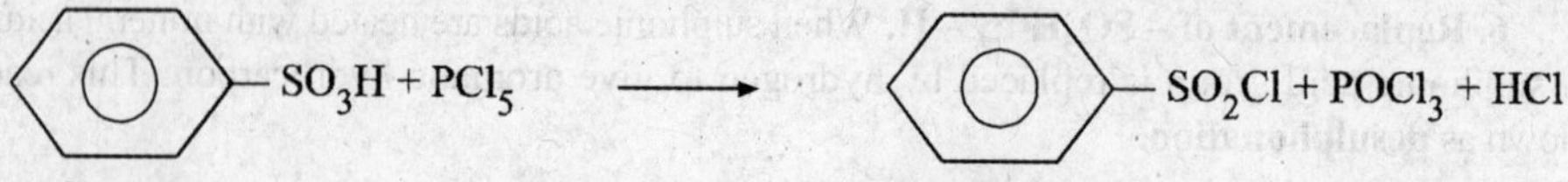

Benzene sulphonic acid → Benzene sulphonyl chloride

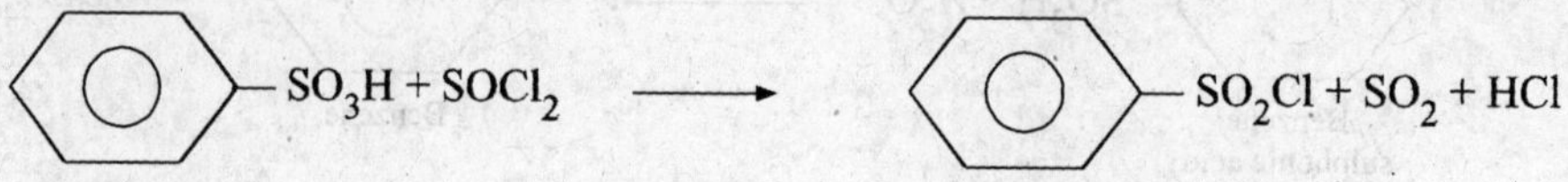

Benzene sulphonic acid → Benzene sulphonyl chloride

2. By the action of chlorosulphonic acid on benzene

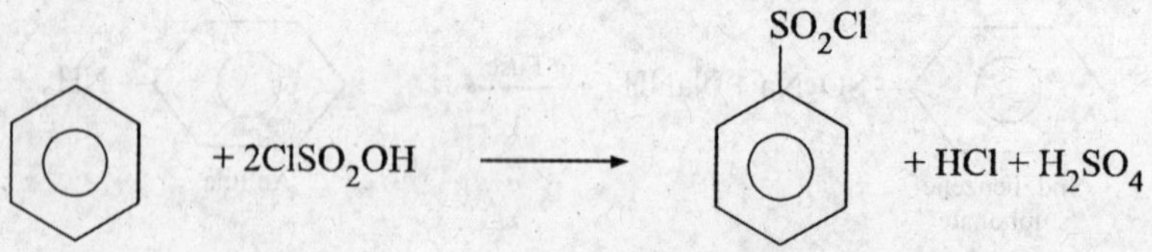

Benzene → Benzene sulphonyl chloride

3.21. PROPERTIES

1. Hydrolysis. On hydrolysis in the presence of alkali, it hydrolyses into benzene sulphonic acid.

Benzene sulphonyl chloride → Benzene sulphonic acid

2. Formation of esters. It reacts with alcohols or phenols to form sulphonic esters.

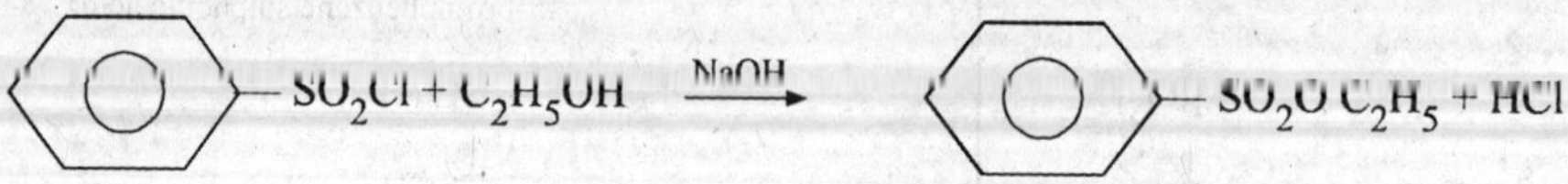

Benzene sulphonyl chloride → Ethyl benzene sulphonate

3. Reaction with ammonia and amines.

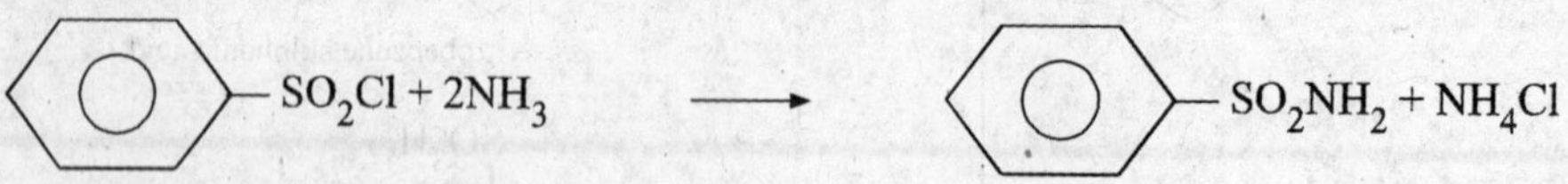

Benzene sulphonyl chloride → Benzene sulphonamide

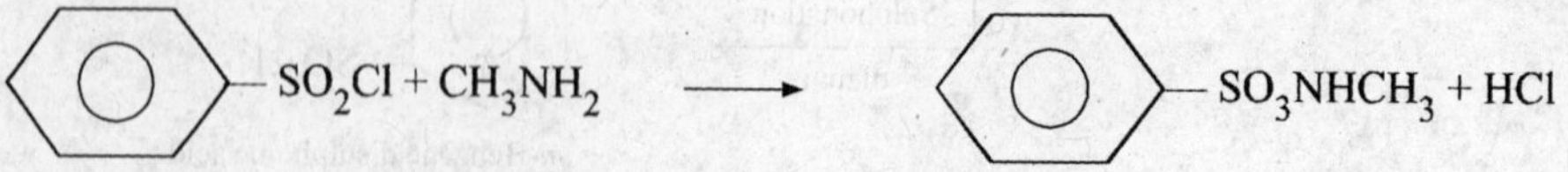

Benzene sulphonyl chloride, Methyl amine → N-methyl benzene sulphonamide

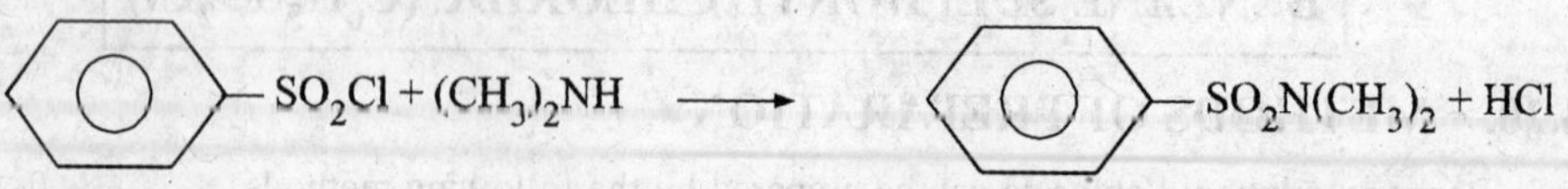

Benzene sulphonyl chloride → N, N Dimethyl benzene sulphonamide

4. Friedel-Crafts Reaction. It undergoes Friedel-Crafts reaction with aromatic hydrocarbons in the presence of anhydrous aluminium chloride to form sulphones.

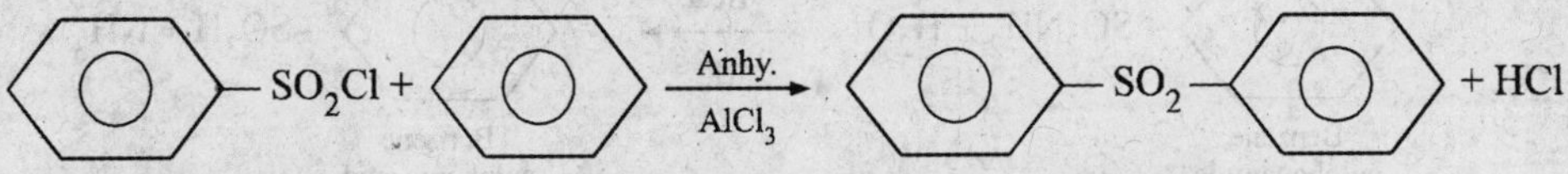

Benzene sulphonyl chloride — Diphenyl sulphone

BENZENE SULPHONAMIDE ($C_6H_5SO_2NH_2$)

3.22. METHODS OF PREPARATION

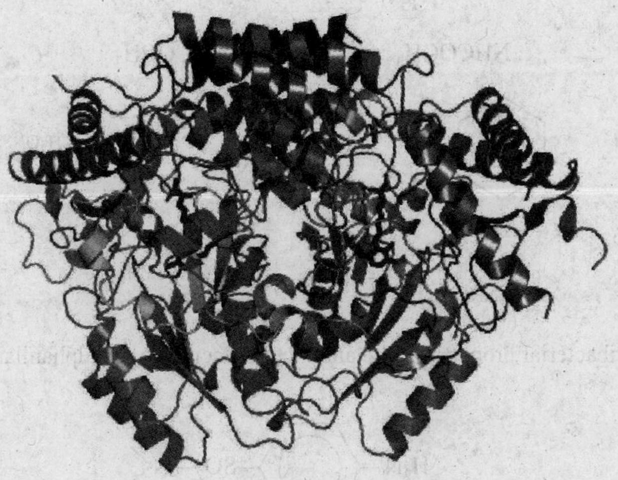

Aminosulphonic acid

Benzene sulphonamide is prepared by the action of ammonia on benzene sulphonyl chloride.

Benzene sulphonyl chloride — Benzene sulphonamide

3.23. PROPERTIES

1. It is weakly acidic in nature and hence reacts with strong alkalis to form water soluble salts

Benzene sulphonamide — Sod. salt

2. On heating with acids, it gets hydrolysed into benzene sulphonic acid.

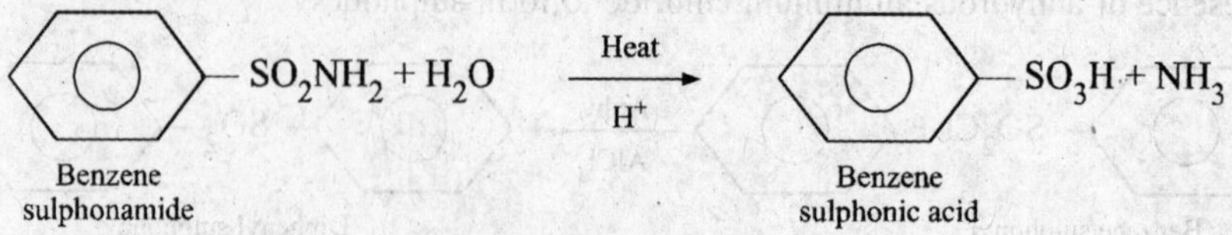

Sulphonamides are well-defined crystialline solids. They are used to characterise sulphonic acids.

3.24. SULPHANILAMIDE (*p*-AMINOBENZENE SULPHONAMIDE)

Synthesis. Acetanilide is treated with chlorosulphonic acid to produce *p*-acetamido benzene sulphonyl chloride which is treated with NH_3 to produce *p*-acetamido benzene sulphonamide. The latter on hydrolysis in the presence of an acid yields sulphanilamide.

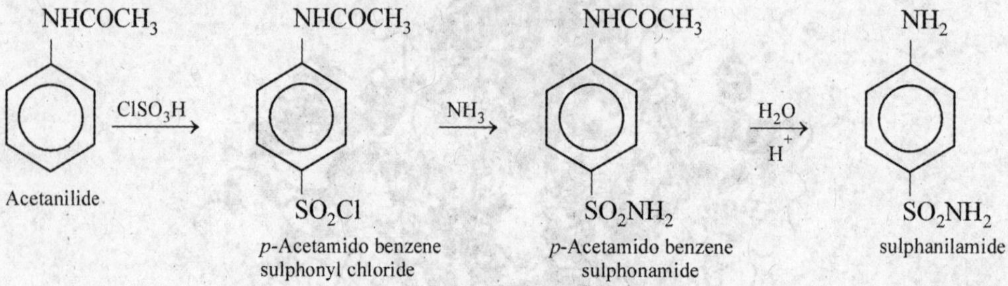

It has got antibacterial properties. The antibacterial acitivity of sulphanilamide is associated with the group.

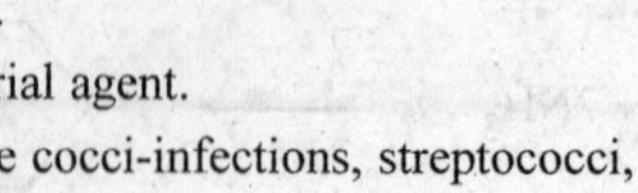

p-aminobenzoic acid is an essential growth factor for most becteria susceptiable to sulphoponamide. The theory of action is, that due to similarity in structure, bacteria absorb sulphonamide by mistake and the bacteria cease to grow in number. Thus sulphonamides are bactericidal as well as bacteriastatic.

Uses. 1. It is used as antibacterial agent.

2. It is used in medicine to cure cocci-infections, streptococci, gonococci and pneumococci.

3.25. SULPHAGUANIDINE

Synthesis. The starting material for sulphaguanidine is the same as for sulphanilamide. *p*-acetamido benzene sulphonyl chloride obtained in the first step is treated with guanidine to obtain sulphaguanidine.

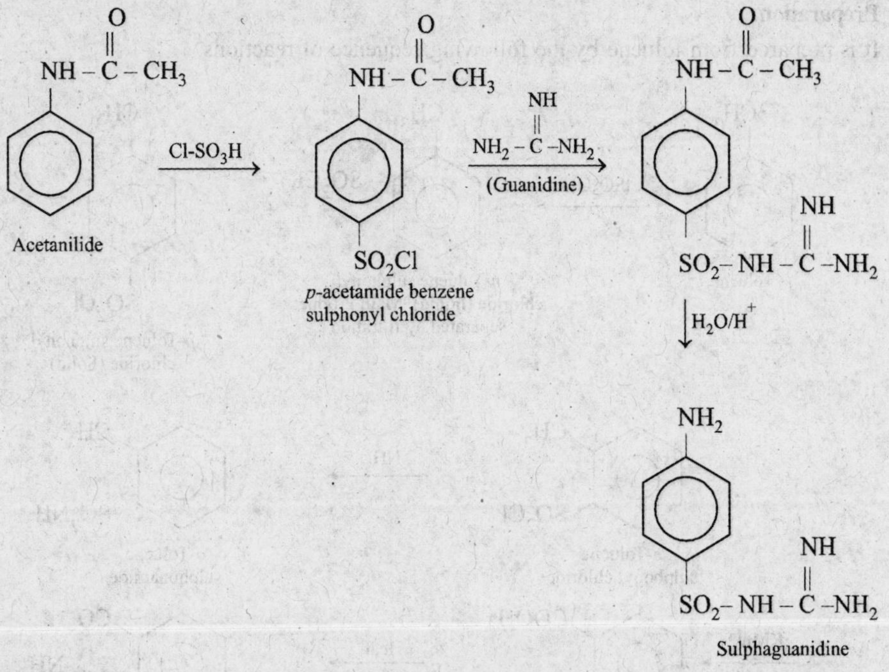

Uses. It is used in the treatment of bacillary dysenetry because it is absorbed only slightly in intestinal tract.

3.26. SACCHARIN

(o-sulphobenzoic imide)

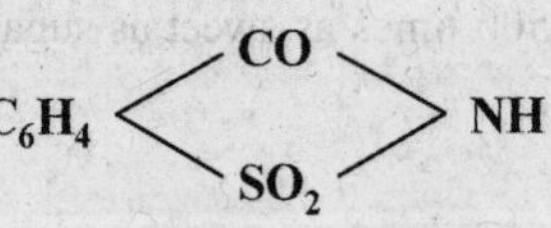

Model of Saccharin

Preparation.

It is prepared from toluene by the following sequence of reactions.

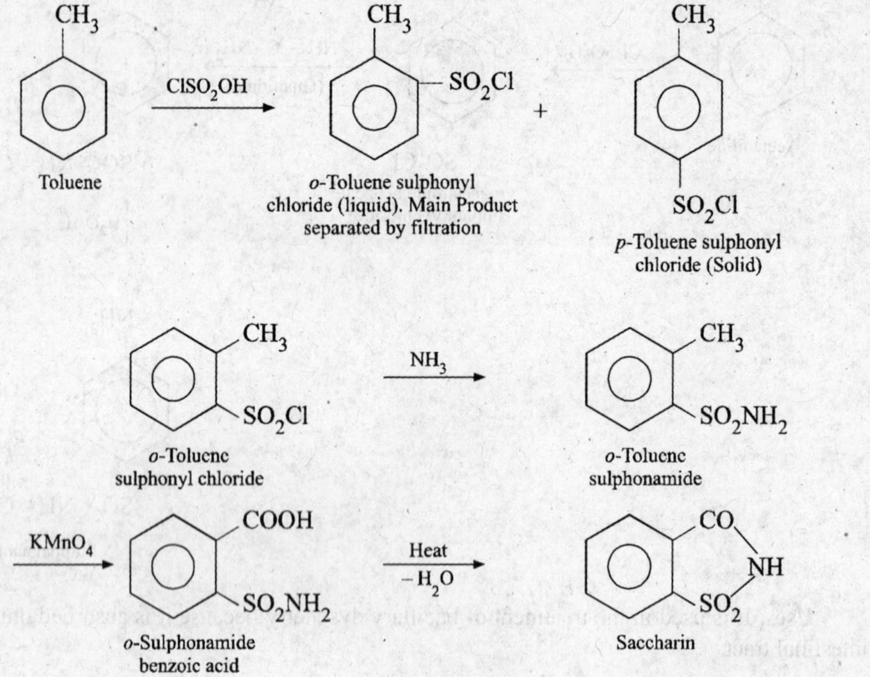

Properties.

1. It is a white solid about 500 times as sweet as sugar.

2. It forms a salt with alkali.

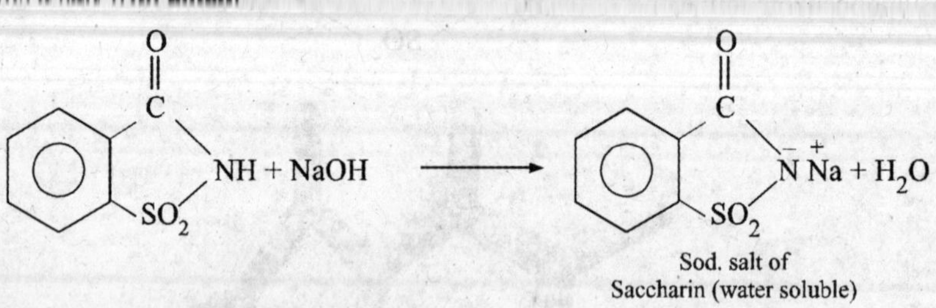

Example 1. Which one is the strongest acid?

(i) C_6H_5COOH **(ii) $OCH_3C_6H_6SO_3H$** **(iii) $NO_2C_6H_4SO_3H$**

(iv) $C_6H_5SO_3H$.

Solution. Benzene sulphonic acid is definitely stronger than benzoic acid because in benzene sulphonate ions, the negative charge is dispersed over three oxygens whereas in carboxylate ion the negative charge is dispersed over two oxygens.

An electron withdrawing group at *o*- or *p*-position in the benzene ring of benzene sulphonic acid increases the acid strength whereas an electron-donating group decreases the acid strengths. Hence the order of acid strength in the compounds is as follows:

o-nitrobenzene sulphonic acid > benzene sulphonic acid > *o*-methoxybenzene sulphonic acid > benzoic acid.

Example 2. Predict the product of monosulphonation of the following

(a) Toluene at 373 K **(b) p-nitrophenol**

(c) nitrobenzene **(d) m-dimethylbenzene.**

Solution.

(*a*)

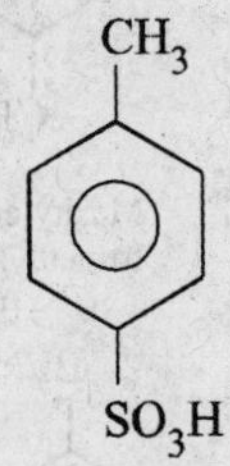

p-Toluene sulphonic acid

(*b*)

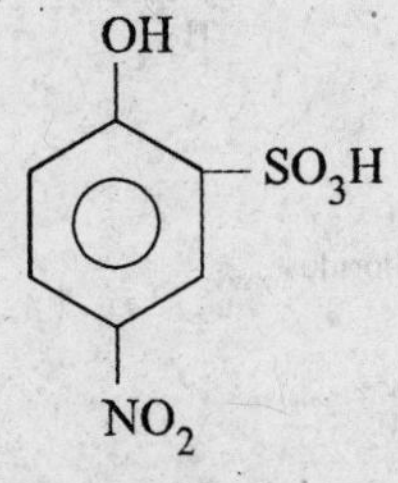

2-Hydroxy-5-nitrobenzene sulphonic acid

(*c*)

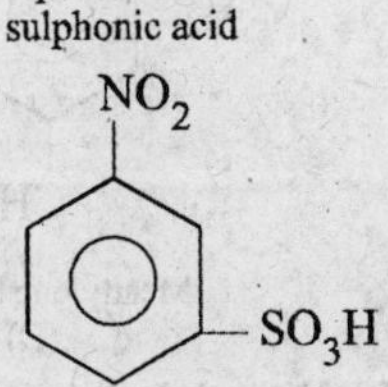

m-Nitrobenzene sulphonic acid

(*d*)

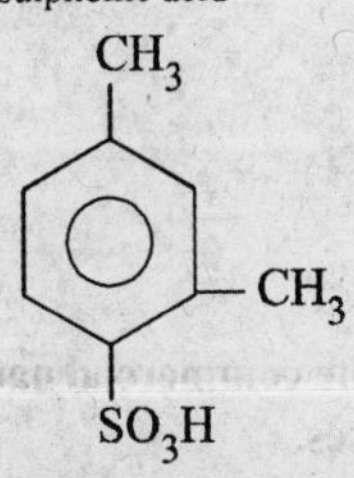

2, 4, Dimethyl benzene sulphonic acid

Example 3. How will you prepare *o*-bromotoluene from toluene without using direct bromination, which gives a large amount of unwanted *p*-isomer?

Solution.

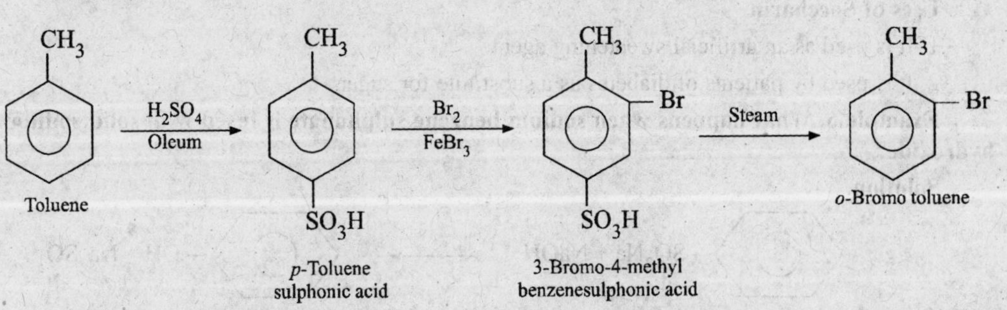

Example 4. Identify the compound $C_9H_{11}O_2SCl$ (A) and the compounds (B), (C) and (D) in the following reactions.

$$\text{AgCl} \xleftarrow[\text{Rapid}]{\text{AgNO}_3} C_9H_{11}O_2SCl\ (A) \xrightarrow[H_3O^+]{\text{NaOH}} \text{Water soluble (B)}$$

$$\text{(B)} \xrightarrow[\text{heat}]{H_3O^+} C_9H_{12}\ (C) \xrightarrow[\text{Fe}]{Br_2} \text{One monobromo derivative (D)}$$

Solution. Compound (A) contains sulphur and gives a precipitate with $AgNO_3$. It indicates an active chlorine and the probable group is – SO_2Cl. Sulphonyl chloride group on treatment with sodium hydroxide gives a sodium salt – SO_2Na and on hydrolysis with an acid is converted into – SO_2OH. On heating, the sulphonic acid group in removed. The following sequence explains the above reactions.

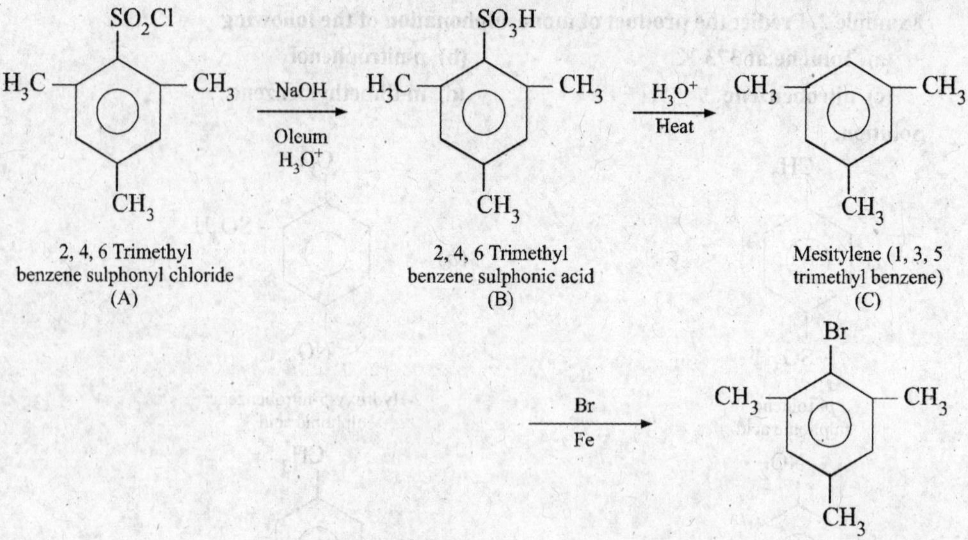

2, 4, 6 Trimethyl benzene sulphonyl chloride (A)

2, 4, 6 Trimethyl benzene sulphonic acid (B)

Mesitylene (1, 3, 5 trimethyl benzene) (C)

Mesitylene bromide (D)

Example 5. Write commercial name and structural formula of ortho sulphobenzoic imide and give its important uses.

Solution. Commercial name of orthosulphobenzoic imide is saccharin and its structural formula is:

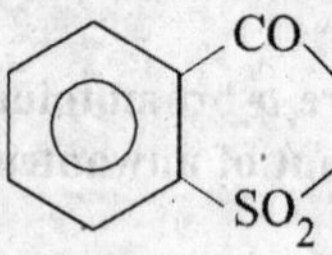

Uses of Saccharin

1. It is used as an artificial sweetening agent.
2. It is used by patients of diabetes as a substitute for sugar.

Example 6. What happens when sodium benzene sulphonate is fused with solid sodium hydroxide?

Solution.

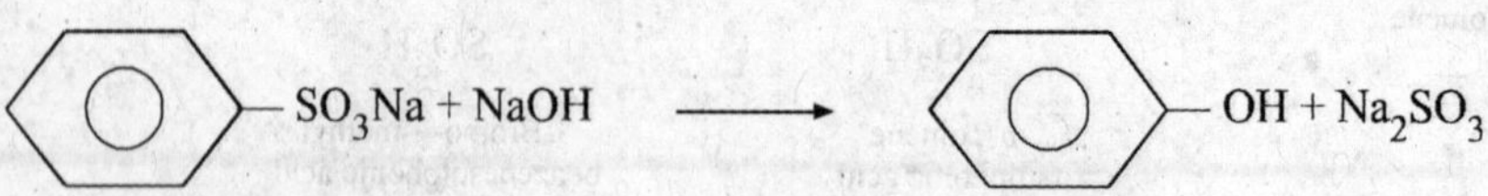

Sodium benzene sulphonate (solid) phenol

Example 7. What happens when benzene sulphonic acid reacts with

(*i*) NaOH Sol (*ii*) Conc HNO_3

Solution.

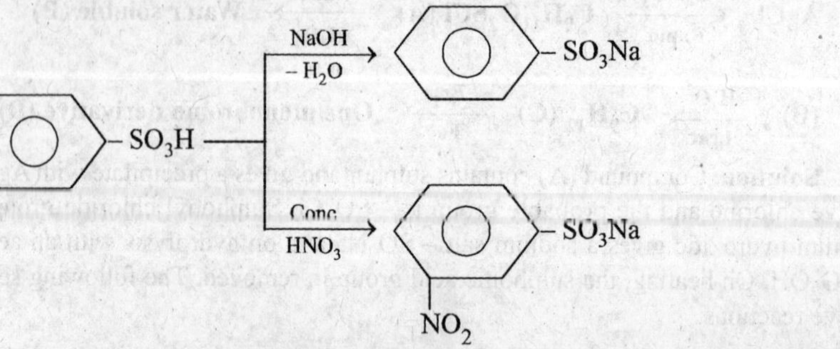

Example 8. How is chloramine – T prepared ? What are its uses ?

Solution. It is prepared as follows :

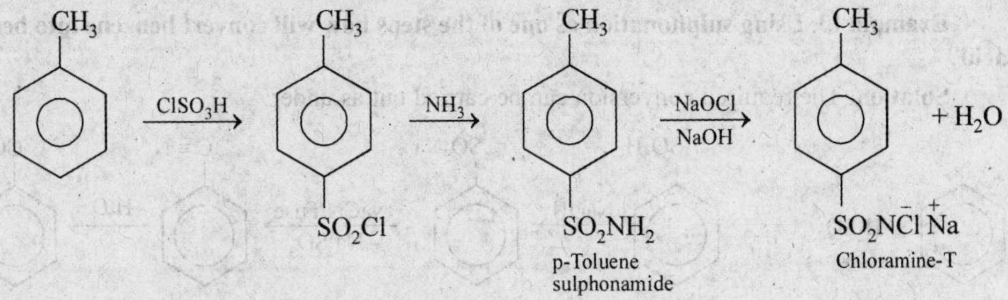

Chloramine – T reacts with water to liberate hypochlorous acid.

It is an effective antiseptic for wounds.

Example 9. How is sulphanilic acid prepared? What are its uses ?

Solution. It is prepared as under :

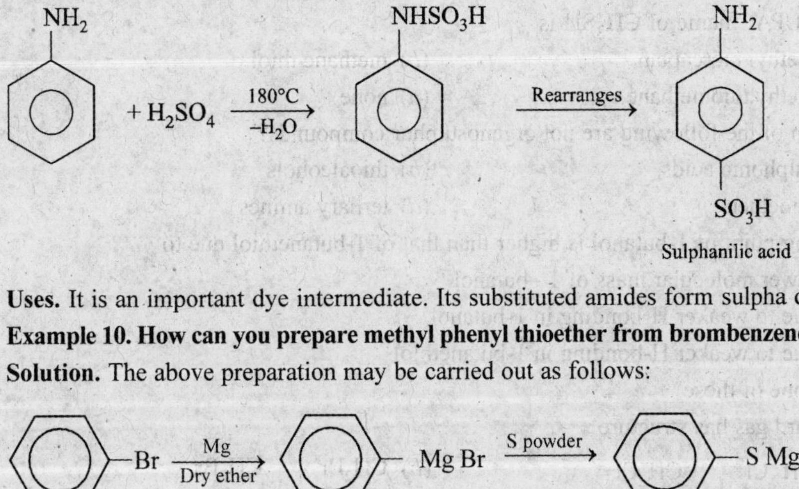

Uses. It is an important dye intermediate. Its substituted amides form sulpha drugs.

Example 10. How can you prepare methyl phenyl thioether from brombenzene?

Solution. The above preparation may be carried out as follows:

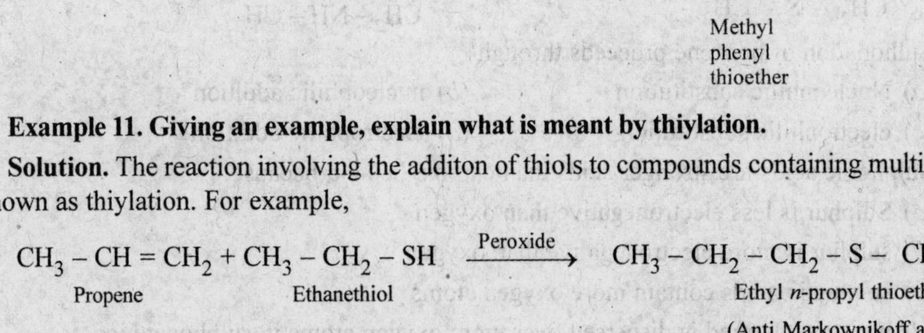

Example 11. Giving an example, explain what is meant by thiylation.

Solution. The reaction involving the additon of thiols to compounds containing multiple bonds is known as thiylation. For example,

$$\underset{\text{Propene}}{CH_3 - CH = CH_2} + \underset{\text{Ethanethiol}}{CH_3 - CH_2 - SH} \xrightarrow{\text{Peroxide}} \underset{\text{Ethyl } n\text{-propyl thioether}}{CH_3 - CH_2 - CH_2 - S - CH_2 - CH_3}$$

(Anti Markownikoff addition)

Example 12. Why do thioethers act as stronger nucleophiles as compared with ethers?

Solution. Sulphur is larger in size and has lower electronegativity as compared with oxygen. As a result, the lone pairs of electrons on sulphur are not tightly held and are easily available for sharing.

Example 13. Using sulphonation as one of the steps how will convert benzene into benzoic acid?

Solution. The required conversion can be carried out as under:

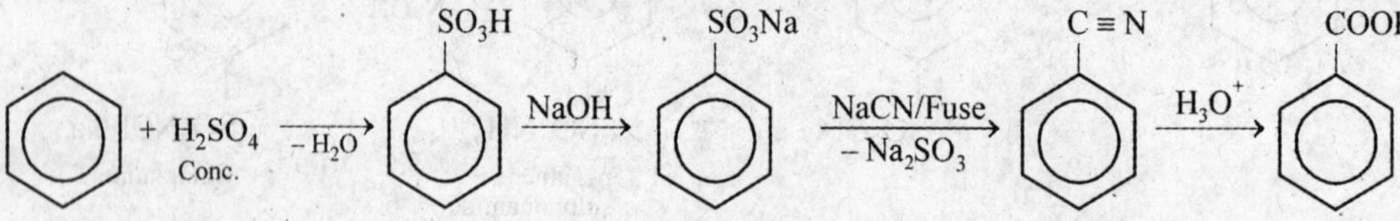

EXERCISES
(Including Questions from Different University Papers)

Multiple choice questions (Choose the correct option)

1. The IUPAC name of CH_3SH is

(*a*) methyl mercaptan (*b*) methane thiol

(*c*) methylthiomethane (*d*) none

2. Which of the following are not organosulphur compounds?

(*a*) sulphonic acids (*b*) thioalcohols

(*c*) thioethers (*d*) tertiary amines

3. Boiling point of 1-butanol is higher than that of 1-butanethiol due to

(*a*) lower molecular mass of 1 –butanol

(*b*) due to weaker H bonding in 1-butanol

(*c*) due to weaker H-bonding in 1-butanethiol

(*d*) none of these

4. Mustard gas has structure

(*a*) CH_2Cl CH_2Cl (each on a CH_2 of) $CH_2 - S - CH_2$

(*b*) CH_2Br CH_2Br (each on a CH_2 of) $CH_2 - S - CH_2$

(*c*) CH_2I CH_2I (each on a CH_2 of) $CH_2 - S - CH_2$

(*d*) CH_2I CH_2I (each on a CH_2 of) $CH_2 - NH - CH_2$

5. Sulhonation of benzene proceeds through

(*a*) Nucleophilic substitution (*b*) nucleophilic addition

(*c*) electrophilic substitution (*d*) electrophilic addition

6. Sulphonic acids are stronger acids than carboxylic acid because

(*a*) Sulphur is less electronegative than oxygen

(*b*) sulphur is more electronegative than oxygen

(*c*) sulphonic acids contain more oxygen atoms

(*d*) negative charged is dispersed over more oxygen atoms in sulphonate ion

7. Commercial name of the compound

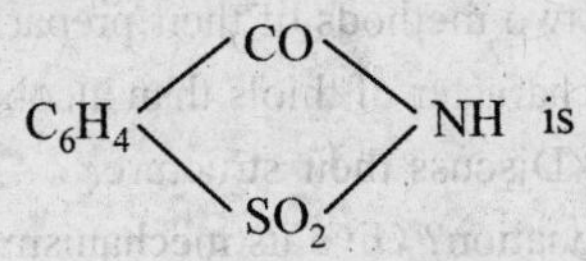

(a) crocin (b) gammaxene
(c) saccharin (d) lindane

8. On boiling with alkali, thioethers are hydrolysed to
(a) alcohols (b) ethers
(c) sulphonic acids (d) sulphanilic acid

9. Thiols can be converted into thioethers by passing vapours of thiols heated to 575 K in the presence of
(a) $Al_2(SO_4)_3$ (b) ZnO
(c) Zn powder (d) Al_2O_3, Zn S

ANSWERS

1. (b)	**2.** (d)	**3.** (c)	**4.** (a)	**5.** (c)	**6.** (d)
7. (c)	**8.** (a)	**9.** (d)			

Short answer questions

1. How will you convert:
   (i) A thiol into a sulphonic acid
   (ii) An alkyl halide into a thioether
   (iii) A thioether into an alcohol
   (iv) Benzene sulphonic acid into phenol.
2. What is meant by sulphonation and sulphonation reagents? Give two example of the latter.
3. Complete the following reactions:

(i) 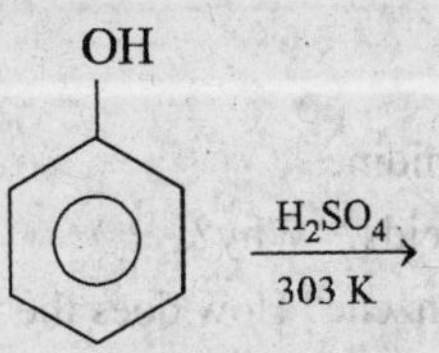

(ii) 2 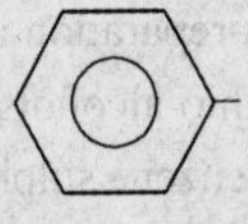 $-SO_3H + Na_2CO_3 \longrightarrow$

4. How does benzenesulphonyl chloride react with ammonia? Give equation.
5. Why are thiols stronger acids than alcohols?
6. How will you convert an alkyl halide into thiether?
7. Write equations for the reactions of ethanethiol with:
   (i) Mercuric oxide (ii) Lead acetate
   (iii) Acetone (iv) I_2.
8. What happens when diethyl thioether is treated with:
   (i) Equimolar amount of H_2O_2 (ii) Excess of H_2O_2?
9. By use of sulphonation as one of the steps, how will you convert benzene into benzoic acids?
10. Give two methods of formation of thioethers.

11. How will you bring about the conversion of benzene to sulphonamide?

12. What are thioethers? Give two methods of their preparation.

13. Explain the greater acidic character of thiols than alcohols.

14. What are sulphonic acids? Discuss their structure.

15. What is meant by desulphonation? Give its mechanism.

16. Write equations for following reactions:

(*i*) $C_2H_5SH + HNO$

(*ii*) $C_2H_5SH + H_2O_2$

(*iii*) $C_2H_5 - S - C_2H_5 + C_2H_5I$

(*iv*)  $-SO_3H + H_2 + H_2O$ (steam) $\longrightarrow$

(*v*) 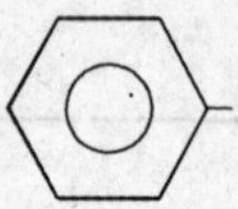 $-SO_3Na + NaHS \longrightarrow$

17. Give two general methods each for the preparation of sulphonic acids and thioethers.

18. Discuss the mechanism of sulphonation of benzene.

19. (*a*) Give mechanism of sulphonation of benzene.

(*b*) Give one methods of preparation and uses of sulphaguanidine.

20. What happens when:

(*i*) Ethyl mercaptan reacts with acetone.

(*ii*) Benzene sulphonyl chloride reacts with NH_3.

21. Predict the product of monobromination of *p*-toluene sulphonic acid followed by treatment with acid and super heated steam.

22. How can you prepare the following from benzene sulphonic acid:

(*i*) Benzene sulphonyl chloride (*ii*) Benzene

(*iii*) Benzoic acid (iv) Phenol?

23. Discuss the mechanism of sulphonation of benzene.

24. Write one method of preparation and uses of sulphaguanidine.

25. Out of alcohols and thio alcohols, which are stronger acids? Why?

26. How can you obtain benzene sulphonyl chloride from benzene? How does the former react with ethyl alcohol?

27. In what respect does sulphonation differ from others electrophitic aromatic substitutions? Illustrate with suitable examples.

28. Write a note on sulphonaton.

General questions

1. Give three methods of preparation of alkane sulphonic acids. Why cannot they be prepared by direct sulphonation of alkanes?

2. Describe the preparation and important reactions of aromatic sulphonic acids.

3. What are mercaptans? Describe their chemical reactions.

4. Outline all steps in the conversions of *p*-toluene sulphonic acid into:
   (*i*) p-Toluenesulphonamide (*ii*) *p*-Cresol
   (*iii*) Toluene (*iv*) Ethyl p-toluenesulphonate
   (*v*) o-Bromotoluene
5. How are thioethers prepared? What are their important properties?
6. Show how benzenesulphonic acid or its sodium salt will react with the following reagents:
   (*i*) Barium hydroxide (*ii*) Fusion with sodium hydroxide
   (*iii*) Fusion with sodamide (*iv*) Fusion with sodium cyanide
   (*v*) Hydrolysis in acid solution.
7. Sketch the following transformations.
   (*i*) Benzene sulphonic acid $\longrightarrow$ Benzene
   (*ii*) *p*-Toluenesulphonic acid $\longrightarrow$ *p*- Toluidine
   (*iii*) Benzene sulphonyl chloride $\longrightarrow$ Benzene sulphonamide
   (*iv*) Toluene $\longrightarrow$ *p*-Toluenesulphonyl chloride.
8. Write an account of the preparation and properties of sulphonamides.
9. Write equations of the following reactions:
   (*i*) Reactions between diethyl sulphide and ethyl iodide
   (*ii*) Reaction between ethanethiol and iodine
   (*iii*) Reaction between ethanethiol and nitric acid
   (*iv*) Hydrolysis of benzenesulphonic acid with steam
   (*v*) Fusion of sodium benzene sulphonate with sodium hydrosulphide.
10. Bring out the following conversions:
    (*i*) Benzene into benzenesulphonic acid
    (*ii*) Benzene into benzenesulphonamide
    (*iii*) Methyl phenyl thioether into methyl phenyl sulphone.

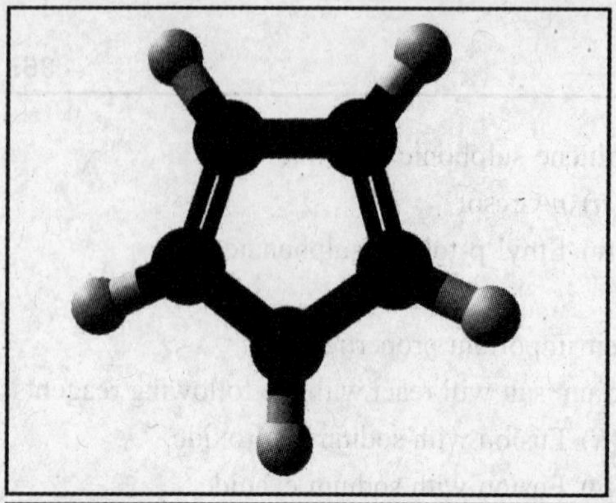

4

Heterocyclic Compounds

4.1. INTRODCUTION

Cyclic compounds having at least one atom other than carbon in the ring are termed as heterocyclic compounds. Although any atom capable of forming two or more covalent bond can be the ring element yet the most commonly found heterocyclic compounds contain only nitrogen, oxygen and sulphur as *hetero atoms*.

Substances such as cyclic anhydrides, alkene oxides and lactones, are heterocyclic in the strict sense of the term but are not generally considered as heterocyclic, because they are highly unstable, also they change into open-chain compounds upon hydrolysis.

Hetero atom in the ring, in place of carbon does not alter the strain relationship significantly. The most stable heterocyclic compounds are those having five or six membered rings.

Thus the heterocyclic compounds may also be defined as *five* or *six membered ring compounds with at least one hetero atom as the ring member. They are relatively stable and exhibit aromatic character.*

Heterocyclic compounds are abundantly available in plant and animal products and form one half of the natural organic compounds. Alkaloids, dyes, drugs, proteins, enzymes, etc. are the important representatives of this class of compounds.

4.2. CLASSIFICATION OF HETEROCYCLIC COMPOUNDS

Heterocyclic compounds may be divided into three categories.

1. Five-membered heterocyclic compounds. These can be considered to be derived from benzene by replacement of a (C = C) by a hetero-atom with an unshared electron pair.

These are further divided into two types.

(*i*) **Compounds having only one hetero atom.** Furan, thiophene, and pyrrole, containing O, S and N respectively as the hetero atoms are examples of this category.

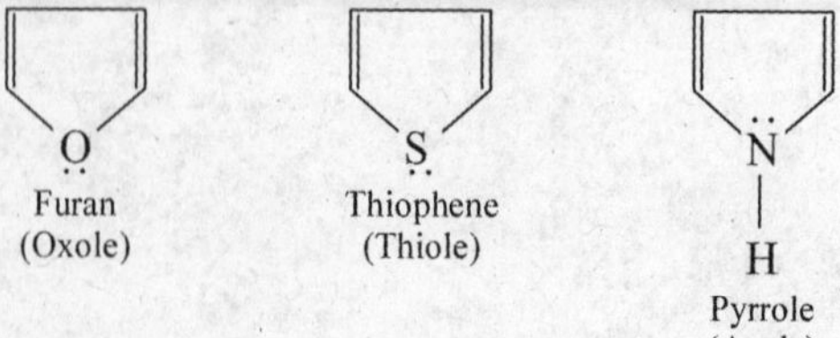

Furan (Oxole) Thiophene (Thiole) Pyrrole (Azole)

(*ii*) **Compounds having more than one hetero-atom.** For example,

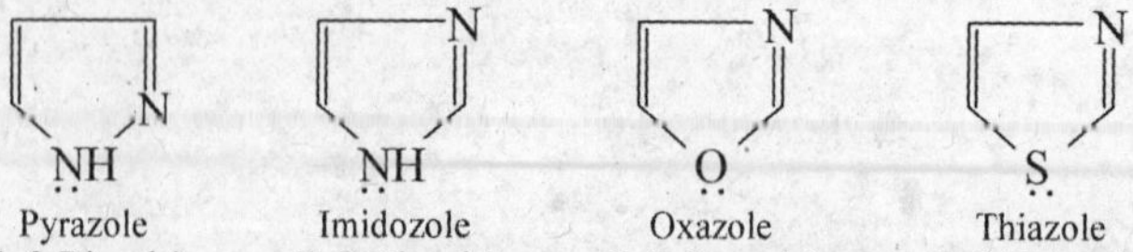

Pyrazole (1, 2-Diazole) Imidozole (1, 3-Diazole) Oxazole (1, 3-Oxazole) Thiazole (1, 3-Thiazole)

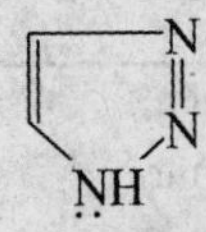

1, 2, 3-Triazole

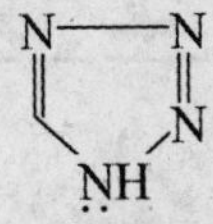

Tetrazole

2. Six-membered heterocyclic compounds. These are obtained by the replacement of a carbon of benzene by an isoelectronic hetero-atom. These are further classified into two types.

(*i*) **Compounds having only one hetero atom.** Pyridine, pyran and thiopyran are some of the examples.

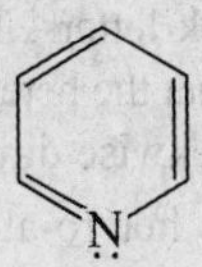

Pyridine

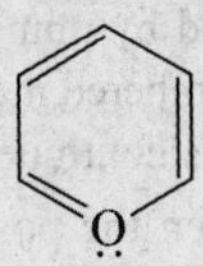

Pyran

Thiopyran

(*ii*) **Compounds having more than one hetero-atom.** Pryrimidine, pyrazine, pyridazine are some of the examples.

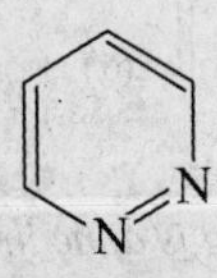

Pyridazine
(1, 2-Diazine)

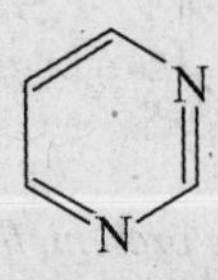

Pyrimidine
(1, 3-Diazine)

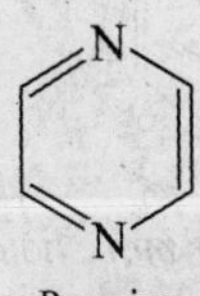

Pyrazine
(1, 4-Diazone)

3. Condensed Heterocyclic compounds. They consist of two or more fused rings which can be partly carbocyclic, and partly heterocyclic, *e.g.*, indole, quinoline, carbazole, etc. or may be fully heterocyclic, *e.g.*, purine, pteridine.

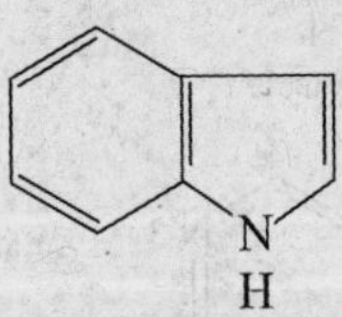

Indole

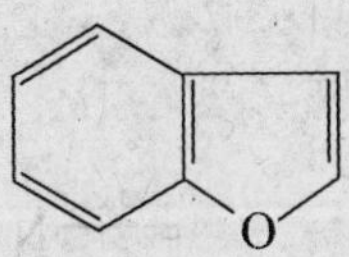

Benzofuran

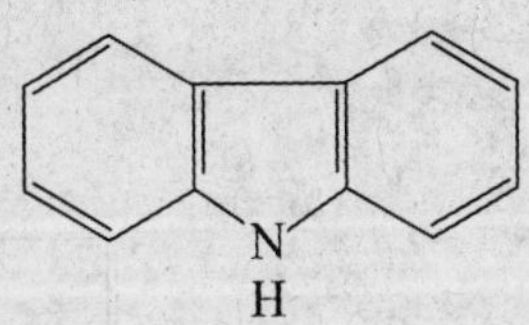

Carbazole

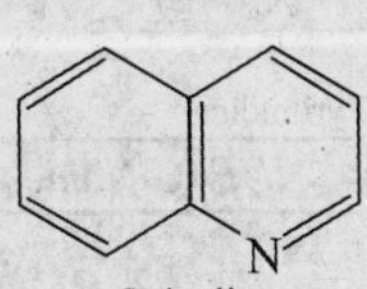

Quinoline

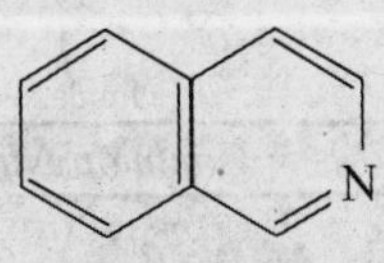

Isoquinoline

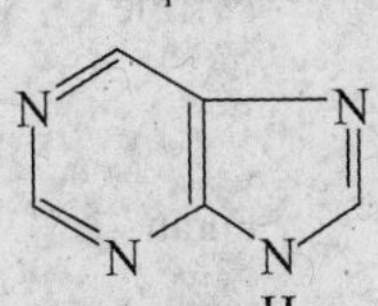

Purine

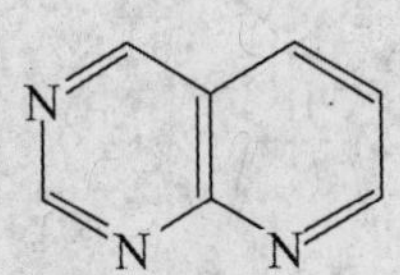

Pteridine

4.3. NOMENCLATURE OF HETEROCYCLIC COMPOUNDS

Heterocyclic compounds are popularly known by their common names. Even the IUPAC system has accepted some of these trivial names as such. IUPAC name of heterocyclic compound is obtained by combining prefixes and suffixes as follows:

Order of Priority	*Prefixes*	*Suffixes*
1.	*oxa*—for oxygen	–*ole* 5-membered ring
2.	thia—for sulphur	–*ine* 6-membered ring
3.	*aza*—for nitrogen	–*epine* 7-membered ring
4.	*phospha*—for phosphorus	

(*a*) The terminal '*a*' of the prefixes is generally removed when combining prefixes and suffixes. In case more than one hetero-atoms are involved, the prefixes are placed in the order of priority *i.e.*, oxa-first, than thio and so on.

(*b*) The ring positions are designated by numerals or Greek letters. The simple heterocyclic compounds having one hetero-atom are numbered in such a way that the hetero-atom gets the number 1 (or lowest number, and the numbering is continued in anti-clockwise direction).

(*c*) When Greek letters are used then the position next to hetero-atom is designated as α- followed by β- and so on.

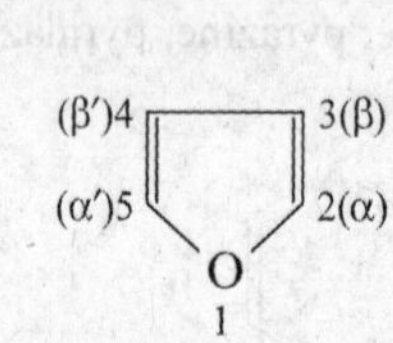
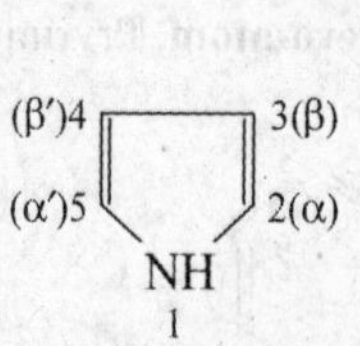
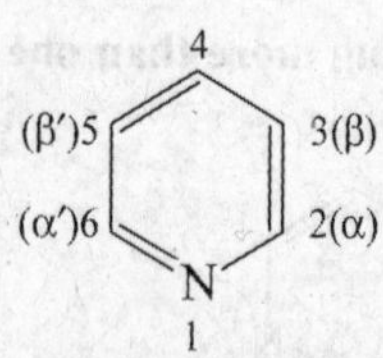
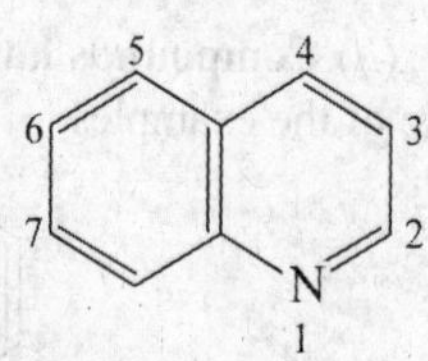

In case two or more hetero-atoms are present, the numbering is done in such a way that hetero-atom highest in priority gets the lowest number i.e. O takes precedence over S and so on.

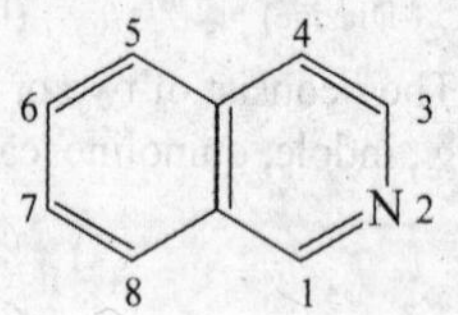
Isoquinoline

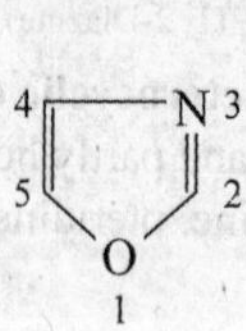
Oxazole

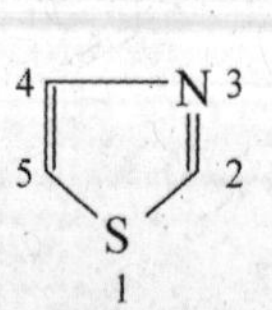
Thiazole

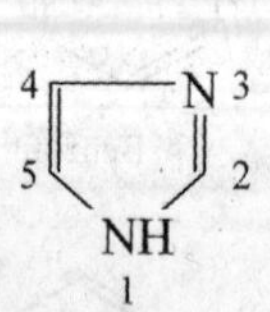
Imidazole

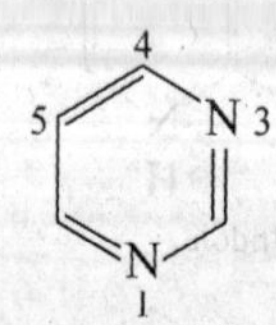
Pyrimidine

Structure	*Common Name*	*IUPAC Name*
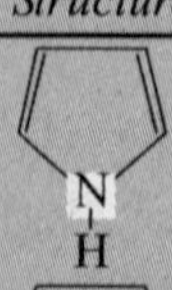	Pyrrole	azole
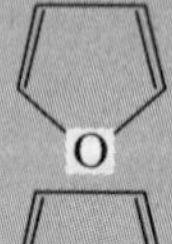	Furan	oxole
	Thiophene	thiole
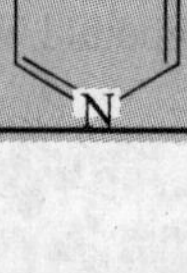	Pyridine	azine

FURAN

4.4. NOMENCLATURE OF FURAN DERIVATIVES

The derivatives of furan are named, giving the substituents and the positions to which they are attached. The numbering system starts with the oxygen in the ring.

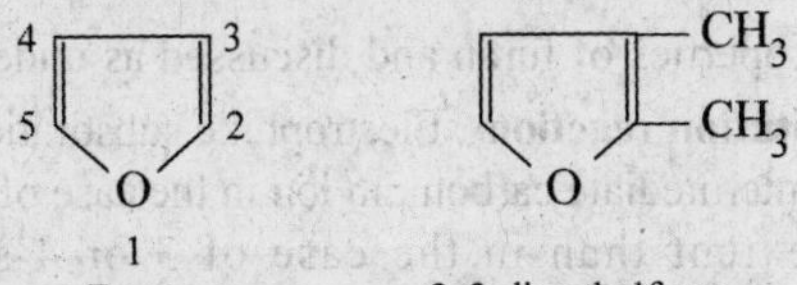

Furan 2, 3-dimethylfuran

The 2, 5 positions are frequently referred to α, α′ and 3, 4 positions as β, β′ positions respectively.

4.5. METHODS OF PREPARATION OF FURAN

Methods of preparation of furan are given as under:

(*i*) **From mucic acid.** *Dry distillation of mucic acid gives furan.*

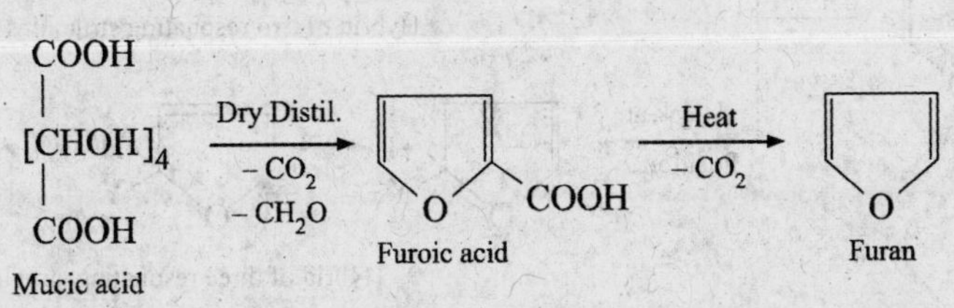

(*ii*) **From furfural.** Furfural and steam are passed over a catalyst consisting of a mixture of zinc and manganese chromites at 673 K, CO is eliminated with formation of furan.

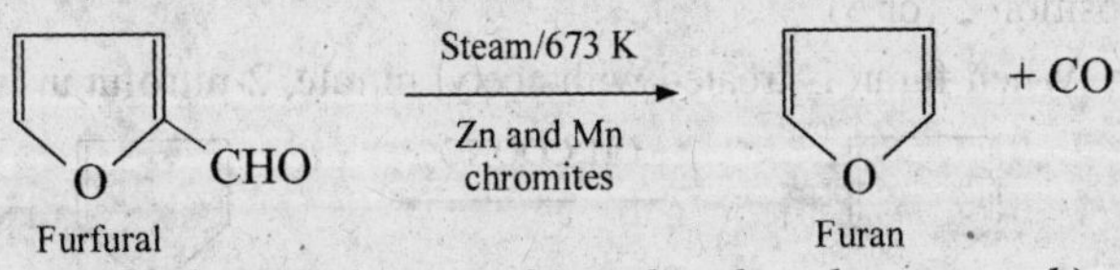

(*iii*) **Paal-Knorr synthesis.** (*Dehydration of 1, 4-dicarbonyl compounds*)

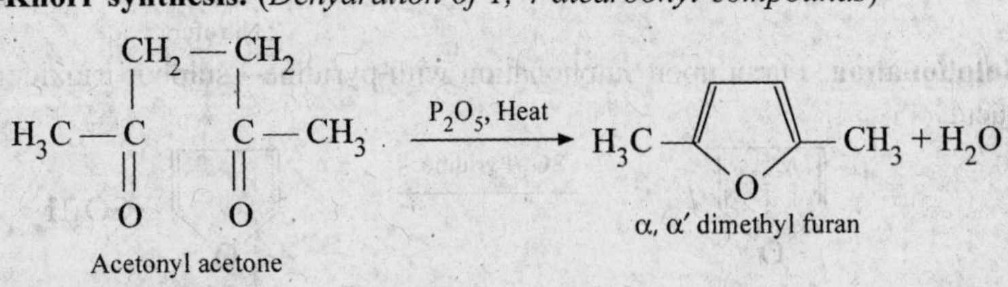

(*iv*) **Fiest-Benary synthesis**

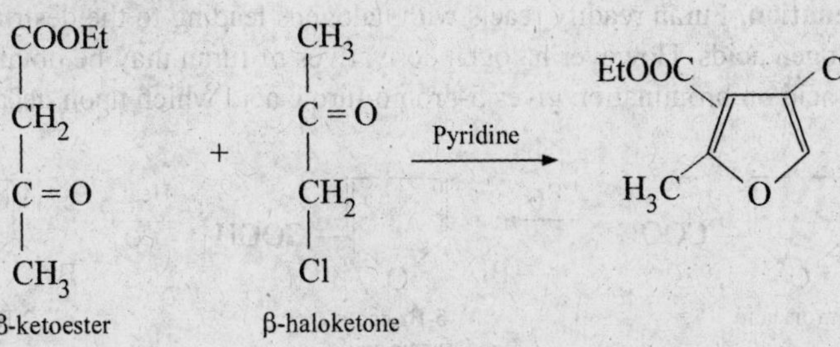

4.6. PHYSICAL PROPERTIES OF FURAN

1. It is a coloured liquid boiling at 305K
2. It is smells like chloroform
3. It is soluble is alcohol and ether but insoluble in water.

4.7. CHEMICAL PROPERTIES OF FURAN

Important chemical properties of furan and discussed as under:

1. Electrophilic substitution reactions. Electrophilic substitutions of furan occurs preferentially at the 2- or 5-position. The intermediate carbonium ion in the case of 2- or 5-substitution is resonance stabilised to a greater extent than in the case of 3 or 4-substitution. Substitution at 3- or 4-position, however, may take place in case 2- and 5-positions of furan are already occupied.

Resonance stabilisation of the intermediate products in substitution at positions 2 and 3 is illustrated below. $Y^{\oplus}$ represents the electrophile.

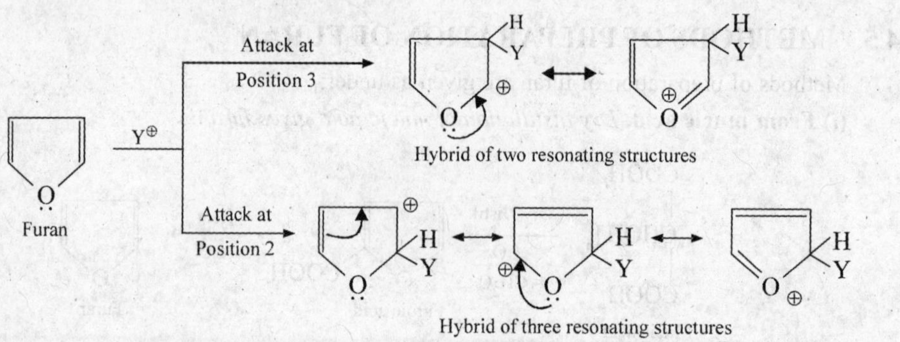

Intermediate product with substitution at position 2 has three resonance structures, while that with substitution at position 3 has two resonating structures. Therefore, substitution takes place preferentially at position 2 (or 5)

(*a*) **Nitration.** When furan is treated with acetyl nitrate, 2-nitrofuran is obtained.

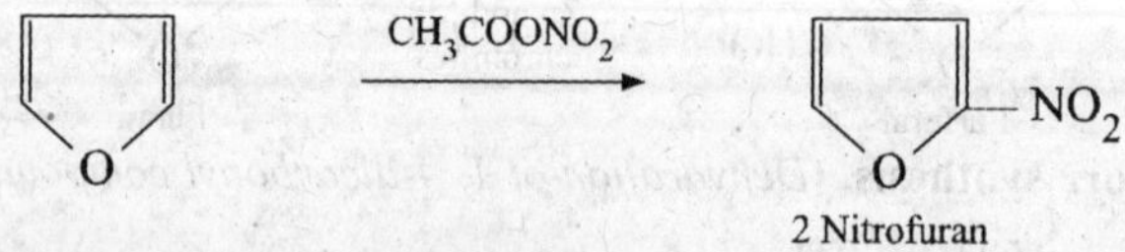

(*b*) **Sulphonation.** Furan upon sulphonation with pyridine—sulphur trioxide forms Furan-2 sulphonic acid.

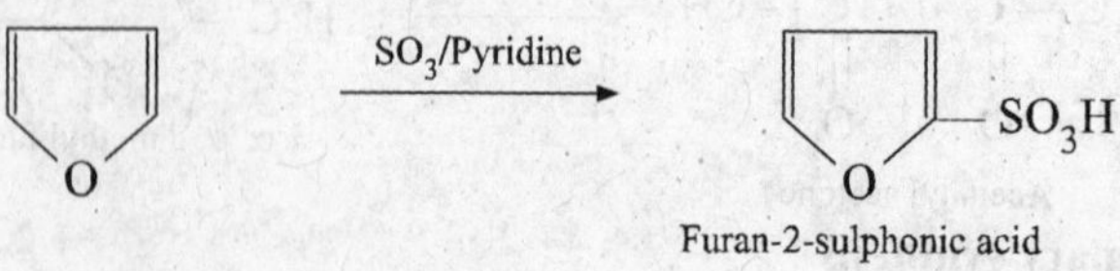

(*c*) **Halogenation.** Furan readily reacts with halogens leading to the destruction of furan ring and forming halogen acids. However halogen derivatives of furan may be obtained indirectly. For example, furoic acid on bromination gives 5-bromo furoic acid which upon decarboxylation yields 2-bromofuran.

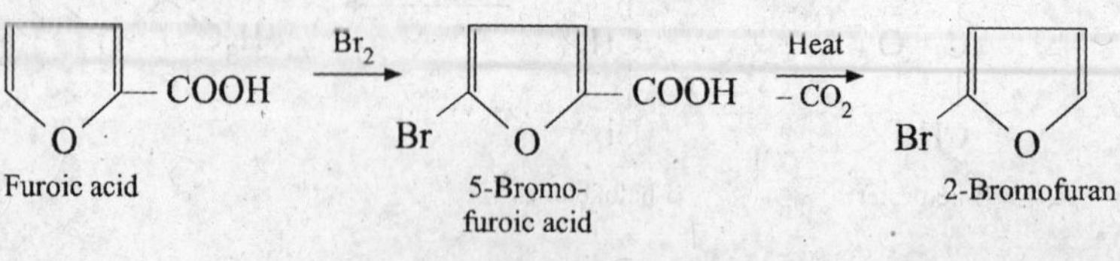

(*d*) **Friedel-Crafts' reaction.** Acylation of furan is done with boron trifluoride in ether.

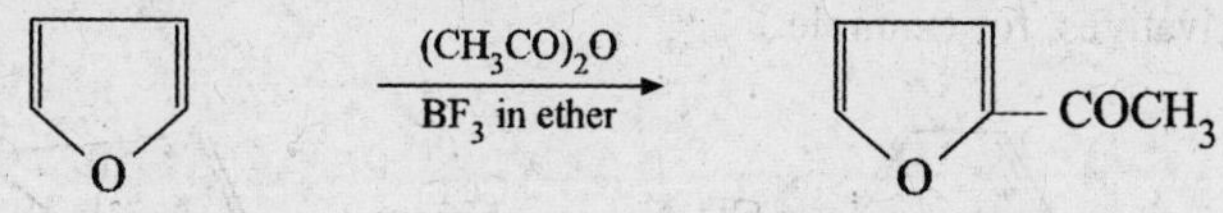

Furan 2-Acetyl Furan

Alkylation reactions with furan results in polymerisation and are not, therefore, possible.

(*e*) **Gattermann Koch Synthesis.** Furan when treated with a mixture of HCN and HCl in the presence of $AlCl_3$ followed by decomposition yields furfural.

$$HCl + HCN \xrightarrow{AlCl_3} HN = CHCl$$

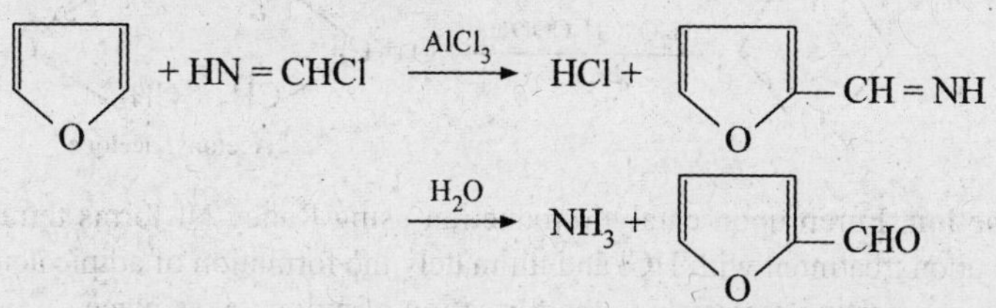

Furfural

(*f*) **Gomberg reaction.** Furan on treatment with diazonium salt in alkaline solution gives aryl furan.

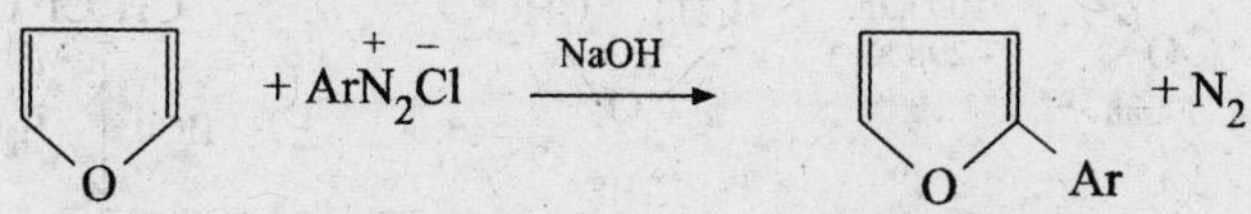

Furan Aryl furan

(*g*) **Reaction with *n*-butyllithium.** Furan gives 2-lithium furan when treated with *n* butyl-lithium.

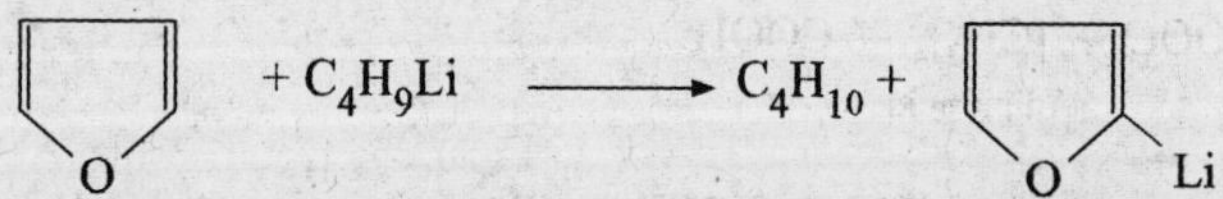

Furan 2-Lithium furan

(*h*) **Formation of 2-chloromercurifuran.** Treatment with a mixture of mercuric chloride and sod. acetate gives 2-chloromercurifuran which is a useful synthetic intermediate as the mercuric group is replaceable.

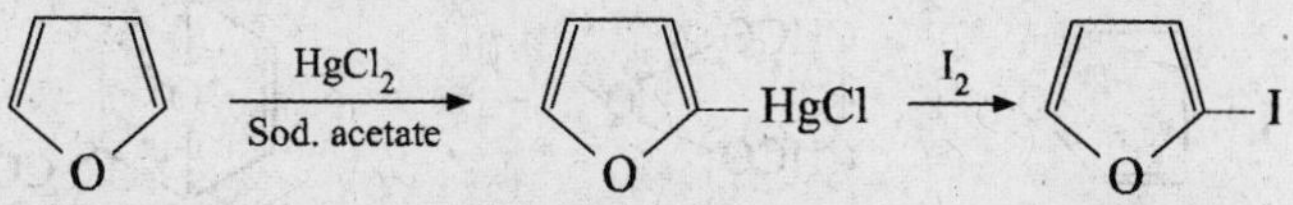

Furan 2-Chloromercuric furan 2-Iodofuran

2. Basic nature on treatment with an acid, furan undergoes protonation at the **ring carbons** rather than oxygen, the protonated product formed undergoes polymerisation.

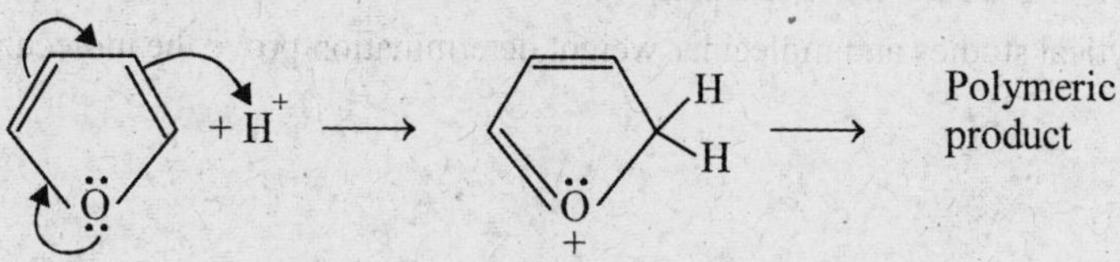

3. Reaction with carbenes. Furan undergoes cycloaddition reaction with carbenes to form cyclopropane derivatives, for example.

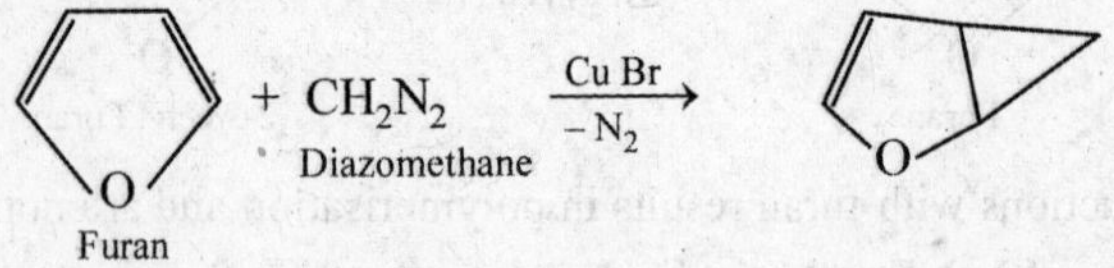

4. Ring cleavage Furan undergoes ring cleavage on treatment with suitable reagents in acidic medium to form dicarbonyl product. For example,

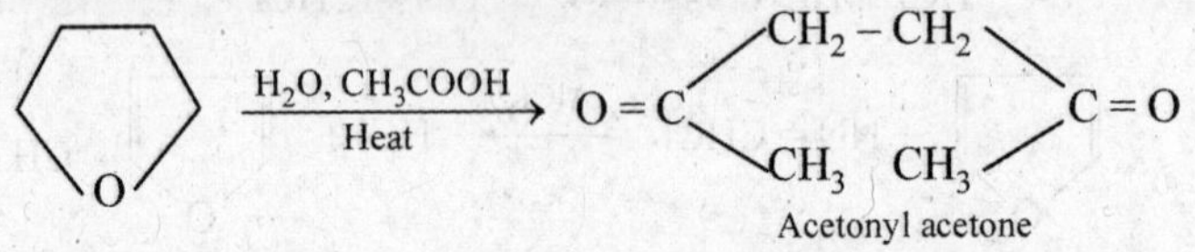

5. Reduction. Furan upon catalytic reduction using Raney Ni forms tetrahydrofuran [THF] which breaks upon treatment with HCl and ultimately the formation of adipic acid and 1, 6 diamino hexane, which are starting materials in the formation of nylon, takes place.

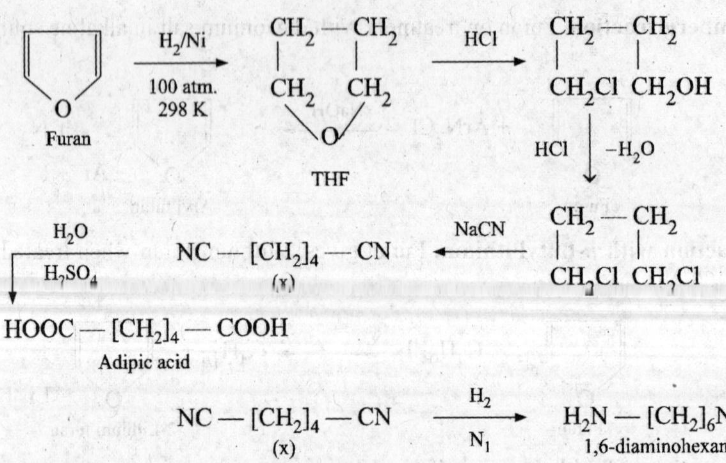

6. Diels Alder Reaction. Furan is the only heterocyclic compound which undergoes Diels Alder reactions.

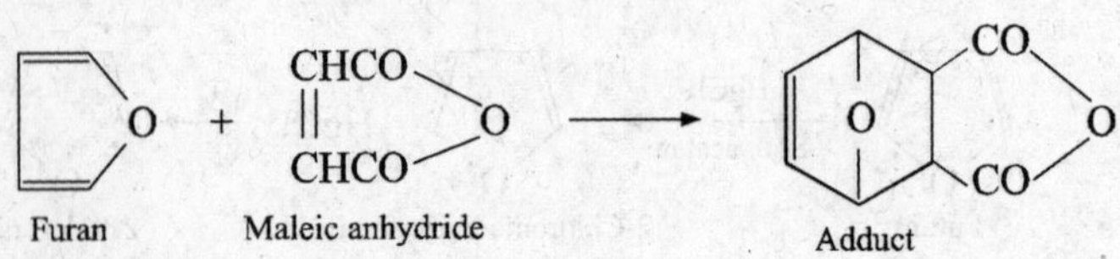

4.8. STRUCTURE OF FURAN

1. Analytical studies and molecular weight determination prove the molecular formula of furan as C_4H_4O.

2. Keeping in view the tetravalency of carbon and divalency of oxygen a possible structure of furan is given below:

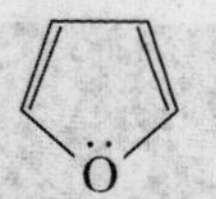

3. The properties of furan are not in keeping with the above structure which has a pair of conjugated double bonds and an ether group. Rather it gives substitution reactions. This is characteristic of aromatic compounds. Its heat of combustion indicates that it is resonance stabilised by an amount of 65 kJ mol^{-1}. Resonance stabilization on account of conjugated double bonds would have provided 12.5 kJ mol^{-1} resonance energy. Therefore the diene structure of furan is ruled out.

4. Resonance structures. To account for extrastability and high value of resonance energy, furan is considered to be resonance hybrid of the following resonating structures.

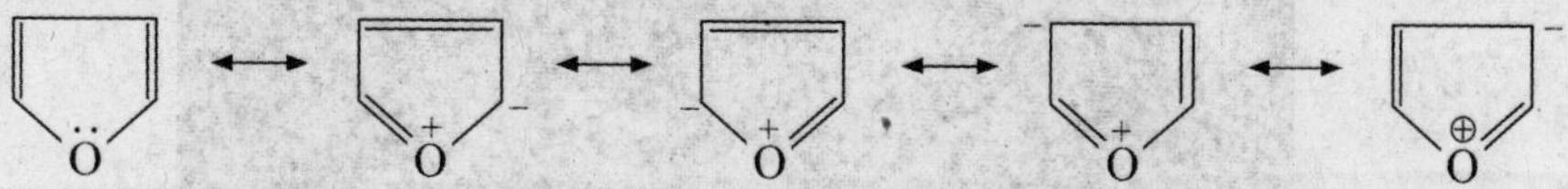

To be aromatic in nature, a monocyclic molecule must be planar and should have $(4n + 2)\pi$ electrons.

Out of the six electrons ($n = 1$) needed for aromaticity, two electrons are provided by oxygen in the form of its lone pair. The resonance hybrid structure of furan has been confirmed by X-ray diffraction studies which shows that C — O bond length in furan is 1.37 Å which is shorter than normal C — O single bond length of 1.43 Å. It establishes that this bond has some double bond character as required by the resonance phenomenon.

5. Orbital structure. Furan has been established to have a flat pentagonal ring involving sp^2 hybridised carbon and oxygen atoms with cyclic π electron cloud containing six electrons each lying above and below the plane of the ring. Oxygen forms two sigma bonds with two adjacent carbon atoms. Of the hybrid orbitals of oxygen, two have got one electron each while the third has a lone pair of electrons.

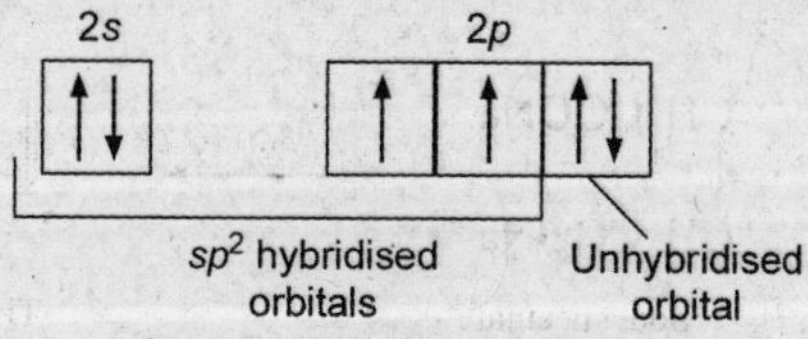

Electronic configuration of oxygen

The unhybridised p-orbital also has a pair of electron. The two hybridised orbitals having single electrons each are involved in the formation of σ-bonds with the two adjacent carbon atoms while the unhybridised p-orbital having two electrons from a π bond by overlapping sideways with the p-orbitals of carbon. The pair of electrons in the third hybrid orbital of oxygen is left unshared. The orbital picture of furan is shown as under:

Orbital structure of furan

THIOPHENE

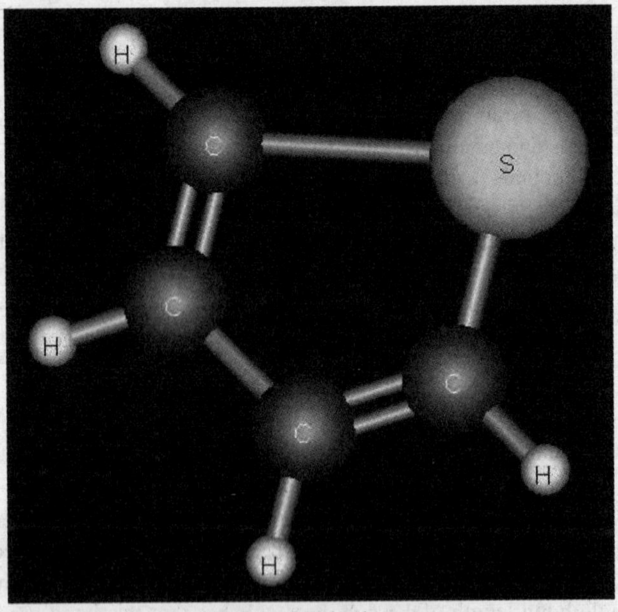

Model of Thiophene

4.9. METHODS OF PREPARATION OF THIOPHENE

Thiophene is prepared by the following methods.

1. Laboratory Method. On heating sodium succinate with phosphorus sulphide, thiophene is obtained.

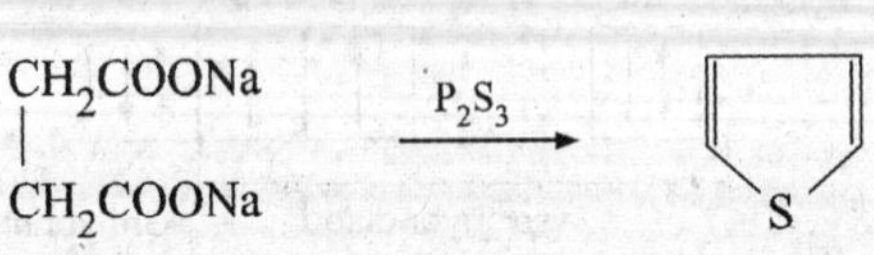

Sod. succinate — Thiophene

2. From *n*-butane. On heating *n*-butane with sulphur at 923 K, thiophene is formed.

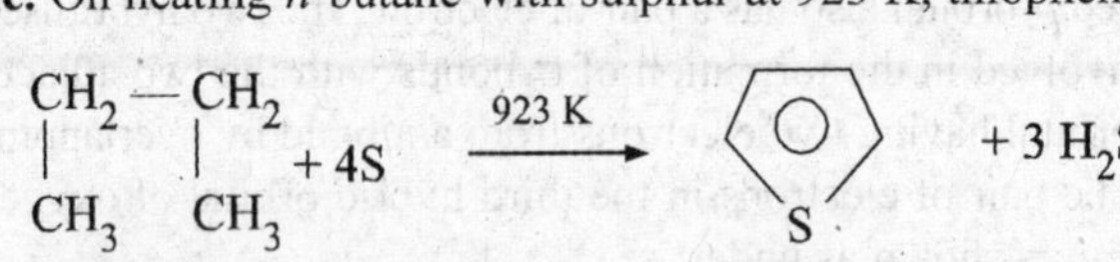

n-Butane — Thiophene

3. Paal-Knorr synthesis of thiophene derivatives. It involves the action of heat on enolisable 1, 4 diketone in the presence of pentasulphide.

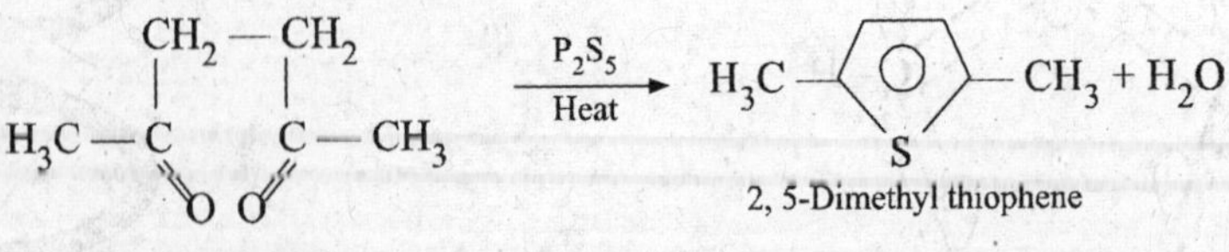

2, 5-Dimethyl thiophene

Acetonyl acetone

4. Manufacture. Thiophene is manufactured by passing a mixture of acetylene and hydrogen sulphide through a tube containing Al_2O_3 at 673 K.

4.10. PHYSICAL PROPERTIES

1. Thiophene is a colourless liquid boiling at 357 K
2. It smells like benzene
3. It is soluble in alcohol and ether but insoluble in water

4.11. CHEMICAL PROPERTIES OF THIOPHENE

Chemical properties of thiophene are discussed an under:

1. Electrophilic Substitution Reactions. Thiophene undergoes electrophilic substitution reactions. On the basis of charge distribution and stability of carbonium ions, the electrophilic substitution would be expected to take place at position 2 [or 5] rather than at position 3 [or 4]. as explained in the case of furan.

(*a*) Nitration. Thiophene when nitrated with fuming HNO_3 in acetic anhydride gives 2-nitro-thiophene.

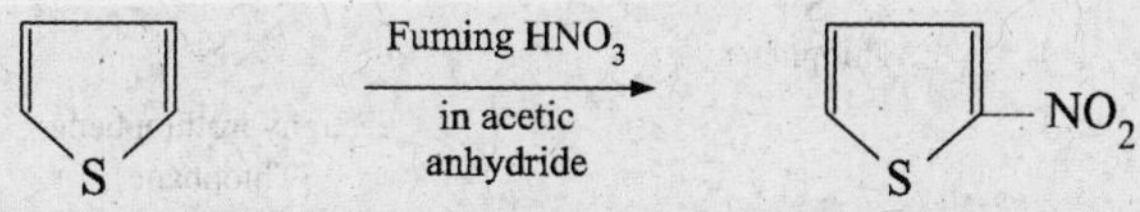

2-Nitrothiophene

(*b*) Sulphonation. Sulphonation of thiophene with cold conc. H_2SO_4 gives 2-thiophene sulphonic acid.

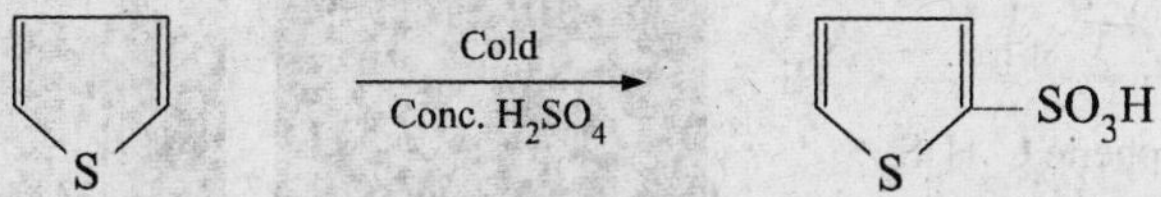

2-Thiophene sulphonic acid

(*c*) Bromination. When thiophene is treated with Br_2 in the absence of any halogen carrier, the formation of 2, 5-dibromothiophene takes place.

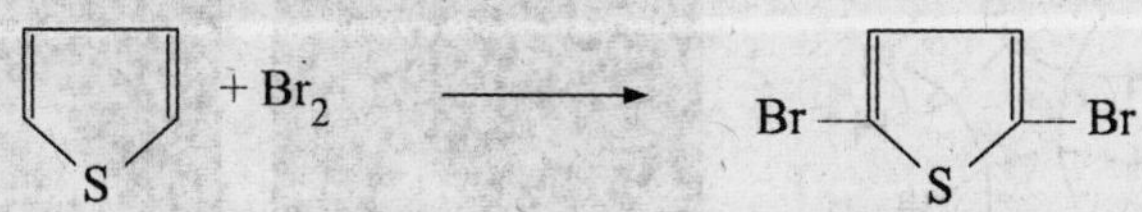

2, 5-Dibromothiophene

(*d*) Iodination. When thiophene is treated with I_2 in the presence of yellow mercuric oxide, 2-Iodothiophene is obtained.

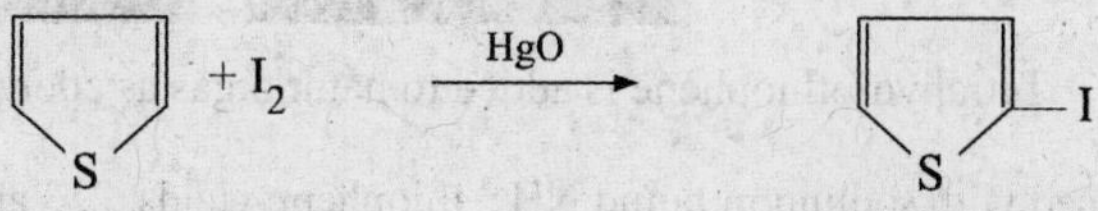

2 Iodothiophene

(*e*) Friedel-Crafts Acylation. Thiophene is acetylated with acetic anhydride in the presence of phosphoric acid.

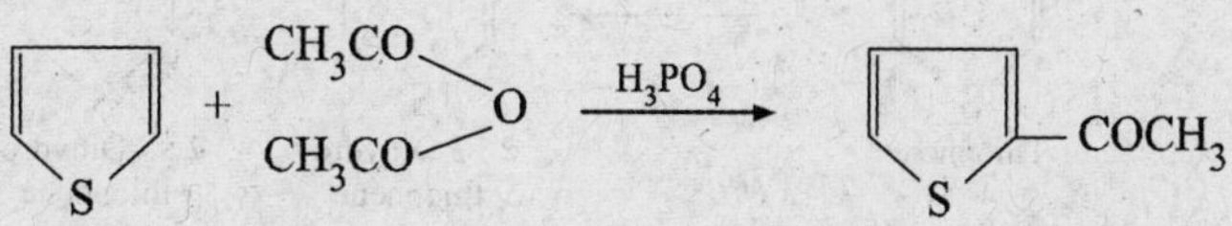

Methyl thiophenyl ketone

(*f*) **Chloromethylation.** Thiophene may be chloromethylated with the mixture of HCHO and HCl.

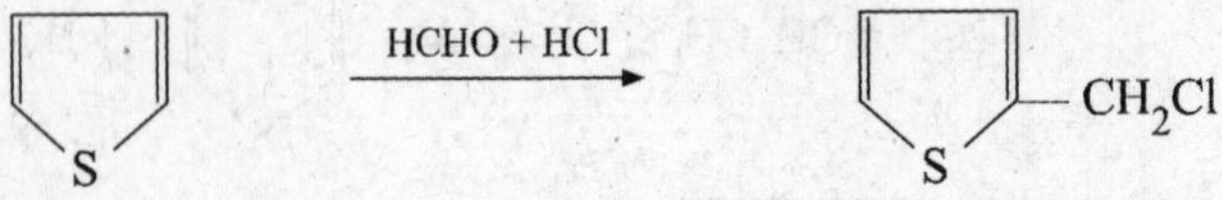

2-Chloromethyl thiophene

(*g*) **Mercuration.** On mercuration with mercuric chloride in the presence of sod. acetate it gives 2-mercuri-chloride.

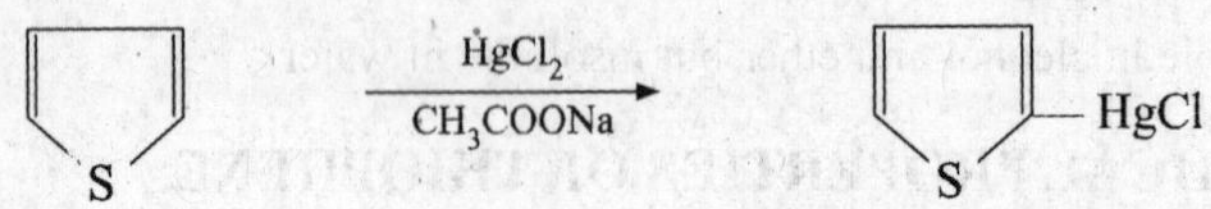

2-Mercuri-chloride thiophene

2. Reduction. (*a*) Catalytic hydrogenation of thiophene using large amounts of catalysts yields tetrahydrothiophene [Thiophan].

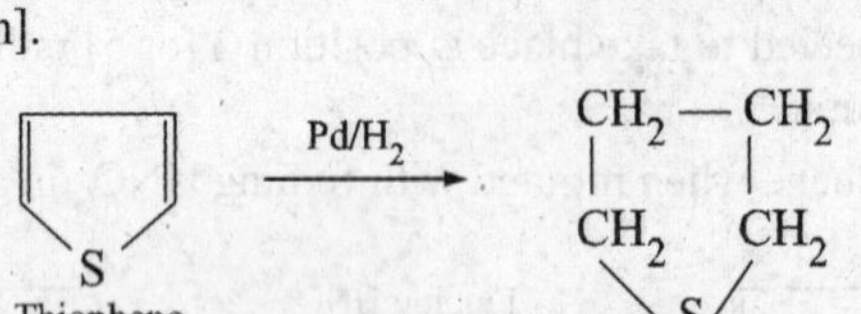

Thiophene

Tetrahydrothiophene [Thiophane]

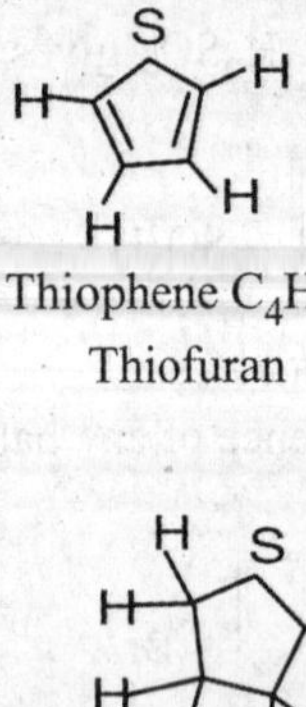

Thiophene C_4H_4S
Thiofuran

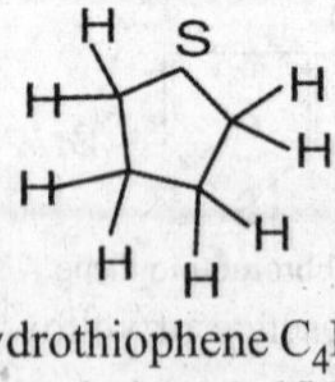

Tetrahydrothiophene C_4H_4S
Tetramethylene Sulfide

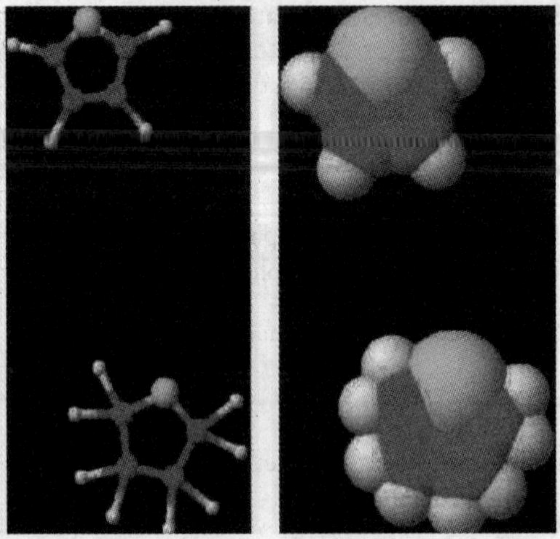

Tetrahydrothiophene is added to natural gas as adorant

(*b*) When reduced with sodium in liquid NH_3, thiophene yields, 2, 3 and 2, 5 dihydrothiophene (Birch reaction).

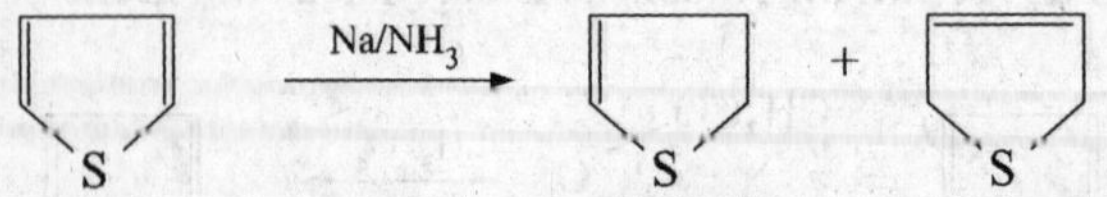

Thiophene 2 : 3 Dihydro-thiophene 2 : 5 Dihydro-thiophene

(*c*) Catalytic reduction of thiophene with Raney Ni as catalyst gives *n*-butane as the main product.

3. Formation of lithium derivative. On treatment with *n*-butyl lithium in ether, thiophene gives 2-lithium thiophene. It is useful in the synthesis of various 2-substituted thiophenes.

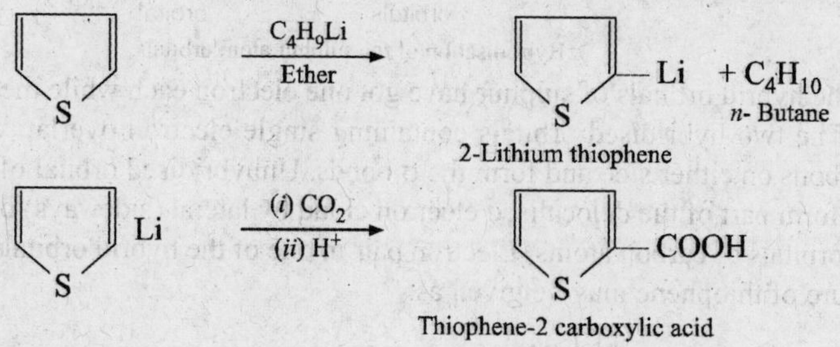

4. Reaction with carbenes. Thiophene react with carbenes to give cyclopropane derivatives. For example:

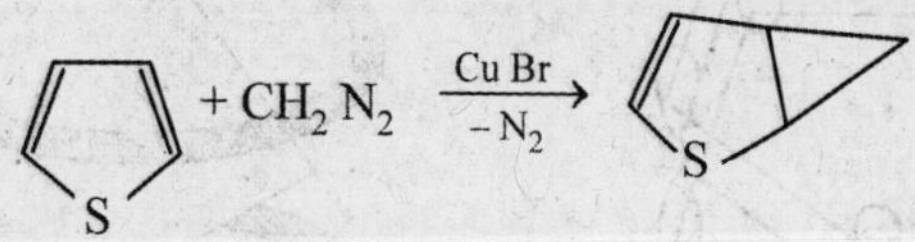

4.12. STRUCTURE OF THIOPHENE

1. Analytical studies and molecular weight determination reveal the formula of thiophene as C_4H_4S.

2. An unsaturated structure is indicated by the formula. A possible structure is shown below:

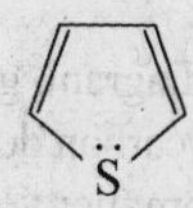

This structure has a pair of conjugated double bonds. But this structure does not fully explain the observed behaviour of thiophene.

3. Thiophene undergoes electrophilic substitution reactions like nitration, sulphonation, halogenation and Friedel-Crafts reactions, quite unexpected from a compound with the above-written unsaturated compound. It is observed that it is stabilised by 125 kJ mol^{-1}, whereas we expect stabilisation to the extent of 12.5 kJ mol^{-1} on account of conjugated double bonds. This is revealed by heat of combustion studies.

4. In the light of 125 kJ mol^{-1} as stabilisation energy, thiophene was considered as a resonance hybrid of the following resonating structures.

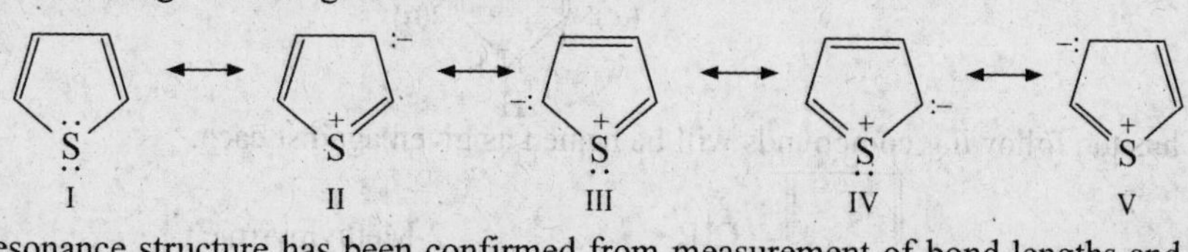

I II III IV V

The resonance structure has been confirmed from measurement of bond lengths and dipole moments.

5. Orbital structure. Although the stabilization energy is explained with the help of resonance phenomenon, electrophilic substitution remains unexplained. This is explained by the orbital structure of thiophene. Thiophene has a flat pentagonal ring of sp^2 hybridised carbon and sulphur atoms. Structure of thiophene is similar to that of furan with the difference that oxygen has been replaced by sulphur.

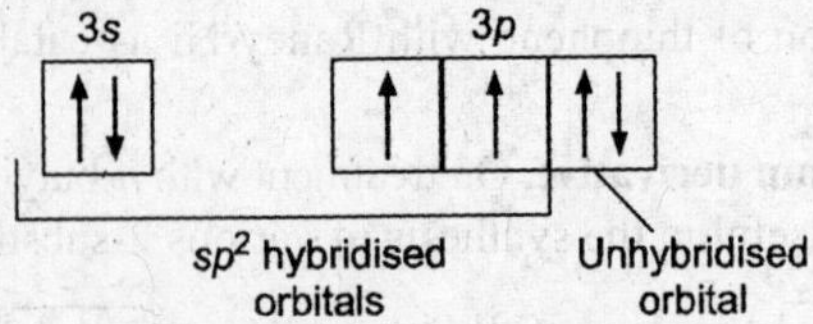

Hybridisation of the sulphur atom orbitals

Two of the hybrid orbitals of sulphur have got one electron each while the third has got a pair of electrons. The two hybridised orbitals containing single electron overlap with the sp^2 hybrid orbitals of carbons on either side and form the σ bonds. Unhybridised orbital of sulphur containing two electrons form part of the delocalised electron cloud by lateral (sideways) overlapping with the unhybridised orbitals of carbon atoms. Electron pair in one of the hybrid orbitals remains unshared. Orbital structure of thiophene may be given as:

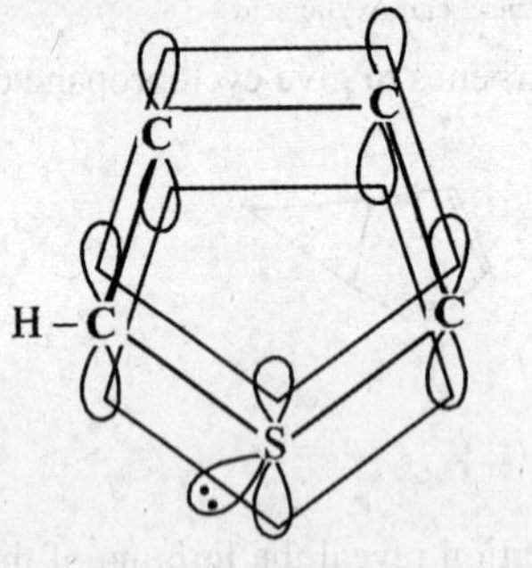

sp^2 hybrid orbital having unshared of electron

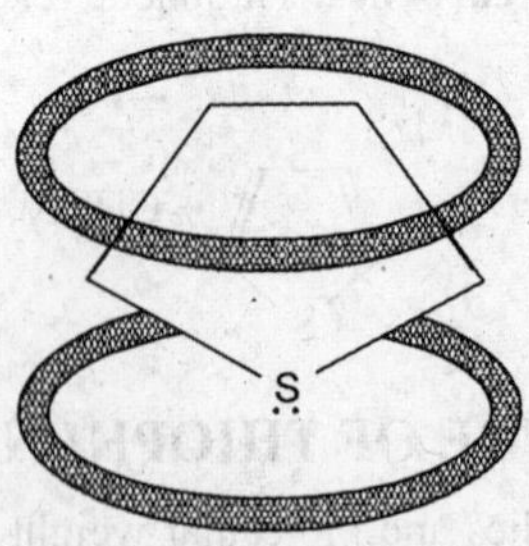

Structure of thiophene

Example 1. How will you show that thiophene is aromatic in nature, giving its molecular orbital diagram?

Solution. Refer to molecular orbital diagram, given above. The no. of π-electrons in thiophene is found to be six *i.e.* four from two carbon-carbon double bonds and two from the hybridised orbitals of nitrogen. Thus the Huckel's rule for aromaticity is satisfied.

PYRROLE

Pyrrole is the nitrogen analogue of furan. It is of interest as it is related to many naturally occurring materials. Pyrrole nucleus is found in many alkaloids and bile pigments, in the chlorophyll of plants and in haemoglobin. It is also found in morphine. It derives its name from the fact that its vapours give a bright red colour on coming in contact with a pure splint moistened with conc. HCl.

The numbering of positions in pyrrole ring is done as under:

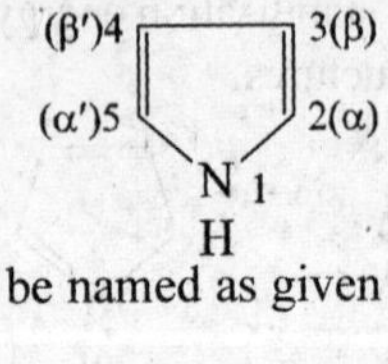

Thus the following compounds will be named as given against each.

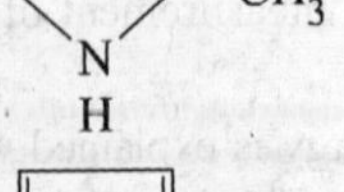

2-Methylpyrrole

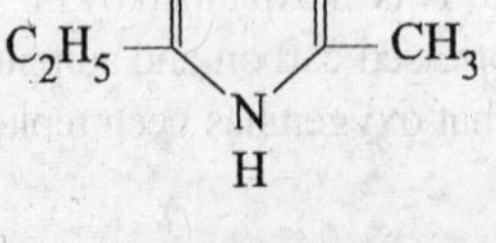

2-Methyl-5-ethylpyrrole

4.13. METHODS OF PREPARATION OF PYRROLE

Following are the general methods of preparation of pyrrole.

***(i)* From Bone oil.** Bone oil is a rich source of pyrrole. Bone oil is thoroughly treated with a solution of dilute alkali to remove acidic impurities and then it is treated with a solution of dilute acid to remove basic impurities. It is then subjected to fractional distillation. The fraction appearing between 373 K and 423 K is collected. It is further purified by fusing with KOH when solid potassiopyrrole is formed. This on steam distillation gives pure pyrrole.

$$C_4H_4NK + H_2O \longrightarrow C_4H_4NH + KOH$$

***(ii)* By distilling succinimide with Zn dust**

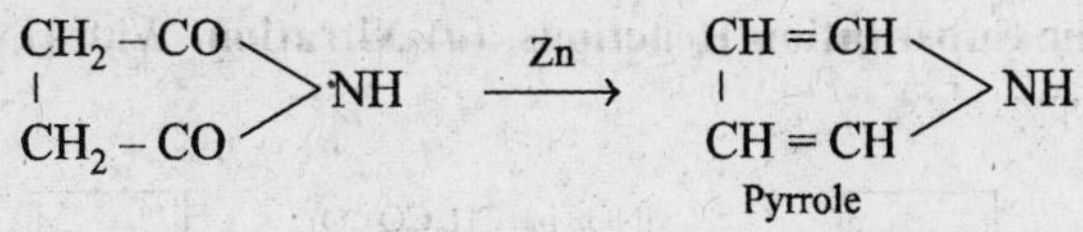

***(iii)* From ammonium mucate.** When ammonium mucate is distilled in the presence of glycerol at 473 K, pyrrole is obtained.

$$\begin{array}{c}(CHOH)_2COONH_4\\|\\(CHOH)_2COONH_4\\\text{Amm. Mucate}\end{array} \xrightarrow[(473\,K)]{\text{Glycerol}} C_4H_5N + 2CO_2 + NH_3 + 4H_2O$$

***(iv)* Synthesis.** By passing a mixture of acetylene and NH_3 through red hot tube.

$$2C_2H_2 + NH_3 \longrightarrow C_4H_5N + H_2$$

(*v*) By the action of acetylene on formaldehyde and then heating with NH_3 under pressure.

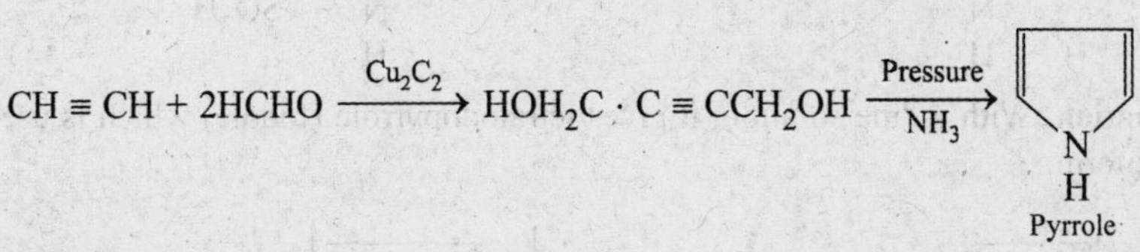

***(vi)* Paal-Knorr synthesis.** In this synthesis, a 1, 4-diketone is heated with ammonia or a primary amine. Pyrrole derivatives are obtained according to the following sequence:

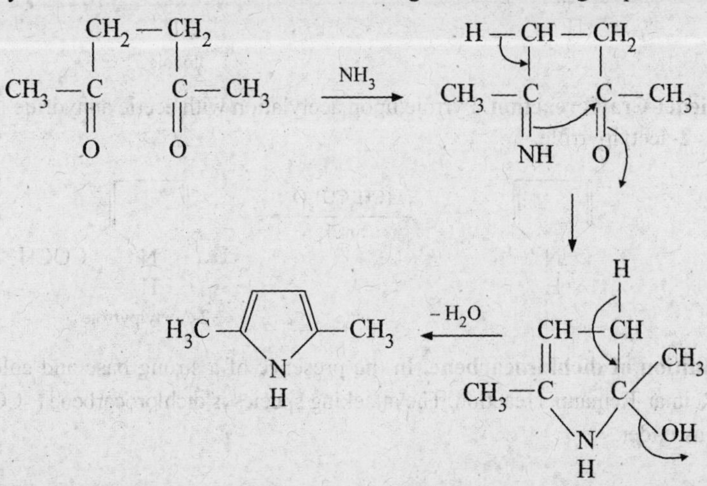

***(vii)* Manufacture.** Pyrrole can be manufactured by passing mixture of furan, ammonia and steam over heated alumina at catalyst.

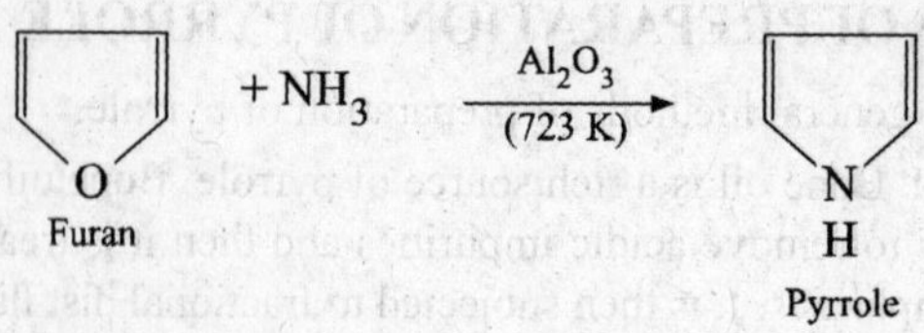

4.14. PROPERTIES OF PYRROLE

Chemically it shows the reactions of aromatic compounds. It is less aromatic than thiophene but more aromatic than furan. Some important reactions of pyrrole are as under:

1. Electrophilic Substitution Reactions. (*a*) Nitration. With HNO_3 in acetic anhydride, pyrrole gives 2-nitropyrrole.

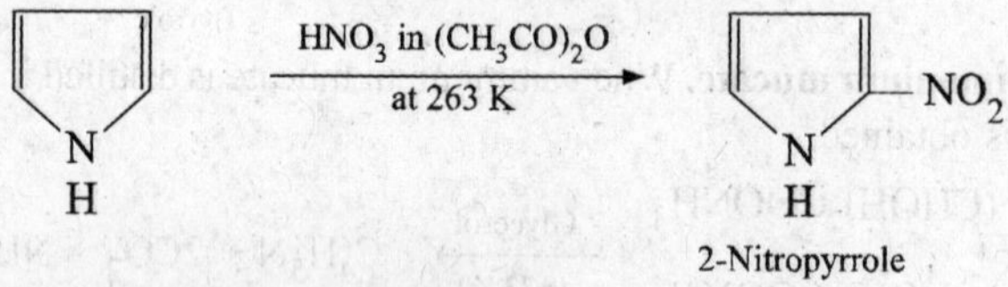

(*b*) Sulphonation. With pyridine – SO_3 mixture in ethylene chloride, pyrrole is sulphonated to give 2-pyrrole sulphonic acid.

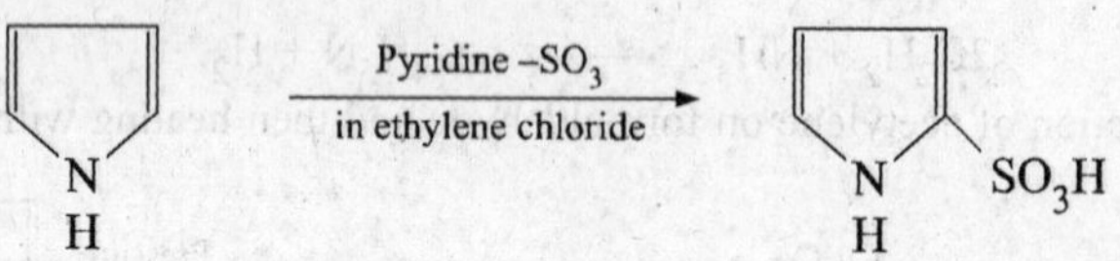

(*c*) Halogenation. With iodine solution, it gives tetraiodopyrrole (Iodole) which is used as a substitute for iodoform.

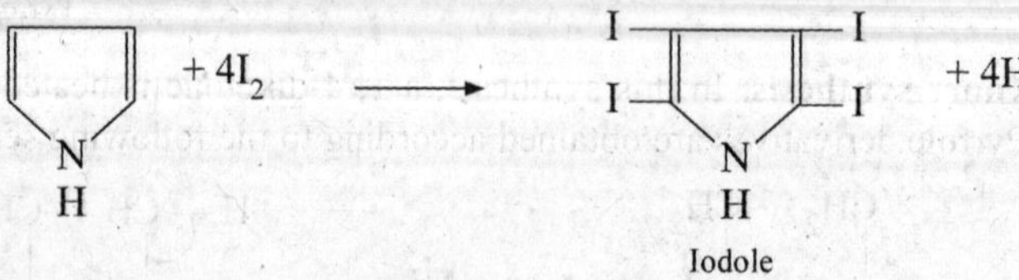

(*d*) Friedel-Crafts reaction. Pyrrole upon acetylation with acetic anhydride in the presence of $SnCl_4$ forms 2-acetylpyrrole.

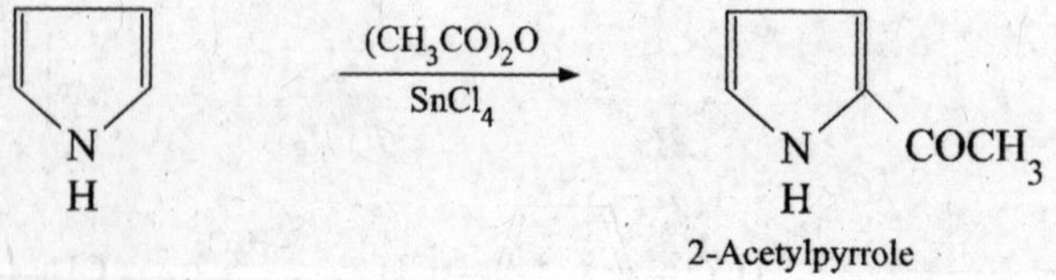

(*e*) Addition of dichlorocarbene. In the presence of a strong base and chloroform, pyrrole undergoes Reimar Tiemann's reaction. The attacking species is dichlorocarbene (: CCl_2). The reaction takes place as under:

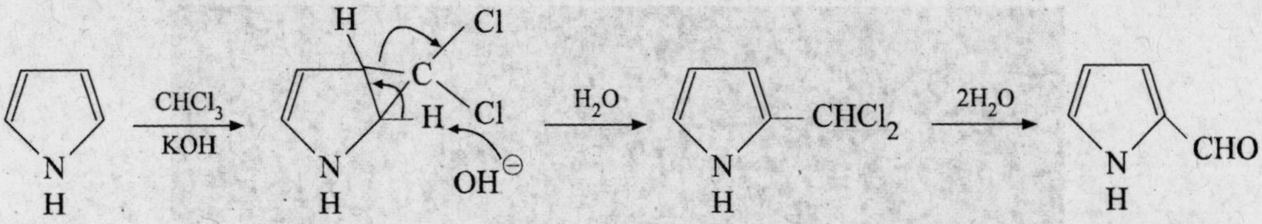

Pyrrole-2-aldehyde

2. Reduction. (*i*) Pyrrole gives pyrroline (2 : 5 dihydropyrrole) on reduction with Zn and acetic acid.

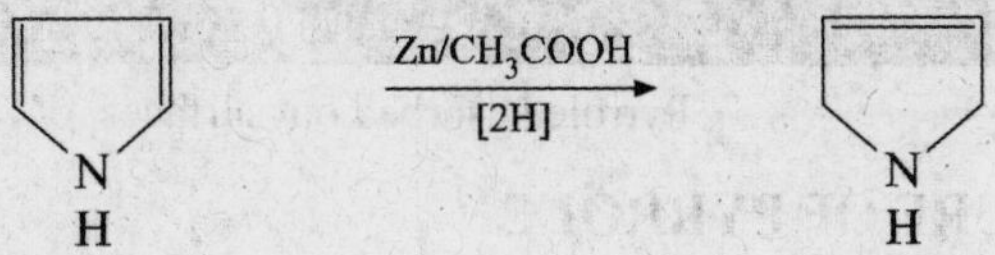

Pyrroline

(*ii*) Pyrrole gives pyrrolidine (tetrahydropyrrole) on hydrogenation in the presence of Ni.

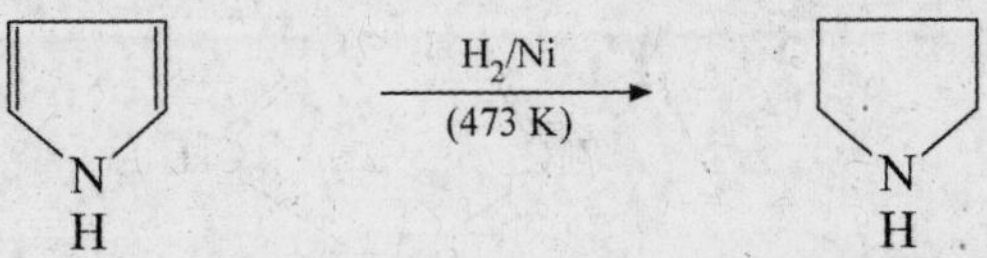

Pyrrolidine

3. Oxidation. When oxidised with chromium trioxide in H_2SO_4, it gives malecimide.

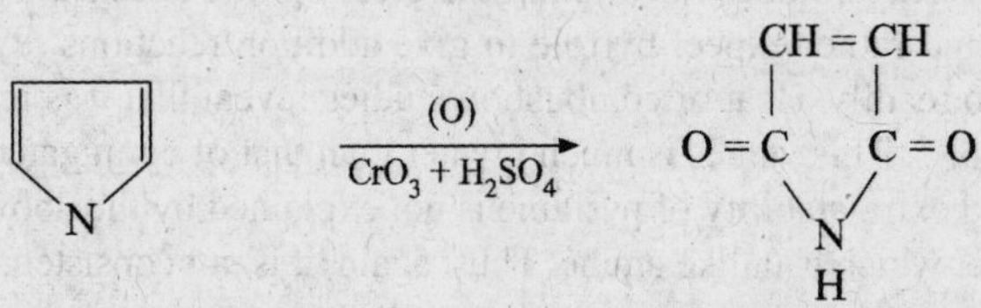

Malecimide

4. Ring expansion. When pyrrole is treated with sodium methoxide and methylene iodide, it gives pyridine (a six membered ring).

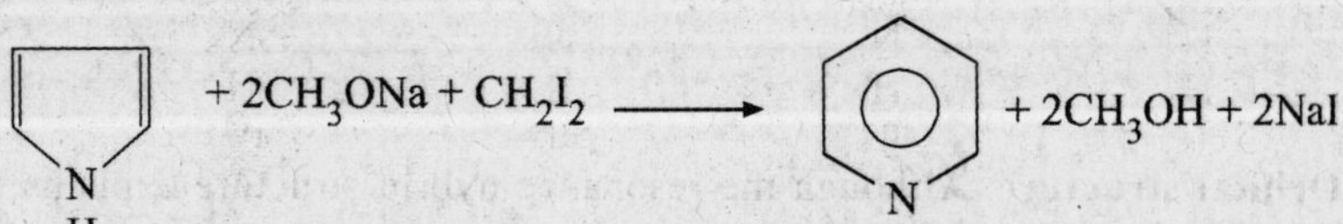

Pyridine

5. Acidic character. Pyrrole unlike furan and thiophene is weakly acidic and forms alkali metal salts.

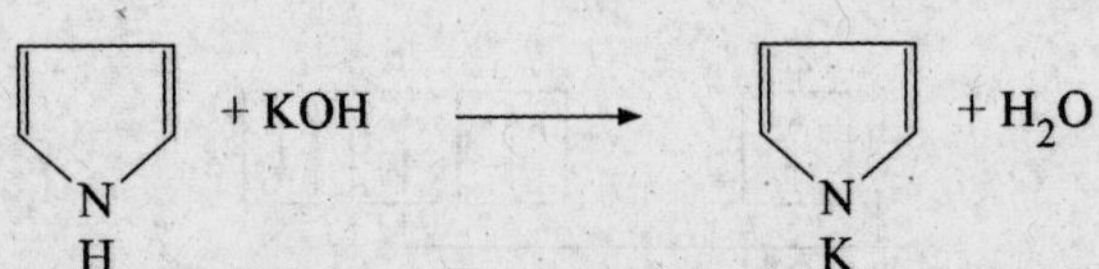

6. Ring fission. When pyrrole is treated with hydroxylamine, the ring is ruptured and succinaldehyde dioxime is formed.

Succinaldehyde dioxime

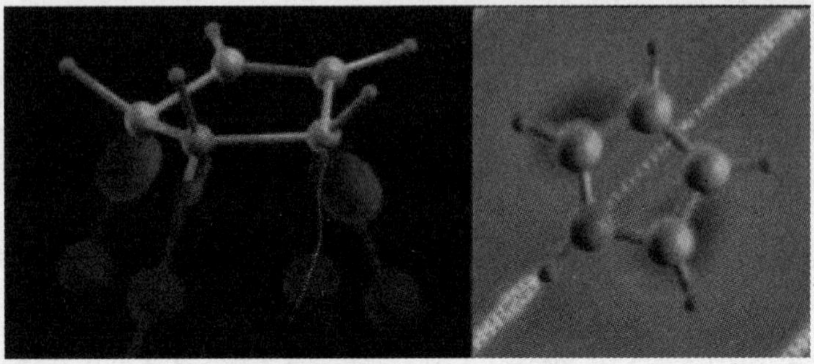

Pyrrole adsorbed on surfaces

4.15. STRUCTURE OF PYRROLE

1. Analytical studies and molecular weight determination establish its molecular formula as C_4H_5H.

2. In terms of tetravalency of carbon and trivalency of nitrogen, a possible structure is

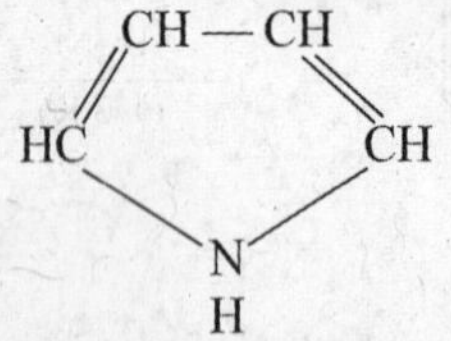

3. The above structure does not explain the electrophilic substitution reactions given readily by pyrrole. We would rather expect pyrrole to give addition reactions. Pyrrole does give addition reactions but not so readily. Heat of combustion studies reveal that it is resonance stabilized to the extent of 100 kJ mol^{-1}. This values is much greater than that of conjugated double bonds. Thus the aromatic nature and extra-stability of pyridine is not explained by the above structure. Also pyrrole lacks basic character which is unlike amine. Thus, again, it is not consistent with the above structure.

4 Pyrrole has been found to be a resonance hybrid of the following resonating structures.

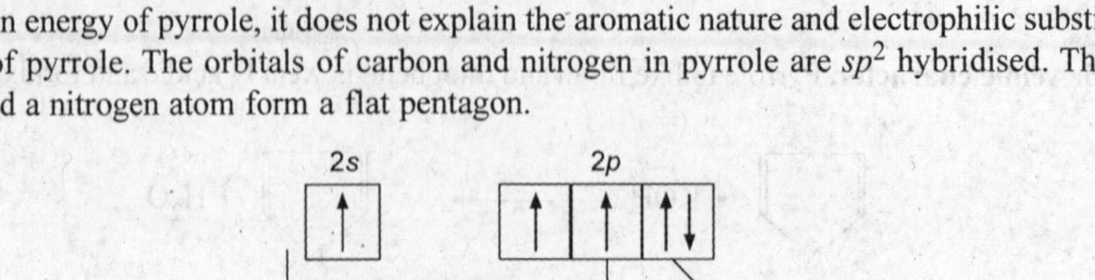

5. Orbital structure. Although the resonance hybrid structure explains the resonance stabilization energy of pyrrole, it does not explain the aromatic nature and electrophilic substitution reactions of pyrrole. The orbitals of carbon and nitrogen in pyrrole are sp^2 hybridised. The four carbons and a nitrogen atom form a flat pentagon.

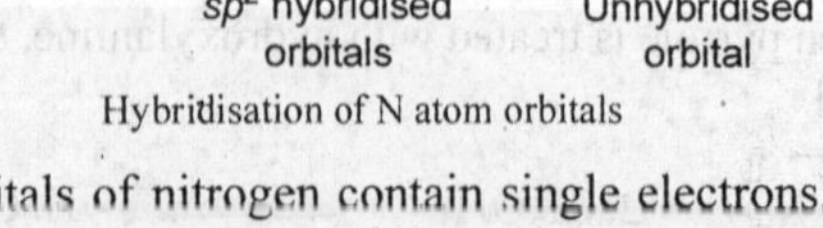

Hybridisation of N atom orbitals

The three sp^2 hybrid orbitals of nitrogen contain single electrons. The unhybridised orbital contains two electrons. The two hybrid orbitals of nitrogen containing single electrons overlap with the hybrid orbitals of two carbon atoms on either side forming σ bonds. The third hybrid orbital overlaps with the orbital of hydrogen and forms N — H bond. Similarly each carbon forms two σ bonds with the neighbouring carbon atoms and one σ bond with hydrogen. Now each carbon has one unhybridised orbitals. These unhybridised orbitals of carbon along with the unhybridised orbital of nitrogen form a delocalised electron cloud above and below the pentagonal ring of pyrrole as shown in the figure below:

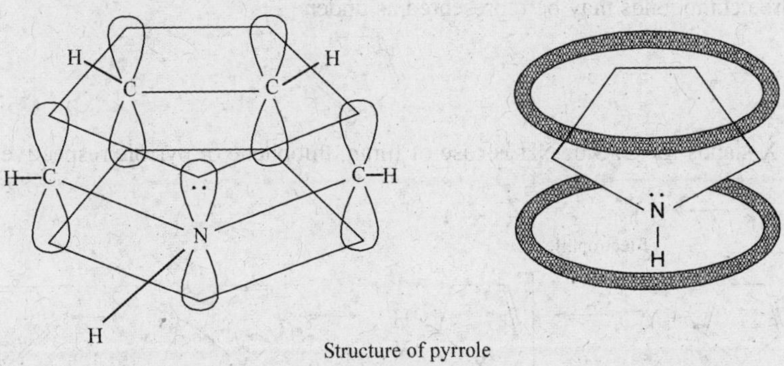

Structure of pyrrole

This accounts for the aromatic nature of pyrrole.

It may be noted that nitrogen contributes both the *p* electrons in the formation of electron cloud. Therefore, nitrogen is not left with any unshared electrons for sharing with acids. As such, pyrrole *has very little basic character*, compared to amines.

It may also be noted that each carbons atom contributes one electrons towards the electron cloud while nitrogen contributes two electrons. Therefore, electron density at the carbon atoms of pyrrole ring is somewhat greater than that in case of benzene. Therefore, *pyrrole is more susceptible to electrophillic attack than benzene*. Thus it undergoes reactions like nitrosation and coupling with diazonium salts which are given only by activated substituated benzens such as phenols and amines.

4.16. RELATIVE AROMATIC CHARACTER OF PYRROLE, THIOPHENE AND FURAN

Molecules of pyrrole, thiophene and furan consist of a flat ring of four carbons atoms and a hetero atom with a *cyclic electron cloud of six delocalised* π*-electrons*. Thus in keeping with Huckel's rule, all these compounds exhibts aromatic character.

The following points may also be noted.

(*i*) Of the five contibuting structures of each of these heterocyclics, there is only one structure in each case which does not involve any charge separation and makes major contribution to the resonace hybrid. As against this, both the major contributing structures of benzene are uncharged and equally stable. As a result, the resonance energies of these heterocylic compounds are considerably lower than that of benzene. In other words. their aromatic character is smaller than that of benzene.

(*ii*) The electronegativities of the hetroatoms present in pyrrole, furan and thiophene are in the order O > N > S. This means that oxygen has the least tendency to release its pair of electrons to contribute to the electrons cloud. Consequently, furan is least aromatic of the three compounds

NMR studies of these compounds show the aromatic character of the compounds in the following order:

Thiophene > Pyrrole > Furan

IN FURAN, THIOPHENE AND PYRROLE

Mechanism

The mechanism of electrophilic substitution in furan, thiophene and pyrrole which occurs preferentially at α-position is represented as under.

The three compounds may be represented as under

where X stands for O, S or NH is case of furan, thiophene or pyrrole respectively.

(*i*) Y : Z ⟶ Y^+ + Z^- :
Electrophile

(*ii*)

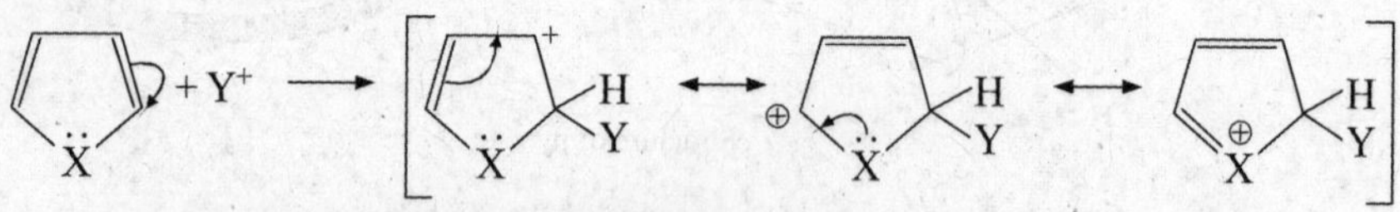

(*iii*)

Orientation

Substitution takes place preferably at that position where the intermediate products is more stabilised. In the case of furan, thiophene and pyrrole, there are two positions 2 or 3 (α or β) where the substitution could take place. Let us see what are the intermediate products in the two cases.

Substitution at α-position (or 2-position)

The intermediate products in case of substitution at α position has three resonating structures as given below :

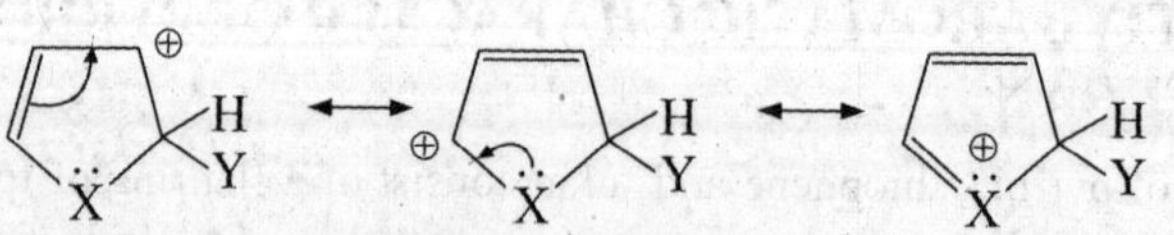

Substitution at β-position (or 3-position)

There are only two resonating structures for intermediate product in case of substitution at β-position.

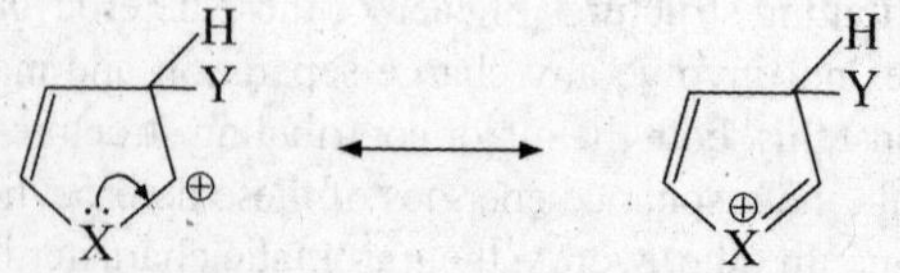

Evidently the carbonium ion intermediate is more stabilised in case of substitution at α-position. Hence substitution is preferred at α-position.

Example 1. Pyrrole is acidic in character like phenol. Explain.

Solution. Both the compounds release the protons and the resulting anions are stabilised by resonance as shown below:

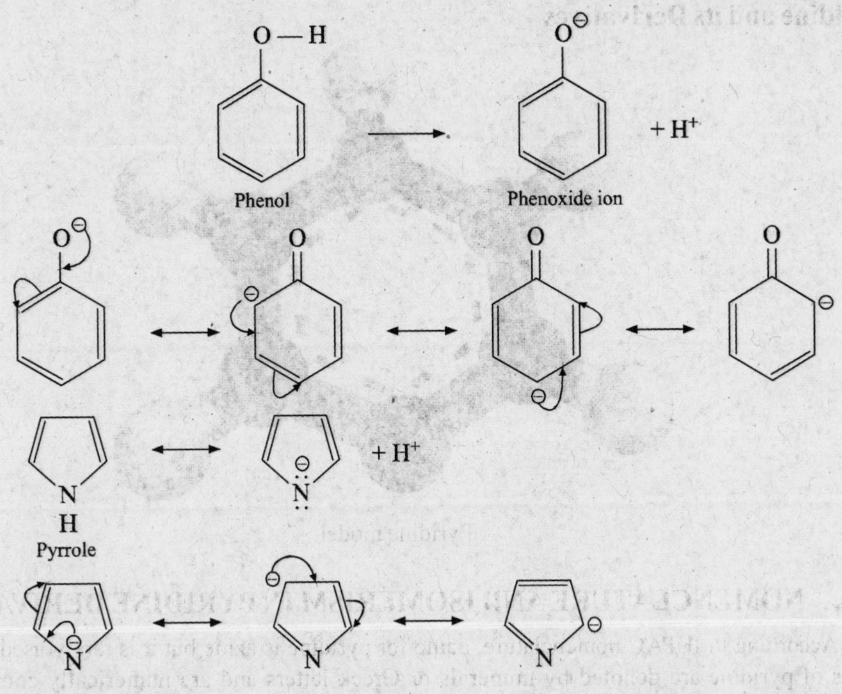

4.18. THE RELATIVE REACTIVITIES OF PYRROLE, FURAN AND THIOPHENE

Like benzene, these compounds show aromatic character and are resonance stabilised. However, their resonance energies are much smaller and hence they are less aromatic or more reactive than benzene.

The order of reactivity is in the order

Pyrrole > Furan > Thiophene > Benzene

The increasing order of aromatic character in pyrrole, furan and thiophene is as follows:

Pyrrole < Furan < Thiophene

(*i*) Thiophene is least reactive or most stable of the three because it can use its *d*-electrons to form a larger no. of resonating structures and hence gets stabilised to a greater extent.

(*ii*) Furan is less reactive than pyrrole because oxygen accommodates a positive charge less readily than nitrogen.

Example 2. Pyrrole, furan and thiophene are more reactive than benzene to electrophilic attack. Explain.

Solution. The above heterocyclics are more reactive than benzene in giving electrophilic reactions because of the formation of intermediate carbonium ions which is specially stabilised by the following resonating structure.

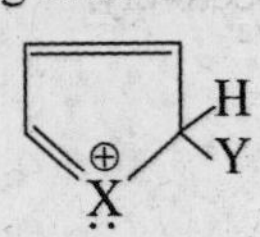

: Y^+ is the electrophile and X is the hetero atom in heterocyclic compound.

This structure is particularly stable because all the atoms have their complete octet.

Pyridine and its Derivatives

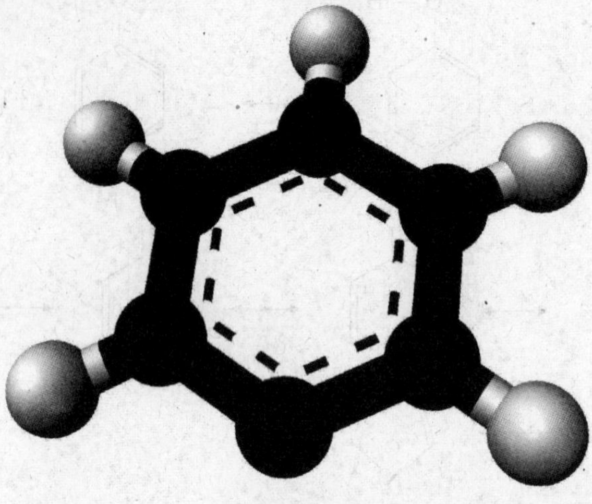

Pyridine model

4.19. NOMENCLATURE AND ISOMERISM IN PYRIDINE DERIVATIVES

According to IUPAC nomenclature, name for pyridine is azine but it is rarely used. The ring atoms of pyridine are denoted by numerals or Greek letters and are numerically counted anti-clockwise with nitrogen occupying position 1.

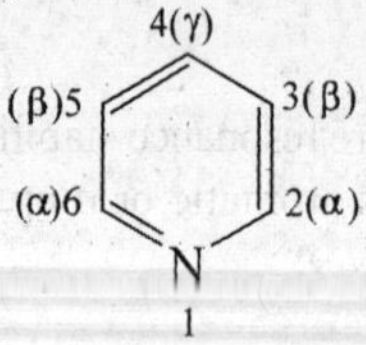

Pyridine

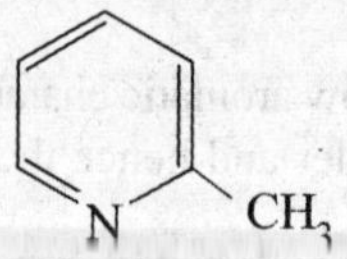

α-Picoline
(2-Picoline)

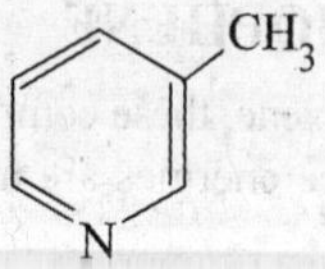

β-Picoline
(3-Picoline)

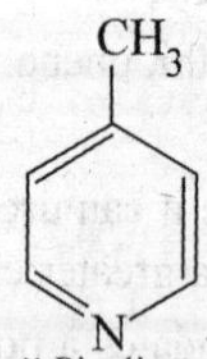

γ-Picoline
(4-Picoline)

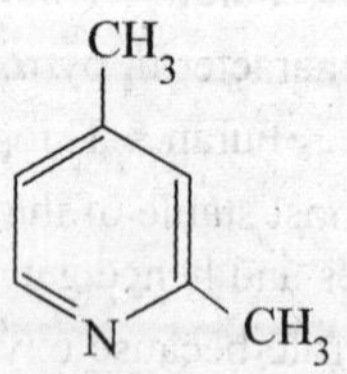

2, 4-Lutidine

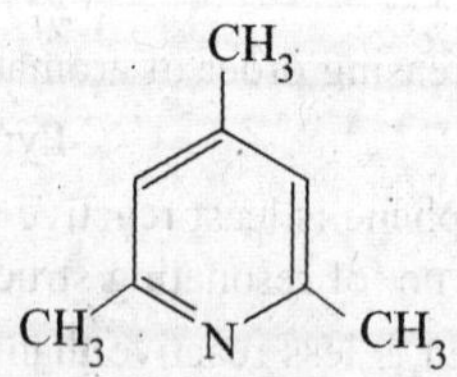

2, 4, 6-Collidine

Pyridine gives *three isomers on mono-substitution*. The nomenclature of pyridine derivative needs special attention because the common names are in general used for methyl pyridine and pyridine carboxylic acids. So mono methyl pyridines are known as *picolines*, the dimethyl pyridines as *lutidines* and trimethyl pyridines as *collidines*. The three isomeric α-, β- and γ- pyridine carboxylic acids are called *picolinic*, *nicotinic* and *isonicotinic* acids respectively.

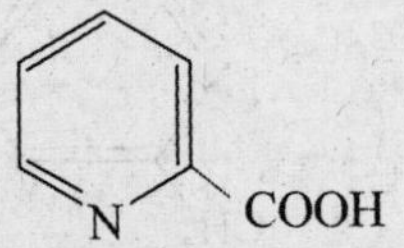

Picolinic acid (Pyridine-2-carboxylic acid)

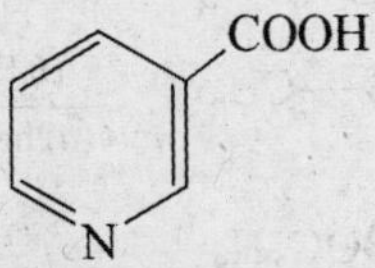

Nicotinic acid (Pyridine-3-carboxylic acid)

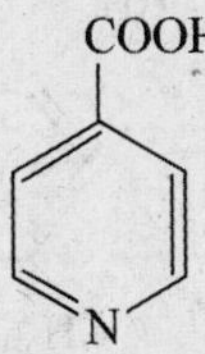

Isonicotinic acid (Pyridine-4-carboxylic acid)

4.20. METHODS OF PREPARATION OF PYRIDINE

Following are the important methods of preparation of pyridine.

1. Isolation from coal-tar. The light oil fraction of coaltar contains aromatic hydrocarbons and phenols besides pyridine and alkyl pyridines. The light oil fraction is thus heated with dil. H_2SO_4 which dissolves pyridine and other basic substances. This aqueous acid layer is neutralised using sodium hydroxide, when bases are liberated as a dark brown oily liquid. This oily layer is separated and pyridine is obtained by fractional distillation.

2. It is prepared by passing a mixture of acetylene and hydrogen cyanide through a red hot tube.

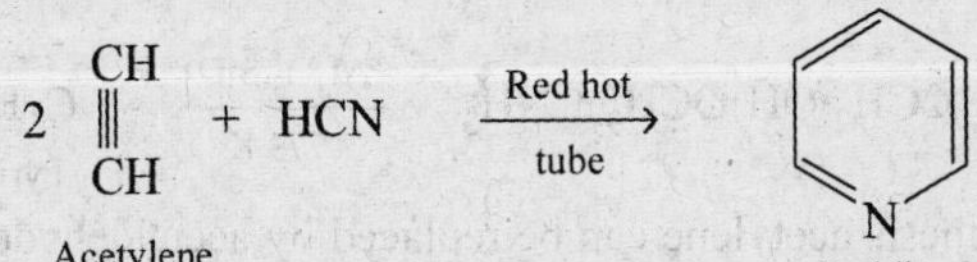

3. It can be prepared by heating pentamethylenediamine—hydrochloride and oxidising the product piperidine with concentrated sulphuric acid at 573 K.

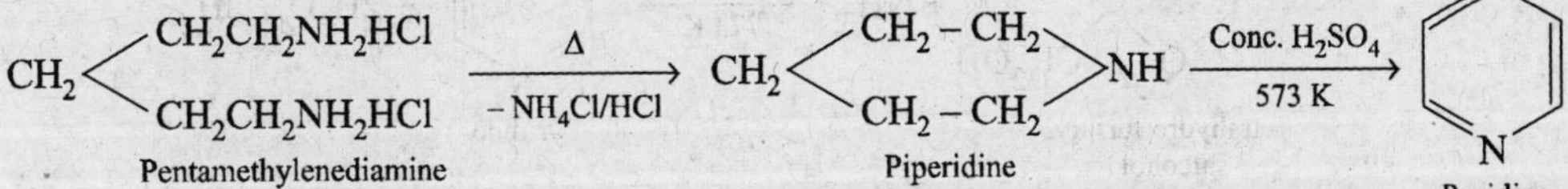

4. Hantzsch Synthesis (1882). It involves the condensation of a β-dicarbonyl compound (2 mole), an aldehyde (1 mole) and ammonia (1 mole). The dihydropyridine derivative formed on oxidation with HNO_3 yields pyridine derivative.

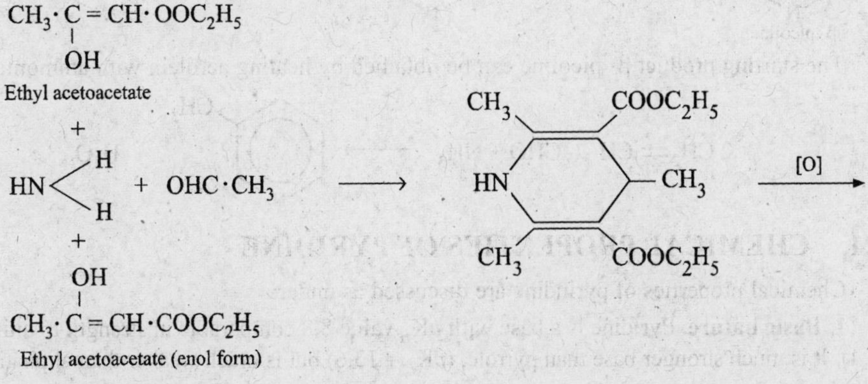

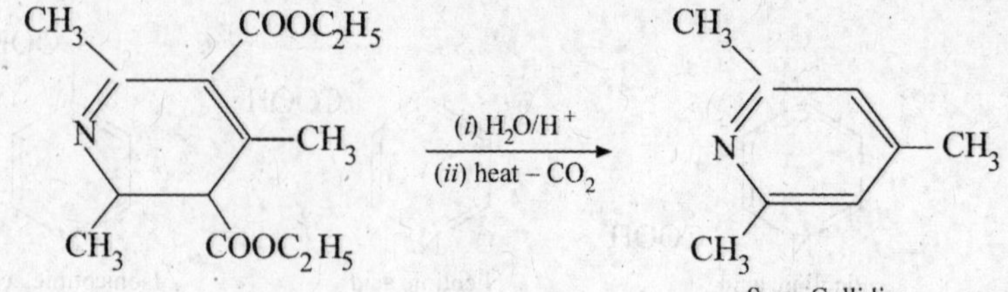

Sym–Collidine
(2, 4, 6-Trimethyl pyridine)

5. From Pyrrole. When heated with methylene dichloride in presence of sod. ethoxide, pyrrole gives pyridine.

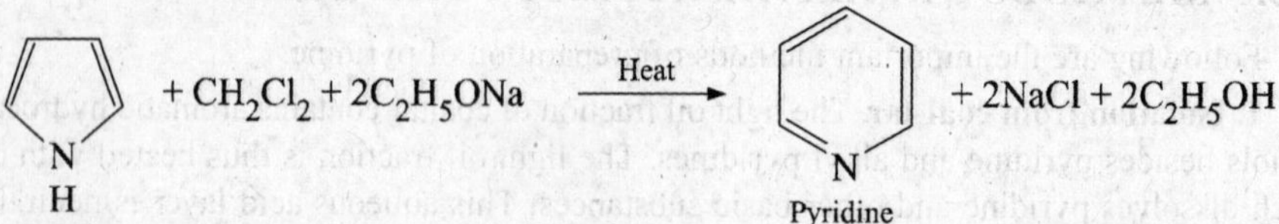

Pyrrole Pyridine

6. Pyridine of very high purity needed for the synthesis of medicinal compounds is obtained on a large scale by passing acetylene, formaldehyde hemi methyl and ammonia over alummia-silica catalyst at 773 K.

$$2\ \begin{matrix} CH \\ \| \! | \\ CH \end{matrix} + 2CH_2(OH)OCH_3 + NH_3 \xrightarrow[773\ K]{Al_2O_3/SiO_2} \underset{\text{Pyridine}}{C_5H_5N} + H_2O + 3CH_3OH$$

Note. In this synthesis acetylene can be replaced by acetaldehyde.

7. A new **industrial method** for the preparation of pyridine is to heat tetrahydro furfuryl alcohol, obtained by the catalytic reduction of furfuryl alcohol, with ammonia at 773 K.

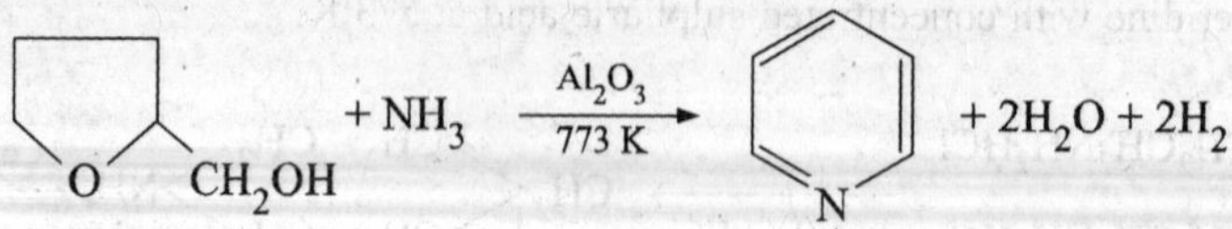

Tetrahydro furfuryl alcohol Pyridine

8. From **picoline.** On oxidation with potassium dichromate and sulphuric acid, β-picoline changes into nicotinic acid. On decarboxylation with calcium oxide, we obtain pyridine.

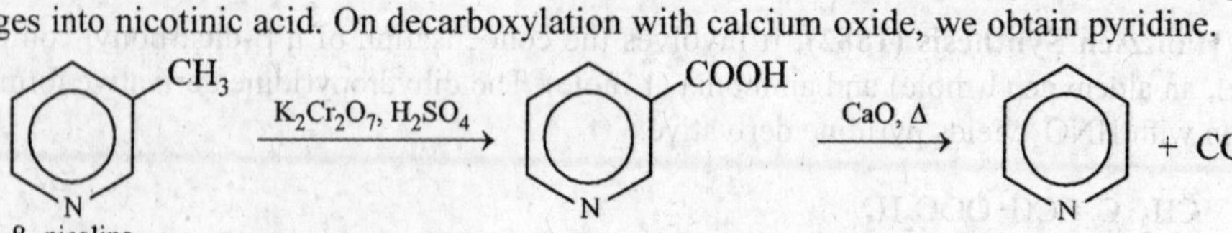

β- picoline

The starting product β- picoline can be obtained by heating acrolein with ammonia.

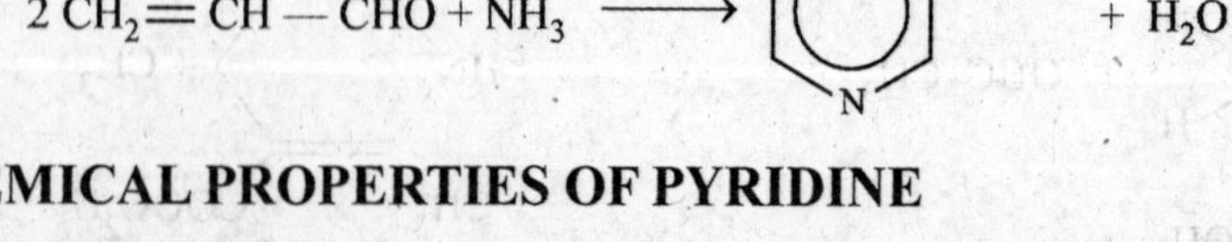

4.21. CHEMICAL PROPERTIES OF PYRIDINE

Chemical properties of pyrindine are discussed as under:

1. Basic nature. Pyridine is a base with pK_b value 8.8 comparable in strength to aniline (pK_b = 9.4). It is much stronger base than pyrrole, (pK_b = 13.6) but is much weaker than aliphatic tertiary amines ($pK_b = 4$). Basic nature of pyridine is due to the presence of the lone pair of electrons on nitrogen atom. It behaves as a mono acid tertiary base and reacts with alkyl halides to give quaternary salts.

$$\underset{\text{Pyridine}}{C_5H_5N:} + CH_3I \longrightarrow \underset{\text{N-methyl pyridinium iodide}}{[C_5H_5\overset{+}{N}CH_3]\,I^-}$$

The fact that pyridine is much weaker base than aliphatic tertiary amines can also be explained in terms of the state of hybridisation of the orbitals having the lone pair of electrons. In case of aliphatic tertiary amines, the unshared pair of electrons is in sp^3 hybrid orbital while in case of pyridine unshared pair is in sp^2 hybrid orbital. Electrons are held more tightly by the nucleus in an sp^2 orbital than in sp^3 orbital. Hence they are less available for sharing with acids. This accounts for low basicity of pyridine.

Why pyrrole is a weaker base than pyridine can be explained in terms of non-avilability of electron pair on nitrogen in pyrrole. All the electrons are involved either in bond formation or in the formation of electron cloud.

2. Reduction. Pyridine undergoes catalytic hydrogenation to form hexahydropyridine, called *piperidine*. Reduction with Na/C_2H_5OH also produces piperidine.

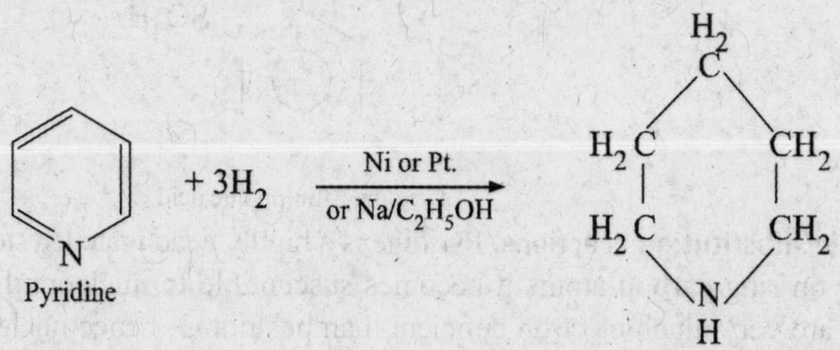

Piperidine

3. Addition of halogen. It adds halogen to yield a dihalide at room temperature and in the absence of catalyst.

$$C_5H_5N + Br_2 \longrightarrow \underset{\text{Dibromopyridine}}{[C_5H_5NBr]\,Br^-}$$

4. Electrophilic substitution reactions. Pyridine behaves as a highly deactivated aromatic nucleus towards electrophilic substitution reactions, and vigorous reaction conditions should be used for these reactions to take place. The low reactivity of pyridine towards electrophilic substitutions is due to the following two reasons.

(*i*) *Greater electronegativity of nitrogen atom decreases electron density* (–I effect) *of the ring thereby deactivating it.*

(*ii*) *In acidic medium it forms a pyridinium cation with a positive charge at nitrogen atom. Reacting electrophile itself (say* E^+*) may also react with pyridine to form pyridinium ion and this decreases the electron density on nitrogen very much,* thus deactivating the ring. However the position – 3 is least affected and is comparatively the position of highest electron density in pyridine.

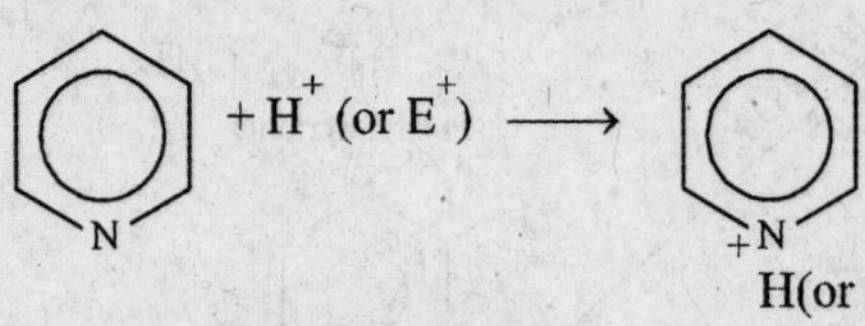

It undergoes electrophilic substitution like halogenation, nitration and sulphonation only under drastic conditions and does not undergo Friedel-Crafts reaction at all. The substitution occurs preferentially at β- or 3-position.

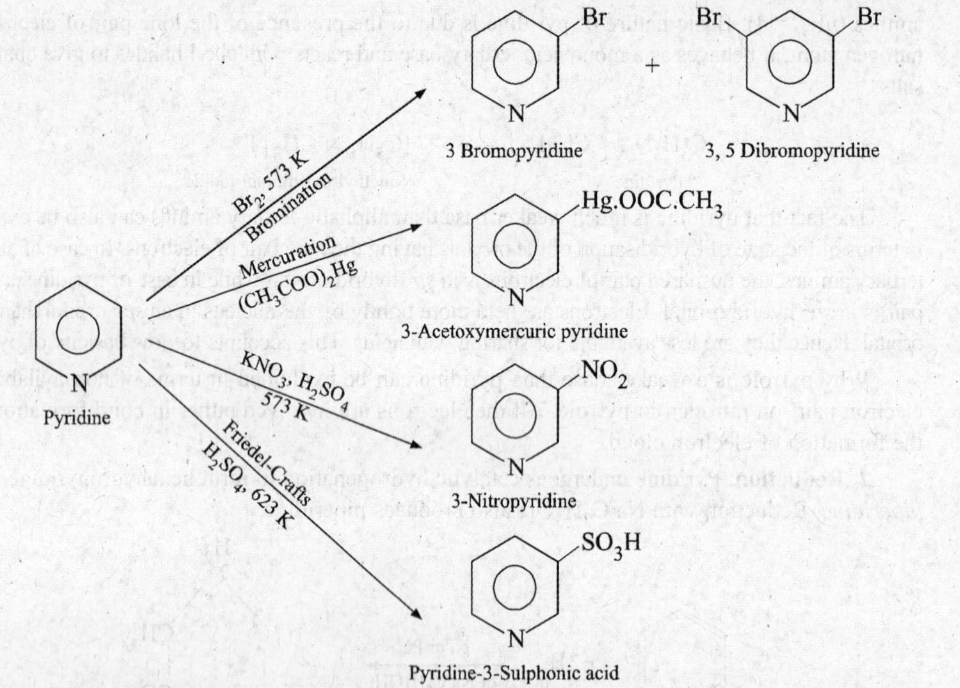

5. Nucleophilic substitution reactions. Pyridine is a highly deactivated system. Due to decrease of electron density on ring carbon atoms it becomes susceptible to nucleophilic attack. Since the positions 2- and 4- are very much electron deficient than position-3, hence nucleophilic substitution occurs readily at positions (2- and 4-), the 2-position being more preferred. The pyridinium ion is still more reactive than pyridine towards nucleophilic substitutions, because of the presence of full positive charge. Some of the important nucleophilic substitution reactions are given as under:

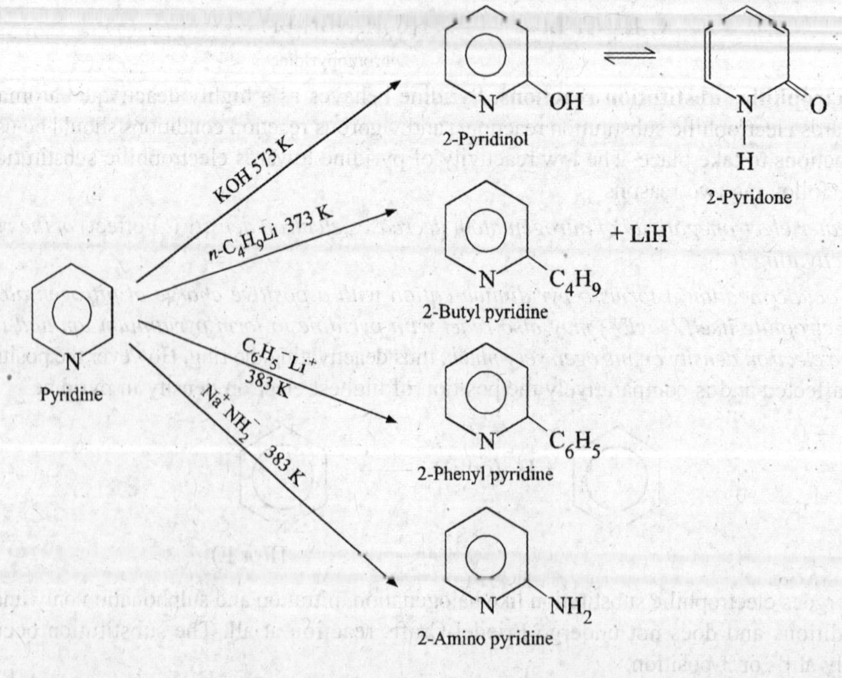

Reaction with sodamide to give 2-aminopyridine is also called as **Chichibabin Reaction.**

6. Oxidation. Like other tertiary amines, pyridine is oxidised by peroxy benzoic acid to yield pyridine N-oxide.

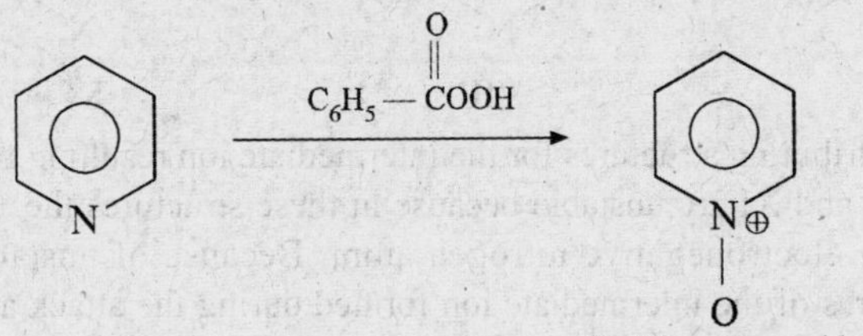

Pyridine – N – oxide

The N-oxide is more reactive than pyridine and finds use is preparing many pyridine derivatives.

4.22. MECHANISM AND ORIENTATION OF ELECTROPHILIC SUBSTIUTION IN PYRIDINE

Mechanism

The mechanism of electrophilic substitution in pyridine which occurs preferentially at position 3 can be described in the following three steps.

(*a*) $Y : Z \longrightarrow Y^{\oplus} + :Z^{-}$

Electrophile

(*b*)

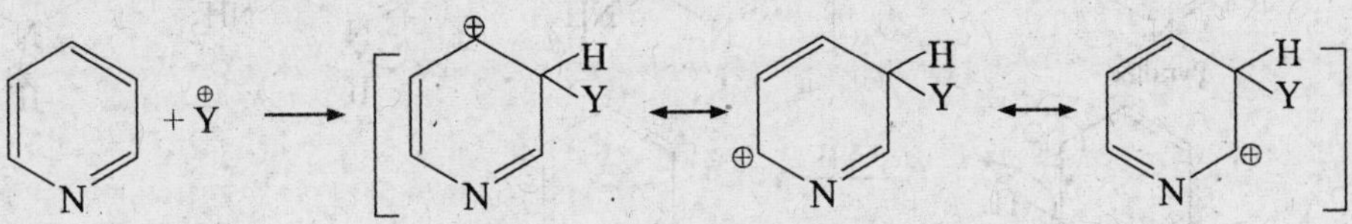

I II III

(*c*)

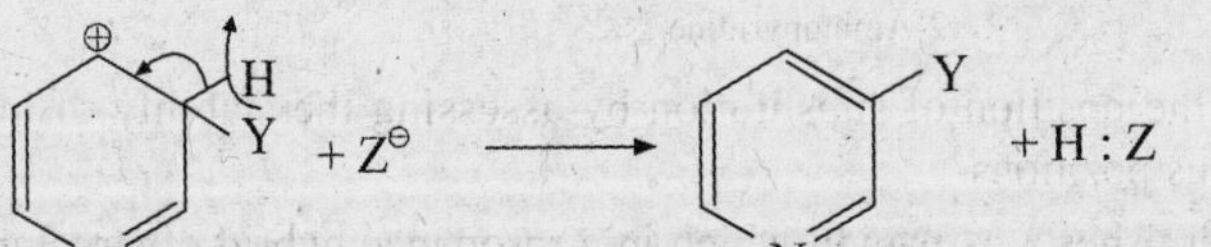

Substituted pyridine

Orientation

Why the substitution takes place preferentially at position 3 and not at 2- or 4-. can be well explained in terms of the contributing structures of carbonium ion intermediate formed in each case. Attack at position 3 furnishes carbonium ion which is a resonance hybrid of structures I, II and III shown above. Attack at position 4 or 2 would give a carbonium ion which is a resonance hybrid of structures IV, V, VI or VII, VIII, IX as given below:

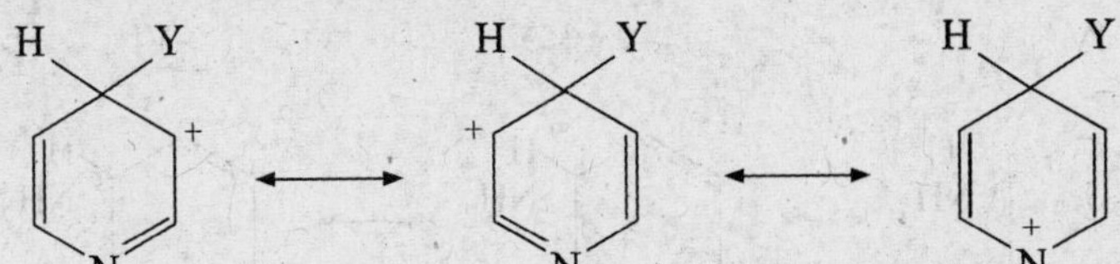

Attack at Position 4

IV V VI

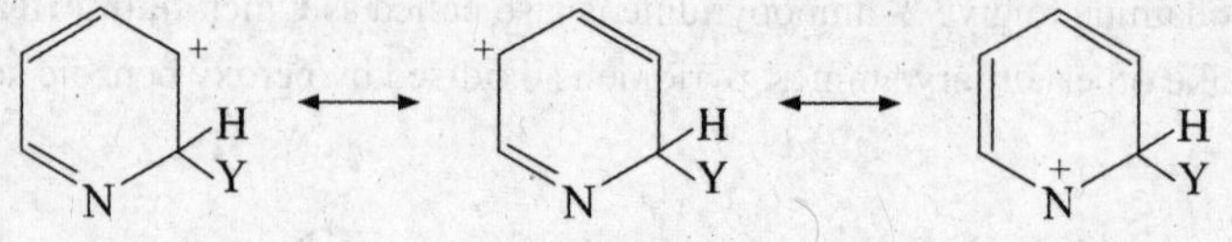

Attack at Position 2

VII VIII IX

Out of the contributing structures for the intermediate ion resulting from the attack at position 4 or 2 structures VI and XI are unstable because in these structures the +ve charge is particularly carried by strongly electronegative nitrogen atom. Because of unstable nature of one of the contributing structures of the intermediate ion formed during the attack at position 4 or position 2 this ion is less stable than that formed during the attack at position 3. Hence substitution at 3-position is favoured.

4.23. MECHANISM AND ORIENTATION OF NUCLEOPHILIC SUBSTI TUTION IN PYRIDINE

The mechanism of nucleophilic substitution in pyridine which usually occurs at position 2 or 4 is outlined as follows:

Reaction with sodamide has been considered.

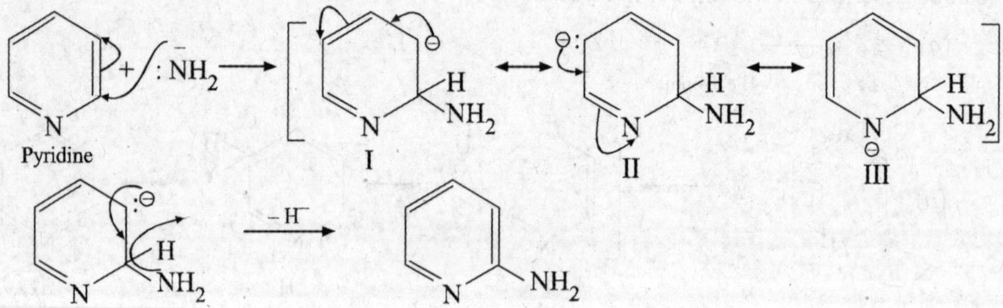

Pyridine I II III

2-Aminopyridine

We can decide about the position of substitution by assessing the stability of intermediate carbanion obtained in various positions.

Attack at position 2 furnishes a carbanion which in a resonance hybrid of structures I, II, III as shown above. Attack at position 4 or 3 would furnish a carbanions which are resonance hybrids of structures IV, V, VI or VII, VIII, IX as shown below:

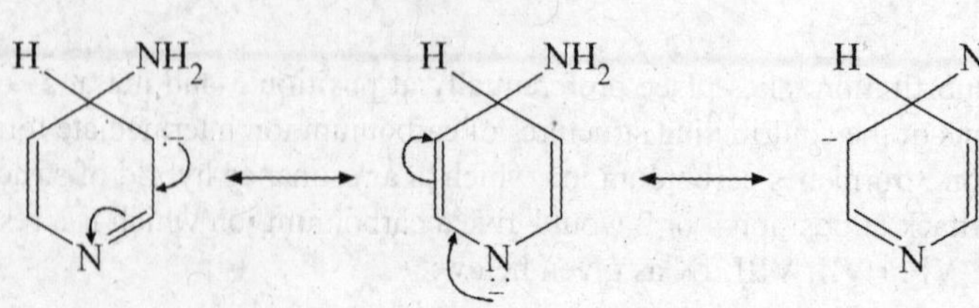

IV V VI

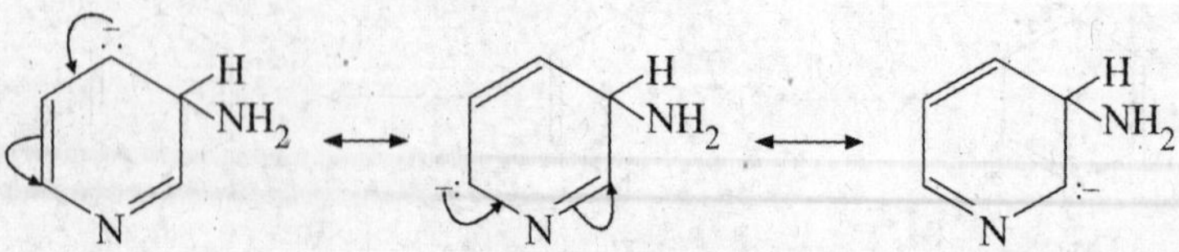

VII VIII IX

It may be noted that when the attack occurs at position 2, the contributing structure III of the *intermediate carbanion is particularly stable, because the negative charge is present on the electronegative nitrogen atom which can be easily accommodated by it. Similarly contributing structure V during attack at position 4 is also quite stable.*

There is no such stable contributing structure of intermediate ion resulting from attack at position 3. This means that the intermediate carbanion resulting from attack at positions 2 or 4 would be more stable, and hence more readily formed than that involved in attack at position 3. That is the reason why nucleophilic substitution is favoured at positions 2 or 4.

4.24. STRUCTURE OF PYRIDINE

1. Molecular formula. On the basis of analytical data and molecular weight determination, the molecular formula of pyridine comes out to be C_5H_5N.

2. Chemical behaviour. (*i*) Pyridine is basic in nature and forms salt with acids

$$\underset{\text{Pyridine}}{C_5H_5N} + HCl \longrightarrow \underset{\text{Pyridine hydrochloride}}{C_5H_5N \cdot HCl}$$

(*ii*) It does not react with acetyl chloride or nitrous acid indicating the absence of primary or secondary amino group. *The above reactions reveal that pyridine is a mono acid tertiary base.*

(*iii*) It reacts with *equimolar* amount of methyl iodide to form a quaternary ammonium salt.

$$\underset{\text{Pyridine}}{C_5H_5N} + CH_3I \longrightarrow \underset{\text{N-Methyl pyridinium iodide}}{[C_5H_5N^+ (CH_3)]\, I^-}$$

(*iv*) It molecular formula indicates that it is a highly unsaturated compound, yet pyridine resists addition reactions and is stable towards usual oxidising agents.

(*v*) Pyridine exhibits aromatic character like benzene and gives electrophilic substitution reactions such as halogenation, nitration and sulphonation.

The last two reactions indicate that pyridine has aromatic character.

3. Korner's formula. On the basis of above evidence Korner assigned the following cyclic structure to pyridine which is quite analogous to Kekule structure of benzene.

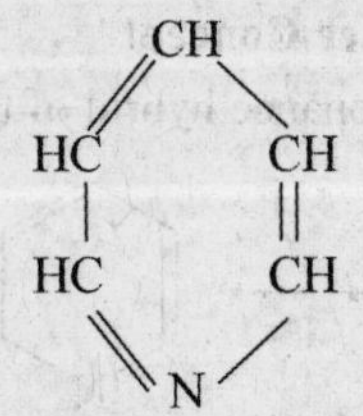

4. Evidence in favour of Korner's formula. (*i*) On reduction, it forms piperidine—a hexahydropyridine—the structure of which is confirmed by its synthesis as given below:

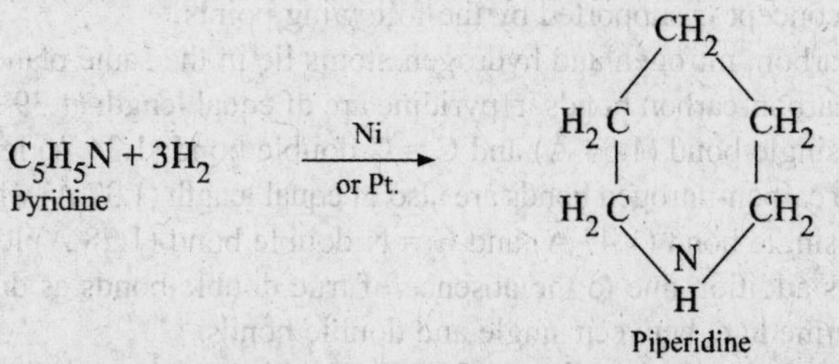

The synthesis of piperidine from penta-methylene-1, 5-diamine hydrochloride and subsequent conversion to pyridine by oxidation with conc. H_2SO_4 further confirms the cyclic structure of piperidine and pyridine both.

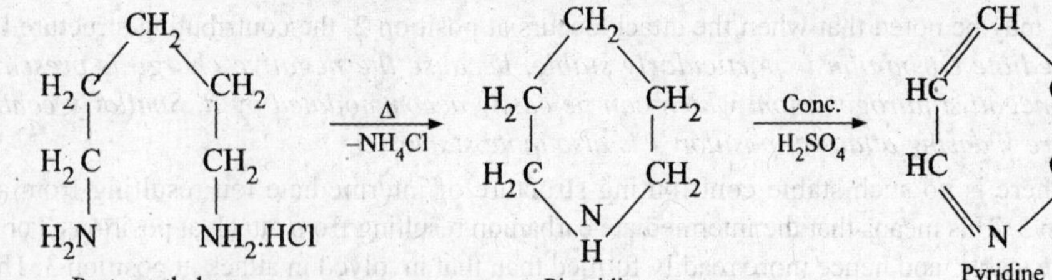

Penta methylene 1, 5-diamine hydrochloride — Piperidine — Pyridine

(*ii*) Pyridine gives three mono substituted products and six disubstitution products (in case the two substituents are identical) quite expected from the Korner's formula. Thus we have:

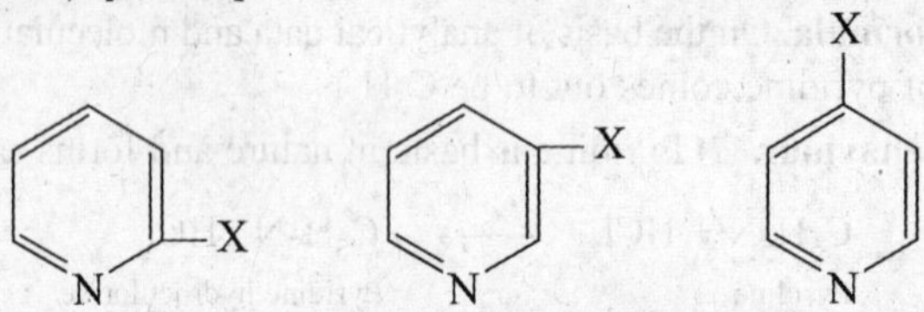

Three mono substitution products of pyridine, are shown above.

5. Objections to Korner formula. It fails to the explain following:

(*i*) its aromatic character.

(*ii*) its resistance to addition reactions.

(*iii*) its electrophilic and nucleophilic substitution reactions.

6. Alternative formulae. Following alternative formulae were suggested for pyridine, but were soon rejected as none was consistent with the behaviour and properties of pyridine.

Riedels formula — Bamberger and Vonpechmann's formula

7. Latest position. (I) Resonance Concept

Pyridine is considered to be resonance hybrid of the following structures:

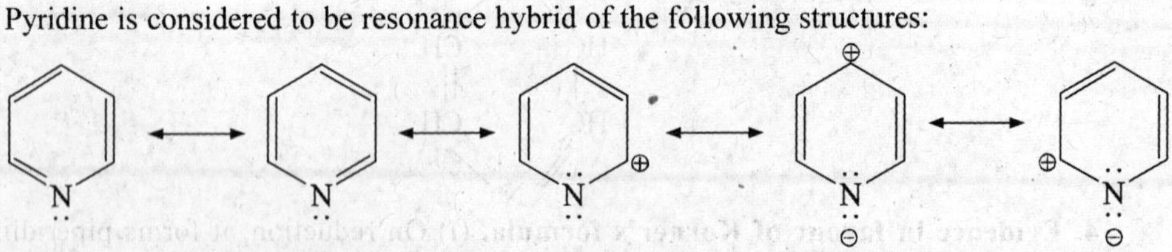

I II III IV V

Resonance concept is supported by the following points:

(*i*) All the carbon, nitrogen and hydrogen atoms lie in the same plane.

(*ii*) All the carbon-carbon bonds in pyridine are of equal length (1.39 Å) which is intermediate between C — C single bond (1.54 Å) and C = C double bond (1.34 Å) length.

(*iii*) The two carbon-nitrogen bonds are also of equal length (1.37 Å) which is again intermediate between C — N single bond (1.47 Å) and C = N double bond (1.28 Å) length.

(*iv*) It resists addition due to the absence of true double bonds as due to resonance all bond distances are intermediate between single and double bonds.

(*v*) Since positive charge is present at 2- and 4-positions in the contributing structures it behaves as deactivated system in electrophilic substitutions—which take place at position 3.

(*vi*) Its resonance energy is 96 kJ mol^{-1}.

II. Molecular Orbital Concept

(*i*) The nitrogen and each of the five carbon atoms in the pyridine are in sp^2 *hybridization* state. These carbon and nitrogen atoms combine with each other to form a ring making use of two of their sp^2 hybrid trigonal orbitals for forming s (sigma) bonds.

(*ii*) The third sp^2 orbital of each the five carbon atoms overlaps with '*s*' orbital of hydrogen to give rise to s bonds whereas *third* sp^2 *orbital of nitrogen possesses the lone pair of electrons*. The nitrogen lone pair electrons in sp^2 hybrid orbital do not interact with π molecular orbital.

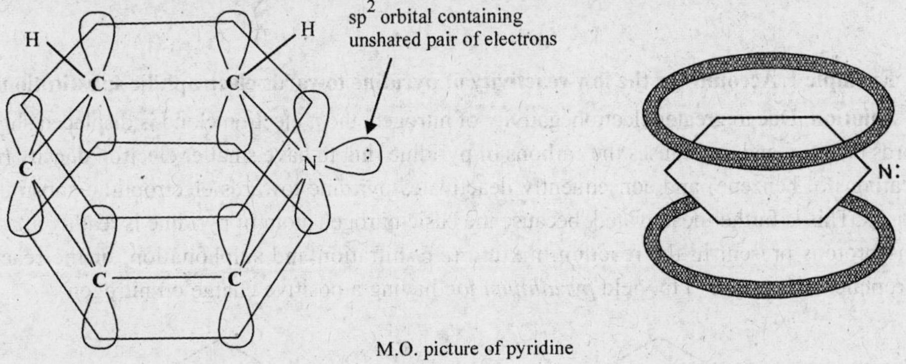

M.O. picture of pyridine

(*iii*) The unhybridised '*p*' orbitals of each carbon atom and nitrogen atom which are perpendicular to the plane of ring atoms, overlap with each other sideway to form a π electron cloud below and above the plane of ring as in the case of benzene.

(*iv*) These delocalised orbitals (having 6 π electrons) which are in the form of π electron clouds above and below the plane of ring are responsible for the stability and aromatic character of pyridine. $(4n + 2)$ Huckel rule where $n = 1$ is satisfied in the case of pyridine.

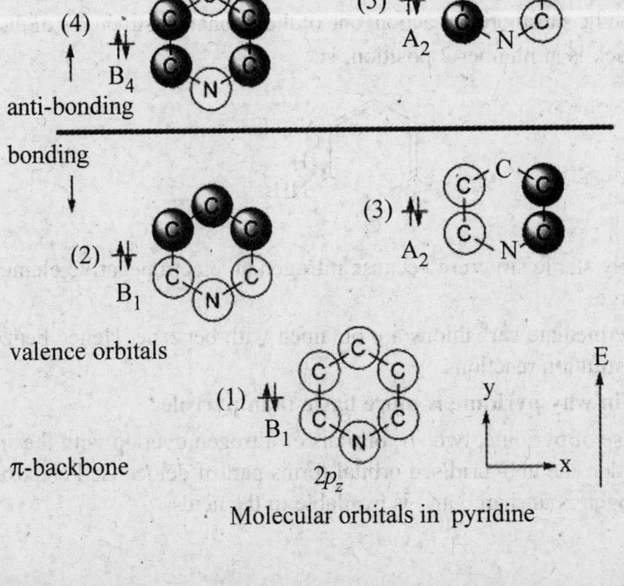

Molecular orbitals in pyridine

The basic nature of pyridine is because of the unshared pair of electron in one of the sp^2 orbitals of nitrogen.

For the sake of convenience, pyridine is generally represented as below:

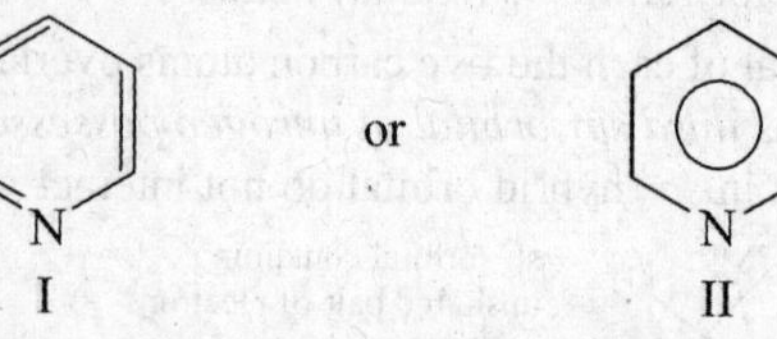

Example 1. Account for the low reactivity of pyridine towards electrophilic substitutions.

Solution. Due to greater electronegativity of nitrogen, the π electron cloud is displaced slightly towards nitrogen and this causes the carbons of pyridine ring to have smaller electron density (than the carbons of benzene) and consequently deactivates pyridine towards electrophilic substitution reactions. This is further deactivated, because the basic nitrogen atom of pyridine is readily attacked by the protons present in the reaction mixture (*e.g.* nitration and sulphonation, or the reacting electrophile itself (say Y^+) to yield *pyridinium ion* having a positive charge on nitrogen.

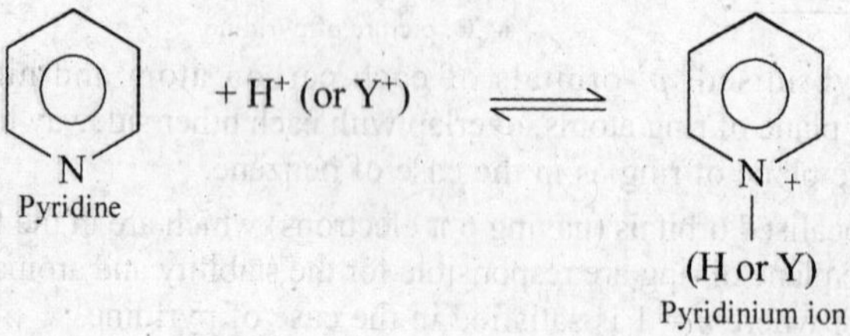

Hence, nitrogen withdraws electrons from the ring quite strongly and makes it very difficult for the electrophilic reagent to attack the ring.

Example 2. While benzene undergoes exclusively electrophilic substitution reactions, pyridine can undergo electrophilic and nucleophilic substitution reactions. Explain.

Solution. Pyridine undergoes nucleophilic addition reactions because the intermediate carbanion gets stabilised due to special features. For example in the reaction of pyridine with sodamide, which is an example of nucleophilic substitution reaction, one of the resonating structures of the intermediate carbanion, when the attack is at number 2 position, is:

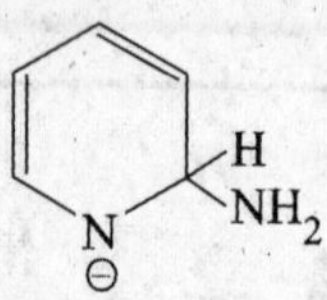

This is a particularly stable structure because nitrogen an electronegative element can easily absorb the negative charge.

No such stable intermediate carbanions are obtained with benzene. Hence benzene does not exhibit nucleophilic substitution reactions.

Example 3. Explain why pyridine is more basic than pyrrole?

Solution. (*i*) In case of pyridine, two sp^2 orbitals of nitrogen overlap with the sp^2 orbitals of two carbons on either side. The unhybridised orbital forms part of delocalised electron cloud. The third sp^2 orbital of nitrogen is unshared and is available to the acids.

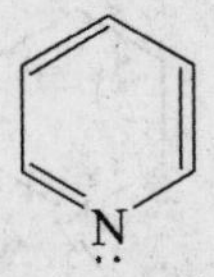

Electrons available for reaction

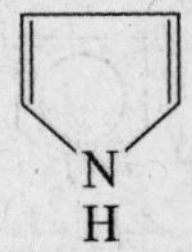

No electrons available

(*ii*) In case of pyrrole, all the electrons are shared. No electrons is free and available to react with acids. Hence pyridine is more basic than pyrrole.

Example 4. Pyrrole is more reactive than pyridine towards electrophilic substitution reactions. Explain.

Solution. Pyrrole is more reactive than pyridine towards electrophilic substitution because the intermediate carbocation formed with pyrrole is more stable as every atom has a complete octet in one of the intermediate structures as shown below:

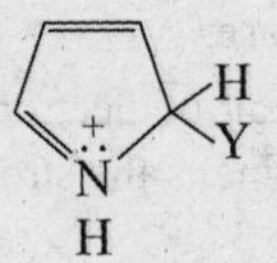

Here Y^+ is the electrophile.

In pyridine, the intermediate carbocation, after attachment of the electrophile has nitrogen with six electrons and hence less stable.

Example 5. How will you synthesise the following compounds?

(*i*) Piperidine **(*ii*) 3- Nitropyridine**

Solution.

(*i*) 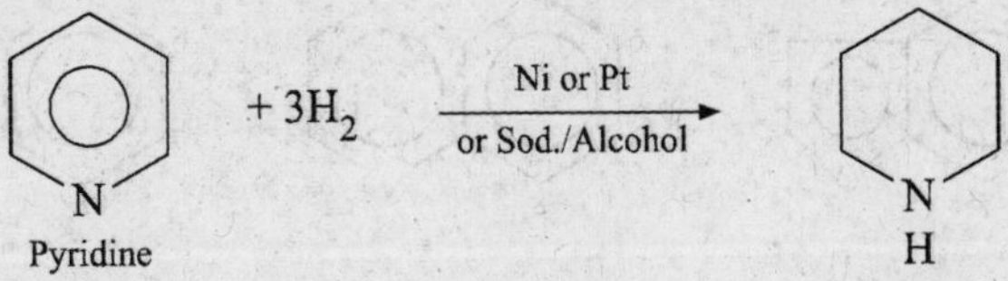

Pyridine — Piperidine

(*ii*) 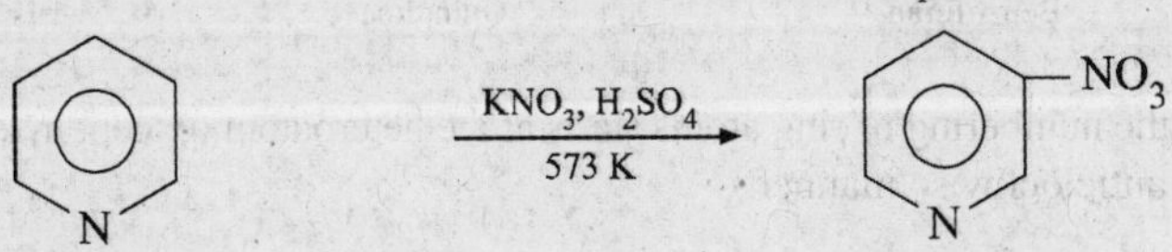

Pyridine — 3-Nitropyridine

4.25. COMARSION OF BASICITY OF PYRROLE, PYRIDINE AND PIPERIDINE

It has been discussed earlier that pyrrole is a very weak base ($pK_b \approx 14$) because, the electrons are not available on nitrogen in pyrrole. Pyridine has basic strength comparable to that of aniline ($pK_b = 8.7$). But, still pyridine is much weaker base than aliphatic amine. The basicity of piperidine is comparable to that of secondary aliphatic amine.

Pyrrole		Pyridine		Piperidine
$pK_b \sim 14$	<	$pK_b \sim 8.7$	<	$pK_b \sim 2.7$
(very poor base)		(Mild base)		(very strong base)

The above order of basic strength of pyrrole, pyridine and piperidine can be justified in terms of the structure of these compounds:

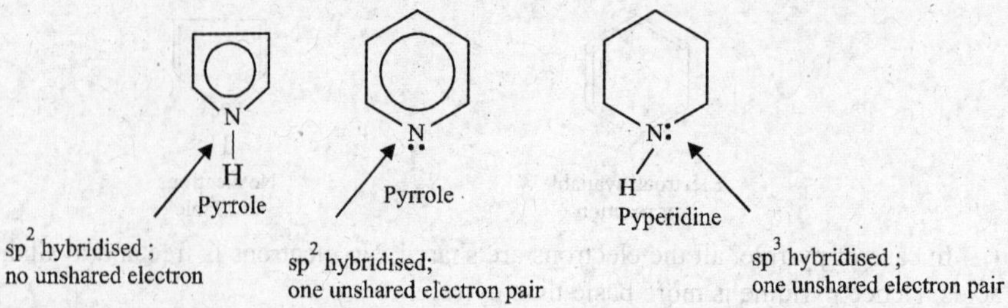

The basicity of nitrogeneous compounds depends upon the avilability of an unshared pair of electrons on nitrogen. In pyrrole, nitrogen does not have any such pair of electrons since two of its electrons have become part of electron cloud. As such the basic character of pyrrole is insignificant.

In pyridine, the nitrgen atom has got a lone pair of electons but it is present in an sp^2 orbital since the electrons in sp^2 orbitals are tightly held by the nucleus, they are less avilable for protonation. As such pyridine acts only as a weak base.

In piperidine, nitrogen has a lone pair of electrons in an sp^3 orbital. These electrons are readily available for protonation. Thus, piperidine is the strongest of the three in basic character.

Condensed five and six membered heterocylics

In addition to the compounds studiedso far a *large number of other heterocylics are known to exist in which a benzene ring is fused to a five or six membered heterocyclic system at positions 2 and 3 of benzene ring*. Such compounds are known as **ring heterocyclics**.

Some wellknown example of such componds are:

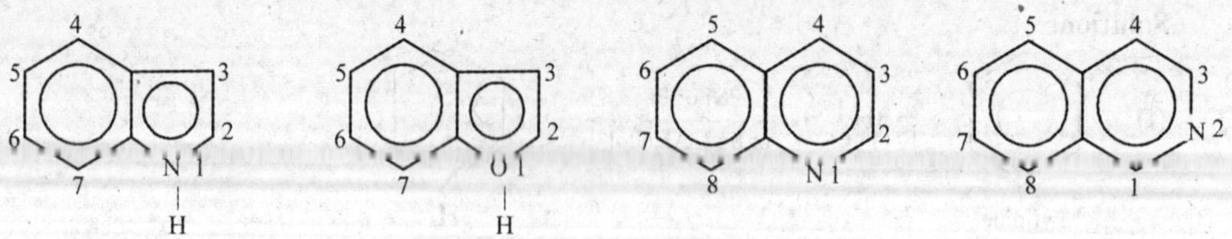

Indole (Benzopyrrole) Benzofuran Quinoline Isoquinoline

It may be seen that the numbering of ring atoms starts at the heteroatom (except in isoquinoline) and then proceeds in an anticlockwise manner.

Indole (Benzopyrrole)

Indole (benzopyrrole) has the following structural formula

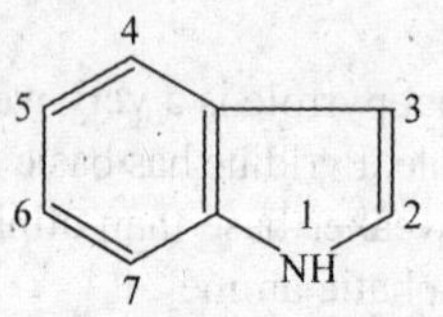

Different positions on the molecule of indole are numbered as shown above.

4.26. SYNTHESIS OF INDOLE

It can be synthesised in a number of ways.

1. Lipp's synthesis. It is carried out by heating *o*-amino-ω-chlorostyrene with sodium ethoxide.

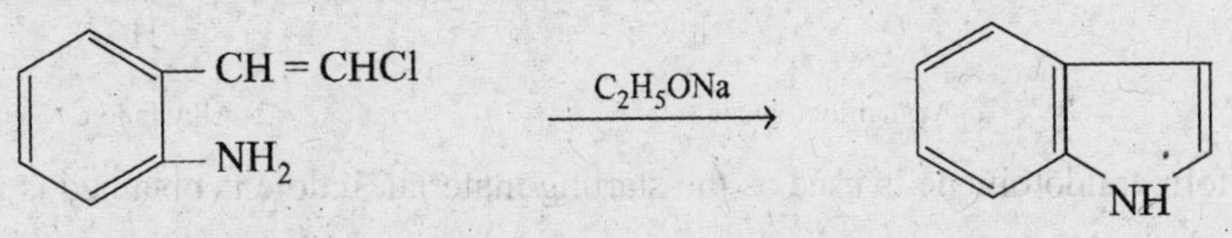

2. Fischer's indole synthesis. It is carried out by heating phenylhydrazone or substituted phenylhydrazone of an appropriate aldehyde or ketone with zinc chloride as catalyst. Synthesis of 2-methyl indole can be achieved by taking acetone phenylhydrazone as shown below:

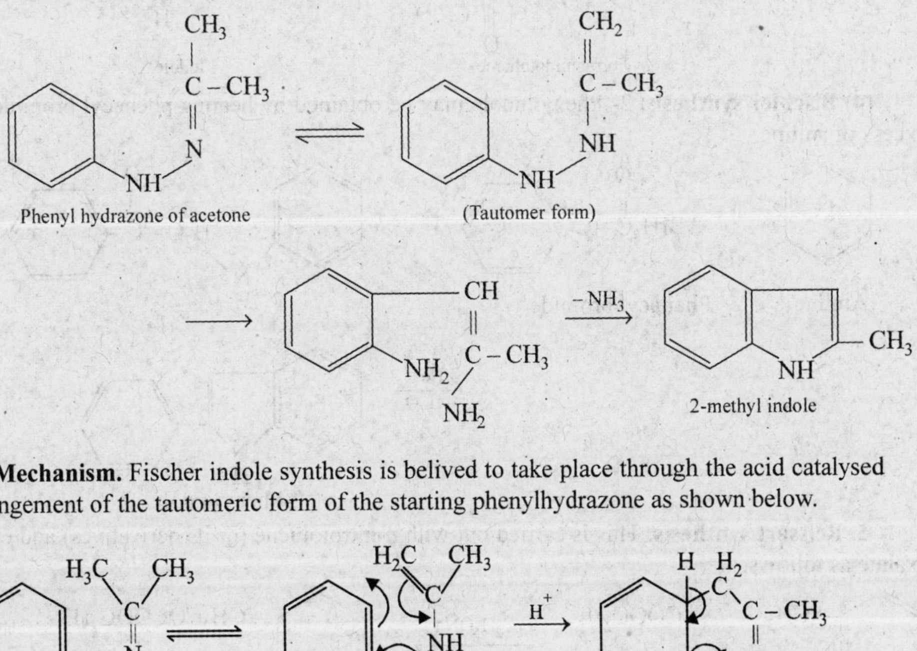

Mechanism. Fischer indole synthesis is belived to take place through the acid catalysed rearrangement of the tautomeric form of the starting phenylhydrazone as shown below.

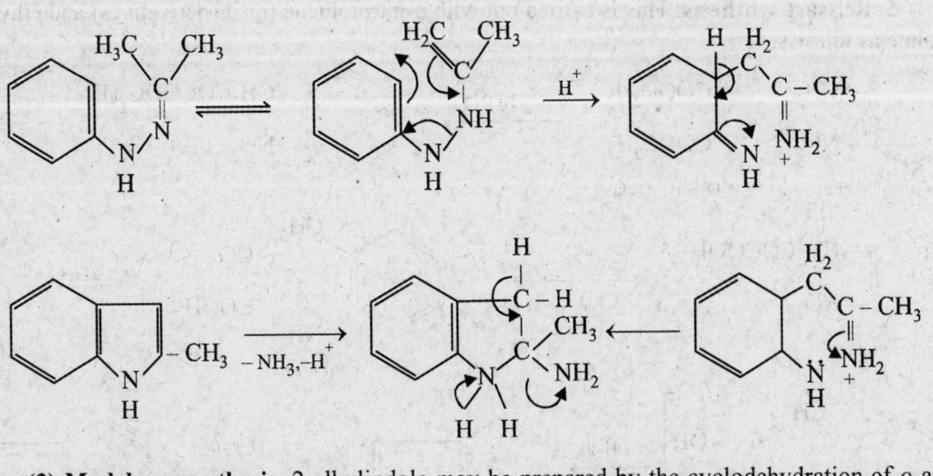

(3) Madelung synthesis. 2-alkylindole may be prepared by the cyclodehydration of *o*-acyl amidotoluene by treatment with a strong base such as potassium tert. butoxide or sodamide.

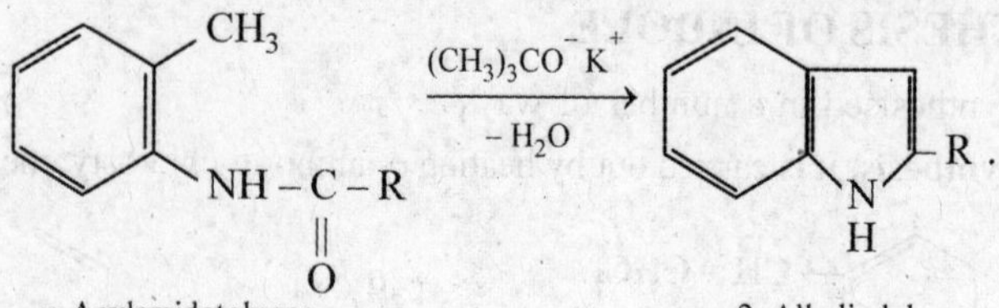

o-Acylamidotoluene 2- Alkylindole

if o- formamidotoluene is used as the starting material, indole is obtained is good yield..

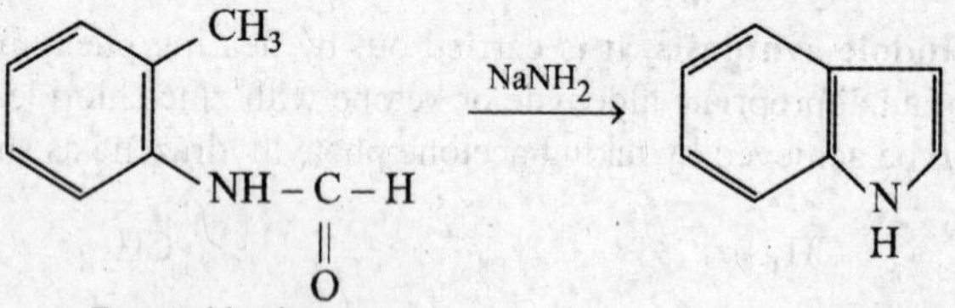

o- Formamidotoluene Iodole

(4) Bischler synthesis. 2- Phenylindole may be obtained by heating phenacyl bromide with excess of aniline.

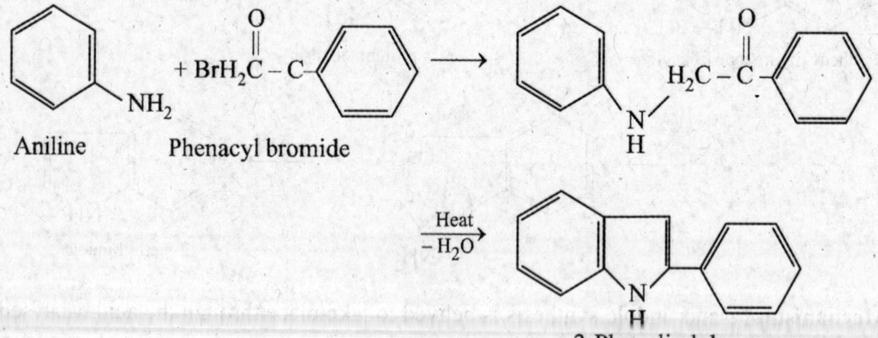

Aniline Phenacyl bromide

2-Phenylindole

5. Reissert synthesis. This is carried out with o-nitrotoluene (or its derivatives) and ethyl oxalate as follows:

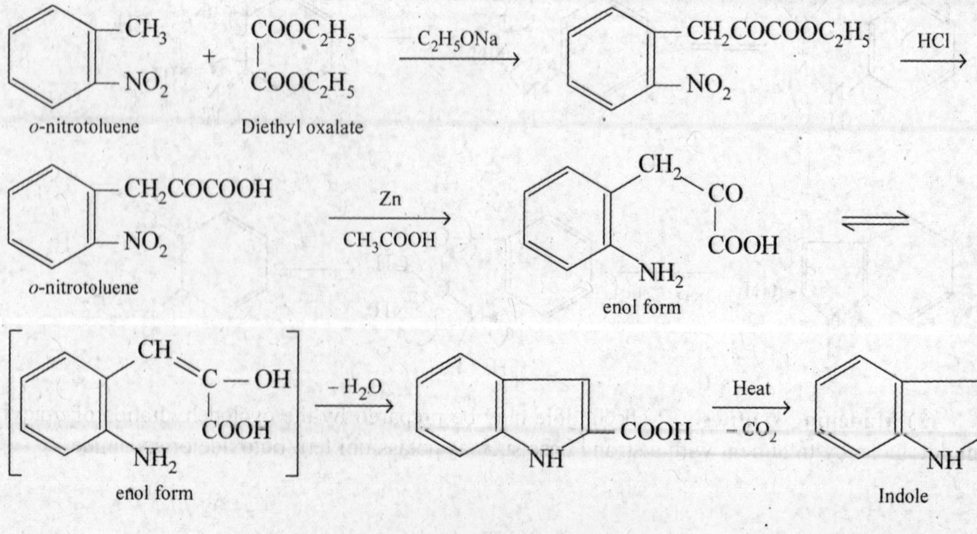

4.27. STRUCTURE OF INDOLE

All the ring atoms in indole are sp^2 hybridised. sp^2 hybrid orbitals overlap with each other and with *s* orbitals of hydrogen to form C – C, C – N, C – H and N – H σ bonds. Each ring atom also possesses a *p*-orbital. These are perpendicular to the plane of the ring. Lateral overlap of these *p* orbitals produces a π molecular orbital containing 10 electrons. Indole is aromatic because it satisfies $4n + 2$ Huckel's rule. Here $n = 2$.

Indole is a hybrid of several resonance structures as shown below:

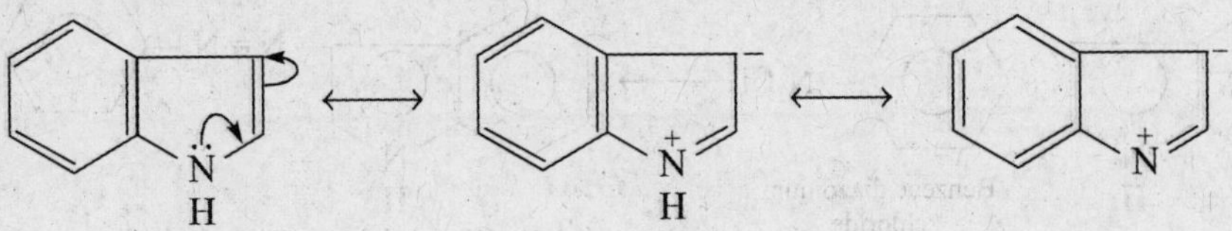

Chemical properties of indole

Important chemical properties of indole are given as under:

1. Nitration

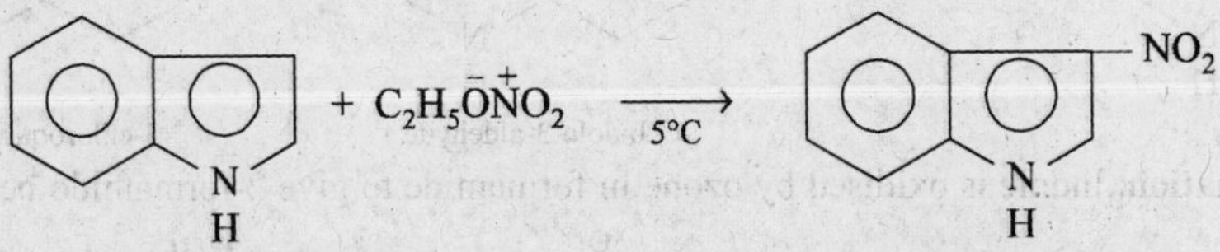

Indole 3-nitroindole

All the electrophilic substitution reactions take place at C – 3 because the intermediate product is stabilized more at this position than at C – 2.

2. Sulphonation

Sulphonation is found to take place at position -2

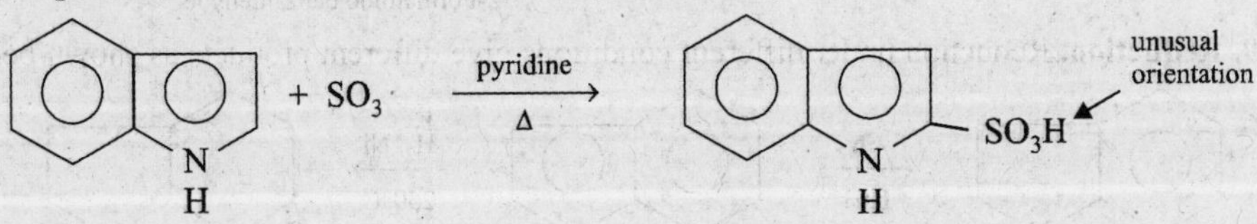

Indole-3-sulphonic acid

3. Bromination

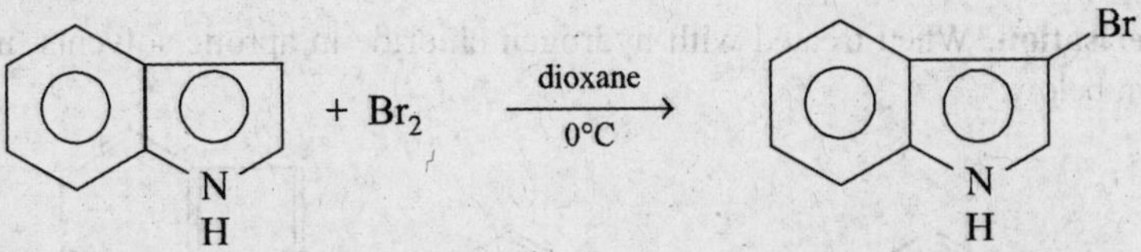

3-bromoindole

4. Friedel crafts alkylation

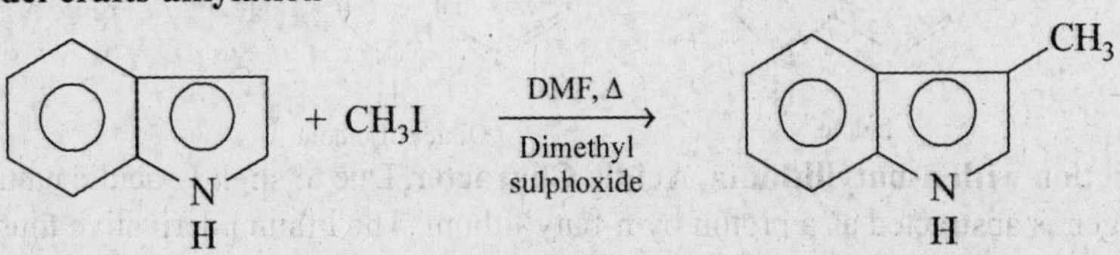

3-methylindole

5. Friedel crafts acylation

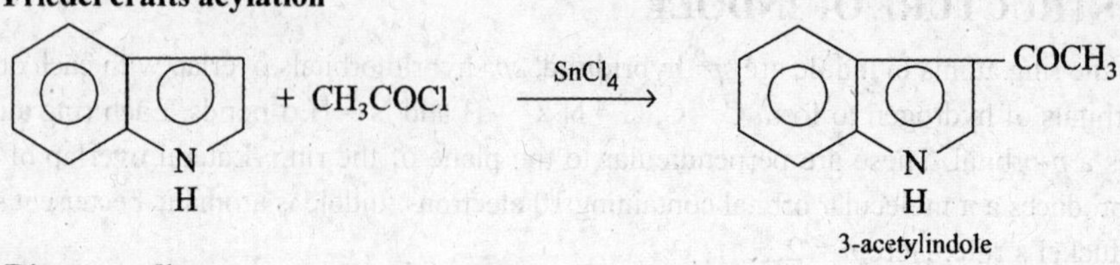

6. Diazocoupling

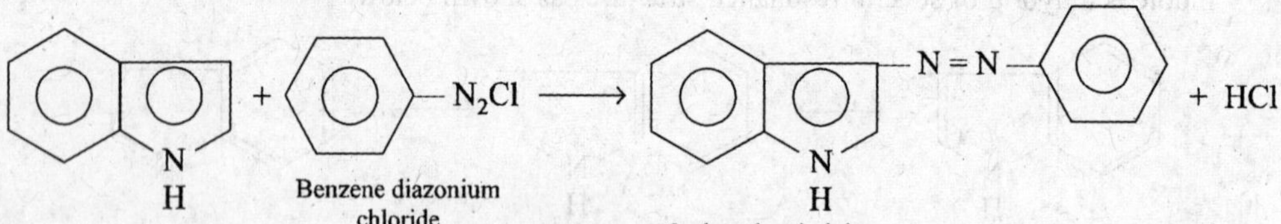

7. Reimer Tiemann formylation

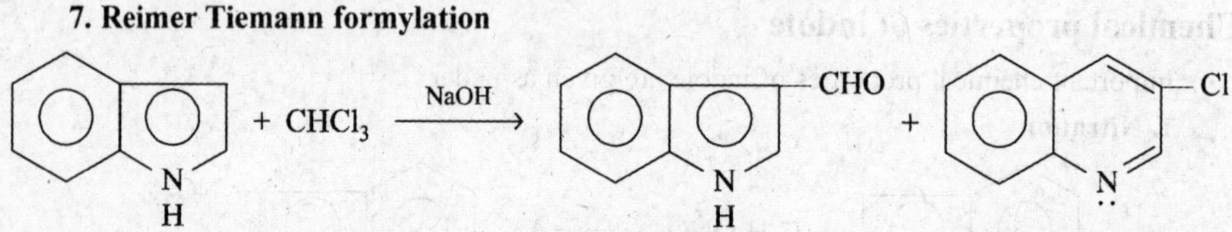

8. Oxidation. Indole is oxidised by ozone in formamide to give 2-formamido benzaldehyde

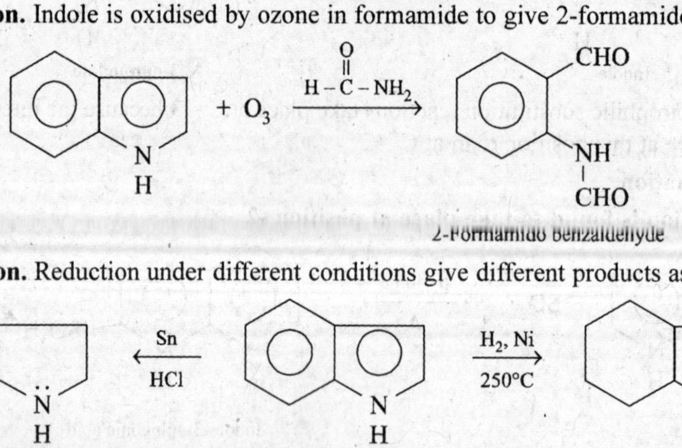

9. Reduction. Reduction under different conditions give different products as shown below:

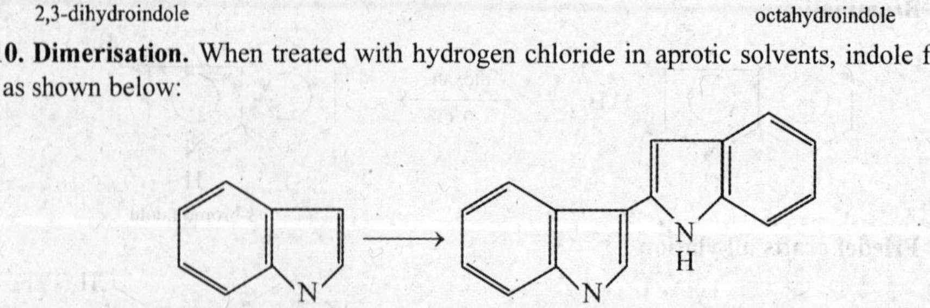

10. Dimerisation. When treated with hydrogen chloride in aprotic solvents, indole forms a dimer as shown below:

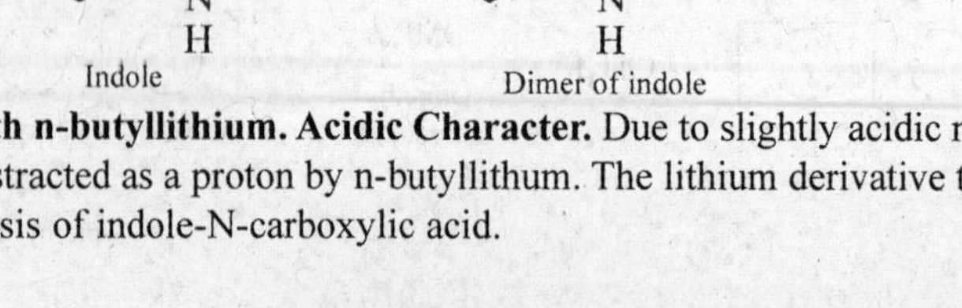

11. Reaction with n-butyllithium. Acidic Character. Due to slightly acidic nature of indole, N – H hydrogen is abstracted as a proton by n-butyllithum. The lithium derivative thus formed can be used for the synthesis of indole-N-carboxylic acid.

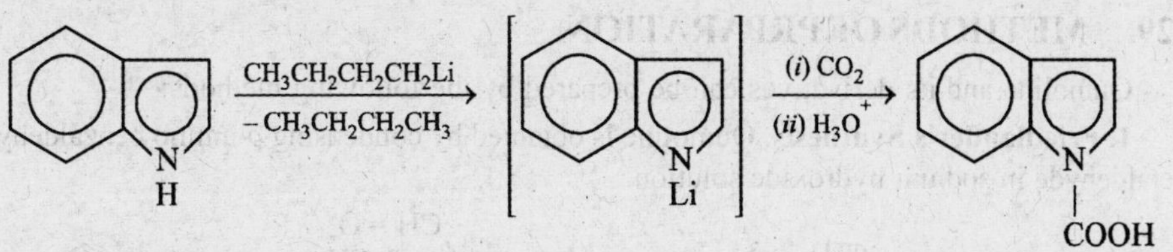

Indole N-carboxylic acid

4.28. MECHANISM OF ELECTROPHILIC SUBSTITUTION

The mechanism of electrophillic substitution of indole is explained as under

(i) Generation of electrophile:

$$E:Z \longrightarrow \underset{\text{Electrophile}}{E^+} + :Z^-$$

(*ii*) Formation of carbocation:

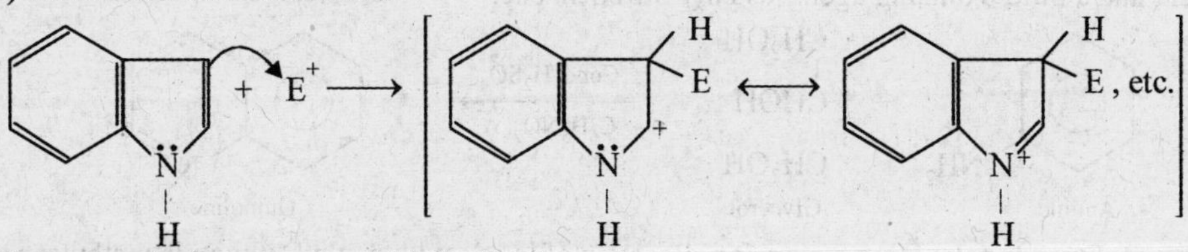

(*iii*) Removal of proton to yield the final product:

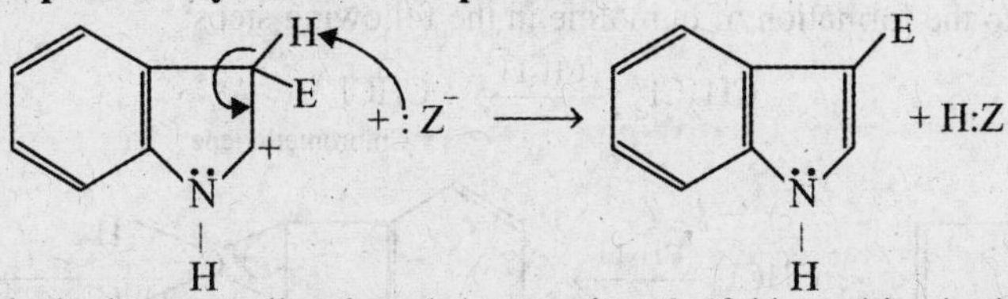

Electrophilic substitution normally takes place at positon-3. If this position is already occupied, substitution takes place at position-2. If both these positions are already occupied, further substitution takes place at the benzene ring at position-6.

QUINOLINE

Structure of quinoline

Quinoline yellow is used as food colour

4.29. METHODS OF PRPARATION

Quinoline and its derivatives can be prepared by the following methods:

1. Friedlander's Synthesis. Quinoline is obtained by condensing *o*-amino benzaldehyde with acetaldehyde in sodium hydroxide solution.

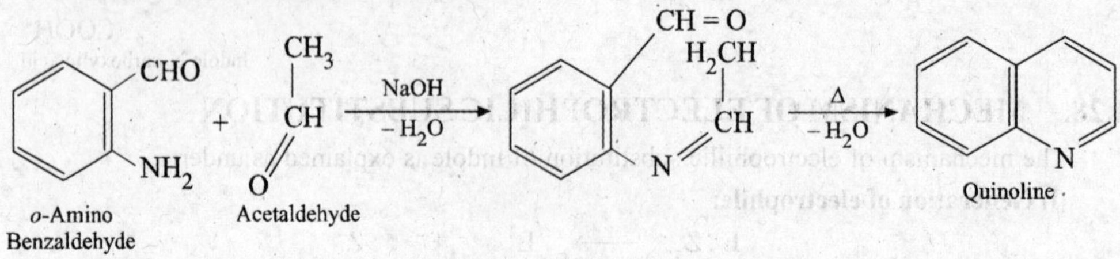

2. Skraup Synthesis. A mixture of aniline and glycerol is heated in the presence of sulphuric acid and a mild oxidising agent, usually nitrobenzene.

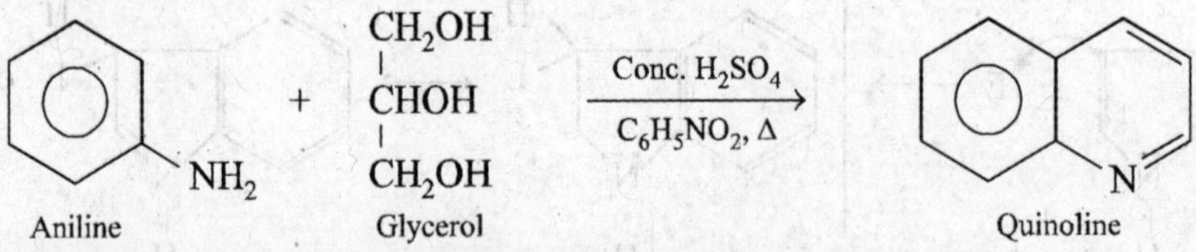

3. From indole: Ring expansion. Reaction of indole with methyl-lithium in methylene chloride solution leads to the formation of quinoline in the following steps:

$$CH_2Cl_2 \xrightarrow{CH_3Li} :CHCl$$

Chloromethylene

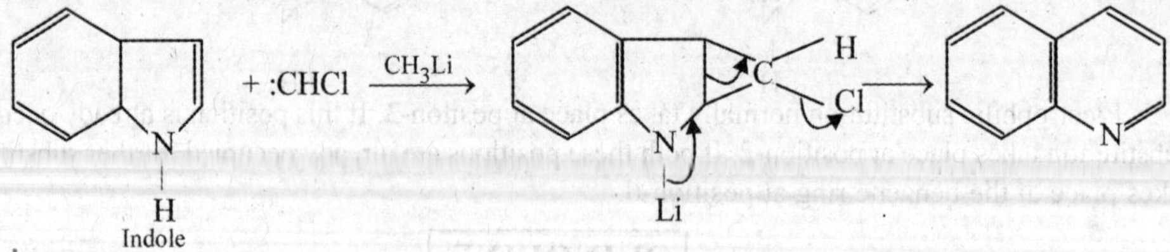

4. The Dobner-Miller synthesis. This is a modified form of Skarup synthesis in which simple aldehydes and ketones act as precurors of α,β-unsaturated carbonyl compounds. The reaction follows the same course as in Skraup synthesis to yield homologues of quinoline. With actaldehyde, 2-methylquinoline is formed as shown below.

$$2CH_3CHO \xrightarrow[\text{(i) Aldol condensation (ii) } -H_2O]{H_3O^+} CH_3-CH=CH-CHO$$

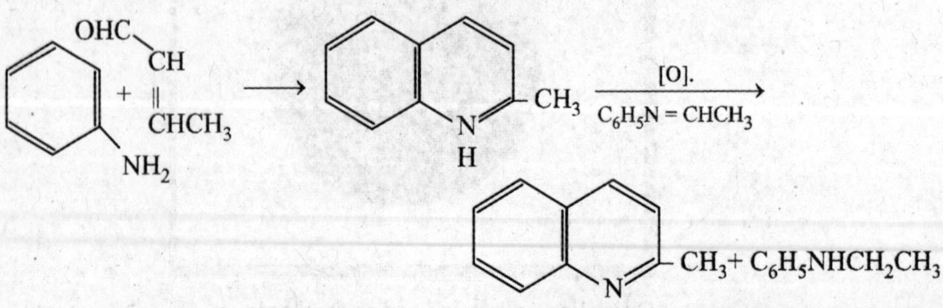

4.30. PROPERTIES OF QUINOLINE

All ring atoms in quinoline are sp^2 hybridised. Like pyridine, the electron pair on nitrogen atom resides in an sp^2 orbital and is not involved in the formation of the delocalised π-molecular orbital. Due to this electron pair, quinoline can readily form an N-oxide and undergoes quaternisation. It satisfies the Huckel's rule as it has 10 delocalised π electrons.

Quinoline exists in the following resonating structures:

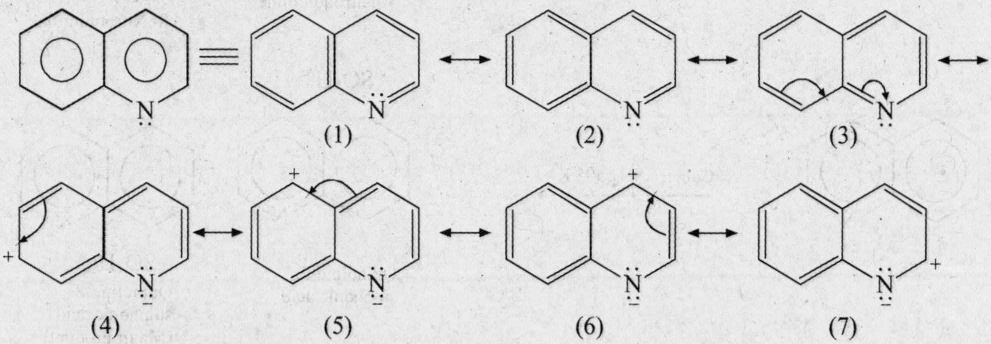

It has a resonance energy of nearly 198 kJ mol^{-1}

The important properties of quinoline are given below:

1. Basicity. Due to the availability of a lone pair of electrons on nitrogen, quinoline acts as a base and forms salts with acids and quaternary salts with alkyl halides.

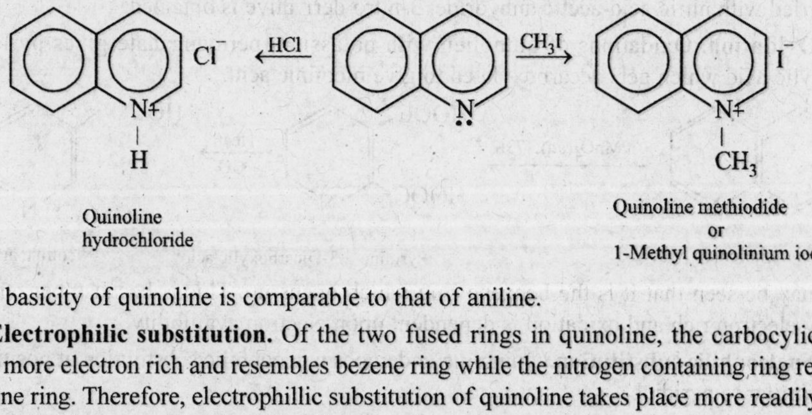

Quinoline hydrochloride

Quinoline methiodide or 1-Methyl quinolinium iodide

The basicity of quinoline is comparable to that of aniline.

2. Electrophilic substitution. Of the two fused rings in quinoline, the carbocylic ring is relatively more electron rich and resembles bezene ring while the nitrogen containing ring resembles the pyridine ring. Therefore, electrophillic substitution of quinoline takes place more readily than in case of pyridine but at position 5 and 8 of benzene ring to give a mixture of products. The important substitution reactions of quinoline are summed up below:

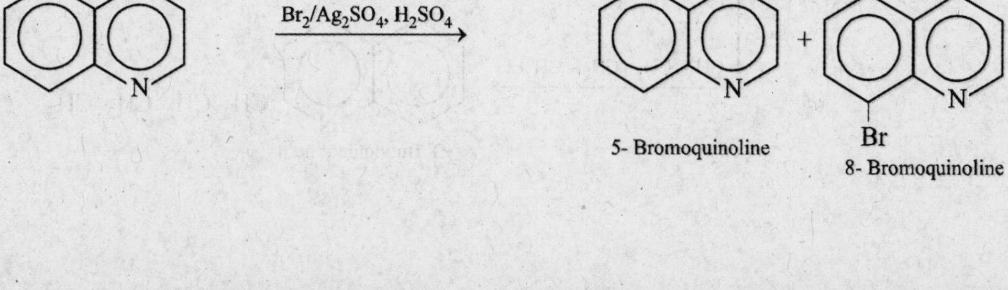

5- Bromoquinoline

8- Bromoquinoline

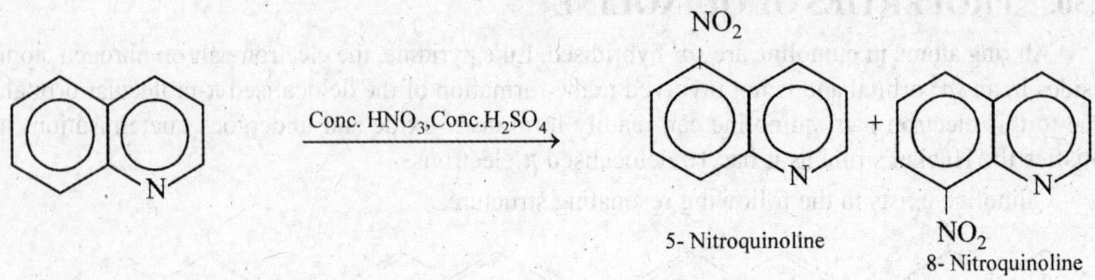

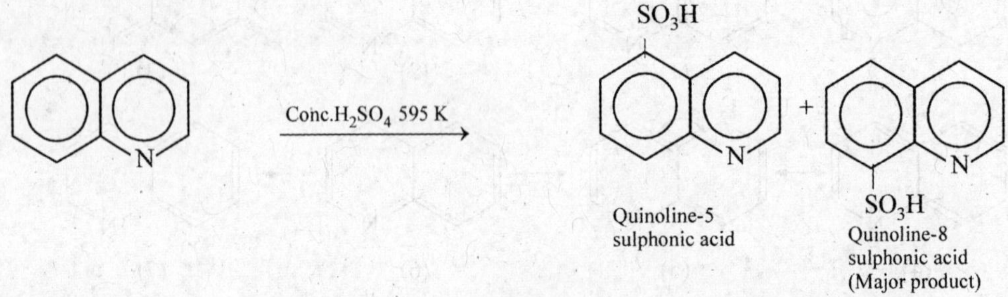

A few important observations about electrophlic substitution reaction of quinoline are:

(*i*) Under highly acidic conditions, bromination takes place at 5- and 8- positions But vapour phase bromination forms 3- bromoqunoline and at 775 K the product is mainly 2- bromoquinoline.

(*ii*) Nitration with conc HNO_3 and conc. H_2SO_4 mixture gives 5- and 8- nitroderivatives but when carried with nitric acid-acetic anhydride, 3-nitro derivative is obtained

3. Oxidation. Oxidations of quinolien with potassium permanganate gives pyridine 2,3-dicarboxylic acid which gets decarboxylated to give nicotinic acid.

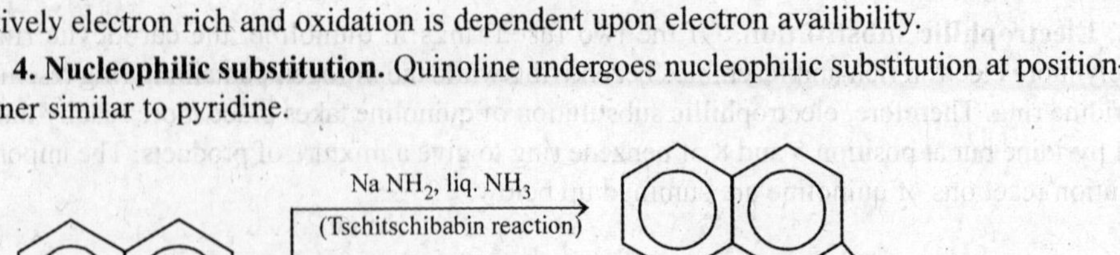

It may be seen that it is the benzene ring which opens up. This is because benzene ring is relatively electron rich and oxidation is dependent upon electron availibility.

4. Nucleophilic substitution. Quinoline undergoes nucleophilic substitution at position-2 in a manner similar to pyridine.

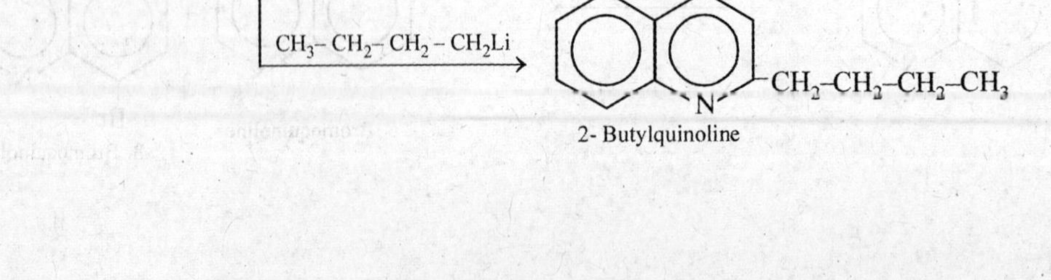

5. Reduction. Reduction of pyridine ring of quinoline takes place relatively more easily but benzene ring is more difficult to reduce. The reduction products formed under different conditions are given below.

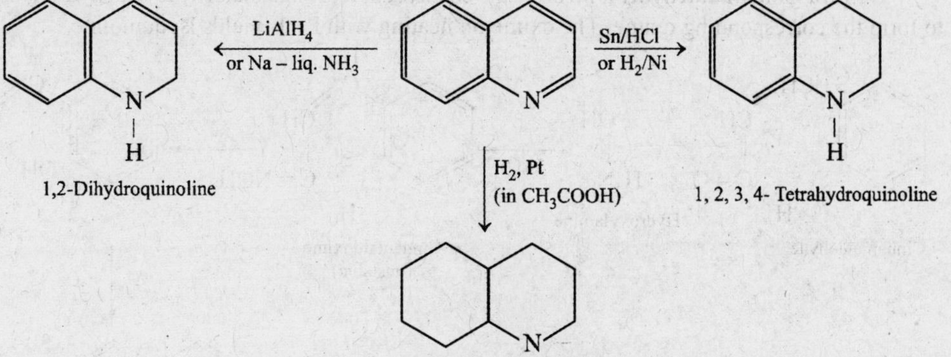

1,2-Dihydroquinoline

1, 2, 3, 4- Tetrahydroquinoline

Decahydroquinoline

4.31. MECHANISM OF ELECTROPHILIC SUBSTITUTION

The mechanism of electrophilic substitution of quinoline at position -5 is described as under:

(*i*) *Generation of electrophile*:

$$E{:}Z \longrightarrow \underset{\text{Electrophile}}{E^+} + {:}Z^-$$

(*ii*) *Formation of carbocation*:

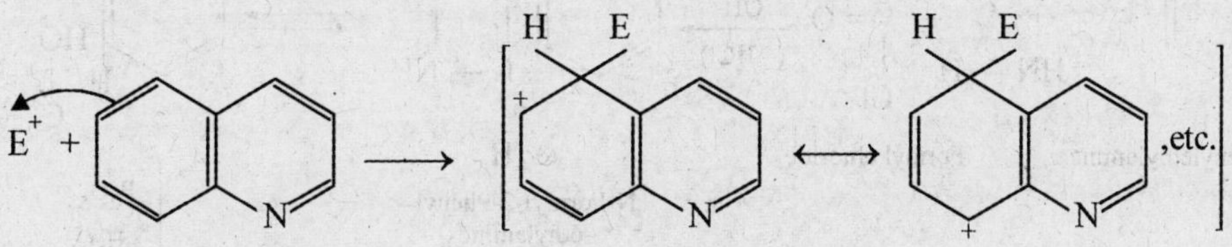

(*iii*) *Removal of proton to form the final product*:

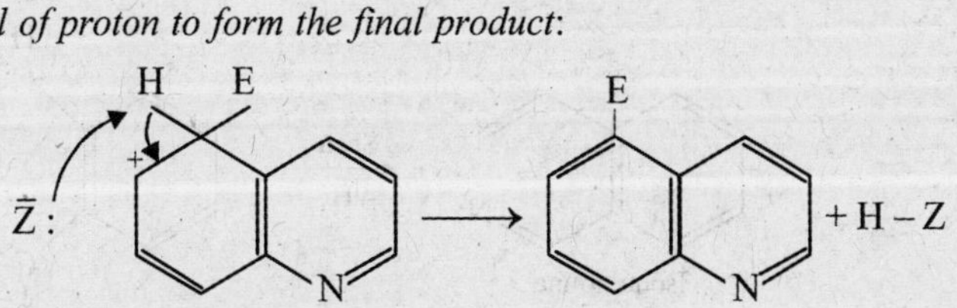

For subtitution at position-8, the formation of carbocation may be represented as:

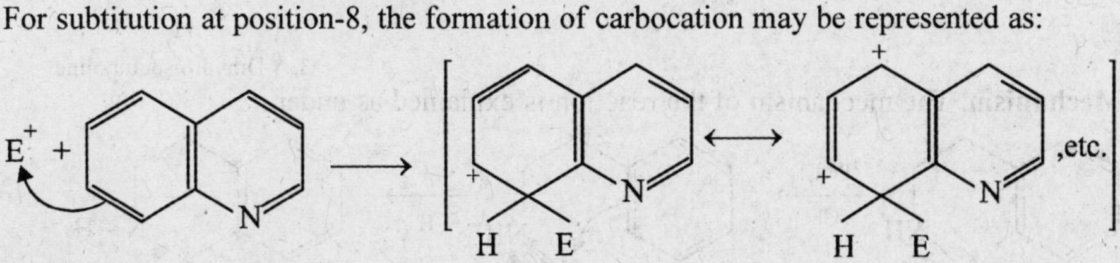

Isoquinoline

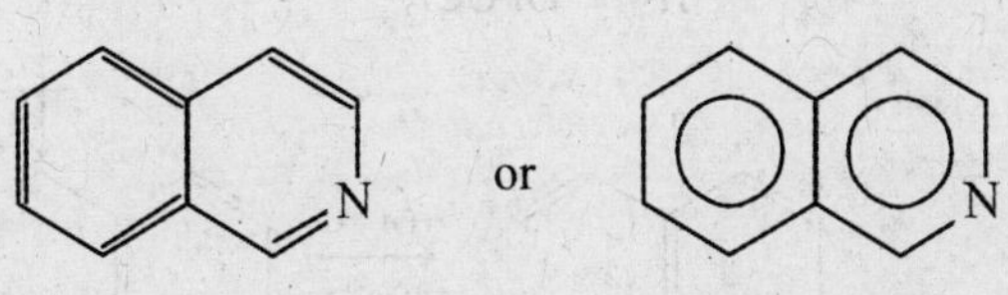

4.32. PREPARATION OF ISOQUINOLINE AND ITS DERIVATIVES

Isoquinoline and its derivatives can be obtained as under:

1. From Cinnamaldehyde. It involves the condensation of cinnamaldehyde with hydroxy-lamine to form the corresponding oxime. The oxime on heating with P_2O_5 yields isoquinoline.

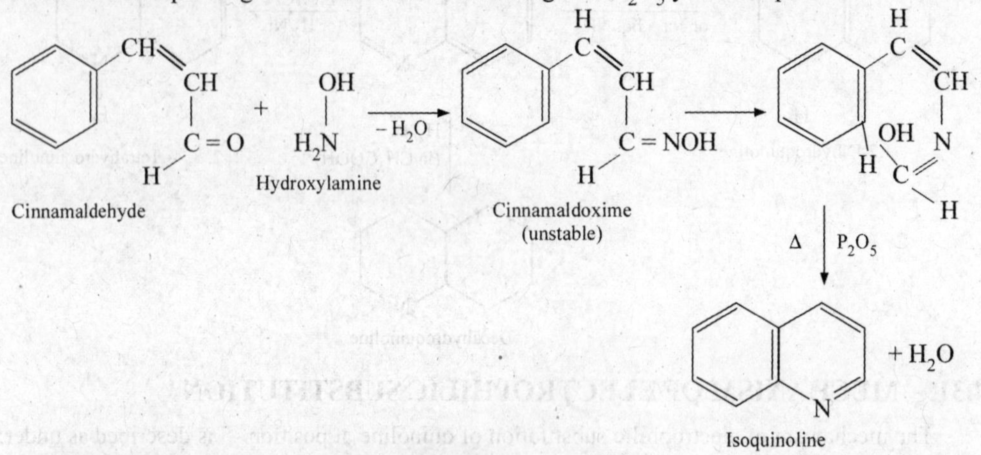

2. Bischler-Napieralski Synthesis. The starting material for the synthesis of isoquinoline is 2-Phenylethylamine.

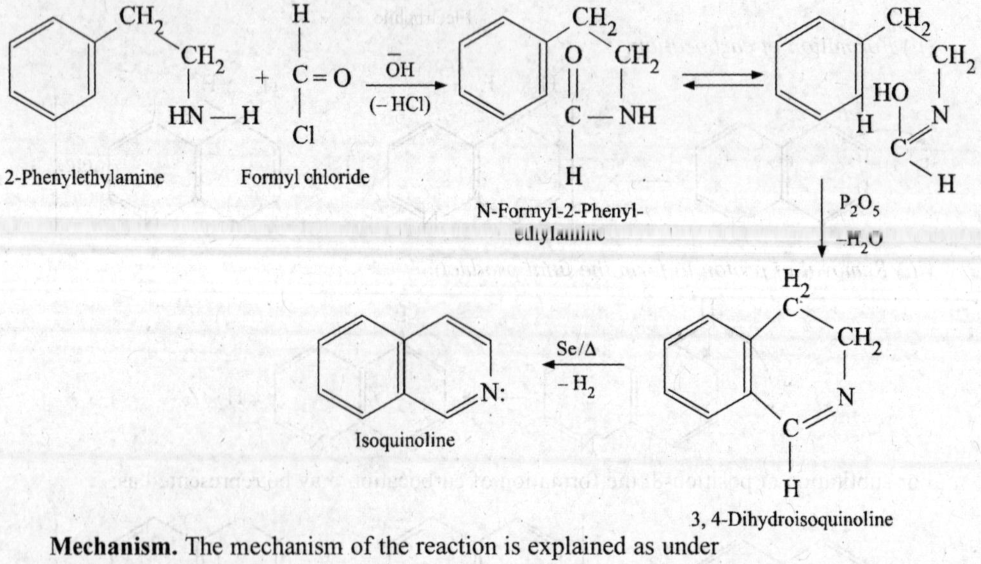

Mechanism. The mechanism of the reaction is explained as under

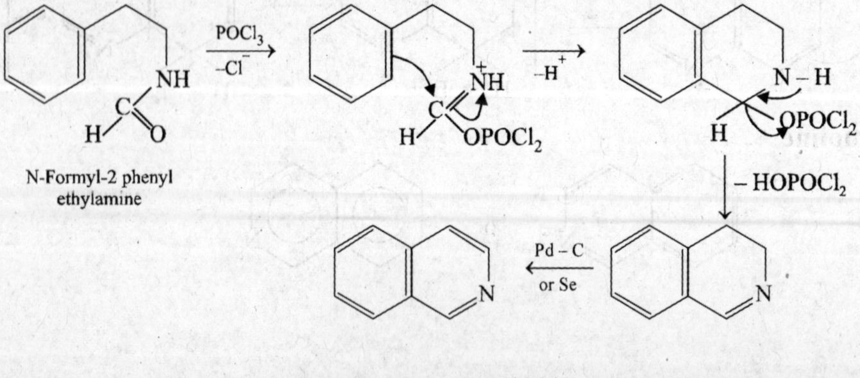

3. The Pomeranz-Fritsch syntheis. In this synthesis, an aromatic aldehyde is condensed with an aminoacetal to form a Schiff's base. This is cyclised in the presence of sulphuric acid to obtain isoquinoline.

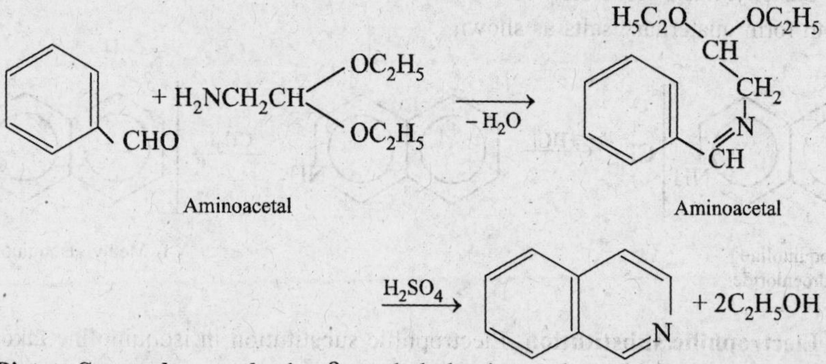

4. The Pictet- Spengaler synthesis.. β- arylethylamine and an aldehyde are condensed the presence of excess of hydrochloric acid at 373 K. As a result, 1,2,3,4-tetrahydroisoquinoline is formed via a protonated imine intermediate. The tetrahydro product is then oxidised by the usual method to form isoquinoline derivative.

Raoul Pierre Pictet

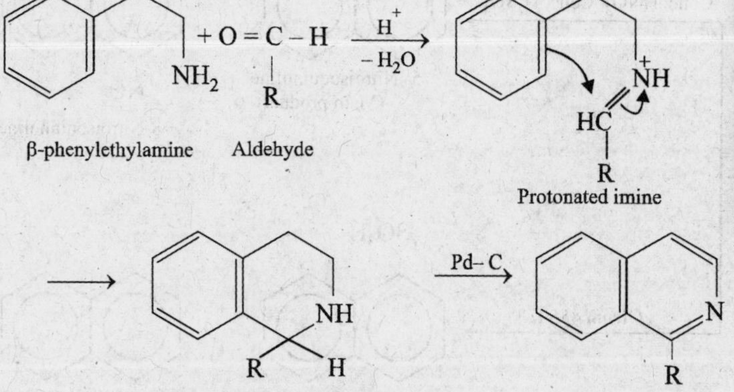

1- Alkyl-1,2,3, 4- tetrahydroisoquinoline

1- Alkylisoquinoline

4.33. PROPERTIES OF ISOQUINOLINE

Isoquinoline is colourless liquid boiling at 516 K.

All the ring atoms in isoquinoline are sp^2 hybridised. The electron pair on nitrogen atom resides in an sp^2 orbital and is not involved in the formation of the delocalised π-molecular orbital. Isoquinoline constitutes delocalised 10 π-electron system and satisfies the Huckel's rule. Isoquinoline can be written in the following resonating structures:

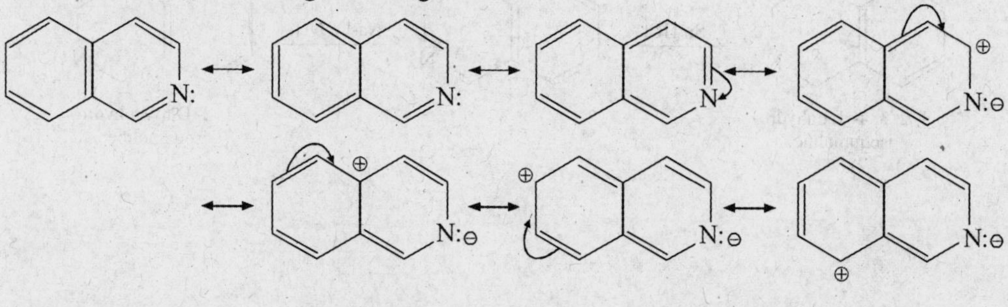

Its important reactions are given below :

1. Basicity. It is a moderately basic compound and reacts with acids to form salts and with alkyl halides to form quaternary salts as shown

Isoquinoline hydrochloride

1- Methyl Isoquinolinium iodide

2. Electrophilic substitution. Electrophilic substitution in isoquinoline takes place at the benzene ring mainly at position -5 and to a smaller extent at position -8.

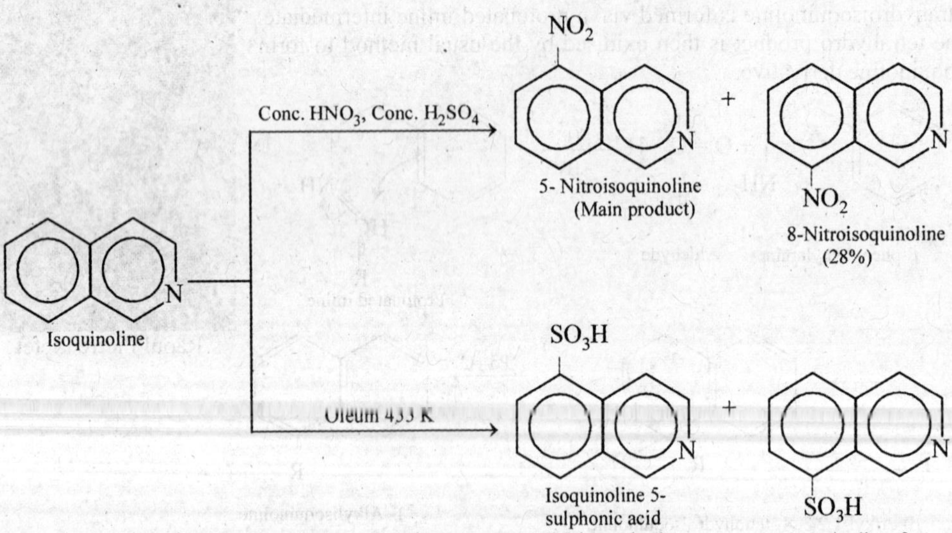

Isoquinoline

5- Nitroisoquinoline (Main product)

8-Nitroisoquinoline (28%)

Isoquinoline 5-sulphonic acid (Main product)

Isoquinoline- 8 sulphonic acid

It is intersting to note that bromination with Br_2/CCl_4 forms 4-bromoisoquinoline

3. Reduction. Pyridine ring in isoquinoline is reduced more easily than benzene ring as is evident form the following reactions.

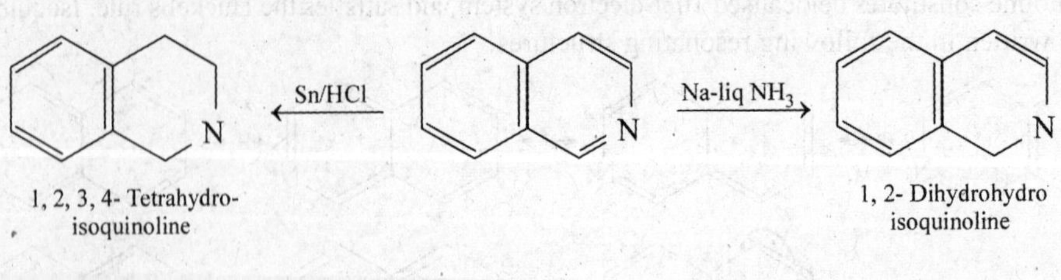

1, 2, 3, 4- Tetrahydro-isoquinoline

1, 2- Dihydrohydro isoquinoline

4. Nucleophillic substitution. Isoquinoline undergoes nucleophilic substitution preferably at position-1.

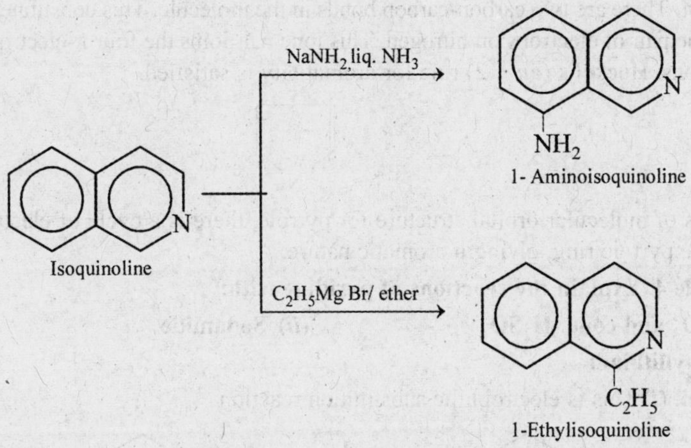

5. Oxidation. Isoquinoline is oxidised by alkaline potassium permanganate to form an equimolar mixture of phthalic acid and cinchomeronic acid.

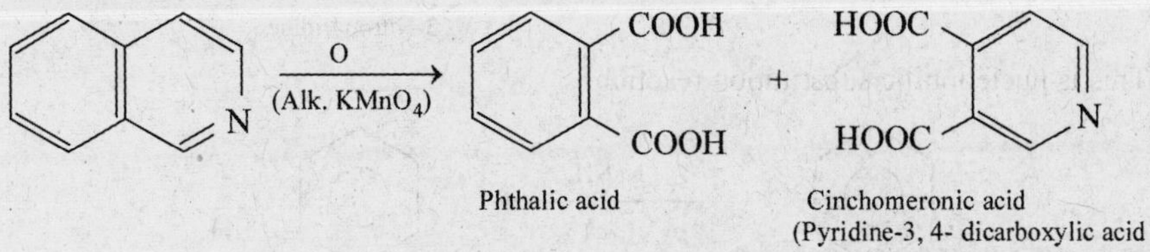

MISCELLANEOUS SOLVED EXAMPLES

Example 1. Electrophilic substitution in 5-membered heterocyclic ring is easier than in pyridine. Why?

Solution. Nitrogen being more electronegative than carbon draws the electrons towards itself in the molecule of pyridine. This is possible because there is a double bonds attached to nitrogen in pyridine. This reduces the availability of π-electrons on the ring which is so essential for electrophilic substitution. However in 5-member heterocyclics (say pyrrole), this is not effectively possible because there is no double bond (or π-electrons) in contact with it.

Example 2. In acidic solutions, pyridine undergoes electrophilic substitution with great difficulty. Explain.

Solution . Nitrogen atom in pyridine gets protonated due to its electronegative nature as shown below:

Protonated nitrogen attracts the π-electrons from the ring. This reduces the electron density on the ring, thereby reducing chances of electrophilic attack.

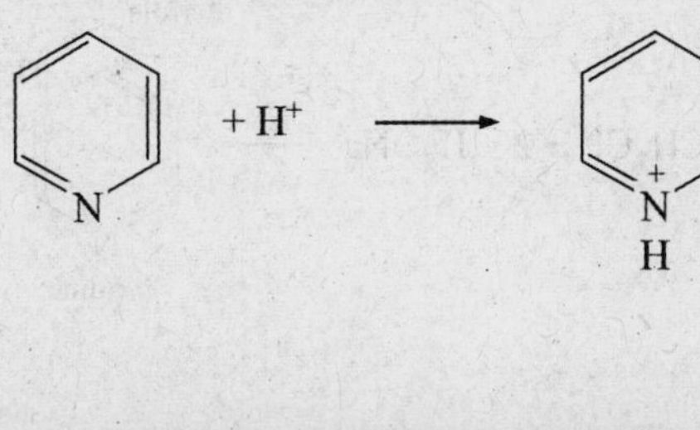

Example 3. Explain the aromatic character of pyrrole on the basis of its resonance and molecular orbital structures.

Solution. There are two carbon-carbon bonds in the molecule. This constitutes four π-electrons. There is a lone pair of electrons on nitrogen. This lone pair joins the four π-electrons, making a total of 6 π-electrons. Huckel's $(4n + 2)$ rule for aromaticity is satisfied.

In terms of molecular orbital structure for pyrrole, there is a cycle of electron clouds, above and below the pyrrole ring, giving it aromatic nature.

Example 4. Explain the reactions of pyridine with:

(*i*) KNO_3 and conc. H_2SO_4 **(*ii*) Sodamide**

(*iii*) Phenyllithium

Solution. (*i*) This is electrophilic substitution reaction

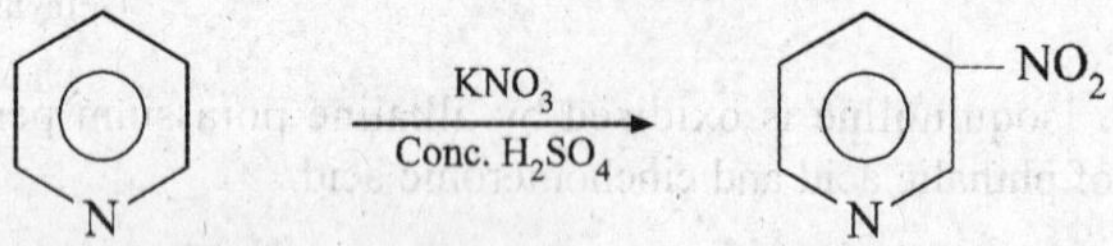

3-Nitropyridine

(*ii*) This is nucleophilic substitution reaction

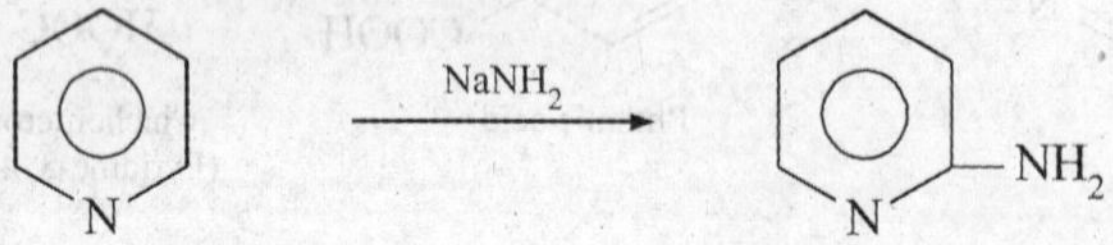

2-Aminopyridine

I his is also nucleophilic substitution reaction.

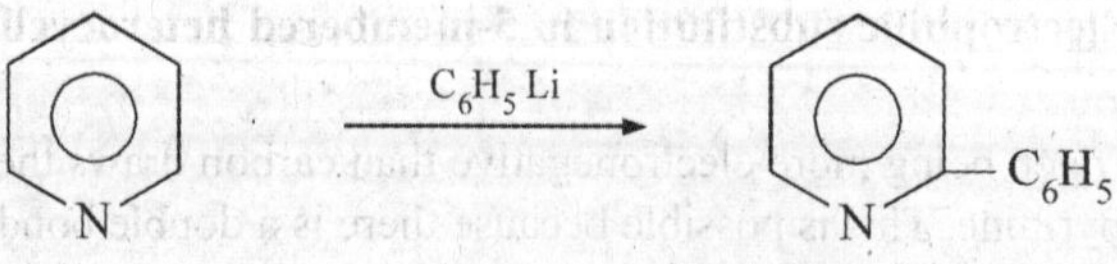

2-Phenylpyridine

Example 5. How will you obtain

(*i*) Pyrrole from furan **(*ii*) Pyridine from pyrrole?**

Solution.

(*i*)

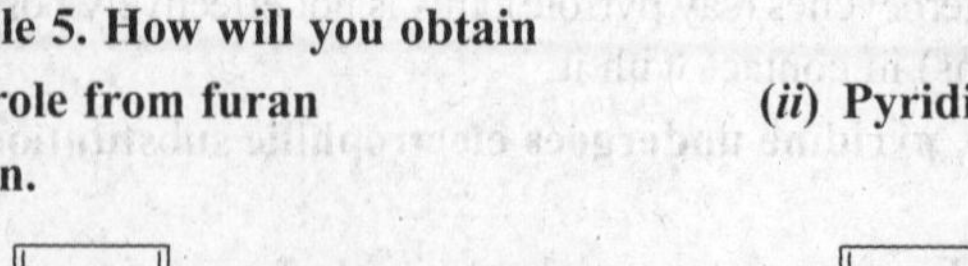

Furan Pyrrole

(*ii*)

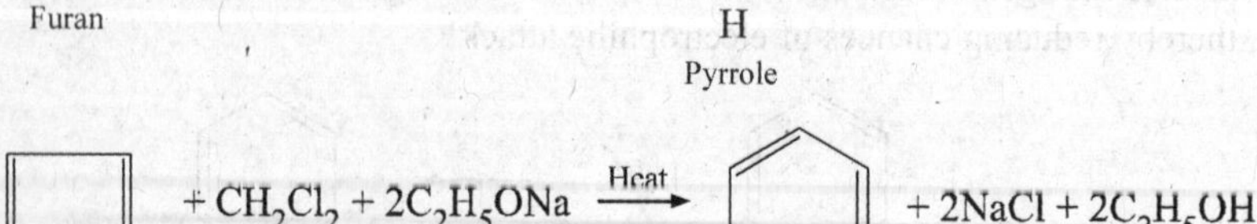

Pyrrole Pyridine

Example 6. Why pyrrole, furan and thiophene are classified as aromatics?

Solution. A compound is called aromatic if it satisfies the $(4n + 2)$ Huckel's rule *i.e.* the no. of π electrons in the molecule is $4n + 2$ where n is an integer including zero. In pyrrole, furan and thiophene molecules, there are two carbon-carbon double bonds *i.e.* 4 π electrons. The lone pair of electrons on nitrogen, oxygen and sulphur also joins the four π electrons to make it $2 + 4 = 6$. Thus Huckel's rule is satisfied.

Example 7. How do you account for the formation of 3-nitropyridine when pyridine is treated with KNO_3/H_2SO_4 at 573 K and 2-aminopyridine when pyridine is treated with $NaNH_2$ at 373 K.

Solution. (*i*) Reaction of pyridine with KNO_3/H_2SO_4 at 573 K is an electrophilic substitution reaction.

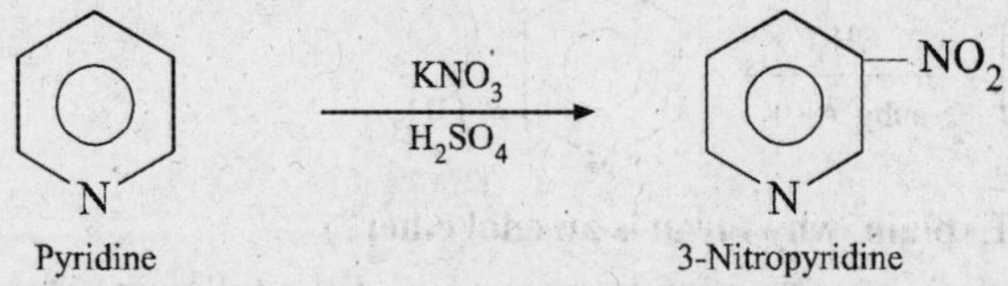

In such reactions, the electrophile preferably attaches itself at position-3. This can be explained in terms of stability of intermediate products.

(*ii*) Reaction of pyridine with $NaNH_2$ at 373 K is a nucleophilic substitution reaction.

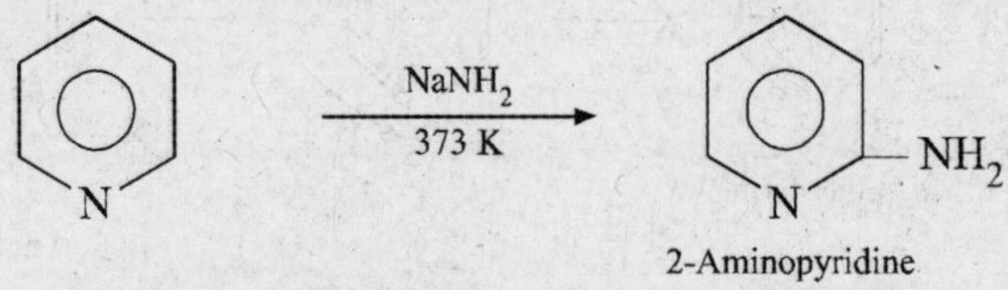

Again this can be explained in terms of stability of intermediate ion.

Example 8. Quinoline undergoes electrophilic attack in the benzene ring while the nucleophilic attack takes place in the pyridine ring. Explain.

Solution. Quinoline molecule consists of two rings, the benzene ring and the pyridine ring. Pyridine ring gives electrophilic reactions less readily because of the formation of pyridinium ion. Pyridine ring, on the other hand, gives nucleophilic reactions readily because of the formation of a stable intermediate anion.

Example 9. Account for the following transformations:

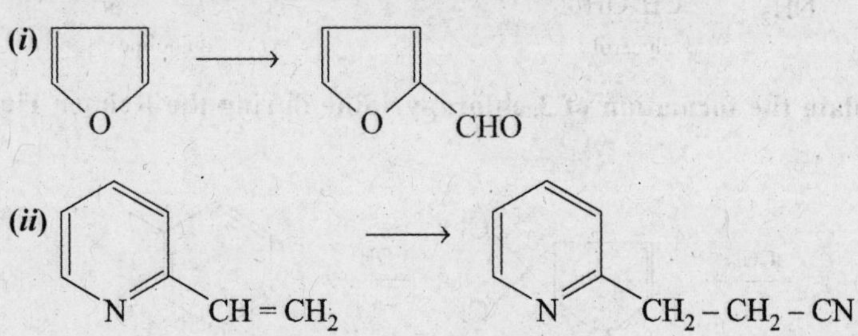

Solution. (*i*) See under properties of furan.

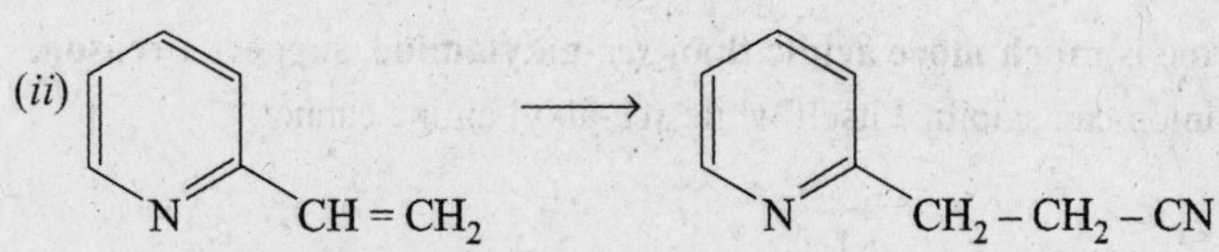

Example 10. Identify compounds A and B in the following reactions:

(*i*) Pyridine + acetyl chloride $\xrightarrow{\text{anhy. AlCl}_3}$ A

(*ii*) Pyridine $\xrightarrow[\text{anhy. AlCl}_3]{\text{CH}_3\text{I}}$

Soluion. (*i*)

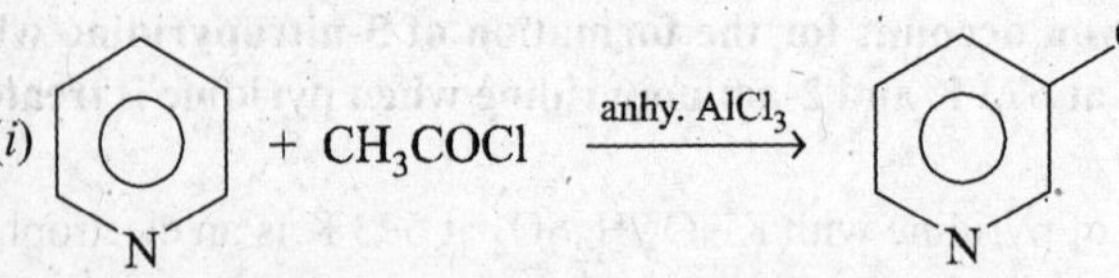

(*ii*)

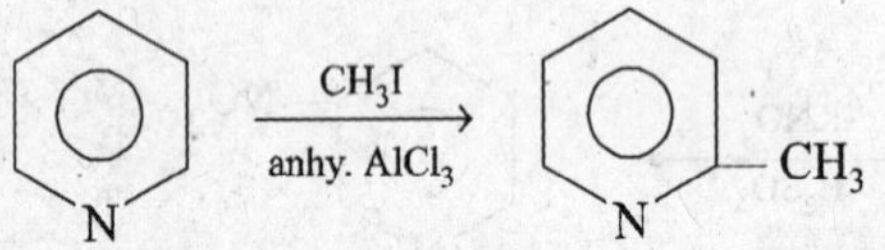

Example 11. Explain, why furan is an enol ether?

Solution. Furan is considered to be enol ether of the following structures:

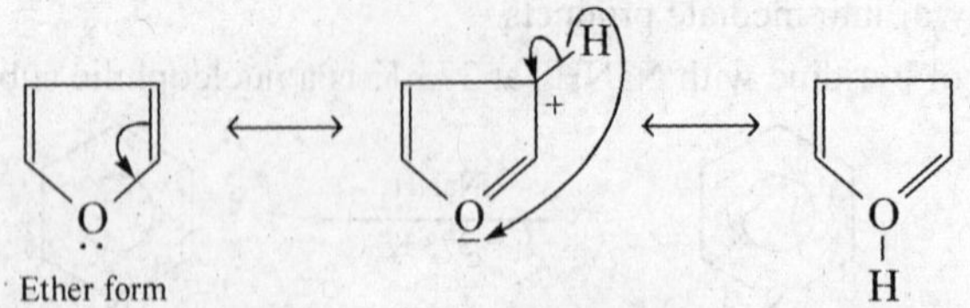

Ether form

Enol form

Example 12. Describe "Skraups synthesis" of quinoline.

Solution. Skraups Synthesis: A mixture of aniline, glycerol and sulphuric acid is heated in the presence of a mild oxidising agent such as nitrobenzene. The reaction being exothermic tends to be violent and ferrous sulphate is added as moderator.

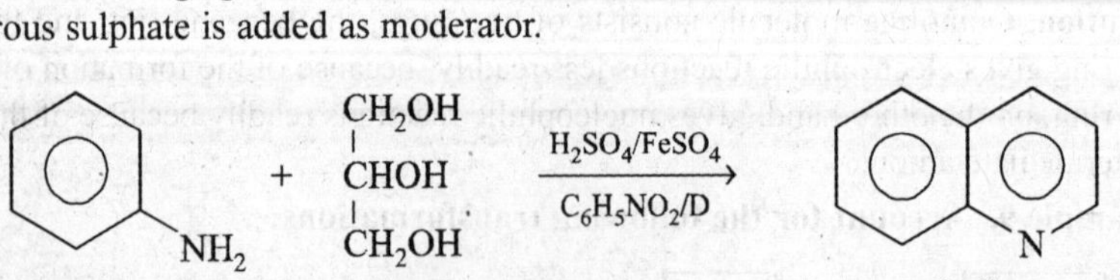

aniline glycerol quinoline

Example 13. Explain the formation of 3-chloropyridine during the Reimer Tiemann reaction of pyrrole.

Solution.

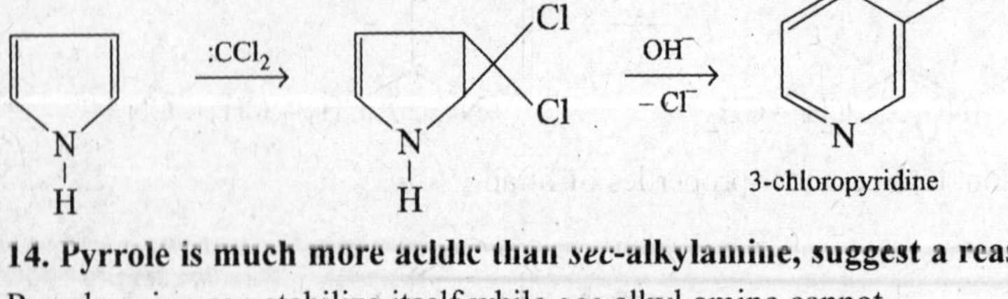

3-chloropyridine

Example 14. Pyrrole is much more acidic than *sec*-alkylamine, suggest a reason.

Solution. Pyrrole anion can stabilize itself while *sec*-alkyl amine cannot.

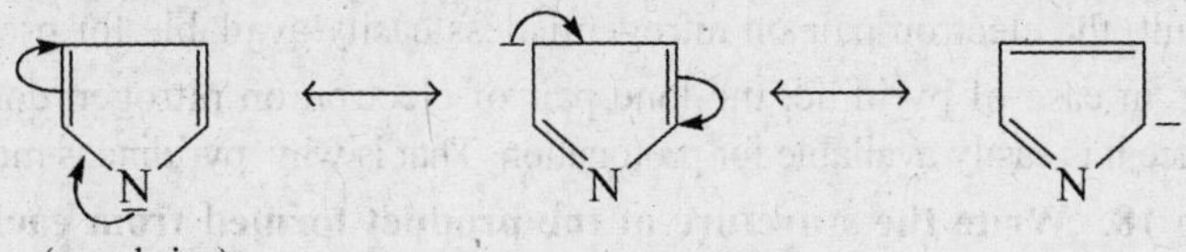

(pyrrole ion)

$$CH_3 - \bar{N} - CH_3$$

Sec-amine anion

Sec-amine anion rather gets destabilised due to the electron repelling effect of — CH_3 groups.

Example 15. Suggest an explanation for the observation that the dipole moment of furan (0.7D) is lower than that of tetrahydrofuran (1.7D).

Solution. In tetrahydrofuran () the highly electronegative oxygen exerts a strong electron attracting inductive effect (– I effect) due to which oxygen gets negatively polarised while carbon atoms of the ring are positively polarised. As a result, tetrahydrofuran shows a dipole moment of 1.7 D.

In case of furan, the –I effect of oxygen is opposed by the resonance effect as shown below:

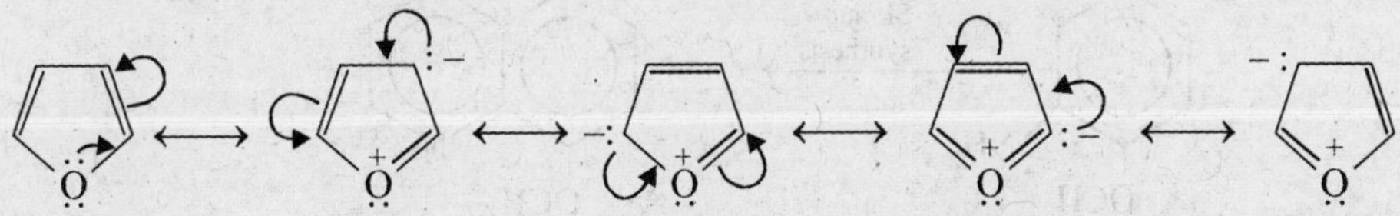

I II III IV V

Due to contribution of structures like II – V, the negative polarisation of oxygen (by – 1 effect) becomes less. Hence furan has lower dipole moment (0.7 D) than tetrahydrofuran.

Example 16. Is it possible to convert a five membered heterocylic compound into a six membered heterocylic compound? If so give an example.

Solution. It is possible to convert a five membered heterocyclic compound into a six membered heterocyclic compound by ring expansion. For example, when lithium salt of pyrrole is heated with chloroform, it undergoes ring expansion to form 3-chloropyrdine is shown below.

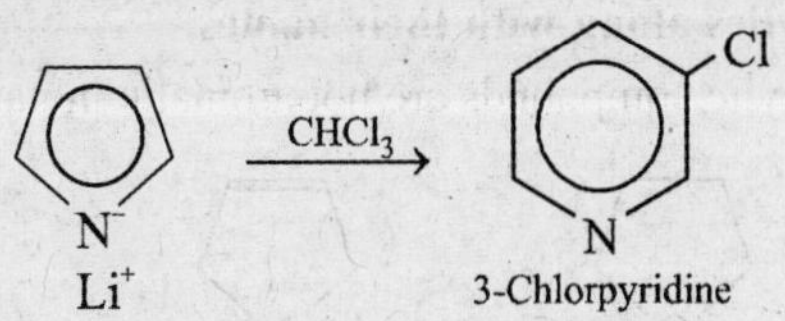

3-Chlorpyridine

Example 17. Out of aniline and pyridine which is less basic and why?

Solution. Aniline is less basic than pyridine. This can be explained as under:

In aniline the lone pair of electrons on nitrogen undergoes resonance interaction with benzene ring

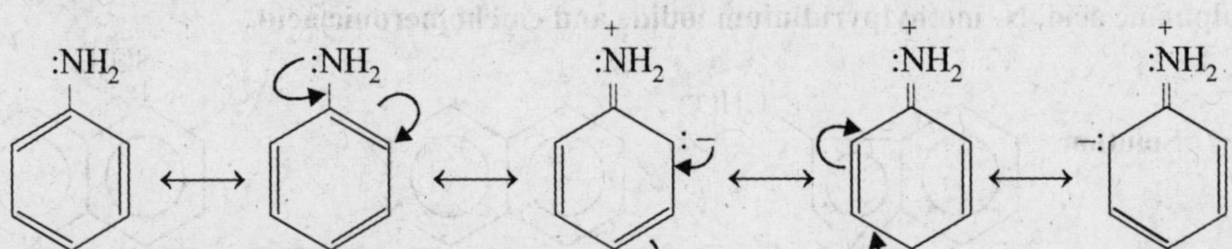

As a result, the electron pair on nitrogen is less easily available for protonation.

However, in case of pyridine, the lone pair of electron on nitrogen does not participate in resonance. Hence it is easily available for protonation. That is why pyridine is more basic than aniline.

Example 18. Write the sturcture of the product formed from each of the subsituted anilines listed below via Skraup synthesis:

(a) 4– toluidine (b) 2-methoxyaniline (c) 3-chloroaniline.

Solution. In Skraup synthesis simple 2 – or 4 – substituted anilines give only one product but 3- substituted anilines give a mixture of 5– and 7 – substituted quinolines

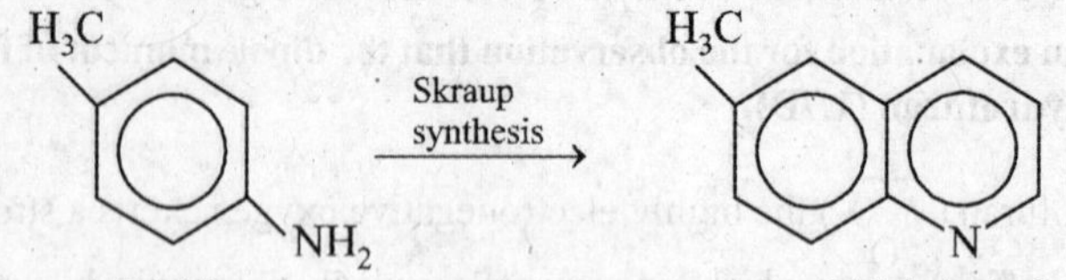

4-Toluidine 6-Methylquinoline

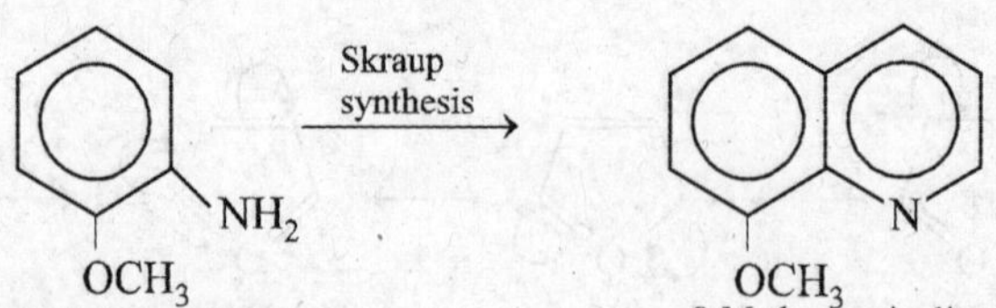

2-Methoxyaniline 8-Methoxyquinoline

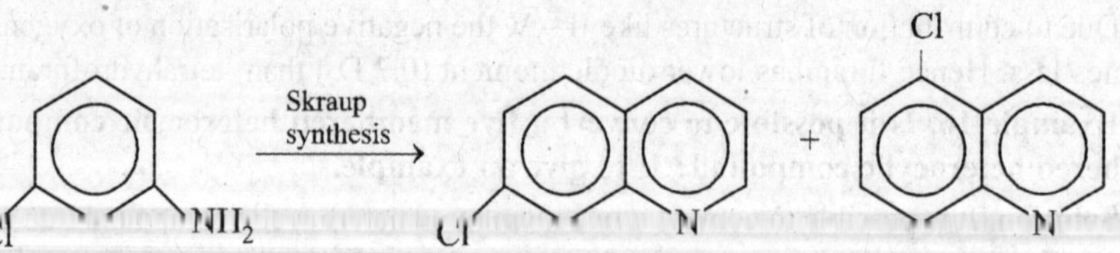

3-Chloroaniline 7-Chloroquinoline 5-Chloroquinoline

Example 19. Are all heterocylic compounds aromatic in nature? If not, give three examples of non-aromatic heterocylics along with their names.

Solution. All heterocylic compounds are not aromatic in nature. For example:

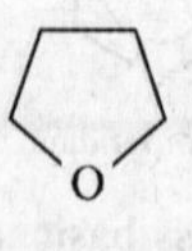

Tetrahydro furan

3-Pyrroline

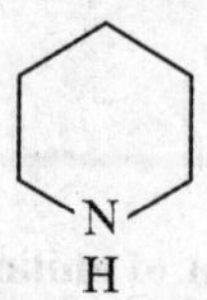

Piperidine

Example 20. Write the formulae of 3- formylindole, 1- methylisoquinoline, quinoline-5-sulphonic acid, N- methylpyridinium iodide and cinchomeronic acid.

Solution.

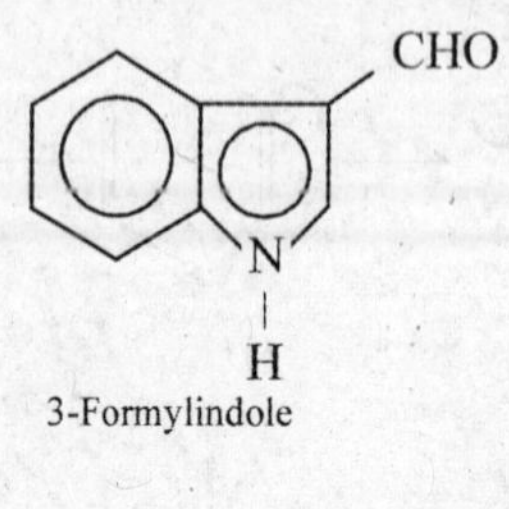

3-Formylindole

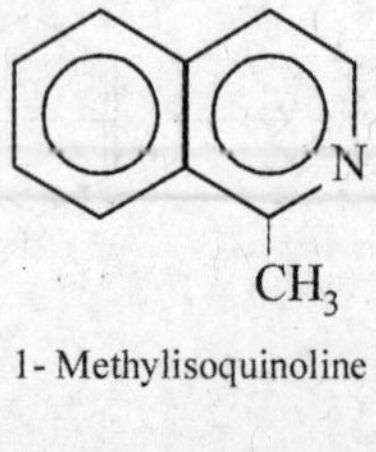

1- Methylisoquinoline

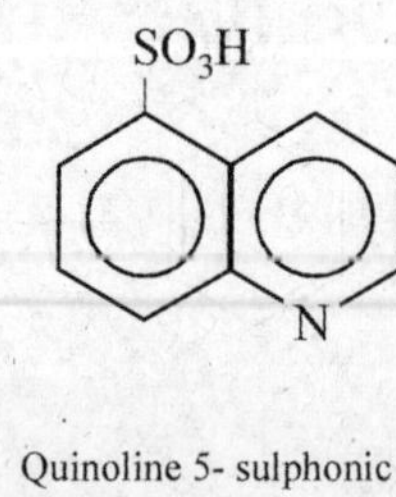

Quinoline 5- sulphonic acid

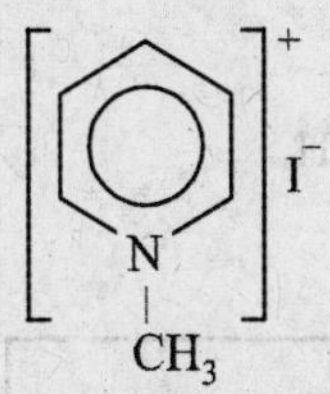

N- Methylpyridinium iodide

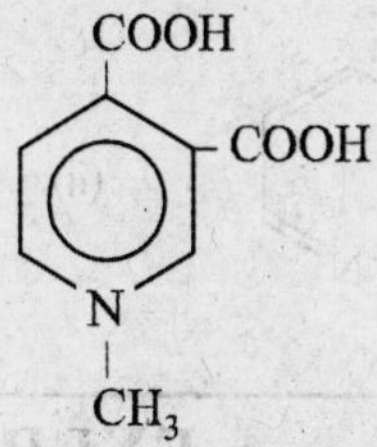

Cinchomeronic acid
(Pyridine-3,4- dicarboxylic acid)

Example 21. Give two examples of condensed heterocyclics. What is the number of delocalised electrons in the two compounds cited?

Solution.

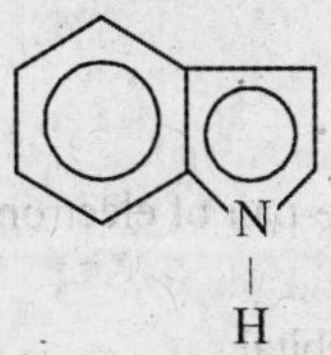

Indole

(No. of delocalised π electrons = 10)

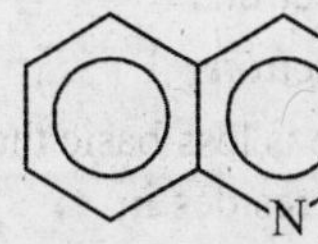

Quinoline

(No. of delocalised π electrons = 10)

Example 22. How can furan, thiophene and pyrrole derivatives be synthesised form 1,4-diketones?

Solution

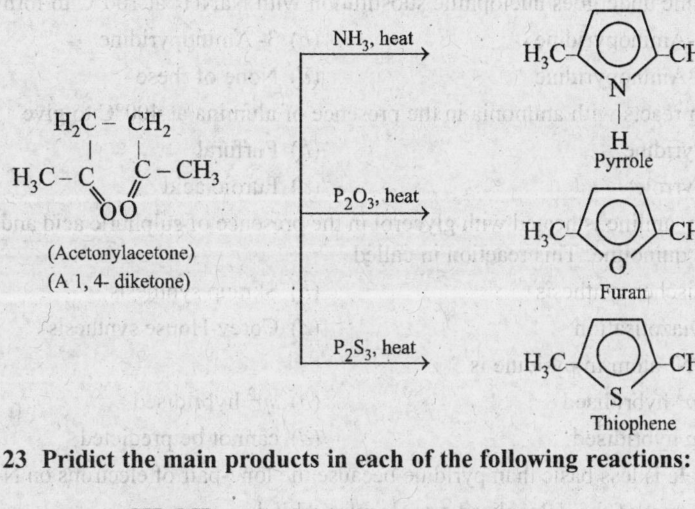

Example 23 Pridict the main products in each of the following reactions:

(i)

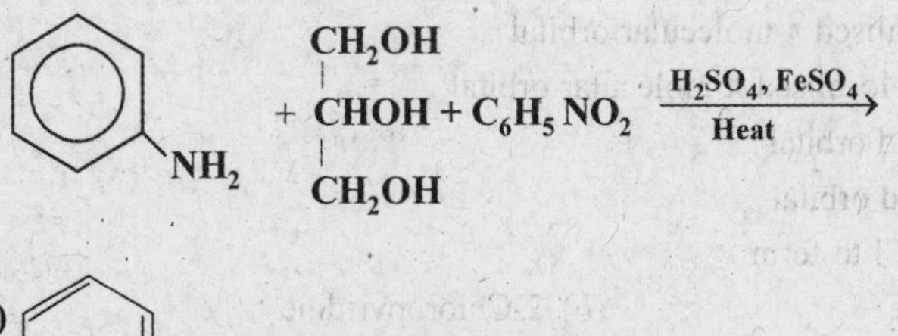

(ii)

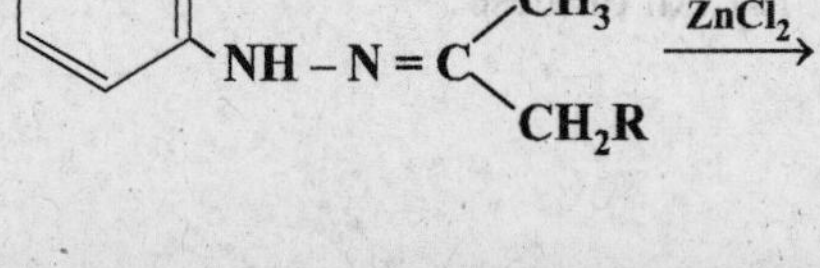

Solution. (*i*) 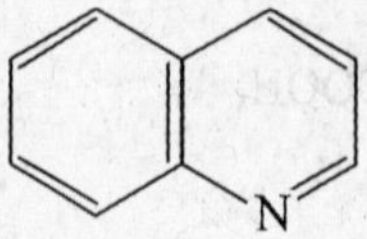 (ii) 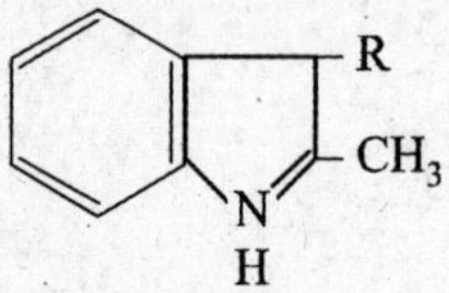

EXERCISES

(Including Questions from Different University Papers)

Multiple choice questions (Choose the correct option)

1. Pyridine has a delocalized π molecular orbital containing

(*a*) 4 electrons (*b*) 6 electrons
(*c*) 8 electrons (*d*) 12 electrons

2. Pyridine is less basic than trimethylamine because the lone-pair of electrons on N-atom in pyridine resides in

(*a*) sp^2 hybrid orbital (*b*) *sp* hybrid orbital
(*c*) sp^3 hybrid orbital (*d*) *p*- orbital

3. Pyridine undergoes electrophilic substitution with fuming H_2SO_4 at 350° C to give

(*a*) 2- Pyridinesulphonic acid (*b*) 4-Pyridinesulphonic acid
(*c*) 3-Pyridinesulphonic acid (*d*) None of these

4. Pyridine undergoes nuclophilic substitution with $NaNH_2$ at 100°C to form

(*a*) 2-Aminopyridine (*b*) 3- Aminopyridine
(*c*) 4-Aminopyridine (*d*) None of these

5. Furan reacts with ammonia in the presence of alumina at 400°C to give

(*a*) Pyridine (*b*) Furfural
(*c*) Pyrrole (*d*) Furoic acid

6. When aniline is heated with glycerol in the presence of sulphuric acid and nitrobenzene, it gives quinoline. This reaction in called

(*a*) Fischer synthesis (*b*) Skraup synthesis
(*c*) Diazotisation (*d*) Corey-House synthesis

7. The 'N' atom in pyridine is

(*a*) sp^3 hybridised (*b*) sp^2 hybridised
(*c*) *sp* hybridised (*d*) cannot be predicted

8. Pyrrole is less basic than pyridine because the lone-pair of electrons on N-atom in pyrrole

(*a*) is part of the delocalised π molecular orbital
(*b*) is not part of the delocalised π molecular orbital
(*c*) resides in sp^2 hybrid orbital
(*d*) resides in *sp* hybrid orbital

9. Pyridine reacts with HCl to form

(*a*) Pyridinium chloride (*b*) 2-Chloropyridine
(*c*) 3-Chlorpyridine (*d*) All of these

10. Pyridine reacts with a mixture of KNO_3 and H_2SO_4 at 300° C to give

(a) 1-Nitropyridine (b) 2-Nitropyridine

(c) 3-Nitropyridine (d) 4-Nitropyridine

11. Which of the following reagents will reacts with furan to form 2- furansulphonic acid?

(a) SO_3 in pyridine at 100°C (b) Dilute H_2SO_4 at 200°C

(c) SO_2 at 100°C (d) Dilute H_2SO_4 at 100°C

12. Which of the following reagents will react with pyrrole to form 2-formylpyrrole?

(a) HCOOH (b) $CHCl_3$/KOH

(c) H_2O_2 (d) $(CH_3CO)_2O/SnCl_4$

ANSWERS

1. (b)	**2.** (a)	**3.** (c)	**4.** (a)	**5.** (c)	**6.** (b)
7. (b)	**8.** (a)	**9.** (a)	**10.** (c)	**11.** (a)	**12.** (b)

Short Answer Questions

1. Give two methods of preparation of pyrrole.
2. Predict the main product in each of the following reactions:

(i)

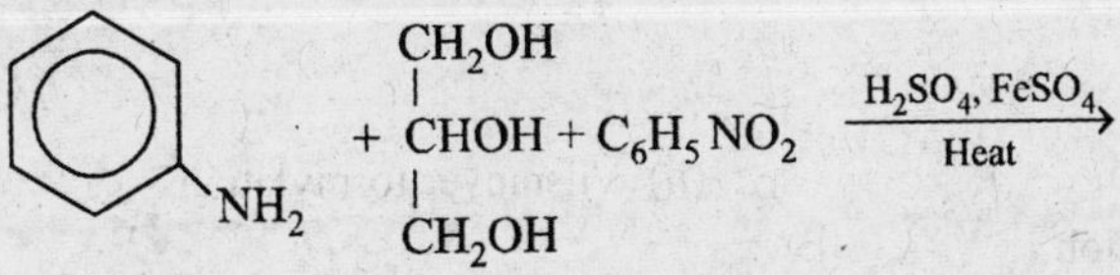

(ii)

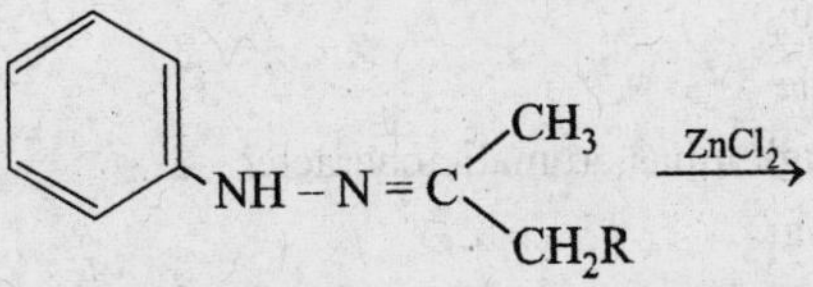

3. What are the important sources of pyridine and its derivatives? How can pyridine be converted into piperidine and 2-aminopyridine?
4. Why is pyridine more basic than pyrrole?
5. Complete the following equations

Furan + Maleic anhydride ⟶

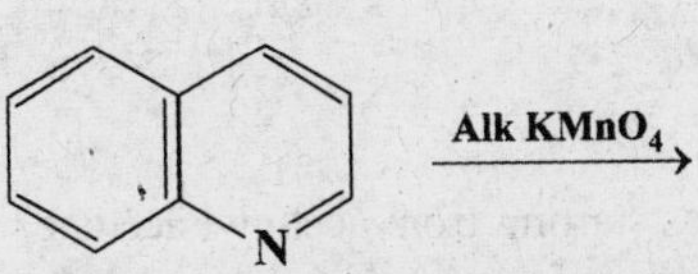

6. Explain Chichibabin reaction.
7. Write any one synthesis of pyrrole.
8. Explain aromatic character of pyrrole on the basis of molecular orbital theory.
9. Explain the following:
   In acidic conditions, pyridine undergoes electrophilic substitution with great difficulty.
10. Write structural formulae of the following:

(i) Piperidine (ii) Isoquinoline

11. Write suitable reasons for the fact that thiophene is more aromatic in nature than furan?

12. Pyrrole is acidic like phenol. Explain.

13. Why pyridine is more aromatic than furan?

14. Explain why piperidine is more basic than pyridine.

15. Starting from 1,4- diketone how can furan, thiophene and pyrrole derivatives be synthesised?

16. Why are pyrrole, thiophene and furan classified as aromatics?

17. How can you bring about the conversion of pyridine to 2- aminopyridine?

18. Why pyrrole is less aromatic than thiophene?

19. Explain why pyridine is more basic than pyrrole?

20. Sketch the structural formulae of quinoline and isoquinoline.

21. Give equations for the following:

(*i*) Feist-Benary synthesis of furan derivatives

(*ii*) Paal-Knorr synthesis of pyrrole derivatives

(*iii*) Chichibabin reactions.

22. Write chemical equations to convert succinic acid into pyrrole and thiophene

23. Which is more basic between pyrrole and pyridine? Give reasons.

24. Explain which one is the most aromatic and which one is the least aromatic of furan, pyrrole and thiophene.

25. Write equations for:

(*i*) Gomberg reaction (*ii*) Vilsmeyer formylation

(*iii*) Chichibabin reaction.

26. Give suitable reasons for the following:

(*i*) Pyridine is more basic than pyrrole

(*ii*) Furan is not stable to acids although it has aromatic character.

27. Give suitable reasons for the following:

(*i*) Thiophene is more aromatic in nature than furan.

(*ii*) Furan is not stable to acids although it has aromatic character.

28. Write suitable reasons for the following:

(*i*) Thiophene is more aromatic in nature than thiophene.

(*ii*) Pyridine is more basic than pyrrole.

29. Give the synthesis of

(*i*) Pyrrole starting from ethyne.

(*ii*) Thiophene starting from succinic acid.

30. How will you prepare the following compounds starting from acetonyl acetone?

(*i*) 2,5-Dimethylfuran (*ii*) 2,5- Dimethylpyrrole.

31. Give equations for paal-Knorr synthesis of pyrrole and thiophene derivatives.

32. Give suitable reasons for the following:

(*i*) Furan is not stable to acids although it has aromatic character.

(*ii*) Pyridine is more basic than pyrrole although it is less basic than aliphatic amines.

33. Discuss the electrophilic substitution reactions of thiophene.

34. Describe the mechanism of electrophilic substitution of pyrrole.

35. Discuss the aromatic character of thiophene and pyridine. How do these compounds differ from benzene in chemical behaviour?

36. Sketch the general mechanism of electrophilic substitution of thiophene.

37. Pyrrole is acidic like phenol. Explain.

38. Describe Chichibabin reaction.

49. Give mechanism of electrophilic substitution of pyrrole for nitration.

40. Explain that pyridine undergoes electrophilic substitution with difficulty. But when it does it does so at β-or 3- position.

General Questions

1. (*a*) What are heterocylic compounds? Justify the statements that pyrrole behaves both as phenol and aniline.

   (*b*) Discuss the orientation of substitution in pyrrole ring.

2. Explain the aromatic character of pyrrole on basis of molecular orbital theory.

3. Draw different resonating structures of pyridine. Explain why electrophilic substitution in pyridine takes place at position-3 and nucleophilic substitution at position -2.

4. What are heterocylic compounds? How are they classified? How do you explain the aromatic character of five membered heterocylics? How do you justify the electrophilic substitution at position 2 in pyrrole and position 3 in pyridine?

5. (*a*) Discuss molecular orbital structure of pyridine

   (*b*) Explain the following:

   (*i*) Pyridine is a weaker base than trimethylamine.

   (*ii*) Pyridine undergoes electrophilic substitution with difficulty. But when it does, it does so at 3 or β-position.

6. Write suitable reasons for the following:

   (*i*) Electrophilic substitution in pyrrole takes place at 2-position; whereas in pyridine at 3-position.

   (*ii*) Pyridine is more basic than pyrrole.

   (*iii*) Electrophilic substitution in five-membered heterocylic ring is easier than in pyridine.

   (*iv*) Thiophene is more aromatic in nature than furan.

7. Give a brief account of the chemistry of furan and pyrrole. How can compounds belonging to each of the ring system be synthesised?

8. Why position-2 is more reactive than position-3 in terms of electrophilic substitution in pyrrole?

9. Starting from 1,4-diketones how can furan, thiophene and pyrrole derivatives be synthesised.

10. (*a*) Explain in detail the orientation observed in electrophilic and nucleophilic substitution. reactions of pyridine.

    (*b*) Give one method of preparation of each of (*i*) of furan (*ii*) pyrrole (*iii*) thiophene.

11. Give the electrophilic substitution reactions of pyridine. Give the mechanism of reaction and justify the position of electrophilic attack.

12. (*a*) What are heterocylic compounds? Describe the molecular orbital structure of pyrrole.

    (*b*) How would you syntheses the following:

    (*i*) 2-Methyl pyrrole (*ii*) Pyrrole 2-carbaldehyde

    (*iii*) 2-Acetyl thiophene (*iv*) 5-Bromofuran.

13. What are heterocyclic compounds and how are they classified? Give one synthesis for each of the following compounds:

(*a*) Pyrrole and (*b*) Pyridine

14. (*a*) What are heterocylic compounds? How are they classified? Give one method of synthesis each of the following:

(*i*) Furan, (*ii*) Pyridine (*iii*) Pyrrole.

(*b*) Why pyridine is more basic than pyrrole. Explain with a diagram.

(*c*) Explain why pyridine is less reactive than benzene in electrophilic substitution with orbital picture.

15. (*a*) Outline the preparation of pyrrole and thiophene giving at least two examples in each case.

(*b*) Why position-2 is more reactive than position-3 in pyrrole.

16. Describe one method of synthesis and four electrophilic substitution reactions of indole.

17. Explain the aromatic character of pyrrole on the basis of its resonance and molecular orbital structures.

18. Draw different resonating structures of pyridine. Explain why electrophilic substitution in pyridine takes place at position-3 and nucleophilic substitution at position-2.

19. Pyridine though aromatic like benzene can undergo both nucleophilic and electrophilic substitution reactions easily while benzene undergoes exclusively electrophilic substitution reactions. Explain.

20. Give orbital picture and resonance structure of furan.

21. Give the mechanism of electrophilic substitution in pyrrole. Why is a particular orientation in this case preferred?

22. Give suitable reasons for the following:

(*i*) Electrophilic substitution in pyrrole takes place at 2-position; whereas in pyridine at 3-position.

(*ii*) Piperidine is more basic than pyridine.

23. Why does electrophilic substitution in pyridine take place preferentially at position -3 and not at 2 or 4?

24. Describe the preparation and properties of indole.

25. Why are electrophilic substitution reactions in five membered heterocylic compounds preferred at position 2 and not a 3?

26. Discuss the electrophilic substitution reactions of thiophene and furan.

27. Give two examples of electrophilic substitution reactions of quinoline where substitution takes place at position -5 and 8 and two examples where it takes place at position -3 and 4.

28. Explain why electrophilic substitution in pyrrole takes place at 3-position, whereas in pyridine it takes place at 3- position.

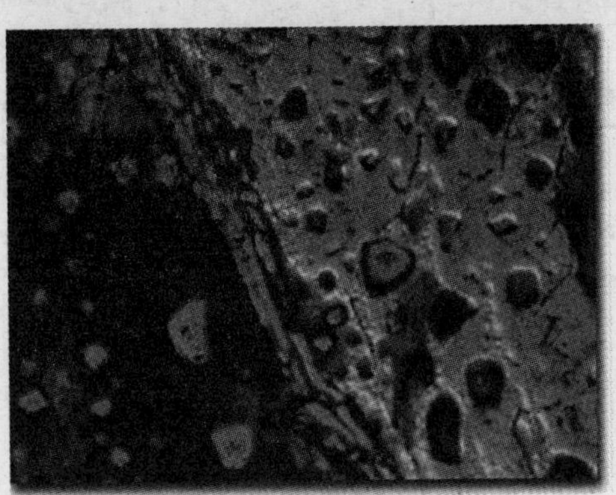

Malonic acid crystallites at 200 X magnification

5

Organic Synthesis Via Enolates

(Acetoacetic and Malonic ester, 1, 3-Dithiane, Enamine)

5.1. INTRODUCTION

It has been shown earlier that hydrogen atoms attached to carbon atoms adjacent to carbonyl group are acidic in nature *i.e.* they have a tendency to be released as protons. This is unlike the hydrogen atoms in hydrocarbon molecules. The pK_a values for the hydrogens in different types of compounds are given below :

Type of Compound		pK_a	K_a
Aldehydes ad ketones		20	10^{-20}
Acetylene	$(CH \equiv CH)$	25	10^{-25}
Vinylene	$(CH_2 = CH - H)$	40	10^{-40}
Alkane	$(CH_3 - CH_2 - H)$	45	10^{-45}

It may be noted that pK_a and K_a are inversely related. K_a stands for dissociation constant. Greater the dissociation of a compound to give H^+, greater value of K_a and smaller the value of pK_a. It may also be noted that, for example, 10^{-20} is greater than 10^{-25}.

The acidic nature of α-hydrogens leads to the formation of species called *enolates* (in simple language, anion). These enolates find use as nucleophile (carrying negative charge) in SN^2 reactions to synthesise a number of organic compounds.

5.2. ACIDITY OF α-HYDROGENS (FORMATION OF ENOLATES)

The electronegativity difference between carbon and hydrogen is small, therefore, C–H bond is normally non-polar. Presence of a carbonyl group, which is strongly electron withdrawing near a C–H bond polarises the electron pair of C–H bond and results in the removal of a proton by a base. Another important fact is that the anion produced after the removal of proton gets stablised by resonance and delocalisation of negative charge

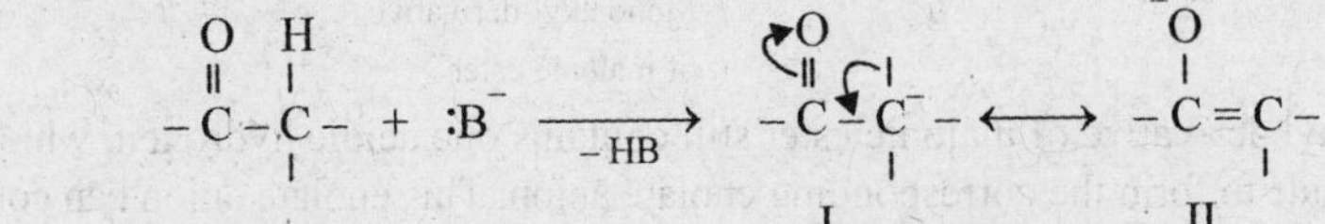

Enolate anion

The resonance stabilised anion is called *enolate anion.* Form II of the enolate ion on combination gives an enol (*en* + *ol*) That is how the anion derives the name enolate anion

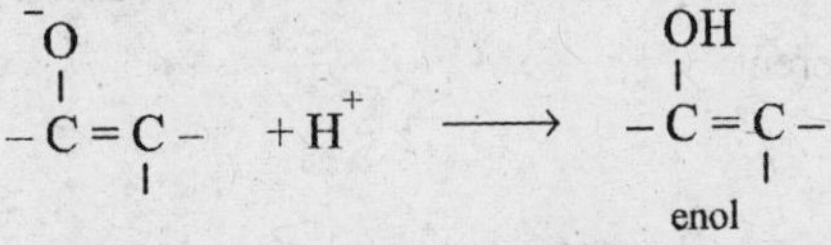

enol

The acidity of the carbonyl compound increases markedly when it contains hydrogen alpha to two carbonyl groups. In such a case, the enolate ion is stabilised to a greater extent by two carbonyl groups. This is illustrated by taking the example of acetonyl acetone.

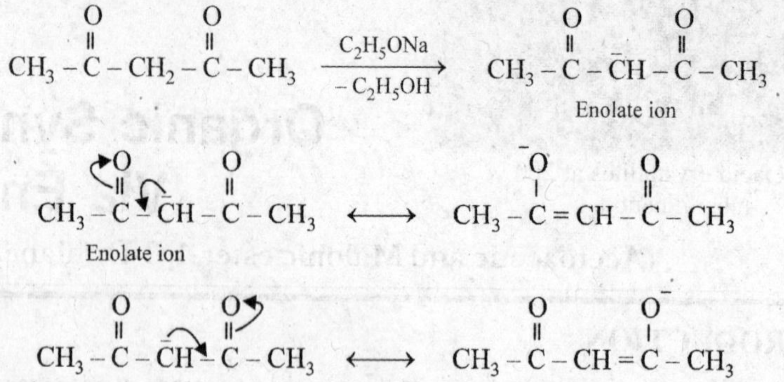

We obtain a greater number of resonating structures, thus increasing the stability of the anion. In other words, the acidity of a compound increases when it contains two carbonyl groups located as above

Enolates obtained from dicarbonyl compounds such as malonic ester and acetoacetic ester, play an important role in organic synthesis. The enolate anions participate as nucleophiles in a large number of SN^2 reactions of synthetic importance.

5.3. ALKYLATION OF MALONIC ESTER

Malonic ester can be converted into its mono and dialkyl derivatives. This process is known as alkylation of malonic ester. This property is responsible for the synthesis of a large number of organic compounds. The two steps in the alkylation of malonic ester are

***(i)* Formation of the enolate**

Malonic ester is a stronger acid than ethanol. Therefore, malonic ester is converted completely into its enolate anion by sodium ethoxide

$$\underset{\text{Malonic ester}}{CH_2(COOC_2H_5)_2} \xrightarrow[-C_2H_5OH]{C_2H_5ONa} \underset{\text{Enolate}}{[CH(COOC_2H_5)_2]^- Na^+}$$

***(ii)* Alkylation of enolate with alkyl halide (SN^2 reaction)**

The enolate anion of malonic ester is a nucleophile and participates in SN^2 reactions with primary and secondary alkyl halides to form alkyl derivatives of malonic esters

$$[CH(COOC_2H_5)_2]^- Na^+ + R-X \longrightarrow \underset{\text{Monoalkyl derivative of malonic ester}}{RCH\,(COOC_2H_5)_2} + Na\,X$$

The monoalkyl derivative of malonic ester still contains one acidic hydrogen, which can react with sodium ethoxide to form the corresponding enolate anion. This enolate anion can combine with another molecule of alkyl halide to form dialkyl derivative of malonic acid

$$RCH(COOC_2H_5)_2 \xrightarrow[-C_2H_5OH]{C_2H_5ONa} [RC(COOC_2H_5)]^- Na^+ \xrightarrow{R'X} \underset{\text{Dialkylmalonic ester}}{RR'C(COOC_2H_5)_2} + Na\,X$$

R and R′ could be same or different

5.4. ALKYLATION OF ACETOACETIC ESTER

Like malonic ester, acetoacetic ester can also be converted into mono and dialkyl derivatives in two steps as shown below :

(*i*) Formation of enolate

On treatment with sodium ethoxide, acetoacetic ester loses a proton quantitatively and is converted into enolate

$$CH_3COCH_2COOC_2H_5 \xrightarrow[-C_2H_5OH]{C_2H_5ONa} \underset{\text{Enolate}}{[CH_3COCHCOOC_2H_5]^- Na^+}$$

(*ii*) Alkylation of enolate with alkyl halide (SN^2 reaction)

Enolate formed above is a strong nucleophile and can displace halide ion from an alkyl halide to form monoalkyl derivative by SN^2 mechanism

$$[CH_3COCHCOOC_2H_5]^- Na^+ + RX \longrightarrow \underset{\text{Monoalkyl acetoacetic ester}}{CH_3CO\underset{\displaystyle R}{\underset{|}{C}}HCOOC_2H_5} + Na\,X$$

The monoalkyl derivative still contains one acidic hydrogen which can react with sodium ethoxide to form corresponding enolate, which will then react with another molecule of alkyl halide to form dialkylacetoacetic ester

$$CH_3CO\underset{\displaystyle R}{\underset{|}{C}}HCOOC_2H_5 \xrightarrow[-C_2H_5OH]{C_2H_5ONa} \left[CH_3CO\underset{\displaystyle R}{\underset{|}{C}}HCOOC_2H_5\right]^- Na^+ \xrightarrow{R'X} \underset{\text{Dialkyl derivative of acetoacetic acid}}{CH_3CO\overset{\displaystyle R'}{\overset{|}{\underset{\displaystyle R}{\underset{|}{C}}}}H - COOC_2H_5} + Na\,X$$

5.5. METHODS OF PREPARATION OF ETHYL ACETOACETATE

The methods for the preparation of ethyl acetoacetate are described as under.

1. Claisen condensation. Two molecules of ethyl acetate undergo self-condensation in the presence of a strong base such as sodium ethoxide, giving ethyl acetoacetate

$$2CH_3COOC_2H_5 \xrightarrow{Na\,/\,Alcohol} CH_3COCH_2COOC_2H_5 + C_2H_5OH$$

Mechanism

1st step. A proton is removed by ethoxide ion from α-carbon of ethyl acetate to form resonance stabilised carbanion.

Ludwig Claisen (1851–1930) German chemist known for his work on condensation of carbonyls

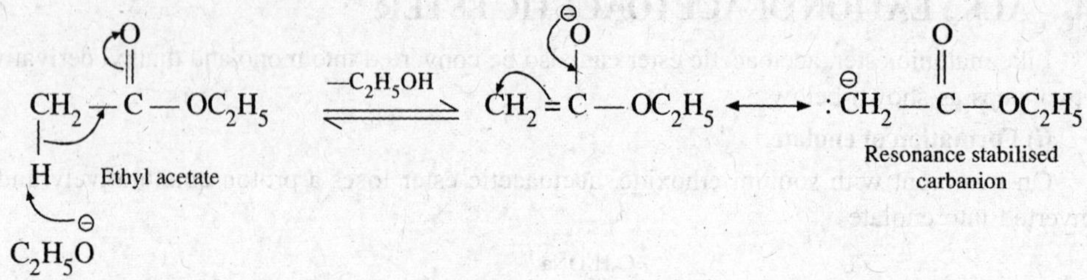

2nd step. The carbanion attacks the carbonyl carbon of the second molecule of ethyl acetate and displaces the ethoxide ion to give ethyl acetoacetate.

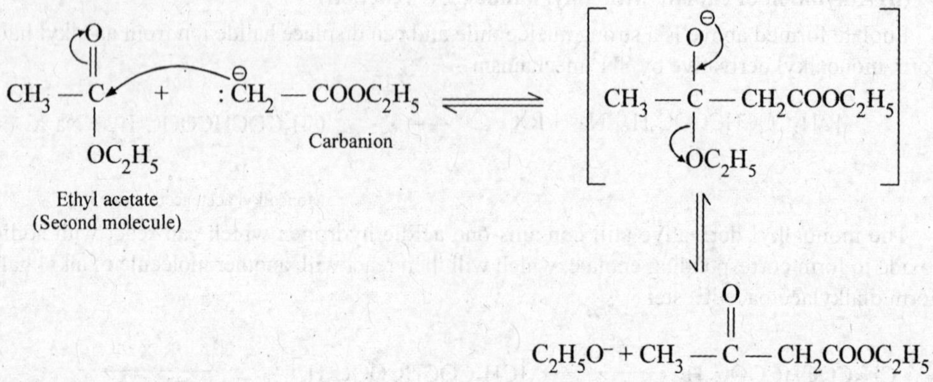

3rd step. Ethoxide ion reacts with ethyl acetoacetate to give ethanol and anion of acetic ester which on acidification with acetic acid yields ethyl acetoacetate in the final step.

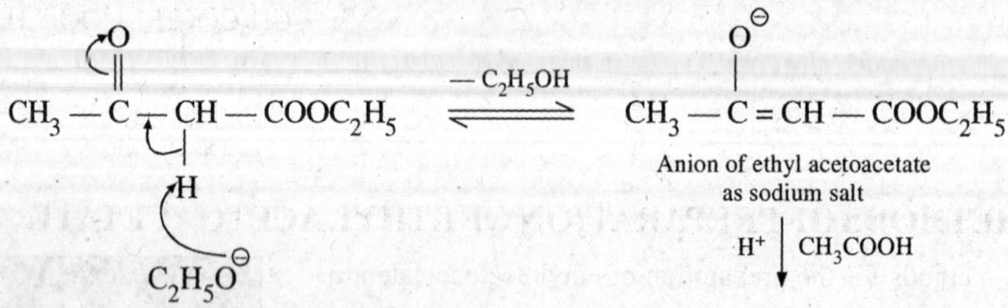

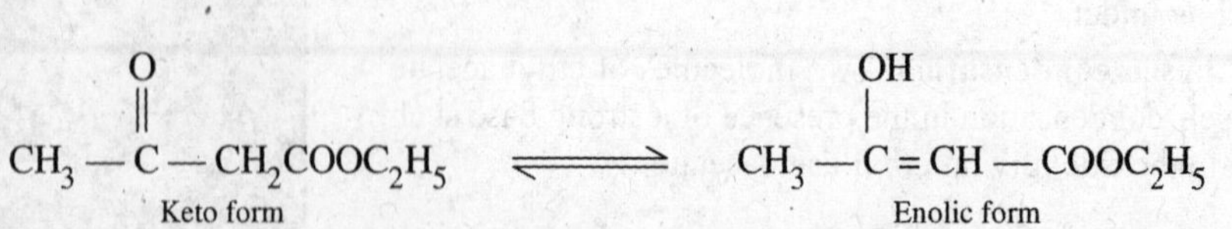

2. From ketene. Commercially, acetoacetic ester is prepared by dimerisation of ketene in acetone solution followed by treatment with alcohol.

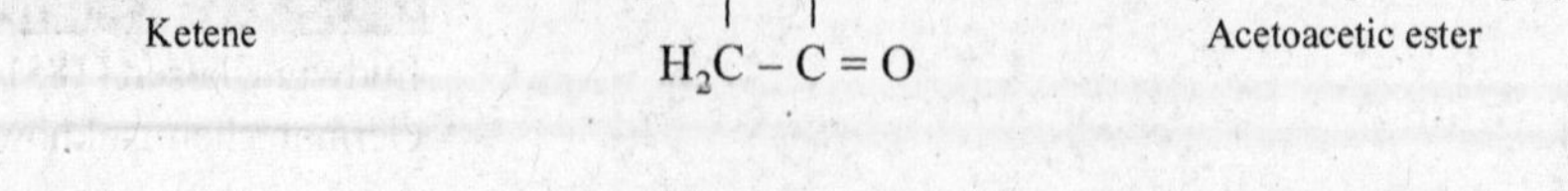

5.6. PROPERTIES OF ETHYLACETOACETATE

Important properties of ethylacetoacetate are described as under :

1. Acidic nature of methylene group. Methylene group in acetoacetic ester is flanked by two electron withdrawing groups, viz., an acetyl group and an ester group. The hydrogens of this methylene group are ionisable because of the electron withdrawing effect of the surrounding groups. Also the negative ion obtained after losing the proton gets stabilised due to resonance as shown below:

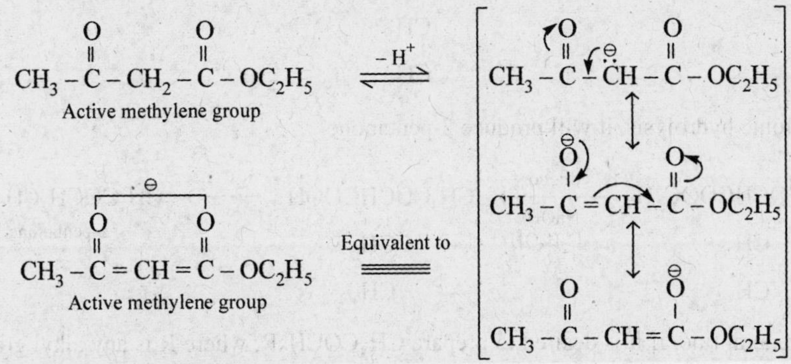

As a consequence of above property, acetoacetic ester reacts with sodium ethoxide to form sodium salt of acetoacetic ester

$$\underset{\text{Acetoacetic acid}}{CH_3COCH_2COOC_2H_5} \xrightarrow{C_2H_5ONa} \underset{\text{Sod. salt}}{[CH_3COCHCOOC_2H_5]^- Na^+}$$

This sodium salt on treatment with an alkyl or acyl halide gives a monoalkyl or acyl derivative of acetoacetic ester

$$[CH_3COCHCOOC_2H_5]^- Na^+ + RX \longrightarrow CH_3CO\underset{\substack{|\\R}}{C}HCOOC_2H_5 + NaX$$

Another such treatment can produce dialkyl derivatives.

2. Ketonic hydrolysis. Acetoacetic ester or its alkyl derivatives on hydrolysis with dil. aqueous alkali form the corresponding acids and then undergo decarboxylation to yield ketones. This process is known as *ketonic hydrolysis*.

$$\underset{\text{Acetoacetic ester}}{CH_3COCH_2COOC_2H_5} \xrightarrow[\text{NaOH}]{\text{dil. aqeous}} \underset{\text{Sodium salt of acetoacetic acid}}{CH_3COCH_2COONa}$$

$$\xrightarrow{H^+} \underset{\text{Acetoacetic acid}}{CH_3COCH_2COOH} \xrightarrow{-CO_2} \underset{\text{Acetone}}{CH_3COCH_3}$$

3. Acid hydrolysis. If hydrolysis of acetoacetic acid or its alkyl derivative is carried out in the presence of conc. alkali, two molecules of carboxylic acid are obtained.

$$CH_3-\overset{\overset{O}{\|}}{C}-CH_2-\overset{\overset{O}{\|}}{C}-OC_2H_5 \xrightarrow{2NaOH} 2CH_3COONa + C_2H_5OH$$

Synthetic Uses of Acetoacetic ester

Based on above three characteristics of the ester, a large variety of compounds can be obtained.

(*i*) **Synthesis of methyl ketones.** As explained above under the heading *ketonic hydrolysis*, a desired ketone may be prepared from the appropriate alkyl derivative of acetoacetic ester. For example, if it is desired to prepare 2-pentanone by ketonic hydrolysis, then the alkyl derivative of acetoacetic ester that needs to be taken is

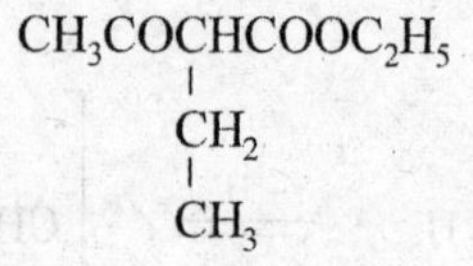

On ketonic hydrolysis, it will produce 2-pentanone

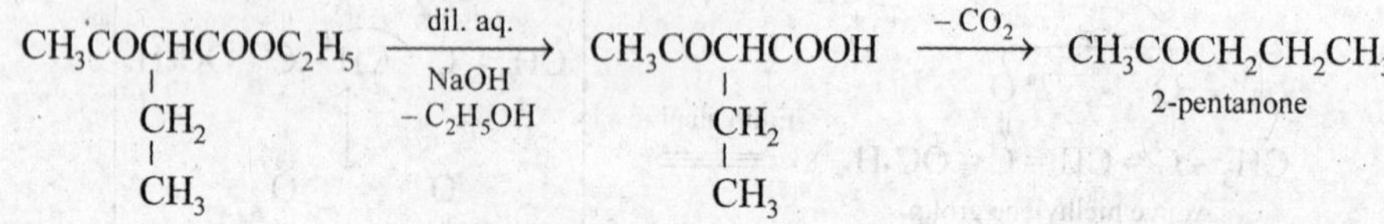

As a general rule, if it is desired to prepare CH_3COCH_2R, where R is any alkyl group, linear or branched, then the starting compound will be

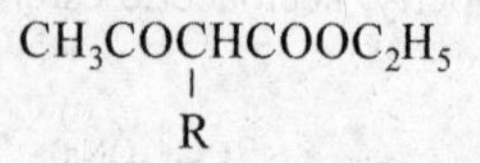

(*ii*) **Synthesis of diketone.** Acyl derivative of acetoacetic ester on ketonic hydrolysis yields diketones. Acetylacetone, for example, may be prepared as under:

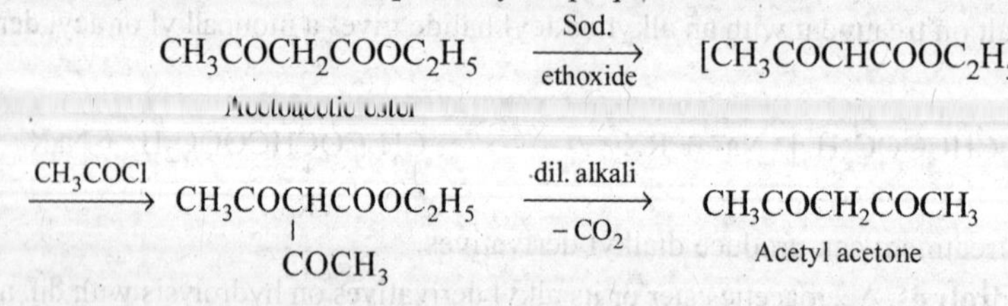

(*iii*) **Synthesis of mono and dialkyl acetic acids.** Acidic hydrolysis of mono alkyl derivative of acetoacetic ester produces monoalkyl acetic acid. Similarly dialkyl derivatives would produce dialkyl acetic acid.

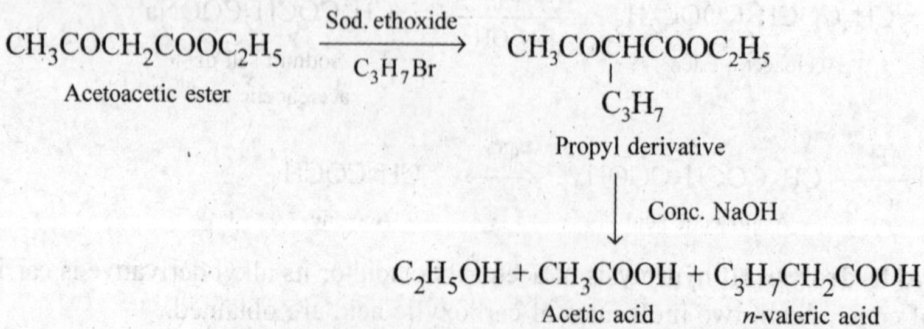

In general to produce RCH_2COOH, the alkyl derivatives required will be

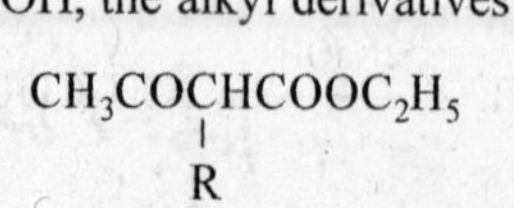

It is also possible to synthesise dialkyl acetic acid *e.g.*

$$\underset{\text{Diethyl derivative}}{CH_3COC(CH_2CH_3)_2\cdot COOC_2H_5} \xrightarrow[-\,C_2H_5OH]{\text{Conc. NaOH}} CH_3COOH + \underset{\text{Diethyl acetic acid}}{(CH_3CH_2)_2CHCOOH}$$

(*iv*) **Synthesis of dicarboxylic acids.** Reaction of sod. salt of acetoacetic ester with halogen derivative of an ester followed by acidic hydrolysis, produces dicarboxylic acid.

$$\underset{\text{Ethyl acetoacetate}}{CH_3COCH_2COOC_2H_5} \xrightarrow[\text{Ethoxide}]{\text{Sod.}} \underset{\text{Sod. salt}}{[CH_3COCHCOOC_2H_5]^- Na^+}$$

$$\xrightarrow{ClCH_2COOC_2H_5} \underset{|\atop CH_2COOC_2H_5}{CH_3COCHCOOC_2H_5} \xrightarrow[\text{NaOH}]{\text{Conc.}} CH_3COOH + \underset{\text{Succinic acid}}{\underset{|\atop CH_2COOH}{CH_2COOH}} + 2C_2H_5OH$$

(*v*) **Synthesis of α, β unsaturated acids.** Acetoacetic ester condenses with aldehydes and ketones in the presence of pyridine (*Knoevenagal Reaction*)

$$CH_3CHO + CH_3COCH_2COOC_2H_5 \xrightarrow{\text{Pyridine}} \underset{\|\atop CHCH_3}{CH_3COCCOOC_2H_5}$$

$$\xrightarrow{\text{Hydrolysis}} CH_3COOH + \underset{\text{Crotonic acid}}{CH_3CH = CHCOOH} + C_2H_5OH$$

(*vi*) **Condensation with urea.** Acetoacetic ester in enol form condenses with urea to form 4-methyl uracil

$$\underset{|\atop OH}{CH_3C} = CHCOOC_2H_5 + \underset{|\atop NH - CO - NH}{H\qquad H} \longrightarrow \underset{|\atop NH - CO - NH}{CH_3C = CH - CO} + C_2H_5OH + H_2O$$

(*vii*) **Condensation with hydroxylamine**

$$\underset{|\atop CH_2 - CO - OC_2H_5}{CH_3 - C = O} + \underset{|\atop HO}{H_2N} \longrightarrow \underset{\text{Methyl iso-oxazolone}}{\underset{|\atop CH_2 - CO}{CH_3 - C = N}\!\!>O} + C_2H_5OH + H_2O$$

(*viii*) **Synthesis of pyrrole derivatives.** Reaction of sod. salt of acetoacetic ester with iodine followed by treatment with ammonia gives pyrrole derivatives

$$2[CH_3COCHCOOC_2H_5]^- Na^+ + I_2 \longrightarrow \begin{array}{c} CH_3COCHCOOC_2H_5 \\ | \\ CH_3COCHCOOC_2H_5 \end{array} + 2NaI$$

$$\xrightarrow{NH_3} \begin{array}{c} CH_3COCHCO \\ | \\ CH_3COCHCO \end{array} \!\!> NH$$

(Pyrrole derivative)

5.7. KETO-ENOL TAUTOMERISM

1. An isomerism in which two forms of the compound exist simultaneously as an equilibrium mixture is called *tautomerism*. This is different from the ordinary isomerism phenomenon in which there are separately different compounds comforming to the same molecular formula.

2. Keto-enol tautomerism is a particular case of tautomerism in which two coexisting substances have the ketonic group and ethylenic double bond.

3. A typical example of a compound exhibiting keto-enol tautomerism is acetoacetic ester. It exists as an equilibrium mixture of two substances one of which has a ketonic group and the other ethylenic bond

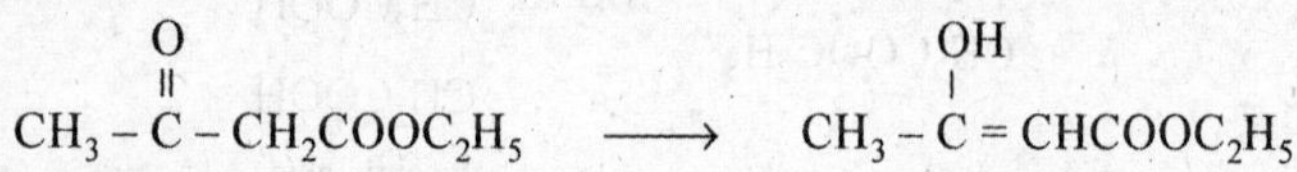

Keto form — Enol form

Enol is a combination of *ene* + *ol i.e.* a double bond and an alcoholic group.

4. This conclusion is based on the observation that acetoacetic ester gives the properties expected of a ketone and those expected of an ethylenic compound containing an alcoholic group. For example, Acetoacetic ester gives reaction with HCN to form cyanohydrin, with hydroxy amine to form oxime and with phenyl hydrazine to form yellow crystalline hydrazone. All these tests point to the presence of a ketonic group.

At the same time, acetoacetic ester liberates hydrogen with metallic sodium, it forms acetyl derivative on treatment with acetyl chloride. These are the characteristic reactions expected of alcohols.

It decolourises bromine solution, which indicates the presence of a double bond. Thus it supports our belief that acetoacetic ester exists as a mixture of two forms one of which is a keto compound and the other an enol *i.e.* ethylenic compound along with an alcoholic group.

5. The two tautomers differ in respect of point of attachment of hydrogen. In the *enol* form, the hydrogen is attached to oxygen whereas it is attached to carbon in the *ketonic* form. The two forms interconvert into each other as follows.

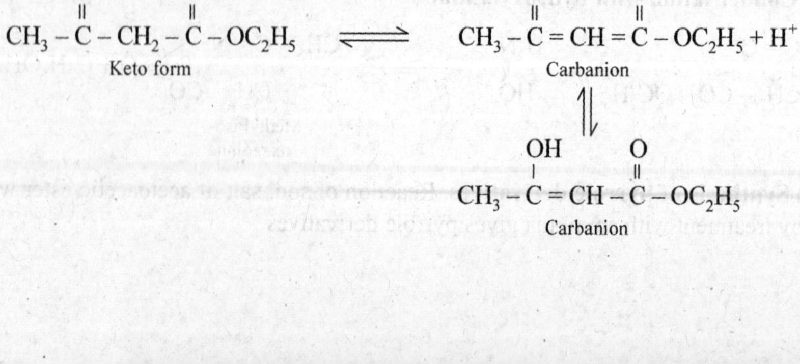

6. Keto form of the ester has a percentage of 93 and enol form has the percentage of 7.

7. Normally it is not possible to isolate either pure keto form or enol form. This is because, as we try to remove keto form from the mixture, some of the enol form in the resulting mixture converts into keto form to maintain 93 : 7 ratio and *vice versa.*

8. However, under conditions of low temp. and using specific solvents, it is possible to affect separation. If a solution of acetoacetic ester in ether and hexane is cooled to 195 K, pure keto form of the ester is obtained. Enol form of the ester can be obtained by treating a suspension of sodium salt of the ester in petroleum ether with dry HCl at 195 K.

Example 1. Starting from ethyl acetoacetate, give a reaction scheme to prepare each of the following:

(A) Methyl ethyl ketone (B) Acetyl acetone (C) 2, 3 dimethyl butanoic acid (D) Succinic acid (F) Crotonic acid.

Solution. (A)

$$\underset{\text{Acetoacetic ester}}{CH_3COCH_2COOC_2H_5} \xrightarrow{C_2H_5ONa} \underset{\text{Sod. salt}}{[CH_3COCHCOOC_2H_5]^- Na^+}$$

$$\xrightarrow{CH_3Br} \underset{\alpha\text{-methyl acetoacetate}}{CH_3CO\underset{\displaystyle CH_3}{\underset{|}{C}H}COOC_2H_5} \xrightarrow[\text{NaOH}]{\text{dil.}} \underset{\text{Methyl ethyl ketone}}{CH_3COCH_2CH_3}$$

(B)

$$CH_3COCH_2COOC_2H_5 \xrightarrow{C_2H_5ONa} [CH_3COCHCOOC_2H_5]^- Na^+$$

$$\xrightarrow{CH_3COCl} CH_3CO\underset{\displaystyle COCH_3}{\underset{|}{C}H}COOC_2H_5 \xrightarrow[\text{NaOH}]{\text{dil.}} \underset{\text{Acetylacetone}}{CH_3COCH_2COCH_3}$$

(C)

$$CH_3COCH_2COOC_2H_5 \xrightarrow[(ii)\ CH_3Br]{(i)\ C_2H_5ONa} CH_3CO\overset{\displaystyle CH_3}{\overset{|}{C}H}COOC_2H_5$$

$$\xrightarrow[(ii)\ \text{Isopropyl bromide}]{(i)\ C_2H_5ONa} CH_3CO-\overset{\displaystyle CH_3}{\overset{|}{C}}\underset{\displaystyle \underset{\displaystyle CH_3}{\underset{|}{CH-CH_3}}}{\underset{|}{H}}-COOC_2H_5 \xrightarrow[\text{NaOH}]{\text{Conc.}}$$

$$CH_3COOH + \underset{\text{2, 3 dimethyl butyric acid}}{CH_3-\overset{\displaystyle CH_3}{\overset{|}{C}H}-\overset{\displaystyle CH_3}{\overset{|}{C}H}-COOH} + C_2H_5OH$$

(D)

$$CH_3COCH_2COOC_2H_5 \xrightarrow{C_2H_5ONa} [CH_3COCHCOOC_2H_5]^- Na^+$$

$$\xrightarrow{ClCH_2COOC_2H_5} CH_3CO\underset{\displaystyle CH_2COOC_2H_5}{\underset{|}{C}H}COOC_2H_5 \xrightarrow[\text{NaOH}]{\text{Conc.}} CH_3COOH + \underset{\displaystyle CH_2COOH}{\underset{|}{CH_2COOH}} + 2C_2H_5OH$$

$$\text{(E) } \underset{\text{Acetoacetic ester}}{CH_3COOCH_2COOC_2H_5} + \underset{\text{Acetaldehyde}}{CH_3CHO} \xrightarrow[-H_2O]{\text{Pyridine}} \underset{\substack{\| \\ CHCH_3}}{CH_3COCCOOC_2H_5}$$

$$\xrightarrow{\text{Acid hydrolysis} \quad \text{Conc. NaOH}} CH_3COOH + \underset{\text{Crotonic acid}}{CH_3CH{=}CHCOOH} + C_2H_5OH$$

Example 2. How will you obtain the following from acetoacetic ester

(*a*) Isobutyric acid

(*b*) Methyl propyl ketone

Solution. (*a*) Isobutyric acid

$$\underset{\text{Acetoacetic ester}}{CH_3COCH_2COOC_2H_5} \xrightarrow[CH_3Br]{C_2H_5ONa} \underset{\text{Methyl derivative}}{\underset{\substack{| \\ CH_3}}{CH_3COCHCOOC_2H_5}}$$

$$\xrightarrow[CH_3Br]{C_2H_5ONa} \underset{CH_3\ CH_3}{CH_3COC - COOC_2H_5} \xrightarrow[H^+]{\text{Conc. NaOH}} CH_3COOH + (CH_3)_2CHCOOH + C_2H_5OH$$

(*b*) Methylpropyl ketone

$$\underset{\text{Acetoacetic ester}}{CH_3COCH_2COOC_2H_5} \xrightarrow[C_2H_5Br]{C_2H_5ONa} \underset{\text{Ethyl derivative}}{\underset{\substack{| \\ C_2H_5}}{CH_3COCHCOOC_2H_5}} \xrightarrow{\text{Dil. NaOH}} \underset{\text{Sod. salt of ethyl acetoacetic acid}}{\underset{\substack{| \\ C_2H_5}}{CH_3COCHCOONa}}$$

$$\xrightarrow[\Delta]{H^+} \underset{\text{Methyl propyl ether}}{CH_3COC_3H_7} + CO_2\uparrow$$

5.8. SYNTHESIS OF DIETHYL MALONATE

Diethyl malonate (malonic ester) can be prepared as under :

1. Malonic ester is prepared by passing dry HCl gas through a solution of pot. cyano acetate in ethanol

$$\underset{\text{Pot. cyanoacetate}}{\underset{\substack{| \\ CN}}{CH_2COOK}} + 2\,C_2H_5OH + 2HCl \longrightarrow \underset{\text{Diethyl malonate}}{\underset{\substack{| \\ COOC_2H_5}}{CH_2COOC_2H_5}} + NH_4Cl + KCl$$

Cyanoacetate itself, is prepared from acetic acid as under:

$$\underset{\text{Acetic acid}}{CH_3COOH} \xrightarrow{Cl_2,\,P} \underset{\text{Chloroacetic acid}}{CH_2\begin{matrix} Cl \\ COOH \end{matrix}} \xrightarrow{K_2CO_3} \underset{\text{Pot. chloroacetate}}{CH_2\begin{matrix} Cl \\ COOK \end{matrix}} \xrightarrow{KCN} CH_2\begin{matrix} CN \\ COOK \end{matrix}$$

2. From malonic acid. Malonic acid is esterified with an alcohol in the presence of conc. H_2SO_4 to yield malonic ester

$$CH_2(COOH)_2 + 2\,C_2H_5OH \xrightarrow{H^+} CH_2(COOC_2H_5)_2 + 2H_2O$$

5.9. PROPERTIES OF MALONIC ESTER

Properties of malonic ester are described as under:

1. Acidic nature of methylene hydrogens

The methylene group in the molecule of malonic ester contains hydrogens which are ionisable and are removed in the form of hydrogen ions. This is because the anion left after releasing the proton is stabilized by resonance as shown below:

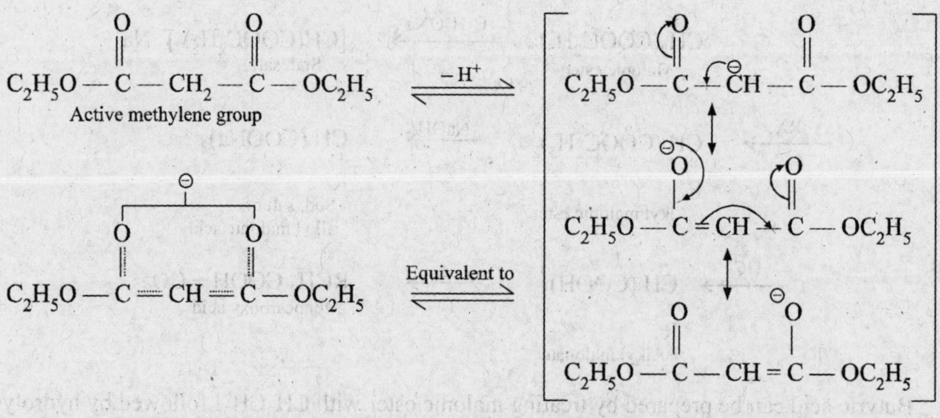

This imparts synthetic importance to malonic ester. When treated with sodium ethoxide, it forms the sodium salt. Sodium salt of malonic ester is capable of reacting with alkyl or acyl halides to give substituted products.

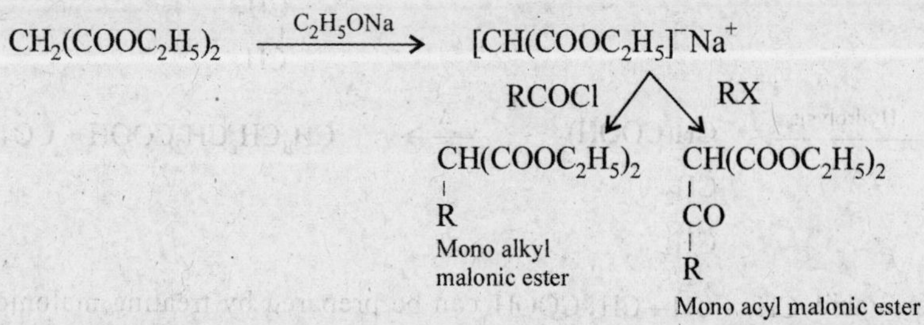

There is a possibility of further substitution as still one more active hydrogen is left on the molecule after monosubstitution. Therefore the monosubstituted product may be subjected to the above treatment once more to obtain disubstituted products.

$$RCH(COOC_2H_5)_2 \xrightarrow[R'X]{C_2H_5ONa} RR'C(COOC_2H_5)_2$$

Disubstituted product

2. Action of heat on substituted malonic acid

Another property of malonic ester and substituted malonic ester is that out of two carboxylic groups, one can be decarboxylated easily. This property is utilized in a number of synthetic reactions.

$$\underset{\text{Alkyl malonic ester}}{RCH(COOC_2H_5)_2} \xrightarrow{NaOH} RCH(COONa)_2$$

$$\xrightarrow{HCl} RCH(COOH)_2 \xrightarrow{\Delta} RCH_2COOH + CO_2$$

Based on the above two properties or principles, following synthetic reactions can be performed.

1. Synthesis of monoalkylacetic acid. Malonic ester is treated with an appropriate alkyl halide after converting the ester into sodium salt. Alkylated malonic ester is then hydrolysed and heated. Heating brings about decarboxylation of one of the carboxyl group. Thus in order to prepare RCH_2COOH, the alkyl halide to be used is RX.

$$\underset{\text{Malonic ester}}{CH_2(COOC_2H_5)_2} \xrightarrow{C_2H_5ONa} \underset{\text{Sod. salt}}{[CH(COOC_2H_5)_2]^- Na^+}$$

$$\xrightarrow{RX} \underset{\text{Alkyl malonic ester}}{\underset{|\atop R}{CH(COOC_2H_5)_2}} \xrightarrow{NaOH} \underset{\text{Sod. salt of alkyl malonic acid}}{\underset{|\atop R}{CH(COONa)_2}}$$

$$\xrightarrow{HCl} \underset{\text{Alkyl malonate}}{\underset{|\atop R}{CH(COOH)_2}} \xrightarrow{\Delta} \underset{\text{Monocarboxy acid}}{RCH_2COOH + CO_2}$$

Butyric acid can be prepared by treating malonic ester with CH_3CH_2I followed by hydrolysis & decarboxylation

$$CH_2(COOC_2H_5)_2 \xrightarrow[CH_3CH_2I]{C_2H_5ONa} \underset{\underset{|\atop CH_3}{CH_2}}{\underset{|}{CH(COOC_2H_5)_2}}$$

$$\xrightarrow{\text{Hydrolysis}} \underset{\underset{|\atop CH_3}{CH_2}}{\underset{|}{CH(COOH)_2}} \xrightarrow{\Delta} CH_3CH_2CH_2COOH + CO_2$$

Isovaleric acid $CH_3-\underset{|\atop CH_3}{CH}-CH_2COOH$ can be prepared by treating malonic ester with isopropyl bromide followed by hydrolysis and decarboxylation.

2. Synthesis of dialkyl acetic acid. Malonic ester is first converted into dialkyl malonic ester. For this sodium ethoxide and alkyl halide are taken in double the quantities of malonic ester (in molar quantities). Dialkyl malonic ester is then hydrolysed and decarboxylated.

$$\underset{\text{Malonic ester}}{CH_2(COOC_2H_5)_2} \xrightarrow[CH_3I]{C_2H_5ONa} \underset{\text{Methyl malonic ester}}{\underset{|\atop CH_3}{CH(COOC_2H_5)_2}}$$

$$\xrightarrow[CH_3CH_2I]{C_2H_5ONa} \underset{\text{Ethyl methyl malonic ester}}{CH_3C(COOC_2H_5)_2(CH_2-CH_3)} \xrightarrow{\text{Hydrolysis}} \underset{\text{Ethyl methyl malonic acid}}{CH_3C(COOH)_2(CH_2-CH_3)} \xrightarrow{\Delta} \underset{\text{2-Methyl butanoic acid}}{CH_3-CH_2-CH(CH_3)COOH}$$

3. Synthesis of-keto acids $RCOCH_2COOH$. Sodium salt of malonic ester is treated with acyl chloride followed by hydrolysis and decarboxylation.

$$CH_2(COOC_2H_5)_2 \xrightarrow{C_2H_5ONa} [CH(COOC_2H_5)_2]^- Na^+$$

$$\xrightarrow{CH_3COCl} CH_3CO-CH(COOC_2H_5)_2 \xrightarrow{\text{Hydrolysis}} CH_3CO-CH(COOH)_2 \xrightarrow[-CO_2]{\Delta} \underset{\text{Acetyl acetic acid}}{CH_3COCH_2COOH}$$

4. Synthesis of dicarboxylic acids.

(*a*) Using α-haloester

$$\underset{\text{Malonic ester}}{CH_2(COOC_2H_5)_2} \xrightarrow{C_2H_5ONa} \underset{\text{Sod. salt of malonic ester}}{[CH(COOC_2H_5)_2]^- Na^+} \xrightarrow[\text{(a-bromoethyl acetate)}]{BrCH_2COOC_2H_5}$$

$$\begin{matrix} CH(COOC_2H_5)_2 \\ | \\ CH_2COOC_2H_5 \end{matrix} \xrightarrow{\text{Hydrolysis}} \begin{matrix} CH_2(COOH)_2 \\ | \\ CH_2COOH \end{matrix} \xrightarrow{\Delta} \underset{\text{Succinic acid}}{\begin{matrix} CH_2COOH \\ | \\ CH_2COOH \end{matrix}} + CO_2$$

(*b*) Using I_2

$$\underset{\text{Malonic ester}}{CH_2(COOC_2H_5)_2} \xrightarrow{C_2H_5ONa} \underset{\text{Sod. salt}}{[CH(COOC_2H_5)_2]^- Na^+}$$

$$\xrightarrow[-2NaI]{I_2} \begin{matrix} CH(COOC_2H_5)_2 \\ | \\ CH(COOC_2H_5)_2 \end{matrix} \xrightarrow{\text{Hydrolysis}} \begin{matrix} CH(COOH)_2 \\ | \\ CH(COOH)_2 \end{matrix} \xrightarrow{\Delta} \underset{\text{Succinic acid}}{\begin{matrix} CH_2COOH \\ | \\ CH_2COOH \end{matrix}}$$

(*c*) Using α, ω dihalide

$$\underset{\text{Ethylene bromide}}{\begin{matrix} CH_2Br \\ | \\ CH_2Br \end{matrix}} + 2\underset{\text{Sod. salt of malonic ester}}{[CH(COOC_2H_5)_2]^- Na^+} \longrightarrow \begin{matrix} CH_2CH(COOC_2H_5)_2 \\ | \\ CH_2CH(COOC_2H_5)_2 \end{matrix}$$

$$\xrightarrow{\text{Hydrolysis}} \begin{matrix} CH_2CH(COOH)_2 \\ | \\ CH_2CH(COOH)_2 \end{matrix} \xrightarrow{\Delta} \underset{\text{Adipic acid}}{\begin{matrix} CH_2CH_2COOH \\ | \\ CH_2CH_2COOH \end{matrix}}$$

5. Synthesis of cycloalkane carboxylic acid

$$\underset{\text{1, 3 dibromo-propane}}{\begin{matrix} CH_2Br \\ | \\ CHCH_2Br \end{matrix}} + 2\underset{\text{Sod. salt of malonic ester}}{[CH(COOC_2H_5)_2]^- Na^+} \longrightarrow \begin{matrix} CH_2-CH(COOC_2H_5)_2 \\ | \\ CH_2CH_2Br \end{matrix}$$

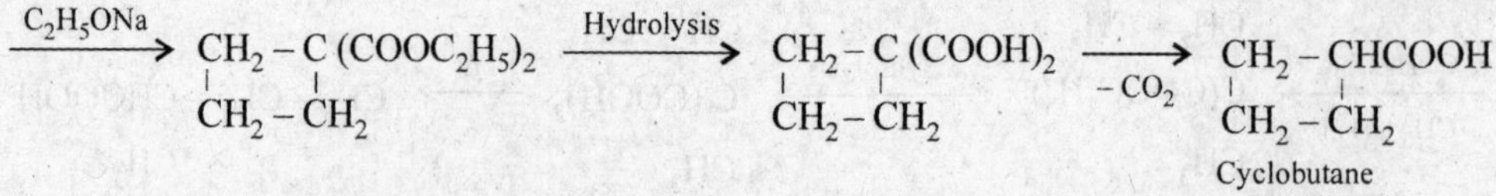

Cyclobutane carboxylic acid

6. Synthesis of α, β unsaturated acids. Malonic ester and its derivatives condense with aldehydes or ketones. Resulting unsaturated esters on hydrolysis and decarboxylation give α, β unsaturated acids.

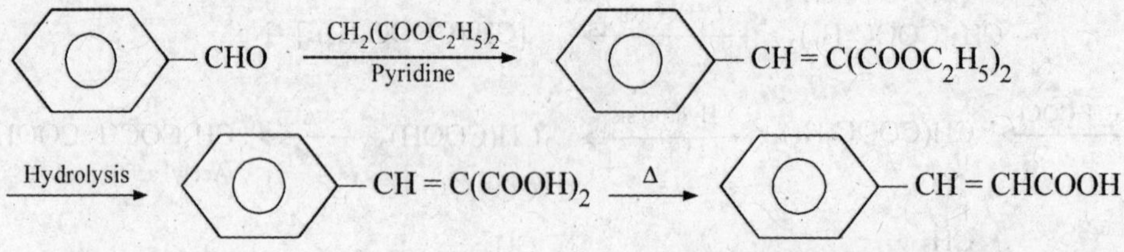

Cinnamic acid

7. Synthesis of amino acids. Malonic ester is treated with nitrous acid. α-oximino malonic ester formed is reduced to give aminomalonic ester. This on hydrolysis and decarboxylation gives amino acetic acid.

$$\underset{\text{Nitrous acid}}{HO-N=O} + \underset{\text{Malonic ester}}{H_2C\,(COOC_2H_5)_2} \longrightarrow \underset{\text{Oximino malonic ester}}{HO-N=C\,(COOC_2H_5)_2}$$

$$\xrightarrow{\text{Zn / Acetic acid}} \underset{\text{Amino malonic ester}}{H_2N-CH(COOC_2H_5)_2} \xrightarrow[-HCl]{CH_3COCl}$$

$$\underset{\text{N-acetyl amino malonic ester}}{CH_3CONH-CH\,(COOC_2H_5)_2} \xrightarrow{\text{Hydrolyis}} H_2N-CH(COOH)_2 \xrightarrow[-CO_2]{\text{Heat}} \underset{\text{Amino acetic acid}}{H_2N-CH_2COOH}$$

8. Synthesis of heterocyclic compounds. Malonic ester undergoes condensation with urea in the presence of sodium ethoxide to form malonyl urea commonly known as barbituric acid

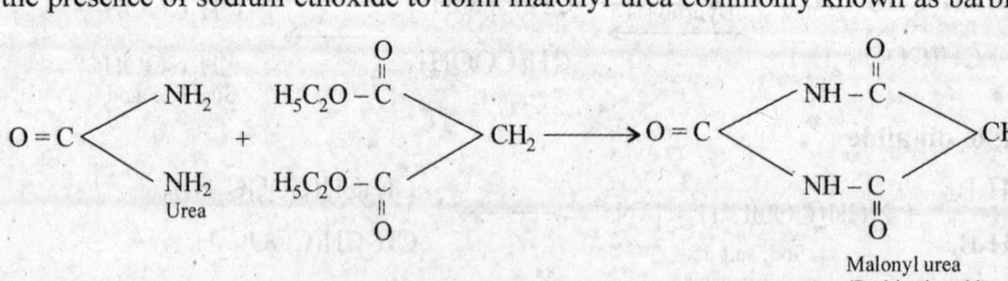

Malonyl urea
(Barbituric acid)

Substituted malonic esters react with urea to give substituted barbituric acids known as barbiturates. Such compounds are used medicinally as sleep-inducing drugs (hypnotics).

1, 3-Dithianes

1, 3-Dithiane is a non-aromatic heterocyclic compound containing two sulphur atoms at positions 1 and 3 in a ring containing only single bonds. That is why the name is *1, 3 dithi* (two sulphur atoms at positions 1 and 3) and *ane* saturated ring :

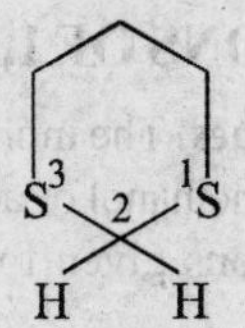

1,3-Dithiane

The hydrogen atoms of carbon 2 which is attached to two sulphur atoms are found to be more acidic ($pK_a \approx 31$) than hydrogens of usual alkyl groups. This is because the sulphur atoms help in stablising the negative charge of the anion formed by the deprotonation of 1, 3-dithiane. The deprotonation is generally carried out with alkylithium reagents.

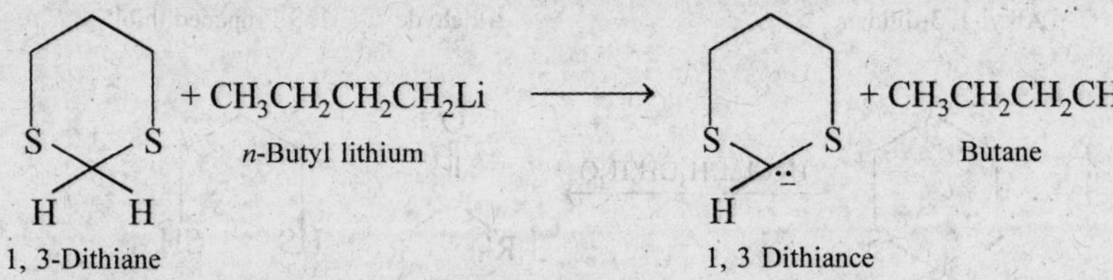

Preparation of 1, 3-dithiane

1, 3-Dithiane and its derivatives can be prepared by treated an aldehyde with 1, 3-Propanedithiol in the presence of a trace of an acid as illustrated below :

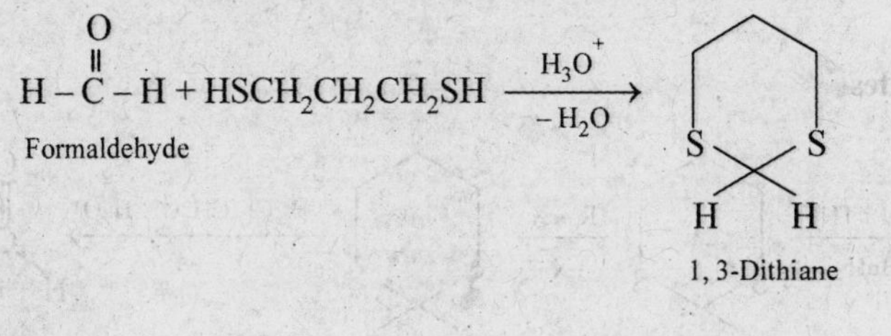

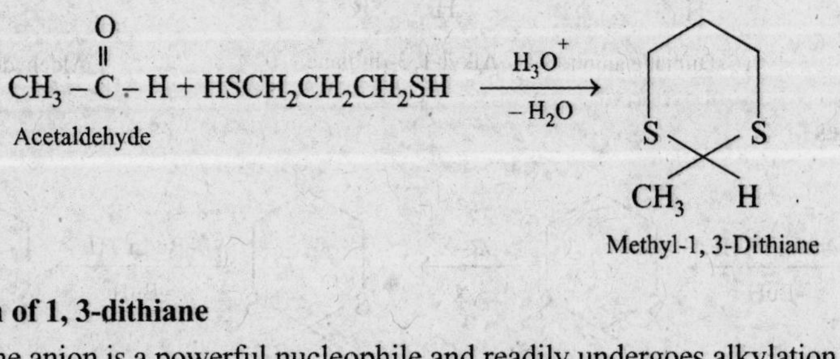

Alkylatioin of 1, 3-dithiane

1, 3-Dithiane anion is a powerful nucleophile and readily undergoes alkylation by S_N2 reaction with alkyl halides to form alkyl derivatives of 1, 3-dithiane.

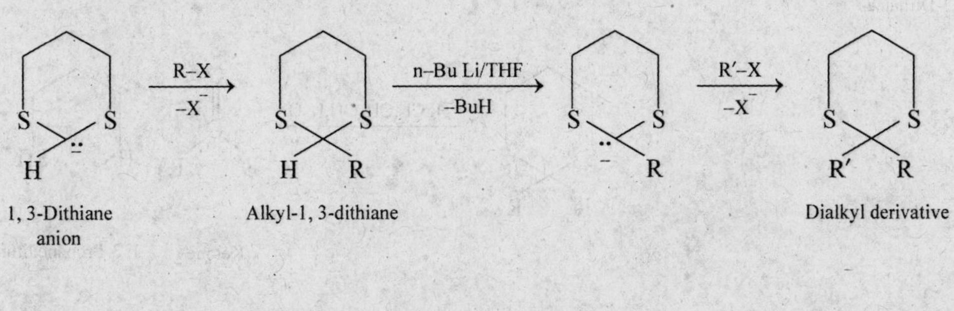

5.10. SYNTHETIC APPLICATIONS OF 1, 3-DITHIANE

1. Preparation of aldehydes and ketones. The mono and dialkyl derivatives of 1, 3-dithianes on hyrolysis with mercuric chloride ($HgCl_2$) in methanol or aqueous acetonitrile (CH_3CN) from aldehydes and ketones respectively, as per the equations given following:

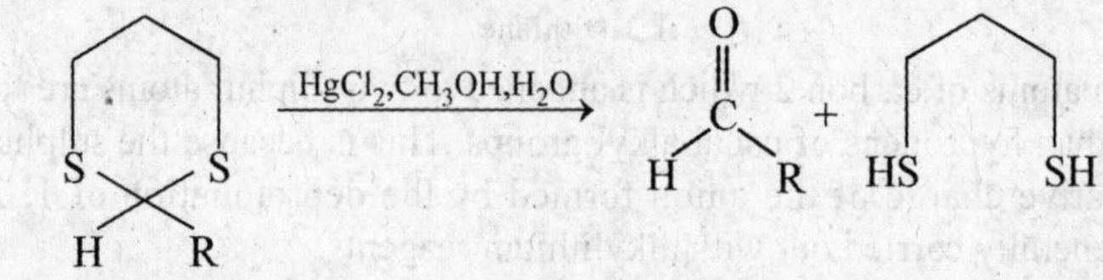

Alkyl-1, 3-dithiane   Aldehyde   1, 3-Propanedithiol

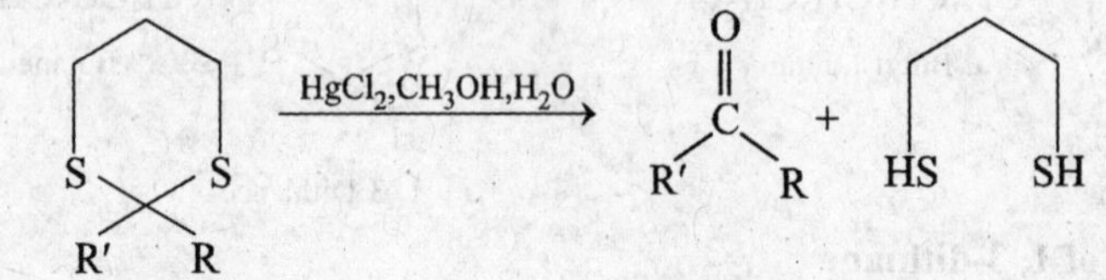

Dialkyl-1, 3-dithiane   Ketone   1, 3-Propanedithiol

Starting from 1, 3-Dithiane the complete sequence of reactions is shown below for aldehydes and ketones.

For aldehydes

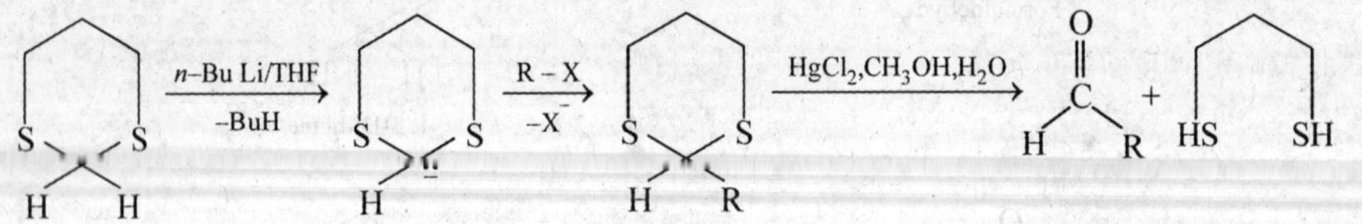

1, 3-Dithiane   1, 3-Dithiane anion   Alkyl-1, 3-dithiane   Aldehyde   1, 3-Propanedithiol

For ketones

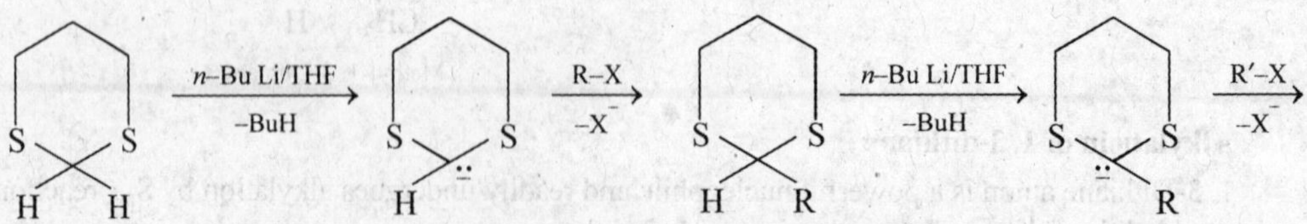

1, 3-Dithiane

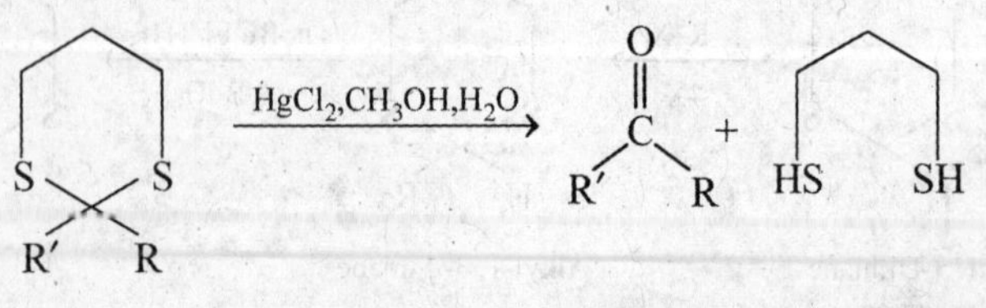

Ketone   1, 3-Propanedithiol

2. Conversion of aldehydes into ketones. Monoalkyl 1, 3-dithianes can be prepared by treating an aldehyde with 1, 3-propanedithiol in the presence of a trace of an acid. The dithiane thus formed on alkylation with alkyl halide followed by hydrolysis yields a ketone as follows:

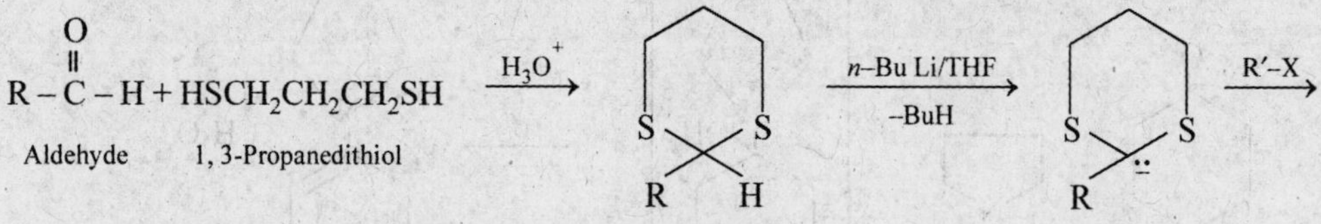

Alkyl-1, 3-diathiane
Alkyl-1, 3-dithiane

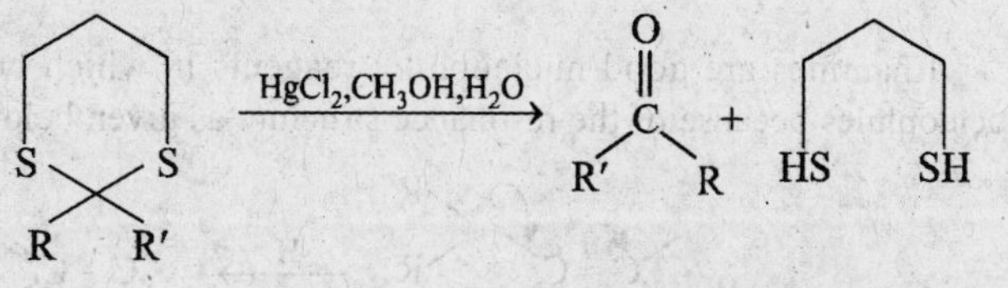

Ketone 1, 3-Propanedithiol

Enamines

Enamines are α, β-unsaturated amines which may be regarded as nitrogen analogs of enols.

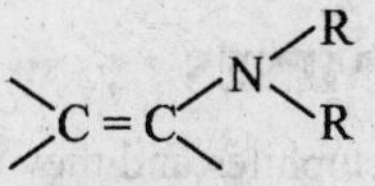

5.11. PREPARATION OF ENAMINES

They are formed by the reaction of an aldehyde or ketone with a secondary amine. The intermediate aminoalcohol initially formed undergoes acid-catalysed dehydration which drives the reversible reaction to completion.

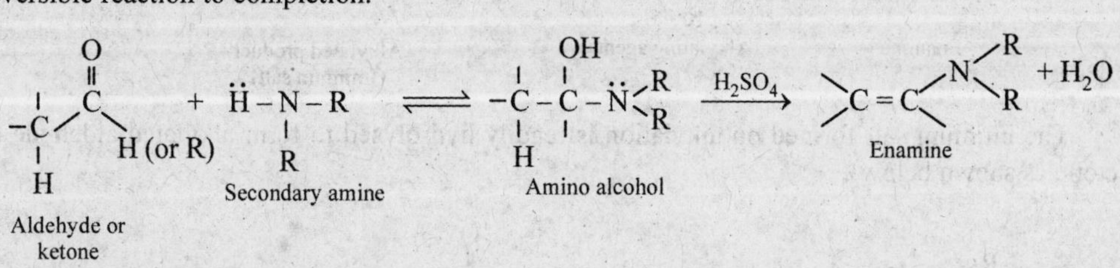

The secondary amines commonly used to prepare enamines are cyclic amines such as pyrrolidine, piperidine and morpholine. However, the reaction above has been illustrated using open-chain secondary amine (R NH R)

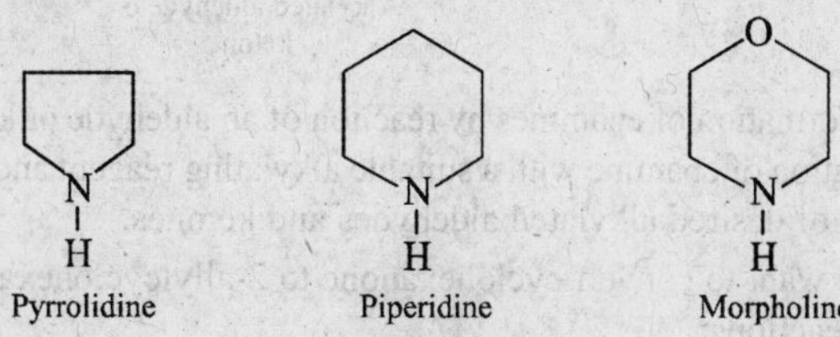

Pyrrolidine Piperidine Morpholine

For example cyclohexanone reacts with pyrrolidine to form the enamine N-(1-cyclohexenyl) pyrrolidine.

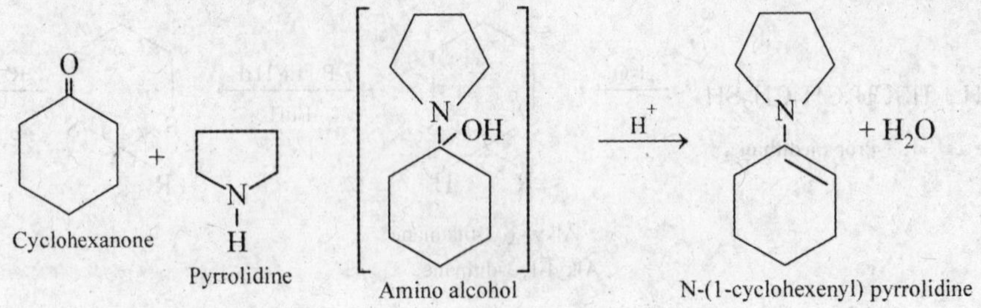

Enamines are good nucleophilic reagents in which both nitrogen and carbon can act as nucleophiles because of the resonance structure as given below :

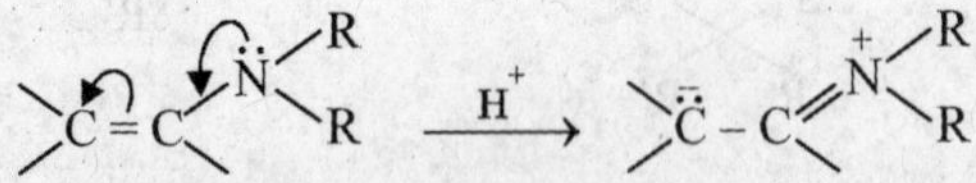

The nucleophilic nature of carbon makes enamines very useful synthetic reagents. They can be easily alkylated, acylated and also participate in many addition reactions. Unlike the basic conditions required for the generation of enolate anions, no base is required to use enamines as nucleophiles.

5.12. ALKYLATION OF ENAMINES

Enamines are not very strong nucelophiles and they undergo alkylation by S_N2 reaction only with methyl, primary alkyl and benzylic halides. For example :

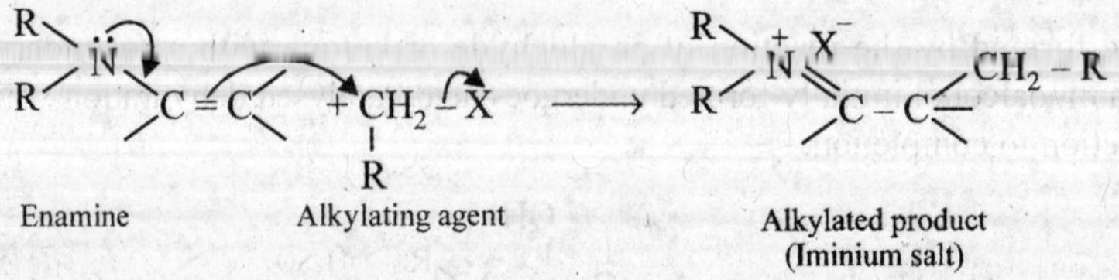

The iminium salt formed on alkylation is readily hydrolysed to form alkylated aldehyde or ketone as shown below :

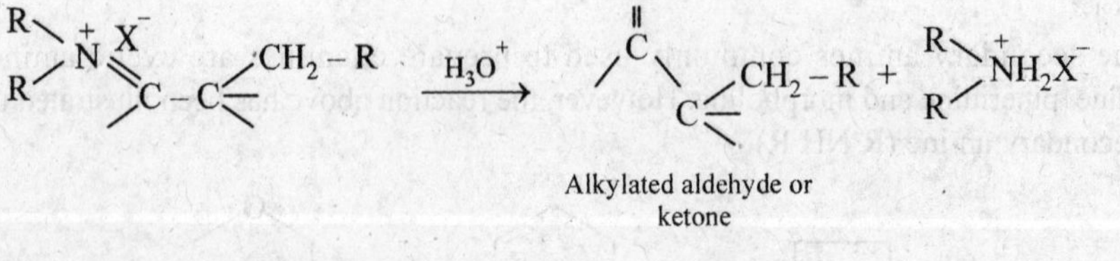

Synthetic utility. Formation of enamines by reaction of an aldehyde or ketone with a secondary amine followed by alkylation of enamine with a suitable alkylating reagent and subsequent hydrolysis is used for the synthesis of desired alkylated aldehydes and ketones.

For example, if we want to convert cyclohexanone to 2-allylcyclohexanone, we carry out the following sequence of reactions:

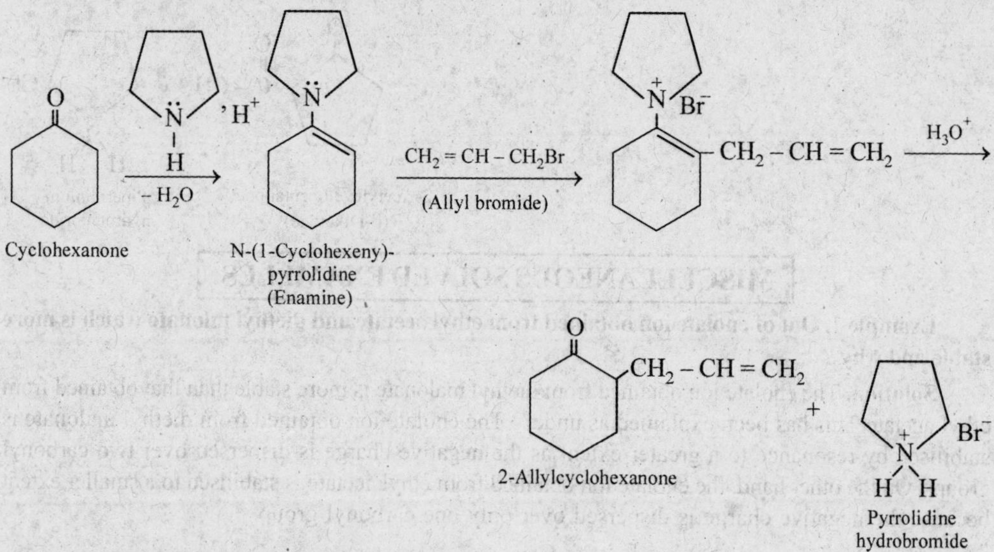

5.13. ACYLATION OF ENAMINES

Enamines undergo acylation on treatment with acid chlorides and acid anhydrides by nucleophilic acyl substitution as shown below :

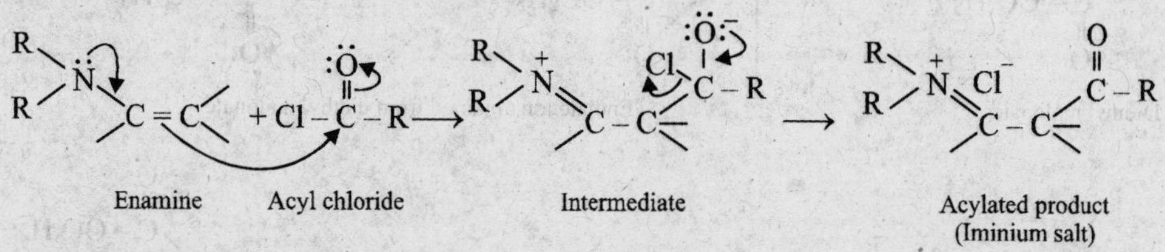

The iminium salt formed above can be hydrolysed to form dicarbonyl compounds as shown below :

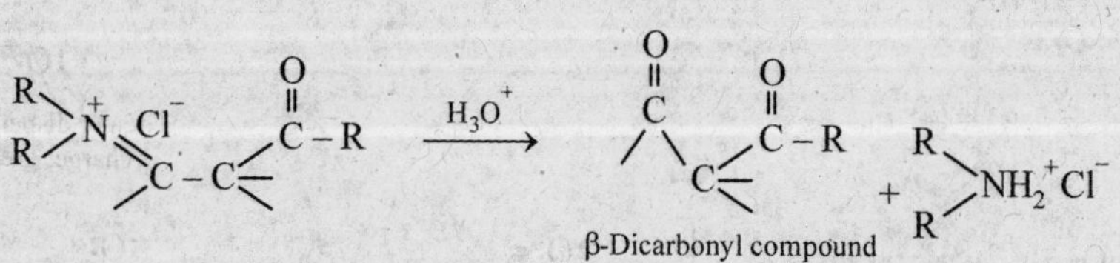

Synthetic utility. Acylation of enamines can be used in the synthesis of dicarbonyl compounds. For example, cyclopentanone can be converted into acetylcyclopentanone as shown below:

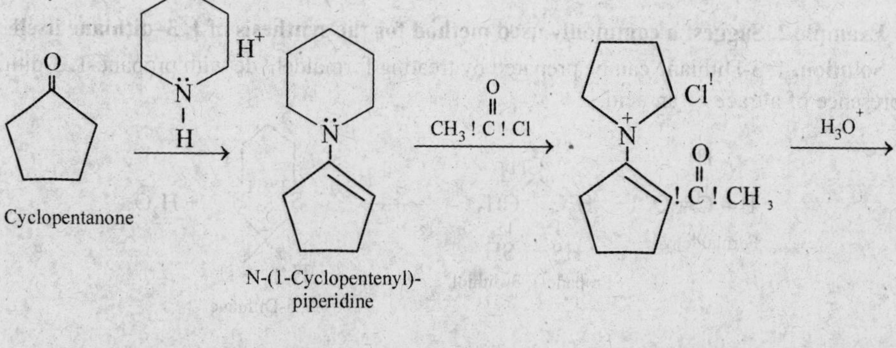

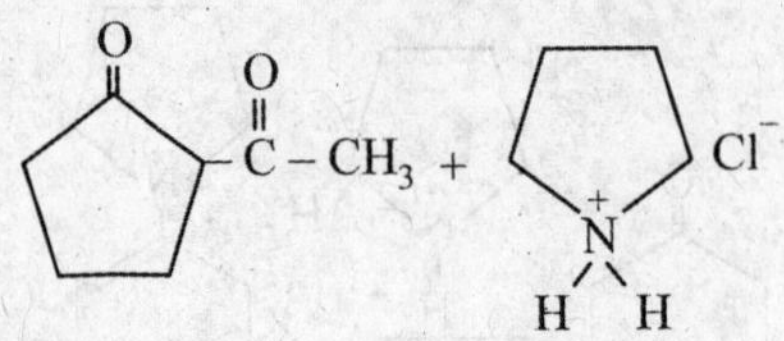

2-Acetylcyclopentanone (β-Diketone) Piperidinium hydrochloride

MISCELLANEOUS SOLVED EXAMPLES

Example 1. Out of enolate ion obtained from ethyl acetate and diethyl mlonate which is more stable and why ?

Solution. The enolate ion obtained from diethyl malonate is more stable than that obtained from ethyl acetate This has been explained as under : The enolate ion obtained from diethyl malonate is stabilised by resonance to a greater extent as the negative charge is dispersed over two carbonyl groups. On the other hand, the enolate ion obtained from ethyl acetate is stabilised to a smaller extent because the negative charge is dispersed over only one carbonyl group.

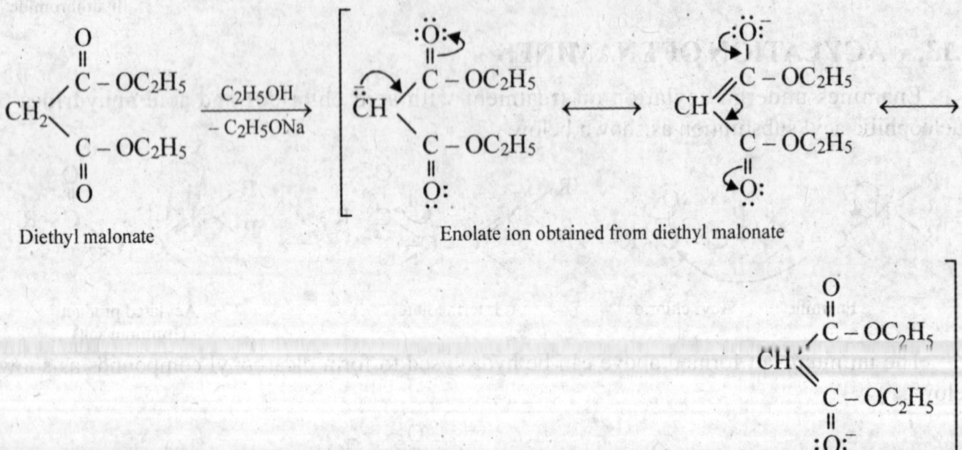

Diethyl malonate Enolate ion obtained from diethyl malonate

Greater dispersal of negative charge; greater stability

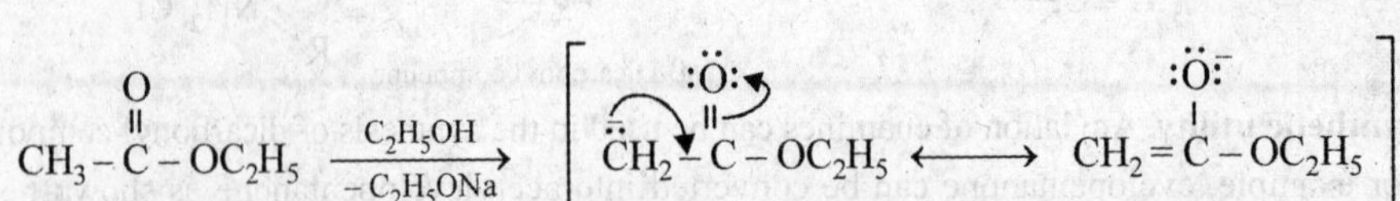

Ethyl acetate Smaller dispersal of negative charge; smaller stability

Example 2. Suggest a commonly used method for the synthesis of 1, 3–dithiane itself.

Solution. 1, 3-Dithiane can be prepared by treating formaldehyde with propane-1, 3-dithiol in the presence of a trace of an acid.

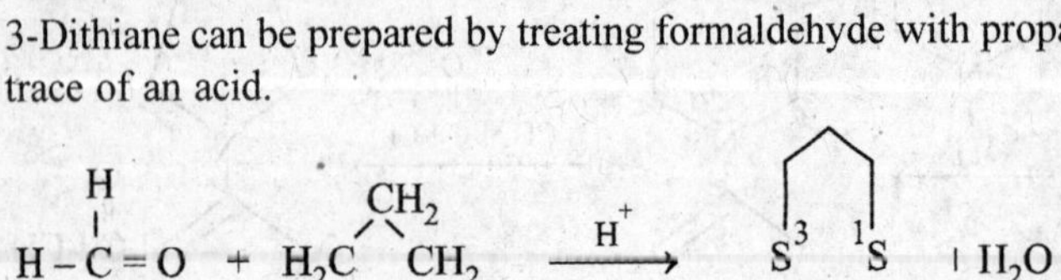

Formaldehyde Propane-1, 3-dithiol 1, 3-Dithiane

Example 3. Which of the following acids can be synthesised from diethyl malonate and which cannot be? Give suitable reasons?

(*i*) $(CH_3)_2CH-CH_2COOH$ (*ii*) $(CH_3)_3C-CH_2COOH$

(*iii*) $CH_3-CH=CHCH_2COOH$ (*iv*) $\begin{array}{l} CH_2-CHCOOH \\ \,|\qquad\;\; | \\ CH_2-CH_2 \end{array}$

Solution. $(CH_3)_2CH-CH_2COOH$ and $\begin{array}{l} CH_2-CHCOOH \\ \,|\qquad\;\; | \\ CH_2-CH_2 \end{array}$ can be synthesised from malonic ester.

Refer to relevant section in the text

$(CH_3)_3C-CH_2COOH$ cannot be synthesised because this would involve alkylation of enolate anion of malonic ester with a tertiary alkyl halide $[(CH_3)_3C-X]$. But tert. alkyl halide undergoes elimination rather than S_N2 reactions.

Also, $CH_3-CH=CHCOOH$ cannot be synthesised because it would involve the reaction between enolate anion and vinylic halide $(CH_3-CH=CHX)$. But vinly halides are highly unreactive and do not undergo SN^2 reactions.

Example 4. How will you convert benzaldehyde into acetophenone via a dithiane?

Solution. The conversion can be carried out as under :

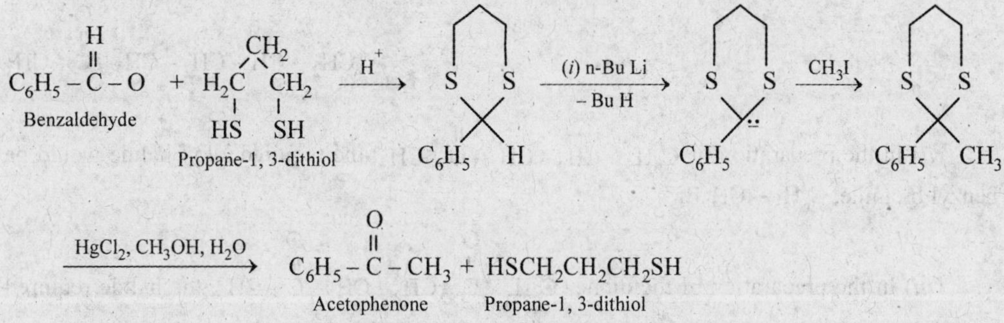

Example 5. Giving suitable explanation arrange the following in increasing order to enolisation :

$$CH_3COCH_2COOC_2H_5,\ \underset{\displaystyle CH_3}{CH_3COCHCOOC_2H_5},\ \underset{\displaystyle CN}{CH_3COCHCOOC_2H_5}$$

Solution. The increasing order of enolisation in the given compounds is :

$$\underset{\displaystyle CH_3}{CH_3COCHCOOC_2H_5} < CH_3COCH_2COOC_2H_5 < \underset{\displaystyle CN}{CH_3COCHCOOC_2H_5}$$

This can be explained by the effect of substituent groups on the negative charge of the anion. Since CH_3 – is an electron releasing group, it intensifies the negative charge of the anion and thus destabilises the anion. On the other hand, – CN being electron attracting halps in the dispersal of negative charge of the anion and thereby stabilises it.

Example 6. Which alkyl halide or other reagent in each case would you use for the alkylation of enolate anion in the synthesis of following ketones from ethyl acetoacetate ?

(*i*) $CH_2 = CH - CH_2 - CH_2 - \overset{\overset{O}{\|}}{C} - CH_3$ (*ii*) $C_6H_5 - CH_2 - CH_2 - \overset{\overset{O}{\|}}{C} - CH_3$

(*iii*) $C_6H_5 - \overset{\overset{O}{\|}}{C} - CH_2 - CH_2 - \overset{\overset{O}{\|}}{C} - CH_3$

Solution. (*i*) In the synthesis of methyl ketones from acetoacetic ester, $- CH_2 - \overset{\overset{O}{\|}}{C} - CH_3$ part is always derived from ethyl acetoacetate. The remaining part of the ketone comes from the alkyl halide. Accordingly, we can derive, which alkyl halide or other reagent will be required to obtain the ketone.

(*i*) For the preparation of $CH_2 = CH - CH_2 - CH_2 - \overset{\overset{O}{\|}}{C} - CH_3$, the required alkyl halide would be alkyl bromide $CH_2 = CH - CH_2Br$.

$$CH_2 = CH - CH_2Br + CH_3 - COCH_2COOC_2H_5 \xrightarrow[(iii)\ H^+/Heat]{(i)\ C_2H_5ONa,\ (ii)\ OH^-/H_2O_2} CH_2 = CH - CH_2 - CH_2 - \overset{\overset{O}{\|}}{C} - CH_3$$

(*ii*) In the preparation of $C_6H_5 - CH_2 - CH_2 - \overset{\overset{O}{\|}}{C} - CH_3$, the required alkyl halide would be benzyl bromide, $C_6H_5 - CH_2Br$.

(*iii*) In the preparation of the dione, $C_6H_5 - \overset{\overset{O}{\|}}{C} - CH_2 - CH_2 - \overset{\overset{O}{\|}}{C} - CH_3$, the halide required would be bromethyl phenyl ketone, $C_6H_5 - \overset{\overset{O}{\|}}{C} - CH_2Br$.

Example 7. Acetylacetone has $pK_a = 9$, while acetone has $pK_a = 20$. How do you justify the observations ?

Solution. $CH_3COCH_2\overset{\overset{O}{\|}}{C} - CH_3$ (Acetylacetone) $CH_3\,CO\,CH_3$ (Acetone)

The enolate formed by acetylacetone stabilises itself to a greater extent due to greater dispersal of negative charge as compared to the enolate of acetone. Thus dissociation of acatylacetone has a greater value of K_a or smaller value of pK_a.

Example 8. Sodium salt of ethyl malonate can be prepared readily by reaction with sodium ethoxide, whereas it is not possible in the case of ethyl acetate. Explain.

Solution.

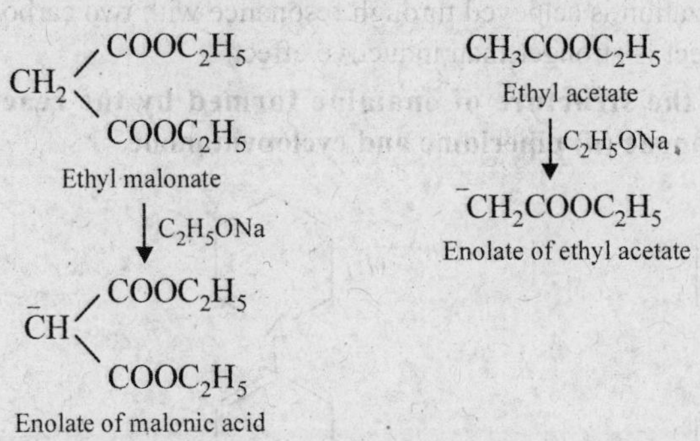

Enolate is the intermediate product in the reaction between ethyl malonate or ethyl acetate and sodium ethoxide. As the enolate of ethyl acetate cannot sufficiently stablise itself, sodium salt of the compound is not formed. On the contrary, enolate of ethyl malonate can sufficiently stabilise itself through two resonating structures. Hence sodium salt in this case is easily formed.

Example 9. Label α-hydrogen atoms in propanal. Explain why these hydrogens are more acidic than the hydrogens of propane.

Solution.

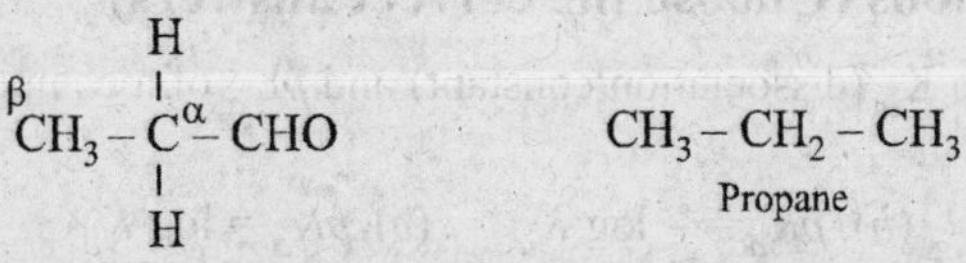

The hydrogens attached to carbon marked α are α-hydrogens. The anion obtained after the removal of H^+ stabilizes through resonance

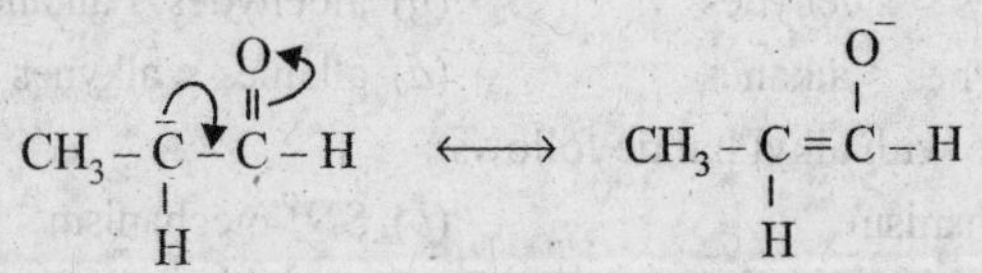

No such resonance stabilization is possible in propane.

Example 10. Arrange the following compounds in increasing order of their expected acidic strength and justify your answer :

Acetone, Chloroacetone, diethylmalonate

Solution. The acidic strength of a compound depends upon the stability of the anion obtained after removal of a proton from the compound. Greater the stability of the anion, greater is the acid strength of the compound. The anions of the three compounds are given below :

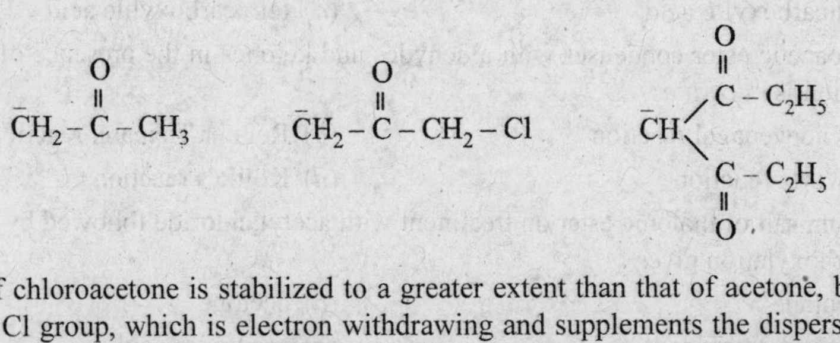

Anion of chloroacetone is stabilized to a greater extent than that of acetone, because of the presence of – Cl group, which is electron withdrawing and supplements the dispersal of negative charge. Anion of diethyl malonate stabilises itself to a still greater extent due to the presence of two

carbonyl group. The stabilization is achieved through resonance with two carbonyl groups. It may be noted that resonance effect is stronger than inductive effect.

Example 11. Draw the structure of enamine formed by the reaction between (*i*) morpholine and cyclohexanone (*ii*) piperidine and cyclopentanone.

Solution. (*i*)

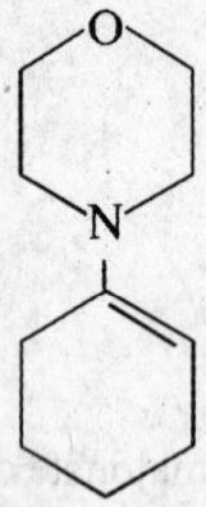

(*ii*)

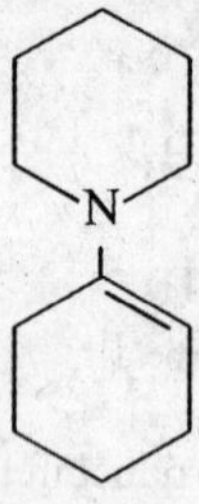

EXERCISES
(Including Questions from Different University Papers)

Multiple Choice Questions (Choose the correct answers)

1. The relation between K_a (dissociation constant) and pK_a of a compound is

(*a*) $K_a = \frac{1}{pK_a}$ (*b*) $pK_a = -\log K_a$ (*c*) $pK_a = \log K_a$ (*d*) None

2. pK_a values for aldehydes, alkynes and alkanes increase in the order :

(*a*) alkynes < alkanes < aldehydes (*b*) aldehydes < alkanes < alkynes

(*c*) aldehydes < alkynes < alkanes (*d*) alkanes < alkynes < aldehydes

3. Alkylation of enolate with alkyl halide follows :

(*a*) free radical mechanism (*b*) SN^1 mechanism

(*c*) elimination reaction (*d*) SN^2 mechanism

4. Ethyl acetoacetate can be prepared by

(*a*) Rosenmund's reaction (*b*)Claisen condensation

(*c*) Kolbe's electrolytic method (*d*) Grignard reagent

5. Reaction of sodium salt of acetoacetic ester with halogen derivative of an ester followed by acidic hydrolysis produces

(*a*) monocarboxylic acid (*b*) tricarboxylic acid

(*c*) dicarboxylic acid (*d*) tetracarboxylilc acid

6. Acetoacetic ester condenses with aldehydes and ketones in the presence of pyridine. This reaction is known as

(*a*) Knoevenagal reaction (*b*) Reimer Tiemann reaction

(*c*) Wurtz reaction (*d*) Kolbe's reaction

7. Sodium salt of malonic ester on treatment with acetyl chloride followed by hydrolysis and decarboxylation gives

(*a*) butane (*b*) acetone

(*c*) acetyl acetic acid (*d*) propionic acid

8. Barbituric acid is obtained by the condensation in the presence of sod. ethoxide between malonic ester and
   (*a*) oxalic acid (*b*) urea (*c*) thiourea (*d*) ethanethiol
9. 1, 3-Dithiane reacts with *n*-butyl lithium to produce
   (*a*) ethane (*b*) pentane (*c*) propane (*d*) butane
10. 1, 3-Dithiane can be synthesised by the reaction between propane-1, 3-dithiol and
   (*a*) formaldehyde (*b*) acetaldehyde (*c*) formic acid (*d*) acetic acid

ANSWERS

1. (*b*)	**2.** (*c*)	**3.** (*d*)	**4.** (*b*)	**5.** (*c*)
6. (*a*)	**7.** (*c*)	**8.** (*b*)	**9.** (*d*)	**10.** (*a*)

Short Answer Questions

1. What are enolates ? Give two examples alongwith their structures.
2. Draw the structures of enolate ions obtained from malonic ester and acetoacetic ester respectively.
3. How can you prepare butanoic acid starting from malonic acid ?
4. Arrange the following in increasing order of enol contents and offer explanation.

$$CH_3COCH(R)COOC_2H_5 \qquad R = H\ ;\ CH_3\ ;\ CN$$

5. A hydrogen α- to two carbonyl groups is found to be much more acidic than a hydrogen α- to only one carbonyl goup. Explain.
6. Mark the most acidic hydrogens in each of the following structures :

(*i*) $C_6H_5-CH(CH_3)COOC_2H_5$ (*ii*) $CH_3-CH_2-C(=O)-CH_2-CN$

(*iii*) $CH_3-CH_2-C(=O)-CH_2COOH$

7. What is dithiane ? How can it be converted in its monoalkyl derivative ?
8. What are enamines ? What is their nucleophilic character due to ?
9. How will you convert malonic ester into its *n*-propyl derivative ? Give equations for the reactions involved.
10. How will you prepare ethyl acetoacetate in the laboratory ? What is the reaction known as ?
11. How will you explain the acidic nature of methylenic hydrogens in diethylmalonate ?
12. Describe keto-enol tautomerism of ethyl acetoacetate.
13. Write the synthesis of 2-methylpentanoic acid and *n*-butyric acid using diethylmalonate.
14. What are 1, 3-dithianes ? How can the formation of 1, 3-dithianes be used for converting acetaldehyde into butan-2-one ?
15. Starting with diethyl malonate, give a reaction scheme for synthesising each of the following :
   (*i*) Succinic acid (*ii*) Ketovaleric acid
   (*iii*) Cyclopropane carboxylic acid.
   Give the reagent used and the conditions employed in each case.

16. Describe the mechanism of formation of ethyl acetoacetate from ethyl acetate and sodium ethoxide. Would you expect a similar reaction to take place with ethyl isobutyrate, $(CH_3)_2CHCOOC_2H_5$?

17. (*a*) How is acetoacetic ester prepared ?

(*b*) Using suitable reagents how the following compounds are prepared from acetoacetic ester :

(*i*) 5-Methyl-2-hexanone (*ii*) 3-Methyl-2-hexanone

18. Describe the mechanism of formation of ethyl acetoacetate from ethyl acetate and sodium ethoxide.

19. Give the reaction and mechanism of synthesis of ethylacetoacetate.

20. What do you mean by (*i*) acid hydrolysis and (*ii*) ketonic hydrolysis of acetoacetic ester and its derivatives ?

21. Explain the acidity of α–hydrogens in ethyl malonate.

22. How can you convert cyclohexanone into 2-hexanoyl cyclohexanone through the intermediate formation of an enamine ?

23. How can diethyl malonate be used for the synthesis of following compounds ?

(*i*) $CH_3-\underset{|}{\overset{CH_3}{CH}}-CH_2-COOH$ (*ii*) $CH_3-\overset{O}{\overset{\|}{C}}-CH_2COOH$

24. Write detailed note on keto-enol tautomerism of acetoacetic ester.

25. How can you synthesis the following compound starting with cyclopentanone and morpholine and employing any other reagents of your choice ?

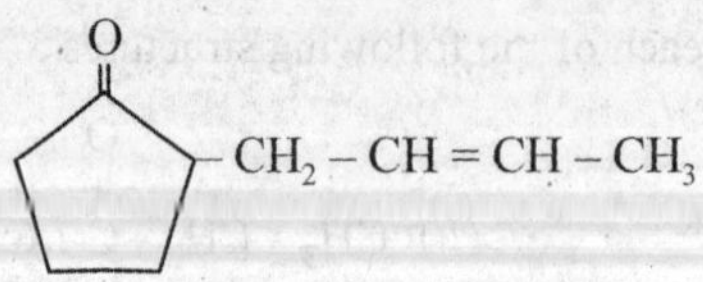

26. Write the Claisen condensation reaction to prepare ethyl acetoacetate. Give its mechanism.

27. How can ethylacetoacetate be used for the synthesis of following compounds ?

(*i*) $CH_3-\overset{O}{\overset{\|}{C}}-CH_2-CH_2-\overset{O}{\overset{\|}{C}}-CH_3$ (*ii*) $CH_3-\overset{O}{\overset{\|}{C}}-CH\begin{matrix}CH_3\\CH_3\end{matrix}$

28. What are 1, 3-dithianes ? How can the formation of 1, 3-dithianes be used for converting acetaldehyde into 2-butanone ?

29. Discuss one method for determining the composition of tautomeric mixture of ethylacetoacetate.

General Questions

1. How do you explain the following :

(*i*) Acidic nature of methylenic hydrogens in ethyl acetoacetate.

(*ii*) Role of acid and base catalysis in keto-enol tautomerism.

(*iii*) The enol content of acetyl acetone is much higher than that of acetone.

2. What are enamines ? How can they be formed ? What is their synthetic utility ?

3. (*a*) How can ethyl acetoacetate be prepared ?
   (*b*) Give the procedure by which the following compounds may be obtained from ethyl acetoacetate :
   (*i*) Methyl ethyl ketone (*ii*) Acetyl acetone
   (*iii*) 2, 3-Dimethylbutyric acid (*iv*) Succinic acid
4. Using the formation of enamines as one of the steps, how will you synthesise the following :

(*i*) 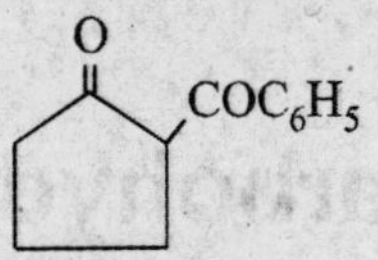
(*ii*) 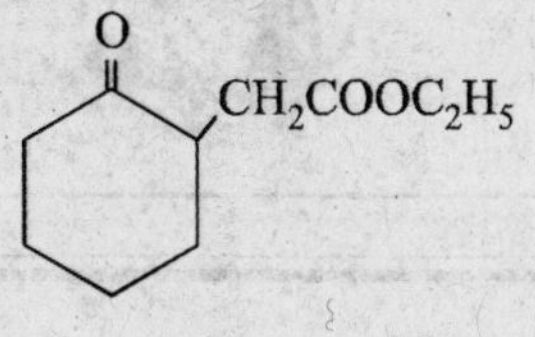

(*iii*) $CH_2=CH-CH_2-C(CH_3)_2-CHO$
(*iv*) 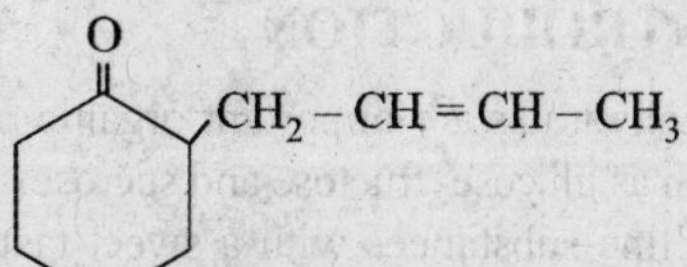

5. Using the formation of enamines as one of the steps, how will you synthesise the following :

(*i*) 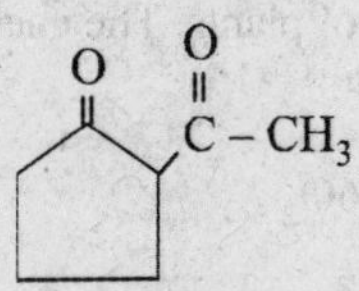
(*ii*) 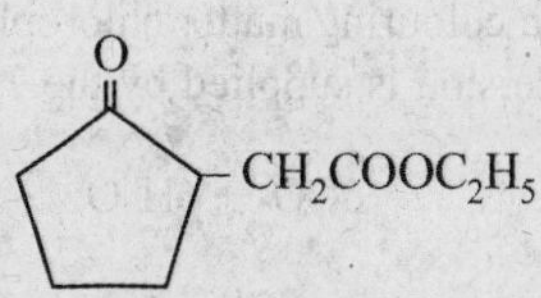

6. How will you carry out ?
   (*i*) Synthesis of 2-ethylpentan-2-one from ethyl acetoacetate.
   (*ii*) Synthesis of 3-pentanone from propanal through the formation of a 1, 3-dithiane.
   (*iii*) Synthesis of cyclobutane carboxylic acid by malonic ester synthesis.
   (*iv*) Synthesis of *n*-valeric acid by malonic ester or acetoacetic ester synthesis.
7. Describe the laboratory preparation of acetoacetic ester. Write the mechanism of the reaction. How acetoacetic ester can be converted to :
   (*i*) Acetic acid (*ii*) Succinic acid
   (*iii*) Crotonic acid (*iv*) 4-Methyluracil ?
8. (*a*) What is Claisen condensation ? Discuss mechanism of the reaction ?
   (*b*) Of the enolate ions obtained from acetone and diethyl malonate which is more stable and why ?

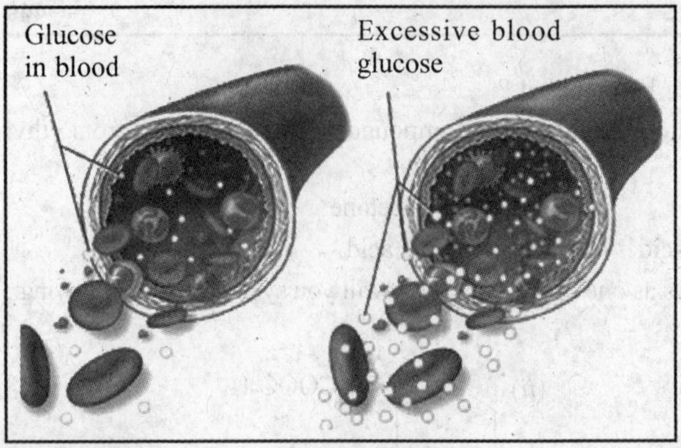

6

Carbohydrates

6.1. INTRODUCTION

Carbohydrates are important organic compounds widely distributed in nature. These include sugars such as glucose, fructose and sucrose as well as non-sugars such as starch and cellulose. Sugars are crystalline substances with a sweet taste and are soluble in water while non-sugars are non-crystalline substances which are not sweet and are insoluble or less soluble in water.

In the green plants, carbohydrates are produced by a process called **photosynthesis.** This process involves the conversion of simple compounds carbon dioxide and water into glucose and is catalysed by green colouring matter chlorophyll present in the leaves of plants. The energy required to effect the conversion is supplied by the sun in the form of sunlight.

$$6CO_2 + 6H_2O \xrightarrow[\text{Chlorophyll}]{\text{Sunlight}} C_6H_{12}O_6 + 6O_2$$

Thousands of glucose molecules can then be combined to form much larger molecules of cellulose which constitutes the supporting framework of the plants.

Carbohydrates are very useful for human beings. They provide us all the three basic necessities of life *i.e., food* (starch containing grain), *clothes* (cellulose in the form of cotton, linen and rayon) and *shelter* (cellulose in the form of wood used for making our houses and furniture etc.). Not only this, our present civilisation depends on cellulose to a surprising degree, particularly in the form of paper. Carbohydrates are also important to the economy of many nations. For example, sugar is one of the most important commercial commodities. Many things of daily use, such as paper, photographic films, plastics, etc. are derived from carbohydrates.

Carbohydrates are so called, because the general formula for most of them could be written as $C_x(H_2O)_y$ and thus they may be regarded as hydrates of carbon. However, this definition was not found to be correct *e.g.,* rhamnose, a carbohydrate, is having the formula $C_6H_{12}O_5$ while acetic acid having formula $C_2H_4O_2$ is not a carbohydrate.

Chemically, carbohydrates contain mainly two functional groups, carbonyl group (aldehyde or ketone) and a number of hydroxyl groups. Accordingly carbohydrates are now defined as *polyhydroxy aldehydes or polyhydroxy ketones or the compound that can be hydrolyzed to either of them.*

The carbonyl group, however does not occur as such but is combined with hydroxyl groups to form intramolecular hemiacetal or acetal linkages. All the carbohydrates contain more than one asymmetric carbon atoms and are optically active.

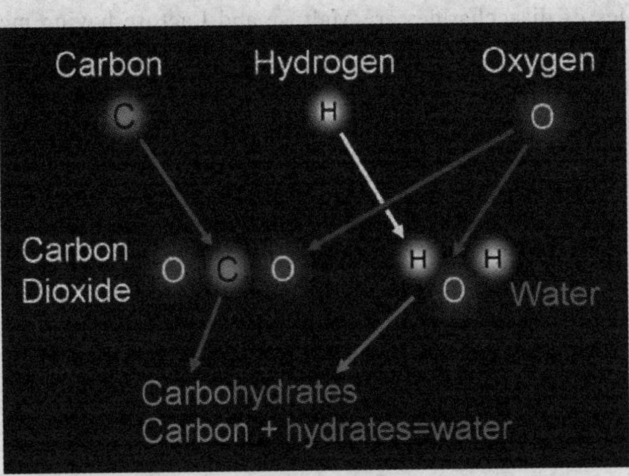

Carbohydrates

6.2. CLASSIFICATION OF CARBOHYDRATES

Carbohydrates are broadly classified into the following types:

(*a*) Monosaccharides. These are the simplest carbohydrates which cannot be hydrolyzed further into smaller units. They contain four, five or six carbon atoms and are known as *tetroses*, *pentoses* or *hexoses* respectively. Depending upon whether they contain an *aldehyde* or *keto* groups, they may be called aldoses or ketoses. For example, a five carbon monosaccharide having aldehyde group is called aldopentose and a six-carbon monosaccharide containing a keto group is called keto-hexose. A few examples of monosaccharides are given below:

Aldotetroses. Erythrose and Threose; $CH_2OH(CHOH)_2$ CHO.

Ketotetroses. Erythrulose, $CH_2OHCOCHOHCH_2OH$.

Aldopentoses. Ribose, Arabinose, Xylose and Lyxose.

$$CH_2OH(CHOH)_3CHO$$

All have a common molecular formula but different structures.

Ketopentoses. Ribulose and xylulose;

$$CH_2OHCO(CHOH)_2CH_2OH$$

Aldohexoses. Glucose, Mannose, Galactose etc;

$$CH_2OH(CHOH)_4CHO$$

Ketohexoses. Fructose, Sorbose etc.

$$CH_2OHCO(CHOH)_3CH_2OH$$

(*b*) Oligosaccharides. These are the carbohydrates which can be hydrolysed into a definite number of monosaccharide molecules. Depending upon the number of monosaccharides that are obtained from them on hydrolysis, they may be called *di-, tri- or tetra-*saccharides. For example:

(*i*) Disaccharides. These give two molecules of monosaccharides on hydrolysis. For example:

$$\underset{\substack{\text{Sucrose} \\ \text{(A disaccharide)}}}{C_{12}H_{22}O_{11}} + H_2O \longrightarrow \underset{\text{Glucose}}{C_6H_{12}O_6} + \underset{\text{Fructose}}{C_6H_{12}O_6}$$

Other examples of disaccharides are Maltose and Lactose having molecular formula, $C_{12}H_{22}O_{11}$.

***(ii)* Trisaccharides.** These give three molecules of monosacharides on hydrolysis. For example:

$$\underset{\substack{\text{Raffinose}\\\text{(A trisaccharide)}}}{C_{18}H_{32}O_{16}} + 2H_2O \longrightarrow \underset{\text{Galactose}}{C_6H_{12}O_6} + \underset{\text{Glucose}}{C_6H_{12}O_6} + \underset{\text{Fructose}}{C_6H_{12}O_6}$$

***(c)* Polysaccharides.** These are high molecular weight carbohydrates yielding a large number of monosaccharide molecules (hundreds or even thousands) on hydrolysis, *e.g.*, Starch and cellulose have molecular formula $(C_6H_{10}O_5)_n$, where is very large number.

$$\underset{\substack{\text{Starch}\\\text{(A polysaccharide)}}}{(C_6H_{10}O_5)_n} + nH_2O \longrightarrow \underset{\text{Glucose}}{nC_6H_{12}O_6}$$

6.3. NOMENCLATURE OF CARBOHYDRATES

Carbohydrates contain hydroxy and aldehydic or ketone groups. They are named according to IUPAC system of nomenclature.

For example

Compound	*Common Name*	*IUPAC Name*
$CH_2OH - CHOH - CHO$	Glyceraldehyde	2, 3-Dihydroxy propanal
$CH_2OH - CO - CH_2OH$	Dihydroxyacetone	1, 3-Dihydroxy propanone
$CH_2OH\ (CHOH)_4\ CHO$	Glucose	2, 3, 4, 5, 6-Pentahydroxyhexanal
$CH_2OH\ (CHOH)_3COCH_2OH$	Fructose	1, 3, 4, 5, 6-Pentahydroxyhexan-2-one

Carbohydrates being among the oldest known compounds continue to be called by their common name. The IUPAC names are hardly referred to in the study of carbohydrates.

6.4. GLUCOSE

Occurrence

Glucose is the most important and abundant sugar and occurs in honey, sweet fruits (ripe grapes and mangoes), blood and urine of animals. It is a normal constituent of blood (0.1%, also known as blood sugar) and is present in the urine (8–10%) of a diabetic person. In the combined form, it is a constituent of many disaccharides such as sucrose and polysaccharides such as starch, cellulose, etc.

Physical Properties of glucose

Some important physical properties of glucose are mentioned as under :

1. It is a colourless sweet crystalline compound having m.p. 419 K.
2. It is readily soluble in water, sparingly soluble in alcohol and insoluble in ether.
3. It forms a monohydrate having m.p. 391 K.
4. It is optically active and its solution is dextrorotatory (hence the name *dextrose*). The specific rotation of fresh solution is + 112°C.
5. It is about three-fourth as sweet as sugarcane *i.e.*, sucrose.

Chemical properties of glucose

Chemical properties of glucose can be studied under the following headings :

(A) Reactions of the aldehydic group

1. Oxidation. Glucose gets oxidised to gluconic acid with mild oxidising agents like bromine water

$$\underset{\text{Glucose}}{CH_2OH\ (CHOH)_4CHO} \xrightarrow[Br_2]{[O]} \underset{\text{Gluconic acid}}{CH_2OH(CHOH)_4COOH}$$

only –CHO group is affected.

(*b*) A strong oxidising agent like nitric acid oxidises both the terminal groups viz. $-CH_2OH$ and –CHO groups and saccharic acid or glucaric acid is obtained.

$$\underset{\text{Glu cos e}}{CH_2OH(CHOH)_4CHO} \xrightarrow[[O]]{HNO_3} \underset{\text{Saccharic or glucaricacid}}{COOH(CHOH)_4\ COOH}$$

(*c*) Glucose gets oxidised to gluconic acid with ammoniacal silver nitrate (*Tollen's reagent*) and alkaline copper sulphate (*Fehling solution*). Tollen's reagent is reduced to metallic silver and Fehling solution to cuprous oxide which is a red precipitate.

(*i*) **With Tollen's reagent**

$$AgNO_3 + NH_4OH \longrightarrow AgOH + NH_4NO_3$$

$$2AgOH \longrightarrow Ag_2O + H_2O$$

$$\underset{\text{Glu cos e}}{CH_2OH(CHOH)_4CHO} + Ag_2O \longrightarrow \underset{\text{Gluconic acid}}{CH_2OH(CHOH)_4COOH} + \underset{\text{Silver mirror}}{2Ag\downarrow}$$

(*ii*) **With Fehling solution**

$$CuSO_4 + 2NaOH \longrightarrow Cu(OH)_2 + Na_2SO_4$$

$$Cu(OH)_2 \longrightarrow CuO + H_2O$$

$$\underset{\text{Glucose}}{CH_2OH(CHOH)_4CHO} + 2CuO \longrightarrow \underset{\text{Gluconic acid}}{CH_2OH(CHOH_4COOH} + \underset{\text{Red ppt}}{Cu_2O}.$$

2. Reduction. (*a*) Glucose is reduced to sorbitol on treatment with sodium amalgam and water.

$$\underset{\text{Glucose}}{CH_2OH(CHOH)_4CHO} + 2[H] \xrightarrow[\text{Water}]{Na/Hg} \underset{\text{Sorbitol}}{CH_2OH(CHOH)_4CH_2OH}$$

(*b*) On reduction with conc. HI and red phosphorus at 373 K glucose gives a mixture of *n*-hexane and 2-iodohexane.

$$\underset{\text{Glucose}}{CH_2OH(CHOH)_4CHO} \xrightarrow{HI/P} \underset{n\text{–hexane}}{CH_3(CH_2)_4CH_3} + \underset{\text{2–Iodohexane}}{CH_3(CH_2)_3CHICH_3}$$

3. Reaction with HCN. An addition compound glucose cyanohydrin is obtained which on hydrolysis gives a hydroxy acid. On treatment with HI, hydroxy acid gives heptanoic acid.

$$CH_2OH(CHOH)_4CHO + HCN \longrightarrow \underset{\text{Glucose cyanohydrin}}{CH_2OH(CHOH)_4CH\begin{matrix} OH \\ CN \end{matrix}}$$

$$\downarrow \text{Hydrolysis}$$

$$\underset{\text{Heptanoic acid}}{CH_3(CH_2)_5COOH} \xleftarrow{HI} \underset{\text{Hydroxy acid}}{CH_2OH(CHOH)_4\underset{\substack{| \\ OH}}{CH} - COOH}$$

4. Reaction with hydroxyl amine. Glucose forms glucoseoxime

$$\underset{\text{Glucose}}{CH_2OH(CHOH)_4CH}\boxed{O + H_2}\underset{\text{Hydroxyl amine}}{NOH} \longrightarrow \underset{\text{Glucose oxime}}{CH_2OH(CHOH)_4CH = NOH} + H_2O$$

5. Reaction with phenyl hydrazine. Glucose reacts with phenyl hydrazine in stages as shown below:

(*i*) One molecule of phenyl hydrazine reacts with one molecule of glucose to form glucose phenyl hydrazone.

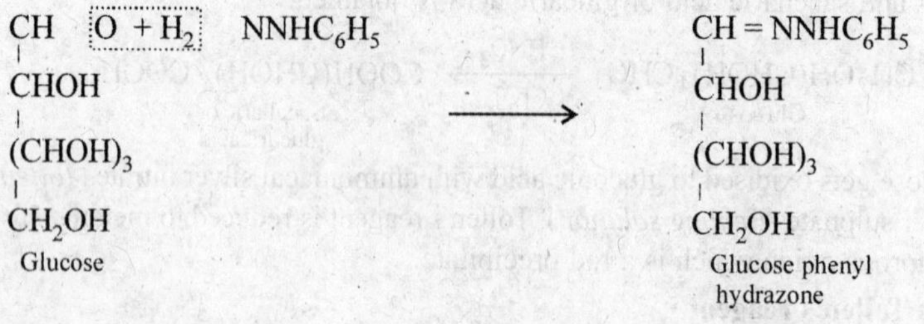

(*ii*) The product obtained above reacts further with a molecule of phenyl hydrazine as under:

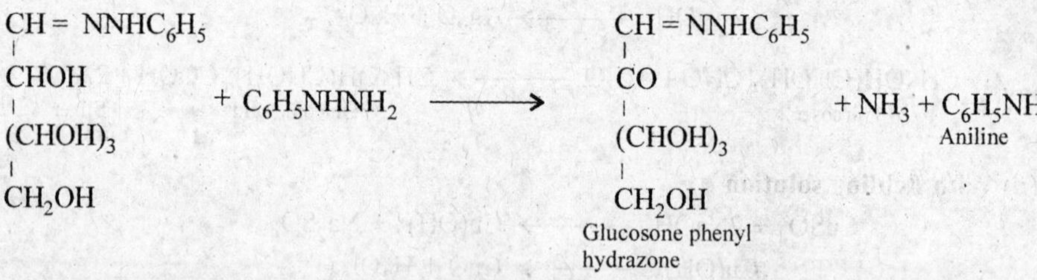

(*iii*) The product in step (*ii*) reacts further with a molecule of phenyl hydrazine to give glucosazone, a yellow solid having a sharp melting point.

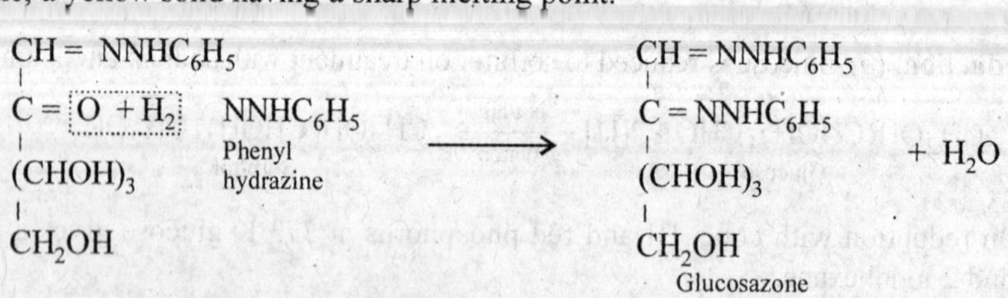

The above steps to explain the formation of glucosazone were given by Fischer.

Lately, however, the above mechanism of glucosazone formation has been superseded by the mechanism of Weygand. It is now explained in terms of *Amadori rearrangement* explained below:

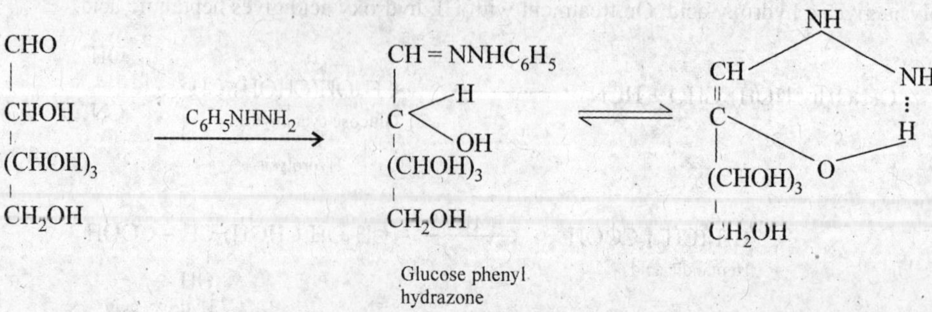

$$\xrightarrow{-C_6H_5NH_2} \begin{array}{c} CH=NH \\ | \\ C=O \\ | \\ (CHOH)_3 \\ | \\ CH_2OH \end{array} \xrightarrow[(-H_2O)]{C_6H_5NHNH_2} \begin{array}{c} CH=NH \\ | \\ C=NNHC_6H_5 \\ | \\ (CHOH)_3 \\ | \\ CH_2OH \end{array} \xrightarrow{C_6H_5NHNH_2} \begin{array}{c} CH=NNHC_6H_5 \\ | \\ C=NNHC_6H_5 \\ | \\ (CHOH)_3 \\ | \\ CH_2OH \end{array} + NH_3$$

Iminoketone Glucosazone

(B) Reactions of the hydroxyl groups

1. Reaction with acetic anhydride or acetyl chloride. Glucose forms penta-acetate with acetic anhydride

$$\begin{array}{c} CHO \\ | \\ (CHOH)_4 \\ | \\ CH_2OH \end{array} + \underset{\text{Acetic anhydride}}{5\,(CH_3CO)_2O} \longrightarrow \begin{array}{c} CHO \\ | \\ (CHOCOCH_3)_4 \\ | \\ CH_2OCOCH_3 \end{array} + 5CH_3COOH$$

Glucose Glucose penta-acetate

2. Reaction with methyl alcohol. Glucose reacts with methyl alcohol in the presence of dry HCl gas to form methyl glucoside.

$$\underset{\text{Glucose}}{C_6H_{11}O_5}\,\boxed{OH + H}\,\underset{\text{Methyl accohol}}{OCH_3} \xrightarrow[\text{Gas}]{\text{HCl}} \underset{\text{Methyl glycoside}}{C_6H_{11}O_5OCH_3} + H_2O$$

3. Reaction with metallic hydroxides. Glucose reacts with calcium hydroxide to form calcium glucosate which is water-soluble.

$$\underset{\text{Glucose}}{C_6H_{11}O_5}\,\boxed{OH + H}\,\underset{\text{Cal.hydroxide}}{OCaOH} \longrightarrow \underset{\text{Cal.glucosate}}{C_6H_{11}O_5OCaOH} + H_2O$$

(C) Miscellaneous Reactions

1. Action of acids. On being heated with conc. HCl, glucose forms 5-hydroxy methyl furfural, which on further reaction gives laevulinic acid.

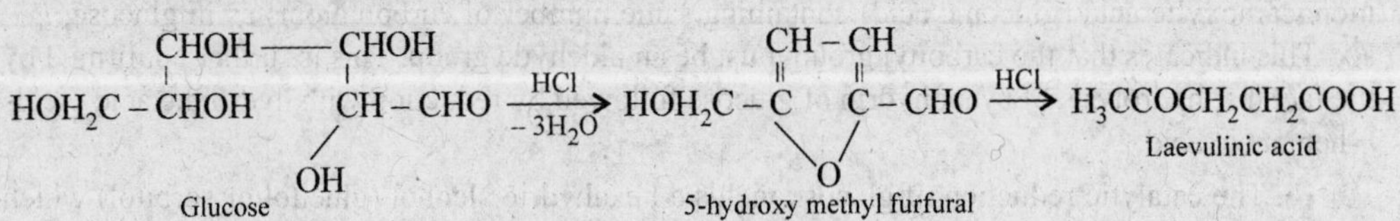

2. Fermentation. Glucose undergoes fermentation into ethyl alcohol in the presence of the enzyme *zymase.*

$$\underset{\text{Glucose}}{C_6H_{12}O_6} \xrightarrow{\text{Zymase}} \underset{\text{Ethyl alcohol}}{2C_2H_5OH} + 2CO_2$$

3. Reaction with alkalis. (*i*) Glucose is turned into brown resins in the presence of conc. alkalis.

(*ii*) In the presence of dilute solutions of alkalis, *Lobry de Bruyn-Van Ekenstein rearrangement* takes place which produces a mixture of (+) glucose, (+) mannose and (–) fructose. The reaction proceeds *via* 1, 2 enolisation.

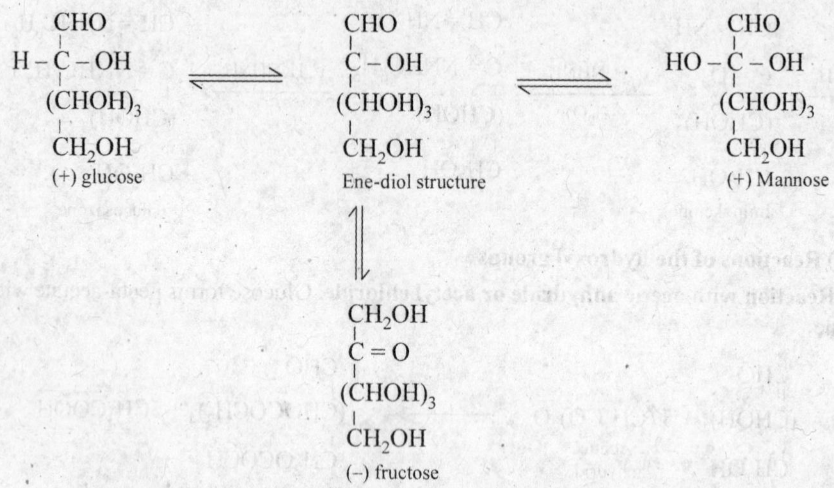

6.5. STRUCTURE OF GLUCOSE

1. On the basis of elemental analysis and molecular weight determination the molecular formula of glucose is $C_6H_{12}O_6$.

2. The reduction of glucose with red phosphorus and HI gives *n*-hexane.

$$\underset{\text{Glucose}}{C_6H_{12}O_5} \xrightarrow{\text{HI}} \underset{n\text{-hexane}}{C_6H_{14}}$$

Therefore, the six carbon atoms of glucoses form a straight chain.

3. It forms a penta acetate on treatment with acetic anhydride which indicates the presence of five hydroxyl groups in the molecule.

4. Glucose reacts with hydroxyl amine to form an oxime and with hydrogen cyanide to form a cyanohydrin. It indicates the presence of a carbonyl group. It also forms phenylhydrazone on treatment with phenylhydrazine.

5. The mild oxidation of glucose with bromine water or sodium hypobromite yields a monocarboxylic acid (gluconic acid) containing same number of carbon atoms as in glucose, *i.e.,* six. This indicates that the carbonyl group must be an aldehyde group. This is further confirmed by the fact that hydrolysis of cyanohydrin of glucose followed by reduction with hydroidic acid yields *n*-heptanoic acid.

6. The catalytic reduction of glucose yields a hexahydric alcohol (glucitol or sorbitol) which gives hexaacetate on treatment with acetic anhydride. The sixth hydroxy group must be obtained by the reduction of aldehyde group, thus further confirming the presence of an aldehyde group and five hydroxyl groups in glucose.

7. Oxidation of gluconic acid with nitric acid yields a dicarboxylic acid (glucaric acid or saccharic acid) with the same number (six) of carbon atoms as in glucose or gluconic acid. Thus besides aldehyde (–CHO) group, glucose must contain a primary alcoholic group ($-CH_2OH$) also, which generates the second carboxylic group on oxidation. The two groups being monovalent must be present on either end of straight chain of six carbon atoms.

8. Glucose is a stable compound and does not undergo dehydration easily, indicating that not more than one hydroxyl group is bonded to a single carbon atom. Thus all the hydroxyl groups are attached to different carbon atoms.

Open-chain structure of glucose

On the basis of the above evidence, the structure of glucose may be written as:

$$CH_2OH - CHOH - CHOH - CHOH - CHOH - CHO$$

This structure explains all the reactions discussed above.

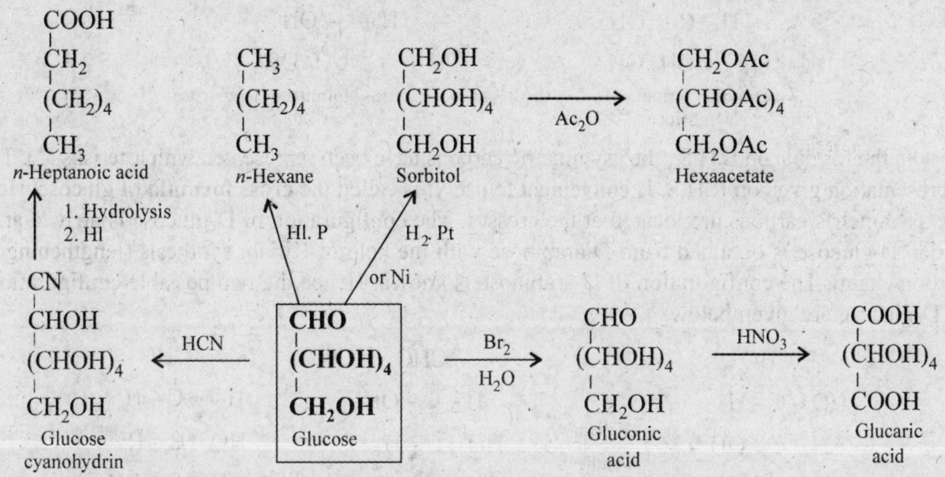

9. The above structure of glucose is also confirmed by the cleavage reaction of glucose with periodic acid. Five moles of periodic acid are consumed by one mole of glucose giving five moles of formic acid and one mole of formaldehyde.

$$\underset{\text{Glucose}}{C_6H_{12}O_6} + \underset{\text{Periodic acid}}{5HIO_4} \longrightarrow HCHO + 5HCOOH + 5HIO_3$$

Stepwise cleavage of glucose molecule is given below:

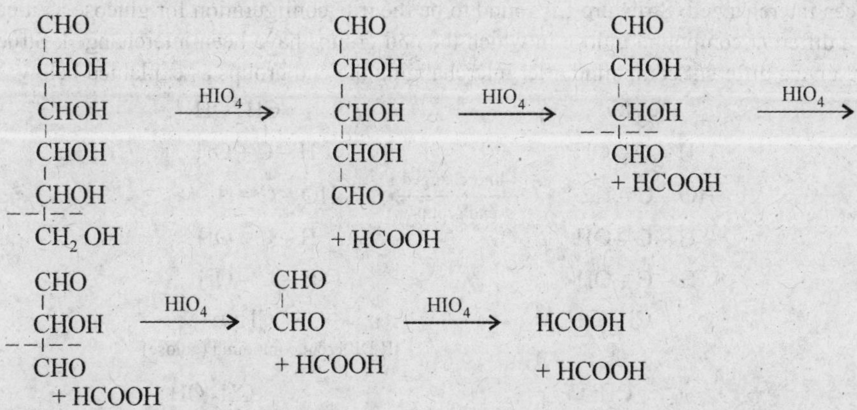

Configuration of (+) glucose

10. *Configuration of a compound means the arrangement of various groups in the molecule.* In a molecule of glucose, there are four asymmetric carbon atoms, i.e., carbon atoms linked to four different groups. This gives rise to a possibility of 2^4, *i.e.*, 16 different configurations for glucose in which the different arrangements of –H and –OH groups exist. All these sixteen compounds having different names have been prepared. Emil Fischer and his coworkers gave the following configuration for D-glucose:

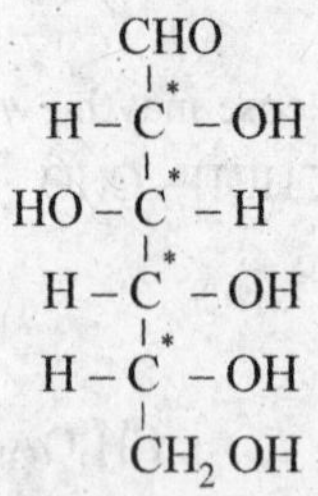

Common formula of D-glucose

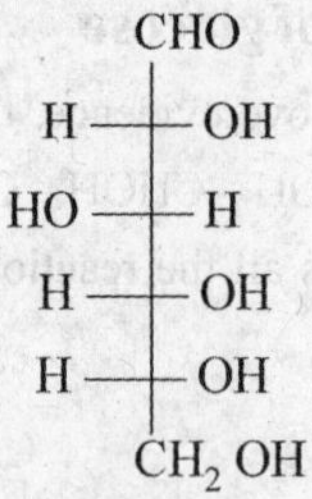

Cross formula of D–glucose

In the formula on L.H.S., the asymmetric carbons have been represented with asterisks (*). The representation given on R.H.S. is convenient to use. It is called the **cross formula** of glucose. Here the asymmetric carbons are located at the crosses. The configuration of D-glucose is arrived at as under. D-glucose is obtained from D-arabinose with the help of Kiliani synthesis (lengthening of carbon chain). The configuration of D-arabinose is known. Hence the two possible configurations of D-glucose are given below:

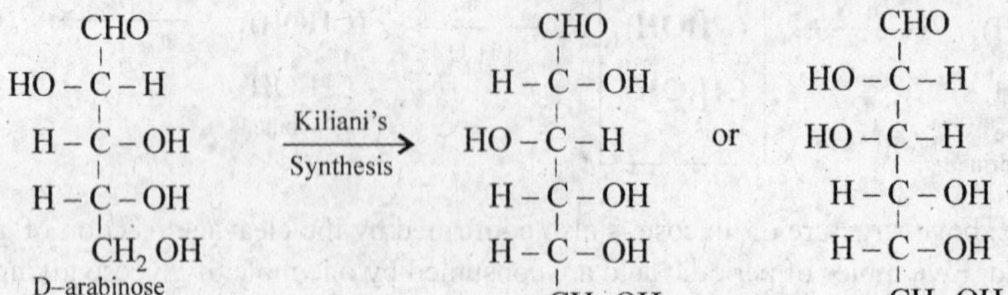

D–arabinose; I; II

11. Two different compounds glucose and gulose on oxidation give saccharic acid in which the end groups –CHO and – CH_2 OH are converted into – COOH groups. This means that the configuration around the asymmetric carbons is the same in glucose and gulose. Only the end groups have been interchanged. Structure I is found to be the true configuration for glucose as it can give rise to a different compound gulose in which the end groups have been interchanged. Structure II will not give a different compound after interchanging the end groups as explained below:

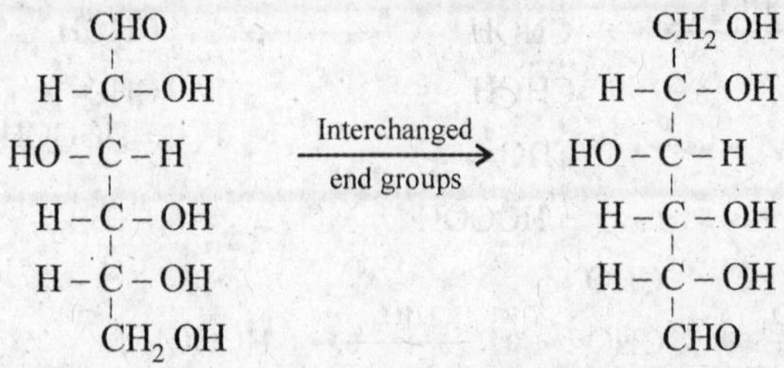

I

III Different compound (gulose)

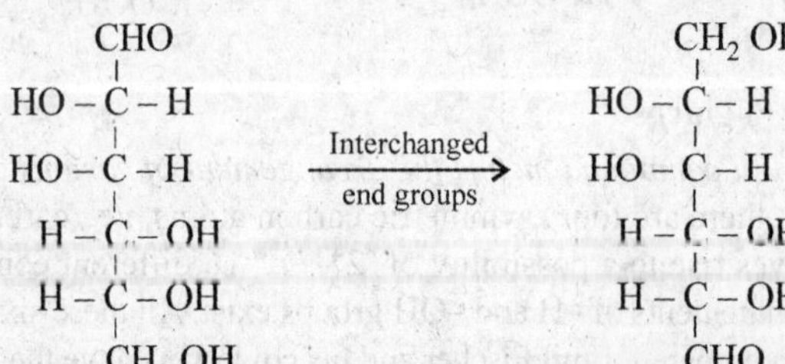

II

IV

In fact there is no difference between structures II and IV. If structure IV is rotated through an angle of 180°, structure II is obtained. This leads us to the conclusion that D-glucose has the configuration I. The configuration II is for D-mannose.

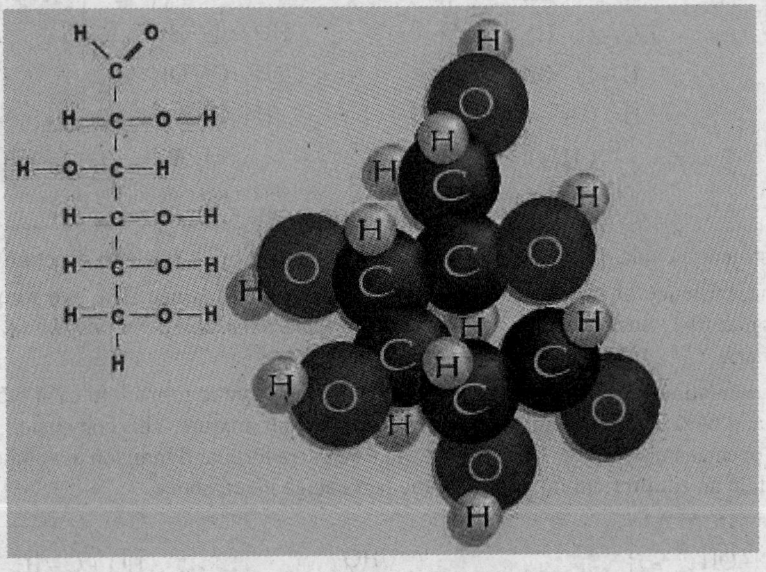

Open chain structure of glucose

Cyclic structure of glucose

The open chain structure for glucose fails to explain some observations given following:

(*i*) Glucose does not restore the pink colour of Schiff's reagent, a characteristic property of aldehydic compounds.

(*ii*) Glucose does not react with ammonia or sodium bisulphite to form addition compounds. It means the aldehydic group in glucose is not free to give these reactions.

(*iii*) D-glucose exists in two isomeric forms which undergo mutarotation. *The phenomenon of change of specific rotation of an optically active compound with time to an equilibrium value of specific rotation is called mutarotation.*

Glucose is an optically active compound because of the presence of asymmetric carbon atoms. It rotates the plane of polarised light. It is observed that when D-glucose having m.p. 419 K is dissolved in water, its specific rotations gradually drops from an initial value of + 112° to a value + 52.7°. On the other hand, when D-glucose having m.p. 423 K is dissolved in water, the specific rotation gradually increases from + 19° to 52.7°. The form of D-glucose having m.p. 419 K and having a higher specific rotation is called α-D-glucose and the other one is called β-D-glucose.

(*iv*) These two forms of glucose form methyl α-D-glucoside and methyl β-D-glucoside on treatment with methyl alcohol in the presence of dry HCl. These glucosides do not exhibit mutarotation and fail to give the Tollen's reagent or Fehling solution tests.

To account for the above, ring structure for glucose was given by Fischer, Tanret and Haworth, as shown below :

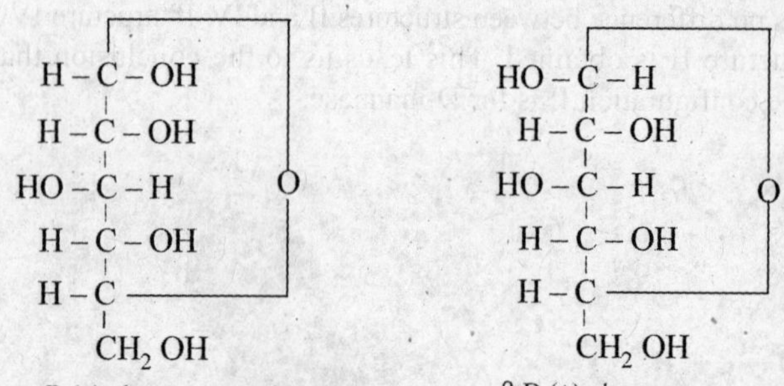

α-D (+) glucose m.p. 419 K, [α] = +112° | β-D (+) glucose m.p. 419 K, [α] = +19°

Ring structure explains the facts which are not explained by open-chain structure as follows:

(*a*) The existence of two forms of glucose, α and β, is explained. The two forms differ in configuration at C–1. Such a carbon atom is called *anomeric carbon atom and the two forms are called anomers.*

(*b*) Mutarotation is explained by the interconversion of cyclic forms into each other till 36% of α-forms and 64% of β-form are present in the equilibrium mixture. This conversion takes place through the open-chain structure. Thus whether we take the α-form or β form, on dissolution in water, we shall obtain an equilibrium mixture with the percentage given above.

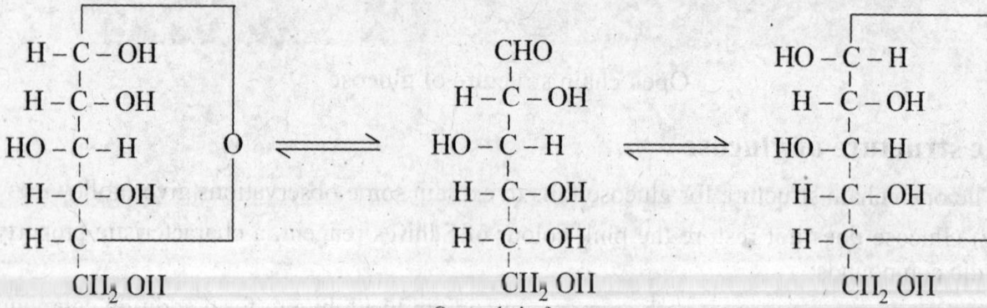

α-D (+) glucose | Open-chain form | β-D (+) glucose

(*c*) α and β methyl glucosides do not exhibit mutarotation, they are not readily hydrolysed to the open-chain form. The structures of methyl glucosides are given below. They are acetals.

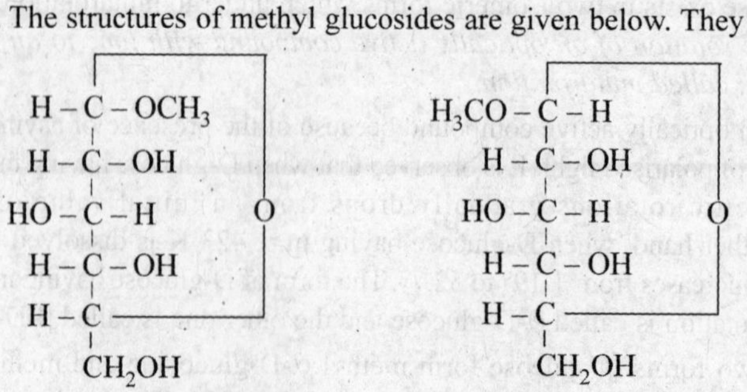

Methyl α-D-glucoside | Methyl β-D-glucoside

(*ii*) Reaction with Tollen's reagent and Fehling solution and osazone formation is explained by the presence of the open-chain form in traces.

Size of the Ring

Heworth and coworkers have shown that D-glucose and glucosides have a six-membered ring and not five-membered ring as proposed earlier. They made these inferences based on the following reactions:

(*i*) Methyl β-D-glucoside (A) on treatment with methyl sulphate and sodium hydroxide is converted to compound (B) as under:

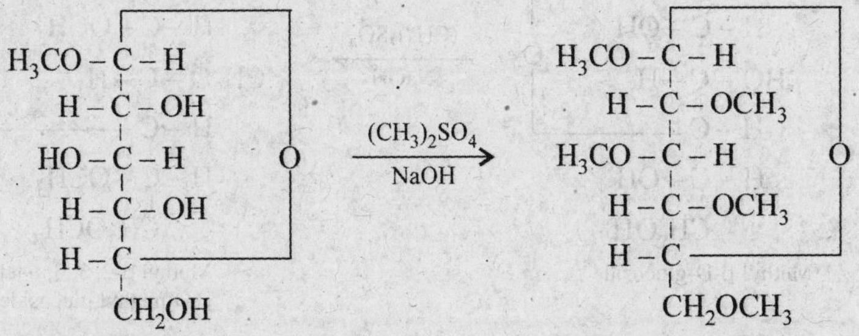

Methyl β-D-glucose (A) — Methyl β-2, 3, 4, 6 tetra O-methyl D-glucoside (B)

The compound (B) is hydrolysed by dil. HCl to compound C which is a cyclic hemi-acetal and is in equilibrium with open-chain structure. Only the $-OCH_3$ at C-1 in the compound B is hydrolysed because acetal is more easily hydrolysed than an ether, ($-OCH_3$ are ether groups at C-2, C-3, C-4 and C-6).

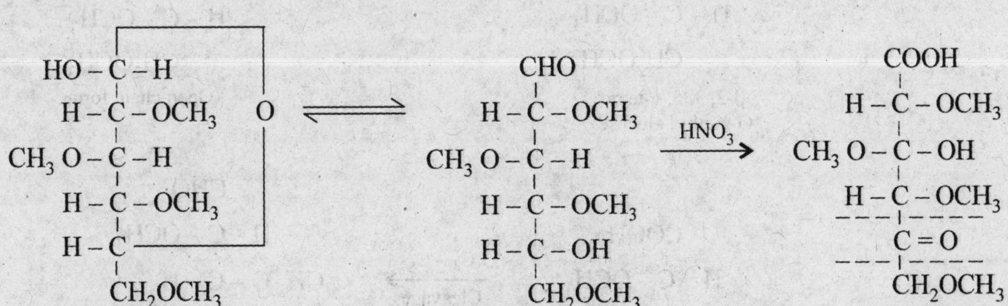

β-2, 3, 4, 6 tetra O-methyl glucose (C) — β-2, 3, 4, 6 tetra O-methyl glucose (Open-chain form) — (keto acid) (D)

Compound (C) on oxidation with HNO_3 gives a mixture of trimethoxy glutaric acid and dimethoxy succinic acid. These may be considered to have been formed via an intermediate keto acid (D). The two acids may be supposed to have been formed by cleavage between $C_5 - C_6$ and $C_4 - C_5$ respectively.

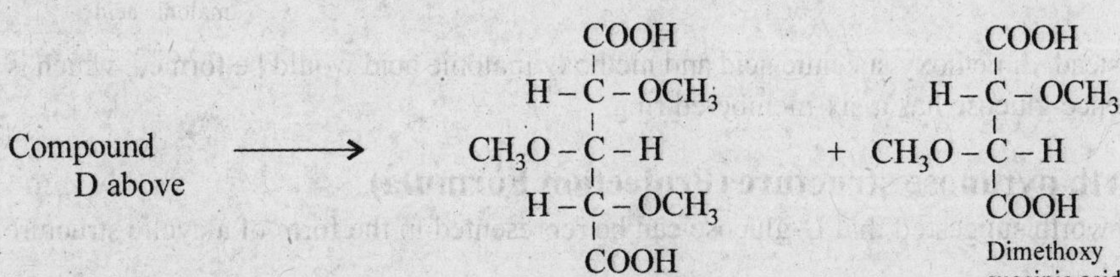

Trimethoxy gluatric acid — Dimethoxy succinic acid

The above reactions are possible if the free –OH group in the open-chain form above is at C-5. Thus C-5 is involved in the ring formation. This supports the six-membered ring for glucose.

If methyl β-D-glucoside had a five-membered ring, it would not have led to the formation of trimethoxy glutaric acid as shown below:

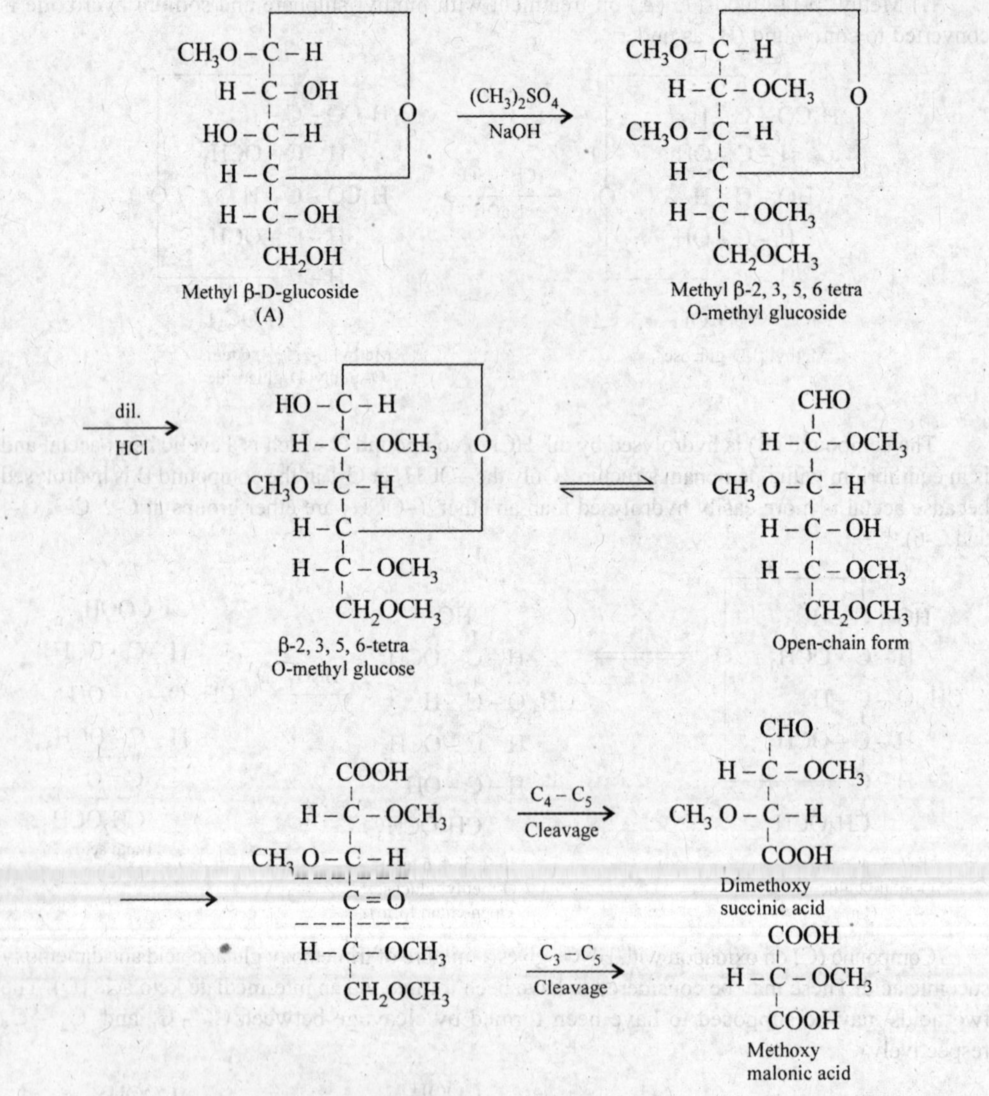

Instead, dimethoxy succinic acid and methoxy malonic acid would be formed, which is not the case. Hence glucose has a six-membered ring.

Haworth pyranose structure (Projection Formula)

Haworth suggested that D-glucose can be represented in the form of a cyclic structure shown below:

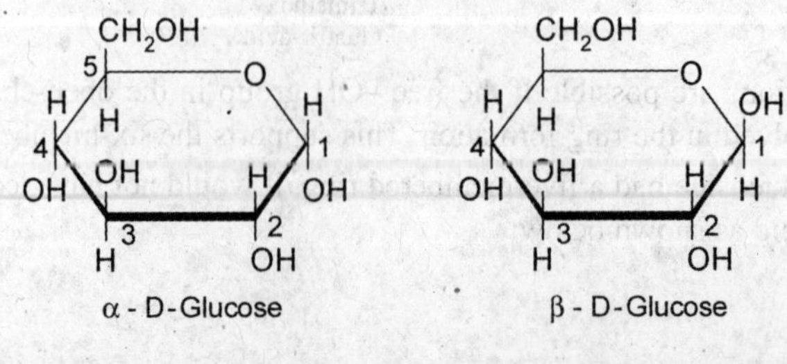

α - D-Glucose β - D-Glucose

The ring is in the same plane. It consists of five carbons (not shown) C_1 to C_5. One position of the ring is occupied by oxygen. At each carbon C_1 to C_4, two groups –H and –OH are attached. These groups are perpendicular to the plane of the ring. The bonds going up are above the plane and going down are below the plane. At C-5, $-CH_2OH$ is above the plane and –H below the plane. α-D-glucose and β-D-glucose differ in the respect that at C-1, –OH is below the plane in α-isomer whereas it is above the plane in β-isomer.

6.6. FRUCTOSE

Fructose is another commonly known monosaccharide having the same molecular formula as glucose. It is laevorotatory because it rotates the plane of polarised light towards the left. It is present abundantly in fruits. That is why it is called *fruit-sugar* also.

Physical properties

1. It is the sweetest of all known sugars.
2. It is readily soluble in water, sparingly soluble in alcohol and insoluble in ether.
3. It is a white crystalline solid with m.p. 375 K.
4. Fresh solution of fructose has a specific rotation –133° which changes to –92° at equilibrium due to mutarotation.

Chemical properties of fructose

Chemical properties of fructose can be studied under the following heads :

[A] Reactions due to ketonic group

1. Reaction with hydrogen cyanide. Fructose reacts with HCN to form cyanohydrin.

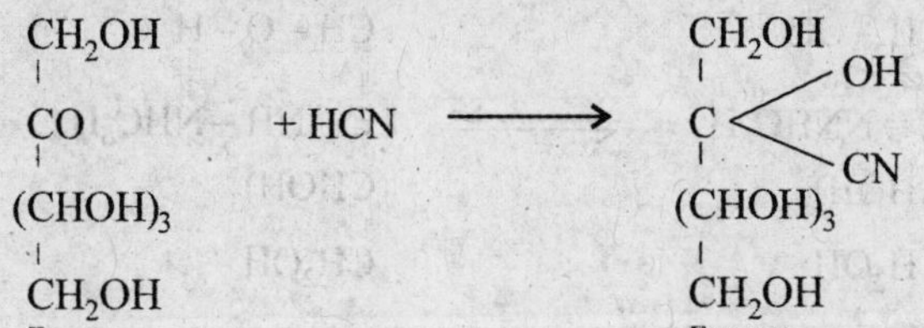

Fructose — Fructose cyanohydrin

2. Reaction with hydroxylamine. Fructose reacts with hydroxylamine to form an oxime

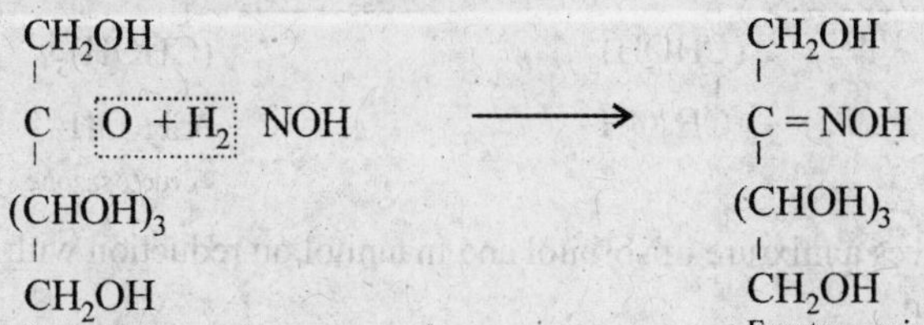

Fructose — Fructose oxime

3. Reaction with phenylhydrazine

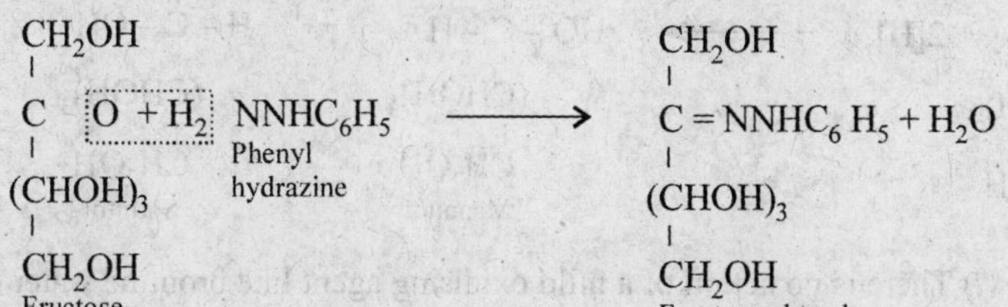

Fructose — Fructose phenyl hydrazone

With another molecule of phenyl hydrazine, the reaction is as under:

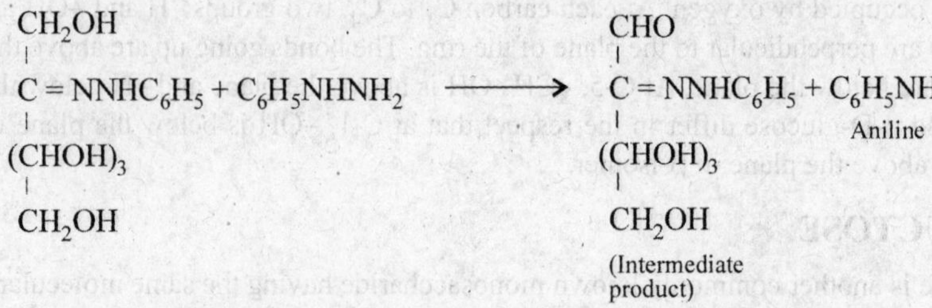

The intermediate product containing the –CHO group finally reacts with the third molecule of phenyl hydrazine to give fructosazone.

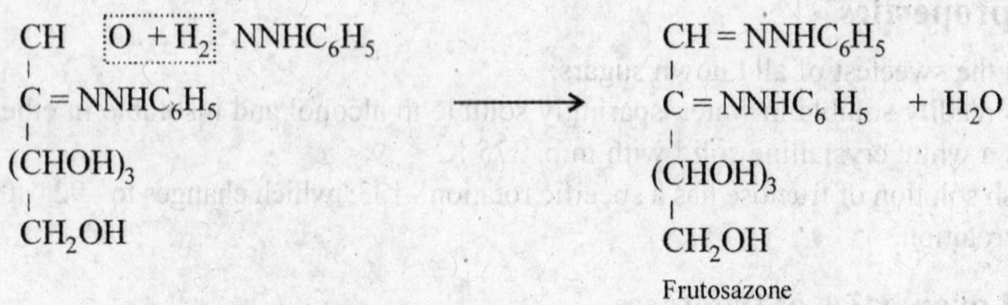

It has been found that fructosazone has the same structure as glucosazone. The m.p. is 478 K.

A modern mechanism due to Weygand explaining the formation of fructosazone is given as follows:

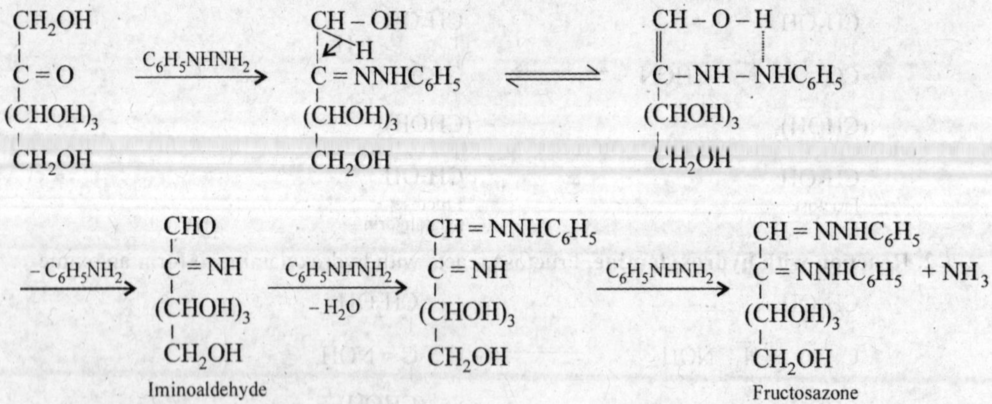

4. Reduction. Fructose gives a mixture of sorbitol and mannitol on reduction with Na–Hg and water or catalytic hydrogenation.

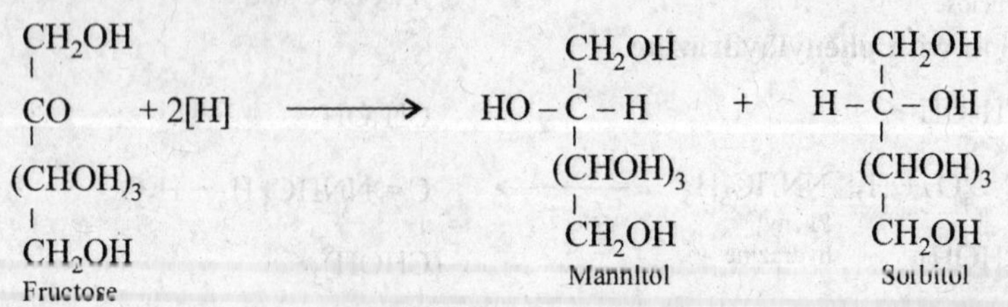

5. Oxidation. (*i*) There is no action of a mild oxidising agent like bromine water on fructose.

(*ii*) Strong oxidising agents like nitric acid oxidise fructose into a mixture of trihydroxy glutaric, glycollic and tartaric acids.

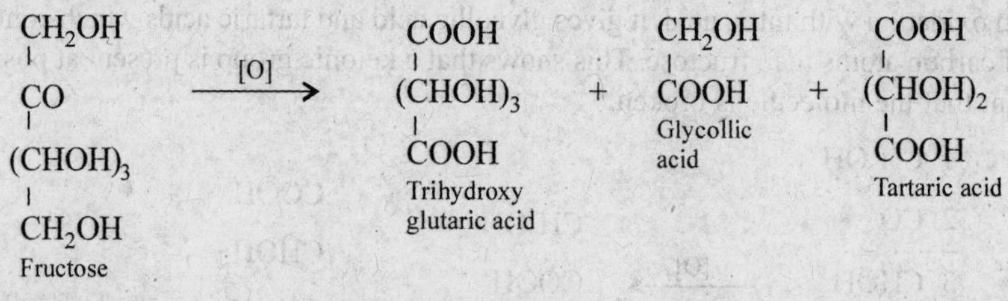

(*iii*) Unlike other ketones, it reduces Tollen's reagent (ammoniacal silver nitrate) and Fehling solution (alkaline copper sulphate). This is due to the presence of traces of glucose in alkaline medium.

[B] Reactions of the alcoholic group

1. Acetylation. With acetic anhydride or acetyl chloride, fructose forms penta-acetate.

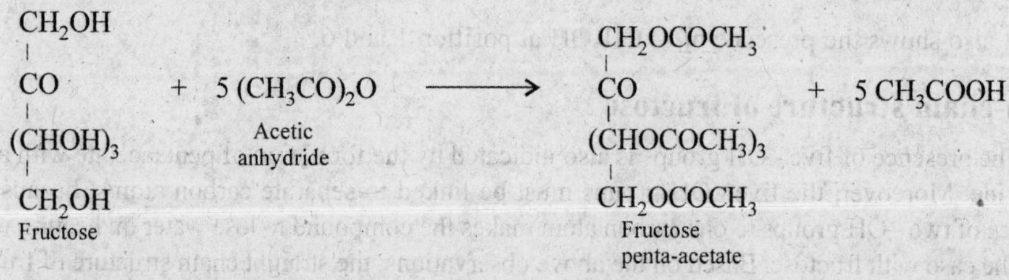

2. Reaction with methyl alcohol (glucoside formation). Fructose reacts with methyl alcohol in the presence of dry HCl gas forming methyl fructoside.

$$\underset{\text{Fructose}}{C_6H_{11}O_5}\ \boxed{OH + H}\ OCH_3 \xrightarrow[\text{gas}]{\text{HCl}} \underset{\text{Methyl fructoside}}{C_6H_{11}O_5OCH_3} + H_2O$$

3. Reaction with metallic hydroxides (fructosate formation)

$$\underset{\text{Fructose}}{C_6H_{11}O_5OH} + \underset{\text{Calcium hydroxide}}{HOCaOH} \longrightarrow \underset{\text{Cal.fructosate}}{C_6H_{11}O_5OCaOH} + H_2O$$

6.7. STRUCTURE OF FRUCTOSE

1. Elemental analysis and molecular weight determination of fructose show that it has the molecular formula $C_6H_{12}O_6$.

2. Fructose on reduction gives sorbitol which on reduction with HI and red phosphorus gives a mixture of *n*-hexane and 2-iodohexane. This reaction indicates that the six carbon atoms in fructose are in a straight chain.

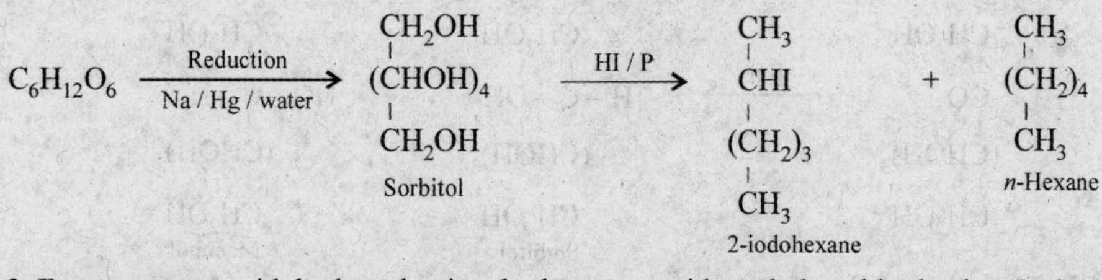

3. Fructose reacts with hydroxylamine, hydrogen cyanide and phenyl hydrazine. It shows the presence of –CHO or > C = O group in the molecule of fructose.

4. On treatment with bromine water, no reaction takes place. This rules out the possibility of presence of –CHO group.

5. On oxidation with nitric acid, it gives glycollic acid and tartaric acids which contain smaller number of carbon atoms than fructose. This shows that a ketonic group is present at position 2. It is at this point that the molecule is broken.

$$\begin{array}{l} 1\ CH_2OH \\ 2\ CO \\ \text{-----------} \\ 3\ CHOH \\ 4\ CHOH \\ 5\ CHOH \\ 6\ CH_2OH \end{array} \xrightarrow{[O]} \begin{array}{c} CH_2OH \\ | \\ COOH \\ \text{Glycollic acid} \end{array} + \begin{array}{c} COOH \\ | \\ CHOH \\ | \\ CHOH \\ | \\ COOH \\ \text{Tartaric acid} \end{array}$$

It also shows the presence of $-CH_2OH$ at position 1 and 6.

Open-chain structure of fructose

The presence of five –OH groups is also indicated by the formation of pentaacetate with acetic anhydride. Moreover, the five –OH groups must be linked to separate carbon atoms, because the presence of two –OH groups to one carbon atom makes the compound to lose water on heating which is not the case with fructose. Based on the above observations, the straight chain structure of fructose is as:

$$\overset{1}{C}H_2OH\overset{2}{C}O\overset{3}{C}HOH\overset{4}{C}HOH\overset{5}{C}HOH\overset{6}{C}H_2OH$$

Facts in support of straight-chain structure

(*i*) Fructose cyanohydrin on hydrolysis gives a mono carboxylic acid which on reduction with HI gives α-methyl caproic acid

$$\begin{array}{c} CH_2OH \\ | \\ C(OH)CN \\ | \\ (CHOH)_3 \\ | \\ CH_2OH \\ \text{Cyanohydrin} \end{array} \xrightarrow{H_2O} \begin{array}{c} CH_2OH \\ | \\ C(OH)COOH \\ | \\ (CHOH)_3 \\ | \\ CH_2OH \\ \text{Hydroxy acid} \end{array} \xrightarrow{H_2O} \begin{array}{c} CH_3 \\ | \\ CHCOOH \\ | \\ (CH_2)_3 \\ | \\ CH_3 \\ \text{α-Methyl caproic acid} \end{array}$$

(*ii*) Fructose gives on reduction, a mixture of mannitol and sorbitol, which confirm the presence of a ketonic group. On reduction the ketonic group is transformed into –CHOH. There are two possible arrangements of –H and –OH groups giving rise to two isomers.

$$\begin{array}{c} CH_2OH \\ | \\ CO \\ | \\ (CHOH)_3 \\ | \\ CH_2OH \end{array} \xrightarrow{[H]} \begin{array}{c} CH_2OH \\ | \\ H-C-OH \\ | \\ (CHOH)_3 \\ | \\ CH_2OH \\ \text{Sorbitol} \end{array} + \begin{array}{c} CH_2OH \\ | \\ HO-C-H \\ | \\ (CHOH)_3 \\ | \\ CH_2OH \\ \text{Mannitol} \end{array}$$

Configuration of D(–) fructose

Glucose and fructose form the same osazone. And in this reaction, only C-1 and C-2 are involved. This means the configuration of rest of the molecule (arrangement of –H and –OH) should be the same as in glucose. In fructose C-3, C-4 and C-5 are chiral carbon atoms, whereas in glucose, C-2, C-3, C-4 and C-5 are chiral atoms. In view of the above, the configuration of fructose is as given below:

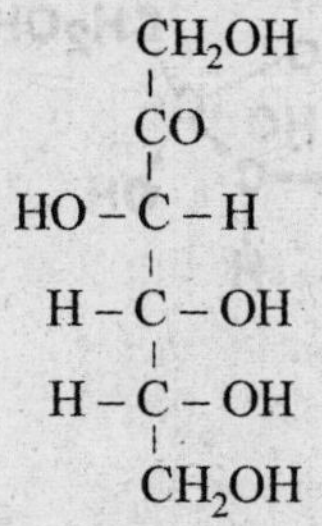

Fructose

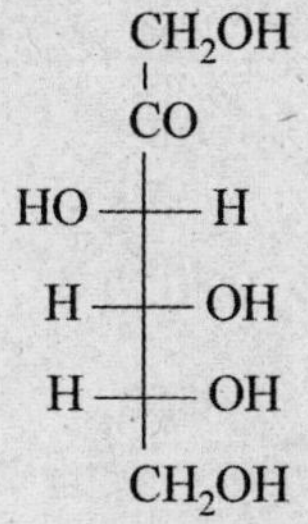

Cross formula of fructose

Ring structure of fructose

Fructose shows the property of mutarotation. This means that it exists in two forms α-fructose and β-fructose which are cyclic in structure and change into each other via the open chain structure. The cyclic and pyranose structures of α-D-fructose and β-D-fructose are represented below:

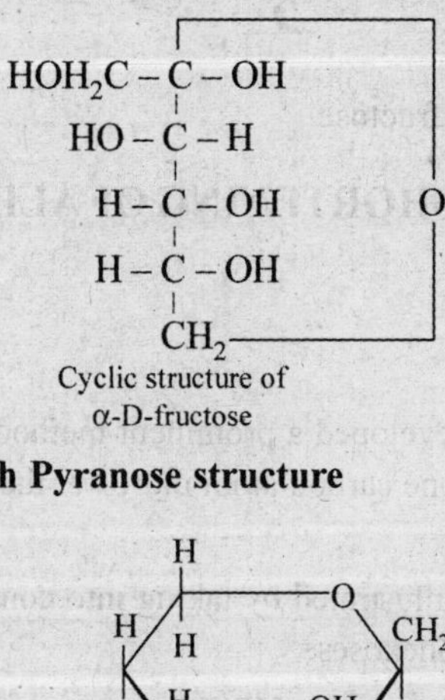

Cyclic structure of α-D-fructose

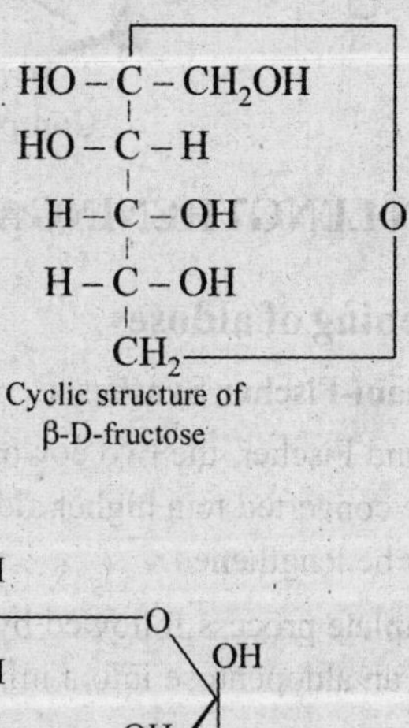

Cyclic structure of β-D-fructose

Haworth Pyranose structure

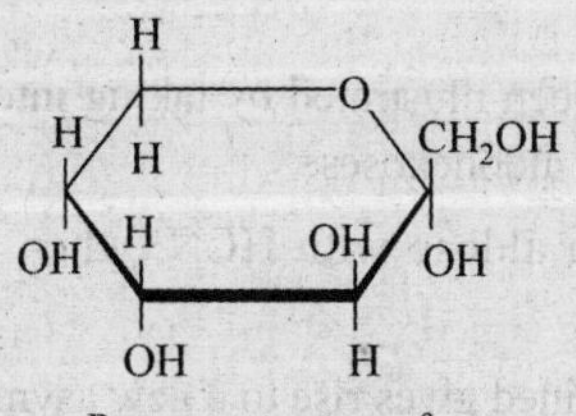

Pyranose structure of α-D-fructose

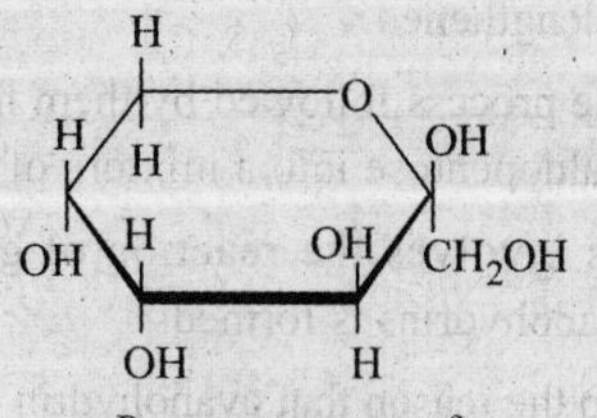

Pyranose structure of β-D-fructose

However, when fructose is linked to glucose in a sucrose molecule, it has the furanose structure as shown below:

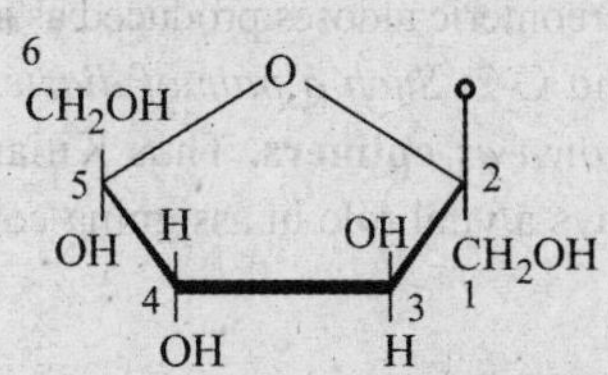

β-D-fructofuranose

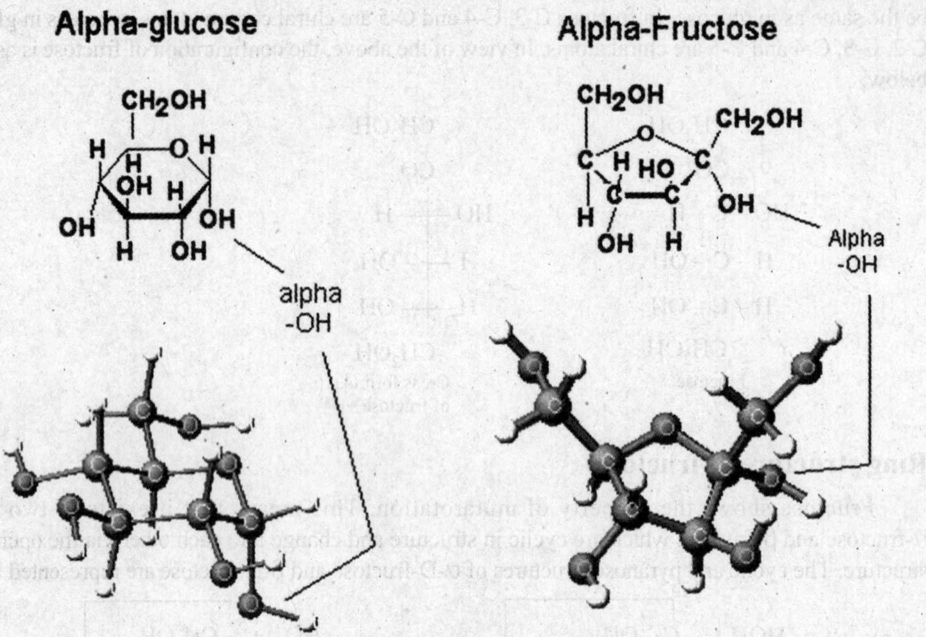

Compare glucose and fructose

6.8. CHAIN LENGTHENING AND CHAIN SHORTENING OF ALDOSES

(*a*) Lengthening of aldoses

The Kiliani-Fischer Synthesis

Kiliani and Fischer, the two coworkers in 1890, developed a prominent method by which an aldose may be converted to a higher aldose containing one carbon atom more *i.e.* the carbon chain of aldose may be lengthened.

The complete process followed by them has been illustrated by taking into consideration the conversion of an aldopentose into a mixture of two aldohexoses.

The process involves the reaction of given aldose with HCN, when a mixture of two diastereomeric cyanohydrins is formed.

This is due to the reason that cyanohydrin obtained gives rise to a new asymmetric centre. The cyanohydrin mixture is hydrolysed to get a mixture of two corresponding acids which are separated from each other. Each of these acids upon heating forms the corresponding lactone which is reduced by sodium amalgam to yield the higher aldoses.

It may be noted that the diastereomeric aldoses produced as a result of Kiliani-Fischer synthesis differ in configuration only around C-2. *Such a pair of diastereomeric aldoses which differ in configuration only at C-2 are known as* **epimers.** Thus Kiliani-Fischer synthesis leads to the formation of epimers. This fact plays a vital role in assigning configurations to aldoses.

Kiliani Fischer Synthesis

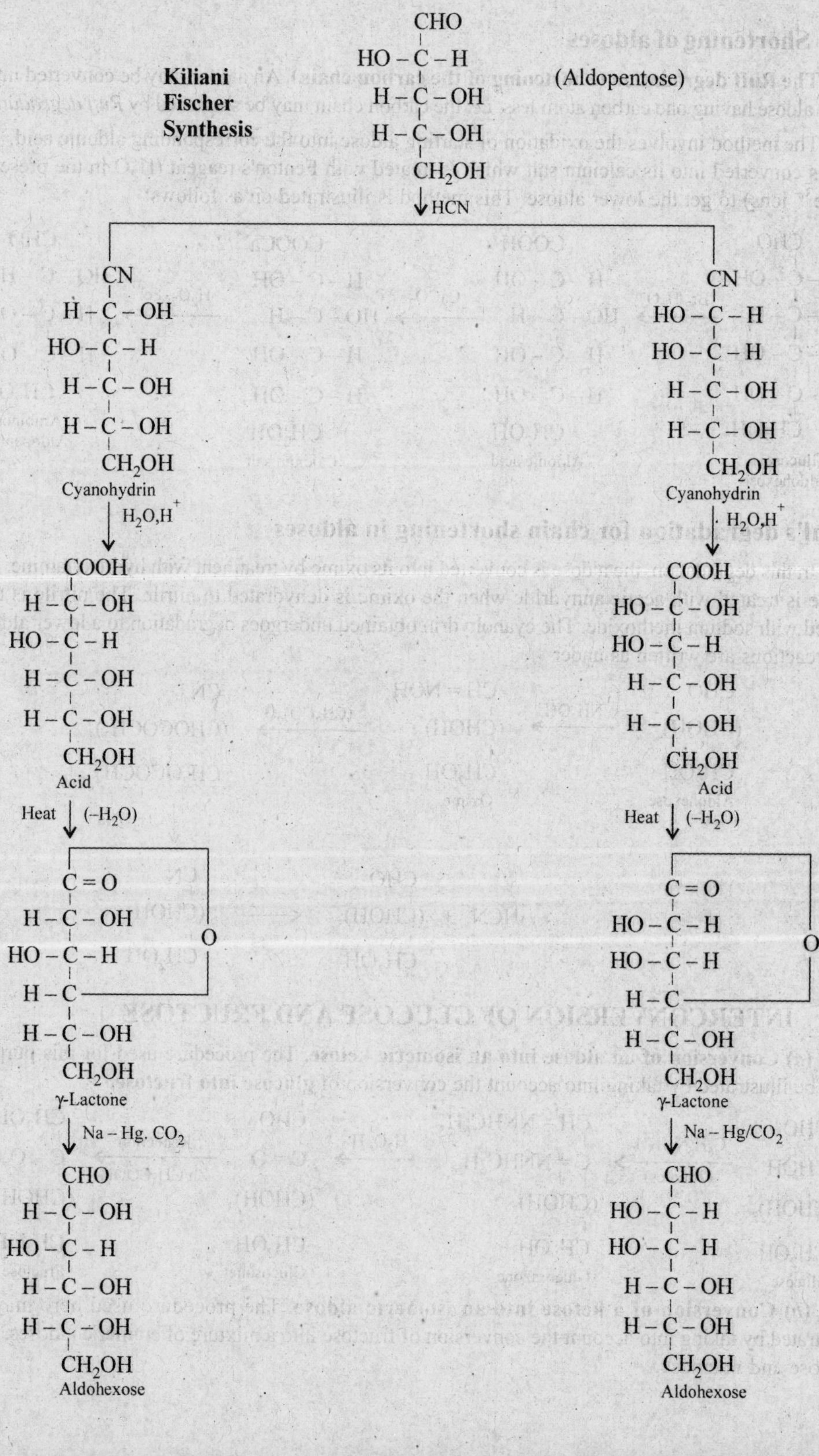

(*b*) Shortening of aldoses

The Ruff degradation (Shortening of the carbon chain). An aldose may be converted into a lower aldose having one carbon atom less, *i.e.* the carbon chain may be shortened by *Ruff degradation.*

The method involves the oxidation of starting aldose into the corresponding aldonic acid. The acid is converted into its calcium salt which is treated with Fenton's reagent (H_2O in the presence of Fe^{3+} ions) to get the lower aldose. This method is illustrated on as follows:

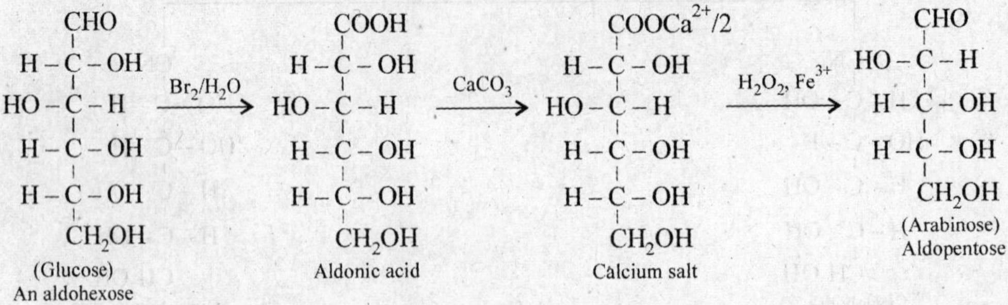

Wohl's degradation for chain shortening in aldoses

In this degradation, the aldose is converted into its oxime by treatment with hydroxylamine. The oxime is treated with acetic anhydride when the oxime is dehydrated to nitrile. The nitrile is then treated with sodium methoxide. The cyanohydrin obtained undergoes degradation to a lower aldose. The reactions are written as under

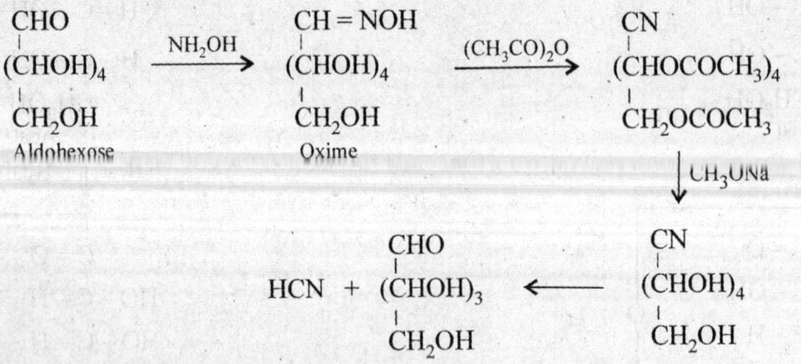

6.9. INTERCONVERSION OF GLUCOSE AND FRUCTOSE

(*a*) Conversion of an aldose into an isomeric ketose. The procedure used for this purpose may be illustrated by taking into account the conversion of **glucose into fructose.**

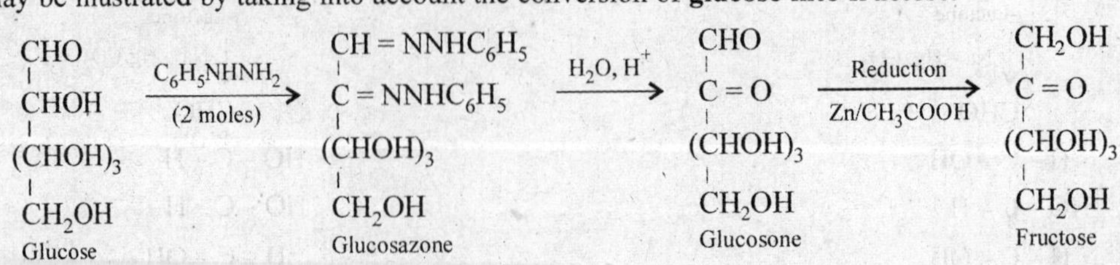

(*b*) Conversion of a ketose into an isomeric aldose. The procedure used here may be illustrated by taking into account the conversion of fructose into a mixture of epimeric aldoses, *viz.*, glucose and mannose.

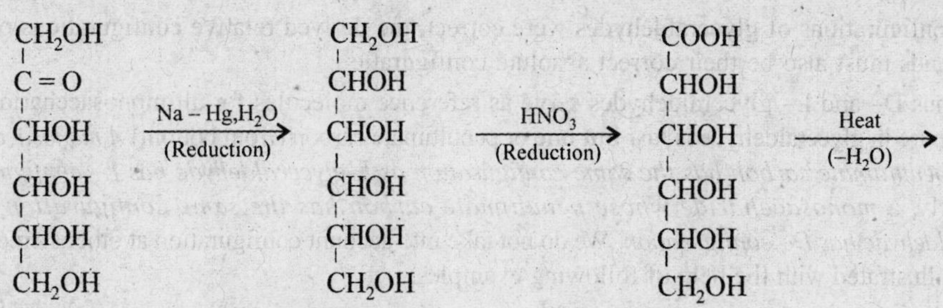

Fructose — Mixture of diastereomeric sorbitol and mannitol — Mixture of diastereomeric gluconic acid and mannonic acid

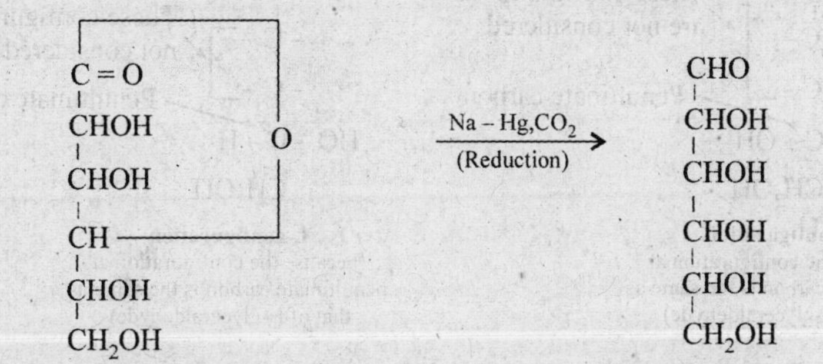

Mixture of diastereomeric γ-lactones — Mixture of epimeric glucose and mannose

6.10. CONFIGURATION OF MONOSACCHARIDES

In early days of development of stereochemistry of organic compounds, it was not possible to determine the **absolute** configurations. The chemists were only interested in knowing the relative configurations [Whether *d*- or *l*- or whether (+) or (–)]. To decide about configurations, Emil Fischer in 1885 chose glyceraldehyde, $CH_2OHCHOHCH_2OH$ as the standard substance and fixed its relative configurations arbitrarily. This compound has one stereocentre and therefore exists in two enantiomeric forms, as given below :

Emil Fischer

D (+)-Glyceraldehyde (I) — L (–)-Glycereldehyde (II)

Compound I was found to be dextrorotatory and compound (II) was found to be laevorotatory. The difference between configuration of the two compounds is that in compound (I), –H is located on the L.H.S. and –OH is located on the R.H.S. of the Fischer projection formula while in compound (II), this is in reverse order.

Configuration of other compounds was then assigned by relating their configuration to that of D– or L– Glyceraldehyde.

In 1951 Bijvoet by using x-ray crystallography established that the arbitrarily assigned configurations of glyceraldehydes actually represented their correct **absolute** configurations. Thus, if the configurations of glyceraldehydes were correct, the derived relative configurations of other compounds must also be their correct absolute configuration.

Thus D– and L– glyceraldehydes serve as reference molecules for all monosaccharides The stereocentre in glyceraldehyde is **last but one** or penultimate carbon (from bottom) *A monosaccharide whose penultimate carbon has the same configuration as L-glyceraldehyde has L– configuration.* Similarly, a *monosaccharide whose penultimate carbon has the same configuration as D–glyceraldehyde has D– configuration.* We do not take into account configuration at other stereocentre. This is illustrated with the help of following examples.

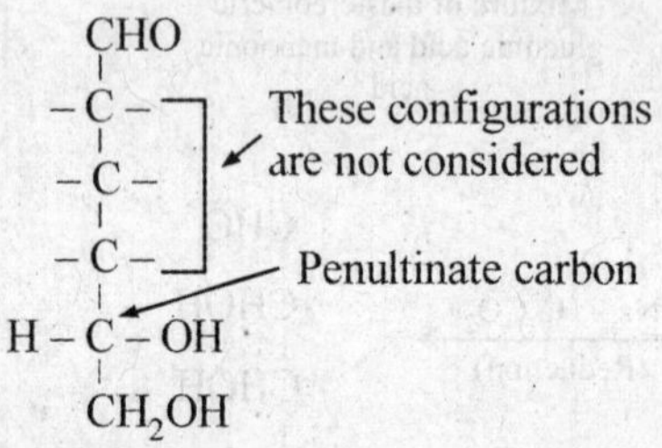

D-configuration
(because the configuration at penultimate carbon is the same as that of D-glyceraldehyde)

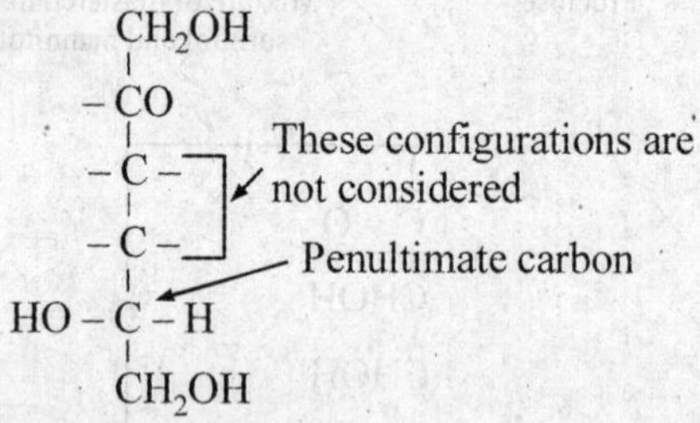

L-configuration
(because the configuration at penultimate carbon is the same as that of L-glyceraldehyde)

6.11. ERYTHRO AND THREO DIASTEREOMERS

Erythrose and Threose are the next homologues of glyceraldehyde.

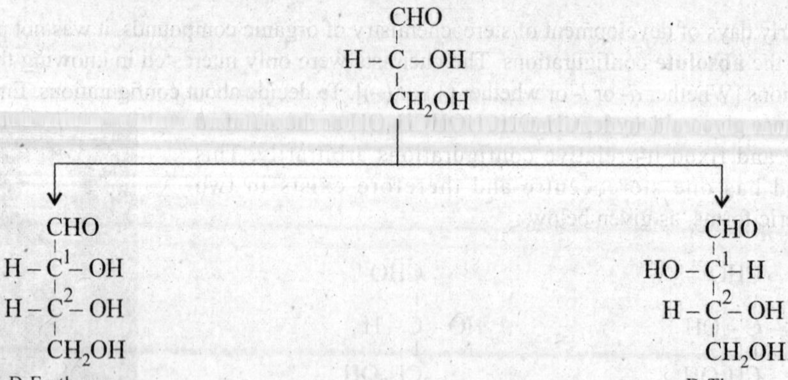

D-Erythrose D-Threose

We can say that erythrose and threose are diastereomers having two stereocentres denoted by 1 and 2. In erythrose, the –OH groups are situated on the same side of the Fischer's projection formula, while in threose, the two –OH groups are oriented on the opposite sides of the formula. We have a system of denoting diastereomers with two adjacent stereocentres (1 and 2 in the above case) in which the two groups attached to one stereocentre are the same as the two groups attached to the other stereocentre.

Thus, if the two similar groups attached to adjacent stereocentres are on the same side in the Fischer formula, the diastereomer is known as **erythro** isomer. This is derived from the word **erythrose.** If the two similar groups are on the opposite side of the Fischer formula, the diastereomer is **threo.** This is derived from the word **threose.** Some examples are given below:

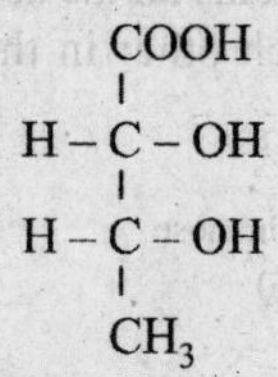

Erythro-2, 3-Dihydroxy butanoic acid

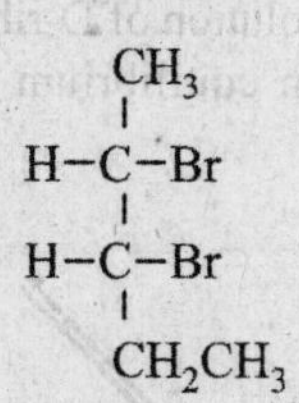

Erythro-2, 3-Dibromo-pentane

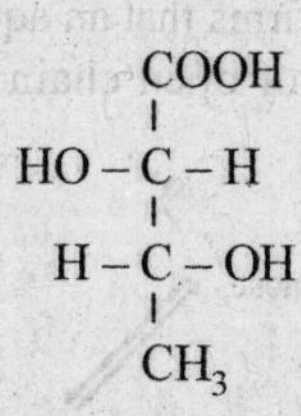

Threo-2, 3-Dihydroxy butanoic acid

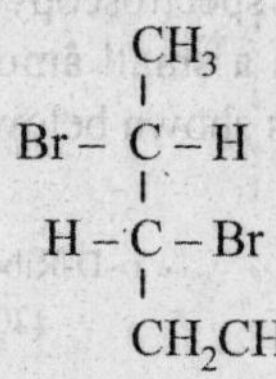

Threo-2, 3-Dibromo pentane

The limitation of this method of notation, however, is that it is applicable to compounds in which the adjacent stereocentres are located next to the bottom carbon atom.

6.12. STRUCTURES OF RIBOSE AND DEOXYRIBOSE

Ribose and deoxyribose are two well-known aldopentoses. Their structures are discussed as under:

Structure of D-(+)-Ribose

D-(+)-Ribose occurs naturally in plant nucleic acids and in liver and pancreas nucleic acids.

D-(+)-ribose gives properties similar to glucose.

Ribose has the molecular formula ($C_5H_{10}O_5$) and shows the presence of an aldehyde group, four hydroxyl groups (one primary and three secondary) and a straight chain of carbon atoms. Therefore, it was assigned an open chain formula as given below:

$$^5CH_2OH - {}^4CHOH - {}^3CHOH - {}^2CHOH - {}^1CHO$$

The configuration of D-ribose has been established as follows.

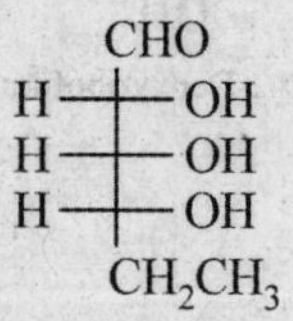

D-(+)-ribose

As in the case of glucose, D-ribose is now assigned a ring structure and is known to exist both in the furanose and pyranose forms as depicted below :

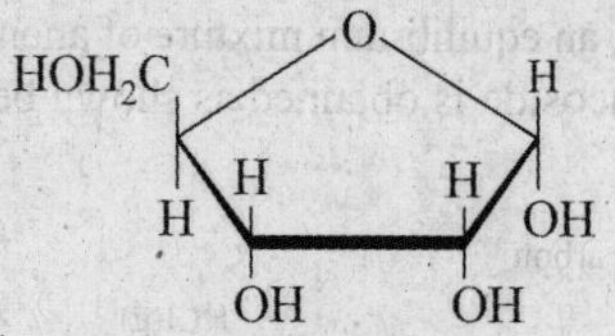

α-D-(–)-Ribofuranose

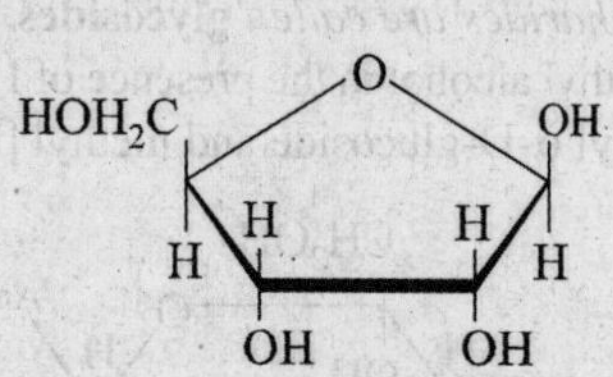

β-D-(+)-Ribofuranose

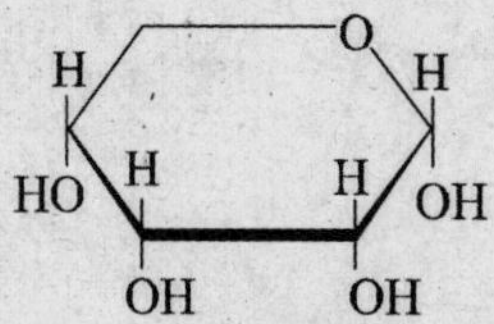

α-D-(–)-Ribopyranose

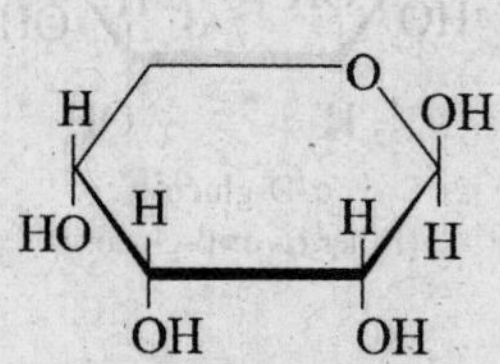

β-D-(–)-Ribopyranose

NMR spectroscopy confirms that an aqueous solution of D-ribose contains all the above forms along with a small amount of open chain form in equilibrium with each other in the relative proportions shown below:

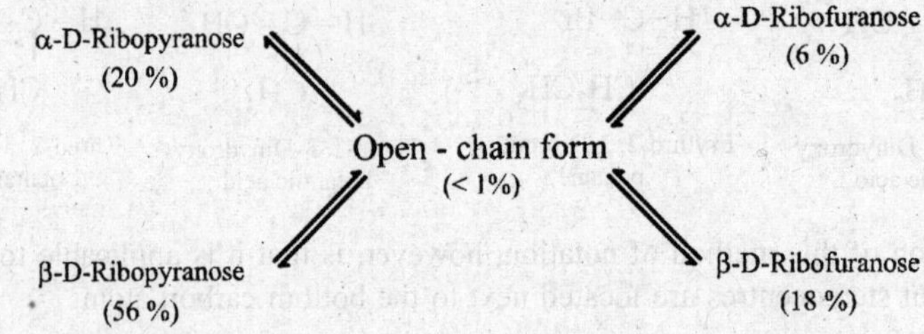

6.13. STRUCTURE OF D-2-DEOXYRIBOSE

In this aldopentose the hydroxyl group at C-2 of ribose has been replaced by hydrogen. That is why it is named as **deoxyribose.** It is fundamental constituent of deoxyribonucleic acid (DNA).

The structure of D-2-deoxyribose is derived from that of D-ribose and may be represented in the open chain and ring forms as follows.

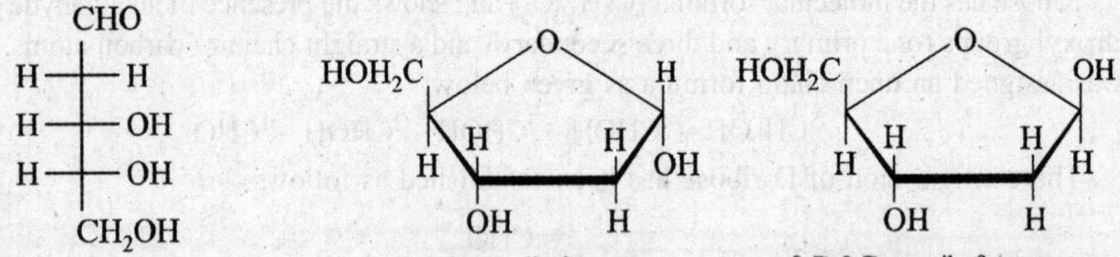

Open-chain form of D-2-Deoxyribose α-D-2-Deoxyribofuranose β-D-2-Deoxyribofuranose

6.14. GLYCOSIDES

Monosaccharides such as glucose, fructose and ribose possess *cyclic hemiacetal* structures. When monosaccharides are treated with alcohols in the presence of traces of acid catalyst, they are converted into cyclic **acetals** by the reaction at the **anomeric carbon.** *Such cyclic acetals derived from monosaccharides are called* **glycosides.** For example, if D-glucose (either α or β anomer) is treated with methyl alcohol in the presence of HCl (g), an equilibrium mixture of anomeric glycosides known as methyl α-D-glucoside and methyl β-D-glucoside is obtained as shown below.

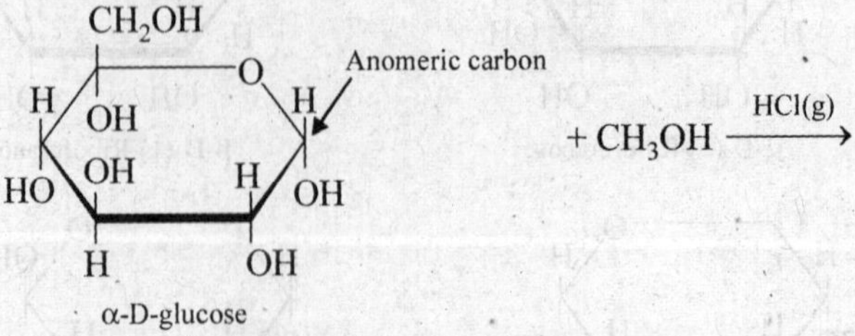

α-D-glucose
(Ether α- or β-glucose)

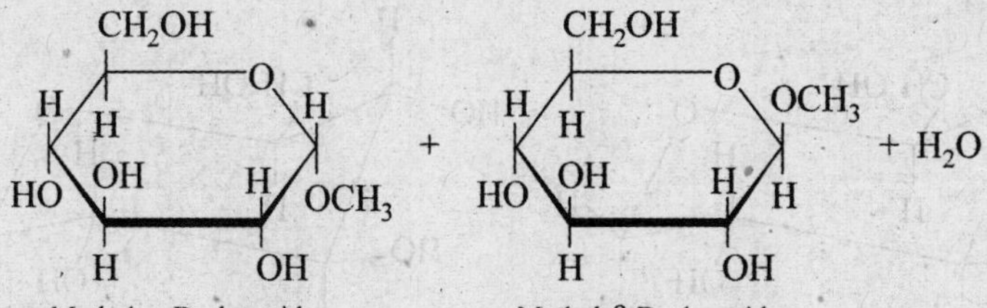

Methyl-α-D-glucoside Methyl-β-D-glucoside

(Mixture of anomeric glycosides)

Whichever anomer of monosaccharide we use as the starting material, both anomers of glycoside are formed as an equilibrium mixture.

Glycosidic linkage and glycosidic hydroxyl group. The bond between the anomeric carbon and the –OR group is known as **glycosidic linkage.**

It should be remembered that only the anomeric hydroxyl group takes part in glycoside formation. This hydroxyl group is known as **glycosidic hydroxyl group.**

The glycosides obtained from different monoaccharides are named by writing the name of group R of the glycosidic linkage followed by the name of the monosaccharide. For example a glycoside of glucose would be named glucoside, a glycoside of mannose would be mannoside and a glycoside of fructose would be called fructoside.

They may be designated as **pyranosides** or **furanosides** according as the ring present is six-membered or five-membered. For example, the two methyl glycosides obtained from glucose are named as **methyl α-D-glucopyranoside and methyl β-D-glucopyranoside.**

Characteristics of glycosides

(*i*) Glycosides are stable in neutral or basic solutions. As is expected from other acetals, glycosides are hydrolysed in acidic solution to form the monosaccharide and the alcohol.

(*ii*) Since glycosides do not split in neutral or basic medium they do not reduce Fehling solution or Tollen's reagent.

(*iii*) For the same reason when either α or β-glycoside is added to water, it remains in the original form and does not get converted into the equilibrium mixture of the two forms. In other words, configuration around anomeric carbon is fixed. Thus, **glycosides do not exhibit mutarotation.**

Interestingly di,tri and higher saccharides are special cases of glycosides where two or more molecules of the same or different monosaccharides are combined to form glycosidic linkages formed between anomeric carbon of one monosaccharide unit and an –OH of another unit. Upon acidic or enzymatic hydrolysis, oligo and polysaccharides change into monosaccharide molecules. For example, sucrose yields an equimolar mixture of α-D-glucose and α-D-fructose.

$$\underset{\text{Sucrose}}{C_{12}H_{22}O_{11}} + H_2O \xrightarrow[\text{Invertase}]{\text{dil. HCl or}} \underset{\alpha\text{-D-Glucose}}{C_6H_{12}O_6} + \underset{\alpha\text{-D-Fructose}}{C_6H_{12}O_6}$$

Conformations of glycosides. Like the monosaccharide from which they are derived, glycosides may also be represented in terms of their conformations. The most stable chair conformations of α- and β-methyl glycosides of glucose and fructose are shown below:

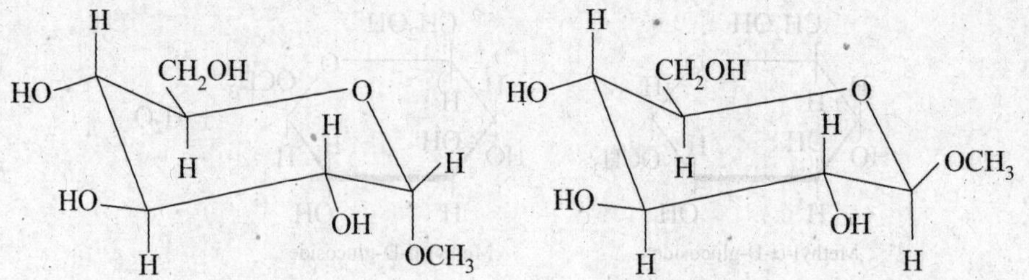

Methyl α-D-glucopyranoside Methyl β-D-glucopyranoside

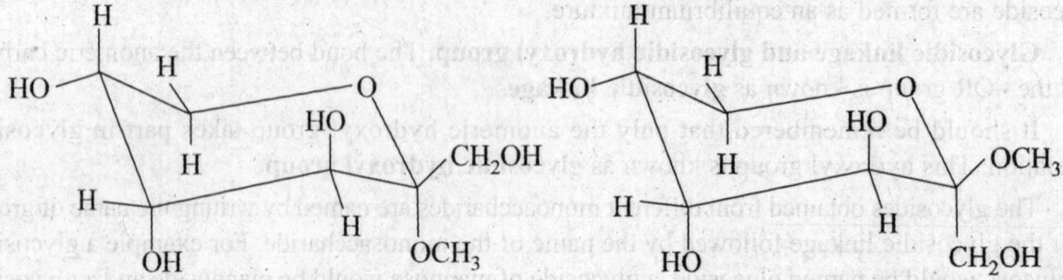

Methyl α-D-fructopyranoside Methyl β-D-fructopyranoside

6.15. FORMATION OF ETHERS

It is possible to convert the –OH groups attached to carbons other than anomeric carbon into alkyl derivatives having ordinary ether C–O–C linkages. For example methyl glucoside can be converted into pentamethyl derivative by treatment with excess dimethyl sulphate in aqueous sodium hydroxide. The function of sodium hydroxide is to convert hydroxyl groups (which are poor nucleophiles) into alkoxide ions (which are good nucleophiles) which then react with dimethyl sulphate by an S_N2 reaction to form methyl ethers.

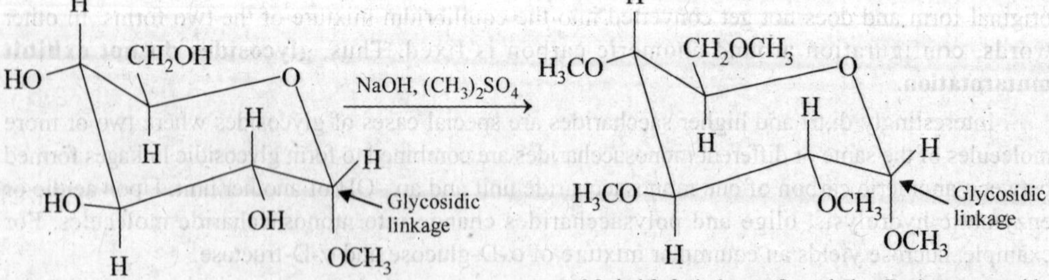

Methyl α-D-glucopyranoside Methyl 2, 3, 4, 6-tetra-O-methyl-α-D-glucopyranoside

Since all the –OH groups are converted into $-OCH_3$ groups, the process is called **exhaustive methylation** or **permethylation.**

For naming these compounds, each $-OCH_3$ group except that of glycosidic linkage is named as an O-methyl group.

When permethylated glycoside is treated with dilute aqueous acid, the methyl glycoside bond gets hydrolysed (since acetals are hydrolysed in acidic solution). But the other methyl groups remain unaffected. This is because ordinary ether groups are stable in dilute aqueous acids. This is shown as under :

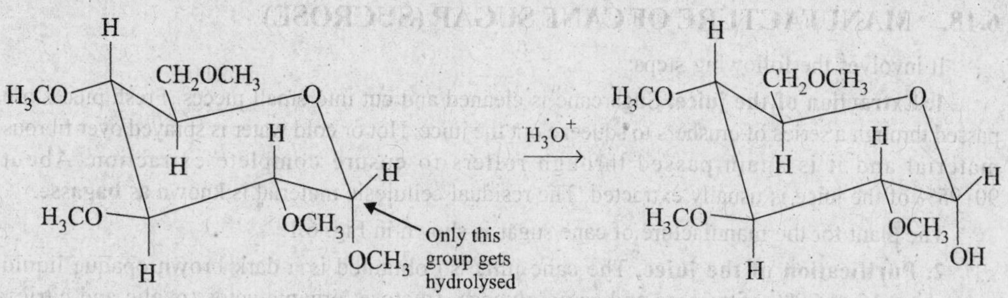

The process of permethylation of glycosides followed by acidic hydrolysis of glycosidic linkage forms an important method for determining the ring size of monosacchrides. This has been illustrated in the case of cyclic structure of glucose.

6.16. FORMATION OF ESTERS

Monosaccharides on treatment with acetic anhydride are converted into ester derivatives which are very useful crystalline compounds. The monosaccharide is treated with acetic anhydride and pyridine (which acts as a mild basic catalyst) when all the hydroxyl groups (including that on anomeric carbon) are converted to ester groups. When carried out at low temperature (273 K), the reaction takes place stereospecifically; α-anomer gives the α-acetate and the β-anomer gives the β-acetate. For example :

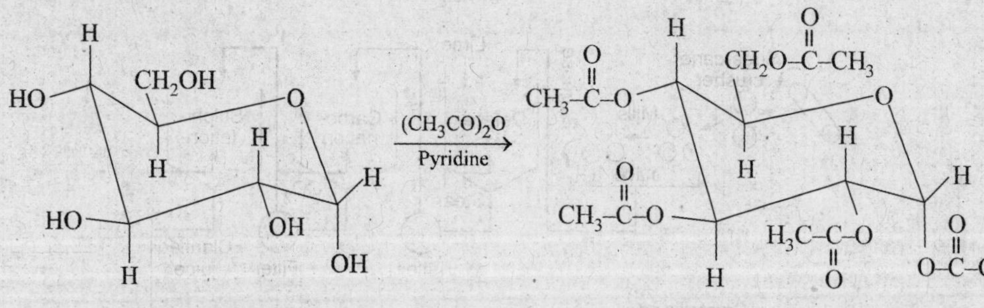

α-D-glucopyranose (α-D-glucose) | Penta-O-acetyl-α-D-glucopyranoside

6.17. HYDROLYSIS OF DISACCARIDES

The disaccharides yield on hydrolysis two monosaccharides. Those disaccharides which yield two hexoses on hydrolysis have a general formula $C_{12}H_{22}O_{11}$. The hexoses obtained on hydrolysis may be same or different *e.g.*,

$$\underset{\text{Sucrose}}{C_{12}H_{22}O_{11}} \longrightarrow \underset{\text{Glucose}}{C_6H_{12}O_6} + \underset{\text{Fructose}}{C_6H_{12}O_6}$$

$$\text{Lactose} \xrightarrow{+H_2O} \text{Glucose + Galactose}$$

$$\text{Maltose} \xrightarrow{+H_2O} \text{Glucose + Glucose.}$$

The disaccharides may be hydrolysed by dil. acids or enzymes. The enzymes which bring about the hydrolysis of sucrose, lactose and maltose are *invertase*, *lactase* and *maltase* respectively.

6.18. MANUFACTURE OF CANE SUGAR (SUCROSE)

It involves the following steps:

1. Extraction of the juice. Sugarcane is cleaned and cut into small pieces. Fresh pieces are passed through a series of crushers to squeeze out the juice. Hot or cold water is sprayed over fibrous material and it is again passed through rollers to ensure complete extraction. About 90–95% of the juice is usually extracted. The residual cellulosic material is known as **bagasse**.

The plant for the manufacture of cane sugar is shown in Fig. 6.1.

2. Purification of the juice. The cane juice so obtained is a dark brown opaque liquid containing 15 to 20% of sucrose and some glucose, fructose, organic acids (oxalic and citric), vegetable proteins, mineral salts, colouring matter, gums and fine particles of bagasse suspended in it in colloidal form. The juice is passed through screens to remove suspended impurities and then subjected to purification as described below:

(*i*) *Defecation.* Fresh juice is run into defecator tanks heated by steam coils, and treated with 2-3% of lime (pH value 7.2) where

(*a*) Proteins and coloidal impurities coagulate out.

(*b*) Free organic acids and phosphates are converted into insoluble calcium salts.

(*c*) Sucrose is partially converted into calcium sucrosate.

The impurities rise to the surface forming a thick scum, which is removed. The precipitated calcium salts are also removed by filtration through canvas. The clear liquid is then pumped to conical tanks for next operation.

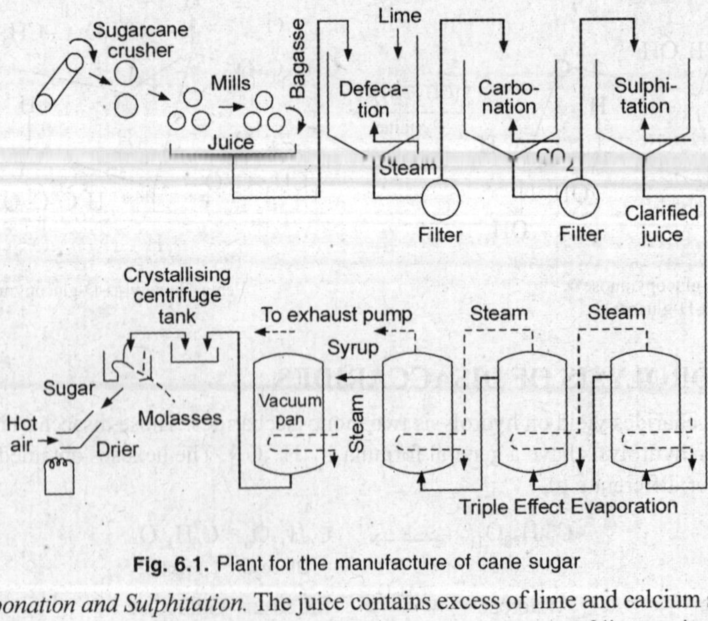

Fig. 6.1. Plant for the manufacture of cane sugar

(*ii*) *Carbonation and Sulphitation.* The juice contains excess of lime and calcium sucrosate. CO_2 is then passed through this juice (*carbonation*) which removes excess of lime as insoluble $CaCO_3$ and decomposes calcium sucrosate to give back sugar.

$$\underset{\text{Cal. sucrosate}}{C_{12}H_{22}O_{11}.3CaO} + 3CO_2 \rightarrow \underset{\text{Cane sugar}}{C_{12}H_{22}O_{11}} + \underset{\text{ppt.}}{3CaCO_3}$$

If SO_2 is used in this operation (*sulphitation*), it being a bleaching agent produces a juice with much lighter colour. At times, carbonation is followed by sulphitation in order to ensure complete neutralisation of lime and decomposition of calcium sucrosate.

$$\underset{\text{Cal.sucrosate}}{C_{12}H_{22}O_{11}.3CaO} + 3SO_2 \rightarrow \underset{\text{Cane sugar}}{C_{12}H_{22}O_{11}} + \underset{\text{ppt.}}{3CaSO_3}$$

3. Concentration and Crystallisation. The clear juice is evaporated to a syrup under reduced pressure in *multiple effect* vacuum pans (Fig. 6.2).

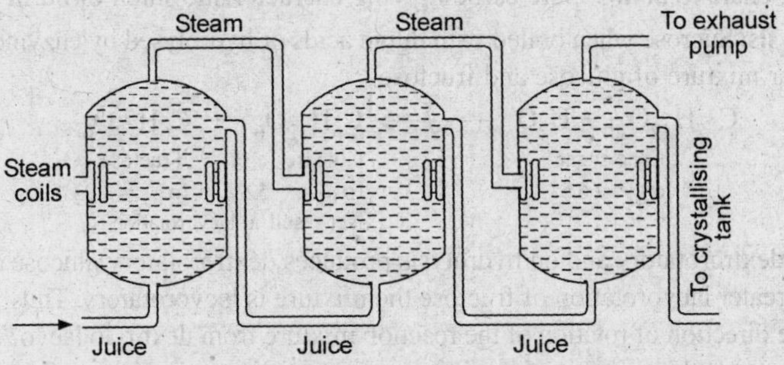

Fig. 6.2. The Multiple effect evaporator

In this arrangement steam from boilers passes through steam coils in the first pan only. The steam produced in first pan is used to boil the juice in second pan at a lower pressure and steam from this pan is sent to the third pan at a still lower temperature and so on. This process of concentration saves a lot of fuel.

The concentrated juice is then sent to vacuum pans where evaporation further reduces water content to 6–8%. On cooling here, partial crystallisation takes place. The contents of this pan are sent to another tank fitted with centrifugal machines.

4. Separation and drying of crystals. Centrifugal machines (Fig. 6.3) separate out sugar crystals from mother liquor known as `molasses'. Sugar crystals are then washed with a little water to remove any impurities sticking to their surface. The crystals are then dried by a current of hot air. In case the sugar crystals so obtained are brown in colour and have slight odour, these are further purified. The crystals are dissolved in water and filtered through animal charcoal or *norit* (activated coconut charcoal). The colourless filtrate is concentrated in vacuum pans and the crystals obtained by centrifuging the syrup.

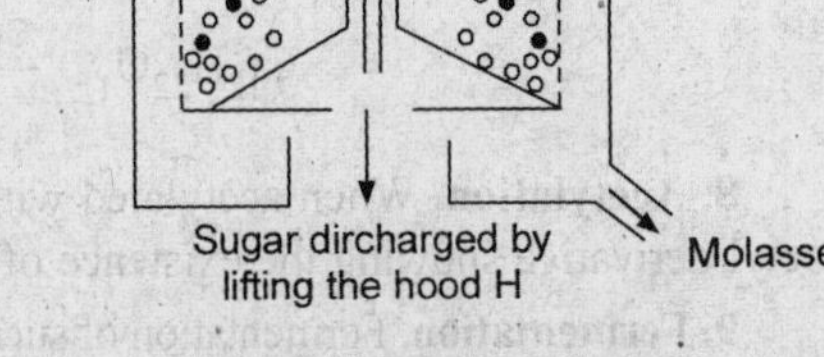

Fig. 6.3. Separation of sugar crystals by centrifugation

By-products of Sugar Industry

(*i*) *Bagasse*. It is the cellulosic material left after the extraction of juice from sugarcane. It is used as a fuel and fodder in our country but in U.S.A., it is used in the manufacture of building and insulating material known as *Celotex*. It is also used in the manufacture of low grade paper.

(*ii*) *Molasses*. It is the mother liquor left after the separation of sugar crystals. It contains nearly 60% fermentable sugar. It is largely used now in the manufacture of ethyl alcohol and also CO_2. It is also used to some extent as a manure and road binder.

6.19. PROPERTIES OF CANE SUGAR, $C_{12}H_{22}O_{11}$

1. It is a colourless, crystalline substance, sweet in taste. It is very soluble in water and the solution is dextrorotatory $[(\alpha)_D = +66.5]$.

2. Effect of heat. Sucrose on heating slowly and carefully melts and then if allowed to cool, it solidifies to pale-yellow glassy mass called '*barley sugar*'.

When heated to 473 K, it loses water to form a brown amorphous mass called *caramel.* On strong heating it chars to almost pure carbon giving characteristic smell of burnt sugar.

3. Hydrolysis. Sucrose when boiled with dilute acids or hydrolysed by enzyme *invertase*, yields an equimolecular mixture of glucose and fructose.

$$\underset{\substack{\text{Cane sugar} \\ [\alpha]_D = 66.5}}{C_{12}H_{22}O_{11}} + H_2O \longrightarrow \underset{\substack{\text{Glucose} \\ [\alpha]_D = +52.7}}{C_6H_{12}O_6} + \underset{\substack{\text{Fructose} \\ [\alpha]_D = -92.4}}{C_6H_{12}O_6}$$

Net result is laevorotation

Sucrose is dextrorotatory and on hydrolysis produces dextrorotatory glucose and laevorotatory fructose. With greater laevorotation of fructose the mixture is laevorotatory. Thus, there is a change (inversion) in the direction of rotation of the reaction mixture from dextro to laevo. This phenomenon is called *inversion* and the enzyme which brings about this inversion is called *invertase*.

4. Formation of Sucrosates. Sucrose solution reacts with calcium, barium and strontium hydroxides to form sucrosates.

$$\underset{\text{Cane sugar}}{C_{12}H_{22}O_{11}} + 3Ca(OH)_2 \longrightarrow \underset{\text{Calcium sucrosate}}{C_{12}H_{22}O_{11}.3(CaO)} + 3H_2O$$

The sucrosate decomposes when carbon dioxide is passed in the solution.

5. Action of sulphuric acid. Concentrated sulphuric acid abstracts water and charring takes place. Carbon produced gets oxidised to carbon dioxide and sulphur dioxide is produced due to reduction of the acid.

$$C_{12}H_{22}O_{11} + H_2SO_4 \longrightarrow 12C + [11H_2O + H_2SO_4]$$

$$C + 2H_2SO_4 \longrightarrow CO_2 + SO_2 + 2H_2O$$

6. Action of nitric acid. Concentrated nitric acid oxidises cane sugar to oxalic acid.

$$C_{12}H_{22}O_{11} + \underset{(\text{From } HNO_3)}{18O} \longrightarrow 6\underset{\text{Oxalic acid}}{\begin{array}{c} COOH \\ | \\ COOH \end{array}} + 5H_2O$$

7. Action of hydrochloric acid. When boiled with concentrated hydrochloric acid, laevulinic acid is obtained.

$$C_{12}H_{22}O_{11} \xrightarrow{\text{Conc. HCl}} \underset{\text{Laevulinic acid}}{CH_3.CO.CH_2.CH_2COOH}$$

8. Acetylation. When acetylated with acetic anhydride and sodium acetate, it gives an octa-acetyl derivative showing the existence of eight –OH groups in sucrose molecule.

9. Fermentation. Fermentation of sucrose is brought about by yeast when the enzymes *invertase* hydrolyses sucrose to glucose and fructose and *zymase* converts them to ethyl alcohol.

$$C_{12}H_{22}O_{11} + H_2O \xrightarrow{\text{Invertase}} \underset{\text{Glucose}}{C_6H_{12}O_6} + \underset{\text{Fructose}}{C_6H_{12}O_6}$$

$$C_6H_{12}O_6 \xrightarrow{\text{Zymase}} 2C_2H_5OH + 2CO_2$$

10. Methylation. When treated with dimethyl sulphate in presence of alkali it gives octamethyl derivative.

Note. *Sucrose does not give positive test with Fehling's solution and Tollen's reagent nor does it react with reagents like HCN, NH_2OH and phenyl hydrazine. This shows the absence of monosaccharide character and absence of aldehydic and ketonic groups. It is stable to alkali and does not show mutarotation.*

6.20. TESTS FOR CANE SUGAR

(1) When heated in a test tube it gives a characteristic odour which is named as 'smell of burnt sugar'.

(2) It gives positive Molische's test. Add alcoholic solution of α-naphthol to cane sugar solution followed by conc. H_2SO_4 along the side of the test tube. A violet ring is obtained.

(3) When it is kept in contact with concentrated sulphuric acid, charring takes place in cold. When warmed a mixture of carbon dioxide and sulphur dioxide are evolved.

(4) Upon adding Fehling's solution to a sucrose solution which has been boiled with dilute hydrochloric acid, a red precipitate of cuprous oxide is obtained. *It is to be noted that no red precipitate is obtained if sucrose solution has not been hydrolysed into glucose and fructose by dilute hydrochloric acid.*

(*Distinction from glucose and fructose*):

It does not turn yellow or brown when heated with caustic soda solution nor gives osazone when treated with phenyl hydrazine.

6.21. STRUCTURE OF SUCROSE (CANE SUGAR)

A brief description of the structure of sucrose is given below :

(1) It has a molecular formula $C_{12}H_{22}O_{11}$.

(2) It does not reduce Fehling's solution or Tollen's reagent. Also it does not react with carbonyl group reagents like hydroxyl amine and phenyl hydrazine. Sucrose does not show mutarotation and forms no methyl glycosides like monosaccharides.

All these negative tests show that sucrose has no free aldehydic or ketonic group.

(3) If hydrolysed, sucrose yields equimolecular mixture of D(+) glucose and D(–) fructose. Since sucrose does not give any indication of the presence of free aldehydic or ketonic group, it is obvious that the union between glucose and fructose must have taken place through C-1 carbon of glucose (carrying aldehydic group) and C-2 carbon of fructose (carrying ketonic group).

(4) It gives octa-acetyl derivative upon acetylation.

(5) Upon complete methylation, octamethyl sucrose is obtained. Formation of octaacetyl and octamethyl derivatives shows *the presence of eight –OH groups in the molecule.*

(6) The hydrolysis of octamethyl sucrose gives 2, 3, 4, 6-tetra-O-methyl derivative of glucose and 1, 3, 4, 6-tetra-O-methyl fructose. Formation of 2, 3, 4, 6, tetra-O-methyl derivative of glucose indicates a ring between C-1 and C-5 of glucose (as shown in structure of glucose) and 1, 3, 4, 6-tetra-O-methyl fructose indicates a ring between C_2 and C_5 of 'fructose'. Hence glucose is pyranoside and fructose is furanoside.

(7) Study of optical rotation and the behaviour of sucrose towards various enzymes have verified that D-glucose unit has α-configuration and D-fructose unit has β-configuration.

With all these facts in view Haworth (1927) suggested the following structure for sucrose:

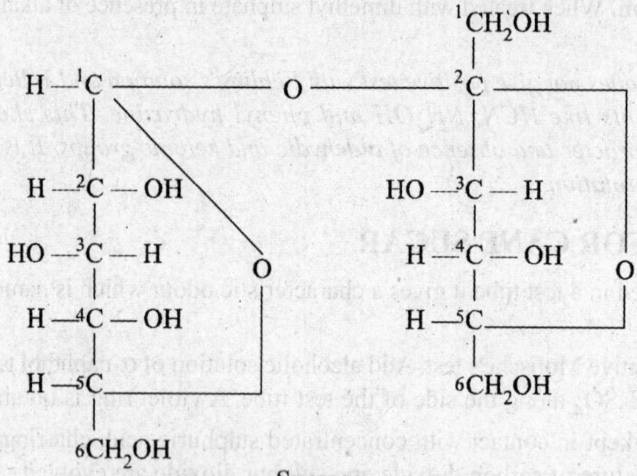

Sucrose

Hudson has designated sucrose as 1-α-D-glucopyranoside-2-β-D-fructofuranoside.

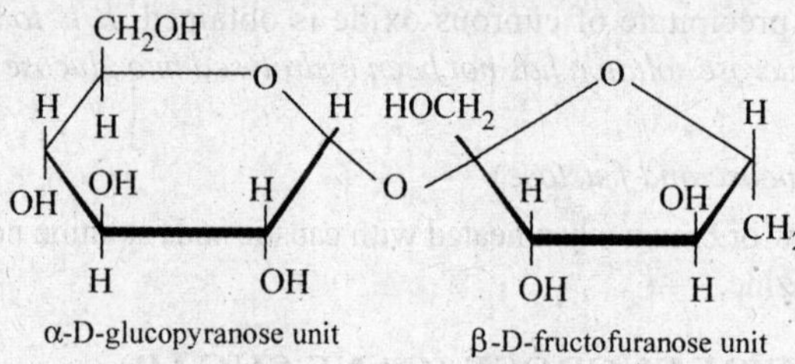

Sucrose

Its conformational structure is written as under:

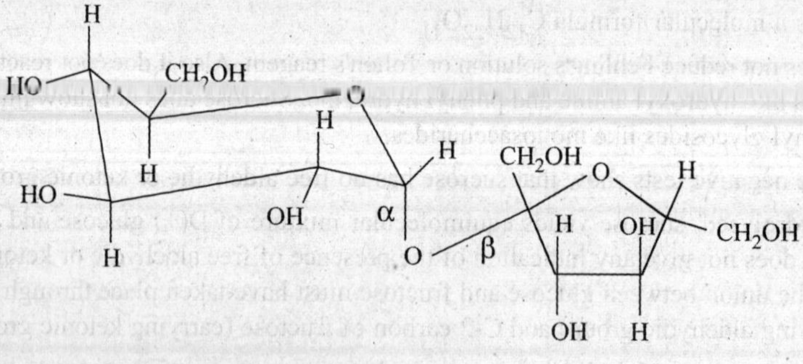

6.22. STRUCTURE OF MALTOSE

Structure of maltose is briefly discussed as under :

Maltose is obtained by hydrolysis of starch using β-amylase.

$$\text{starch} \xrightarrow{\beta\text{-amylase}} \text{Maltose}$$

1. Molecular formula of maltose as determined by usual analytical techniques comes out to be $C_{12}H_{22}O_{11}$.

2. Maltose on heating with dil HCl yields two equivalents of D(+) glucose. It is thus made up of two units of glucose.

$$\underset{\text{Maltose}}{C_{12}H_{22}O_{11}} + H_2O \xrightarrow{H^+} \underset{\text{Glucose}}{2C_6H_{12}O_6}$$

3. Maltose reduces Tollen's reagent and Fehling solution. This means it is a reducing sugar. It also forms an osazone and shows the phenomenon of mutarotation.

$$\alpha\text{-form} \rightleftharpoons \text{Equilibrium Mixture} \rightleftharpoons \beta\text{-form}$$

4. Maltose on treatment with bromine water gives a mono-carboxylic acid. This shows that maltose contains a carbonyl group present in a reactive hemiacetal form as in case of glucose.

5. The following structure has been proposed for maltose. This structure explains the observations described above

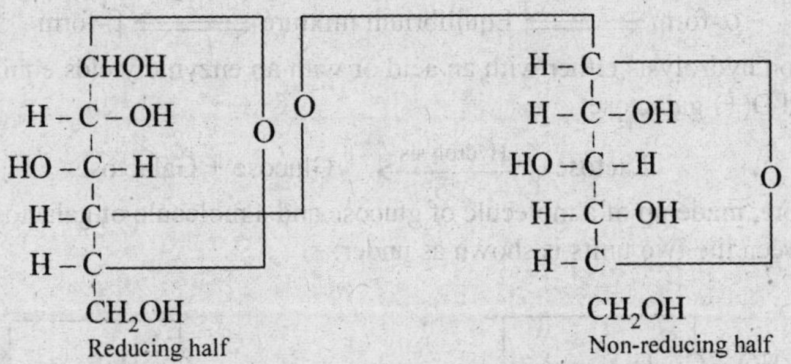

The structure can also be represented as Haworth projection formula

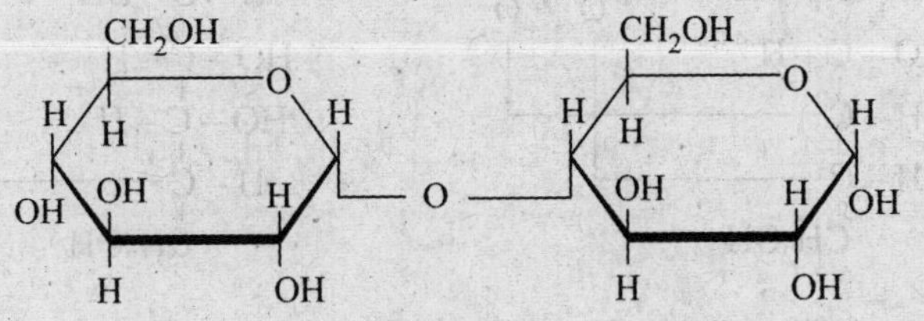

Maltose (α-form)

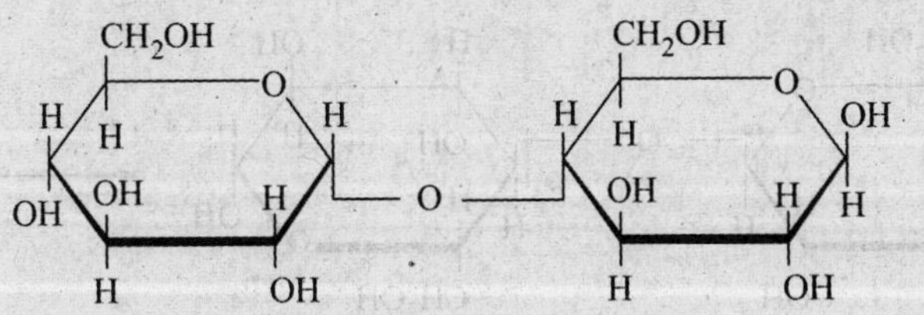

Maltose (β - form)

Conformational formula of α-form of maltose may be represented as

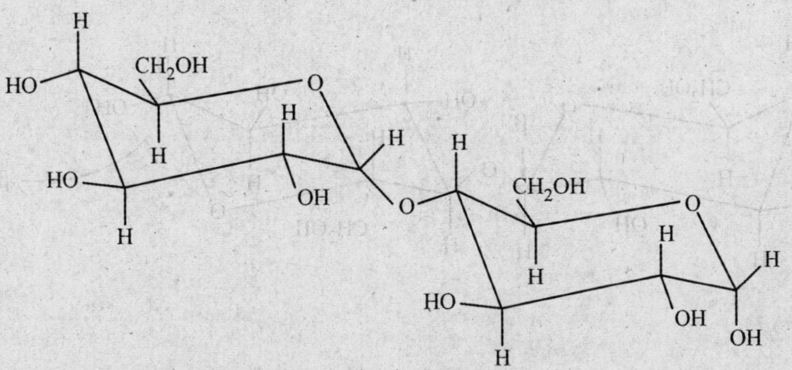

6.23. STRUCTURE OF LACTOSE

Lactose is found in the milk of mammal and in *whey*, a bye-product in the manufacture of cheese.

A brief description of the structure of lactose is given below :

1. Its molecular formula is determined to be $C_{12}H_{22}O_{11}$.

2. It reduces Fehling solution and forms an osazone. It means its is a reducing sugar. It also exhibits mutarotation.

$$\alpha\text{-form} \rightleftharpoons \text{Equilibrium mixture} \rightleftharpoons \beta\text{-form}$$

3. Lactose on hydrolysis either with an acid or with an enzyme yields equivalent amounts of D(+) glucose and D(+) galactose.

$$\text{Lactose} \xrightarrow{\text{Hydrolysis}} \text{Glucose + Galactose}$$

It is, therefore, made up of a molecule of glucose and a molecule of galactose linked together. The linkage between the two units is shown as under:

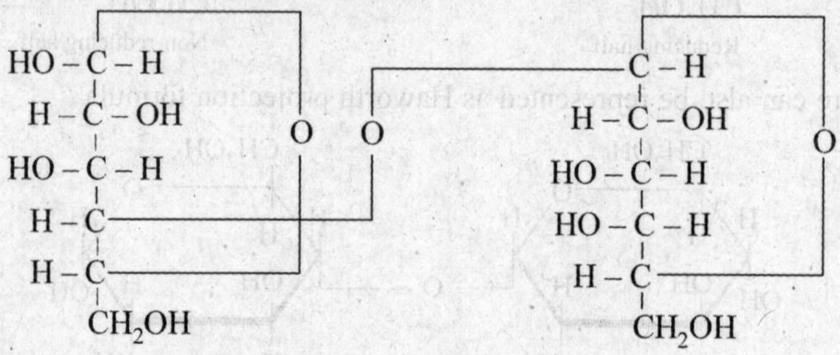

Haworth projection formula of lactose can be represented as

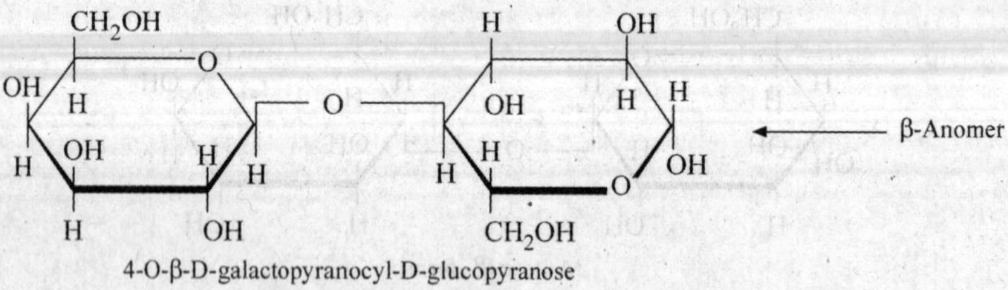

4-O-β-D-galactopyranocyl-D-glucopyranose

Its conformational formula may be written as

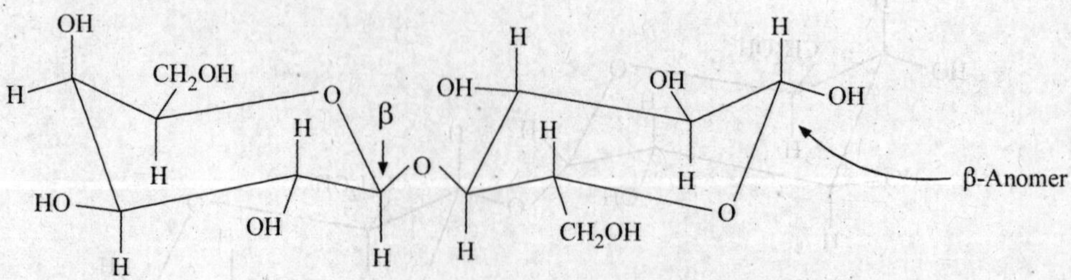

POLYSACCARIDES

6.24. STARCH

Starch is most widely distributed in vegetable kingdom. In nature, it is transformed into complex polysaccharides like gum and cellulose and into simpler mono and disaccharides by enzymes working in vegetable kingdom. Its rich sources are potatoes, wheat, maize, rice, barley and arrow root. It is interesting to note that no two sources give identical starch.

Physical properties

It is a white, amorphous substance with no taste or smell. It is insoluble in water but when starch is added to boiling water the granules swell and burst forming colloidal, translucent suspension.

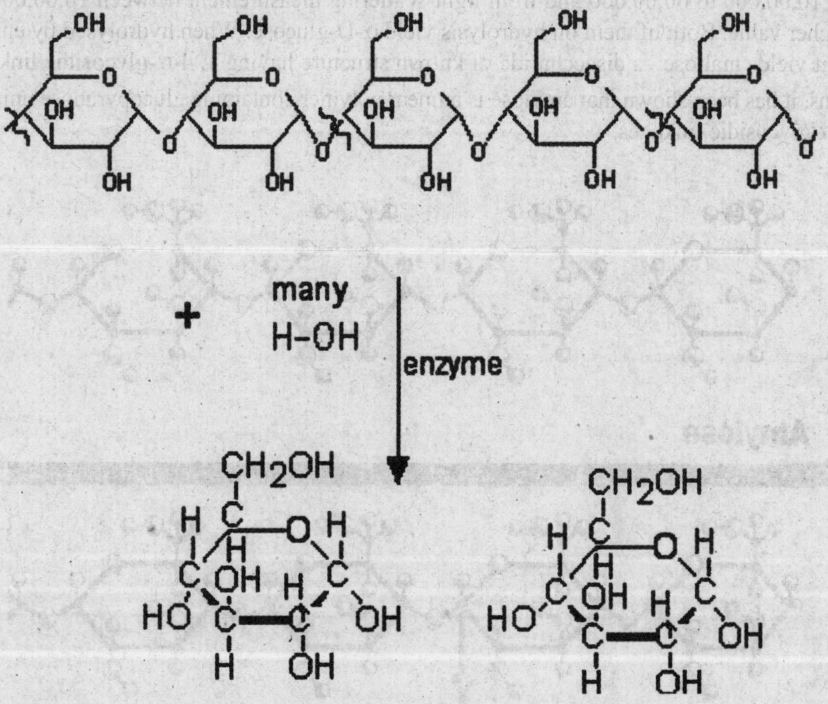

Hydrolysis of starch

Chemical properties of starch

(*i*) When heated to a temperature between 200 to 250°C it changes into dextrin. At higher temperature charring takes place.

(*ii*) Starch, when boiled with dilute acid, yields ultimately glucose.

$$\underset{\text{Starch}}{(C_6H_{10}O_5)_n} \longrightarrow \underset{\text{Dextrin}}{(C_6H_{10}O_5)_{n1}} \longrightarrow \underset{\text{Maltose}}{C_{12}H_{22}O_{11}} \longrightarrow \underset{\text{Glucose}}{C_6H_{12}O_6}$$

When hydrolysed with enzyme diastase, maltose is obtained.

$$\underset{\text{Starch}}{2(C_6H_{10}O_5)_n} + nH_2O \xrightarrow{\text{Diastase}} \underset{\text{Maltose}}{nC_{12}H_{22}O_{11}}$$

(*iii*) Starch solution gives a blue colour with a drop of iodine solution. The blue colour disappears on heating and reappears on cooling. *In fact it is the amylose that gives a blue colour with iodine; the amylopectin gives a red brown colour with iodine.*

(*iv*) When heated with a mixture of concentrated sulphuric and nitric acids it gives nitro-starch.

6.25. STRUCTURE OF STARCH

The exact chemical nature of starch varies from source to source. Even the starch obtained from same source consists of two fractions (*i*) *amylose*, and (*ii*) *amylopectin*, present in 1 : 3 or 1 : 4 ratio.

Molecular weight of amylose from ultracentrifuge method has been shown to lie between 15,000 to 225,000 whereas osmotic pressure measurement suggests that it may have value between 1,00,000 to 20,00,000. Molecular weight of amylopectin from osmotic pressure data has been shown to be between 10,00,000 to 60,00,000 and from light-scattering measurement between 10,00,000 to and much higher value. Both of them on hydrolysis yield α-D-glucose. When hydrolysed by enzyme β-*maltase* it yields maltose –a disaccharide of known structure having 1, 4-α-glycosidic linkage.

Thus, it has been shown that amylose is a linear polymer containing glucopyranose units joined by 1, 4-α-glycosidic linkages.

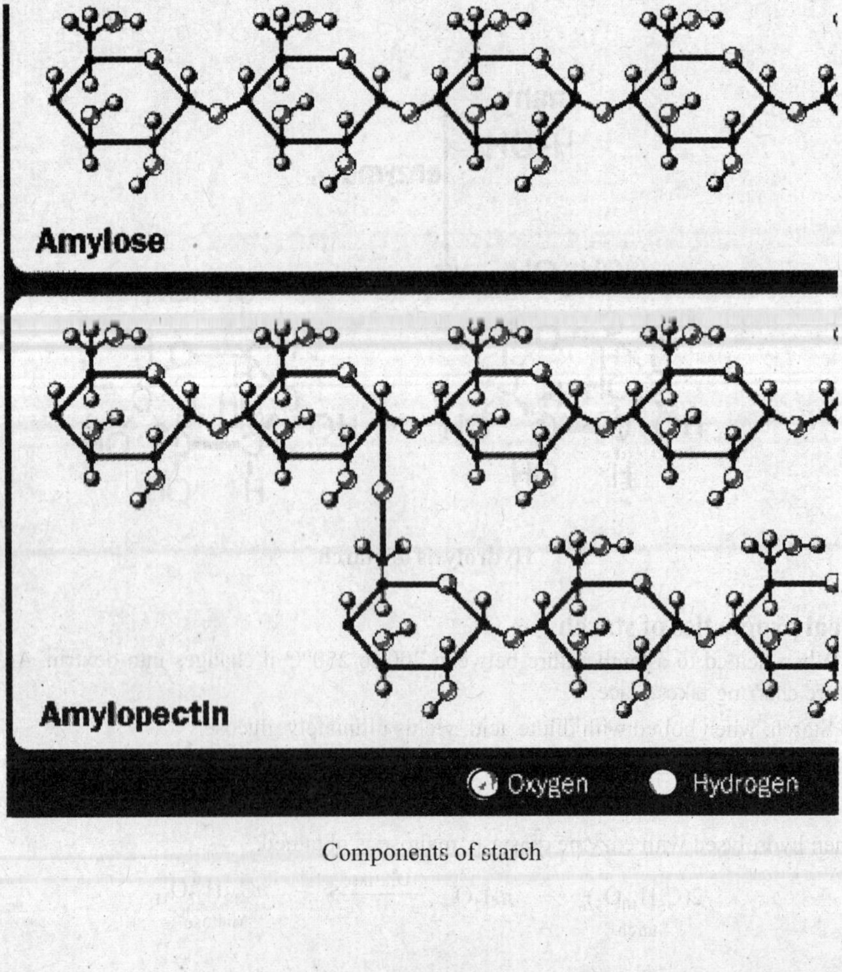

Components of starch

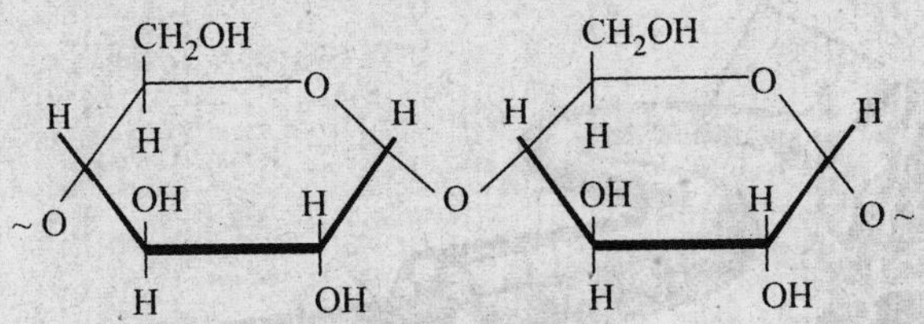

Amylose

Amylopectin is a highly branched polymer. The branches consist of 20 to 25 glucose units joined by 1, 4-α-linkages and are joined to each other by 1, 6-α-glycosidic linkages.

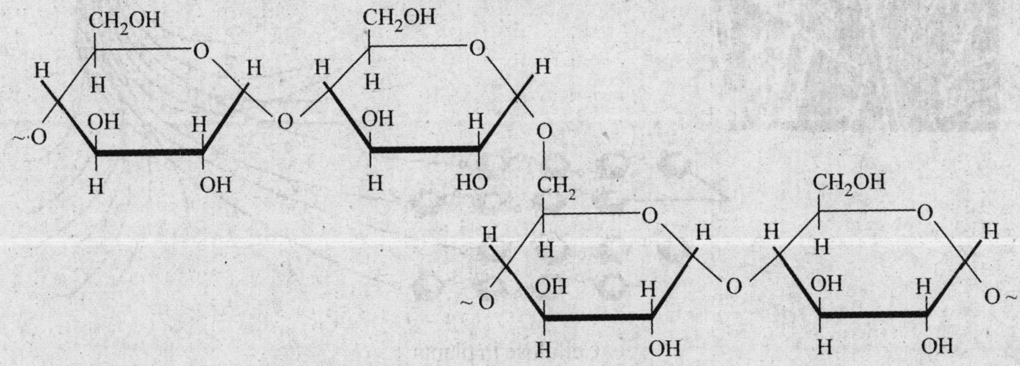

Amylopectin

6.26. CELLULOSE

Cellulose $(C_6H_{10}O_5)_n$ is widely distributed in plant kingdom and plants maintain their structure due to the support of fibrous material which is cellulose. Cotton is almost pure cellulose and jute, hemp, wood, paper, etc. are all different forms of cellulose.

Preparation

(1) *From Cotton.* Cotton is treated with organic solvents to remove fats and waxes. It is then treated with hydrofluoric acid to remove mineral matter. The crude cellulose thus obtained is bleached by alkali and sodium hypochlorite to get pure cellulose as white amorphous mass.

(2) *From Wood.* Cellulose can also obtained by taking wood shavings or wood chips and boiling them with sodium bisulphite or caustic soda to remove lignin and resins. It is then treated successively with dilute acid, water and organic solvents to remove other substances. Lastly it is bleached with sodium or calcium hypochlorite to get white amorphous cellullose.

Physical properties

Cellulose is a colourless, amorphous substance with organised fibrous structure which is characteristic of the source. It is insoluble in water but dissolves in ammoniacal solution of cupric hydroxide (Schweitzer's reagent). Cellulose also dissolves in a solution of zinc chloride in hydrochloric acid.

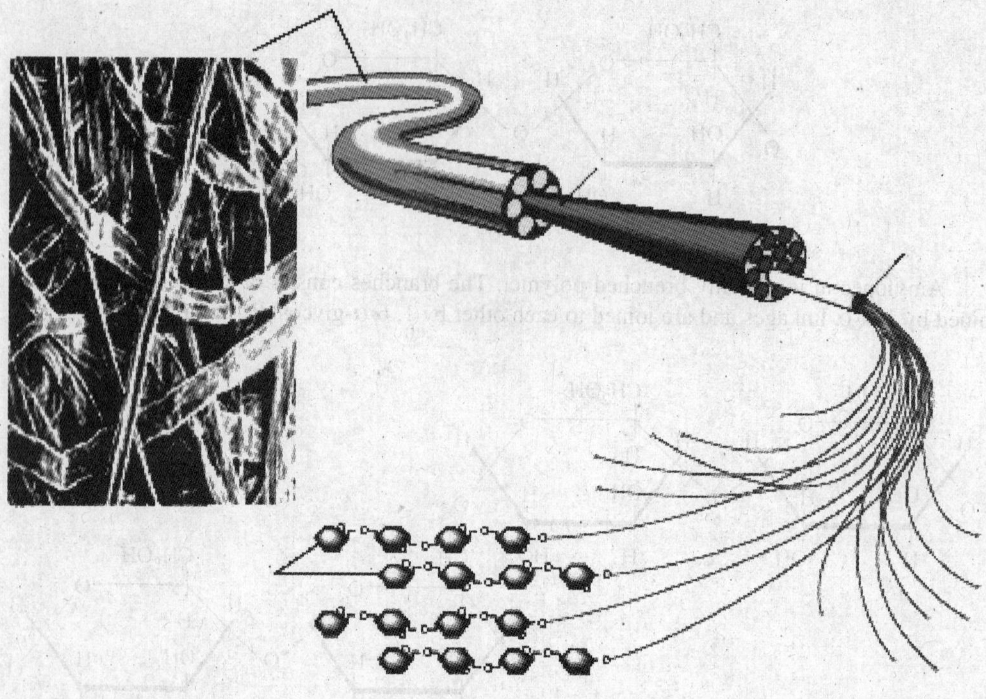

Cellulose in plants

Chemical properties

(1) When cellulose is treated with concentrated sulphuric acid in cold it slowly passes into solution. The solution when diluted with water, precipitates a starch like substances *amyloid*.

(2) When boiled with dilute sulphuric acid it is completely hydrolysed into D-glucose.

$$\underset{\text{Cellulose}}{(C_6H_{10}O_5)_n} + nH_2O \longrightarrow \underset{\text{Glucose}}{nC_6H_{12}O_6}$$

(3) It forms *hydrocellulose* with dilute acids and water on mechanical communication.

(4) When treated with 20% caustic soda solution its appearance becomes smooth and lustrous. This property was noted by John Mercer* in 1884.

(5) Cellulose has D-glucose unit in its molecule which has 3 hydroxyl groups free for esterification. Thus when treated with a mixture of concentrated nitric and sulphuric acids it forms mono, di and tri nitrates of cellulose. Cellulose nitrates are used in preparation of explosives.

Similarly when treated with a mixture of glacial acetic acid and acetic anhydride, cellulose yields a mixture of di and tri acetates. Cellulose acetate is used in the manufacture of synthetic fibres and paints.

---

* A cotton fibre or cloth is immersed in strong alkali solution and stretched to neutralise the contraction due to swelling of the cotton. This gives it a lustrous appearance. The process is called mercerisation and is of industrial importance in textile industry.

6.27. STRUCTURE OF CELLULOSE

The structure of cellulose is briefly discussed as under :

Like starch the molecular weight of cellulose also varies with the source. Its molecular weight as determined by ultracentrifugal method lies between 1–2 million.

On complete hydrolysis it also yields glucose, like starch, but glucose is present in cellulose as β-D-glucopyranose unit. Enzymatic hydrolysis of cellulose yields a disaccharide cellobiose having the glucose units linked through 1, 4-β-glycosidic linkages.

Cellulose, therefore, is considered to be a linear polymer having β-D-glucopyranose units joined by 1, 4-β-linkages.

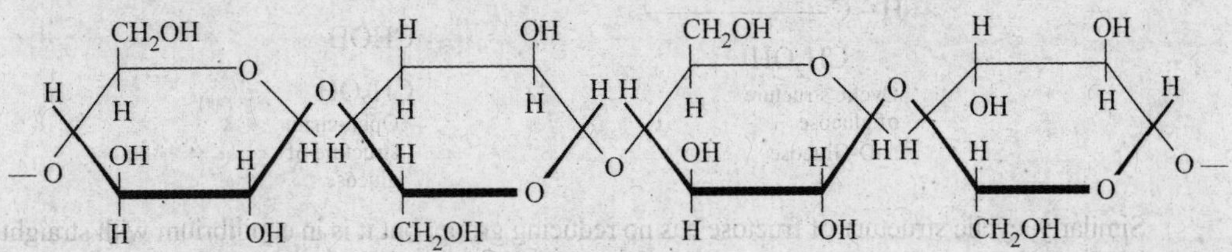

Cellulose

Uses of cellulose

(1) It is used as such in the manufacture of paper and cloth.

(2) Cellulose nitrates are used in the manufacture of explosives, medicines, paints and lacquers.

(3) Cellulose nitrate with camphor yields celluloid which is used in manufacturing toys, decorative articles and photographic films.

(4) Cellulose acetate is used in rayon manufacture and in plastics.

6.28. REDUCING AND NON-REDUCING SUGARS

Reducing sugars are those carbohydrates which reduce the solution of alkaline copper sulphate (Fehling solution) into a red precipitate of Cu_2O and reduce Tollen's reagent into silver mirror. Non-reducing sugars are those which don't respond to these tests. In reducing sugars, there is an aldehydic group or an alcoholic group adjacent to a ketonic group. Such groups are easily oxidised and hence compounds containing such group act as reducing agents.

(1) Fehling solution test

$$CuSO_4 + 2NaOH \longrightarrow Cu(OH)_2 + Na_2SO_4$$

$$Cu(OH)_2 \longrightarrow CuO + H_2O$$

$$\underset{\text{(from sugar)}}{-CHO + 2CuO} \longrightarrow -COOH + \underset{\text{Red ppt.}}{Cu_2O}$$

(2) Tollen's reagent test

$$AgNO_3 + NH_4OH \longrightarrow AgOH + NH_4NO_3$$

$$2AgOH \longrightarrow Ag_2O + H_2O$$

$$\underset{\text{(from carbohydrate)}}{-CHO + Ag_2O} \longrightarrow -COOH + \underset{\text{Silver mirror}}{2Ag\downarrow}$$

As non-reducing sugars do not contain these groups in the free state, they do not respond to these tests.

Glucose and fructose are reducing sugars because they contains either an aldehydic group or an alcoholic group adjacent to a ketonic group.

Take the case of glucose. Although there is no free aldehydic group in the cyclic structure, but as the cyclic form is in equilibrium with straight-chain structure, there is always some concentration of aldehydic species and hence it gives the Fehling solution and Tollen's reagent tests.

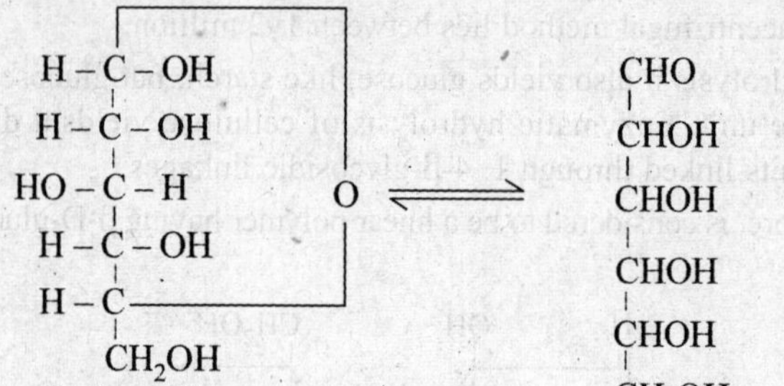

Cyclic structure of glucose α-D-Glucose

Open-chain structure of glucose

Similarly, cyclic structure of fructose has no reducing group, but it is in equilibrium with straight chain structure which has alcoholic groups adjacent to a ketonic group, because of which it gives reducing properties.

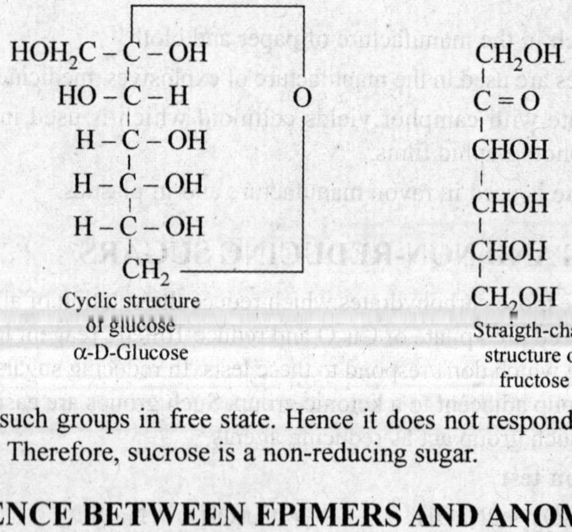

Cyclic structure of glucose α-D-Glucose

Straigth-chain structure of fructose

Sucrose has no such groups in free state. Hence it does not respond to Fehling solution and Tollen's reagent tests. Therefore, sucrose is a non-reducing sugar.

6.29. DIFFERENCE BETWEEN EPIMERS AND ANOMERS

Epimers

Sugars having a common configuration (arrangement of –H and –OH around carbon atoms), except at α-carbon or carbon no. 2 are called epimers. Examples are glucose and mannose. They are epimers of each other.

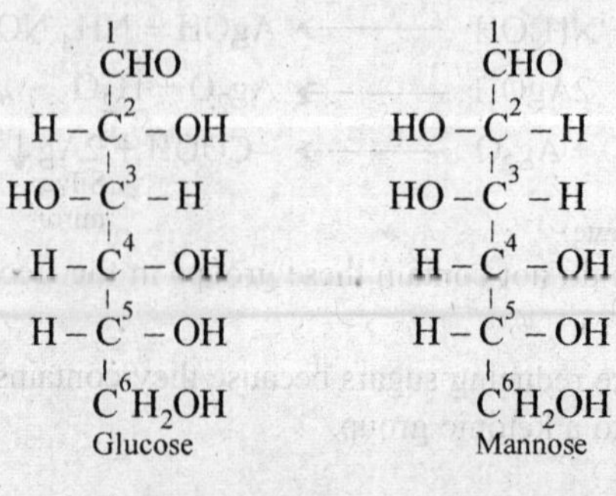

Glucose Mannose

We observe that glucose and mannose differ in their structures only at carbon no. 2.

Both the compounds give the same osazone with phenyl hydrazine. This reactions is completed in 3 steps. (Refer to the chemical properties of glucose in this chapter). In the first step, one molecule of phenyl hydrazine reacts with the aldehydic group (carbon no. 1). This step will be common to both the above compounds, *viz*., glucose and mannose. In the second step, two hydrogens are taken out by the phenyl hydrazine molecule from carbon no. 2 to convert it into > C = O. This step will also be given by both compounds in spite of different configuration at C-2. Third step is again common to both compounds. Therefore, the osazone that will be obtained from the two compounds will be the same.

Anomers

Sugars having a common configuration except at carbon no. 1 are called anomers. α-D-glucose and β-D-glucose are enomers because they have the same configuration except at carbon no. 1 in the cyclic structure.

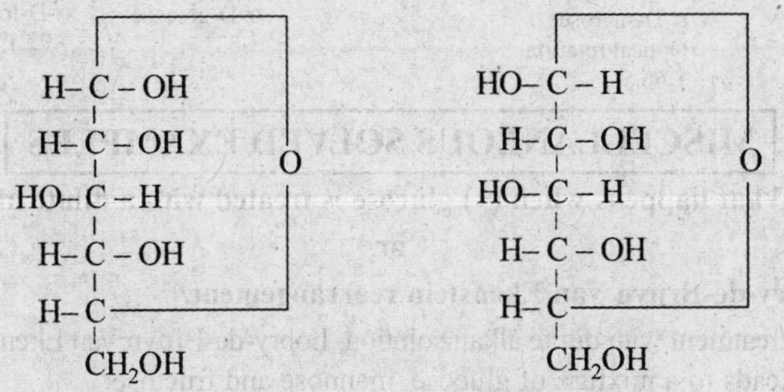

α-D-glucose β-D-glucose

6.30. MUTAROTATION

Glucose contains asymmetric carbon atoms, therefore it exhibits optical properties like optical rotation i.e. rotation of plane polarised light when it passes through glucose solution. It is observed that when D-glucose having m.p. 419 K is dissolved in water, it shows an optical rotation of + 112°. Gradually it drops to + 52.7°.

On the other hand, when D-glucose having m.p. 423 K is dissolved in water, it shows a value of + 19° for optical rotation. Gradually the value rises to + 52.7°. This happens because they attain a state of equilibrium having a fixed composition after sometime thereby giving the same value of optical rotation *i.e.* 52.7°. The compound having m.p. 419 K is α-D-glucose with optical rotation value as + 112° and the compound having m.p. equal to 423 K is β-D-glucose and has optical rotation + 19°. On dissolving either α or β-D-glucose in water, we shall obtain, after sometime, an equilibrium mixture of the two forms having 36% of α-form and 64% of β-form having optical rotation value 52.7°. This can be calculated mathematically also.

$$\text{Optical rotation of mixture} = \frac{\text{Percentage of } \alpha\text{-form} \times \text{opt. rotation of } \alpha\text{-form} + \text{Percentage of } \beta\text{-form} \times \text{opt. rotation of } \beta\text{-form}}{100}$$

$$= \frac{36 \times 112 + 64 \times 19}{100} = 52.48$$

The slight difference is due to approximation that has been made in the percentage of the two isomers.

Thus mutarotation may be defined as the phenomenon of change in specific rotation of an optically active compound with time to an eqiuilibrium value of specific rotation. (For reason of mutarotation, see question on ring structure of glucose.)

6.31. INVERSION OF SUGAR

A change in the sign of specific rotation of a sugar solution, after hydrolysis, is called inversion of sugar.

Sucrose in an optically active compound having sp. rotation of + 66.5°. When sucrose solution is subjected to hydrolysis in the presence of an acid or an enzyme, we observe that the sp. rotation has reversed its sign from positive to negative or from dextrorotatory to laevorotatory. This is because sucrose gets hydrolysed to glucose and fructose. Out of the two substances, glucose and fructose, the latter has a negative sp. rotation and to a larger magnitude. As a consequence, the resulting mixture is laevorotatory.

$$C_{12}H_{22}O_{11} + H_2O \longrightarrow C_6H_{12}O_6 + C_6H_{12}O_6$$

α-D-sucrose, Optical rotation + 66.5° → α-D-glucose + 53° + α-D-fructose – 92.3°

MISCELLANEOUS SOLVED EXAMPLES

Example 1. What happens when (+) glucose is treated with a dilute alkali solution?

or

What is Lobry-de-Bruyn Van Ekenstein rearrangement.

Solution. On treatment with dilute alkali solution, Lobry-de-Bruyn Van Ekenstein rearrangement takes place which leads to a mixture of glucose, mannose and fructose.

For complete reaction, see the section on properties of (+) glucose.

Example 2. Glucose and fructose have the same formula. Glucose has an aldehydic group, while fructose has a ketonic group but both reduce Fehling solution. Explain.

Solution. This is because Fehling solution is alkaline in nature. In alkaline medium, fructose undergoes Lobry-de-Bruyn Van Ekenstein rearrangement when an equilibrium mixture of glucose, mannose and fructose is obtained (for complete reaction, see the section on properties of glucose).

It is because of the glucose fraction in the solution that fructose reduces Fehling solution.

Example 3. Why does D-glucose not give Schiff's test?

Solution. D-glucose does not give Schiff's test because the aldehydic group in glucose is not completely free. It is involved in acetal formation with the hydroxyl group at carbon no. 5 to give the cyclic structure.

Example 4. Draw the chair conformations of α-D(+) glucose and β-D(+) glucose. Which is more stable and why?

Solution. The chair conformations of the two forms of glucose are given below:

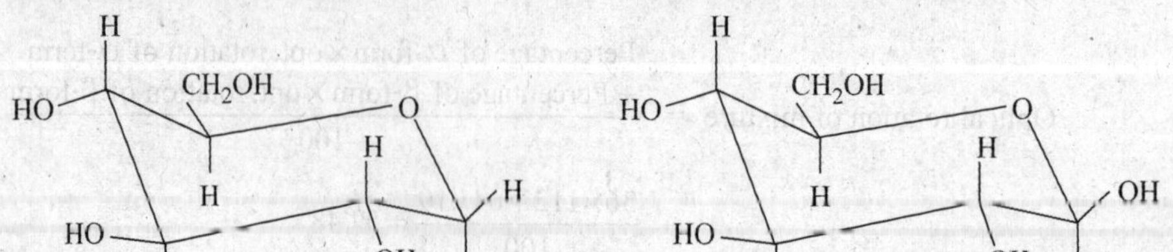

Out of α- and β-forms of glucose, β-form is more stable. This is because all bulky groups like –OH and – CH_2 OH are in equitorial positions in β-form while the bulky group –OH at carbon no.1 in α-form is in axial position. In general, there is less crowding and hence greater stability if the groups are oriented in equatorial positions. Hence, β-form is more stable than α-form.

Example 5. Explain why unlike glucose, neither α-nor β-methyl glucoside reduce Tollen's reagent or Fehling solution.

Solution. Tollen's reagent and Fehling solution tests are given by glucose because its cyclic structure is converted into linear structure which contains –CHO group. Aldehydic group gives Tollen's and Fehling solution tests. But methyl glucosides are not hydrolysed readily to obtain cyclic and open-chain structure of glucose. In the absence of –CHO groups, these tests are not given.

Example 6. What is amylose? Give its structure.

Solution. Amylose is a fraction of starch. It has the structure as given below:

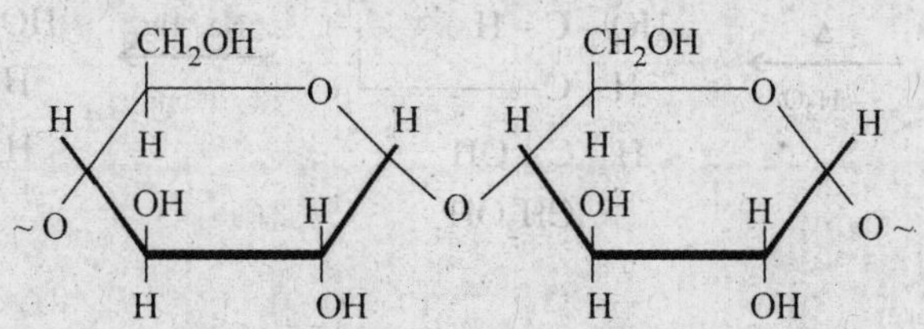

Example 7. Why sucrose is known as invert sugar? How do you account for the fact that sucrose does not reduce Fehling solution or form oxime? Write down the structure of sucrose molecule.

Solution. Sucrose is an optically active compound. It is dextrorotatory. But on hydrolysis, it produces glucose and fructose in equivalent amounts. Glucose is dextrorotatory and fructose is laevorotatory. But the laevorotation of fructose is much more in magnitude than the dextrorotation of glucose, with the result, that solution of sucrose exhibits laevorotation. As inversion in optical rotation takes place, it is called invert sugar.

Sucrose does not reduce Fehling solution nor does it form oxime. These reactions are give by a compound containing a carbonyl group. Although the constituents of sugar, *viz.*, glucose and fructose both contain carbonyl group but sucrose does not. This is because the two units are linked to each other through their carbonyl group. Thus carbonyl groups are not free in the molecule of sugar.

Example 8. Write a note on epimerization (Conversion of glucose into mannose)

Solution. Aldoses which differ in their configuration only at C-2 are called epimers and the interconversion of such compounds is called epimerisation. An aldose can be converted into its epimer as follows:

1. Oxidise the aldose into aldonic acid by treatment with bromine water.

2. Treat the aldonic acid with pyridine. An equilibrium mixture of aldonic acid and its epimer will be obtained.

3. Separate the apimeric aldonic acid from the aldonic acid.

4. Reduce the epimeric aldonic acid to obtain the epimeric aldose.

The reactions given below show how glucose is converted into its epimer mannose.

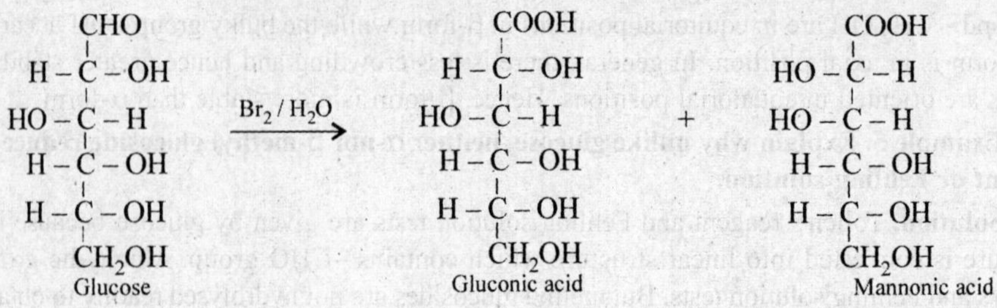

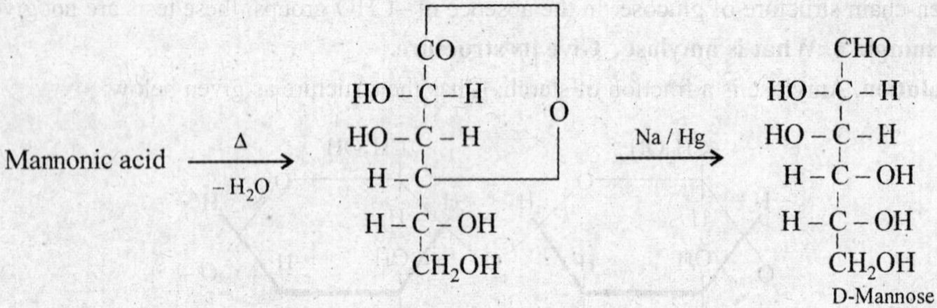

Example 9. Why do glucose and fructose form the same osazone.

Solution. This can be understood in terms of mechanism of reactions of glucose and fructose with phenyl hydrazine. Refer to "Reaction with phenylhydrazine".

Example 10. Fructose is laevorotatory whereas it is written as D(–) fructose. Explain.

Solution. Terms leavorotatory or dextrorotatory denote the sign of rotation of plane-polarised light when passed through an optically active substance. For *leavo*, the sign (–) is used. Capital letters D and L stand for absolute configuration of the substance i.e. relative arrangement of –H and –OH groups around different carbon atoms. Fructose has the absolute D-configuration and rotates the plane of polarised light in the anticlockwise (or left) direction. Hence it is written as D( ) fructose.

Example 11. How will you distinguish between glucose and fructose by a chemical test ?

Solution. Glucose gives a silver mirror with ammoniacal solution of silver nitrate (Tollen's test) and a red precipitate with alkaline solution of copper sulphate (*Fehling solution test*). Fructose does not give these tests.

Example 12. Glucose does not react with sodium bisulphite even though it has an aldehydic group. Explain.

Solution. This is because the aldehydic group is not completely free. It is involved in acetal formation with hydroxyl group at C–5.

Example 13. Explain the high solubility of sugars in water.

Solution. There is extensive hydrogen bonding in water because of which it is highly soluble in water. This is explained by takes the example of glucose.

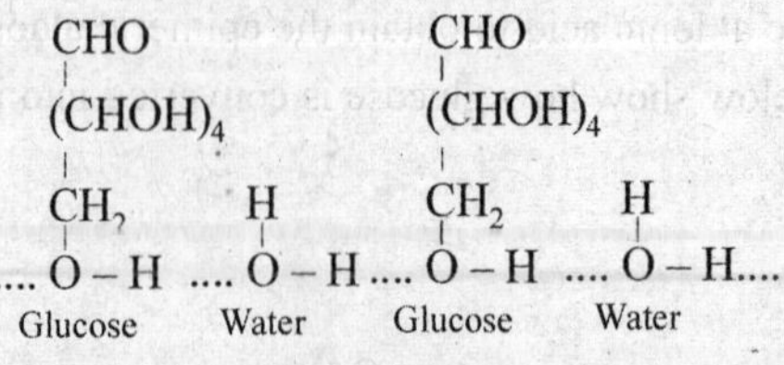

Example 14. Give the products of periodic acid oxidative cleavage of the following. How many moles of the reagent will be consumed in each case ?

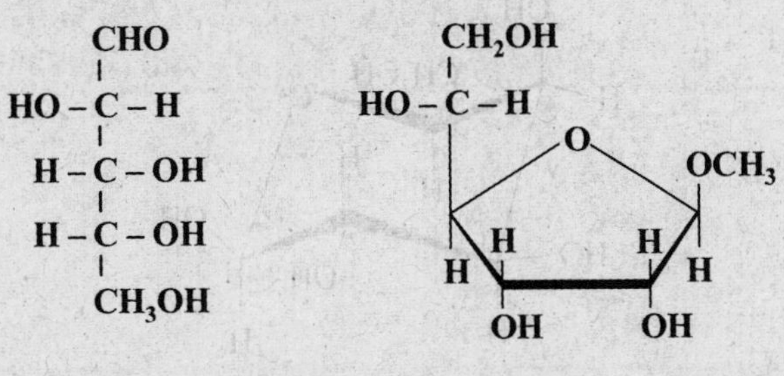

(a) **(b)**

Solution. (*a*) The molecule will require four moles of periodic acid to form four moles of formic acid and 1 mole of formaldehyde as shown following:

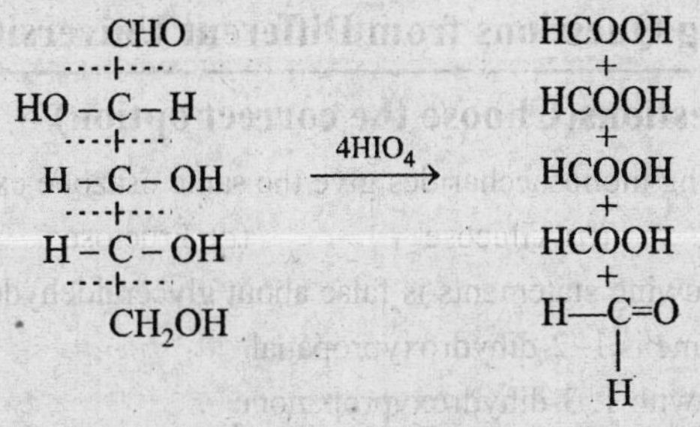

(*b*) The molecule will require two moles of periodic acid to form the following products.

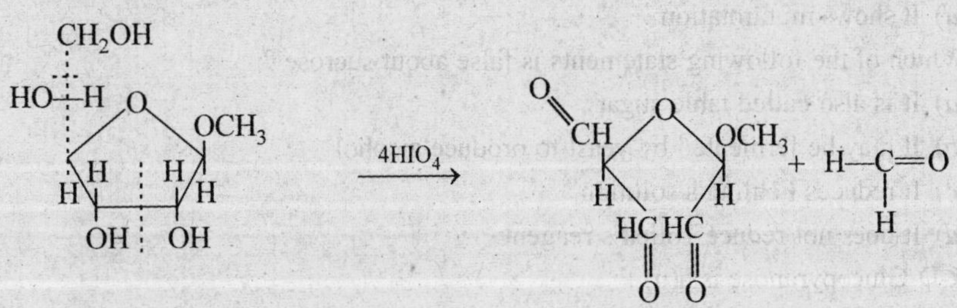

Example 15. Draw the most stable conformation of β–D-galactopyranose.

Solution. D-galactose is an epimer of D-glucose which differs from the former only in configuration at C – 4. Therefore open chain formula of galactose given below can be converted into hemi-acetal form and then into Haworth projection formula as follows :

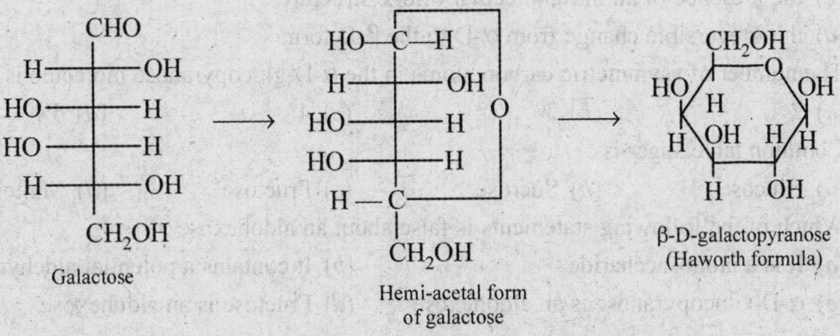

Galactose

Hemi-acetal form of galactose

β-D-galactopyranose (Haworth formula)

Now the Haworth form is converted to the chair conformation in which $-CH_2OH$ group is equatorial.

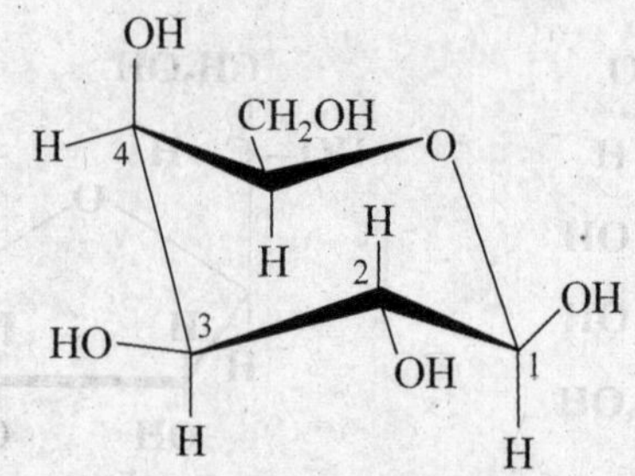

It may be noted that C – 4 hydroxyl group is axial in β-D-galactopyranose but equatorial in β-D-Glucopyranose.

EXERCISES
(Including Questions from Different University Papers)

Multiple Choice Questions(Choose the correct option)

1. All of the following monosaccharides give the same osazone except

(*a*) Galactose (*b*) Glucose (*c*) Fructose (*d*) Mannose

2. Which of the following statements is false about glyceraldehyde ?

(*a*) Its IUPAC name is 1, 2-dihydroxypropanal

(*b*) It is isomeric with 1, 3-dihydroxypropanone

(*c*) It is optically active

(*d*) It shows mutarotation

3. Which of the following statements is false about sucrose ?

(*a*) It is also called table sugar

(*b*) It may be fermented by yeast to produce alcohol

(*c*) It reduces Fehling's solution

(*d*) It does not reduce Tollen's reagent

4. α-D-Glucopyranose is a(n) :

(*a*) hemiacetal (*b*) hemiketal (*c*) acetal (*d*) ketal

5. The mutarotation of glucose is characterised by :

(*a*) a change from an aldehyde to ketone structure

(*b*) a change of specific rotation from a (+) to a (–) value

(*c*) the presence of an intramolecular bridge structure

(*d*) the irreversible change from α-D to the β-D form

6. The number of asymmetric carbon atoms in the α-D-glucopyranose molecule is :

(*a*) 2 (*b*) 3 (*c*) 4 (*d*) 5

7. Common table sugar is

(*a*) Glucose (*b*) Sucrose (*c*) Fructose (*d*) Maltose

8. Which of the following statements is false about an aldohexose ?

(*a*) It is a monosaccharide (*b*) It contains a potential aldehyde group

(*c*) α-D-Glucopyranose is an aldohexose (*d*) Fructose is an aldohexose

9. The sugar that yields only glucose on hydrolysis is

(*a*) Lactose (*b*) Sucrose (*c*) Maltose (*d*) Fructose

10. Which of the following compounds reduces Tollen's reagent ?

(*a*) Glucose (*b*) Sucrose (*c*) Methanol (*d*) Acetic acid

11. Which of the following products is not derived from cellulose ?

(*a*) Rayon (*b*) Insulin (*c*) Gun cotton (*d*) Paper

12. Starch

(*a*) is a trisaccharide (*b*) is also called amylose

(*c*) is also called amylopectin (*d*) is a mixture of amylose + amylopectin

13. Which of the following carbohydrates will not give a red precipitate of Cu_2O when heated with Benedict's solution ?

(*a*) Maltose (*b*) Glucose (*c*) Sucrose (*d*) Fructose

14. A reducing sugar will

(*a*) react with Fehling's test (*b*) not react with Fehling's test

(*c*) have fewer calories (*d*) always be a ketone

15. The reagent that can be used to differentiate an aldose and a ketose is :

(*a*) Bromine water (*b*) Fehling's solution (*c*) Tollen's reagent (*d*) None of these

16. Which of the following carbohydrates is not a reducing sugar ?

(*a*) Glucose (*b*) Sucrose (*c*) Fructose (*d*) Lactose

17. The monosaccharide obtained by hydrolysis of starch is :

(*a*) D-Glucose (*b*) Maltose (*c*) D-Galactose (*d*) D-Ribose

ANSWERS

1. (*a*)	**2.** (*d*)	**3.** (*c*)	**4.** (*a*)	**5.** (*c*)	**6.** (*d*)
7. (*b*)	**8.** (*d*)	**9.** (*c*)	**10.** (*a*)	**11.** (*b*)	**12.** (*d*)
13. (*c*)	**14.** (*a*)	**15.** (*a*)	**16.** (*b*)	**17.** (*a*)	

Short Answer Questions

1. Explain :

(*i*) Prefix D- is given to fructose even though it is dextrorotatory.

(*ii*) Glucose does not react with sodium bisulphite even though it contains an aldehyde group.

2. (*a*) What are disaccharides ? Write the ring structure of lactose.

(*b*) Define carbohydrates.

3. Draw the structures of starch and cellulose indicating their point of difference.

4. Write a short note on mutarotation.

5. Explain, why do both glucose and fructose form the same osazone.

6. How will you convert an aldohexose into a ketohexose ?

7. Name the enantiomer of α-D-(+)-glucose.

8. How will you convert an aldohexose into ketohexose ?

9. Describe Ruff's degradation for the conversion of an aldohexose into an aldopentose.

10. How would you explain that fructose, without containing any reducing group, is able to reduce Tollen's reagent and Fehling solution ?

11. What are disaccharides ? Write the ring structure of sucrose.

12. Complete the reaction :

$$\text{Glucose} \xrightarrow{\text{HI/P}} \text{A + B}$$

13. What are reducing and non-reducing sugars ? Give an example in each case.

14. Give the main points of difference in the structure of starch and cellulose.

15. Draw Haworth's formula of

(*i*) α-D-Glucose (*ii*) β-D-Fructose

(*iii*) α-D (+) glucopyranose (*iv*) β-D (–) Fructofuranose

16. What is the difference between anomers and epimers ? Give one example in each case.

17. How do you explain that glucose does not react with $NaHSO_3$ even though it contains an aldehyde group ?

18. Explain Lobry de Bruyn-Van Ekenstein rearrangement.

19. Write down the structure of sucrose molecule indicating the monosaccharides and the linkages involved.

20. (*a*) How will you convert aldopentose to aldohexose ?

(*b*) Give evidence in favour of ring structure of fructose.

21. Write notes on the following :

(*i*) Anomers and epimers (*ii*) Glycosides and glucosides.

22. (*a*) Write various steps involved in Ruff's degradation.

(*b*) Give evidences to show that fructose is keto-hexose.

23. What happens when fructose is heated with excess of phenylhydrazine ? Give its modern mechanism.

24. (*a*) How will you convert an aldohexose to ketohexose ?

(*b*) Draw conformational formulae of β-D-Glucose and α-D-Fructose.

25. (*a*) Draw the Haworth formulae of

(*i*) Sucrose (*ii*) β-D-(–)-fructofuranose

(*b*) Indicating clearly the linkage involved, write down the structure of

(*i*) Amylose (*ii*) Cellulose.

26. Why does D-(+)-glucose show the phenomenon of mutarotation ?

27. (*a*) Explain Lobry de Bruyn van-Ekenstein rearrangement.

(*b*) Convert fructose to glucose.

28. Identify compounds A, B, C, D and E in the following sequence of reactions :

$$\text{Methyl β-D-glucoside} \xrightarrow[\text{NaOH}]{(CH_3)_2SO_4} \text{A} \xrightarrow{\text{Dil HCl}} \text{B} \xrightarrow{HNO_3} \text{C (Intermediate Keto acid)} \longrightarrow \text{D + E}$$

29. How will you convert glucose into fructose and fructose into glucose ?

30. Write a note on Kiliani-Fischer synthesis.

31. Give the conversion of an aldopentose to aldohexose and *vice-versa.*

32. Justify the existence of two types of methyl glycosides. Explain why they do not reduce Tollen's reagent.

33. What is an anomeric carbon ? Give suitable example.
34. Differentiate :
(*i*) Starch and Cellulose (*ii*) Sucrose and maltose.
35. How is glucose converted into fructose and mannose ?
36. What is mutarotation ? Give its mechanism.
37. How will you convert aldohexose into ketohexose and ketohexose into an aldohexose ?
38. Why do glucose and fructose give the same osazone ?
39. Why does D-(+)-glucose show the phenomenon of mutarotation ?
40. While glucose is a reducing sugar but neither α- nor β-glucoside reduces either Tollen's reagent or Fehling solution. Why is it so ? Discuss.
41. Why sucrose is known as invert sugar ? How do you account for the fact that sucrose does not reduces Fehling solution ? Write Haworth projection formula for sucrose.
42. Write the modern mechanism for the osazone formation of Glucose.

General Questions

1. How do you explain the following :
(*i*) Glucose and fructose give the same osazone.
(*ii*) A freshly prepared aqueous solution of glucose has an optical rotation of + 112°. On standing at room temperature the rotation gradually changes to + 52.5° and no further.
(*iii*) Unlike glucose, neither α- nor β-methyl glucoside reduces Tollen's reagent or Fehling solution.
(*iv*) Fructose is a pentahydroxy ketone, still it reduces Fehling solution and Tollen's reagent.
2. Why is sucrose known as invert sugar ? How do you account for the fact that sucrose does not reduce Fehling solution or form oxime ? Write down the structure of sucrose molecule.
3. Explain the ring size of D-glucose.
4. (*a*) Explain the mechanism of conversion of glucose to glucosazone.
(*b*) How will you convert an aldohexose into ketohexose and ketohexose into aldohexose ?
5. How could ring structure of glucose explain the limitations of open-chain structure ? How is ring size determined in case of glucose ?
6. Give the evidence(s) leading to the cyclic structure of D-(+)-glucose.
7. Describe Ruff's degradation for conversion of aldohexose into aldopentose.
8. (*a*) Explain the limitations of open chain D-(+)-glucose structure.
(*b*) Explain Lobry de-Bruyn-Van Ekenstein rearrangement.
9. Give evidence to show that fructose is a keto-hexose.
10. Explain the following :
(*i*) Mutarotation (*ii*) Epimerisation
(*iii*) Ruff's degradation in aldoses.
11. Write notes on
(*i*) Ruff's degradation (*ii*) Epimerisation
(*iii*) Kiliani-Fischer synthesis (*iv*) Formation of glucosides.
12. Give evidence to show that fructose is a ketohexose.
13. Explain the following :
(*i*) Prefix D is given to fructose even though it is laevorotatory.

(*ii*) Glucose does not react with $NaHSO_3$ even though it contains an aldehyde group.
(*iii*) Fructose contains a keto group, still it reduces Fehling's solution and Tollen's reagent.

14. (*a*) Discuss the evidence(s) leading to the cyclic structure of D-(+) glucose.
(*b*) Explain the following :
(*i*) Mutarotation (*ii*) Wohl degradation

15. Draw the Haworth formulae of :
(*i*) α-D-glucose (*ii*) β-D-fructose
(*iii*) Sucrose (*iv*) Lactose

16. Explain Kiliani-Fischer synthesis.

17. Write notes on the following :
(*i*) Kiliani-Fischer synthesis (*ii*) Ruff's degradation.

18. (*a*) Giving suitable examples, write an account of the classification of carbohydrates.
(*b*) Discuss the cyclic structure of D (+) glucose.

19. Give Kiliani-Fischer synthesis of carbohydrates.

20. Explain :
(*i*) Glucose does not react with sodium bisulphite even though it contains an aldehyde group.
(*ii*) Prefix-D is given to fructose even though it is laevorotatory.
(*iii*) Fructose is a pentahydroxy ketone, still it reduces Fehling solution and Tollen's reagent.

21. Give evidence to prove that D (+) – Glucose has a cyclic structure.

7

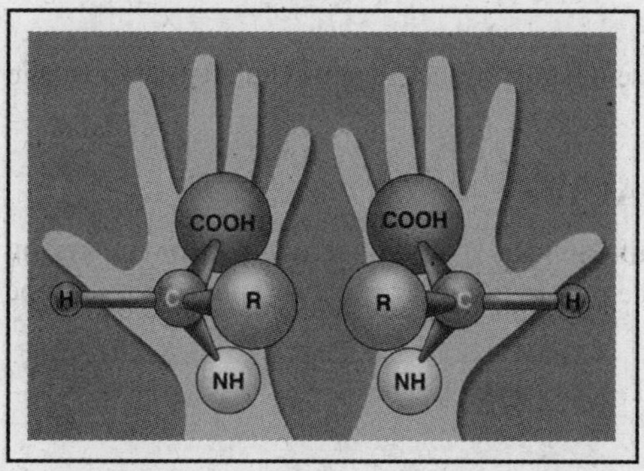

Amino Acids, Peptides, Proteins and Nucleic Acids

7.1. INTRODUCTION

Amino acids are compounds which contain at least one amino group and one carboxylic group. The amino group could be linked to carbon just next to carboxylic group or with a gap of one or two carbons. They are called α, β or γ acids respectively. Some examples of amino acids are:

Structure	Name
$CH_2(NH_2)—COOH$	α-amino acetic acid (or 2-aminoethanoic acid)
$CH_2(NH_2)—CH_2—COOH$	β-amino propionic acid (or 3-aminopropanoic acid)
$CH_2(NH_2)—CH_2—CH_2—COOH$	γ-amino butyric acid (or 4-aminobutanoic acid)

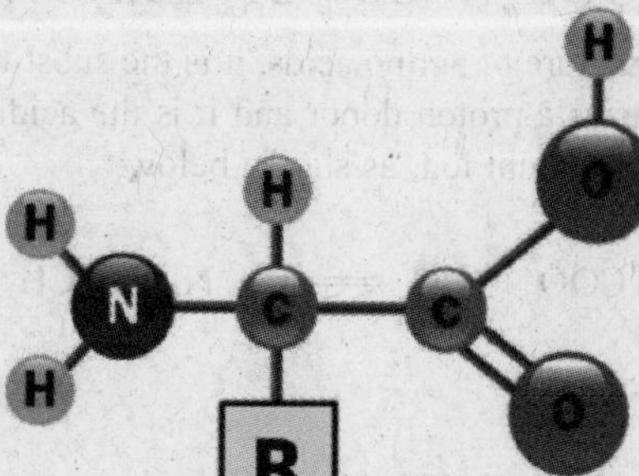

Model of amino acid

α-amino acids have the maximum importance because they are the constituent units of proteins. Common system of nomenclature is still more popular in naming amino acids and proteins. It is mostly α-amino acids, which are involved in the building up of protein molecules. The simplest α-amino acid is α-amino acetic acid also called *glycine*. In general, an α-amino acid can be represented as

$$\underset{\displaystyle NH_2}{R-\underset{|}{CH}-COOH}$$

where R is an alkyl or aryl group. It could also represent highly branched or unsaturated carbon chain or heterocyclic ring.

7.2. DIPOLAR NATURE OF AMINO ACIDS

It has been found that an amino acid molecule appears as a dipole, one part of it carrying positive charge and the second negative charge. The dipolar ionic structure of amino acids can be represented as:

$$\underset{\displaystyle \overset{+}{N}H_3}{R-\underset{|}{CH}-COO^-}$$

This is also called a *Zwitter ion* or *Internal salt*. There is no free amino or carboxylic group present in the molecule.

Evidence in support of dipolar nature

1. Spectroscopic studies of amino acids do not show bands characteristics of $-NH_2$ and $-COOH$ groups

2. Amino acids are insoluble in non-polar solvents and soluble in polar solvents like water. This behaviour can be expected of the polar substances.

3. Amino acids are non-volatile crystalline solids, which melt at high temperature. This is quite like ionic substances which have high melting points and unlike amines and carboxylic acids which have low melting points.

4. They have high dipole moments indicating polar nature of the molecule.

5. Dissociation constant K_a and K_b give us an idea about the acid and base strengths. Amino acids have very low values of K_a and K_b indicating that the molecule does not possess these groups in the normal forms.

7.3. LOW VALUES OF K_a AND K_b OF AMINO ACIDS

Because of the dipolar nature of amino acids, it is the substituted ammonium ion, and not the carboxylic group, which acts as a proton donor and it is the acidic centre. K_a, therefore, refers to the acidity of substituted ammonium ion, as shown below:

$$\overset{+}{N}H_3-\underset{\displaystyle R}{\underset{|}{CH}}-COO^- + H_2O \rightleftharpoons NH_2-\underset{\displaystyle R}{\underset{|}{CH}}-COO^- + H_3\overset{+}{O}$$

$$K_b = \frac{\left[NH_2-\underset{\displaystyle R}{\underset{|}{CH}}-COO^-\right][H_3O^+]}{\left[\overset{+}{N}H_2-\underset{\displaystyle R}{\underset{|}{CH}}-COO^-\right]}$$

The concentration of H_2O remains constant and hence it is not taken up in the denominator.

Similarly it is the carboxylate ion, and not the amino group, which acts as the proton acceptor and it is the basic centre. K_b, therefore, refers to the basicity of carboxylate ion as given below:

$$\overset{+}{N}H_3-\underset{\displaystyle R}{\underset{|}{CH}}-COO^- + H_2O \rightleftharpoons \overset{+}{N}H_3-\underset{\displaystyle R}{\underset{|}{CH}}-COOH + OH^-$$

or

$$K_a = \frac{\left[\overset{+}{N}H_3-\underset{\displaystyle R}{\underset{|}{CH}}-COOH\right][OH^-]}{\left[\overset{+}{N}H_3-\underset{\displaystyle R}{\underset{|}{CH}}-COO^-\right]}$$

The measured values of K_a and K_b of amino acids, of say, glycine (α-amino acetic acid), can be justified, in terms of dipolar structure, by calculating the values for conjugate bases or conjugate acids and comparing them with observed values.

Let us take into consideration K_a and K_b values for glycine $H_3\overset{+}{N}CH_2COO^-$. It shows K_a values equal to 1.6×10^{-10}. K_b of its conjugate base $NH_2-CH_2-COO^-$ can be calculated by using the generalisation that the product of K_a and K_b in an aqueous solution is 1×10^{-14}. Thus,

$$K_a \times K_b = 1.0 \times 10^{-14}$$

$$1.6 \times 10^{-10} \times K_b = 1.0 \times 10^{-14}$$

or

$$K_b = \frac{1.0\times10^{-14}}{1.6\times10^{-10}} = 6.3 \times 10^{-3}$$

This value is quite reasonable for an aliphatic amine. Basicity constant K_b of glycine is measured as 2.5×10^{-12}. Acidity constant K_a of its conjugate acid $\overset{+}{N}H_3\ CH_2COOH$ can be calculated as per the above criteria.

$$K_a \times K_b = 1.0 \times 10^{-14}$$

$$K_a \times 2.5 \times 10^{-12} = 1.0 \times 10^{-14}$$

or

$$K_a = \frac{1.0\times10^{-14}}{2.5\times10^{-12}} = 4 \times 10^{-3}$$

This value is quite reasonable for an aliphatic acid with an electron-withdrawing group.

Effect of pH on the structure of amino acids

1. When the solution of an amino acid is made acidic, the dipolar ion (A) gets converted into a cation as shown below. This is because the stronger acid H_3O^+ releases a proton to the carboxylate ion and displaces a weaker acid.

$$\underset{\text{(A)}}{\overset{+}{N}H_3-\underset{\displaystyle R}{\underset{|}{CH}}-CO\bar{O}} + \underset{\text{Stronger acid}}{H_3O^+} \rightleftharpoons \underset{\text{(Weaker acid)}\ \text{(I)}}{\overset{+}{N}H_3-\underset{\displaystyle R}{\underset{|}{CH}}-COOH} + H_2O$$

2. Similarly when the solution of amino acid is made alkaline with NaOH, the dipolar ion is converted into an anion as shown below

$$\underset{\underset{(A)}{R}}{\overset{+}{N}H_3 - \underset{|}{CH} - CO\bar{O}} + \underset{\text{Stronger base}}{OH^-} \rightleftharpoons \underset{\underset{(II)}{R}}{NH_2 - \underset{|}{CH} - CO\bar{O}} + \underset{\text{Weaker base}}{H_2O}$$

The stronger base OH^- abstracts a proton from the amino acid to give a weaker base.

It is to be borne in mind that the ions I and II shown above are in equilibrium with the dipolar ion (A) such that amino acids can still exhibit properties of amines as well as carboxylic acid.

$$\underset{\underset{(I)}{R}}{\overset{+}{N}H_3 - \underset{|}{CH} - COOH} \underset{H^+}{\overset{OH^-}{\rightleftharpoons}} \underset{\underset{(A)}{R}}{\overset{+}{N}H_3 - \underset{|}{CH} - COO^-} \underset{H^+}{\overset{OH^-}{\rightleftharpoons}} \underset{\underset{(II)}{R}}{NH_2 - \underset{|}{CH} - CO\bar{O}}$$

7.4. CLASSIFICATION OF AMINO ACIDS DERIVED FROM PROTEINS

Amino acids can be classified in different ways:

(a) As essential or non-essential amino acids:

Amino acids which are very important or essential for the growth of humans and animals are called essential amino acids. These amino acids cannot be synthesised by the body and must be supplied in the diet as such. Lack of these amino acids in the diet may cause the disease *Kwashiorkar*. Amino acids which can be synthesised by our body are called non-essential amino acids. Essential amino acids are indicated by * sign in Table 7.1

(*b*) As neutral, basic or acidic amino acids

Neutral Amino Acids

Amino acids having one amino and one carboxylic groups are called neutral amino acids. Examples are glycine and alanine.

Basic Amino Acids

Amino acids containing two amino (or imino) and one carboxylic groups are called basic amino acids. Examples: lysine and arginine.

Acidic Amino Acids

Amino acids having one amino (or imino) and two carboxylic groups are called acidic amino acids. Examples are aspartic acid and glutamic acid.

Table 7.1: List of amino acids

S.No.	*Name*	*Abbreviation and one letter code*	*Structure*
	Neutral		
1.	Glycine	Gly (G)	$\underset{^+NH_3}{\underset{\mid}{CH_2}}\,COO^-$
2.	(+) Alanine	Ala (A)	$CH_3\,\underset{^+NH_3}{\underset{\mid}{CH}}\,COO^-$

S.No.	*Name*	*Abbreviation and one letter code*	*Structure*
3.	(+) Valine*	Val (V)	$(CH_3)_2CHCH(^+NH_3)COO^-$
4.	(–) Leucine*	Leu (L)	$(CH_3)_2CHCH_2CH(^+NH_3)COO^-$
5.	(+) Isoleucine*	Ile (I)	$CH_3CH_2CH(CH_3)—CH(^+NH_3)—COO^-$
6.	(–) Phenylalanine*	Phe (F)	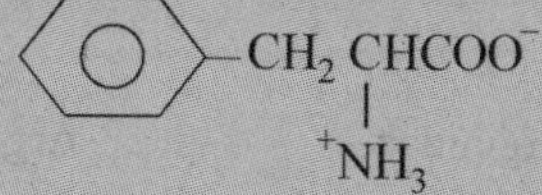
7.	(–) Serine	Ser (S)	$HOCH_2CH(^+NH_3)COO^-$
8.	(–) Threonine*	Thr (T)	$CH_3CHOHCH(^+NH_3)COO^-$
9.	(–) Cysteine	Cys (C)	$HSCH_2CH(^+NH_3)COO^-$
10.	(–) Cystine	Cys-Cys	$^-OOCCH(^+NH_3)CH_2S—SCH_2CH(^+NH_3)COO^-$
11.	(–) Methionine*	Met (M)	$CH_3SCH_2CH_2CH(^+NH_3)COO^-$
12.	(–) Tyrosine	Tyr (Y)	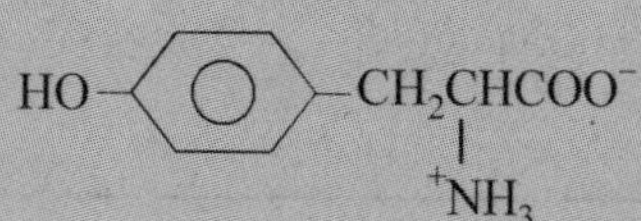
13.	(–) Proline	Pro (P)	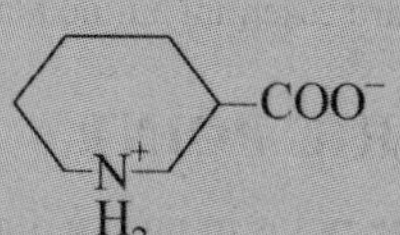
14.	(–) Hydroxyproline	Hyp	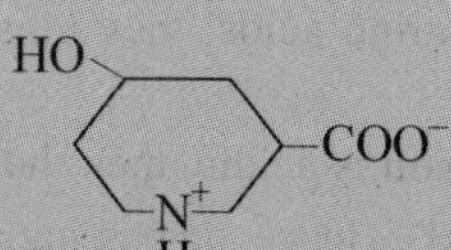

S.No.	*Name*	*Abbreviation and one letter code*	*Structure*
15.	(–) Tryptophan*	Try (W)	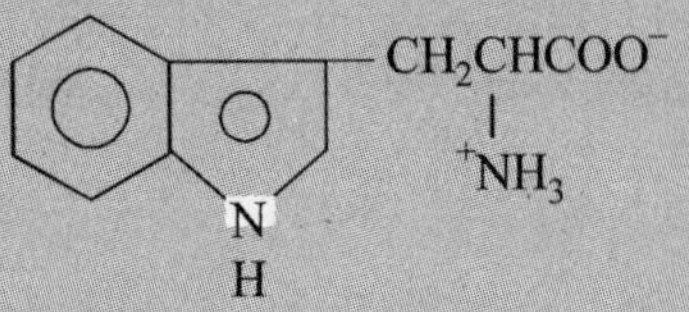
	Basic		
16.	(+) Lysine*	Lys (k)	$H_3N^+ CH_2 CH_2 CH_2 CH_2 CH(NH_2) COO^-$
17.	(–) Hydroxylysine	Hyl	$H_3N^+ CH_2 CHOHCH_2 CH_2 CH(NH_3)COO^-$
18.	(+) Arginine*	Arg (R)	$H_2NCH(^+NH_3) — NH — CH_2 CH_2 CH_2 CH(NH_2)COO^-$
19.	(–) Histidine*	His (H)	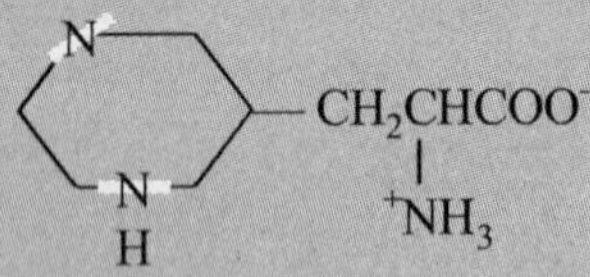
	Acidic		
20.	(+) Aspartic acid	Asp (D)	$HOOCCH_2 CH(^+NH_3)COO^-$
21.	(–) Asparagine	Asn (N)	$H_2NCOCH_2 CH(^+NH_3)COO^-$
22.	(+) Glutamic Acid	Glu (E)	$HOOCCH_2CH_2 CH(^+NH_3)COO^-$
23.	(+) Glutamine	Gln (O)	$H_2NCOCH_2CH_2 CH(^+NH_3)COO^-$

* Amino acids marked * are essential amino acids.

7.5. ISOELECTRIC POINT

As amino acids are polar in nature, they show electrical properties. On applying electrical field to the solution of amino acids, they migrate to one or the other electrode depending upon the following factors:

(*a*) If the solution is acidic, then the equilibrium lies towards positively charged amino acid (NH_3^+ CHR – COOH). Hence, on passing an electric current through the solution of the amino acid, it moves towards the cathode.

(*b*) If the solution is alkaline, then the equilibrium is predominantly lying towards the negatively charged amino acid ($NH_2CHR - COO^-$). Hence on passing electricity, amino acid molecule which is in the form of anion, moves towards the anode.

(*c*) At a certain pH of the solution, the anionic and cationic structures will be in equal concentrations. On passing electricity we shall observe that there is no movement of the amino acid.

The pH at which a particular amino acid does not migrate under the influence of the electrical field is called isoelectric point. Every amino acid has a characteristic isoelectric point. Glycine has an isoelectric point at pH 6.1. It may be noted that amino acids have the minimum solubility at the isoelectric point. This is because at isoelectric point, there is maximum concentration of dipolar ions which are relatively less soluble.

Isoelectric points of some amino acids are given in table 7.2.

Table 7.2: Isoelectric points of some α-amino acids.

Amino acid		Isoelectric point
Alanine	Neutral	6.02
Valine	Neutral	5.97
Leucine	Neutral	5.98
Serine	Neutral	5.70
Threonine	Neutral	5.60
Aspartic acid	Acidic	2.87
Glutamic acid	Acidic	3.22
Lysine	Basic	9.74
Arginine	Basic	10.70

7.6. ELECTROPHORESIS (Separation of amino acids)

Electrophoresis is a process of separation and purification of compounds on the basis of movement of charged particles in an electric field. Separation of mixture of a amino acids can be brought about by electrophoresis.

To carry out electrophoresis, a strip of paper, a suitable plastic of cellulose acetate is used as solid support. A solution of mixture of different amino acids which is to be separated from each other is placed near the centre of strip of paper (Fig 7.1). The strip is then moistened with an aqueous buffer having a particular predetermined pH which depends upon the isoelectric points of the acids to be separated. The ends of the strip are connected to the electrodes on applying an electric field, we observe that

(*i*) The amino acids whose isoelectric points is below the pH of the buffer start migrating slowly towards the positive electrode because they exist mainly in the anionic form.

(*ii*) The amino acids having isoelectric point higher than the pH of the buffer migrate towards the negative electrode because they exist mainly in the cationic form.

(*iii*) The amino acids whose isoelectric points corresponds to the pH of the buffer do not migrate from the origin because they have no net charge.

Different amino acids migrate at different rates depending upon the isoelectric point of the acid. Thus, different amino acids get separated from each other. After the separation is complete, the strip is dried and sprayed with a dye such as ninhydrin when the separated components become visible.

Thus if a mixture of alanine (isoelectric point = 6.02) aspartic acid (isoelectric point = 2.87) and lysine (isoelectric point = 9.74) is to be separated, it is subjected to electrophoresis in a buffer having pH = 6.0. At this pH, aspartic acid would migrate towards the positive electrode, alanine would remain at the origin while lysine would move towards the negative electrode seen in the figure

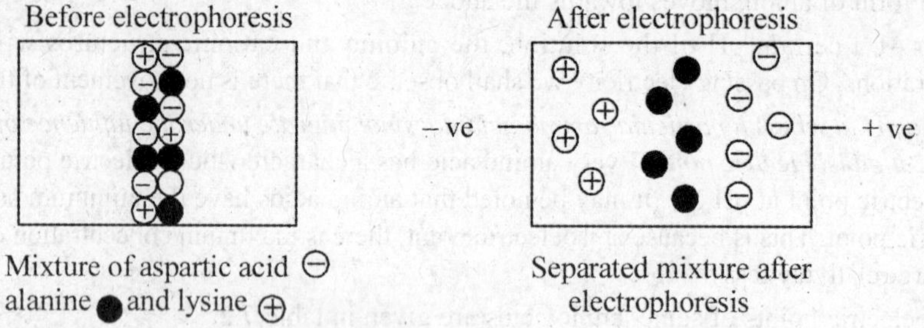

Fig. 7.1 Separation of amino acid mixture by electrophoresis

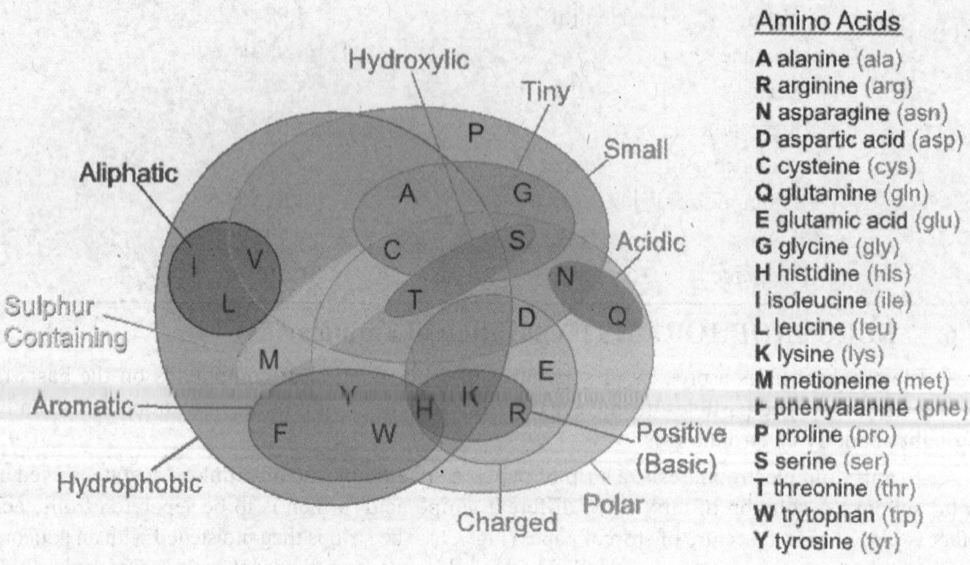

Amino acid grouping

7.7. SOME CHARACTERISTICS OF NATURAL AMINO ACIDS

1. Some of the amino acids are acidic, some neutral and some basic. It may be noted that neutral amino acids are those which have one amino and one carboxy group thus neutralizing each other. Cystine has two amino and carboxy groups each. Compounds containing more amino groups than carboxy groups are expected to show basic behaviour. Similarly compounds having more carboxy groups (or acidic groups) than amino groups are supposed to exhibit acidic properties.

2. Valine, leucine, isoleucine, phenylalanine, threonine, methionine, tryptophan, lysine, histidine, asparagine cannot be synthesised by animals. And these are very important for the growth of animals. Hence these must be supplied in the diet as such.

3. All these acids except glycine have four different groups attached to the α-carbon atom. Thus the α-carbon is asymmetric and hence there is a possibility of optical isomerism. However they belong to the same stereochemical series. They have the same configuration as that of L (–) glyceraldehyde which is shown below (Fischer Projection formula)

$$\begin{array}{c} CHO \\ | \\ HO - C - H \\ | \\ CH_2OH \end{array} \qquad\qquad \begin{array}{c} COO^- \\ | \\ H_3N^+ - C - H \\ | \\ R \end{array}$$

L (–) Glyceraldehyde — L - Amino acid

They are optically active compounds and rotate the plane of polarised light. It may be mentioned that the sign L refers to the *configuration* whereas the (+) or (–) sign refers to the *direction* of rotation of plane-polarised light.

4. The amino acids isolated from the proteins rotate the plane of polarised light i.e. they are *optically active* whereas the same amino acids if synthesised are *optically inactive*. It is because during the synthesis, both dextro and laevo forms are obtained in equal forms which counterbalance the rotation of each form. Such mixtures are called *racemic mixtures*.

5. Another interesting observation may be made. All the proteins which are resynthesised by human body have L-configuration. This happens so in all humans. That is why all human beings have almost similar characteristics. Had different persons produced different types (L and D) of proteins, then perhaps there would have been vast differences in the characteristics of different persons. This would have resulted into two different genetic breeds of persons (L type and D type).

7.8. METHODS OF PREPARATION OF AMINO ACIDS

Methods commonly used for preparation of amino acids are described below:

1. From α-halogeno acids. Treatment of α-halogeno acids with ammonia or ammonium hydroxide gives α-amino acids.

$$\underset{\alpha\text{-Chloroacetic acid}}{\begin{array}{l} CH_2COOH \\ | \\ Cl \end{array}} + 2NH_4OH \longrightarrow \underset{\text{Glycine}}{\begin{array}{l} CH_2 - COO^- \\ | \\ {}^+NH_3 \end{array}} + NH_4Cl + H_2O$$

A conc. solution of ammonia is needed to obtain a good yield of the amino acid. The α-halogen used as a starting substance is obtained by Hell-Vohlard-Zelinsky method as follows:

$$\underset{\text{Acetic acid}}{CH_3COOH} + Cl_2 \xrightarrow[P]{Red} \underset{\text{Chloroacetic acid}}{\begin{array}{l} CH_2COOH \\ | \\ Cl \end{array}} + HCl$$

2. From sodium salt of malonic ester

$$\underset{\text{Monosodium diethyl malonate}}{Na^+ \, CH\,(COOC_2H_5)_2} + \underset{\text{Isobutyl bromide}}{(CH_3)_2\,CHCH_2\,Br}$$

$$\xrightarrow{-NaBr} (CH_3)_2CHCH_2CH(COOC_2H_5)_2 \xrightarrow{\text{Hydrolysis}} (CH_3)_2CHCH_2CH(COOH)_2$$

$$\xrightarrow{Br_2} (CH_3)_2CHCH_2CBr(COOH)_2 \xrightarrow[-CO_2]{\text{Heat}} (CH_3)_2CHCH_2CH(Br)COOH$$

$$\xrightarrow{NH_4OH} (CH_3)_2CHCH_2CH(\overset{+}{N}H_3)COO^-$$

Leucine

3. Phthalimido malonic ester synthesis. It is a modification of the Gabriel phthalimide synthesis.

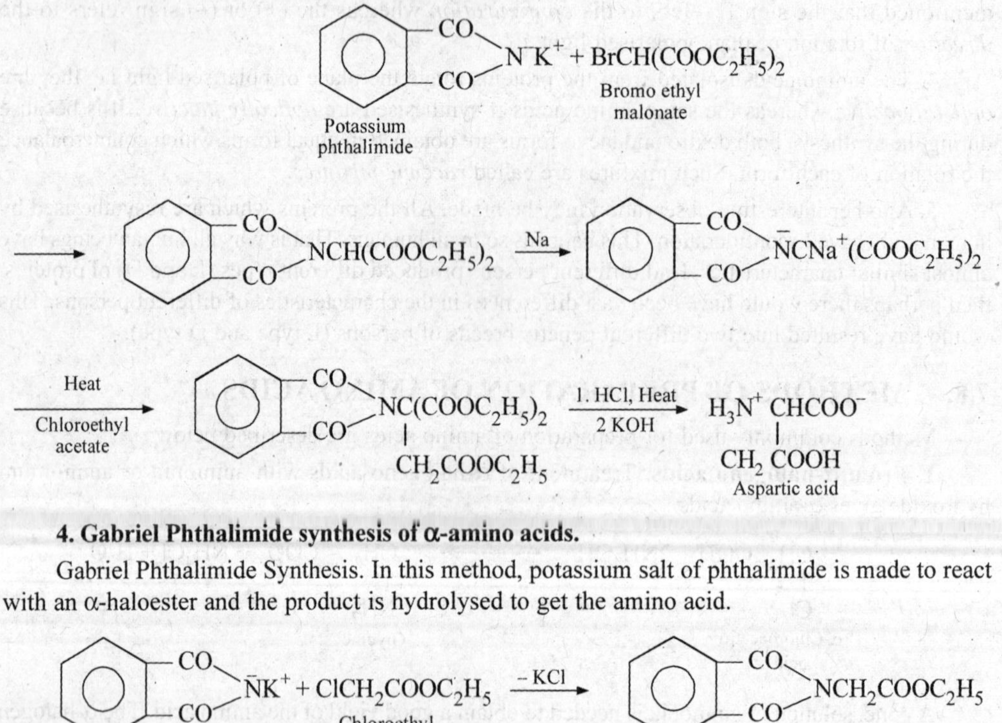

4. Gabriel Phthalimide synthesis of α-amino acids.

Gabriel Phthalimide Synthesis. In this method, potassium salt of phthalimide is made to react with an α-haloester and the product is hydrolysed to get the amino acid.

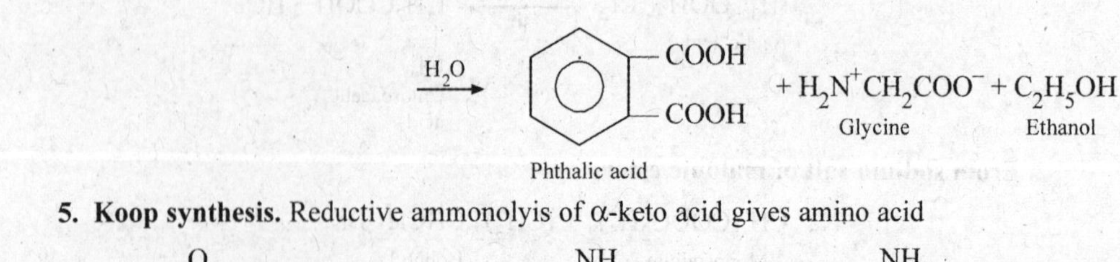

5. Koop synthesis. Reductive ammonolyis of α-keto acid gives amino acid

$$\underset{\alpha\text{-Keto acid}}{R-\overset{O}{\overset{\|}{C}}-COOH} + NH_3 \longrightarrow R-\overset{NH}{\overset{\|}{C}}-COOH \xrightarrow{H_2/Pd} \underset{\alpha\text{-amino acid}}{R-\overset{NH_2}{\overset{|}{C}H}-COOH}$$

6. Strecker synthesis of α-amino acid.

Strecker's synthesis. Here we treat an aldehyde with ammonia and hydrogen cyanide followed by hydrolysis. Thus

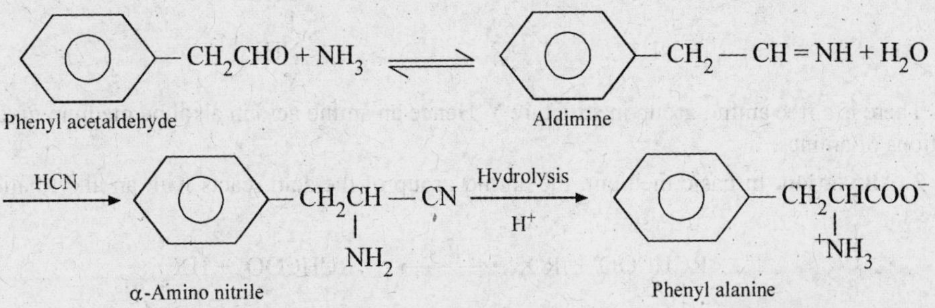

7. Erlenmeyer azlactone synthesis of α-amino acids.

This method is used for the preparation of aromatic amino acid. The sequence of reactions given below give the synthesis of phenylalanine from glycine.

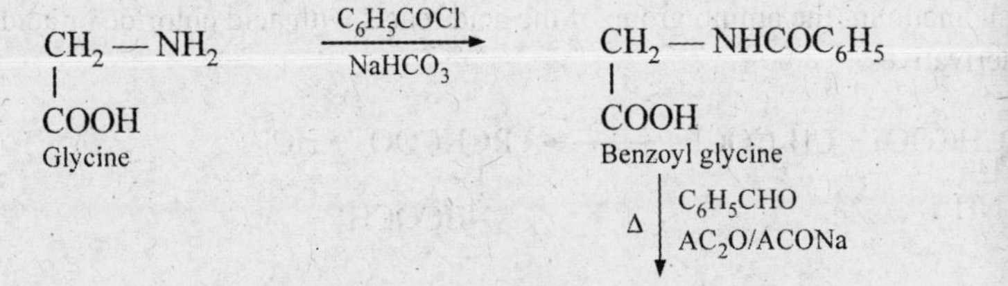

Richard Erlenmeyer

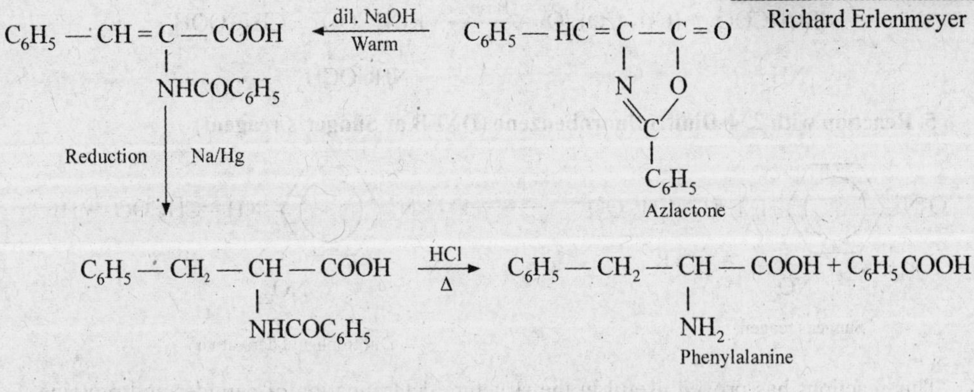

7.9. CHEMICAL PROPERTIES OF α-AMINO ACIDS

Some important properties of α-amino acids are given below:

1. With acids. The dipolar ionic structure of amino acid exists in equilibrium with structure X in acidic medium as follows :

$$\underset{\overset{|}{{}^{+}NH_3}}{R-CHCOO^-} \xrightleftharpoons{H^+} \underset{\underset{X}{\overset{|}{{}^{+}NH_3}}}{RCHCOOH}$$

There is free carboxy group in structure X. Hence it gives the reactions of carboxy group.

2. With alkalis. The dipolar ionic structure Y exists in alkaline medium as follows:

2. With alkalis. The dipolar ionic structure Y exists in alkaline medium as follows:

$$\underset{\overset{|}{^{+}NH_3}}{R-CHCOO^-} \xrightleftharpoons{OH^-} \underset{\underset{Y}{\overset{|}{NH_2}}}{R-CHCOO^-} + H_2O$$

There is a free amino group in structure Y. Hence an amino acid in alkaline medium gives the reactions of amines.

3. Alkylation. In basic medium, the amino group of the acid reacts with an alkyl halide as follows.

$$\underset{\overset{|}{NH_3}}{RCHCOO^-} + R'X \xrightarrow{Base} \underset{\underset{\text{N-Alkyl amino acid}}{\overset{|}{NHR'}}}{RCHCOO^-} + HX$$

$$\underset{\underset{\text{N-Methyl glycine}}{\overset{|}{NH_2}}}{CH_2COO^-} + CH_3Cl \xrightarrow{Base} \underset{\overset{|}{NHCH_3}}{CH_2COO^-} + HCl$$

4. Acetylation. In basic medium, the amino group of the acid reacts with acid chloride or acid anhydride to form acetyl derivatives.

$$\underset{\overset{|}{NH_2}}{RCHCOO^-} + CH_3COCl \xrightarrow{Base} \underset{\overset{|}{NHCOCH_3}}{RCHCOO^-} + HCl$$

$$\underset{\overset{|}{NH_2}}{RCHCOO^-} + (CH_3\,CO)_2\,O \xrightarrow{Base} \underset{\overset{|}{NHCOCH_3}}{RCHCOO^-} + CH_3COOH$$

5. Reaction with 2, 4-Dinitrofluorobenzene (DNFB or Sanger's reagent)

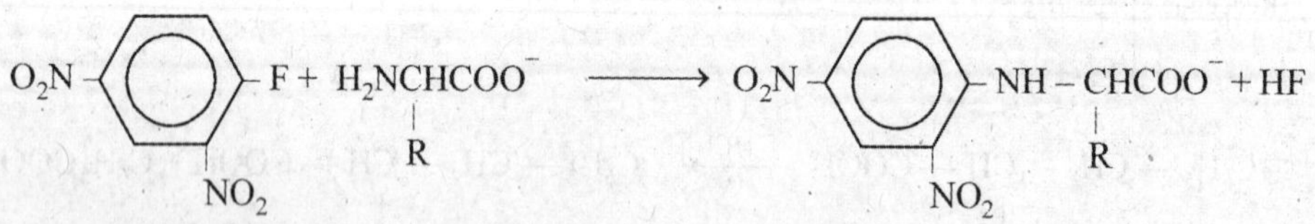

Sanger's reagent   Dinitrophenyl derivative

This reactions has proved useful in the structure determination of peptides and proteins.

6. Esterification. In acidic medium, the amino acids give the reactions of carboxy group like formation of esters, acid chlorides and acid anhydrides.

$$\underset{\overset{|}{^{+}NH_3}}{RCHCOOH} + C_2H_5OH \xrightarrow[HCl]{Anhy.} \underset{\underset{\text{Ester}}{\overset{|}{^{+}NH_3}}}{RCHCOOC_2H_5} + H_2O$$

$$R \quad \underset{\overset{|}{^{+}NH_3}}{CH} \quad COOH + PCl_5 \longrightarrow R-\underset{\underset{\text{Acid chloride}}{\overset{|}{^{+}NH_3}}}{CH}-COCl + POCl_3 + HCl$$

7. Reaction with nitrous acid. Nitrous acid reacts with amino acids to liberate nitrogen gas. This method is used to analyse amino acids and is known as **Van Slyke's method**. The volume of nitrogen evolved is measured. It may be noted that half of nitrogen comes from the amino acid and half of it comes from nitrous acid.

$$\underset{\stackrel{|}{^{+}NH_3}}{RCHCOO^-} + HNO_2 \longrightarrow N_2 + RCH(OH)COOH + H_2O$$

8. Reaction with formaldehyde. Amino acids with formaldehyde to form N-methylene amino acids.

$$HCHO + H_3\overset{+}{N} - \underset{\stackrel{|}{R}}{CH} - COO^- \longrightarrow CH_2 = N - \underset{\stackrel{|}{R}}{CHCOOH} + H_2O$$

N-methylene amino acid

As a result of this change, the amino group of the amino acid gets blocked and the resulting product is acidic in nature. It can be titrated with alkali and forms the basis of **Sorenson formol titration** method for the estimation of proteins.

9. Reduction. Amino acids are converted into amino alcohols on reduction with lithium aluminium hydride.

$$H_3\overset{+}{N}\underset{\stackrel{|}{R}}{CH}COO^- \xrightarrow[\text{[H]}]{LiAlH_4} H_2N\underset{\stackrel{|}{R}}{CH}CH_2OH$$

Amino alcohol

10. Decarboxylation. Amino acids undergo decarboxylation in the presence of acids or bases or enzymes.

$$H_3\overset{+}{N}\underset{\stackrel{|}{R}}{CH}COO^- + Ba(OH)_2 \xrightarrow{\Delta} \underset{\text{Primary amine}}{RCH_2NH_2} + BaCO_3 + H_2O$$

11. Action of heat. (*a*) α-amino acids. On heating, α-amino acids undergo dehydration by interaction between two amino acid molecules.

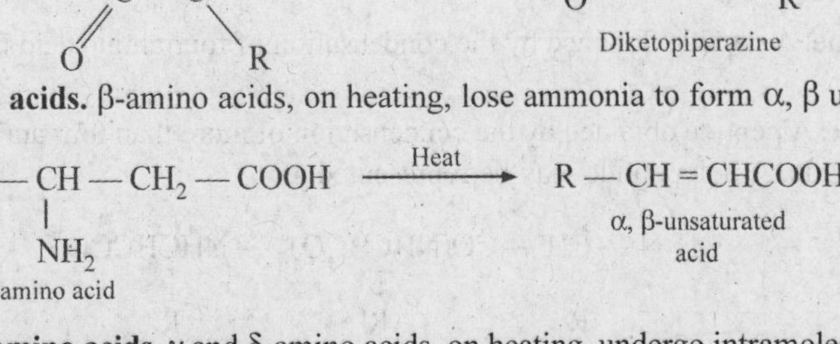

Diketopiperazine

(*b*) β-amino acids. β-amino acids, on heating, lose ammonia to form α, β unsaturated acids.

$$\underset{\beta\text{-amino acid}}{R - \underset{\stackrel{|}{NH_2}}{CH} - CH_2 - COOH} \xrightarrow{\text{Heat}} \underset{\alpha,\ \beta\text{-unsaturated acid}}{R - CH = CHCOOH} + NH_3$$

(*c*) γ and δ-amino acids. γ and δ-amino acids, on heating, undergo intramolecular dehydration to form cyclic amides called *lactums*.

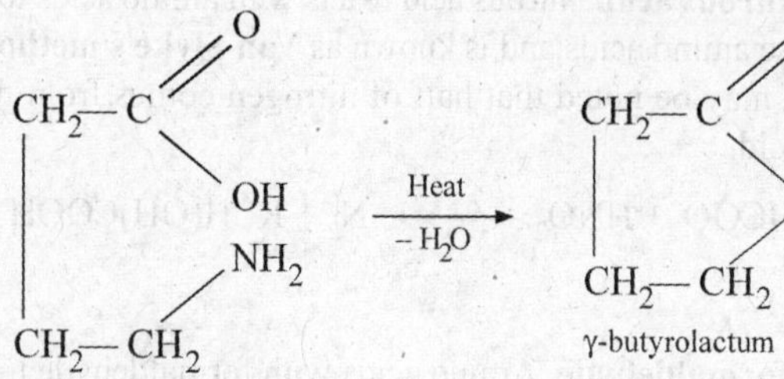

γ-amino butyric acid γ-butyrolactum

12. Reaction with ninhydrin

Amino acids react with ninhydrin (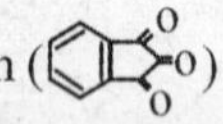) to form a dark blue or violet complex. This reaction is used as a test for α-amino acids. However, proline and hydroxyproline give a yellow colour.

13. Reaction with metallic ions. Amino acids react with heavy metal ions in aqueous solution to form deep coloured complexes.

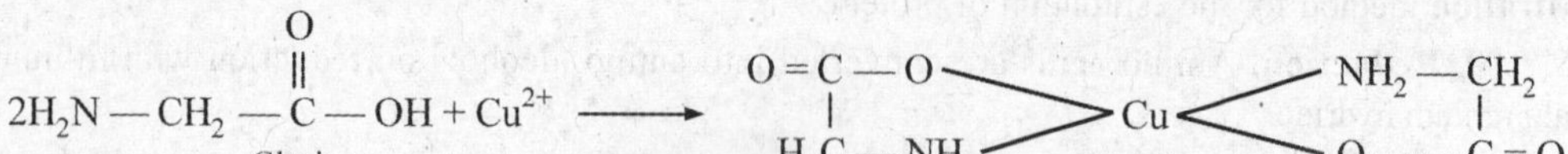

Glycine Complex (Deep blue colour)

PEPTIDES

Peptides are amides obtained by interaction between the amino and carboxylic groups of two or more amino acid molecules. Two molecules of glycine, for example, combine to form amide substance known as glycyl glycine.

$$\underset{\text{2 molecules of glycine}}{\overset{+}{N}H_3CH_2COO^- + \overset{+}{N}H_3CH_2COO^-} \xrightarrow{-H_2O} \underset{\text{Glycyl glycine}}{H_3\overset{+}{N}CH_2CONHCH_2COO^-}$$

The amide group — CO—NH— in the peptides is called peptide linkage.

7.10. CLASSIFICATION OF PEPTIDES

Peptides are classified as under:

Dipeptide. A peptide obtained by the condensation of two amino acid molecules is called dipeptide.

Tripeptide. A peptide obtained by the condensation of three amino acid molecules is called tripeptide.

Tetrapeptide. A peptide obtained by the condensation of four amino acid molecules is called tetrapeptide.

Polypeptide. A peptide obtained by the condensation of more than four amino acid molecules is called polypeptide. A polypeptide may be represented as:

$$H_3\overset{+}{N} - \underset{\displaystyle R}{\underset{|}{CH}} - CO(\underset{\displaystyle R'}{\underset{|}{NHCHCO}})_n - \underset{\displaystyle R''}{\underset{|}{NHCHCOO^-}}$$

It may be noted that peptides have a free – NH_3^+ on one end and a free – COO^- group on the other. The amino acid having the free – NH_3^+ group is called N– terminal **amino acid residue** while the amino acid having the free – COO^- group is called **C– terminal amino acid residue.** In writing the formula of peptides, we start from the left with the N-terminal amino acid residue and proceed to the right towards the C-terminal amino acid residue.

Peptides of molecular weight upto 10000 are known as **polypeptides** whereas peptides of higher molecular weight are the **proteins**. Conventionally, N-terminal amino acid residue is written at the left end while the C-terminal amino acid residue is written at the right end.

7.11. NOMENCLATURE

While naming a peptide, the names of the constituent amino acids are written from N-terminal (L.H.S.) to C–terminal (R.H.S)

The suffix "ine" of the names of all amino acids except the C-terminus acid is replaced by "yl"

Sometimes the name of polypeptides are abbreviated by using three letter abbreviations for constituent amino acids. The examples given below will illustrate.

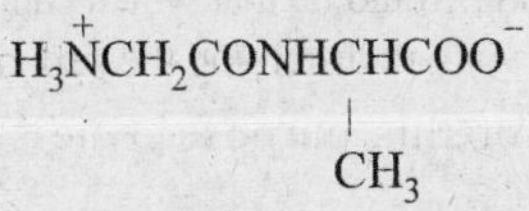

Glycylalanine
(Gly-Ala)

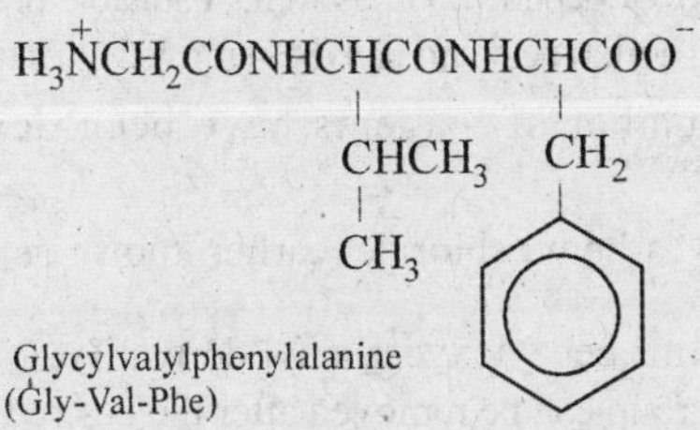

Glycylvalylphenylalanine
(Gly-Val-Phe)

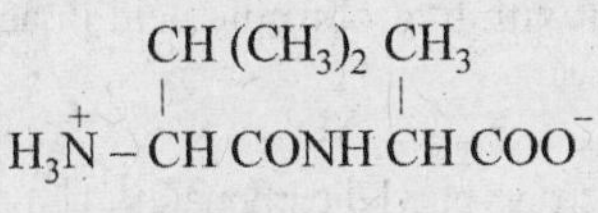

Valylalanine

Geometry of peptide linkage

X-ray diffraction studies reveal that the peptide linkage is flat *i.e.* carbonyl carbon, nitrogen and atoms attached to them lie in the same plane. The C—N bond distance comes out to be 1.32Å compared to usual C—N single bond distance of 1.47 Å, indicating that C—N bond has 50% double bond character. Further, measurement of bond angles shows that bonds to nitrogen are similar to those about trigonal carbon atom as shown below:

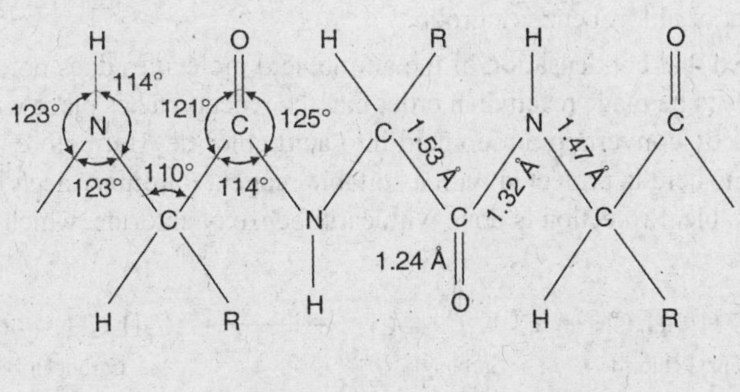

SYNTHESIS OF PEPTIDES FROM AMINO ACIDS

7.12. CLASSICAL PEPTIDE SYNTHESIS

The condensation reaction between two amino acid molecules to form peptide linkage is an endothermic reaction and does not take place easily. Therefore, in actual practice the carboxylic group of one amino acid molecule is activated by conversion of the amino acid into acid chloride. This is then reacted with the amino group of another amino acid molecules.

But there is another problem caused by the presence of both the acid group and the amino group in the same molecules. Consider the synthesis of a simple dipeptide, alanoylglycine (Ala – Gly). Let us say, we first activate the carboxyl group of alanine by converting it into acid chloride and then allow it to react with glycine. In practice alanoyl chloride reacts not only with glycine but also with another molecule of alanoyl chloride. Therefore, the reaction would yield not only Ala-Gly but also Ala–Ala. The different products formed would have to the separated which is a tedious process.

To overcome this problem we protect the amino group of the first amino acid before the acid is converted into the chloride. This involves the conversion of amino group into some other group which would not react with the acid chloride formed. Protecting the amino group by acetylation or benzoylation is not feasible here. This is because removing the protecting acetyl or benzoyl group from the acetylated peptide formed leads to the cleavage of peptide bond as well. A suitable protecting group would be that which can be introdcued in the starting amino acid and can be removed after the reaction without disturbing the peptide bond formed.

A number of reagents have been developed for this purpose. One such reagent is benzyloxycarbonyl chloride (earlier known as carbobenzoxy chloride), $C_6H_5CH_2-O-\overset{\overset{O}{\|}}{C}-Cl$.

It forms benzyloxycarbonyl derivatives of the amino group of the amino acid. The benzyloxy carbonyl group can be removed after the desired reaction, without disturbing the peptide bond formed, either on catalytic reduction or treatment with hydrobromic acid in acetic acid.

Another amino protecting group is ditertiarybutyl dicarbonate $(CH_3)_3\,CO\overset{\overset{O}{\|}}{C}-O-\overset{\overset{O}{\|}}{C}-OC\,(CH_3)_3$.

It is used for the introduction of butoxycarbonyl group $(CH_3)_3\,CO\overset{\overset{O}{\|}}{C}-$ (BOC group). This group can be eliminated by treatment with HBr, after the peptide has been formed.

Synthesis of a peptide from amino acids involves the following steps:

(*a*) Protection of the —NH_2 group of the amino acid with carbobenzoxy chloride (also known as benzyloxy carbonyl chloride).

(*b*) Conversion of the carboxyl group into acid chloride.

(*c*) Formation of peptide linkage.

(*d*) Removal of chlorobenzoxy group.

It is observed that condensation of the amino acid molecules does not take place rapidly. The carboxy group has to be made reactive in order that the reaction takes place with appropriate kinetics. This can be done by converting amino acid into acid chloride. But before this is done, the amino group of the amino acid is protected with a suitable reagent. Ordinary acetylation or benzylation is found to be untenable. Protection is done with carbobenzoxy chloride, which is prepared as follows:

$$\underset{\text{Benzyl chloride}}{C_6H_5CH_2Cl} + \underset{\text{Carbonyl chloride}}{Cl-\overset{\overset{O}{\|}}{C}-Cl} \longrightarrow \underset{\text{Carbobenzoxy chloride}}{C_6H_5CH_2-O-\overset{\overset{O}{\|}}{C}-Cl}$$

This group is easy to remove at the completion of peptide synthesis. Stepwise synthesis of glycylalanine (dipeptide) is illustrated as under:

(*i*) Protection of amino group

$$\underset{\text{Glycine}}{H_3\overset{+}{N}CH_2COO^-} + \underset{\text{Carboxybenzoxy chloride}}{C_6H_5CH_2OCOCl} \xrightarrow{-HCl} \underset{\text{Carbobenzoxy glycine}}{C_6H_5CH_2OCONHCH_2COOH}$$

(*ii*) Formation of acid chloride

$$C_6H_5 - CH_2OCONHCH_2COOH + SOCl_2 \xrightarrow[-HCl]{-SO_2} \underset{\text{Acid chloride of carbobenzoxy glycine}}{C_6H_5CH_2OCONHCH_2COCl}$$

(*iii*) Formation of peptide linkage

$$C_6H_5CH_2OCONHCH_2COCl + \underset{\text{Alanine}}{H_3\overset{+}{N}\underset{\underset{CH_3}{|}}{C}HCOO^-} \xrightarrow{-HCl} \underset{\text{Carbobenzoxy glycylalanine}}{C_6H_5CH_2OCONHCH_2CONH\underset{\underset{CH_3}{|}}{C}HCOOH}$$

(*iv*) Removal of protecting group

$$C_6H_5CH_2OCONHCH_2CONH\underset{\underset{CH_3}{|}}{C}HCOOH \xrightarrow{Hg^{\cdot}-Pd} \underset{\text{Glycylalanine}}{H_3\overset{+}{N}CH_2CONH\underset{\underset{CH_3}{|}}{C}HCOO^-} + \underset{\text{Toluene}}{C_6H_5CH_3} + CO_2$$

Repeating the above steps could yield tri, tetra and polypeptides.

Carboxyl protecting groups

Carboxyl group is generally protected by conversion to methyl, ethyl, or benzyl esters (whenever required). After the peptide formation, methyl and ethyl esters are hydrolysed to free carboxylic acid. In case of benzeyl esters, catalytic reduction is carried out to obtain free carboxylic acid.

Use of DCC as reagent for peptide bond formation. The method of peptide bond formation discussed above has certain drawbacks. A serious drawback is that it leads to racemisation of the product at the stereocentre alpha to the acid chloride. Therefore, attempts were made to discover better methods of peptide bond formation. It was found that treatment of a solution containing an amino protected amino acid and a carbonyl-protected amino acid with dicylclohexyl carbodiimide (DCC) leads directly to peptide bond formation

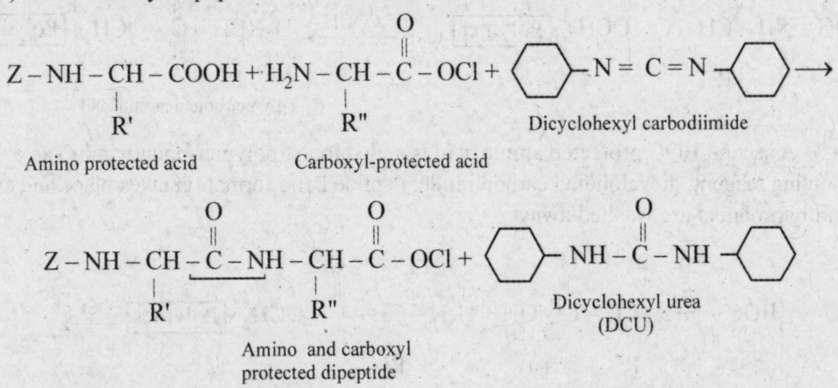

Peptide bond formation is a dehydration reaction and DCC is a strong dehydrating agent. It removes – OH from the carboxyl group of amino acid and – H from the amino group of the other acid to form a peptide bond and it itself converted into dicyclohexylurea (DCU). Removal of protecting groups from the products gives the dipeptide.

7.13. SOLID-PHASE PEPTIDE SYNTHESIS

R.B Merrifield discovered a unique method of peptide synthesis known as solid–phase peptide synthesis. Here a solid support is used to carry out the synthesis.

The solid support used is a stryrene polymer cross-linked with about 2% *p*-divinylbenzene. It is further chloromehylated in such a way that about 5% benzene rings carry chloromethyl ($-CH_2Cl$) substituents.

The advantage of this method is that the excess reagents, impurities and by-products formed can be just washed away with suitable solvents while the growing peptide chain remains bound to the polymer.

We illustrate the procedure followed in this method by considering a generalised example. The chloromethylated polystyrene polymer has been shown as $\boxed{\text{Polymer}}-CH_2Cl$. The amino protecting group used is *tert*-butoxycarbonyl abbreviated as BOC. The carboxyl group is protected by formation of a benzyl ester with the solid polymer via – CH_2Cl substituent on the benzene ring. The process is completed in five steps as shown below:

Step 1. The amino group of the amino acid which is to be C-terminal end of the peptide to be formed is protected by BOC. The BOC protected amino acid is attached to the polystyrene polymer by formation of a benzyl ester linkage by reaction between carboxylate part of the acid and chloromethyl group on the polymer.

$$\underset{\text{Amino-protected amino acid}}{BOC-NH-\underset{\underset{R'}{|}}{CH}-\overset{\overset{O}{\|}}{C}-OH} + ClCH_2-\boxed{\text{Polymer}} \xrightarrow{\text{Base}} \underset{\text{Ester linkage with the polymer}}{BOC-NH\underset{\underset{R'}{|}}{CH}-\overset{\overset{O}{\|}}{C}-OCH_2-\boxed{\text{Polymer}}}$$

The excess reagents are removed by washing with suitable solvents.

Step 2. The BOC protecting group is removed by treatment with trifluoroacetic acid. The polymer bonded amino acid is purified by washing.

$$BOC-NH-\underset{\underset{R'}{|}}{CH}-\overset{\overset{O}{\|}}{C}-OCH_2-\boxed{\text{Polymer}} \xrightarrow{CF_3COOH} \underset{\text{Polymer bound amino acid}}{H_2N\underset{\underset{R'}{|}}{CH}-\overset{\overset{O}{\|}}{C}-OCH_2-\boxed{\text{Polymer}}}$$

Steps 3. A second BOC protected amino acid is added to the polymer bound amino acid along with the coupling reagent, dicyclohexyl carbodiimide. Peptide bond formation takes place and excess reagents and byproducts are washed away.

$$BOC-NH-\underset{\underset{R''}{|}}{CH}-\overset{\overset{O}{\|}}{C}-OH + H_2N-\underset{\underset{R'}{|}}{CH}-\overset{\overset{O}{\|}}{C}-OCH_2-\boxed{\text{Polymer}} \xrightarrow{DCC}$$

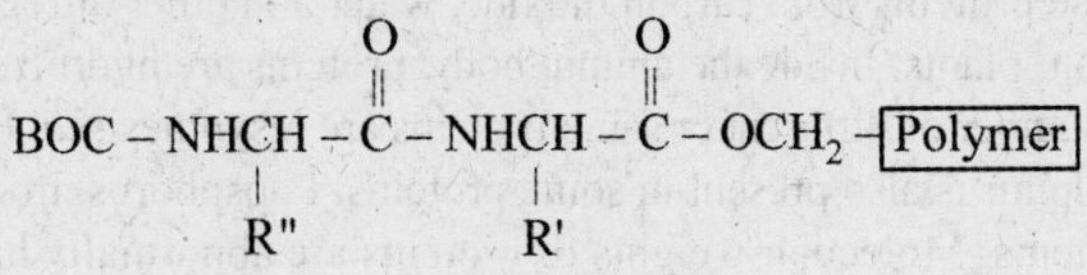

Peptide bond formation takes place

Steps 4. The BOC protecting group is removed as in step 2.

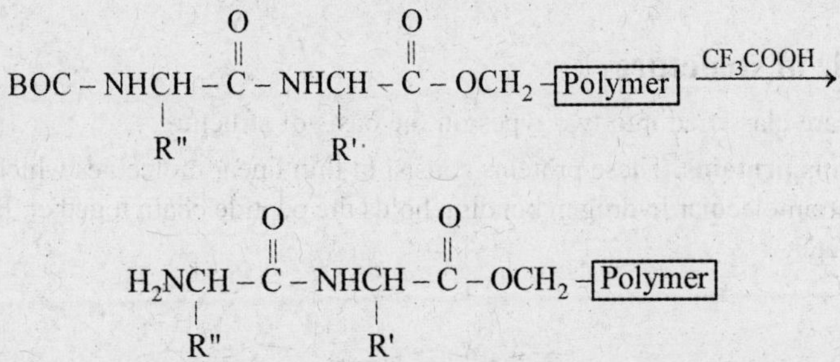

Step 3 and 4 are repeated to introduce as many amino units as required to build the polypeptide chain. The last acid to be attached is the N-terminal amino acid.

Steps 5. The completed polypeptide is removed from the polymer by treatment with anhydrous hydrogen fluoride.

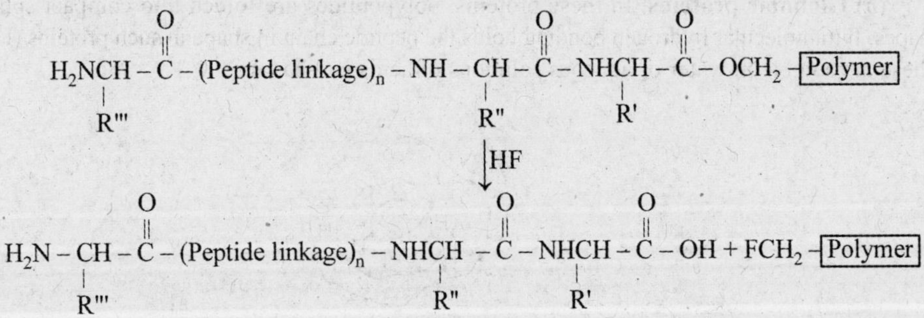

Free peptide

Ribonuclease which contains a sequence of 124 amino acids and involves 369 chemical reactions and 11931 steps, has been successfully synthesised by this process.

Proteins

Proteins are complex organic compounds essential for growth and maintenance of life. These are nitrogenous compounds obtained from α-amino acids. They are the constituents of all living organisms and found in every part of plants and animals. They are present in muscles, skin, hair, nails, blood, tendons and arteries.

Proteins present in different plants and animals differ from one another in composition and biological action. Proteins perform diverse functions in life processes. Some proteins are responsible only for structural shapes of parts of the body *e.g.*, keratin in hair, some are responsible for regulating metabolic processes *e.g.*, insulin in blood sugar level whereas some act as catalysts for biological reactions *e.g.*, enzymes.

Plants synthesise proteins from carbon dioxide, water and other nitrogenous materials. Animals consume proteins from plants. Inside the animal body, proteins are hydrolysed to a mixture of amino acids from which a number of different or same proteins are resynthesised. Besides carbon, hydrogen, oxygen, nitrogen, sulphur is also present in some proteins. Phosphorus, iron and magnesium may be present in select proteins. Molecular weights of proteins are abnormally high as a protein molecule is constituted of several thousands of amino acid molecules. In all there are 26 different natural amino acids which are the building blocks of different proteins.

7.14. CLASSIFICATION OF PROTEINS

On the basic of structure

Proteins are classified into two types on the basis of structure:

(a) Fibrous proteins. These proteins consist of thin linear molecules which lie side by side to form fibres. Intramolecular hydrogen bonding holds the peptide chain together. Fibrous proteins are insoluble in water.

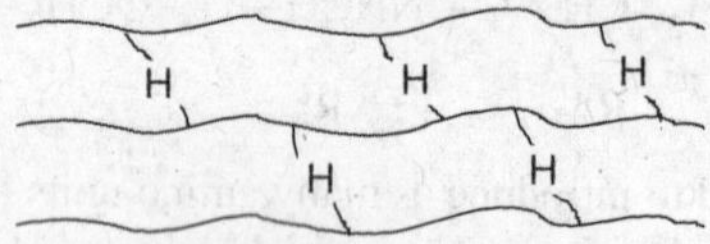

Structure of fibrous proteins

Fibrous proteins are the main structural materials of tissues. Examples of important fibrous proteins are *keratin* in skin and hair, *callagen* in tendon, *fibroin* in silk and *myosin* in muscles.

(*b*) Globular proteins. In these proteins, polypeptides are folded into compact spheroidal shapes. Intramolecular hydrogen bonding holds the peptide chain in shape in such proteins (Fig 7.2). These are soluble in water or aqueous solutions of acids and bases.

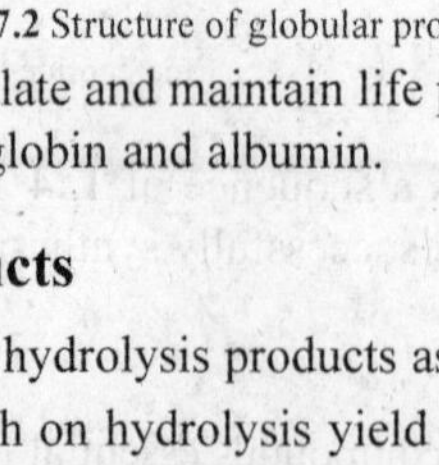

Fig. 7.2 Structure of globular proteins

These proteins have the role to regulate and maintain life processes. Examples of this class of proteins are enzymes, hormones, haemoglobin and albumin.

On the basis of hydrolysis products

Proteins are classified according to hydrolysis products as follows:

(*a*) Simple proteins. Proteins which on hydrolysis yield only amino acids are called simple proteins. Examples of this class of proteins are albumins (such as egg albumin, serum albumin), globulins (such as tissue globulin) and glutelins (such as wheat gluteline).

(*b*) Conjugated proteins. Proteins which are a combination of two parts, a proteinous part and non-proteinous part, are called conjugate proteins. The non-proteinous part is called prosthetic group. Prosthetic group plays its part in biological function of the protein.

Conjugated proteins are further classified as nucleoproteins, glycoproteins and chromoproteins. The prosthetic groups in such proteins are nucleic acid, carbohydrate and haemoglobin (or chlorophyll) respectively.

7.15. GENERAL CHARACTERISTICS OF PROTEINS

1. Composition. The elements generally present in proteins are carbon, hydrogen, oxygen and nitrogen.

2. High molecular weights. As protein molecules are obtained from hundreds and thousands of amino acid molecules, their molecular masses run into several thousands and sometimes into several lakhs.

3. Physical state. Generally speaking, proteins are colourless, tasteless, amorphous solids having no sharp melting points. They are colloidal size particles. This property is used to separate proteins from crystalline salts by the process of dialysis.

4. Optical activity. Because of the presence of asymmetric carbon atoms, proteins show optical activity. However, all naturally occurring proteins have the same configuration *viz*. the L-glyceraldehyde configuration.

5. Amphoteric nature. Protein molecules exist as dipolar ions. Hence they react with both acids and alkalis and are thus amphoteric.

6. Hydrolysis. As amino acids are the constituents of proteins, the latter on hydrolysis in the presence of acids, bases or enzymes give back amino acids but in steps. The steps involved during hydrolysis are represented as under:

$$\begin{array}{ccccc} \text{Proteins} & \longrightarrow & \text{Proteoses} & \longrightarrow & \text{Peptones} \\ & & & & \downarrow \\ \text{Amino acids} & \longleftarrow & \text{Simple Peptides} & \longleftarrow & \text{Polypeptides} \end{array}$$

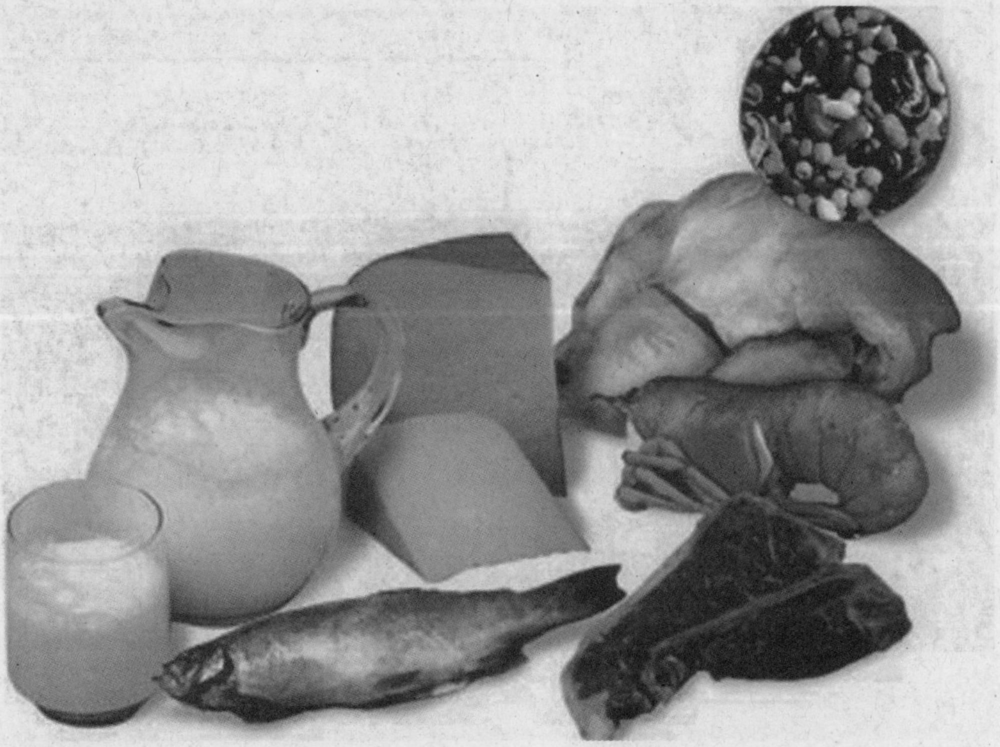

Sources of protein

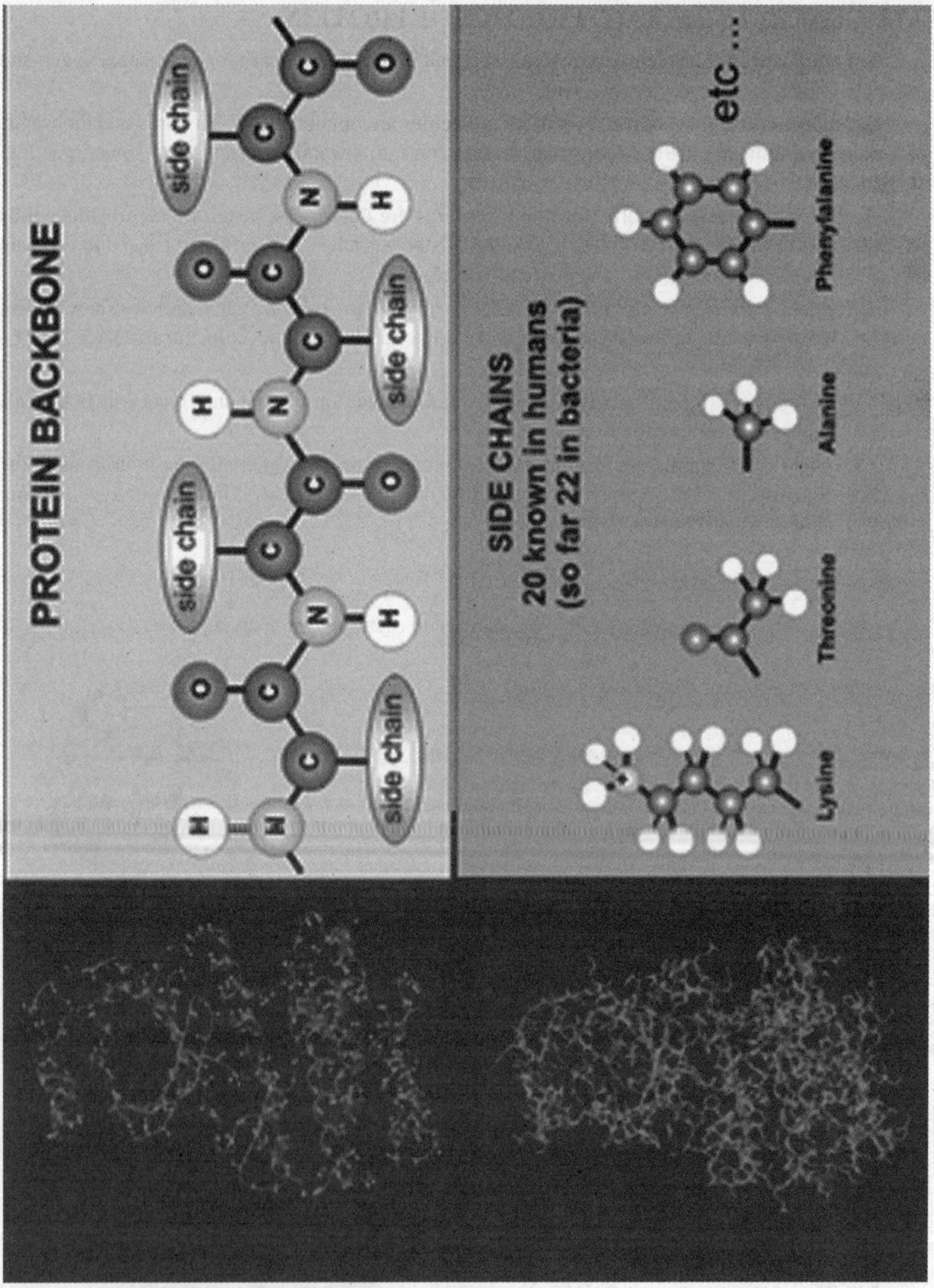
PROTEIN BACKBONE
side chain
side chain
side chain
side chain
SIDE CHAINS
20 known in humans
(so far 22 in bacteria)
Lysine
Threonine
Alanine
Phenylalanine
etc ...

7.16. DENATURATION AND RENATURATION OF PROTEINS

Proteins are very tender and delicate substances. When subjected to heat or action of acids or alkalis, they lose their biological activity. They are said to be **denatured**. This phenomenon in which the proteins lose their biological activity and other characteristics under the effect of temperature, is called **denaturation**. The denatured proteins can be brought back to its original state by cooling the protein solution very slowly. This process is called **renaturation**.

During denaturation, there is a rearrangement in the secondary and tertiary structure of the protein but the primary structure remains unchanged. Coagulation of egg-white by the action of heat is an example of irreversible denaturation of proteins.

7.17. TESTS FOR PROTEINS

Following tests are performed to identify proteins:

1. Biuret test. A drop of copper sulphate is added to an alkaline solution of protein. A bluish colour develops. This test is also given by proteoses and peptones which are the hydrolytic products of proteins.

2. Ninhydrin test. Ninhydrin is triketo hydrindene hydrate. A blue to red-violet is obtained when a protein is treated with ninhydrin.

3. Xanthoproteic test. This tests is given by proteins containing tyrosine or tryptophane. When warmed with conc. HNO_3, such proteins give a yellow colour.

4. Millon's test. Millon's reagent is a mixture of mercurous and mercuric nitrates. Proteins containing tyrosine give a white precipitate turning red when treated with Millon's reagent.

5. Heller's test. This test is commonly employed for detecting albumin in urine. When conc. HNO_3 is poured along the side of a test tube containing protein solution, a white precipitate is obtained.

7.18. BIOLOGICAL IMPORTANCE OF PROTEINS

Proteins are vital to animal life. They perform a number of biological processes and play a pivotal role in the running and maintenance of body. They are hydrolysed inside the body into its constituent amino acids before performing various functions.

In the stomach, the proteins are hydrolysed, in the presence of hydrochloric acid and enzyme pepsin, into lower molecular weight polypeptides.

In the intestine, proteins are hydrolysed in the presence of enzymes trypsin and pepsin. The simple amino acids are then assimilated in the blood stream and transported to various cells of the body. It is interesting to note that some of the amino acids are reconverted into some specific proteins which are needed by the body for specific requirements and other amino acids are oxidised to produce energy. Thus there is a continuous cycle of decomposition and synthesis of proteins in the body. Functions of different kinds of proteins are discussed as under:

***(a)* Enzymes.** These are the proteins which catalyse biological reactions in the body. Every enzyme is very specific in its action. They are able to perform reactions without requiring any change in temperature, pressure or pH. They carry out degradation and synthetic reaction in the body with much greater speed and efficiency than would be achieved in the laboratory. For example, enzyme carbonic anhydrase present in the blood catalyses the decomposition of over 30 million molecules of carbonic acid into carbon dioxide in just a minute, thereby maintaining carbon dioxide level in the body fluids.

Enzymes pepsin and trypsin catalyse the hydrolysis of peptide linkage in protein molecules in digestive tract.

The enzyme ptyalum, which is present in saliva catalyses the conversion of starch into maltose during chewing of food.

The enzymes maltase, lactase, sucrase catalyse the decomposition of disaccharides into monosaccharides.

***(b)* Antibodies.** These proteins defend the body against attack from foreign organism. These proteins are produced by the body during attack by foreign infectious species called *antigens*. The newly produced proteins combine with these antigens and thus protect the body against the destructive action of antigens. Gamma globulins present in blood are examples of antibodies.

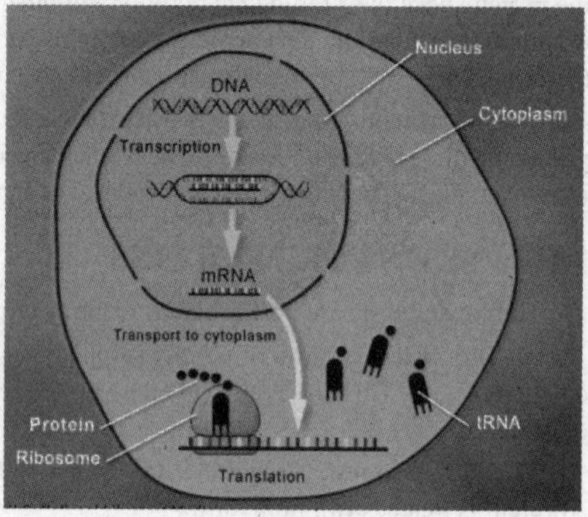

From genes to proteins

***(c)* Haemoglobin.** This is an example of transporting proteins. Haemoglobin present in blood transports oxygen from lungs to all parts of the body. Haemoglobin consists of two parts. The protein part is *globin* and the non-protein part is *haemo*. It is haemo which is responsible for transporting oxygen and is also responsible for the red colour of blood.

***(d)* Structural proteins.** Structural proteins are responsible for providing distinct structure to various parts of the body. Keratin is a structural protein present in skin, hair, nails, horn and features.

Myosin is a structural protein present in muscles and *collagen* in tendons.

7.19. PRIMARY, SECONDARY AND TERTIARY STRUCTURE OF PROTEINS

Studying structure of a protein is a very complex exercise. If we divide the work into smaller steps, the task becomes easier. For structure elucidation of protein, the work is divided under four headings as listed below:

(*a*) Primary structure (*b*) Secondary structure
(*c*) Tertiary structure (*d*) Quaternary structure

Primary Structure

It refers to the number, nature and sequence of the amino acids in polypeptide chains. Primary structure determination of protein involves the following steps:

***(i)* Determination of amino acid composition.** The protein under investigation is hydrolysed by means of acid, alkali or enzyme to its constituent amino acids. The amino acids are separated and identified by means of ion-exchange chromatography. The weights of amino acids produced are noted. Thus the number of moles of each amino acid and the relative number of various amino acid residues in the protein are estimated. Knowing the molecular weights of the protein, which is determined by a suitable method, the actual number of amino acid residues in the molecules is calculated.

***(ii)* Sequence of arrangement of amino acids.** The next job, and a difficult job is to arrive at the sequence of arrangement of amino acids. For this, a controlled hydrolysis of protein is carried out to produce small polypeptides. The sequence of arrangement of amino acids is determined by terminal and residue analysis.

Terminal residue analysis. The amino acid residues at the two extremes of a peptide chain are different from all other amino acid residue and also from each other. Amino acid present at one end of the chain has free — NH_2 group. It is called as *N-terminal amino acid*. The amino acid present on the other end has a free — COOH group. It is called as *C-terminal amino acid.*

$$\underset{\text{N-terminal residue}}{NH_2-\overset{\overset{R}{|}}{C}HCO}\left[NH\overset{\overset{R'}{|}}{C}HCO\right]_n\underset{\text{C-terminal residue}}{NH\overset{\overset{R''}{|}}{C}HCOOH}$$

The two terminal amino acid residues are identified as follows:

N-terminal residue analysis

Frederick Sanger

(*i*) Sanger's method. The polypeptide is treated with 2, 4-dinitrofluorobenzene (DNFB) commonly known as Sanger's reagent, in presence of sod. bicarbonate. The reagent reacts with free amino group of N-terminal amino acid residue to form a 2, 4-dinitrophenyl (DNP) derivative of the polypeptide. The product obtained is hydrolysed by an acid to form dinitrophenyl (DNP) derivative of N-terminal amino acid and a smaller polypeptide molecule. The DNP derivative of the amino acid is isolated and analysed chromatographically or by thin layer chromatography. The reactions are given as under:

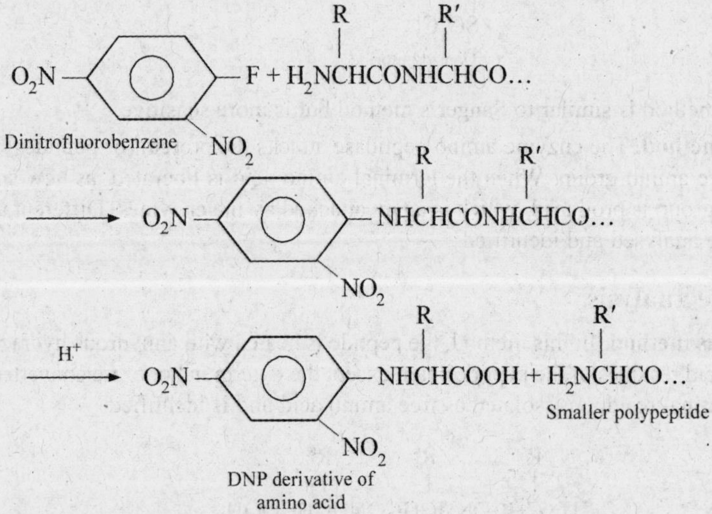

(*ii*) Edman's method. This method involves the reaction between peptide and phenylisothiocyanate, $C_6H_5N = C = S$ (commonly termed as Edman reagent) in the presence of a dilute alkali. N-phenylthiacarbomyl peptide thus formed is treated, with a mild acid when the N-terminal amino acid residue gets removed as phenylthiohydantoin (PTH) while the rest of peptide chain remains intact. This phenylthiohydantoin is separated from the reaction mixture and identified by comparison with phenylhydantoins obtained from known amino acids. This helps in the identification N-terminal amino acid. For example

$$\underset{\text{Phenylisothiocynate}}{C_6H_5-N=C=S} + \underset{\text{Peptide}}{H_2N\underset{|}{\overset{}{C}}HCO-NH\,CHCO-}\ \ (R,\ R') \xrightarrow{OH^-} \underset{\text{N-Phenylthiocarbamyl peptide}}{C_6H_5-NH-\overset{S}{\overset{\|}{C}}-NHCHCO-NH\,CHCO-}\ \ (R,\ R')$$

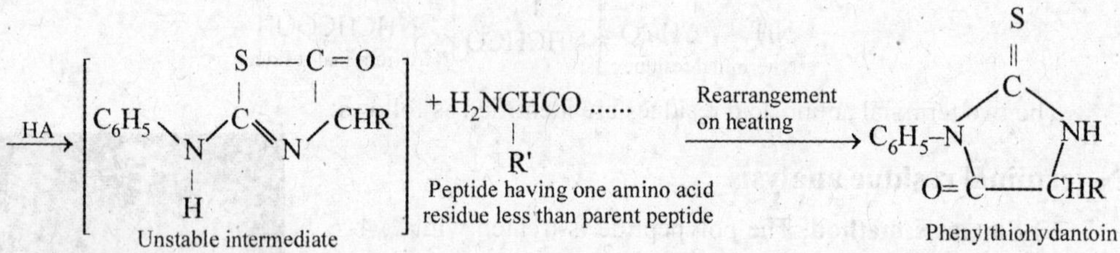

The complete process can be repeated with the newly formed degraded peptide to identify the second end amino acid and so on.

***(iii)* Dansyl method.** In this method the reagent used for N-terminal residue analysis is 1-dimethylamino-naphthalene -5-sulphonyl chloride (*called dansyl chloride*)

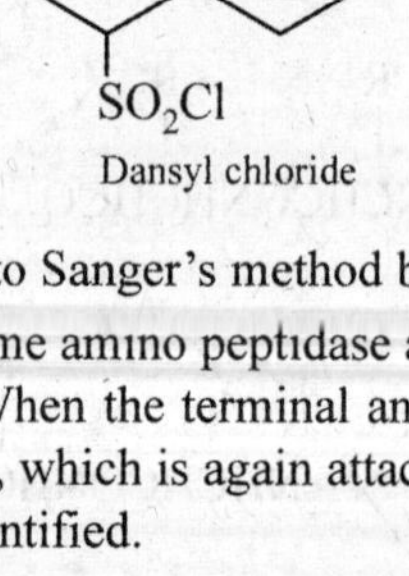

Dansyl chloride

Dansyl chloride method is similar to Sanger's method but is more sensitive.

***(iv)* Enzymatic method.** The enzyme amino peptidase attacks the protein (or peptides) at the terminal which has free amino group. When the terminal amino acid is liberated, as new smaller peptide having amino group is produced, which is again attacked by the enzymes. Different amino acids liberated thus are analysed and identified.

C-terminal residue analysis

***(i)* Hydrazinolysis method.** In this method, the peptide is heated with anhydrous hydrazine at 373 K. All the amino acid residues of the peptide chain except the C-terminal one, are converted into hydrazides. The C-terminal residue is isolated as free amino acid and is identified.

$$\ldots HN\overset{R}{\overset{|}{C}}HCONH\overset{R'}{\overset{|}{C}}HCONH\overset{R''}{\overset{|}{C}}HCOOH$$

$$\Big\downarrow \text{Heat} \mid NH_2NH_2$$

$$\underbrace{H_2N\overset{R}{\overset{|}{C}}HCONHNH_2 + H_2N\overset{R'}{\overset{|}{C}}HCONHNH_2}_{\text{Hydrazides}} + \underset{\text{Free C-terminal amino acid}}{H_2N\overset{R''}{\overset{|}{C}}HCOOH}$$

***(ii)* Enzymatic method.** The enzyme carboxypeptidase specifically hydrolyses the peptide chain adjacent to the free carboxy group. The C-terminal amino acid is thus set free and identified. The process is repeated on smaller peptide chains to set free amino acids one by one from the side of the free carboxy group.

Secondary Structure

Secondary structure tells us about the shape and conformation of the peptide chains in protein molecule. X-ray studies show that proteins have two different conformations viz. α-helix and β-structure or pleated sheet structure.

The β-structure. Peptide chains in silk fibroin are fully extended to form flat zig-zags. These chains of polypeptides lie side by side and are held to one another by hydrogen bonds as shown in Fig 7.3.

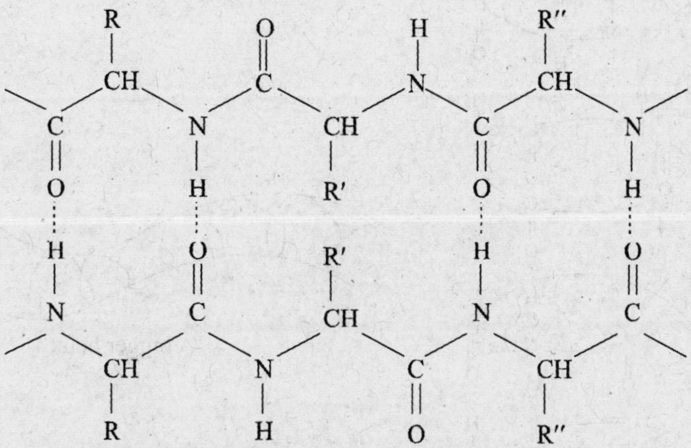

Fig. 7.3 β-Flat sheet structure of a protein

Due to steric hindrance between R, R', R'' side chains, the flat sheet structure contracts and adopts pleated sheet structure (Fig 7.4). The exact contraction depends upon the size of the side chains.

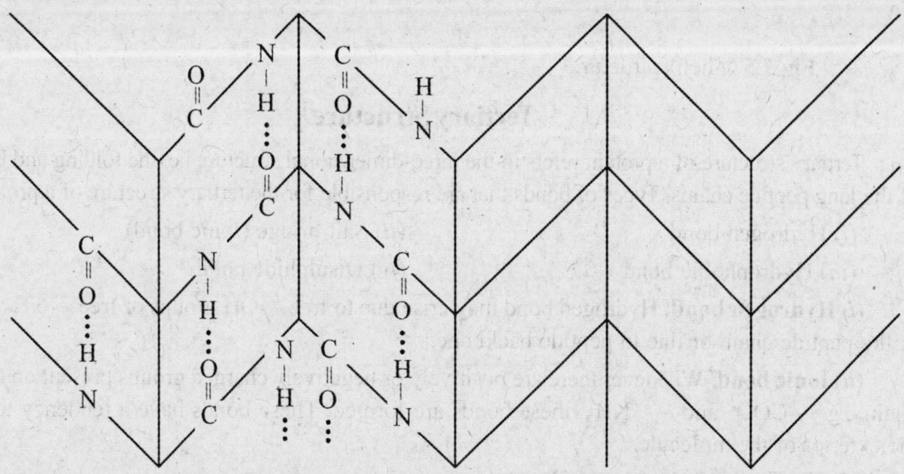

Fig. 7.4 β-pleated sheet structure of protein

The α-helix. When the side chains in a polypeptide are bulky, the β-structure is not feasible. In such a case, α-helix structure is adopted.

Pauling proposed α-helix structure for α-keratin, which is the constituent structural protein in hair, nails etc. In α-helix structure, each peptide chain is coiled to form helix as shown in Fig. 7.5 below. Each turn of the helix has about 3-7 amino acid residues and the distance between two helices is 5.4 Å. Two adjacent turns are linked by means of hydrogen bonds which involve NH— group of one amino group and the carbonyl oxygen of the fourth residue in the chain. Hydrogen bonding keeps the structure tight and prevents free rotation keeping the helix intact. α-helix may be left or right-handed. But it is observed that the right-handed helix is more stable. Diameter of the helix is about 10 Å.

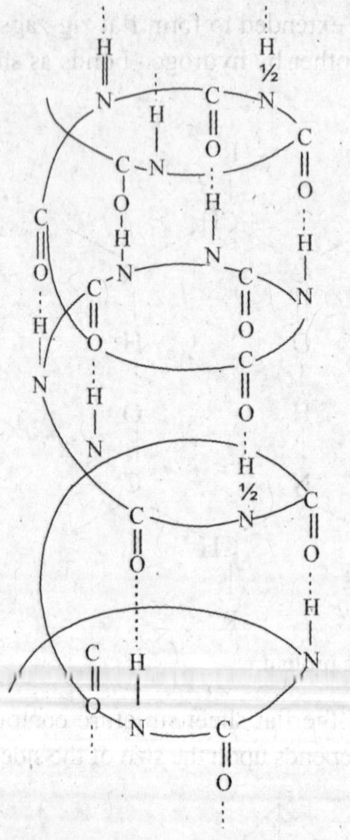

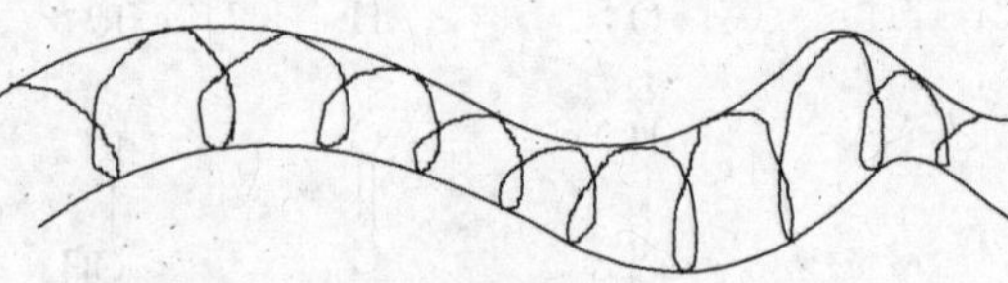

A bigger helix

Fig 7.5 α-helix structure

Tertiary Structure

Tertiary structure of a protein refers to the three-dimensional structure i.e. the folding and bending of the long peptide chains. Types of bonds that are responsible for the tertiary structure of a protein are:

(*i*) Hydrogen bond (*ii*) Salt bridge (Ionic bond)

(*iii*) Hydrophobic bond (*iv*) Disulphide bond

***(i)* Hydrogen bond.** Hydrogen bond may arise due to free —OH groups or free —NH_2 groups in the peptide chain or due to peptide backbone.

***(ii)* Ionic bond.** Whenever there are positively or negatively charged groups present on the side chain e.g. —COO^- and — $^+NH_3$, these bonds are formed. These bonds have a tendency to be on the exterior of the molecule.

***(iii)* Hydrophobic bond.** Such bonds are formed between the two methyl or two phenyl groups in the side chain. As they are hydrophobic groups, they prefer to remain in the interior where there is less water.

(*iv*) Some portions of the chain are linked by means of disulphide (–S–S–) bond. This is illustrated in the Fig. 7.6.

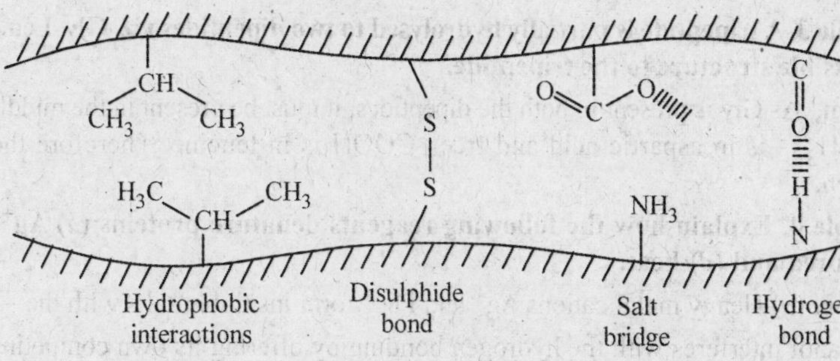

Fig 7.6 Tertiary structure of proteins

Quaternary Structures

There are certain proteins which have more than one polypeptide chain and also more than one independently folded unit (sub units) each of at least one chain. Thus they are aggregates of polypeptide sub units. The *quaternary structure* of a protein includes any protein in which the native molecule is made up of several different sub-units and it refers to the manner in which these units are grouped together. Each sub-unit contains its own independent three dimensional conformation. The quaternary structure is stabilized by the same non-covalent forces as in tertiary structure.

The Fig. 7.7 gives the representation of primary, secondary, tertiary and quaternary structures of proteins.

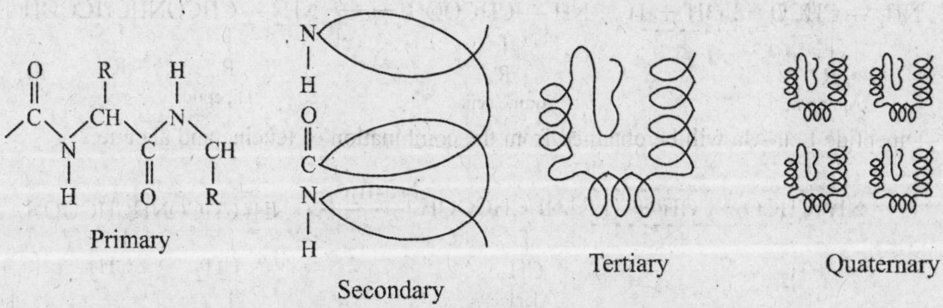

Fig. 7.7. Diagrammatic representation of primary, secondary, tertiary and quaternary structure of protein

SOME SOLVED EXAMPLES

Example 1. How can lysine and glycine be separated from each other? The isoelectric points are pH 9.6 for lysine and pH 5.97 for glycine.

Solution. An aqueous solution of the mixture is placed between two electrodes. The pH is adjusted to 9.6 and an electric current is passed. Lysine does not migrate at this pH. Glycine can be collected at the anode. Now the pH is adjusted to 5.97. At this pH, glycine does not migrate. Lysine can be collected at the cathode.

Example 2. How many different peptides can be synthesised from (*a*) glycine and alaine and (*b*) gly. ala$_2$?

Solution. (*a*) Four products can be obtained:

(*i*) Gly. Gly. (*ii*) Ala. Gly. (*iii*) Ala. Ala. (*iv*) Gly. Ala.

(*b*) Three products are possible:

(*i*) Gly. Ala. Ala. (*ii*) Ala. Gly. Ala. (*iii*) Ala. Ala. Gly.

Example 3. A tripeptide is partially hydrolysed to two dipeptides viz. Gly. Leu. and Asp. Gly. Assign a possible structure to the tripeptide.

Solution. As Gly is present in both the dipeptides, it must be present in the middle of tripeptide. The free $^{+}NH_3$— is in aspartic acid and free –COOH is in leucine. Therefore the tripeptide is Asp. Gly. Leu.

Example 4. Explain how the following reagents denature proteins (*a*) Ag^+ and Pb^{2+} (*b*) ethanol (*c*) urea and (*d*) heat.

Solution. (*a*) Heavy metal cations Ag^+ and Pb^{2+} form insoluble salts with the —COO^- group.

(*b*) Ethanol interferes with the hydrogen bonding by offering its own competing hydrogens.

(*c*) Urea can form a good hydrogen bond and thus interferes with the original hydrogen bonding in the protein.

(*d*) Heat produces more random conformations in the protein and thus hampers the formation of helix.

Due to these reasons, the denaturation of proteins takes place.

Example 5. Write the structure of dipeptide Leu-Ala.

Solution. Dipeptides are compounds obtained by the condensation of two amino acid (same or different) molecules. Hydroxy from the carboxy group of one molecule and hydrogen from amino group of second molecule are removed as water, resulting in the formation of peptide linkage as given below:

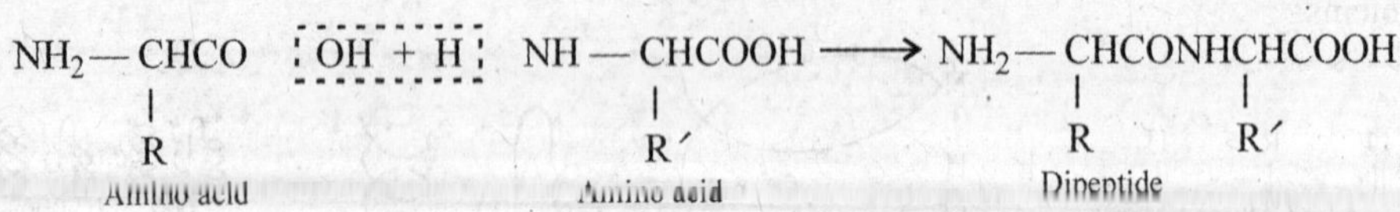

Dipeptide Leu-Ala will be obtained from the combination of leucine and alanine.

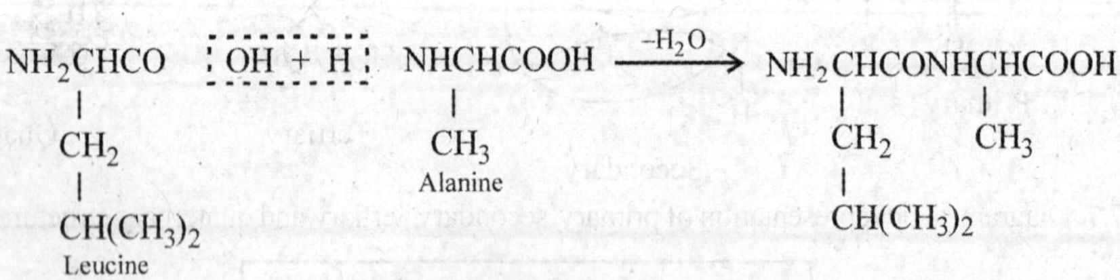

Since peptides exist as dipolar compounds, the above compound Leu-Ala dipeptide may be correctly written as

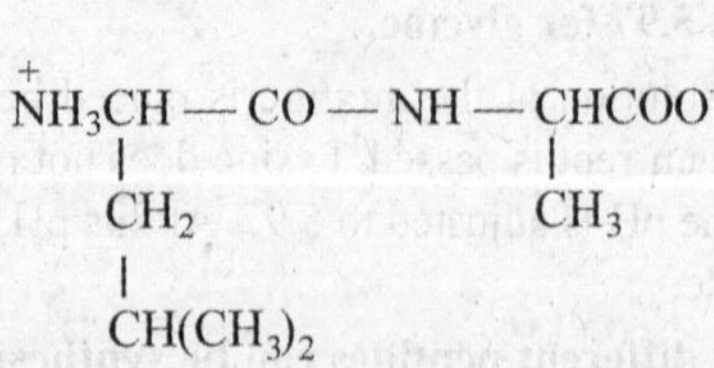

NUCLEIC ACIDS, RNA AND DNA

Nucleic acids form an important class of compounds, having high molecular weights. They play an important role in the development and reproduction of all forms of life. Living cells contains nucleic acids in the form of nucleoproteins. Nucleoproteins consists of a protein and a nucleic acid.

There are two types of nucleic acids

(*i*) Deoxyribonucleic acid or DNA (*ii*) Ribonucleic acid or RNA

These names for nucleic acids are derived from their hydrolysis products.

A nucleic acid on hydrolysis gives a nucleotide which on further hydrolysis gives a nucleoside and phosphoric acid. A nucleoside on hydrolysis in the presence of inorganic acid gives a mixture of sugar, purines and pyrimidines.

$$\text{Nucleic acid} \xrightarrow{Ba(OH)_2} \text{Nucleotide} \xrightarrow[NH_3]{\text{Aqueous}} \text{Nucleoside} \xrightarrow{\text{Inorganic acid}} \text{Sugar + Purines + Pyrimidines}$$

Sugars present in nucleic acids

Only two sugars have been isolated from the hydrolysis products of nucleic acids. They are D-Ribose and D-2-Deoxyribose.

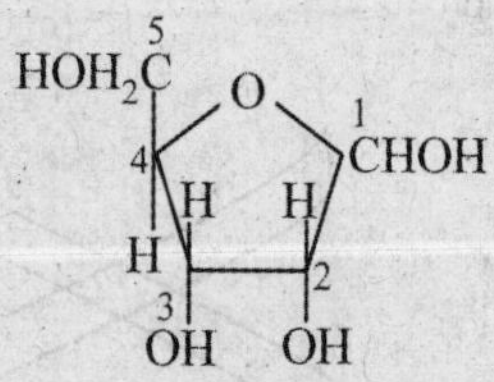

D-Ribose

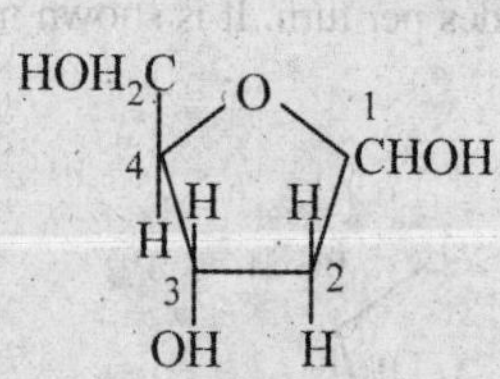

D-2-Deoxyribose

Bases present in nucleic acids

The bases present in nucleic acids are: adenine, guanine, cytosine, thymine and uracil

Adenine and guanine belong to the heterocyclic bases, purines.

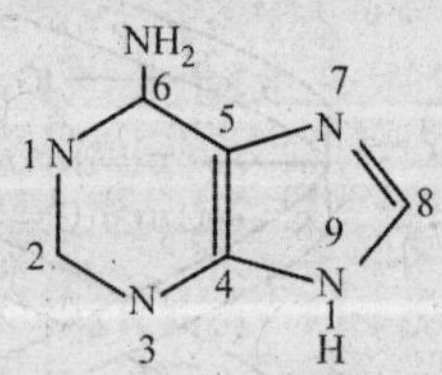

Adinine (A)

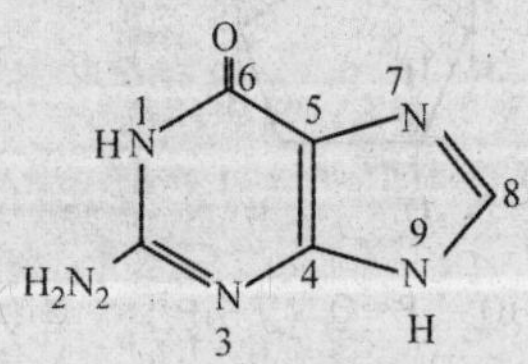

Guanine (G)

Cytosine, thymine and uracil belong to the class of **pyrimidines**

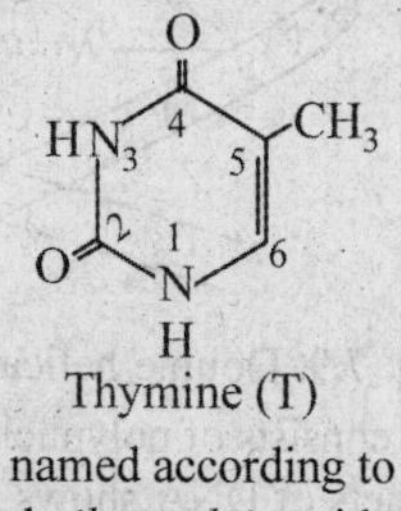

Thymine (T)

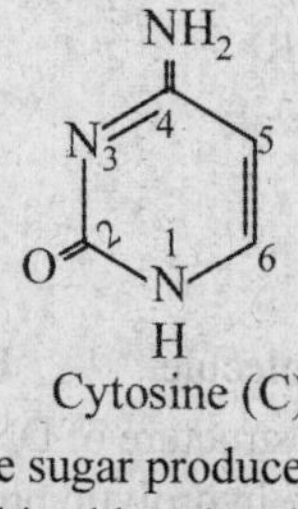

Cytosine (C)

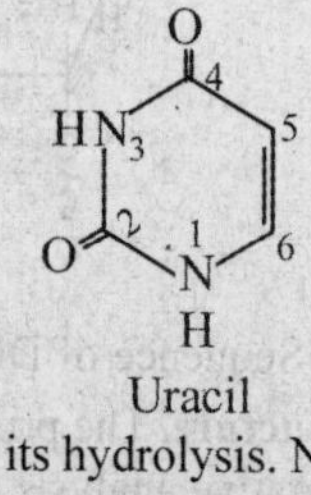

Uracil

Nucleic acids are named according to the sugar produced by its hydrolysis. Nucleic acid which contains ribose is called *ribonucleic* acid and is abbreviated as RNA. Nucleic acid which contains deoxyribose as a component is called *deoxyribonucleic* acid and is abbreviated as DNA. Correspondingly, there are ribonucleoproteins and deoxyribonucleo proteins. Nucleic acids are derived from sugars, phosphoric acid and heterocyclic bases (purines and pyrimidines). The linkage between them is shown as under:

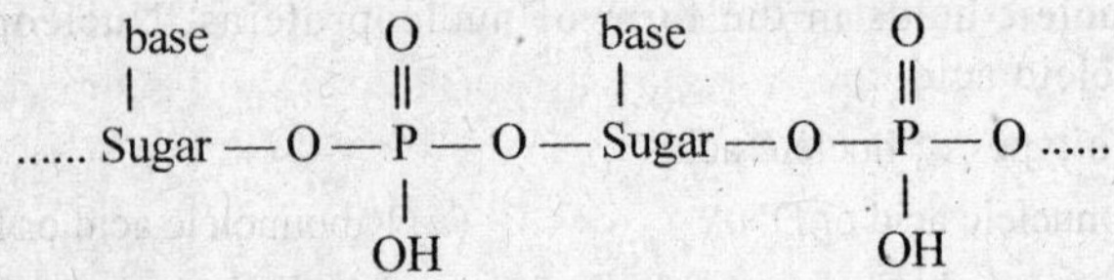

7.20. STRUCTURE OF DNA

DNA is a huge molecule with mol. mass ranging between 10^6 and 10^9. On hydrolysis of DNA, we obtain 2-deoxyribose, phosphoric acid and the bases adenine, guanine, cytosine, 2-methyl cytosine and thymine. Different DNA molecules contain the bases in different order. It is this different sequence of bases in the DNA molecule that is responsible for genetic variations. A general sequence of DNA molecule is shown in Fig. 7.8.

Watson and Crick in 1953 gave the secondary structure of DNA. They described DNA molecule as two identical polynucleotide chains twisted about each other to form double helix with a diameter of 18 Å in which the heads of two chains are in opposite directions. Both helices are right handed and have ten nucleotide resides per turn. It is shown in Fig. 7.9.

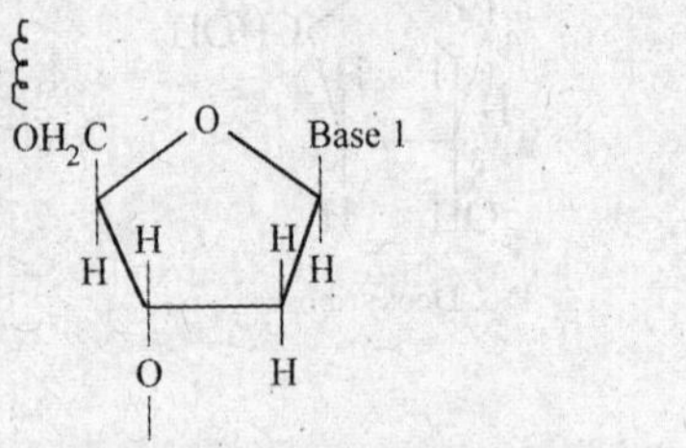

Fig. 7.8. Sequence of DNA mlecule

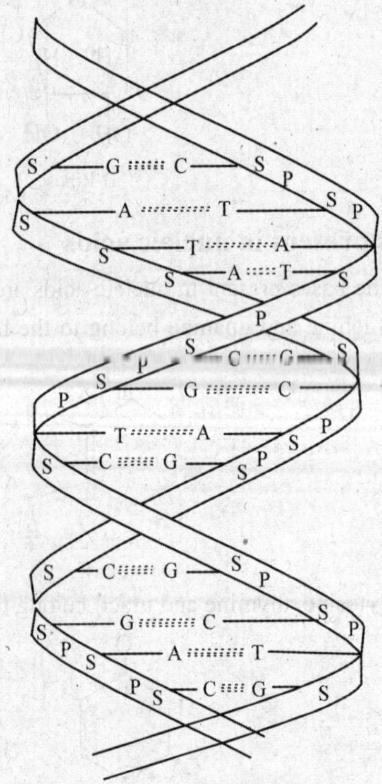

Fig. 7.9. Double helical structure of DNA

Primary structure. The primary structure of DNA consists of polynucleotide chains as shown in Fig. 7.9 Quantitative analysis of the hydrolytic products of DNA shows that in DNA the ratios of adenine to thymine and guanine to cytosine are 1:1.

Secondary structure. Watson and Crick (1953) proposed that DNA consists of two identical polynucleotide chains or strands twisted about each other to form a double right-handed helix in which the heads of the two chains are in opposite directions (Fig. 7.9)

The two chains of the double helix are held in position of hydrogen bonding between base part of one chain and an appropriate base part of the other chain. In other words, hydrogen bonding can occur only between specific bases, one of which is always a **purine** base while the other is a **pyrimidine** base.

It has found that guanine (G) part of one chain is always linked to cytosine (C) part of the other chain by three hydrogen bonds.

DNA Structures

Similarly adenine (A) is linked to thymine (T) by two hydrogen bonds.

These base pairings may be denoted as G :::: C and A ::: T. Thus the base sequence in one chain must have a specific and complementary base sequence in the other chain. This is known as complementary base pairing.

Structure of RNA

The difference between DNA and RNA is that the former contains the sugar 2-deoxyribose whereas the latter contains the sugar ribose. There is also some difference between the base content of the two nucleic acids. RNA contains the bases adenine, guanine, cytosine and uracil. DNA has thymine in place of uracil. RNA is lighter than DNA in weight by 33%.

In general, the polynucleotide chain of RNA may be represented in the same way as that of DNA except that the 2-deoxyribose sugar unit is replaced by ribose unit.

Functions of RNA and DNA

Functions of RNA

1. There are three types of RNA in the living cell and they perform different roles as given below:

(*i*) Messenger RNA (m-RNA) carries the message of DNA for performing a particular synthesis.

(*ii*) Ribosomal RNA (*r*-RNA) makes provision for a site for performing synthesis of protein.

(*iii*) Transfer RNA (*t*-RNA) directs the amino acids to the site for the purpose of synthesis.

2. Processes of learning and memory storage are believed to be controlled by RNA.

Functions of DNA

1. It has a property of self replication and is responsible for maintaining hereditary traits from one generation to the next.

2. It controls the synthesis of RNA which guides the synthesis of proteins.

3. The base sequence of DNA gets altered by the action of UV light, X-rays and certain chemicals. This process is called *mutation*. These mutations are responsible for changing hereditary traits of animals and their off-springs.

MISCELLANEOUS SOLVED EXAMPLES

Example 1. If one chain of a DNA double helix has the sequence ATGCTTGA, what is the sequence in the complementary chain?

Solution. Guanine (G) in one chain is always linked to cytosine (C) in the complementary chain and vice versa. Similarly adenine (A) in one chain is always linked to thymine (T) in the complementary chain and vice versa. Therefore the sequence in the complementary chain is TACGAACT

Example 2. Assign absolute configuration (R or S) to the following L-amino acids?

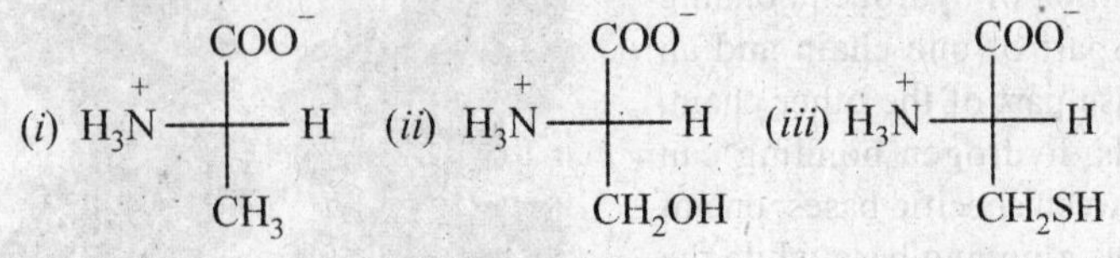

L-Alanine L-Serine L-Cysteine

Solution. (*i*) The decreasing order of priority of the groups attached to chiral carbon is $H_3N^+ > -COO^- > -CH_3 > H$

Make two interchanges in the given structure so that the group of least priority (*i.e.* H) points away from the observer.

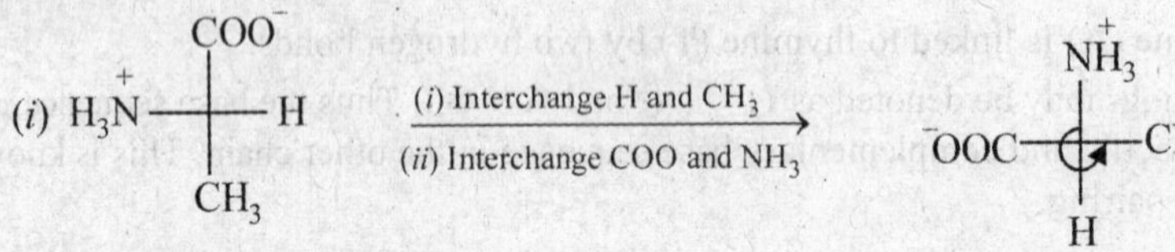

Thus, the configuration is S.

(*ii*) The decreasing order of priorities of the groups in this case : $H_3N^+ > -COO^- > -CH_2OH > H$

Make two interchanges to get an equivalent structure in which H is bonded vertically and points away from the observer

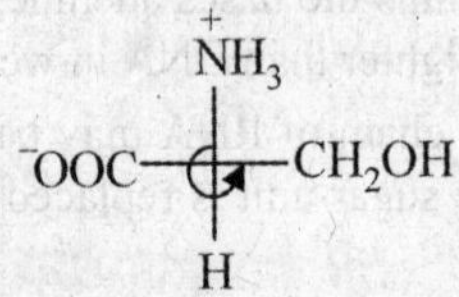

Thus the configuration is S.

(*iii*) The decreasing order of priorities is

$$H_3N^+ > -CH_2SH > -COO^- > H$$

Following the rules as above, we get the structure:

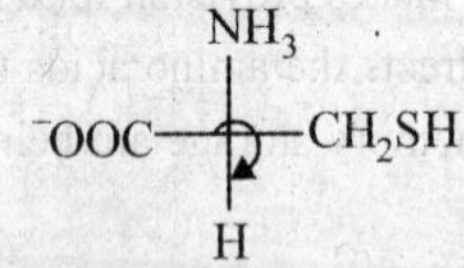

Thus configuration is R.

Example 3. How will you separate a mixture of leucine (pl = 5.98) and lycine (pl = 9.6) by electrophoresis?

Solution. An aqueous solution of the mixture is applied in the middle of a strip of filter paper and the pH is adjusted at 5.98. The ends of the strip are connected to the electrodes and electric field is applied. Leucine does not migrate while lycine moves to the cathode.

Example 4. The isoelectric point of glycine is 6.1 while that of glutamic acid $\left(HOOC-CH_2-CH_2-\underset{NH_2}{\underset{|}{CH}}-COOH\right)$ is as low as 3.22. How do you explain this?

Solution. This is because glycine is a monobasic amino acid while glutamic acid as dibasic amino acid. Due to the presence of two carboxylic groups, glutamic acid exists in the following equilibrium in aqueous solution:

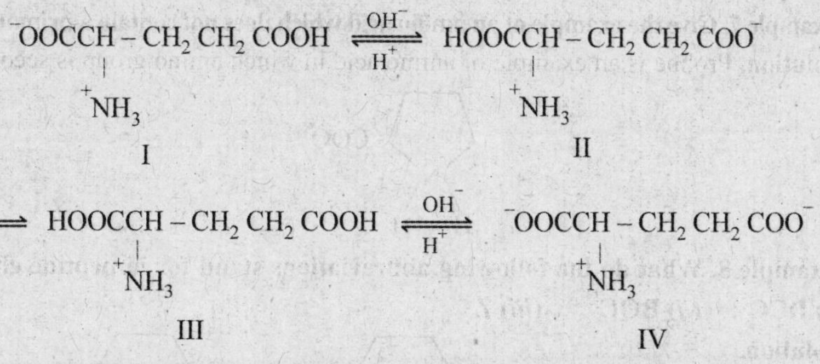

In the above equilibrium I and II are neutral species, III is a cation and IV is an anion. The anionic species IV arises due to the presence of second carboxyl group in glutamic acid. No such species is possible in case of glycine or other monbasic amino acids. The aqueous solution contains considerable concentration of IV. To suppress the formation of this species, sufficient acid has to be added so that the concentration of cationic species (III) becomes equal to the anionic species (IV). Therefore, isoelectric point of glutamic acid is much lower than that of glycine.

Example 5. Write structural formula to show the constitution of each of the following dipeptides.

(*i*) Leu-gly (*ii*) Gly-Ala (*iii*) Ala-Phe.

Solution. (*i*) Leu-gly is a dipeptide in which leucine is the N-terminal amino acid while glycine is the C-terminal amino acid as shown below:

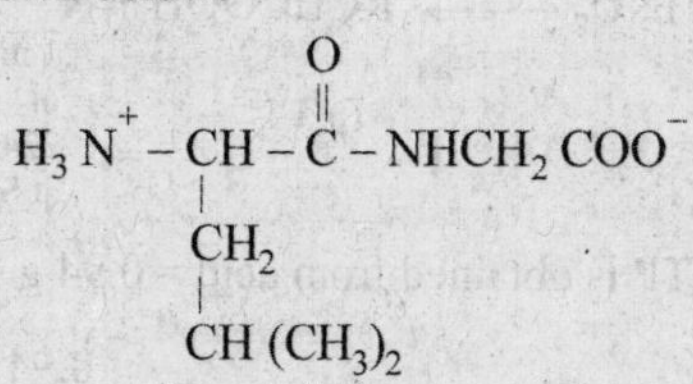

Leucyl glycine

(*ii*) Gly-Ala is a dipeptide in which glycine is N-terminal residue while alanine is C-terminal residue as shown below:

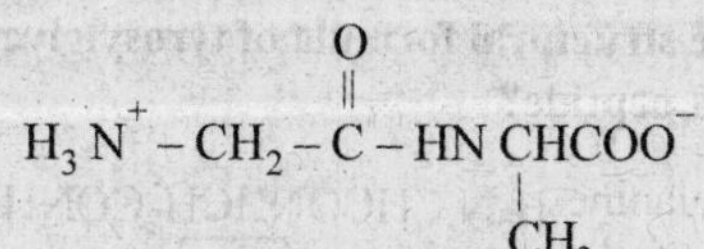

Glycylalanine

(*iii*) Ala-Phl is a dipeptide in which alamine is N-terminal residue while phenyl alanine is C-terminal residue.

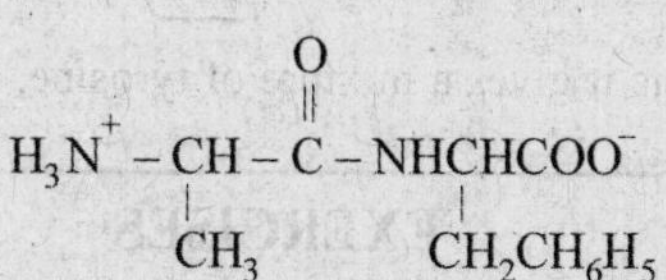

Alanylphenylalanine

Example 6. In an aqueous solution of alanine, which group acts as the acidic group and which as the basic group?

Solution. In aqueous solution of alanine $H_3\overset{+}{N}-CH(CH_3)-COO^-$, $-\overset{+}{N}H_3$ acts as the acid while $-COO^-$ acts as base.

Example 7. Give the example of an amino acid which does not contain a primary amino group.

Solution. Proline is an example of amino acid in which amino group is secondary in nature,

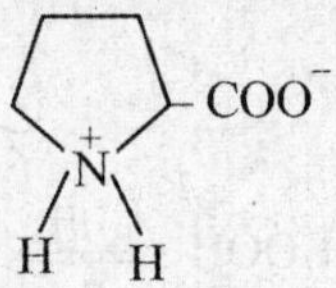

Example 8. What do the following abbreviations stand for in peptide chemistry?

***(i)* DCC** ***(ii)* BOC** ***(iii)* Z**

Solution.

DCC = Dicyclohexylcarbodiimide, 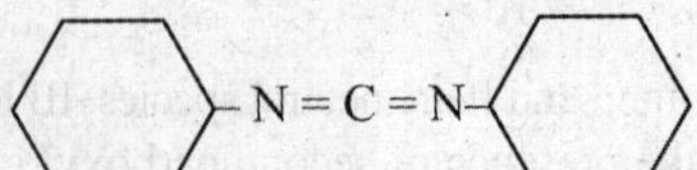

BOC = Ditertiarybutyl dicarbonate, $(CH_3)_3\,CO-\overset{\overset{O}{\|}}{C}-O-\overset{\overset{O}{\|}}{C}-OC\,(CH_3)_3$

Z = Benzyloxycarbonyl chloride, $(C_6H_5\,CH_2O-\overset{\overset{O}{\|}}{C}-Cl$

Example 9. When a solution of 0.94 g of an unknown amino acid was treated with nitrous acid, 10.8 ml of nitrogen at STP was evolved. What is the minimum molecular mass of the acid?

Solution. $\underset{\underset{1\text{ mol}}{\overset{+}{N}H_3}}{RCHCOO^-} + HNO_2 \longrightarrow \underset{OH}{RCHCOOH} + \underset{\substack{1\text{ mol}\\ \text{or}\\ 22400\text{ ml}\\ \text{at STP}}}{N_2} + H_2O$

10.8 ml of N_2 at STP is obtained from acid = 0.94 g

∴ 22400 ml of N_2 at STP is obtained from acid = $\frac{0.94}{10.8}\times 22400\text{ g} \approx 195\text{ g}$

∴ Molecular mass of amino acid = 195 g mol^{-1}

Example 10. What is the structural formula of tyrosylglycylalanine? What are the products os complete hydrolysis of this peptide?

Solution. Tyrosylglycylalanine 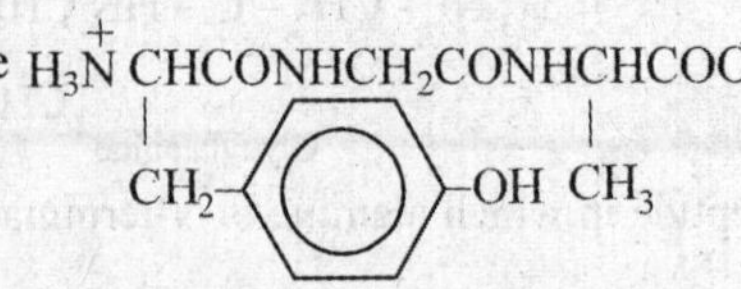

Upon complete hydrolysis it gives a mixture of tyrosine, glycine and alanine.

EXERCISES
(Including Questions from Different University Papers)

Multiple Choice Questions (Choose the correct option)

1. An aqueous solution of glycine is neutral because of the formation of:

(a) carbanion (*b*) zwitterion

(*c*) carbonium ions (*d*) free radicals

2. The isoelectric point of a protein is

(a) the pH at which the protein molecule has no charges on its surface.

(*b*) the pH at which a protein in solution has an equal number of positive and negative charge

(*c*) the electric charge under isothermal conditions

(*d*) none of these.

3. Glycine is a unique amino acid because it

(a) has no chiral carbon (*b*) has a sulphur containing *R* group

(*c*) cannot form a peptide bond (*d*) is an essential amino acid

4. Glycine is

(a) NH_2CH_2COOH (*b*) $NH_2CH_2CH_2CH_2CH_2NH_2$

(*c*) $NO_2CH_2CH_2COOH$ (*d*) $BrCH_2COOH$

5. A zwitterion is

(a) an ion that is positively charged in solution

(*b*) an ion that is negatively charged in solution

(*c*) a compound that can ionise both as a base and an acid.

(*d*) a carbohydrate with an electrical charge.

6. The pH at which the amino acid shows no tendency to migrate when placed in an electric field known as its:

(a) isoelectric point (*b*) dipole moment

(*c*) iodine number (*d*) wavelength

7. Glycine reacts with nitrous acid to form:

(*a*) glycollic acid (*b*) diketopiperazine

(*c*) methylamine (*d*) ethyl alcohol

8. Which one of the following compounds form zwitterions?

(a) carbonyl compounds (*b*) amino acids

(*c*) phenols (*d*) heterocyclic compounds

9. Which of the following organic ions results when glycine is treated with concentrated HCl?

(a) $\overset{+}{N}H_3CH_2COOH$ (*b*) $NH_2CH_2C\bar{O}$

(*c*) $\overset{+}{N}H_3CH_2CO\bar{O}$ (*d*) $HOCH_2C\bar{O}$

10. The five elements present in most naturally occurring proteins are:

(a) C, H, O, P, and S (*b*) N, C, H, O, and I

(*c*) N, S, C, H, and O (*d*) C, H, O, S, and I

11. Complete hydrolysis of proteins produces:

(a) ammonia and carbon dioxide (*b*) urea and uric acid

(*c*) a mixture of amino acids (*d*) glycogen and a fatty acid

12. The α-Helix is held in a coiled conformation partially because of:

(a) optical activity (*b*) hydrogen bonding

(*c*) resonance (*d*) delocalization

13. The primary structure of a proteins refers to:

(a) whether the protein is fibrous or globular

(*b*) the amino acid sequence in the polypeptide chain
(*c*) the orientation of the amino acid side chains in space
(*d*) the presence or absence of an α-helix.

14. Upon hydrolysis, proteins give

(a) amino acids (*b*) hydroxy acids
(*c*) fatty acids (*d*) alcohols

15. Proteins are

(a) polyamides (*b*) polymers of ethylene
(*c*) α-Aminocarboxylic acids (*d*) polymers of propylene

16. The linear arrangement of amino acid units in proteins is called:

(a) primary structure (*b*) secondary structure
(*c*) tertiary structure (*d*) quaternary structure

17. The double helical structure of DNA is held together by

(a) sulphur-sulphur linkages (*b*) peptide bonding
(*c*) hydrogen bonding (*d*) glycosidic bonds

18. Which of the following does not belong to the class of pyramidine?

(a) Cytosine (*b*) Guanine
(*c*) Thymine (*d*) Uracil

ANSWERS

1. (*b*)	**2.** (*b*)	**3.** (*a*)	**4.** (*a*)	**5.** (*c*)	**6.** (*a*)
7. (*a*)	**8.** (*b*)	**9.** (*a*)	**10.** (*c*)	**11.** (*c*)	**12.** (*b*)
13. (*b*)	**14.** (*a*)	**15.** (*a*)	**16.** (*a*)	**17.** (*c*)	**18.** (*b*)

Short Answer Questions

1. What is meant by amino protecting group? Give two examples of such groups.
2. Give two methods of synthesis of amino acids.
3. Explain isoelectric point with reference to α-amino acids. Do all α-amino acids have the same isoelectric point?
4. How will you synthesise Glycine from potassium phthalimide?
5. What colour tests are used for the detection of proteins?
6. Why amino acids are called amphoteric compounds?
7. What is isoelectric point?
8. Explain denaturation of proteins.
9. How will you synthesise phenylalanine from potassium phthalimide?
10. What is an isoelectric point? Discuss.
11. Describe Gabriel phthalimide synthesis of amino acids.
12. Give equations for the reactions of amino acids with the following:
(*i*) 2,4-Dinitrofluorobenzene (*ii*) Nitrosyl chloride
13. How will you convert glycine into diketopiperazine?
14. Explain denaturation and renaturation of proteins.
15. What do you understand by isoelectric point? Explain.

16. Write Erlenmeyer azlactone synthesis of Phenylalanine.
17. What are α-amino acids? Give evidence supporting their dipolar ionic structure.
18. Discuss the isoelectric point of amino acids.
19. Give two methods of synthesis of α-amino acids.
20. Why are amino acids called amphoteric compounds?
21. What are proteins? Discuss the secondary structure of proteins.
22. Write a brief note on solid peptide synthesis.
23. Explain isoelectric point with reference to α-amino acids. Do all α-amino acids have the same isoelectric point?
24. Explain the terms:

    (*i*) Denaturation of proteins (*ii*) Isoelectric points.
25. In amino acids $-\overset{+}{N}H_3$ is the acidic group and $-COO^-$ is the basic group. Explain.
26. Give two methods to synthesis α-amino acids.
27. Give the Zwitterion structure of amino acids and justify the same in terms of any two expected properties shown by them.
28. Write a note on Zwitterion structure of amino acids. Give evidence in favour of Zwitterion structure.
29. What are conjugated proteins? How are they classified?
30. What are proteins? Discuss secondary structure of proteins.
31. Explain fibrous proteins and globular proteins.
32. Write a short note on primary structure of proteins.
33. Using carbobenzoxy chloride, sketch the synthesis of a dipeptide.
34. What is the configuration of natural amino acids? Give evidence in support of Zwitterion structure.
35. Describe Gabriel phthalimide synthesis of α-amino acids. Explain why $-\overset{+}{N}H_3$ group is acidic and $-COO^-$ group in basic in them.
36. Biologically proteins are very important. Comment.
37. White a short note on primary structure of proteins.
38. Write a note on tertiary structure of proteins. Name the different kinds of bonds responsible for the formation of tertiary structure of proteins.
39. Write a note on Strecker synthesis.

General Questions

1. What are peptides? What are the difficulties encountered in their synthesis? Illustrate the various methods used for protecting amino and carboxyl groups.
2. What is the geometry of peptide bond? Discuss the two main types of secondary structure of proteins.
3. Explain the terms:

   (*i*) Zwitterion (*ii*) Denaturation of proteins.
4. Explain the terms with examples:

   (*i*) Nucleoside (*ii*) Nucleotide (*iii*) Polynucleotides.
5. Define the term protein. Write what you know about the structure of proteins.

6. What are peptides? Using carbobenzoxy chloride as the N-Protecting agent, sketch the synthesis of a dipeptide.

7. Write short notes on the following:

(*i*) Fibrous proteins and globular proteins (*ii*) Zwitterions.

8. Discuss the following

(*i*) Zwitterion structure of amino acids.

(*ii*) Gabriel phthalimide synthesis of α-amino acids.

9. Explain the following terms giving suitable examples:

(*i*) Denaturation of proteins (*ii*) Isoelectric point of amino-acids.

10. How are proteins related to amino acids ? What do you understand by isoelectric point? Describe the primary, secondary and tertiary structure of proteins.

11. What are nucleic acids ? What are they made up off ? Discuss the structure of deoxyribonucleic acids.

12. What are proteins? Discuss, with diagram, β-pleated sheet structure of proteins.

13. How are α-amino acids, peptides and proteins related to each other? How are α-amino acids and dipeptides synthesised ?

14. (a) How are amino acids prepared in the laboratory? Discuss the stereochemistry of natural amino acids.

(*b*) How can we determine the structure of polypeptide chain ?

15. Explain the various steps in classical peptide synthesis. What are its drawbacks ?

16. Discuss the secondary and tertiary structure of proteins. What are all the forces of interaction present in tertiary structure?

17. What are peptides? Using a suitable protecting group, sketch the synthesis of a dipeptide.

18. How is a peptide bond formed? Discuss its hydrolysis. Give the products formed from (*i*) complete (*ii*) partial hydrolysis of pentapeptide: Gly. Ala. Met. Val. Gly.

19. Give the evidences in favour of dipolar ionic structure of amino acids.

20. What are polypeptides? Using benzyloxycarbonyl chloride, sketch the synthesis of a dipeptide.

21. Classify the proteins based on their shapes and proteins.

8

Fats, Oils, Soaps and Detergents

8.1. INTRODUCTION

Fats and oils are found in abundance in plants and animals. Seeds store the fats and oils in plants whereas in animals, they are located under the skin and in muscles.

Fats and oils are triesters of glycerol (triglycerides) and their general structure can be written as follows:

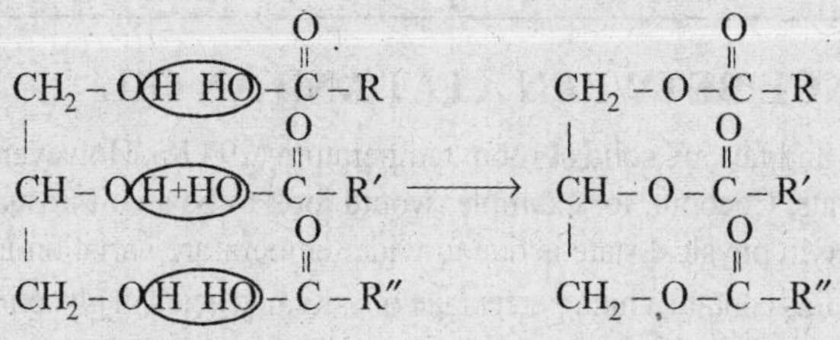

Glycerol Higher fatty acids — A triglyceride

where R, R′, R″ are parts of the higher fatty acids forming ester with glycerol. The three acids could be same, two of them same, or all different from one another. Accordingly, they are called simple or mixed glycerides.

$CH_2OCOC_{15}H_{31}$	$CH_2OCOC_{15}H_{31}$
$CHOCOC_{15}H_{31}$	$CHOCOC_{17}H_{35}$
$CH_2OCOC_{15}H_{31}$	$CH_2OCOC_{15}H_{31}$
Glyceryl tripalmitate (Tripalmitin) Simple glyceride	Glyceryl stearato dipalmitate (Mixed glyceride)

Natural fats and oils contain a mixture of mixed and simple glycerides.

8.2. INDUSTRIAL OILS OF VEGETABLE ORIGIN

Oils as mentioned above find the following industrial uses :

1. Many vegetable oils such as tripalmitin and tristearin are used to make soaps, skin products, candles, perfumes and other cosmetics.
2. Some oils containing oleic and linoleic acids are suitable as drying agents and are used in making paints and other wood treatment products.
3. Vegetable oils are finding increasing use as insulators in electrical industry as they are non-toxic to environment and biodegradable.

4. Vegetable based oils such as castor oil has been used in medicine and as lubricant for a long time. Castor oil has numerous industrial applications primary due to the presence of hyroxyl groups on the fatty acid chains. Oils which have been chemically modified to contain hydroxy groups are becoming increasingly important in the production of polyurethane plastic for many applications.
5. Vegetable oils are also used to make biodiesel, which can be used like conventional (mineral) diesel. Lately, jatrofa oil is being used on a large commercial scale as a substitute for diesel.

The carboxylic acids that form ester chains in natural triglycerides (oils or fats) may be saturated or unsaturated. Some of these are listed in the table below.

Table 8.1. Some Common Fatty Acids

Name	*Formula*	*mp°C*
Myristic acid	$CH_3(CH_2)_{12}COOH$	58
Palmitic acid	$CH_3(CH_2)_{14}COOH$	63
Stearic acid	$CH_3(CH_2)_{16}COOH$	70
Oleic acid	$CH_3(CH_2)_7\ \underset{10}{C}H{=}\underset{9}{C}H(CH_2)_7\underset{1}{C}OOH$	4
Linoleic acid	$CH_3(CH_2)_4\ \underset{13}{C}H{=}\underset{12}{C}HCH_2\underset{10}{C}H{=}\underset{9}{C}H(CH_2)_7COOH$	– 5

8.3. DIFFERENCE BETWEEN A FAT AND AN OIL

1. Oils are liquid and fats are solid at room temperature (293 K). However, we cannot rely on this definition of oils and fats. Coconut, for example, would look to be an oil in summer and fat in winter, although this difference in physical state is due to wide temperature variation in summer and winter.

2. It is found that fats contain a high percentage of saturated acids in glycerides whereas oils contain a high percentage of unsaturated acids in glycerides. Thus fats are saturated and oils are unsaturated in nature.

The melting point of a fat or oil depends on its structure. If all the fatty acids forming a triglyceride are saturated, their carbon-chains can align themselves in a regular pattern. Such molecules can pack well in a crystal have higher melting points and thus form solids at room temperature. On the other hand, if some of the ester chains are formed by unsaturated acids, there are kinks in their carbon-chains at the double bonds that have *cis* configuration. These kinks make the carbon-chains irregular and do not allow close-packing of the molecules. There will be weaker molecular interactions resulting in lower melting points. Thus triglycerides of unsaturated acids are liquids at room temperature.

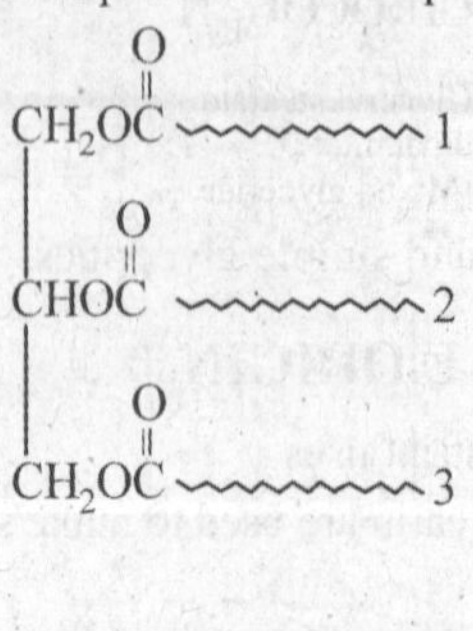

A saturated glyceride (solid)

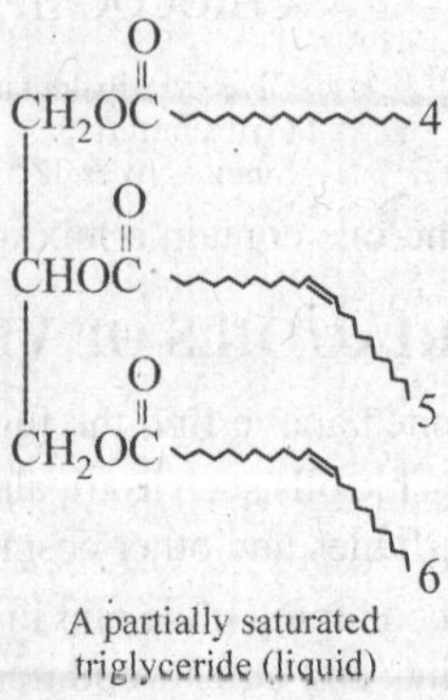

A partially saturated triglyceride (liquid)

Substances which like or behave like fats and oils but are not accepted as such.

Following substances are not accepted in the category of oils and fats:

Mineral oils. Kerosene oil, diesel oil etc. have mineral origin. They are mixtures of hydrocarbons. Hence they are not acceptable as fats and oils.

Essential oils. Pleasant-smelling liquids found in plants, clove oil, lemon oil, turpentine oil, although resembling fats and oils in physical properties, are again not accepted in this category as such.

8.4. EXTRACTION OF FATS AND OILS FROM ANIMALS AND PLANTS

Following steps are employed for the extraction of fats and oils from animal or plant sources:

1. Chopping or rendering. This method is employed for extraction of fats and oils from animals. Animal tissues richer in fats are chopped off and subjected to heat-treatment either as such or in combination with water till the fat melts and form a separate layer.

2. Crushing. This method is employed for extraction of fats or oils from plant sources. Oil-seeds like cotton-seed, castor seeds, mustard-seeds and linseed are crushed between steel rollers and crushed seeds are pressed in a hydraulic press. The oil is extracted and the residue called **deoiled-cake** is left.

3. Solvent extraction. In another process, the crushed animal tissues and seeds are extracted with suitable solvents like benzene, petroleum, ether etc. to take out the maximum amount of fats and oils. The solvent is distilled off leaving behind the fat or oil. The same solvents is recycled. This method gives better yield of fats and oils than conventional crushing and passing through rollers.

4. Refining. The fat or oil obtained above may not be in pure state. It is purified as follows:

(*a*) **Neutralisation.** Some free fatty acid might be produced as a result of heating or crushing or hydrolysis. This is neutralised with small amounts of alkali.

(*b*) The fat or oil is decolourised by warming with animal charcoal.

(*c*) Odour is removed from the oil or fat by passing superheated steam through it and then separating the aqueous layer from the oily layer.

8.5. PHYSICAL PROPERTIES OF OILS AND FATS

1. Oils and fats are liquids or solids having a greasy touch. In pure state, they are colourless, odourless and tasteless.

2. They are lighter than water (specific gravity less than 1) and thus float on the surface of water.

3. Oils and fats are insoluble in water but soluble in organic solvents like chloroform, benzene and ether.

4. They form emulsions with water in the presence of soap or gelatin.

8.6. CHEMICAL PROPERTIES OF FATS AND OILS

Following are some of the important chemical properties of oils and fats:

1. Hydrolysis. Fats and oils are hydrolysed by the action of acids, alkalis or superheated steam.

$$\begin{array}{l} CH_2OCOR \\ | \\ CHOCOR' + 3H_2O \\ | \\ CH_2OCOR'' \end{array} \longrightarrow \begin{array}{l} CH_2OH \\ | \\ CHOH \\ | \\ CH_2OH \\ \text{Glycerol} \end{array} \begin{array}{r} + RCOOH + R'COOH \\ \\ + R''COOH \\ \text{Mixture of fatty acids} \end{array}$$

It hydrolysis is carried out by alkali, a mixture of alkali salts of fatty acids and glycerol is obtained. Salts of fatty acids are used as soaps and therefore, this hydrolysis using alkali is also called **saponification.**

$$\begin{array}{l} CH_2OCOR \\ | \\ CHOCOR' + 3NaOH \\ | \\ CH_2OCOR' \end{array} \longrightarrow \begin{array}{l} CH_2OH \\ | \\ CHOH \\ | \\ CH_2OH \\ \text{Glycerol} \end{array} \begin{array}{l} RCOO\,Na + R'COONa \\ \\ \quad + R''COONa \\ \quad\quad \text{(Soap)} \end{array}$$

This reaction forms the basis of soap industry. Luxury soaps, beauty soaps and shaving soaps are prepared by using potassium hydroxides instead of sodium hydroxide.

2. Hydrogenation. Oils contain unsaturated hydrocarbon chain as part of the constituent fatty acid of the glyceride. If hydrogen gas is passed through such oils in the presence of finely divided nickel, the double bonds in the hydrocarbon chain is saturated and we obtain hydrogenated oil which is a solid (fat).

$$\begin{array}{l} CH_2OCOC_{17}H_{33} \\ | \\ CHOCOC_{17}H_{33} + 3H_2 \\ | \\ CH_2OCOC_{17}H_{33} \\ \text{Triolein (Oil)} \end{array} \xrightarrow[450\text{ K}]{\text{Ni}} \begin{array}{l} CH_2OCOC_{17}H_{35} \\ | \\ CHOCOC_{17}H_{35} \\ | \\ CH_2OCOC_{17}H_{35} \\ \text{Tristearin (Oil)} \end{array}$$

The hydrogenated fat has a longer shelf life and does not become rancid easily.

3. Hydrogenolysis. If hydrogen gas is passed in excess through a fat or oil at high temperature and under high pressure, glycerol and long chain aliphatic alcohols are obtained.

$$\begin{array}{l} CH_2OCOC_{15}H_{31} \\ | \\ CHOCOC_{15}H_{31} + 6H_2 \\ | \\ CH_2OCOC_{15}H_{31} \\ \text{Tripalmitin} \end{array} \xrightarrow[\text{Chromite}]{\text{Copper}} \begin{array}{l} CH_2OH \\ | \\ CHOH \\ | \\ CH_2OH \\ \text{Glycerol} \end{array} \begin{array}{l} \\ \\ + 3C_{15}H_{31}CH_2OH \\ \quad \text{Hexadecyl alcohol} \end{array}$$

4. Trans-esterification. Simple esters can be prepared by treating oils or fats with lower alcohols in the presence of sod. alkoxides

$$\begin{array}{l} CH_2OCOC_{17}H_{35} \\ | \\ CHOCOC_{17}H_{35} + 3CH_3OH \\ | \\ CH_2OCOC_{17}H_{35} \\ \text{Stearin} \end{array} \xrightarrow[\text{Chromite}]{\text{Copper}} \begin{array}{l} CH_2OH \\ | \\ CHOH \\ | \\ CH_2OH \end{array} \begin{array}{l} \\ \\ + 3C_{17}H_{35}COOCH_3 \\ \quad \text{Methyl stearate} \end{array}$$

5. Rancidification. Fats and oils, on exposure to air and light, start giving a foul smell and taste. We say that the fat or oil has become **rancid.** This phenomenon is called rancidification. This change takes place due to the formation of fatty acids and carbonyl compounds, which are foul smelling. The low hydrolysis of the fats or oils under the effect of atmospheric moisture produces these fatty acids. Similarly attack of oxygen at the double bonds in oil produces foul-smelling carbonyl compounds. These reasons contribute together towards the phenomenon of rancidification.

8.7. MANUFACTURE OF VANASPATI (HYDROGENATION OF OIL)

Hydrogenation is commercially used in the manufacture of vegetable ghee (vanaspati). Groundnut oil, cotton seed oil, coconut oil etc. are used as raw materials for the preparation of vanaspati. The hydrogenator which is used for this purpose is shown in Fig. 8.1.

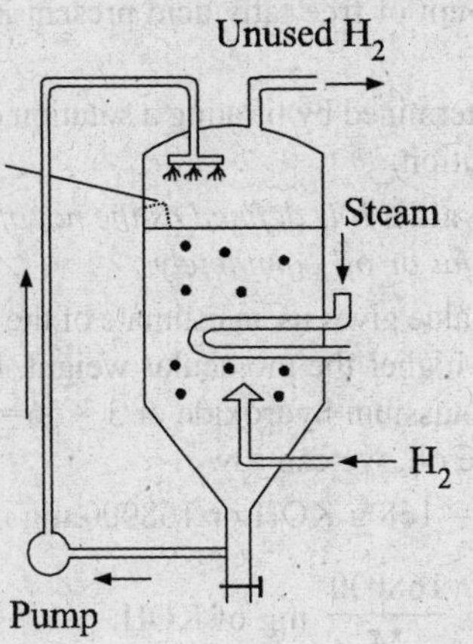

Fig. 8.1. Manufacture of vegetable ghee.

The oil is filtered and refined and taken in the hydrogenator fitted with steam coils for heating. Finely divided nickel is added to the oil. It is heated to 450 K and a current of hydrogen gas is passed through the oil. The oil is cycled again and again into the hydrogenator to ensure complete saturation of double bonds in the oil molecules. After the complete hydrogenation has taken place, the molten fat is taken out and filtered to remove nickel catalyst.

8.8. ISOLATION OF CARBOXY ACIDS AND ALCOHOLS FROM FATS AND OILS

Isolation of carboxylic acids

Fats or oils are subjected to acidic hydrolysis *i.e.* hydrolysis in the presence of a mineral acid like HCl. Ester linkage is broken and glycerol and fatty acids are produced. Fractional distillation of these acids is carried out to separate them.

In an alternative process, alkaline hydrolysis of the ester is carried out. We obtain glycerol and a mixture of sodium salts of the fatty acids. The sodium salts are treated with a mineral acid to liberate a mixture of carboxylic acids which are separated by fractional distillation. In still another alternative method, trans-esterification is carried out with methyl alcohol. A mixture of methyl esters is obtained. It is separated into individual esters by fractional distillation. Esters are hydrolysed separately to obtain carboxylic acids.

Isolation of Alcohols

The fats or oils are subjected to hydrogenation at high temperature and under high pressure in the presence of copper chromite. A mixture of long-chain alcohols is obtained which is separated into individual alcohols by fractional distillation.

8.9. ANALYSIS OF FATS AND OILS

Oils and fats are used as raw materials for the manufacture of soaps, drying oils and other products. Therefore, their quality control is essential. Some of the parameters of quality of oils and fats are described below :

(*a*) **Acid value.** *It is defined as the no. of milligrams of potassium hydroxide required to neutralise one gram of oil or fat.*

It indicates the amount of free fatty acid present in an oil or fat. A high acid value indicates a stale oil.

The acid value is determined by titrating a solution of oil or fat in pure alcohol against standard potassium hydroxide solution.

(*b*) **Saponification value.** *It is defined as the no. of miligram of potassium hydroxide required to saponify one gram of fat or oil completely.*

The saponification value gives us an estimate of the molecular mass of the fat or oil, the smaller the saponification value, higher the molecular weight. It is made clear like this. One mole of the oil requires 3 moles of potassium hydroxide or 3 × 56 = 168 g of KOH for saponification. If M is the molecular mass of the oil, we can say

M g of oil requires = 168 g KOH or 168000 mg of KOH.

$$1 \text{ g of oil requires} = \frac{168000}{M} \text{ mg of KOH.}$$

It is evident from the equation that greater the value of M, smaller will be the saponification value.

Saponification value of an oil or fat is determined by refluxing a known amount of a sample with excess of standard alcoholic potassium hydroxide solution and titrating the unused alkali against a standard acid solution.

A lower value of saponification number, therefore, shows the abundance of high-molecular-weight fatty acid residues in the given fat. A higher saponification number, on the other hand, indicates the abundance of low-molecular-weight fatty acid residues. The saponification number is characteristic of a particular fat or oil and serves for its identification.

(*c*) **Iodine value.** *It is the no. of grams of iodine that combine with 100 g of oil or fat.*

It gives a measure of unsaturation in an oil or fat. Iodine value is determined by adding a known excess of Wij's solution, which is iodine monochloride in glacial acetic acid, to a solution of known weight of the oil or fat in carbon tetrachloride. Unused iodine is estimated by titration with standard hypo solution. We use Wij's solution because iodine as such does not react with the unsaturated oils.

Table 8.2. Analysis of Some Fats and Oils

Fat or Oil	*Saponification Number*	*Iodine Number*
Coconut fat	250-260	8-10
Butter fat	210-230	26-28
Tallow	190-200	30-48
Lard	193-200	46-70
Olive oil	187-196	79-90
Cottonseed oil	190-198	105-114
Sunflower oil	188-194	140-156
Linseed oil	187-195	170-185

Table 8.2 shows the saponification and iodine numbers of different fats and oils. We notice the high saponification number of butter fat, which is an indication that it contains some short carbon-chain acid residues. Also we notice the low iodine number of butter fat, which has few double bonds and is therefore a solid at room temperature. Linseed oil, on the other hand has a higher iodine number which shows that it is highly unsaturated. Such highly unsaturated oils find use as drying oil in point industry.

(*d*) **Reichert-Meissl Value (R.M. Value).** *It is the number of millilitre of N/10 potassium hydroxide solution required to neutralise the distillate of 5 g of hydrolysed fat or oil.*

It is a measure of steam volatile fatty acids present as esters in fats or oils. It is used for checking the purity of butter or ghee.

To determine Reichert-Meissl value, 5 g of the sample is hydrolysed with NaOH and the mixture is acidified with dil. H_2SO_4 and then steam distilled. The distillate containing acids upto C_{10} is cooled, filtered and titrated against N/10 alkali.

Example 1. Calculate saponification value of the following ester:

$$\begin{array}{l} CH_2OCOC_{15}H_{31} \\ | \\ CHOCOC_{15}H_{31} \\ | \\ CH_2OCOC_{15}H_{31} \end{array}$$

Solution. Saponification value of an oil or fat is the no. of milligram of potassium hydroxide required to saponify one gram of fat or oil completely. The above ester (fat) requires three moles of KOH per mole of fat for saponification.

$$\underset{\substack{\text{Tripalmitin} \\ (806\text{ g})}}{\begin{array}{l} CH_2OCOC_{15}H_{31} \\ | \\ CHOCOC_{15}H_{31} \\ | \\ CH_2OCOC_{15}H_{31} \end{array}} + \underset{(168\text{ g})}{3KOH} \longrightarrow \begin{array}{l} CH_2OH \\ | \\ CHOH \\ | \\ CH_2OH \end{array} + \underset{\text{Pot. palmitate}}{3C_{15}H_{31}COOK}$$

806 g of ester requires for saponification = 168000 mg KOH

$$1 \text{ g of ester requires for saponification} = \frac{168000}{806} \text{ mg} = 208.4 \text{ mg}$$

Hence saponification value of the ester = 208.4.

8.10. DRYING OF OILS (HARDENING OF OIL)

Some oils which contain glycerides of unsaturated acids, like linoleic acid or linolenic acid, become thick and harden to resin-like solid, when they are exposed to air. *This phenomenon of hardening of oil on exposure to air and light is called drying.* It is believed that this hardening process takes place due to the oxidation of double bonds in the carboxylic acid component of the glyceride (oil). There is also some possibility of polymerisation taking place. This drying process is quickened or catalysed by certain manganese or lead salts.

Based on behaviour on exposure to air, the oils can be divided into three categories:

(*a*) **Drying oils.** These oils show strong tendency towards drying and within 4-5 hours of exposure to air, harden to a resin-like solid. It is found that such oils are constituted of linoleic and linolenic acids. Examples of oils of this category are linseed oil and hemp seed oil. Such materials are used in paints and varnish industries. They leave a shining surface in the coating of paint or varnish. Iodine value of such oils lies above 130.

(*b*) **Semi-drying oils.** These oils do have some tendency towards drying although they take a much longer time. This is because the proportion of linoleic and linolenic acid component is much less in the oils. Examples of oils of this category are sunflower and cotton seed oils. They absorb oxygen slowly and gradually convert into a solid mass. However, they are not commercially exploited for use in paints and varnishes because of poor results. Their iodine value ranges between 100 and 130.

(*c*) **Non-drying oils.** These oils remain stable to air. They don't dry on exposure for any long period. Examples of this type of oils are olive oil and almond oil. Their iodine value is below 100.

8.11. SOAPS

Soaps are sodium salts of higher fatty acids. A typical example of soap is sodium stearate $C_{17}H_{35}COONa$. Soaps are obtained by alkaline hydrolysis of oils and fats which are glyceryl esters of higher fatty acids. Upon hydrolysis, sodium salt of the higher fatty acid *i.e.* soap and glycerol are obtained.

$$\underset{\text{Oil or fat}}{\begin{array}{l} CH_2OCOC_{17}H_{35} \\ | \\ CHOCOC_{17}H_{35} \\ | \\ CH_2OCOC_{17}H_{35} \end{array}} + 3\,NaOH \longrightarrow \underset{\text{Glycerol}}{\begin{array}{l} CH_2OH \\ | \\ CHOH \\ | \\ CH_2OH \end{array}} + \underset{\text{Soap}}{3C_{17}H_{35}COONa}$$

This process is called *saponification.*

A higher proportion of salts of saturated acids (palmitic, stearic) gives **hard soaps,** while a higher proportion of salts of unsaturated acids (oleic acid) yields **soft soaps.**

The potassium soaps, produced by the saponification of fats with KOH, are usually softer and more soluble than the sodium soaps. Therefore, potassium soaps are used mainly in shaving creams and liquid soaps (*Shampoo*).

8.12. MANUFACTURE OF SOAP

Two methods are being used for the manufacture of soap.

(*a*) Kettle Process

(*b*) Hydrolyser Process

The *Kettle Process* is the old process while hydrolyser process is a modern process.

Kettle Process

This process is carried in a steel tank or Kettle from which it gets the name. The following steps are involved :

(1) **Boiling.** The fat and NaOH solution are fed into the kettle and boiled with steam entering from a perforated coil at the bottom (Fig. 8.2).

$$\underset{\text{(Fat)}}{C_3H_5(COOC_{17}H_{35})_3} + 3NaOH \xrightarrow[\text{steam}]{\Delta} \underset{\substack{\text{Sod stearate} \\ \text{(Soap)}}}{3C_{17}H_{35}CO\overset{-}{O}\overset{+}{Na}} + \underset{\text{Glycerol}}{C_3H_5(OH)_3}$$

The boiling is continued till saponification is nearly 80% complete.

(2) **Salting Out.** Common salt is then added and boiling resumed till soap has separated by common ion effect. The soap being lighter floats to the surface as curdy mass. The lower layer containing glycerol and salt is drawn off, leaving soap plus unreacted fat in the kettle.

(3) **Addition of Fresh Lye.** The soap from step (2) contains 20% fat which escaped esterification earlier. It is boiled with more of NaOH solution. More soap is formed. The lower layer is drawn off. The soap layer left in the kettle is boiled with water to wash away excess of lye (NaOH).

(4) **Finishing.** The soap in the kettle while still molten, is pumped into the *crutcher* by means of swing pipe. The *crutcher* is a steam jacketed tank fitted with a blade-stirrer. Here the soap is mixed with colour and perfumes, till it becomes a homogeneous mass. The crutched soap is poured

into frames and on solidification cut into cakes. For making toilet soap, the neat crutched soap is shreded, dried, and stamped into cakes.

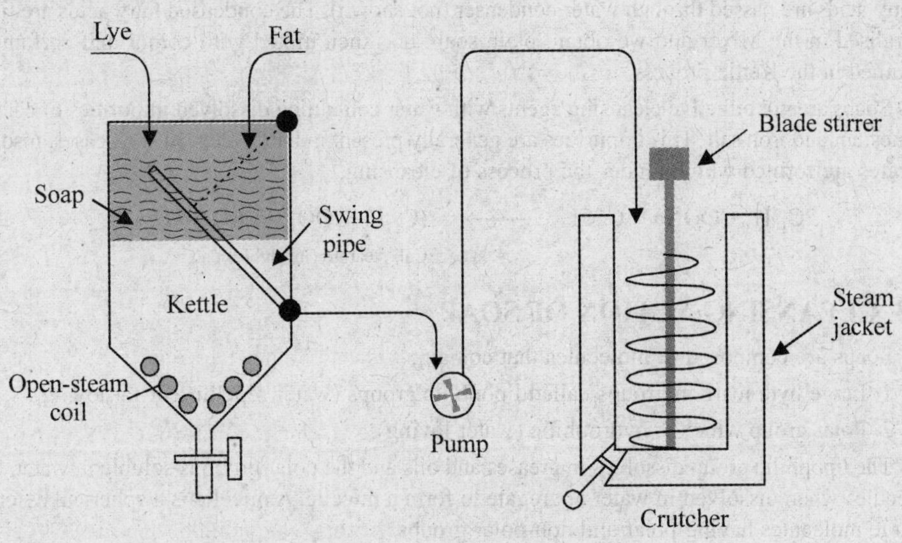

Fig. 8.2. Manufacture of soap by the Kettle Process

Hydrolyser Process

This is the modern continuous process for soap manufacture. It is more economical than the *Kettle Process* and gives better quality product. The following steps are involved here :

(*a*) Hydrolysis of fat with water in the presence of zinc oxide at high temperature and pressure.

(*b*) Distillation of fatty acids under vacuum.

(*c*) Neutralization of the condensed fatty acids with alkali.

$$\underset{\text{Fat}}{C_3H_5(COOR)_3} + 3H_2O \xrightarrow[\Delta,\ \text{pressure}]{ZnO} \underset{\text{Fatty acids}}{3RCOOH} + \underset{\text{Glycerol}}{C_3H_5(OH)_3}$$

$$RCOOH + NaOH \longrightarrow \underset{\text{Soap}}{RCO\overset{-}{O}\overset{+}{Na}} + H_2O$$

The flowsheet of the *Hydrolyser Process* is shown in Fig. 8.3. The fat mixed with zinc oxide catalyst, and water are heated to 230-250° C under 40-45 atm pressure in the *Hydrolyser.* The fatty

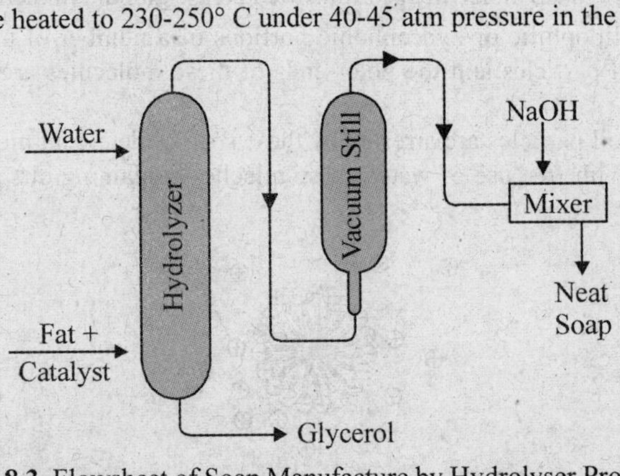

Fig. 8.3. Flowsheet of Soap Manufacture by Hydrolyser Process

acids mixed with water are discharged at the top into a steam flash-tank (not shown). The water vaporises, cooling the fatty acids which are then vacuum-distilled in the *Vacuum Still.* The vapours of fatty acids are passed through water condenser (not shown). The condensed fatty acids are finally neutralised in the Mixer and we obtain plain soap. It is then mixed with colour and perfumes as explained in the Kettle process.

Soaps are not effective cleansing agents with water containing dissolved impurities of calcium, magnesium and iron salt. This impurities are generally present in hard water. In such cases, insoluble stearates are formed which hinder the process of cleansing.

$$2C_{17}H_{35}COONa + CaCl_2 \longrightarrow \underset{\text{Cal. stearate (insoluble)}}{(C_{17}H_{35}COO)_2\,Ca} + 2NaCl$$

8.13 CLEANSING ACTION OF SOAP

Soaps are composed of molecules that contain:

1. Large hydrocarbon groups called lipophilic groups (water repelling or fat loving).
2. Polar group which is hydrophilic (water loving).

The lipophilic group dissolves in greases and oils and the polar group is soluble in water. Soap molecules when dissolved in water aggregate to form a micelle. A micelle is a spherical assembly of large molecules having polar and non-polar groups.

The washing or cleansing action of soaps or detergents can be visualised as shown in Fig. 8.4.

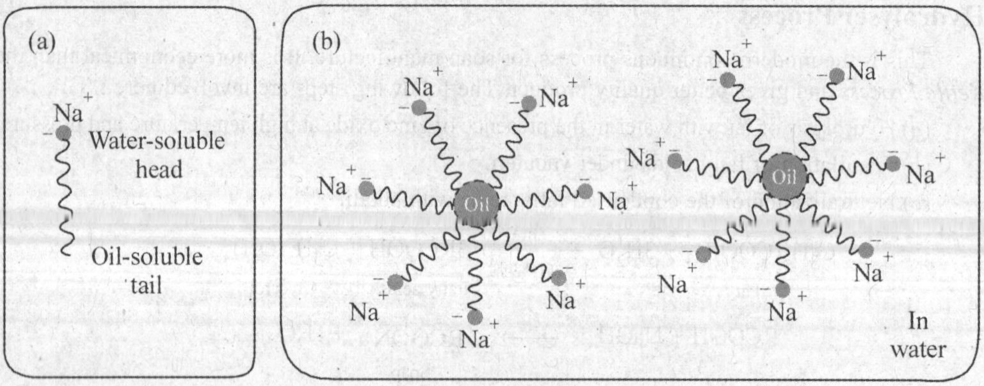

Fig. 8.4. (*a*) A soap molecule (*b*) Emulsified grease globule (micelle formation)

The non-polar lipophilic or hydrophobic portions of a number of molecules are directed towards grease or oil particles and the polar ends of these molecules are surrounded by water molecules.

Thus grease or oil particles are arrested by these soap or detergent molecules in the form of micelles (Fig. 8.5). With free use of water, these micelles containing dirt particles in grease are washed away.

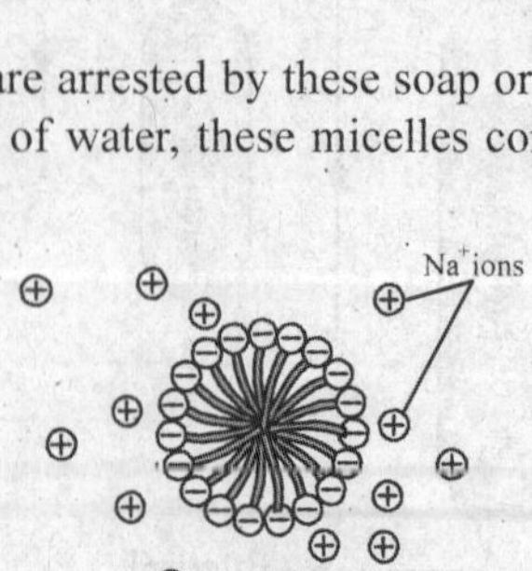

Fig. 8.5. Cross section of a soap micelle in water

8.14. SYNTHETIC DETERGENTS

Synthetic detergents also known as syndets, are synthetic substances that are being increasingly employed as cleansing agents these days. Unlike soap, detergents can be used satisfactorily even in hard water since they do not form precipitate in such water. Huge quantities of different types of detergents are now being manufactured and used all over the world. However, all these detergents are similar to soap because their molecules also have a large non-polar hydrocarbon end that is oil soluble and an ionic end that is water-soluble.

Synthetic detergents are either salts of alkyl sulphates or alkylbenzene sulphonates as given below.

(1) Sodium alkyl sulphates. These detergents consist of sodium alkyl sulphates obtained from long chain alcohols. The first such detergent was made from 1-dodecanol (or lauryl alcohol). The alcohol was treated with concentrated sulphuric acid to form a sulphate ester which was neutralised with sodium hudroxide to obtain a detergent known as sodium dodecyl sulphate (SDS) as shown below :

$$\underset{\substack{\text{1-Dodecanol}\\\text{(Lauryl alcohol)}}}{CH_3(CH_2)_{10}CH_2OH} \xrightarrow[-\,H_2O]{+\,H_2SO_4} \underset{\substack{\text{Dodecyl hydrogen sulphate}\\\text{(Lauryl hydrogen sulphate)}}}{CH_3(CH_2)_{10}CH_2OSO_3H} \xrightarrow[-H_2O]{NaOH} \underset{\substack{\text{Sodium dodecyl sulphate}\\\text{(Sodium lauryl sulphate)}}}{CH_3(CH_2)_{10}CH_2OSO_3^- Na^+}$$

The long hydrocarbon chain is evidently the non-polar end while the polar end is $—OSO_3^-Na^+$.

SDS is mainly used in shampoos and cosmetics.

(2) Alkyl benzenesulphonates. The most commonly used detergents these days are linear alkylbenzenesulphonates (**LAS**) such as sodium dodecylbenzenesulphonates. They are anionic detergents and are obtained by conversion of benzene into alkyl benzene followed by sulphonation and treatment with alkali as shown below:

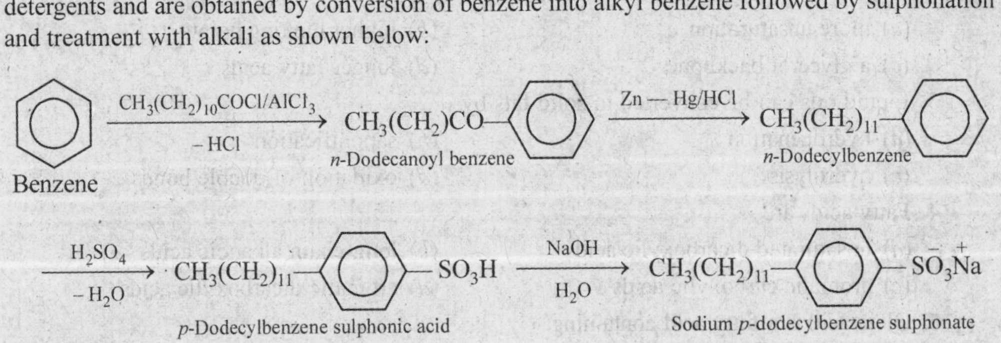

The cleansing action of these detergents depends upon the presence in their molecules of a large non-polar hydrocarbon group which is oil soluble and an ionic group which is water soluble.

Synthetic Detergents Versus Soaps

Synthetic detergents have a long nonpolar hydrocarbon chain and a highly polar group at the end of the molecule like soaps. Thus they have cleansing power as good or even better than ordinary soap. The synthetic detergents are superior to soaps because they do not form insoluble salts with Ca^{2+}, Mg^{2+} ions present in hard water.

$$\underset{\text{Soap}}{2RCO\overset{-}{O}\overset{+}{Na}} + MgSO_4 \longrightarrow \underset{\text{(Insoluble)}}{(RCO\overset{-}{O})_2Mg^{2+}} + Na_2SO_4$$

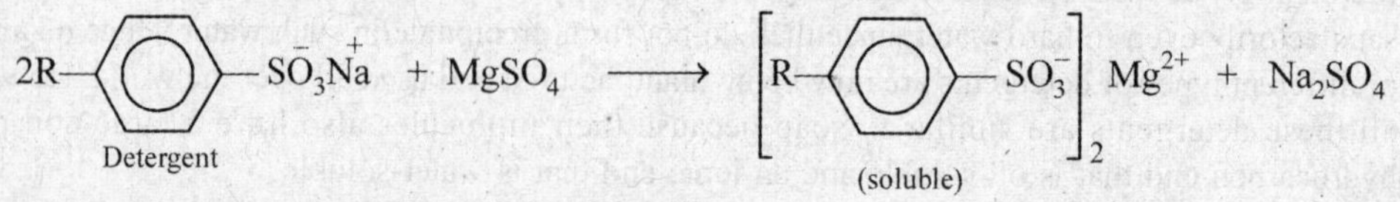

Thus detergents can be used in both soft water and hard water, while ordinary soaps are precipitated in hard water and go waste.

The earlier detergents containing branched chaing in the hydrocarbon tails, were not readily biodegradable. They were not broken down by bacteria in the sewage treatment plants. Thus, they caused water pollution. All the new LAS detergents now are biodegragable, and the problem of water pollution has now been eliminated. It may be mentioned that soaps although not working efficiently in hard water are completely biodegradable.

EXERCISES
(Including Questions from Different University Papers)

Multiple Choice Questions (Choose the correct option)

1. Soap is
   (*a*) a mixture of salts of fatty acids (*b*) a salt of glycerol
   (*c*) a mixture of ethers (*d*) a mixture of aromatic ethers
2. Fats differ from waxes in that fats have :
   (*a*) more unsaturation (*b*) higher melting points
   (*c*) a glycerol backbone (*d*) longer fatty acids
3. Liquid oils can be converted to solid fats by
   (*a*) hydrogenation (*b*) saponification
   (*c*) hydrolysis (*d*) oxidation of double bonds
4. Fatty acids are
   (*a*) unsaturated dicarboxylic acids (*b*) long-chain alkanoic acids
   (*c*) aromatic carboxylic acids (*d*) aromatic dicarboxylic acids
5. Oleic acid is a fatty acid containing
   (*a*) 12 carbons (*b*) 14 carbons (*c*) 16 carbons (*d*) 18 carbons
6. Sodium or potassium salts of fatty acids are called
   (*a*) proteins (*b*) terpenes (*c*) carbohydrates (*d*) soaps
7. Fats and oils are
   (*a*) monoesters of glycerol (*b*) diesters of glycerol
   (*c*) triesters of glycerol (*d*) diesters of glycol
8. Both stearic acid and linoleic acid have 18 carbons. Linoleic acid is unsaturated, while stearic acid is saturated. The melting point of stearic acid :
   (*a*) is higher than linolcic acid
   (*b*) is lower than linoleic acid

(*c*) is same as linoleic acid

(*d*) can not predict, insufficient information

9. Alkaline hydrolysis of oils (or fats) is called :

(*a*) saponification (*b*) fermentation (*c*) diazotisation (*d*) rancidification

10. Synthetic detergents can be represented by the following general formula

(*a*) RONa (*b*) $ROSO_3Na$ (*c*) RCOONa (*d*) RCOOH

11. The degree of unsaturation of a fat can be determined by means of its

(*a*) iodine number (*b*) octane number

(*c*) saponification number (*d*) melting point

12. Saponification of a fat

(*a*) always results in the formation of insoluble soaps

(*b*) produces glycerol and soap

(*c*) is used in the production of detergents

(*d*) is used in the production of lactic acid

ANSWERS

1. (*a*)	**2.** (*c*)	**3.** (*a*)	**4.** (*b*)	**5.** (*d*)	**6.** (*d*)
7. (*c*)	**8.** (*a*)	**9.** (*a*)	**10.** (*b*)	**11.** (*a*)	**12.** (*b*)

Short Answer Questions

1. Explain how oils and fats are formed from glycerol and higher fatty acids. Give the structure of glyceryl tripalmitate.

2. Draw a labelled diagram for the manufacture of vegetable ghee.

3. Name the different steps involved in the extraction of fats and oils from animals and plants.

4. Write the following properties of oils and fats :

(*a*) Hydrogenolysis (*b*) Hydrolysis (*c*) Trans-esterification

5. What is meant by rancidification? How is it caused ?

6. What is the significance of saponification value and iodine value ?

7. Which kind of oils are used in points? Give some examples.

8. How is soap obtained from oil or fat and caustic soda ? Give the chemical reaction.

9. With the help of a neat labelled diagram, explain the cleansing action of soap.

10. Which problems are faced when we use soaps for bathing and washing clothes ?

General Questions

1. What are Oils and Fats and how they differ from each other.

2. What is Soap ? How is Soap manufactured ?

3. What are Detergents ? How do they differ from Soaps ?

4. Write notes on :

(*a*) Hydrogenation of oils (*b*) Saponification value

(*c*) Iodine value (*d*) Cleansing action of soap

5. A triglyceride has a molecular weight of 790 and contains three double bonds. Calculate (*a*) the saponification number and (*b*) the iodine number.

Hint. (*a*) *Saponification Number:* Three moles of KOH are required to saponify one mole of triglyceride. 3 moles of KOH *i.e.* 168 g KOH are needed to saponify one mole or 790 g of triglyceride or 168000 mg of KOH are need to saponify 790 g triglyceride.

Saponification number is defined as the number of milligrams of KOH required to saponify 1 g of a fat or an oil.

$$\therefore \quad \text{Saponification number} = \frac{168000}{790}$$

(*b*) *Iodine Number:* The triglyceride contains three double bonds; therefore each mole of triglyceride will require three moles of I_2 to saturate the double bonds.

Thus, 762 g I_2 are needed to saturate 790 g triglyceride. Iodine number is defined as the number of grams of iodine that will add to 100 g of a fat or an oil.

$$\therefore \quad \text{Iodine number} = \frac{762}{790} \times 100$$

$$= 96 \text{ (Rounded)}$$

6. What are the advantages of detergents over soap ? What were the limitations of earlier detergents ?

7. Explain the cleansing action of soap with the help of a neat labelled diagram.

9

Synthetic Polymers

9.1. INTRODUCTION

Polymers may be defined as the substances made up of giant molecules of high molecular mass, each molecule of which consists of a very large number of simple molecules joined together through covalent bonds in a regular manner.

(*i*) **Monomers.** The simple molecules which combine with one another repetitively to form the polymers are called monomers.

(*ii*) **Repeat Unit.** Consider the case of polyethylene polymer. It is made from polymerizing ethylene as shown below :

$$nCH_2 = CH_2 \xrightarrow{\text{Polymerization}} (-CH_2 - CH_2 -)_n$$

Thus, we find that ethylene ($CH_2 = CH_2$) is not the repeat unit in the polymer. Repeat unit is $-CH_2-CH_2-$ which is repeated over and over again to produce the polymer. *Thus, repeat unit is the smallest chain which is repeated to provide the polymer.*

(*iii*) **Degree of Polymerisation.** It is the number of times a repeat unit is contained in the polymer molecule. In the above example of polyethylene, n denotes the degree of polymerisation. Usually, n is very large, of the order of a few thousands or tens of thousands.

9.2. HOMOPOLYMERS AND COPOLYMERS

Homopolymers. *Polymers whose repeat-unit is derived from only one type of monomer are called* **homopolymers.** *For example, in case of polyethylene, polymer which is obtained by polymerization of ethylene molecules, the repeat unit, i.e.,* $-CH_2-CH_2-$ is derived from only one type of monomer, *i.e.*, ethylene.

$$\underset{\substack{\text{Ethylene}\\(\text{Monomer})}}{nCH_2 = CH_2} \xrightarrow{\text{Polymerization}} \underset{\substack{\text{Polyethylene}\\(\text{Polymer})}}{(-CH_2 - CH_2 -)_n}$$

Polypropylene, polyvinyl chloride (PVC), polyisoprene, neoprene (polychloroprene), polyacrylonitrile (PAN), nylon-6, polybutadiene, teflon (polytetrafluoroethylene), etc., are other examples of homopolymers.

Copolymers. Polymers whose repeat units are derived from two or more types of monomers an called **copolymers.** For example, in case of nylon-66, the repeat-unit, *i.e.*,

$-NH-(CH_2)_6-NH-CO-(CH_2)_4-CO-$ is derived from two monomer, *i.e.*, hexamethylenediamine and adipic acid.

$$nH_2N-(CH_2)_6-NH_2 + nHOOC-(CH_2)_4-COOH \xrightarrow{\text{Polymerization}}$$

Hexamethylenidiamine (Monomer) Adipic acid (Monomer)

$$[-NH-(CH_2)_6-NH-CO-(CH_2)_4-CO-]_n + nH_2O$$

Nylon-66 (Polymer)

Buna-S, polyesters, bakelite, melamine-formaldehyde polymer, etc., are other examples of polymers.

9.3. CLASSIFICATION OF POLYMERS

Based on the source

Depending upon the source from which they are obtained, polymers are broadly divided into two classes:

(1) *Natural polymers* **and** **(2) *Synthetic polymers.***

1. Natural Polymers. Polymers which are obtained from animals and plants are called **natural polymers**. Starch cellulose, proteins, nucleic acids and natural rubber are some examples of this type of polymers. A brief description of these natural polymers is given below :

Starch. It is a polymer of α-glucose. It is the chief food reserve of the plants and is made up of two fraction — amylose and amylopectin. Amylose is a *linear polymer* of α-glucose and amylopectin is a *branched polymer* of α-glucose.

Cellulose. It is a polymer of β-glucose. It is the chief structural material of the plants and is obtained from wood and cotton. About 50% of wood is cellulose. Cotton contains about 90–95% cellulose.

Both starch and cellulose are made by plants from glucose produced during *photosynthesis.*

Proteins. Proteins are polypeptides or polyamides. These polymers contain a large number of α-amino acids joined together through peptide (NH–CO) bonds in a particular sequence. These are either long chain or crossed linked polymers.

Natural rubber. It is prepared from latex which, in turn, is obtained from rubber trees. It is a polymer of the hydrocarbon isoprene (2-methyl-1, 3-butadiene). Polymerization is illustrated as under:

$$nCH_2=\underset{\underset{CH_3}{|}}{C}-CH=CH_2 \xrightarrow{\text{Polymerization}} (-CH_2-\underset{\underset{CH_3}{|}}{C}=CH-CH_2-)_n$$

Isoprene (2-Methyl-1, 3-butadiene) Polyisoprene (Natural rubber)

2. Sythetic polymers. Polymers which are made in the laboratory or industry are called synthetic polymers. Natural rubber swells and loses elasticity after prolonged exposure to petrol and motor oil. Stability and melting points of many natural polymers are such that they cannot be melted and cast into desired shapes. Further, natural fibres such as cotton, wool, silk, etc., cannot meet the increasing demands. All these needs led man to synthesize polymers in the laboratory and industry. *These man-made polymers are called* **synthetic polymers.** Some important synthetic polymers are : polyethylene, polystyrene, polyvinyl chloride (PVC), bakelite, nylon and dacron.

Based upon structure

On the basis of structure, polymers are classified into three types :

1. Linear Polymers. In such polymers, the monomers are joined together to form long straight chains of polymer molecules. The various polymeric chains are then stacked over one another to give a well packed structure (Fig. 9.1a). *Linear polymers have high melting points, high densities and high tensile strength because of close packing of chains.* Nylon and polyesters are examples of linear polymers.

2. Branched Chain Polymers. In such polymers, the monomer units not only combine to form the linear chain but also form branches along the main chain (Fig. 9.1b).

These polymer molecules do not pack well because of branches. As a result, branched chain polymers have lower melting points, densities and tensile strength compared to linear polymers. An important example of a branched chain polymer is low density polythene.

3. Three-dimensional Network Polymers. In such polymers, the linear polymer chains formed initially are joined together to form a three-dimensional network structure (Fig. 9.1c). Because of the presence of crosslinks, these polymers are also called **cross-linked polymers.** These polymers are hard, rigid and brittle. Bakelite, urea-formaldehyde polymer and melamine-formaldehyde polymer are some examples of this class.

(a) (b) (c)

Fig. 9.1. Different structures of polymers (*a*) linear structure. (*b*) branched chain structure and (*c*) three-dimensional network structure.

Based on synthesis

Polymerization mainly occurs by two modes :

1. *Addition polymerization* *and* 2. *Condensation polymerization*

Corresponding to these two modes of synthesis, polymers have been classified into two types:

1. *Addition polymers* *and* 2. *Condensation polymers.*

1. Addition Polymerization. In this type of polymerization, the molecules of the same or different monomers simply add to one another in repetition leading to the formation of a polymer in which the molecular formula of the repeat unit is the same as that of the monomer. The polymers, thus, formed are called **addition polymers.**

Addition polymerization generally occurs among molecules containing double bonds.

$$\underset{\text{Ethylene}}{nCH_2 = CH_2} \longrightarrow \underset{\text{Polyethylene}}{(-CH_2-CH_2-)_n}$$

$$\underset{\text{Propylene}}{nCH_3-CH=CH_2} \longrightarrow \underset{\text{Polypropylene}}{(-CH_2-\overset{\overset{\large CH_3}{|}}{C}H-)_n}$$

2. Condensation Polymerization. In this type of polymerization, a large number of monomer molecules combine together usually with the loss of simple molecules like water, alcohol, ammonia, carbon dioxide, hydrogen chloride, etc., to form a polymer in which the molecular formula of the repeat unit is not the same as that of the monomer. The polymers thus formed are called **condensation polymers.** Condensation polymerization generally occurs between bifunctional compounds.

Nylon-66, obtained by the condensation between two monomers, *viz.* hexamethylenediamine and adipic acid, each containing two functional groups with the loss of water molecules, is an example of condensation polymer.

$$\underset{\text{Hexamethylenediamine}}{nH_2N-(CH_2)_6-NH_2} + \underset{\text{Adipic acid}}{nHOOC-(CH_2)_4-COOH} \xrightarrow{525\,K} \underset{\text{Nylon-66}}{[-NH-(CH_2)_6-NH-\overset{\overset{O}{\|}}{C}-(CH_2)_4-\overset{\overset{O}{\|}}{C}-]_n} + nH_2O$$

Terylene and bakelite are other examples of condensation polymers.

Based on Mode of Addition of Monomer Units

A more rational method of classification based upon the *mode of addition of the monomer units to the growing chain has been proposed.* According to this method of classification, there are two types of polymers:

(*i*) *Chain-growth polymers* and (*ii*) *Step-growth polymers.*

(*i*) **Chain-growth polymers** (*also called as addition polymers*). These polymers are formed by the successive addition of monomer units to the growing chain carrying a reactive intermediate such as a free-radical, a carbocation or a carbanion.

Monomer	Polymer
Ethylene	Polyethylene
Propylene	Polypropylene
Butadiene	Polybutadiene
Tetrafluoroethylene	Polytetrafluoroethylene (PTFE) or Teflon
Vinyl chloride	Polyvinyl chloride (PVC)
Isoprene	*cis*-Polyisoprene (natural rubber)

Thus, chain-growth polymers are formed by a process which involves chain reactions.

The most commonly used radical initiator is benzoyl peroxide.

Examples of chain growth polymers are given in the table above :

All the examples given in the table are those of homopolymers. An important example of a chain-growth copolymer is Buna-S involving 1, 3-butadiene and styrene as monomers.

(*ii*) **Step-growth polymers** (*also called as condensation polymers*). These polymers are formed through a series of reactions. Each such reaction involves the condensation (bond formation) between two difunctional monomer molecules to produce dimers which, in turn, produce tetramers and so on with the loss of simple molecules like H_2O, NH_3, HCl, etc. Since in this process, *the polymer is formed in a step-wise manner, it is called* **step-growth polymer** *and the process is called* **step growth polymerization.**

In contrast to chain growth polymers, the formation of step-growth polymers does not occur through chain reactions involving free radicals, carbanions or carbocations as reactive chemical species. Some typical examples of step-growth polymers are given in the table below :

S.No.	Monomers	Polymer
(*i*)	Hexamethylenediamine and adipic acid.	Nylon-66
(*ii*)	Phenol and formaldehyde.	Bakelite
(*iii*)	Terephthalic acid or its methyl ester and ethylene glycol.	Polyester (Terylene)

9.4. DIFFERENT TYPES OF POLYMERS

Polymer is a wide term involving a wide variety of materials having different characteristics. Some of them are discussed as under :

(1) Elastomers. Elastomers are those polymers in which the intermolecular forces of attraction between the polymer chains are the weakest.

Elastomers are amorphous polymers having high degree of elasticity. They have the ability to stretch out many times their normal length and return to original position when the force is withdrawn. These polymers consist of randomly coiled molecular chains of irregular shape having a few cross links (Fig. 9.2). When the force applied, these coiled chains open out and the polymer is stretched. Since the *van der Waals'* forces of attraction between the polymer chains are very weak, these cannot maintain this stretched form. Therefore, as soon as the force is withdrawn, the polymer chains return to its original coiled state. Thus, we observe that weak *van der Waals' forces of attraction allow the polymer chains to be stretched on applying the force but the cross links bring the polymer back to the original position when the force is withdrawn.* The most important example of an elastomer is natural rubber.

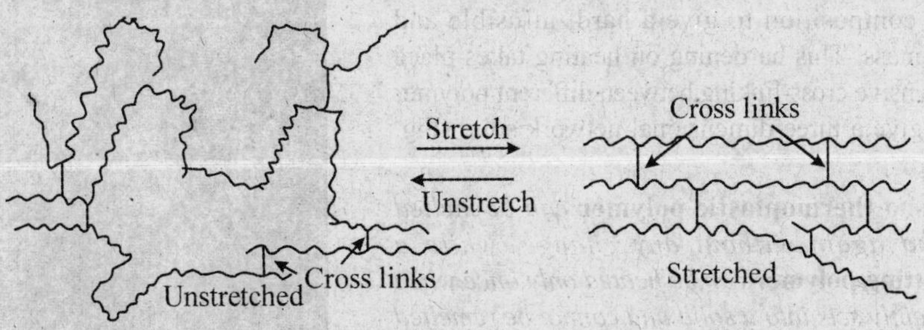

Fig. 9.2. Unstretched and stretched forms of an elastomer.

(2) Fibres. Polymers having the strongest intermolecular forces of attraction are called **fibres.** These forces are either due to H-bonding or dipole-dipole interactions. In case of nylons, the intermolecular forces are due to H-bonding (Fig. 9.3) while in polyesters and polyacrylonitrile they are due to powerful dipole-dipole interactions between the polar carbonyl (C = O) groups and between carbonyl and cyano (–C ≡ N) groups respectively.

Fibres show high tensile strength and minimum elasticity due to strong intermolecular forces.

The molecules of these polymers are long, thin and thread-like and, hence, polymer chains can be easily packed over one another. As a result, they have high melting points and low solubility.

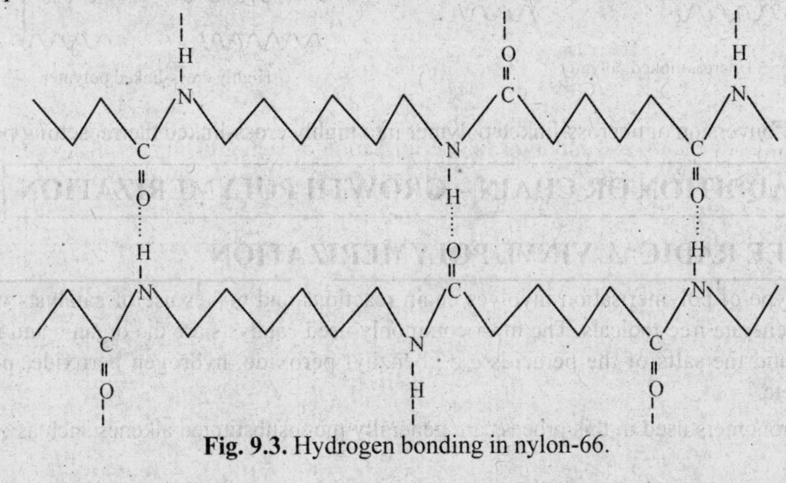

Fig. 9.3. Hydrogen bonding in nylon-66.

(3) Thermoplastics. Thermoplastics are those polymers in which the intermolecular forces of attraction are in between those of elastomers and fibres. These polymers which are linear in shape and hard at room temperature, become soft and viscous on heating and again become rigid on cooling. The process of softening and cooling can be repeated as many times as desired without any change in properties of the plastic. As a result, these plastics can be moulded into toys, buckets, telephone and television cases, etc. Thermoplastics have little or no cross linking and, hence, the individual polymer chains can slip past one another on heating. Some common examples of thermoplastics are polythene, polystyrene, polyvinyl chloride, teflon, polyvinyl acetate and polyacrylonitrile.

Plasticizers. Those plastics which do not soften easily on heating can be made soft by the addition of certain organic compound called *plasticizers.* Dialkylphthalates or cresyl phosphates are the commonly used plasticizers.

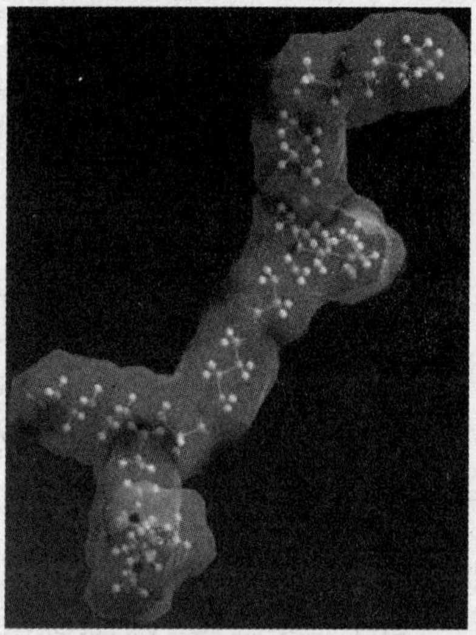

Biodegradable polymers are capable of gene delivery

4. Thermosetting polymers. Thermosetting polymers are semi-fluid substances with low molecular weights which on heating in a mould, undergo change in chemical composition to give a hard, infusible and insoluble mass. This hardening on heating takes place due to extensive cross-linking between different polymer chains to give a three-dimensional network solid (Fig. 9.4).

Thus, a **thermoplastic polymer** *can be melted again and again without any change, while a* **thermosetting polymer** *can be heated only once when it permanently sets into a solid and cannot be remelted and reworked.* Some examples of thermosetting polymers are phenol-formaldehyde (bakelite), urea-formaldehyde and melamine-formaldehyde.

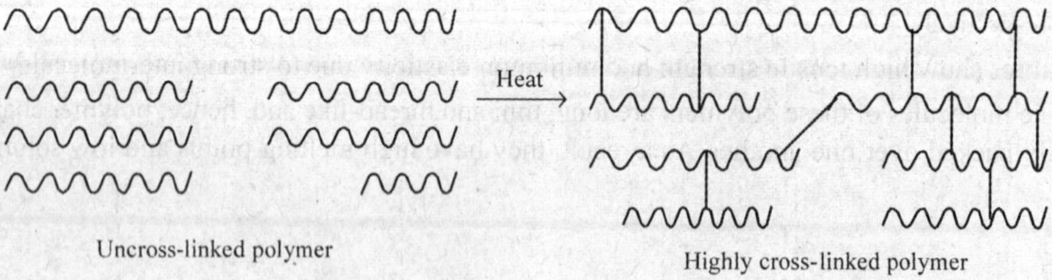

Fig. 9.4. Conversion of uncross-linked polymer into highly cross-linked thermosetting polymer.

ADDITION OR CHAIN – GROWTH POLYMERIZATION

9.5. FREE RADICAL VINYL POLYMERIZATION

This type of polymerisation involves chain reactions and makes use of catalysts which are known to generate free radicals. The most commonly used catalysts are the organic and inorganic peroxides and the salts of the peracids *e.g.,* benzoyl peroxide, hydrogen peroxide, potassium perborate, etc.

The monomers used in this process are generally monosubstituted alkenes such as :

vinyl chloride ($CH_2 = CHCl$), styrene ($CH_2 = CH -$ ) and propylene ($CH_2 = CHCH_3$).

They contain a **vinyl** group.

The mechanism of free radical vinyl polymerisation may be described as follows by considering the polymerisation of an ethylenic compound, $CH_2 = CHG$, in the presence of an organic peroxide. Here G stands for some group like $-Cl$, $-C_6H_5$ or $-CH_3$, etc.

(1) Chain initiation step : $\underset{\text{Peroxide}}{(RCOO)_2} \longrightarrow RCOO^{\bullet} \longrightarrow R^{\bullet} + CO_2$

$$R^{\bullet} + CH_2 = \underset{\underset{G}{|}}{CH} \longrightarrow RCH_2 - \underset{\underset{G}{|}}{CH^{\bullet}}$$

(2) Chain propagation step :

$$RCH_2 - \underset{\underset{G}{|}}{CH^{\bullet}} + CH_2 = \underset{\underset{G}{|}}{CH} \longrightarrow RCH_2 - \underset{\underset{G}{|}}{CH} - CH_2 - \underset{\underset{G}{|}}{CH^{\bullet}}$$

and repetition of the steps similar to (2) a number of times.

(3) Chain termination step : Chain termination could take place in one of the following ways :

***(i)* Coupling :** Collision between growing chains or a growing chain and a catalyst radical to form a deactivated molecule could take place as follows :

$$2R(CH_2\underset{\underset{G}{|}}{CH})_nCH_2 - \underset{\underset{G}{|}}{CH^{\bullet}} \longrightarrow R(CH_2\underset{\underset{G}{|}}{CH})_n CH_2 \underset{\underset{G}{|}}{CH} - \underset{\underset{G}{|}}{CH} - CH_2 (\underset{\underset{G}{|}}{CH} - CH_2)_n R$$

$$R(CH_2\underset{\underset{G}{|}}{CH})_n CH_2 \underset{\underset{G}{|}}{CH^{\bullet}} + R^{\bullet} \longrightarrow R(CH_2\underset{\underset{G}{|}}{CH})_n CH_2 \underset{\underset{G}{|}}{CHR}$$

***(ii)* Disproportionation :** One free radical acquiring hydrogen from the other leading to their deactivation could take place.

$$2R(CH_2\underset{\underset{G}{|}}{CH})_n CH_2 \underset{\underset{G}{|}}{CH^{\bullet}} \longrightarrow R(CH_2\underset{\underset{G}{|}}{CH})_n CH_2 \underset{\underset{G}{|}}{CH_2} + R(CH_2\underset{\underset{G}{|}}{CH})_n CH = \underset{\underset{G}{|}}{CH}$$

***(iii)* Chain transfer reaction :** Sometimes a growing polymer chain abstracts an atom from some impurity in the monomer. This leads to termination of the original chain but gives to a new radical which sets up a new polymerisation chain. For example,

$$R(CH_2\underset{\underset{G}{|}}{CH})_n CH_2 \underset{\underset{G}{|}}{CH^{\bullet}} + ASH \longrightarrow R(CH_2 CH)_n \underset{\underset{G}{|}}{CH_2} CH_2 + \underset{\underset{G}{|}}{AS^{\bullet}}$$

ASH stands for the impurity in the monomer.

Thus, the free radical vinyl polymerisation is generally initiated by the decomposition of the catalyst to generate a suitable free radical which adds to a multiple bond to generate a new free radical. The new free radical, in turn, adds to a second alkene molecule to form a bigger new free radical. This process of addition of each newly formed free radical to a molecule of alkene continues to form a large polymeric molecule.

Some examples of free radical vinyl polymerisation :

(i) $\underset{\text{Acrylonitrile}}{CH_2=\underset{|}{CH}(CN)} \xrightarrow{\text{Peroxide}} \underset{\text{Polyacrylonitrile or Orlon}}{-CH_2-\underset{CN}{CH}-CH_2-\underset{CN}{CH}-CH_2-\underset{CN}{CH}-}$

(ii) $\underset{\text{Chloroprene}}{CH_2=\underset{Cl}{C}-CH=CH_2} \xrightarrow[\text{persulphate}]{\text{Potassium}} \underset{\text{Polychloroprene}}{-CH_2-\underset{Cl}{C}=CH-CH_2}$

(iii) $\underset{\text{Styrene}}{CH_2=\underset{C_6H_5}{CH}} \xrightarrow{\text{Peroxide}} \underset{\text{Polystyrene}}{-CH_2-\underset{C_6H_5}{CH}-CH_2-\underset{C_6H_5}{CH}-CH_2-\underset{C_6H_5}{CH}-}$

(iv) $\underset{\text{Styrene}}{CH_2=\underset{C_6H_5}{CH}} + \underset{\text{Butadiene}}{CH_2=CH-CH=CH_2} \xrightarrow[\text{persulphate}]{\text{Potassium}}$

$$\underset{\text{Styrene butadiene rubber (A copolymer)}}{-CH_2-\underset{C_6H_5}{CH}-CH_2-CH=CH-CH_2-}$$

9.6. IONIC VINYL POLYMERISATION

This type of polymerization takes place through ionic species instead of free radicals. If the ionic species are cations, it is called cationic polymerisation. If the ionic species anions, it is called anionic polymerization.

Cationic vinyl polymerisation

This type of polymerisatioin is initiated by acids. The commonly used acid catalysts being H_2SO_4, HF and Lewis acids such as $AlCl_3$, $SnCl_2$ or BF_3.

Mechanism of cationic addition polymerisation may be explained as follows by considering the polymerisation of a substituted alkene $CH_2 = CHG$, in the presence of an acid A. Here G stands for an alkyl or aryl or halogen atom.

(1) Chain initiation step. The process is initiated by the reaction of an acid with the monomer to form a carbocation.

$$\underset{\text{Acid}}{\overset{+}{A}} + CH_2=\underset{G}{CH} \longrightarrow \underset{\text{Carbocation}}{A-CH_2-\underset{G}{\overset{+}{CH}}}$$

(2) Chain propagation step. The carbocation formed above being electrophilic adds to another molecule of the monomer to yield a second carbocation. This process continues till a large polymer molecule is formed.

$$A-CH_2-\underset{G}{\overset{+}{CH}} + CH_2=\underset{G}{CH} \longrightarrow A-CH_2-\underset{G}{CH}-CH_2-\underset{G}{\overset{+}{CH}}$$

(3) Chain terminating step. The chain reaction is usually terminated by the loss of a hydrogen ion from the growing carbocation.

$$A-CH_2-\underset{G}{CH}-\left[CH_2-\underset{G}{CH}\right]_n-CH=\underset{G}{CH} \xrightarrow{-H^+} A-CH_2-\underset{G}{CH}-\left[CH_2-\underset{G}{CH}\right]_n-CH=\underset{G}{CH}$$

Polymerisation of isobutylene in the presence of BF_3 and a trace of water to form butyl rubber as shown below, is an example of cationic addition polymerisation.

$$BF_3 + H_2O \rightleftharpoons H^+ + BF_3OH^-$$

$$\overset{+}{H} + H_2C=C(CH_3)_2 \longrightarrow H_3C-\overset{+}{C}(CH_3)_2 \xrightarrow{CH_2=C(CH_3)-CH_3} H_3C-C(CH_3)_2-CH_2-\overset{+}{C}(CH_3)_2$$

and so on to yield ultimately a large molecule of butyl rubber, *i.e.*, $\left[-CH_2-C(CH_3)_2-\right]_n$.

Anionic vinyl polymerisation

This type of polymerisation is initiated by bases. The catalysts usually employed for this purpose includes the alkali metals, alkali metal alkyls, alkali metal amides and Grignard reagents. The monomers which generally undergo this type of polymerisation are alkenes carrying electron withdrawing substituents.

The mechanism of anionic addition polymerisation may be explained as follows by considering polymerisation of a substituted alkene $CH_2=\underset{G}{CH}$ in the presence of a base, B:, G is an electron withdrawing group like $-C_6H_5$.

(1) Chain initiation step. The reaction is initiated by some base with the formation of carbanion intermediate.

$$\underset{\text{Base}}{B:} + CH_2=\underset{G}{CH} \longrightarrow \underset{\text{Base}}{B}-CH_2-\underset{G}{\bar{C}H}$$

(2) Chain propagation step. The carbanion formed above reacts with another monomer molecule and this step is repeated unitl a polymeric molecule is formed.

$$B-CH_2-\underset{G}{\bar{C}H} + CH_2=\underset{G}{CH} \longrightarrow B-CH_2-\underset{G}{CH}-CH_2-\underset{G}{\bar{C}H}$$

(3) Chain termination step. The chain reaction gets terminated by combination with H^+ ion or some Lewis acid present in the reaction mixture:

$$B-CH_2-\underset{G}{CH}\left[-CH_2-\underset{G}{CH}\right]_n-CH_2-\underset{G}{\bar{C}H}+\overset{+}{H} \longrightarrow B-CH_2-\underset{G}{CH}\left[-CH_2-\underset{G}{CH}\right]_n-CH_2-\underset{G}{CH_2}$$

It is noteworthy that anionic addition polymerisation takes place with the formation of carbanion intermediates. That is why this type of polymerisation occurs only if the carbanions are stabilised by electron withdrawing substituents.

Another feature of the mechanism is that chain termination can take place only if there is some source of H^+ ions or some Lewis acid in the reaction mixture. Otherwise, the reaction goes on till all monomer molecule are used up and a **living** polymer which is capable of reacting further is obtained.

The polymerisation of styrene in the presence of the base $K^+NH_2^-$ to form polystyrene is an example of anionic addition polymerisation.

$$\overset{-}{NH_2} + CH_2{=}\underset{C_6H_5}{CH} \longrightarrow NH_2-CH_2-\underset{C_6H_5}{\overset{-}{CH}} \xrightarrow{CH_2=CH(C_6H_5)} NH_2-CH_2-\underset{C_6H_5}{CH}-CH_2-\underset{C_6H_5}{\overset{-}{CH}}$$

and so on till polymeric molecule of polystyrene, $(-CH_2-\underset{C_6H_5}{CH}-)_n$ is formed.

9.7. ZIEGLER-NATTA POLYMERISATION

Polymerisation, making use of Ziegler-Natta catalysts is called Ziegler-Natta polymerization.

Ziegler-Natta catalysts are a mixture of alkyl aluminium and titanium halides for example $(C_2H_5)_3Al/TiCl_4$ in an inert solvent such as heptane.

Karl Ziegler

Giulio Natta

Free-radical polymerisation of isoprene produces non-stereospecific rubbers, which do not possess the desired qualities of natural rubber since they contain both *cis* and *trans* isomers.

To overcome this problem, we carry out polymerisation by ionic mechanism producing stereospecific polymers. This is done by using Ziegler-Natta catalysts, which enables us to obtain polyisoprene with all *cis* configuration. We obtain the product with properties comparable to those of natural rubber. The exact mechanism of Ziegler-Natta polymerization is not certain. It is believed that the monomer gets successively inserted between titanium and alkyl groups of the growing chain through π bonding.

9.8. VULCANISATION

Properties like tensile strength, elasticity and resistance to abrasion of natural rubber can be improved by a process called vulcanization. It consists of heating rubber with 3-5% sulphur. During vulcanization, sulphur cross-links between polymer chains are introduced (Fig. 9.5). The process of vulcanization was discovered Charles Goodyear in 1839. Goodyear is now a famous brand of automobile tyres and tubes. Fig. 9.5 below demonstrates how sulphur cross-links between the polymer chains are formed.

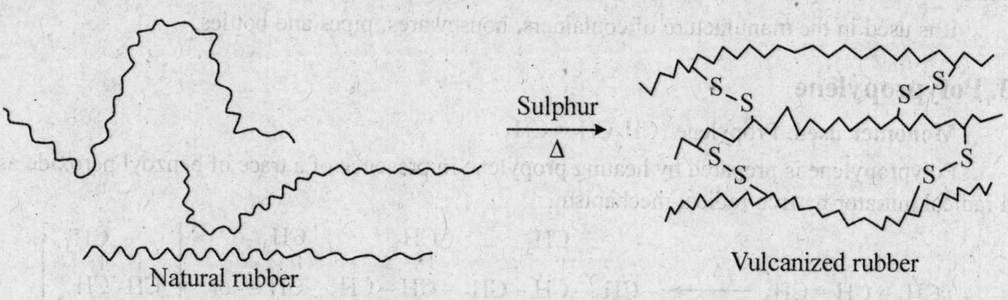

Fig. 9.5. In vulcanized rubber, the polymer chains are held together by polysulphide bridges or cross links.

9.9. SOME IMPORTANT VINYL POLYMERS

1. Low density polythene

In this polymer, the repeat unit is $-CH_2-CH_2-$. It is manufactured by heating ethylene to 473 K under a pressure of 1500 atmospheres and in the presence of a trace of oxygen. This polymerization occurs by a free radical mechanism initiated by oxygen.

$$nCH_2 = CH_2 \xrightarrow[\text{Traces of oxygen}]{473\text{ K, }1500\text{ atm}} \underset{\text{Polythene}}{+CH_2-CH_2+_n}$$

The polythene, thus, produced has a molecular mass of about 20,000 and *has a branched structure.* There branched polythene molecules do not pack well and, hence, this type of polythene has a low density (0.92 g/cm^3) and a low melting point (384 K). That is why polythene prepared by free radical polymerization is called low density polythene.

Properties and uses. Low density polythene is a *transparent polymer* of moderate tensile strength and high toughness. It is chemically inert, and is a poor conductor of electricity.

It is widely used as a packaging material (in the form of thin plastic films, bags etc.) and as insulation for electrical wires and cables.

Charles Goodyear

2. High density polythene

It is prepared by *co-ordinaion polymerization* of ethylene. In this process, enthylene in a solvent is heated to 333-343 K under a pressure of 6-7 atmospheres in the presence of a catalyst consisting of triethylaluminium and titanium tetrachloride called *Ziegler-Natta catalyst.*

$$\underset{\text{Ethylene}}{n(CH_2 = CH_2)} \xrightarrow[\text{Ziegler- Natta catalyst}]{333\text{ - }343\text{ K, }6\text{ - }7\text{ atm.}} \underset{\text{Polythene}}{+CH_2-CH_2+_n}$$

This polythene consists of linear chains of polymer molecules. These polymer molecules pack well and hence, this polythene has higher density (0.97 g/cm^3) and higher melting point (403 K) than the polymer produced by *free-radical* polymerisation. That is why polythene prepared by coordination polymerisation is called high density polythene.

Properties and Uses. High density polythene is a *translucent polymer.* It is also chemically inert and has greater toughness, hardness and tensile strength compared to low density polythene.

It is used in the manufacture of containers, housewares, pipes and bottles.

3. Polypropylene

Monomer used. Propylene ($CH_3CH = CH_2$)

Polypropylene is prepared by heating propylene in presence of a trace of benzoyl peroxide as a radical initiator by free-radical mechanism.

$$\underset{\text{Propylene}}{nCH_3 - CH = CH_2} \xrightarrow[\text{Peroxide}]{\Delta} -CH_2 - \underset{}{\overset{CH_3}{\overset{|}{C}H}} - CH_2 - \overset{CH_3}{\overset{|}{C}H} - CH_2 - \overset{CH_3}{\overset{|}{C}H} - \text{ or } \underset{\text{Polypropylene}}{\left[CH_2\,\overset{CH_3}{\overset{|}{C}H} \right]_n}$$

The monomer of polypropylene is $CH_3 - CH = CH_2$ and the repeat unit is $-CH_2\,\overset{CH_3}{\overset{|}{C}H}-$

Uses. It is harder and stronger than polythene. It is used :

(*i*) in the manufacture of stronger pipes and bottles.

(*ii*) for packing of textiles and foods.

(*iii*) for making liners for bags and heat shrinkable wraps for records.

(*iv*) for making automotive mouldings, seat covers, carpet fibres and ropes.

4. Polystyrene or Styron

Monomer used. Styrene ($C_6H_5CH = CH_2$).

In the presence of peroxides, styrene polymerises to form polystyrene.

$$\underset{\text{Styrene}}{nCH_2 = \underset{\underset{C_6H_5}{|}}{CH}} \xrightarrow{\text{Peroxides}} \left[CH_2 - \underset{\underset{C_6H_5}{|}}{CH} \right]_n$$

Uses. It is a transparent polymer and is used for making plastic toys, household wares, radio and television bodies.

5. Neoprene

It is a polymer of chloroprene and is a synthetic rubber.

Monomer used. Chloroprene ($CH_2 = CH - \overset{Cl}{\overset{|}{C}} = CH_2$)

Chloroprene is prepared by the Markovnikov's addition of HCl to vinylacetylene at the triple bond.

$$\underset{\text{Vinylacetylene}}{CH_2 = CH - C \equiv CH} + HCl \longrightarrow \underset{\text{Chloroprene}}{CH_2 = CH - \overset{Cl}{\overset{|}{C}} = CH_2}$$

Vinylacetylene needed for the purpose is prepared by dimerization of acetylene as given below :

$$\underset{\text{Acetylene (2 molecules)}}{CH \equiv CH + HC \equiv CH} \xrightarrow[343\,K]{NH_4Cl,\ CuCl} \underset{\text{Vinylacetylene}}{CH_2 = CH - C \equiv CH}$$

Chloroprene polymerises very fast. No specific catalysts are needed but the polymerisation is slower in absence of oxygen. The reaction occurs by 1, 4-addition of one chloroprene molecule to the other as shown below :

$$\underset{\text{Chloroprene}}{\overset{1}{C}H_2 = \overset{2}{C}(Cl) - \overset{3}{C}H = \overset{4}{C}H_2} + \underset{\text{Chloroprene}}{\overset{1}{C}H_2 = \overset{2}{C}(Cl) - \overset{3}{C}H = \overset{4}{C}H_2} + \ldots \xrightarrow{O_2 \text{ or Peroxide}} \underset{\text{Polychloroprene or Neoprene}}{- CH_2 - C(Cl) = CH - CH_2 - CH_2 - C(Cl) = CH - CH_2 -}$$

The polymer may be represented as $\underset{\text{Neoprene}}{-\!\!\left(CH_2 - \underset{\displaystyle Cl}{\underset{|}{C}} = CH - CH_2\right)\!\!-_n}$

It has excellent rubber like properties.

Uses. Neoprene is inferior to natural rubber in some properties but is quite stable to aerial oxidation and resistant to oils, gasoline and other solvents. It is, therefore, used in the manufacture of hoses, shoe heels, stoppers, etc.

6. Buna-S

It is a synthetic rubber and is a copolymer of 1, 3-butadiene and styrene. **Bu** stands for 1, 3-butadiene, **na** for sodium which is used as the polymerizing agent and **S** stands for styrene. It is also called SBR (Styrene, Butadiene, Rubber).

Materials used. 1,3-Butadiene ($CH_2 = CH - CH = CH_2$) and styrene ($C_6H_5CH = CH_2$).

It is obtained by copolymerization of 1, 3-butadiene and styrene in the ratio 3:1 in the presence of sodium.

$$\underset{\text{1, 3-Butadiene}}{nCH_2 = CH - CH = CH_2} + \underset{\text{Styrene}}{n\underset{\displaystyle C_6H_5}{\underset{|}{C}H} = CH_2} \xrightarrow{\text{Na, heat}} \underset{\text{Buna-S or SBR}}{\left[- CH_2 - CH = CH - CH_2 - \underset{\displaystyle C_6H_5}{\underset{|}{C}H} - CH_2 - \right]_n}$$

Uses. It is used in the manufacture of tyres, rubber soles, water-proof shoes, etc.

7. Poly (vinyl acetate), PVA, $-\!\!\left[CH_2 - \underset{\displaystyle Cl}{\underset{|}{C}H}\right]\!\!-_n$

Preparation. This homopolymer prepared by free radical addition polymerisation of vinyl acetate ($CH_2 = CHOCOCH_3$) in the presence of an initiator such as benzoyl peroxide.

$$\underset{\text{Vinyl acetate}}{nCH_2 = CHOCOCH_3} \xrightarrow{\text{Benzoyl peroxide}} \underset{\text{Poly (vinyl acetate)}}{-\!\!\left[CH_2 - \underset{\displaystyle OCOCH_3}{\underset{|}{C}H}\right]\!\!-_n}$$

Vinyl acetate required in this preparation is obtained by passing a mixture of ethylene, acetic acid vapours and oxygen over a heated catalyst consisting of Pd Cl_2 and Cu Cl_2.

$$CH_2 = CH_2 + CH_3COOH + \frac{1}{2}O_2 \xrightarrow{\text{Heated catalyst}} \underset{\text{Vinyl acetate}}{CH_2 = CHOCOCH_3} + H_2O$$

Uses. It is a soft rubbery polymer and is used :

(1) in making plastic emulsion paints.

(2) in making paper grease-proof.

8. Poly(methyl methacrylate), PMMA

Monomer used. Methyl methacrylate $(CH_2 = \overset{\overset{\displaystyle CH_3}{|}}{C}H - COOCH_3)$

The monomer methyl methacrylate is obtained by treating acetone cyanohydrin with $CH_3OH - H_2SO_4$ which brings about simultaneous hydrolysis, dehydration and esterification. This is polymerized in the prese.nce of a radical initiator to give poly (methyl methacrylate).

$$\underset{\text{Acetone}}{CH_3 - \overset{\overset{\displaystyle CH_3}{|}}{C} = O} \xrightarrow{HCN} \underset{\text{Acetone cyanohydrin}}{CH_3 - \overset{\overset{\displaystyle CH_3}{|}}{\underset{\underset{\displaystyle OH}{|}}{C}} - CN} \xrightarrow{CH_3OH - H_2SO_4} \underset{\text{Methyl methacrylate}}{CH_2 = \overset{\overset{\displaystyle CH_3}{|}}{C} - COOCH_3}$$

$$\underset{\text{Methyl methacrylate}}{nCH_2 = \overset{\overset{\displaystyle CH_3}{|}}{C} - COOCH_3} \xrightarrow{\text{Peroxides}} \underset{\text{Poly (methyl methacrylate)}}{\left[CH_2 - \overset{\overset{\displaystyle CH_3}{|}}{\underset{\underset{\displaystyle COOCH_3}{|}}{C}} \right]_n}$$

Poly (methyl methacrylate) is a hard and transparent substance

The most important property of poly (polymethyl methacrylate) is its clearness and excellent light transmission even better than glass.

Uses. It is used in the manufacture of lenses, light covers, light shades, transparent domes and skylights, aircraft windows, dentures and plastic jewellery.

9. Poly (ethyl acrylate)

Monomer used. Ethyl acrylate $(CH_2 = CH - COOC_2H_5)$.

Ethyl acrylate on polymerization in the presence of peroxides gives poly (ethyl acrylate).

$$\underset{\text{Ethyl acrylate}}{nCH_2 = \underset{\underset{\displaystyle COOC_2H_5}{|}}{CH}} \xrightarrow{\text{Peroxides}} \underset{\text{Poly (ethylacrylate)}}{\left[CH_2 - \underset{\underset{\displaystyle COOC_2H_5}{|}}{CH} \right]_n}$$

Poly (ethyl acrylate) is tough but with somewhat rubber-like properties.

10. Polyacrylonitrile (PAN), Acrilan, Orlon

Monomer used. Acrylonitrile $(CH_2 = CH - CN)$.

Acrylonitrile polymerizes in presence of peroxides to give polyacrylonitrile.

$$\underset{\text{Acrylonitrile}}{nCH_2 = \underset{\underset{\displaystyle CN}{|}}{CH}} \xrightarrow{\text{Peroxides}} \underset{\text{Polyacrylonitrile}}{\left[CH_2 \quad \underset{\underset{\displaystyle CN}{|}}{CH} \right]_n}$$

The monomer acrylonitrile is manufactured by either of the following reactions :

$$\underset{\text{Acetylene}}{CH \equiv CH} + HCN \xrightarrow{CuCl-HCl} \underset{\text{Acrylonitrile}}{CH_2 = CH - CN}$$

$$\underset{\text{Propylene}}{2CH_3CH = CH_2} + \underset{\text{(From air)}}{3O_2} + 2NH_3 \xrightarrow[\text{Mo, Co and Al, 723 K}]{\text{Oxides of}} \underset{\text{Acrylonitrile}}{2CH_2 = CH - CN} + 6H_2O$$

Uses. Polyacrylonitrile is a hard and high melting material.

(*i*) It is used in the manufacture of *Orlon* and *Acrilan* fibres used for making clothes, carpets and blankets.

(*ii*) It is blended with other polymers to improve their qualities.

11. Polyvinyl chloride (PVC)

Monomer used. Vinyl chloride ($CH_2 = CH - Cl$).

Vinyl chloride polymerises in the presence of peroxides to form polyvinyl chloride.

$$\underset{\text{Vinyl chloride}}{nCH_2 = \underset{|}{CH} \atop \quad Cl} \xrightarrow{\text{Peroxides}} \underset{\text{Polyvinyl chloride}}{\left(CH_2 - \underset{\substack{| \\ Cl}}{CH} \right)_n}$$

The monomer vinyl chloride is itself manufactured by the addition of HCl to acetylene in the presence of mercury salts as catalyst or by dehydrochlorination of ethylene dichloride.

$$\underset{\text{Acetylene}}{CH \equiv CH} + HCl \xrightarrow{Hg^{2+}} \underset{\text{Vinylchloride}}{CH_2 = CH - Cl}$$

$$CH_2Cl - CH_2Cl \xrightarrow{873-923\ K} \underset{\text{Vinyl chloride}}{CH_2 = CH - Cl} + HCl$$

Uses. (*i*) It is a good electrical insulator and hence is used for coating wires and cables.

(*ii*) It is also used in making gramophone records and pipes.

(*iii*) It is used for making raincoats, hand bags, plastic dolls, upholstery, shoe soles and vinyl flooring.

12. Polytetrafluoroethylene (PTFE) or Teflon

Monomer used. Tetrafluoroethylene ($F_2C = CF_2$).

Tetrafluoroethylene polymerises in the presence of oxygen to give polytetrafluoroethylene popularly called **Teflon**.

$$\underset{\text{Tetrafluoroethylene}}{nCF_2 = CF_2} \xrightarrow{O_2} \underset{\text{Polytetrafluoroethylene or Teflon}}{\left(CF_2 - CF_2 \right)_n}$$

Uses. Teflon is unaffected by solvents, boiling acids and aqua regia upto 598 K.

(*i*) Because of its great chemical inertness and high thermal stability, teflon is used for making non-sticks utensils.

(*ii*) It is also used for making gaskets, pump packings, valves, seals, non-lubricated bearings, etc.

9.10. CONDENSATION OR STEP-GROWTH POLYMERIZATION

In this type of polymerization, the monomer molecules combine together in a step-wise manner with the elimination of some simple molecule such as water or methyl alcohol.

This type of polymerization generally takes place between difunctional monomers *i.e.* monomers containing two functional groups. Some condensation polymers are described as under:

Polyesters

1. Terylene or Dacron. It is prepared by condensation polymerization of ethylene glycol and terephthalic acid with elimination of water at 425-475 K.

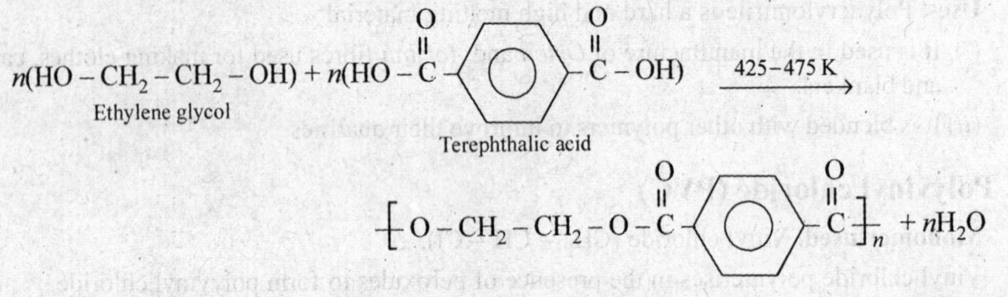

Terylene or Dacron

Uses. (*i*) The fibre of terylene is highly crease-resistant, durable and has low moisture content. It is, therefore used for the manufacture of wash and wear fabrics, tyre cords, seat belts and sails. It is also blended with cotton and wool to increase their durability.

(*ii*) The Mylar film made from dacron is flexible and resistant to ultraviolet degradation. It is, therefore, used for making magnetic recording tapes.

2. Glyptal or Alkyd resin. Glyptal, *i.e.*, poly (ethylene phthalate) is formed by the condensation of ethylene glycol and phthalic acid.

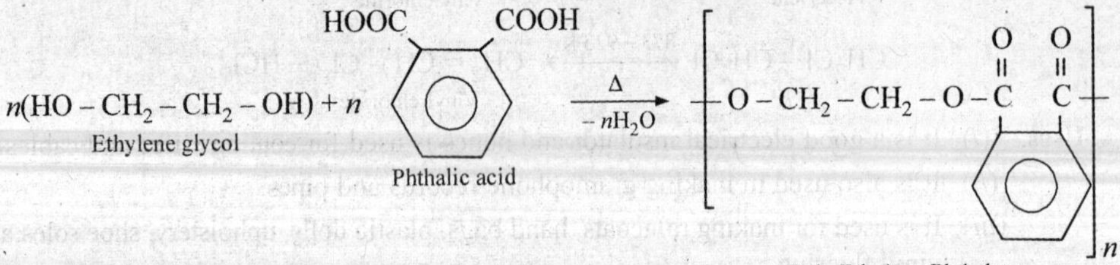

Poly Ethylene Phthalate

Uses. Poly (ethylene phthalate) is a thermoplastic. Its solution on evaporation leaves a tough and inflexible film. It is, used in the manufacture of paints and lacquers.

Polyamides

1. Nylon-6, 6. * It is manufactured by the condensation polymerization of adipic acid and hexamethylenediamine at about 525 K when water is lost as steam and the nylon is produced in the molten state. It is then cast into a sheet or fibres by passing through a spinneret.

$$\underset{\text{Adipic acid}}{n\mathrm{HO}-\overset{\overset{\large O}{\|}}{C}-(\mathrm{CH_2})_4-\overset{\overset{\large O}{\|}}{C}-\mathrm{OH}} + \underset{\text{Hexamethylenediamine}}{n\mathrm{H_2N}-(\mathrm{CH_2})_6-\mathrm{NH_2}} \xrightarrow[\text{Polymerization}]{525\ \mathrm{K}}$$

$$\underset{\text{Nylon-6, 6}}{\left[-\overset{\overset{\large O}{\|}}{C}-(\mathrm{CH_2})_4-\overset{\overset{\large O}{\|}}{C}-\mathrm{NH}-(\mathrm{CH_2})_6-\mathrm{NH}-\right]_n} + n(\mathrm{H_2O})$$

* Nylon was originally prepared in New York and London simultaneously. (*Ny* – New York, *Lon* – London). This is how the polymer got its name.

It is called nylon-6, 6 (read as nylon six, six) since both adipic acid and hexamethylenediamine contain six carbon atoms each.

2. Nylon-6, 10. Another commonly used nylon is nylon-6, 10 (read as nylon six, ten) which is obtained by the condensation of hexamethylenediamine (containing six carbon atoms) and sebacic acid [$HOOC(CH_2)_8COOH$], dibasic acid containing ten carbon atoms.

$$\underset{\text{Hexamethylenediamine}}{nH_2N-(CH_2)_6-NH_2} + \underset{\text{Sebacic acid}}{nHOOC(CH_2)_8COOH} \xrightarrow{-nH_2O} \underset{\text{Nylon-6, 10}}{\left[-HN-(CH_2)_6-NH-\overset{O}{\overset{\|}{C}}-(CH_2)_8-\overset{O}{\overset{\|}{C}}- \right]_n}$$

Nylon fibres possess high tensile strength and are abrasion resistant. They also possess some elasticity.

Uses. (*i*) These are used in the manufacture of carpets, textile fibres and bristles for brushes.

(*ii*) Crinkled nylon fibres are used for making elastic hosiery.

(*iii*) Being tough nylon is used as a substitute for metals in bearings and gears.

3. Nylon-6 or Perlon. Monomer *caprolactam* on polymerization gives nylon-6.

Caprolactam needed for the purpose is manufactured from cyclohexane as described below :

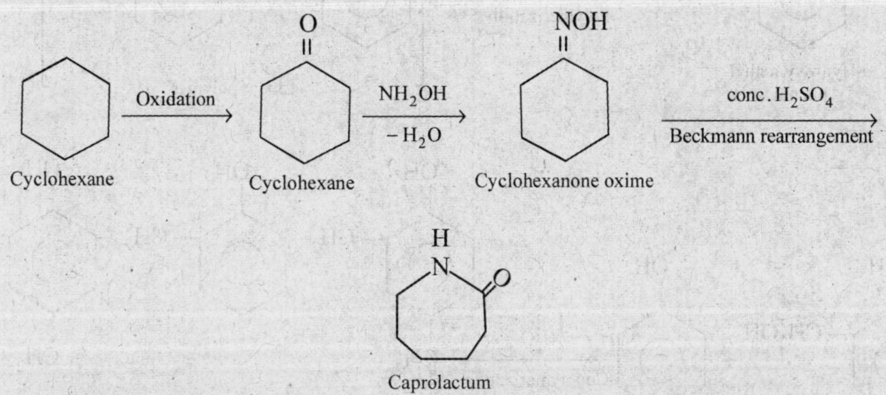

Caprolactam is heated with a trace of water, it hydrolyses to ε-aminocaproic acid which upon continued heating undergoes polymerization to give nylon-6. It is called as nylon-6 (read as nylon six) since the monomer (caprolactam) contains six carbon atoms.

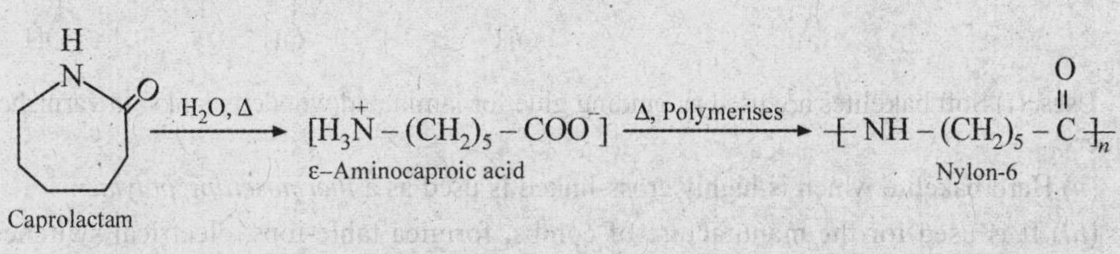

The fibres of nylon-6 are obtained when molten polymer is forced through a spinneret.

Uses. It is used for the manufacture of tyre cords, fabrics and mountaineering ropes.

Phenol-formaldehyde Resins (Bakelite)

Phenol on treatment with formaldehyde in the presence of a basic catalyst undergoes condensation polymerization to form either a linear or a cross-linked polymer called **phenol-formaldehyde resin or bakelite.** Methylene bridges are formed either at *ortho* or *para*-position or both at *ortho* and *para*-positions with respect to the phenolic group.

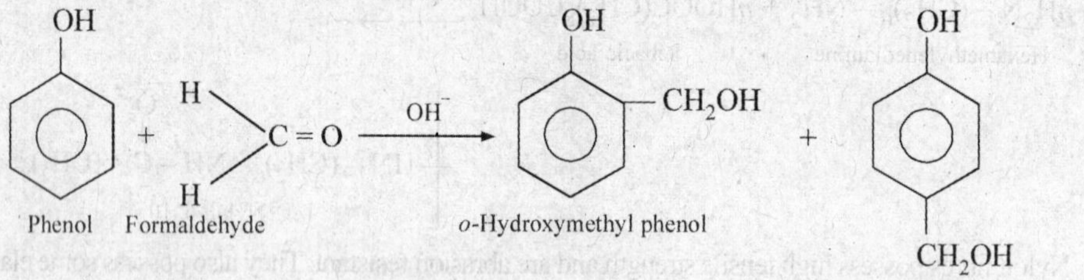

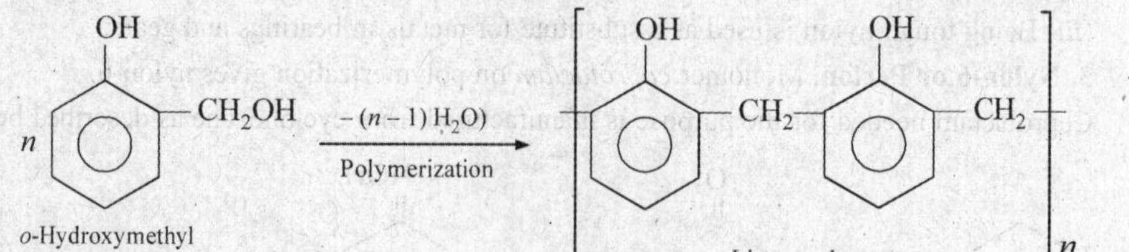

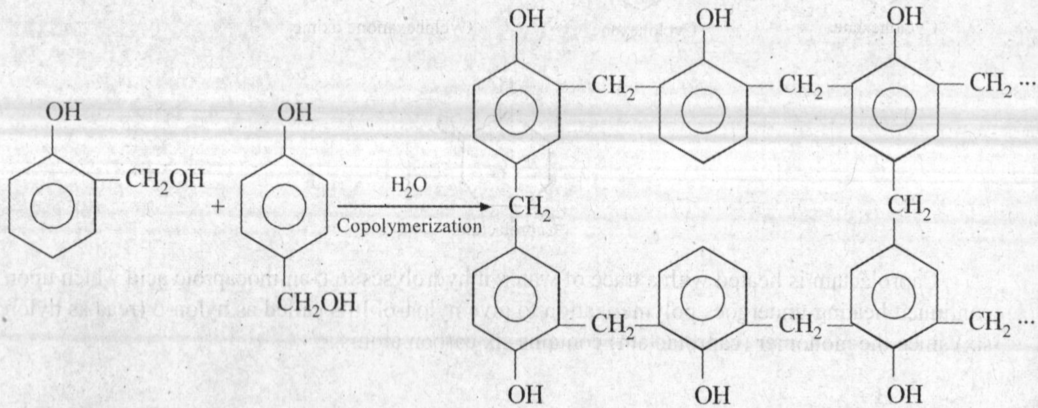

Uses. (*i*) Soft bakelites are used as binding glue for laminated wooden planks, in varnishes and lacquers.

(*ii*) Hard bakelite which is highly cross-linked is used as a *thermosetting polymer.*

(*iii*) It is used for the manufacture of combs, formica table-tops, electrical switches and gramophone records, etc.

Bracelets made from phenol-formaldehyde resin

Melamine-formaldehyde resin

Melamine and formaldehyde on heating undergo copolymerization to for melamine-formaldehyde resin.

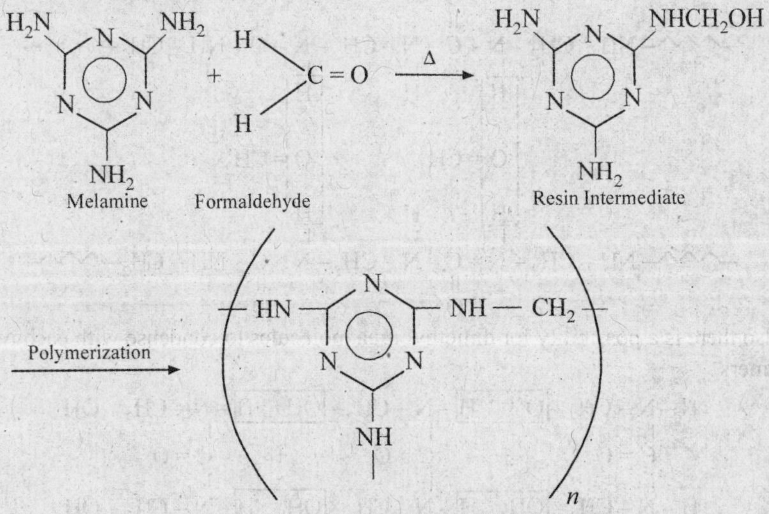

MELAMINE-FORMALDEHYDE POLYMER

Use. It is used for making unbreakable crockery.

Urea-formaldehyde resin

Preparation. (*i*) This polymer is obtained by a reaction between urea and formaldehyde in the presence of a base or an acid

$$H_2N-\overset{\overset{\displaystyle O}{\|}}{C}-NH_2 + HCHO \longrightarrow \underset{\text{Methyl urea}}{H_2N-\overset{\overset{\displaystyle O}{\|}}{C}-NH-CH_2-OH}$$

(*ii*) Methyl urea obtained in step (*i*) reacts with another molecule of formaldehyde to produce dimethylurea

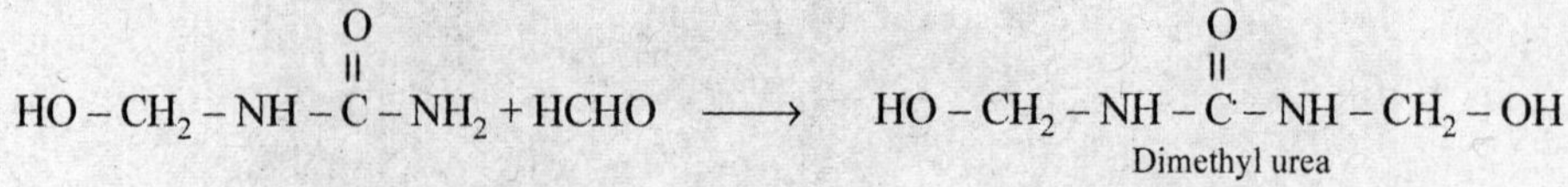

(*iii*) Dimethyl urea molecules condense with more urea molecules to form linear polymers

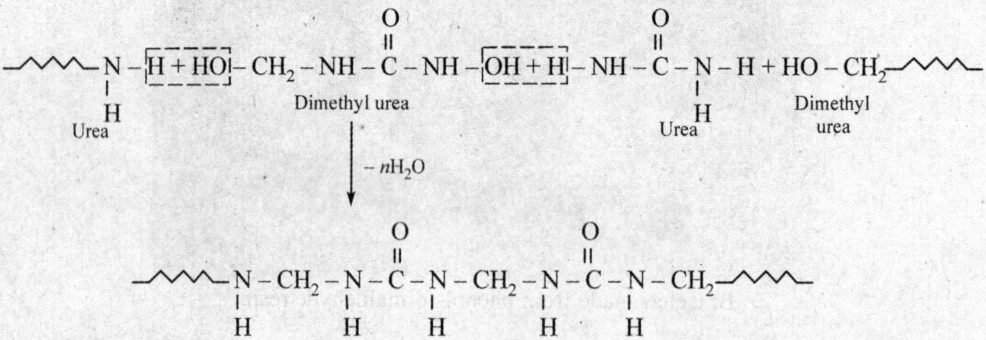

(*iv*) The linear molecules can condense with more formaldehyde molecules to cross-linked structures

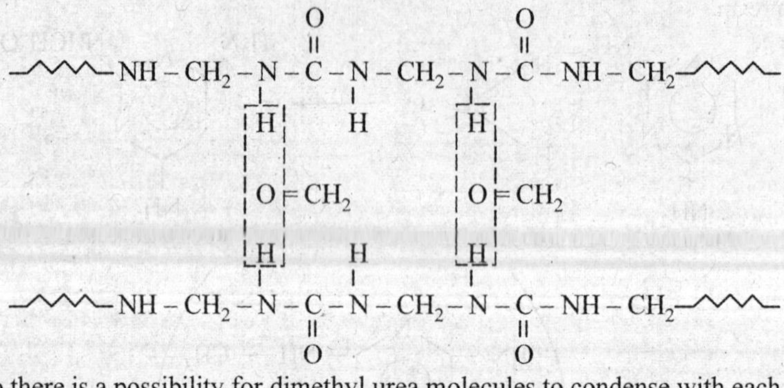

(*v*) Also there is a possibility for dimethyl urea molecules to condense with each other to form linear polymers

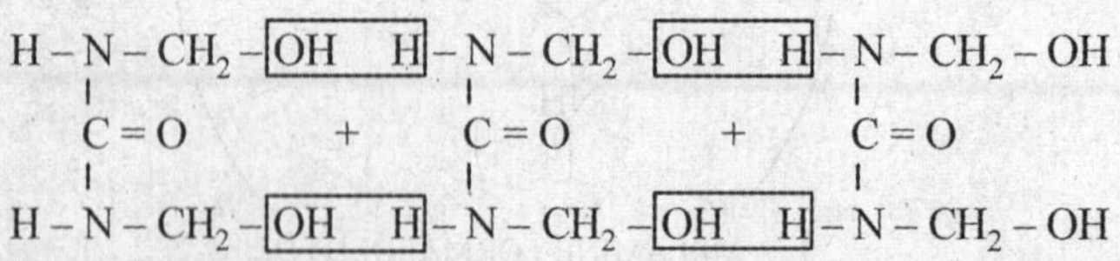

Depending upon the reaction conditions, resins of different types are obtained.

Uses. 1. These polymers find use as ingradients of varnishes and lacques

2. These are used in the manufacture of electrical fittings, radio and TV cabinets, telephones, table tops and toys

3. These are used for gluing and impregnating timber

Epoxy Resins

Preparation. Epoxy resins are prepared by the condensation of excess of chloroepoxy alkanes with dihydric phenols in the presence of NaOH catalyst at 50-60°C.

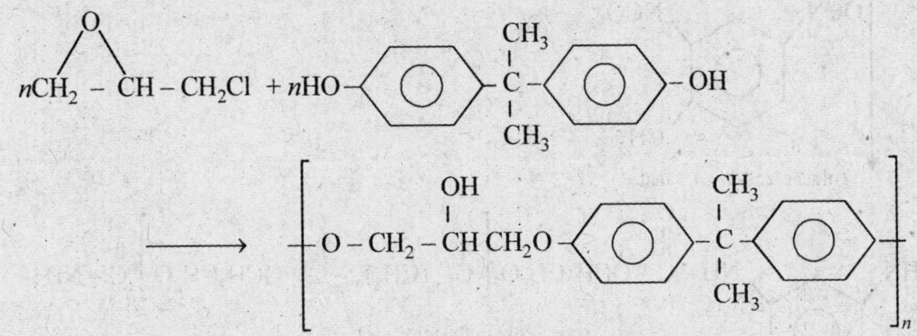

Cross-linking of chains is achieved by heating the above product with amines, when terminal epoxy groups of the chain react with amine.

Uses 1. Epoxy resins are known for their bonding properties and as such are used in adhesives such as araldite. Epoxy resins are used to bind glass, porcelain, metal and wood.

2. Epoxy resins possess the properties of inertness, handness and flexibility and as such are used as protective coating. Fibreglass parts of boats and automobiles have a metal frame coated with a layer of a mix of span glass and epoxy resin.

Polyurethanes

Urethane, the monomer of polyurethane may be considered to be obtained as under :

$$\underset{\text{Isocyanate}}{R-N=C=O} + \underset{\text{Alcohol}}{HO-R'} \longrightarrow \underset{\text{A urethane}}{R-NH-\overset{\overset{O}{\|}}{C}-OR'}$$

Polyurethanes are polymers containing urethane linkages.

Preparation. Polyurethanes are usually prepared by reaction between a **diol** and **di-isocyanate.** The diol generally employed is a polyester having CH_2OH end groups and with a molecular mass of 1000 to 2000. It is referred to as *prepolymer.* The di-isocyanate commonly used is toluene-2, 4-di-isocyanate. The di-isocyanate reacts with the prepolymer to give polyurethane. Synthesis of a polyurethane has been worked out as under :

Polyurethane enameled copper wire

$$\underset{\substack{\text{Glycol}\\\text{(Excess)}}}{HOCH_2CH_2OH} + \underset{\text{Adipic acid}}{nHO-\overset{\overset{O}{\|}}{C}-(CH_2)_4-\overset{\overset{O}{\|}}{C}-OH}$$

$$\downarrow$$

$$\underset{\text{Prepolymer}}{HOCH_2CH_2O\left[\overset{\overset{O}{\|}}{C}-(CH_2)_4-\overset{\overset{O}{\|}}{C}-OCH_2CH_2O\right]_n H}$$

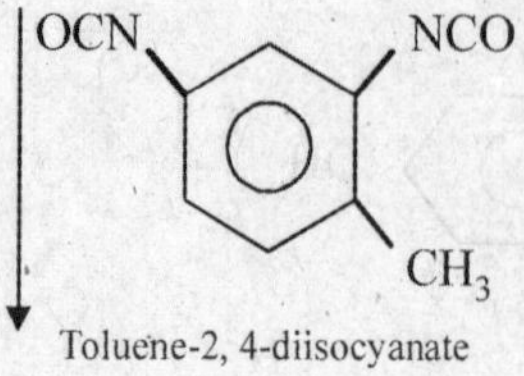

Toluene-2, 4-diisocyanate

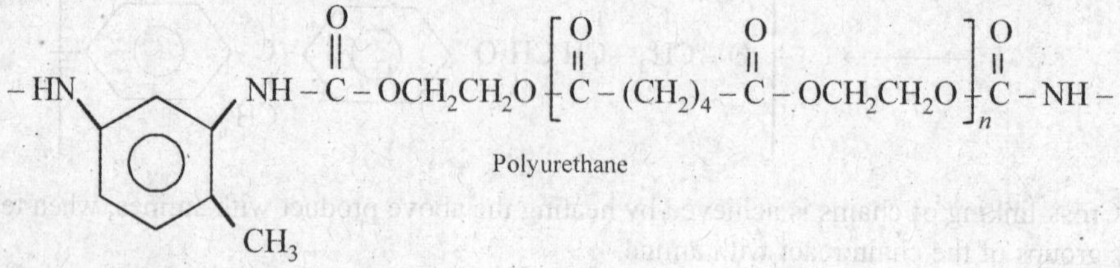

Polyurethane

Characteristics of polyurethanes

Polyurethanes can be spun into elastic fibres or they can be obtained in the form of foam. For the formation of foams some water is added to the reaction mixture during polymerisation. This reacts with some of the isocyanate groups to form amino groups and liberate carbon dioxide which acts as the foaming agent.

$$-N=C=O + H_2O \longrightarrow -NH_2 + CO_2$$

Polyurethanes consist of long straight chain molecules without having any interlinkages between the individual strands.

Uses. Polyurethane foams are used in pillows, from backed fabrics, car seats etc.

Polyurethane foams are used in packing, construction and interior decoration.

9.11. NATURAL AND SYNTHETIC RUBBERS

Over the past years, the demand for rubber has increased tremendously. Therefore, apart from natural rubber, a number of synthetic varieties of rubber have been developed.

A brief description of natural and synthetic rubbers is given below.

Natural Rubber

It is a natural polymer which is obtained from *latex*. Latex is a milky liquid which comes out from rubber tree (*hevea brasiliensis*) when an incision is made in the bark of the tree. Latex contains 30-40% rubber which is coagulated by adding salt and acetic acid to latex. The coagulation takes about 3-4 hours. The rubber thus obtained is converted into sheets and processed further.

Natural rubber has been found to be a polymer of isoprene. There may be 10,000 to 20,000 isoprene units in the polymer chain of natural rubber. A characteristics feature of natural rubber which is responsible for its unique properties is that it has got *cis-configuration* at almost every double bond. It may be represented as :

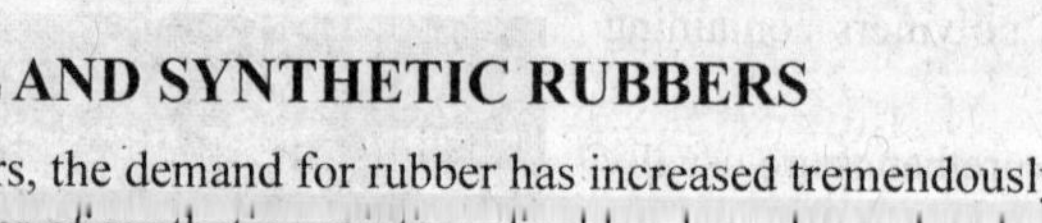

Natural rubber (all cis)
(cis-1, 4-polyisoprene)

Latex being collected from rubber tree

Natural rubber has a number of unique properties particularly its *elasticity.* It finds use in the manufacture of many articles. But the products made from natural rubber become sticky in hot weather and stiff in cold weather. This drawback can be removed by heating natural rubber with sulphur. This process is known as **vulcanisation** (Charles Goodyear, 1839). During vulcanisation, cross links are produced between different polymeric chains through disulphide linkages as shown below. This improves the elasticity of rubber and makes it tough and resistant to heat.

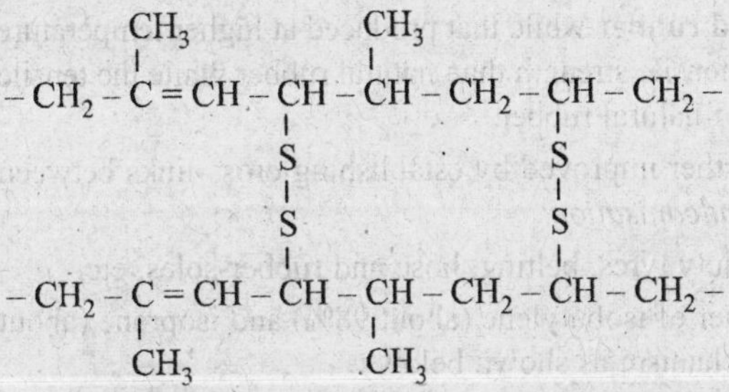

It is possible to prepare rubber with desired physical properties by varying the amount of sulphur used in vulcanisation. If the amount of sulphur used is 1-3%, rubber obtained is soft and stretchy. It can be used for rubber bands and tubes used in tyres. Use of 3-10% sulphur in vulcanisation produces some what harder but flexible rubber which can be used in the manufacture of tyres. Vulcanised rubber containing 20-30% sulphur is very hard and can be used like a hard synthetic plastic.

Natural rubber inner-tube

Synthetic Rubber

Synthetic rubber is a general name used for synthetic polymeric materials having rubber-like properties.

The *cis*-polyisoprene having the properties of natural rubber was synthesised in 1955 by the polymerisation of isoprene,

$$n\,CH_2=\underset{}{\overset{CH_3}{\overset{|}{C}}}-CH=CH_2 \xrightarrow{Li} \left[CH_2-\overset{CH_3}{\overset{|}{C}}=CH-CH_2\right]_n$$

in the presence of finely divided lithium. Some of the important synthetic rubbers are described below.

1. Styrene butadiene rubber, SBR. It is obtained by the free radical addition copolymerisation between one part by weight of styrene and three parts of butadiene.

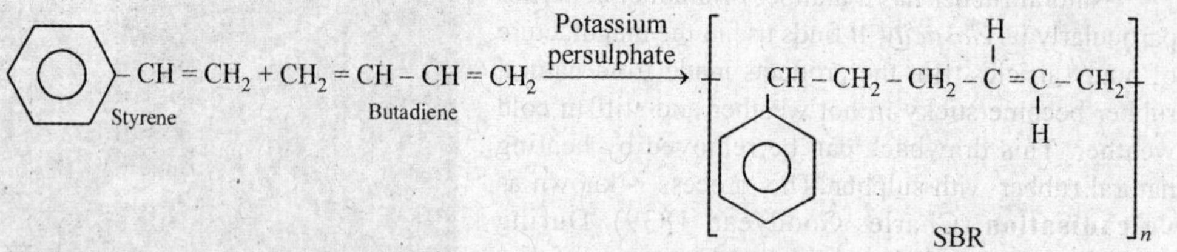

The steric configuration about the double bond may be *cis* or *trans* depending on conditions. If the polymerisation is done at 323 K, 60% of the butadiene units have *trans*-configuration and this percentage increases if polymerisation is performed at lower temperature. The configuration around the double bond plays an important role in determining the properties of the product.

SBR made at 255 to 278 K is known as **cold rubber** while that produced at higher temperature is termed **regular** SBR. Regular SBR has lower tensile strength than natural rubber while the tensile strength of cold rubber is almost equal to that of natural rubber.

The physical properties of SBR can be further improved by establishing cross-links between the polymer molecules through the process of *vulcanisation.*

Uses. SBR is now widely used in lighter duty tyres, belting, hose and rubber-soles, etc.

2. Butyl rubber. It is an addition copolymer of isobutylene (about 98%) and isoprene (about 2%). The reaction takes place by a cationic mechanism as shown below :

$$CH_3-\overset{\overset{CH_3}{|}}{C}=CH_2 + CH_2=\overset{\overset{CH_3}{|}}{C}-CH=CH_2$$

$$\Big\downarrow \; AlCl_3,\ 173\text{ K}$$

$$\left[CH_2-\underset{\underset{CH_3}{|}}{\overset{\overset{CH_3}{|}}{C}}-CH_2-\overset{\overset{CH_3}{|}}{C}=CHCH_2-CH_2-\underset{\underset{CH_3}{|}}{\overset{\overset{CH_3}{|}}{C}}-CH_2-\underset{\underset{CH_3}{|}}{\overset{\overset{CH_3}{|}}{C}} \right]_n$$

The polymeric chains which are practically linear may be cross-linked through vulcanisation.

3. Polyisoprene. It is homopolymer of 2-methyl-1, 3-butadiene or isoprene obtained through free radical polymerisation as shown below.

$$\underset{\text{Isoprene}}{n\,CH_2=\overset{\overset{CH_3}{|}}{C}-CH=CH_2} \longrightarrow \underset{\text{Polyisoprene}}{\left[CH_2-\overset{\overset{CH_3}{|}}{C}=CH-CH_2 \right]_n}$$

The properties of synthetic rubber made from isoprene are somewhat different from natural rubber even though natural rubber is also a polymer of isoprene formed in nature. The difference may be attributed to: the difference in the stereochemistry of synthetic and natural products.

4. Neoprene or Polychloroprene. This is a polymer of chloroprene,. *i.e.*, 2-chloro-1, 3-butadiene, $CH_2=CClCH=CH_2$. The polymerisation proceeds by a free radical addition mechanism and the product has *trans* configuration around the double bond.

$$n\,CH_2 = CCl - CH = CH_2 \xrightarrow[\text{persulphate}]{\text{Potassium}} \left[-CH_2 - \underset{\underset{H}{|}}{\overset{\overset{Cl}{|}}{C}} = C - CH_2 - \right]_n$$

Chloroprene → Neoprene

Neoprene is superior to natural rubber in certain, respects. It resists the action of oil and other natural solvents. It has also high tensile strength and good weathering resistance.

Neoprene is vulcanised by heating alone without using sulphur.

Uses. It is used in the manufacture of hoses, belting, shoe.heels and stoppers.

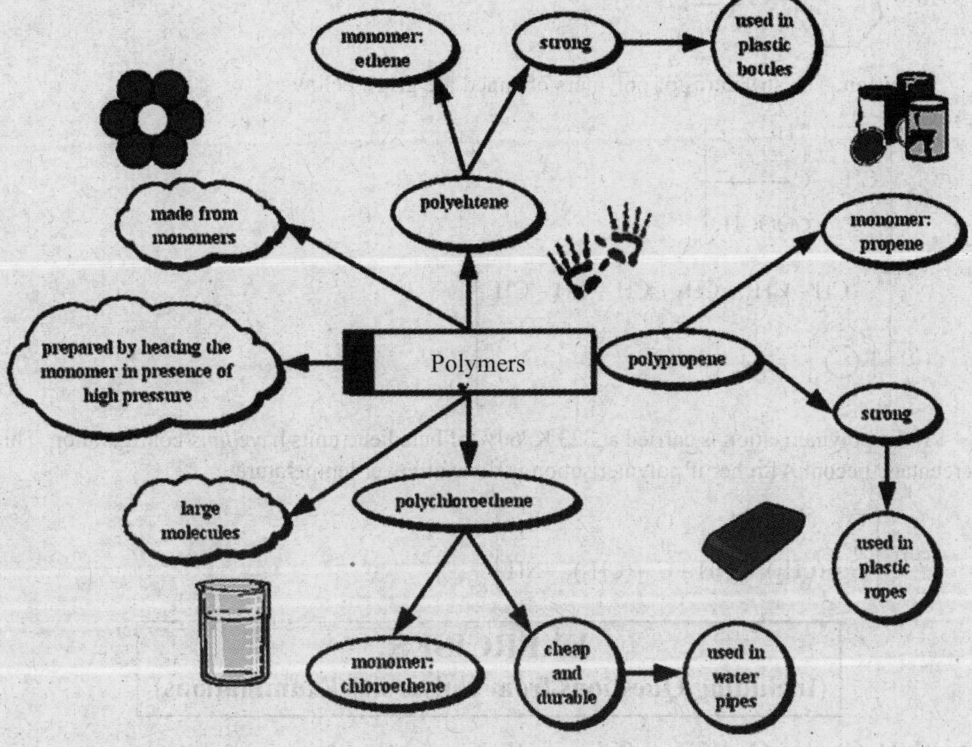

Polymers-Revision

SOLVED EXAMPLE

Example 1. Give the likely structure of the polymers obtained in each of the following cases :

(*i*) $CH_2 = \underset{|}{\overset{CH_3}{\overset{|}{C}}} - COOCH_3 \xrightarrow{\text{organic peroxide}}$

(*ii*) 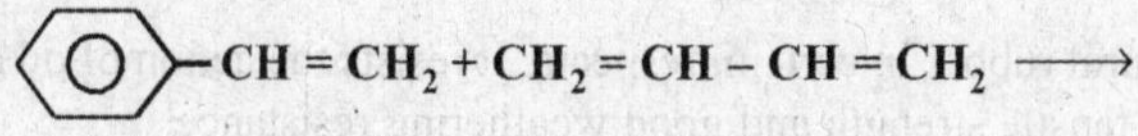

(*iii*) 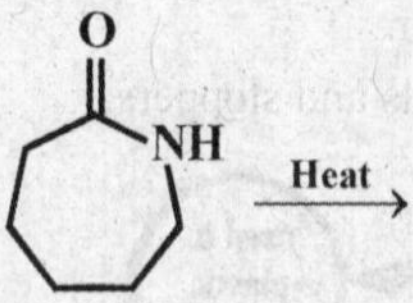

Solution. The structures of polymers obtained are given below :

(*i*) $$\left[CH_2 - \underset{\underset{COOCH_3}{|}}{\overset{\overset{CH_3}{|}}{C}} \right]_n$$

(*ii*) 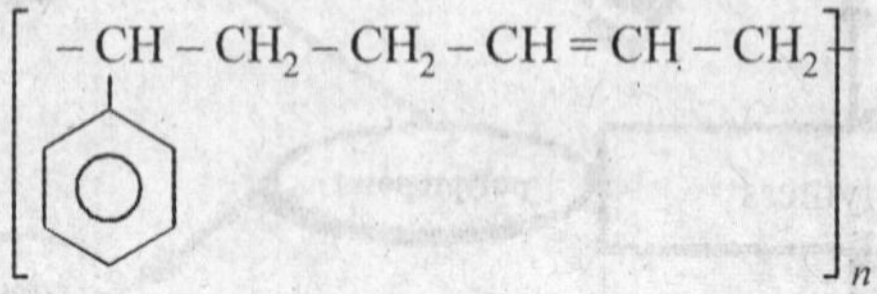

If the polymerisation is carried at 323 K, 60% of butadiene units have *tans*-configuration. This percentage becomes higher if polymerisation is done at lower temperature.

(*iii*) $$\left[\overset{\overset{O}{||}}{C} - (CH_2)_5 - NH - \overset{\overset{O}{||}}{C} - (CH_2)_5 - NH \right]_n$$

EXERCISES
(Including Questions from University Examinations)

Multiple-Choice Questions (Choose the correct option)

1. Bakelite is obtained from :

(*a*) phenol and formaldehyde

(*b*) adipic acid and hexamethylene diamine

(*c*) dimethyl terephthalate and ethylene glycol

(*d*) neoprene

2. Neoprene is a polymer of the following monomer:

(*a*) Chloroprene (*b*) Isoprene (*c*) Isobutane (*d*) Isopentene

3. Which of the following is an example of a condensation polymer ?

(*a*) Nylon-6, 6 (*b*) Teflon (*c*) Polypropylene (*d*) Orlon

4. Orlon is prepared by the polymerisation of
(*a*) vinyl cyanide (*b*) allyl alcohol (*c*) vinyl chloride (*d*) allyl chloride

5. Nylon-6, 6 is obtained from :
(*a*) adipic acid and hexamethyl diamine (*b*) tetrafluoroethylene
(*c*) vinyl cyanide (*d*) vinylbenzene

6. Teflon is prepared by the polymerisation of
(*a*) butadiene (*b*) vinyl cyanide
(*c*) vinyl chloride (*d*) tetrafluoroethylene

7. Which of the following is a thermosetting polymer ?
(*a*) Bakelite (*b*) Nylon-6, 6 (*c*) Polyethylene (*d*) Teflon

8. Adipic acid reacts with hexamethylene diamine to form
(*a*) bakelite (*b*) nylon-6, 6 (*c*) terylene (*d*) nylon-6, 10

9. The monomers for *Buna-S* are 1, 3-butadiene and
(*a*) ethylene glycol (*b*) adipic acid (*c*) styrene (*d*) caprolactum

10. Natural rubber is a polymer of
(*a*) propene (*b*) isoprene (*c*) formaldehyde (*d*) phenol

11. Which of the following polymers contain nitrogen?
(*a*) PVC (*b*) Teflon (*c*) Nylon (*d*) Terylene

12. Ethylene glycol reacts with dimethyl terephthalate to form
(*a*) nylon-6, 6 (*b*) teflon (*c*) dacron (*d*) orion

ANSWERS

1. (*a*)	**2.** (*a*)	**3.** (*a*)	**4.** (*a*)	**5.** (*a*)	**6.** (*d*)
7. (*a*)	**8.** (*b*)	**9.** (*c*)	**10.** (*b*)	**11.** (*c*)	**12.** (*c*)

Short Answer Questions

1. What kind of polymerisation cationic or anionic or vinyl polymerisation would be suited for the preparation of :
(*i*) polypropylene (*ii*) polyacrylonitrile (*iii*) polystyrene

2. Give likely structures of polymers obtained in each of the following cases :

(*i*) 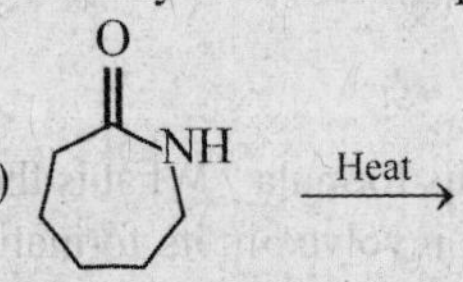

(*ii*) 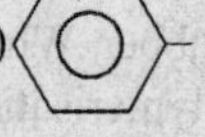 $-CH=CH_2 + CH_2=CH-CH=CH_2 \longrightarrow$

3. Outline the synthesis of Bakelite. Give its two important uses.

4. What is SBR ? How is it synthesised ?

5. What is mean by Ziegler-Natta polymerisation ? How is it brought about ?

6. What do you understand by chain growth and step growth polymerisation ? Give one example of the polymer in each case.

7. What do you mean by the term "addition polymers" ? Give two examples of such polymers.

8. What is a polyamide ? Give one example alongwith its method of preparation.

9. Give one example for each of the following types of polymers :
(*i*) Polyamides (*ii*) Polyesters (*iii*) Polyolefins (*iv*) Polyacrylate

10. Write the synthesis and uses of the following polymers:

(*i*) Nylon 6, 6 (*ii*) Polyacrylonitrile

11. Give one example for each of the following types of polymers : (only name is to be given)

(*i*) Polyester (*ii*) Polyacrylate

(*iii*) Polyamide (*iv*) Phenol-formaldehyde resin

12. Give the information asked for the following polymers :

(*i*) Bakelite : materials used for preparation

(*ii*) Neoprene : material (s) required for preparation

(*iii*) High density polythene : mode of preparation of free radical or ionic vinyl polymerisation or Ziegler-Natta polymerisation.

13. What is meant by vulcanisation ? How is it carried out ?

14. Write the formulae, type of polymer (addition or condensation), one method of preparation of the following :

(*i*) Polymethylmethacrylate (*ii*) Poly(ethyleneterephthalate)

(*iii*) Styrene-butadiene rubber.

15. How will you synthesise Nylon-66 and Teflon ? Identify the type of polymerisation involved in each case in terms of chain growth and step growth polymerisation.

16. Give the preparation and uses of orlon.

17. Give the possible chain terminating steps of free radical vinyl polymerisation.

18. Write a brief note on polyurethanes.

19. Give the structure of butyl rubber. What are the starting materials required for its preparation ? Is it an addition or condensation polymer? If the answer is addition polymer, give the conditions (free radical, anionic or cationic) most appropriate for its polymerisation.

20. Write a short note on condensation polymerisation.

21. What are epoxyresins ? Give the preparation of one epoxy resin alongwith its characteristics and uses.

22. Give the preparation and uses of :

(*i*) urea-formaldehyde resin (*ii*) polyurethanes (*iii*) polystyrene.

23. Write a note on natural and synthetic Rubbers.

24. Give a brief account of Urea-formaldehyde resins.

25. Give mechanism of free radical addition polymerisation.

26. Neoprene is a commonly used synthetic rubber. What is its formula ? What is the monomer used in its preparation ? What kind of polymerisation is involved in its formation ?

27. What is Ziegler-Natta Polymerisation ? What are its advantages over free radical vinyl polymerisation ?

General Questions

1. Write short notes on :

(*i*) Homopolymer (*ii*) Copolymer

(*iii*) Condensation polymerisation (*iv*) Styrene-butadiene rubber

(*v*) Vulcanisation.

2. Give the likely structures of the polymers obtained in each of the following cases :

(*i*) $CH_2=\underset{}{\overset{CH_3}{\overset{|}{C}}}-COOCH_3 \xrightarrow{\text{organic peroxide}}$

(*ii*) 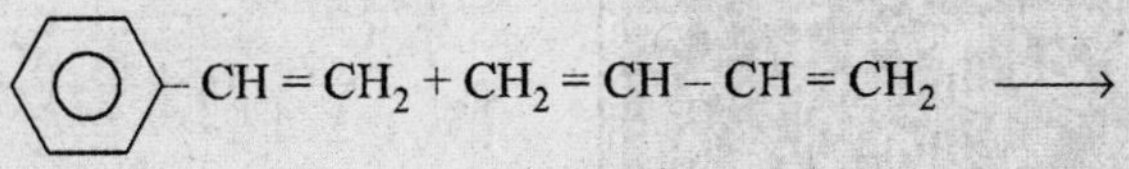

(*iii*) 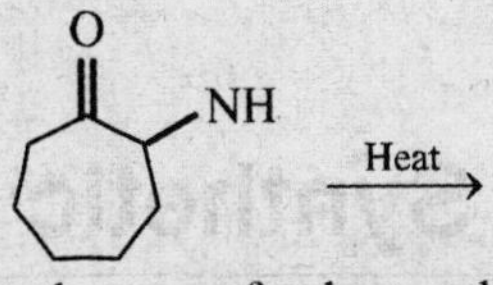

Give the uses of polymer obtained in each case.

3. What do you understand by chain growth and step growth polymerisation ? What are the various types of chain growth polymerisation ? Discuss the mechanism of one of the type of chain growth polymerisation.

4. Discuss the mechanisms of the following :

(*i*) Free-radical addition polymerisation (*ii*) Cationic addition polymerisation

(*iii*) Anionic addition polymerisation.

5. Define the term addition polymerisation. Write the method of preparation and uses of the following :

(*i*) Teflon (*ii*) Acrilon (*iii*) Styron.

6. (*a*) What is meant by Ziegler-Natta polymerisation ? What are its advantages over free radical vinly polymerisation ?

(*b*) Write the formulae, type of polymer (addition or condensation), one method of preparation of the following :

(*i*) Poly (methylmethacrylate) (*ii*) Poly (ethylene terephthalate).

7. Give preparation and important uses of the following :

(*i*) Teflon (*ii*) Nylon 66 (*iii*) Neoprene.

8. Name one polymer in each of the following cases. Give its preparation and important uses?

(*i*) Polyester (*ii*) Polyurethane (*iii*) Synthetic rubber (*iv*) Polyolefin.

10

Synthetic Dyes

10.1. INTRODUCTION

A dye may be defined as a coloured substance which when applied to the fabrics imparts a permanent colour and the colour is not removed by washing with water, soap or on exposure to light. All coloured organic substances are not necessarily dyes. For instance, both picric acid and trinitrotoluene are yellow in occur, but only the former can fix to a cloth and is a dye, while, the latter does not fix to a cloth and is not a dye.

Characteristics of a dye

1. A dye should be a coloured substance
2. It should be able to fix itself to the material from solution or be capable of being fixed on it.
3. It should resist the action of water, soap and light.

Various dyeing processes

The following four methods are usually employed for dyeing, depending upon the nature of the cloth (fabric) and dye itself.

(*a*) *Substantive dyeing method,* (*b*) *Adjective dyeing (mordant) method,*

(*c*) *Developed dyeing (ingrain) method,* (*d*) *Vat dyeing method.*

These are explained further as follows :

(*a*) Substantive dyeing method. This is also called *direct dyeing* method. In this method, a cloth is dyed directly by immersing it in a solution of dye. It is largely used for dyeing of animal fabrics (wool and silk), but not for vegetable fabrics (cotton and artificial silk).

(*b*) Adjective dyeing method. This is also known as *mordant dyeing* method. In this method, the fabric is first impregnated with certain substances referred to as mordants which combine with the dye. The mordants used are basic substances for acidic dyes and acidic substances for basic dyes. The impregnated fabric is then immersed in the solution of a dye to form insoluble *lakes* on the cloth. Dyeing with alizarin is done by this method.

(*c*) Developed dyeing method. This is also known as *ingrain dyeing method*. It consists in preparing and applying the dye on the cloth itself. For example for azo-dyes the cloth to be dyed is first impregnated with a phenol solution and then treated with a solution of diazotised amine to produce azo-dye on the cloth.

(*d*) Vat dyeing method. The water insoluble coloured compound which are used in their reduced state (leuco-compounds) for dyeing purpose are called **Vat dyes.**

In this method, the cloth to be dyed is soaked in the solution of reduced dye and then spread in the air so that the leuco-compound is oxidised back to generate the original dye.

10.2. CLASSIFICATION OF DYES

Dyes may be classified on the basis of chemical constitution or on the basis of application to fibre.

Based on chemical constitution

The chemical classification is based on the common parent structure of the dye. The number of dyes based on this classification is fairly large. We shall give here an account of important classes of dyes.

(*i*) **Azo-Dyes.** These dyes contain one or more azo groups (–N = N–) which form bridges between two or more aromatic rings. The azo-dyes form the largest and the most important group of dyes. Azo group is the chromophore. Aniline yellow, methyl orange, congo red, and bismark brown are some examples of this class.

(*ii*) **Nitroso dyes.** These dyes contain a nitroso group (–NO) as the chromophore and hydroxyl group (–OH) as the auxo chrome. Fast green O (Dinitrosoresorcinol) and Gambine Y (α-nitroso-β-naphthol) are some examples of this class.

(*iii*) **Nitro dyes.** These dyes contain a nitro group (–NO_2) as the chromphore and hydroxyl group (–OH) as the auxo-chrome. They are generally nitro-derivatives of phenols containing at least one nitro group in ortho or para position to the hydroxyl group. Pictric acid (2, 4, 6-trinitrophenol) and martius yellow (2, 4-dinitro-l-naphthol) belong to this category.

(*iv*) **Phthaleins.** These dyes are obtained by the condensation of phthalic anhydride with phenols in the presence of some dehydrating agent like conc. H_2SO_4 or anhydrous $ZnCl_2$. An important example of this class is phenolphthalein.

(*v*) **Xanthene dyes.** These dyes are usually derived from Xanthen. These dyes have a common structural feature and are obtained by the condensation of phenols with phthalic anhydride in the presence of a dehydrating agent like anhydrous $ZnCl_2$, H_2SO_4, etc. Examples of this class are fluorescein and eosin.

(*vi*) **Triphenyl methane dyes.** The nucleus of this group is triphenyl methane. These dyes are usually obtained by introducing groups like NH_2, NR_2 or OH into the rings of this nucleus which contains three phenyl radicals. The typical instances of triphenyl-methane dyes are rosaniline, malachite green and crystal violet.

Classification of dyes based on application to fibre

Dyes can be classified as follows based on application to fibre:

(*i*) **Acid dyes.** These dyes are essentially the sodium salts of acids which may contain sulphonic acid or phenolic acid group. The colour of an acid dye is in its negative ion. These dyes give very bright colour and have a wide range of fastness. These dyes are also known as *anionic dyes.*

The acid dyes are always used in acidic solution. The fabric is stirred in the hot solution of the dye in the presence of either an acid or salt till it is smoothly dyed. It is then removed and dried. Some typical instances of acid dyes are picric acid, orange-II, naphthol yellow, etc.

(*ii*) **Basic dyes.** These are also called *cationic dyes*. The basic dyes are those which contain a basic amino group and it is protonated under the acid conditions of fibres by formation of salt linkages with anionic or acidic group in the fibres. Some examples of basic dyes are crystal violet, methylene blue and methyl violet.

(*iii*) **Direct dyes.** These dyes are also known as *substantive dyes*.

They dye cotton as well as wool and silk. The dye is applied to the fabric by immersing it in its hot boiling solution, removing and then drying the fabric. Some typical members of direct dyes are congo red, naphthol yellow S and martius yellow.

(*iv*) **Developed dyes.** These dyes are also called *azotic* or *ingrain dyes*. These are the dyes which are produced within the cloth itself as a result of chemical action between the two reactants producing the dye. For example, the cloth is first dipped in an alkaline solution of phenol, resorcinol or β-naphthol and then immersed in an alkaline solution of diazo compound. The coupling reaction takes place between the phenols and diazo compound within the textile fibres giving rise to the formation of a dye.

(*v*) **Mordant dyes.** These dyes are also called *adjective dyes.* They cannot directly dye cotton, silk or wool but require the help of a mordant. A mordant is a substance which is taken up by the fibres and which in turn takes up the dye. There are basic and acidic mordants. If the dye is acidic, the mordant must be basic, *e.g.*, salts of Cr, Al, Sn and Fe. On the other hand, if the dye is basic, the mordant must be acidic, *e.g.*, tannin or tannic acid containing some amount of tartar emetic.

The mordanted cloth is dipped into the solution of the dye. The dye is absorbed by the mordant forming an insoluble mordant dye compound which gets firmly fixed within the fibres. Certain dyes give different colours with different mordants.

Colour lakes formed by alizarin with different mordants are given in Table 10.1.

Table 10.1: Coloured lakes formed by alizarin

Ion used (mordant)	*Colour*	*Ion used*	*Colour*
Al^{3+}	Rose red	Ba^{2+}	Blue
Cr^{3+}	Brownish red	Mg^{2+}	Violet
Sn^{2+}	Red	Fe^{2+}	Black violet
Sr^{2+}	Red violet	Fe^{3+}	Brown black

The metal ions first get attached to the fabric and then the dye molecules get coordinated to the metal forming *lakes* on the fabric. Thus, the mordant forms an insoluble complex compound (lake) between the fabric and the dye and binds the two. This is represented as under: (lake formation).

Aluminium mordant dyed object

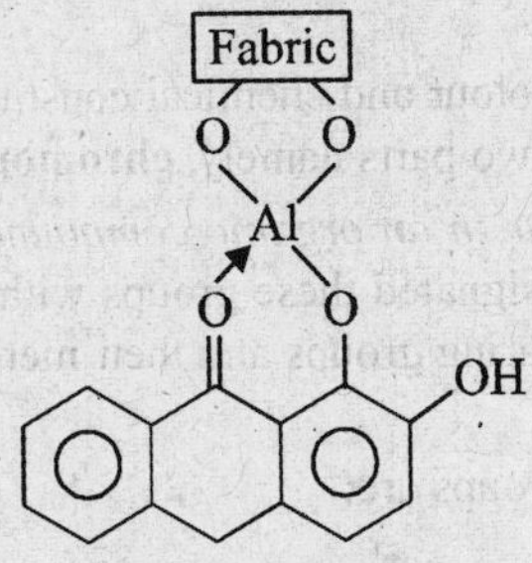

Lake formation

(*vi*) **Vat dyes.** These dyes are insoluble in water but their reduced forms are soluble. On reduction with alkaline sodium bisulphite, the vat dyes are converted into water soluble compounds called *leuco-compounds*. They dye both vegetable and animal fibres directly. The vat dyes are mostly used to dye cotton. Some typical examples of vat dyes are indigo and anthraquinone dyes.

10.3 THEORIES OF COLOUR AND CONSTITUTION

White light is the combination of light of all wavelengths in the visible range 400-750 nm. The reason why different substances show different colours is because they absorb and reflect different wavelengths from the white light that falls on them. The following generalisations can be made in respect of colours.

(*i*) A substance which totally reflects the white light appears white.

(*ii*) A substance which absorb all the wavelengths of white light appears black.

(*iii*) A substance which reflects a narrow band of wavelength of one colour and absorbs all the other wavelengths has the colour of the reflected light. Thus a substance appears blue because it absorbs all the wavelengths of visible light except a narrow band corresponding to blue (around 450 nm) which it reflects.

(*iv*) A substance which absorbs a single narrow band of one particular colour and reflects the remaining wavelengths has the colour due to the combination of remaining wavelengths. For example, if a substance absorbs in the wavelength region corresponding to blue and reflects the remaining wavelengths, it will appear yellow and *vice-versa*.

The complementary colours are related to each other as the colour absorbed and the colour observed. The relationship between the complementary colours is shown below:

Table 10.2: Complementary Colours

Wavelength (nm) of colour absorbed	*Colour absorbed*	*Complementary colour (i.e. colour observed)*
400–435	Violet	Yellowish green
435–480	Blue	Yellow
480–490	Greenish blue	Orange
490–500	Bluish green	Red
500–560	Green	Purple
560–580	Yellow green	Violet
580–595	Yellow	Blue
595–605	Orange	Greenish blue
605–750	Red	Bluish green

Witt's Theory

According to Witt's theory of colour and chemical constitution of dyes, a coloured substance or a dye is essentially composed of two parts namely, **chromophores** and **auxochromes**.

(1) Chromophores. *The colour in an organic compound is due to the presence of certain groups with multiple bonds.* Witt designated these groups with multiple bonds as **chromophores.** The chromophores are the colour bearing groups and their mere presence produces a colour in the molecule of an organic compound.

The important chromophoric groups are:

1. Nitro, 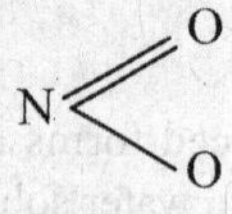   2. Nitroso, $-N=O$

3. Azo, $-N=N-$ 4. Azoxy, $-N=\underset{\substack{\downarrow \\ O}}{N}N-$

5. Carbonyl, 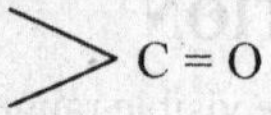   6. *p*-quinonoid, 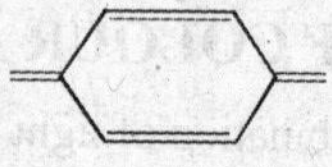

7. *o*-quinonoid, 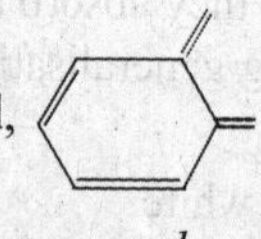

The organic compound containing a chromophoric group in its molecule is referred to as **chromogen.** The presence of any one of the above stated chromophoric groups in the molecule is sufficient to impart colour to an organic compound. For instance,

	Compound	Colour
1.	Benzene, C_6H_6	Colourless
2.	Nitrobenzene, $C_6H_5.NO_2$	Pale-yellow
3.	Azobenzene, $C_6H_5-N=N-C_6H_5$	Red
4.	*p*-Benzoquinone, $O=C_6H_4=O$	Yellow
5.	*o*-Benzoquinone, $O=C_6H_4=O$	Orange-red

It has been observed that the chromogen containing only one chromophoric group is usually coloured (yellow). The intensity of colour generally increases with number of chromophoric groups. A single $C=C$ group as in ethene $CH_2=CH_2$ does not produce any colour, but if a number of these groups are present in conjugation, the colour may develop. For instance, $CH_3-(CH=CH)_6-CH_3$ is yellow in colour.

In case of weaker chromophoric groups, more than one group is needed to develop a visible colour.

(2) **Auxochromes.** *Certain groups (which are not chromophores) when present in the chromogen tend to intensify its colour.* Such groups are referred to as **auxochromes.** The auxochromes may be acidic or basic in character.

The most effective auxochromes are as under:

1. Hydroxyl	$-OH$	5. Sulphonic acid	$-SO_3H$
2. Alkoxy	$-OR$	6. Carboxyl	$-COOH$
3. Amino	$-NH_2$	7. Phenolic	$-OH$

4. Alkylated amino – NHR or – NR_2

Auxochromes are salt-forming groups and perform two functions:

(*i*) *Auxochromes deepen the colour of chromogen.*

(*ii*) *The presence of auxochromes is essential to make the chromogen a dye.*

Further, the groups which deepen the colour are referred to as **bathochromic groups.** The groups which bring about the opposite effect are known as **hypsochromic groups.** The replacement of H in NH_2 group by R or Ar has bathochromic effect while replacement of H in OH group by acetyl group has a hypsochromic effect. Often a colourless chromogen becomes coloured when an auxochrome is introduced in the molecules. For example, benzophenone (colourless) becomes yellow when an amino (– NH_2) group is introduced into it. Nitroaniline is deeper in colour than nitrobenzene.

Quinonoid theory of colour and chemical constitution

According to this theory, all colouring substances may be represented by quinonoid structures (*o* – or – *p*). If a particular substance can be formulated in a quinonoid form, it is coloured otherwise it is colourless.

(*i*) According to the quinonoid theory, benzene is colourless while benzoquinone is coloured.

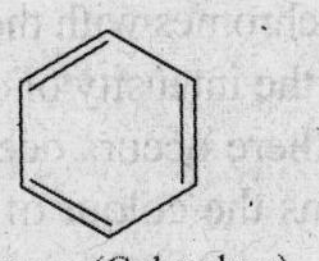

Benzene (Colourless)

Benzoquinone (Yellow)

(*ii*) On the basis of this theory, the dye like phenolphthalein is coloured when present in *p*-quinonoid structure but is colourless when *p*-quinonoid structure is absent.

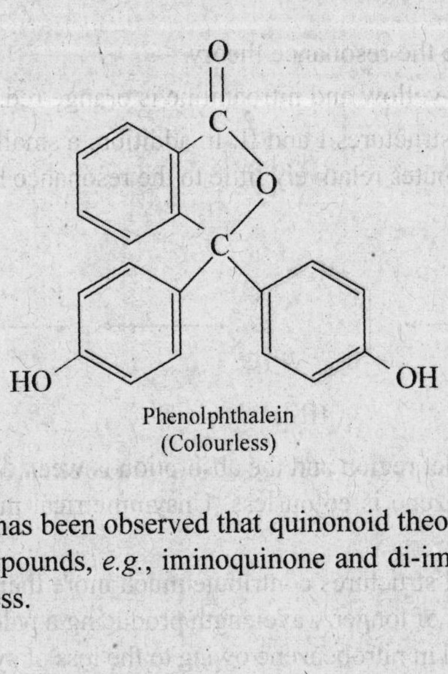

Phenolphthalein (Colourless)

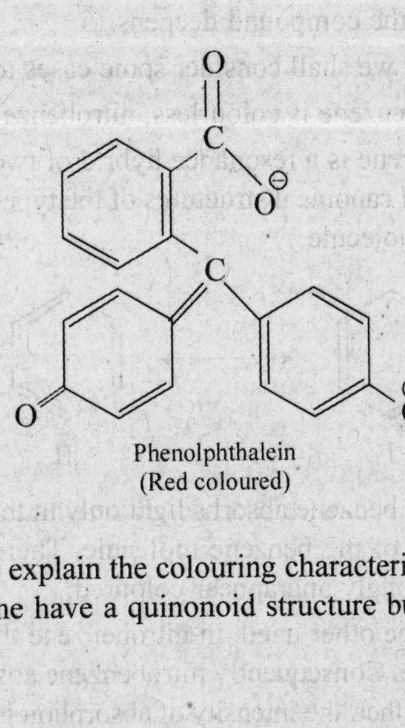

Phenolphthalein (Red coloured)

It has been observed that quinonoid theory fails to explain the colouring characteristics of all the compounds, *e.g.*, iminoquinone and di-iminoquinone have a quinonoid structure but they are colourless.

Iminoquinone Di-iminoquinone

Similarly, many coloured compounds like diacetyl, and azobenzene are coloured but they cannot be represented by quinonoid structures.

Valence bond theory of colour and constitution

This theory is also called **resonance theory.** The various important postulates of valence bond theory are as under:

(*a*) Chromophores are the groups of atoms in which the π-electrons may get transferred from ground state to excited state by the absorption of radiations. This absorption gives the colour.

(*b*) Auxochromes are the groups which tend to increase resonance by interaction of the unshared pair of electrons on nitrogen or oxygen atoms of the auxochromes with the π-electrons of the aromatic rings. This increase in resonance results in an increase of the intensity of absorption of light and shifts the absorption band to longer wavelength. As a result, there occurs deepening of the colour. It has been observed that increase in resonance always deepens the colour of a compound.

(*c*) The dipole moment changes as a consequence of oscillation of electron pairs. The ease of excitation of different groups follows the order:

$$N=O \; > \; C=S \; > \; N=N \; > \; C=O \; > \; C=C$$

(*d*) Valence bond theory explains the relation between colour and the symmetry of the molecule or transition dipole of the molecule. As the number of charged canonical structures increases, the colour of the compound deepens.

Now we shall consider some cases to illustrate the resonance theory.

(*i*) Benzene is colourless, nitrobenzene is pale-yellow and nitroaniline is orange red.

Benzene is a resonance hybrid of two Kekule structures I and II. In addition, a small number of charged canonical structures of the type III contributes relatively little to the resonance hybrid of benzene molecule.

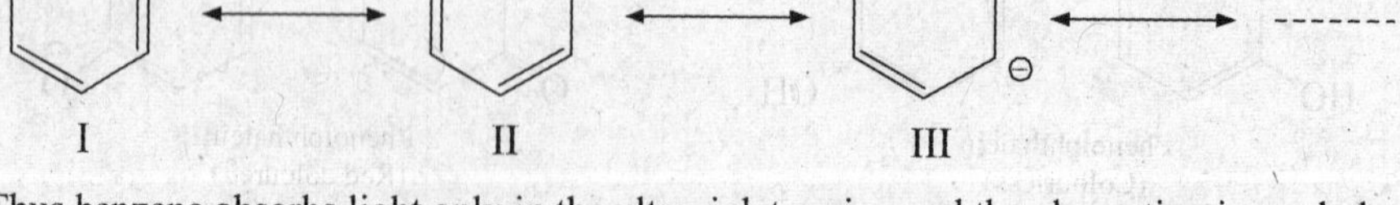

I II III

Thus benzene absorbs light only in the ultraviolet region and the absorption is weak due to the symmetry of the benzene molecule. Therefore, benzene is colourless. Unsymmetrical molecules absorb strongly and appear coloured.

On the other hand, in nitrobenzene the charged structures contribute much more than in case of benzene. Consequently nitrobenzene absorbs light of longer wavelength producing a pale yellow colour. Further, the intensity of absorption is increased in nitrobenzene owing to the loss of symmetry of the molecule.

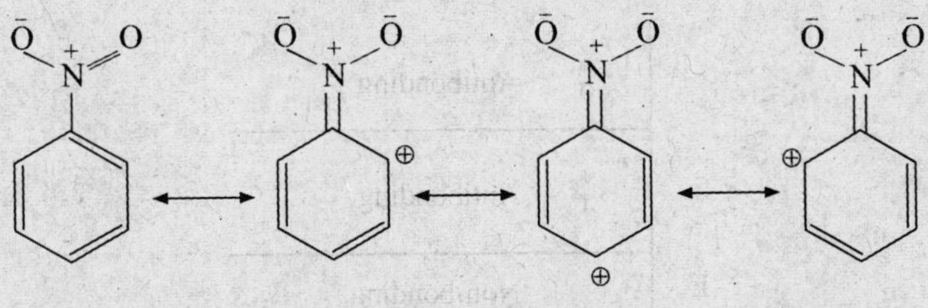

In a similar manner, in *p*-nitroaniline, the contribution of the charged structure is still larger and hence the light of longer wavelength is absorbed and further, the deepening of the colour to orange red is produced.

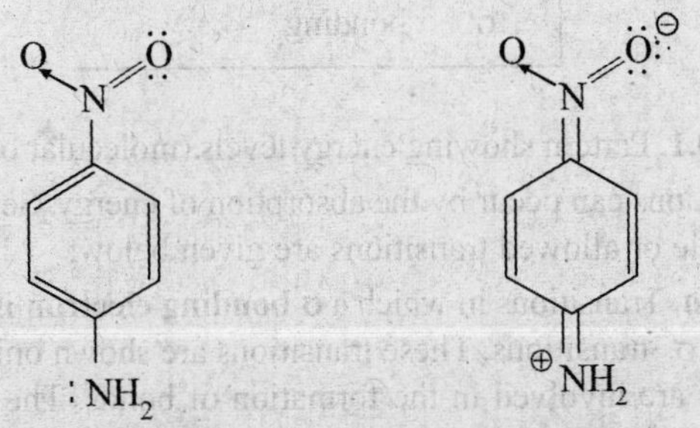

(*ii*) Valence bond or resonance theory also explains why pure *p*-nitrophenol is colourless but has yellow colour in alkaline solution. This is explained on the basis that in alkaline solution, *p*-nitrophenol exists as nitrophenoxide ion in which only the charged structures are contributing to the resonance hybrid and hence the compound absorbs light of higher wavelength.

(*iii*) The resonance theory also explains why *p*-aminobenzene is yellow but in acidic medium, it becomes violet. The deepening of colour in acidic solution is due to the fact that in the yellow form it is a resonance hybrid of the two structures out of which only one is charged while in the acidic solution (violet form) both the contributing structures are charged. The absorption band is shifted to longer wavelength and the colour gets deepened in the acidic solution.

(*iv*) As the conjugated system of double bonds provides a long path for resonance, it plays an extremely important role in producing deep colour. The longer the conjugation in a molecule, the deeper will be the colour.

Molecular orbital theory of colour and constitution

According to molecular orbital theory, the excitation of an atom or a molecule involves the transference of one electron from an orbital of lower energy (a bonding orbital) to that of higher energy (an antibonding orbital). The electrons involved may be σ (sigma), π (pi) or *n* (non-bonding) electrons. The higher energy states are usually known as *antibonding orbitals* while the lower energy states are called *bonding orbitals*. The antibonding orbitals associated with σ and π bond are represented by σ* and π* orbitals, respectively. There are, however, no antibonding orbitals associated with *n* (non-bonding) electrons as they do not form bonds. Energy levels of various orbitals are shown in Fig. 10.1

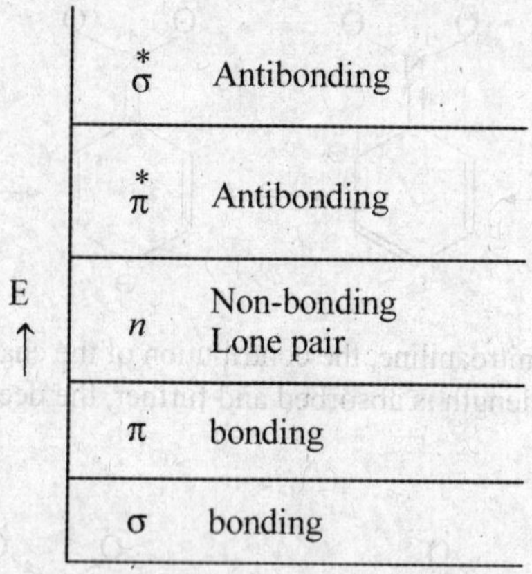

Fig. 10.1. Pattern showing energy levels (molecular orbitals)

The electronic transitions can occur by the absorption of energy. Several electronic transitions are possible but permissible or allowed transitions are given below:

(*i*) **$\sigma \rightarrow \sigma^*$ Transition.** Transitions in which a **σ bonding electron** is excited to an antibonding **σ^* orbital** are called $\sigma \rightarrow \sigma^*$ transitions. These transitions are shown only by compounds in which all valence shell electrons are involved in the formation of bonds. The energy required for these transitions is very high and consequently these occur at very short wavelengths. Saturated hydrocarbons show this type of transition.

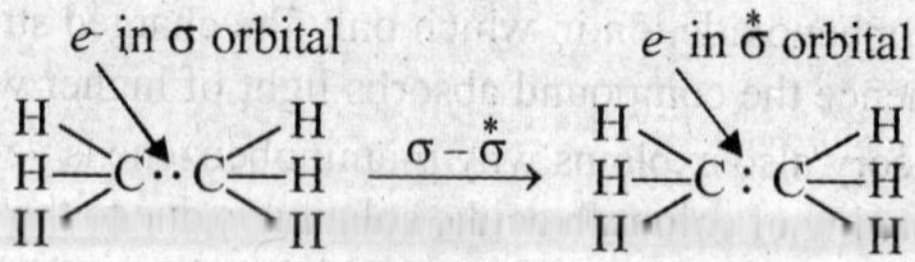

(*ii*) ***n* $\rightarrow \sigma^*$ Transitions.** These are the electronic excitations from a **non-bonding atomic orbital to an antibonding σ^* orbital.** Compounds having non-bonding electrons on hetero atoms such as nitrogen, sulphur or halogens can show $n \rightarrow \sigma^*$ transitions. These transitions are of lower energy and hence occur at longer wavelength than $\sigma \rightarrow \sigma^*$ transitions.

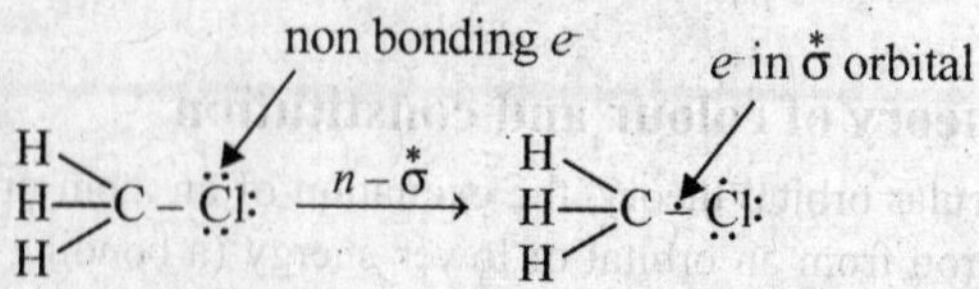

(*iii*) ***n* $\rightarrow \pi^*$ Transitions.** These are the transitions in which an electron in a **non-bonding** atomic **orbitals** is promoted to an **antibonding π* orbital.** Compounds having double bonds between hetero atoms, *e.g.*, C = O, C = S and N = O, undergo these transitions.

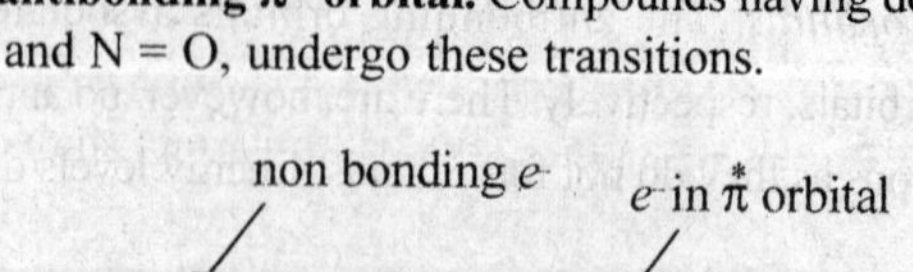

These transitions require small amounts of energy and take place at relatively longer wavelengths.

(*iv*) **$\pi \rightarrow \pi^*$ Transitions.** The transitions in which a **π electron** is excited to an **antibonding π^* orbital** are called $\pi \rightarrow \pi^*$ transitions. These have relatively higher energy requirement than $n \rightarrow \pi^*$ transitions and occur at short wavelengths for simple molecules. The $\pi - \pi^*$ transition for ethylene is shown as under :

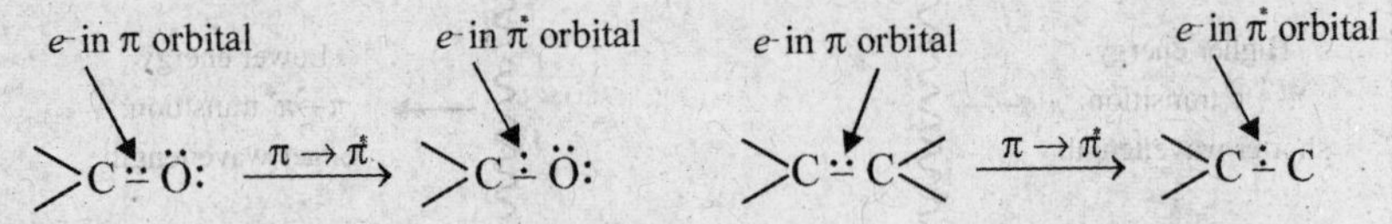

In general, the order of energy difference is :

$$\sigma \rightarrow \sigma^* > n \rightarrow \sigma^* > \pi \rightarrow \pi^* > n \rightarrow \pi^*$$

Energy changes occurring in various transitions follow the sequence

$$\sigma \rightarrow \overset{*}{\sigma} > n \rightarrow \overset{*}{\sigma} > \pi \rightarrow \overset{*}{\pi} > n \rightarrow \overset{*}{\pi}$$

We shall consider some examples of electronic transitions in organic compounds as under:

1. Alkanes contain C – H and C – C bonds and can show only $\sigma - \overset{*}{\sigma}$ transitions. The energy required to bring about this transition is very high and is available in the ultraviolet or far ultraviolet region. For example, ethane has a absorption maximum λ_{max} at 135 nm (C – C) which lies in the far ultraviolet region and hence appears colourless.

2. In ethylene, 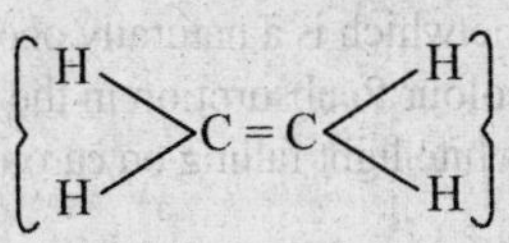 , two types of electronic transitions $\sigma \rightarrow \overset{*}{\sigma}$ and $\pi \rightarrow \overset{*}{\pi}$ are possible. Ethylene has λ_{max} 175 nm and this absorption band is due to a $\pi \rightarrow \overset{*}{\pi}$ transition in the far ultraviolet region. For this reason, ethylene is colourless. Comparatively the absorption band in ethylene is in the region of longer wavelength due to the presence of a double bond in it.

3. Butadiene shows λ_{max} 217 nm for a $\pi \rightarrow \overset{*}{\pi}$ transition. The conjugation in butadiene shifts the absorption band to a longer wavelength (217 nm) although still not into the visible region. This compound also appears colourless.

Effect of conjugation. Conjugation of double bonds lowers the energy required for the electronic transitions. As a result, molecule containing conjugated systems such as $>C = \overset{|}{C} - \overset{|}{C} = C<$ and $>C = \overset{|}{C} - \overset{|}{C} = O$ absorb the radiations of longer wavelengths than in case of non-conjugated systems. For instance 1, 3-butadiene ($CH_2 = CH - CH = CH_2$) has absorption wavelength 217 nm as compared to 175 nm of ethylene.

The double bond in ethylene consists of a sigma bond formed by the combination of sp^2 hybrid orbitals of two carbon atoms and a pi bond formed by the combination of unhybridised $2p$ orbitals. Due to the large difference in the energies of π and π^* orbitals, $\pi \rightarrow \pi^*$ transition takes place with the absorption of radiation at 175 nm (Fig 10.2)

In a conjugated system such as 1, 3-butadiene, there are two ethylene units in conjugation with each other. Due to conjugation, the π-bonding orbitals of the two ethylene units interact with each other to form two new bonding orbitals π_1 and π_2 of different energies. Similarly two π^* orbitals (*i.e.* pi antibonding orbitals), π_3^* and π_4^* of different energies from the two ethylene units are formed. The energy difference between highest occupied pi molecular orbital (HOMO) and the lowest unoccupied molecular orbital (LUMO) for 1, 3-butadiene is less than it is for ethylene. As the HOMO-LUMO energy gap becomes smaller, absorption takes place at longer wavelength.

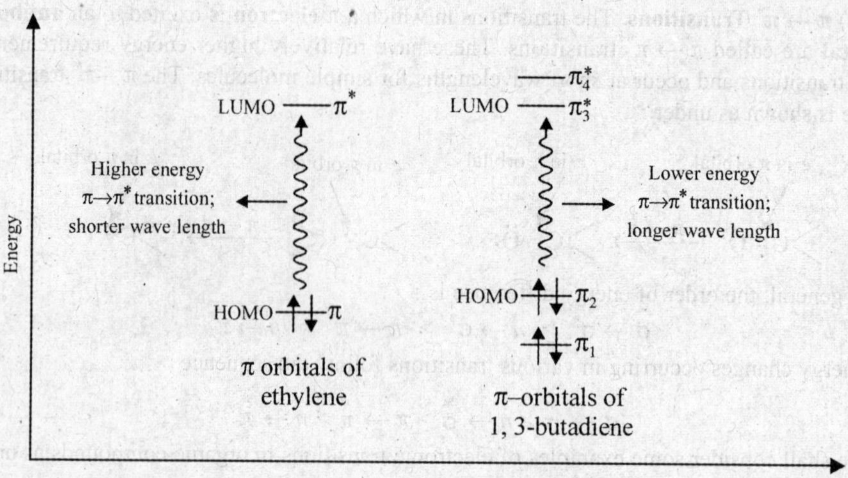

Fig. 10.2. Comparison of $\pi \rightarrow \pi^*$ transition energy on ethylene and 1,3-butadiene

With greater number of double bonds in conjugation, the energy difference between HOMO and LUMO further decreases so that absorption moves to still longer wavelengths. Thus in 1, 3, 5, 7-octatetraene ($CH_2 = CH - CH = CH - CH = CH - CH = CH_2$) with four double bonds in conjugation, absorption takes place at 290 nm. Carotene which is a naturally occuring yellow pigment containing 11 double bonds in conjugation, owes its colour to absorption in the blue region (455 nm) of visible light. As the blue light is removed from the white light falling on carotene, it appears yellow.

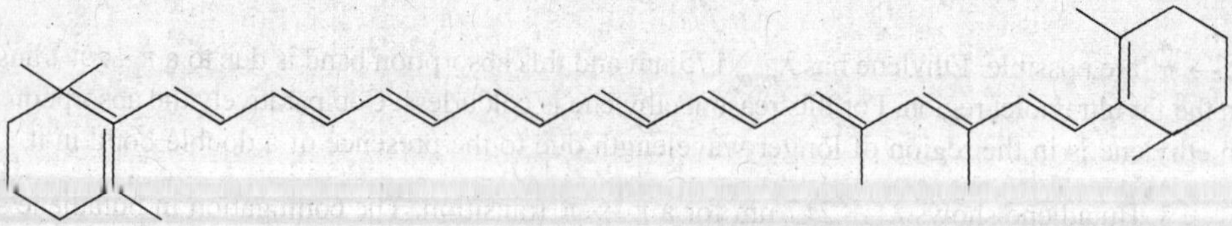

Extended system of conjugation in carotene :
(Absorbs in blue region, appear yellow)

The effect of conjugation in raising the wavelength of absorbed radiation is explained in the case of benzene ring compounds.

Benzene has a system of 6π electrons but its absorption takes place in the ultraviolet region (254 nm) so that it is colourless in nature. When a nitro group is introduced as a substituent, the resulting compound (*i.e.* nitrobenzene) has a pale yellow colour. This is because nitro group shifts the wavelength of absorption to longer region by extending the conjugation in the system.

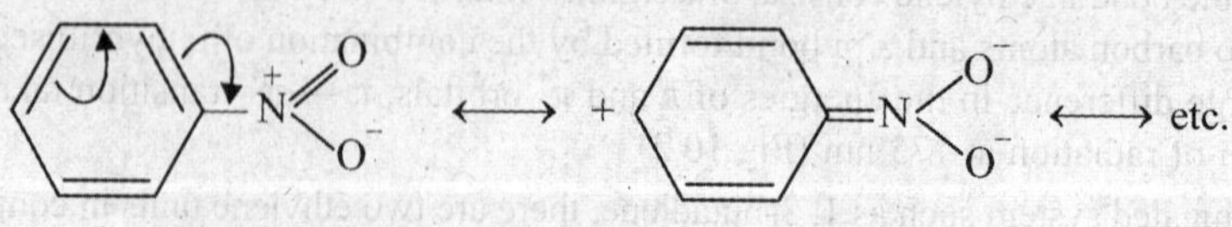

Azobenzene, 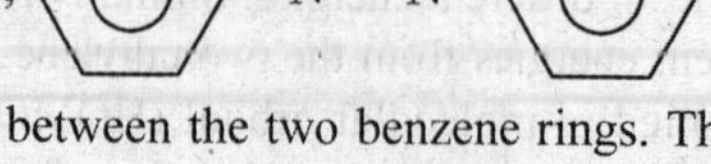 owes its yellow colour to the presence of azo group, –N = N–, between the two benzene rings. The azo group acts as a 'delocalisation' bridge between the two rings to form an extended delocalised system which acts as the chromophore.

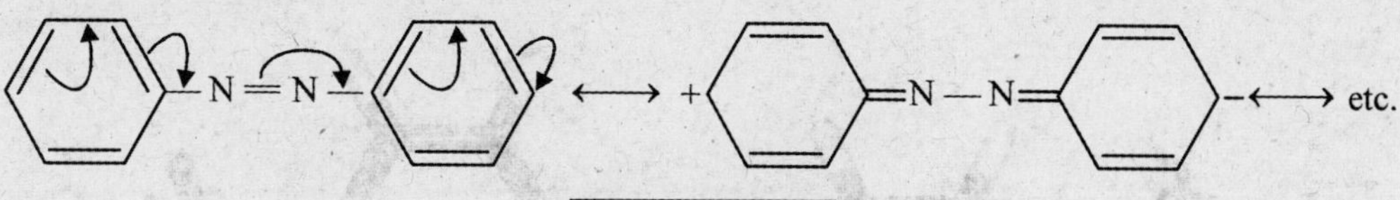

AZO DYES

In azo dyes, the chromophore is an aromatic system joined to the azo-group, and the common auxochromes are NH_2, NR_2, OH. They are generally prepared by direct coupling between a diazonium salt and a phenol or an amine.

Azo-dyes are classified as (*i*) Cationic dyes (*ii*) Anionic dyes

Cationic dyes

Chrysoidine is a cationic dye and is prepared by coupling benzenediazonium chloride with *m*-phenylenediamine. It is used for dyeing paper, leather, and jute.

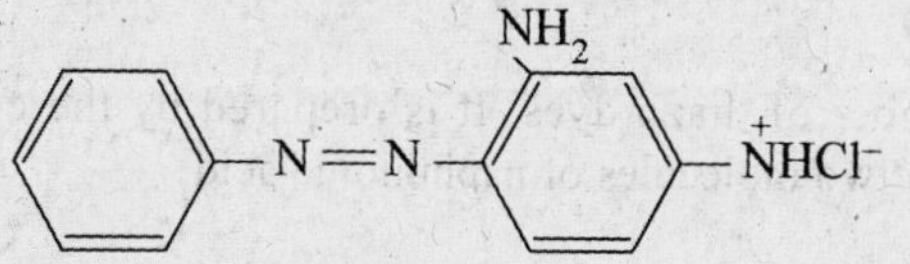

Chrysoidine

Anionic dyes

These dyes contain a sulphonic or a carboxylic group. They belong to the class of *soluble dyes*.

10.4. METHYL ORANGE (HELIANTHIN)

Methyl Orange (*Helianthin*), an anionic dye is prepared by coupling diazotised sulphanilic acid with dimethylaniline. It is used as an indicator, being orange in alkaline solution and red in acid solution. This colour change takes place in the pH range 3.1–4.5.

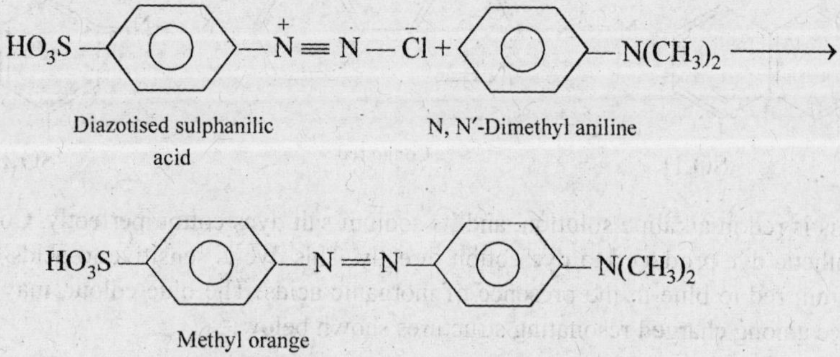

Diazotised sulphanilic acid N, N′-Dimethyl aniline

Methyl orange

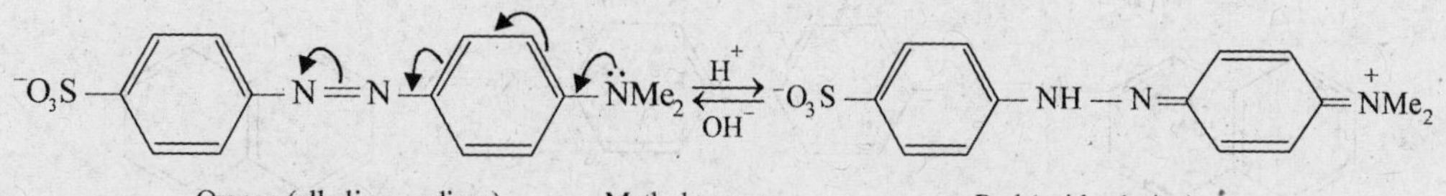

Orange (alkaline medium) Methyl orange Red (acid solution)

Methyl orange is not being used as a dye because it is not sufficiently stable to soap and light. However, it acts as a good acid-base indicator

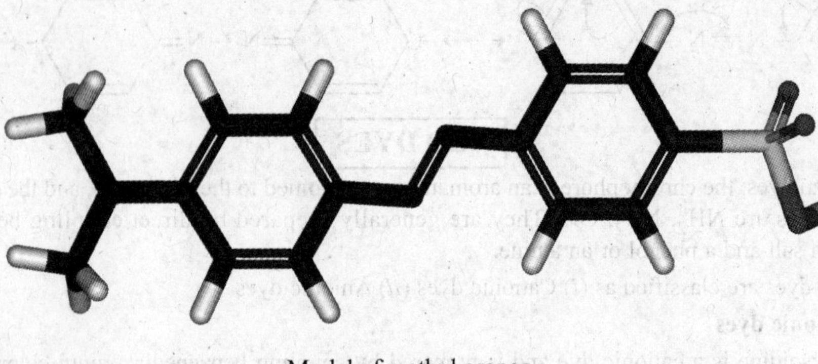

Model of methyl orange

10.5. CONGO RED

Congo red is a member of diazo dyes. It is prepared by the coupling reaction between tetrazotised benzidine with two molecules of naphthionic acid.

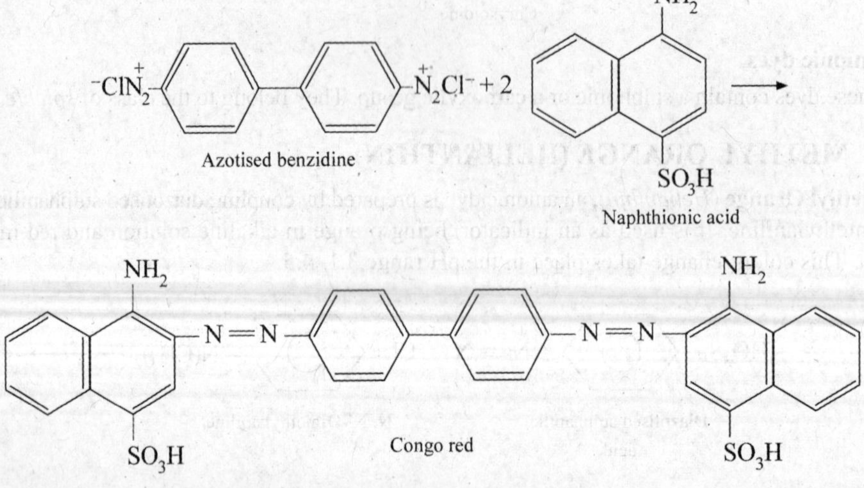

This is red in alkaline solution, and its sodium salt dyes cotton perfectly. Congo red was the first synthetic dye produced to dye cotton directly. This dye is sensitive to acids and changes the colour from red to blue in the presence of inorganic acids. The blue colour, may be attributed to resonance among charged resonating structures shown below:

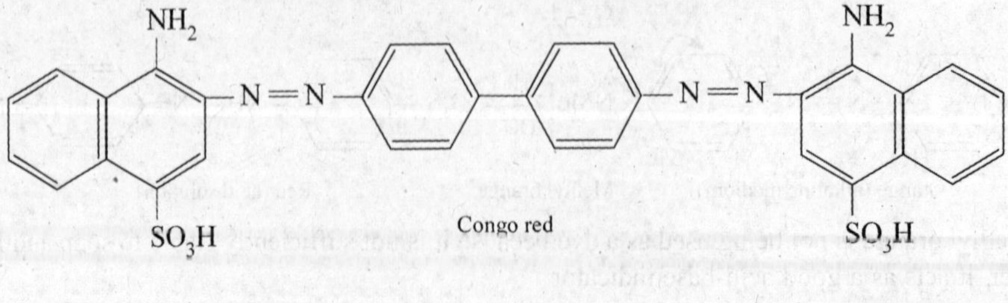

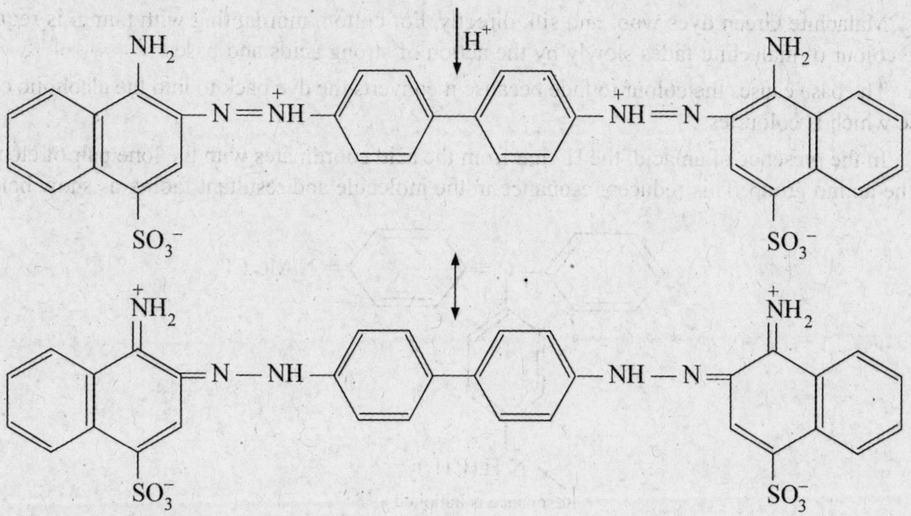

10.6. MALACHITE GREEN

Compounds obtained by introducing NH_2, NR_2 or OH groups into the rings are called **leuco-compounds**. These are colourless compounds but on oxidation, are converted into the corresponding tertiary alcohols, the **colour bases,** which readily change from the colourless benzenoid forms to the quinonoid dyes in the presence of acid, due to salt formation. The sequence of changes is given below:

$$\underset{\text{(colourless)}}{\text{leuco-base}} \underset{\text{reduction}}{\overset{\text{oxidation}}{\rightleftharpoons}} \underset{\text{(colourless)}}{\text{colour base}} \underset{\text{alkali}}{\overset{\text{acid}}{\rightleftharpoons}} \underset{\text{(coloured)}}{\text{dye}}$$

Malachite Green is prepared by condensing dimethylaniline (2 molecules) with benzaldehyde (1 molecule) at 100°C in the presence of concentrated sulphuric acid. The leuco-base produced is oxidised with lead dioxide to produce Malachite Green with excess of hydrochloric acid:

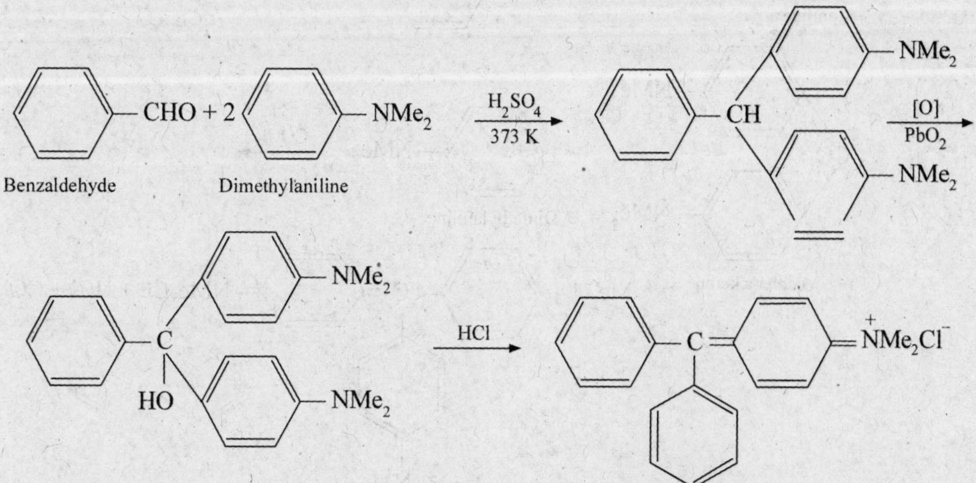

Malachite Green

Malachite Green dyes wool and silk directly. For cotton, mordanting with tannin is required. The colour of malachite fades slowly by the action of strong acids and bases.

The base causes the colour to fade because it converts the dye back to into the alcoholic colour base which is colourless.

In the presence of an acid, the H^+ ion from the acid coordinates with the lone pair of electrons of the amino group. This reduces resonance in the molecule and resultant fading as show below :

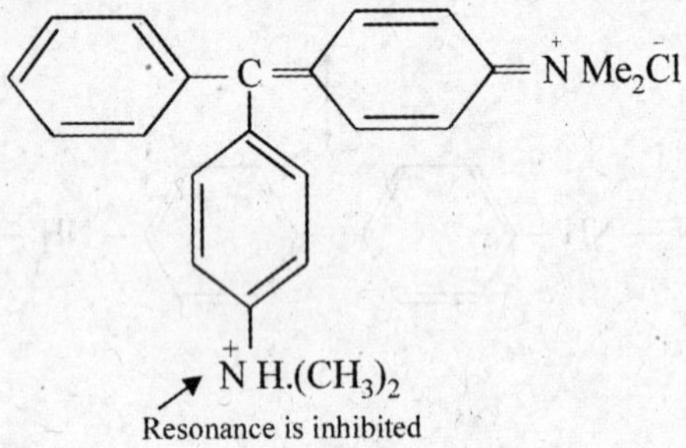

10.7. CRYSTAL VIOLET

Crystal violet may be prepared by heating *Michler's ketone* with dimethylaniline in the presence of phosphoryl chloride or carbonyl chloride. It belongs to the family of triphenylmethane dyes. It can be used as a direct dye for animal fibres like silk and wool.

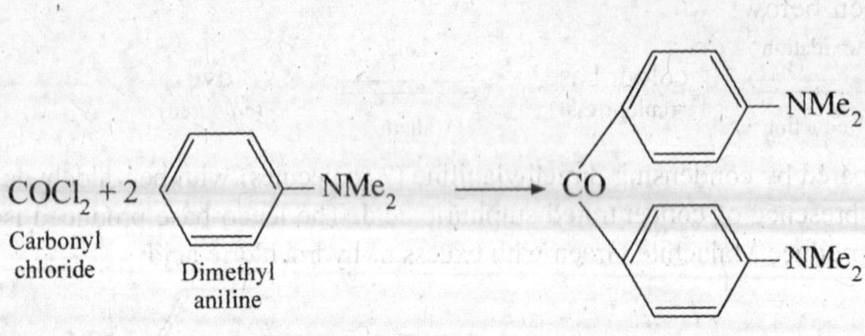

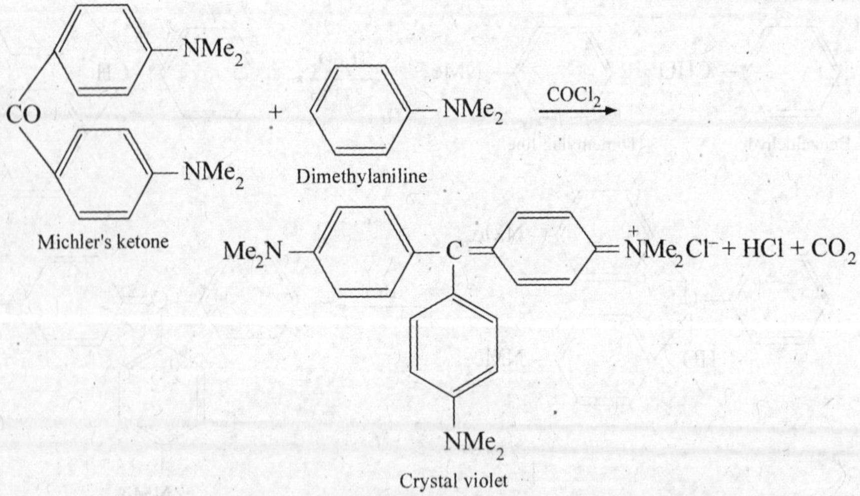

A weakly acid solution of crystal violet gives a purple colour. The colour changes with the acid strength of the solution. Thus the colour changes to green in strongly acidic medium, whereas a yellow colour is obtained in still stronger acidic medium.

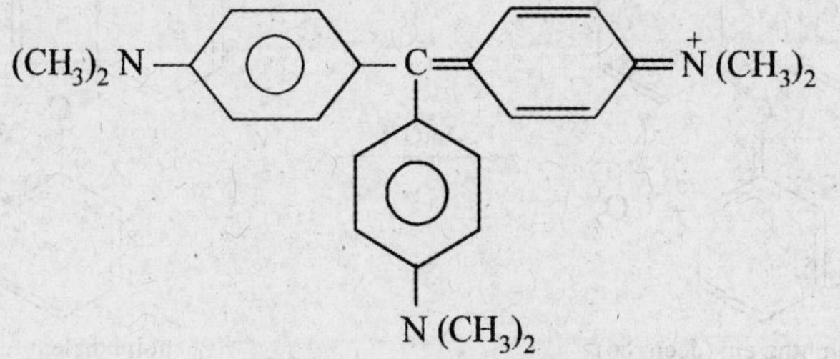

Weakly acidic solution (purple)

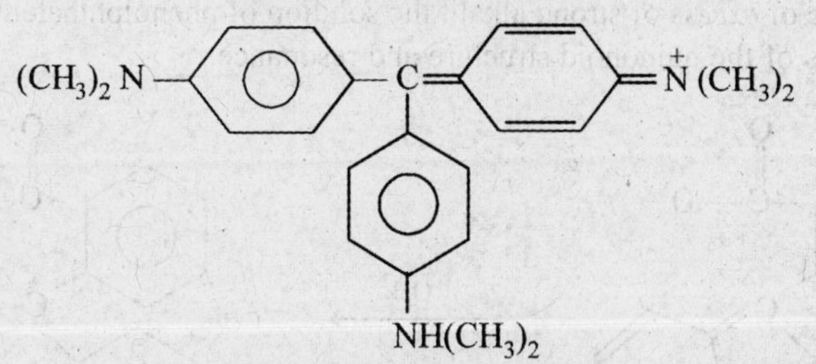

Strongly acidic solution (green)

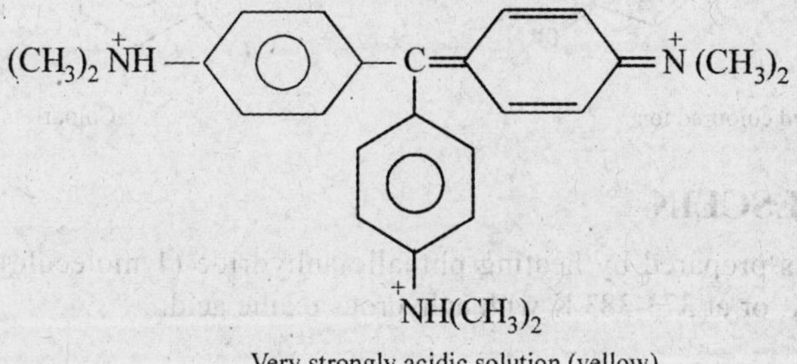

Very strongly acidic solution (yellow)

10.8. PHENOLPHTHALEIN

(*i*) **Phenolphthalein** is prepared by heating phthalic anhydride (1 molecule) with phenol (2 molecule) in the presence of concentrated sulphuric acid. It belongs to the phthalein group of dyes. It is primarily used as an indicator rather than a dye.

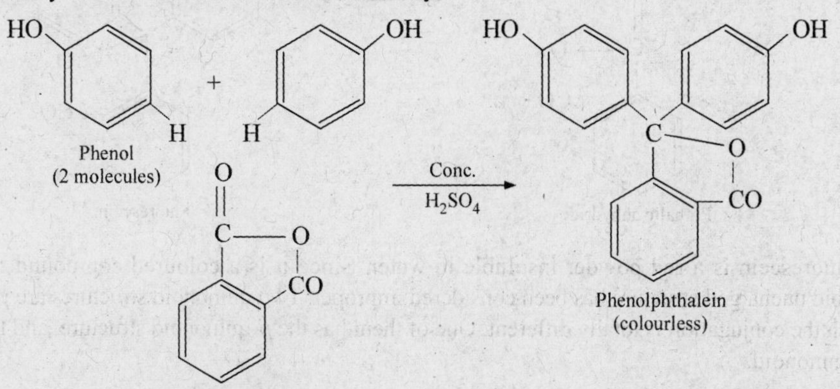

It is a white crystalline solid, insoluble in water, but soluble in alkalis to form deep red solutions:

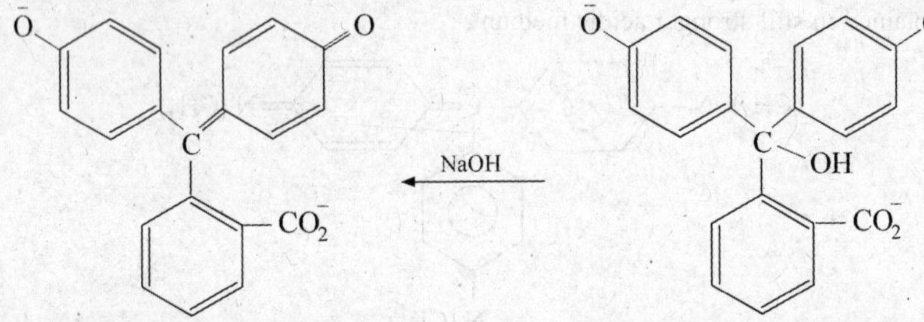

Phenolphthalein (deep red) Phenolphthalein (colourless)

In the presence of *excess* of strong alkali, the solution of phenolphthalein becomes colourless again due to the loss of the quinonoid structure and resonance.

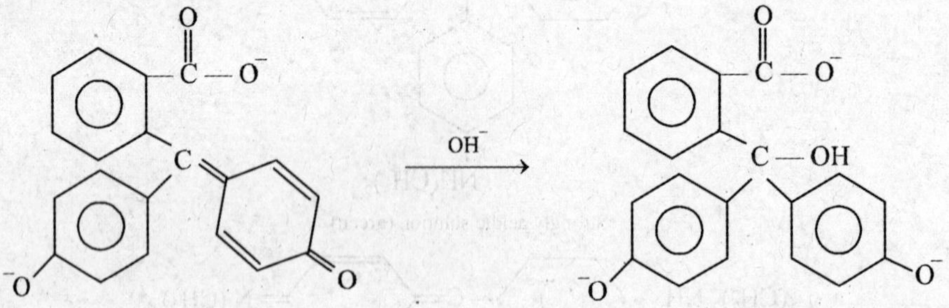

Red coloured ion Colourless ion

10.9. FLUORESCEIN

Fluorescein is prepared by heating phthalic anhydride (1 molecule) with resorcinol (2 molecules) at 473 K, or at 373-383 K with anhydrous oxalic acid:

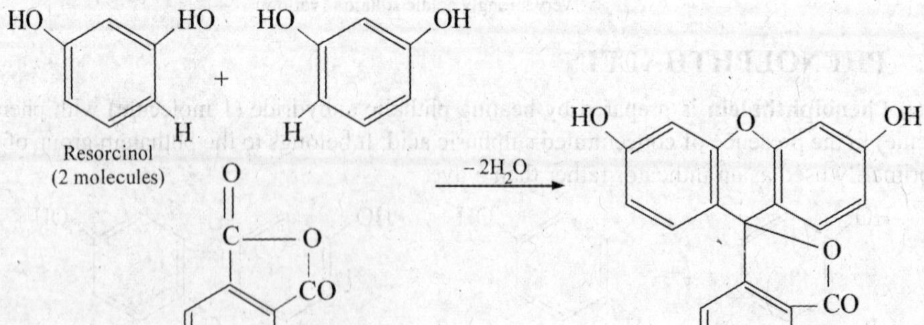

Phthalic anhydride Fluorescein

Fluorescein is a red powder insoluble in water. Since it is a coloured compound the non-quinonoid uncharged structure has been considered improper. Two quinonoid structures are proposed in which the conjugation is totally different. One of them has the *p*-quinonoid structure, and the other the *o*-quinonoid.

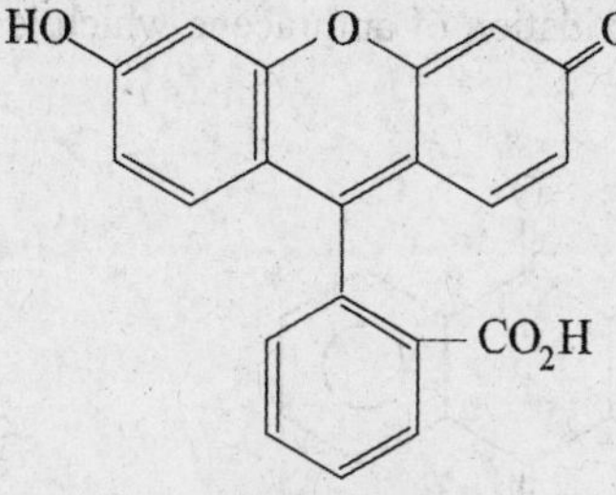

p-quinonoid structure of fluorescein

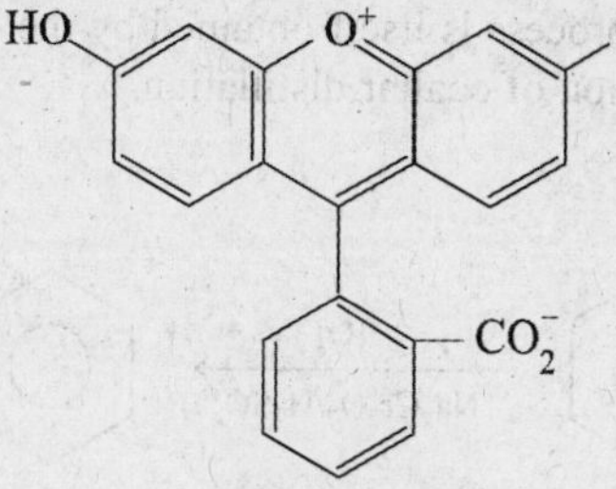

o-quinonoid structure of fluorescein

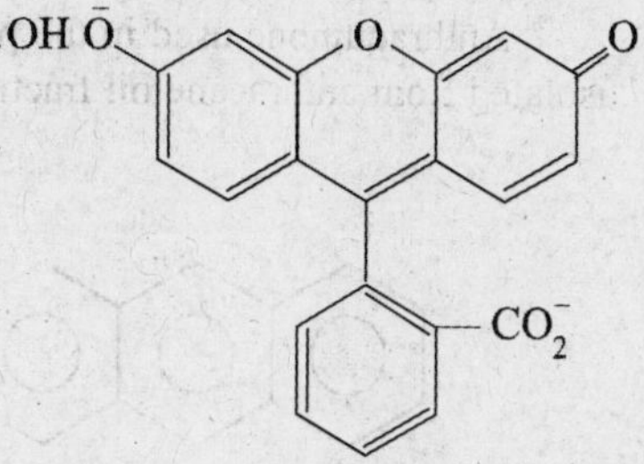

Structure of fluorescein anion

Fluorescein dissolves in alkalis to give a reddish-brown solution which gives a strong yellowish-green fluorescence on dilution. The structure of the fluorescein anion is given above. The sodium salt of fluorescein is known as **Uranine**, and is used to dye wool and silk yellow from an acid bath.

10.10 ALIZARIN

Synthesis of Alizarin

Alizarin is one of the most important anthraquinoid dyes. It is a dye of splendid colour. Alizarin is the chief constituent of the madder root wherein it is present as the glucoside, *ruberythric acid.* It can be prepared by the following methods:

1. **From madder root.** Alizarin is extracted from the root of the madder plant. It contains alizarin as glucoside, ruberythric acid, which forms alizarin on hydrolysis in the presence of enzyme or dilute acid.

$$\underset{\text{Ruberythric acid}}{C_{26}H_{28}O_{14}} + 2H_2O \xrightarrow[\text{(or Enzyme)}]{H^+} \underset{\text{Glucose}}{2C_6H_{12}O_6} + \underset{\text{Alizarin}}{C_{14}H_6O_2(OH)_2}$$

2. **From anthraquinone.** Anthraquinone on sulphonation with fuming H_2SO_4 at 140°C yields anthraquinone-2-sulphonic acid which is converted into its sodium salt by treatment with NaOH. This sodium salt on fusion with NaOH in the presence of $KClO_3$ at 200°C under pressure produces alizarin (1, 2, dihydroxyquinone).

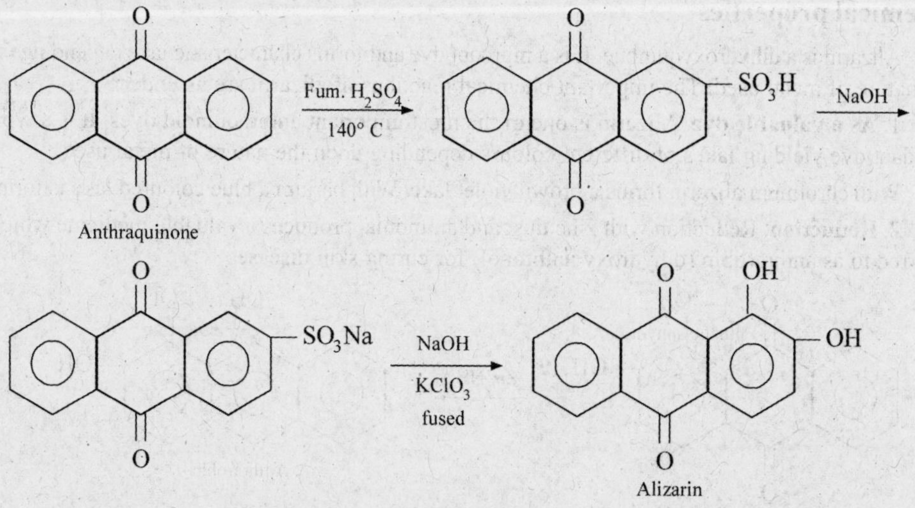

Anthraquinone used in the process is itself obtained by the oxidation of anthracene which is isolated from anthracene oil fraction of coaltar distillation.

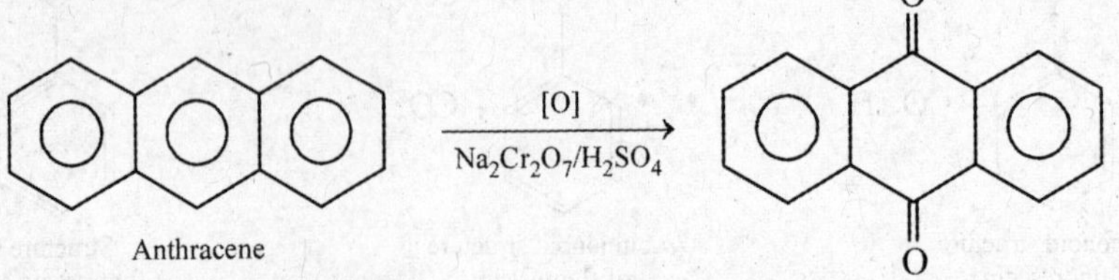

3. From Phthalic anhydride. Phthalic anhydride on condensation with catechol in the presence of H_2SO_4 at 180°C gives rise to the formation of alizarin.

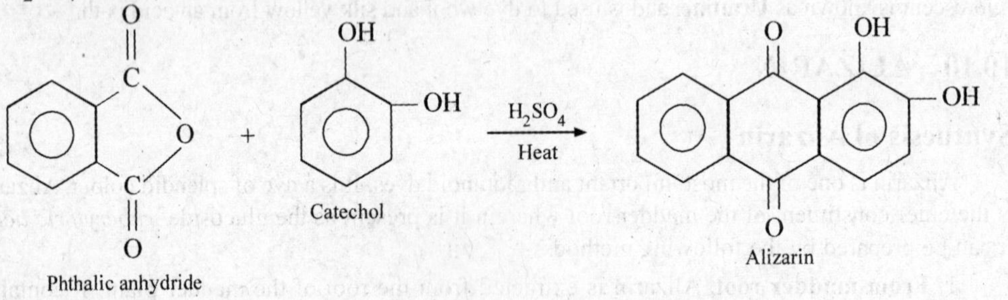

Properties of Alizarin

Physical properties

1. It melts at 290°C.
2. It sublimes on heating.
3. Alizarin forms ruby red crystals.
4. Alizarin dissolves in caustic alkalis to give a purple coloured solution.
5. It is insoluble in water and sparingly soluble in ethanol.

Chemical properties

Alizarin is a dihydroxyquinone. It is a mordant dye and forms characteristic lake depending upon the nature of metal used. The important chemical reactions of alizarin are as under:

1. **As a valuable dye**. Alizarin is one of the most important anthraquinoid dyes. It is a typical mordant dye yielding lakes of different colours depending upon the nature of metal used.

With chromium alizarin forms a brown violet lake. With barium a blue coloured lake is formed.

2. **Reduction.** Reduction with zinc dust and ammonia, produces a valuable medicine which is referred to as anthrarobin (dihydroxy anthranol) for curing skin diseases

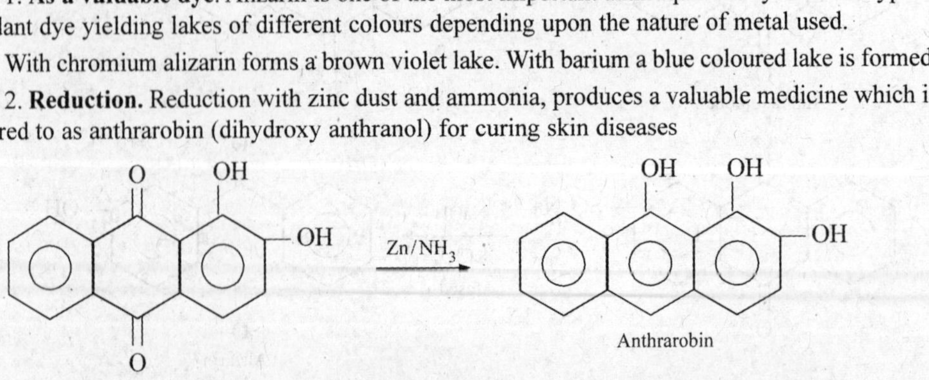

3. **Oxidation.** On mild oxidation with MnO_2 and H_2SO_4, alizarin gives rise to the formation of a dye having an additional hydroxyl group, referred to as *purpurin*.

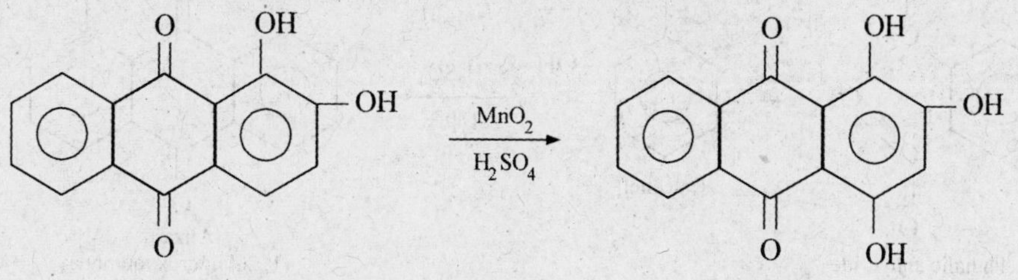

Purpurin

Vigorous oxidation of alization yields phthalic acid.

4. **Action of alkalis.** Alizarin dissolves in caustic alkalis forming purple solution of alizarate.

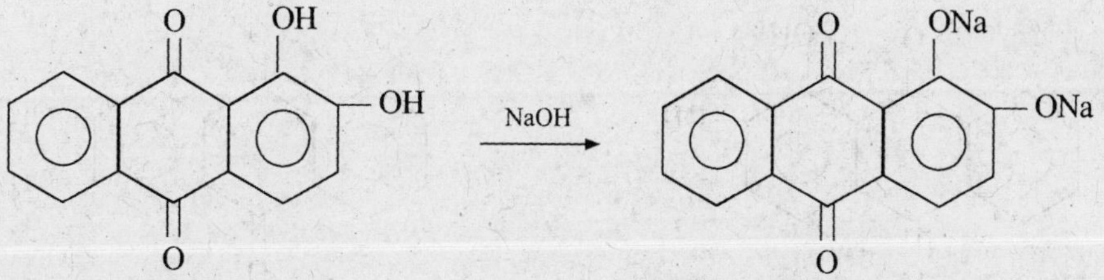

Alizarin Sodium alizarate

Uses. Alizarin is used:

1. As a mordant dye.
2. In the preparation of anthrarobin.
3. In the manufacture of printing inks.
4. As a purgative in medicine.

Structure (constitution) of alizarin

Structure of alizarin is established as under :

1. Molecular formula of alizarin. The elemental analysis and molecular weight determination show that the molecular formula of alizarin is $C_{14}H_6O_4$.

2. Carbon skeleton of alizarin. On distillation with zinc dust, alizarin is converted into anthracene. This indicates that alizarin has the same carbon skeleton as that of anthracene. Thus, it has been derived from anthracene.

3. Presence of two –OH groups. (*a*) Alizarin on treatment with acetic anhydride forms a diacetate showing the presence of two –OH groups in the molecule of alizarin.

(*b*) Anthraquinone on heating with a calculated amount of bromine yields dibromoanthraquinone which on fusion with KOH gives rise to the formation of *alizarin*. This indicates that alizarin is dihydroxy-anthraquinone.

4. Possible positions of two –OH groups. (*a*) Alizarin on vigorous oxidation gives rise to the formation of phthalic acid. This indicates that the two OH groups are in the same ring – the ring which is destroyed during the vigorous oxidation, otherwise either hydroxyphthalic acid would have been formed or the entire molecule would have been broken down completely.

5. Possible formulae of alizarin. Alizarin is obtained by condensing phthalic anhydride and catechol in the presence of H_2SO_4 at 180°C in the same way as anthraquinone is formed by heating phthalic anhydride with benzene.

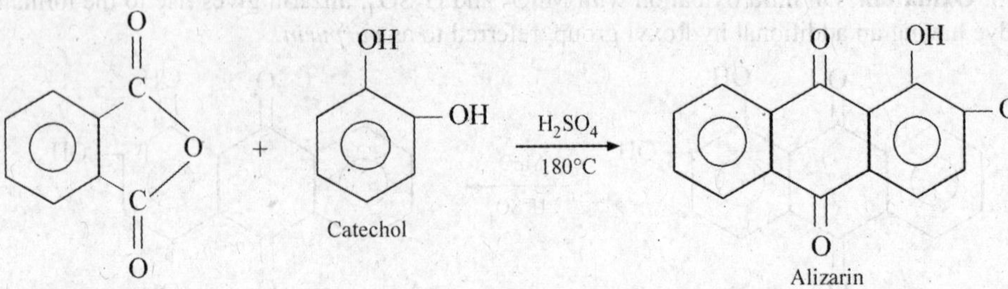

Phthalic anhydride   Catechol   Alizarin (1, 2 dihydroxyquinone)

This shows that the two –OH groups in alizarin must be in the same ring and in the ortho-position to each other (*i.e.* 1 : 2 or 2 : 3 positions).

Keeping in view the above, the following two structural formulae for alizarin are possible.

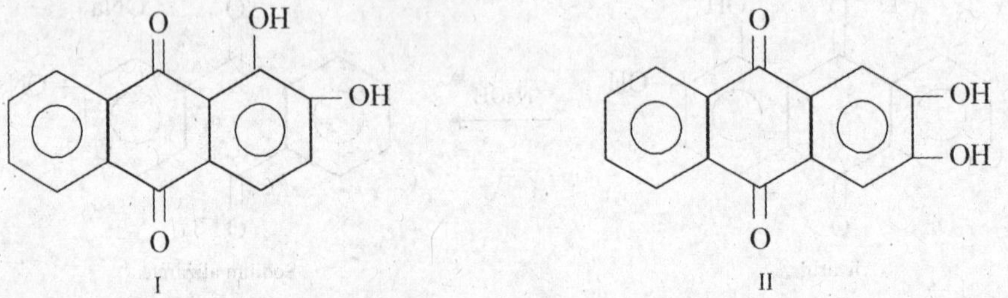

I   II

6. Confirmation of structural formula. (*a*) On nitration, alizarin gives rise to the formation of two isomeric mononitro derivatives having –OH and –NO_2 groups in the same ring. Further, both these derivatives on oxidation yield phthalic acid. This indicates that in each of these derivatives, the –NO_2 group is inducted on the benzene ring containing –OH groups.

The possible mononitro derivatives from structure I are as under:

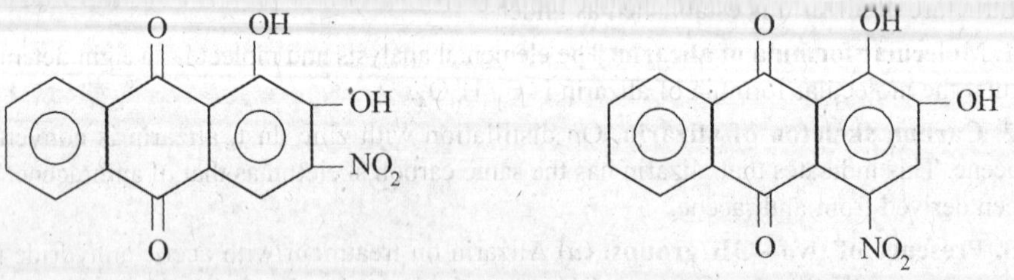

Different

The possible mono-nitro derivatives from structure II are as under:

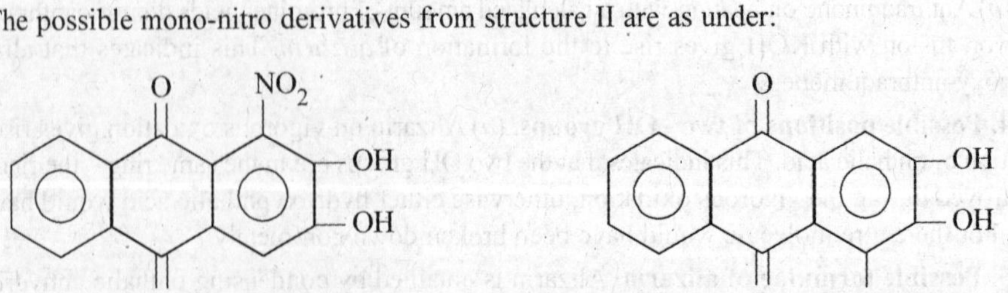

Identical

Only structure I permits the formation of two isomeric mononitro-derivatives and hence, structure I gives the true representation of alizarin.

(*b*) Final confirmation of structure is provided by the synthesis of alizarin from 1 : 2 dibromoanthraquinone. Alizarin is formed by fusing 1 : 2 dibromoanthraquinone with NaOH.

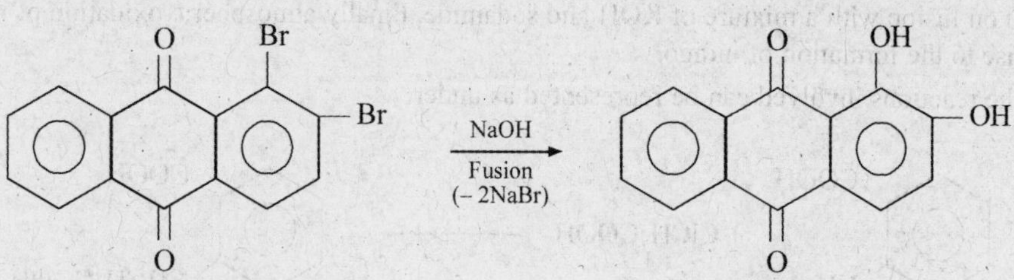

(1, 2, Dihydroxyanthraquinone) Alizarin (1, 2, Dihydroxyanthraquinone)

10.11 INDIGO

Natural indigo from the plants of *indigofera* group is the oldest known dye. It is present in the plants as β-glucoside of indoxyl– known as indican. On acidic or enzymatic hydrolysis indican gives glucose and indoxyl. Indoxyl on oxidation gives indigo

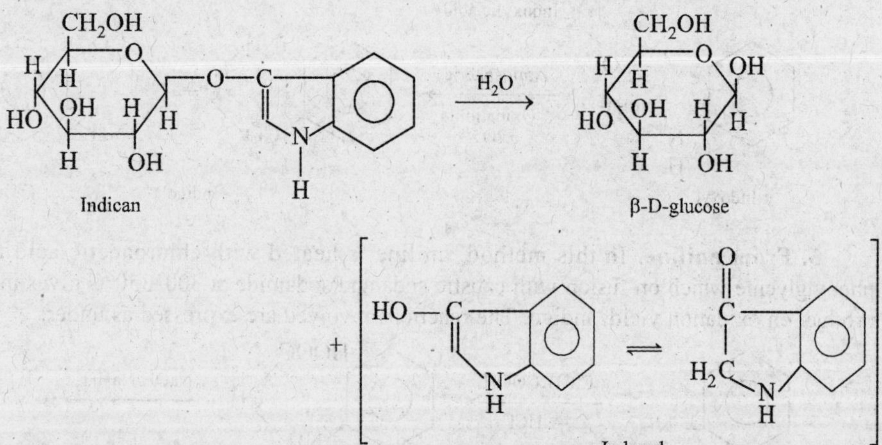

Indican β-D-glucose Indoxyl

Synthesis of Indigo

Indigo is prepared by the following methods:

1. From isatin chloride. Isatin chloride on reduction with zinc dust in glacial acetic acid yields indoxyl which on oxidation in air forms indigo.

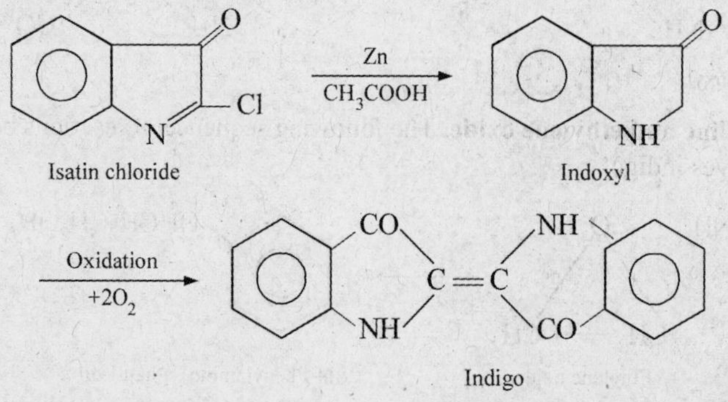

Isatin chloride Indoxyl Indigo

2. From anthranilic acid. In this method, anthranilic acid is treated with chloroacetic acid to give phenyl glycine-*o*-carboxylic acid which undergoes ring closure and decarboxylation to give indoxyl on fusion with a mixture of KOH and sodamide. Finally atmospheric oxidation of indoxyl gives rise to the formation of indigo.

The reactions involved can be represented as under:

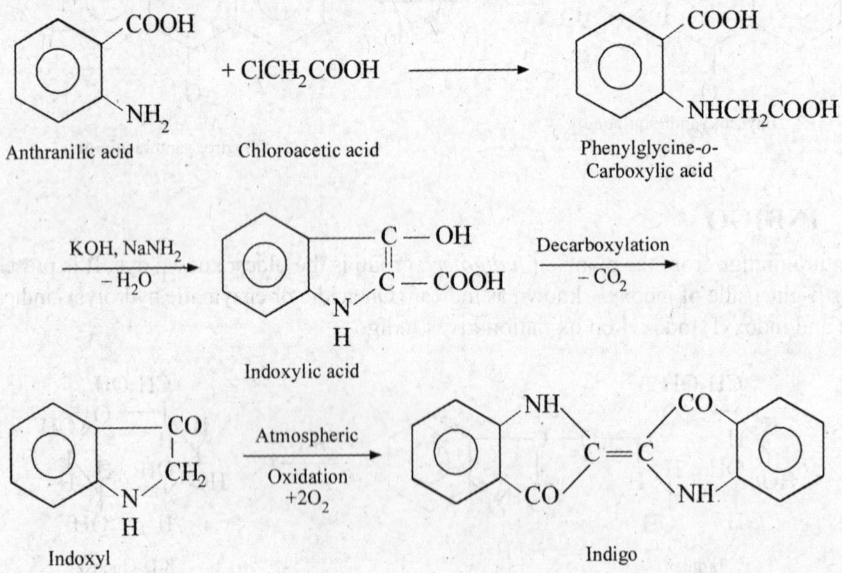

3. From aniline. In this method, aniline is heated with chloroacetic acid to form N-phenylglycine which on fusion with caustic soda and sodamide at 300–350°C gives indoxyl. This product on oxidation yields indigo. The reactions involved are expressed as under:

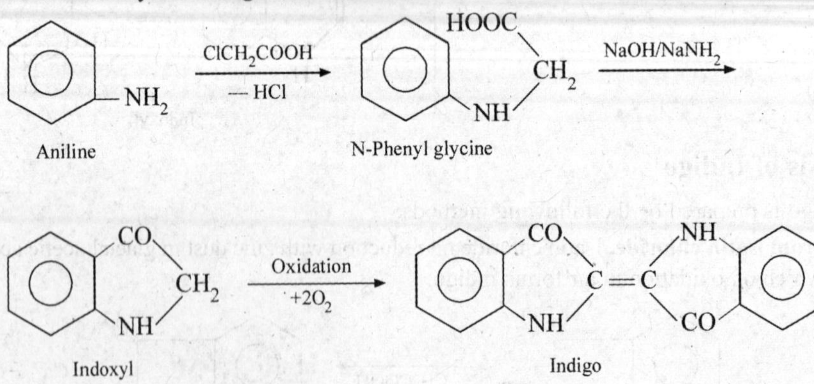

4. From aniline and ethylene oxide. The following sequence of reactions between aniline and ethylene oxide gives indigo

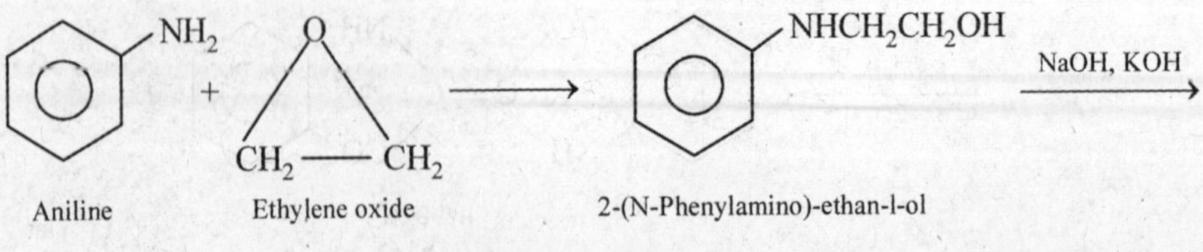

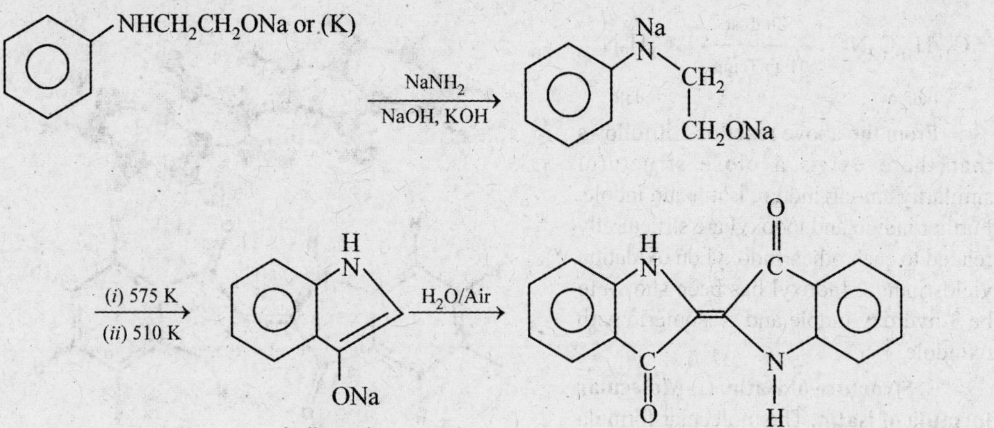

Sodium salt of indoxyl

Indigo

For dyeing a fabric with indigo, a paste of the dye is stirred with a reducing agent such as sodium hydrosulphite ($Na_2S_2O_4$) in large vats. As a result, indigo is reduced to a soluble leucobase *indigo white.*

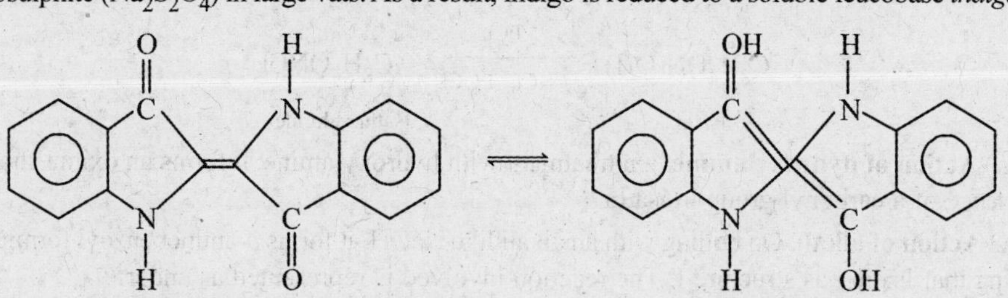

Indigo (blue)

Indigo white (leucobase)

The fabric to be dyed is soaked in the solution of the leucobase and then exposed to air when the leucobase is oxidised back to the original blue dye which gets bound to the fabric and is extremely stable to washing.

Properties. Indigo is a dark blue solid. Its melting point is 390–392°C. It is insoluble in water. It dissolves in aniline, nitrobenzene and chloroform. It sublimes under reduced pressure to give deep red vapours.

Structure of indigo or indigotin

Structure of indigo is established as under :

1. Empirical and molecular formulae. From the elemental analysis, the empirical formula of indigo is found to be C_8H_5ON. Vapour density determination reveals that its molecular formula is $C_{16}H_{10}O_2N_2$.

2. Degradation of indigo. (*i*) Vigorous oxidation of indigo with HNO_3 forms two molecules of isatin, showing the presence of two similar units joined by a double bond (point of attack). Each unit on oxidation yields one molecule of isatin.

$$\underset{\text{Indigo}}{C_{16}H_{10}O_2N_2} \xrightarrow{2(O)} 2C_8H_5O_2N$$

(*ii*) On distillation with zinc dust at high temperature, indigo gives a new product known as indole C_8H_7N.

$$\underset{\text{Indigo}}{C_{16}H_{10}O_2N_2} \xrightarrow[\text{High Temp.}]{\text{Zn dust}} \underset{\text{Indole}}{C_8H_7N}$$

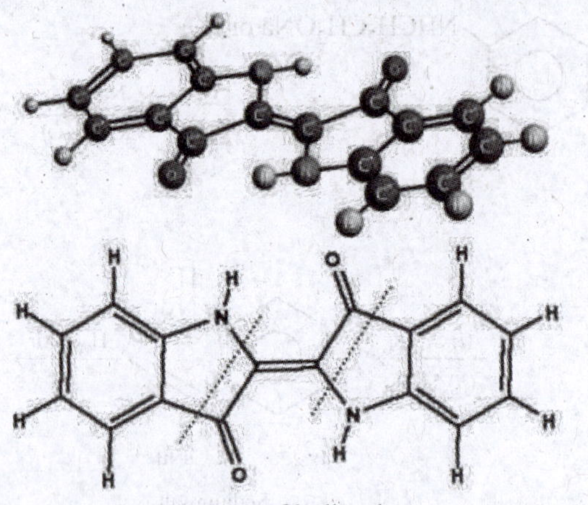

Structure of indigotin

From the above reactions, it follows that there exists a close structural similarity among indigo, isatin and indole. Further indigo and indoxyl are structurally related to each other. Indoxyl on oxidation yields indigo. Indoxyl has been shown to be 3-hydroxy indole and is isomeric with oxindole.

3. Structure of isatin. (*i*) **Molecular formula of isatin.** The molecular formula of isatin has been found to be $C_8H_5O_2N$.

(*ii*) **Action of phosphorus pentachloride.** On treatment with PCl_5, it forms isatin chloride indicating the presence of a hydroxyl group in isatin.

$$\underset{\text{Isatin}}{C_8H_4ON(OH)} \xrightarrow[-HCl,\ -POCl_3]{PCl_5} \underset{\text{Isatin chloride}}{C_8H_4ONCl}$$

(*iii*) **Action of hydroxylamine.** On treatment with hydroxylamine, it forms an oxime showing the presence of a carbonyl group in isatin.

(*iv*) **Action of alkali.** On boiling with an alkali like NaOH, it forms *o*-aminobenzoyl formic acid indicating that Isatin has structure I. The reaction involved is represented as under:

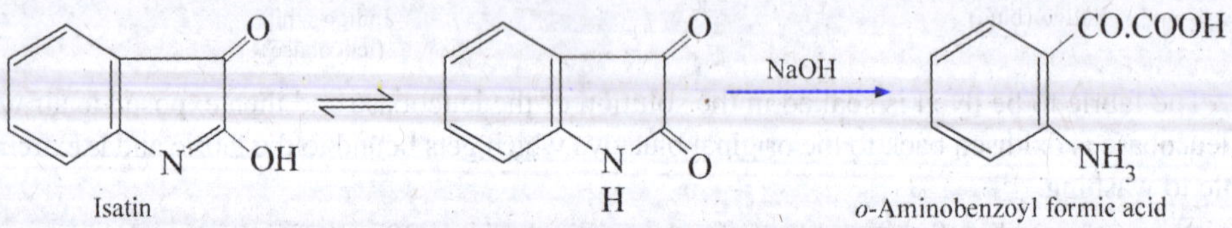

Isatin (I) *o*-Aminobenzoyl formic acid

(*v*) **Synthesis of isatin.** The structure (I) for isatin has further been confirmed by its synthesis from o-Nitrobenzoyl chloride. The reactions involved in the synthesis of isatin are as under:

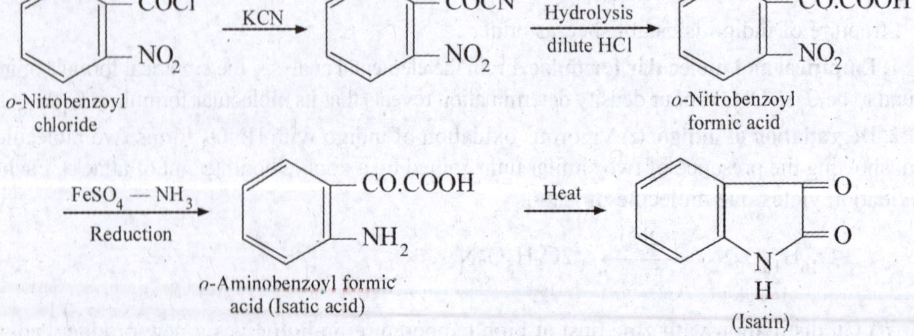

o-Nitrobenzoyl chloride; *o*-Nitrobenzoyl formic acid; *o*-Aminobenzoyl formic acid (Isatic acid); (Isatin)

4. Structure of indigo. On the basis of the above structure of isatin, indigo may possess either of the following structures (*i.e.* II, III and IV). All of them when oxidised yield two molecules of isatin.

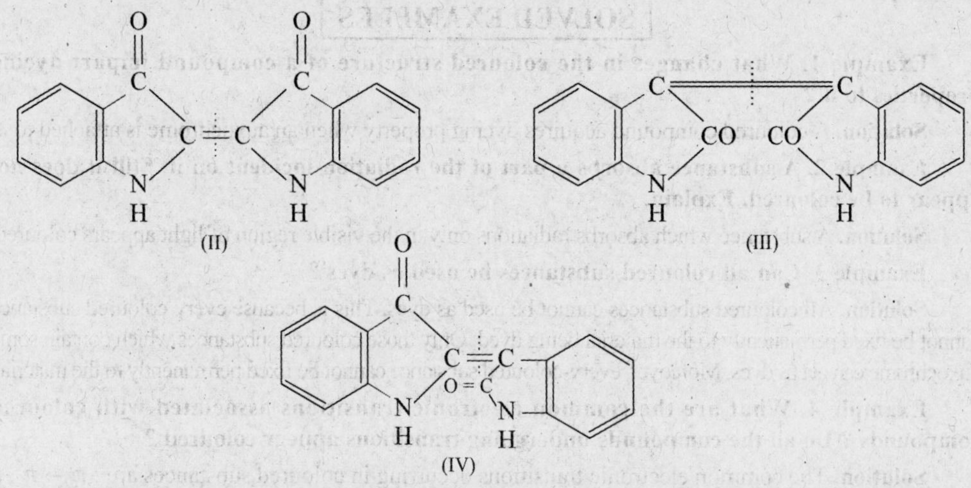

Out of these structures, the structure (II) is found to be correct because indigo when hydrolysed with dilute alkali like NaOH yields anthranilic acid and indoxyl-2-aldehyde.

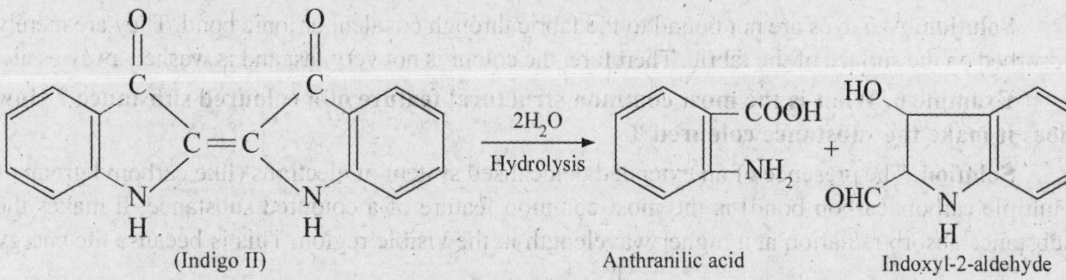

5. Synthesis of indigo. The structure of indigo has been confirmed by the various syntheses of indigo. Baeyer in 1872 synthesised indigo from isatin.

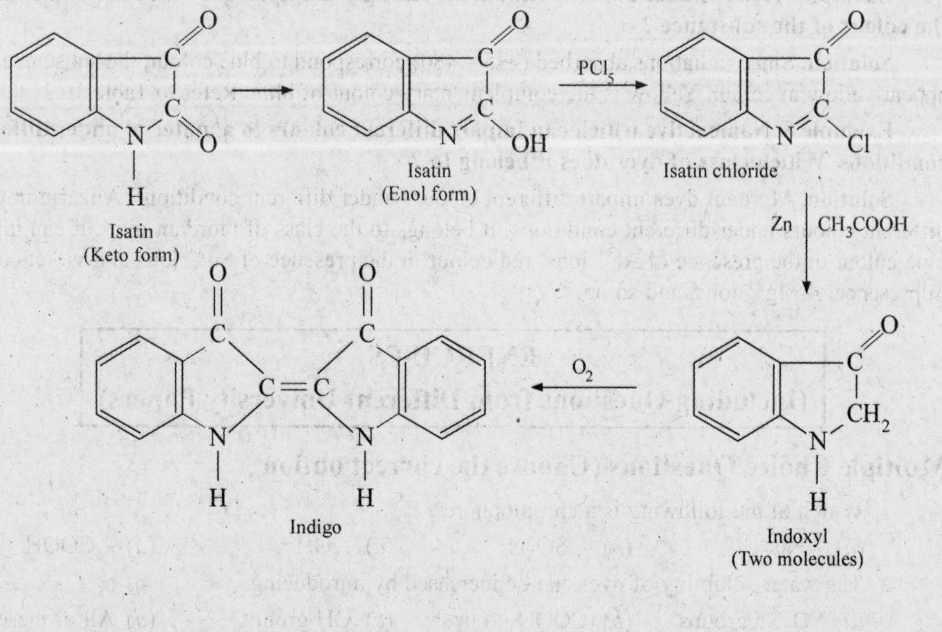

SOLVED EXAMPLES

Example 1. What changes in the coloured structure of a compound impart dyeing properties to it ?

Solution. A coloured compound acquires dyeing property when an auxochrome is attached to it.

Example 2. A substance absorbs a part of the radiation incident on it. Still it does not appear to be coloured. Explain.

Solution. A substance which absorbs radiations only in the visible region of light appears coloured.

Example 3. Can all coloured substances be used as dyes ?

Solution. All coloured substances cannot be used as dyes. This is because every coloured substance cannot be fixed permanently to the material being dyed. Only those coloured substances which contain some auxochrome can act as dyes. Moreover, every-coloured substance cannot be fixed permanently to the material.

Example 4. What are the common electronic transitions associated with coloured compounds ? Do all the compounds undergoing transitions appear coloured ?

Solution. The common electronic transitions occurring in coloured substances are : $\pi \rightarrow \pi^*$, $n \rightarrow \pi^*$. Only such compounds which absorb radiations in the visible range are coloured.

Example 5. Azo dyes do not impart fast colour to the fabric. Why ?

Solution. Azo dyes are not bound to the fabric through covalent or ionic bond. They are merely adsorbed on the surface of the fabric. Therefore, the colour is not very fast and is washed away easily.

Example 6. What is the most common structural feature of a coloured substance ? How does it make the substance coloured ?

Solution. The presence of an extended delocalised system of electrons (like carbonyl group or multiple carbon–carbon bond) is the most common feature of a coloured substance. It makes the substance absorb radiation at a higher wavelength in the visible region. This is because the energy for electronic transition decreases ($E \propto \frac{1}{\lambda}$).

Example 7. A substance absorbs radiations strongly in region 435–480 nm. What will be the colour of the substance ?

Solution. Since radiations absorbed (435 – 480) correspond to blue colour, the substance will appear yellow in colour. Yellow is the complementary colour of blue. Refer to Table 10.2

Example 8. Name a dye which can impart different colours to a material under different conditions. Which class of dyes does it belong to ?

Solution. Mordant dyes impart different colours under different conditions. Alizarin imparts different colours under different conditions. It belongs to the class of mordant dyes. It can impart blue colour in the presence of Al^{3+} ions, red colour in the presence of Sn^{2+} ions and violet colour in presence of Mg^{2+} ions, and so on.

EXERCISES
(Including Questions from Different University Papers)

Multiple Choice Questions (Choose the correct option)

1. Which of the following is a chromophore ?

(*a*) $-NO_2$ (*b*) $-SO_3H$ (*c*) $-OH$ (*d*) $-COOH$

2. The water-solubility of dyes can be increased by introducing

(*a*) SO_3Na groups (*b*) COOH groups (*c*) OH groups (*d*) All of these

3. The reason why materials appear coloured is
   (*a*) The selective absorption of spectral colours
   (*b*) The interaction between the light and the electrons of the dye molecules
   (*c*) The composition of the white light from various spectral colours
   (*d*) All of these
4. Which dyes become linked to the fibre by chemical reaction ?
   (*a*) Acid dyes (*b*) Direct dyes (*c*) Disperse dyes (*d*) None of these
5. Which of the following is an auxochrome ?
   (*a*) – N= O (*b*) $-NO_2$ (*c*) – N = N – (*d*) – OH
6. A substance which appears orange absorbs the radiation in the range (nm)
   (*a*) 400–435 (*b*) 480–490 (*c*) 500–560 (*d*) 605–750
7. Colour lake formed by alizarin with Mg^{2+} ions (mordant) is
   (*a*) brown black (*b*) rose red (*c*) violet (*d*) red
8. Saturated hydrocarbons give the transitions :
   (*a*) $\pi \rightarrow \pi^*$ (*b*) $n \rightarrow \pi^*$ (*c*) $n \rightarrow \sigma^*$ (*d*) $\sigma \rightarrow \sigma^*$
9. Colour of a compound is best explained by
   (*a*) Molecular orbital theory (*b*) Valence bond theory
   (*c*) Witt's theory (*d*) None of these
10. Malachite green is obtained by the condensation of
   (*a*) benzaldehyde and dimethylaniline (*b*) Formaldehyde and dimethyl aniline
   (*c*) benzaldehyde and N–methylaniline (*d*) Formaldehyde and N–methyl aniline

ANSWERS

1. (*a*)	**2.** (*d*)	**3.** (*d*)	**4.** (*a*)	**5.** (*d*)	**6.** (*b*)
7. (*c*)	**8.** (*d*)	**9.** (*a*)	**10.** (*a*)		

Short Answer Questions

1. Write the name and structure of each of the following dyes :
   (*i*) Vat dye (*ii*) Mordant.
2. Classify the following dyes on the basis of their structures and methods of application.
   (*i*) Congo red (*ii*) Malachite green (*iii*) Alizarin (*iv*) Crystal violet.
3. How are the dyes generally classified according to their constitution. Give at least two examples each of any three such classes of dyes ?
4. Give an example of each of the following dyes ?
   (*i*) Acid dye (*ii*) Diazo dye
   (*iii*) Triphenylmethane dye (*iv*) Vat dye.
5. What is meant by complementary colours ? What are the complementary colours of
   (*i*) green and (*ii*) yellow respectively
6. Relate the colour change of methyl orange with pH with its structure.
7. What is a chromophore ? How does it help in the development of colour ?
8. Enumerate the various kinds of electronic transitions taking place in organic compounds. Which of these transitions are generally associated with coloured substances ?

9. Discuss the process of vat dyeing with the help of a suitable example.

10. Give six important classes of dyes according to their method of application. Give at least two examples of each.

11. Suggest two methods of preparation of indigo and describe its method of application.

12. What is meant by mordant dyes and mordant ? Name one such dye and give its method of preparation.

13. Give the formulae and important features of :

(*i*) fluorescein (*ii*) malachite green

14. Draw the structural formulae of the following dyes. To which type according to chemical constitution does each of them belong to ? Which is the auxochrome in each ?

(*i*) Martius yellow (*ii*) Chrysoidine (*iii*) Fast green

15. Write a note on electronic concept of colour and constitution.

16. Giving at least one example of each type, name six different classes of dyes according to their chemical constitution.

17. What are mordant dyes ? Giving suitable examples explain how colour is dependent on mordant used.

18. Give one method of preparation of each of the following dyes :

(*i*) Crystal violet (*ii*) Alizarin (*iii*) Congo red

19. Draw structural formulae of the following dyes. To which type according to chemical constitution does each of these belong to ?

(*i*) Martius yellow (*ii*) Fast green (*iii*) Crystal violet

20. Write an account of the preparation of alizarin.

General Questions

1. Name the various classes of dyes according to (*i*) chemical constitution and (*ii*) method of application. Give the method of preparation of one dye belonging to each type.

2. Write the structures, synthesis and chemistry of the following dyes :

(*i*) Methyl orange (*ii*) Phenolphthalein (*iii*) Indigo

3. Explain the meanings of the terms of chromophore and auxochrome. Give two examples in each case. How is a dye distinguished from other coloured substances ?

4. Explain each of the following terms as used in the study of dyes and illustrate with specific examples :

(*i*) Direct dye (*ii*) Complementary colour

(*iii*) Mordant (*iv*) Leuco base (*v*) Deepening of colour.

5. Give one method of preparation of each of the following dyes :

(*i*) Malachite green (*ii*) Fluorescein or Alizarin

Explain following terms with suitable examples

(*i*) Chromophore (*ii*) Auxochrome.

6. Explain the meaning of the terms chromophore and auxochrome. Give at least three examples of each. How is a dye distinguished from other coloured substances ?

7. Discuss, giving examples, the process of dyeing with respect to

(*i*) vat dyes (*ii*) mordant dyes (*iii*) ingrain dyes

8. Discuss the role of extended delocalised electron system in making a compound appear coloured.

PHYSICAL CHEMISTRY

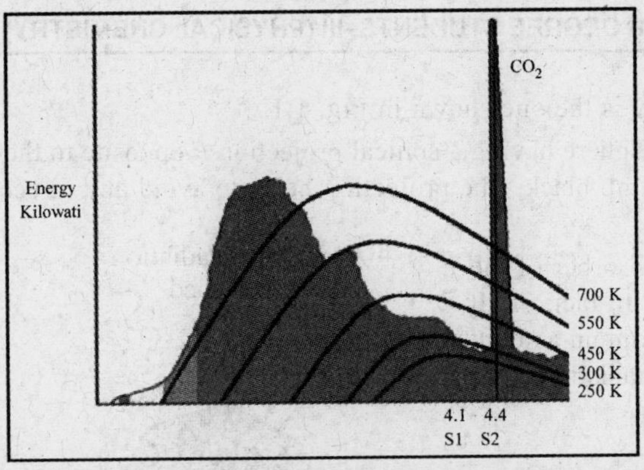

1

Elementary Quantum Mechanics

1.1. INTRODUCTION

The branch of science based on Newton's laws of motion and Maxwell's electromagnetic wave theory to explain phenomenon related to motion and energy is known as classical mechanics or Newtonian mechanics. According to classical mechanics, it should be possible to determine simultaneously the position and velocity of a moving particle. But Heisenberg uncertainty principle contradicts it. Classical mechanics assumes that energy is emitted or absorbed in a continuous manner but Planck's quantum dictates that emission or absorption of energy takes place in an uncontinuous manner in the form of packets of energy called *quanta*.

Limitations of classical mechanics

Classical mechanics does not provide satisfactory explanation for the following phenomena:

(*i*) Blackbody radiation

(*ii*) Photoelectric effect

(*iii*) Atomic and molecular spectra

(*iv*) Heat capacities of solids.

The classical point of view is adequate for objects of appreciable size but is quite unsatisfactory for describing the behaviour of particles of atomic dimensions. It has been necessary therefore to devise a new mechanics for the treatment of electrons and atomic nuclei. In this new mechanics, which takes into account the de-Broglie's dual nature of matter and Heisenberg's uncertainty principle, the exact position of the moving object such as the orbit of an electron around the nucleus of an atom, is replaced by a function which determines the probability of the object being in the particular position. These probability functions satisfy differential equations which are analogous to those representing the variation of the amplitude of a wave. This new atomic mechanics has been referred to as the **wave mechanics.** This new approach to the study of small particles provides a satisfactory basis for many of the quantum postulates and the term quantum mechanics is generally used for this new approach.

1.2. BLACK BODY AND BLACK BODY RADIATIONS

If the radiant energy is allowed to fall on a blackened metallic surface or carbon black, it is found that the energy is almost completely absorbed. *A body which completely absorbs the radiant energy falling on it is called a perfectly black body.* The absorption is found to be more perfect if we take a hollow sphere blackened on the inside and having a small hole for the entry of the radiation. This is so because any radiation that enters through the hole is reflected over and again by the walls of the sphere till finally it is completely absorbed.

The most commonly used black body is the one shown in Fig. 1.1.

It is a hollow double walled metallic sphere having a conical projection P opposite to the hole H and is coated on the inside with lamp black. The projection helps to avoid any direct reflection.

A black body is not only a perfect absorber of the radiant energy, but also a perfect radiator. In fact, of all the bodies, the black body radiates the maximum amount of energy for the given temperature. The radiations thus emitted are called *black body radiation.*

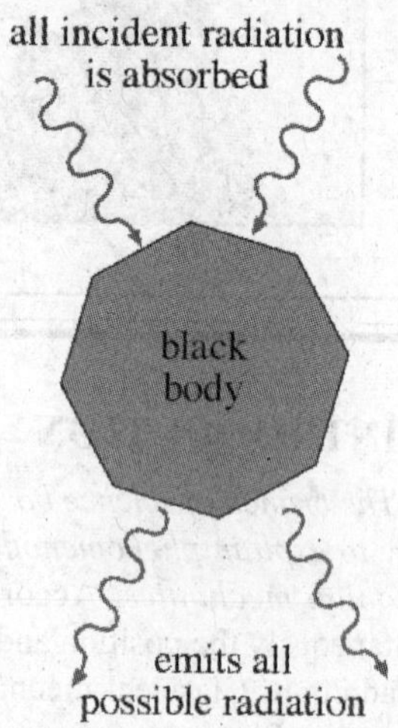

Black body

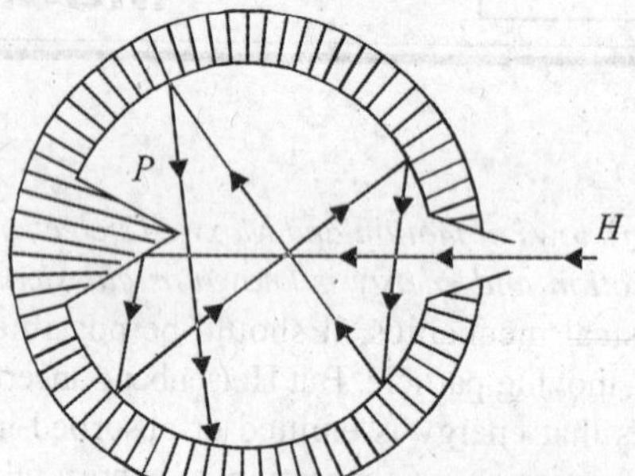

Fig. 1.1. Black body due to Fermy.

1.3. KIRCHOFF'S LAW OF BLACK BODY RADIATIONS

The *quantitative* relationship between the energy absorbed and the energy emitted by a body was put forward by Kirchoff in 1859. It is called Kirchoff's law and it states as follows:

At any temperature, the ratio of the emissive power (or emittance) of a body to the absorptive power (or absorbance or absorptivity) is constant, independent of the nature of the surface and equal to the emissive power (emittance) of a perfectly black body.

The term *emissive power* or *emittance* means the energy emitted by the surface per unit time per unit area whereas the term *absorptive power* or *absorbance* or *absorptivity* is the fraction of the incident energy absorbed by the surface per unit time per unit area.

If E_S and A_S represent the emissive power and absorptive power of any substance and E_B represents the emissive power of a perfectly black body, then according to Kirchoff's law,

$$\frac{E_S}{A_S} = E_B$$

Derivation of Kirchoff's law : Kirchoff's law can be derived in a simple manner as follows:

Suppose Q is the amount of radiation incident per unit area per second on a body. If A, is the absorptive power of the surface, then the amount of radiation absorbed by the body per unit area per second = $A_s \times Q$.

If E_S represents the emissive power of the surface, then the amount of radiation emitted by the surface per unit area per second = E_S

When the body and the enclosure are in *thermal equilibrium,*

Amount of radiation emitted by the surface per unit area per second = Amount of radiation absorbed by the body per unit area per second.

Gustav Robert Kirchoff

Thus $E_S = A_S \times Q$ ... (*i*)

For a perfectly black body, $A_S = A_B = 1$ and E_S may be replaced by E_B where A_B and E_B represent the absorptive power and the emissive power of a perfectly black body. Thus we have

$$E_B = Q \quad \text{... } (ii)$$

From equations (*i*) and (*ii*), we have

$$E_S = A_S \times E_B$$

or

$$\boxed{\frac{E_S}{A_S} = E_B} \quad \text{... } (iii)$$

As E_B is constant, the equation gives the expression for the Kirchoff s law.

1.3.1. Stefan's fourth power law

On the basis of experimental studies, Stefan in 1879 showed that the *total amount of energy E radiated by a perfectly black body per unit area per unit time is directly proportional to the fourth power of its absolute temperature T, i.e.*

$$E \propto T^4$$

or

$$E = \sigma T^4$$

where σ is a universal constant called Stefan's constant. Its value is 5.6697×10^{-5} if energy is expressed in ergs, time in seconds and area in square centimeters. The above statement is called **Stefan's fourth power law or Stefan-Boltzmann fourth power law.** The theoretical proof of this law was given by Boltzmann by using the laws of thermodynamics.

Joseph Stefen

1.4. SPECTRAL DISTRIBUTION OF BLACK BODY RADIATION

The energy emitted by a black body at any temperature does not consist of a single frequency and is also not uniformly distributed along the spectrum. Lummer and Pringsheim in 1899 studied the distribution of energy amongst different wavelengths of the black body spectrum as described below.

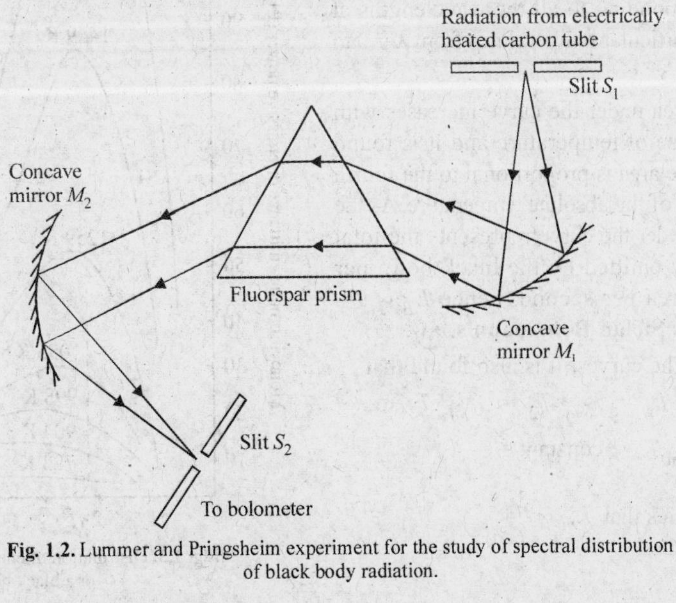

Fig. 1.2. Lummer and Pringsheim experiment for the study of spectral distribution of black body radiation.

The black body radiations were obtained by heating a carbon tube electrically. After passing through the slit S_1 these were made parallel with the help of a concave reflector mirror M_1 and then allowed to fall on a fluorspar prism to get the spectrum. The radiation after refraction through the prism were focussed by another concave mirror M_2 and slit S_2 on to a linear bolometer as shown in Fig. 1.2.

Bolometer is a sensitive instrument for measuring heat radiations. It consists of two very thin blackened platinum gratings, forming two arms of a Wheatstone bridge circuit. Radiant heat falling on one of gratings raises its electrical resistance, thus indicating a deflection of the needle of a galvanometer in the circuit.

By gradually rotating the prism, the radiations corresponding to different wavelengths are obtained and focussed on to the bolometer one by one and their intensities measured. By keeping the carbon tube (black body) at different temperatures a number of observations were made between 621 K to 1646 K. The curves were then plotted between the monochromatic emittance E_λ along the Y-axis and the wavelengths along the X-axis for each of the temperatures. These curves are shown in Fig. 1.3. Following observations can be made in respect of these curves.

(*i*) At a particular temperature, the distribution of energy is not uniform among the various wavelengths of the radiation emitted by the black body.

(*ii*) For each temperature, there is a wavelength (λ_m) at which the energy radiated is maximum ($= E_m$).

(*iii*) With increase in temperature, the maximum shifts higher but towards lower wavelengths. This means that with increase in temperature E_m increases but the corresponding λ_m decreases.

(*iv*) Higher the temperature, more pronounced is the maximum.

(*v*) The area under the curve for a particular temperature gives the total energy emitted by the black body per unit area per second for the complete spectrum *i.e.* corresponding to all the wavelengths at that particular temperature (from $\lambda = 0$ to $\lambda = \infty$).

(*vi*) The area under the curve increases with increase of temperature and it is found that the area is proportional to the fourth power of the absolute temperature. As the area under the curve represents the total energy emitted by the black body per unit area per second, hence $E_B \propto T^4$. This is **Stefan Boltzmann's law.**

(*vii*) From the curves, it is also found that

$$\lambda_{m_1} T_1 = \lambda_{m_2} T_2 = \lambda_{m_3} T_3 \ldots\ldots$$

i.e.
$$\lambda_m T = \text{constant}$$

This implies that $\lambda_m \propto \dfrac{1}{T}$

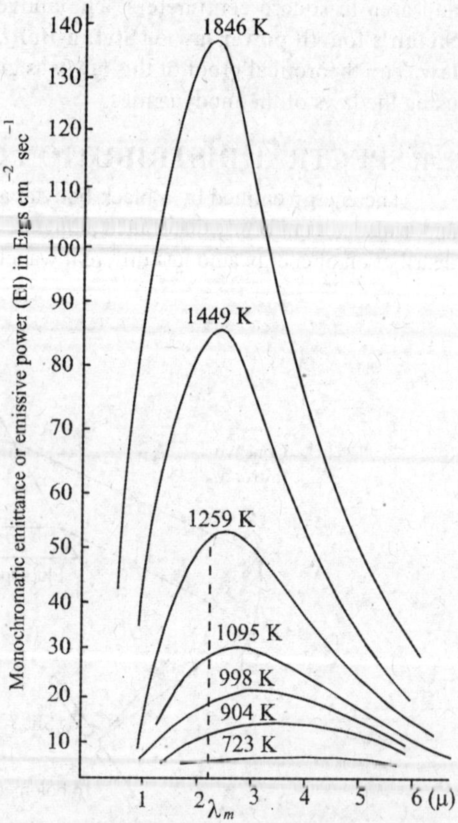

Fig. 1.3. Distribution of energy in the spectrum or a black body.

Wien's displacement law

The wavelength λ_m for which the emittance of a black body is maximum is inversely proportional to its absolute temperature. This is called Wien's displacement law. Mathematically, the law may be represented as :

$$\lambda_m \propto \frac{1}{T}$$

where λ_m = Wavelength corresponding to maximum radiation of energy

T = Temperature on Kelvin scale.

This law is able to explain why the colour of the visible light radiation changes from red to yellow as the temperature of the hot body is increased.

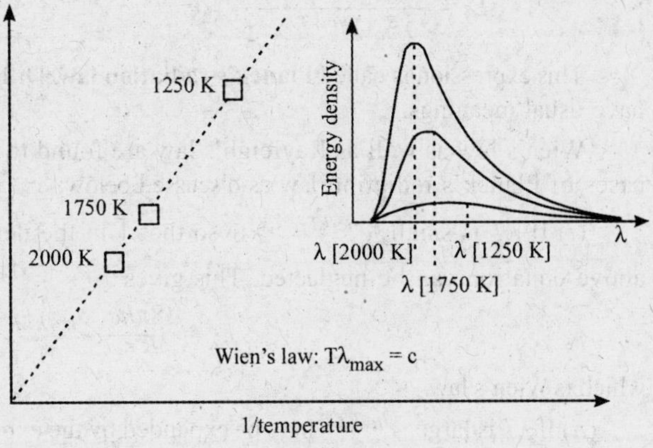

Wien's law

1.4.1. Failure of classical mechanics to explain the spectral distribution of black body radiation

According to Maxwell electromagnetic wave theory of classical mechanics, the source of radiant energy are the vibrating particles called oscillators and these oscillators could radiate or absorb any amount of energy. The frequency of the energy (or the wave) emitted is equal to the frequency of the oscillator. Based on these classical concepts, Wien in 1896 and Lord Rayleigh and Jeans in 1900 explained the distribution of black body radiation using the classical concepts of continuous emission of radiation. Wien's equation was able to explain the distribution only at lower wavelengths whereas Rayleigh-Jeans equation was able to do so only at higher wavelengths. However neither was able to explain the complete distribution of black body radiation. Later, Planck in 1900 was able to provide explanation for the complete distribution of black body radiation by propounding his theory of emission and absorption of energy in quantas.

Wien's radiation law equation is

$$E_\lambda = \frac{8\pi hc}{\lambda^5} e^{-hc/kT\lambda}$$

Rayleigh-Jeans radiation law equation is

$$E_\lambda = \frac{8\pi}{\lambda^4} kT$$

where E_λ is the emissive power of the black body corresponding to wavelength λ, c is velocity of light, h is Planck's constant, k is Boltzmann's constant and T is the absolute temperature.

1.5. PLANCK'S RADIATION LAW

Max Planck in 1900 put forward a theory that the oscillator of black body cannot have *just any* amount of energy but can have only a *discrete* amount of energy depending upon the frequency of the oscillator. Planck summed up his result as under :

Energy is emitted or absorbed not continuously but discontinuously in the form of packets of energy called quanta. The energy of each quantum is given by the relation, E= hv, where v is the frequency of the radiation and h is called Planck's constant. Thus the total energy emitted or absorbed is either unit quantum i.e. hv or a whole number multiple of hv i.e. equal to nhv.

Max Planck

Based upon these concepts, Planck deduced expression for the energy E_λ radiated by a black body at wavelength λ, which is as under:

$$E_\lambda = \frac{8\pi hc}{\lambda^5}\frac{1}{e^{hc/kT\lambda}-1} \quad ... (i)$$

This expression is called **Planck's radiation law.** Different symbols have usual meanings.

Wien's law as well as Rayleigh's law are found to be particular cases of Planck's radiation law as discussed below:

(*i*) If λT is small, $e^{hc/kT\lambda} >> 1$ so that 1 in the denominator in above equation can be neglected. This gives

$$E_\lambda = \frac{8\pi hc}{\lambda^5}e^{-hc/kT\lambda}$$

which is Wien's law.

(*ii*) If λT is large, $e^{hc/kT\lambda}$ may be expanded by the exponential theorem

$$e^x = 1 + x + x^2 + \cdots$$

we get $$e^{hc/kT\lambda} = 1 + \frac{hc}{kT\lambda} + \cdots \quad \text{(Higher terms may be neglected)}$$

So that eqn. (*i*) reduces to the form

$$E_\lambda = \frac{8\pi hc}{\lambda^5} \times \frac{kT\lambda}{hc} = \frac{8\pi}{\lambda^4}.kT$$

which is Rayleigh-Jeans law.

1.5.1. Derivation of Planck's radiation law

Planck considered the black body radiations to consist of linear oscillators of molecular dimensions and that the energy of a linear oscillator can assume only the discrete values

$$0, h\nu, 2h\nu, 3h\nu.....nh\nu$$

If N_0, N_1, N_2 ...... are the number of oscillators per unit volume possessing energies 0, $h\nu$, $2h\nu$ respectively, then the total number of oscillator N per unit volume will be

$$N = N_0 + N_1 + N_2 + ... \quad ... (i)$$

Number of oscillators, N_r having energy E_r is given by (Maxwell's formula)

$$N_r = N_0 e^{-E_r/kT} \quad ... (ii)$$

Substituting these values in eqn. (*i*), we get

$$N_r = N_0 + N_0e^{-h\nu/kT} + N_0e^{-2h\nu/kT} + ... + N_0e^{-nh\nu/kT}$$
$$= N_0(1 + e^{-h\nu/kT} + e^{-2h\nu/kT} + \cdots)$$
$$= \frac{N_0}{1-e^{-h\nu/kT}} \quad [\because (1-e^{-x})^{-1} = 1 + e^{-x} + e^{-2x} + \cdots] \quad ... (iii)$$

The total energy of N oscillators will be given by

$$E = N_1 . h\nu + N_2 . 2h\nu + N_3 . 3h\nu + ...$$
$$= h\nu (N_1 + 2N_2 + 3N_3 + ...)$$
$$= h\nu (N_0e^{-h\nu/kT} + 2N_0e^{-2h\nu/kT} + 3N_0e^{-3h\nu/kT} + ...)$$
$$= N_0 h\nu (e^{-h\nu/kT} + 2e^{-2h\nu/kT} + 3e^{-3h\nu/kT} + ...)$$
$$= N_0 h\nu e^{-h\nu/kT} (1 + 2e^{-h\nu/kT} + 3e^{-2h\nu/kT} + ...)$$
$$= N_0 h\nu e^{-h\nu/kT} (1 + 2x + 3x^2 + ...) \quad \text{(Putting } e^{-h\nu/kT} = x\text{)}$$

$$= N_0\, h\nu\, e^{-h\nu/kT}\,(1-x)^{-2}$$

$$= \frac{N_0 h\nu e^{-h\nu/kT}}{(1-x)^2}$$

$$= N_0 h\nu \frac{e^{-h\nu/kT}}{(1-e^{-h\nu/kT})^2} \qquad \text{... (iv)}$$

Hence the average energy per oscillator is given by

$$\bar{E} = \frac{E}{N} = N_0 h\nu \frac{e^{-h\nu/kt}}{(1-e^{-h\nu/kT})^2} \times \frac{1-e^{-h\nu/kT}}{N_0}$$

$$\frac{h\nu e^{-h\nu/kt}}{1-e^{-h\nu/kT}} = \frac{h\nu}{e^{h\nu/kT}-1} \quad \left(\begin{array}{l}\text{Dividing numerator and}\\ \text{denominator by } e^{-h\nu/kT}\end{array}\right) \qquad \text{... (v)}$$

Thus we find that the average energy of the oscillator is not kT (as given by classical theory) but equal to $(e^{h\nu/kT} - 1)$ according to Planck's quantum theory.

Further it can be deduced that the number of oscillators per unit volume having frequency in the range of ν and $\nu + d\nu$ is equal to

$$\frac{8\pi\nu^2 d\nu}{c^3} \qquad \text{... (vi)}$$

The average energy per unit volume (*i.e.* energy density) inside the enclosure can be obtained by multiplying (*v*) with (*vi*).

$$E_\nu d\nu = \frac{8\pi\nu^2 d\nu}{c^3} \times \frac{h\nu}{e^{h\nu/kT}-1} \qquad \text{... (vii)}$$

Putting $\nu = \frac{c}{\lambda}$ and $d\nu = -\frac{c}{\lambda^2} d\lambda$, the average energy per unit volume in the enclosure for the wavelengths between λ and $\lambda + d\lambda$ is given by

$$E_\nu d\lambda = \frac{8\pi hc}{\lambda^5} \times \frac{1}{e^{hc/\lambda kT}-1} d\lambda \qquad \text{... (viii)}$$

or the energy radiated by the black body corresponding to wavelength λ is

$$\boxed{E_\lambda = \frac{8\pi hc}{\lambda^5} \frac{1}{e^{hc/\lambda kT}-1}} \qquad \text{... (ix)}$$

which is Planck's radiation law or Planck's distribution law.

Planck's radiation law is able to explain the spectral distribution curves of black body radiations. That proves the validity of Planck's distribution law.

1.6. PHOTOELECTRIC EFFECT

When a beam of light with frequency equal to or greater than a particular value (called threshold frequency) is allowed to strike the surface of a metal, electrons are ejected instantaneously from the surface of the metal. This effect is called 'photoelectric effect'.

Some characteristics of the photoelectric effect are given below:

(*i*) The electrons are ejected only if the frequency of the incident light is equal to or greater than a minimum value, called *threshold frequency* (ν_0) (Fig. 1.4).

(*ii*) The electrons are ejected *instantaneously.* There is no time lag between striking of the metal surface by the light and emission of electrons.

(*iii*) The kinetic energy of the emitted electrons depends upon the *frequency* of the incident light.

(*iv*) Einstein explained different observations about photoelectric effect by applying Planck's quantum theory. According to this theory, each quantum of light called *photon* has energy to $h\nu$. When the photon hits the metal atom, it transfers its energy to the electron. Energy equal to threshold value is used up in the release of electron and the remaining energy is stored as the kinetic energy of the electron. (The quantity $h\nu_0$ is called **work function** and is equal to the ionization energy of the metal atom *i.e.* $h\nu_0 = IE$). Thus

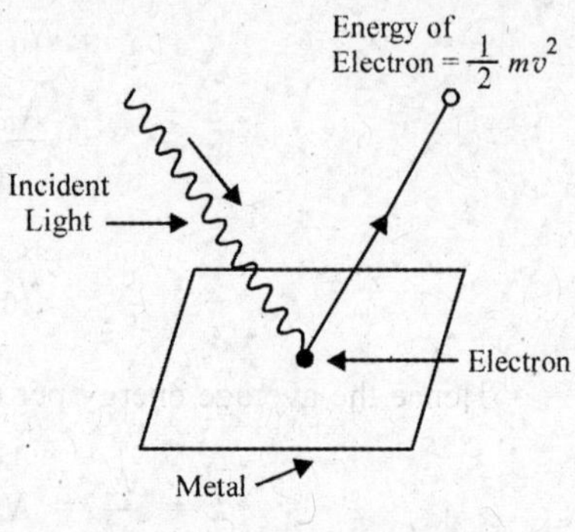

Fig. 1.4. Photoelectric effect

$$h\nu = h\nu_0 + 1/2\, mv^2 \quad \text{... }(i)$$

Thus, if the frequency of the incident light is equal to the threshold value, electron emitted will not possess any kinetic energy.

Further, intensity of light means the number of photons hitting the metal surface per unit time. An increase in intensity can increase the number of electrons emitted but will have no effect on their kinetic energy.

Eqn. (*i*) can be rewritten in the form

$$KE = h\nu - h\nu_0 \quad \text{... }(ii)$$

This equation shows that kinetic energy of the electrons emitted varies linearly with frequency of the incident radiation. A plot of kinetic energy of the emitted electrons versus frequency of the incident radiation will be a straight line with slope equal to the Planck's constant, h (Fig. 1.5). This provides a method to find the value of the Planck's constant.

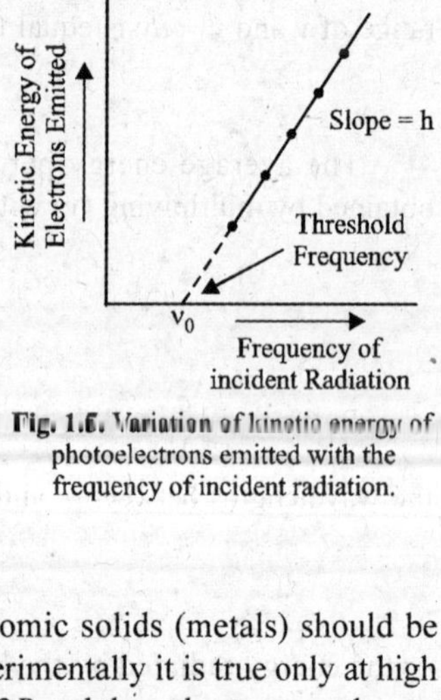

Fig. 1.5. Variation of kinetic energy of photoelectrons emitted with the frequency of incident radiation.

1.7. HEAT CAPACITY OF SOLIDS

Normally it is expected that the heat capacity of all monatomic solids (metals) should be constant and equal to $3R$ (**Dulong and Petit's law**). However, experimentally it is true only at high temperature. At low temperature, the value is found to be less than $3R$ and the values approach zero as T tends to zero degree.

The variation of molar heat capacity of a few elements with temperature is shown in Fig. 1.6.

Einstein (in 1905) explained the variation of heat capacity with temperature again by using Planck's theory of quantisation.

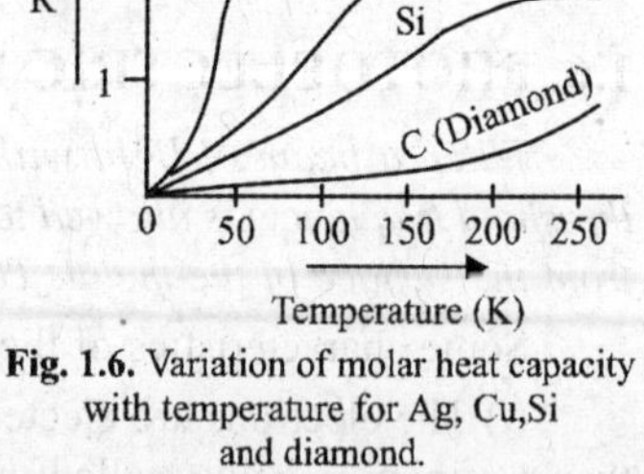

Fig. 1.6. Variation of molar heat capacity with temperature for Ag, Cu,Si and diamond.

Explanation. A monatomic solid can be considered as a collection of oscillators each having three vibrational degrees of freedom. According to law of equipartition of energy, each atom (oscillator) in the solid has mean vibrational energy = $3kT$. Therefore, one mole of monatomic solid (having Avogadro's number of atoms, N_A), will have total vibrational energy *i.e.*, molar internal energy equal to

$$E = N_A(3\,kT) = 3\,RT \qquad (N_A k = R, \text{ the gas constant})$$

or
$$C_v = \left(\frac{\partial U}{\partial T}\right)_v = 3R \qquad \text{... } (i)$$

Using the concepts of Planck's quantum theory, Einstein explained that all the oscillators do not vibrate with the same frequency v *i.e.,* they do not have the same vibrational energy (hv) but may have energies which are integral multiples of some minimum value, hv_0 (= nhv_0). The mean energy of the oscillators is then given by

$$\bar{E} = \frac{hv_0}{e^{hv_0/kT} - 1} \qquad \text{... } (ii)$$

Molar energy of the solid will be

$$E = N_A \times 3\,\bar{E} = \frac{3N_A hv_0}{e^{hv_0/kT} - 1} \qquad \text{(in place of classical value of } 3RT)$$

$$\therefore \quad C_v = \left(\frac{\partial U}{\partial T}\right)_v = 3N_A k\left(\frac{hv_0}{kT}\right)^2 \frac{e^{hv_0/kT}}{(e^{hv_0/kT} - 1)^2} \qquad \text{... } (iii)$$

Stepwise detailed solution of eq. (*iii*) (optional for medical students)

$$C_v = \left(\frac{\partial U}{\partial T}\right)_v = 3\,N_A\,hv_0 \frac{\partial}{\partial T}(e^{hv_0/kT} - 1)^{-1} = 3\,N_A\,hv_0(-1)\,(e^{hv_0/kT} - 1)^{-2}\ \frac{\partial}{\partial T}(e^{hv_0/kT} - 1)$$

$$= 3\,N_A\,hv_0(-1)\,(e^{hv_0/kT} - 1)^{-2}\,e^{hv_0/kT}\,\frac{\partial}{\partial T}\left(\frac{hv_0}{kT}\right)$$

$$= 3\,N_A\,hv_0(-1) \times \frac{e^{hv_0/kT}}{(e^{hv_0/kT} - 1)^2} \times \frac{hv_0}{k}\frac{\partial}{\partial T}(T^{-1})$$

$$= 3\,N_A\,hv_0(-1) \times \frac{e^{hv_0/kT}}{(e^{hv_0/kT} - 1)^2} \times \frac{hv_0}{k}(-1)\,T^{-2}$$

$$= 3\,N_A\,hv_0\,\frac{hv_0}{kT^2} \times \frac{e^{hv_0/kT}}{(e^{hv_0/kT} - 1)^2}$$

$$= 3\,N_A\,k\left(\frac{hv_0}{kT}\right)^2 \frac{e^{hv_0/kT}}{(e^{hv_0/kT} - 1)^2}$$

At low temperature, $hv_0/kT >> 1$ and hence $e^{hv_0/kT} >> 1$. The eq. (*iii*) then reduces to

$$C_v = 3N_A k\left(\frac{hv_0}{kT}\right)^2 e^{-hv_0/kT} \qquad \text{... } (iv)$$

On decreasing the temperature, the exponential factor decreases much more rapidly than the corresponding increase in the factor $(hv_0/kT)^2$. Hence C_v decreases with decrease in temperature.

At high temperature, it can be seen from eq. (*iii*) that the value of C_v reduces to the classical value of $3R$. This is because at high temperature, $hv_0/kT << 1$. Applying expansion series $e^x = 1 + x + x^2 + ...$, eqn. (*iii*) becomes

$$C_v = 3N_A k\left(\frac{h\nu_0}{kT}\right)^2 \frac{1+(h\nu_0/kT)+...}{\left\{1+(h\nu_0/kT)+\frac{1}{2}(h\nu_0/kT)^2+...-1\right\}^2}$$

$$= 3N_A k\left(\frac{h\nu_0}{kT}\right)^2 \frac{1+(h\nu_0/kT)+...}{\left\{(h\nu_0/kT)+\frac{1}{2}(h\nu_0/kT)^2+...\right\}^2}$$

$$= 3N_A k \frac{(h\nu_0/kT)^2+(h\nu_0/kT)^3+...}{\left\{(h\nu_0/kT)+\frac{1}{2}(h\nu_0/kT)^2+...\right\}^2}$$

At high temperatures, $h\nu_0/kT$ is very small and hence the terms with powers 2 and above can be neglected. Thus the above equation reduces to the form

$$C_v = 3N_A K = 3\,R \text{ (classical value)}$$

1.8. QUANTUM MECHANICS

The *wave mechanics* put forward by Schrodinger in 1926 is based upon de Broglie concepts of dual character of matter and thus takes into account the particle as well as wave nature of the material particles. It also called '*particle mechanics*' or '*quantum mechanics*' because it deals with the problems that arise when particles such as electrons, nuclei, atoms, molecules etc., are subjected to a force.

The branch of science which takes into consideration de Broglie concept of dual nature of matter and Planck's quantum theory and is able to explain the phenomena related to small particles is known as quantum mechanics.

The heart of the quantum mechanics is an equation called *Schrodinger wave equation.* It can be derived directly or on the basis of certain postulates of quantum mechanics. We shall study certain results that follow from the Schrodinger wave equation and some applications of the Schrodinger wave equation.

1.8.1. Comparison of classical mechanics with quantum mechanics

Classical Mechanics	*Quantum Mechanics*
(*i*) It deals with macroscopic (big) particles.	(*i*) It deals with microscopic (small) particles.
(*ii*) It is based upon Newton's laws of motion.	(*ii*) It takes into account Heisenberg's uncertainity principle and de Broglie concept of dual nature of matter
(*iii*) It is based on Maxwell's electromagnetic wave theory according to which any amount of energy may be emitted or absorbed continuously.	(*iii*) It is based on Planck's quantum theory according to which only discrete values of energy are emitted or absorbed.
(*iv*) The state of a system is defined by specifying all the forces acting on the particles as well as their positions and velocities (momenta).	(*iv*) It gives probabilities of finding the particles at various locations in space.

1.9. BOHR'S MODEL OF ATOM

Neils Bohr, a Danish physicist in 1913 proposed a model of atom based upon Planck's quantum theory. This new model is called Bohr's model of atom.

The main postulates of Bohr's model of atom are as follows :–

(*i*) An atom consists of a small, heavy positively charged nucleus in the centre and the electrons revolve around in circular orbits.

(*ii*) Out of a large number of circular orbits theoretically possible around the nucleus, the electrons revolve only in those orbits which have a *fixed* value of *energy*. Hence these orbits are called **energy levels** or **stationary states.** The word stationary means that the energy of the electron revolving in a particular orbit is fixed and does not change with time. The different energy levels are numbered as 1, 2, 3, 4.... etc. or designated as K, L, M, N, O, P.....etc. starting from the nucleus side (Fig. 1.7).

The energies of the different stationary states in case of hydrogen atom are given by the expression.

$$E_n = -\frac{2\pi^2 me^4}{n^2 h^2}$$

Substituting the values of m (mass of the electron), e (charge on the electron) and h (Planck's constant), we obtain the equation for the energy associated with the nth orbit

$$E_n = \frac{21.8\times10^{-19}}{n^2}\ \text{J/atom}$$

$$= -\frac{1312}{n^2}\ \text{kJ mol}^{-1}$$

$$= \frac{13.6}{n^2}\ e\text{V/atom}$$

$$(\because 1\ e\text{V} = 1.602\times10^{-19}\ \text{J})$$

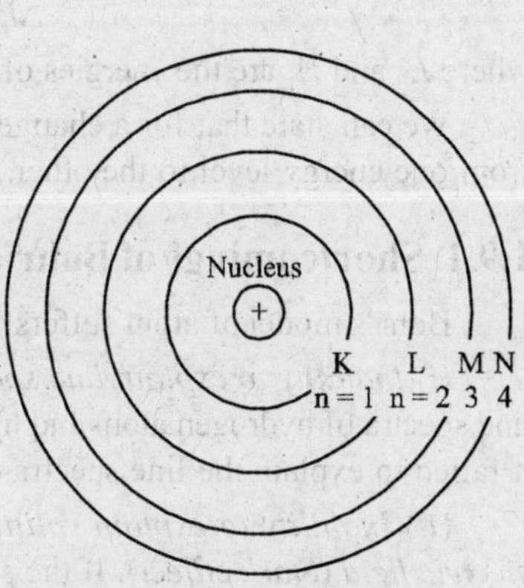

Fig. 1.7. Circular orbits (energy levels/ stationary states) around the nucleus.

where n = 1, 2, 3.....etc. stand for 1st, 2nd, 3rd....etc. level respectively. We find that the 1st energy level (n = 1) which is closest to the nucleus has lowest energy. The energy of the levels increases as we move outwards starting from the 1st level (K-level).

For H-like particles *e.g.* He^+, Li^{2+} etc. (containing one electron only), the expression for energy is

$$E_n = -\frac{2\pi^2 mZ^2e^4}{n^2h^2} = -\frac{1312\,Z^2}{n^2}\ \text{kJ mol}^{-1}$$

where Z is the atomic number of the element (For He^+, Z = 2; for Li^{2+}, Z = 3).

(*iii*) As the electrons revolve only in those orbits which have fixed values of energy, electrons in an atom can have only certain definite or discrete values of energy and not just any value. This is expressed by saying that the energy of an electron is **quantized.**

(*iv*) Also the angular momentum of an electron in an atom can have certain definite or discrete values and not just any value. The permissible values of angular momentum are given by the expression.

$$mvr = \frac{nh}{2\pi}$$

This is expressed by saying that *angular momentum of the electron is an integral multiple of $h/2\pi$*. Here m is the mass of the electron, v is the tangential velocity of the revolving electron, r is the radius of the orbit, h is the Planck's constant and n is any integer. Thus, the angular momentum of the electron can be out of $h/2\pi$, $2h/2\pi$, $3h/2\pi$ .....etc. This means that *like energy, the angular momentum of an electron in an atom is also* **quantized.**

(*v*) In their lowest (normal) energy state, electrons keep on revolving in their respective orbits without losing energy because energy can neither be lost nor gained continuously. This state of atom is called **normal** or **ground state.**

(*vi*) Energy is emitted or absorbed only when the electrons jump from one orbit to the other. When energy is supplied to an atom in a reaction etc., an electron in the atom may jump from its normal energy level (ground state) to some higher energy level by absorbing a definite amount of energy. This state of atom is called **excited** state. Since the life time of the electron in the excited state is short, it starts jumping back to the lower energy level by emitting energy in the form of light of suitable frequency or wavelength. The amount of energy emitted or absorbed is given by the difference of energies (ΔE) of the two energy levels concerned, *i.e.*,

$$\Delta E = E_2 - E_1$$

where E_2 and E_1 are the energies of the electron in the higher and lower energy levels respectively.

We can state that for a change of electronic energy, the electron has to jump and not to flow from one energy level to the other.

1.9.1. Shortcomings of Bohr's Model of Atom

Bohr's model of atom suffers from the following limitations:

(*i*) *Inability to explain line spectra of multi-electron atoms.* Bohr's theory could explain the line spectra of hydrogen atom and hydrogen like particles containing single electron only. However, it failed to explain the line spectra of multi-electron atoms.

(*ii*) *Inability to explain splitting of lines in the magnetic field* (*Zeeman effect*) *and in the electric field* (*Stark efffect*). If the source emitting the radiation is placed in a magnetic field or in an electric field, it is observed that each spectral line splits up into a number of lines. The splitting of spectral lines in the magnetic field is called *Zeeman effect* while the splitting of spectral lines in the electric field is called *Stark effect.* Bohr's model of atom could not explain this splitting.

(*iii*) *Inability to explain the three dimensional model of atom.* According to Bohr's model of atom, the electrons move along certain circular paths in one plane. Thus *it envisages a flat model of atom.* But now it is well established that the atom is three dimensional and not flat, as suggested by Bohr.

(*iv*) *Inability to explain the shapes of molecules.* We know that in covalent molecules, the bonds have directional characteristics (*i.e.*, atoms are linked to each other in particular directions) and hence they possess definite shapes. Bohr's model is not able to explain it.

1.10. DE BROGLIE HYPOTHESIS (DUAL NATURE OF MATTER AND RADIATION)

Earlier, it was thought that light is a stream of particles which are photons (or called corpuscles). However, this concept failed to explain the phenomena of interference and diffraction which could be explained only if light is considered to have wave nature. But at the same time, it was observed that the phenomena of black body radiation and photoelectric effect could be explained only if light is considered to have particle character. Therefore, it was concluded that light has a particle nature as well as wave nature *i.e.*, it has a dual nature.

Louis de Broglie, a French physicist, in 1924, advanced *the idea that like photons, all material particles such as electron, proton, atom, molecule, a piece of chalk, a piece of stone or an iron ball* (*i.e. microscopic as well as macroscopic objects*) *also possessed dual character.* The wave associated with a particle is called a **matter wave** or **de Broglie wave.**

The de Broglie Relation. The wavelength of the wave associated with any material particle was calculated by analogy with photon as follows :—

In case of a photon, if it is assumed to have wave character, its energy is given by

$$E = h\nu \qquad ... (i)$$

(according to the *Planck's quantum theory*)

where ν is the frequency of the wave and h is Planck's constant.

If the photon is supposed to have particle character, its energy is given by

$$E = mc^2 \qquad ... (ii)$$

(according to *Einstein equation*)

where m is the mass of photon and c is the velocity of light.

From equations (*i*) and (*ii*), we get

$$h\nu = mc^2$$

But $$\nu = c/\lambda$$

$\therefore$ $$h \cdot c/\lambda = mc^2 \qquad ... (iii)$$

or $$\lambda = h/mc$$

de Broglie pointed out that the above equation is applicable to any material particle big or small. The mass of the photon is replaced by the mass of the material particle and the velocity c of the photon is replaced by the velocity v of the moving particle. Thus, for any material particle like electron, we may write

$$\lambda = h / mv$$

or $$\lambda = h / p \qquad ... (iv)$$

where $mv = p$ is the momentum of the particle.

The above equation is called **de Broglie equation.**

Significance of de Broglie equation. *Although the Broglie equation is applicable to all material objects but it has significance only in case of microscopic particles.* This is because the wavelength produced by a bigger particle (say a ball) using eq. (*iv*) come out to be too small to be observed. Only particles like electrons, atoms, etc. give an observable value of λ according to eq. (*iv*).

1.11. HEISENBERG'S UNCERTAINTY PRINCIPLE

Werner Heisenberg, a German physicist, in 1927 gave a principle about the uncertainties in simultaneous measurements of position and momentum of small particles. It is known as Heisenberg's uncertainty principle and it states as follows :

It is impossible to measure simultaneously the position and momentum of a small particle with absolute accuracy or certainty. If an attempt is made to measure any one of these two quantities with greater accuracy, the other becomes less accurate. The product of the uncertainty in the position (Δx) *and the uncertainty in the momentum* ($\Delta p = m.\ \Delta v$ *where m is the mass of the particle and* Δv *is the uncertainty in velocity*) *is always constant and is equal to* $h/4\pi$, *where h is Planck's constant i.e.*

$$\Delta x \cdot \Delta p \approx \frac{h}{4\pi} \qquad ... (i)$$

Putting $\Delta p = m \times \Delta v$ eqn. (i) becomes

$$\Delta x \cdot (m\, \Delta v) = \frac{h}{4\pi} \qquad \text{... }(ii)$$

or

$$\Delta x \cdot \Delta v = \frac{h}{4\pi m} \qquad \text{... }(iii)$$

This implies that the *position and velocity of a particle cannot be measured simultaneously with certainty.*

Explanation of Heisenberg's uncertainty principle. Suppose we attempt to measure both the position and momentum of an electron. To locate the position of the electron, we have to use light so that the photon of light strikes the electron and the reflected photon is seen in the microscope (Fig. 1.8). As a result of the hitting by the photon, the position as well as the velocity of the electron are altered.

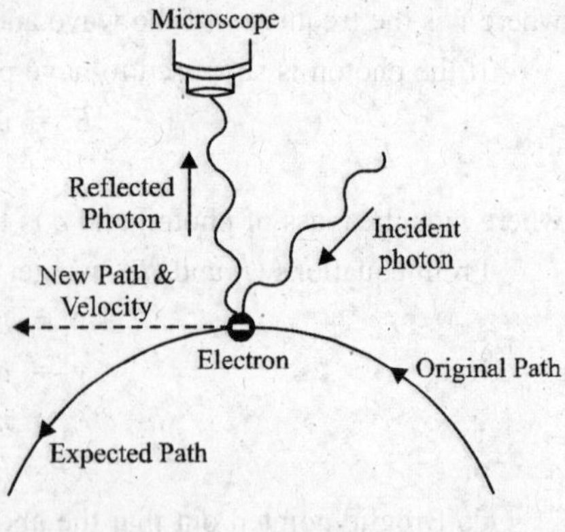

Fig. 1.8. Change of momentum and position of electron on impact with a photon

Significance of Heisenberg's uncertainty principle. Although Heisenberg's uncertainty principle holds good for all objects but *it is of significance only for microscopic particles.* Obviously the energy of the photon is insufficient to change the position and velocity of bigger bodies when it collides with them. For example, the light from a torch falling on a running ball in a dark room neither changes the speed of the ball nor its direction *i.e.* position.

This may be further illustrated by comparing $\Delta x \cdot \Delta v$ values for a particle of mass 1 g and a microscopic particle.

For a particle of mass 1g, we have

$$\Delta x \cdot \Delta v = \frac{h}{4\pi m} = \frac{6.626 \times 10^{-34}\ \text{kg}\, m^2 s^{-1}}{4 \times 3.1416 \times (10^{-3}\ \text{kg})} \simeq 10^{-25}\ \text{m}^2\text{s}^{-1}$$

Thus the product of Δx and Δv is extremely small. Hence these values are negligible.

For a microscopic particle like an electron, we have

$$\Delta x \cdot \Delta v = \frac{h}{4\pi m} = \frac{6.626 \times 10^{-34}\ \text{kg}\, m^2 s^{-1}}{4 \times 3.1416 \times (9.11 \times 10^{-31}\ \text{kg})} \simeq 10^{-4}\ \text{m}^2\text{s}^{-1}.$$

Thus if uncertainty in position is 10^{-8} m, uncertainty in velocity will be = 10^4 ms^{-1} which is quite significant.

1.12. THE COMPTON EFFECT

A.H. Compton

A.H. Compton in 1923 performed experiments to establish the particle nature of the radiation. He found that if *monochromatic X-rays (i.e. X-rays consisting of a single wavelength) are allowed to hit a material of low atomic weight e.g, graphite, the scattered radiations contained not only wavelengths of that of the incident X-rays but also contained radiations of higher wavelength* (*or lower frequency and hence lower energy*). This effect is called "Compton effect."

As the scattering is produced by electrons, it was thought that it must be due to collision between the X-ray photon and the individual electron of the target that must have resulted in the increase of the wavelength of the scattered X-rays.

Suppose λ is the wavelength of the incident X-rays or $h\nu$ is the energy of the incident X-ray photon. By de Broglie relation, its momentum will be h/λ or $h\nu/c$ where c is the velocity of light. Suppose λ' is the wavelength of the scattered X-ray or $h\nu'$ is the energy of the scattered X-ray photon, then its momentum will be h/λ' or $h\nu'/c$.

If mv is the momentum of the electron after collision (m is the rest mass of the electron and v is its velocity) then by applying the laws of conservation of energy and momentum, we can say that the increase in wavelength ($\Delta\lambda = \lambda' - \lambda$) is given by

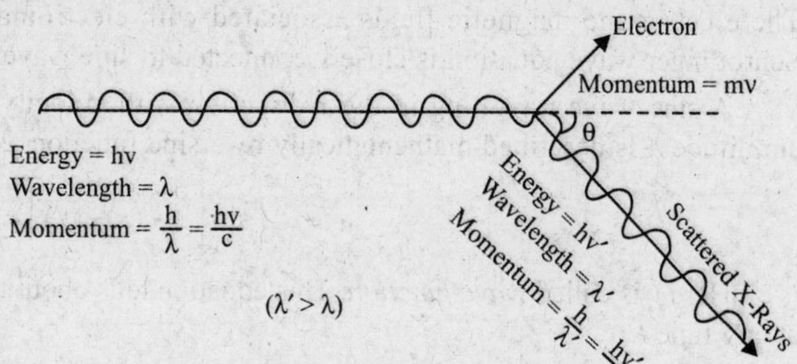

Fig. 1.9. Compton effect

$$\Delta\lambda = \lambda' - \lambda = \frac{2h}{mc}\sin^2\frac{\theta}{2} \qquad \text{... } (i)$$

where θ is the angle between the incident and the scattered X-rays (Fig. 1.9) · $\Delta\lambda$ is called **Compton shift.**

Equation (*i*), indicates that the Compton shift of wavelength is independent of the wavelength of the incident X-rays. This has been confirmed by comparing the experimental value of $\Delta\lambda$ with the theoretically calculated value by substituting the values of m, c, h and θ in the term on right hand side of eqn (*i*). The two values are found to be in reasonable agreement.

Compton effect also supports the uncertainty principle. If X-rays are used to determine the position and momentum of the electron, after the collision, the wavelength of the scattered X-rays photon is found to be more. That is, the scattered X-ray photon has lower energy. Thus some energy of the X-ray photon might have been transferred to the electron whose energy will, therefore, change and so the momentum will also change. Consequently, the momentum of the electron cannot be determined with certainty, which is uncertainty principle.

Compton shift does not depend upon the nature of the material used as target. It depends only on angle θ (Refer to eq. (*i*)). To interpret results for different values of θ, we write eq. (*i*) in the form

$$\Delta\lambda = \lambda' - \lambda = \frac{h}{mc}(1 - \cos\theta) \qquad \text{... } (ii)$$

$$[\because 2\sin^2 x = 1 - \cos 2x]$$

(*i*) When $\theta = 0$ so that $\cos\theta = 1$, $\Delta\lambda = 0$ *i.e.* no change in Compton shift takes place.

(*ii*) When $\theta = 90°$ so that $\cos\theta = 0$, $\Delta\lambda = \frac{h}{mc} = 2.42$ pm (on substituting the values of h, m and c)

(*iii*) When $\theta = 180°$ so that $\cos\theta = -1$, $\Delta\lambda = \frac{2h}{mc} = 4.84$ pm *i.e.* double the value when $\theta = 90°$.

1.13. SINUSOIDAL WAVE EQUATION

Consider the motion of a wave (say ocean wave) in a particular direction, say *x*-axis. At any time *t*, the upward displacement of the wave, represented by *y*, is a function of position *x* and time *t i.e.* $y = f(x, t)$.

Waves of many shapes occur in nature. One of the simplest and most common is the *sine wave.* The electric and magnetic fields associated with electromagnetic waves are sine waves. The Schrodinger wave equation is closely connected to sine waves.

A sine wave travelling in the *x*-direction with velocity v, wavelength λ, frequency ν and amplitude A is described mathematically by a sine function as follows :

$$y = A \sin\left[\frac{2\pi}{\lambda}(x - vt)\right] = A \sin\left[2\pi\left(\frac{x}{\lambda} - \nu t\right)\right] \qquad (\because v = \nu\lambda)$$

Here y is called *wave function*. This equation tells about the displacement y at any point x and at any time t.

Any equation having the characteristics of a sine function is called **sinusoidal wave function.**

1.13.1. Stationary or standing waves

A stationary or standing wave can be explained by taking the example of a string stretched between two fixed points. Suppose a string is stretched between two fixed points and a wave is allowed to travel along the string in one direction. When the wave reaches the other end, it is diverted in the opposite direction with the same wavelength, speed and amplitude. In this process the string is divided into a number of vibrating segments. Such a vibrating motion is called *stationary* or *standing wave* as shown in Fig. 1.10.

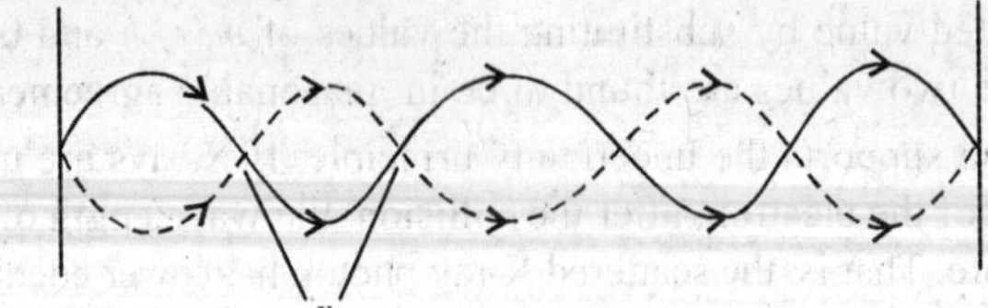

Fig. 1.10. Stationary waves in a string

Points x which represent zero amplitude are called *nodes.*

1.14. SCHRODINGER WAVE EQUATION

Consider the simplest type of wave motion like that of the vibration of a stretched string travelling along the *x*-axis with a velocity v (Fig. 1.11). If ω is the amplitude of the wave at any point whose co-ordinate is x, at any time t, then the equation for such a wave motion is

$$\frac{\partial^2 \omega}{\partial x^2} = \frac{1}{v^2}\frac{\partial^2 \omega}{\partial t^2} \qquad ...(i)$$

This differential equation indicates that the amplitude ω of the wave at any time travelling with a particular velocity depends upon the displacement x and the time t. In other words, ω is a function of x and t. Hence we may write

$$\omega = f(x) f'(t) \qquad ...(ii)$$

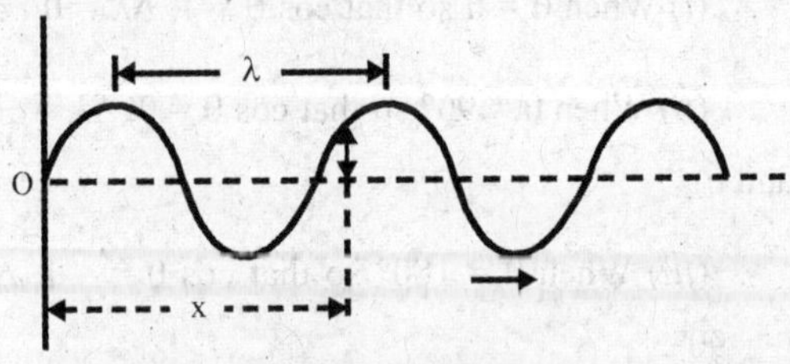

Fig. 1.11. Vibration in a stretched string.

where $f(x)$ is a function of the co-ordinate x only and $f'(t)$ is a function of the time t only.
But for the stationary waves, we have

$$f'(t) = A \sin 2\pi\nu t \qquad \text{... (iii)}$$

where ν is the frequency of vibration and A is a constant, equal to the maximum amplitude of the wave.

Substituting the value of $f'(t)$ from eqn. (*iii*), in eqn. (*ii*), we get

$$\omega = f(x) A \sin 2\pi\nu t \qquad \text{...(iv)}$$

Differentiating this equation twice w.r.t. t, we get

$$\frac{\partial^2 \omega}{\partial t^2} = -f(x) 4\pi^2\nu^2 (A \sin 2\pi\nu t)$$

$$= -4\pi^2\nu^2 f(x) f'(t) \qquad \text{...(v)}$$

Differentiating eqn. (*ii*) twice w.r.t x, we get

$$\frac{\partial^2 \omega}{\partial x^2} = \frac{\partial^2 f(x)}{\partial x^2} f'(t) \qquad \text{...(vi)}$$

Substituting the values of $\partial^2\omega/\partial t^2$ and $\partial^2\omega/\partial x^2$ from equations (*v*) and (*vi*) into the equation (*i*), we get

$$\frac{\partial^2 f(x)}{\partial x^2} f'(t) = -\frac{1}{v^2} 4\pi^2\nu^2 f(x) f'(t)$$

or

$$\frac{\partial^2 f(x)}{\partial x^2} = -\frac{4\pi^2\nu^2}{v^2} f(x) \qquad \text{...(vii)}$$

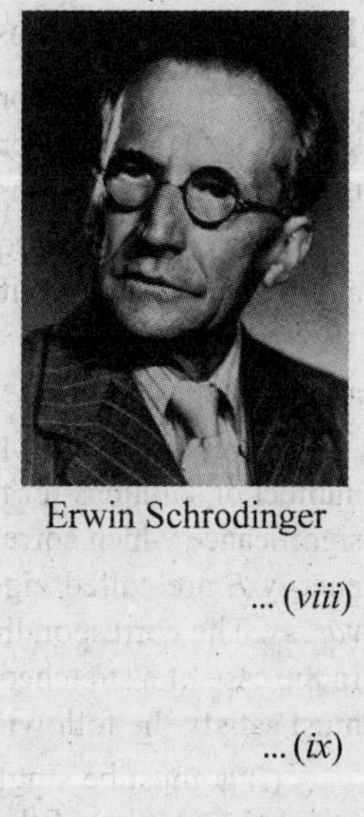

Erwin Schrodinger

This equation has become **time-independent** and hence represents the variation of the amplitude function $f(x)$ with x only.

The velocity v is related to the frequency ν and the wavelength λ of the wave by the expansion

$$v = \nu\lambda \qquad \text{... (viii)}$$

Substituting this value in equation (*vii*), we obtain the result

$$\frac{\partial^2 f(x)}{\partial x^2} = -\frac{4\pi^2}{\lambda^2} f(x) \qquad \text{... (ix)}$$

This equation has been derived for the wave motion in one direction only. For extending this equation to the wave motion in three dimensions, the amplitude function $f(x)$ for one co-ordinate may be replaced by ψ (*x, y, z*) or simply written as ψ for the sake of brevity which represents the *amplitude function* for three co-ordinates. Thus the equation for wave motion in three dimensions becomes

$$\frac{\partial^2 \psi}{\partial x^2} + \frac{\partial^2 \psi}{\partial y^2} + \frac{\partial^2 \psi}{\partial z^2} = -\frac{4\pi^2}{\lambda^2}\psi \qquad \text{... (x)}$$

This equation is applicable to all particles like electrons, nuclei, atoms, molecules, photons etc. Further applying de Broglie equation, according to which $\lambda = h/mv$ where m is the mass of the particle, v is its velocity and h is Planck's constant, eqn. (x) becomes

$$\frac{\partial^2 \psi}{\partial x^2} + \frac{\partial^2 \psi}{\partial y^2} + \frac{\partial^2 \psi}{\partial z^2} = -\frac{4\pi^2 m^2 v^2}{h^2}\psi \qquad \text{... (xi)}$$

Also

Total energy (E) of the particle = Kinetic energy + Potential energy

i.e.,
$$E = \frac{1}{2} mv^2 + V$$

or
$$\frac{1}{2} mv^2 = E - V \qquad \text{... } (xii)$$

where V represents the potential energy of the particle.

Substituting the value of mv^2 from eqn. (*xii*) in eqn. (*xi*), we have

$$\frac{\partial^2 \psi}{\partial x^2} + \frac{\partial^2 \psi}{\partial y^2} + \frac{\partial^2 \psi}{\partial z^2} + \frac{8\pi^2 m(E-V)}{h^2}\psi = 0 \qquad \text{... } (xiii)$$

This equation which represents the wave motion of the particle in three dimensions is called *Schrodinger wave equation.*

1.14.1. Eigen values and Eigen functions (Wave functions)

For a stationary (or standing) wave in a stretched string, the amplitude function $f(x)$ can have significance only for certain definite values of λ. These functions will be those which must satisfy the following conditions :

(*i*) $f(x)$ must be equal to zero at each end of the string, as the string is fixed at these points and the amplitude of vibration is zero.

(*ii*) $f(x)$ must be single valued and finite *i.e.*, at every point on the vibrating string, the amplitude has a definite value at any given instant.

Similarly, Schrodinger wave equation, being a second order differential equation, can have a number of solutions for the amplitude function ψ but only those values of ψ are acceptable and have significance which correspond to some definite values of the total energy E. Such values of the total energy E are called **eigen values.** These values are also called *proper values or characteristic values.* The corresponding values of the function ψ are called **eigen functions** or **wave functions.** In the case of a stretched string, the eigen functions of the Schrodinger equation will be those which must satisfy the following conditions :

(*i*) ψ must be single valued and finite *i.e.*, for each value of the variable x, y, z there is only one definite value of the function ψ.

(*ii*) ψ must be continuous *i.e.*, there must not exist any sudden changes in ψ as its variables are changed.

(*iii*) ψ must become zero at infinity.

We define a class of functions as all those functions which obey certain specified conditions. It may happen that in a given class *there are functions which when operated on by an operator,* α *are merely multiplied by some constant a, or in symbols*

$$\alpha f(x) = a\, f(x) \qquad \text{... (1)}$$

Those members of the class which obey such a relation are known as characteristic or eigen functions of the operator α. The various possible values of a are called the characteristic values or eigen values of the operator.

For example, when function $\Psi = \sin kx$ on which the operator $\delta = \frac{d}{dx}$ operates, we get

$$\frac{d\psi}{dx} = \frac{d}{dx}(\sin kx) = k \cos kx$$

Since $\cos kx \neq \psi$, $\frac{d}{dx}$ is not an eigen operator for the function $\psi = \sin kx$. Now let us consider $\frac{d^2}{dx^2}$ as the operator for the same functions $\psi = \sin kx$, then upon operating, we get

$$\frac{d^2\Psi}{dx^2} = \frac{d}{dx}\left[\frac{d}{dx}\sin kx\right]$$

$$= \frac{d}{dx}[k\cos kx]$$

$$= -k^2 \sin kx = -k^2\psi.$$

Here $\frac{d^2}{dx^2}$ is an eigen operator for $\psi = \sin kx$. Thus $\psi = \sin kx$ is an eigen function for operator, $\frac{d}{dx^2}$ having eigen value $-k^2$.

Let us take another example when function $\psi = e^{imx}$, operators are $\frac{d}{dx}$ and $\frac{d^2}{dx^2}$.

Then $\frac{d}{dx}(\psi) = ime^{imx} = im\psi$

Here $\frac{d}{dx}$ is an eigen operator, im is the eigen value for the eigen function $\psi = e^{imx}$.

Also, $\frac{d^2\psi}{dx^2} = im\ im\ e^{imx} = -m^2\psi$.

Here, $\frac{d^2}{dx^2}$ is an eigen operator with eigen value $-m^2$ for eigen function $\psi = e^{imx}$. As is clear different eigen operators have different eigen values for the same function.

1.15. BORN'S INTERPRETATION AND SIGNIFICANCE OF WAVE FUNCTION (Ψ)

For a particle like an electron, the wave function ψ represents the amplitude of the electron wave at any instant of time when the co-ordinates of the electron are x, y, z. However, it was suggested by Max Born that, just as in the case of light or sound, the square of the amplitude of the wave at any point gives the intensity of the light or sound at that point, the square of the amplitude of the electron wave *i.e.*, ψ^2 at any point gives the intensity of the electron wave at that point. However, keeping in view the Heisenberg's uncertainty principle, the intensity of the electron at any point may be interpreted as the probability of finding the particle electron at that point (Fig. 1.12). Thus ψ^2 *at any point gives the probability of finding the electron at that point.* That is why the wave function, ψ^2 is also called **probability amplitude.** But the probability is directly

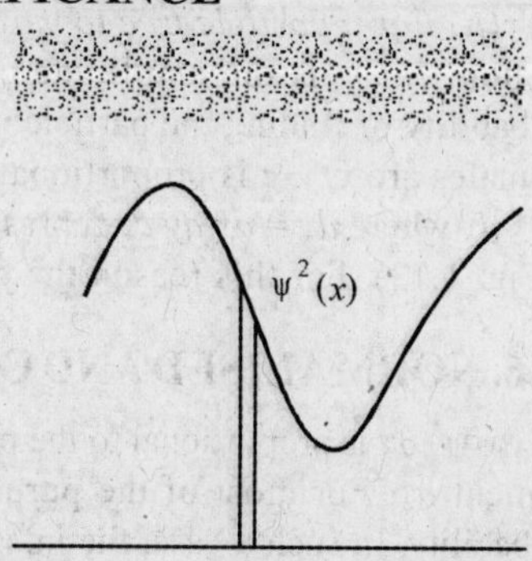

Fig. 1.12. Born interpretation of wave function in one dimensional system. Probability of finding the particle at points along x-axis. $\psi^2(x)$ is represented by the density of shading in the upper half of the diagram.

related to the density (concentration) of the electron cloud. Thus ψ^2 gives the electron density at any given point. Since the region around the nucleus which represents the electron density at different points is called an orbital, that is why the wave function for an electron in an atom is called *orbital wave function* or simply atomic orbital.

Max Born

Further, it may be noted that ψ may not always come out to be real. In some cases it may come out to be imaginary *i.e.*, may include the imaginary quantity $i(=\sqrt{-1})$. Then ψ^2 may be real or imaginary depending upon the nature of the expression for ψ. However, since the probability of finding a material particle at any point in space must always be real, the value taken is $\psi\psi^*$ in place of ψ^2 where, ψ^* is called the *complex conjugate* of ψ and is obtained by replacing i by $-i$. This product (*i.e.*, $\psi\psi^*$) is always real irrespective of whether ψ is real or imaginary. For example, suppose $\psi = a + ib$. Then the complex conjugate will be $\psi^* = a - ib$. The product $\psi\psi^* = (a + ib)\ (a - ib) = a^2 + b^2$ which is real. The product $\psi\psi^*$ is usually represented by $|\psi|^2$ where, $|\psi|$ represents *modulus* or *absolute* value of ψ. $|\psi|^2$ is called the **square modulus** of ψ. If ψ is real, then $\psi = \psi^*$ and the product will automatically be equal to ψ^2.

Born interpretation of the wave function may be summed up as follows :

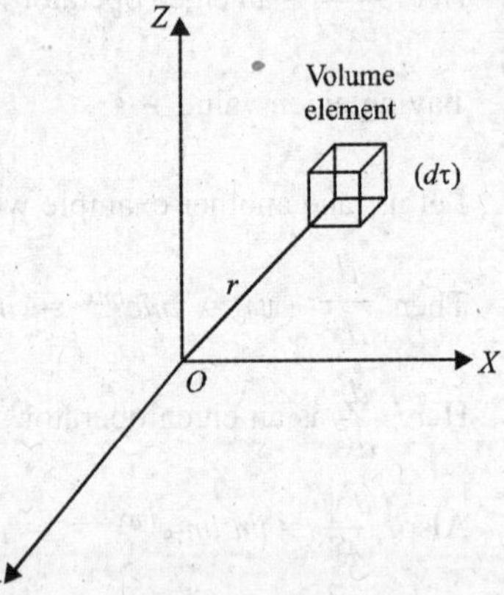

Fig. 1.13. Born interpretation of the wave function in three dimensional space.

(*i*) *For a one-dimensional system,* if the amplitude of the wave function of a particle is ψ at some point x, then the probability of finding the particle between x and $x + dx$ is proportional to $\psi(x)\psi^*(x)\ dx$ or in short it is written as $\psi\psi^*\ dx$.

(*ii*) *For a particle free to move in three-dimensions,* (*e.g.* an electron near the nucleus of an atom), if ψ is the amplitude of the wave function at some point at distance r, then the probability of finding the particle in the small volume element $dx\ dy\ dz$ at that point whose co-ordinates are x, y, z is proportional to $\psi\ (x, y, z)\ \psi^*\ (x, y, z)\ dx\ dy\ dz$ or in short it is written as $\psi\psi^*\ d\tau$ where $d\tau = dx\ dy\ dz$ represents the small volume element of the space at some distance r (Fig. 1.13). For this reason the product $\psi\psi^*$ is called the *probability distribution function.*

1.16. NORMALISED AND ORTHOGONAL WAVE FUNCTIONS

$\psi\psi^*\ \partial\tau$ is proportional to the probability of finding the particular system in the small volume element $\partial\tau$. For most of the purposes it is taken as equal to rather than proportional to the probability. In such a case, the integration of $\psi\psi^*\ \partial\tau$ over the whole of the configuration space, which gives the total probability, must be equal to unity *i.e.*,

$$\oint \psi\psi^*\ d\tau = 1 \qquad \text{... } (i)$$

where $\oint$ represents the integration over the whole space. In place of $\oint$, we can also use the symbol $\int_{-\infty}^{+\infty}$

A wave function which satisfies the above equation is said to be normalised.

In some cases, the result of integration is found to be equal to some constant, say *N*. Then we have

$$\oint \psi\psi^* d\tau = N \qquad \text{... (ii)}$$

To make the result equal to unity, we have

$$\frac{1}{N}\oint \psi\psi^* d\tau = 1 \qquad \text{... (iii)}$$

or $$\oint \left(\frac{1}{N^{1/2}}\psi\right)\left(\frac{1}{N^{1/2}}\psi^*\right)d\tau = 1 \qquad \text{... (iv)}$$

The factor $1/N^{1/2}$ is called the *normalisation constant* and the function $\left(\frac{1}{N^{1/2}}\psi\right)$ is called the *normalised function.*

If ψ_i and ψ_j are two different eigen functions obtained as satisfactory solutions of the wave equation, then these functions will be normalized if

$$\oint \psi_i\psi_i^* \, d\tau = 1 \quad \text{and} \quad \oint \psi_j\psi_j^* \, d\tau = 1 \qquad \text{... (v)}$$

Further if they satisfy the following conditions

$$\oint \psi_i\psi_j^* \, d\tau = 0 \quad \textbf{and} \quad \oint \psi_i^* \psi_j \, d\tau = 0 \qquad \text{... (vi)}$$

they are said to be mutually *orthogonal.*

Wave functions that are both orthogonal and normalized are called *orthonormal.*

1.17. OPERATORS

An operator is a mathematical instruction or procedure to be carried out on a function so as to get another function.

The result of an 'operator' on a function gives another function.

(Operator) · (function) = (Another function)

The function on which the operation is carried out is called an *operand.* The above equation does not mean that the function is multiplied with the operator. An operator written alone has no significance.

A few examples are given below :

(*i*) $\frac{d}{dx}(x^4) = 4x^3$. Hence $\frac{d}{dx}$ which stands for differentiation w.r.t. x is the operator, x^4 is the operand and $4x^3$ is the result of the operation.

(*ii*) $\int x^3 \, dx = x^4/4 + C$. Here $\int dx$ which stands for integration w.r.t. x is the operator, x^3 is the operand and $x^4/4 + C$ is the result of the operation.

Similarly, *taking the square root, cube root or multiplication by a constant k* etc. are different operations which can be carried on any function.

In case the symbol used for the operator is not self-explanatory, a suitable letter or some symbol for the operator is used with the symbol (^) over it *i.e.* a carat is put over the symbol.

Algebra of Operators. The operators follow certain rules similar to those of algebra, as given below :

(1) Addition and Subtraction of Operators. If $\hat{A}$ and $\hat{B}$ are two different operators, and f is the operand, then

$$(\hat{A}+\hat{B})f = \hat{A}f+\hat{B}f \quad ...(i)$$

$$(\hat{A}-\hat{B})f = \hat{A}f-\hat{B}f \quad ...(ii)$$

(2) Multiplication of Operators. If $\hat{A}$ and $\hat{B}$ are two different operators and f is the operand, then expression $\hat{A}\,\hat{B}\,f$ implies that f is operated first by the operator $\hat{B}$ to get the result, say f', and then f' is operated by the operator $\hat{A}$ to get the final result say f'', *i.e.* $\hat{A}\,\hat{B}\,f$ implies that $\hat{B}\,f = f'$ and then $\hat{A}\,f' = f''$ so that we have

$$\hat{A}\,\hat{B}\,f = f'' \quad ...(iii)$$

Thus the order of using the operators is from **right to left** as written in the given expression.

If the same operation is to be done a number of times in continuation, it is shown by a power of operator. For example, $\hat{A}\,\hat{A}\,f$ is written as $\hat{A}^2 f$

or $$\hat{A}\,\hat{A}\,f = \hat{A}^2 f \quad ...(iv)$$

It may be noted that usually

$$\hat{A}\,\hat{B}\,f \neq \hat{B}\,\hat{A}\,f \quad ...(v)$$

Suppose $\hat{A} = x$, $\hat{B} = \dfrac{d}{dx}$ and $f = f(x) = x^2$. Then

$$\hat{A}\,\hat{B}\,f = x\frac{d}{dx}(x^2) = x(2x) = 2x^2$$

$$\hat{B}\,\hat{A}\,f = \frac{d}{dx}x\,(x^2) = \frac{d}{dx}x^3 = 3x^2$$

In case

$$\hat{A}\,\hat{B}\,f = \hat{B}\,\hat{A}\,f \quad ...(vi)$$

i.e., the change of order of the operators gives the same result, the operators are said to *commute.*

The **commutator** of the two operators $\hat{A}$ and $\hat{B}$, represented as $[\hat{A}, \hat{B}]$, is defined as

$$[\hat{A}, \hat{B}] = \hat{A}\,\hat{B} - \hat{B}\,\hat{A}$$

As stated above, if $\hat{A}\,\hat{B} = \hat{B}\,\hat{A}$, the operators are said to commute and hence $[\hat{A}, \hat{B}] = 0$.

(3) Linear operators. An operator $\hat{A}$ is said to be linear if for the two functions f and g.

$$\hat{A}\,(f+g) = \hat{A}\,f+\hat{A}\,g \quad ...(vii)$$

i.e., the operator on the sum of two functions gives the same result as the sum of two results obtained by carrying out the same operation on the two functions separately. For example, d/dx, d^2/dx^2 etc. are linear operators. Taking the square or taking the square root etc. are non-linear, because we may not get the same result with the operator on the sum of two functions and the sum of the operations on two functions separately.

Laplacian operator. This is a very common operator used in quantum mechanics. It is represented by ∇^2 (pronounced as '*del squared*) and is defined as

$$\nabla^2 = \frac{\partial^2}{\partial x^2}+\frac{\partial^2}{\partial y^2}+\frac{\partial^2}{\partial z^2} \quad \text{... (viii)}$$

Schrodinger wave equation *viz.*,

$$\frac{\partial^2\psi}{\partial x^2}+\frac{\partial^2\psi}{\partial y^2}+\frac{\partial^2\psi}{\partial z^2}+\frac{8\pi^2 m(E-V)}{h^2}\psi=0$$

may be written in terms of Laplacian operator in the form

$$\nabla^2\psi+\frac{8\pi^2 m(E-V)}{h^2}\psi=0 \quad \text{... (ix)}$$

Hamiltonian operator. Schrodinger wave equation may also be written in the form

$$\nabla^2\psi = -\frac{8\pi^2 m(E-V)}{h^2}\psi$$

or $$\nabla^2\psi = -\frac{8\pi^2 m}{h^2}(E\psi - V\psi)$$

or $$-\frac{h^2}{8\pi^2 m}\nabla^2\psi+V\psi = E\psi$$

or $$\left(-\frac{h^2}{8\pi^2 m}\nabla^2+V\right)\psi = E\psi \quad \text{... (x)}$$

This equation implies that the operation $\left(-\frac{h^2}{8\pi^2 m}\nabla^2+V\right)$ carried on the function ψ, is equal to the total energy multiplied with the function ψ. The operator $\left(-\frac{h^2}{8\pi^2 m}\nabla^2+V\right)$ is called Hamiltonian operator and is represented as $\hat{H}$. Thus

$$\hat{H} = -\frac{h^2}{8\pi^2 m}\nabla^2+V \quad \text{... (xi)}$$

Eqn. (*x*) may, therefore, be written as

$$\hat{H}\psi = E\psi \quad \text{... (xii)}$$

This is a brief format of writting the Schrodinger wave equation.

In this equation, ψ is called the eigen function and E is called eigen value.

Such an equation is, therefore, called **eigen value equation.** Thus for Schrodinger wave equation, we can write

(Energy operator) (Wave function) = (Energy) × (Wave function)

This is a general result as it is equally applicable to all other **observables** *i.e.* measurable properties of a system. Thus, in general, we can write

(Operator corresponding to an observable) (Wave function) = (Value of observable) × (Wave function)

SOLVED EXAMPLES ON OPERATOR

Example 1. Find the commutator $\left[x, \dfrac{d}{dx}\right]$.

Solution. By definition of commutator,

$$\left[x, \frac{d}{dx}\right] = \left[x\frac{d}{dx} - \frac{d}{dx}x\right]$$

Let us operate this on some arbitrary function $\psi(x)$.

$$\left[x\frac{d}{dx} - \frac{d}{dx}x\right]\psi = x\frac{d\psi}{dx} - \frac{d}{dx}(x\psi)$$

$$= x\frac{d\psi}{dx} - x\frac{d\psi}{dx} - \psi\frac{dx}{dx}$$

$$= (-1)\,\psi$$

Thus $$\left[x, \frac{d}{dx}\right] = -1.$$

Example 2. Find the commutator of the operators for momentum and position, the two conjugate properties of Heisenberg's uncertainty principle.

Solution. Operator for momentum $= \hat{p} = \dfrac{h}{2\pi i}\dfrac{\partial}{\partial x}$

Operator for position $= \hat{x}$ = Multiplication by x

$$[\hat{p}, \hat{x}]\psi = (\hat{p}\,\hat{x} - \hat{x}\,\hat{p})\psi = \hat{p}\,\hat{x}\,\psi - \hat{x}\,\hat{p}\,\psi \quad \text{... }(i)$$

$$\hat{p}\,\hat{x}\,\psi = \frac{h}{2\pi i}\frac{\partial}{\partial x}(\hat{x}\,\psi) = \frac{h}{2\pi i}\frac{\partial}{\partial x}(x\psi)$$

$$= \frac{h}{2\pi i}\left(x\frac{\partial\psi}{\partial x} + \psi\right) = \frac{h}{2\pi i}x\frac{\partial\psi}{\partial x} + \frac{h}{2\pi i}\psi \quad \text{... }(ii)$$

$$\hat{x}\,\hat{p}\,\psi = \hat{x}\,\frac{h}{2\pi i}\frac{\partial\psi}{\partial x} = x\frac{h}{2\pi i}\frac{\partial\psi}{\partial x} \quad \text{... }(iii)$$

Substituting the values from eqns. (*ii*) and (*iii*) in eqn. (*i*), we have

$$[\hat{p}, \hat{x}]\psi = \frac{h}{2\pi i}\psi \quad \text{or} \quad [\hat{p}, \hat{x}] = \frac{h}{2\pi i} \simeq h$$

PROBLEMS FOR PRACTICE

Problem 1. Evaluate the commutator $\left[\dfrac{d}{dx}, \dfrac{d^2}{dx^2}\right]$ **(Ans. 0)**

Problem 2. Prove that $[\hat{A}^2, \hat{B}] = \hat{A}[\hat{A}, \hat{B}] + [\hat{A}, \hat{B}]\hat{A}$

Solution. R.H.S. $= \hat{A}(\hat{A}\hat{B} - \hat{B}\hat{A}) + (\hat{A}\hat{B} - \hat{B}\hat{A})\,\hat{A}$

$$= \hat{A}^2\hat{B} - \hat{A}\hat{B}\hat{A} + \hat{A}\hat{B}\hat{A} - \hat{B}\hat{A}^2$$

$$= \hat{A}^2\hat{B} - \hat{B}\hat{A}^2 = [\hat{A}^2, \hat{B}]$$

Expressions for operators. The expression for an operator can be obtained by the following procedure.

Step 1. *The operator is allowed to operate on the operand.*

Step 2. *The results of calculation are arranged such that the operand appears on both sides in a way that it can be cancelled from both sides.*

Example 1. Derive expression for the following operator

$$\left(\frac{d}{dx}x\right)^2$$

Solution. Taking the operand as $\psi(x)$. Then

$$\left(\frac{d}{dx}x\right)^2 \psi(x) = \left(\frac{d}{dx}x\right)\left(\frac{d}{dx}x\right)\psi = \left(\frac{d}{dx}x\right)\left(\frac{d}{dx}x\psi\right)$$

$$= \frac{d}{dx}x\left(x\frac{d\psi}{dx}+\psi\frac{dx}{dx}\right) = \frac{d}{dx}x\left(x\frac{d\psi}{dx}+\psi\right)$$

$$= \frac{d}{dx}\left(x^2\frac{d\psi}{dx}+x\psi\right) = \left(x^2\frac{d^2\psi}{dx^2}+\frac{d\psi}{dx}(2x)\right)+\left(x\frac{d\psi}{dx}+\psi\frac{dx}{dx}\right)$$

$$= x^2\frac{d^2\psi}{dx^2}+2x\frac{d\psi}{dx}+x\frac{d\psi}{dx}+\psi = x^2\frac{d^2\psi}{dx^2}+3x\frac{d\psi}{dx}+\psi$$

$$= \left(x^2\frac{d^2}{dx^2}+3x\frac{d}{dx}+1\right)\psi$$

Removing ψ from both sides, we get

$$\left(\frac{d}{dx}x\right)^2 = x^2\frac{d^2}{dx^2}+3x\frac{d}{dx}+1.$$

Example 2. Derive expression for the following operator $\left(\frac{d}{dx}+x\right)^2$.

Solution. Taking the operand as $\psi(x)$. Then

$$\left(\frac{d}{dx}+x\right)^2\psi(x) = \left(\frac{d}{dx}+x\right)\left(\frac{d}{dx}+x\right)\psi = \left(\frac{d}{dx}+x\right)\left(\frac{d\psi}{dx}+x\psi\right)$$

$$= \frac{d}{dx}\left(\frac{d\psi}{dx}+x\psi\right)+x\left(\frac{d\psi}{dx}+x\psi\right)$$

$$= \frac{d^2\psi}{dx^2}+\left(x\frac{d\psi}{dx}+\psi\right)+x\frac{d\psi}{dx}+x^2\psi$$

$$= \frac{d^2\psi}{dx^2}+2x\frac{d\psi}{dx}+x^2\psi+\psi$$

$$= \left(\frac{d^2}{dx^2} + 2x\frac{d}{dx} + x^2 + 1\right)\psi$$

Removing ψ from both sides, we get

$$\left(\frac{d}{dx} + x\right)^2 = \frac{d^2}{dx^2} + 2x\frac{d}{dx} + x^2 + 1.$$

1.18. POSTULATES OF QUANTUM MECHANICS

Postulates of quantum mechanics which can lead to the derivation of Schrodinger equation are given as under :

(1) *For every time independent state of system, a function ψ of the coordinates can be written which is single valued, continuous and finite throughout the configuration space. This function describes completely the state of the system.*

(2) *To each observable quantity in classical mechanics, like position, velocity, momentum, energy etc., there corresponds a certain mathematical operator in quantum mechanics, the nature of which depends upon the classical expression for the observable quantity.* For example,

(*i*) The operator corresponding to a position co-ordinate is multiplied by the value of that coordinate *i.e.*, operator for a position coordinate x is the multiplier x.

(*ii*) The operation representing the momentum (p) in the direction of any co-ordinate q is the differential operator

$$\frac{h}{2\pi i}\frac{\partial}{\partial q} \quad \text{or} \quad -\frac{ih}{2\pi}\frac{\partial}{\partial q}$$

where h is Planck's constant and $i = \sqrt{-1}$.

For example, the operator for linear momentum parallel to x-axis is

$$\hat{p} = \frac{h}{2\pi i}\frac{\partial}{\partial x}$$

This means that to find the linear momentum of a particle parallel to the x-axis, we use the eigen value equation

$$\hat{p}\,\psi = p\psi$$

i.e. we differentiate the wave function ψ with respect to x and multiply the result with $h/2\pi i$ and then from the eigen value equation, we can know the value of the momentum p.

For three-dimensional systems, the operators are obtained by summation of the corresponding operators. For example,

(*i*) For momentum $\hat{p} = \frac{h}{2\pi i}\left(\frac{\partial}{\partial x} + \frac{\partial}{\partial y} + \frac{\partial}{\partial z}\right)$

(*ii*) For kinetic energy, $\hat{T} = \frac{-h^2}{8\pi^2 m}\left(\frac{\partial^2}{\partial x^2} + \frac{\partial^2}{\partial y^2} + \frac{\partial^2}{\partial z^2}\right)$

(*iii*) For total energy, $\hat{H} = \frac{-h^2}{8\pi^2 m}\nabla^2 + V(x, y, z)$ $\quad (\because E = T + V)$

(3) *If ψ is a well behaved function for the given state of a system and $\hat{A}$ is a suitable operator for the observable quantity or property, then the operation on ψ by the operator $\hat{A}$ gives ψ multiplied by a constant value (say, a) of the observable property i.e.,*

$$\hat{A}\psi = a\,\psi \quad \text{... } (i)$$

The given state is called the eigen state of the system, ψ is called the eigen function and a is called the eigen value. The above equation is a general expression for **eigen value equation.**

(4) *The only possible measured values of an observable are the eigen values obtained from the eigen value equation given in postulate* (3) *above. Thus if the wave function ψ of the system is an eigen function of the operator for the observable, then measurement of that observable gives eigen value. If the wave function ψ of the system is not an eigen function, then the value of the observable obtained on measurement has a probability that can be calculated from the wave function.*

(5) *If a number of measurements are made over the configuration space, then the average value of the quantity (represented by $\bar{a}$) is given by*

$$\bar{a} = \frac{\oint \psi^* \hat{A} \psi \, d\tau}{\oint \psi^* \psi \, d\tau} \quad \text{... } (ii)$$

where $\oint$ represents integration over the whole of the configuration space.

This postulate can be understood from eq. (*i*)

$$a = \frac{\hat{A}\psi}{\psi}$$

But this expression does not give a constant value of a because it involves the function of the variable coordinates. By multiplying the numerator and the denominator by the complex conjugate ψ^* and integrating over the whole of the configuration space, the result obtained is no longer a function of coordinates and gives an average value of the constant a.

How to test whether the given function is an eigen function of the given operator?

Operate on the function with the operator and check whether the result is a constant factor times the original function. If so, the given function is an eigen function of the given operator.

Example 1. Find out whether the function cos *ax* is an eigen function of (*a*) *d/dx* (*b*) d^2/dx^2. What is the corresponding eigen value if any ?

Solution. (*a*) Here $\psi = \cos ax$

$$\therefore \quad \frac{d\psi}{dx} = \frac{d}{dx}(\cos ax) = -a \sin ax$$

Hence cos *ax* is not an eigen function of *d/dx*.

(*b*)

$$\frac{d^2\psi}{dx^2} = \frac{d^2}{dx^2}(\cos ax) = \frac{d}{dx}\left(\frac{d}{dx}\cos ax\right)$$

$$= \frac{d}{dx}(-a \sin ax) = -a^2 \cos ax = -a^2\psi$$

Hence cos *ax* is an eigen function of d^2/dx^2. Eigen value $= -a^2$.

Example 2. Determine whether e^{ax} is an eigen function of the operator *d/dx* and find the corresponding eigen value if any. Show that e^{ax^2} is not an eigen function of *d/dx*.

Solution. In the first case, $\psi = e^{ax}$

$$\therefore \quad \frac{d\psi}{dx} = \frac{d}{dx}(e^{ax}) = a\,e^{ax} = a\psi$$

Hence e^{ax} is an eigen function of *d/dx* and its eigen value = *a*.

In the second case, $\psi = e^{ax^2}$

$$\therefore \quad \frac{d\psi}{dx} = \frac{d}{dx}(e^{ax^2}) = 2axe^{ax^2} = 2\,ax\psi$$

Thus the result obtained is not a constant factor multiplied with ψ but a variable factor, 2*ax* multiplied with ψ. Hence e^{ax^2} is not an eigen function of *d/dx*.

Example 3. If all the eigen functions of two operators $\hat{A}$ and $\hat{B}$ are the same functions, prove that $\hat{A}$ and $\hat{B}$ commute with each other.

Solution. Let us suppose the eigen functions of the operators $\hat{A}$ and $\hat{B}$ are ψ_i and the corresponding eigen values are a_i and b_i respectively. Then eigen value equations will be $\hat{A}\psi_i = a_i\psi_i$ and $\hat{B}\psi_i = b_i\psi_i$.

The eigen functions of the operator $\hat{A}\,\hat{B}$ are obtained as follows :

$$\hat{A}\,\hat{B}\,\psi_i = \hat{A}(\hat{B}\psi_i) = \hat{A}(b_i\psi_i) = b_i\hat{A}\psi_i = b_i a_i \psi_i$$

The eigen functions of the operator $\hat{B}\,\hat{A}$ are obtained as follows :

$$\hat{B}\,\hat{A}\,\psi_i = \hat{B}(\hat{A}\psi_i) = \hat{B}(a_i\psi_i) = a_i\hat{B}\psi_i = a_i b_i \psi_i$$

As $b_i a_i = a_i b_i$, this means $\hat{A}\,\hat{B}\,\psi_i = \hat{B}\,\hat{A}\,\psi_i$ or $\hat{A}\,\hat{B} = \hat{B}\,\hat{A}$

Hence they commute with each other.

Example 4. Show that the function ψ = cos *ax* cos *by* cos *cz* is an eigen function of the Laplacian operator. What is the corresponding eigen value?

Solution. Laplacian operator $\nabla^2 = \frac{\partial^2}{\partial x^2} + \frac{\partial^2}{\partial y^2} + \frac{\partial^2}{\partial z^2}$

Hence,

$$\nabla^2\psi = \left(\frac{\partial^2}{\partial x^2} + \frac{\partial^2}{\partial y^2} + \frac{\partial^2}{\partial z^2}\right)(\cos ax \cos by \cos cz)$$

$$= \cos by \cos cz \frac{\partial^2}{\partial x^2}(\cos ax) + \cos ax \cos cz \frac{\partial^2}{\partial y^2}(\cos by) + \cos ax \cos by \frac{\partial^2}{\partial z^2}(\cos cz)$$

$$= \cos by \cos cz\,(-a^2 \cos ax) + \cos ax \cos cz\,(-b^2 \cos by) + \cos ax \cos by\,(-c^2 \cos cz)$$

$$\left[\because \frac{\partial^2}{\partial x^2}(\cos ax) = \frac{\partial}{\partial x}(-a \sin ax) = -a^2 \cos ax \text{ and so on}\right]$$

$$= -(a^2 + b^2 + c^2) \cos ax \cos by \cos cz = -(a^2 + b^2 + c^2)\,\psi$$

As $(a^2 + b^2 + c^2)$ is constant, hence the given function is an eigen function of Laplacian operator. Eigen value $= -(a^2 + b^2 + c^2)$.

Example 5. If ψ_1 and ψ_2 are the proper functions of an atom, show that any linear combination of these wave functions will also be a proper wave function of the Schrodinger equation.

Solution. Consideration for proper or eigen wave functions for a system of constant energy (E) is given by

$$\hat{H}\psi = E\psi$$

where $\hat{H}$ is Hamiltonian operator.

Thus, $\hat{H}\psi_1 = E\psi_1$, $\hat{H}\psi_2 = E\psi_2$.

Linear combination of ψ_1 and ψ_2 may be represented by ψ as

$$\psi = c_1\psi_1 + c_2\psi_2$$

Where c_1 and c_2 are constants.

$$\therefore \quad \hat{H}\psi = \hat{H}[c_1\psi_1 + c_2\psi_2]$$
$$= c_1\hat{H}\psi_1 + c_2\hat{H}\psi_2$$
$$= c_1 E\psi_1 + c_2 E\psi_2$$
$$= E(c_1\psi_1 + c_2\psi_2)$$
$$= E\psi$$

Hence, ψ is also proper function of the Schrodinger equation.

Example 6. What of the following wave functions are acceptable in quantum mechanics over the range when x goes from 0 to 2.

(i)* sin *x ***(ii)* tan *x*** ***(iii)* cos *x* + sin *x***

Solution.

(*i*) sin x is continuous, single valued and finite in the given range of x. It is acceptable. (sin x varies between –1 and 1 in the given range).

(*ii*) tan x is infinite for $x = \frac{\pi}{2}$. It is not acceptable.

(*iii*) cos x + sin x is continuous, single valued and finite in the given range. It is acceptable. (cos x + sin x) varies between – 1 to + 1 in the given range).

Example 7. If two operators $\hat{A}$ and $\hat{B}$ commute then they have the same set of eigen functions. Justify it.

Solution. Suppose the operator $\hat{A}$ has eigen function ψ, then

$$\hat{A}\psi = a\psi$$

where "*a*" is the eigen value since $\hat{A}$ and $\hat{B}$ commute

$$\hat{A}\hat{B} = \hat{B}\hat{A}\psi = \hat{B}a\psi = a\hat{B}\psi$$

This shows that $\hat{B}\psi$ is an eigen function of the operator $\hat{A}$ with eigen value a.

This is possible only if $(\hat{B}\psi)$ is a multiple of ψ *i.e.*

$$\hat{B}\psi = b\psi$$

where "*b*" is constant.

In other words, "ψ" is also an eigen function of $\hat{B}$.

1.19. DERIVATION OF SCHRODINGER WAVE EQUATION BASED ON THE POSTULATES OF QUANTUM MECHANICS

Using the postulates of quantum mechanics, Schrodinger wave equation can be derived in a simple manner as follows :

Consider a single particle *e.g.* an electron of mass m moving with a velocity v. Its total energy E is the sum of its kinetic energy (T) and potential energy (V), *i.e.*,

$$E = T + V \qquad \text{... (i)}$$

Put $$T = \frac{1}{2}mv^2 = \frac{p^2}{2m} \qquad \text{... (ii)}$$

where p represents the total momentum of the particle.

Further we know that

$$p^2 = p_x^2 + p_y^2 + p_z^2$$

where p_x, p_y and p_z are the momenta of the electron along the three mutually perpendicular directions along the three axes

$$T = \frac{p_x^2 + p_y^2 + p_z^2}{2m} \qquad \text{... (iv)}$$

Substitute this value in equation (*i*), we get

$$E = \frac{p_x^2 + p_y^2 + p_z^2}{2m} + V \qquad \text{... (v)}$$

According to the second postulate of quantum mechanics, the operators for the momenta p_x, p_y and p_z are

$$p_x = \frac{h}{2\pi i}\frac{\partial}{\partial x}$$

$$p_y = \frac{h}{2\pi i}\frac{\partial}{\partial y}$$

$$p_z = \frac{h}{2\pi i}\frac{\partial}{\partial z}$$

V is a function of position co-ordinates and hence the operator for V is V itself.

The operator for the energy E is the Hamiltonian operator, $\hat{H}$.

Hence equation (v) takes the form (taking into consideration the algebraic rules of the operators)

$$\hat{H} = \frac{1}{2m}\left[\left(\frac{h}{2\pi i}\frac{\partial}{\partial x}\right)^2 + \left(\frac{h}{2\pi i}\frac{\partial}{\partial y}\right)^2 + \left(\frac{h}{2\pi i}\frac{\partial}{\partial z}\right)^2\right] + V$$

$$= -\frac{h^2}{8\pi^2 m}\left(\frac{\partial^2}{\partial x^2} + \frac{\partial^2}{\partial y^2} + \frac{\partial^2}{\partial z^2}\right) + V$$

$$= -\frac{h^2}{8\pi^2 m}\nabla^2 + V \qquad \text{... (vii)}$$

where $\nabla^2 = \frac{\partial^2}{\partial x^2} + \frac{\partial^2}{\partial y^2} + \frac{\partial^2}{\partial z^2}$ represents the Laplacian operator.

According to the third postulate, we have

$$\hat{H}\psi = E\psi$$

$$\hat{H}\psi - E\psi = 0 \quad \text{... (viii)}$$

where E is the eigen value of the energy for the given state of the system.

Substituting the value of $\hat{H}$ from equation (*vii*) into equation (*viii*) we get

$$\left(-\frac{h^2}{8\pi^2 m}\nabla^2 + V\right)\psi - E\psi = 0 \quad \text{... (ix)}$$

which can be rearranged and written in the form

$$\nabla^2\psi + \frac{8\pi^2 m}{h^2}(E - V)\psi = 0 \quad \text{... (x)}$$

which is the required Schrodinger wave equation in one of the common forms.

1.20. PARTICLE IN A ONE DIMENSIONAL BOX

Particle in a box provides us the application of Schrodinger wave equation to the translational motion of a particle like electron, atom, etc. It also explains as to why the energies associated are quantised.

The motion of a particle in a one dimensional box is like the flow of an electron in a wire but still it is called a "particle in a box" and a general case can be considered in which we may assume that a single particle *e.g.*, a gas molecule of mass m is restricted to move in a region of space from $x = 0$ to $x = a$ and that its potential energy within the box is constant and taken as equal to zero for the sake of convenience. (Fig. 1.14). In order that the particle may remain within the box, it is essential to assume that the potential energy on or outside the walls is very high (= ∞) so that as soon as the particle reaches the walls, it is reflected back into the box.

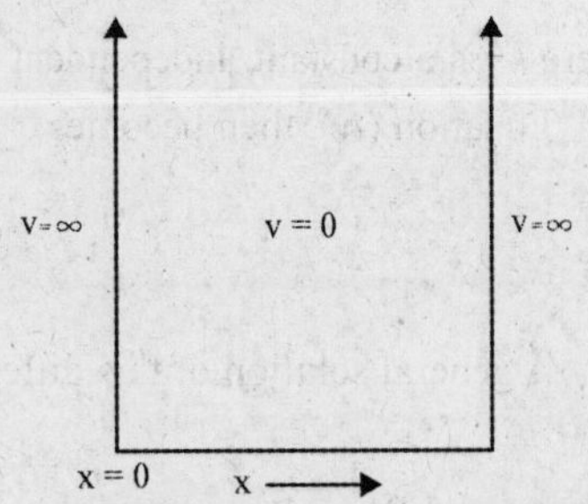

Fig. 1.14. A particle in a one dimensional box with P.E. = 0 inside the box and P.E. = ∞ on the walls of the box and outside the box

Schrodinger wave equation for one dimension is

$$\frac{d^2\psi}{dx^2} + \frac{8\pi^2 m}{h^2}(E - V)\psi = 0 \quad \text{... (i)}$$

where ψ has been taken as the function of x co-ordinate only.

Outside the box, $V = \infty$, therefore for outside the box, equation (*i*) becomes

$$\frac{d^2\psi}{dx^2} + \frac{8\pi^2 m}{h^2}(E - \infty)\psi = 0 \quad \text{... (ii)}$$

Neglecting E in comparison to ∞, equation (*ii*) reduces to

$$\frac{d^2\psi}{dx^2} - \infty\,\psi = 0 \qquad \left(\frac{8\pi^2 m}{h^2} \times \text{infinity} = \text{infinity}\right)$$

$$\frac{d^2\psi}{dx^2} = \infty\ \psi$$

$$\psi = \frac{1}{\infty}\frac{d^2\psi}{dx^2} = \text{zero} \qquad \text{... (iii)}$$

This proves that outside the box, $\psi = 0$ which implies that the particle cannot go outside the box.

For the particle within the box, $V = 0$, therefore the Schrodinger wave equation (*i*) takes the form

$$\frac{d^2\psi}{dx^2} + \frac{8\pi^2 m}{h^2}E\psi = 0 \qquad \text{... (iv)}$$

For the given state of the system, the energy E is constant which is one of the postulates of the quantum mechanics. Therefore we put

$$\frac{8\pi^2 m}{h^2}E = k^2 \qquad \text{... (v)}$$

where k^2 is a constant, independent of x.

Equation (*iv*), then becomes

$$\frac{d^2\psi}{dx^2} + k^2\psi = 0 \qquad \text{... (vi)}$$

A general solution of this differential equation is given by

$$\psi = A \sin kx + B \cos kx \qquad \text{... (vii)}$$

where A and B are constants.

Depending upon the value of A, B and k, ψ can have many values. But all the values are not acceptable. Only those values of ψ are acceptable which satisfy the boundary conditions, *viz.*,

$$\psi = 0 \text{ at } x = 0 \text{ and } x = a$$

Putting $\psi = 0$ when $x = 0$, equation (*vii*) becomes

$$0 = B \sin 0 + B \cos 0$$

$$= 0 + B \qquad (\because \sin 0 = 0 \text{ and } \cos 0 = 1)$$

i.e., $B = 0$

Thus when $x = 0$, equation (*vii*) becomes (by putting $B = 0$),

$$\psi = A \sin kx \qquad \text{... (viii)}$$

Putting $\psi = 0$ when $x = a$, equation (*viii*) becomes

$$0 = A \sin ka$$

i.e., $\sin ka = 0$ ... (*ix*)

This equation holds good only when the value of ka are integral multiplies of π *i.e.*,

$$ka = n\pi \qquad \text{... (x)}$$

where n is an integer *i.e.*, $n = 0, 1, 2, 3$..... However the value $n = 0$ may be excluded because it makes $k = 0$ and hence $\psi = 0$ for any value of a between 0 and a *i.e.,* within the box. This is not true because the particle is always assumed to be present within the box,

From equation (*x*)

$$k = \frac{n\pi}{a} \qquad \text{... (xi)}$$

Substituting this value in equation (*viii*), we get

$$\boldsymbol{\psi = A\sin\left(\frac{n\pi}{a}x\right)} \qquad \text{... (xii)}$$

This gives the expression for the eigen function ψ.

The expression for the eigen value of the energy may be obtained as follows :

From equation (*v*)

$$E = \frac{k^2h^2}{8\pi^2 m} \qquad \text{... (xiii)}$$

Substituting the value of k from equation (*xi*), we get

$$E = \frac{(n\pi/a)^2\, h^2}{8\pi^2 m}$$

i.e.

$$\boldsymbol{E = \frac{n^2h^2}{8ma^2}} \qquad \text{... (xiv)}$$

Equations (*xii*) and (*xiv*) are the solutions of the Schrodinger wave equation for a particle in one dimensional box.

The value of A in eq. (*xii*) can be obtained by normalisation of wave function.

The process of *normalisation of the wave function* can be carried out as follows :

$$\int_0^a \psi\psi^* d\tau = 1$$

Substituting the value of $\psi(=\psi^*)$ from equation (*xii*), we get

$$A^2\int_0^a \sin^2\frac{n\pi}{a}x\,dx = 1$$

or

$$A^2\cdot\frac{a}{2} = 1 \qquad \left[\text{as} \int_0^a \sin^2\left(\frac{n\pi}{a}x\right)dx = \frac{a}{2}\right]$$

or

$$A = \sqrt{\frac{2}{a}}$$

Trigonometric relation

$$2\sin^2 x = 1 - \cos 2x$$

$$\sin^2 x = \frac{1-\cos 2x}{2}$$

Hence the normalised wave function (which will also be a solution of the Schrodinger equation) is

$$\boxed{\boldsymbol{\psi = \sqrt{\frac{2}{a}}\sin\left(\frac{n\pi}{a}x\right)}}$$

If there are any two eigen functions ψ_m and ψ_n corresponding to two different values of n then ψ_m and ψ_n will be orthogonal to each other. This can be derived as under :

$$\int_0^a \psi_m \psi_n dx = \frac{2}{a}\int_0^a \sin\left(\frac{m\pi x}{a}\right)\sin\left(\frac{n\pi x}{a}\right)dx$$

$$= \frac{2}{a}\int_0^a \frac{1}{2}\left[\cos(m-n)\frac{\pi x}{a} - \cos(m+n)\frac{\pi x}{a}\right]dx$$

$[\because 2 \sin A \sin B = \cos(A - B) - \cos(A + B)]$

$$= \frac{1}{a}\left[\frac{\sin(m-n)\frac{\pi x}{a}}{(m-n)\frac{\pi}{a}} - \frac{\sin(m+n)\frac{\pi x}{a}}{(m+n)\frac{\pi}{a}}\right]_0^a$$

$$= \frac{1}{\pi}\left[\frac{\sin(m-n)\frac{\pi x}{a}}{(m-n)} - \frac{\sin(m+n)\frac{\pi x}{a}}{(m+n)}\right]_0^a$$

$$= \frac{1}{\pi}\left[\frac{\sin(m-n)\pi}{(m-n)} - \frac{\sin(m+n)\pi}{(m+n)}\right] = 0$$

[$\because$ m and n are integers and $m \# n$, therefore $(m - n)$ and $(m + n)$ are also integers and $\sin N\pi = 0$, where N is an integer]

Results from the study of a particle in a box

Expression for energy for a particle in one dimensional box is

$$E = \frac{n^2h^2}{8ma^2} \qquad \text{... }(i)$$

where m is the mass of the particle and a is the length of the box ($x = 0$ to $x = a$)

***(i)* Quantization of energy:** Since n can have only integral values equal to 1, 2, 3 etc., therefore from equation (*i*), it follows that the energy E associated with the motion of a particle in a box can have only discrete values *i.e.*, the energy is quantized.

The integer n is called the *quantum number* of the particle.

Further putting $n = 1, 2, 3$.... etc., the discrete energy levels obtained for the particle of mass m confined in the box of length a are shown in Fig. 1.15. It is important to note that as the quantum number n increases, the separation between them increases. It may also be noted that energy levels also depends upon the box length a. As a increases, *i.e.*, the space available to a particle increases, energy quanta become smaller and energy levels move closer together. If the box length becomes very large, quantization disappears"and there is a smooth transition from quantum behaviour to classical behaviour. (Continuous change).

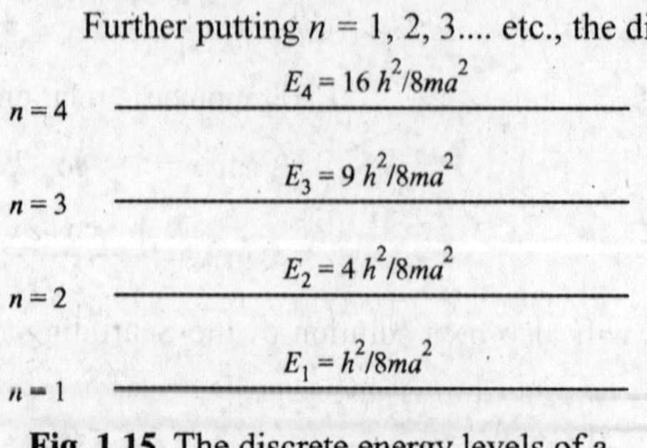

Fig. 1.15. The discrete energy levels of a particle of mass m confined in a box of lengh *a*.

***(ii)* Existence of zero point energy :** Minimum value of $n = 1$ ($n = 0$ is ruled out, as discussed earlier). Substituting this value in equation (*i*), we get

$$E = \frac{h^2}{8ma^2} \qquad \text{... } (ii)$$

This implies that the minimum energy possessed by the particle is not zero but has a definite value. This is called *zero point energy.*

***(iii)* Non-quantisation of energy for the particle.**

If the walls of the box are removed, so that the particle is free to move in a field whose potential energy may be supposed to be zero, then the boundary conditions are no longer applicable and so the constants *A, B* and *k* can have any value. The energy of the particle is then given by equation

$$E = \frac{k^2h^2}{8\pi^2 m}$$

As the constant *k* can have any value, the energy of the free particle is not quantised.

1.21. CALCULATION OF EXPECTATION VALUES USING WAVE FUNCTION

Physical property of a particle moving in one dimension which does not depend upon time can be calculated from the time-independent wave function $\psi(x)$ of the system. In these calculations, the probability of finding the particle in any region of the one-dimensional space is calculated using Born's postulate, according to which the probability $P(x)dx = \psi(x)\,\psi^*(x)\,dx$. Thus, the wave function helps us to calculate observables like position, momentum etc. However, if we carry out an experiment to measure the position *x* of the particle, $\psi(x)$ may not give the result of that particular experiment. But if we repeat the experiment a number of times, the value of the position *x* calculated from $\psi(x)$ is the same as the mean value of the different experiments. *This mean (or average) value, denoted by* $<x>$, *is called the expectation value of the observable x.*

(*i*) *Expectation value of position x*

According to one of the postulates of quantum mechanics, as operator for the position co-ordinate *x* is the multiplier *x*, therefore the expectation value $<x>$ will be given by

$$<x> = \int \psi^* x \psi \, dx = \int x\, \psi^* \psi \, dx = \int x\, P(x)\, dx \qquad \text{... } (i)$$

where $P(x)\,dx$ is the probability calculated by Born's postulate.

(*ii*) *Expectation value for any observable quantity, G*

To calculate the expectation value of any other physical quantity (or observable), *G*, remember that the operator for *G* is represented by $\hat{G}$ and apply the basic postulates of quantum mechanics. The expectation value $<G>$ is given by

$$<G> = \int_{-\infty}^{\infty} \psi^* \hat{G} \psi dx \qquad \text{... } (ii)$$

which would be same as the mean value of the results of the repeated experiments for the measurement of *G* in the state $\psi(x)$.

It may be noted that $\hat{G}$ is not a simple factor like *x*. It is an operator for the quantity *G*.

Hence $\int \psi^* \hat{G} \psi \, dx$ cannot be rearranged to give $\int \hat{G} \psi^* \psi \, dx$. Thus $\hat{G}$ must be allowed to operate on ψ. For example, according to one of the postulates of quantum mechanics, for the momentum, p_x, the operator p_x is $-\frac{ih}{2\pi}\frac{d}{dx}$ *i.e.* take derivative with respect to x and multiply the result by $-ih/2\pi$.

1.21.1. Expectation values of x, x^2, p_x and p_x^2 for a particle in one dimensional box of length a

For a particle in one dimensional box, we have the relation

$$\psi_n = \sqrt{\frac{2}{a}}\sin\left(\frac{n\pi x}{a}\right)$$

ψ_n is a normalized wave function, therefore

$$(i)\qquad \langle x\rangle = \int_0^a \psi\, x\, \psi dx = \int_0^a x\psi^2 dx$$

$$= \int_0^a x.\frac{2}{a}\sin^2\left(\frac{n\pi x}{a}\right)dx$$

$$= \frac{2}{a}\int_0^a x.\sin^2\left(\frac{n\pi x}{a}\right)dx \qquad \text{Trigonometric relation } 2\sin^2 A = 1-\cos 2A$$

$$= \frac{2}{a}\int_0^a x.\left[\frac{1-\cos\left(\frac{2n\pi x}{a}\right)}{2}\right]dx$$

Integrate the definite integral by parts using the formula $\int (uv)dx = u\int vdx - \int[\frac{du}{dx}\int vdx]\, dx$

The value of the integral comes out to be zero

$$= \frac{2}{a}\left(\frac{a^2}{4}\right) = \frac{a}{2} \qquad \text{... } (i)$$

$$(ii)\qquad \langle x^2\rangle = \int_0^a \psi x^2 dx = \int_0^a x^2\psi^2 dx = \frac{2}{a}\int_0^a x^2\sin^2\left(\frac{n\pi x}{a}\right)dx$$

$$= \frac{2}{a}\int_0^a x^2\left[\frac{1-\cos\left(\frac{2n\pi x}{a}\right)}{2}\right]dx$$

$$= \frac{2}{a}\left[\frac{a^3}{6} - \frac{a^3}{4n^2\pi^2}\right] = \frac{1}{a}\left[\frac{a^3}{3} - \frac{a^3}{2n^2\pi^2}\right]$$

$$= \left(\frac{a^2}{3} - \frac{a^2}{2n^2\pi^2}\right) \qquad \text{... } (ii)$$

(*iii*) $$< p_x >_n = \int_0^a \psi\left(\frac{-ih}{2\pi}\frac{d}{dx}\right)\psi dx$$

$$= \int_0^a \left[\left(\frac{2}{a}\right)^{1/2} \sin\frac{n\pi x}{a}\right]\left(\frac{-ih}{2\pi}\frac{d}{dx}\right)\left[\left(\frac{2}{a}\right)^{1/2} \sin\frac{n\pi x}{a}\right] dx$$

$$= -\frac{ih}{2\pi}\left(\frac{2}{a}\right)\int_0^a \sin\left(\frac{n\pi x}{a}\right)\left(\frac{n\pi}{a}\right)\cos\left(\frac{n\pi x}{a}\right) dx$$

$$= -\frac{ih}{2\pi}\left(\frac{2}{a}\right)\frac{n\pi}{a}\int_0^a \sin\left(\frac{n\pi x}{a}\right)\cos\left(\frac{n\pi x}{a}\right) dx$$

$$= 0 \text{ (on integration by parts)} \qquad \text{... } (iii)$$

This result is justified, because the average momentum has to be zero as the particle in a box cannot continue to travel in one direction only. Also note that the limits used are 0 and a in place of $-\infty$ to $+\infty$ because the wave function is known to be zero outside the box.

(*iv*) $$< p_x^2 >_n = \int_0^a \left[\left(\frac{2}{a}\right)^{1/2} \sin\frac{n\pi x}{a}\right]\left(\frac{-ih}{2\pi}\frac{d}{dx}\right)^2 \left[\left(\frac{2}{a}\right)^{1/2} \sin\frac{n\pi x}{a}\right] dx^*$$

$$= -\frac{h^2}{4\pi^2}\left(\frac{2}{a}\right)\int_0^a \sin\left(\frac{n\pi x}{a}\right)\frac{d^2}{dx^2}\left(\sin\frac{n\pi x}{a}\right) dx$$

$$= -\frac{h^2}{4\pi^2}\left(\frac{2}{a}\right)\int_0^a \sin\left(\frac{n\pi x}{a}\right)\left[(-)\left(\frac{n\pi}{a}\right)^2 \sin\left(\frac{n\pi x}{a}\right)\right] dx^*$$

$$= \left(\frac{h^2}{4\pi^2}\right)\left(\frac{2}{a}\right)\left(\frac{n\pi}{a}\right)^2 \int_0^a \sin^2\left(\frac{n\pi x}{a}\right) dx$$

$$= \frac{n^2h^2}{2a^3}\int_0^a \sin^2\left(\frac{n\pi x}{a}\right) dx$$

$$= \frac{n^2h^2}{2a^3}\int_0^a \frac{1-\cos\frac{(2n\pi x)}{a}}{2} dx \qquad [\because 2\sin^2 A = 1 - \cos 2A]$$

---

* The operator for p_x^2 is $(\hat{p}_x)\ (\hat{p}_x)$

$$\hat{p}_x^2 = (\hat{p}_x)(\hat{p}_x) = \left(-\frac{ih}{2\pi}\frac{d}{dx}\right)\left(-\frac{ih}{2\pi}\frac{d}{dx}\right) = -\frac{h^2}{4\pi^2}\frac{d^2}{dx^2}$$

* $\int \sin x dx = -\cos x + C$, $\int \cos x dx = \sin x + C$.

$\frac{d}{dx}(\sin x) = \cos x$, $\frac{d}{dx}(\cos x) = -\sin x$. That is why (–) sign appears.

$$= \frac{n^2h^2}{4a^3}\left[\frac{x - \sin\frac{(2n\pi x)}{a}}{\frac{2n\pi x}{a}}\right]_0^a$$

$$= \frac{n^2h^2}{4a^3}\cdot a = \frac{n^2h^2}{4a^2} = 2mE \qquad \left[\because E = \frac{n^2h^2}{8ma^2}\right]$$

1.21.2. Calculation of standard deviation in x and p_x *i.e.* Δx and Δp_x and derivation of Heisenberg uncertainty principle

The standard deviation in the position x for the particle in an eigen state of a one dimensional box is given by

$$\Delta x = (< x^2 > - < x >^2)^{1/2} \qquad \text{... }(i)$$

Substituting the values derived in sec. 1.21.1, we get

$$\Delta x = \left[\left(\frac{a^2}{3} - \frac{a^2}{2n^2\pi^2}\right) - \left[\frac{a}{2}\right]^2\right]^{1/2} = \left[\left(\frac{a^2}{12} - \frac{a^2}{2n^2\pi^2}\right)\right]^{1/2}$$

$$= a\left(\frac{1}{12} - \frac{1}{2n^2\pi^2}\right)^{1/2} \qquad \text{... }(ii)$$

The standard deviation in the momentum, p_x for the particle is given by

$$\Delta p_x = (< p^2 > - < p >^2)^{1/2} \qquad \text{... }(iii)$$

Substituting the values derived in Sec. 1.21.1, we get

$$\Delta p_x = \left(< p^2 >\right)^{1/2} = \left(\frac{n^2h^2}{4a^2}\right)^{1/2} = \frac{nh}{2a} \qquad \text{... }(iv)$$

$$\therefore \quad \Delta x \cdot \Delta p = a\left(\frac{1}{12} - \frac{1}{2n^2\pi^2}\right)^{1/2} \cdot \frac{nh}{2a}$$

$$= \frac{nh}{2}\left(\frac{1}{12} - \frac{1}{2n^2\pi^2}\right)^{1/2}$$

$$= \frac{nh}{2} \times \frac{1}{2n\pi}\left(\frac{4n^2\pi^2}{12} - \frac{4n^2\pi^2}{2n^2\pi^2}\right)^{1/2}$$

$$= \frac{h}{4\pi}\left(\frac{n^2\pi^2}{3} - 2\right)^{1/2} \qquad \text{... }(v)$$

The quantity on the right hand side is $> \frac{h}{4\pi}$ because $(n^2\pi^2/3) > 2$ always. Hence

$$\Delta x \Delta p > \frac{h}{4\pi} \qquad \text{... }(vi)$$

which is Heisenberg's uncertainty principle.

Example 1. Calculate the probability of finding the particle between 0.49 *a* and 0.51 *a* for the states ψ_1 and ψ_2.

Solution. Probability $= \int_{0.49a}^{0.51a} \psi^2 dx$

$$= \int_{0.49a}^{0.51a} \frac{2}{a}\sin^2\left(\frac{n\pi x}{a}\right)dx$$

$$= \frac{2}{a}\int_{0.49a}^{0.51a} \sin^2\left(\frac{n\pi x}{a}\right)dx$$

$$= \frac{2}{a}\int_{0.49a}^{0.51a} \frac{1-\cos\left(\frac{2n\pi x}{a}\right)}{a}dx$$

$$= \frac{1}{a}\int_{0.49a}^{0.51a} \left[1-\cos\left(\frac{2n\pi x}{a}\right)\right]dx$$

$$= \frac{1}{a}\left[x-\frac{\sin\left(\frac{2n\pi x}{a}\right)}{\frac{2n\pi}{a}}\right]_{0.49a}^{0.51a}$$

$$= \frac{1}{n\pi}\left[\frac{n\pi x}{a}-\frac{1}{2}\sin\frac{2n\pi x}{a}\right]_{0.49a}^{0.51a}$$

Putting $n = 1$ and reading the value of sines of the angles from the tables, Probability = **0.0399**

Similarly, for $n = 2$, Probability = **0.0001**

Example 2. What is the ground state energy for an electron which is confined to a potential well (one-dimensional box) having a width of 0.2 nm ?

Solution. The energy for a particle in a one-dimensional box is given by

$$E = \frac{n^2h^2}{8ma^2}$$

For the ground state, $n = 1$

For the electron, $m = 9.11 \times 10^{-31}$ kg

We are given $a = 0.2 \text{ nm} = 0.2 \times 10^{-9}$ m

Also we know that $h = 6.626 \times 10^{-34}$ Js

$$\therefore \quad E = \frac{(1)^2(6.626\times10^{-34})^2}{8\times9.11\times10^{-31}\times(0.2\times10^{-9})^2}$$

$$= 1.506 \times 10^{-18}\text{J}$$

$\therefore$ For 1 mole of electrons, the energy will be

$$E = \frac{(1.506\times10^{-18})\times(6.022\times10^{23})}{10^3}\text{ kJ mol}^{-1}$$

$$= \mathbf{907\ kJ\ mol^{-1}}$$

1.22. PARTICLE IN A THREE-DIMENSIONAL BOX

Consider a single particle *e.g.*, a gas molecule of mass m restricted to move within a rectangular box of dimensions a, b and c along the X-axis, Y-axis and Z-axis respectively (Fig 1.16). As assumed earlier that the potential energy of the particle within the box is

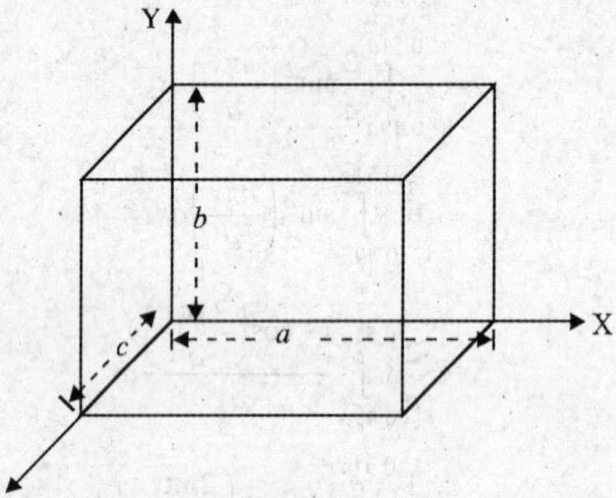

Fig. 1.16. Particle in a three dimensional box.

constant (taken as equal to zero) and on the walls or outside, it suddenly rises to a very large value (= ∞).

Now, the Schrodinger wave equation for three dimensions is

$$\frac{\partial^2 \psi}{\partial x^2} + \frac{\partial^2 \psi}{\partial y^2} + \frac{\partial^2 \psi}{\partial z^2} + \frac{8\pi^2 m}{h^2}(E - V)\psi = 0 \qquad \dots (i)$$

where E represents the total energy of the particle, V its potential energy and ψ is a function of the three co-ordinates x, y and z.

For the particle within the box, $V = 0$. Hence equation (i) takes the form

$$\frac{\partial^2 \psi}{\partial x^2} + \frac{\partial^2 \psi}{\partial y^2} + \frac{\partial^2 \psi}{\partial z^2} + \frac{8\pi^2 m}{h^2} E\psi = 0 \qquad \dots (ii)$$

To solve this differential equation it is convenient to separate the variables. For this purpose, ψ, which is a function of x, y and z is taken as equal to the product of three functions $X(x)$, $Y(y)$ and $Z(z)$, each in one variable, *i.e.,*

$$\phi = X(x).\ Y(y).\ Z(z) \qquad \dots (iii)$$

or for the sake of simplicity, we may write it as

$$\phi = XYZ \qquad \dots (iv)$$

Substituting this value of ϕ in equation (ii), we get

$$\frac{\partial^2 (XYZ)}{\partial x^2} + \frac{\partial^2 (XYZ)}{\partial y^2} + \frac{\partial^2 (XYZ)}{\partial z^2} + \frac{8\pi^2 m}{h^2} EXYZ = 0$$

or

$$YZ\frac{\partial^2 X}{\partial x^2} + XZ\frac{\partial^2 Y}{\partial y^2} + XY\frac{\partial^2 Z}{\partial z^2} + \frac{8\pi^2 m}{h^2} EXYZ = 0$$

Dividing throughout by XYZ, we get

$$\frac{1}{X}\frac{\partial^2 X}{\partial x^2}+\frac{1}{Y}\frac{\partial^2 Y}{\partial y^2}+\frac{1}{Z}\frac{\partial^2 Z}{\partial z^2}+\frac{8\pi^2 m}{h^2}E=0$$

$$\frac{1}{X}\frac{\partial^2 X}{\partial x^2}+\frac{1}{Y}\frac{\partial^2 Y}{\partial y^2}+\frac{1}{Z}\frac{\partial^2 Z}{\partial z^2}=-\frac{8\pi^2 m}{h^2}E \quad \text{... }(v)$$

Total energy, E of the particle is constant. Therefore R.H.S. of equation (v) is constant and we may put $\frac{8\pi^2 mE}{h^2}=k^2$. So we may write equation (v) as

$$\frac{1}{X}\frac{\partial^2 X}{\partial x^2}+\frac{1}{Y}\frac{\partial^2 Y}{\partial y^2}+\frac{1}{Z}\frac{\partial^2 Z}{\partial z^2}=-k^2 \quad \text{... }(vi)$$

Thus the sum of the three terms is constant. Suppose we change the value of x, keeping y and z constant, the sum of three terms should be again equal to the same constant, *viz.* k^2. This indicates that the first term on L.H.S. of equation (vi) is not only independent of y and z but is also independent of x. Likewise the second and the third terms must also be independent of x and z. In other words, each term on the L.H.S. of equation (vi) must be constant and the sum of the three constants should be equal to k^2. Hence we may write

$$\frac{1}{X}\frac{\partial^2 X}{\partial x^2}=-k_x^2 \quad \text{... }(vii)$$

$$\frac{1}{Y}\frac{\partial^2 Y}{\partial y^2}=-k_y^2 \quad \text{... }(viii)$$

$$\frac{1}{Z}\frac{\partial^2 Z}{\partial z^2}=-k_z^2 \quad \text{... }(ix)$$

where $k_x{}^2$ and $k_y{}^2$ and $k_z{}^2$ are constants such that

$$k_x^2+k_y^2+k_z^2=k^2 \quad \text{... }(x)$$

and the values of these constants are as follows :

$$k_x^2=\frac{8\pi^2 mE_x}{h^2} \quad \text{... }(xi)$$

$$k_y^2=\frac{8\pi^2 mE_y}{h^2} \quad \text{... }(xii)$$

$$k_z^2=\frac{8\pi^2 mE_z}{h^2} \quad \text{... }(xiii)$$

where E_x, E_y and E_z represent the components of the energy in the directions parallel to X-axis, Y-axis and Z-axis respectively such that

$$E=E_x+E_y+E_z \quad \text{... }(xiv)$$

Equations (*vii*) to (*ix*) may be rewritten as

$$\frac{\partial^2 X}{\partial x^2} + k_x^2 X == 0, \frac{\partial^2 Y}{\partial y^2} + k_y^2 Y == 0 \text{ and } \frac{\partial^2 Z}{\partial z^2} = k_z^2 Z = 0$$

Each of these equations is of the same form as for particle in one-dimensional box.

The solutions of these equations are :

$$X = \sqrt{\frac{2}{a}} \sin \frac{n_x \pi}{a} x, Y = \sqrt{\frac{2}{b}} \sin \frac{n_y \pi}{b} y$$

and $$Z = \sqrt{\frac{2}{c}} \sin \frac{n_z \pi}{c} z \qquad ... (xv)$$

and $$E_x = \frac{n_x^2 h^2}{8ma^2}, E_y = \frac{n_y^2 h^2}{8mb^2} \text{ and } E_z = \frac{n_z^2 h^2}{8mc^2} \qquad ... (xvi)$$

where n_x, n_y and n_z having integral values represent three quantum numbers in directions parallel to the three axes.

The normalised wave function for a particle in three dimensional box is

$$\psi = XYZ$$

$$= \left(\sqrt{\frac{2}{a}} \sin \frac{n_x \pi}{a} x\right)\left(\sqrt{\frac{2}{b}} \sin \frac{n_y \pi}{b} y\right)\left(\sqrt{\frac{2}{c}} \sin \frac{n_z \pi}{c} z\right)$$

$$= \sqrt{\frac{8}{abc}} \sin\left(\frac{n_x \pi}{a} x\right) \sin\left(\frac{n_y \pi}{b} y\right) \sin\left(\frac{n_z \pi}{c} z\right) \qquad ... (xvii)$$

The total energy (E) will be given by

$$E = E_x + E_y + E_z$$

$$= \frac{n_x^2 h^2}{8ma^2} + \frac{n_y^2 h^2}{8mb^2} + \frac{n_z^2 h^2}{8mc^2}$$

or $$\boxed{E = \frac{h^2}{8m}\left(\frac{n_x^2}{a^2} + \frac{n_y^2}{b^2} + \frac{n_z^2}{c^2}\right)} \qquad ... (xviii)$$

Thus three quantum numbers follow from the solution of the Schrodinger wave equation in three dimensions.

1.23. CONCEPT OF DEGENERACY

When the sides of the three dimensional box are equal

$$a = b = c$$

The equation for the energy of a particle in three dimensional box

$$E = \frac{h^2}{8m}\left(\frac{n_x^2}{a^2} + \frac{n_y^2}{b^2} + \frac{n_z^2}{c^2}\right) \text{ reduces to}$$

$$E = \frac{h^2}{8ma^2}\left(n_x^2 + n_y^2 + n_z^2\right) \qquad ... (i)$$

The different combinations of the values of quantum number n_x, n_y and n_z may give the same value of the energy E. For example, the following six combinations of the values 1, 2 and 3 give the same value of E.

n_x	1	1	2	2	3	3
n_y	2	3	1	3	1	2
n_z	3	2	3	1	2	1

Such an energy level having different states of the system (Different wave functions) but having the same energy is said to be **degenerate** and the number of independent wave functions associated with the given energy level is called its **degeneracy.**

The degeneracies of a few energy levels with the different quantum numbers are given below:

$n_x n_y n_z$	111	211	221	311	222	321	322
Degeneracy	1	3	3	3	1	6	3

As an example, the three states corresponding to the quantum numbers 211 will have the quantum numbers, 211, 121 and 112. All these states will have energy $= \dfrac{h^2}{8ma^2}\,(4 + 1 + 1)$

$\dfrac{3h^2}{4ma^2}$ [obtained by putting the values of n_x, n_y and n_z in eqn. (*i*)]

This is illustrated in Fig. 1.17

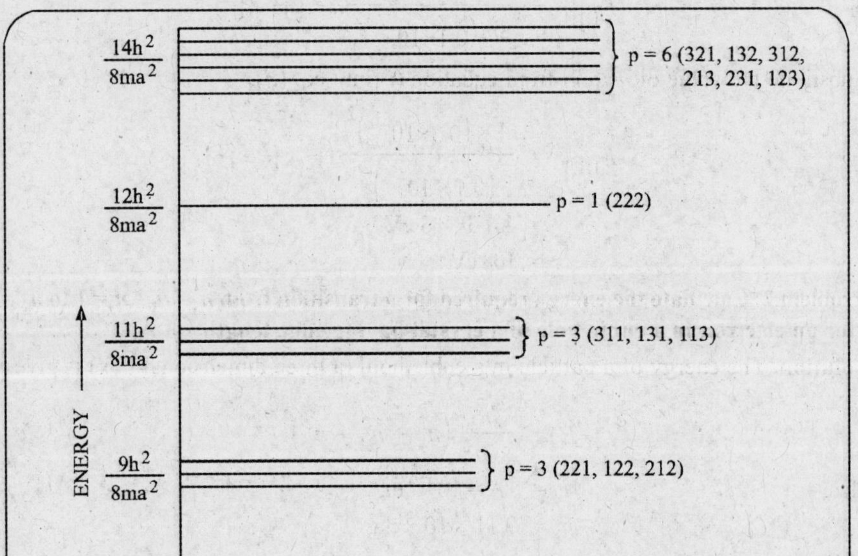

Fig. 1.17. Representation of degeneracy level, for a particle in cubical box.

It may be mentioned here that degeneracies arise in quantum mechanics when there is some element of symmetry. If the symmetry is broken *e.g.* by taking a box whose sides have different lengths, the degeneracy disappears.

SOLVED PROBLEMS ON PARTICLE IN A BOX

Problem 1. Calculate the expected ground state energy of a hydrogen atom electron assumed to be present in a three dimensional cubical box of length 0.1 nm if the ground state energy of the electron in a one dimensional box of length 0.3 nm is 4 eV.

Solution. For one-dimensional box, for ground state $n = 1$ so that

$$E_1 = \frac{n^2h^2}{8ma^2} = \frac{(1)^2(h)^2}{8m(0.3\times10^{-9})^2} = 4eV$$

or,
$$\frac{h^2}{8m} = 4\times(0.3\times10^{-9})^2 \quad \text{... (i)}$$

For three dimensional box, for ground state,

$n_x = n_y = n_z = 1$ so that

$$E_{1,1,1} = \frac{h^2}{8ma^2}\left(n_x^2 + n_y^2 + n_z^2\right)$$

or
$$E_{1,1,1} = \frac{h^2}{8m(0.1\times10^{-9})^2}(1^2+1^2+1^2) \quad \text{... (ii)}$$

Substitute the value of $h^2/8\,m$ from equation (*i*) into eq. (*ii*)

$$E_{1,1,1} = \frac{4\times(0.3\times10^{-9})^2}{(0.1\times10^{-9})^2}(1^2+1^2+1^2)$$

$$= 4\times3^2\times3 \text{ eV}$$

$$= 108 \text{ eV}$$

Problem 2. Calculate the energy required for a transition from $n_x = n_y = n_z = 1$ to $n_x = n_y = 1$, $n_z = 2$ for an electron in a cubic hole of a crystal having edge length = 1 Å.

Solution. The energy of a particle in a cubic hold (a three dimensional box) is given by

$$E = \frac{h^2}{8ml^2}\left(n_x^2 + n_y^2 + n_z^2\right)$$

Here
$$l = 1\text{Å} = 10^{-10}\text{ m},$$
$$m = 9.11\times10^{-31}\text{ kg}$$

∴ For $n_x = n_y = 1,\ n_z = 1$

$$E_{1,1,1} = \frac{(6.626\times10^{-34})^2}{8(9.11\times10^{-31})(10^{-10})^2}(1^2+1^2+1^2)$$
$$= 1.8072\times10^{-17}\text{ J}$$

For $n_x = n_y = n_z = 2$

$$E_{1,1,2} = \frac{(6.626\times10^{-34})^2}{8(9.11\times10^{-31})(10^{-10})^2}(1^2+1^2+2^2)$$
$$= 3.6144\times10^{-17}\text{J}.$$

HenceΔE $=(3.6144 \times 10^{-17} - 1.8072 \times 10^{-17})$ J

$= \mathbf{1.8072 \times 10^{-17}}$ **J**

Example 3. An electron is confined to an infinite one dimensional box of width 4Å. Calculate its energy (in eV) in the fourth energy level.

Solution. For one dimensional box,

$$E = \frac{n^2h^2}{8mL^2}$$

Here, $n = 4, m = 9.1 \times 10^{-31}$ kg

$L = 4 \times 10^{-10}\, m, h = 6.626 \times 10^{-34}$ Js

$$E = \frac{(4)^2 \times (6{\cdot}26 \times 10^{-31}\text{ Js})^2}{8\times(9{\cdot}1\times10^{-31}\text{ kg})\times(4\times10^{-10}\, m)^2}$$

$$= 6.03 \times 10^{-18}\text{ J} \qquad (1\text{ J} = 1\text{ kg } m^2\, s^{-1})$$

$$= \frac{6{\cdot}03 \times 10^{-18}}{1{\cdot}6 \times 10^{-19}}\text{ eV} \qquad (1\text{ eV} = 1.6 \times 10^{-19}\text{ J})$$

$$= 37.68\text{ eV.}$$

Example 4. An electron is confined to a molecule of length 1nm. What is its minimum energy? What is the probability of finding it is the region of the molecule lying between $x = 0$ and $x = 0.2$ nm?

Solution. The system of an electron confined to length of 1 nm is just like a particle in one dimensional box; so energy is

$$E = \frac{n^2h^2}{8mL^2} \text{ and wave function } \psi = \sqrt{\frac{2}{L}} \sin \frac{n\pi}{L} x.$$

For minimum energy $n = 1$, we get

$$E = \frac{h^2}{8mL^2} \text{ and } \psi = \sqrt{\frac{2}{L}} \sin \frac{\pi}{L} x.$$

Now $$E = \frac{h^2}{8mL^2} = \frac{(6{\cdot}626 \times 10^{-34}\text{ Js})^2}{8\times(9{\cdot}1 \times 10^{-31}\text{ kg}) \times (1 \times 10^{-9}\, m)^2}$$

$$= 6.025 \times 10^{-20}\text{ J.}$$

The probability of finding the electron in the given region is:

$$= \int_0^{0{\cdot}2\,nm} \psi\psi^*\, dx = \int_0^{0{\cdot}2\,nm} \psi^2\, dx$$

$$= \frac{2}{L} \int_0^{0{\cdot}2\,nm} \sin^2 \frac{\pi}{L} x\, dx$$

$$= \frac{2}{L} \int_0^{0{\cdot}2\,nm} \frac{1-\cos\frac{2\pi}{L}x}{2}\, dx$$

$$= \frac{2}{L} \int_0^{0\cdot 2\,nm} \frac{1}{2} dx - \int_0^{0\cdot 2\,nm} \frac{1}{2} \cos \frac{2\pi}{L} x\, dx$$

$$= \frac{2}{L} \left[\left| \frac{x}{2} \right|_0^{0\cdot 2\,nm} - \frac{1}{2} \left| \frac{\sin \frac{2\pi}{L}}{\frac{2\pi}{L}} \right|_0^{0\cdot 2\,nm} \right]$$

$$= \frac{2}{1nm} \left[\frac{0\cdot 2\,nm}{2} - \frac{1}{2} \times \frac{1nm}{2\pi} \sin \frac{2\pi \times 0\cdot 2\,nm}{1\,nm} \right]$$

$$= \left[0\cdot 2 - \frac{1}{2\pi} \sin 0\cdot 4\pi \right]$$

$$= \left[0\cdot 2 - \frac{1}{2\pi} \times 0\cdot 9511 \right]$$

$$= [0.2 - 0.1514]$$

$$= 0.0486.$$

Problem 5. What will happen if the walls of the one dimensional box are suddenly removed?

Solution. When the walls of a one-dimensional box are suddenly removed, the particle in it becomes free to move without any restriction on the value of potential energy.

Accordingly, it can possess any value of energy *i.e.*, the energy of the particle is no longer quantized, rather it becomes continuous.

1.24. SCHRODINGER WAVE EQUATION IN TERMS OF POLAR (SPHERICAL) CO-ORDINATES

The Schrodinger equation in three dimensions is

$$\frac{\partial^2 \psi}{\partial x^2} + \frac{\partial^2 \psi}{\partial y^2} + \frac{\partial^2 \psi}{\partial z^2} + \frac{8\pi^2 m}{h^2} (E - V)\, \psi = 0 \quad ...(i)$$

In order to apply it to hydrogen and hydrogen-like particles, it is convenient to express the above equation in terms of polar co-ordinates (r, θ, ϕ), (Fig. 1.18) for which the transformations are :

$$x = r \sin \theta \cos \phi$$

$$y = r \sin \theta \sin \phi$$

$$z = r \cos \theta$$

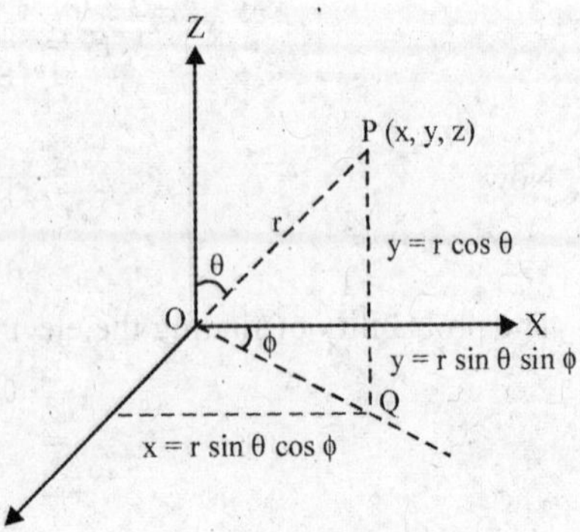

Fig. 1.18. Cartesian and spherical co-ordinates (PQ is perpendicular on the plane XOY)

Substituting these values for x, y and z in equation (i) and simplifying, the Schrodinger equation is obtained in spherical co-ordinates as follows :

$$\frac{1}{r^2} \frac{\partial}{\partial r} \left(r^2 \frac{\partial \psi}{\partial r} \right) + \frac{1}{r^2 \sin \theta} \frac{\partial}{\partial \theta} \left(\sin \theta \frac{\partial \psi}{\partial \theta} \right) + \frac{1}{r^2 \sin^2 \theta} \frac{\partial^2 \psi}{\partial \phi^2} + \frac{8\pi^2 m}{h^2} (E - V)\, \psi = 0 \quad ... (ii)$$

Schrodinger wave equation for hydrogen-like particles

Schrodinger wave equation for the motion of a single particle is given by

$$\frac{\partial^2 \psi}{\partial x^2} + \frac{\partial^2 \psi}{\partial y^2} + \frac{\partial^2 \psi}{\partial z^2} + \frac{8\pi^2 m}{h^2}(E - V)\,\psi = 0 \qquad \ldots (i)$$

where m is the mass of the particle, E is its total energy, V is its potential energy, and x, y, z are the coordinates of the particle.

For hydrogen-like particles, the mass m is replaced by the reduced mass (μ) of the system given by the equation

$$\mu = \frac{m_e \times m_n}{m_e + m_n} \qquad \ldots (ii)$$

where m_e and m_n represent the mass of the electron and the nucleus respectively. (As $m_e < m_n$, in fact $\mu \approx m_e$)

A hydrogen like atom can have two types of motion, the revolution of the electron round the nucleus and the translational motion of the centre of mass. Taking only the revolution of the electron around the nucleus (assuming that the nucleus does not move), E will represent the electronic energy.

Thus equation (*i*) for a hydrogen-like atom becomes

$$\frac{\partial^2 \psi}{\partial x^2} + \frac{\partial^2 \psi}{\partial y^2} + \frac{\partial^2 \psi}{\partial z^2} + \frac{8\pi^2 \mu}{h^2}(E - V)\,\psi = 0 \qquad \ldots (iii)$$

Here x, y, z respresent the co-ordinates of the centre of mass of the system.

This is the same equation as that of a single particle of mass μ (which is nearly equal to that of the electron) moving in a field of potential V.

For the hydrogen-like particles, the electron is moving under the central force field of the nucleus. Hence if $-e$ is the charge on the electron and $+Ze$ is the charge on the nucleus (Z being the atomic number of the element) and r is the distance between the electron and the nucleus, the value of V will be given by

$$V = -\frac{Ze^{2*}}{r} \qquad \ldots (iv)$$

Substituting this value of V in eqn. (*iii*), Schrodinger wave equation for hydrogen like particles in terms of spherical coordinates, can be written using eq. (*ii*) sec. 1.24

$$\frac{1}{r^2}\frac{\partial}{\partial r}\left(r^2 \frac{\partial \psi}{\partial r}\right) + \frac{1}{r^2 \sin\theta}\frac{\partial}{\partial \theta}\left(\sin\theta \frac{\partial \psi}{\partial \theta}\right) + \frac{1}{r^2 \sin^2\theta}\frac{\partial^2 \psi}{\partial \phi^2} + \frac{8\pi^2 \mu}{h^2}\left(E + \frac{Ze^2}{r}\right)\psi = 0 \qquad \ldots (v)$$

or

$$\frac{1}{r^2}\left(\frac{\partial}{\partial r}\left(r^2 \frac{\partial \psi}{\partial r}\right) + \frac{1}{\sin\theta}\frac{\partial}{\partial \theta}\left(\sin\theta \frac{\partial \psi}{\partial \theta}\right) + \frac{1}{\sin^2\theta}\frac{\partial^2 \psi}{\partial \phi^2}\right) + \frac{8\pi^2 \mu}{h^2}\left(E + \frac{Ze^2}{r}\right)\psi = 0 \qquad \ldots (vi)$$

*Force acting between the particles is given by

$$F = \frac{-e \cdot Ze}{r^2} = -\frac{Ze^2}{r^2} \quad \text{(Coulomb's law)}$$

$$\therefore \quad V = \int_{\infty}^{r} F\,dr = \int_{\infty}^{r} -\frac{Ze^2}{r^2}\,dr = -\frac{Ze^2}{r}$$

1.25. SEPARATION OF VARIABLES

ψ is a function of three variables r, θ and φ see Eq. (*vi*) above. To separate the variables, as before we suppose that ψ (r, θ, φ) is a product of three functions $R(r)$, $\Theta(\theta)$ and $\Phi(\phi)$, each being a function of one variable only, as indicated *i.e.*,

$$\psi(r, \theta, \phi) = R(r), \Theta(\theta), \Phi(\phi)$$

or For the sake of simplicity, we may write

$$\psi = R.\Theta.\Phi$$

Differentiating this equation to get different derivatives required in equation (*vi*) above and substituting these values in that equation

$$\frac{1}{r^2}\left[\Theta\Phi\frac{d}{dr}\left(r^2\frac{dR}{dr}\right)+\frac{R\Phi}{\sin\theta}\frac{d}{d\theta}\left(\sin\theta\frac{d\Theta}{d\theta}\right)+\frac{R\Theta}{\sin^2\theta}\frac{d^2\Phi}{d\phi^2}\right]+\frac{8\pi^2\mu}{h^3}\left(E+\frac{Ze^2}{r}\right)R\Theta\Phi = 0 \quad ...(i)$$

Multiplying throughout by r^2 and dividing by $R\,\Theta\,\phi$, we get

$$\frac{1}{R}\frac{d}{dr}\left(r^2\frac{dR}{dr}\right)=\frac{1}{\Theta\sin\theta}\frac{d}{d\theta}\left(\sin\theta\frac{d\Theta}{d\theta}\right)+\frac{1}{\Phi\sin^2\theta}\frac{d^2\Phi}{d\phi^2}+\frac{8\pi^2\mu r^2}{h^3}\left(E+\frac{Ze^2}{r}\right)=0 \quad ...(ii)$$

Writting the terms of the radial variables r on L.H.S. and those of the angular variables θ and φ on R.H.S., equation (*ii*) takes the form

$$\frac{1}{R}\frac{d}{dr}\left(r^2\frac{dR}{dr}\right)+\frac{8\pi^2\mu r^2}{h^2}\left(E+\frac{Ze^2}{r}\right)=-\frac{1}{\Theta\sin\theta}\frac{d}{d\theta}\left(\sin\theta\frac{d\Theta}{d\theta}\right)-\frac{1}{\Phi\sin^2\theta}\frac{d^2\Phi}{d\phi^2} \quad ...(iii)$$

As r, θ and φ are independent variables, equation (*iii*) holds good if each side is equal to the same constant, say equal to β. Thus

$$\frac{1}{R}\frac{d}{dr}\left(r^2\frac{dR}{dr}\right)+\frac{8\pi^2\mu r^2}{h^2}\left(E+\frac{Ze^2}{r}\right)=\beta \quad ...(iv)$$

(called the *radial equation*)

and

$$\frac{1}{\Theta\sin\theta}\frac{d}{d\theta}\left(\sin\theta\frac{d\Theta}{d\theta}\right)+\frac{1}{\Phi\sin^2\theta}\frac{d^2\Phi}{d\phi^2}=-\beta \quad ...(v)$$

(called the *angular equation*)

To further separate the terms of θ and φ, multiply equation (*v*) throughout by $\sin^2\theta$. Thus

$$\frac{\sin\theta}{\Theta}\frac{d}{d\theta}\left(\sin\theta\frac{d\Theta}{d\theta}\right)+\frac{1}{\Phi}\frac{d^2\Phi}{d\phi^2}=-\beta\sin^2\theta$$

or

$$\frac{\sin\theta}{\Theta}\frac{d}{d\theta}\left(\sin\theta\frac{d\Theta}{d\theta}\right)+\beta\sin^2\theta=-\frac{1}{\Phi}\frac{d^2\Phi}{d\phi^2} \quad ...(vi)$$

Again this equation holds good if each side is equal to a constant. Using the constant as m^2, we get

$$\frac{\sin\theta}{\Theta}\frac{d}{d\theta}\left(\sin\theta\frac{d\Theta}{d\theta}\right)+\beta\sin^2\theta=m^2 \quad ...(vii)$$

and $$\frac{1}{\Phi}\frac{d^2\Phi}{d\phi^2} = -m^2 \qquad \text{... }(viii)$$

Thus the Schrodinger wave equation for the hydrogen-like atoms has been separated into three equations *viz.* (*iv*), (*vii*) and (*viii*) in r, θ and ϕ respectively.

For hydrogen like atom, it is found that $\beta = l\,(l+1)$ where l is zero or an integer (*i.e.* $l = 0$, 1, 2, 3.........)

1.26. EXPRESSIONS FOR THE ANGULAR SPHERICAL WAVE FUNCTIONS

The solutions of equations in θ and ϕ give the expressions for the angular wave function Θ and Φ.

Equation in the variable ϕ (eqn. *viii* sec. 1.25) is

$$\frac{1}{\Phi}\frac{d^2\Phi}{d\phi^2} = -m^2$$

Its solution to provide the normalized eigen function is

$$\boxed{\Phi_m = \frac{1}{\sqrt{2\pi}}e^{im\phi}} \qquad \text{... }(i)$$

where $m = 0, \pm 1, \pm 2$ .... $\pm\, l$. The positive and negative values represent different possible solutions.

Equation in the variable θ (eqn. (*vii*), sec. 1.25) is

$$\frac{\sin\theta}{\Theta}\frac{d}{d\theta}\left(\sin\theta\frac{d\Theta}{d\theta}\right) + \beta\sin^2\theta = m^2$$

Its solution, putting $\beta = l\,(l+1)$ is

$$\Theta_{l,m}^{|m|}\sqrt{\frac{2l+1\,(l-|m|)!}{2\,(l+|m|)!}}\,P_l^{|m|}\,(\cos\theta) \qquad \text{... }(ii)$$

where $P_l^{|m|}$ represents *associated Legendre polynomial* of degree l and order m defined by the expression.

$$P_l^{|m|}\,(x) = (1-x^2)^{|m|/2}\,\frac{d^{|m|}P_l(x)}{dx^{|m|}} \qquad \text{... }(iii)$$

in which $P_l\,(x)$ is defined by the expression

$$P_1\,(x) = \frac{1}{2^l\,l!}\frac{d^l\,(x^2-1)^l}{dx^l} \qquad \text{... }(iv)$$

The solutions of the angular equation (*v*) *sec. 1.25, given by* $\Theta\,\Phi$, *are called* **spherical harmonics.** Their values depend upon the quantum numbers l and m. The first few spherical harmonics (angular wave functions) are given in Table 1.1.

Table 1.1. First few spherical harmonics (angular wave functions for hydrogen like atoms)

l	m	$\Theta_{l,m}\,\Phi_m$
0	0	$\Theta_{0,0}\,\Phi_0 = \left(\frac{1}{4\pi}\right)^{1/2}$
1	0	$\Theta_{1,0}\,\Phi_0 = \left(\frac{3}{4\pi}\right)^{1/2} \cos\theta$
1	1	$\Theta_{1,1}\,\Phi_1 = \left(\frac{3}{4\pi}\right)^{1/2} \sin\theta\cos\phi$
1	–1	$\Theta_{1,-1}\,\Phi_{-1} = \left(\frac{3}{4\pi}\right)^{1/2} \sin\theta\sin\phi$
2	0	$\Theta_{2,0}\,\Phi_0 = \left(\frac{5}{16\pi}\right)^{1/2} (3\cos^2\theta - 1)$
2	1	$\Theta_{2,1}\,\Phi_1 = \left(\frac{15}{4\pi}\right)^{1/2} \sin\theta\cos\theta\cos\phi$
2	–1	$\Theta_{2,-1}\,\Phi_{-1} = \left(\frac{15}{4\pi}\right)^{1/2} \sin\theta\cos\theta\sin\phi$
2	2	$\Theta_{2,2}\,\Phi_2 = \left(\frac{15}{16\pi}\right)^{1/2} \sin^2\theta\cos 2\phi$
2	–2	$\Theta_{2,-2}\,\Phi_{-2} = \left(\frac{15}{16\pi}\right)^{1/2} \sin^2\theta\sin 2\phi$

1.27. EXPRESSION FOR THE RADIAL WAVE FUNCTION

The expression for the radial wave function (R) is obtained by solving the equation (*iv*), sec. 1.25 in the variable r. Complete solution of the equation being complicated, the expression for R is directly given as under :

$$R_{n,l} = Ne^{-l/2}\rho^l\, L_{n+l}^{2l+1}(\rho) \qquad \text{... (i)}$$

where N is the normalization constant, given by

$$N = \left(\frac{2Z}{na_0}\right)^{3/2} \left\{\frac{(n-l-1)!}{2n[(n+l)!]^3}\right\}^{1/2} \qquad \text{... (ii)}$$

ρ is new variable in place of r, defined by

$$\rho = \frac{2Z}{na_0} r \qquad \text{... (iii)}$$

in which n is a new parameter defined by

$$n^2 = -\frac{2\pi^2 \mu Z^2 e^4}{h^2 E} \quad ...(iv)$$

and a_0 is a constant defined by

$$a_0 = \frac{h^2}{4\pi^2 \mu e^4} \quad ...(v)$$

and $L_{n+l}^{2l+1}(\rho)$ is called *associated Legendre polynomial* in ρ of degree $(n + l) - (2l + 1)$ and order $(2l + 1)$. It is defined by the expression

$$L_{n+l}^{2l+1} = \frac{d^{2l+1}}{d\rho^{2l+1}} L_{n+l} \quad ...(vi)$$

in which L_{n+l} is called *Legendre polynomial* of degree $n + l$ and is defined by

$$L_{n+l} = e^{\rho} \frac{d^{n+l}}{d\rho^{n+l}} (\rho^{n+l} e^{-\rho}) \quad ...(vii)$$

The above equations being quite complex for the level, the book is meant for, the values for the normalized radial wave functions R for the first three values of n and the corresponding possible values of l are directly given in Table 1.2.

Table 1.2. A few normalized radial wave functions for Hydrogen-like atoms

n	l	*Radial wave function* (R_n)
1	0	$R_{1,0} = 2\left(\frac{Z}{a_0}\right)^{3/2} e^{-zr/a_0}$
2	0	$R_{2,0} = \frac{1}{2\sqrt{2}}\left(\frac{Z}{a_0}\right)^{3/2}\left(2 - \frac{Zr}{a_0}\right) e^{-zr/2a_0}$
2	1	$R_{2,1} = \frac{1}{2\sqrt{6}}\left(\frac{Z}{a_0}\right)^{3/2}\left(\frac{Zr}{a_0}\right) e^{-zr/2a_0}$
3	0	$R_{3,0} = \frac{2}{81\sqrt{3}}\left(\frac{Z}{a_0}\right)^{3/2}\left(27 - 18\frac{Zr}{a_0} + 2\left(\frac{Zr}{a_0}\right)^2\right) e^{-zr/3a_0}$
3	1	$R_{3,1} = \frac{4}{81\sqrt{6}}\left(\frac{Z}{a_0}\right)^{3/2}\left(6\left(\frac{Zr}{a_0}\right) - \left(\frac{Zr}{a_0}\right)^2\right) e^{-zr/3a_0}$
3	2	$R_{3,2} = \frac{1}{81\sqrt{30}}\left(\frac{Z}{a_0}\right)^{3/2}\left[\frac{Zr}{a_0}\right]^{3/2}\left[\frac{Zr}{a_0}\right]^2 e^{-zr/3a_0}$

1.28. EXPRESSION FOR THE ENERGY FOR HYDROGEN–LIKE PARTICLES

Rearranging eq. (*iv*), sec. 1.27, we get

$$E = -\frac{2\pi^2\mu Z^2 e^4}{n^2 h^2} \quad \text{... } (i)$$

This expression gives the values of the energy corresponding to different values of *n*. Replacing μ by m_e (mass of the electron), we get an equation which was put forward by Bohr in which *n* corresponds to *principal quantum number* and can have the integral values *i.e.*,

$$n = 1, 2, 3.......$$

From equation (*i*), it may be noted that the energy of the electron depends only on the value of the principal quantum number *n* and is independent of the values of *l* and *m*. Now as the quantum number *n* appears only in the radial wave function *R* (*r*), it may be concluded that the energy of the electron depends only on its distance from the nucleus and not upon angular value.

Substituting the values of the various constants in eqn. (*i*), we get the value of *E* for hydrogen like particles

$$E_n = -\frac{21.8\times10^{-19} Z^2}{n^2} \text{ J atom}^{-1} = -\frac{13.6\, Z^2}{n^2}\, eV$$

For H-atom, $Z = 1$ and for the ground state, $n = 1$. Hence energy of the electron in the first Bohr orbit $(E_1) = -13.6$ eV. The energies of the electron in different excited states can be found by substituting the approriate value of *n*. Similarly, the energies of electron in ground state and different excited states for H-like particles can be calculated by substituting the corresponding values of *Z* and *n*.

1.29. QUANTUM NUMBERS FROM SCHRODINGER WAVE EQUATION

To solve Schrodinger wave equation, it is first separated into three equations, each in one variable *i.e.*, *r*, θ and φ. Each of these equations is then solved to obtain the expression for the functions *R* (*r*), Θ (θ) and Φ (φ). The expression for these functions involve *n*, *l* and *m*. For example, expression for *R* involves *n* and *l*, that for Θ involves *l* and *m*, and that for Φ involves *m* only. Hence we may write for the eigen function

$$\psi_{n,\,l,\,m} = R_{n,\,l} \times \Theta_{l,\,m} \times \Phi_m$$

It is found that the solutions of the Schrodinger wave equation obtained in terms of expressions for *R*, Θ and Φ are acceptable solutions only if they follow the following restrictions about *n*, *l* and *m* :

$$n = 1, 2, 3, ...$$
$$l = 0, 1, 2, ... (n-1)$$
$$m = 0, 1, 2, ... n \pm 1, \pm 2 .... \pm l$$

i.e., – *l* to + *l* including zero.

These values of *n*, *l* and *m* are the same as those for the principal quantum number, azimuthal quantum number and magnetic quantum number respectively. Thus these three quantum numbers can be obtained from the solutions of the Schrodinger wave equation.

1.30. CONCEPT OF ORBITAL

Heisenberg's uncertainty principle does not approve the Bohr's concept of definite orbits. The quantum mechanical approach helps us to calculate the value of ψ for an electron of known values of *n*, *l* and *m*. ψ^2 gives the probability of finding the electron at a point.

We can visualise three dimensional space around the nucleus within which the probability of finding the electron under consideration is maximum. Such a three dimensional space is called an *orbital.* This term is the quantum mechanical analogue of the term *orbit*. Depending upon the values of l, these orbitals or electrons are given the following symbols :

Value of l	*Symbols for the orbital*
0	*s*
1	*p*
2	*d*
3	*f*

Further, depending upon the value of l for a given value of n, the different orbitals or electrons are given different names as follows :

Values of n	*Value of l*	*Designation of orbital*
1	0	1*s*
2	0	2*s*
2	1	2*p*
3	0	3*s*
3	1	3*p*
3	2	3*d*

1.31. SHAPES OF ORBITALS

The shape of an orbital can be found from the probabilities at different points around the nucleus. These can be obtained by finding the values of the wave function ψ (r, θ, ϕ) (because ψ^2 gives the probability) which in turn is the product of the radial wave function R (r) and the angular wave function Θ (θ) Φ (ϕ). Representing their dependence on the quantum numbers n, l and m (*i.e.*, the quantum numbers which they give on their solutions), we can write

$$\psi_{n,\,l,\,m} = R_{n,\,l} \times \Theta_{l,\,m} \times \Phi_m.$$

The radial wave function R gives the energy and size of the orbitals whereas the angular wave function, Θ, Φ, *gives the shape and orientation of the orbitals.*

Before we take up the shapes of different orbitals, we shall observe the types of curves that are obtained when we plot ψ^2 versus distance (r) from the nucleus. Such plots are known as **probability distribution curves.** However since only R (r) varies with r and the product, $\Theta \cdot \Phi$ does not depend upon r, therefore plots of ψ^2 versus r will be similar to those for R (r). The values of the radial wave function R, for the electron in first two orbitals (Table 1.2.) are as follows :

$$R_{1s} \text{ i.e., } R_{1,0} = 2\left(\frac{Z}{a_0}\right)^{3/2} e^{-zr/a_0} \quad \text{... } (i)$$

$$R_{2s} \text{ i.e., } R_{2,0} = \frac{1}{2\sqrt{2}}\left(\frac{Z}{a_0}\right)^{3/2}\left(2-\frac{Zr}{a_0}\right) e^{-zr/2a_0} \quad \text{... } (ii)$$

The plot of R^2 vs r obtained for the above two obritals are as shown in Fig. 1.19.

These plots, however, lead to unbelievable results. From these plots, we observe that $R_{n,l}^2$ should be maximum at $r = 0$. If $R_{n,l}^2$ represents the probability, this means that the probability of finding the electron should be maximum at the nucleus. But this cannot be true. This anomaly is solved by dividing the space around the nucleus into different concentric shells of small thickness *dr* and finding the probability in these shells. This probability is called **radial probability.**

It is equal to product of $R_{n,l}^2$ and volume of the shell of thickness *dr*.

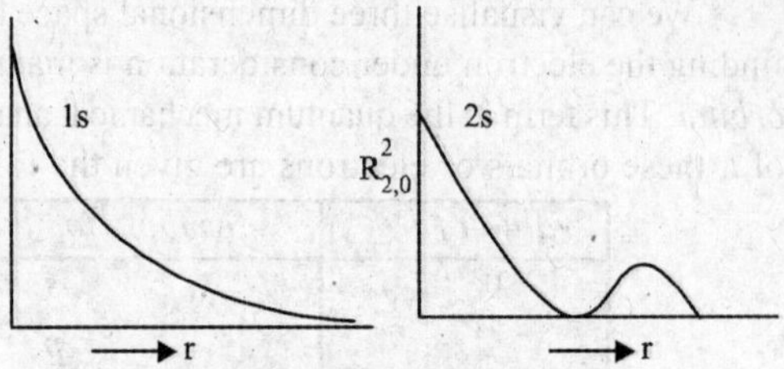

Fig. 1.19. Plots of $R_{1,0}^2$ versus *r* for electrons of 1s and 2s orbitals.

Volume of a shell of thickness *dr* is given by (Fig. 1.20)

$$dV = \frac{4}{3}\pi (r + dr)^3 - \frac{4}{3}\pi r^3$$

$$= \frac{4}{3}\pi (r^3 + dr^3 + 3r^2 dr + 3r dr^2) - \frac{4}{3}\pi r^3$$

As *dr* is very small, dr^2, dr^3 etc. can be neglected. Hence

$$dV = 4\pi r^2 dr$$

$$\therefore \quad \text{Radial probability} = R_{n,l}^2 \times dV$$

$$= R_{n,l}^2 \times 4\pi r^2 \, dr$$

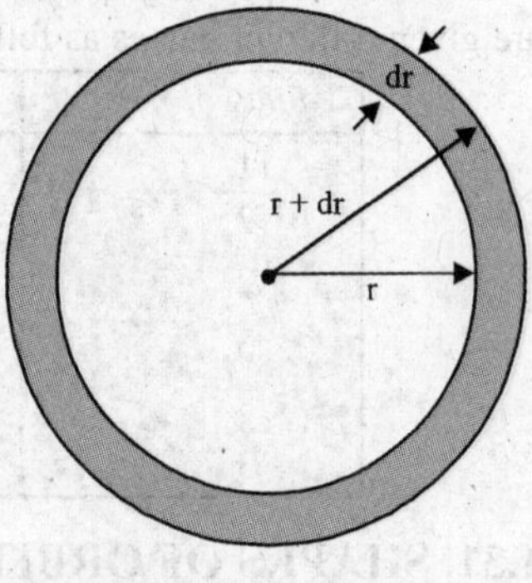

Fig. 1.20. Shell of thickness for around the nucleus.

Thus radial probability distribution can be obtained by plotting $4\pi r^2 R_{n,l}^2$ versus *r*. Such plots are called **radial probability distribution curves.** For 1s and 2s electrons, these plots are shown in Fig. 1.21.

Radial probability is the product of two terms *viz.* $4\pi r^2$ and $R_{n,l}^2$. At the nucleus, though R^2 is large but $r = 0$. Hence radial probability at the nucleus is zero. As *r* increases, $4\pi r^2$ increases but $R_{n,l}^2$ decreases. Hence the plots of the type shown in Fig. 1.21 are obtained.

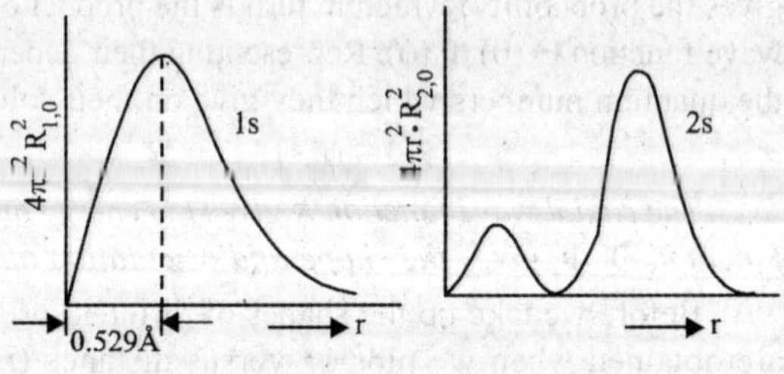

Fig. 1.21. Radial probability distribution curves for 1s and 2s electrons.

The radial probability distribution curves for 1*s*, 2*s*, 2*p*, 3*s*, 3*p* and 3*d* orbitals are given in Fig. 1.22 for comparison.

The number of nodes *i.e.*, the regions at which the radial probability is zero = $n - l - 1$. The distance of the highest peak gives the *radius of maximum probability.* In case of hydrogen atom, it is 0.529 Å (Fig. 1.21) which is same as Bohr's radius. While Bohr's model postulates that the electron in hydrogen atom revolves in a fixed orbit at a distance of 0.529 Å from the nucleus, the quantum mechanical model says that probability of finding the electron is maximum at a radius of 0.529 Å but the electron also keeps on moving closer or farther than this distance from the nucleus.

Another important feature may be observed by comparing the radial probability distribution curves of 3*s*, 3*p* and 3*d* electrons. Their radii of maximum probability are in the order 3*s* > 3*p* > 3*d*. In other words, 3*s* is more penetrating than 3*p* or 3*d* electron. Thus their probability of occurrence closer to the nucleus is in the order 3*s* > 3*p* > 3*d*.

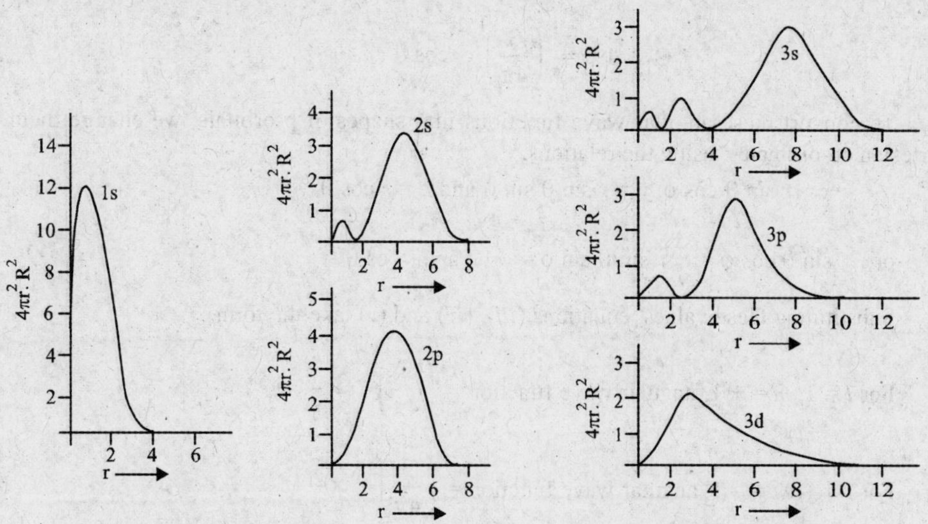

Fig. 1.22. Radial probability distribution curves for 1s, 2s, 2p; 3s, 3p and 3d orbitals.

Shape of *s*-orbitals. For an *s*-orbital, $l = 0$. Hence $m = 0$. The angular wave function is given by

$$\Theta_{0,0}\ \Phi_0 = \sqrt{\frac{1}{4\pi}} = \frac{1}{2\sqrt{\pi}}$$

Thus the angular wave function for the *s*-orbital is independent of the angles θ and ϕ. This means that it will have symmetrical geometry in all directions. Hence an *s*-orbital has a **spherical** shape. However 1s has no radial node, 2s has one and 3s has two radial nodes, they will have the shapes as shown in Fig. 1.23.

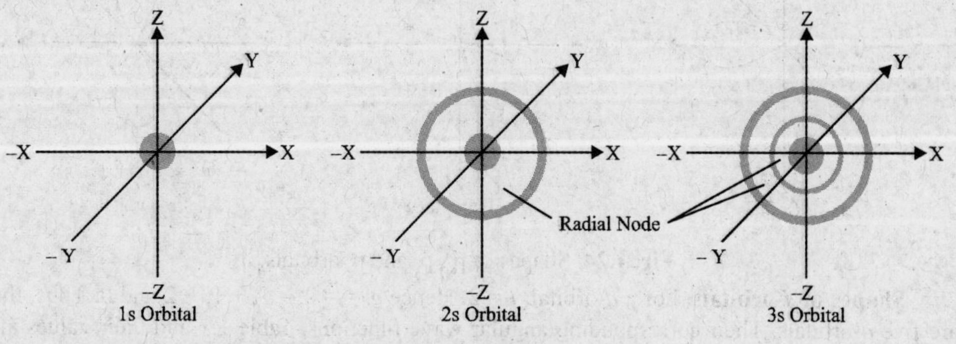

Fig. 1.23. Shapes of 1s, 2s and 3s orbitals.

Shapes of *p*-orbitals. For any *p*-orbital, $l = 1$. Hence $m = +1, -1, 0$. Their angular wave functions as given in Table1.1 are represented as under :

$$\Theta_{1,+1}\ \Phi_{+1} = \left(\frac{3}{4\pi}\right)^{1/2} \sin\theta\cos\phi \qquad \text{... (iii)}$$

$$\Theta_{1,-1}\ \Phi_{-1} = \left(\frac{3}{4\pi}\right)^{1/2} \sin\theta\sin\phi \qquad \text{... (iv)}$$

$$\Theta_{1,0}\,\Phi_0 = \left(\frac{3}{4\pi}\right)^{1/2} \cos\theta \qquad ...(v)$$

To convert these angular wave functions into shapes of p-orbitals, we change them into cartesian co-ordinates, using the relations.

$$x = r\sin\theta\cos\phi,\ y = r\sin\theta\sin\phi \text{ and } z = r\cos\theta$$

or $$\sin\theta\cos\phi = \frac{x}{r},\ \sin\theta\sin\phi = \frac{y}{r} \text{ and } \cos\theta = \frac{z}{r}$$

Substituting these values, equations (*iii*), (*iv*) and (*v*) take the form

For $l = 1$, $m = +1$, angular wave function $= \left(\frac{3}{4\pi}\right)^{1/2} \times \frac{x}{r}$

For $l = 1$, $m = -1$, angular wave function $= \left(\frac{3}{4\pi}\right)^{1/2} \times \frac{y}{r}$

For $l = 1$, $m = 0$, angular wave function $= \left(\frac{3}{4\pi}\right)^{1/2} \times \frac{z}{r}$

Thus it may be seen that these wave functions are functions of x, y and z respectively. Hence they are called p_x, p_y and p_z wave functions. Any of these wave functions when plotted in the three dimensional space gives a **dumb-bell** shape as shown in Fig. 1.24. Plots show that when x, y, z are positive, these functions are also positive and when x, y, z are negative, these functions are negative.

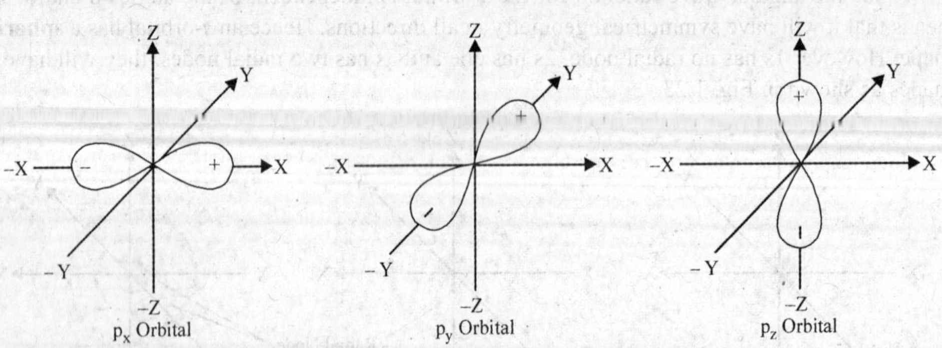

Fig. 1.24. Shapes of p_x, p_y and p_z orbitals.

Shapes of *d*-orbitals. For a *d*-orbital, $l = 2$. Hence $m = -2, -1, +1, +2$ and 0. Thus, there are five *d*-orbitals. Their corresponding angular wave functions, Table 1.1 and their values after changing into cartersian co-ordinates (as done in case of *p*-orbitals) are given below:

1. $$\Theta_{2,-2}\,\Phi_{-2} = \left(\frac{15}{16\pi}\right)^{1/2} \sin^2\theta\sin 2\phi = \left(\frac{15}{16\pi}\right)^{1/2} \sin^2\theta \cdot 2\sin\phi\cos\phi$$

$$= \left(\frac{15}{4\pi}\right)^{1/2} \sin\theta\sin\phi \cdot \sin\theta\cos\phi = \left(\frac{15}{4\pi}\right)^{1/2} \times \frac{xy}{r^2}$$

2. $$\Theta_{2,-1}\,\Phi_{-1} = \left(\frac{15}{4\pi}\right)^{1/2} \sin\theta\cos\theta\cdot\sin\phi = \left(\frac{15}{4\pi}\right)^{1/2} \times \frac{yz}{r^2}$$

3. $$\Theta_{2,+1}\,\Phi_{+1} = \left(\frac{15}{4\pi}\right)^{1/2} \sin\theta\cos\theta\cos\phi = \left(\frac{15}{4\pi}\right)^{1/2} \times \frac{xz}{r^2}$$

4. $$\Theta_{2,+2}\,\Phi_{+2} = \left(\frac{15}{16\pi}\right)^{1/2} \sin^2\theta\cos 2\phi = \left(\frac{15}{16\pi}\right)^{1/2} \sin^2\theta\,(\cos^2\phi - \sin^2\phi)$$

$$= \left(\frac{15}{16\pi}\right)^{1/2} (\sin^2\theta\cos^2\phi - \sin^2\theta\sin^2\phi) = \left(\frac{15}{16\pi}\right)^{1/2} \frac{x^2-y^2}{r^2}$$

5. $$\Theta_{2,0}\,\Phi_0 = \left(\frac{5}{16\pi}\right)^{1/2} (3\cos^2\theta - 1) = \left(\frac{5}{16\pi}\right)^{1/2} (3z^2 - r^2)\times\frac{1}{r^2}$$

Thus it may be seen that these wave functions vary with xy, yz, xz, $x^2 - y^2$ and z^2 respectively. Hence they are represented by d_{xy}, d_{xz}, d_{yz}, $d_{x^2-y^2}$ and d_{z^2} respectively. When we plot the wave functions in the three-dimensional space, we get the shapes shown in Fig. 1.25 :

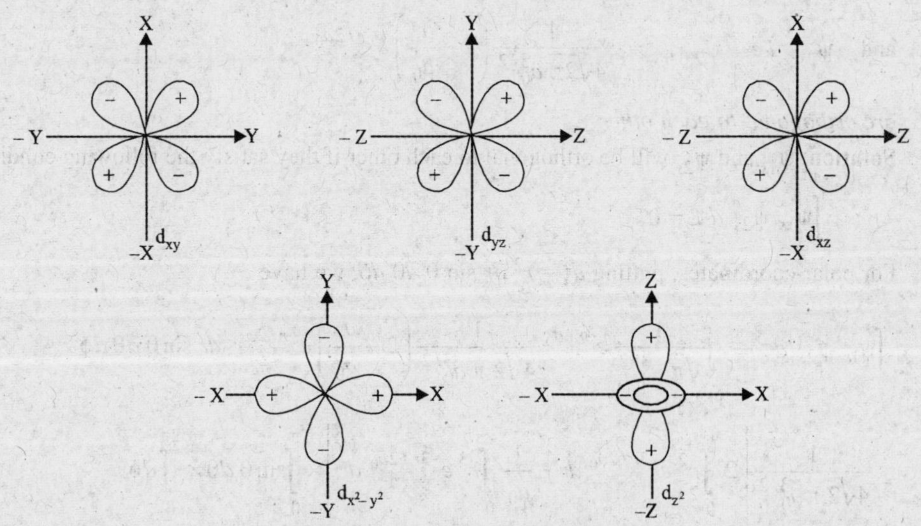

Fig. 1.25. Shapes of *d*-orbitals

The first three shapes are **double dumb-bell** shapes lying *between the axes.* The fourth is also double dumb-bell shape but *lying along the axes.* The fifth (d_{z^2}) is dumb-bell shape *symmetrical about z-axis* **but** with a **dough-nut** shape in the centre in the *xy* plane. The signs of the wave functions may be explained by taking the example of the wave function (1). If x and y both are positive or both are negative, then as xy will be positive, the wave function will be positive. If either x or y is negative then as xy will be negative, the wave function will also be negative. The signs of the wave functions (2) and (3) can be explained in a similar manner. In the wave function (4), x may be positive or negative, x^2 will always be positive whereas y may be positive or negative,

$-y^2$ will always be negative. Hence lobes along x-axis are always positive while those along y-axis are always negative. In wave function (5), as z^2 is always positive, lobes along z-axis are always positive but the dought-nut (in the xy plane) has a negative sign.

Further the shapes of $4d$ orbitals will be similar to those of $3d$ orbitals. They will differ in two aspects, *size* of $4d$ orbitals will be bigger than that of $3d$ orbitals, and $4d$ orbitals will have one node whereas $3d$ orbitals have no node. Number of radial nodes in any orbital = $n - l - 1$.

SOLVED PROBLEMS ON QUANTUM MECHANICS

The following two results may be kept in mind while solving problems on quantum mechanics.

1. Just as in cartesian coordinates, volume element $d\tau = dx\, dy\, dz$, in polar coordinates, $d\tau = r^2\, dr \sin\theta\, d\theta\, d\phi$

2. $\int_0^\infty x^n e^{-ax}\, dx = \frac{n!}{a^{n+1}}$ (n = +ve integer, $a > 0$)

Example 1. Show that 1s and 2s wave functions of hydrogen atom given by

$$\psi_{1s} \quad i.e. \quad \psi_{1,0,0} = \frac{1}{\sqrt{\pi}\, a_0^{3/2}} e^{-r/a_0}$$

and $$\psi_{2s} \quad i.e. \quad \psi_{2,0,0} = \frac{1}{4\sqrt{2\pi}\, a_0^{3/2}} \left(2 - \frac{r}{a_0}\right) e^{-r/2a_0}$$

are orthogonal to each other.

Solution. ψ_{1s} and ψ_{2s} will be orthogonal to each other if they satisfy the following condition

$$\int \psi_{1s}\, \psi_{2s}\, d\tau = 0$$

For polar coordinates, putting $d\tau = r^2\, dr \sin\theta\, d\theta\, d\phi$, we have

$$\int \psi_{1s}\, \psi_{2s}\, d\tau = \int \frac{1}{\sqrt{\pi}\, a_0^{3/2}} e^{-r/a_0} \frac{1}{4\sqrt{2}\, \pi\, a_0^{3/2}} \left(2 - \frac{r}{a_0}\right) e^{-r/2a} r^2 dr \sin\theta\, d\theta\, d\phi$$

$$= \frac{1}{4\sqrt{2}\, \pi a_0^{3/2}} \left[2 \int_0^\infty r^2 e^{-3r/2a_0}\, dr - \frac{1}{a_0} \int_0^\infty r^3 e^{-3r/2a_0}\, dr\right] \times \int_0^\pi \sin\theta\, d\theta \times \int_0^{2\pi} d\phi$$

The first two integrals may be solved by applying the formula

$$\int_0^\infty x^n e^{-ax}\, dx = \frac{n!}{a^{n+1}}$$

Thus $$\int_0^\infty r^2 e^{-3r/2a_0}\, dr = \frac{2!}{(3/2a_0)^3} = \frac{16 a_0^3}{27}$$

$$\int_0^\infty r^3 e^{-3r/2a_0} dr = \frac{3!}{(3/2a_0)^4} = \frac{32a_0^4}{27}$$

Solution of the other two integrals is given below :

$$\int_0^\pi \sin\theta \, d\theta = [-\cos\theta]_0^\pi = -(-1)-(-1) = 2$$

and
$$\int_0^\pi d\phi = [\phi]_0^{2\pi} = 2\pi$$

$$\int \psi_{1s}\, \psi_{2s}\, d\tau = \frac{1}{4\sqrt{2}\,\pi a_0^3}\left[2\times\frac{16a_0^3}{27} - \frac{1}{a_0}.\frac{32\,a_0^4}{27}\right]\times 2\times 2\pi = 0$$

Thus the condition of orthogonality is satisfied. We can say that 1s and 2s wave functions of hydrogen atom are orthogonal to each other.

Example 2. Prove that 1*s* wave function of hydrogen atom given by

$$\psi_{1s} \quad i.e. \quad \psi_{1,0,0} = \frac{1}{\sqrt{\pi}\, a_0^{3/2}} e^{-r/a_0}$$

is a normalised wave function where a_0 represents Bohr radius.

Solution. The wave function ψ_{1s} will be a normalized wave function if it satisfies the following condition :

$$\int \psi_{1s}\, \psi_{1s}\, d\tau = 1 \quad \text{or} \quad \int \psi_{1s}^2\, d\tau = 1$$

In polar coordinates, volume element $d\tau = r^2\, dr \sin\theta\, d\theta\, d\phi$

Substituting this value of $d\tau$ and given value of ψ_{1s},

$$\int \psi_{1s}^2\, d\tau = \int \frac{1}{\pi a_0^3} e^{-2r/a_0}\, r^2\, dr \sin\theta\, d\theta\, d\phi \qquad ...(i)$$

$$= \frac{1}{\pi a_0^3} \int_0^\infty r^2 e^{-2r/a_0}\, dr \int_0^\pi \sin\theta\, d\theta \int_0^{2\pi} d\phi$$

Each of the integrals may be solved separately as follows :

Evaluate the first integral using the formula $\int_0^\infty x^n e^{-ax}\, dx = \frac{n!}{a^{n+1}}$

$$\therefore \quad \int_0^\infty r^2 e^{-2r/a_0}\, dr = \frac{2}{(2/a_0)^3} = \frac{a_0^3}{4}$$

$$\text{Second integral} = \int_0^{\pi} \sin\theta\, d\theta = [-\cos\theta]_0^{\pi} = -(-1)-(-1) = 2$$

$$\text{Third integral} = \int_0^{2\pi} d\phi = [\phi]_0^{2\pi} = 2\pi$$

$\therefore$ Eq. (1) reduces to

$$\int \psi_{1s}^2\, d\tau = \frac{1}{\pi a_0^3} \times \frac{a_0^3}{4} \times 2 \times 2\pi = 1$$

Thus the condition of the normalized wave function is satisfied. Hence it is a normalised wave function.

Example 3. Calculate the average distance of the electron from the nucleus in the ground state of hydrogen atom. Given that the normalized wave function for 1*s* electron of hydrogen atom is

$$\psi_{1s} = \frac{1}{\sqrt{\pi}\, a_0^{3/2}} e^{-r/a_0}$$

Solution. The average value of the distance *i.e.* the variable r is equal to the expectation value which is calculated by using the following formula

$$< r > = \int \psi^* r \psi\, d\tau$$

In the present case

$$< r > = \int \psi_{1s}\, r\, \psi_{1s}\, d\tau$$

$$= \int \frac{1}{\sqrt{\pi a_0^{3/2}}} e^{-r/a_0} \cdot r \frac{1}{\sqrt{\pi a_0^{3/2}}} \cdot r^2 dr \sin\theta\, d\theta\, d\phi$$

$$= \frac{1}{\pi a_0^3} \int_0^{\infty} r^3 e^{-2r/a_0}\, dr \int_0^{\pi} \sin\theta\, d\theta \int_0^{2\pi} d\phi$$

$$= \frac{1}{\pi a_0^3} \frac{3!}{(2/a_0)^4} (2)(2\pi)$$ (For solutions of different integrals see Example 2)

$$= \frac{1}{\pi a_0^3} \frac{3 \times 2 \times 1 \times a_0^4}{16} \times 2 \times 2\pi = \frac{3}{2} a_0 .$$

THEORIES OF BONDING

1.32. INTRODUCTION

There are two theories to explain the nature of chemical bonding and to predict the properties of molecules such as bond distance, bond dissociation energy, dipole moment etc. These are Valence Bond Theory (VBT) and Molecular Orbital Theory (MOT). According to VBT, atoms are considered to retain their *identity* and the bond is formed as a result of interaction of only the *valence electrons* of the atoms, (*i.e.*, half-filled orbitals overlap to form a covalent bond). Fully-filled orbitals do not participate in bonding. Molecular orbital theory considers the molecule as a collection of nuclei and all the electrons of the molecule are distributed in different energy levels around the group of nuclei just as the electrons of an atom are distributed in different energy levels of a single nucleus. Energy levels of an atom are called atomic orbitals while the energy levels of the molecule are called molecular orbitals. According to MOT, all the atomic orbitals of an atom overlap with atomic orbitals of comparable energy of the other atoms. These theories are being discussed separately.

1.33. VALENCE BOND THEORY

It was put forward by Heitler and London in 1927 and was further extended by Slater and Pauling. Some basic quantum mechanical principles involved in the theory are given below :

(*i*) The total wave function (ψ) of two independent systems with wave functions ψ_A and ψ_B is equal to their product *i.e.* it is given by

$$\psi = \psi_A \psi_B \qquad ...(i)$$

Walter Heitler

Fritz London

(*ii*) The total energy (E) of the two independent systems with energies E_A and E_B is equal to their sum.

$$E = E_A + E_B \qquad ...(ii)$$

(*iii*) The total hamiltonian is also equal to the sum of their values for independent systems.

$$\hat{H} = \hat{H}_A + \hat{H}_B \qquad ...(iii)$$

(*iv*) For a many-electron system, the total wave function is a linear combination of the different wave functions involved *i.e.*

$$\psi = c_1 \psi_1 + c_2 \psi_2 + ... \qquad ...(iv)$$

where coefficients c_1, c_2 etc. are found by variation method so that ψ is closest to the true wave function corresponding to which the energy of the system is minimum.

(*v*) The square of the coefficient of any function gives the contribution of that wave function towards total wave function and is usually called the *weight* of that wave function. For the function ψ to be normalized, the following condition must be satisfied

$$c_1^2 + c_2^2 + c_3^2 + ... = 1 \quad ...(v)$$

Application of Valence Bond Theory to the study of H_2 molecule. We start with two hydrogen atoms separated by a large distance so that there are no forces of interaction between them and the energy of the system is taken as zero. Let the two H atoms be represented by H_A and H_B and the associated electrons by e_1 and e_2 respectively. Thus, the two H-atoms may be represented as $H_A(e_1)$ and $H_B(e_2)$. Also suppose the wave functions associated with these electrons are represented by $\psi_A(1)$ and $\psi_B(2)$ respectively. The total wave function for the two separated hydrogen atoms will be given by

$$\psi = \psi_A(1)\,\psi_B(2) \quad ...(vi)$$

If we plot the value of E as a function of intermolecular distance, curve *I* is obtained as shown in Fig. 1.26. It may be seen that energy of the system decreases with distance till a minimum is reached. This represents the formation of H_2 molecule. The decrease in energy corresponding to this minimum represents the bond energy and the distance represents the bond length. This curve has been obtained by using the wave function as represented by eqn. (*vi*). The bond energy is found to be only 24 kJ mol^{-1} while the actual (experimental) value is 458 kJ mol^{-1}. Similarly, bond distance is found to be 90 pm which is much different from the actual bond length of H_2 molecule *viz* 74 pm. This shows that the wave function used in the above calculations *viz.* $\psi_A(1)\,\psi_B(2)$ is not accurate. Hence some other interactions between the two hydrogen atoms are taking place and need to be taken into consideration to arrive at a better wave function. The different possible interactions are discussed as under :

(1) Exchange of electrons. In writing the above wave function, it was assumed that electron e_1 belongs to H_A and e_2 belongs to H_B. However as the electrons are actually indistinguishable after the overlap of the two atomic orbitals, electron e_1 may be under the influence of nucleus of H_B and also electron e_2 may be under the influence of nucleus of H_A. Thus the system may exist in two different states as follows :

$$\text{I. } H_A(e_1)\,H_B(e_2) \qquad \text{II. } H_A(e_2)\,H_B(e_1) \quad ...(vii)$$

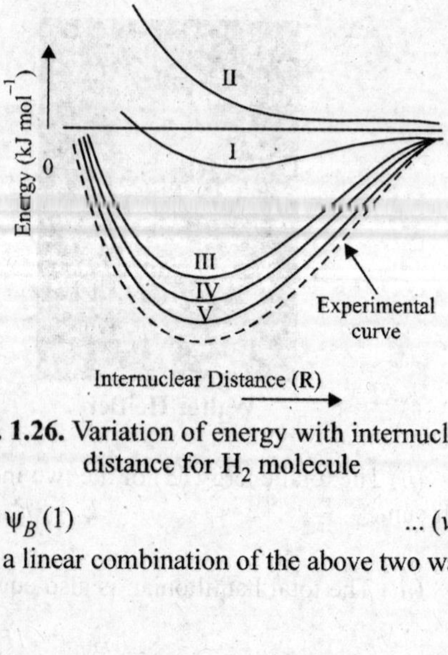

Fig. 1.26. Variation of energy with internuclear distance for H_2 molecule

The wave functions corresponding to these two states may be written as

$$\psi_I = \psi_A(1)\,\psi_B(2), \quad \psi_{II} = \psi_A(2)\,\psi_B(1) \quad ...(viii)$$

The total wave function of the system will then be a linear combination of the above two wave functions *i.e.*

$$\psi_{VB} = c_1\,\psi_I \pm c_2\,\psi_{II}$$
$$= c_1\,\psi_A(1)\,\psi_B(2) \pm c_2\,\psi_A(2)\,\psi_B(1) \quad (ix)$$

For H_2 molecules, as both atoms are same, put $c_1 = c_2 = 1$. Hence

$$\psi_{VB} = \psi_A(1)\,\psi_B(2) \pm \psi_A(2)\,\psi_B(1) \quad ...(x)$$

Thus there are two possible linear combinations of the wave functions corresponding to wave functions represented by ψ_+ and ψ_- *i.e.*

$$\psi_+ = \psi_A(1)\,\psi_B(2) + \psi_A(2)\,\psi_B(1) \quad ...(xi)$$
$$\psi_- = \psi_A(1)\,\psi_B(2) - \psi_A(2)\,\psi_B(1) \quad ...(xii)$$

If the energies are calculated as a function of intermolecular distance by employing the wave function ψ_-, the curve II is obtained which shows that energy increases when the atoms are brought close together. This represents a *non-bonding state.* On the other hand, if energies are calculated as a function of intermolecular distance by using the wave function ψ_+, curve III is obtained. It represents decrease of energy with minimum at a particular distance and hence represents a *bonding state.*

Using this wave function, the minimum in the curve is found to be at 303 kJ mol^{-1} at an internuclear distance of 80 pm. Thus both the values have come closer to the experimental values. *The extra lowering of energy is due to the exchange of electrons between the two hydrogen atoms and is called* **exchange energy**. Thus exchange energy for H_2 molecule = 303 – 24 = 279 kJ mol^{-1}.

(2) Screening effect of electrons. When the atoms come close together, the electron of one atom shields the electron of the other atom from the nucleus. Thus the electron does not feel full attraction by nucleus. Thus, its effective nuclear charge is smaller than the actual nuclear charge. Taking this fact into consideration, the wave function was further improved. Using this wave function, energy versus internuclear distance curve IV is obtained as shown in Fig. 2.4. The minimum in this curve lies at 365 kJ mol^{-1} which shows a further improvement in the value.

(3) Ionic structure for H_2 molecule. Since the calculated value of 365 kJ mol^{-1} is still much lower than the experimental value of 458 kJ mol^{-1} which requires further improvement in the wave function. It was suggested that when the atoms come closer together, both the electrons may lie close to nucleus H_A or close to nucleus H_B, giving rise to two possible ionic structures as represented below:

$$\text{III} \quad {}^{(e_1)}_{(e_2)}H_A\,H_B \qquad\qquad \text{IV} \quad H_A\,H_B{}^{(e_1)}_{(e_2)} \qquad \textit{(xiii)}$$

$$(H_A^-\,H_B^+) \qquad\qquad (H_A^+\,H_B^-)$$

Though the possibility of existence of these states is very little because repulsions between the electrons will be much larger than the attractions between H_A^- and H_B^+ or H_A^+ and H_B^-, yet by taking these structures into consideration, wave function could be further improved. The wave functions corresponding to states III and IV may be written as

$$\psi_{III} = \psi_A(1)\,\psi_A(2),$$
$$\psi_{IV} = \psi_B(1)\,\psi_B(2) \qquad \text{... }(xiv)$$

The total wave function due to these ionic structures (represented by ψ_{ionic}) may be obtained by linear combination of ψ_{III} and ψ_{IV}. Thus

$$\psi_{ionic} = \psi_{III} + \psi_{IV} = \psi_A(1)\,\psi_A(2) + \psi_B(1)\,\psi_B(2) \qquad \text{... }(xv)$$

Now combining the covalent and ionic wave functions, the complete wave function for the system may be written as

$$\psi = \psi_{covalent} + \lambda\,\psi_{ionic}$$
$$= [\psi_A(1)\,\psi_B(2) + \psi_A(2)\,\psi_B(1)] + \lambda\,[\psi_A(1)\,\psi_A(2) + \psi_B(1)\,\psi_B(2)] \qquad \text{... }(xvi)$$

where λ gives the extent of contribution of the ionic structures towards bonding and is usually called **mixing coefficient.** Its value for H_2 is found to be 0.17.

Using the above wave function, the minimum in the curve (shown at V) is obtained at 388 kJ mol^{-1}, thus bringing the value still closer to the experimental value of bond energy of 458 kJ mol^{-1}. The bond distance is found to be 75 pm which agrees fairly well with the experimental value of 74 pm.

Further improvement in the wave function have been done but these will not be discussed here.

It may be mentioned that Valence Bond Theory led to the concept that a molecule or ion may exist in different structures, none of which represents the actual structure and the actual structure is in between the different structures. This phenomenon is called **Resonance** and the different structures are called **Resonating structures.** Thus the concept of resonance follows from VBT.

1.34. MOLECULAR ORBITAL THEORY

This theory developed by Hund, Mulliken, Huckel and others differs from Valence Bond Theory mainly in the fact that the electron in a molecule is not considered to be under the influence of a particular nucleus but is considered to be under the influence of all the nuclei present. The total wave function is the product of the wave functions for various electrons involved in the bonding. Main points of the theory are listed below:

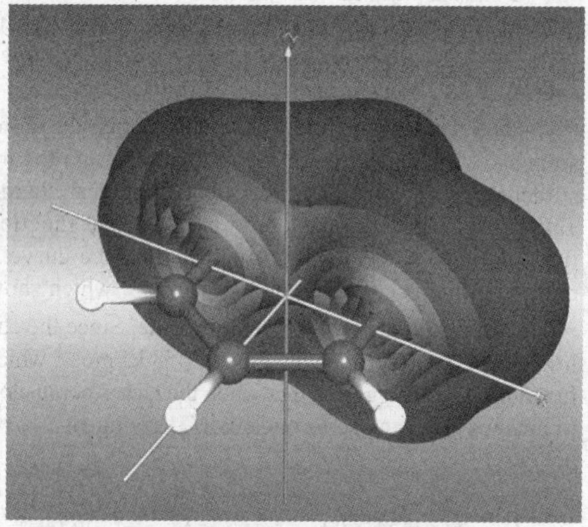

Molecular orbitals for benzene

(*i*) A molecule is considered as a group of nuclei and all the electrons are distributed around this group in different energy levels like in an atom the electrons are distributed in different energy levels around a single nucleus. The energy levels of the molecule are called molecular orbitals. While the atomic orbitals are *monocentric*, the molecular orbitals are *polycentric.*

(*ii*) Each electron or each molecular orbital in the molecule is described by a wave function (called molecular orbital wave function) the correct value of which can be found by a method called Linear Combination of Atomic Orbitals (LCAO method).

(*iii*) Each molecular orbital is associated with a definite amount of energy, the correct value of which can be calculated by applying Schrodinger wave equation.

(*iv*) The significance of molecular orbital wave function is same as that of the atomic orbital wave function. Thus just as in an atom $\psi^2\, d\tau$ gives the probability of finding the electron in a volume element $d\tau$, in a molecule, $\psi^2\, d\tau$ gives the *probability density* or *electron charge density* in the volume element $d\tau$.

(*v*) The size shape and energy of the molecular orbitals depend upon the size, shape and energy of the combining atomic orbitals. The atomic orbitals combining together lose their identity to form molecular orbitals.

(*vi*) Only those atomic orbitals combine to form molecular orbitals which have comparable energies and same symmetry with respect to the molecular axis such that sufficient overlapping is possible.

(*vii*) The filling of electrons into molecular orbitals takes place according to the same rules as followed by atomic orbitals *viz*. Aufbau principle, Pauli exclusion principle and Hund's rule of maximum multiplicity.

1.34.1. Linear combination of atomic orbitals (LCAO)–H_2^+ ion

We can construct a molecular orbital wave function by an approximation of Linear Combination of Atomic Orbitals or in short called as *LCAO approximation*. To understand this method, let us consider the simplest case of hydrogen molecular ion (H_2^+).

H_2^+ ion consists of two protons H_A and H_B separated by a distance R and only one electron. The electron is considered to be moving in a molecular orbital under the influence of both the nuclei. When at some point of time the electron is in the neighbourhood of A (state I, shown in Fig. 1.27), it will behave as if it were moving in the atomic orbital of H_A and the effect of the nucleus of H_B will be negligible. Similarly, at some other point of time, the electron may be in the neighbourhood of H_B so that the influence of the nucleus of H_A on the electron is negligible and the electron will behave as if it were moving in the atomic orbital of H_B (state II).

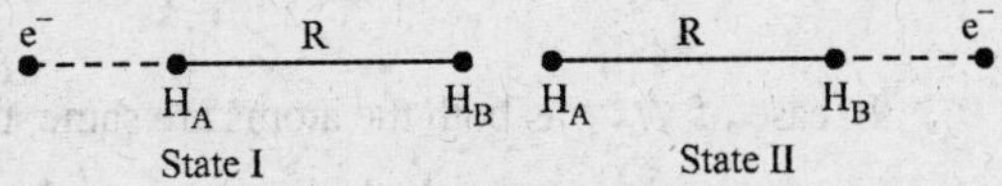

Fig. 1.27. Two extreme in H_2^+ ion

In the first case, the wave function of the electron will be ψ_A (*i.e.* like that of the electron in atom H_A) and in the second case, the wave function of the electron will be ψ_B (*i.e.* like that of the electron in atom H_B). In between these two extremes, the electron may be under the influence of both the nuclei and hence may have characteristics of both the wave functions ψ_A and ψ_B. The wave function of the system, called *molecular orbital wave function* (ψ_{MO}) is then obtained as linear combination of ψ_A and ψ_B according to the expression.

$$\psi_{MO} = c_1 \psi_A + c_2 \psi_B \quad \text{... }(i)$$

where c_1 and c_2 are coefficients of the wave functions ψ_A and ψ_B such that their squares *i.e.* c_1^2 and c_2^2 give their contribution (weightage) towards the molecular orbital wave function.

The coefficients c_1 and c_2 are found by variation method which give minimum value of the energy E. The condition for getting minimum energy with respect to coefficients c_1 and c_2 is that we put $\partial E / \partial c_1 = 0$ and $\partial E / \partial c_2 = 0$. Combining results of different equations, we get a quadratic equation in E which also contains the coefficients c_1 and c_2. The solution of the quadratic equation gives two values of E. These are represented by E_+ and E_-. The symbol E_+ represents molecular orbital with energy lower than that of the combining atomic orbitals (E_A or E_B) and E_-represents the molecular orbital with energy higher than of the combining atomic orbitals. These values are then substituted separately in the quadratic equation in energy obtained above and we get two equations in terms of c_1 and c_2. Applying the normalization condition for the molecular orbital wave function, we can calculate the values of c_1 and c_2 for the two molecular orbitals (one with energy E_+ and the other with E_-) as explained below :

As one of the basic principles of quantum mechanics, the function ψ is acceptable solution of Schrodinger wave equation only if it is normalized *i.e.* it must satisfy the condition.

$$\int \psi_{MO}^2 \, d\tau = 1 \quad \text{... }(ii)$$

i.e.

$$\int (c_1\psi_A + c_2\psi_B)^2 \, d\tau = 1 \quad \text{... }(iii)$$

or

$$\int c_1^2\psi_A^2 \, d\tau + \int c_2^2\psi_B^2 \, d\tau + \int 2c_1c_2\psi_A\psi_B \, d\tau = 1$$

or

$$c_1^2 \int \psi_A^2 \, d\tau + c_2^2 \int \psi_B^2 \, d\tau + 2\, c_1c_2 \int \psi_A \psi_B \, d\tau = 1 \quad \text{... }(iv)$$

As the atomic orbital wave functions ψ_A and ψ_B combining to form molecular orbital wave function are also normalized, we have

$$\int \psi_A^2\, d\tau = 1 \quad \text{and} \quad \int \psi_B^2\, d\tau = 1 \qquad \text{... (v)}$$

Substituting these values in eq. (*iv*) and representing the integral $\int \psi_A \psi_B\, d\tau$ by S (called overlap *integral* between the orbitals ψ_A and ψ_B), we have

$$c_1^2 + c_2^2 + 2c_1c_2 S = 1 \qquad \text{... (vi)}$$

For zero-overlap approximation, we put $S = 0$. Hence the normalization condition becomes

$$c_1^2 + c_2^2 = 1 \qquad \text{... (vii)}$$

In case of H_2^+, as both the atoms are same, there will be equal contribution by the atomic wave functions ψ_A and ψ_B towards the molecular orbital wave function. Hence we will have

$$c_1^2 = c_2^2 \qquad \text{... (viii)}$$

From eqs. (*vii*) and (*viii*)

$$c_1^2 = c_2^2 = \frac{1}{2}$$

or

$$c_1 = \pm c_2 = \pm \frac{1}{\sqrt{2}} \qquad \text{... (ix)}$$

Thus two sets of values are possible for c_1 and c_2 *i.e.*, $+\frac{1}{\sqrt{2}}, +\frac{1}{\sqrt{2}}$ and $+\frac{1}{\sqrt{2}}, -\frac{1}{\sqrt{2}}$.

Substituting these values in eqn. (*i*), two linear combination are possible, one by using first set of values of c_1 and c_2 is represented by $\psi_{+(MO)}$ and the other by using second set of values represented by $\psi_{-(MO)}$.

$$\psi_{+(MO)} = +\frac{1}{\sqrt{2}}\psi_A + \frac{1}{\sqrt{2}}\psi_B = +\frac{1}{\sqrt{2}}(\psi_A + \psi_B) \qquad \text{... (x)}$$

$$\psi_{-(MO)} = +\frac{1}{\sqrt{2}}\psi_A - \frac{1}{\sqrt{2}}\psi_B = \frac{1}{\sqrt{2}}(\psi_A - \psi_B) \qquad \text{... (xi)}$$

The wave function $\psi_{+(MO)}$ corresponds to the energy E_+ whereas $\psi_{-(MO)}$ corresponds to the energy E_-.

Equations (*x*) and (*xi*) give the values of the wave functions according to LCAO–MO treatment of H_2^+ ion. Using these wave functions, energies (E_+ and E_-) are calculated for different internuclear distance *R*. A graph is then plotted between energy versus internuclear distance. The potential energy curves thus obtained for H_2^+ ion are shown in Fig. 1.28. The potential energy curve of $\psi_{+(MO)}$ shows a minimum at 171 kJ mol^{-1} and internuclear distance of 132 pm. The actual (observed) values are 268 kJ mol^{-1} and 106 pm. There is no satisfactory agreement between the two. Hence a number of further improvements have been suggested, discussion of which is beyond the scope of this level.

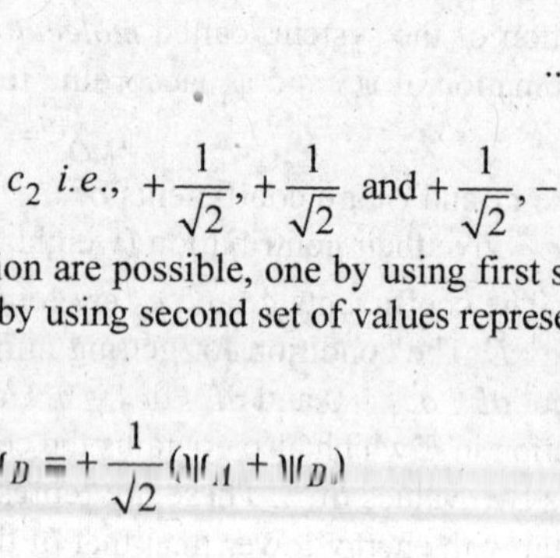

Fig. 1.28. Plots showing variation of energy (calculated from $\psi_{-(MO)}$ and $\psi_{-(MO)}$) versus internuclear distance.

1.35. PHYSICAL PICTURE OF BONDING AND ANTIBONDING WAVE FUNCTIONS

We have seen from the LCAO–MO treatment of H_2^+ ion that on calculation, two values of energy are obtained for this system, represented by E_+ and E_-. Their corresponding wave functions, represented by $\psi_{+(MO)}$ and $\psi_{-(MO)}$ are given by eqs. (*x*) and (*xi*), section 1.34.1.

The value of E_+ is found to be lower than the energies of the combining atomic orbitals *i.e.* E_A and E_B of the isolated atoms H_A and H_B. Thus it leads to a greater stability and hence is responsible for the bonding between H_A and H_B. The molecular orbital formed is, therefore, called **Bonding Molecular Orbital** (BMO). On the other hand, the value of E_- is higher than that of the combining atomic orbitals. The molecular orbital thus formed has lower stability and hence is called **Antibonding Molecular Orbitals** (ABMO). Further, it is found that the extent of lowering of energy (for bonding orbital) is almost equal to the extent of increase in energy for anti-bonding orbital. Hence a physical picture of the bonding and antibonding molecular orbitals formed from atomic orbitals may be represented as shown in Fig. 1.29.

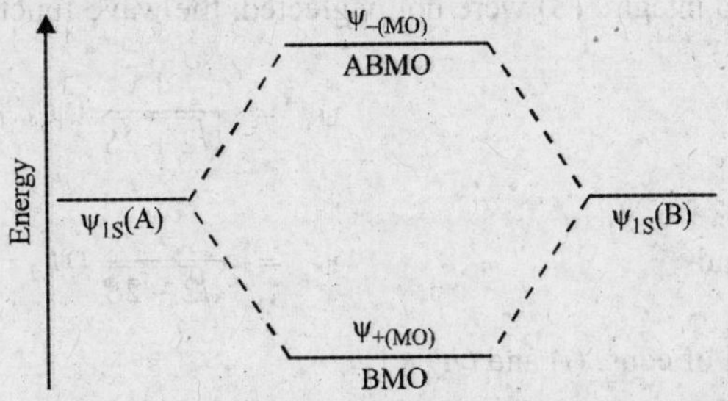

Fig. 1.29. Representing the energy of bonding and antibonding molecular orbitals with respect to those of combining atomic orbitals.

Rewritting expressions (*x*) and (*xi*), sec. 1.34.1 for the wave functions ψ_+ and ψ_-, we have

$$\psi_{+(MO)} = \frac{1}{\sqrt{2}}(\psi_A + \psi_B) \quad \text{... } (i)$$

$$\psi_{-(MO)} = \frac{1}{\sqrt{2}}(\psi_A - \psi_B) \quad \text{... } (ii)$$

Squaring both sides of these equations, we have

$$\psi_+^{\,2} = \frac{1}{2}(\psi_A^2 + \psi_B^2 + 2\psi_A\psi_B) \quad \text{... } (iii)$$

$$\psi_-^{\,2} = \frac{1}{2}(\psi_A^2 + \psi_B^2 - 2\psi_A\psi_B) \quad \text{... } (iv)$$

ψ^2 gives the probability of finding the electron and $\psi_+^2 > \frac{1}{2}(\psi_A^2 + \psi_B^2)$ because there is an extra positive term in eqn. (*iii*) and $\psi_-^2 < \frac{1}{2}(\psi_A^2 + \psi_B^2)$, (because there is a negative term in eqn. (*iv*).

It may be noted that the electron charge density in the region between the nuclei of the uncombined atomic orbitals, ψ_A and ψ_B is $\frac{1}{2}\psi_A^2 + \frac{1}{2}\psi_B^2$ and not $\psi_A^2 + \psi_B^2$.

This shows that probability of finding the electron in the region between the nuclei increases in case of bonding molecular orbitals and it decreases in case of antibonding molecular orbitals compared to their probabilities in the separated atoms. The presence of greater electron density between the nuclei results in greater attraction by both the nuclei and hence decrease in energy and thus result in bonding.

As seen in Fig. 1.29, the antibonding molecular orbital has been raised by the same energy by which the bonding molecular orbital has been lowered. Strictly speaking, if more precise calculations are carried out, the ABMO is raised slightly more in energy than the energy by which BMO is lowered, although the general pattern of the diagram remains the same. For example, if overlap integral (S) were not neglected, the wave functions ψ_+ and ψ_- would be given by

$$\psi_+ = \frac{1}{\sqrt{2+2S}}(\psi_A + \psi_B)$$

and

$$\psi_- = \frac{1}{\sqrt{2-2S}}(\psi_A - \psi_B)$$

instead of eqns. (*i*) and (*ii*).

Main points about bonding and antibonding molecular orbitals

(*i*) Whenever two atomic orbitals Ψ_A and Ψ_B combine together, their linear combination gives rise to two new molecular orbitals Ψ_+ and Ψ_-.

(*ii*) The energy E_+ corresponding to molecular orbitals Ψ_+ is lower than that of the combining atomic orbitals. Hence it is more stable than the combining atomic orbitals and is, therefore, called *Bonding Molecular Orbital* (BMO).

(*iii*) The energy E_- corresponding to molecular orbital Ψ_- is higher than that of the combining atomic orbitals. Hence it is less stable than the combining atomic orbitals and is, therefore, called *Antibonding Molecular Orbital* (ABMO).

(*iv*) The amount of lowering in energy of the BMO is almost equal to the increase in energy of the ABMO compared to the combining atomic orbitals.

(*v*) The electrons belonging to atomic orbitals leave the atomic orbitals and enter the bonding and antibonding molecular orbitals in order of their increasing energies, the molecular orbitals with lower energy being filled first (according to Aufbau principle).

(*vi*) On the basis of LCAO calculations, is found that greater the extent of overlap between the combining atomic orbitals, greater is the lowering in energy of the bonding molecular orbitals.

The main points of difference between BMO and ABMO are given in the Table 1.3 below:

Table 1.3. Comparison of Bonding and Antibonding Molecular Orbitals

Bonding Molecular Orbitals (BMO)	*Antibonding Molecular Orbitals (ABMO)*
1. They are formed by the additive effect of the atomic orbitals.	**1.** They are formed by the *subtractive effect* of the atomic orbitals.
2. Their energy is lower than that of combining atomic orbitals.	**2.** Their energy is higher than that of the combining atomic orbitals.
3. The electron charge density in the internuclear region is very high. As a result, the repulsion between the nuclei decreases and hence stability increases *i.e.* it favours the formation of bonding.	**3.** The electron charge density in the internuclear region decreases. Hence there is repulsion between the nuclei resulting in increase in energy and decrease in stability.
4. They are represented by σ, π δ etc.	**4.** They are represented by $\sigma*$, $\pi*$, $\delta*$ etc.

1.36. VARIATION OF ELECTRON PROBABILITY DENSITY ALONG INTERNUCLEAR AXIS IN BONDING AND ANTIBONDING MOLECULAR ORBITALS

Graphical Representation

(*a*) **Bonding Molecular Orbitals.** The bonding molecular orbitals wave function of the two atomic orbitals wave functions (Ψ_A and Ψ_B) is given by

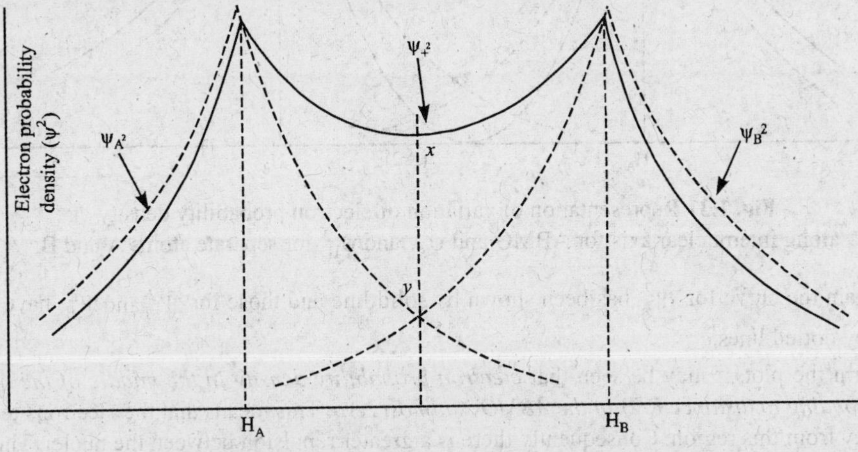

Fig. 1.30 Representation of variation of electron probability density for BMO and ψ_A^2 and ψ_A^2 for separate atoms A and B along the internuclear axis

$\Psi_+ = \frac{1}{\sqrt{2}}(\psi_A + \psi_B)$ (eq. (*i*) sec. 1.35) Hence the electron probability density in the bonding molecular orbitals will be given by

$\Psi_+^2 = \frac{1}{2}(\psi_A + \psi_B)^2 = \frac{1}{2}\Psi_A^2 + \frac{1}{2}\psi_B^2 + \Psi_A\Psi_B$. The sum $\frac{1}{2}\Psi_A^2 + \frac{1}{2}\psi_B^2$ gives the electron probability density between the nuclei of the two atoms in the *uncombined state* (Note that it is not equal to $\Psi_A^2 + \Psi_B^2$). However, Ψ_A^2 and Ψ_B^2 give the electron probability densities of the electrons in separate atoms *A* and *B* respectively. Plotting the values of Ψ_A^2 (solid line along internuclear axis and also plotting the values of Ψ_A^2 and Ψ_B^2 (dotted lines) for the individual atoms *A* and *B*, the curves obtained are shown in Fig. 1.30 For Ψ_+^2.

It may be seen from the plot that there is a *greater electron probability density between the nuclei in BMO than in the uncombined atomic orbitals* (look at points *x* and *y*).

(*b*) **Antibonding Molecular Orbitals.** For the antibonding molecular orbitals, the wave function is $\Psi_- = \frac{1}{\sqrt{2}}(\psi_A - \psi_B)$.

Hence electron probability density will be given by

$$\Psi_-^2 = \frac{1}{2}\Psi_A^2 + \frac{1}{2}\Psi_B^2 - \psi_A\psi_B.$$

Plotting the values as before, the curves obtained are shown in Fig. 1.31.

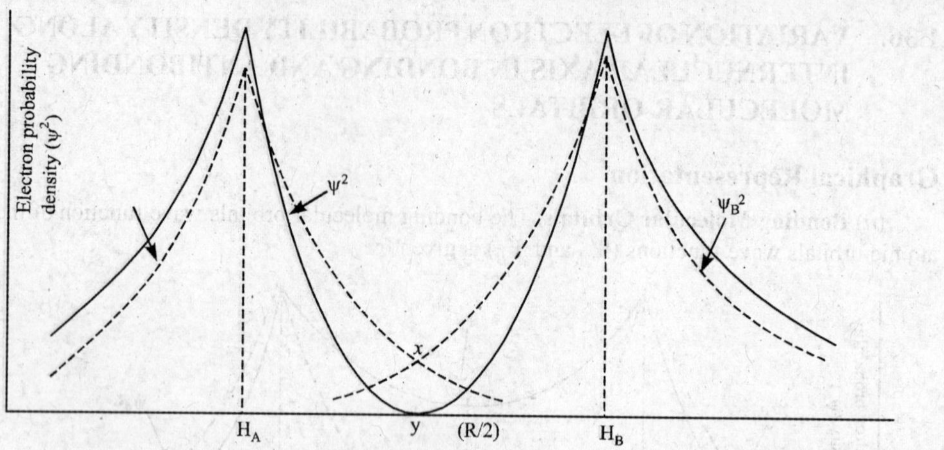

Fig. 1.31 Representation of variation of electron probability density along internuclear axis for ABMO and ψ_A^2 and ψ_B^2 for separate atoms A and B.

Again the curve for Ψ_-^2 has been shown by solid line and those for Ψ_A^2 and Ψ_B^2 have been shown by dotted lines.

Form the plot, it may be seen that *electron probability density in the middle of the nuclei (corresponding to distance R/2) in the ABMO is almost zero*. This means that the electrons tend to stay away from this region. Consequently there is a greater repulsion between the nuclei. There is a plane passing through the point exactly in the centre between the two nuclei and perpendicular to the internuclear axis on which the electron probability density in AMBO is zero. This plane is called the *nodal plane*.

Three-dimensional contour diagrams

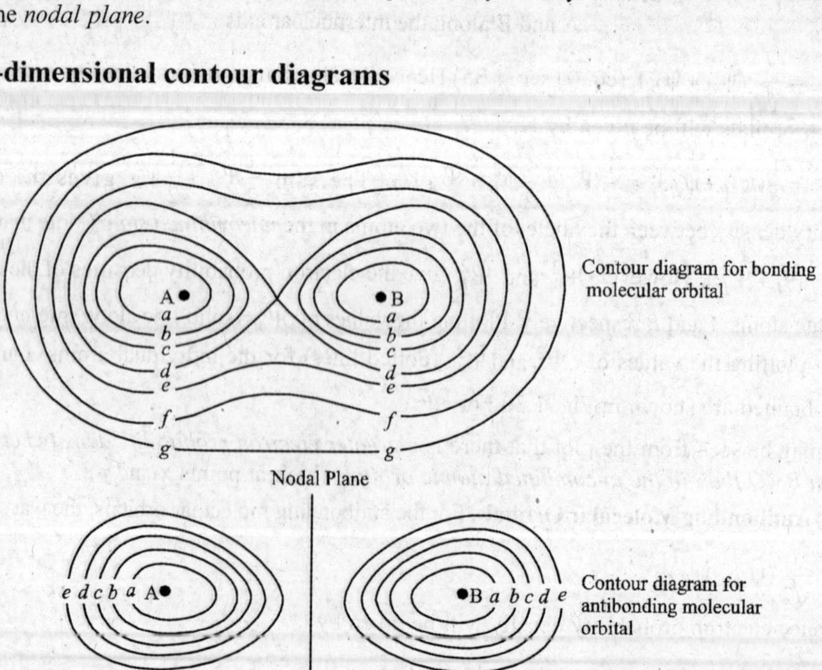

Fig. 1.32 Electron probability density contour diagrams for bonding and antibonding molecular orbitals

Electron probability densities are calculated at different points around both the nuclei. Regions showing the equal probability density are joined together. The resulting surface that we obtain is called a **contour**.

A number of contours are obtained around both the nuclei for bonding and antibonding molecular orbitals, as shown in Fig. 1.32.

We observe that in BMO the electron probability density tends to concentrate towards the internuclear region whereas in the AMBO, it tends to shift away from the internuclear region.

1.37. FORMATION OF MOLECULAR ORBITALS FROM ATOMIC ORBITALS

Just any two atomic orbitals cannot combine to form molecular orbitals. They combine to form molecular orbitals only if the following conditions are satisfied:

1. *The combining atomic orbitals must possess comparable energies*. Thus in the formation of a homonuclear diatomic molecules A_2 (like H_2, O_2 etc), 1s orbitals of one atom can combine with 1s of the other, 2s of one atom with 2s of the other and so on. 1s orbitals of one atom cannot combine with 2s of the other because of a big difference of energy. However, such combinations may be displayed by heteronuclear diatomic molecules (A – B).

We can explain it by saying that if the combining atomic orbitals (Ψ_A and Ψ_B) have comparable energies, their coefficients c_A and c_B have comparable values. As a matter of fact, if two combining atoms are identical, they will have the same coefficients *i.e.* $c_A = c_B$.

2. *The combining atomic orbitals must overlap to a considerable extent.* Greater the extent of overlap, greater is the electrons density in the internuclear region and hence greater is the attraction on the electron density by both nuclei and smaller is the repulsion between the nuclei.

3. *The combining atomic orbitals must be symmetrical with respect to their molecular axis*. For example, taking Z-axis as the internuclear axis, an s-orbital can overlap with p_z-orbitals or p_z orbital can overlap with p_z-orbitals but *s*-orbital cannot overlap with p_x or p_y orbital. Similarly p_x can overlap with p_x or p_y can overlap with p_y but p_x cannot overlap with p_y as illustrated in Fig. 1.33

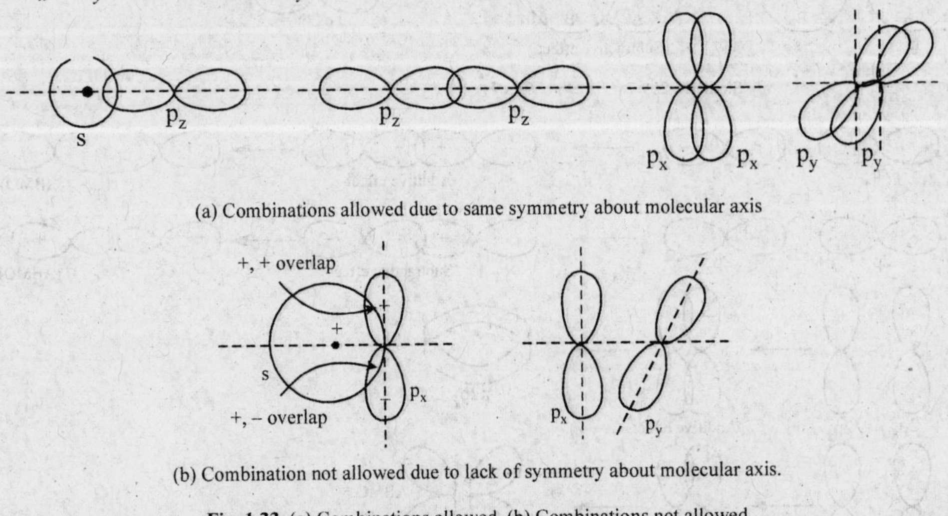

Fig. 1.33. (a) Combinations allowed (b) Combinations not allowed

This may also be explained in terms of the signs of the wave functions. In $s - p_x$ overlap, the area of +, + overlap is equal to the area of +, – overlap. Thus the two areas are equal but opposite in sign. Hence +, + overlap is cancelled by the +, – overlap and there is no net overlap.

1.38. CONCEPT OF σ, σ*, π, π* ORBITALS AND THEIR CHARACTERISTICS

For homonuclear diatomic molecules, (H_2, O_2 etc.) only similar atomic orbitals have comparable energies, hence only similar atomic orbitals of the two atoms can combine to form molecular orbitals *i.e.* 1*s* of one atom with 1*s* of the other, 2*s* with 2*s* and 2*p* with 2*p* and so on. Representing the combining atomic orbitals and the molecular orbitals formed with proper signs of the wave function, the formation of molecular orbitals is represented in Fig. 1.34. It may be noticed that when 1*s* combines with 1*s* or 2*s* with 2*s* or $2p_z$ with $2p_z$ (taking Z-axis as the internuclear axis), the atomic orbitals overlap along the internuclear axis resulting into formation of molecular orbitals whose electron charge density is symmetrical about the molecular axis. In each case, two molecular orbitals are formed, the bonding molecular orbital (formed by additive effect) is called σ-molecular orbitals which has high electron charge density in the internuclear region whereas the corresponding antibonding molecular orbitals (formed by subtractive effect) is called σ^*- molecular orbital in which case the electron density in the internuclear region decreases and hence the repulsion between the nuclei increases.

Additive effect implies +ve wave function combining with +ve wave function. The result is (+) × (+) = + ve i.e. electron density in the internuclear region increases. Subtractive effect means +ve wave function combining with –ve wave function . The result is (+) × (–) = –ve *i.e.* there is no electron density in the internuclear region. To represent the particular atomic orbitals combining to form molecular orbitals, symbols of the atomic orbitals are written alongwith the molecular orbitals.

σ and σ^* *e.g.* σ_{1s}, σ^*_{1s}, σ_{2s}, σ^*_{2s}, σ_{2p_z}, $\sigma^*_{2p_z}$ etc.

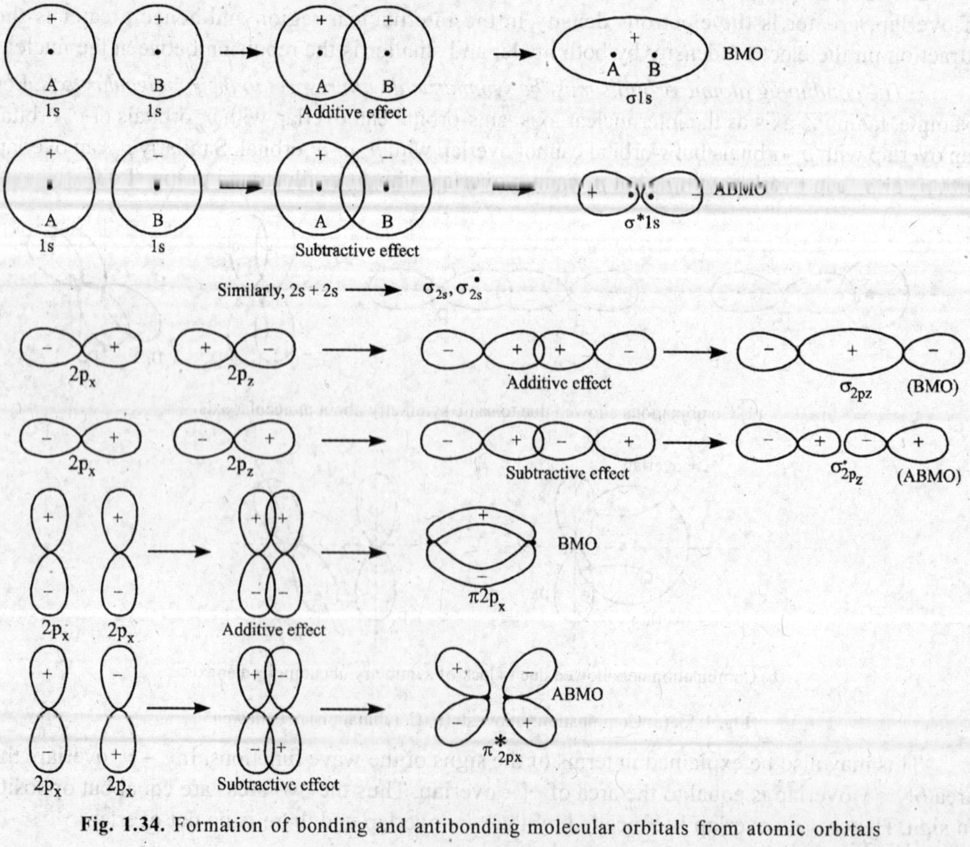

Fig. 1.34. Formation of bonding and antibonding molecular orbitals from atomic orbitals

If we take Z-axis as the internuclear axis, p_x orbital of one atom combines with p_x orbital of the other atom to give sideway overlapping. The molecular orbitals formed in such a case do not have symmetrical electron density about the molecular axis. The bonding molecular orbital in this case is called π-molecular orbitals and the corresponding antibonding molecular orbitals is called π*-molecular orbital. And to indicate that $2p_x$ orbitals are involved in the formation of molecular orbitals, we represent them by π_{2p_x} and $\pi^*_{2p_x}$. Similarly, if $2p_y$ orbital of one atom combines with $2p_y$ orbital of the other atom, molecular orbitals formed are unsymmetrical and are represented by π_{2p_y} and $\pi^*_{2p_y}$.

In summary, molecular orbitals which are formed by the overlap of the atomic orbitals along the internuclear axis and which have symmetrical electron charge density about the internuclear axis are called σ-molecular orbitals if they are bonding molecular orbitals and are called σ-molecular orbitals if they are antibonding. Similarly, molecular orbitals which are formed by the sideway overlap of atomic orbitals in a direction perpendicular to the internuclear axis and have unsymmetrical electron charge density about the internuclear axis are called π-molecular orbitals if they are bonding and are called π* molecular orbitals if they are antibonding.*

For **heteronuclear diatomic molecules,** the contribution by the two wave functions Ψ_A and Ψ_B is not equal towards the molecular orbitals wave function. In such cases different atomic orbitals of the two atoms may combine to form molecular orbitals. For example, consider the formation of HCl molecule. The electronic configuration of $_{17}Cl$ is $1s^2\ 2s^2\ 2p^6\ 3s^2\ 3p_x^2\ 3p_y^2\ 3p_z^1$. In the formation of HCl, $1s$ orbitals of H atom combines with $3p_z$ orbitals of Cl atom to form σ and σ* molecular orbitals, although, the two atomic orbitals involved possess much different energies.

1.39. FORMATION OF H_2 MOLECULE

In this case both the atomic orbitals are identical, their wave functions will contribute equally towards the molecular orbitals wave function. Hence applying LCAO approximation, the molecular orbitals wave function can be taken as the sum of the atomic orbitals wave functions (*i.e.* coefficients c_1 and c_2 can be ignored).

In H_2 molecule, there are two nuclei (protons) H_A and H_B. The two electrons may be written as 1 and 2. Let us consider first electron 1. The situation will be like that of H_2^+ ion. As the electron is under the influence of both the nuclei (H_A and H_B), therefore, if atomic orbital wave functions of atoms H_A and H_B are Ψ_A and Ψ_B respectively, then by LCAO approximation, the molecular orbital wave function for the electron 1 can be written as

$$\Psi_1 = \Psi_A(1) + \Psi_B(1) \quad \text{... } (i)$$

Similarly, as electron 2 is also under the influence of both the nuclei, by LCAO approximation, the molecular orbital wave function of electron 2 can be written as

$$\Psi_2 = \Psi_A(2) + \Psi_B(2) \quad \text{... } (i)$$

Thus total molecular orbital wave function will be given by

$$\Psi_{MO} = \Psi_1\Psi_2 = [(\Psi_A(1) + \Psi_B(1)]\ [\Psi_A(2) + \Psi_B(2)] \quad \text{... } (iii)$$

Using the molecular orbital wave function given by eqn. (*iii*), the energy is calculated for different internuclear distances and a graph is plotted between energy versus internuclear distance as usual and the values corresponding to the minimum in the curve are found and compared with the experimental values. The calculated value of bond dissociation energy and the equilibrium internuclear distance are 258.6 kJ mol^{-1} and 85 pm respectively. The experimental values are found to be 458.1 kJ mol^{-1} and 74.1 pm. Thus, the result obtained is worse than that obtained by the valence bond method.

A number of modifications have been done both in the valence bond method as well as molecular orbital method to obtain a wave function which give results in close agreement with the experimental values. However, these are beyond the scope of this book.

1.40. HYBRIDIZATION

"*Hybridization is a process of mixing of different atomic orbitals of the same atom which may have slightly different energies and different shapes such that a redistribution of energy takes place between them to form new orbitals of exactly same shape and energy.*" The new orbitals formed are known as hybrid orbitals . The number of hybrid orbitals formed is equal to the number of atomic orbitals mixed together. For example, if one 2*s* and one 2*p* orbital mix, the hybridization is called *sp* hybridization and the two orbitals formed are called *sp* hybrid orbitals. If one 2*s* and two 2*p* orbitals mix, hybridization is called sp^2 and the three orbitals formed are called sp^2 hybrid orbitals. Similarly, if one 2*s* and three 2*p* orbitals mix, the hybridization is called sp^3 and the four orbitals formed are called sp^3 hybrid orbitals. Hybrid orbitals has one lobe much bigger in size than the other on the opposite side. Due to bigger size of the lobe, they undergo better overlapping with the hybridized or unhybridized orbitals of the other atom and hence form a comparatively stronger bond. *sp*, sp^2 and sp^3 hybridization is illustrated in Fig. 1.35.

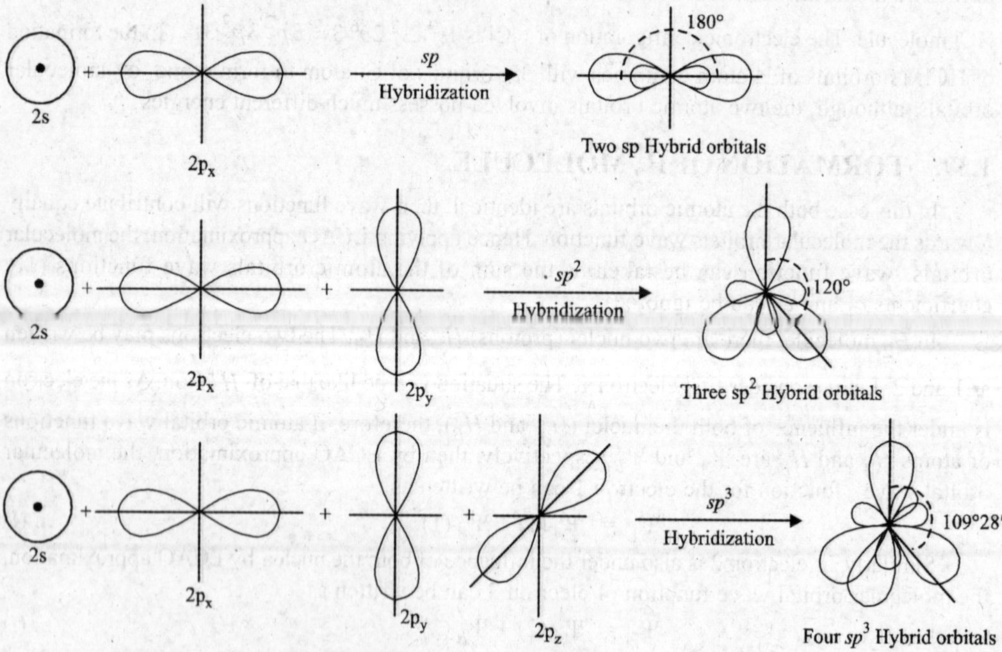

Fig. 1.35. Formation of *sp*, sp^2 and sp^3 hybrid orbitals.

1.41. QUANTUM MECHANICAL PRINCIPLES OF HYBRIDIZATION

The following quantum mechanical principles are involved in the formation of hybrid orbitals from atomic orbitals.

1. The hybrid orbitals are formed by linear combination of atomic orbitals belonging to the same atom. Thus.

$$\Psi_1 = a_1\Psi_s + b_1\,\Psi_{px} + c_1\Psi_{py} + d_1\,\Psi_{pz}$$
$$\Psi_2 = a_2\,\Psi_s + b_2\,\Psi_{px} + c_2\,\Psi_{py} + d_2\,\Psi_{pz} \text{ and so on.}$$

2. As the s-orbitals is spherically symmetrical, its charge density will be equally distributed among the n possible hybrid orbitals *i.e.* square of the coefficient of s-orbital will be equal in all the n hybrid orbitals which means that s-orbital in each hybrid orbitals will have square of coefficient $= 1/n$.

3. Each wave function (hybrid or atomic) is normalized. Thus

$$\int \Psi_1^2 \, d\tau = 1$$

$$\int \Psi_2^2 \, d\tau = 1$$

$$\int \Psi_s^2 \, d\tau = 1$$

$$\int \Psi_p^2 \, d\tau = 1 \qquad \text{and so on.}$$

4. The wave function of the hybrid orbitals as well atomic orbitals are orthogonal to each other.

Thus, $\int \Psi_1 \Psi_2 \, d\tau = 0$ and $\int \Psi_s \Psi_p \, d\tau = 0$.

5. In the formation of sp^2 hybrid orbitals, the first hybrid orbital may be consider to have maximum charge density along X-axis. Then p_y and p_z will not contribute towards this hybrid orbital.

6. In the formation of sp^3 hybrid orbitals, the second orbital may be considered to be in a plane, say xz plane, p_y will not contribute towards this hybrid orbitals.

1.42. CALCULATION OF THE COEFFICIENTS OF ATOMIC ORBITALS IN DIFFERENT HYBRID ORBITALS

Calculation have been done separately for sp, sp^2 and sp^3 hybrid orbitals as described below.

1. Calculation of coefficients of atomic orbitals in *sp* hybrid orbitals. The wave functions for the two hybrid orbitals may be written as

$$\Psi_1 = a_1 \Psi_s + b_1 \Psi_p \qquad \text{... } (i)$$

$$\Psi_2 = a_2 \Psi_s + b_2 \Psi_p \qquad \text{... } (ii)$$

(*i*) As the charge density is distributed equally among the two hybrid orbitals (two sp hybrid orbitals)

$$a_1^2 = a_2^2 = \frac{1}{2} \qquad \text{... } (iii)$$

$$\therefore \quad a_1 = a_2 = \frac{1}{\sqrt{2}} \qquad \text{... } (iv)$$

(*ii*) As Ψ_1 is a normalized wave function

$$\int \Psi_1^2 \, d\tau = 1 \quad \text{or} \quad \int (a_1 \Psi_s + b_1 \Psi_p)^2 \, d\tau = 1 \qquad \text{... } (v)$$

or

$$a_1^2 \int \Psi_s^2 \, d\tau + b_1^2 \int \Psi_p^2 \, d\tau + 2a_1 b_1 \int \Psi_s \Psi_p \, d\tau = 1 \qquad \text{... } (vi)$$

As atomic wave functions Ψ_s and Ψ_p are nomalized and they are also orthogonal to each other.

$$\int \Psi_s^2 \, d\tau = 1, \int \Psi_p^2 \, d\tau = 1, \int \Psi_s \Psi_p \, d\tau = 0 \qquad \text{... } (vii)$$

From eqs. (*vi*) and (*vii*), we have

$$a_1^2(1) + b_1^2(1) + 2a_1b_1(0) = 1$$

i.e. $$a_1^2 + b_1^2 = 1 \qquad \text{... (viii)}$$

But $a_1 = \dfrac{1}{\sqrt{2}}$ (see eq. (*iii*))

$$\therefore \quad \frac{1}{2} + b_1^2 = 1 \quad \text{or} \quad b_1^2 = \frac{1}{2} \quad \text{or} \quad b_1 = \frac{1}{\sqrt{2}} \qquad \text{... (ix)}$$

(*iii*) As Ψ_1 and Ψ_2 are orthogonal to each other, we have

$$\int \Psi_1 \Psi_2 \, d\tau = 0$$

This means

$$a_1 a_2 + b_1 b_2 = 0 \qquad \text{... (x)}$$

Substituting the values of a_1, a_2 and b_1, we get

$$\left(\frac{1}{\sqrt{2}}\right)\left(\frac{1}{\sqrt{2}}\right) + \left(\frac{1}{\sqrt{2}}\right) b_2 = 0$$

i.e. $$\frac{1}{2} + \frac{1}{\sqrt{2}} b_2 = 0 \quad \text{or} \quad b_2 = -\frac{1}{2} \times \sqrt{2} = -\frac{1}{\sqrt{2}} \qquad \text{... (xi)}$$

Thus, the wave function for the two *sp* hybrid orbitals are:

$$\boxed{\begin{aligned} \Psi_1 &= \frac{1}{\sqrt{2}}(\Psi_s + \Psi_p) \\ \Psi_2 &= \frac{1}{\sqrt{2}}(\Psi_s - \Psi_p) \end{aligned}} \qquad \begin{aligned} &\text{... (xii)} \\ &\text{... (xiii)} \end{aligned}$$

2. Calculation of coefficients of atomic orbitals in sp^2 hybrid orbitals. The wave functions for the three sp^2 hybrid orbitals may be written as

$$\Psi_1 = a_1 \Psi_s + b_1 \Psi_{px} + c_1 \Psi_{py} \qquad \text{... (xiv)}$$
$$\Psi_2 = a_2 \Psi_s + b_2 \Psi_{px} + c_2 \Psi_{py} \qquad \text{... (xv)}$$
$$\Psi_3 = a_3 \Psi_s + b_3 \Psi_{px} + c_3 \Psi_{py} \qquad \text{... (xiv)}$$

(*i*) As the charge density of *s*-orbital is equally distribute among the three hybrid orbitals

$$a_1^2 = a_2^2 = a_3^2 = \frac{1}{3}$$

or $$a_1 = a_2 = a_3 = \frac{1}{\sqrt{3}}.$$

(*ii*) Let us assume that Ψ_1 has maximum charge density along *X*-axis, it will have contribution only from *s* and p_x orbitals and the contribution form p_y orbital will be zero i.e. $c_1 = 0$.

(*iii*) As Ψ_1 is normalized wave function, we have

$$a_1^2 + b_1^2 = 1$$

But $$a_1 = \frac{1}{\sqrt{3}} \quad \therefore \quad \frac{1}{3} + b_1^2 = 1 \quad \text{or} \quad b_1^2 = \frac{2}{3} \quad \text{or} \quad b_1 = \sqrt{\frac{2}{3}}.$$

(*iv*) Ψ_1 and Ψ_2 are orthogonal to each other and also Ψ_1 and Ψ_3 are orthogonal to each other. That means

$$\int \Psi_1 \Psi_2 d\tau = 0 \text{ and } \int \Psi_1 \Psi_3 d\tau = 0$$

We can write the above results as

$$a_1 a_2 + b_1 b_2 = 0 \text{ and } a_1 a_3 + b_1 b_3 = 0$$

or
$$b_2 = -\frac{a_1 a_2}{b_1} = -\frac{(1/\sqrt{3})(1/\sqrt{3})}{\sqrt{2/3}} = -\frac{1}{\sqrt{6}}$$

and
$$b_3 = -\frac{a_1 a_3}{b_1} = -\frac{(1/\sqrt{3})(1/\sqrt{3})}{\sqrt{2/3}} = -\frac{1}{\sqrt{6}}$$

(*v*) As Ψ_2 is a normalized wave function, we have

$$\int \Psi_2^2 d\tau = 1 \text{ or } a_2^2 + b_2^2 + c_2^2 = 1$$

or
$$c_2^2 = 1 - (a_2^2 + b_2^2) = 1 - \left(\frac{1}{3} - \frac{1}{6}\right) = \frac{1}{2}$$

or
$$c_2 = \frac{1}{\sqrt{2}}.$$

(*vi*) As Ψ_3 is a normalized wave function, we have

$$\int \Psi_3^2 d\tau = 1 \text{ or } a_3^2 + b_3^2 + c_3^2 = 1$$

or
$$c_3^2 = 1 - (a_2^2 + b_2^2) = 1 - \left(\frac{1}{3} + \frac{1}{6}\right) = 1 - \frac{1}{2} = \frac{1}{2}$$

or
$$c_3 = \pm\frac{1}{\sqrt{2}}$$

c_2 and c_3 cannot be identical. If we take +ve sign for c_2 [see under (*v*) above], we have to take –ve sign for c_3 i.e. we take $c_3 = -\frac{1}{\sqrt{2}}$. (Only then the condition of orthogonality between Ψ_2 and Ψ_3 will be satisfied i.e. $a_2 a_3 + b_2 b_3 + c_2 c_3 = 0$).

Thus the wave functions for the three sp^2 hybrid orbitals are:

$$\Psi_1 = \frac{1}{\sqrt{3}}\Psi_s + \sqrt{\frac{2}{3}}\Psi_{px}$$

$$\Psi_2 = \frac{1}{\sqrt{3}}\Psi_s - \frac{1}{\sqrt{6}}\Psi_{px} + \frac{1}{\sqrt{2}}\psi_{py}$$

$$\Psi_3 = \frac{1}{\sqrt{3}}\Psi_s - \frac{1}{\sqrt{6}}\Psi_{px} - \frac{1}{\sqrt{2}}\psi_{py}$$

3. **Calculation of coefficients of atomic orbitals of sp^3 hybrid orbitals.** The wave functions for the four sp^3 hybrid orbitals can be written in terms of the following four equation

$$\Psi_1 = a_1 \Psi_s + b_1 \Psi_{px} + c_1 \Psi_{py} + d_1 \Psi_{pz}$$
$$\Psi_2 = a_2 \Psi_s + b_2 \Psi_{px} + c_2 \Psi_{py} + d_2 \Psi_{pz}$$
$$\Psi_3 = a_3 \Psi_s + b_3 \Psi_{px} + c_3 \Psi_{py} + d_3 \Psi_{pz}$$
$$\Psi_4 = a_4 \Psi_s + b_4 \Psi_{px} + c_4 \Psi_{py} + d_4 \Psi_{pz}$$

(*i*). As the charge density of *s*-orbital is distributed equally among the four sp^3 hybrid orbitals, we have

$$a_1^2 + a_2^2 + a_3^2 + a_4^2 = 1$$

or $$a_1^2 = a_2^2 = a_3^2 = a_4^2 = \frac{1}{4}$$

$$a_1 = a_2 = a_3 = a_4 = \frac{1}{2}$$

(*ii*) Assuming that Ψ_1 is aligned along the X-axis, contribution from Ψ_{py} and Ψ_{pz} will be taken as zero *i.e.* $c_1 = 0$ and $d_1 = 0$.

(*iii*) As Ψ_1 is a normalized wave function, we have

$$\int \psi_1^2 dt = 1 \quad \text{or} \quad a_1^2 + b_1^2 + c_1^2 + d_1^2 = 1 \quad \text{or} \quad a_1^2 + b_1^2 = 1 \qquad (\because c_1 = d_1 = 0)$$

$$\therefore \quad b_1^2 = 1 - a_1^2 = 1 - \left(\frac{1}{2}\right)^2 = \frac{3}{4}$$

or $$b_1 = \sqrt{\frac{3}{4}} = \frac{\sqrt{3}}{2}$$

(*iv*) As ψ_1 is orthogonal with ψ_2, ψ_3 and ψ_4 separately, we have the following three relations:

$$\int \Psi_1 \Psi_2 \, d\tau = 0, \int \Psi_1 \Psi_3 \, d\tau = 0, \int \Psi_1 \Psi_4 \, d\tau = 0.$$

$$a_1 a_2 + b_1 b_2 = 0, a_1 a_3 + b_1 b_3 = 0, a_1 a_4 + b_1 b_4 = 0$$

or $$b_2 = -\frac{a_1 a_2}{b_1},$$

$$b_3 = -\frac{a_1 a_3}{b_1},$$

$$b_4 = -\frac{a_1 a_4}{b_1},$$

But $$a_1 = a_2 = a_3 = a_4 = \frac{1}{2} \quad \text{(Derived above)}$$

$$\therefore \quad b_2 = b_3 = b_4 = -\frac{\frac{1}{2} \cdot \frac{1}{2}}{\sqrt{3}/2} = -\frac{1}{2\sqrt{3}}.$$

(*v*) Let us assume that the second hybrid orbital Ψ_2 lies in the *xz* plane, then p_y will not contribute towards Ψ_2 *i.e* $c_2 = 0$.

(*vi*) As Ψ_2 is normalized, $\int \Psi_2^2 d\tau = 1$. This means

$$a_2^2 + b_2^2 + c_2^2 + d_2^2 = 1$$

$$a_2^2 + b_2^2 + d_2^2 = 1 \qquad (\because c_2 = 0)$$

or $$d_2^2 = 1-(a_2^2+b_2^2)$$

$$= 1-\left(\frac{1}{4}+\frac{1}{12}\right)=1-\frac{1}{3}=\frac{2}{3}$$

or $$d_2 = \sqrt{\frac{2}{3}}.$$

(*vii*) As Ψ_2 is orthogonal with Ψ_3 and Ψ_4 separately, we have the following two relations.

$$\int\Psi_2\,\Psi_3\,d\tau=0,\ \int\Psi_2\,\Psi_4\,d\tau=0$$

i.e. $$a_2\,a_3+b_2\,b_3+d_2\,d_3 = 0,\quad a_2\,a_4+b_2\,b_4+d_2\,d_4=0$$

or $$d_3 = -\frac{a_2a_3+b_2b_3}{d_2}\ \text{and}\ d_4=-\frac{a_2\,a_4+b_2\,b_4}{d_2}$$

or $$d_3 = -\frac{1/4+1/12}{\sqrt{2/3}}=-\frac{1}{\sqrt{6}}\ \text{and}\ d_4=-\frac{1/4+1/12}{\sqrt{21/3}}=-\frac{1}{\sqrt{6}}$$

Thus, $$d_3 = d_4=-\frac{1}{\sqrt{6}}.$$

(*viii*) As Ψ_3 is normalised , $\int\Psi_3^2 d\tau=1.$ This means

$$a_3^2+b_3^2+c_3^2+d_3^2 = 1$$

or $$c_3^2 = 1-(a_3^2+b_3^2+d_3^2) = 1-\left(\frac{1}{4}+\frac{1}{12}+\frac{1}{6}\right)=\frac{1}{2}$$

$\therefore$ $$c_3 = \pm\frac{1}{\sqrt{2}}$$

To satisfy the condition of orthogonality between Ψ_3 and Ψ_4 and the normalization condition of Ψ_4 we take $c_4=-\frac{1}{\sqrt{2}}$.

Thus, the wave functions for the four sp^3 hybrid orbitals are:

$$\Psi_1 = \frac{1}{2}\Psi_s+\frac{\sqrt{3}}{2}\Psi_{px}$$

$$\Psi_2 = \frac{1}{2}\Psi_s-\frac{1}{2\sqrt{3}}\Psi_{px}+\sqrt{\frac{2}{3}}\Psi_{pz}$$

$$\Psi_3 = \frac{1}{2}\Psi_s-\frac{1}{2\sqrt{3}}\Psi_{px}+\frac{1}{\sqrt{2}}\Psi_{p_y}-\frac{1}{\sqrt{6}}\Psi_{p_z}$$

$$\Psi_4 = \frac{1}{2}\Psi_s-\frac{1}{2\sqrt{3}}\Psi_{px}-\frac{1}{\sqrt{2}}\Psi_{p_y}-\frac{1}{\sqrt{6}}\Psi_{p_z}$$

1.43. COMPARISON OF VALENCE BOND (VB) AND MOLECULAR ORBITAL (MO) MODELS

Points of Similarities

(1) Both the VB and MO models make use of atomic orbital wave functions ψ_A and ψ_B of the electrons involved.

(2) The bond is formed due to greater electron charge density in the internuclear region according to both models.

(3) Both the models postulate that bond is formed between the atoms due to overlap of their atomic orbitals.

(4) Both the models assume that the combining atomic orbitals must have comparable energies and same symmetry with respect to the internuclear axis.

Points of Differences

(1) According to VB Theory, only the valence electrons take part in the bonding and hence atoms do not lose their identity even after the overlap. According to the MO Theory, all the electrons are distributed around all the nuclei taken together and hence atoms lose their identity. The energy levels in which the electrons of the molecule are distributed are called molecular orbitals.

(2) According to VB theory only the half-filled orbitals of the valence shell take part in the bonding but according to MO theory all the atomic orbitals *i.e.* half-filled as well as fully-filled combine together provided they have comparable energies and same symmetries.

(3) According to VB Model the total wave function (ψ_{VB}) obtained as a linear combination of the two wave functions $\psi_I = \psi_A(1)\,\psi_B(2)$ and $\psi_{II} = \psi_A(2)\,\psi_B(1)$ is given by

$$\psi_{VB} = \psi_A(1)\,\psi_B(2) + \psi_A(2)\,\psi_B(1) \quad ...(i)$$

According to MO Model total molecular orbital wave function is given by

$$\psi_{MO} = [\psi_A(1) + \psi_B(1)]\,[\psi_A(2) + \psi_B(2)] \quad ...(ii)$$

On multiplying out this equation becomes

$$\psi_{MO} = \psi_A(1)\,\psi_B(2) + \psi_A(2)\,\psi_B(1) + \psi_A(1)\,\psi_A(2) + \psi_B(1)\,\psi_B(2) \quad ...(iii)$$

In this equation, the first two terms are same as those of ψ_{VB} in eqn. (*i*). Hence eqn. (*iii*) becomes

$$\psi_{MO} = \psi_{VB} + \psi_A(1)\,\psi_A(2) + \psi_B(1)\,\psi_B(2) \quad ...(iv)$$

The last two terms represent the ionic structures *i.e.* when both the electrons are on *A* or both the electrons are on *B*.

Thus, MO Model justifies the contribution of ionic structure whereas in VB Model these terms were included later but were not present in the original VBT.

(4) VB Theory leads to the concept of resonance whereas MO Theory has nothing to do with resonance.

PROBLEMS BASED ON LCAO, MOT AND VBT

Problem 1. Using LCAO approximation, write down the complete wave function for a heteronuclear diatomic molecule AB assuming that it has 85% covalent character and 15% ionic character.

Solution. If *AB* were purely covalent, its wave function will be

$$\psi_{\text{covalent}} = \psi_A(1)\,\psi_B(2) + \psi_A(2)\,\psi_B(1)$$

If *AB* were purely ionic, both electrons will be on more electronegative atom. Hence its wave function will be

$$\psi_{\text{ionic}} = \psi_B(1)\,\psi_B(2)$$

When *AB* has both covalent and ionic character, its wave function will be

$$\psi = c_1\,\psi_{\text{covalent}} + c_2\,\psi_{\text{ionic}}$$

Such that $c_1^2 + c_2^2 = 1$

Here, we are given that $c_1^2 = 85\% = 0.85$ and $c_2^2 = 15\% = 0.15$

$\therefore \quad c_1 = \sqrt{0.85} = 0.92, \quad c_2 = \sqrt{0.15} = 0.39$

Hence $\quad \psi = 0.92\,[\psi_A(1)\,\psi_B(2) + \psi_A(2)\,\psi_B(1)] + 0.39\,[\psi_B(1)\,\psi_B(2)].$

Problem 2. What will be the wave function for BMO of a hetero nuclear diatomic molecule AB, given that the electron spends 70% of its time on the nucleus of A and 30% of its time on the nucleus of B.

Solution. According to LCAO – MO method

$$\psi_{MO} = C_A\psi_A + C_B\psi_B$$

The significance of the coefficients C_A and C_B is that their squares *i.e.*, $|C_A|^2$ and $|C_B|^2$ represent the weightages of ψ_A and ψ_B towards the total wave function. This means

$$|C_A|^2 = 70\% = 0.70 \text{ and } |C_B|^2 = 30\% = 0.30$$

$\therefore \quad C_A = \pm\sqrt{0.70} = \pm 0.84$

and $\quad C_B = \pm\sqrt{0.30} = \pm 0.55$

Hence, $\quad \psi_{MO} = 0.84\ \psi_A + 0.55\ \psi_B$

SOME SOLVED CONCEPTUAL PROBLEMS

Problem 1. How do spectral distribution curves of black body radiation prove (i) Stefan-Boltzmanns' law (ii) Wien's displacement law?

Solution. (*i*) The area under the curve at any temperature gives the total energy emitted by the black body per second. The area is found to be proportional to 4th power of temperature. Hence $E_B \propto T^4$. This is Stefan – Boltzmann's law.

(*ii*) Product of the wavelength (λ_m) corresponding to maximum emittance and temperature is constant that is

$$\lambda_{m_1} \times T_1 = \lambda_{m_2} \times T_2 = \lambda_{m_3} \times T_3 = ...$$

or $\quad \lambda_m \times T = \text{constant}$

or $\quad \lambda_m \propto \dfrac{1}{T}$

This is Wien's displacement law.

Problem 2. Name the basic principles on which (a) classical mechanics is based (b) quantum mechanics is based. Name the phenomena which the classical mechanics fails to explain but quantum mechanics can explain them.

Solution. Classical mechanics is based on Newton's laws of motion and electromagnetic wave theory. Quantum mechanics is based on Heisenberg's uncertainty principle, de Broglie concept of dual nature of matter, Schrodinger wave equation and Planck's Quantum theory. Classical mechanics fails to explain Black body radiation, Photoelectric effect, Heat capacities of solids and Atomic and molecular spectra, while a satisfactory explanation of these phenomena is provided by quantum mechanics.

Problem 3. What is the difference in heat capacities of solids when classical mechanics is applied and when quantum mechanics is applied?

Solution. When classical mechanics is applied, the heat capacity of all monatomic solids comes out to be constant and equal to 3 R (Dulong and Petit's law). Experimentally, it is found to be true only at high temperature. At low temperature, the value is found to be less than $3R$ and decreases with decrease of temperature. Quantum mechanics is able to explain the anomalies.

Problem 4. What is Compton effect? Write expression for Compton shift. Prove that Compton shift is doubled if the angle between the incident and the scattered rays is doubled from 90° to 180°.

Solution. Compton effect states: *"If monochromatic X-rays are allowed to hit a material of low atomic weight e.g. graphite, the scattered radiations contain not only wavelengths of that of the incident X-rays but also contain radiations of higher wavelength.*

Expression for Compton shift is given by the relation

$$\Delta\lambda = \lambda' - \lambda = \frac{h}{mc}(1 - \cos\theta)$$

where θ is the angle between the incident and the scattered X-rays

When $\theta = 90°$, $\cos\theta = 0$, $\Delta\lambda_1 = \dfrac{h}{mc}$

When $\theta = 180°$, $\cos\theta = -1$, $\Delta\lambda_2 = \dfrac{2h}{mc}$ so that $\dfrac{\Delta\lambda_2}{\Delta\lambda_1} = \dfrac{2h}{mc} \times \dfrac{mc}{h} = 2.$

Problem 5. What are the conditions which an eigen function (ψ) must satisfy?

Solution. (*i*) ψ must be single valued and finite.

(*ii*) ψ must be continuous

(*iii*) ψ must become zero at infinity.

Problem 6. What is the commutator of two operators $\hat{A}$ and $\hat{B}$? When are the operators said to commute?

Solution. Commutator of two operators $\hat{A}$ and $\hat{B}$, represented as $\left[\hat{A}, \hat{B}\right]$ is given by

$$\left[\hat{A}, \hat{B}\right] = \hat{A}\hat{B} - \hat{B}\hat{A}$$

If $\hat{A}\hat{B} = \hat{B}\hat{A}$ so that $\left[\hat{A}, \hat{B}\right] = 0$, the operators are said to commute.

Problem 7. What is sinusoidal wave equation?

Solution. The upward displacement (*y*) of a wave at any time (*t*) travelling in the *x*-direction with velocity *v*, wavelength λ, frequency ν and amplitude *A* is represented as

$$y = A \sin\left[\frac{2\pi}{\lambda}(x - vt)\right] = A \sin\left[2\pi\left(\frac{x}{\lambda} - \nu t\right)\right]$$

This is called sinusoidal wave equation.

Problem 8. Write expressions for the following:

(*i*) Probability of finding electron in a small volume element in terms of (*a*) cartesian coordinates (*b*) polar coordinates.

(*ii*) Condition for a normalized wave function.

(*iii*) Condition for orthogonality of two wave functions.

Solution. (*i*) (*a*) Probability = $\psi\psi^* \, d\tau = \psi\psi^* \, dx\, dy\, dz$ (in cartesian coordinates)

(*b*) Probability = $\psi\psi^* \, r^2 \, dr \sin\theta \, d\theta \, d\phi$ (in polar coordinates)

(*ii*) For a wave function ψ to be a normalized wave function, the following condition is to be satisfied.

$$\int_{-\infty}^{+\infty} \psi\psi^* \, d\tau = 1$$

(*iii*) For two wave functions ψ_m and ψ_n to be orthogonal to each other, the following condition is to be satisfied.

$$\int_{-\infty}^{+\infty} \psi_m \psi_n^* d\tau = 0, \qquad \int_{-\infty}^{+\infty} \psi_m^* \psi_n d\tau = 0.$$

Problem 9. The angular wave function ($\Theta\Phi$) for one of the d-orbitals is given by

$$\Theta_{2,-2}\phi_{-2} = \left(\frac{15}{16\pi}\right)\sin^2\theta \ \sin 2\phi$$

Prove that opposite lobes of the d-orbital have the same sign.

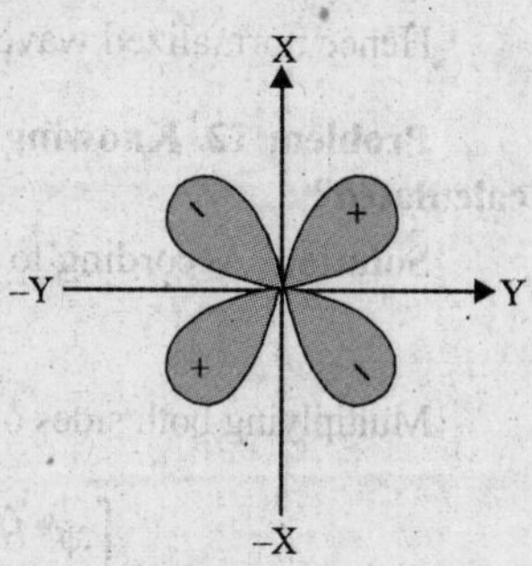

Solution.

$$\Theta_{2,-2}\phi_{-2} = \left(\frac{15}{16\pi}\right)^{1/2} \sin^2\theta \sin 2\phi$$

$$= \left(\frac{15}{16\pi}\right)^{1/2} \sin^2\theta . 2 \sin\phi \cos\phi$$

$$= \left(\frac{15}{16\pi}\right)^{1/2} \sin\theta \sin\phi \cdot \sin\theta \cos\phi$$

$$= \left(\frac{15}{16\pi}\right)^{1/2} \frac{y}{r} \cdot \frac{x}{r} \qquad (\because x = r \sin\theta \cos\phi, y = r \sin\theta \sin\phi)$$

$$= \left(\frac{15}{16\pi}\right)^{1/2} \frac{xy}{r^2}$$

If x and y both are positive or both are negative xy will be positive Hence wave function will be positive. If either x or y is negative then xy will be *negative* and hence wave function will be negative.

Problem 10. What happens to the energy of the particle in one dimensional box if the length of the box is made larger?

Solution. Energy of the particle in one-dimensional box is given by

$$E = \frac{n^2 h^2}{8ma^2}$$

If 'a' is made larger, values of E becomes smaller. Consequently, energy levels move close together. If the box length becomes very large, quantization disappears and there is a smooth transition from quantum behaviour to classical behaviour. This is called Bohr's **correspondence principle**.

Problem 11. Normalize the wave function $\psi = A \sin\left(\frac{n\pi}{a}x\right)$ by finding the value of the constant A when the particle is restricted to move in one dimensional box of width 'a'.

Solution. For normalized wave function

$$\int_0^a \psi\psi^* dx = 1$$

or $$A^2 \int_0^a \sin^2 \frac{n\pi}{a} x\, dx = 1$$

or $$A^2 \cdot \frac{a}{2} = 1 \quad \text{or} \quad A = \sqrt{\frac{2}{a}}$$

Hence normalized wave function $= \sqrt{\frac{2}{a}} \sin\left(\frac{n\pi}{a} x\right)$

Problem 12. Knowing the wave function of a system, how is the energy of the system calculated?

Solution. According to Schrodinger wave equation

$$\hat{H}\psi = E\psi$$

Multiplying both sides of the equation with ψ^* and integrating over the whole space ($-\infty$ to $+\infty$)

$$\int_{-\infty}^{+\infty} \psi^* \hat{H} \psi\, d\tau = \int_{-\infty}^{+\infty} \psi^* E \psi\, d\tau$$

As E is constant, this equation can be written as

$$\int_{-\infty}^{+\infty} \psi^* \hat{H} \psi\, d\tau = E \int_{-\infty}^{+\infty} \psi^* \psi\, d\tau$$

or $$E = \int_{-\infty}^{+\infty} \psi^* \hat{H} \psi\, d\tau \Big/ \int_{-\infty}^{+\infty} \psi^* \psi\, d\tau$$

If ψ is a real function, the above equation can be written as

$$E = \int_{-\infty}^{+\infty} \psi \hat{H} \psi\, d\tau \Big/ \int_{-\infty}^{+\infty} \psi^2\, d\tau$$

where Hamiltonian operator $\hat{H} = -\frac{h^2}{8\pi^2 m}\nabla^2 + V$ (∇^2 = Laplacian operator).

EXERCISES

(Based on Question Papers of Different Universities)

Multiple Choice Questions (Choose the correct option)

Wave mechanics

1. Classical mechanics does not provide satisfactory explanation for the following

(*a*) Blackbody radiation (*b*) Photoelectric effect

(*c*) Heat capacities of solids (*d*) All the above

2. A quantitative relationship between the energy absorbed and energy emitted by a body was put forward by

(*a*) Einstein (*b*) Kirchoff

(*c*) Hess (*d*) Fermi

3. Total amount of energy emitted by a perfectly black body per unit area per unit time is directly proportional to

(*a*) T (*b*) T^2

(*c*) T^4 (*d*) T^3

4. Schrodinger wave equation in three dimensions is given by

(*a*) $\frac{\partial^2\psi}{\partial x^2}+\frac{\partial^2\psi}{\partial y^2}+\frac{\partial^2\psi}{\partial z^2}+\frac{8\pi^2 m(E-V)}{h^2}\psi=0$

(*b*) $\frac{\partial^2\psi}{\partial x^2}+\frac{\partial^2\psi}{\partial y^2}+\frac{\partial^2\psi}{\partial z^2}-\frac{8\pi^2 m(E-V)}{h^2}=0$

(*c*) $\frac{\partial^2\psi}{\partial x^2}+\frac{\partial^2\psi}{\partial y^2}+\frac{\partial^2\psi}{\partial z^2}+\frac{8\pi^2 m(E+V)}{h^2}\psi=0$

(*d*) $\frac{\partial^2\psi}{\partial x^2}+\frac{\partial^2\psi}{\partial y^2}+\frac{\partial^2\psi}{\partial z^2}-\frac{8\pi^2 m(E+V)}{h^2}\psi=0$

5. Expression for energy for a particle in one dimensional box is

(*a*) $E=\frac{n^2h^2}{8ma^2}$ (*b*) $E=\frac{n^2h^2}{4ma^2}$

(*c*) $E=\frac{n^2h}{8ma^2}$ (*d*) $E=\frac{nh^2}{8ma^2}$

6. Compton shift in the scattered radiations, $\Delta\lambda$ is given by

(*a*) $\Delta\lambda=\frac{2h}{mc}\cos^2\frac{\theta}{2}$ (*b*) $\Delta\lambda=\frac{2h}{mc}\tan^2\frac{\theta}{2}$

(*c*) $\Delta\lambda=\frac{2h}{mc}\sin^2\frac{\theta}{2}$ (*d*) $\Delta\lambda=\frac{2h}{mc}\cot^2\frac{\theta}{2}$

Theories of chemical bonding

7. Valence bond theory of chemical bonding was put forward initially

(*a*) Heisenberg (*b*) Einstein

(*c*) Heitler and London (*d*) de Broglie

8. The phenomenon of resonance is explained by

(*a*) Molecular orbital theory (*b*) Valence bond theory

(*c*) Neither of the two (*d*) Both the above

9. Choose the incorrect answer

(*a*) Bonding M.O. are formed by the additive effect of the atomic orbitals

(*b*) The electron density in the internuclear region in bonding M.O. is very low

(*c*) Energy of the antibonding M.O. is higher than that of the combining atomic orbitals.

(*d*) Antibonding M.O. are represented by σ^*, π^*, δ^*, etc.

10. The condition to be satisfied for the atomic orbitals to form molecule orbitals is

(*a*) They must possess comparable energies

(*b*) They must overlap to a considerable extent

(*c*) They must be symmetrical with respect to the molecular axis
(*d*) all the above

11. The coefficient of atomic orbitals in sp^2 hybrid orbitals is

(*a*) 0.5 (*b*) $\frac{1}{\sqrt{2}}$

(*c*) $\frac{1}{\sqrt{3}}$ (*d*) 0.33

12. The wave functions corresponding to ionic structures of hydrogen molecule, according to valence bond theory are

(*a*) $\psi_A(1)\,\psi_B(1),\ \psi_A(2)\,\psi_B(2)$
(*b*) $\psi_A(1)\,\psi_A(1),\ \psi_B(2)\,\psi_B(2)$
(*c*) $\psi_A(2)\,\psi_B(2),\ \psi_A(1)\,\psi_B(1)$
(*d*) $\psi_A(1)\,\psi_A(2),\ \psi_B(1)\,\psi_B(2)$

ANSWERS

1. (*d*)	**2.** (*b*)	**3.** (*c*)	**4.** (*a*)	**5.** (*a*)	**6.** (*c*)
7. (*c*)	**8.** (*b*)	**9.** (*b*)	**10.** (*d*)	**11.** (*c*)	**12.** (*d*)

Short Answer Questions

Wave mechanics

1. State and explain Kirchoff's law about black body radiation.

2. List the main points of difference between Classical mechanics and Quantum mechanics. What is the relation between the two?

3. What do you understand by a 'normalised eigen function'? How do we normalise a wave function?

4. How can you explain that the energy of a free particle is not quantised?

5. How does the study of particle in three dimensional box in quantum mechanics lead to the concept of degeneracy?

6. What are probability distribution curves? What are their shortcoming? How is it overcome by radial probability distribution curves?

7. With the help of the phenomenon of black body radiation, explain how classical mechanics failed and it led to the origin of quantum mechanics.

8. What is sine wave? Give examples of sine waves. Write expression for it. What type of plots we get keeping time or position constant? What is sinusoidal wave equation?

9. Write the expression for energy for the particle in a one-dimensional box. How can you justify
(*i*) quantisation of energy?
(*ii*) existence of zero point energy?

10. What do you understand by expectation value of an observable in quantum mechanics? Write expressions for the expectation values of (*i*) position *x* (*ii*) any other observable *G* (e.g. momentum).

11. How do the three quantum numbers follow from the solutions of Schrodinger wave equation?

12. Show that the normalised wave functions of a particle in one-dimensional box are orthogonal for any pair of different values of n.
13. What do you understand by eigen value and eigen function? What are the conditions which eigen functions must satisfy?
14. Write expression for the energy of hydrogen-like atoms as obtained by quantum mechanics. How is it identical with the equation obtained by Bohr?
15. How does classical mechanics fail to explain the spectral distribution of black body radiation? What is Planck's radiation law? How do Wien's law and Rayleigh Jean's law follow from it.
16. How do we proceed to find out the probability if ψ does not come out to be real?
17. When are the two eigen functions said to be (*i*) mutually orthogonal, (*ii*) orthonormal?
18. How does quantum mechanics lead to the concept of orbital?
19. Write the expression for the angular and the radial wave function. What do different symbols signify?

Theories of chemical bonding

20. Taking the example of H_2 molecule, explain how the energy is calculated for this molecule from wave function.
21. Write expressions for ψ_{VB} and ψ_{MO} for H_2 molecule. How do they differ? What conclusion do you draw from it? Can this conclusion be generalized?
22. What do you understand by gerade and ungerade molecular orbitals? What is the criterion to check whether a molecular orbital is gerade or ungerade? Explain with suitable examples.
23. What is Born-Oppenheimer approximation in Quantum Mechanics? How is it applied in the study of potential energy curve of H_2^+ ion?
24. How Ritz linear combination method is employed to arrive at the correct wave function? What is the advantage of this method?
25. Write expressions for the wave functions of BMO and ABMO. Using these expressions, how can you explain the electron probability in the internuclear region in the two cases?
26. How is the total molecular orbital wave function obtained for H_2 molecule according to molecular orbital theory?
27. Apply quantum mechanics to obtain the wave functions of the two *sp* hybrid orbitals.
28. How are three dimensional contour diagrams used to represent the electron probability density for bonding and antibonding molecular orbitals?
29. Taking exchange of electrons into consideration and ionic structures of H_2 molecules, write expressions for ψ_{covalent} and ψ_{ionic} and then for the complete wave function ψ.
30. Write expression for ψ_{MO} for H_2^+ ion according to LCAO method. Starting from it, how do you arrive at the expressions for $\psi_{+(\text{MO})}$ and $\psi_{-(\text{MO})}$?
31. Derive the values of the coefficients of atomic orbitals in the three sp^2 hybrid orbitals.
32. Compare the important characteristics of σ and π molecular orbitals.
33. Outline the main points of variation method to arrive at the correct wave function.
34. Taking the example of H_2^+ ion, explain how the energy is calculated for this species from wave function.
35. Write expressions for bonding and antibonding molecular orbital wave functions in terms of the two combining atomic orbital wave functions ψ_A and ψ_B. Discuss graphically the variation of electron probability density for BMO and ψ_A^2 and ψ_B^2 along the internuclear axis.

36. What is hybridization? What is meant by *sp*, sp^2 and sp^3 hybridization? Why the bond formed by a hybrid orbital is stronger than that formed by unhybridized orbital?

37. What are the criteria or the conditions for the formation of molecular orbitals from atomic orbitals?

38. List the basic ideas of Molecular Orbital Theory.

General Questions

1. Derive Planck's radiation law. How can it be verified experimentally?

2. What is 'classical mechanics'? What are its limitations? Name the two mechanics that have been put forward to overcome the limitations of classical mechanics. Name the scientists responsible for each of them. What is the basic difference between the two mechanics?

3. Briefly explain how classical mechanics fails when applied to the following:

(*i*) Photoelectric effect

(*ii*) Heat capacity of solids.

How could these phenomena be explained by Planck's quantum theory?

4. Briefly describe the spectral distribution of black body radiation. How do the following laws follow from it?

(*i*) Stefan Boltzmann's law (*ii*) Wien's displacement law

5. Briefly explain de Broglie hypothesis related to dual nature of matter. Derive de Broglie relation. How has the dual nature of electrons been verified experimentally?

6. What is Compton effect? What is 'Compton shift'? Write expression for Compton shift and explain the results obtained for scattering angles of 0°, 90° and 180°. How does it explain the results of Heisenberg's uncertainty principle?

7. Applying de Broglie relationship, derive Schrodinger wave equation.

8. What is the concept of particle in one-dimensional box? What is the Schrodinger wave equatin for such a case? How can this equation be solved for ψ and E?

9. Derive expressions for the expectation values of x, x^2, p_x and p_x^2.

10. Write Schrodinger wave equation for hydrogen-like atom in spherical polar co-ordinates. How can you separate the variables of this equation to get expressions each containing one variable only.

11. Apply Schrodinger wave equation to a particle in one-dimensional box and obtain the expression for the eigen function and eigen value of the energy.

12. Derive expressions for standard deviations in x and p_x and hence deduce Heisenberg's uncertainty principle.

13. What are the main postulates of Bohr model of atom. What are its defects?

14. State and explain Heisenberg's uncertainty principle.

15. What is the basic equation on which quantum mechanics is based? Derive the equation in the usual form.

16. What is an operator? When are the operators said to commute? Explain with an example that the operators usually do not commute. What is the commutator of the two operators $\hat{A}$ and $\hat{B}$? What is its value when the operators commute?

17. What are the postulates of quantum mechanics? Based on the postulates of quantum mechanics, derive Schrodinger wave equation.

18. Derive expression for the eigen function for a particle in one-dimensional box. How can this function be normalized? Write expression for the normalised wave function.
19. Describe expression for the normalised eigen function for a particle in a three-dimensional box.
20. Briefly explain Born's interpretation of wave function or Explain the significance of wave function (ψ).
21. Derive expressions for the total energy for a particle in a three-dimensional box.
22. Using the concepts of quantum mechanics describe the shapes of *s*, *p* and *d*-orbitals.
23. On the basis of angular wave functions, justify the different designations given to three *p*-orbitals and five *d*-orbitals.
24. What do you understand by Radial probability? Draw and discuss radial probability distribution curves for 1*s*, 2*s*, 3*s*, 3*p* and 3*d* orbitals.

Theories of chemical bonding

25. Discuss the application of Valence Bond Theory to study the bond dissociation energy and equilibrium bond distance of H_2 molecule. Explain at least three improvements that are made in the wave function to get better agreement of calculated value with the experimental value.
26. How LCAO – MO treatment of H_2^+ ion leads to the concept of Bonding and Antibonding Molecular Orbitals? Represent them diagrammatically.
27. What are gerade and ungerade molecular orbitals? Explain with suitable examples.
28. What are the main points of similarities and differences between VBT and MOT?
29. Using LCAO – MO method, derive expressions for molecular orbital wave functions. Compare the calculated value of energy with the experimental value on the energy versus internuclear distance diagram.
30. How electron probability density varies with internuclear distance? Represent these graphically as well as by contour diagrams.
31. Applying quantum mechanical principles, derive expressions for the wave functions of the two *sp* hybrid orbitals, three sp^2 hybrid orbitals and four sp^3 hybrid orbitals.
32. Using LCAO for wave function for H_2^+, obtain the normalized wave function for BMO and ABMO without neglecting the overlap integral.
33. What do you understand by Linear Combination of Atomic Orbitals (LCAO). How can it be applied to H_2^+ ion to calculate its energy? Comment on the values of the energy obtained.
34. Compare the main features of the Valence Bond Model with those of the Molecular Orbital Model.
35. Represent diagrammatically the formation of bonding and antibonding molecular orbitals formed by combination of 2*s* with 2*s* and 2*p* with 2*p* orbitals. How are they designated?
36. Derive expressions for Hamiltonian operator for (*i*) H_2^+ ion (*ii*) H_2 molecule. How is the correct (true) wave function obtained by (*i*) variation method (*ii*) Ritz linear combination method? How are bond energy and bond length of these species calculated from suitable plots?
37. Discuss the application of LCAO–MO method to the study of H_2 molecule. Compare the results obtained with the experimental values.
38. Apply quantum mechanical principles to calculate the coefficients of atomic orbitals in *sp*, sp^2 and sp^3 hybrid orbitals.

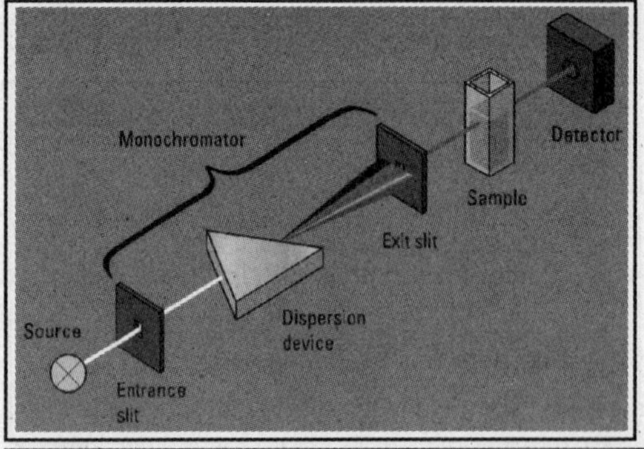

2

Spectroscopy

2.1. INTRODUCTION

Spectro means radiations and *scopy* means measurement. Thus spectroscopy means measurement of radiations.

However, *spectroscopy is defined as the branch of science which is associated with the interaction of radiations of different wavelengths with matter.*

Electromagnetic radiations

An electromagnetic radiations is the radiant energy emitted from any source in the form of heat or light or sound. Some of the characteristics of radiations are listed below.

1. Electromagnetic radiations possess the particle nature as well as the wave nature. Thus they have the dual nature. A light ray consists of a stream of particles (photons) moving in the form of waves.

2. Electromagnetic radiations travel with the velocity of visible light i.e, 2.998×10^{0} ms ^{1}

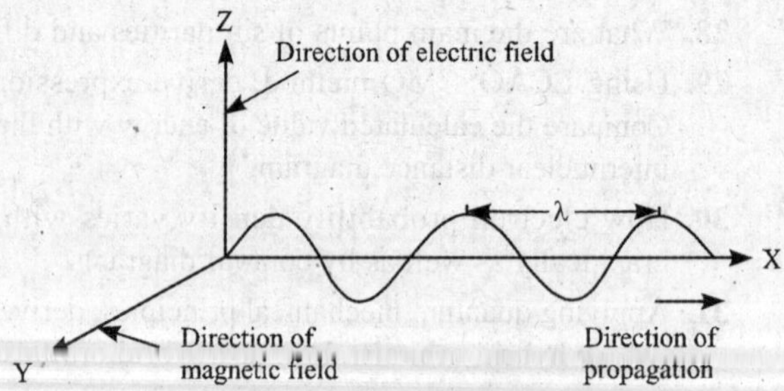

Fig 2.1. Electromagnetic radiation travelling in the form of waves and associated with electric and magnetic fields at right angles to each other.

3. The electromagnetic waves have electric and magnetic fields associated with them at right angles to one another as shown in Fig 2.1. Thus, if the radiation is moving along X-axis, the magnetic field is oriented along Y-axis and electric field is oriented along Z-axis.

Visible light, X-rays, microwaves, radiowaves, etc., are all electromagnetic radiations. Collectively, they make up the electromagnetic spectrum (Fig. 2.2).

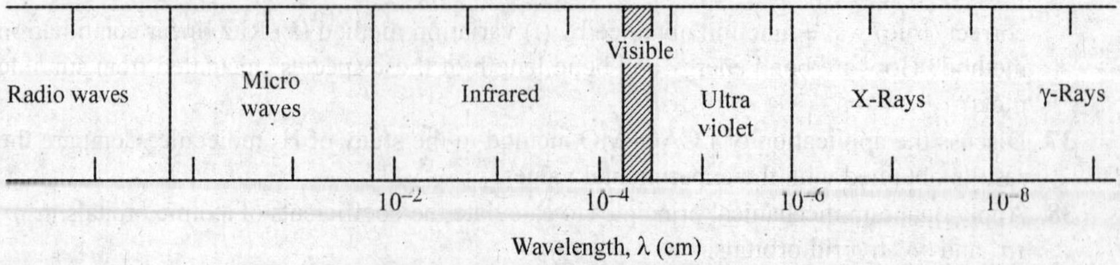

Fig. 2.2. The Electromagnetic Spectrum.

Electromagnetic radiations consist of electrical and magnetic waves oscillating at right angles to each other. They travel away from the source with the velocity of light (c) in vacuum. Different terms used in the study of electromagnetic waves are discussed as under :

1. Frequency (ν) : It is the number of successive crests (or troughs) which pass a stationary point in one second. The unit is **hertz;** $1\text{Hz} = 1\text{s}^{-1}$.

2. Wavelength (λ) : It is the distance between successive crests (or troughs). λ is expressed in centimeters (cm), meters (m), or nanometers ($1 \text{ nm} = 10^{-9}\text{m}$).

3. Wave number ($\bar{\nu}$) : It is the reciprocal to wavelength. Its unit is cm^{-1}.

$$\bar{\nu} = \frac{1}{\lambda}$$

Relation between Frequency, Wavelength and Wave number

Frequency and wavelength of an electromagnetic radiation are related by the equation

$$\nu\lambda = c \qquad \ldots(i)$$

Or

$$\nu = \frac{c}{\lambda} \qquad \ldots(ii)$$

where c is the velocity of light. It may be noted that wavelength and frequency are inversely proportional. That is, **higher the wavelength, lower is the frequency; lower the wavelength higher is the frequency.**

Further

$$\bar{\nu} = \frac{1}{\lambda}\text{ cm}^{-1} \qquad \ldots(iii)$$

From (ii) and (iii)

$$\nu = c\bar{\nu}$$

Energy of Electromagnetic Radiation

Electromagnetic radiation behaves as consisting of discrete wave-like particles called **quanta** or **photons.** Photons possess the characteristics of a wave and travel with the velocity of light in the direction of the beam. The amount of energy corresponding to 1 photon is expressed by **Planck's equation.**

$$E = h\nu = hc/\lambda$$

where E is the energy of 1 photon (or quantum), h is Planck's constant (6.62×10^{-27} ergs-sec.); ν is frequency in hertz; and λ is wavelength in centimeters.

If N is the Avogadro number, the energy of 1 mole photons can be expressed as

$$E = \frac{Nhc}{\lambda} = \frac{2.85\times10^{-3}}{\lambda}\text{ kcal/mol}$$

2.2. ROTATIONAL, VIBRATIONAL AND ELECTRONIC ENERGIES OF MOLECULES

Rotational Energy

It involves the rotation of molecules or of parts of molecules about the centre of gravity (Fig. 2.3).

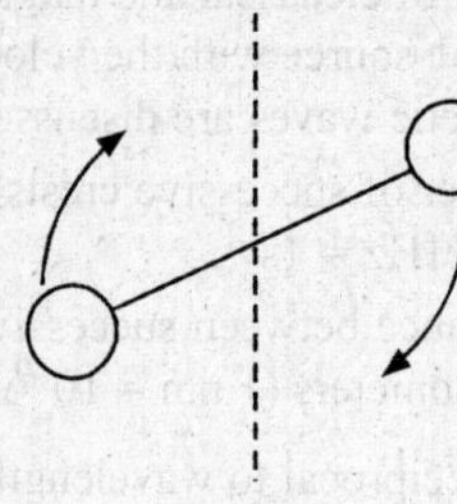

Fig. 2.3. Molecular rotation in molecules

Vibrational Energy

It is associated with stretching, contracting or bending of covalent bonds in molecules. The bonds behave as spirals made of wire (Fig. 2.4).

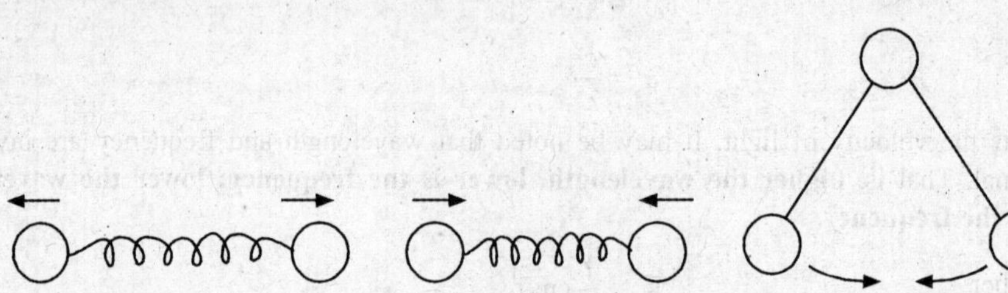

Fig. 2.4. Vibrations within molecules

Electronic Energy

It involves changes in the distribution of electrons by splitting of bonds or the promotion of electrons into higher energy levels (Fig. 2.5).

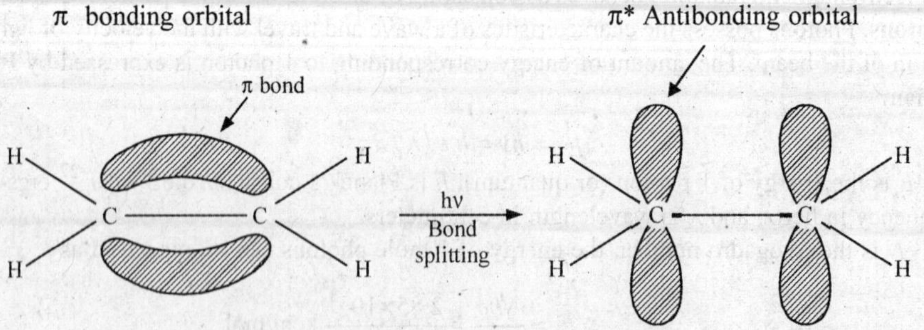

Fig. 2.5. Splitting of π bond by absorption of energy.

2.2.1. Experimental set-up of absorption spectrophotometer

Absorption spectrum of a given sample is obtained experimentally with the help of an apparatus called **absorption spectrophotometer** (Fig. 2.6). Light of a range of wavelengths from the source is passed through the sample. The wavelengths corresponding to allowed molecular transitions are absorbed. The transmitted light passes through a prism which resolves it into various wavelengths. It is then reflected from the mirror onto a detector. The prism is rotated so that light of each given wavelength is focussed on the detector. The pen recorder records the intensity of radiation as a function of frequency and gives the absorption spectrum of the sample.

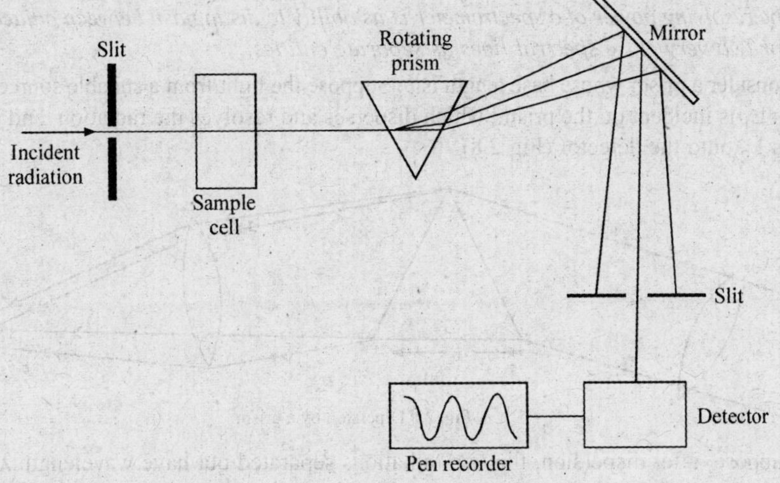

Fig. 2.6. Working of absorption spectrophotometer

2.2.2. Signal to noise ratio

The signals produced in the analysis of a sample using a spectrophotometer consist of two parts:

(*i*) signals produced due to the sample

(*ii*) signals produced due to other components (solvent etc.) and the instrument used.

The signals produced as in (*ii*) are called **noise**. The noise signals are unwanted and hinder sometimes the interpretation of signals. *Noise may be defined as the random electronic signals that are usually visible as fluctuation of base line in the signal* (Fig 2.7). It is therefore always designed to root out the signals or at least to minimize them so that a correct interpretation of the signals in the spectrum can be made.

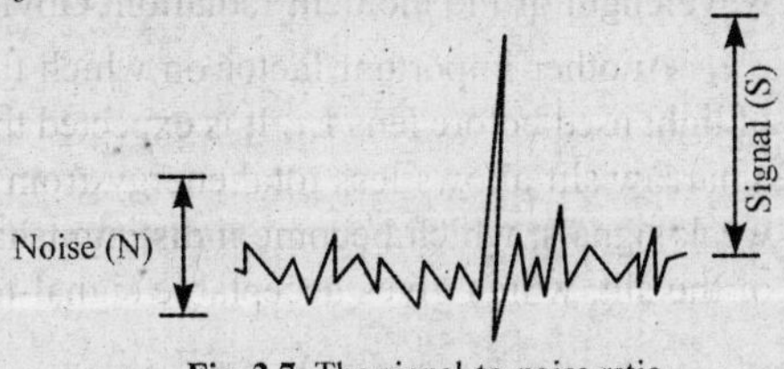

Fig. 2.7. The signal-to-noise ratio

The amount of radiation absorbed by the sample depends upon the concentration. Therefore the amount of noise obtained using an instrument determines the smallest concentration of the sample that can be measured accurately. With decreasing concentration of the sample, we observe greater difficulty in distinguishing noise from the signal. This leads to decreased precision in the measurements The ability of a instrument to distinguish between the noise and the signal is expressed as signal-to-noise ratio.

$$\frac{\text{Average signal amplitude}}{\text{Average noise amplitude}} = \frac{S}{N}$$

Obviously a greater value S/N indicates greater noise reduction or greater precision in measurement. It is important to understand that S/N ratio cannot be increased to simple amplification (increase of current), because the magnitude of both noise and signal increases. Thus this does not solve the problem. Some techniques, for example, hardware techniques using filters and lock-in-amplifiers and software techniques using software algorithm of ensemble averaging and Fourier transformation have been used with success to increase S/N ratio.

2.2.3 Resolving power

The resolving power of a spectrometer is its ability to distinguish between adjacent absorption bands or two very close spectral lines as separate entities.

Consider a prism whose base length is *b*. Suppose the light from a suitable source after crossing the lens L_1 is incident on the prism, which disperses and resolves the radiation, and is focussed by the lens L_2 onto the detector (Fig 2.8)

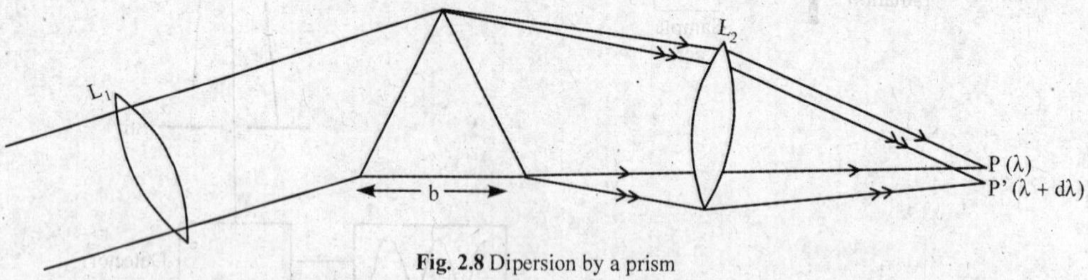

Fig. 2.8 Dipersion by a prism

Suppose, after dispersion, the two radiations separated out have wavelength λ and $\lambda + d\lambda$, then the intervel $d\lambda$ gives the value of **resolution.** Smaller the intervals, higher the resolution of the prism. The resolving power *R* of a dispersing element is given by

$$R = \frac{\lambda}{d\lambda}$$

If *n* is the refractive index and *b* is the base length of the prism, then resolving power of the prism is given by

$$R = b\frac{dn}{d\lambda}$$

$dn/d\lambda$ represents the change in refractive index of the material of the prism with change in wavelength of the incident radiation. Obviously for high resolving power, $dn/d\lambda$ should be large.

Another important factor on which the resolution depends is the width of the slit (aperture) of light used before lens L_1. It is expected that a narrower slit would give better resolution, however, a narrow slit allows less total energy from the beam to reach the detector. As a result, we receive weak signals, which become indistinguishable from the noise. Thus an optimum minimum width of the slit, which gives acceptable signal-to-noise ratio has to be decided.

2.3. MOLECULAR SPECTROSCOPY VERSUS ATOMIC SPECTROSCOPY

In the case of atoms, electrons jump from lower orbit to higher orbit. When they start coming back to the lower orbit, emission of energy equal to the difference in the energies of the two orbitals takes place in the form of atomic spectrum.

However, in case of molecules, when the energy is absorbed, it may result into rotation, vibration or electronic transition depending upon the amount of energy absorbed. Just as electronic energy is quantized *i.e.*, there are only discrete electronic energy levels in an atom or a molecule, the rotational and vibrational energies are also quantized *i.e.*, there are only discrete rotational and vibrational energy levels in a molecule. The rotational, vibrational and electronic energy levels of a molecule are collectively called **molecular energy levels.** The transitions of energies can take place only between these levels. The result is a **molecular spectra.** Just as the study of atomic spectra is helpful in the study of structure of atoms, the study of molecular spectra is one of the best experimental methods of studying the structure of molecules.

When the energy absorbed by a molecule is so large that an electron can jump from one electronic energy level to another (say $n = 1$ to $n = 2$) then as the energy required for transition from one vibrational energy level to another is much less and for one rotational level to another is still less, hence the electronic transition is also accompanied by the transition between the vibrational levels as well as the rotational levels. The complete picture of the various molecular energy levels for the first two electronic levels is shown in Fig. 2.9.

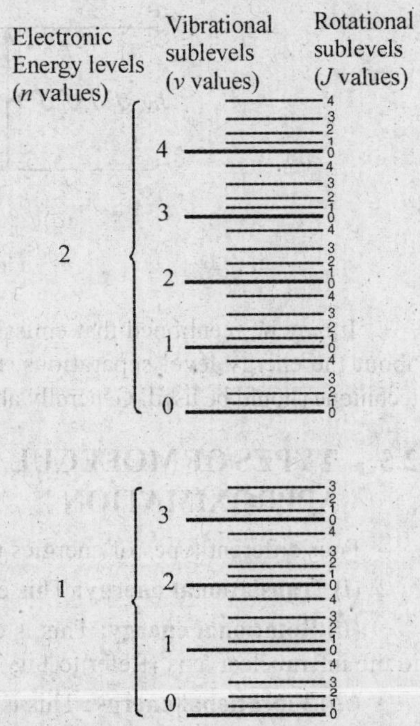

Fig. 2.9. Molecular energy levels.

It is evident that each electronic level consists of a number of vibrational sublevels represented by $v = 0, 1, 2, 3$ etc. and further each vibrational level consists of a number of rotational sublevels represented by $J = 0, 1, 2, 3$ etc. v and J are called *vibrational and rotational quantum numbers respectively.*

Due to a large number of energy level involved in transition, the molecular spectra are much more complex than atomic spectra. However, just as atomic spectra help in the study of the structure of the atoms, the study of molecular spectra gives even more information about the structure and properties of molecules such as bond lengths, bond angles, bond strength, shapes, dipole moments *etc.*

2.4. ABSORPTION AND EMISSION SPECTROSCOPY

An atom or a molecule has a number of energy levels which are quantized. The transitions take place only between these energy levels according to certain rules, called *selection rules.* Transition between any two energy levels can take place in either of the following two ways:

(*i*) The transition may take place from lower energy level to higher energy level by absorbing energy Fig. 2.10(*a*). It is then called **absorption spectroscopy** and the result obtained as a result of number of such transitions is called "absorption spectrum". Absorption spectrum consists of **dark** lines.

(*ii*) The transition may take place from higher energy level to a lower energy level Fig. 2.10(*b*) thereby emitting the excess energy as a photon. It is then called **emission spectroscopy** and the result obtained as a result of number of such transitions is called "emission spectrum". Emission spectrum consists of **bright** lines.

In fact, a number of groups of closely spaced lines are observed. Each such group of closely spaced lines is called a **band**. Thus a number of bands are observed. We can, therefore, conclude that whereas atoms give line spectra, molecules give band spectra.

In either case, the energy of the photon $h\nu$ emitted or absorbed is given by Bohr's frequency formula

$$E_2 - E_1 = h\nu$$

or in terms of wave length, $\nu = \dfrac{c}{\lambda}$

or in terms of wave number, $\bar{\nu} = \dfrac{1}{\lambda} = \dfrac{\nu}{c}$ (in cm^{-1} if λ is in cm or velocity c is in cm s^{-1})

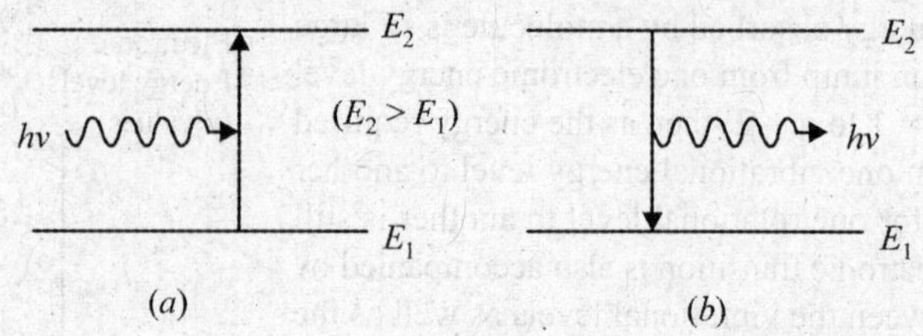

Fig. 2.10. (*a*) Transition involving absorption of photon.
(*b*) Transition involving emission of photon.

It may be mentioned that emission and absorption spectroscopy give the same information about the energy level separations. However, the practical considerations decide as to which technique should be used. Generally absorption spectra are easier to interpret than emission spectra.

2.5. TYPES OF MOLECULAR ENERGIES AND BORN OPPENHEIMER APPROXIMATION

Four different types of energies possessed by a molecule are:

(*i*) **Translational energy:** This energy is due to translational linear motion of the molecule.

(*ii*) **Rotational energy:** This is due to rotation of the molecule about an axis perpendicular to the internuclear axis (Refer to Fig 2.3)

(*iii*) **Vibrational energy:** This is due to the to-and-fro motion of the nuclear of the molecule without allowing any change in the centre of gravity of the molecule (see Fig 2.4).

(*iv*) **Electronic energy:** This is due to absorption of energy by the electron resulting in excitation to higher energy levels (see Fig 2.5)

Born-Oppenheimer approximation

According to the above postulate, **the total energy of a molecule is the sum of translational, rotational, vibration and electronic energies.**

or $\quad E = E_t + E_r + E_v + E_e$

It has been found that translational energy E_t is negligible as compared to E_r, E_v and E_e. Therefore

$$E = E_r + E_v + E_e$$

2.6. TYPES OF MOLECULAR SPECTRA

A molecule possesses quantised translational, rotational, vibrational and electronic energy levels. Within any two successive electronic energy levels, there are a number of different vibrational levels and within any two successive vibrational levels, there are a number of different rotational levels. The energy difference between any two successive electronic levels is more than that between any two successive vibrational levels which in turn is more than that between any two successive rotational levels. It can be represented as

$$E_r << E_v << E_e$$

We are not taking into consideration translational energies E_t. The difference between two successive translational levels is so small that it cannot be determined experimentally. For practical purposes, translational energy is considered as continuous and we do not observe any translational spectrum.

Different types of molecular spectra are described as under:

(*i*) **Pure rotational (Microwave) spectra:** If the energy absorbed by the molecules is such that it can cause transition only from one rotational level to another within the same vibrational level, the result obtained is called the rotational spectrum. These spectra are, therefore, observed in the

far infra-red region or in **the microwave region** whose energies are extremely small ($\bar{\nu}$ = 1 – 100 cm^{-1}). The spectra obtained is, therefore, also called **micro-wave spectra.**

(*ii*) **Vibrational rotational spectra:** If the incident energy is sufficiently large so that it can cause a transition from one vibrational level to another within the same electronic level then as the energies required for the transitions between the rotational levels are still smaller, both types of transitions *viz.,* vibrational and rotational will take place. The result is, therefore, a vibration-rotation spectrum. Since such energies are available in the near infra-red region, these spectra are observed in this region ($\bar{\nu}$ = 500 – 4000 cm^{-1}) and are called **infra-red spectra.**

(*iii*) **Electronic Band spectra:** If the incident energy is higher such that it can result in a transition from one electronic level to another, then this will also be accompanied by vibrational level changes and each of these is further accompanied by rotational level changes. For each vibrational change, a set of closely spaced lines is observed due to rotational level changes. Such a group of closely spaced rotational lines is called a **band.** Thus for a given electronic transition, a set of bands is observed. This set of bands is called a **band group** or a **band system.** Each electronic transition gives a band system. The complete set of band systems obtained due to different electronic transitions gives the electronic band spectrum of the gaseous molecule. As such high excitation energies are obtained from the visible and ultraviolet regions, these spectra are observed in the **visible region** (12,500 – 25,000 cm^{-1}) and **ultraviolet region** (25,000 – 70,000 cm^{-1}).

(*iv*) **Raman spectra:** This is also a type of vibrational-rotational spectrum but is based on scattering of radiation and not on the absorption of radiation by the sample. It is based upon the principle that when a sample is hit by monochromatic radiation of the visible region and scattering is observed at *right angles to the direction of the incident beam,* the scattered radiation has frequency equal to that of the incident beam (called *Rayleigh scattering*) as well as frequencies different (higher as well as lower) from that of the incident beam (called *Raman scattering*). The difference in the frequencies of the incident beam and that of the scattered beam (called *Raman frequencies*) are similar to those observed for the vibrational and rotational transitions. However, by suitably adjusting the frequency of the incident radiation, Raman spectra are observed in the visible region (12,500 – 25,000 cm^{-1}).

(*v*) **Nuclear Magnetic Resonance (NMR) spectra:** This type of spectrum arises from the transitions between the *nuclear spin energy levels* of the molecule when an external magnetic field is applied on it. The energies involved in these transitions are very high which lie in the *radio frequency region* (5–100 MHz). The method is based upon applying such frequencies on the sample so that it resonates with the applied frequency.

(*vi*) **Electron Spin Resonance (ESR) spectra:** This type of spectrum arises from the transitions between the *electron spin energy levels* of the molecule when an external magnetic field is applied on it. These involve frequencies corresponding to microwave region (2000 – 9600 MHz). Frequencies of this range are applied on the sample to bring the sample in resonance condition.

ROTATIONAL SPECTRA

2.7. SELECTION RULES FOR ROTATIONAL AND VIBRATIONAL SPECTRA

Atomic spectrum as well as molecular spectrum are obtained due to transition taking place between energy levels. However, such transitions can take place only between definite energy levels not between just any two energy levels. *The restrictions thus applied on the transition are governed by certain rules which are called the selection rules.* If these rules are followed, the transition can take place and it is called an **allowed transition.** If these are not followed, the transition cannot take place and it is called a **forbidden transition.** The selection rules to be followed depend upon the type of transition.

The **selection rules** are generally expressed in terms of changes in quantum numbers for the allowed transitions. For example, for the pure rotational transition, the selection rule is $\Delta J = \pm 1$ where J represents rotational quantum number ($\Delta J = +1$ corresponds to **absorption** and $\Delta J = -1$ corresponds to **emission**). Similarly, for pure vibrational transition, the selection rule is $\Delta v = \pm 1$ where v represents the vibrational quantum number.

The fundamental principle which forms the basis of selection rules is that for a molecule to absorb or emit a radiation of frequency ν, it must possess a dipole at that frequency. Thus a transition is allowed only if transition dipole moment, μ_{tr} # 0. μ_{tr} will be zero for molecule which has no dipoles. Hence only the molecule like NO, CO, etc which have permanent dipole, give pure rotational spectra.

2.8. WIDTH AND INTENSITIES OF THE SPECTRAL LINES

Width of the spectral lines

If the spectral line is sharp, it will appear in the spectrum as a vertical line with no width. On the contrary, if the line is not sharp it will have certain width as shown in Fig. 2.11.

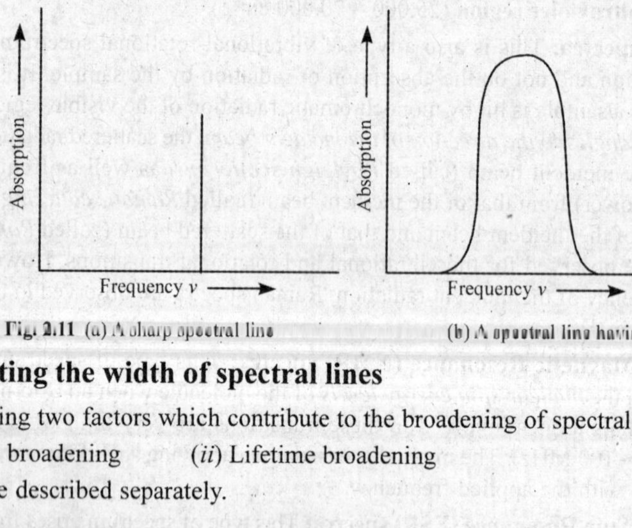

Fig. 2.11 (a) A sharp spectral line (b) A spectral line having width.

Factors affecting the width of spectral lines

The following two factors which contribute to the broadening of spectral line

(*i*) Doppler broadening (*ii*) Lifetime broadening

The two are described separately.

Doppler broadening

This type of broadening occurs due to Doppler effect. Doppler effect which is applicable to gaseous samples states:

Frequency of the radiation emitted or absorbed changes when the molecule is moving towards or away from the observer.

If the molecule (source) emitting the frequency ν is moving away from the observe with a velocity v, observer records the frequency of radiation as

$$\nu' = \frac{\nu}{1+\frac{v}{c}} \qquad ...(i)$$

If the molecule (source) is approaching the observer with a velocity v, the observer records the frequency of radiation as

$$\nu' = \frac{\nu}{1-\dfrac{v}{c}} \quad \text{...(ii)}$$

It is found by calculation based on Maxwell distribution of velocities that if m is the mass of the gas molecule, then at a temperature T, the width of the line at half the height is

$$\delta\nu = \frac{2\nu}{c}\left(\frac{2kT\, ln2}{m}\right)^{1/2} \quad \text{...(iii)}$$

Doppler broadening increases with temperature because of increase in molecular velocities with temperature. It is therefore best to work at low temperatures to obtain spectra of maximum sharpness.

Lifetime broadening

This broadening applies to samples in gases, liquid and solids including solutions. When Schrodinger equation is solved for a system that is changing with time, it is not possible to specify energies of levels exactly. Suppose, on an average, a system remains in a state for the time τ (lifetime of the state), then energy level become uncertain to the extent δE given by the equation

$$\delta E = \frac{h}{2\pi\tau} \quad \text{...(iv)}$$

The quantity δE is called lifetime broadening. Equation (*iv*) is analogous to Heisenberg uncertainty principle and the quantity δE may also be called as *uncertainty broadening.*

We can see from eq. (*iv*) that δE and τ are inversely related. Thus, shorter the lifetime of the state involved in the transition, greater will be the broadening of the spectral line. No excited state lines in infinite lifetime. Therefore, all states are subject to some lifetime broadening.

Collision between the molecules or with the walls of the container is the main reason for the finite lifetimes of the excited states. If τ_{coll} is the mean time between collisions, then the width of the resulting line is given by the relation.

$$\delta E_{coll} = \frac{h}{2\pi\tau_{coll}} \quad \text{...(v)}$$

Broadening can be minimised by working at low pressures with a gaseous sample.

2.8.1. Intensity of the spectral lines

Factors which determine the intensity of the spectral lines are:

(*i*) Population density of a state

(*ii*) Strength of the incident radiation

(*iii*) Probability of transition taking place between the energy levels

Greater the intensity of radiation, greater is the rate at which transitions take place resulting in stronger absorption by the sample. Similarily transition probability is decided by selection rules. If a transition is allowed, we expect a strong spectral line. If it is a forbidden transition, we obtain a weak spectral line (or no spectral line). Einstein explained how the population density of the states involved in the transition affects the intensity of the spectral lines. He suggested that the net rate of absorption in the presence of an electromagnetic field (i.e. incident radiation) is the outcome of the following processes:

1. Stimulated (by the incident radiation) absorption
2. Stimulated emission (emission of photon)
3. Spontaneous emission

The net rate of absorption as a result of the above three processes is given by

$$R_{net} = (N - N')\,B\rho \qquad ...(vi)$$

Where N = Number of molecules present in the lower state

N' = Number of molecule present in the upper state

B = Einstein coefficient of stimulated absorption

ρ = Energy density of the radiation at the frequency of transition.

At any temperature T, the ratio of the population of the two states with energy E and E′ can be obtained from Boltzmann distribution law

$$\frac{N'}{N} = e^{-h\nu/kT} \qquad ...(vii)$$

But $h\nu = E' - E = \Delta E$

If ΔE is large, from eq. (*vii*), N'/N will be small. This means population in the upper state is small. Hence, the only absorption will be from the lower state. The stimulated emission from the upper state will also be small.

2.9. DEGREES OF FREEDOM

Consider a molecule made up of N atoms. We know all the mass of the atom is concentrated in its nucleus and the nucleus is very small in size. Therefore the atoms may be considered as mass points. To represent each mass point, we require three co-ordinates. Hence to represent the instantaneous position of N mass points in space, we require $3N$ co-ordinates. This is how we introduce the term degrees of freedom.

The number of co-ordinates required to specify the position of all the mass points i.e, atoms in a molecule is called the number of **degrees of freedom.**

Thus a molecule made up of N atoms has 3 N degrees of freedom.

When thermal energy is absorbed by a molecule, it is stored within the molecule in the form of:

(*i*) translational motion of the molecule

(*ii*) internal movement of the atoms within the molecule, *i.e.*, rotational motion and vibrational motion.

The translational motion of the molecule means the motion of the centre of mass of the molecule as a whole. The centre of mass of the molecule can be represented by three co-ordinates. Thus there are *three translational degrees of freedom*. The remaining (3N – 3) co-ordinates represent the *internal degrees of freedom*.

The internal degrees of freedom may be subdivided into

(*i*) rotational degrees of freedom, and (*ii*) vibrational degrees of freedom

For a rotational motion, there are two degrees of freedom for a linear molecule and three for a non-linear molecule as shown in Fig. 2.12.

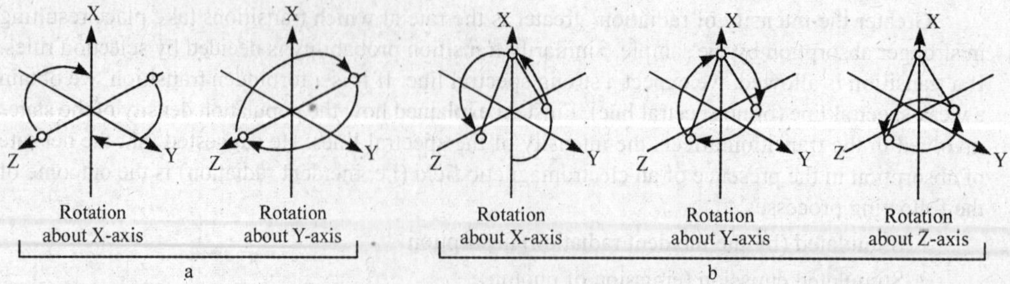

Fig. 2.12. (a) Rotations of a linear molecule about two mutually perpendicular axes.
(b) Rotations of a non-linear molecule about three mutually perpendicular axes.

Remaining degrees of freedom for the *linear* molecules are equal to (3*N*–5) and for the *non-linear* molecules, they are equal to (3*N*–6). These degrees of freedom represent the number of *vibrational degrees* of freedom. Hence

Vibrational degrees of freedom of a linear molecule containing *N* atoms = $3N - 5$.

Vibrational degrees of freedom of non-linear molecule containing *N* atoms = $3N - 6$.

A few example on the calculation of vibrational degrees of freedom are given below:

(*i*) As a diatomic molecule ($N = 2$) is always linear, its vibrational degrees of freedom $= 3N - 5 = 3 \times 2 - 5 = 1$.

(*ii*) For the linear polyatomic molecule CO_2, $N = 3$. Hence vibrational degrees of freedom = $3N - 5 = 3 \times 3 - 5 = 4$. Similarly for $HC \equiv CH$, $N = 4$. Hence the vibrational degrees of freedom $= 3N - 5 = 3 \times 4 - 5 = 7$.

(*iii*) For the non-linear polyatomic molecule H_2O, $N = 3$. Hence vibrational degrees of freedom $= 3N - 6 = 3 \times 3 - 6 = 3$. Similarly, for HCHO, $N = 4$ the vibrational degrees of freedom $= 3N - 6 = 3 \times 4 - 6 = 6$.

The vibrational degree of freedom represent the interatomic distances and angles needed to specify the geometry of the atomic framework within the molecule.

Also, the vibrational degrees of freedom represent the number of independent *vibrational modes* that can occur in the molecule, the vibrational energy being associated with each mode.

The three modes of vibration of a diatomic molecule like CO, four modes of vibration of linear triatomic CO_2 molecule and three modes vibration of non-linear triatomic H_2O molecule are illustrated in Fig 2.13.

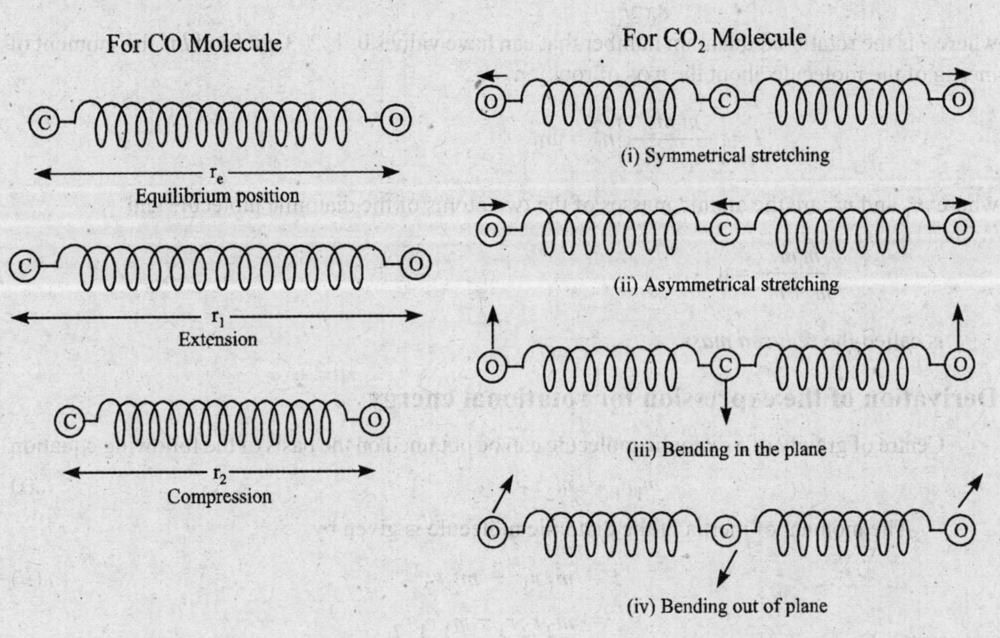

For H_2O Molecule

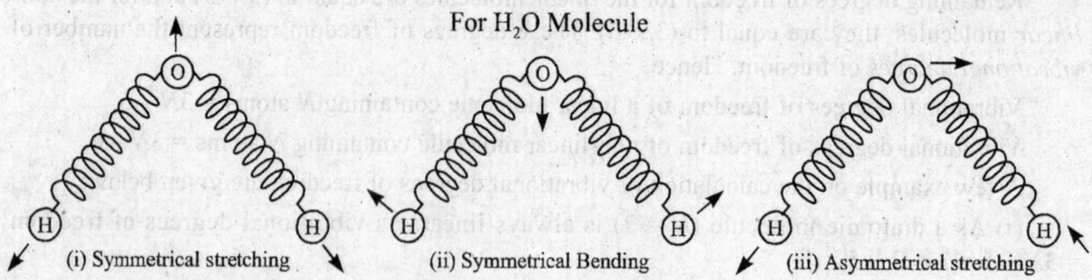

Fig 2.13. Normal modes of vibration of (*i*) diatomic molecule, CO, (*ii*) linear triatomic molecule, CO_2 and (*iii*) non-linear triatomic molecule, H_2O.

2.10. PURE ROTATIONAL SPECTRA OF DIATOMIC MOLECULES

Energy levels of Rigid Rotator

If *diatomic* molecule is considered to be a rigid rotator, *i.e.*, like a rigid dumbbell joined along its line of centres by a bond equal in length to the distance r_0 between the two nuclei (Fig. 2.14) then the allowed rotational energies of the molecule around the axis passing through the centre of gravity and perpendicular to the line joined the nuclei are given by

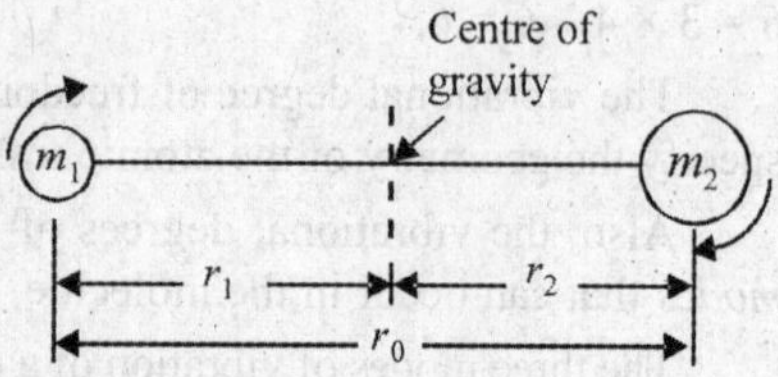

Fig. 2.14. Rigid rotator.

$$E_r = \frac{h^2}{8\pi^2 I} J(J+1)$$

where J is the rotational quantum number that can have values 0, 1, 2, 3 etc., and I is the moment of inertia of the molecule about the axis of rotation, *i.e.*,

$$I = \left(\frac{m_1 m_2}{m_1 + m_2}\right) r_0^2 = \mu r_0^2$$

where m_1 and m_2 are the atomic masses of the two atoms of the diatomic molecule and

$$\frac{m_1 m_2}{m_1 + m_2} = \mu$$

is called the *reduced mass.*

Derivation of the expression for rotational energy

Centre of gravity of a diatomic molecule can be obtained on the basis of the following equation

$$m_1 r_1 = m_2 r_2 \qquad ...(i)$$

The moment of inertia of the diatomic molecule is given by

$$I = m_1 r_1^2 + m_2 r_2^2 \qquad ...(ii)$$

$$= m_2 r_2 r_1 + m_1 r_1 r_2$$

$$= r_1 r_2 (m_1 + m_2) \qquad ...(iii)$$

But $\quad r_1 + r_2 = r_0 \qquad ...(iv)$

$\therefore$ From eqn. (i), $\quad m_1 r_1 = m_2 (r_0 - r_1)$

or
$$r_1 = \frac{m_2 r_0}{m_1 + m_2} \qquad ...(v)$$

Similarly,
$$r_2 = \frac{m_1 r_0}{m_1 + m_2} \qquad ...(vi)$$

Substituting these values in eqn. (*ii*), we get

$$I = \frac{m_1 m_2^2}{(m_1 + m_2)^2} r_0^2 + \frac{m_1^2 m_2}{(m_1 + m_2)^2} r_0^2$$

$$= \frac{m_1 m_2 (m_1 + m_2)}{(m_1 + m_2)^2} r_0^2 = \frac{m_1 m_2}{m_1 + m_2} r_0^2$$

$$= \mu r_0^2 \qquad ...(vii)$$

$$\left(\text{where } \mu = \frac{m_1 m_2}{m_1 + m_2} \text{ is the reduced mass} \right)$$

By definition, the angular momentum of a rotating molecule is given by

$$L = I\omega \qquad ...(viii)$$

where ω is the angular velocity (just as linear momentum is mass × velocity).

Angular momentum is quantized whose values are given by

$$L = \sqrt{J(J+1)} \frac{h}{2\pi} \qquad ...(ix)$$

where J = 0, 1, 2, 3..., called the *rotational quantum numbers.*

The energy of a rotating molecule is given by

$$E = \frac{1}{2} I\omega^2$$

∴ The quantized value of the rotational energy is given by

$$E_r = \frac{1}{2} I\omega^2 = \frac{(I\omega)^2}{2I} = \frac{L^2}{2I} \qquad ...(x)$$

Substituting the value of L From eqn. (*ix*), we get

$$E_r = J(J+1) \frac{h^2}{4\pi^2} \times \frac{1}{2I} \text{ or } E_r = \frac{h^2}{8\pi^2 I} J(J+1) \qquad ...(xi)$$

Putting J = 0, 1, 2, 3 etc. in equation (*xi*), pattern of the rotational energy levels obtained will be as shown in Fig. 2.15.

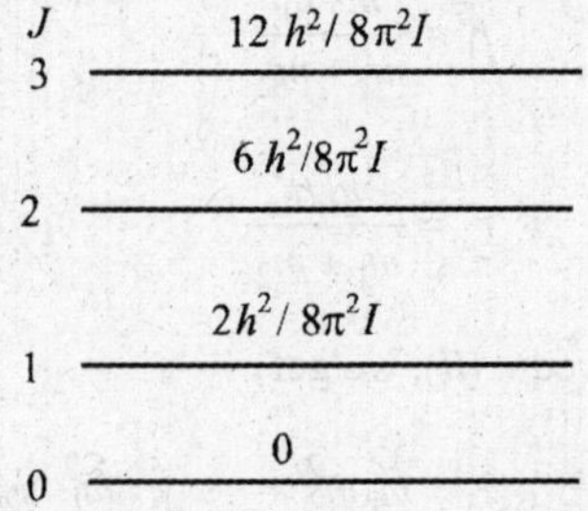

Fig. 2.15. Rotational energy levels of a diatomic molecule treating it as a rigid rotator.

As can be seen above, the spacing between the energy levels increases as J increases, because of the factor $J(J + 1)$ in eq. (*xi*).

2.10.1. Frequency and wave number of lines in the rotational spectrum

Allowed rotational energies are given by the expression

$$E_r = \frac{h^2}{8\pi^2 I} J(J+1) \qquad ...(i)$$

As $E = h\nu$, therefore, in terms of frequency, we can write eq. (*i*) as

$$\nu = \frac{h}{8\pi^2 I} J(J+1) \qquad ...(ii)$$

Also $c = \nu\lambda = \nu / \bar{\nu}$ (because wave number $\bar{\nu} = 1/\lambda$). In term of wave numbers, eqn. (*ii*) can be written as

$$\bar{\nu} = \frac{h}{8\pi^2 Ic} J(J+1) \qquad ...(iii)$$

Putting $\frac{h}{8\pi^2 Ic}$ equal to B, equal (*iii*) reduces to

$$\bar{\nu} = BJ(J+1) \qquad ...(iv)$$

where B is called *rotational constant.*

Putting $J = 0, 1, 2, 3$ etc in eqn. (*iv*), *the wave numbers of the different rotational levels* will be

0, $2B$, $6B$, $12B$, $20B$, $30B$,...... so on.

Suppose a transition takes place from a lower rotational level with rotational quantum number J to a higher rotational level with rotational quantum number J'.

The energy absorbed will be given by

$$\Delta E_r = E_r' - E_r \qquad (\text{or } E_{j'} - E_j)$$

$$= \frac{h^2}{8\pi^2 I} J'(J'+1) - \frac{h^2}{8\pi^2 I} J(J+1)$$

$$= \frac{h^2}{8\pi^2 I} [J'(J'+1) - J(J+1)] \qquad ...(v)$$

According to the rotational selection rules, only those rotational transitions are allowed for which $\Delta J = \pm 1$. Hence in the present case $J' = J + 1$.

Substituting this value in eqn. (*v*), we get

$$\Delta E_r = \frac{h^2}{8\pi^2 I}[(J+1)(J+2) - J(J+1)]$$

$$= \frac{h^2}{8\pi^2 I} 2(J+1) \quad ...(vi)$$

To express in terms of wave numbers, we put

$$\Delta E_r = h\nu = h\frac{c}{\lambda} = hc\bar{\nu}$$

Hence $$hc\bar{\nu} = \frac{h^2}{8\pi^2 I} \times 2(J+1)$$

or $$\bar{\nu} = \frac{h}{8\pi^2 Ic} \times 2(J+1)$$

$$= 2B(J+1) \quad ...(vii)$$

where $B = \dfrac{h}{8\pi^2 Ic}$ is rotational constant (as already explained).

$\bar{\nu}$ in eqn. (*vii*) represents the wave numbers of the spectral lines which will be obtained as a result of the transitions between the rotational levels. Putting $J = 0, 1, 2, 3$ etc. (*i.e.*, for the transitions $J = 0$ to $J' = 1$; $J = 1$ to $J' = 2$ etc.) or in general from J to $J + 1$ (involving absorption of energy), the wave numbers of the lines obtained will be

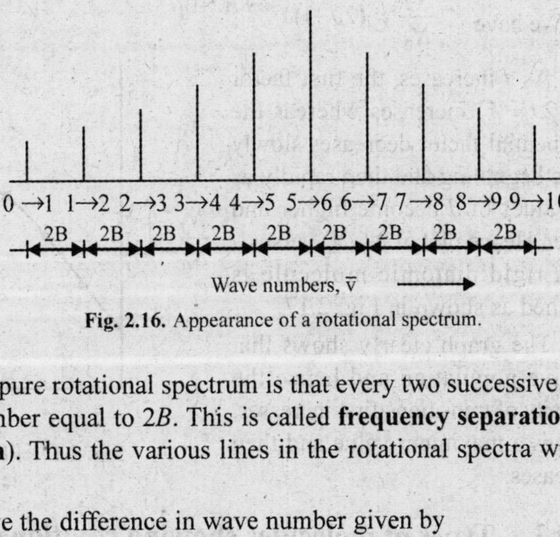

Fig. 2.16. Appearance of a rotational spectrum.

$2B, 4B, 6B, 8B,$..... and so on.

The most important feature of the pure rotational spectrum is that every two successive lines have a constant difference of wave number equal to $2B$. This is called **frequency separation** (or more strictly **wave number separation**). Thus the various lines in the rotational spectra will be equally spaced as shown in Fig. 2.16.

Any two successive lines will have the difference in wave number given by

$$\Delta\bar{\nu} = 2B = \frac{h}{4\pi^2 Ic}$$

We observe in this spectra that the intensities of different transitions are not equal. The intensities increase with increasing J, pass through a maximum and then decrease as J increases further.

2.10.2. Intensities of rotational spectra lines

As explained in Sec. 2.8.1 the relative intensities of the spectral lines will depend upon the relative populations of the energy levels. Greater the population of an energy level, greater is the number of molecules that can be promoted to the next higher level and hence greater is the intensity of absorption. But the population of a rotational energy level with quantum number J relative to that of the ground level for which $J = 0$ is given according to Boltzmann's distribution law by equation.

$$\frac{N_J}{N_0} = e^{-E_J/kT} \qquad ...(i)$$

However, rotational energy levels are degenerate and the degeneracy of any rotational level with quantum number J is given by

$$g_j = 2J + 1 \qquad ...(ii)$$

It we take into consideration degeneracy factor, equation (i) should be replaced by

$$\frac{N_J}{N_0} = g_j\, e^{-E_J/kT} \qquad ...(iii)$$

From (ii) and (iii)

$$\frac{N_J}{N_0} = (2J+1)e^{-E_J/kT} \qquad ...(iv)$$

$$\text{Intensity} \propto \frac{N_J}{N_0} = (2J+1)e^{-E_J/kT}$$

Putting $E_j = h\nu = h\frac{c}{\lambda} = hc\bar{\nu} = hcBJ(J+1)$ (Refer to eq. (iv), sec. 2.10.1)

We have $\dfrac{N_J}{N_0} = (2J+1)e^{-hcBJ(J+1)/kT}$

As J increases, the first factor *viz.* $(2J + 1)$ increases whereas the exponential factor decreases slowly in the beginning and then rapidly as the values of J become higher and higher. Hence plot of N_J/N_0 versus J for a rigid diatomic molecule is obtained as shown in Fig. 2.17.

The graph clearly shows that relative population and hence the intensity of transitions first increases, reaches a maximum value and then decreases.

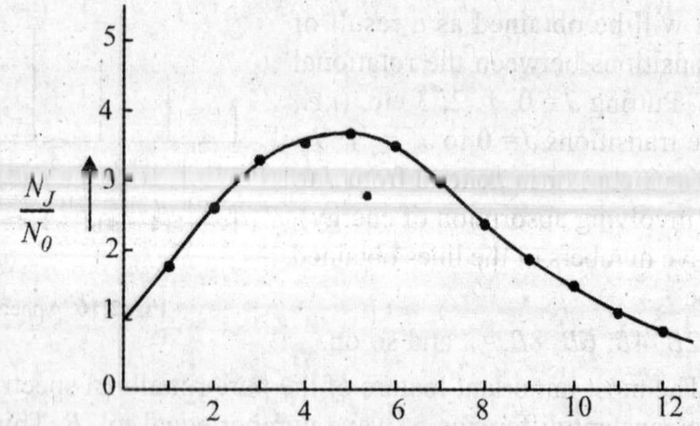

Fig. 2.17. Plot of relative population of rotational energy levels versus rotational quantum number.

2.10.3. Types of molecules showing rotational spectra

The condition for obtaining a pure rotational spectrum is that the molecule must have a permanent dipole moment so that it can interact with the electromagnetic radiation and absorb or emit a photon. Thus *pure rotational spectrum is given only by polar molecules.* Consequently, homonuclear diatomic molecules (like H_2, N_2 etc.) and symmetrical linear molecules like CO_2 (or symmetrical molecules like C_6H_6) do not give rotational spectra. Polar molecules like H_2O, NO, N_2O *etc.*, give pure rotational spectra.

Example 1. Which of the following molecules show rotational spectra? HCl, CO, H_2 and O_2.

Solution. Out of HCl, CO, H_2 and O_2, the first two namely HCl and CO will give rotational spectra as they possess a permanent dipole moment. This is the necessary condition for a molecule to show rotational spectra.

2.10.4. Applications of rotational spectra

By determining the frequency separation, $\Delta\bar{v}$, it is possible to calculate the moment of inertia I of the molecule, using equation $B = h/8\pi^2 Ic$. Then knowing the masses of the atoms present in the diatomic molecule, it is possible to calculate the distance between the atoms, r_0 (*i.e.*, internuclear distance or bond length) using equation

$$I = \left(\frac{m_1 m_2}{m_1 + m_2}\right) r_0^2$$

Working of the spectrometer

The source of radiations is a special type of electronic oscillator (called klystron). It produces radiations of frequency corresponding to wavelength range of 0.1 to a few hundred centimetres. The energies corresponding to these wavelengths are same as required for rotational transitions. The procedure consists in passing the monochromatic energy of the microwaves through the gaseous sample of the substance being studied and measuring the intensity of the transmitted radiation. By varying the frequency of the oscillator and observing the intensity of the transmitted beams, the frequencies at which absorption takes place can be determined from the plot of transmitted intensity versus frequency.

SOVED PROBLEMS ON ROTATIONAL SPECTRA

Example 1. What is the moment of inertia of a diatomic molecule whose internuclear distance is 150 pm and the reduced mass is 1.5×10^{-27} kg.

Solution. Moment of inertia I of a diatomic molecule is given by the relation

$$I = \frac{m_1 m_2}{m_1 + m_2} \cdot r_0^2$$

where $\frac{m_1 m_2}{m_1 + m_2}$ is the reduced mass and r_0 is the internuclear distance.

In S.I. units

$$\text{Reduces mass} = 1.5 \times 10^{-27} \text{ kg}$$

and

$$r_0 = 150 \text{ pm} = 150 \times 10^{-12} \text{ m}$$

$$\begin{aligned} I &= 1.5 \times 10^{-27} \times (150 \times 10^{-12})^2 \text{ kg m}^2 \\ &= 1.5 \times 150 \times 150 \times 10^{-27} \times 10^{-24} \\ &= 33750 \times 10^{-51} \\ &= 3.375 \times 10^{-47} \text{ kg m}^2 \end{aligned}$$

Example 2. The far infrared spectrum of HI consists of series of equally spaced lines with $\Delta\bar{v} = 12.8 \text{ cm}^{-1}$. What is (*a*) the moment of inertia and (*b*) the internuclear distance?

Solution. In C.G.S. system

(*a*) Here we are given

$$\Delta\bar{v} = 12.8 \text{ cm}^{-1}$$

$$\therefore \quad I = \frac{h}{4\pi^2 c.\Delta\bar{v}} = \frac{6.62 \times 10^{-27}}{4 \times \left[\frac{22}{7}\right]^2 \times (3 \times 10^{10})(12.8)}$$

$$= \mathbf{4.36 \times 10^{-40} \text{ g cm}^2}$$

(*b*) $$I = \left(\frac{m_1 m_2}{m_1 + m_2}\right) r_0^2$$

Here m_1 (for H-atom) $= \dfrac{1.008}{6.022\times10^{23}} g = 1.6739\times10^{-24}$ g

m_2 (for I-atom) $= \dfrac{127}{6.022\times10^{23}} g = 2.1089\times10^{-22}$ g

$$\therefore \quad r_0^2 = \left(\frac{m_1 + m_2}{m_1 m_2}\right) \times I$$

$$= \frac{1.6739\times10^{-24} + 2.1089\times10^{-22}\, g}{1.6739\times10^{-24} \times 2.1089\times10^{-22}} \times 4.36\times10^{-40}$$

$$= \frac{(1.6739 + 210.89)\times10^{-24} \times 4.36\times10^{-40}}{1.6739 \times 2.1089 \times 10^{-46}}$$

$$= 2.6254 \times 10^{-16}\ \text{cm}^2$$

$$\therefore \quad r_0 = 1.62 \times 10^{-8}\ \text{cm}^2 = 1.62\text{Å}$$

In S.I. units

(*a*) $$\Delta\bar{\nu} = 12{\cdot}8\ \text{cm}^{-1} = 1280\ \text{m}^{-1}$$

$$\therefore \quad I = \frac{h}{4\pi^2 c\cdot\Delta\bar{\nu}}$$

$$= \frac{6.62\times10^{-34}}{4\times\left(\frac{22}{7}\right)^2 \times(3\times10^8)\times1280}$$

$$= \mathbf{4.36\times10^{-47}\ kg\ m^2}$$

(*b*) $m_1 = 1.6739 \times 10^{-27}$ kg

$m_2 = 2.1089 \times 10^{-25}$ kg

$$r_0^2 = \left(\frac{m_1 + m_2}{m_1 m_2}\right) \times I$$

$$= \frac{1.6739\times10^{-27} + 2.1089\times10^{-25}}{1.6739\times10^{-27} \times 2.1089\times10^{-25}} \times 4.36\times10^{-47} = 2.6254\times10^{-20}$$

$\therefore \quad r = 1.62 \times 10^{-10}$ m $=$ **0.162 nm.**

Example 3. The internuclear distance of carbonmonoxide molecule is 1.13 ×10^{-10} m. Calculate the energy (in joules and in eV) of this molecule in the first excited rotational level. The atomic masses of carbon and oxygen are 1.99 × 10^{-26} kg and 2.66 × 10^{-26} kg respectively.

Solution We are given, $r = 1.13\times 10^{-10}$ m

The reduced mass μ of CO is given by the equation

$$\mu = \frac{m_1 m_2}{m_1 + m_2}$$

$$= \frac{(1.99 \times 10^{-26} \text{kg})\ (2.66 \times 10^{-26} \text{kg})}{(1.99 + 2.66) \times 10^{-26} \text{kg}} = 1.14 \times 10^{-26} \text{kg}$$

Moment of inertia, $I = \mu r^2$

$$= (1.14 \times 10^{-26}\text{kg}) \times (1.13 \times 10^{-10}\text{m})^2$$

$$= 1.46 \times 10^{-46}\ \text{kgm}^2$$

The rotational energy levels of a rigid diatomic molecule are given by

$$E_J = \frac{h^2}{8\pi^2 I} J(J+1)\ \text{joule}$$

For the first excited rotational level, $J = 1$. Here

$$E_1 = \frac{h^2 \times 1(1+1)}{8\pi^2 I} = \frac{h^2}{8\pi^2 I}\ \text{joule}$$

$$= \frac{(6.626 \times 10^{-34}\ \text{Js})^2}{4\pi(1.46 \times 10^{-46}\text{kg m}^2)} = 7.61 \times\times 10^{-23}\ \text{joule}$$

Since, $\quad 1\text{eV} = 1.602 \times 10^{-19}\ \text{J}$

$$E_1 = \frac{7.61 \times 10^{-23}\ \text{J}}{1.6023 \times 10^{-19}\ \text{J (eV)}^{-1}} = 4.76 \times 10^{-4}\ \text{eV}$$

Example 4. The pure rotational spectrum of gaseous HCl contains a series of equally spaced lines separated by 20.80 cm^{-1}. Calculate the internuclear distance of the molecule. The atomic masses of H and Cl are 1.673 × 10^{-27} kg and 58.06 × 10^{-27} kg respectively.

Solution. The spacing between the lines $(2B) = 20.80\ \text{cm}^{-1}$

$$\therefore \quad B = 10.40\ \text{cm}^{-1}$$

$$B = \frac{h}{8\pi^2 Ic}\text{cm}^{-1} \ \text{or}\ I = \frac{h}{8\pi^2 Bc}$$

substituting the values

$$I = \frac{6.626 \times 10^{-34}\ \text{Js}}{(8\pi^2)(10.40\text{cm}^{-1})(3 \times 10^{10}\ \text{cms}^{-1})}$$

$$= 0.2689 \times 10^{-46}\ \text{kg}^2\text{m}^2 \qquad (\text{J} = \text{kgm}^2\ \text{s}^{-2})$$

The reduced mass is given by the equation

$$\mu = \frac{m_1 m_2}{m_1 + m_2}$$

where m_1 and m_2 are the atomic mass of H and Cl respectively.

Thus, $$\mu = \frac{(1.673 \times 10^{-27}\ \text{kg})(58.06 \times 10^{-27}\ \text{kg})}{(1.673 + 58.06) \times 10^{-27}\ \text{kg}}$$

$$= \frac{(1.673 \times 58.06) \times 10^{-27}\ \text{kg}}{59.74}$$

$$= 1.626 \times 10^{-27}\ \text{kg}$$

Since, $$I = \mu r^2$$

$$r = \left(\frac{I}{\mu}\right)^{1/2} = \left(\frac{0.2689\times10^{-46}\,\text{kg m}^2}{1.626\times10^{-27}\,\text{kg}}\right)^{1/2}$$

$$r = 1.29 \times 10^{-10}\,\text{m} = 129\,\text{pm}$$

Example 5. The pure rotational spectrum of the gaseous molecule CN has a series of equally spaced lines separated by 3.7978 cm^{-1}. Calculate the internuclear distance of the molecule. The molar masses of C and N are 12.011 and 14.007 g/mol respectively.

Solution. The spacing between the lines $(2B) = 3.7978\ \text{cm}^{-1}$

$$B = \frac{3.7978}{2} = 1.8989\ \text{cm}^{-1} = 189.89\ \text{m}^{-1}$$

$$B = \frac{h}{8\pi^2 Ic}\,\text{cm}^{-1} \qquad \text{[Refer to eq. (}iv\text{) Sec. 2.10.1]}$$

$$I = \frac{h}{8\pi^2 Bc} = \frac{6.626\times10^{-34}\,\text{Js}}{8\pi^2(189.89\ \text{m}^{-1})\,(3\times10^8\,\text{m/s})}$$

$$= 1.4742\times10^{-46}\ \text{kg m}^2 \qquad (\text{J} = \text{kg m}^2\,\text{s}^{-2})$$

The reduced mass is given by the equation

$$\mu = \frac{m_1 m_2}{(m_1 + m_2)}$$

where N_A is the Avogadro's no. and m_1, m_2 are the atomic masses of C and N, respectively.

$$\mu = \frac{(12.011\text{g mol}^{-1})\,(14.007\ \text{g mol}^{-1})}{(26.018\ \text{g mol}^{-1})\,(6.022\times10^{23}\ \text{mol}^{-1})} = 1.0737\times10^{-23}\,\text{g}$$

$$= 1.073710\times10^{-26}\ \text{kg}$$

Thus, $$I = \mu r^2$$

$$r = \left(\frac{I}{\mu}\right)^{1/2} = \left(\frac{1.4742\times10^{-46}\,\text{kgm}^2}{1.0737\times10^{-26}\,\text{kg}}\right)^{1/2}$$

$$\therefore \quad r = 1.1717\times10^{-10}\,\text{m} = 117\ \text{pm}$$

2.11. NON-RIGID ROTOR (QUALITATIVE DESCRIPTION)

Frequency of rotation of a rigid rotor is given by

$$\nu = \frac{h}{8\pi^2 I}J(J+1) \qquad (i)$$

We can say from the above equation that the frequency of rotation increases with increase in rotational quantum number J. However the increase in the frequency of rotation increases the centrifugal force which tends to move the atoms apart. This implies that the bond length of the diatomic molecule increases with increase in the value of J. Thus the molecule will no longer behave as a rigid rotor. Instead, it will behave like a non-rigid rotor.

Wave numbers of the rotational level of a rigid rotor as explained earlier are given by

$$\bar{\nu} = BJ\,(J+1) \qquad ...(ii)$$

For a molecule behaving like non-rigid rotor, the wave numbers of the rotational levels are given by

$$\bar{\nu} = BJ\,(J+1) - DJ^2\,(J+1)^2 \qquad ...(iii)$$

where D is called *centrifugal distortion constant*. When a transition takes place from rotational level J to $J+1$, the wave number of the line produced, for the molecule behaving like a non-rigid rotor, will be given by

$$\begin{aligned}\Delta\bar{\nu} &= \Delta\bar{\nu}_{J+1} - \bar{\nu}_J \\ &= [B\,(J+1)\,(J+2) - D(J+1)^2\,(J+2)^2] - [BJ\,(J+1) - DJ^2\,(J+1)^2] \\ &= 2B\,(J+1) - 4\,D\,(J+1)^3 \\ &= 2\,(J+1)\,\{B - 2D\,(J+1)^2\} \qquad ...(iv)\end{aligned}$$

As J increases, the factor $\{B - 2\,D\,(J+1)^2\}$ decreases. Hence $\Delta\bar{\nu}$ decreases with increase in the value of J.

2.12. ISOTOPIC EFFECT

If an atom in a molecule is replaced by its isotope, chemically the new molecule is not different from the original molecule in respect of electron distribution, internuclear distance and dipole moment. But, because of the difference in mass of the isotope, the moment of inertia (I) of the molecule will change. This means that if I increases, B will decrease ($B = h/8\pi^2Ic$). As the energies or the wave numbers of the rotational levels are directly related to the value of B, the energies of the rotational levels will decrease *i.e.* the energy levels will come closer. As the spacing between the adjacent energy levels is equal to $2B$, decrease in the value of B means that the energy levels will come closer.

By studying the rotational spectra with isotopic substitution in molecules it is possible to calculate not only the atomic masses of the isotopes but also their relative abundances by comparing the absorption intensities.

PROBLEMS FOR PRACTICE

1. Calculate the reduced mass and moment of inertia of DCl^{35}, given that internuclear bond distance is 0.1275 nm.

(**Ans.** 3.162×10^{-27} kg, 5.141×10^{-47} kgm^2)

2. The far infrared spectrum of HI consists of a series of equally spaced lines with spacing equal to 12.8 cm^{-1}. Calculate the moment of inertia and the internuclear distance.

(**Ans.** 4.37×10^{-47} kgm^2, 0.163 nm)

3. Calculate the energy in joules per quantum and joules per mole of photons of wavelength 30 nm. (**Ans.** 6.62×10^{-19} J, 3.98×10^5 J mol^{-1})

4. Calculate the relative Boltzmann population of the first four rotational levels of the ground state of HCl^{35} at 300 K. Given the rotational constant is 10.4398 cm^{-1}.

(**Ans.** N_J/N_O for J = 0, 1, 2, 3, 4 are 1.00, 2.71, 3.70, 3.84, 3.31)

5. Calculate the frequency in cm^{-1} and wave length in cm of the first rotational transition J = $0 \rightarrow 1$ for DCl^{35}. (**Ans.** 10.89 cm^{-1}, 0.9183 cm)

VIBRATIONAL SPECTRA

2.13. VIBRATIONAL SPECTRA OF DIATOMIC MOLECULES (INFRARED SPECTRA)

2.13.1. Vibrational energy levels of a simple harmonic oscillator

A simple harmonic oscillator is the one in which the restoring force is proportional to the displacement in accordance with Hook's law *i.e.*,

$$F = -kx \quad ...(i)$$

For a diatomic molecule, $x = R - R_e$, where R is distance to which the atoms have been stretched and R_e is the equilibrium distance between the two atoms. k is called the force constant.

For such an oscillator, the potential energy is given by

$$V = \frac{1}{2}kx^2 \quad ...(ii)$$

This is the equation for a parabola. Hence a *parabolic potential energy curve* is obtained as shown in Fig. 2.18.

Using the concepts of quantum mechanics, it can be shown that if the vibratory motion of the nuclei of diatomic molecule (Fig. 2.19) is taken as equivalent to that of **simple harmonic oscillator,** then the vibrational energy is obtained using quantum mechanics in the form of eqn. (*iii*)

$$E_v = \left(v + \frac{1}{2}\right) h\nu_0 \quad ...(iii)$$

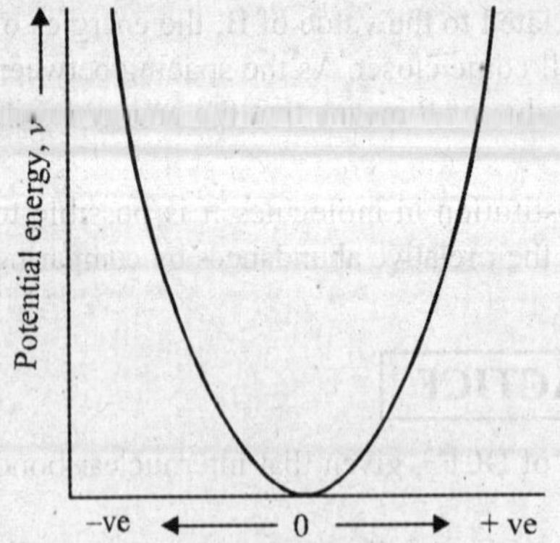

Fig. 2.18. Parabolic potential energy curve of a harmonic oscillator.

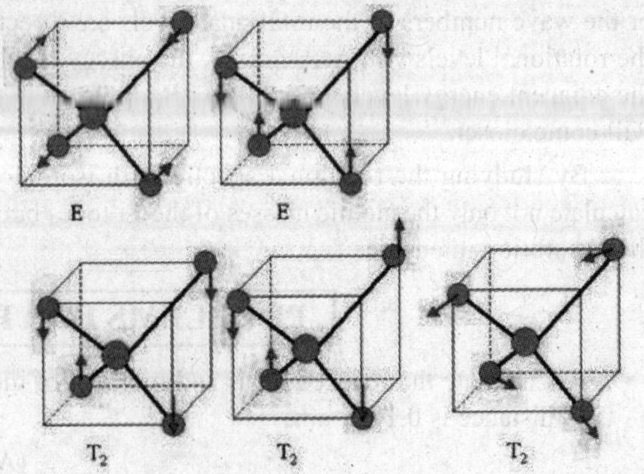

Vibrational modes for CCl_4

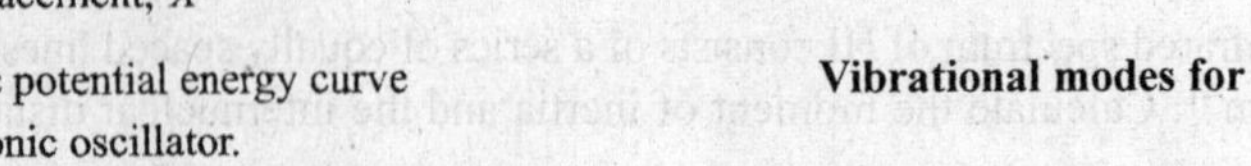

Fig. 2.19. Vibratory motion of a diatomic molecule.

where ν_0 is the frequency of vibration and v is the vibrational quantum number with allowed values of 0, 1, 2, 3, etc.

Putting $\nu_0 = \frac{c}{\lambda} = c\bar{\nu}_e$ or $c\,\omega_e$ where $\bar{\nu}_e$ or ω_e, represents the equilibrium vibrational frequency in terms of wave numbers, (cm^{-1}) eqn. (*iii*) can be written as

$$E_v = \left(v + \frac{1}{2}\right) hc\omega_e \quad ...(iv)$$

Putting $v = 0, 1, 2, 3$ etc. in equation (*iv*), it may be seen that the vibrational energy levels of a harmonic oscillator are equally spaced. These are shown diagrammatically in Fig. 2.20.

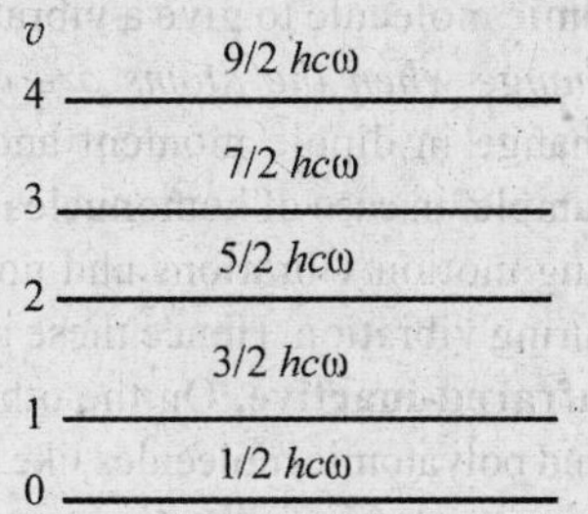

Fig. 2.20. Equally spread vibrational energy levels of a harmonic oscillator.

For the lowest vibrational level, $v = 0$. The energy for the level will, therefore, be

$$E_0 = \frac{1}{2} hc\omega_e \quad ...(v)$$

This is called **zero point energy.** It implies that even *at absolute zero when all translational and rotational motion ceases in a crystal, the residual energy of vibration E_0 still remains, i.e., the vibrational motion still exists.*

2.13.2. Selection rules for vibrational transitions in a simple harmonic oscillator

The selection rule for a molecular vibration is that the *electric dipole moment must change when the atoms are displaced.* Here note that the molecule need not have a permanent dipole. Based on this, the specific selection rule for vibrational transitions is

$$\Delta v = \pm 1$$

i.e., change is vibrational quantum number should be unity. Transitions for which $\Delta v = +1$ correspond to absorption and those with $\Delta v = -1$ correspond to emission.

2.13.3. Vibrational spectrum for simple harmonic oscillator

As the transitions can take place between vibrational levels whose vibrational quantum numbers differ by unity (*i.e.* $\Delta v = \pm 1$), the energy absorbed by a diatomic molecule, when the transition takes place from v to $v+1$ level will be given by

$$\Delta E_v = \left(v + 1 + \frac{1}{2}\right) hc\omega_e - \left(v + \frac{1}{2}\right) hc\omega_e = hc\omega_e$$

or $$\frac{\Delta E}{hc} = \omega_e$$

i. e., $$\Delta \overline{v}_e = \omega_e$$

Thus *only one absorption line* will be obtained in the vibrational spectrum whose wave number is equal to the equilibrium vibrational frequency of the diatomic molecule. Otherwise also, as the vibrational levels are equally spaced (Fig. 2.20), any transition from any v to $v + 1$ will give rise to the same energy change and hence only one absorption line is expected. Further, at room temperature, as most of the molecules are in the lowest (ground state) vibrational level with $v = 0$, therefore, the transitions take place from $v = 0$ to $v = 1$. The frequency thus observed is called the **fundamental frequency.** These spectra are observed in the infrared (IR) region.

2.13.4. Types of molecules showing vibrational spectra

The selection rule for a diatomic molecule to give a vibrational spectrum is that the *dipole moment of the molecule must change when the atoms are displaced due to vibration.* The condition required is only the change in dipole moment and the molecule need not have a permanent dipole moment. For example, in case of homonuclear diatomic molecules like H_2, O_2, N_2 etc., which have only stretching motion/vibrations and no bending motion/vibrations, the dipole moment does not change during vibration. Hence these molecules do not give vibrational spectra *i.e.*, they are said to be **infrared-inactive.** On the other hand, heteronuclear diatomic molecules like HCl, CO, NO etc., and polyatomic molecules like CO_2, H_2O, CH_4, C_2H_4 etc., which show change in dipole moment in some mode of vibration give vibrational spectra and are said to be **infrared-active.**

2.14. VIBRATIONAL ENERGY LEVELS OF AN ANHARMONIC OSCILLATOR (MORSE EQUATION)

When a molecule is undergoing simple harmonic oscillations, the restoring force of the harmonic vibrations is directly proportional to the displacement and the energy of the harmonic oscillator is given by the relation

$$E_v = \left(v + \frac{1}{2}\right) hc\omega_e \qquad ...(i)$$

where v is the vibrational quantum number, which may be zero or an integer and ω_e is the equilibrium frequency of vibration of the oscillator *i.e.*, for small displacements.

Strictly speaking the movement of a real oscillator is not perfectly harmonic. This is because as the displacement increases, the restoring force becomes weaker and for large amplitude of vibration, the atoms of the molecule must fall apart *i.e.*, the molecule must dissociate into atoms. Such a real oscillator is said to be an *anharmonic* oscillator. P.M. Morse in 1929 suggested that the equation (*i*) has to be modified for an anharmonic oscillator by including one additional term as follows:

$$E_v = \left(v + \frac{1}{2}\right) hc\omega_e - \left(v + \frac{1}{2}\right)^2 hcx_e\omega_e \qquad ...(ii)$$

where x_e is called the *anharmonicity constant.*

Putting $v = 0, 1, 2....$ in equations (*i*) and (*ii*), the energy levels of the harmonic oscillator and the anharmonic oscillator may be obtained. We observe that whereas the energy levels of a harmonic oscillator are equally spaced, those of the anharmonic oscillator are not equally spaced but fall more closely together as the quantum number increases. Also, while the potential energy curve for a harmonic oscillator is a parabola 2.21 (*a*), that of the anharmonic oscillator is different as shown in Fig. 2.21 (*b*).

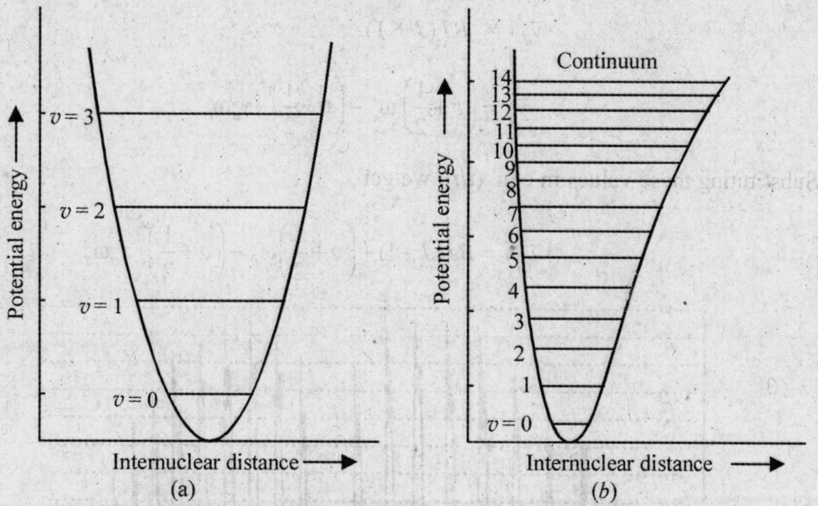

Fig. 2.21. Potential energy curve and energy levels for (*a*) harmonic oscillator (b) anharmonic oscillator.

2.14.1. Selection rules for vibrational transition of an anharmonic oscillator

Selection rules for the anharmonic oscillator are

$$\Delta v = \pm 1, \pm 2, \pm 3 \text{ and so on.}$$

Thus all types of vibrational transitions are possible. However, the transition from $v = 0$ to $v = 1$ is found to be most intense (due to greater population of these energy levels) and is called the **fundamental absorption.** The transition from $v = 0$ to $v = 2$ has a very weak intensity and is called the **first overtone.** The transition from $v = 0$ to $v = 3$ has still weaker intensity and is called the **second overtone** and so on.

2.15. VIBRATIONAL-ROTATIONAL SPECTRA OBTAINED FOR A DIATOMIC MOLECULE TAKING IT AS ANHARMONIC OSCILLATOR

As the energy required for rotation is much less than that required for vibration, the vibrational motion is always accompanied by rotational motion and hence we do not have a pure vibrational spectra; instead we have a vibrational-rotational spectra.

As rotational motion and vibrational motion take place independent of each other, the total energy of the molecule may be taken as the sum of the rotational energy and the vibrational energy. Thus

$$E_{\text{total}} = E_r + E_v \qquad ...(i)$$

We may write it as

$$E_{v,J} = E_J + E_v \qquad ...(ii)$$

Expressing in terms of wave numbers, it can be written as

$$\overline{v}_{v,J} = \overline{v}_J + \overline{v}_v \qquad ...(iii)$$

But the rotational energy in terms of wave number $\overline{v}_J$ and vibrational energy $\overline{v}_v$ are given by the following two equations

$$\overline{\nu}_J = BJ(J+1)$$

$$\overline{\nu}_v = \left(v+\frac{1}{2}\right)\omega_e - \left(v+\frac{1}{2}\right)^2 x_e\omega_e$$

Substituting these values in eqn. (*iii*), we get

$$\overline{\nu}_{v,J} = BJ(J+1) + \left(v+\frac{1}{2}\right)\omega_e - \left(v+\frac{1}{2}\right)^2 x_e\omega_e \quad ...(iv)$$

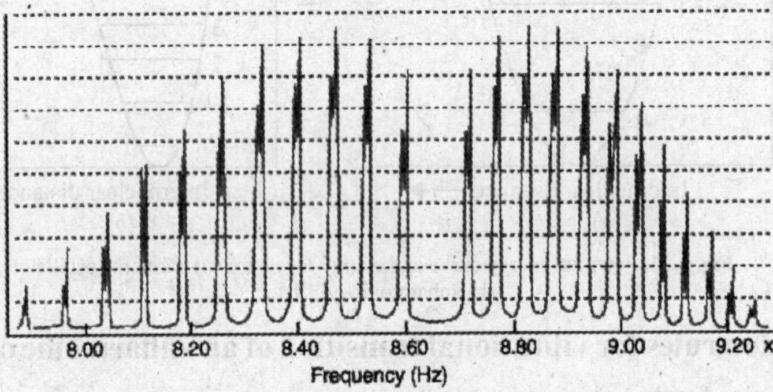

Rotational-vibrational spectra of HCl

Here $\overline{\nu}_{v,J}$ represents the wave number of a rotational level with quantum number J in the vibrational level with quantum number v.

From eqn. (iv), it is clear that in the same vibrational level (so that v = constant), there will be a number of rotational levels corresponding to $J = 0, 1, 2, 3,.....$

When a transition takes place from a level with quantum number v and J to a level with quantum numbers v' and J', the energy change expressed in terms of wave numbers will be given by

$$\Delta\overline{\nu} = \left[BJ'(J'+1) + \left(v'+\frac{1}{2}\right)\omega_e - \left(v'+\frac{1}{2}\right)^2 x_e\omega_e\right]$$
$$- \left[BJ(J+1)\left(v+\frac{1}{2}\right)\omega_e - \left(v+\frac{1}{2}\right)^2 x_e\omega_e\right] \quad ...(v)$$

Restricting to the transition from $v = 0$ to $v = 1$, we get

$$\Delta\overline{\nu} = \left[BJ'(J'+1) + \frac{3}{2}\omega_e - \frac{9}{4}x_e\omega_e\right] - \left[BJ(J+1) + \frac{1}{2}\omega_e - \frac{1}{4}x_e\omega_e\right]$$
$$= B\,[J'(J'+1) - J(J+1)] + \omega_e - 2x_e\,\omega_e$$
$$= B\,[(J'-J)(J+J'+1)] + (1-2x_e)\,\omega_e \quad ...(vi)$$

Putting $(1-2x_e)\,\omega_e = \omega_0$, we can write

$$\Delta\overline{\nu} = \omega_0 + B\,(J'-J)(J+J'+1) \quad ...(vii)$$

ω_0 represents the fundamental frequency. This is called the centre or origin of the fundamental band.

(*i*) For transition, $\Delta J = +1$, *i.e.*,

$$J' - J = +1 \text{ or } J' = J + 1$$

$$\Delta\bar{\nu} = \omega_0 + 2B(J+1) \text{ where } J = 0, 1, 2, 3.... \quad ...(viii)$$

(*ii*) For transition, $\Delta J = -1$, *i.e.*,

$$J' - J = -1 \text{ or } J = J' + 1$$

$$\Delta\bar{\nu} = \omega_0 - 2B(J' + 1) \text{ where } J' = 0, 1, 2, 3... \quad ...(ix)$$

Combining equations (*viii*) and (*ix*), we get

$$\Delta\bar{\nu} = \omega_0 \pm 2Bm \quad ...(x)$$

where $m = J + 1$ in eqn. (*viii*) and $m = -(J' + 1)$ in eqn. (*ix*). As $J = 0, 1, 2, 3,....$, this means that in eqn. (*x*), $m = \pm 1, \pm 2, \pm 3,...$ (+ sign for $\Delta J = +1$ and – sign for $\Delta J = -1$).

The nature of the vibration-rotation spectrum can be predicted from eqn. (*x*) as follows:

(*i*) As $m \# 0$, therefore, line corresponding to the frequency ω_0 will not appear in the spectrum. This wave number is called the **band centre** or **band origin.**

(*ii*) As $m = \pm 1, \pm 2, \pm 3,....$, it follows that the spectrum will consist of equally spaced lines on each side of the band centre, with spacing equal to $2B$ between any two adjacent lines.

Keeping in view the selection rule, $\Delta J = \pm 1$, the absorption transitions that can take place from level $v = 0$ to $v = 1$ and the corresponding lines obtained in the spectrum may be represented as shown in Fig. 2.22.

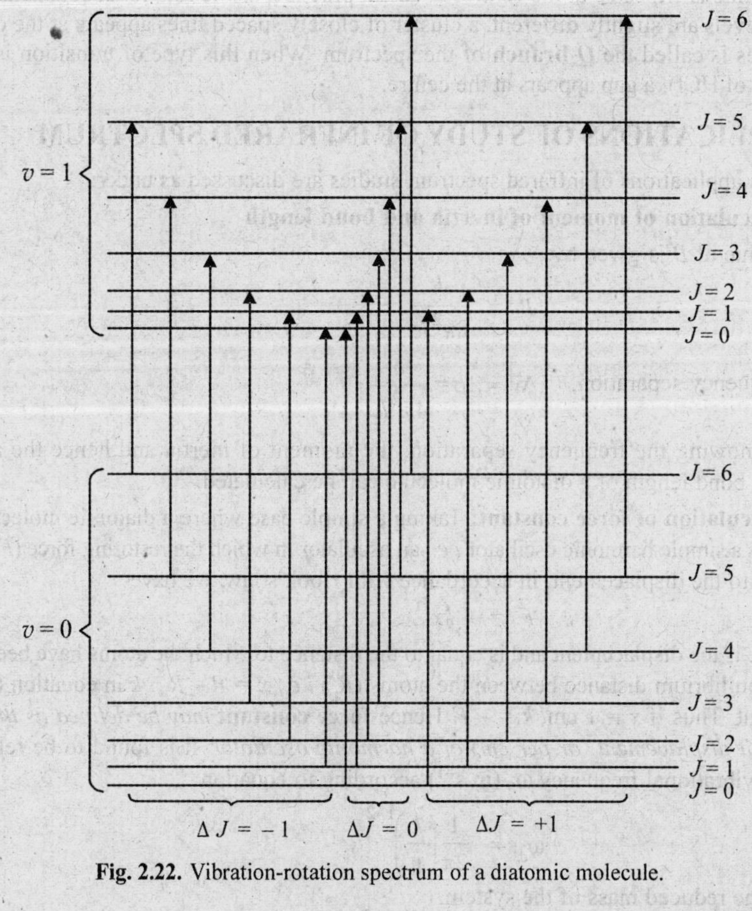

Fig. 2.22. Vibration-rotation spectrum of a diatomic molecule.

P, Q and *R* branches of Rotation-Vibration Spectrum

In the case of rotation-vibration spectrum:

(*i*) For $\Delta J = -1$, lines with frequency lower than the fundamental frequency are obtained. These lines are called ***P*-branch** of the spectrum (Fig. 2.23).

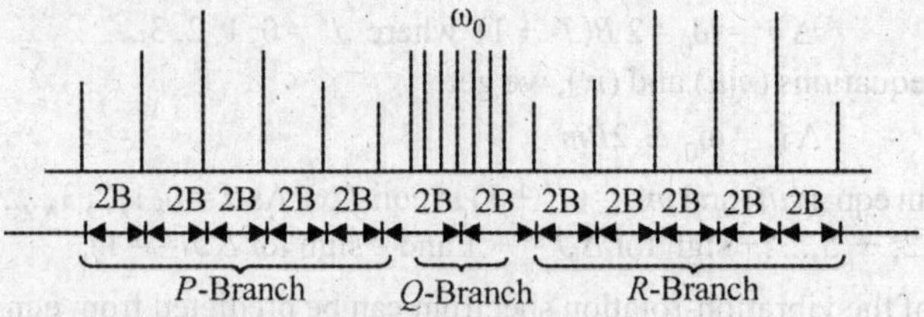

Fig. 2.23. *P*, *Q* and *R* branches of rotation-vibration spectrum.

(*ii*) For $\Delta J = +1$, lines with frequency greater than the fundamental frequency are obtained. These lines are called the ***R*-branch** of the spectrum.

(*iii*) For $\Delta J = 0$ *i.e.*, when a vibrational transition occurs without being accompanied by rotational transition (*e.g.*, in case of NO),

$$\Delta\bar{\nu} = \omega_0$$

i.e., a single line is expected at the centre. In actual practice, as the rotational constants of the two vibrational levels are slightly different, a cluster of closely spaced lines appears at the centre. This group of lines is called the ***Q*-branch** of the spectrum. When this type of transition is forbidden (*e.g.*, in case of HCl), a gap appears at the centre.

2.16. APPLICATIONS OF STUDY OF INFRARED SPECTRUM

Various applications of infrared spectrum studies are discussed as under:

(1) **Calculation of moment of inertia and bond length**

The value of *B* is given by

$$B = \frac{h}{8\pi^2 Ic}$$

$\therefore$ Frequency separation, $\quad \Delta\bar{\nu} = 2B = \dfrac{h}{4\pi^2 Ic}$

Thus knowing the frequency separation, the moment of inertia and hence the antinuclear distance *i.e.*, bond length of a diatomic molecule can be calculated.

(2) **Calculation of force constant:** Taking a simple case where a diatomic molecule may be considered as a simple harmonic oscillator *i.e.*, an oscillator in which the restoring force (*F*) is directly proportional to the displacement, in accordance with Hook's law, we have

$$F = -kx \qquad ...(i)$$

where *x* is the displacement and is equal to the distance to which the atoms have been stretched (*R*) minus equilibrium distance between the atoms (R_e) *i.e.*, $x = R - R_e$. *k* in equation (*i*) is called force constant. Thus if $x = 1$ cm, $k = -F$. Hence **force constant** *may be defined as the restoring force per unit displacement* (*or per cm*) *of a harmonic oscillator.* It is found to be related to the equilibrium vibrational frequency ω_e (in s^{-1}) according to equation

$$\omega_e = \frac{1}{2\pi}\left(\frac{k}{\mu}\right)^{1/2} s^{-1} \qquad ...(ii)$$

where μ is the reduced mass of the system.

To convert the frequency ω_e from s^{-1} to cm^{-1}, we divide by velocity of light, *c*. Thus,

$$\omega_e = \frac{1}{2\pi c}\left(\frac{k}{\mu}\right)^{\frac{1}{2}} \text{ cm}^{-1}. \qquad ...(iii)$$

To calculate *k*, equation (*iii*) can be rewritten as follows :

$$k = 4\pi^2 \omega_e^2 \mu$$

or

$$\boxed{k = 4\pi^2 \omega_e^2 \frac{m_1 m_2}{m_1 + m_2}} \qquad ...(iv)$$

where m_1 and m_2 are the masses of the oscillating atoms.

Thus knowing vibration frequency w_e and the masses of the atoms of the diatomic molecule, the force constant *k* can be calculated. Note that w_e is in s^{-1}.

It is found that force constant, increases almost directly with the multiplicity of the bond is given below:

Bond	*Force constant (dynes/cm)*
C – C	4.6×10^5
C = C	9.5×10^5
C ≡ C	15.8×10^5

Thus multiplicity of the bond can be predicted from the value of the force constant. For example, the force constant for C ≡ O in carbon monoxide is 18.6×10^5 dynes/cm, whereas the force constant for C = O in carbon dioxide is 15.2×10^5 dynes/cm. This result confirms that CO_2 has resonating structures containing both carbon oxygen double and triple bonds *i.e.*, we write

$$O^+ — C \equiv O^- \leftrightarrow O = C = O \leftrightarrow O^- \equiv C — O^+$$

(3) **Calculation of equilibrium dissociation energy:** The equilibrium dissociation energy of a diatomic molecule is given by

$$D_e = \frac{\omega_e}{4x_e} \qquad ...(v)$$

This is the dissociation energy when it is measured from the bottom of the Morse potential energy curve. The true dissociation energy is, however, measured from the ground level ($v = 0$) and is given by

$$D_0 = \frac{\omega_e}{4x_e} - \frac{1}{2}\omega_e\left(1 - \frac{1}{2}x_e\right) \qquad ...(vi)$$

2.17. VIBRATIONAL FREQUENCIES OF FUNCTIONAL GROUPS

It is observed that skeletal vibrations take place when the absorption of energy is upto 1500 cm^{-1}. They are characteristic of the molecule as a whole whether it is linear, branched or has benzenoid structure. This part of the spectrum is called *fingerprint region*. But, certain groups of atoms and functional groups irrespective of the molecule in which they are present, absorb at a particular wave number (> 1500 cm^{-1}) to exhibit group vibrations. They give rise to characteristic absorption bands in the spectrum.

The frequencies (in terms of wave numbers) at which different functional groups show absorption are listed in the Table below for some very common functional groups or groups of atoms.

Table 2.1. Characteristic Infrared Absorption Frequencies

Bond	Compound (Group)	Type of Vibration	Absorption frequency (cm^{-1})
C—H	Alkane	Stretching	2850 – 2960
	— CH_3	Bending	1350 – 1470
	— CH_2	Bending	1465
	Alkenes	Stretching	3000 – 3100
		Bending out of plane	650 – 1000
	Alkynes	Stretching	3300
	Aromatics	Stretching	3050 – 3150
		Bending out of plane	690–900
C — C	Alkane, $(CH_3)_3$ C—	Skeletal	1250
	—$(CH_2)_n$ (n >3)	Skeletal	725
C = C	Alkenes	Stretching	1600 – 1680
	Aromatics		1475 & 1600
C ≡ C	Alkynes	Stretching	2100 – 2250
C = O	Aldehydes	....	1720 – 1740
	Ketones	....	1705 – 1725
	Carboxylic acids	....	1700 – 1725
	Esters	....	1730 – 1750
	Amides	....	1630 – 1680
	Anhydride	....	1760 & 1810
	Acid chloride	....	1800
C — O	Alcohols, ethers	....	1000 – 1300
	esters, carboxylic	....	
	acids, anhydrides	....	
O — H	Alcohols and phenols	....	
	Free	....	3600 – 3650
	H-bonded	....	3200 – 3400
	Carboxylic acids	....	2400 – 3400
N — H	Primary amines, sec-amines, amides	Stretching	3100 – 3500
		Bending	1550 – 1640
C — N	Amines	....	1000 – 1350
C ≡ N	Nitriles	....	2240 – 2260
N = O	Nitro (R — NO_2)	....	1350 – 1550
C – X	Fluoride	....	1000 – 1400
	Chloride	....	540 – 785
	Bromide, Iodide	....	< 667

The group absorption frequency ($\Delta\bar{\nu}$) is related to the force constant (k) of the bond and the reduced mass (μ) of the group according to the approximate relation.

$$\Delta\bar{v} = \frac{1}{2\pi c}\sqrt{\frac{k}{\mu}}$$

i.e. $\Delta\bar{v}$ increases with increase in force constant and decrease with increase in reduced mass.

SOLVED EXAMPLES ON INFRARED SPECTROSCOPY

Example. Calculate the force constant for the bond in HCl from the fact that the fundamental vibration frequency is $8.667 \times 10^{13}s^{-1}$.

Solution. m_1 (for H-atom) $= \dfrac{1.008}{6.022\times10^{23}} g = 1.6739\times10^{-27}$ kg

m_1 (for Cl-atom) $= \dfrac{35.5}{6.022\times10^{23}} g = 5.8951\times10^{-26}$ kg

$\therefore$ Reduced mass $\mu = \dfrac{m_1 \times m_2}{m_1 + m_2} = 1.6277\times10^{-27}$ kg

Also it is given that

$$\omega = 8.667 \times 10^{13}\ s^{-1}$$

$$k = 4\pi^2\omega^2\mu$$

$$= 4 \times \left(\frac{22}{7}\right)^2 \times (8.667 \times 10^{13})^2 \times (1.6277 \times 10^{-22}\ kg)$$

$$= \mathbf{483.1\ Nm^{-1}}.$$

PROBLEMS FOR PRACTICE

1. The force constant of HF molecule in 970 Nm^{-1}. Calculate the fundamental vibrational frequency and zero point energy. (**Ans.** $1.247 \times 10^{-14}s^{-1}$, $4.129 \times 10^{-20}J$)
2. The hydrogen halides have the following fundamental vibrational frequencies: HF(4141.3cm^{-1}), HCl^{35} (2988.9 cm^{-1}), HBr^{81} (2649.7 cm^{-1}), HI (2309.5 cm^{-1}). Calculate the force constant of the hydrogen-halogen bonds. (**Ans.** 967 × 516, 412 × 314 Nm^{-1})
3. Calculate the relative Boltzmann population of the $v = 1$ and $v = 0$ vibrational energy levels of a diatomic molecule at 25°C if they are separated by 1000 cm^{-1}. (**Ans.** $N_{v=1}/N_{v=0} \approx 0008$)
4. The antiharmonicity constant of $Cl^{35}F^{19}$ is $\omega_e x_e = 9.9\ cm^{-1}$ and the fundamental vibrational frequency is 793.2 cm^{-1}. Calculate the energies of the first three vibrational energy levels. (**Ans.** 1167.52 cm^{-1}, 1921.13 cm^{-1}, 2654.23cm^{-1}.)
5. If the observed fundamental frequency of H_2 is 4159 cm^{-1}, what fraction of the molecules are in $v = 1$ state at room temperature? (**Ans.** 6.84×10^{-10})

RAMAN SPECTRA

2.18. INTRODUCTION

C.V. Raman

Prof C.V. Raman of Calcutta University, observed in 1928 that when a substance (gaseous, liquid or solid) is irradiated with monochromatic light of a definite frequency ν, the light scattered at right angles to the incident light contained lines not only of the incident frequency but also of lower frequency and sometimes of higher frequency as well. The lines with *lower frequency* are called **Stokes' lines** whereas lines with *higher frequency* are called **anti-Stokes' lines.** Line with the same frequency as the incident light is called **Rayleigh line.** Raman further observed that the difference between the frequency of the incident light and that of a particular scattered line was constant depending only **upon the nature** of the substance being irradiated and was completely independent of the frequency of the incident light. If ν_i is the frequency of the incident light and ν_s, that of a particular scattered line, the difference $\Delta\nu = \nu_i - \nu_s$ is called **Raman frequency** or **Raman shift.** Thus the Raman frequencies observed for a particular substance are characteristic of that substance. The various observations made by Raman are called **Raman effect** and the spectrum observed is called **Raman spectrum.** Raman spectrum is represented by Fig. 2.24.

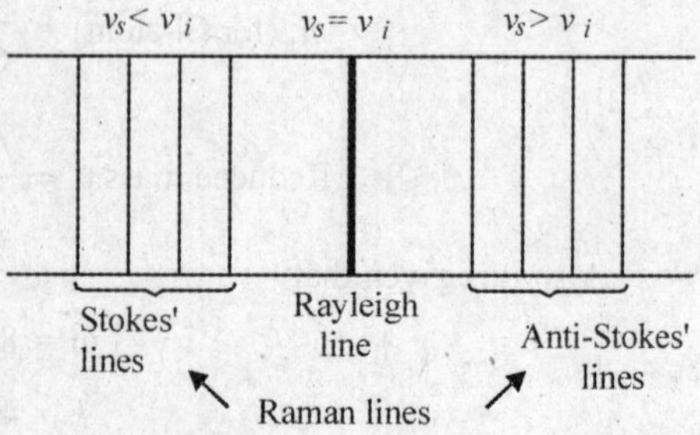

Fig. 2.24. A simplified reprensentation of Raman spectrum.

2.18.1. Explanation of Rayleigh's line, Stokes' lines and anti-Stokes' lines in Raman spectra

When a photon is incident on the molecule, the energy is absorbed by the molecule and it gets excited to some higher energy level. Now if it returns to the original level, it will emit the same energy as absorbed and thus we have Rayleigh scattering. However, in most of the cases, the excited molecule does not return to the original level. It may return to a level higher than the original level, thereby emitting less energy than absorbed. This explains the occurrence of Stokes's lines. Thus a part of the energy of the incident photon remains absorbed by the molecule (so that molecule has higher energy than before). Alternatively, the excited molecule may return to a level lower than the original level. Thus more energy is emitted than absorbed. This explain the occurrence of anti-Stokes' lines. In this case, the molecule has less energy than before. The different cases may be represented diagrammatically by Fig. 2.25.

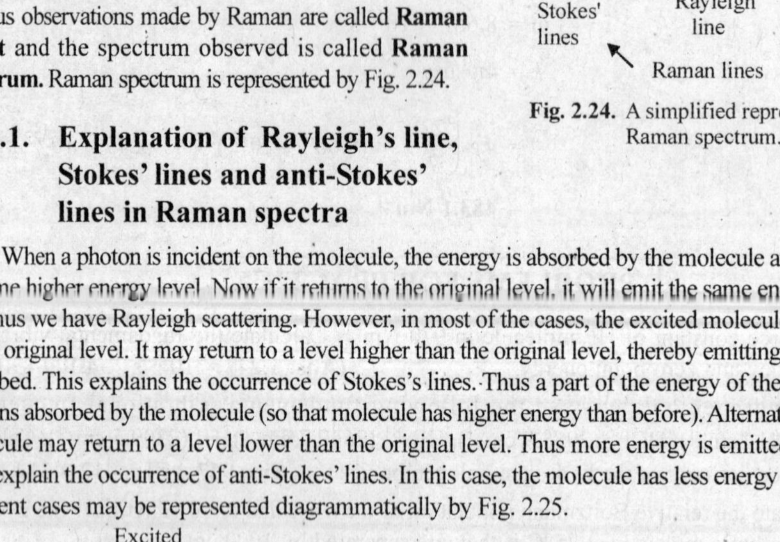

Fig. 2.25 (*a*) Energy absorbed by the molecule. (*b*) Rayleigh scattering. (*c*) Formation of Stokes's lines (*d*) Formation of anti-Stokes' line

2.18.2. Pure rotational Raman spectrum expected for diatomic molecule

The selection rules for pure rotational Raman spectra of diatomic molecules are

$$\Delta J = 0, \pm 2$$

The selection rule $\Delta J = 0$ corresponds to Rayleigh scattering whereas selection rule $\Delta J = \pm 2$ gives rise to Raman lines as explained below :

Energy of a rotational level (in terms of wave number) with quantum number J is given by

$$\bar{\nu} = BJ(J+1) \qquad ...(i)$$

When a transition takes place from a lower rotational level with quantum number J to a higher rotational level with quantum number $(J+1)$ the energy absorbed in terms of wave numbers will be

$$\Delta\bar{\nu} = BJ'(J'+1) - BJ(J+1) \qquad ...(ii)$$

For the selection rule, $\Delta J = +2$ *i.e.*, $J' - J = 2$, we get

$$\Delta\bar{\nu} = B(J+2)(J+3) - BJ(J+1)$$
$$= B(4J+6) \text{ where } J = 0, 1, 2, 3 \qquad ...(iii)$$

Again for the selection rule, $\Delta J = -2$ *i.e.*, $J' - J = -2$, we get

$$\Delta\bar{\nu} = BJ'(J'+1) - B(J'+2)(J'+3)$$
$$= -B(4J'+6) \text{ where } J' = 0, 1, 2, 3... \qquad ...(iv)$$

Combining the results of eqns. (*iii*) and (*iv*), the wave numbers of the lines obtained in the Raman will be given by

$$\bar{\nu} = \bar{\nu}_i \pm \Delta\bar{\nu}$$

or

$$\boxed{\bar{\nu} = \bar{\nu}_i \pm B(4J+6)} \qquad ...(v)$$

where $J = 0, 1, 2, 3,....$

Here $\bar{\nu}_i$ represents the wave number of the Rayleigh line, plus sign gives lines with higher wave numbers called anti-Stokes' lines and minus sign gives lines with lower wave numbers, called Stokes' lines.

From eqn. (*v*), it may be seen that for $J = 0$, $\bar{\nu} = \bar{\nu}_i + 6B$ *i.e.*, the first Stokes' and anti-Stokes' line will be at a separation of $6B$ from the Rayleigh line. Further putting $J = 1, 2, 3, ...$, the separation between any two adjacent Stokes' lines or anti-Stokes' lines will be $4B$. Thus, the pure rotational Raman spectrum expected for a diatomic molecule may be represented by Fig. 2.26.

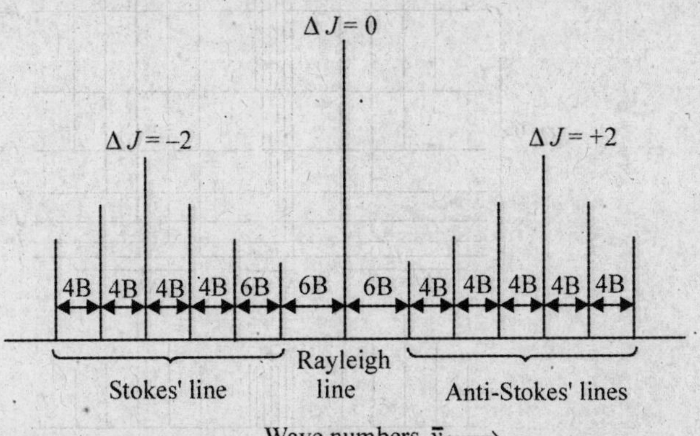

Fig. 2.26. Pure rotational Raman spectrum expected for a diatomic molecule.

2.19. ROTATIONAL–VIBRATIONAL RAMAN SPETRA OF DIATOMIC MOLECULES

Diatomic gaseous molecules give rotational-vibrational Raman spectra which are governed by the following selection rules:

$$\Delta v = \pm 1$$

$$\Delta J = 0, \pm 2$$

However at room temperature, as most of the molecules are in the lowest vibrational level ($v = 0$), therefore, the significant vibrational transition is from $v = 0$ to $v = 1$ (called fundamental vibration transition). Restricting to this vibrational transition only, the results obtained are as follows:

For $\Delta J = 0$, $\Delta \bar{v} = \omega_e (1 - 2x_e)$ (called ***Q*-branch**).

For $\Delta J = +2$, $\Delta \bar{v} = \omega_e(1 - 2x_e) + B(4J + 6)$ (called ***S*-branch**).

For $\Delta J = -2$, $\Delta \bar{v} = \omega_e(1 - 2x_e) - B(4J' + 6)$ (called ***O*-branch**).

Representing the value for *Q*, *S* and *O* branch by $\Delta\bar{v}_Q, \Delta\bar{v}_S$ and $\Delta\bar{v}_O$ respectively, and the wave number of the exciting radiation by $\bar{v}_i$ the wave numbers of the Stokes lines will be as follows:

$$\bar{v}_Q = \bar{v}_i - \Delta\bar{v}_Q \; ; \quad \bar{v}_S = \bar{v}_i - \Delta\bar{v}_S \; ;$$

$$\bar{v}_O = \bar{v}_i - \Delta\bar{v}_O$$

We have considered the case for $\Delta v = +1$ which represents Stokes lines. The anti-Stokes lines are those for which $\Delta v = -1$. They are usually weak because very few molecules are in the excited vibrational state initially.

The rotational transitions accompanying the vibrational transition from $v = 0$ to $v = 1$ and the rotation-vibration Raman spectrum obtained may be represented as shown in Fig. 2.27.

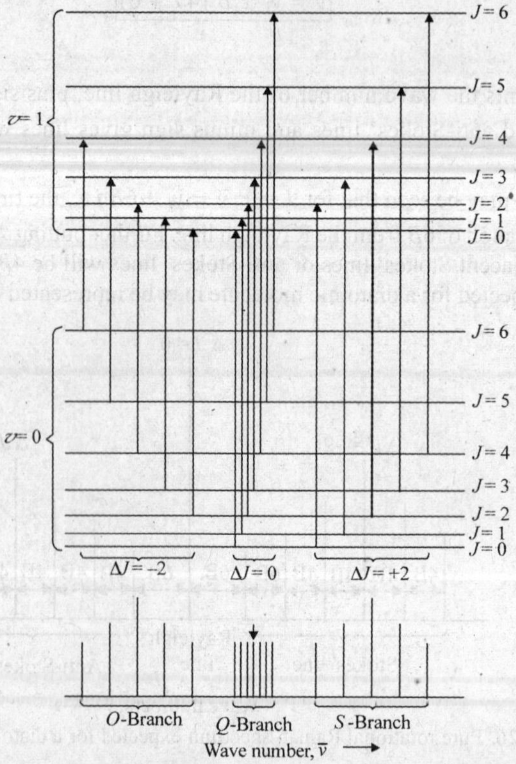

Fig. 2.27. Rotation-vibration Raman spectrum of a diatomic molecule.

2.19.1. Type of molecules that give rotation-vibration Raman spectra

The selection rule for a molecule to give rotation-vibration Raman spectrum is that the *polarization of the molecule must change as the molecule vibrates.* In case of both types of diatomic molecules *i.e.*, homonuclear as well as heteronuclear, as the molecule vibrates, the control of the nuclei over the electrons varies and hence there is a change in polarizability. That is why both types of diatomic molecules give rotation-vibration Raman spectra *i.e.*, they are *vibrational Raman-active.* This is one of the main advantages of Raman spectra over infrared spectra because homonuclear diatomic molecules do not give pure vibrational or rotational spectra because they do not possess a permanent dipole moment.

2.19.2. Advantages of Raman spectroscopy over infrared spectroscopy

Raman spectroscopy has a number of advantages over infrared spectroscopy which are discussed hereunder:

(*i*) Raman spectra can be obtained not only for gases but even for liquids and solids whereas infrared spectra for liquids and solids are quite diffused.

(*ii*) Since Raman frequencies are independent of the frequency of the incident radiation, hence by suitably adjusting the frequency of the incident radiation, Raman spectra can be obtained in the visible spectrum range where they can be easily observed rather than the more difficult infrared range.

(*iii*) Raman spectra can be obtained even for molecules such as O_2, N_2, Cl_2 etc., which have no permanent dipole moment. Such a study is not possible by infra-red spectroscopy.

2.19.3. Experimental set-up of Raman spectrophotometer

The experimental arrangement for observing Raman spectra is shown in Fig. 2.28. The fraction of the light scattered is so low that normally very weak Raman lines are observed. Hence to detect these lines, it is necessary to use a powerful source of incident light. Earlier mercury vapour lamps were used which gave an intense monochromatic radiation of wavelength 435.8 nm. However, with the development of lasers, which are very powerful sources of monochromatic radiation, Raman spectroscopy has undergone radical changes.

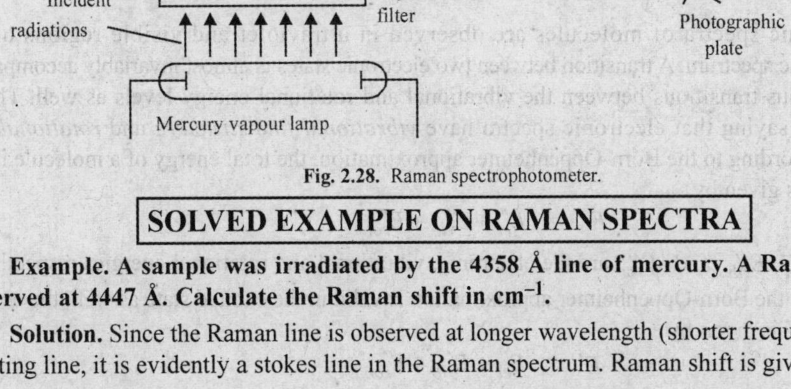

Fig. 2.28. Raman spectrophotometer.

SOLVED EXAMPLE ON RAMAN SPECTRA

Example. A sample was irradiated by the 4358 Å line of mercury. A Raman line was observed at 4447 Å. Calculate the Raman shift in cm^{-1}.

Solution. Since the Raman line is observed at longer wavelength (shorter frequency) than the exciting line, it is evidently a stokes line in the Raman spectrum. Raman shift is given by the eqn.

$$\Delta\bar{V}_{Raman}(cm^{-1}) = \frac{10^8}{\lambda_{exc.}(Å)} - \frac{10^8}{\lambda_{Raman.}(Å)}$$

$$= \frac{10^8}{4.358\times10^3} - \frac{10^8}{4.447\times10^3} = (2.295 - 2.249)\times10^4$$

$$= 0.046 \times 10^4$$

$$\Delta\bar{V}_{Raman}(cm^{-1}) = 460\,cm^{-1}$$

PROBLEMS FOR PRACTICES

1. The triatomic molecule AB_2 shows three strong infrared absorption bands. What is the structure of the molecule? (**Ans.** ABB bent or BAB bent)
2. Calculate the Raman line in nm if HCl^{35} is irradiated with 435.8 nm mercury line. Given that the fundamental vibrational frequency of HCl^{35} is $8.667 \times 10^{13} s^{-1}$. (**Ans.** 498.57 nm)
3. The infrared and Raman spectra of a triatomic molecule of the type MX_2 show two infrared frequencies and one Raman frequency. Determine the shape of the molecule, whether linear or bent.
4. The fundamental vibrational frequencies of SO_2 are 1151.38, 517.69 and 1361.76 cm^{-1}. Account for the absorption bands at 1875.55, 2295.88 and 2499.55 cm^{-1}.

(**Ans.** $\bar{v}_2 + \bar{v}_3, 2\bar{v}_1, \bar{v}_1 + \bar{v}_3$)

ELECTRONIC SPECTRA

2.20. INTRODUCTION

Spectra which arises due to electronic transition in a molecule by absorption of radiation falling in the visible and UV range is called electronic spectra. The range of visible region is 12500–25000 cm^{-1} while that of UV region is 25000–70000 cm^{-1}. It is observed that electronic transitions in a molecule are invariably accompanied by vibrational and rotational transitions, thus, making electronic spectra quite complex. Electronic spectra help in the determination of ionization energies of molecules.

It the light radiation incident on a molecule possesses very high energy, it causes removal of electron from the molecule *i.e.* it causes ionization. A beam of photons of known energy is allowed to fall on a sample and the kinetic energy of the ejected electron is measured. The difference between the photon energy and the excess kinetic energy gives the binding energy of the molecule.

2.20.1. BORN-OPPENHEIMER APPROXIMATION IN RELATION TO ELECTRONIC SPECTRA OF MOLECULES

Electronic spectra of molecules are observed in ultraviolet and visible regions of the electromagnetic spectrum. A transition between two electronic states is almost invariably accompanied by simultaneous transitions between the vibrational and rotational energy levels as well. This is expressed by saying that electronic spectra have *vibrational fine structure* and *rotational fine structure*. According to the Born-Oppenheimer approximation, the total energy of a molecule in the ground state is given by

$$E'' = E''_{el} + E''_{vib} + E''_{rot}$$

Here, E''_{el}, E''_{vib} and E''_{rot} are the electronic, vibrational and rotational energies respectively. Assuming that the Born-Oppenheimer approximation is valid in the excited state as well, the excited state energy E' is given by

$$E' = E'_{el} + E'_{vib} + E'_{rot}$$

The energy change for an electronic transition is given by

$$\Delta E = E' - E'' = (E'_{el} - E''_{el}) + (E'_{vib} - E''_{vib}) + (E'_{rot} - E''_{rot})$$
$$= \Delta E_{el} + \Delta E_{vib} + \Delta E_{rot}$$

It may be understood that

$$\Delta E_{el} >> \Delta E_{vib} >> \Delta E_{rot}$$

The frequency for the electronic transition is given by Bohr-Planck relation.

$$\therefore \quad \bar{v} = \frac{\Delta E}{hc} = \frac{\Delta E_{el} + \Delta E_{vib} + \Delta E_{rot}}{hc} \text{ cm}^{-1}$$

2.20.2. Potential energy curve and Franck-Condon principle

A very useful principle for investigating the vibrational structure of electronic spectra is provided by **Franck-Condon principle.** It states that *an electronic transition takes place so rapidly that a vibrating molecule does not change its internuclear distance appreciably during the transition.* This principle is true since the electrons move so much faster than the nuclei that during the electronic transition the nuclei do not change their position. *An electronic transition may be represented by a vertical line on a plot of potential energy versus the internuclear distance.*

Consider Fig. 2.29 where we have shown two potential energy curves for the molecule in the ground electronic state (E_0) and in the first excited electronic state (E_1). Since the bonding in the excited state is weaker than in the ground state, the minimum in the potential energy curve for the excited state occurs at a slightly greater internuclear distance than the corresponding minimum in the ground electronic state. Also quantum mechanically it is known that the molecule is in the *centre* of the ground vibrational level of the ground electronic state. When a photon falls on the molecule, the most probable electronic transition, according to the Franck-Condon principle, takes place from $v'' = 0$ to $v' = 2$ (written schematically as $0 \rightarrow 2$). Transitions to other vibrational levels of the excited electronic state occur with smaller probabilities so that their relative intensities are smaller than the intensity of the $0 \rightarrow 2$ transition, as shown in Fig. 2.30.

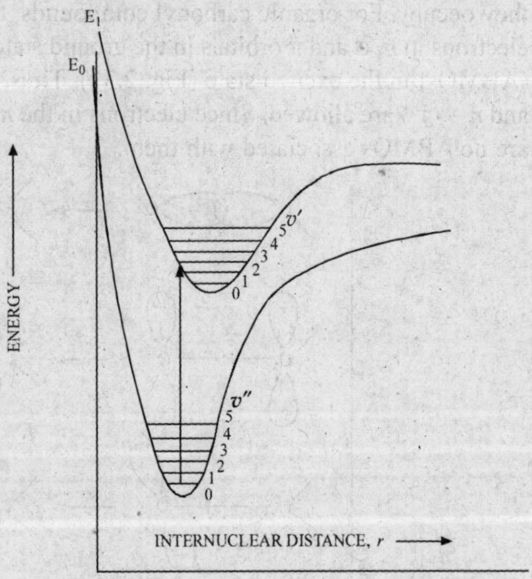

Fig. 2.29. Electronic transition in a diatomic molecule

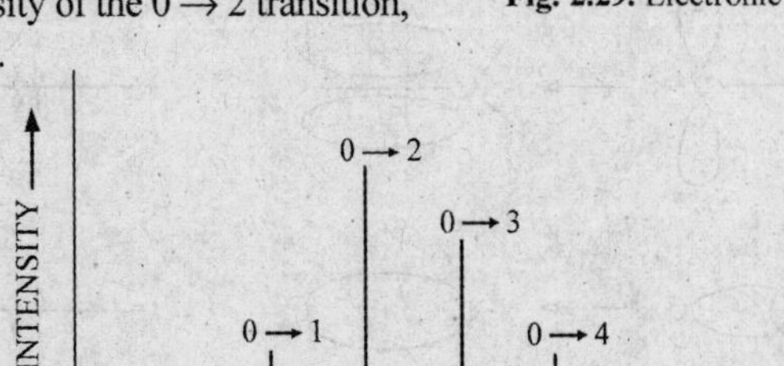

Fig. 2.30. Electronic spectrum of a diatomic molecule

2.20.3. Electronic transition in homonuclear molecules

In molecules such as H_2 and N_2, highest occupied molecular orbital (HOMO) in the ground state is a bonding molecular orbital (BMO) whereas the lowest unoccupied molecular orbital (LUMO) is an antibonding molecular orbital (ABMO). The HOMO and LUMO orbitals are collectively referred to as *Frontier Molecular Orbitals* (FMOs). The electronic transition, HOMO $\rightarrow$ LUMO takes place, *when the electron absorbs a photon.*. This lies in the UV range. Oxygen molecule O_2 is an exception. It has two unpaired electrons in the ground state which is thus a *triplet state.* The electronic transition occurs from the triplet ground state to the triplet excited state (rather than to the singlet excited state). The electronic spectrum of O_2 molecule is, of course, complex.

2.21. ELECTRONIC TRANSITION IN σ, π AND *n* MOLECULAR ORBITALS

Organic compounds, particularly those containing groups like C = C, C = O, – N = N– and extensively conjugated systems, form a special class of polyatomic molecules whose electronic spectra can be easily interpreted even though the investigation of their detailed spectral features may require knowledge of quantum mechanics and group theory. On the basis of the molecular orbital theory (MOT), the electrons can be classified as σ, π or *n* (non-bonding) depending upon the MOs they occupy. For organic carbonyl compounds, the electronic transitions involve promotion of the electrons in *n*, σ and π orbitals in the ground state to the σ* and π* anti-bonding molecular orbitals. (ABMOs) in the excited state (Fig. 2.31). Thus, only the transitions of the type $\sigma \rightarrow \sigma^*$, $\pi \rightarrow \pi^*$, and $n \rightarrow \pi^*$ are allowed. Since electrons in the *n* orbitals are not involved in bond formation, there are no ABMOs associated with them.

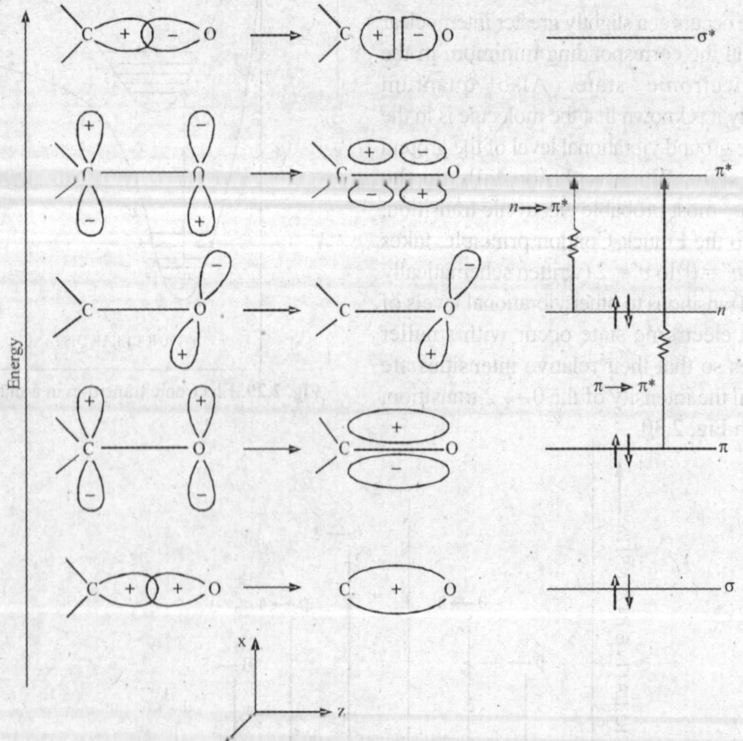

Fig. 2.31. Localized molecular orbitals and electronic transitions for the carbonyl group

The $\sigma \rightarrow \sigma^*$ transitions occurring in saturated hydrocarbons and other types of compounds in which all valence shell electrons are involved in single bonds, are found in the far ultraviolet region since they involve very high energy. The $\pi \rightarrow \pi^*$ and the $n \rightarrow \pi^*$ transitions, on the other hand, are found either in the UV or visible regions. The unsaturated molecules containing C = C and C = O groups (such as aldehydes and ketones) show $n \rightarrow \pi^*$ and $\pi \rightarrow \pi^*$ transitions. For aldehydes and ketones the more intense band near 180 nm is due to $\pi \rightarrow \pi^*$ transition and the weaker band around 285 nm is due to $n \rightarrow \pi^*$ transition. Olefinic hydrocarbons show $\pi \rightarrow \pi^*$ transition in the wave length 160 – 170 nm. Acetylene shows an absorption near 180 nm.

Molecules such as methylamine and methyl iodide which do not contain a π-orbital, show $n \rightarrow \sigma^*$ transitions.

In highly conjugated systems, the π-electrons are delocalized over the entire skeletal framework. It is found that *the absorption bands shift to longer wave lengths as the extent of conjugation increases.* Thus, in the compound $C_6H_5 - (CH = CH)_n - C_6H_5$, the $\pi \rightarrow \pi^*$ transition lies in the UV region when n = 1 or 2. As n increases, the electronic transition shifts to the visible region. This phenomenon is shown schematically in Fig. 2.32.

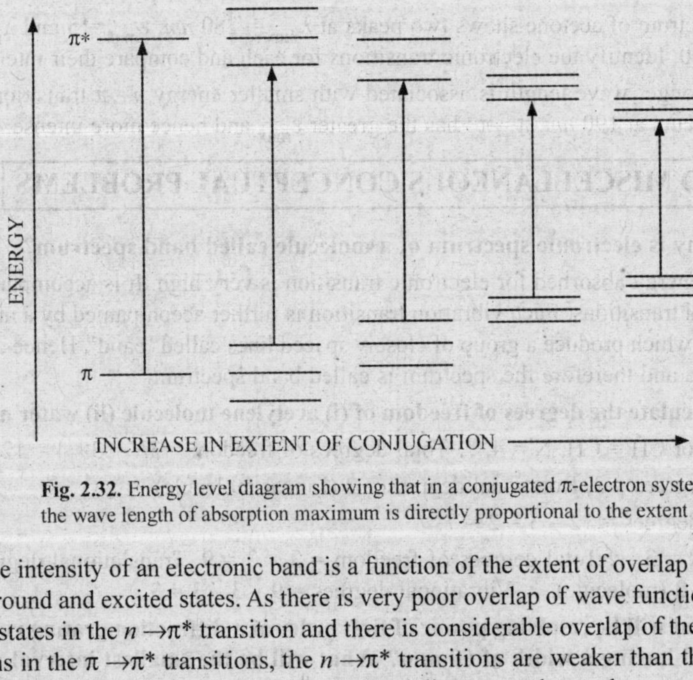

Fig. 2.32. Energy level diagram showing that in a conjugated π-electron system, the wave length of absorption maximum is directly proportional to the extent of conjugation

The intensity of an electronic band is a function of the extent of overlap of the wave functions in the ground and excited states. As there is very poor overlap of wave functions of the ground and excited states in the $n \rightarrow \pi^*$ transition and there is considerable overlap of the corresponding wave functions in the $\pi \rightarrow \pi^*$ transitions, the $n \rightarrow \pi^*$ transitions are weaker than the $\pi \rightarrow \pi^*$ transitions. Again in strongly acidic media, the $n \rightarrow \pi^*$ band disappears due to the protonation of lone pair of electrons. In fact, the protonation increases the excitation energy to such an extent that the $n \rightarrow \pi^*$ transition shifts into the UV region and may not be observed.

The protonation of a functional group introduces major changes in the spectra. The spectrum in such cases is strongly dependent upon pH. Solvent effects are useful in identifying the nature of these transition. The $n \rightarrow \pi^*$ transitions are altered by the solvent effects in cases where the lone pair electrons in oxygen or nitrogen-containing systems interact with polar solvents. The shifts in absorption bands and their intensity changes are presented as follows :

1. *Bathochromic shift* (or, the *red shift*): a shift of λ_{max} to longer wave lengths.
2. *Hypochromic shift* (or, the *blue shift*): a shift of λ_{max} to shorter wave lengths.

3. *Hyperchromic shift*: an increase in the intensity of an absorption band, usually with reference to its molar extinction coefficient ε_{max}.

4. *Hypochromic shift*: a decrease in the intensity of an absorption band with reference to ε_{max}.

PROBLEMS FOR PRACTICES

1. List all the electronic transitions possible for (a) CH_4 (b) $H_2C = O$

Ans. [(a) $\sigma \rightarrow \sigma^*$ (b) $\sigma \rightarrow \sigma^*$, $\sigma \rightarrow \pi$, $\pi \rightarrow \sigma^*$, $n \rightarrow \sigma^*$, $\pi \rightarrow \pi^*$, and $n \rightarrow \pi^*$]

2. Identify the two geometrical isomers of stilbene $C_6H_5CH = CHC_6H_5$ from their λ_{max} value 294 *nm* and 278 *nm*.

Ans. *cis*-isomer has a higher energy (due to steric hindrance of phenyl groups) and therefore shorter λ_{max}. Thus the value 278 *nm* corresponds to *cis*-stilbene.

3. Predict the kind of electronic transitions and intensities in (a) Cl_2 and (b) $C = O$ groups

Ans. (*a*) $\sigma \rightarrow \sigma^*$ (allowed transition, strong), $n \rightarrow \sigma^*$ (forbidden transition, weak)

(*b*) $\pi \rightarrow \pi^*$ and $\sigma \rightarrow \sigma^*$ (both allowed transitions, strong) $n \rightarrow \pi^*$ and $n \rightarrow \sigma^*$ (both forbidden transition, weak)

4. The UV spectrum of acetone shows two peaks at λ_{max} = 280 *nm*, ε_{max}=15 and λ_{max} =190 *nm*, ε_{max}=100. Identify the electronic transitions for each and compare their intensities.

Ans. The longer wave length is associated with smaller energy $n \rightarrow \pi$ transition, $\pi \rightarrow \pi^*$ transition occurs at 190 *nm*. $\pi \rightarrow \pi^*$ has the greater ε_{max} and hence more intense.

SOLVED MISCELLANEOUS CONCEPTUAL PROBLEMS

Problem 1. Why is electronic spectrum of a molecule called band spectrum?

Solution. The energy absorbed for electronic transition is very high. It is accompanied by a number of vibrational transitions. Each vibration transition is further accompanied by a number of rotational transitions which produce a group of closely spaced lines called 'band'. Hence a number of bands are obtained and therefore the spectrum is called band spectrum.

Problem 2. Calculate the degrees of freedom of (i) acetylene molecule (ii) water molecule.

Solution. (*i*) For CH ≡ CH, N = 4, ∴ Total degrees of freedom = $3N = 3 \times 4 = 12$.

Translational degrees = 3, Rotational degrees = 2 (Linear molecule)

∴ Vibrational degrees = 12 – 3 – 2 = 7.

(*ii*) For H_2O, N = 3, ∴ Total degrees of freedom = 3 × 3 = 9. Translational degrees = 3, Rotational degrees = 3 (molecule), ∴ Vibrational degrees = 9 – 3 – 3 = 3.

Problem 3. What will be resolving power of first order of a diffraction grating ruled with 500 grooves/mm and having a width of 60 mm? What will be the smallest interval resolved by this grating for sodium light with wavelength 5890 Å ?

Solution. Resolving power of grating (R) = $m \times N$

For resolving power of 1st order, $m = 1$

N = Total no. of grooves = 500 × 60 = 30,000

Smallest wavelength interval resolved = 5890/30,000 = 0.2 Å.

Problem 4. How the study of pure rotational spectrum of a diatomic molecule helps in the calculation of bond length?

Solution. The spacing between any two adjacent lines has difference in wave number given by the relation

$$\Delta\bar{\nu} = 2B = 2\times\frac{h}{8\pi^2 Ic} = \frac{h}{4\pi^2 Ic}$$

$I = \left(\frac{m_1 m_2}{m_1 + m_2}\right) r_0^2$. Knowing the masses m_1 and m_2 of the atoms of the diatomic molecules, r_0 *i.e.* bond length can be calculated.

Problem 5. What is the effect on rotational energy levels of a molecule if an atom is replaced by its heavier isotope (isotopic substitution)?

Solution. $I = \left(\frac{m_1 m_2}{m_1 + m_2}\right) r_0^2$ and spacing between the adjacent lines $= 2B = \frac{h}{8\pi^2 Ic}$. Here I appears in the denominator. Thus if I increases, the spacing between the adjacent energy levels will decrease i.e. energy levels will come closer.

Problem 6. Why only one absorption line is expected in the vibrational spectrum of a diatomic molecule, treating it is a simple harmonic oscillator?

Solution. For the transition taking place from v to $v + 1$ level, we have the equation

$$\Delta E_v = \left(v+1+\frac{1}{2}\right)hc\omega_e - \left(v+\frac{1}{2}\right)hc\omega_e = hc\omega_e$$

or $$\frac{\Delta E}{hc} = \omega_e \text{ i.e } \Delta\bar{\nu} = \omega_e$$

Thus only one absorption line will be obtained.

Alternatively as the vibrational levels are equally spaced, any transition from any v to v +1 will give rise to same energy change.

Problem 7. What types of molecules exhibit rotational spectra? Out of CO_2, C_6H_6, H_2O, NO, N_2O, which will exhibit pure rotational spectra?

Solution. Pure-rotational spectrum is obtained for those molecules which have permanent dipole moment so that the molecule can interact with electromagnetic radiation and absorb or emit a photon. Thus it is given only by *polar* molecules. As CO_2 and C_6H_6 are symmetrical molecules and hence non-polar, they do not give rotational spectra. H_2O, NO, N_2O are polar and hence give rotational spectra.

Problem 8. Write expression for the energy of vibrational energy levels. Comment on the spacing between the levels. What is zero point energy?

Solution. $E_v = \left(v+\frac{1}{2}\right) hc\omega_e$ (v = vibrational quantum no.)

Putting v = 0, 1, 2 etc., we find that vibrational energy levels are equally spaced. For the lowest vibrational level, v = 0. The energy for this level will be $E_0 + \frac{1}{2}hc\omega_e$. This is called zero point energy. It means that even at absolute zero when all translational and rotational motion ceases in a crystal, the residual energy of vibration E_0 or vibrational motion still exists.

Problem 9. In the vibration-rotation spectrum of a diatomic molecule, why a line does not appear at the band centre?

Solution. The wave numbers of the lines obtained in vibration-rotation spectrum are given by $\Delta\bar{\nu} = \omega_0 + 2Bm$ where ω_0 represents the wave number of the line at the band centre and $m = \pm$ 1, $\pm$ 2 etc. As m # 0, therefore line corresponding to ω_0 is not obtained.

Problem 10. Arrange the following groups in order of their absorption frequencies:
***(i)* CF, CCl, CBr, CH *(ii)* C — C, C = C, C ≡ C. Give justification.**

Solution. Group absorption frequency is related to the force constant (k) of the bond and the reduced mass (μ) of the group by the following equation

$$\Delta\nu = \frac{1}{2\pi c}\sqrt{\frac{k}{\mu}}$$

The equation shows that $\Delta\bar{\nu}$ increases with increase in force constant and decreases with increase in reduced mass

(*i*) Reduced mass varies in the order : CH < CF < CCl < CBr

∴ Absorption frequencies vary in the order : CH > CF > CCl > CBr

(*ii*) Force constants vary as : C ≡ C > C = C > C – C

∴ Absorption frequencies vary in the order : C ≡ C > C = C > C – C.

Problem 11. In the electronic band spectrum, why there are no simple selection rules for transitions among vibrational levels?

Solution. According to Franck-Condon principle, electronic transition takes place much faster than the nuclei can oscillate to change their internuclear distance. Representing the transition by a vertical line, transition can take place from lowest vibrational level to any vibrational level of higher electronic state. That explains why there are no simple selection rules.

Problem 12. In the vibration-rotation spectrum of HCl, why each individual line of the spectrum is found to consist of doublets?

Solution. The splitting takes place due to presence of two isotopes of Cl viz. ^{35}Cl and ^{37}Cl in HCl. These isotopes with different masses result into different moments of inertia (I) and fundamental frequency of vibration (ω_e) for HCl^{35} and HCl^{37}.

Problem 13. Are the lines in the Rotational Raman spectra equally spaced? Explain.

Solution. The lines are not equally spaced. This is because wave numbers of the lines are given by

$$\bar{\nu} = \bar{\nu}_i \pm B(4J + 6)$$

where $\bar{\nu}_i$ represents the wave number of Rayleigh line (*i.e.* equal to that of the incident radiation). Putting $J = 0$, $\bar{\nu} = \bar{\nu}_i \pm 6B$ *i.e.* first Strokes' and anti-Strokes' line will be separated by $6B$ from the Rayleigh line Further putting J = 1, 2, 3,...., the separation between any two adjacent Strokes' lines or anti-strokes lines comes out to $4B$.

Problem 14. Why does the rotational constant B of a diatomic molecule depend upon the electronic state of the molecule?

Solution. This is because B depends upon the moment of inertia, which in turn depends upon the internuclear distance, r. Internuclear distance, r is different for different electronic states of a molecule.

Problem 15. Why is it that the vibrational frequency ν is smaller in the excited state than in the ground state of a molecule?

Solution. This is because the vibrational frequency is proportional to force constant, k. Force constant, k is smaller in the excited state.

Problem 16. The selection rules for spectral transitions in atomic spectra are: (*i*) Δn = 1, 2, 3, 4.... and (*ii*) $\Delta l = \pm 1$. Using these selection rules, determine which of the following transitions are allowed and which are forbidden:

(*a*) 1s → 2p (*b*) 2s → 3s (*c*) 2p → 3s (*d*) 4p→ 5f

Solution. For allowed transition, both the selection rules should be obeyed transitions (*a*) and (*c*) obey the selection rules. Hence these are allowed transitions.

Transitions (*b*) and (*d*) do not obey the selection rules. Hence these are forbidden transitions.

Problem 17. Using Heisenberg uncertainty principle, estimate the life time of an energy state which gives rise to a line width of (*a*) 0.1 cm (*b*) 60 MHz

Solution. $$\tau = \frac{1}{4\pi\Delta\nu} = \frac{1}{4\pi c\Delta\bar{\nu}} \quad (\text{As } \nu = c\bar{\nu})$$

(*a*) $\Delta\bar{\nu} = 0.1 \text{ cm}^{-1}$ $$\therefore \tau = \frac{1}{4\pi \times (3\times10^{10} \text{ cm s}^{-1}) \times (0.1\text{cm}^{-1})} = 2.5\times10^{-11}\text{s}$$

(*b*) $\Delta\bar{\nu} = 60\text{MHz}$ $$\therefore \tau = \frac{1}{4\pi \times (60\times10^{6} s^{-1})} = 1.35\times10^{-9}\text{s}$$

Problem 18. Which of the following molecules will show rotational Raman spectra : HCl, H_2, CO, CH_4, H_2O, NH_3, F_6? Explain.

Solution. All the molecules except CH_4 and SF_6 give rotational Raman spectra.

Rotation Raman spectra are given by a molecule in which the rotation motion takes place with a change in the polarizability.

EXERCISES
(Based on Question-Papers of Different Universities)

Multiple Choice Questions (choose the correct answers)

1. A molecule can be excited to only the next higher rotational level by

(*a*) absorption of energy (*b*) release of energy

(*c*) the electric current (*d*) applying magnetic field

2. The IR spectra of a compound helps in

(*a*) proving the identity of compounds

(*b*) showing the presence of certain functional groups in the molecule

(*c*) neither of the above

(*d*) both of the above

3. The Raman and IR spectra can tell us whether

(*a*) a molecule is linear or non-linear (*b*) a molecule is symmetrical or asymmetrical

(*c*) neither of the above (*d*) both of the above

4. The wave numbers are expressed in

(*a*) sec^{-1} (*b*) cm sec^{-1}

(*c*) cm^{-1} (*d*) $\text{cm}^2 \text{ sec}^{-1}$

5. In the Raman spectrum, the middle line is called

(*a*) Raman line (*b*) Rayleigh line

(*c*) functional group line (*d*) none of these

6. The rotational spectra involve

(*a*) a very high energy changes (*b*) small energy changes

(*c*) no energy change (*d*) none of these

7. The spectra caused in the infrared region by the transition in vibrational levels in different modes of vibrations are called

(*a*) rotational spectra (*b*) electronic spectra
(*c*) vibrational spectra (*d*) none of these

8. The electronic spectra consists of

(*a*) a large number of absorption bands (*b*) a large number of closely packed lines
(*c*) a large number of peaks (*d*) none of these

9. The change in frequency by scattering (Raman Effect) occurs due to _____ of energy between the incident photon and the scattering molecule.

(*a*) release (*b*) absorption
(*c*) exchange (*d*) none of these

10. The electromagnetic radiating of higher wavelengths has ______ energy

(*a*) higher (*b*) lower
(*c*) intermediate (*d*) zero

ANSWERS

1. (*a*)	**2.** (*d*)	**3.** (*d*)	**4.** (*c*)	**5.** (*b*)	**6.** (*b*)
7. (*c*)	**8.** (*b*)	**9.** (*c*)	**10.** (*b*)		

Short Answer Questions

1. What is molecular spectroscopy? How does it differ from atomic spectroscopy? Draw the molecular energy levels of the first two electronic levels. What are vibrational and rotational quantum numbers?

2. What are basic components of a spectrometer? Show them diagrammatically.

3. What is Rayleigh's criterion for resolution of two wavelengths λ and $\lambda + d\lambda$?

4. Explain the selection rules followed in atomic spectroscopy. Give suitable examples.

5. Explain the relative intensities of the lines obtained in a pure rotational spectrum.

6. What is the effect of isotopic substitution on rotational spectra?

7. What are electromagnetic radiations? List their important characteristics.

8. Why do we get a band spectrum in case of molecules?

9. What do you mean by electromagnetic spectrum? List the different types of electromagnetic radiations alongwith with frequency ranges and sources.

10. What do you understand by resolving power of a spectrometer? What is the formula for the resolving power of a prism?

11. What are the different types of energies possessed by a molecule? What is Born-Oppenheimer approximation?

12. What do you understand by Doppler broadening (Doppler effect) and lifetime broadening?

13. How is the rotational spectrum observed experimentally?

14. What type of potential energy curve is obtained for a simple harmonic oscillator and why?

15. Briefly explain the term – Absorption and Emission spectroscopy. Why absorption spectroscopy is preferred ? Give its experimental set-up.

16. What do you understand by the terms: Resolution, Linear dispersion and Angular dispersion? Explain how the resolution depends upon the slit width?

17. What do you mean by selection rules in spectroscopy? What are gross and specific selection rules?
18. What are the factors that affect the width of a spectral line?
19. Write the expression for the rotational energy of a diatomic molecule, taking it as a rigid rotator. Draw the rotational energy level diagram for such a molecule.
20. Explain why molecules behave as non-rigid rotors. Write expression for the wave numbers of rotational levels of a non-rigid rotor and hence derive expression for the wave numbers of the lines produced.
21. What are the factors which affect the intensity of a spectral line? Explain briefly.
22. What do you mean by 'zero point energy'?
23. What types of molecules exhibit vibrational spectra? Out of H_2, O_2, N_2, HCl, CO, NO, CO_2, H_2O and CH_4, which will give pure vibrational spectra?
24. What is isotopic effect on vibration-rotation spectrum? Explain with a suitable example.
25. On the basis of polarizability, explain which type of molecules will be rotationally Raman active and which will be inactive.
26. What do you understand by symmetric, asymmetric, bending, parallel and perpendicular modes of vibration? Explain with suitable examples.
27. What type of vibrational spectrum is expected for a diatomic molecule taking it as a simple harmonic oscillator?
28. Write expression for the vibrational energy of a diatomic molecule, taking it as a simple harmonic oscillator. Represent the vibrational energy level of such a molecule diagrammatically.
29. What are *P*, *Q* and *R* branches of the vibration-rotation spectrum?
30. How are infrared spectra helpful in the identification of organic compounds?
31. Explain with suitable mathematical equations the type of pure rotational Raman spectrum expected for a diatomic molecule. Mark clearly the separation between the different lines.
32. Why a diatomic molecule should be considered as an anharmonic oscillator? Write Morse equation for the energy of the vibrational levels of an anharmonic oscillator. Compare the potential energy curve of an anharmonic oscillator with that of a harmonic oscillator.
33. What are the selection rules for the vibrational transitions in a diatomic molecule, taking it as a simple harmonic oscillator?
34. What structural information is obtained from the study of infrared spectra?
35. What type of vibration-rotation spectrum is obtained if the resolving power of the spectrometer is not so high?
36. What is the most important use of studying pure rotational Raman spectrum?

Long Answer Questions

1. What do you understand by signal-to-noise (S/N) ratio? How can it be enhanced? Briefly explain Fourier Transformation for enhancement of S/N ratio. Why is it considered better than the conventional spectroscopy?
2. Name the different types of molecular spectra. Explain each of them briefly. In which region of the electromagnetic spectrum are they obtained?
3. What do you understand by degrees of freedom of motion of a molecule? Briefly explain the different types of degrees of freedom possessed by linear and non-linear molecules.
4. Explain with suitable derivations what type of rotation-vibration spectrum is obtained for a diatomic molecule, taking it as an anharmonic oscillator.

5. What is Raman spectrum? Name the different types of lines present in it and explain the reasons for observing these lines? Give the experimental set-up for observing Raman spectra.

6. Taking the example of carbonyl compounds, represent molecular orbitals and explain the electronic transitions taking place between them.

7. What are the selection rules for rotation-vibration Raman spectra of diatomic molecules? Applying these rules explain what type of rotation-vibration Raman spectrum is obtained for a diatomic molecule?

8. Write short notes on the following:

(*a*) Electromagnetic spectrum (*b*) Molecular energy levels

(*c*) Absorption and emission spectroscopy.

9. What do you understand by resolving power of a spectrometer? Explain taking the example of a prism as the dispersing element. Also explain how the resolution depends upon the slit width.

10. Briefly explain terms – stimulated absorption, stimulated emission and spontaneous emission. Prove that the net absorption is proportional to the population difference of the two states involved in the transition.

11. Derive expression for the frequency (or wave number) of the rotational lines in a pure rotational spectrum. What type of pure rotational spectrum will be observed? How do the intensities of these lines vary?

12. What do you understand by normal modes of vibration of a polyatomic molecule? Show diagrammatically the different normal modes of vibration of CO_2 and H_2O molecules. Give their names. Which of them are inactive in the infrared spectra?

13. Explain Raman effect on the basis of polarizability of molecules.

14. Explain the formation of electronic band spectrum on the basis of potential energy curves. How do these curves help in the calculation of the dissociation energy of the molecules?

15. Considering a diatomic molecule as a rigid rotator, explain the type of rotational (microwave) spectra obtained after deriving the expressions required.

16. Briefly explain why the electronic band spectrum obtained is very complex. Inspite of its complexity, why is it preferred? How does it help in the calculation of dissociation energy of a molecule?

17. What do you understand by 'width' and 'intensity' of a spectral line? Briefly explain the factors on which each of these depend.

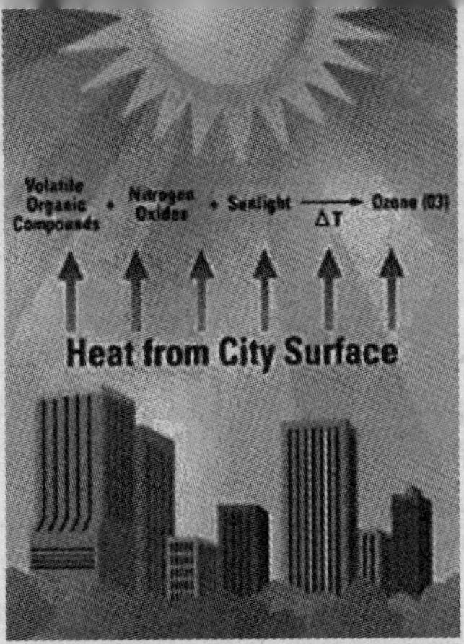

3

Photochemistry

3.1. INTRODUCTION

Photochemistry is defined as that branch of chemistry which deals with the process involving emission or absorption of radiation. Two types of reactions are studied under photochemistry.

Photophysical processes: These processes take place in the presence of light but do not involve any chemical reaction. These processes take place due to absorption of light by the substances followed by the emission of the absorbed light. If there is instantaneous emission of the absorbed light, the process is called *fluorescence.*

If a high amount of energy is absorbed, the electrons may leave the atoms completely. Such a process is called *photoelectric effect.* Photophysical processes include fluorescence, phosphorescence and photoelectric effect.

Photochemical reactions: These are the reactions that take place by absorption of radiations of suitable wavelength. In such processes, the absorbed light energy is first stored in the molecule and then it is further used to bring about the reaction. Examples of photochemical reactions are :

(*i*) *Reaction between hydrogen and chlorine*

$$H_2(g) + Cl_2(g) \xrightarrow{\text{Light}} 2HCl(g)$$

(*ii*) *Photosynthesis of carbohydrates in plants taking place in presence of chlorophyll, the green colouring matter, present in the leaves.*

$$\underset{\text{From air}}{\underline{6CO_2 + 6H_2O}} \xrightarrow[\text{(chlorophyll)}]{\text{Light}} \underset{\text{Glucose}}{C_6H_{12}O_2} + 6O_2$$

An experiment to explain a photochemical reaction.

A mixture of hydrogen and chlorine remains unchanged with lapse of time. But when exposed to light, the reaction occurs with a loud explosion.

$$H_2 + Cl_2 \xrightarrow{\text{Dark}} \text{No reaction}$$

$$H_2 + Cl_2 \xrightarrow{\text{Light}} 2HCl$$

A bottle is filled with equimolar amounts of hydrogen and chlorine (Fig. 3.1). It is tightly stoppered with a handball. When the lamp is turned on, a beam of light falls on the mixture through the bottom of the bottle. The reaction occurs chlorophyll absorbs energy from the sum and passes on to the mixture of $CO_2 + H_2O$

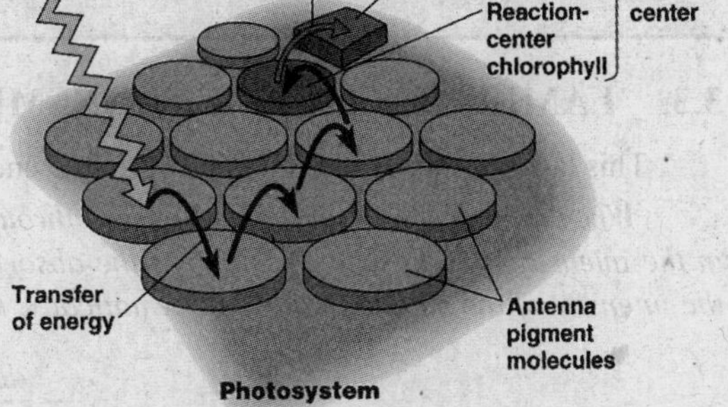

Photosynthesis

to form carbohydrate with an explosion. The ball is expelled with high velocity so that it strikes the opposite wall of the room.

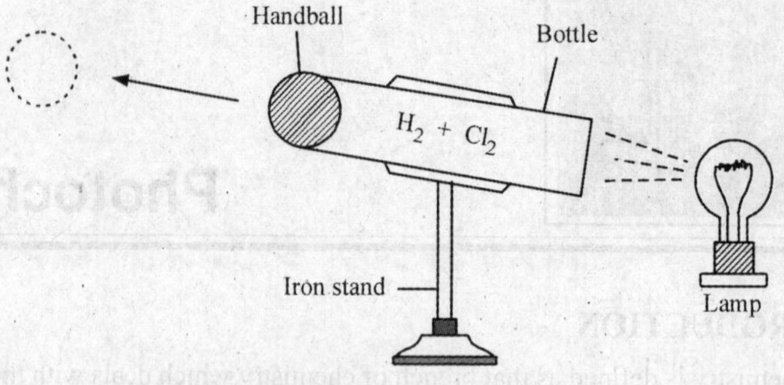

Fig. 3.1. The 'HCl-cannon' experiment.

3.2. THE POINTS OF DIFFERENCE BETWEEN THERMOCHEMICAL REACTIONS AND PHOTOCHEMICAL REACTIONS

The main points of difference between the two types of reactions are given in the tabular form as under:

Thermochemical reactions	*Photochemical reactions*
(*i*) These reactions involve absorption of heat	(*i*) These reactions involve absorption or evolution light.
(*ii*) They can take place even in the dark	(*ii*) The presence of light is the primary requisite for the reaction to take place
(*iii*) The free energy change (ΔG) of a thermochemical reaction is always negative.	(*iii*) The free energy change (ΔG) of a photochemical reactions may not be negative. A few examples of photochemical reactions for which ΔG is positive and still they are spontaneous are synthesis of carbohydrates in plants and decomposition of HC1 into H_2 and $C1_2$.
(*iv*) Temperature has significant effect on the rate of a thermochemical reaction.	(*iv*) Temperature has very little effect on the rate of photochemical reaction. Instead, the intensity of light has a marked effect on the rate of a photochemical reaction.

3.3. LAMBERT'S LAW OF TRANSMISSION OF LIGHT

This law put forward by Lambert states as under :

When a monochromatic light is passed through a pure homogeneous medium, the decrease in the intensity of light with thickness of the absorbing medium at any point X is proportional to the intensity of the incident light. Mathematically, the law can be put as

$$-\frac{dI}{dx} \propto I$$

or $$-\frac{dI}{dx} = kI \qquad \text{... }(i)$$

where dI is the small decrease in intensity of the light on passing through a small thickness dx, I is the intensity of the incident light just before entering the thickness dx and k is the constant of proportionality called the *absorption coefficient.* It depends upon the nature of the absorbing medium.

The intensity I at any point X at a distance x from the start of the medium, can be found in terms of the original intensity I_0 as follows:

Equation (i) can be rewritten as

$$\frac{dI}{I} = -\,kdx \qquad \text{... }(ii)$$

When $x = 0, I = I_0$.

Integrating equation (ii) between the limits $x = 0$ to x and $I = I_0$ to I, we get

$$\int_{I_0}^{I} \frac{dI}{I} = \int_{x=0}^{x=x} -k\,dx$$

or $$\ln \frac{I}{I_0} = -\,kx \qquad \text{... }(iii)$$

or $$\frac{I}{I_0} = e^{-kx}$$

or $$\mathbf{I = I_0 e^{-kx}} \qquad \text{... }(iv)$$

This equation expresses how the original intensity I_0 is reduced to intensity I after passing through a thickness x of the medium.

Equation (iii) can be written as

$$2.303 \log \frac{I}{I_0} = -\,kx$$

or $$\log \frac{I}{I_0} = -\frac{k}{2.303}x$$

or $$\log \frac{I}{I_0} = -\,k'x \qquad \text{... }(v)$$

or $$\frac{I}{I_0} = I_0^{-k'x}$$

or $$\mathbf{I = I_0 10^{-k'x}} \qquad \text{... }(vi)$$

where $k' = \frac{k}{2.303}$ *is* called **extinction coefficient** of the substance *i.e.* the absorbing medium.

This quantity is now called **absorption coefficient** or **adsorptivity** of the substance.

$-\lambda \log I/I_0$ is called **absorbance** of the medium, denoted by A.

The physical significance of the absorption coefficient or absorptivity follows more clearly from equation (*v*), which can be rewritten as

$$k' = -\frac{1}{x}\log\frac{I}{I_0} = \frac{1}{x}\log\frac{I_0}{I}.$$

If $\log\frac{I_0}{I} = 1, k' = \frac{1}{x}$

But $\log\frac{I_0}{I} = 1$, means $\frac{I_0}{I} = 10$ *i.e.* $I = \frac{1}{10} I_0$.

Hence extinction coefficient or absorptivity may be defined as follows :

The extinction coefficient or absorptivity is the reciprocal of the thickness (expressed in cm) at which the intensity of the light falls to one-tenth of its original value.

3.4. BEER'S LAW

Beer's law states as :

When a monochromatic light is passed through a solution, the decrease in the intensity of light with the thickness of the solution is directly proportional not only to the intensity of the incident light but also to the concentration c of the solution. Mathematically, we have

$$-\frac{dI}{dx} \propto I \times c \quad \text{... (i)}$$

or $$-\frac{dI}{dx} = \varepsilon Ic$$ (ε is pronounced as epsilon)

where ε is a constant of proportionality and is called **molar absorption coefficient.** Its value depends upon the nature of the absorbing solute and the wavelength of the light used.

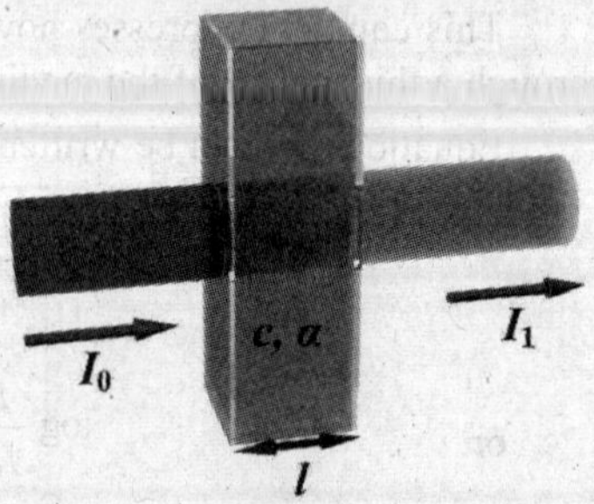

Beer-Lambert's law

Equation (*i*) can be rewritten as

$$\frac{dI}{I} = -\varepsilon c\, dx.$$

Integrating this equation between the limits $x = 0$ to x and $I = I_0$ to I, we get

$$\int_{I_0}^{I}\frac{dI}{I} = \int_{x=0}^{x=x} -\varepsilon c\, dx$$

or $$\ln\frac{I}{I_0} = -\varepsilon cx \quad \text{... (ii)}$$

or $$\frac{I}{I_0} = e^{-\varepsilon cx} \quad \text{... (iii)}$$

or $$\boldsymbol{I = I_0 e^{-\varepsilon cx}} \quad \text{... (iv)}$$

This equation expresses how the intensity of a monochromatic light falls from I_0 to I on passing through a thickness x of a solution of concentration c.

Equation (*ii*) can be expressed in another form.

$$2.303 \log \frac{I}{I_0} = -\varepsilon c x$$

or
$$\log \frac{I}{I_0} = -\frac{\varepsilon}{2.303} c x$$

or
$$\log \frac{I}{I_0} = -\varepsilon' c x \qquad ...(v)$$

or
$$\frac{I}{I_0} = 10^{-\varepsilon' c x}$$

or
$$\mathbf{I = I_0 \times 10^{-\varepsilon' c x}} \qquad ...(vi)$$

$\varepsilon' = \dfrac{\varepsilon}{2.303}$ was earlier called **molar extinction coefficient** of the absorbing solution. Now this quantity is called molar absorption coefficient or molar absorptivity of the absorbing solution.

The physical significance of molar extinction coefficient or **molar absorptivity** follows from equation (*v*), which can be rewritten as

$$\varepsilon' = -\frac{1}{cx} \log \frac{I}{I_0} = \frac{1}{cx} \log \frac{I_0}{I}$$

If
$$c = 1M \text{ and } \log \frac{I_0}{I} = 1$$

then
$$\varepsilon' = \frac{1}{x}$$

$$\log \frac{I_0}{I} = 1 \text{ means } \frac{I_0}{I} = 10 \;\; i.e.\; I = \frac{1}{10} I_0$$

Hence *molar extinction coefficient or molar absorptivity may be defined as the reciprocal of that thickness of the solution layer of 1 molar concentration which reduces the intensity of the light passing through it to one-tenth of its original value.*

Additional Information for Solving Numerical Problems. In Art 3.3 and 3.4, we have assumed that I is the intensity of light after crossing the distance x which has have any numerical value. In fact, this is the intensity of transmitted light after absorption has taken place from the incident light with intensity I_0. This may also be written as It. In fact I and I_t represent the same quantity you may come across the term I_t in the numerical problems. We have some more relations:

Intensity of incident light = Incident of absorbed light + Incident of transmitted light

or $I_0 = I_{abs} + I_t$

or $I_{abs} = I_0 - I_t$

Also $I_t = I_0 - I_{abs}$

We also have some other terms

Transmittance, $T = \dfrac{I_t}{I_0}$

Absorbance, $A = -\log \dfrac{I_t}{I_0} = \log \dfrac{I_0}{I_t}$

SOME SOLVED EXAMPLES ON BEER-LAMBERT'S LAW

Example 1. Calculate the transmittance, absorbance and absorption co-efficient of a solution which absorbs 90% of a certain wavelength of light beam passed through a 1 cm cell containing 0.25 M solution.

Solution. It is given that

$$I_{abs} = 0.90\ I_0\ ;\ x = 1\ \text{cm};\ c = 0.25\ \text{M}$$

$$\therefore \quad I_t = I_0 - I_{abs} = I_0 - 0.90\ I_0 = 0.10\ I_0$$

$$\therefore \quad \text{Transmittance} \quad T = \frac{I_t}{I_0} = \mathbf{0.10}$$

$$\text{Absorbance} \quad A = -\log\frac{I}{I_0} = \log\frac{I_0}{I} = \log\frac{1}{0.1} = 1$$

Applying Beer's law, we have

$$\log\frac{I_t}{I_0} = -\varepsilon' cx$$

i.e.
$$\log 0.10 = -\ \varepsilon' \times 0.25 \times 1$$
$$-1 = -\varepsilon' \times 0.25$$

or
$$\varepsilon' = 0.4\ \text{L mol}^{-1}\text{cm}^{-1}$$

Example 2. A certain substance in a cell of length *l* absorbs 10% of the incident light. What fraction of the incident light will be absorbed in a cell which is five times longer ?

Solution. In the first case, $I_{abs} = 0.10\ I_0$ so that $I_t = 0.90\ I_0$

or
$$\frac{I_t}{I_0} = 0.90$$

We have the relation : $\log\frac{I_t}{I_0} = -\ k'x$. *Put* $x = l$

$$\log\frac{I_t}{I_0} = -\ k'l \text{ i.e. } \log 0.90 = -\ k'l$$

or
$$k' = \frac{0.0458}{l}$$

In the second case, $x = 5l$ so that

$$\log\frac{I_t}{I_0} = -\ k' \times 5l = -\frac{0.0458}{l} \times 5l = -0.2290$$

or
$$\log\frac{I_t}{I_0} = \bar{1}.7710$$

i.e.
$$\frac{I_t}{I_0} = 0.5902 \text{ or } I_t = 0.5902 I_0$$

$$\therefore \quad I_{abs} = I_0 - I_t = I_0 - 0.5902\ I_0 = 0.4098\ I_0$$
$$= 40.98\ \%\ \text{of}\ I_0$$

Q. 9. When an incident beam of wavelength 3000 Å was allowed to pass through 2 mm thick pyrex glass, the intensity of the radiation was reduced to one-tenth of its initial value. What part of the same radiation will be absorbed by 1 mm thick glass ?

Solution. *In the first case,* it is given that

$$\frac{I_t}{I_0} = \frac{1}{10} \quad \text{(It denotes transmitted light)}$$

$$x = 2\ \text{mm} = 0.2\text{cm}$$

Applying Lambert's law, we have

$$\log \frac{I_t}{I_0} = -k'x$$

i.e. $$\log \frac{1}{10} = -k'(0.2)$$

or $$k' = \frac{1}{0.2} = 5$$

or *In the second case,* it is given that

$$x = 1 \text{ mm} = 0.1 \text{ cm}$$

As the medium is the same, k' will have the same value *viz.* 5

Applying Lambert's law, we have

$$\log \frac{I_t}{I_0} = -k'x = -5 \times 0.1 = -0.5$$

$$\therefore \quad \frac{I_t}{I_0} = \text{antilog}(-0.5)$$

$$= \text{antilog}(\bar{1}.5) = 0.3162$$

i.e., $$I_t = 0.3162\, I_0$$

$$I_{abs} = I_0 - I_t = I_0 - 0.3162\, I_0 = 0.6838 I_0$$

PROBLEMS FOR PRACTICE

1. The percentage transmittance of an aqueous solution of disodium fumarate at 250 mm and 298 K is 19.2% for a 5×10^{-4} molar solution in a 1 cm cell. Calculate the absorbance A and the molar absorption coefficient ε. What will be the percentage transmittace of a 1.75×10^{-5} molar solution in a 10 cm cell?

   [**Ans.** $A = 0.717$, $\varepsilon = 1.43 \times 10^3$ L mol^{-1} cm^{-1}, $T = 56.1\%$]

2. Light of wavelength 256 nm passing through a 1.0 mm thick cell containing a sample solution reduced to 16% of its initial intensity. Calculate the absorbance and the molar absorption coefficient of the sample. What would be the transmittance through a 2.0 mm cell?

   [**Ans.** $\varepsilon = 160$ M cm^{-1}, $A = 0.80$ $T = 0.025 = 2.5\%$]

3. If 2 mm thick plate of a material transmits 70% of the incident light, what percentage will be transmitted by 0.5 mm thick plate? [**Ans.** 91.47%]

4. A 1 molar solution in a cell of 0.200 mm thickness transmits 10% of a beam of light. Neglecting the absorption by solvent and cell windows, what percentage of the initial intensity will be transmitted by a molar solution in a cell 0.150 mm thick? [**Ans.** 3.16%]

   [**Hint.** From the first data, calculate ε'. Using the value of ε', calculate I_t/I_0 from the second data.]

3.5. LAWS OF PHOTOCHEMISTRY

1. Grotthus law (first law of photochemistry).

When light falls on a body, a part of it is reflected, a part of it is transmitted and the rest of it is absorbed. It is only the absorbed light which is effective in bringing about a chemical reaction.

This law, however, does not imply that the absorbed light must always result into chemical reaction. The absorbed light may simply bring about phenomena such as fluorescence, phosphorescence etc. Similarly, the absorbed light energy may be simply converted into thermal energy *e.g.* in case of potassium permanganate solution, the light energy is absorbed strongly but no chemical effect is produced. Further in some cases it is observed that light energy may not be absorbed by the reacting substance directly but may be absorbed by some other substance present along with the reacting substance. The energy thus absorbed is then passed onto the reacting substance which then starts reacting. This process is called *photosensitization.* Photosynthesis of carbohydrates in plants where chlorophyll acts as a photosensitizer is a prominent example of photosensitization.

2. Stark-Einstein law of photochemical equivalence (Second law of photochemistry).

The law states:

Every atom or molecule that takes part in a photochemical reaction absorbs one quantum of the radiation to which the substance is exposed.

Albert Einstein

If v is the frequency of the absorbed radiation, then the energy absorbed by each reacting atom or molecule is one quantum *i.e.* hv where h is Planck's constant. The energy absorbed by one mole of the reacting molecules is given by

$$E = Nhv \quad ...(i)$$

where N is Avogadro's number

Putting $v = \frac{c}{\lambda}$, we can write $E = Nh\frac{c}{\lambda}$ ... (*ii*)

where c is the velocity of light and λ is the wavelength of the absorbed radiation.

The energy possessed by one mole of photons or the energy absorbed by one mole of the reaching molecules is called one Einstein

Numerical value of Einstein

Substituting $N = 6.022 \times 10^{23}\ \text{mol}^{-1}$

$h = 6.626 \times 10^{-27}$ erg sec and $c = 3.0 \times 10^{10}$ cm./sec

we get from eq. (*ii*) above

$$E = \frac{(6.022 \times 10^{23})\,(6.626 \times 10^{-27})\,(3.0 \times 10^{10})}{\lambda}$$

$$= \frac{119.7 \times 10^{6}}{\lambda} \text{ ergs per mole or } \frac{119.7 \times 10^{6}}{4.184 \times 10^{7}\lambda} \text{ cal per mole}$$

or $E = \frac{2.86}{\lambda}$ cal per mole (in CGS units) ... (*iii*)

In this equation, λ is expressed in cm. However as λ is usually expressed in Angstrom units (Å) and we know that 1Å = 10^{-8} cm, therefore equation (*iii*) can be written as

$$E = \frac{2.86}{\lambda} \times 10^{8} \text{ cal per mole} \quad ... (iv)$$

where λ is in Angstrom units

Further putting 10^3 cal = 1 kcal, equation (*iv*) can be written as

$$\mathbf{E = \frac{2.86 \times 10^{5}}{\lambda} \textbf{ kcal per mole}} \quad ... (v)$$

This equation gives *the energy possessed by one mole of photons and is called* **one Einstein.** Thus the *energy per Einstein is inversely proportional to the wavelength of the radiation.* Shorter the wavelength of a radiation, greater is the energy per Einstein.

The unit of energy used most frequently in photochemistry is electron volt (eV).

$$1 \text{ eV} = 23.06 \text{ kcal/mole.}$$

It may be mentioned that when a potential of one volt is applied to an electron, it acquires one eV of energy.

To obtain the equation in **SI units,** we substitute

$$N = 6.022 \times 10^{23} \text{ mol}^{-1}, h = 6.626 \times 10^{-34} \text{ J s}, c = 3{\cdot}0 \times 10^{8} \text{ ms}^{-1}$$

we get $$E = \frac{\mathbf{0.1197}}{\lambda} \textbf{ Jmol}^{-1} = \frac{\mathbf{11.97 \times 10^{-5}}}{\lambda} \textbf{ kJ mol}^{-1}$$

In this equation λ is in metre.

SOLVED PROBLEMS ON CALCULATION OF EINSTEIN OF ENERGY

Exampl 1. If the value of an Einstein is 72 kcal, calculate the wavelength of the light.

Ans. Here $$E = 72 \text{ kcal}$$

From the equation, $$E = \frac{2.86 \times 10^5}{\lambda} \text{ kcal/mole}$$

$$\lambda = \frac{2.86 \times 10^5}{E} \text{Å} = \frac{2.86 \times 10^5}{72} = \mathbf{3972.2Å}$$

Example 2. Calculate the value of Einstein of energy for radiation of wavelength 4000 Å.

Solution. $$E = \frac{2.86 \times 10^5}{\lambda} \text{ kcal /mole } (\lambda \text{ in Å units})$$

$$= \frac{2.86 \times 10^5}{4000} \text{ kcal /mole}$$

$$= \mathbf{71.5 \text{ kcal/mole}}$$

Example 3. Calculate the value of an Einstein of energy in electron volts for radiation of frequency 3×10^{13} s^{-1}.

Solution. Here we are given that

$$\nu = 3 \times 10^{13} \text{ s}^{-1}$$

$$\therefore \quad \lambda = \frac{c}{\nu} = \frac{3 \times 10^{10} \text{ cm s}^{-1}}{3 \times 10^{13} \text{ s}^{-1}} = 10^{-3} \text{ cm} = 10^5 \text{Å}$$

$$\therefore \quad E = \frac{\mathbf{2.86 \times 10^5}}{\lambda} \textbf{ kcal/mole } (\lambda \text{ in Å units}) = \frac{2.86 \times 10^5}{10^5} = 2.86 \text{ kcal/mole}$$

$$= \frac{2.86}{23.06} \text{ eV} = 0.124 \text{ eV}$$

Example 4. Calculate the energy of one photon of light of wavelength 2450 Å. Will it be able to dissociate a bond in diatomic molecule which absorbs this photon and has a bond energy equal to 95 kcal per mole.

Planck's constant (h) = 6.626×10^{-27} erg sec molecule^{-1}

Velocity of light (c) = 3×10^{10} cm sec^{-1}

Avogadro's number (N) = 6.02×10^{23}.

Solution. Energy of one photon $= hv = h\frac{c}{\lambda}$

$$= \frac{(6.026\times10^{-27})\times(3\times10^{10})}{2450\times10^{-8}} \text{ ergs}$$

$$= 8.11\times 10^{-12} \text{ ergs}$$

$$= \frac{8.11\times10^{-12}}{4.184\times10^{7}} \text{ cal} = 1.938\times10^{-19}\text{cal}$$

Energy required for dissociation of one bond

$$= \frac{95000}{6.02\times10^{23}} \text{cal}$$

$$= \mathbf{1.578 \times 10^{-19} \ cal}$$

Thus energy of photon is greater than the energy required to break the bond. Hence the bond will be dissociated.

PROBLEMS FOR PRACTICE

1. Calculate the value of an einstein of energy in electron volts for radiation of frequency 3×10^{15} s^{-1} **[Ans.** 12.4 eV]

2. What is the energy in kcals of one mole of photons of wavelength 2573 Å? **[Ans.** 111.15 kcal]

3. Find the value of an Einstein of energy for the radiation of wavelength 4240 Å. **[Ans.** 67.45 kcal/mol]

4. Calculate the energy of one photon of light of wavelength 2450 Å. Will it be able to dissociate a bond in diatomic molecules which absorbs this photon and has a bond energy equal to 95 kcal per mole.

Planck's constant (h) $= 6.626 \times 10^{-27}$ erg sec molecule^{-1}
Velocity of light (c) $= 3 \times 10^{10}$ cm sec^{-1}
Avogadro's number (N) $= 6.02 \times 10^{23}$

Hint. Calculate the energy of one proton. If this energy is greater than that required to break the bond (bond energy), dissociation of the bond will happen

[Ans. 1.938×10^{-19} cal, bond will dissociate]

5. Calculate the value of the einstein corresponding to radiation of wavelength 6000 Å. **[Ans.** 47.66 kcal/mol]

3.6. QUANTUM EFFICIENCY OR QUANTUM YIELD

To relate the number of quanta absorbed with the number of reacting molecules, a term called **quantum efficiency** or **quantum yield** (ϕ) has been introduced. It is expressed as

$$\phi = \frac{\text{Number of molecules reacting in a given time}}{\text{Number of quanta of light absorbed in the same time}}$$

$$= \frac{\text{Number of moles reacting in a given time}}{\text{Number of Einsteins of light absorbed in the same time}}$$

Hence quantum efficiency may be defined as the number of moles reacting per einstein of the light absorbed.

If Stark-Einstein's law is strictly obeyed in the form already stated, ϕ should be equal to unity. However, it has been experienced that the law is applicable only to the primary processes. Hence it is sometimes preferred to state it as follows :

Each molecules that gets activated to initiate the reaction absorbs one quantum of light.

3.6.1. Determination of Quantum yield of a photochemical reaction

To determine the quantum yield of any photochemical reaction, we have to measure

(*i*) the rate of the chemical reaction in terms of molecules per sec.

(*ii*) the quanta absorbed per sec.

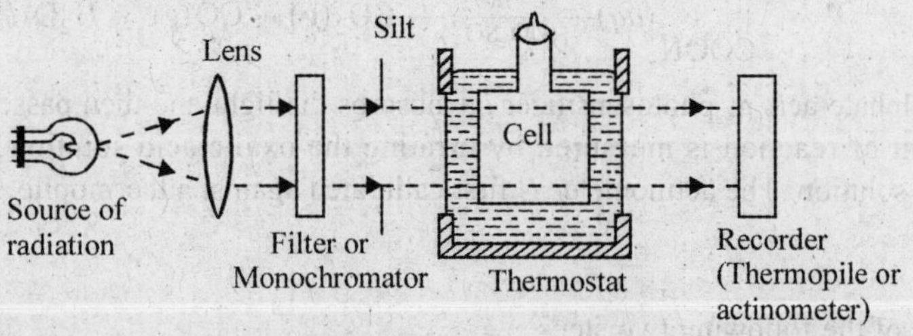

Fig. 3.2. Apparatus for the determination of quantum yeild of photochemical reaction.

Appratus: The apparatus used for measuring the above two quantities is shown in Fig.3.2

Various components of the apparatus are described as under:

(*i*) *A source of radiation* emitting radiation of suitable intensity in the desired spectral range. The commonly used sources are filament lamps, carbon and metal arcs and various gas discharge tubes.

(*ii*) *Lens* of suitable focal length.

(*iii*) *Monochromator or filter* which cuts off all radiations except the radiations of the desired wavelength.

(*iv*) *Slit.*

(*v*) *A cell placed in a thermostat* and containing the reaction mixture. The cell is made of glass or quartz and has optically plane windows for the entrance and exit of light. Glass is used only if the wavelength of the light used lies in the visible range. For radiations with wavelength below 3500 Å the cell completely made of quartz is used.

(*vi*) *Recorder* is usually a *thermopile* or an *actinometer* which is used to measure intensity of the radiation. The principle of a thermopile and that of an actinometer are briefly explained below :

A *thermopile* is a set of thermocouples. Rods of two different metals (*e.g.* Ag and Bi) are joined alternately as shown in Fig. 3.3. One set of junctions is soldered to metal strip blackened with platinum or lamp black and

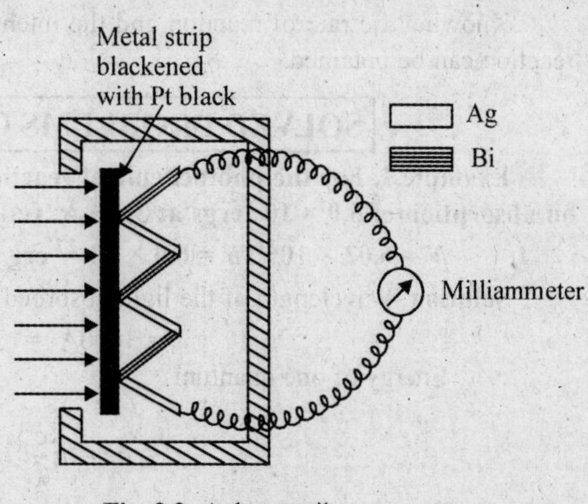

Fig. 3.3. A thermopile.

the other set of junctions is protected completely from radiation by placing the system in a box. The radiations falling on the blackened metal strip are almost completely absorbed by it. As a result, this set of junctions becomes hot. The temperature difference between the hot end and the cold end produces a current which can be measured by connecting a milliammeter to the thermopile. By calibrating the thermopile with radiations of known intensity, the intensity of the desired radiation can be measured.

A *chemical actinometer* is a device in which solutions sensitive to light are used. The working of this device is based upon the fact that a definite amount of the radiation absorbed brings about a definite amount of chemical reaction. The most common actinometer is the uranyl oxalate actinometer which consists of 0.05 molar oxalic acid and 0.01 molar uranyl sulphate (UO_2SO_4) solutions in water. When exposed to light, the following reaction takes place:

$$\begin{matrix} COOH \\ | \\ COOH \end{matrix}(aq) \xrightarrow[(UO_2SO_4)]{h\nu} CO_2(g) + CO(g) + H_2O(l)$$

Uranyl sulphate acts as photosensitizer *i.e.* absorbs the light and then passes it on to oxalic acid. The extent of reaction is measured by titrating the oxalic acid solution with potassium permanaganate solution. The actinometer is first calibrated against a thermopile.

Procedure

It consists of the following two steps.

(*i*) *Measurement of the intensity of light absorbed.* First the empty cell or the cell filled with the solvent alone (in case of solutions), is placed in the thermostat. The monochromatic radiation is allowed to pass through the cell for a definite time. The reading is taken on the recorder. From this we get the energy of the incident radiations. Now the reactants are taken in the cell and the radiations are allowed to pass through it for definite time and the reading is taken again. The difference between the two readings gives the total energy absorbed by the reactants in a given time. From this the intensity of the radiation absorbed I_{abs} is calculated as follows :

$$I_{abs} = \frac{\text{Total energy absorbed}}{\text{Volume of the reaction mixture} \times \text{Time in sec}}$$

(*ii*) *Measurement of the rate of reaction.* The rate of the reaction can be studied by noting the change in some physical property. Alternatively, samples of the reaction mixtures are removed periodically from the cell and analysed.

Knowing the rate of reaction and the intensity of light absorbed, the quantum yield of the reaction can be obtained.

SOLVED PROBLEMS ON QUANTUM YIELD

Example 1. For the photochemical reaction $A \rightarrow B$, 1.0 × 10^{-5} moles of B were formed on absorption of 6.0 × 10^{7} ergs at 3600 Å. Calculate the quantum efficiency.

$N = 6.02 \times 10^{23}$, $h = 6.0 \times 10^{-27}$ erg sec, $c = 3 \times 10^{10}$ cm /sec.

Solution. Wavelength of the light absorbed

$$= 3600 \text{Å} = 3600 \times 10^{-8} \text{cm}$$

∴ Energy of one quantum

$$= h\nu = h\frac{c}{\lambda} = \frac{(6.0\times10^{-27})(3\times10^{10})}{3600\times10^{-8}} \text{ergs}$$

$$= 5.0\times10^{-12} \text{ ergs}$$

Total energy absorbed

$$= 6.0 \times 10^7 \text{ ergs (Given)}$$

∴ Number of quanta absorbed

$$= \frac{6.0 \times 10^7}{5.0 \times 10^{-12}} = 1.2 \times 10^{19}$$

Number of *moles* reacting $= 1.0 \times 10^{-5}$

Number of *molecules* reacting

$$= 1.0 \times 10^{-5} \times 6.02 \times 10^{23} = 6.02 \times 10^{18}$$

$$\text{Quantum efficiency } (\phi) = \frac{\text{Number of molecules reacting}}{\text{Number of quanta absorbed}}$$

$$= \frac{6.02 \times 10^{18}}{1.2 \times 10^{19}} \cong \mathbf{0.5}$$

Example 2. A certain system absorbs 3 × 10^18 quanta of light per second. On irradiation for 20 minutes 0.003 mole of the reactant was found to have reacted. Calculate the quantum yield for the process (Avogadro's number = 6.02 × 10^23)

Solution. Number of quanta absorbed per second

$$= 3 \times 10^{18}$$

Number of *moles* reacting in 20 minutes

$$= 0.003$$

∴ Number of *molecules* reacting in 20 minutes

$$= 0.003 \times 6.022 \times 10^{23}$$

Number of *molecules* reacting per second

$$= \frac{0.0303 \times 6.022 \times 10^{23}}{20 \times 60} = 1.506 \times 10^{18}$$

$$\therefore \text{ Quantum yield } (\phi) = \frac{\text{Number of molecules reacting per second}}{\text{Number of quanta absorbed per second}}$$

$$= \frac{1.506 \times 10^{18}}{3 \times 10^{18}} = \mathbf{0.5}$$

Example 3. In the photobromination of cinnamic acid using blue light of 4358 Å at 30.6°C an intensity of 14,000 ergs per second produced a decrease of 0.075 millimole of Br_2 during an exposure of 1105 seconds. The solution absorbed 80.1% of the light passing through it. Calculate the quantum yield for the photoreaction of bromine.

Solution. Intensity of incident light

$$= 14{,}000 \text{ ergs per sec.}$$

Energy of one quantum

$$= \frac{hc}{\lambda} = \frac{(6.62 \times 10^{-27}) \times (3 \times 10^{10})}{4358 \times 10^{-8}} \text{erg}$$

$$= 4.56 \times 10^{-12} \text{ erg}$$

No. of quanta/sec $= \dfrac{14,000}{4.56 \times 10^{-12}}$

$$= 3.07 \times 10^{-15}$$

As the light absorbed is 80.1%

Number of quanta absorbed/sec $= 3.07 \times 10^{15} \times \dfrac{80.1}{100} = 2.46 \times 10^{15}$

Amount of substance reacting $=$ 0.075 millimole $=$ 0.000075 moles

Number of molecules reacting/sec $= \dfrac{0.000075 \times 6.023 \times 10^{23}}{1105}$

$$= 4.09 \times 10^{16}$$

Quantum yield $(\phi) = \dfrac{4.09 \times 10^{16}}{2.46 \times 10^{15}}$ = 16.6 molecules/quantum

Example 4. A uranyl oxalate actinometer is irradiated for 15 minutes with light of 4350 Å. At the end of this time, it is found that oxalic acid equivalent to 12.0 ml of 0.001 molar $KMnO_4$ solution has been decomposed by the light. At this wavelength, the quantum efficiency of the actinometer is 0.58. Find the average intensity of the light used in (*a*) ergs per sec and (*b*) in quanta per sec.

Solution. Oxalic acid decomposed in 15 minutes

$$= 12.0 \text{ ml of } 0.001 \text{ molar} = \frac{0.001}{1000} \times 12 \text{ moles}$$

$$= \frac{0.001}{1000} \times 12 \times 6.022 \times 10^{23} \text{ molecules}$$

$$= 7.23 \times 10^{18} \text{ molecules}$$

∴ Number of molecules decomposed per sec

$$= \frac{7.23 \times 10^{18}}{15 \times 60} = 8 \times 10^{15}$$

∴ Number of quanta absorbed per sec

$$= \frac{\text{Number of molecules reacting/sec}}{\text{Quantum efficiency}}$$

$$= \frac{8 \times 10^{15}}{0.58} = \mathbf{1.38 \times 10^{16}}$$

Further, wavelength of the light absorbed $=$ 4350 Å $= 4350 \times 10^{-8}$ cm

∴ Energy of one quantum $= h\nu = \dfrac{hc}{\lambda}$

$$= \frac{(6.62\times10^{-27})\times(3\times10^{10})}{4350\times10^{-8}}$$

$$= 4.57 \times 10^{-12} \text{ ergs}$$

∴ Energy absorbed per second

$$= (1.38 \times 10^{16}) \times (4.57 \times 10^{-12}) \text{ ergs}$$

$$= \mathbf{6.31 \times 10^4 \text{ ergs}}$$

PROBLMS FOR PRACTICE

1. A certain system absorbs 3.0×10^{16} quanta of light per second. On irradiation for 20 minutes, 0.002 mole of the reactant was found to have reacted. Calculate the quantum efficiency of the process. **[Ans.** 0.669]
2. When gaseous *HI* is irradiated with radiation of wavelength 2530 Å, it is observed that 1.85 $\times 10^{-2}$ mole decomposes per 1000 cal of radiant energy absorbed. Calculate the quantum efficiency of the reaction. **[Ans.** 1.9]

Example. What are the reasons for very high and very low quantum yields in photochemical reactions?

Solution. The reason for very high or very low quantum yield is found in the fact that a photochemical reaction takes place in two steps, which are :

(*i*) **Primary process:** This involves the activation of certain molecules by absorption of light (one quantum per molecule) *i.e.*

$$AB + h\nu \longrightarrow \underset{\substack{\text{Activated}\\\text{molecule}}}{AB^*}$$

or the dissociation of certain molecules to produce active atoms or free radical *i.e.*

$$AB + h\nu \longrightarrow A + \underset{\substack{\text{Active or}\\\text{excited atom}}}{B^*}$$

(*ii*) **Secondary process:** This involves the reaction of the activated molecules or the products of the primary process with other molecules or the deactivation of the activated molecules. Actually a number of reactions are involved in the secondary process.

Some of these are endothermic reactions and thus take place slowly thereby affecting the overall reaction. Some of these are exothermic reactions and the heat evolved decomposes the products again affecting overall reaction. The effect of these factors brings about sudden rise or fall of quantum yield.

3.7. SOME EXAMPLES OF PHOTOCHEMICAL REACTIONS

1. Photochemical decomposition of hydrogen iodide.

The photolysis of HI to give H_2 and I_2 has been studied in the wavelength range 2070 –2820Å and the quantum yield of this reaction has been found to be approximately 2. Mechanism for the photolysis of HI has been proposed as under.

(*a*) *Primary process:*

$$HI + h\nu \longrightarrow H + I^*$$

i.e. a molecules of HI absorbs one photon of light and dissociates to give a normal hydrogen atom and an excited iodine atom.

(*b*) *Secondary processes:* The products of the primary process then take part in the secondary processes as follows :

(*i*) $\dot{H} + HI \longrightarrow H_2 + \dot{I}$ (*ii*) $I^* + HI \xrightarrow{\text{Endo}} \dot{H} + I_2$

(*iii*) $\dot{H} + \dot{H} \longrightarrow H_2$ (*iv*) $\dot{I} + \dot{I} \longrightarrow I_2$

(*v*) $\dot{H} + \dot{I} \xrightarrow{\text{Exo}} HI$

All these secondary processes may not be important. The reactions (*iii*) and (*v*) are highly exothermic and the heat produced causes the dissociation of the products of these reactions. Reaction (*ii*) is endothermic and hence takes place slowly. Thus the only secondary processes of importance are the reaction (*i*) and (*iv*). Rewriting the reaction of the primary process and the two important reactions *viz.* (*i*) and (*iv*) of the secondary processes, we have

$$HI + h\nu \longrightarrow \dot{H} + I^* \} \text{ Reaction of the primary process}$$

$$\left.\begin{array}{l} \dot{H} + HI \longrightarrow H_2 + \dot{I} \\ \dot{I} + \dot{I} \longrightarrow I_2 \end{array}\right\} \text{The only important reactions of the secondary processes}$$

Adding these three equations, we get the overall reaction

$$2HI + h\nu \longrightarrow H_2 + I_2$$

Thus for every one quantum of light absorbed, two molecules of HI are decomposed. Hence the quantum yield value of 2 is explained.

2. Photolysis (photochemical decomposition) of HBr

The quantum yield of this reaction is found to be 2. Mechanism of this reaction is suggested to be similar to that of photolysis of HI. Use HBr in place of HI, Br_2 in place of I_2 and Br in place of I.

3. Photosynthesis of HBr from hydrogen and bromine

The reaction may be represented as

$$H_2 + Br_2 \xrightarrow{h\nu} 2HBr$$

The quantum efficiency of this reaction is very low *i.e.* about 0.01 at ordinary temperatures. This is explained by proposing the following mechanism for the above reaction

(*a*) *Primary process :*

$$Br_2 + h\nu \longrightarrow 2Br$$

(*b*) *Secondary processes :*

(*i*) $\dot{Br} + H_2 \xrightarrow{\text{Endo}} HBr + \dot{H}$ (*ii*) $\dot{H} + Br_2 \longrightarrow HBr + \dot{Br}$

(*iii*) $\dot{H} + HBr \longrightarrow H_2 + \dot{Br}$ (*iv*) $\dot{Br} + \dot{Br} \longrightarrow Br_2$

The H-atom produced in reaction (*i*) attacks Br_2 according to reaction (*ii*) producing HBr and Br atom. The Br-atom thus produced attacks more of H_2 according to reaction (*i*) producing H-atom again. As a result, reactions (*i*) and (*ii*) should repeat over and again *i.e.* a chain reaction should be set up and the quantum yield of the reaction should be very high. But actually the quantum yield is very low. This is explained as follows :

Reaction (*i*) is endothermic and hence takes place slowly or almost does not tend to take place. The reactions (*ii*) and (*iii*) depend upon the formation of H atoms in reaction (*i*); consequently, reactions (*ii*) and (*iii*) are very slow. So the only important reaction in the secondary processes is the reaction (*iv*). This is just the reverse of the primary process. Thus the Br atom formed in the primary process do not combine to form HBr; instead, they combine together to give back Br_2. As a result, the quantum yield of the reaction is very low at ordinary temperatures.

The quantum yield of the above reaction increases with increase of temperature. This is obviously because of the fact that the reaction (*i*) which is endothermic and requires high energy of activation becomes faster as high temperatures because the required energy of activation becomes available at higher temperatures. This reaction produces more H-atoms thereby increasing the rate of reactions (*ii*) and (*iii*) also.

4. Photosynthesis of HCl from hydrogen and chlorine.

This reaction whose quantum yield is very high *i.e.* 10^4 to 10^6 may be represented as

$$H_2 + Cl_2 \xrightarrow{h\nu} 2HCl$$

The high quantum yied of this reaction is explained by the chain mechanism. The different steps involved are as follows:

(*a*) *Primary process:*

$$Cl_2 + h\nu \longrightarrow 2\dot{C}l \qquad \textit{chain initiating step}$$

Thus a chlorine molecule absorbs one quantum of light and dissociates to give Cl atoms.

(*b*) *Secondary processes :*

(*i*) $\dot{C}l + H_2 \xrightarrow{\text{Exo}} HCl + \dot{H}$

Chain propagating steps

(*ii*) $\dot{H} + Cl_2 \longrightarrow HCl + \dot{C}l$

(*iii*) $\dot{C}l + \dot{C}l \longrightarrow Cl_2$

Chain terminating step

The H-atoms produced in reaction (*i*) attacks Cl_2 according to reaction (*ii*) producing HC1 and Cl atom. The Cl atom thus produced attacks more of H_2 according to reaction (*i*) producing H-atom again. In this way, the reactions (*i*) and (*ii*) repeat over and again till almost whole of H_2 and Cl_2 have reacted to form HC1. The reaction stops when the Cl atoms left off combine with each other to form $C1_2$, according to reaction (*iii*). *This reaction takes place on the walls of the reaction vessel.*

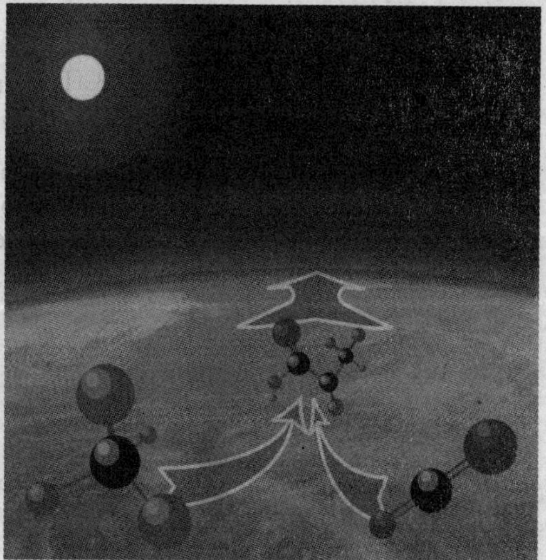

Photochemistry of methane and carbondioxide during the formation of earth

In this reaction, the reaction (*i*) of the secondary processes which immediately follows the primary process is *exothermic* and therefore, takes place very easily and consequently the chain reaction is set up easily. In the photosynthesis of HBr, the reaction (*i*) of the secondary process is *endothermic* and thus has very little tendency to take place.

The quantum yield of the above reaction decreases if the reaction vessel contains small traces of oxygen. This explained on the basis that oxygen brings about chain termination by the following reactions :

$$\dot{H} + O_2 \longrightarrow HO_2$$
$$\dot{Cl} + O_2 + HCl \longrightarrow HO_2 + Cl_2$$

5. Photolysis of ammonia

This is an example of a photochemical reaction whose quantum yield is very low *i.e.* about 0·25 at ordinary temperatures and pressures. The probable mechanism is as follows:

(*a*) *Primary process* :

$$NH_3 + h\nu \longrightarrow \dot{N}H_2 + \dot{H}$$

The presence of H-atoms has been established.

(*b*) *Secondary processes:* The ultimate products of the photolysis of ammonia have been found to be nitrogen, hydrogen and hydrazine. Hence the following probable secondary reactions have been proposed.

(*i*) $\dot{N}H_2 + \dot{N}H_2 \longrightarrow N_2 + 2H_2$ (*ii*) $\dot{N}H_2 + \dot{N}H_2 \longrightarrow N_2H_4$

(*iii*) $\dot{H} + \dot{H} \longrightarrow H_2$ (*iv*) $\dot{N}H_2 + \dot{H} \longrightarrow NH_3$

Since the quantum yield of the reaction is very low, the reaction (*i*) to (*iii*) take place only to a very small extent. The important secondary reaction is the reaction (*iv*) only which involves the reversal of the primary process leading to decline in the value of (ϕ).

6. Photolysis of acetone.

The reaction may be represented as

$$\underset{\text{Acetone}}{CH_3COCH_2} \xrightarrow{h\nu} \underset{\text{Ethane}}{C_2H_6} + \underset{\text{Carbon monoxide}}{CO}$$

The major products of the photolysis of acetone are ethane and carbon monoxide but small amounts of methane (CH_4) and appreciable amounts of diacetyl, $(CH_3CO)_2$ are also found to be present. The quantum yield of the reaction is found to be nearly unity. All these observations suggest the following probable mechanism for the photolysis of acetone.

(*a*) *Primary process* :

$$\underset{\text{Acetone}}{CH_3COCH_3} \xrightarrow{h\nu} \underset{\text{Methyl free radical}}{\dot{C}H_3} + \underset{\text{Acetyl free radical}}{CH_3\dot{C}O}$$

(*b*) *Secondary processes* : A number of secondary reactions are possible. A few of these are as under :

(*i*) $CH_3\dot{C}O \longrightarrow \dot{C}H_3 + CO$

(*ii*) $\dot{C}H_3 + CH_3\dot{C}O \longrightarrow C_2H_6 + CO$

(*iii*) $CH_3\dot{C}O + CH_3\dot{C}O \longrightarrow (CH_3CO)_2$

Thus no chain reaction is set up. The free radicals formed in the primary process simply react with each other. Hence the quantum efficiency of the reaction is unity.

3.8. LUMINESCENCE

The glow produced by a body, without the involvement of heat is called luminescence. Luminescence is also called **cold light**. We know when a block of iron is heated, it first becomes red hot and eventually becomes white and begins to glow. But this will not be called luminescence. Luminiscence does not occur by the action of heat. It can occur, for example when the electrons excited to higher levels by the absorption of radiation return to original levels. Luminescence is of the following three types:

1. Chemiluminescence: When a photochemical reaction takes place, light is absorbed. However, there are certain reactions in which light is produced. *The emission of light in chemical reactions at ordinary temperatures is called chemiluminescence.* Thus chemiluminescence is just the reverse of a photochemical reaction.

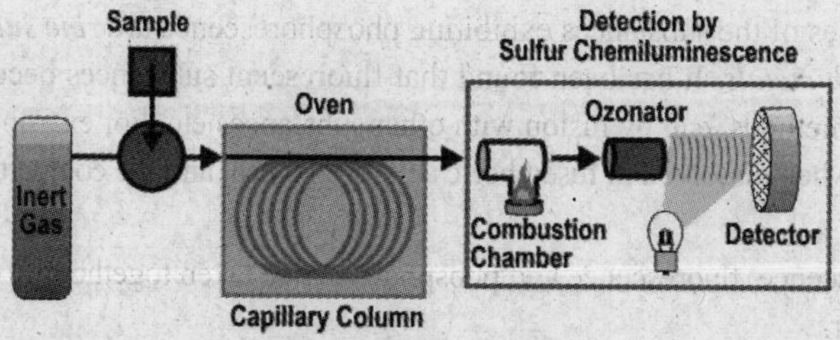

Detection by sulphur chemiluminiscence

A few examples of chemiluminescence are as given below:

(*a*) The light emitted by glow-worms is due to the oxidation of the protein, *luciferin, present in the glow-worms.*

(*b*) The oxidation of yellow phosphorus in oxygen or air to give P_2O_5 at ordinary temperatures (–10 to 40°C) is accompanied by the emission of visible greenish-white luminescence.

In the above cases, a part or whole of the energy emitted during the reaction, is used up for the excitation of electrons. When they jump back to the inner orbits, the emission of light takes place.

2. Fluorescence: There are certain substances which when exposed to light or certain radiations absorb the energy and then immediately or instantaneously start re-emitting the energy. Such substances are called fluorescent substances and the phenomenon is called *fluorescence.* Obviously the absorption of energy results into the excitation of the electrons followed immediately by the jumping back of the excited electrons to the lower levels. As a result, the absorbed energy is emitted back.

Fluorescence starts as soon as the substance is exposed to light and the fluorescence stops as soon as the light is cut off.

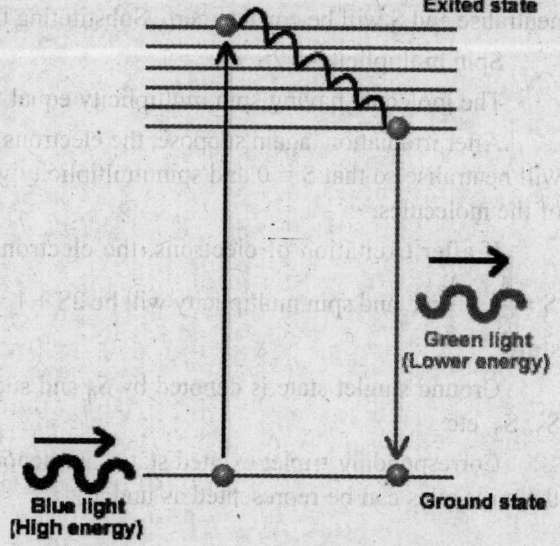

Fluorescence phenomenon

A few examples of the substances showing the phenomenon of fluorescence are given below:

(*a*) fluorite, CaF_2

(*b*) certain organic dyes such as eosin, fluorescein etc.;

(*c*) certain inorganic compounds such as uranyl sulphate, UO_2SO_4.

Notably in the phenomenon of fluorescence, the wavelength of the emitted light is usually greater than that of the absorbed light. This is explained on the basis that the energy absorbed raises the electron to a sufficiently higher level but the return of the excited electron to the original level takes place in steps, through the intermediate levels. The energy thus produced in every jump is smaller and hence the wavelength of the light emitted is larger than that of the absorbed light.

3. Phosphorescence: There are certain substances which continue to glow for some time even after the external light is cut off. Such substances are called *phosphors or phosphorescent substances* and the phenomenon is called *phosphorescence.* Thus in a way phosphorescence is a slow fluorescence. It is found mostly in solids, because the molecules have least freedom of motion. The excited electrons thus keep on jumping back slowly for quite sometime.

Some examples of the substances exhibiting phosphorescence are *zinc sulphide* and *sulphides* of the *alkaline earth metals.* It has been found that fluorescent substances become phosphorescent if fixed by suitable methods, *e.g.* by fusion with other substances etc. For example, many dyes which show fluorescence when dissolved in fused boric acid or glycerol and the cooled to rigid mass become phosphorescent.

Chemiluminescence, fluorescence and phosphorescence taken together are called *luminescence.* as mentioned earlier.

3.9. EXCITATION OF ELECTRONS (JABLONSKI DIAGRAM)

When radiations are incident on certain substances, some of the electrons may absorb energy and jump to higher levels. The electrons may retrurn to the ground state or may bring about a chemical change.

The electronic spins are expressed in terms of **spin multiplicity** which is equal to 2S + 1, where S is the total electron spin. Taking $+\frac{1}{2}$ and $-\frac{1}{2}$ as spin values of electrons spinning clockwise and anti-clockwise, we can say that if a molecule contains all the paired electrons, their spins will neutralise and S will be equal to zero. Substituting the value of S = 0 in the spin multiplicity relation

Spin multiplicity = 2S + 1 = 1

The molecule having spin multiplicity equal to 1 is said to be in **singlet ground state**

After irradiation, again suppose, the electrons in the excited state have opposite spin, the spins will neutralise so that S = 0 and spin multiplicity will be 2S + 1 i.e 1. This is **singlet excited state** of the molecules.

If after excitation of electrons, the electrons occupy different orbitals with similar spins, $S = \frac{1}{2} + \frac{1}{2} = 1$ and spin multiplicity will be 2S + 1 = 3. The molecule is said to be in **triplet excited state.**

Ground singlet state is denoted by S_0 and successive singlet excited state are denoted by S_1, S_2, S_2, etc.

Corresponding triplet excited states are denoted by T_1, T_2, T_3, etc. Excitation of electrons in the molecules can be represented as under.

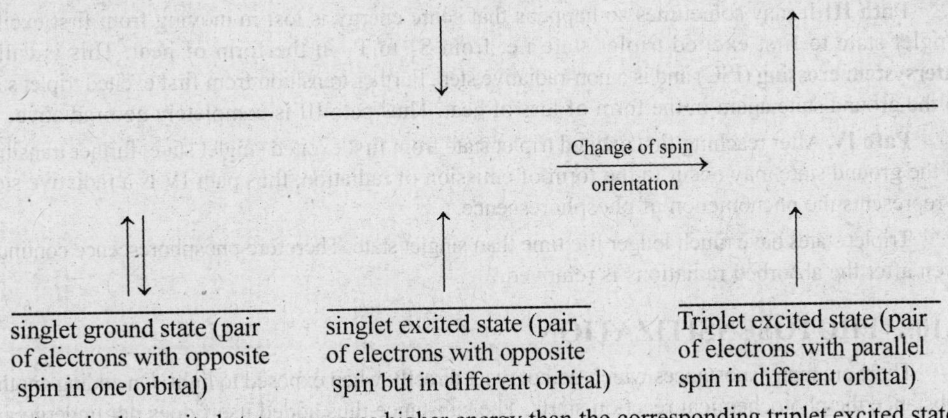

Any singlet excited state possesses higher energy than the corresponding triplet excited state.

Thus $E_{S_1} > E_{T_1}, E_{S_2} > E_{T_2}$ and so on.

Excited molecules retrurns to the original (ground) state by emitting energy in a number of ways as shown in the Fig 4.5. This is known as **Jablonski diagram.**

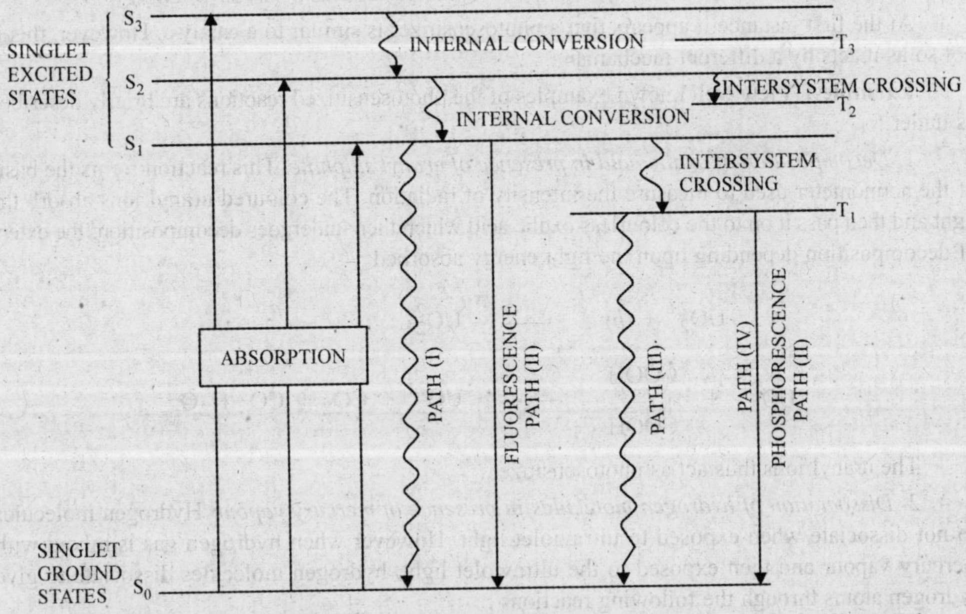

Jablonski diagram showing excitation of molecules and their return to the ground state by different paths.

In the first step of reverting to ground state, transition from higher excited singlet states (S_2, S_3, etc) to lowest excited singlet state state S_1 occurs. This is called internal conversion (IC). Loss of energy takes place only in the form of heat, no radiations are emitted. Therefore this step is called **non-radiative** or **radiationless** transition. From the first excited singlet state, the molecules return to the ground state through one of the following paths.

Path I If the molecule emits rest of the energy also in the form of heat, the whole process becomes non-radiative. This is shown by a wavy line in the figure.

Path II If the molecule loses energy in the form of radiation to reach the ground state, it is radiative transition. This transition represents **fluorescence** and is shown by a straight line in the figure.

Path III It may sometimes so happens that some energy is lost in moving from first excited singlet state to first excited triplet state i.e. from S_1 to T_1 in the form of heat. This is called **intersystem crossing** (ISC) and is a non-radiative step. Further transition from first excited triplet state to the ground state again in the form of loss of heat. Thus path III is completely non-radiative.

Path IV. After reaching first excited triplet state from first excited singlet state, further transition to the ground state may occur in the form of emission of radiation, thus path IV is a radiative step. It represents the phenomenon of phosphorescence.

Triplet states have much longer life time than singlet state. Therefore phosphorescence continues even after the absorbed radiations is removed.

3.10.. PHOTOSENSITIZATION

There are many substances which do not react directly when exposed to light. On adding another substance the photochemical reaction starts. The substance thus added itself does not undergo any chemical change.

The substances added merely absorbs the light energy and then passes it on to one of the reactants. *Such a substance which when added to a reaction mixture helps to start the photochemical reaction but itself does not undergo any chemical change is called a photosensitizer and the process is called photosensitization.* Thus a photosensitizer simply acts as a *carrier of energy.*

At the first instance it appears that a photosensitizer is similar to a catalyst. However, this is not so as it acts by a different mechanism.

Examples: A few well known examples of the photosensitized reactions are briefly described as under :

1. *Decomposition of oxalic acid in presence of uranyl sulphate.* This reaction forms the basis of the actinometer used to measure the intensity of radiation. The coloured uranyl ions absorb the light and then pass it on to the colourless oxalic acid which then undergoes decomposition; the extent of decomposition depending upon the light energy absorbed.

$$UO_2^{2+} + h\nu \longrightarrow UO_2^{2+*}$$

$$UO_2^{2+*} + \begin{array}{c} COOH \\ | \\ COOH \end{array} \longrightarrow UO_2^{2+} + CO_2 + CO + H_2O$$

The uranyl ions thus act as photosensitizer.

2. *Dissociation of hydrogen molecules in presence of mercury vapour.* Hydrogen molecules do not dissociate when exposed to ultraviolet light. However when hydrogen gas is mixed with mercury vapour and then exposed to the ultraviolet light, hydrogen molecules dissociate to give hydrogen atoms through the following reactions :

$$Hg + h\nu \longrightarrow Hg^*$$

$$Hg^* + H_2 \longrightarrow Hg + 2H$$

where Hg^* represents the activated mercury atom. Thus in the above reaction mercury acts a photosensitizer.

The H-atom being highly reactive can easily reduce metallic oxides, nitrous oxide, carbon monoxide etc.

3. *Photosynthesis of carbohydrates in plants.* Carbon dioxide and water vapour present in the air do not absorb the visible light emitted by the sun. However, the green colouring matter, namely *chlorophyll* present in the plants can absorb the visible light. After absorption, it passes on the energy

to the carbon dioxide and water molecules which then combine to form carbohydrates alongwith the evolution of oxygen.

$$\underset{\text{From air}}{\underline{CO_2 + H_2O}} + \underset{\text{From Sunlight}}{\underline{h\nu}} \xrightarrow{\text{Chlorophyll}} \frac{1}{6}\underset{\text{Glucose}}{(C_6H_{12}O_6)} + O_2$$

Thus chlorophyll acts as a photosensitizer in the above reaction.

3.10.1. Photo-inhibitors

Presence of certain substances considerably reduces the quantum yield of some photochemical reactions. For example, in the photosynthesis of HCl, the presence of traces of oxygen reduces the quantum yield of this reaction. Substances like nitric oxide, sulphur dioxide, propylene etc. also show similar effects. *Such substances which reduce the quantum yield of photochemical reactions are called photo-inhibitors.* It is believed that such substances react with the chain propagating atoms or radicals resulting into the chain termination. The chain termination in the photosynthesis of HCl by the presence of traces of oxygen takes place as follows:

$$\dot{H} + O_2 \longrightarrow HO_2$$

$$\dot{Cl} + O_2 + HCl \longrightarrow HO_2 + Cl_2$$

3.10.2. Photochemical equilibrium

Suppose a substance A changes into a substance B by the absorption of light. If the reverse reaction can also occur either as a *photochemical reaction* or *as thermal reaction* (*dark reaction*), the situation may be represented as

$$A \underset{\text{Light}}{\overset{\text{Light}}{\rightleftharpoons}} B$$

or

$$A \underset{\text{Thermal}}{\overset{\text{Light}}{\rightleftharpoons}} B$$

A stage may reach when the rate of forward reaction may become equal to the rate of backward reaction. Now further absorption of light produces no chemical change. The reaction is then said to have attained a *photochemical equilibrium.*

Examples: (*i*) *Photochemical decomposition of nitrogen dioxide*

$$2NO_2 \underset{\text{Thermal}}{\overset{\text{Light}}{\rightleftharpoons}} 2NO + O_2$$

(*ii*) *Photochemical decomposition of sulphur trioxide*

$$2SO_3 \underset{\text{Light}}{\overset{\text{Light}}{\rightleftharpoons}} 2SO_2 + O_2$$

Equilibrium constant: The equilibrium constant of a photochemical equilibrium is different from that of the ordinary chemical equilibrium. This is because the rate of a photochemical reaction does not depend upon the concentration of the reactants but it depends upon the intensity of the light absorbed. Thus for the reaction

$$A \underset{\text{Thermal}}{\overset{\text{Light}}{\rightleftharpoons}} B$$

Rate of forward reaction $\propto I_{abs} = k_1 I_{abs}$

Rate of backward reaction $\propto [B] = k_2$ [B]

When the reaction is in equilibrium,

Rate of forward reaction = Rate of backward reaction

i.e. $$k_1 I_{abs} = k_2 [B]$$

or $$\frac{k_2}{k_1} = \frac{I_{abs}}{[B]} \quad \text{... } (i)$$

or $$K = \frac{I_{abs}}{[B]} \quad \text{... } (ii)$$

where $K = \frac{k_2}{k_1}$ is the **photochemical equilibrium constant.**

From eqn (*ii*), it is clear that photochemical equilibrium constant depends upon the intensity of the light absorbed.

Further eqn (*i*) may be written as

$$[B] = \frac{k_2}{k_1} I_{abs}$$

It follows that the concentration of the products formed at equilibrium is directly proportional to the intensity of the light absorbed.

SOME SOLVED CONCEPTUAL PROBLEMS

Problem 1. Why does quantum yield of photosynthesis of HCl decrease if (*i*) the reaction is carried out in capillary tubes (*ii*) small traces of oxygen are present?

Solution. (*i*) This is because capillary tube is very narrow. The chain reaction taking place through Cl and H-atoms gets terminated on the walls of the capillary tube.

(*ii*) This is because oxygen leads to chain termination through the following reactions

$$H + O_2 \longrightarrow HO_2; \; Cl + O_2 + HCl \longrightarrow HO_2 + Cl_2$$

Problem 2. Comment on the free energy change of photochemical reaction giving reasons.

Slution. The free energy change (ΔG) of photochemical reaction may not be negative. There are many photochemical reactions for which ΔG is positive (e.g. photosynthesis of carbohydrates in plants ozonization of oxygen etc.) The reason for increase of free energy is that a part of the light energy absorbed by the reactants is converted into free energy of products.

Problem 3. What types of transitions are involved in fluorescence and in phosphoresence? Comment on the wavelength of radiation emitted in these phenomena as compared to that of the light absorbed for excitation.

Solution. Fluorescence involves a radiative transition between two states of same multiplicity (usually two singlets) whereas phosphorescence is due to radiative transition between two states of different multiplicity (usually triplet to singlet). The energy emitted during these phenomena is less than the energy absorbed because a part of the internal energy is lost as heat in internal conversion or intersystem crossing. Hence frequencies emitted are smaller or wavelengths are longer.

Problem 4. What is the difference between fluorescence and phosphorecence?

Solution. When certain substances like calcium fluorite are exposed to light, they immediately start emitting the light. The phenomenon is called fluorescence. The fluorescence stops as soon as the source of external light is removed. However, there are certain substances (e.g. ZnS) which continue to glow for some time even after the external light is cut off. This phenomenon is called phosphorescence.

Problem 5. What type of plot is obtained when optical density (D) or absorbance (A) is plotted against the concentration of the solution?

Solution. According to Beer's law, $\log\frac{I}{I_0} = -\log T = D \text{ (or } A) = \varepsilon\, cx$

Hence a plot of optical density (D) or absorbance (A) versus concentration will be a straight line passing through the origin.

Problem 6. What is the physical significance of extinction coefficient or absorptivity?

Solution. According to Lambert's law $I = I_0\, e^{-kx}$ or $\ln\frac{I}{I_0} = -kx$

or $\log\frac{I}{I_0} = -\frac{k}{2.303}x = -k'x$ where $k' = \frac{k}{2.303}$ is called extinction coefficient. We can rewrite it as

$$\log\frac{I_0}{I} = k'x \quad \text{or} \quad k' = \frac{1}{x}\log\frac{I_0}{I}. \text{ If } \frac{I_0}{I} = 10 \text{ i.e. } I = \frac{1}{10}I_0, k' = \frac{1}{x}.$$

Hence extinction coefficient is the reciprocal of that layer thickness at which the intensity of light falls to 1/10th of its original value.

Problem 7. What do you understand by the following terms?

(*i*) Radiative and non-radiative (radiationless) transitions.

(*ii*) Internal conversion and inter system crossing.

Solution. (*i*) In a transition when energy is lost only in the form of heat due to collision with other molecules, it is called non-radiative (radiationless) transition. When energy is lost in the form of light or ultraviolet radiation etc., it is called radiative transition.

(*ii*) The transition from higher excited singlet states (S_2, S_3 etc.) to lowest singlet excited state (S_1) accompanied by loss of energy in the form of heat is called *internal conversion*. The transition from an excited singlet state to the corresponding excited triplet state ($S_1 \rightarrow T_1$), ($S_2 \rightarrow T_2$) etc., accompanied by loss of energy in the form of heat is called *intersystem crossing*.

EXERCISES
(Based on Questions Papers of Different Universities)

Multiple choice questions (Choose the correct option)

1. The equation for the Lambert's law is

(*a*) $\ln\left(\frac{I_0}{I}\right) = -bx$ (*b*) $\ln\left(\frac{I}{I_0}\right) = -bx$

(*c*) $\ln\left(\frac{I}{I_0}\right) = -\varepsilon cx$ (*d*) $\ln\left(\frac{I}{I_0}\right) = \varepsilon cx$

2. Which of the following statement is true?

(*a*) It is the secondary reaction in which absorption of radiation takes place

(*b*) It is the primary reaction in which absorption of radiation takes place

(*c*) The absorption of radiation takes place in both-the primary and secondary reactions

(*d*) None of the above

3. One einstein energy is

(*a*) $E = \dfrac{2.859}{\lambda} \times 10^5 \text{ cal mol}^{-1}$ (*b*) $E = \dfrac{2.859}{\lambda} \times 10^5 \text{ k cal mol}^{-1}$

(*c*) $E = \dfrac{2.859}{\lambda} \times 10^5 \text{ J mol}^{-1}$ (*d*) $E = \dfrac{2.859}{\lambda} \times 10^5 \text{ kJ mol}^{-1}$

4. "It is only the absorbed light radiations that are effective in producing a chemical reactions." This is the statement of

(*a*) Lambert law (*b*) Lambert-Beer law

(*c*) Grothus-Draper law (*d*) Stark-Einstein law

5. The number of molecules reacted or formed per photon of light absorbed is called

(*a*) yield of the reaction (*b*) quantum efficiency

(*c*) quantum yield (*d*) quantum productivity

6. "In a photochemical reaction each molecule of the reacting substance absorbs a single photon of radiation causing the reaction and is activated to form the products." This is the statement of

(*a*) Lambert-Beer's law (*b*) Grothus-Draper law

(*c*) Stark-Einstein law (*d*) Lambert's law

7. In some photochemical reactions low quantum yield is obtained. It is due to

(*a*) deactivation of reacting molecules

(*b*) occurrence of reverse primary reaction

(*c*) recombination of dissociated fragments

(*d*) all of these

8. A species which can both absorb and transfer radiant energy for activation of the reactant molecules is called

(*a*) radioactive substance (*b*) an ionisor

(*c*) a photochemical substance (*d*) a photosensitizer

9. One einstein is given by (N is Avogadro's number)

(*a*) $E = \dfrac{Nhc^2}{\lambda}$ (*b*) $E = \dfrac{Nhc}{\lambda^2}$

(*c*) $E = \dfrac{Nhc}{\lambda}$ (*d*) $E = \dfrac{Nh}{c\lambda}$

10. Photochemical decomposition of a substance is called

(*a*) thermal dissociation (*b*) thermolysis

(*c*) photolysis (*d*) none of the above

11. The light emitted in a chemiluminescent reaction is also called

(*a*) cold light (*b*) hot light

(*c*) bright light (*d*) none of these

12. "Only the fraction of incident light that is absorbed by the substance can bring about a chemical change". is

(*a*) First law of photochemistry (*b*) Second law of photochemistry

(*c*) Third law of photochemistry (*d*) none of these

ANSWERS

1. (*b*)	**2.** (*b*)	**3.** (*b*)	**4.** (*c*)	**5.** (*c*)	**6.** (*c*)
7. (*d*)	**8.** (*d*)	**9.** (*c*)	**10.** (*c*)	**11.** (*a*)	**12.** (*a*)

Short Answer Questions

1. State and explain Lambert's law.
2. Briefly explain the following terms:
   Transmittance, absorbance or optical density, molar extinction coefficient.
3. State and explain first and second law of photochemistry.
4. Give at least three points in which photochemical reactions differ from thermochemical reactions.
5. What is a chemical actinometer? Briefly explain the working of uranyl oxalate actinometer.
6. Define 'Photochemistry'. Give two examples of photophysical processes and two examples of photochemical reactions.
7. Give one example of a photochemical reaction in which the quantum yield is very high. Briefly explain the reasons for the same.
8. What mechanism has been proposed for the photolysis of ammonia?
9. Write short notes on 'fluorescence', and 'phosphorescene'. What is the difference between them?
10. Briefly explain fluorescence and phosphorescence using Jablonski diagram.
11. Write a short note on 'photosensitization'.
12. What are photo-inhibitors? How do they work?
13. What is 'resonance fluorescence'? Give one example.
14. What mechanism has been proposed to explain photolysis of acetone?
15. What is photosensitization? Explain with two suitable examples.
16. Why is it that in certain cases, quantum yield is lower than that expected from Einstein law of photochemical equivalence? Give two examples to support your point.
17. State and explain Beer's law.
18. Define Stark-Einstein's law of photochemical equivalence. What is one einstein of energy? How is it calculated in kJ mol^{-1}?
19. What is meant by primary and secondary process in photochemistry?
20. Give the photolysis of ammonia. What is the quantum yield of the process? Give reasons for your answer.
21. On the basis of mechanism, how can you justify that quantum yield of photolysis of HI is 2.
22. How do you explain that in fluorescence, the wavelength of the light emitted is usually greater than that of the light absorbed?
23. What is quenching in fluorescence? What do you mean by internal quenching and external quenching. Write the various steps from activation to deactivation in the process.
24. What do you understand by quantum yield of a photo-chemical reaction? Why some reactions have high quantum yield whereas some others have very low value? What is the modified definition of Stark-Einstein law?
25. How do you explain that photochemical yield of the reaction between hydrogen and chlorine is very high?

General Questions

1. What are photochemical reactions. Give at least five examples of photochemical reactions. List the main points of difference between a photochemical reaction and thermochemical reaction.
2. State and explain Lambert's law and Beer's law governing absorption of light by a sample. What are the limitations of Beer's law?
3. State and explain the first and second law of photochemistry (Grotthus-Draper law and Stark-Einstein's law of photochemical equivalence). What do you understand by 'one einstein' of energy? How is its value calculated is CGS units and in SI units?
4. What do you understand by 'quantum yield' or quantum efficiency' of a photochemical reaction? How is it determined experimentally?
5. What do you mean by quantum yield of a photochemical reaction? Explain why photosynthesis of HCl has very high quantum yield while that of photosynthesis of HBr is very low. What happens to the quantum yield of photosynthesis of HCl if the vessel contains small traces of oxygen? Explain with reasons.
6. What do you understand by 'luminescence'? Briefly explain the different types of luminescence.
7. What do you understand by the terms spin multiplicity, singlet states and triplet states. Explain the phenomenon of fluorescence and phosphorescence using Jablonski diagram.
8. Draw Jablonski diagram. Depict the non-radiative (radiationless) and radiative transitions, internal conversion and inter system crossing fluorescence phosphorescence. What should be the type of multiplicity for fluorescence and for phosphorescence? What should be the type of excited state for a chemical reaction to occur and why?
9. What do you understand by quenching of fluorescence? Derive Stern-Volmer equation.
10. Explain the term "photosensitization" with at least three examples along with their mechanism.
11. What do you understand by photochemical equilibrium? Give two examples. Show that for the reaction.

$$A \underset{\text{Thermal}}{\overset{\text{Light}}{\rightleftharpoons}} B$$

the concentration of the product at equilibrium is proportional to the intensity of the light absorbed.

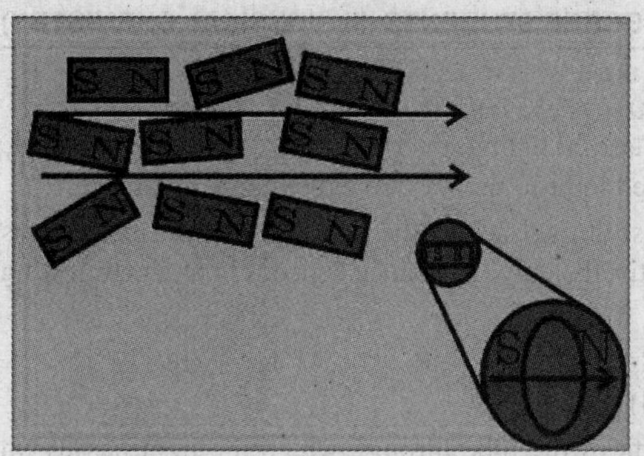

4

Physical Properties and Molecular Structure

4.1. INTRODUCTION

Various physical properties that support the determination of molecular structure are:

(*i*) Molar refraction

(*ii*) Parachor (Surface Tension)

(*iii*) Optical rotation

(*iv*) Dipole moment

(*v*) Magnetic susceptibility.

It is found that these properties are additive as well as constitutive. This fact makes the study of the above physical properties useful for the determination of structures of compounds.

An *additive property* is defined as the property, the aggregate value of which is the sum of values of its constituents.

A *constitutive property* is defined as the property which depends upon the constitution or structure of the molecule like arrangement of atoms within the molecule, multiple bonds and rings etc.

4.2. OPTICAL ACTIVITY

4.2.1. Explanation of certain terms

1. Plane polarized light. If a ray of light travels in any direction, it has vibrations in all directions at right angles to the path of propagation (Fig. 4.1). If this light is passed through Nicol prism, then the light which emerges out of the prism is found to have vibrations only in one plane. This light is called *'plane polarized light'*. The Nicol prism thus used is called a *polarizer*. In place of nicol prism, a polaroid consisting of a film of cellulose acetate mounted between two glass plates can also be used.

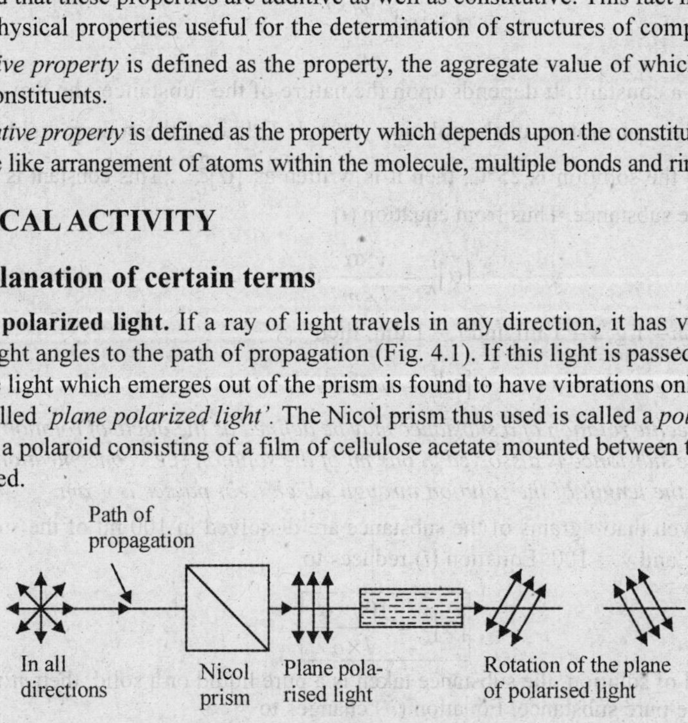

Fig. 4.1. Plane polarized light and its rotation.

2. Optical activity. If the plane polarized light is passed through a solution of sugar, it is found that the plane of the polarized light gets rotated when it comes out of the solution. *The substances like quartz, sugar etc., which rotate the plane of the polarized light are called optically active substances* *and the property of a substance to rotate the plane of the polarized light is called optical activity.* The substances which rotate the plane of the polarized light to the right are called dextero-rotatory and those which rotate the plane of the polarized light to the left are called laevo-rotatory.

3. Specific rotation. The angle (in degrees) through which the plane of the polarized light is rotated on passing through the *solution* of an optically active substance depends upon the following factors:

(*i*) Nature of the optically active substance.

(*ii*) Concentration of the solution (in g/ml).

(*iii*) Length of the solution *i.e.*, the length of the tube through which the light passes (in decimeters).

(*iv*) Wavelength of the light used.

(*v*) Temperature of the solution.

Thus if *m* grams of the substance are dissolved in *v* ml of the solution. (so that the concentration of the solution is *m/v* g/ml), *l* is the length of the tube (in decimeters) and α is the angle of rotation (in degrees), then it is found that

$$\alpha \propto (\text{concentration in g/ml}) \times (\text{length of the tube in dm})$$

i.e.,
$$\alpha \propto \frac{m}{v} \times l$$

or
$$\alpha = [\alpha]_\lambda^t \frac{m \times l}{v} \quad ...(i)$$

where $[\alpha]_\lambda^t$ is a constant. It depends upon the nature of the substance, the wavelength of light used (λ) and the temperature of the solution (*t*°*C*). If *D* line of the sodium light is used and the temperature of the solution is 25°C, then it is written as $[\alpha]_D^{25°}$. This constant is called **'specific rotation'** of the substance. Thus from equation (*i*)

$$[\alpha]_\lambda^t = \frac{v \times \alpha}{l \times m} \quad ...(ii)$$

Now, if *m*= 1 g, *v*= 1 ml and *l* = 1 dm, then

$$[\alpha]_\lambda^t = \alpha$$

Hence *specific rotation of a substance may be defined as the angle of rotation produced when one gram of the substance is dissolved in one ml of the solution (i.e., concentration of the solution is 1 g/ml) and the length of the solution through which light passes is 1 dm.*

If it is given that *c* grams of the substance are dissolved in 100 ml of the solution, then we may put *m* = *c* and *v* = 100. Equation (*i*) reduces to

$$\boxed{[\alpha]_\lambda^t = \frac{100 \times \alpha}{l \times c}} \quad ...(iii)$$

If instead of solution, the substance taken is a pure liquid or a solid, then *m/v* is replaced by density *d* of the pure substance. Equation (*ii*) changes to

$$\boxed{[\alpha]_\lambda^t = \frac{\alpha}{l \times d}}, \text{ for pure liquids or solids} \quad ...(iv)$$

4. Molar rotation. If we multiply the specific rotation with the molecular mass of the substance and divide the result by 100, we obtain molar rotation. It is represented by $[\mathrm{M}]_\lambda^t$

Thus $$[M]_{\lambda}^{t} = \frac{M\,[\alpha]_{\lambda}^{t}}{100}$$

4.3. MEASUREMENT OF OPTICAL ACTIVITY

The instrument used for the measurement of the amount of rotation caused by an optically active substance to the plane of the polarized light is called a *polarimeter.* It consists of two Nicol prisms, one of which is called the *polarizer* and the other is called the *analyser.* Both of these are fitted on the opposite ends of the same metal tube. However, the polarizer is fixed but the analyser can be rotated about the axis of the tube. In between the polarizer and the analyser, there is space for keeping the glass tube containing the solution. The glass tube is usually 10 or 25 cm in length and has glass windows on both ends. A circular scale is fixed on the metal tube near the analyser, as shown in Fig. 4.2.

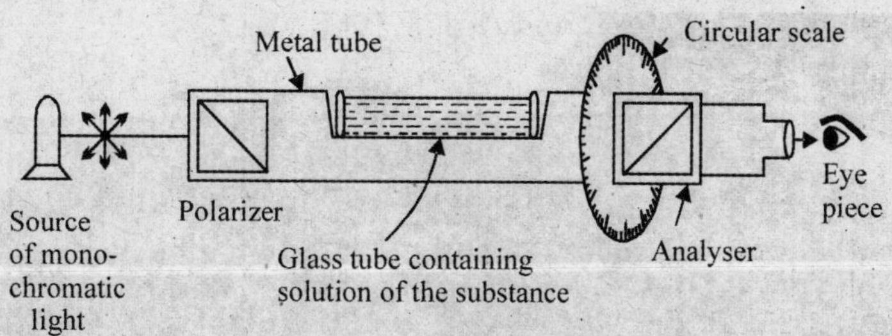

Fig. 4.2. Working of a polarimeter.

Procedure

First of all monochromatic light from a suitable source (usually sodium vapour lamp) is passed through the instrument, without the liquid or the solution. The light emerging out of the analyser is viewed through the eye-piece. The analyser is then rotated slowly till the field of view is completely dark. This gives the initial reading of the instrument. Next, the tube is fitted with the liquid or the solution and the shutter is closed. On seeing through the eye-piece, the field of view is no longer found to be dark. The analyser is rotated slowly till the field of view becomes completely dark again. This gives the final reading. The difference between the *final reading* and the *initial reading* gives the angle of rotation 'α'. If it comes out to be positive the substance is *dextro-rotatory* and if negative, it is *laevo-rotatory*.

4.4. CAUSE OF OPTICAL ACTIVITY IN A COMPOUND

As mentioned earlier optical activity is a *constitutive* property *i.e.*, it depends upon the arrangement of atoms within the molecule. Organic compounds which are found to be optically active contain at least one carbon atom to which four different atoms or groups are attached. Such a carbon atom is called an *asymmetric carbon* atom and is usually shown by putting an asterisk over it. A very simple example of such a compound is lactic acid, as represented below:

```
        CH3
        |
  H — *C — OH
        |
       COOH
```

Such a compound exists in two forms, one of which rotates the plane of the polarized light to the right and is called *dextero-rotatory* (*d*-form or '+' form) and the other which rotates the plane of the polarized light to the left and is called *laevo-rotatory* (*l*-form or '–' form). The two forms have the same constitution but differ in the arrangement of atoms or groups in space. Thus the two forms are called *optical isomers.*

Le Bel and van't Hoff explained the differences in the two forms by suggesting that in case of such molecules, the asymmetric carbon atom is present in the centre of a regular tetrahedron and the four atoms or groups attached to it are present at the corners of the tetra-hedron. This gives rise to two different spatial arrangement which are mirror images of each other, as shown in Fig. 4.3.

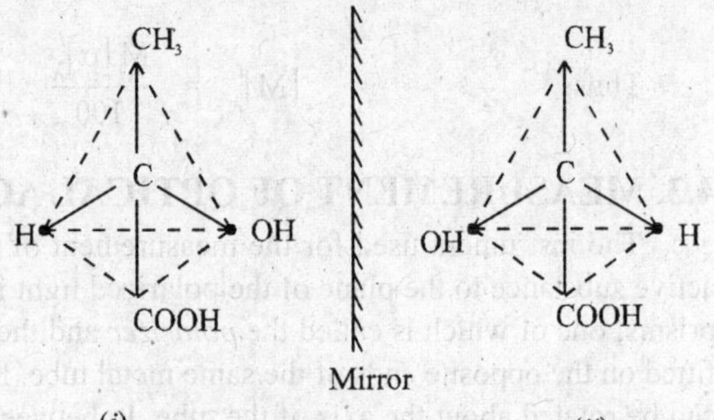

(*i*) (*ii*)

Fig. 4.3. Spatial difference in *d* and *l* forms of lactic acid.

Mirror images in nature

The two configuration are different in the sense that they are not superimposable over each other. The above two forms may be represented in a simplified way as follows:

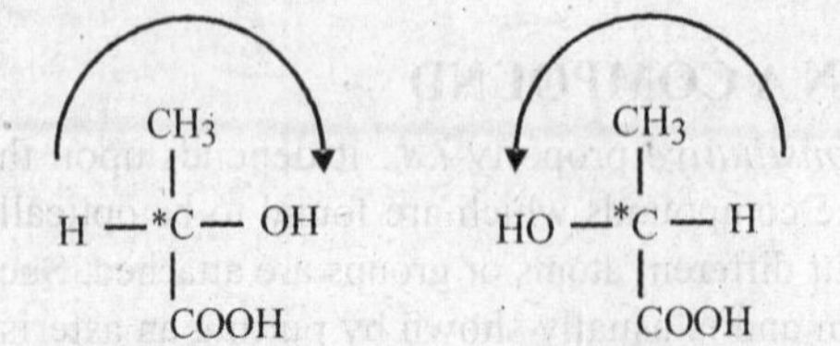

(I) (II)

In form I, the arrangement from H to OH is clockwise and in form II, the arrangement from H to OH is anticlockwise. Hence if one of them is dextero-rotatory, the other is laevo-rotatory. However, by merely looking at the configurations, it is not possible to predict as to which form is dextero or laevo. The two forms are also said to be *enantiomorphs.*

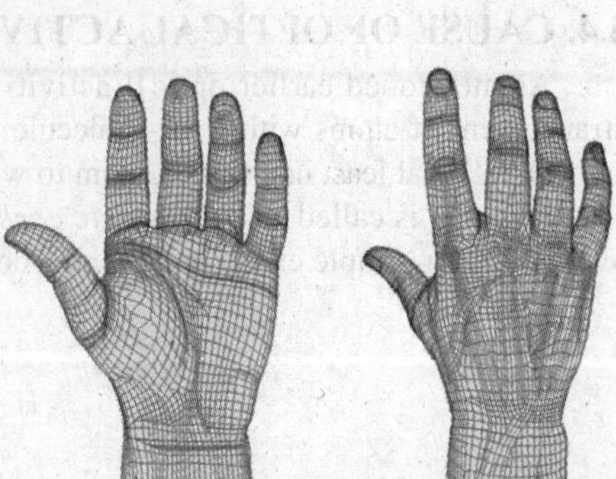

LEFT HAND **RIGHT HAND**

Left hand is the mirror image of right hand

External compensation and Internal compensation

External compensation: If equimolar quantities of *d*- and *l*-forms are mixed together, the mixture obtained is optically inactive. This is obviously due to the fact that the rotation caused by one form to the right is compensated by the equal rotation caused by the second form to the left. This process is called *external compensation* and the mixtures thus obtained are called *racemic mixtures* (or *dl*-form or ± form). This form can be separated into *d* and *l* forms by suitable methods.

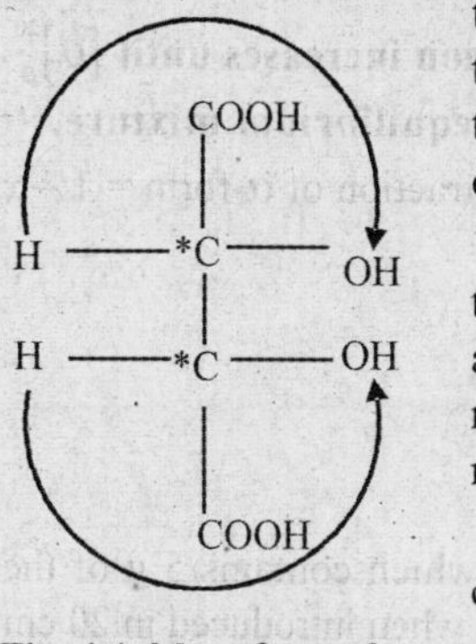

Fig. 4.4. Meso form of tartaric acid.

Internal Compensation: In case of compounds containing two asymmetric carbon atoms (*e.g.*, tartaric acid, Fig. 4.4), another form is also found to exist. In this form, the rotation caused by half of the molecule is compensated by the other half within the molecule. Thus the molecule is optically inactive. This is a case of *internal compensation.*

This form which is optically inactive due to *internal compensation* is called *meso form.*

It may be mentioned that certain compounds may not contain asymmetric atoms but still show optical activity. This is because such a molecule as a whole is asymmetric. In other words, the molecule cannot be divided into two equal halves with a plane. We can therefore conclude that *optical activity is a property related to the asymmetry of the molecule.*

SOLVED EXAMPLES ON OPTICAL ACTIVITY

Example 1. The value of $[\alpha]_D^{20}$ for lactose is 55.4°. What is the concentration in grams per litre of a solution of lactose which gives a rotation of 7.24° in a 10 cm cell at 20° with sodium *D* light?

Solution. Here, we are given

$$[\alpha]_D^{20} = 55.4^\circ$$
$$\alpha = 7.24^\circ$$
$$l = 10 \text{ cm} = 1 \text{ dm}$$

Putting the values in the formula

$$[\alpha]_\lambda^t = \frac{100 \times \alpha}{l \times c} \quad \text{or} \quad 55.4 = \frac{100 \times 7.24}{l \times c} = \frac{100 \times 7.24}{1 \times c}$$

or $$c = \frac{100 \times 7.24}{55.4 \times 1}$$

or $$c = 13.07 \text{ g /100 ml} = \mathbf{130.7 \text{ g/litre}}$$

Example 2. A solution of a certain optically active substance in water containing 1.56 g in 100 ml rotated polarised light by 4.91° in a polarimeter which had a cell 20 cm long. The *D* line of sodium was used as a light source. Calculate the specific rotation.

Solution. Here, we are given

$$c = 1.56 \text{ g/100 ml}$$
$$\alpha = 4.91^\circ$$
$$l = 20 \text{ cm} = 2 \text{ dm}$$

$$\therefore \quad [\alpha]_\lambda^t = \frac{100 \times \alpha}{l \times c} = \frac{100 \times 4.91}{2 \times 1.56} = \mathbf{157.37^\circ}$$

Example 3. When α-D glucose $\left([\alpha]_D^{20} = +112.2°\right)$ is dissolved in water the optical rotation decreases as β-D glucose is formed until at equilibrium $[\alpha]_D^{20} = +52.7°$. As expected, when β-D glucose $\left([\alpha]_D^{20} = +18.7°\right)$ is dissolved in water, the optical rotation increases until $[\alpha]_D^{20} = +52.7°$ is obtained. Calculate the percentage of β-form in the equilibrium mixture.

Solution. Suppose x is the fraction of D-glucose in β form. Then fraction of α-form = $1 - x$

$$\therefore \quad x(+18.7) + (1-x)(+112.2) = 52.7$$

or

$$x = 0.636 = 63.6\%\ \beta$$

$$1 - x = 0.364 = 36.4\%\ \alpha$$

PROBLEMS FOR PRACTICE

1. Calculate the specific rotation of a substance the solution of which contains 5 g of the substance dissolved in 25 ml of water and shows a rotation of 5° when introduced in 20 cm long polarimeter tube. **[Ans.** 12.5°]
2. A 13 per cent aqueous solution of maltose showed a rotation = 17° in 10 cm tube for *D*-line of sodium at 25°C. Another solution of maltose under identical conditions gave a rotation of 35°. Calculate the strength of this maltose solution in grams per litre. **[Ans.** 267.6 g/litre]
3. When α-*D*-mannose $\left([\alpha]_D^{20} = +29.3°\right)$ is dissolved in water, the optical rotation decreases as β-*D*-mannose is formed until at equilibrium $[\alpha]_D^{20} = +14.2°$. As expected, when β-*D*-mannose $\left([\alpha]_D^{20} = -17.0°\right)$ is dissolved in water the optical rotation increases until $[\alpha]_D^{20} = +14.2°$ is obtained. Calculate the percentage of α-form in the equilibrium mixture. **[Ans.** 67.4%]

4.5. DIPOLE MOMENT

Explanation of certain terms

1. Polar and non-polar covalent bonds. The pair of electrons shared by two identical atoms in a bond is situated exactly in the middle of the two atoms. Such a bond has no separation of charges and the bond formed between the two atoms is called **non-polar** covalent bond.

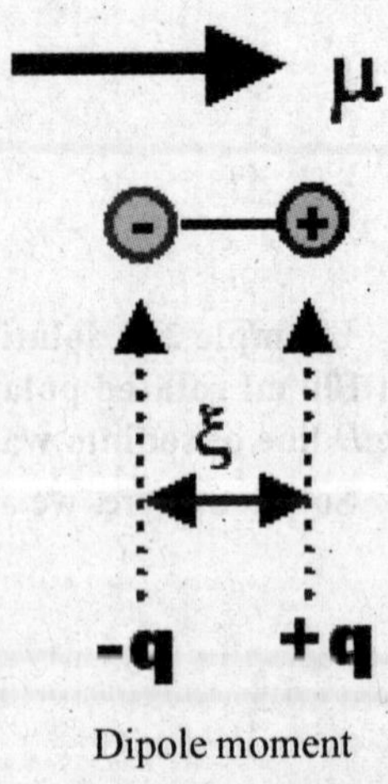

Dipole moment

However, if the atoms involved in a covalent bond are different having different electronegativity, the pair of electrons is not situated in the middle. It is tilted towards more electronegative atom. This causes separation of charges, one atom gaining partial negative charge and the other atom gaining partial positive charge. This is illustrated as under for the two types of covalent bonds.

H ẋ H	H ẋ Cl
↑ Exactly in the centre (non-polar covalent bond)	↑ Tilted towards chlorine (polar covalent bond)

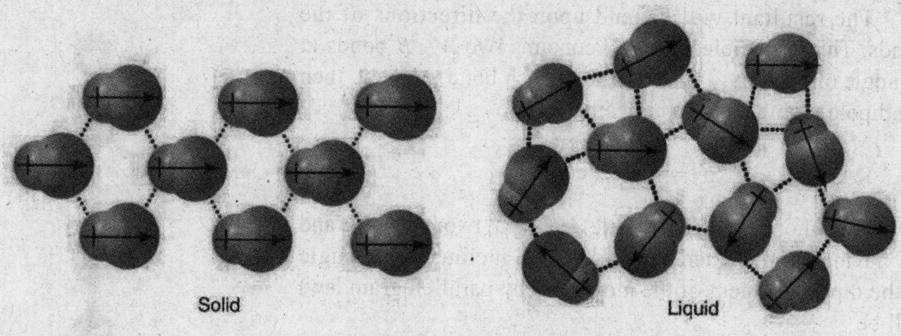

Dipole moment in solids and liquids

The molecule containing polar covalent bond is called a dipole (two poles) having a positive pole and a negative pole. In general a dipole AB can be represented as

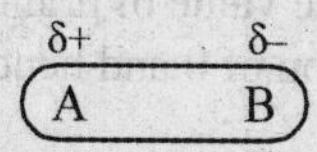

It may be mentioned that the charges on the poles are not unit charges (+ 1 and – 1) because complete transference of electrons form one atom to the other does not take place. Only a partial polarity is produced on the two poles

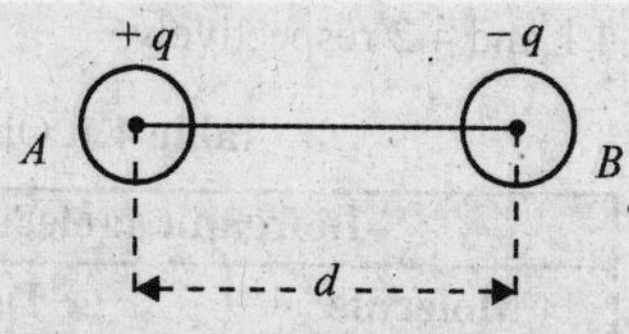

2. Dipole moment of a diatomic molecule. The extent of polarity of a molecule is expressed in terms of dipole moment which is defined as follows:

The dipole moment of a polar diatomic molecule is the product of the charge on each end of the molecule and the average distance between the centres of their nuclei i.e., the bond length.

In general, for a polar diatomic molecule AB in which atom B is more electronegative than A, if $+q$ is the charge on the end A and $-q$ on the end B of the molecule and d is the average distance between the centres of their nuclei, then the dipole moment of the molecule AB is given by

$$\mu = q \times d$$

Obviously if the molecule is non-polar, $q = 0$ and hence $\mu = 0$.

The charge present on the ends of the molecule is usually of the order of 10^{-10} esu and the distance between the atoms (*i.e.*, bond length) is usually of the order of 10^{-8} cm (*i.e.* Å units), therefore, μ must be of the order of $10^{-10} \times 10^{-8} = 10^{-18}$ esu cm. This value is called 1 Debye. Thus

$$1 \text{ Debye} = 10^{-18} \text{ esu cm.}$$

The dipole moment is a vector quantity *i.e.*, it has both magnitude as well as direction. It is, therefore, represented by an arrow pointing from the +ve end to the –ve end of the molecule *i.e.*,

$$\overset{+ \quad -}{\longrightarrow}$$

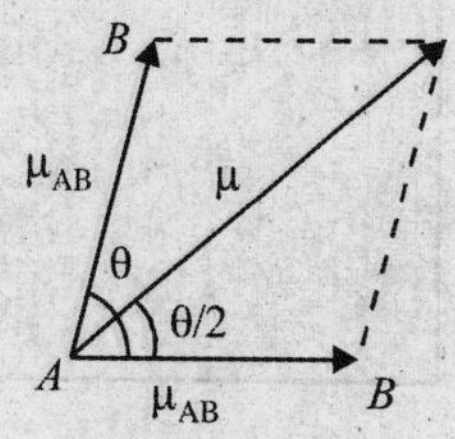

3. Dipole moment of polyatomic molecule. In a polyatomic molecule every two bonded atoms have their dipole moment. This is called the **bond moment.** The *net dipole moment of the polyatomic molecule is the resultant of the different bond moments.*

The resultant will depend upon the directions of the bonds. Thus if a molecule, AB_2 contains two $A - B$ bonds at an angle of θ and μ_{AB} is the value of each bond moment, then the dipole moment of the molecule will be

$$\mu = 2\mu_{AB} \cos \frac{\theta}{2}$$

For a more general molecule ABC with two bonds AB and AC having bond moments μ_1 and μ_2 and inclined at an angle θ, the dipole moment of the molecule (by parallelogram law) will be

$$\mu = \sqrt{\mu_1^2 + \mu_2^2 + 2\mu_1\mu_2 \cos\theta}$$

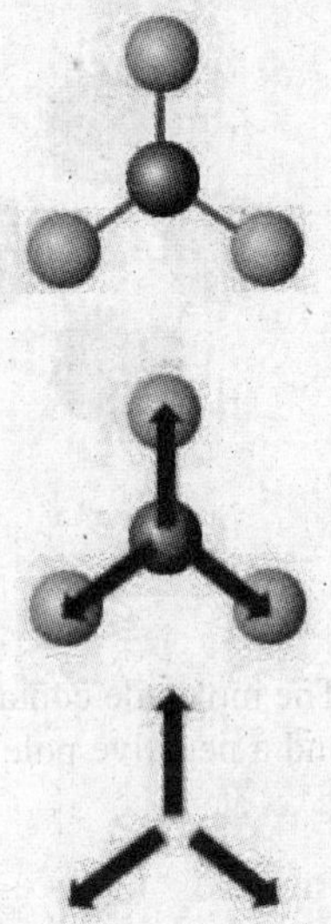

BF$_3$ dipole

As μ is known from experiments, therefore, knowing the value of θ, the method is usually used for the calculation of bond moments. Alternatively, knowing the value of μ and μ_{AB}, the method can be used for the calculation of θ and hence for predicting the shapes of molecules.

Observed values of dipole moments of certain substances and calculated moments of certain bonds are given in Tables 4.1 and 4.2 respectively.

Table 4.1. Observed values of dipole moments some substances

Inorganic molecules		**Organic molecule**	
Molecule	**Dipole moment (Debye)**	**Molecule**	**Dipole moment (Debye)**
H_2, Cl_2, Br_2, I_2, N_2	0	CH_4, C_2H_6, C_2H_4, C_2H_2	0
CO_2, CS_2, $SnCl_4$, SnI_4	0	CCl_4, CBr_4	0
HCl	1.03	CH_3Cl	1.86
HBr	0.78	CH_3Br	1.80
HI	0.38	C_2H_5Br	2.03
H_2O	1.84	CH_3OH	1.78
NH_3	1.46	C_2H_5OH	1.85
H_2S	1.10	CH_3NH_2	1.24
SO_2	1.63	CH_3COOH	1.74
N_2O	0.17	C_6H_5Cl	1.73
CO	0.12	C_6H_5Br	1.71
PH_3	0.55	C_6H_5I	1.20
PCl_3	0.78	$C_6H_5NO_2$	4.23
HCN	2.93	$C_6H_5NH_2$	1.56
		C_6H_5OH	1.70
		$C_6H_4Cl_2$	0
		m-$C_6H_4Cl_2$	1.72
		o-$C_6H_4Cl_2$	2.50

Table 4.2. Bond moments (in Debye units) of some common bonds
(Atoms on the right hand side are more electronegative than on the left hand side)

Bond	Bond moment	Bond	Bond moment	Bond	Bond moment
H—C	0.2	H—F	(2.0)*	C—O	0.9
H—N	1.5	H—Cl	1.03	C=O	2.5
H—O	1.6	H—Br	0.78	C≡O	5.3
H—F	(2.0)*	H—I	0.38	C—N	0.4
H—P	0.55	C—Cl	1.7	C—S	1.2
H—S	0.8	C—Br	1.6	C=S	3.2
H—Cl	1.03	C—I	1.4	N—O	1.9
		P—Cl	1.8		
		P—Br	1.6		

* Arbitrarily fixed value

The dipole moments of some groups like—NO_2, —NH_2, —CH_3 etc. have also been reported. These are called group moments. The value depends upon whether the group is present in an aliphatic compound or aromatic compound. The values of a few groups are given below:

Group (X)	—NO_2	—CN	—CH_3	—NH_2
Group moment (in Alk—X) :	−3.05	−3.46	0	+ 1.23
Group moment (in Ar—X) :	−3.93	−3.99	+ 0.45	+ 1.55

A negative value of group moment means that the group is net electron-withdrawing *i.e.* more electronegative and a positive value means that the group is net electron repelling or less electronegative than the alkyl or the aryl radical. Based on these values, we can perdict the course of reaction involving compounds with such groups and propose a mechanism of the reaction.

4.6. POLARIZATION OF MOLECULES (INDUCED POLARIZATION)

Every molecule is made up of positively charged nuclei and negatively charged electron cloud as shown in Fig. 4.5 (*a*). The centre of positive charge and centre of negative charge coincide. When such a molecule is introduced between the two plates of an electric field, the positively charged nuclei and hence the centre of positive charge is attracted towards the negative plate, whereas the negatively charged electron cloud is attracted towards the positive plate of the electric field. As a result, the molecule gets distorted, as shown in Fig. 4.5 (*b*). This is called *polarization* of the molecule and the distorted molecule with positive and negative ends is called an *electric dipole*. The molecule remains polarized so long as it is under the influence of electric field but goes back to the original state as soon as the electric field is switched off. That is why this type of polarization is called *induced polarization* and the electric dipole formed is called *induced dipole.*

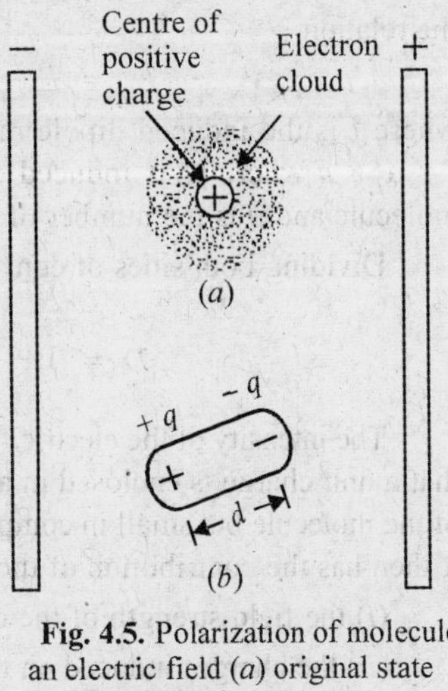

Fig. 4.5. Polarization of molecule an electric field (*a*) original state (*b*) polarized state.

Since even after polarization, the molecule is neutral as a whole, this means that the positive charge on one end must be equal to the negative charge on the other. Suppose this charge is *q*. If

d is the distance of separation between the charges, then dipole moment of the induced dipole μ_i will be given by

$$\mu_i = q \times d \qquad ...(i)$$

Evidently, the value of μ_i depend upon the nature of the molecule and the strength of the electric field, say X. Thus

$$\mu_i = \alpha \times X \qquad ...(ii)$$

where α is a constant called the *polarizability* of the given molecule.

α is found to be related to the dielectric constant, D, of the medium present between the plates *i.e.*, the molecules being studied, according to the equation given by Clausius and Mossotti. (derivation given in the next section)

$$\left(\frac{D-1}{D+2}\right)\frac{M}{\rho} = \frac{4}{3}\pi N\alpha \qquad ...(iii)$$

where M is molecular mass of the molecules

ρ is the density of the molecules

N is Avogadro's number.

4.7. CLAUSIUS-MOSOTTI EQUATION

Suppose an electrostatic field is applied to two plates so that they are charged. Suppose the strength of the uniform electric field produced between them = E_0.

Now if a non-polar substance is placed between the plates, the strength of the electric field will be reduced to E because the induced dipole (called *dielectrics*) acts against the applied field and hence partially neutralizes the charges of the plates as shown in Fig. 4.6. The ratio E_0/E is called the *dielectric constant* D of the medium. From electrostatics, have the relation

$$E_0 = E + 4\pi I \qquad ... (1)$$

where I is the induced dipole moment per unit volume *i.e.* $I = \mu_i \times n$ where μ_i is induced dipole moments in a single molecule and n is the number of molecules per cc.

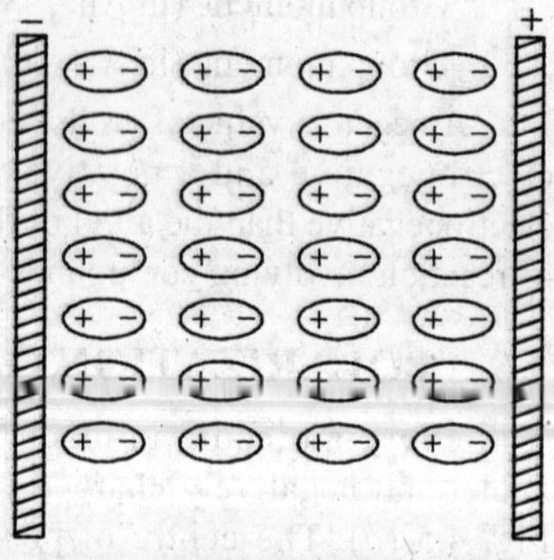

Fig. 4.6. Partial neutralisation of the charges on the plates by the induced electric dipole (dielectrics)

Dividing both sides of eqn. (1) by E and putting $E_0/E = D$, we get

$$D = 1 + \frac{4\pi I}{E} \quad \text{or} \quad (D-1)E = 4\pi I \qquad ... (2)$$

The intensity of the electric field strength X acting on each molecule is calculated by supposing that a unit charge is enclosed in a small spherical cavity which is large as compared with the size of the molecule but small in comparison with the distance between the plates. The electric intensity X then has the contribution of the following:

(*i*) the field strength of the charges on the plates = E_0.

(*ii*) the charges induced on the surfaces of the dielectrics in contact with the plates = $-4\pi I$.

(*iii*) the charge induced on the surface of the spherical cavity = $+\frac{4}{3}\pi I$

$$\therefore \qquad X = E_0 + \frac{4}{3}\pi I - 4\pi I \qquad ...(3)$$

Substituting the value of E_0 from eqn. (1), we get

$$X = (E + 4\pi I) + \frac{4}{3}\pi I - 4\pi I = E + \frac{4}{3}\pi I \qquad ...(4)$$

Substituting the value of $4\pi I$ from eqn. (2), it becomes

$$X = E + \left(\frac{D-1}{3}\right)E = \left(1 + \frac{D-1}{3}\right)E = \left(\frac{D+2}{3}\right)E \qquad ...(5)$$

From eqns. (4) and (5), we have

$$E + \frac{4}{3}\pi I = \left(\frac{D+2}{3}\right)E$$

or

$$\left(\frac{D+2}{3} - 1\right)E = \frac{4}{3}\pi I$$

or

$$\left(\frac{D-1}{3}\right)E = \frac{4}{3}\pi I \qquad ...(6)$$

Putting

$$I = \mu_i \times n$$

$$= \alpha X \times n \qquad (\because \mu_i = \alpha X)$$

$$= \alpha\left(\frac{D+2}{3}\right)E \times n \qquad \because X = \left(\frac{D+2}{3}\right)E \text{ from eqn. (5)}$$

Eqn. (6) becomes

$$\left(\frac{D-1}{3}\right)E = \frac{4}{3}\pi\alpha\left(\frac{D+2}{3}\right)E \times n$$

or

$$\left(\frac{D-1}{D+2}\right) = \frac{4}{3}\pi n \alpha \qquad ...(7)$$

If ρ is the density of the substance placed between the charged plates and M is its molecular mass, then molar volume of the substance = $\frac{M}{\rho}$. If N represents Avogadro's number, number of molecules per unit volume will be

$$n = \frac{N}{M/\rho} = \frac{N\rho}{M} \qquad ...(8)$$

Substituting this value in eqn. (7), we get

$$\left(\frac{D-1}{D+2}\right) = \frac{4}{3}\pi\frac{N\rho}{M}\alpha$$

or

$$\left(\frac{D-1}{D+2}\right)\frac{M}{\rho} = \frac{4}{3}\pi N \alpha \qquad ...(9)$$

All quantities on the right hand side of eqn. (9) are constant (α being constant for a substance) independent of temperature, therefore, the quantity on the left hand side of eqn. (9) must also be constant, independent of temperature and depending only upon the nature of the molecules. This quantity is called *induced molar polarization* and is represented by P_i. Thus

$$P_i = \left(\frac{D-1}{D+2}\right)\frac{M}{\rho} \quad \text{... (10)}$$

Induced molar polarization is defined as the amount of polarization produced in 1 mole of the substance when placed between the plates of an electric field of unit strength.

Expression (10) gives the relationship between dielectric constant and polarizability of a substance and is called Mosotti-Clausius equation.

4.8. STUDY OF CERTAIN SUBSTANCES

Using eqn. (10), the values of molar polarization have been determined for a number of substances. It is found that whereas molar polarization values of a number of substances such as O_2, CO_2, N_2 and CH_4 are constant and independent of temperature, the polarization values for some other substances such as HCl, $CHCl_3$, $C_6H_5NO_2$ and CH_3Cl do not come out to be constant, but decrease with increase of temperature.

In case of molecules such as HCl, $CHCl_3$ etc., their centres of positive and negative charges do not coincide with each other. Thus they possess some dipole moment even in the absence of electric field. The dipole moment thus possessed by the molecules is called *permanent dipole moment.* In the absence of electric field these permanent dipoles are oriented in a random manner

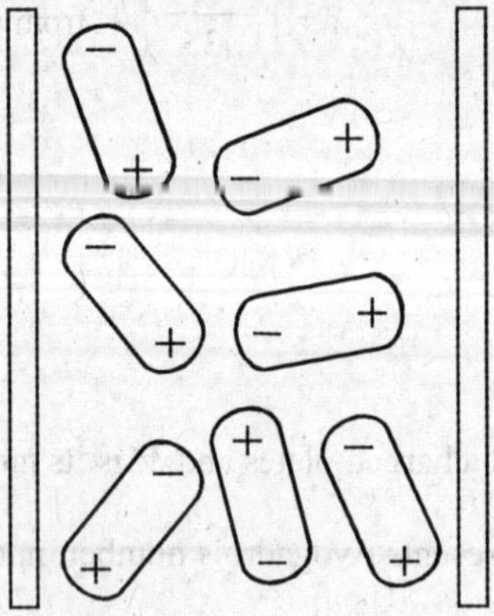

Fig. 4.7. Permanent dipoles oriented in a random manner in the absence of an electric field.

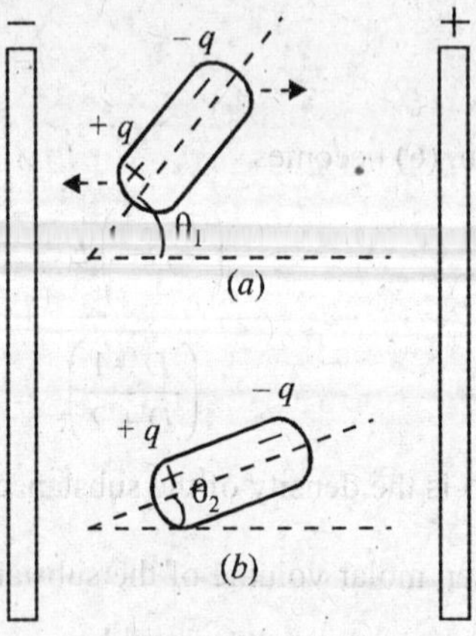

Fig. 4.8. (*a*) Original state of a permanent dipole: (*b*) Final state after polarization and orientation.

as shown in Fig. 4.7. However in the presence of an electric field (Fig. 4.8), two disturbing effects take place. These are:

(*i*) The electric field will tend to rotate and orient these dipoles in the direction of the field.

(*ii*) The electric field will tend to polarize the molecules.

If the molecules were stationary, the dipoles would have been oriented perfectly parallel to the electric field *i.e.*, at right angles to the plates of the electric field. However, due to thermal agitation, they take up position in between that of the original state and the perfectly parallel state, as shown at (*b*) in Fig. 4.8 for any one permanent dipole.

Thus in case of molecules possessing permanent dipoles, the right-hand side of eqn. (10) will give the total polarization (P_t) which will be the sum of the induced molar polarization (P_t) and the *orientation molar polarization* (P_o)

i.e. $$P_t = \left(\frac{D-1}{D+2}\right)\frac{M}{\rho} = P_i + P_o \quad ...(1)$$

And, the induced polarization is given by [Refer eq. (*iii*) Section 4.6]

$$P_i = \frac{4}{3}\pi N\alpha \quad ...(2)$$

The orientation molar polarization is given by the equation put forward by Debye, *viz.*,

$$P_o = \frac{4}{3}\pi N\left(\frac{\mu^2}{3kT}\right) \quad ...(3)$$

where μ is the *permanent dipole moment* of the molecule, *k* is Boltzmann's constant (=*R*/*N*) and *T* is the temperature.

Substituting the values of P_i and P_o from equations (2) and (3) in eqn. (1), we get

$$P_t = \left(\frac{D-1}{D+2}\right)\frac{M}{\rho} = \frac{4}{3}\pi N\alpha + \frac{4}{3}\pi N\left(\frac{\mu^2}{3kT}\right) \quad ...(4)$$

4.9. MEASUREMENT OF DIPOLE MOMENT

(1) Vapour–Temperature Method

The equation (4) can be written in the form

$$P_i = A + \frac{B}{T} \quad ...(5)$$

where $A = \frac{4}{3}\pi N\alpha$ and $B = \frac{4\pi N\mu^2}{9k}$ are constant for the given substance.

Thus, if P_t is plotted against $1/T$, a straight line graph will be obtained, if the molecules possessed permanent dipoles. The slope of the plot will be given by

$$B = \frac{4\pi N\mu^2}{9k} \quad ...(6)$$

From this eqn. the permanent dipole moment of the molecule can be calculated, as all other quantities are constants whose values are known.

Solving eqn. (6) by putting the values of the various constants, we get

$$\mu = 0.0128\sqrt{B} \times 10^{-18}$$

Fig. 4.9 gives the plots for a few polar and non polar substances. We observe that P_t varies with temperature for substances with permanent dipoles (for example, HCl, CH_3Cl, etc.). But there is no

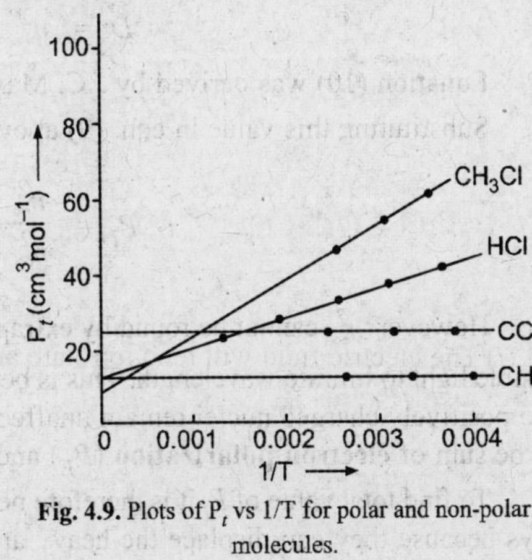

Fig. 4.9. Plots of P_t vs 1/T for polar and non-polar molecules.

effect of temperature on non polar substances such as CH_4, CCl_4, etc. not having permanent dipoles. These substances possess only induced molar polarization. The basic equation for the determination of dipole moment is

$$P_t = \left(\frac{D-1}{D+2}\right)\frac{M}{\rho} = A + \frac{B}{T} \qquad \text{... (7)}$$

where A and B are constants with

$$B = \frac{4\pi N \mu^2}{9k}$$

Thus the experiment involves the determination of the dielectric constant D and the density ρ at different temperatures T. The values of P_t are then calculated and plotted against $1/T$. If P_t is found to vary linearly with $1/T$, the substance possesses a permanent dipole moment and the value of the dipole moment (μ) is calculated from the slope of the line. On the other hand, if P_t is found to be independent of temperature, the substance does not possess any permanent dipole moment (*i.e.*, it is non-polar).

The dielectric constant required in the above equation is determined by first measuring the capacity of a condenser in vacuum between the plates (C_o) and then measuring the capacity of the same condenser when filled with the given substance (C). The dielectric constant of the given substance is then given by the relation, as mentioned earlier also.

$$D = \frac{C}{C_o} \qquad \text{... (8)}$$

(2) Refraction Method. According to Clausius-Mosotti equation, the induced molar polarization (P_i) of the molecules is given by

$$P_i = \frac{D-1}{D+2}\cdot\frac{M}{\rho} = \frac{4}{3}\pi N \alpha \qquad \text{... (9)}$$

As D is a pure number, molar polarisation P is simply molar volume of the material. Further, it has already been proved that it is independent of temperature if the substance is non-polar and does not have a permanent dipole moment. For such molecules, the dielectric constant of a substance is related to its refractive index for light of long wavelength according to the equation

$$D = n_\infty^2 \qquad \text{... (10)}$$

Equation (10) was derived by J.C. Maxwell,

Substituting this value in eqn. (9) above, we get

$$P_i = \frac{n_\infty^2 - 1}{n_\infty^2 + 2}\,\frac{M}{\rho} = \frac{4}{3}\pi N \alpha \qquad \text{... (11)}$$

However n_∞^2 cannot be found by extrapolating the values of refractive indices measured with visible light to infinite wavelength. This is because visible light can displace only electrons whereas the positively charged nuclei remain unaffected. Thus the induced polarization can be considered to be sum of **electron polarization** (P_E) and **atom polarization** (P_A) *i.e.* $P_i = P_E + P_A$.

To find total value of P_i it is therefore necessary to measure the referactive index using infrared rays because they can displace the heavy atomic nuclei as well as electrons but measurement of refractive index with infrared radiations is difficult. Thus problem is solved by finding electron polarization by measuring the refractive index (n) using visible light (D-line) *i.e.*

$$P_E = \frac{n^2 - 1}{n^2 + 2}\frac{M}{\rho} \qquad \text{... (12)}$$

5% of this value is added to it to accound for P_A to get the total value of P_i. This serves as satisfactory aproximation.

Total polarization (P_t) of a molecule possessing a permanent dipole moment is the sum of two terms, the first called *induced polarization* (P_i) and the second called the *orientation polarization* (P_o) due to permanent moment possessed by the molecule. Hence we can write

$$P_t = P_i + P_o \qquad \text{... (13)}$$

Putting the value of P_o from Debye equation viz.

$$P_o = \frac{4}{3}\pi N\left(\frac{\mu^2}{3\,kT}\right) \qquad \text{... (14)}$$

Eqn. (13) becomes

$$P_t = P_i + \frac{4}{3}\pi N\left(\frac{\mu^2}{3\,kT}\right) \qquad \text{... (15)}$$

Putting the values of the constants and simplifying, we get

$$\mu = 0.0128\sqrt{(P_t - P_i)\,T} \qquad \text{... (16)}$$

P_t can be determined by finding the dielectric constant of the vapour at temperature T and P_i can be found either by measuring the refractive index using infrared radiation or by approximation method explained above.

(3) Dilute Solution Method. The equation $P_t = \left(\frac{D-1}{D+2}\right)\frac{M}{\rho}$ is applicable to gases and vapours only. For substances which are not gaseous the following procedure is followed:

(*i*) A *dilute* solution of the given substance is prepared in a *non-polar* solvent such as benzene, CCl_4, CS_2 etc. The density and dielectric constant of the solution are determined at different temperatures.

(*ii*) For each temperature, the molar polarization of the solution P_s is calculated using the relation

$$P_s = \left(\frac{D-1}{D+2}\right)\left(\frac{x_1M_1 + x_2M_2}{\rho}\right) \qquad \text{... (17)}$$

where x_1 and x_2 are the mole fractions of the solvent and the solute in the solution, M_1 and M_2 are their respective molecular masses and ρ is the density of the solution.

(*iii*) The molar polarization of the solution is related to the molar polarization of the solvent (x_1) and that of the solute (x_2) as (it is an additive property)

$$P_s = x_1P_1 + x_2P_2 \qquad \text{... (18)}$$

Thus knowing the values of x_1, x_2 and P_1, the value of P_2 *i.e.* molar polarization of the substance can be calculated. The values of P_2 obtained for different temperatures are plotted against $1/T$. The value of μ is obtained from the slope of the line.

4.10. DIPOLE MOMENT AND STRUCTURE OF MOLECULES

Dipole moment studies help in elucidating the structure of compounds as follows:

1. In comparing the relative polarities of molecules: The dipole moments of HF, HCl, HBr, HI are 2.0, 1.03, 0.78 and 0.38 Debye respectively. This shows that the polar character of the molecules decreases as we move from HF to HI.

2. In calculating the percentage ionic character of a polar covalent bond: For example, the dipole moment of HCl is 1.03 D. If the bond were 100% ionic, each end would acquire full one unit charge *i.e.* 4.8×10^{-10} esu. The bond distance in HCl is taken to be 1.275 Å *i.e.*, 1.275×10^{-8} cm. Hence if HCl were 100% ionic, the dipole moment would have been

$$= (48 \times 10^{-10}) \times (1.275 \times 10^{-8}) \text{ esu cm.}$$

$$= 6.11 \times 10^{-18} \text{ esu cm} = 6.11 \text{ D.}$$

Actual value of dipole moment of HCl = 1.03 D.

$$\therefore \text{ percentage ionic character H–Cl bond} = \frac{1.03}{6.11} \times 100 = 17\% \text{ approx.}$$

3. In predicting the shapes of molecules: The fact that the dipole moment is a vector quantity and that a particular bond has almost a constant value of bond moment is very useful in predicting the shapes of molecules. A few cases are discussed below:

(*a*) **Triatomic molecules:** In triatomic molecules of the type AB_2, there are two possibilities.

(*i*) *The dipole moment of the molecule may be zero as* in case of CO_2 and CS_2. This shows that in these molecules, one bond moment is equal and opposite to the other. Hence the molecules must be linear, as shown below:

$$\delta^- \leftarrow \delta^+ \rightarrow \delta^- \qquad\qquad \delta^- \leftarrow \delta^+ \rightarrow \delta^-$$
$$O = C = O \qquad\qquad S = C = S$$
$$\mu = 0 \qquad\qquad \mu = 0$$

(*ii*) The molecule may have a definite value of dipole moment (*e.g.*, μ for H_2O = 1.84 D, μ for SO_2= 1.63 D). Obviously, in these cases, the possibility of linear shape is ruled out. The two bonds must have a definite angle between them which can be calculated, if we know the bond moments. For example, the bond moment for H—O is 1.6 D, the bond angle is calculated to be nearly 105°. Thus H_2O may be represented as shown in the figure. Such molecules are called *bent molecules* and the shape is called *V-shape.*

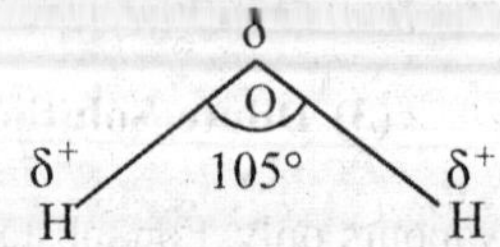

(*b*) **Tetratomic molecules:** In tetratomic molecules of the type AB_3, the two simple possibilities are:

(*i*) *The dipole moment of the molecule may be zero as in* BF_3. In such a case, all the four atoms must be in one plane and the angles between the bonds must be equal. Such a shape is called **planar trigonal.**

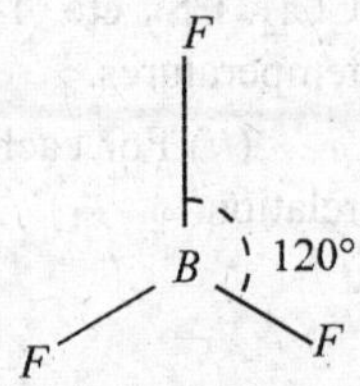

(*ii*) *The molecule may have some definite value of dipole.* (*e.g.*, NH_3, PF_3, PCl_3). This indicates that one atom lies out of plane of rest of the atoms. For example, the molecule of NH_3 may be represented as shown in the figure.

Such a shape is called *pyramidal shape.*

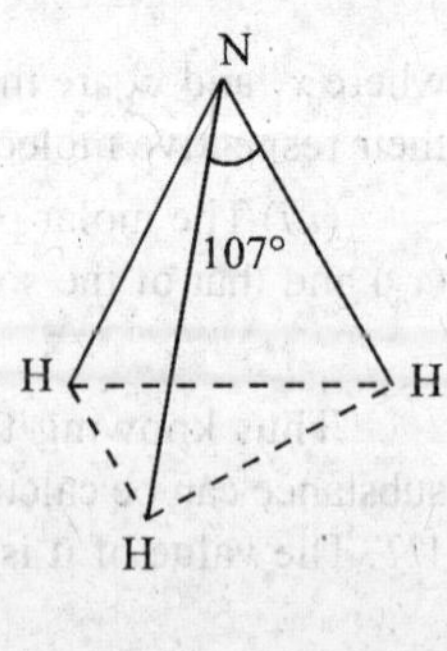

(*c*) **Pentaatomic molecules:** The pentaatomic molecules of the type AB_4 which we commonly come across are CH_4, CCl_4, $SiCl_4$, $SnCl_4$ etc. The dipole moments of all these molecules is found to be zero. This is possible only if these molecules have symmetrical or regular geometry so that the different bond moments cancel out one another. The experimental facts confirm that these molecule have **tetrahedral** shape as represented below:

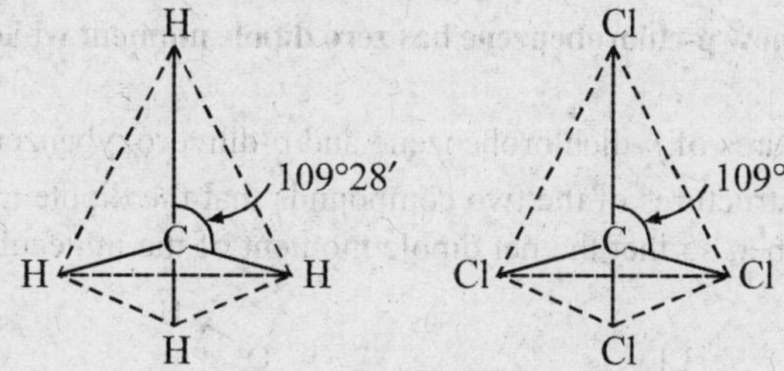

4. In differentiating between the *cis*- and *trans*-isomers: The *cis*- and *trans*- isomers are represented as:

cis- *trans*-

The bond moments will cancel each other in *trans* isomer. Hence *trans* isomer will show zero value for dipole moment. On the other hand, the *cis* isomer will show some definite value of dipole moment.

For example, in case of dichloroethylene (putting X = Cl), *cis*-form has been found to have a dipole moment of 1.9 D, while the *trans*- form is found to have zero dipole moment.

5. In differentiating between *o, m* and *p* isomers. As benzene is a regular hexagon its dipole moment is zero, therefore, on substituting two atoms or groups in the ortho, meta or para positions, the disubstituted derivative will have different dipole moment for the *o, m* and *p* positions, as shown below:

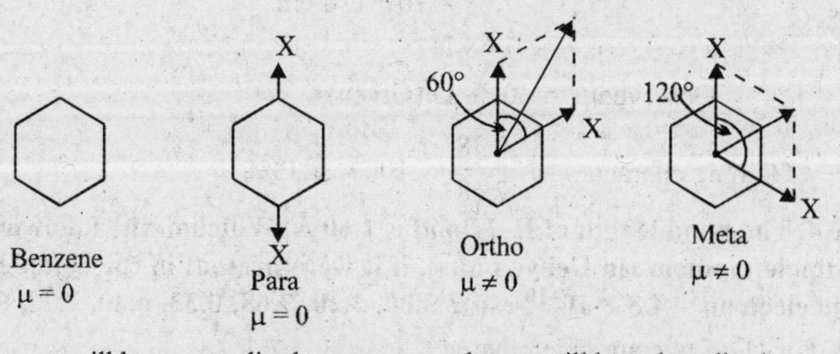

Obviously, para will have zero dipole moment, and meta will have less dipole moment than ortho.

SOME SOLVED EXAMPLES

Example 1. How can you establish the planar nature of benzene using dipole moment measurement?

Solution. Dipole moment of benzene molecule is zero *i.e.*, $\mu = 0$

This is possible only if benzene has planar structure as represented in the figure. In this structure, dipole moments of six C—H bonds cancel one another and the net dipole moment of the benzene molecule is zero.

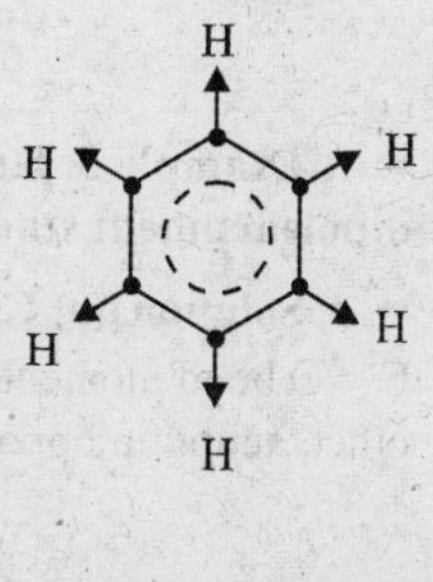

Had the structure not been planar, molecule would have shown a net dipole moment which is not the case.

Example 2. Explain how p-chlorobenzene has zero dipole moment while p-dihydroxy benzene has a definite value.

Solution. The structures of *p*-dichlorobenzene and *p*-dihydroxybenzene are given below:

It is clear from the structures of the two compounds that the dipole moments of two C—Cl bonds will cancel each other, so that the net dipole moment of the molecule will be zero.

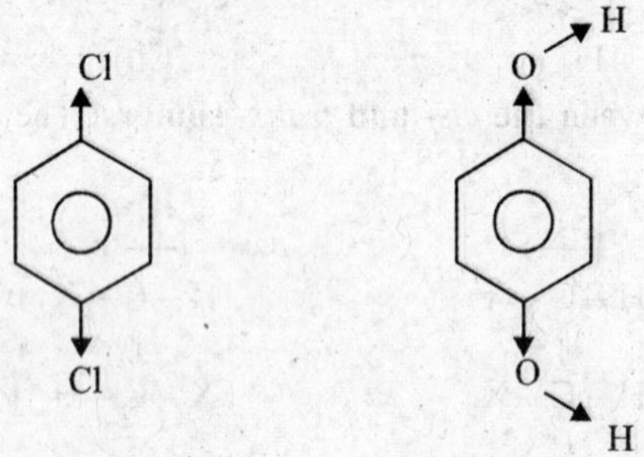

p-dichlorobenzene　　*p*-dihydroxybenzene

This cancellation of dipole moments of two C—O—H groups in *p*-dihydroxybenzene is not possible. Hence it shows a definite value of dipole moment.

Example 3. The bond length of H-I bond is 1.60 Å and its dipole moment is 0.38 D. Calculate the percentage ionic character of H–I bond.

Solution. Assuming H–I to be 100% ionic bond, each end would acquire full one unit charge *i.e.*, $q = 4.8 \times 10^{-10}$ esu

$$\mu = q \times d$$
$$= (4.8 \times 10^{-10} \text{ esu})\ (1.60 \times 10^{-8} \text{ cm})$$
$$= 7.68 \times 10^{-18} \text{ esu cm}$$
$$= 7.68 \text{ D}$$
$$\text{Observed } \mu = 0.38 \text{ D (Given)}$$
$$\therefore \quad \% \text{ ionic character} = \frac{0.38}{7.68} \times 100 = \mathbf{4.95\%}$$

Example 4. The bond length of H–I bond is 1.60 Å. Which of the following is the correct value for its dipole moment (in Debye units), if it were present in the completely ionic form (charge on the electron = 4.8 × 10⁻¹⁰ esu)? 3.00, 3.20, 7.68, 0.33, 6.40, 2.77, 9.28.

Solution. For H—I, we are given that

$$d = 1.60 \text{ Å} = 1.60 \times 10^{-8} \text{ cm}$$
$$q = 4.8 \times 10^{-10} \text{ esu}$$
$$\therefore \quad \mu = q \times d = (4.8 \times 10^{-10}) \times (1.60 \times 10^{-8})$$
$$= 7.68 \times 10^{-18} \text{ esu cm} = \mathbf{7.68\ D.}$$

Example 5. How will you arrive at the structure of (*a*) Carbon dioxide (*b*) Water from dipole moment studies ?

Solution. (*a*) Carbon dioxide. The dipole moment of carbon dioxide (μ_{CO_2}) is zero, although C = O bond moment ($\mu_{C=O}$) is 2.5 D. This implies that the two C = O bond moments nullify each other, acting in opposite directions.

$$:\ddot{O}=C=\ddot{O}:$$

Using the parallelogram law of bond moments, the bond angle θ for a triatomic molecule is calculated as follows :

$$\mu^2 = \mu_1^2 + \mu_2^2 + 2\mu_1\mu_2 \cos\theta$$

or
$$\cos\theta = \frac{\mu^2 - (\mu_1^2 + \mu_2^2)}{2\mu_1\mu_2}$$

where, μ_1, μ_2 are the bond moments of bonds and μ is the dipole moment of the molecule. For carbon dioxide the above equation becomes

$$\cos\theta = \frac{\mu_{CO_2}^2 - 2\mu_{CO}^2}{2\mu_{CO}^2} \qquad (\because \mu_1 = \mu_2)$$

$$\cos\theta = \frac{0 - 2\mu_{CO}^2}{2\mu_{CO}^2} = -1$$

$$\theta = \cos^{-1}(-1) = 180°$$

That is, carbon dioxide molecule is linear.

(*b*) Water. The bond moment, μ_{O-H} and the dipole moment μ_{H_2O} are found to be 1.505 D and 1.84 D respectively.

Again, using parallelogram law of bond moments, we can calculate the bond angle θ.

$$\cos\theta = \frac{\mu^2_{H_2O}}{2\mu^2_{O-H}} - 1$$

$$\cos\theta = \frac{(1.84)^2}{2(1.505)^2} - 1 = -0.2520$$

$$\theta = \cos^{-1}(-0.2520)$$

$$= 104°36'$$

Thus, water molecule has *bent* (*or V*) shape.

4.11. MAGNETIC SUSCEPTIBILITY

4.11.1. Various terms involved

Magnetic permeability. If two magnetic poles of strength m_1 and m_2 are separated by a distance *r*, then the force *F* acting between them is given by

$$F = \frac{1}{\mu}\frac{m_1 \times m_2}{r^2}$$

The quantity μ in the above expression is called the magnetic permeability of the medium between the poles. *It is defined as the relative tendency of the magnetic lines of force to pass through the medium as compared to that in the vacuum.* For vacuum, μ is taken as equal to 1. Similarly, for air, μ is taken as approximately equal to 1.

Diamagnetic, paramagnetic and ferromagnetic substances. For media other than vacuum or air, μ can be less than or greater than 1. *If μ is less than 1, then the magnetic lines of force prefer* *to pass through the vacuum rather than through the substance comprising the medium.* In other words, *the magnetic lines of force are repelled away from the substance. Such substances are said to be* **diamagnetic**. On the other hand, *if* μ *is greater than 1, then the magnetic lines of force have greater tendency to pass through the substance rather than through the vacuum.* In other words, *the magnetic lines of force are attracted towards the substance. Such substances are called* **paramagnetic**. If μ *is very high* (*i.e.* of the order of 10^3 or more), *the tendency of the substance to attract the magnetic lines of force is very high. Such substances are said to be* **ferromagnetic**. Ferromagnetism is exhibited mainly by iron, cobalt, nickel and their alloys. The behaviour of diamagnetic and paramagnetic substances in a magnetic field is shown diagrammatically in Fig. 4.10.

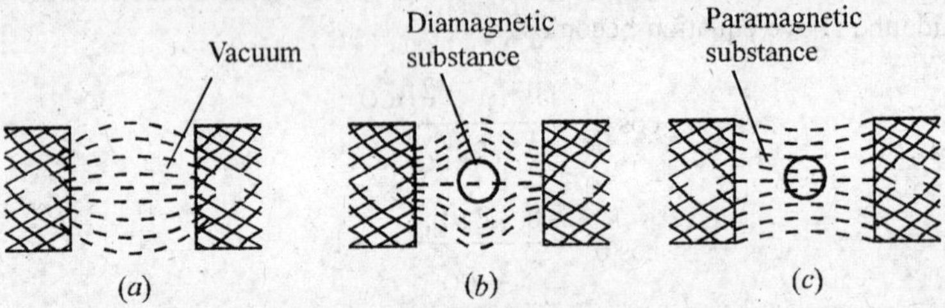

Fig. 4.10. (*a*) Magnetic lines of force passing through vacuum. (*b*) Magnetic lines of force pushed away by the diamagnetic substance. (*c*) Magnetic lines of force drawn in by the paramagnetic.

Magnetic susceptibility. If a material is placed in a magnetic field, the strength of the magnetic field in the material is different from that in the vacuum. The strength of the magnetic field thus present in the material is called **magnetic induction.** Mathematically.

If H is the strength of the applied magnetic field and B is the magnetic induction in the given material, then

$$B = H + 4\pi I \quad \text{... }(i)$$

where I represents the *magnetic moment per volume* in the material and is called **intensity of magnetisation.** It may be positive or negative depending upon whether the material is paramagnetic or diamagnetic.

The intensity of magnetisation produced per unit strength of the applied magnetic field is called **specific magnetic susceptibility**. It is usually represented by χ. Thus mathematically,

$$\chi = \frac{I}{H} \quad \text{...}(ii)$$

It is this quantity which is usually measured experimentally in the study of magnetic properties of different materials.

Multiplying the specific magnetic susceptibility with the molar volume (M/ρ) of the material, we get *magnetic susceptibility per mole* or **molar magnetic susceptibility.** It is usually represented by χ_M. Mathematically,

$$\chi_M = \frac{M}{\rho} \cdot \frac{1}{H} \quad \text{...}(iii)$$

Thus **molar magnetic susceptibility** *may be defined as the intensity of magnetization induced per mole of the material by the unit field strength.*

Quite often it is convenient to deal with the quantity χ_M in place of χ.

Further, the magnetic induction B is related to the strength of the applied magnetic field H according to the equation

$$B = \mu H \quad \text{...}(iv)$$

where μ is the magnetic permeability of the medium (*i.e.*, the material placed between the poles).

Also, from equation (*ii*)

$$I = \chi H \quad ...(v)$$

Substituting the values of B and I from equations (*iv*) and (*v*) in equation (*i*), we get

$$\mu H = H + 4\pi \chi H$$

or

$$\mu = 1 + 4\pi\chi \quad ...(vi)$$

4.11.2. Determination of magnetic susceptibility (Gouy's method)

Gouy method for magnetic susceptibility measurement is described below:

The material under examination is taken in a thin cylinder or a tube which is then suspended vertically with a wire from one arm of the balance between the two poles of an electromagnet such that the lower end of the cylinder is nearly in the middle of the poles of the electromagnet (Fig. 4.11). The cylinder is now counter-balanced in this position by placing weights in the other pan of the balance. The current is now switched on into the electromagnet. If the substance is paramagnetic, the cylinder will be pulled down (so that more magnetic lines of force can pass through it). On the other hand if the substance is diamagnetic, the cylinder will be pushed upward (so that less magnetic lines of force may pass through it). In the first case, some extra weights will have to be added into the pan, whereas in the second case, some weights will have to be removed from the pan to restore the balance.

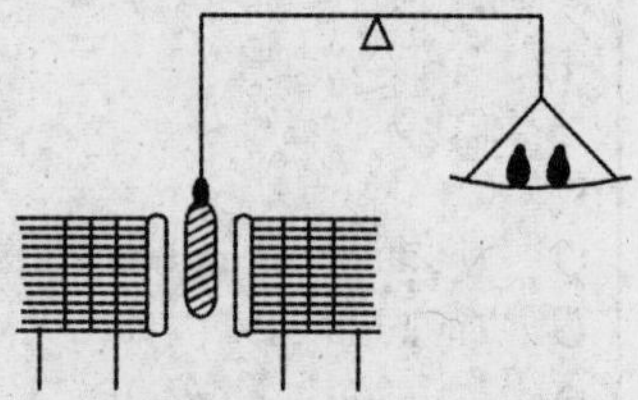

Fig. 4.11. Gouy's method for measurement of magnetic susceptibilities.

The change in weight is equal to the magnetic force acting on the material. Therefore if Δm is the change in mass, then we have

$$\Delta mg = \frac{1}{2}(\chi - \chi_a)AH^2 \quad ...(i)$$

where g is the acceleration due to gravity so that Δmg gives the change in weight. The right hand side is the magnetic force acting on the material of volume susceptibility χ when A is the area of cross-section of the specimen, H is the field strength and χ_a is the volume susceptibility of the air.

Louis Georges Guoy

Value of $\chi_a = 0.03 \times 10^{-6}$. Thus using equation (*i*), the value of χ can be calculated.

In the above calculations, the absolute value of the field strength H is also required. However, the necessity of measuring H and A can be eliminated if the experiment is first performed with substance of known susceptibility using the same magnetic field and the same cylinder for taking this substance. Usually water is taken as the standard substance for this purpose. Then evidently we will have

$$\frac{\Delta m_s}{\Delta m_w} = \frac{\chi_s - \chi_a}{\chi_w - \chi_a} \quad ...(ii)$$

where Δm_s is the change in mass when the sample under examination is placed in the magnetic field and Δm_w is the change in mass when water is taken in the same cylinder and placed in the same magnetic field. χ_s is the magnetic susceptibility (volume susceptibility) of the sample and χ_w and χ_a are those of the water and air respectively. We can thus eliminate the usage of H.

4.12. MAGNETIC SUSCEPTIBILITY AND MOLECULAR STRUCTURE

It has been revealed that molar magnetic susceptibility (χ_M) is an **additive** as well as a **constitutive** property. The values assigned to the various atoms and to the structural factors (*i.e.*, double bond, triple bond, rings etc.) can be obtained from the experimentally determined values for a large number of elements and compounds. The values of magnetic susceptibility for some of the atoms and structural groups are given in Table 4.1 below.

Table. 4.1. Atomic and structural susceptibilities

Atom	Magnetic *susceptibility*	*Structural factor*	*Magnetic susceptibility*
	$\times 10^{-6}$		$\times 10^{-6}$
H	–2.93	C = C double bond	+ 5.4
C	–6.00	N = N double bond	+ 1.9
O (in alcohols and ethers)	–4.61	N = N double bond	+ 1.9
O (in ketones)	+1.73	C = N double bond	+ 8.2
O (in esters and acids)	–3.36	C ≡ C triple bond	+ 0.8
N (in open chain)	–5.57	C ≡ N triple bond	– 0.8
N (in ring)	–4.61	Benzene ring	– 1.5
N (in amines)	–1.54	**Note:** For an aromatic compound, the value of benzene ring has been added, no value has been added for C = C double bond.	
F	–11.5		
Cl	–20.1		
Br	–30.6		

The calculation of molar magnetic susceptibility of a compound for the atomic and structural values help us to decide between the possible structures for a compound or to confirm the structure of the compound by comparing the calculated value with the experimental value.

SOLVED EXAMPLES ON MAGNETIC SUSCEPTIBILITY

Example 1. Establish that the structure of methyl acetate is

$$CH_3-\overset{\overset{\displaystyle O}{\|}}{C}-O-CH_3$$

Given that the magnetic susceptibility of the compound is found to be -70.3×10^{-6}

Soluton. The value for χ_M for this structure can be calculated as follows:

$$\begin{aligned} 3\text{ C atom} = 3 \times (-6.00 \times 10^{-6}) &= -18.00 \times 10^{-6} \\ 6\text{ H atom} = 6 \times (-2.93 \times 10^{-6}) &= -17.58 \times 10^{-6} \\ 1\text{ O atom} = 1 \times (-4.61 \times 10^{-6}) &= -4.61 \times 10^{-6} \\ 1\text{O (ester)} = 1 \times (-3.36 \times 10^{-6}) &= -3.36 \times 10^{-6} \\ \hline \text{Total calculated value} &= -43.55 \times 10^{-6} \\ \text{Observed value} &= -43.6 \times 10^{-6} \end{aligned}$$

Calculated value agrees well with the observed value. Hence the given structure is correct.

Example 2. Confirm that the structure of benzoic acid is

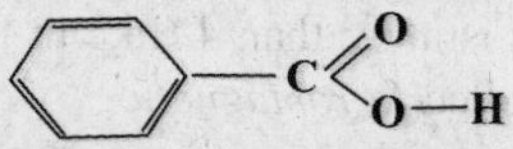

Given that the magnetic susceptibility of the compound is measured as -70.3×10^{-6}.

Solution. The magnetic susceptibility contributions due to various atoms and structural factors and hence the total value for χ_M can be calculated as follows:

$$7\text{ C atom} = 7 \times (-6.00 \times 10^{-6}) = -42.00 \times 10^{-6}$$

$$6\text{ H atom} = 6 \times (-2.93 \times 10^{-6}) = -17.58 \times 10^{-6}$$

$$1\text{ O atom (alcohol)} = 1 \times (-4.61 \times 10^{-6}) = -4.61 \times 10^{-6}$$

$$1\text{O (acid)} = 1 \times (-3.36 \times 10^{-6}) = -3.36 \times 10^{-6}$$

$$1\text{ benzene ring} = 1 \times (-1.5 \times 10^{-6}) = -1.50 \times 10^{-6}$$

$$\text{Total calculated value} = -69.05 \times 10^{-6}$$

$$\text{Observed value} = -70.3 \times 10^{-6}$$

As the calculated value is very close to the observed value, the given structure is correct.

4.13. EXPLANATION OF DIAMAGNETISM AND PARAMAGNETISM

Study of atomic structure supports that an electron is associated with orbital motion and spin motion. Movement of a charged particle produces a small magnetic field and we obtain a definite value of magnetic moment. An election spinning clockwise will give magnetic moment in a direction opposite to that spinning anticlockwise.

Thus, if all the electrons in a molecule are paired, they will give the net value of magnetic moment as zero. If such a substance is placed in a magnetic field, it is *repelled* by the magnetic field because a small magnetic moment is induced in the substance which works against the external magnetic field.

We can generalise that *diamagnetism is the property of substances containing paired electrons only.*

On the contrary, if the molecule (substance) contains some electrons which are unpaired. Each one of them will possess magnetic moment which will not get cancelled. Hence such a substance will possess a definite value of magnetic moment. If such a substance is placed in a magnetic field, we shall find that the molecules will arrange themselves parallel to the external magnetic field. In such a case, the induced magnetic moment of the substance is too small to overcome the permanent magnetic moment of the substance. Such substances get attracted by the magnetic field, and are known as paramagnetic substances.

We can say that *paramagnetism is the property of substances containing a certain number of unpaired electrons.*

Table 4.2. Expected magnetic moments and molar magnetic susceptibilities for different number of unpaired electrons

No. of unpaired electrons (n)	Spin magnetic moment (B.M.) $\mu_M = \sqrt{n(n+2)}$	Molar magnetic susceptibility at 298 K $\chi_M = \frac{N}{3kT}[n(n+2)]$
1	1.73	1260×10^{-6}
2	2.83	3360×10^{-6}
3	3.83	6290×10^{-6}
4	4.90	10100×10^{-6}
5	5.92	14700×10^{-6}

Further, greater the number of unpaired electron, greater are the magnetic properties associated. If the number of unpaired electrons is more than 4 (like in Fe, Co, Ni), the substance has a high value of magnetic moment and is called *ferromagnetic*.

4.13.1. Molecular interpretation of diamagnetism and paramagnetism

Molecular Interpretation of Diamagnetism

The classical theory of diamagnetism and paramagnetism was put forward Langevin in 1905.

According to Langevin, *diamagnetism results from the orbital motion of the atoms and molecules.* According to Lenz's law which states that when a current flows through a wire in a magnetic field, the magnetic field induces current in the coil in such a manner that it opposes the applied field. Similarly, the applied magnetic field perturbs the orbital motion thereby inducing a magnetic field which opposes its own direction. In multi-electron atom, the orbits are spatially oriented at random relative to the field.

The molar diamagnetic susceptibility χ_M^d is given by

$$\chi_M^d = -\frac{Z\,Ne^2}{6\,mc^2}\sum \overline{r_i^2}$$

$$= -2.832 \times 10^{10}\,\overline{r_i^2} \quad \text{(After substituting for various quantities)}$$

Where $\overline{r_i^2}$ ≡ the mean square distance of the *i*th orbit from the nucleus, Z = atomic number , e = charge on the electron, m = mass of electron and c = velocity of light.

Diamagnetism is a common property of all atoms of all matter whether it possesses a permanent dipole moment or not.

Molecular Interpretation of Paramagnetism

According to Langevino's theory of paramagnetism each atom or molecule behaves as a small bar magnet with magnetic moment μ. The magnetic moment may either be due to orbital motion of an electron or due to unpaired electron spin. The *molar paramagnetic susceptibility* due to permanent magnetic moment is given by

$$\chi_m^p = \frac{N\mu_M^2}{3kT} = \frac{\text{constant}}{T} \qquad (\mu_M = \text{Magnetic moment})$$

i.e. χ_m^p varies inversely with absolute temperature T. This is known as *Curie's law*, which holds over a limited range of temperature.

In case, the system is distorted by the external field, the molar paramagnetic susceptibility is more correctly given by

$$\chi_m^p = \frac{N\mu_M^2}{3kT} = \alpha_M$$

where α_M is magnetic polarizability of the system.

As all substances exhibit diamagnetic behaviour, the total molar magnetic susceptibility is the sum of χ_m^d and χ_m^p. However, the diamagnetic contribution is negligibly small as compared to paramagnetic contribution.

4.14. DETERMINATION OF THE MAGNETIC MOMENT OF A PARAMAGNETIC SUBSTANCE

Each atom, ion or molecule of a paramagnetic substance behaves like a micromagnet with a definite inherent magnetic moment of magnitude μ_m. When such a substance is placed in the external magnetic field, micromagnets tend to arrange themselves parallel to the applied field. However, this tendency is opposed by the thermal motion of the particles. Hence the actual arrangement taken up will depend upon the temperature. Langevin in 1905 put forward the following relationship between molar magnetic susceptibility χ_M, magnetic moment μ_m and temperature T :

$$\chi_M = \alpha_m \times \frac{\mu_m^2 N}{3kT} \quad \text{... }(i)$$

where N is Avogadro's number, k is Boltzmann's constant and α_m is a constant for the given substance. It represents the diamagnetic molar susceptibility of the substance induced by the applied magnetic field.

In eqn. (i), all quantities are constant except χ_M and T. Thus determining χ_M for the given substance at different temperatures T, and then plotting χ_M *vs* $\frac{1}{T}$, a straight line graph will be obtained, the slope of which will be equal to $\frac{\mu_m^2 N}{3k}$. Hence μ_m can be calculated.

Magnetic moment is usually expressed in Bohr-Magneton (B.M.)

B.M. = $eh/4\pi\, m\, c$

where h is Planck's constant, c is velocity of light, e is the charge on electron and m is its mass. Substituting various values, we have

$$1 \text{ B.M.} = 9.2732 \times 10^{-21} \text{ ergs gauss}^{-1}$$

4.15. RELATION BETWEEN MAGNETIC MOMENT AND NUMBER OF UNPAIRED ELECTRONS

The total magnetic moment of any atom, ion or molecule is the sum of the following two values:

(i) Magnetic moment due to the *orbital motion* of the electron. It is equal to $\sqrt{l(l+1)}$ B.M. where l is the azimuthal or orbital quantum number for that electron.

(ii) Magnetic moment due to the *spin motion* of the electron. It is equal to $2\sqrt{s(s+1)}$ B.M. where s is the spin quantum number having value equal to $\frac{1}{2}$.

If ions or molecules are considered instead of free atoms, it is expected that the orbital motions of the electrons are tied into the nuclear configuration of the molecule or the ion so tightly that they are unaffected by applied magnetic field. Hence the magnetic moment contribution due to the orbital motion is negligible. Thus the magnetic moment of the molecule or the ion is mainly due to the spin motion of the electrons. (Hence it is called spin magnetic moment). Thus we have

$$\mu_m = 2\sqrt{s(s+1)} \text{ B.M.} \quad \text{...}(i)$$

Here s represents spin quantum number for one unpaired electron. For a molecule or ion containing a number of unpaired electrons, the expression is usually written as

$$\mu_m = 2\sqrt{S(S+1)} \text{ B.M.} \quad \text{...}(ii)$$

where S represents the total spin quantum number.

Thus for one unpaired electron, $S = \frac{1}{2}$

for two unpaired electrons, $S = \frac{2}{2}$

for three unpaired electron, $S = \frac{3}{2}$ and so on.

In general, for n unpaired electrons, $S = \frac{n}{2}$, hence eqn. (*ii*) becomes

$$\mu_m = 2\sqrt{\frac{n}{2}\left(\frac{n}{2}+1\right)}$$

$$\boxed{\mu_m = \sqrt{n(n+2)}} \text{ B.M.} \qquad ...(iii)$$

Thus knowing the number of unpaired electrons present, the magnetic moment (μ_m) can be calculated or *vice versa.*

4.15.1. Relation between molar susceptibility and the number of unpaired electrons

Taking α_m as constant, eqn. (*i*) of section 4.14 can be written as:

$$\chi_M = \frac{\mu_m^2 N}{3kT} \qquad ... (i)$$

where χ_M and μ_m are magnetic susceptibility and magnetic moment respectively.

Substituting the value of μ_m in eqn. (*i*), we get

$$\boxed{\chi_M = \frac{N}{3kT}[n(n+2)]} \qquad ... (ii)$$

The expected values of spin magnetic moments and magnetic susceptibilities for different values of n are given in Table 4.2.

4.16. APPLICATIONS OF MAGNETIC SUSCEPTIBILITY MEASUREMENT

Some important applications of magnetic susceptibility measurement are given below:

(1) **In confirming the structure of a given molecule:** The use of molar magnetic susceptibilities in confirming the structure of a given molecule is based upon the fact that magnetic susceptibility is additive as well constitutive property.

(2) **In calculation of the number of unpaired electrons in a molecule or ion:** Knowing the molar magnetic susceptibility χ_M which can be easily determined experimentally, the number of unpaired electrons present in a molecule or an ion can be calculated. This has helped in understanding the electronic structures of a number of molecules. For example, the molar magnetic susceptibility of oxygen has been found to be 3400 × 10^{-6} which shows that it must contain two unpaired electrons. Thus the structure of oxygen molecule should be represented as

:Ȯ : Ȯ:

It is a paramagnetic molecule.

(3) **In the study of coordination compounds:** An important application of magnetic susceptibility in chemistry has been in the study of coordination compounds of transition elements. Measurement of magnetic susceptibility helps us to know the number of unpaired electrons present in the complex which in turn helps to assign the correct structure to the complex ion in some cases. For example, the magnetic susceptibility values of the complexes $[Fe(CN)_6]^{3-}$ and $[Fe\,F_6]^{3-}$ (though each has the same coordination number *i.e.*, 6) suggest that there is one unpaired electron present in the former whereas there are five unpaired electrons present in the latter. These are explained by suggesting the following electronic structures to these complexes (Fig. 4.12).

$$Fe = 26 = 1s^2\, 2s^2\, 2p^6\, 3s^2\, 3p^6\, 3d^6\, 4s^2$$

$$Fe^{3+} = 1s^2\, 2s^2\, 2p^6\, 3s^2\, 3p^6\, 3d^5$$

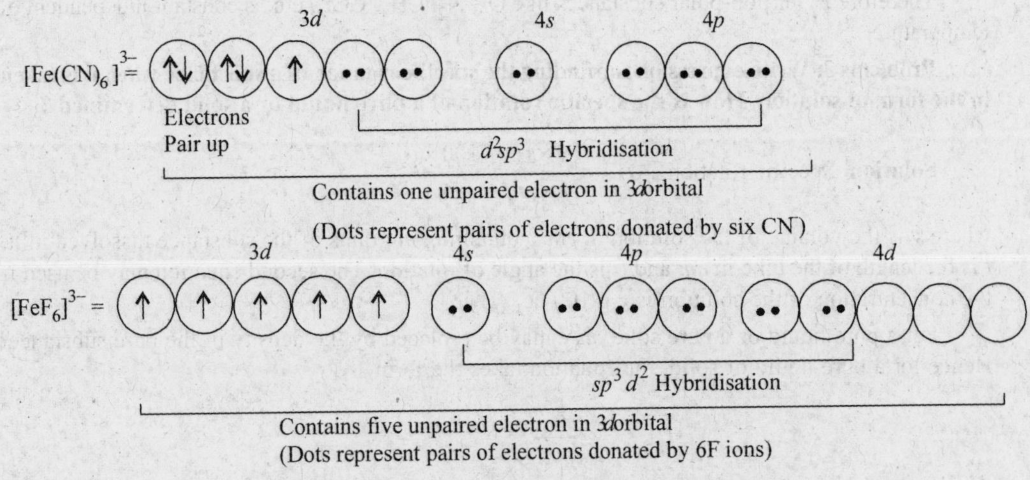

Fig. 4.12. Structure of (*i*) $[Fe(CN)_6]^{3-}$ and (*ii*) $[FeF_6]^{3-}$

(4) **In establishing the formulas of certain compounds:** For example, it has been shown that the formula of hypophosphoric acid is $H_4P_2O_6$ and not H_2PO_3. This is because if the latter formula were correct, sodium and silver salts, $NaHPO_3$ and Ag_2PO_3, would contain an odd number of electrons and should be paramagnetic while actually they are diamagnetic. Hence their salts must have the formulas $Na_2H_2P_2O_6$ and $Ag_4P_2O_6$ respectively. Similarly the formula of mercurous chloride is Hg_2Cl_2 and not HgCl. Dimerisation leads to spin neutralisation of electrons.

(5) **In explaining the existence of odd electron molecules:** The compounds containing odd electrons are expected to be paramagnetic. This is actually found to be so for a few gaseous compounds, namely nitric oxide, nitrogen dioxide and chlorine dioxide. Moreover, their magnetic susceptibilities show that they contain one unpaired electron in each case.

SOME SOLVED CONCEPTUAL PROBLEMS

Problem 1. How do meso form and racemic mixture of tartaric acid differ from each other ?

Solution. Both are optically inactive but for different reasons. In case of meso form, it is due to *internal compensation i.e.* the molecule has a plane of symmetry so that rotation caused by one half of the molecule is compensated by other half within the molecule. In case of racemic mixture, optical inactivity is due to *external compensation.* It is a mixture of *d* and *l*-forms such that the rotation to the right by *d*-form is neutralised by equal rotation to the left by l-form.

Problem 2. Write Clausius-Mosotti equation. What important result do you draw from it for non-polar molecules ?

Solution. The induced polarization P_i in a non-polar molecule is given by

$$P_i = \left(\frac{D-1}{D+2}\right)\frac{M}{\rho} = \frac{4}{3}\pi N \alpha$$

where D = dielectric constant of the medium (substance)

M = molecular mass, ρ = density, N = Avogadro's number

α = constant for the given substance.

All quantities on the R.H.S of the equation are constants and also independent of temperature.

Therefore P_i for non-polar substances like O_2, N_2, CH_4, CCl_4, etc. is constant independent of temperature.

Problems 3. Write expression for finding the specific rotation of a substance when it is taken in the form of solution. How is the specific rotation of a pure liquid or a solid determined ?

Solution. Specific rotation, $[\alpha]_\lambda^t = \dfrac{v \times \alpha}{l \times m} = \dfrac{100 \times \alpha}{l \times c}$

where v is the volume of the solution in cm^3 containing m grams of the substance dissolved in it, l is the length of the tube in *dm* and α is the angle of rotation. The second equation may be used if the concentration of the solution is c g/100 cc.

For a pure liquid or a pure solid, m/v may be replaced by the density of the pure substance. Hence for a pure liquid or solid, the equation takes the form

$$[\alpha]_\lambda^t = \frac{\alpha}{l \times d}$$

Problem 4. For polar molecules with permanent dipole moment, write the expression for the total polarization. What is the most important use of this expression ?

Solution. For polar molecules with permanent dipole moment, total polarization is given by the sum of induced polarization and orientation polarization *i.e.*

$$P_t = P_i + P_o$$

$$= \frac{4}{3}\pi N \alpha + \frac{4}{3}\pi N \left(\frac{\mu^2}{3\,kT}\right)$$

It can be written in the form

$$P_t = A + \frac{B}{T}$$

where $A = \dfrac{4}{3}\pi N \alpha$ and $B = \dfrac{4\pi N \mu^2}{9\,k}$

Thus a plot of P_t versus $1/T$ will give a straight line whose slope is

$$B = \frac{4\pi N \mu^2}{9\,k}$$

or
$$\mu = \sqrt{\frac{9k}{4\pi N}B} = 0.0128\sqrt{B}\times 10^{-18} \text{ esu cm.}$$

Thus permanent dipole moment μ of the molecule can be calculated.

Problem 5. How is the dipole moment calculated for a molecule AB_2 with bond angle θ ?

Solution. In general for a triatomic molecule with bond moments μ_1 and μ_2, net dipole moment μ of the molecule is given by

$$\mu = \sqrt{\mu_1^2 + \mu_2^2 + 2\mu_1\mu_2\cos\theta}$$

For AB_2, $\mu_1 = \mu_2 = \mu_{AB}$ so that $\mu = 2\mu_{AB}\cos\frac{\theta}{2}$. $\left(\because 1+\cos\theta = 2\cos^2\frac{\theta}{2}\right)$

Problem 6. Comment on the value of magnetic permeability μ for diamagnetic, paramagnetic and ferromagnetic substances.

Solution. For diamagnetic substance $\mu < 1$

For paramagnetic substances $\mu > 1$

For ferromagnetic substances $\mu >> 1$ ($> 10^3$)

Problem 7. How are magnetic moment and molar magnetic susceptibility calculated from the number of unpaired electrons ?

Solution. $\mu_m = \sqrt{n(n+2)}$ BM, $\chi_M = \frac{\mu_m^2 N}{3kT} = \frac{N}{3kT}[n(n+2)]$

Problem 8. How is magnetic moment of a substace determined from molar magnetic susceptibility ?

Solution. Molar magnetic susceptibiltiy (χ_M) is related to magnetic moment (μ_m) according to the equation

$$\chi_M = \alpha_M \times \frac{\mu_m^2 N}{3kT}$$

A plot of χ_M vs $1/T$ gives a straight line with slope $= \frac{\mu_m^2 N}{3k}$ from which μ_m can be calculated.

Problem 9. How is magnetic susceptibility (χ) related to magnetic permeability (μ) ? Comment on its value for diamagnetic, paramagnetic and ferromagnetic substances.

Solution. $\chi = \frac{\mu - 1}{4\pi\rho}$ (ρ = density of the substance)

For diamagnetic substances, $\mu < 1$, therefore χ is –ve.

For paramagnetic substances, $\mu > 1$, therefore χ is +ve but small.

For ferromagnetic substances, $\mu >> 1$, χ is +ve and very large.

EXERCISES
(Based on Question Papers of Different Universities)

Multiple Choice Questions (Choose the correct option)

1. A Chiral molecule has
   (*a*) no plane of symmetry (*b*) one plane of symmetry
   (*c*) infinite planes of symmetry (*d*) none of these
2. An equimolar mixture of (+)-tartaric acid and (–)-tartaric acid is called
   (*a*) enantiomers (*b*) optically active mixture
   (*c*) racemic mixture (*d*) asymmetric mixture
3. The paramagnetism is due to the presence of
   (*a*) paired electrons (*b*) unpaired electrons
   (*c*) both paired as well as unpaired electrons
   (*d*) none of these
4. Dichlorobenzene exists in three isomers-*ortho*, *meta* and *para* isomers. Out of these three isomers, one with highest dipole moment will be
   (*a*) *ortho*-isomer (*b*) *meta*-isomer
   (*c*) *para*-isomer (*d*) all will have the same dipole moment
5. Optical activity is
   (*a*) an additive property (*b*) a constitutive property
   (*c*) both an additive and constitutive property
   (*d*) none of these
6. A racemic mixture has
   (*a*) zero optical rotation (*b*) positive optical rotation
   (*c*) negative optical rotation (*d*) infinite optical rotation
7. The magnetic moment, μ, is given by the formula in which *n* is the number of unpaired electrons
   (*a*) $\mu = n \times (n + 2)$ (*b*) $\mu = \sqrt{n \times (n+1)}$
   (*c*) $\mu = \sqrt{n \times (n+2)}$ (*d*) $\mu = n \times (n + 1)$
8. The net dipole moment of the molecule is
   (*a*) sum of all individual bond moments (*b*) product of all individual bond moments
   (*c*) vector resultant of all the individual bond moments
   (*d*) none of the above
9. The necessary condition for a compound to be optically active is
   (*a*) the presence of no chiral atoms
   (*b*) the presence of at least one asymmetric atoms
   (*c*) the presence of chirality in the molecule
   (*d*) none of these
10. The substances which retain their magnetic field when removed from the magnetic field are called
    (*a*) paramagnetic (*b*) diamagnetic
    (*c*) ferrimagnetic (*d*) ferromagnetic

11. The dipole moment of a polar substance is given by the formula, where q is charge at one end and r is the distance between the opposite charges

(*a*) $\mu = q + r$ (*b*) $\mu = q - r$

(*c*) $\mu = q \times r$ (*d*) $\mu = q \div r$

12. Iron, cobalt and nickel are examples of

(*a*) diamagnetic substances (*b*) paramagnetic substances

(*c*) ferromagnetic substances

(*d*) sometimes diamagnetic & sometimes paramagnetic substances

13. The molecules that are non-superimposable mirror images are called

(*a*) optical isomers (*b*) racemic isomers

(*c*) enantiomers (*d*) none of these

14. A diamagnetic substance is ______ by / in the magnetic field

(*a*) attracted (*b*) repelled

(*c*) rotated (*d*) revolved

ANSWERS

1. (*b*)	**2.** (*c*)	**3.** (*b*)	**4.** (*a*)	**5.** (*b*)	**6.** (*a*)
7. (*c*)	**8.** (*c*)	**9.** (*c*)	**10.** (*d*)	**11.** (*c*)	**12.** (*c*)
13. (*c*)	**14.** (*b*)				

Short Answer Questions

1. What are the factors on which optical rotation depends ? Derive expression for specific rotation.

2. What information can dipole moment give about the structure of a molecule ? Give examples.

3. What is 'magnetic permeability' ? Differentiate between diamagnetic, paramagnetic and ferromagnetic substances in terms of magnetic permeability.

4. How can you justify the following ?

(*i*) O_2 has two unpaired electrons

(*ii*) $[Fe(CN)_6]^{3-}$ has less magnetic moment than $[FeF_6]^{3-}$ though both are octahedral.

(*iii*) Formula of hypophosphoric acid is $H_4P_2O_6$ and not H_2PO_3.

5. Write expression for the total polarization of a molecule. How does it help in determining the dipole moment of a substance ?

6. How can you differentiate between the following by dipole moment measurements ?

(*i*) *cis* and *trans* isomers of dichloroethene

(*ii*) *o*, *m* and *p*-isomers of dichlorobenzene.

7. What do you understand by internal compensation and external compensation in optical activity ? Explain with suitable examples.

8. What do you understand by additive and constitutive properties ? Name some physical properties of molecules which are additive as well as constitutive.

9. Briefly explain temperature method for measurement of dipole moment. How the dielectric constant required is determined ?

10. Derive the relationship between

(*i*) magnetic moment of a substance and the number of unpaired electrons

(*ii*) molar magnetic susceptibility and the number of unpaired electrons.

11. What is cause of optical activity ? How is measurement of optical rotation helpful in deciding chemical constitution of a substance ?

12. How do you explain the action of magnetic field on diamagnetic and paramagnetic substances ?

13. How is the dipole moment of a solid determined from molar polarization studies ?

14. Briefly explain the terms 'optical activity' and 'specific rotation'.

15. What is dipole moment ? What are its units ? How is the dipole moment of a triatomic molecule calculated ?

16. How molar magnetic susceptibility can be used to find the magnetic moment of paramagnetic substance ?

17. "Optical activity is a constitutive property". Exemplify.

18. Write Clausius-Mosotti equation giving relationship between distortion (or induced) polarization and dielectric constant of the medium. What important result do you draw from it ?

19. Briefly explain the principle of refraction method for determining dipole moment of a substance.

20. What do you understand by magnetic induction, specific magnetic susceptibility and molar magnetic susceptibility ?

General Questions

1. What is optical activity ? What is the cause of optical activity? Briefly explain with suitable examples. How is measurement of optical rotation helpful in elucidating the chemical constitution of a substance ?

2. Explain the effect of temperature on molar polarization of molecules. How does it help in finding the dipole moment of a substance ?

3. Briefly explain the following terms:

(*i*) Magnetic permeability (*ii*) Magnetic induction

(*iii*) Specific magnetic susceptibility (*iv*) Molar magnetic

4. Define dipole moment. How is it determined by

(*i*) Vapour-Temperature method ?

(*ii*) Refraction method ?

5. What do you understand by magnetic susceptibility ? Briefly describe Gouy's method for its measurement. How does it help

(*i*) in deciding the molecular structure of a substance

(*ii*) in the study of co-ordination compounds

(*iii*) in explaining the existence of odd electron molecules ?

6. What do you understand by 'optical rotation' and 'specific rotation' ? How will you determine the specific rotation of a substance ? Explain the principle of the instrument used.

7. What do you understand by electrical polarization of molecules ? Derive Clausius-Mosotti equation.

8. Briefly describe at least five important applications of dipole moment in deciding chemical constitution.

9. Briefly explain the following terms :

(*i*) Induced dipole moment (*ii*) Induced or Distortion polarization

(*iii*) Orientation polarization.

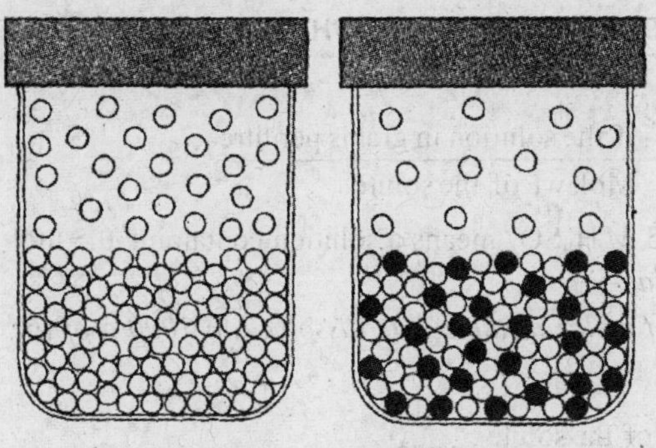

Pure solvent   Solution with a nonvolatile solute

5

Solutions, Dilute Solutions and Colligative Properties

5.1. INTRODUCTION

A **solution** is defined as a homogeneous mixture of two or more non-reacting components whose relative amounts can be varied upto certain limits. Salt-water and alcohol-water are the examples of solutions.

A solution has two components. The substance which is present relatively in smaller proportion in the solution is called **solute**. For example, in a sugar solution in water, sugar is the solute.

The substance present in larger proportion in the solution is called **solvent**. For example in sugar solution in water, water is the solvent. However in alcohol-water mixture, alcohol can be present as a solute as well as solvent, depending upon whether it is present in smaller or greater amount compared to water.

5.2. METHODS OF EXPRESSING THE CONCENTRATION OF A SOLUTION

A number of methods are used to express the concentration of a solution, as given below.

(*i*) **Percentage by mass:** *It is defined as the amount of solute in grams present in 100 grams of the solution.*

$$\text{Thus \% by mass} = \frac{\text{Mass of the solute}}{\text{Mass of the solution}} \times 100$$

A 5 % solution of NaCl in water means, 5 g of NaCl is present in 95 g (or 95 cc) water.

(*ii*) **Percentage by volume:** It is the mass of the solute (if solid) in grams or the volume of the solute (if liquid) in cc present in 100 cc of the solution. Thus 5 % sodium chloride solution by volume means 5 g of sodium chloride are present in 100 cc of the solution, while 5 % ethanol solution in water means 100 cc of the solution contains 5 c.c. of ethanol.

(*iii*) **Strength:** *The strength of the solution is defined as the number of grams of the solute dissolved per litre of the solution.*

$$\text{Thus strength} = \frac{\text{Weight of the solute in grams}}{\text{Volume of the solution in litres}}$$

(*iv*) **Molarity :** *It is defined as the number of moles of the solute dissolved per litre of the solution.*

$$\text{Thus molarity} = \frac{\text{Number of moles of the solute}}{\text{Volume of the solution in litres}}$$

$$= \frac{\text{Strength of the solution in grams per litre}}{\text{Mol. wt. of the solute}}$$

It is denoted by symbol *M*. For example, 0.5 *M* H_2SO_4 means a solution containing 0.5 mol of H_2SO_4 present per litre and is read as 0.5 *molar*.

(*v*) **Molality:** *It is defined as the number of moles of the solute dissolved in 1000 g of the solvent.*

$$\text{Molality} = \frac{\text{Number of moles of the solute}}{\text{Weight of the solvent in grams}} \times 1000$$

$$= \frac{\text{Number of grams of the solute per 1000 g of the solvent}}{\text{Mol. wt. of the solute}}$$

It is denoted by *m*. For example, 0.1 molal solution means a solution containing 0.1 mol of the solute dissolved per 1000 g of the solvent and is written as 0.1 m.

(*vi*) **Normality:** *The normality of a solution is defined as the number of gram equivalents of the solute dissolved per litre of the solution.* Thus

$$\text{Normality} = \frac{\text{Number of gram equivalents of the solute}}{\text{Volume of the solution in litres}}$$

$$= \frac{\text{Strength of the solution in grams per litre}}{\text{Equivalent wt. of the solute}}$$

It is represented by *N*. For example, 1*N* Na_2CO_3 solution means a solution containing one gram equivalent (*i.e.*, 53 g) of Na_2CO_3 dissolved per litre of the solution. It is read as *one normal.*

(*vii*) **Mole Fraction:** *The mole fraction of a component in a solution is defined as the number of moles of that component divided by the total number of moles of all the components.*

Mole fraction of the solute in the solution (x_2)

$$= \frac{\text{No. of moles of the solute}}{\text{No. of moles of solute + No. of moles of solvent}}$$

Mole fraction of solvent in the solution (x_1)

$$= \frac{\text{No. of moles of the solvent}}{\text{No. of moles of solute + No. of moles of solvent}}$$

For example if n_2 moles of the solute are dissolved in n_1 moles of the solvent, we may write

$$x_2 = \frac{n_2}{n_1 + n_2} \quad \text{and} \quad x_1 = \frac{n_1}{n_1 + n_2}$$

Evidently, $x_1 + x_2 = 1$

(*viii*) **Parts per million (ppm):** It is the number of parts of the solute present in one million parts of the solution.

SOLVED PROBLEMS ON CONCENTRATION OF SOLUTIONS

Example 1. What is the molarity of a solution of sodium chloride (at wt of Na = 23 and Cl = 35.5) which contains 60 g of sodium chloride in 2000 ml of solution?

Solution. 2000 ml = 2 litre

No. of moles of sodium chloride $= \dfrac{60}{23+35.5} = \dfrac{60}{58.5}$

2 litre of the solution contains $= \dfrac{60}{58.5}$ moles of the solute

1 litre of the solution contains $= \dfrac{60}{58.5} \times \dfrac{1}{2} = \dfrac{30}{58.5} = 0.513$ moles

Hence molarity of the solution $= 0.513$ M.

Example 2. 0.212 g of Na_2CO_3 with molecular mass 106 is dissolved in 250 ml of the solution. Calculate the molarity of Na_2CO_3 in the solution.

Solution. No. of moles of $Na_2CO_3 = \dfrac{0.212}{106}$ moles

250 ml of the solution contains $= \dfrac{0.212}{106}$ moles

1000 ml of the solution contains $= \dfrac{0.212 \times 1000}{106 \times 250}$ moles $= 0.008$

Hence molarity of the solution $= 0.008$ M.

Example 3. 4.0 g of NaOH is contained in a deci-litre of a solution. Calculate the following.

(*i*) Mole fraction of NaOH

(*ii*) Molarity of NaOH

(*iii*) Molality of NaOH

Density of the solution is 1.038 g/ml.

Solution. (*i*) **Mole fraction of NaOH**

Volume of the solution = 100 ml, Density = 1.038 g/ml

Mass of 100 ml of the solution = $1.038 \times 100 = 103.8$ g

Mass of NaOH = 4.0 g

$\therefore$ Mass of water = $(103.8 - 4)$ g $= 99.8$ g

No. of moles of NaOH $= \dfrac{4}{40} = 0.1$

No. of moles of water $= \dfrac{99.8}{18} = 5.54$

Mole fraction of NaOH $= \dfrac{0.1}{0.1+5.54} = 0.018$

(*ii*) **Molarity of NaOH *(M)***

100 ml of the solution contains = 0.1 moles of NaOH

1000 ml of the solution contains $= \dfrac{0.1 \times 1000}{100} = 1$ mole

$\therefore$ Molarity of the solution = 1 *M*

(*iii*) **Molality of NaOH (*m*)**

99.8 g water contains = 0.1 mole of NaOH

1000 g water contains $= \frac{0.1}{99.8} \times 1000 = 1.002$ moles

Hence molality of the solution = 1.002 m

Example 4. A solution contains 25% water, 25% ethanol and 50% acetic acid by mass. Calculate mole fraction of each component.

Solution. Let the total mass of the solution be 100 g.

(*i*) Mass of water = 25 g

No. of moles of water $= \frac{25}{18} = 1.39$

(*ii*) Mass of ethanol = 25 g

No. of moles of ethanol $= \frac{25}{46} = 0.544$

(*iii*) Mass of acetic acid = 50 g

No. of moles of acetic acid $= \frac{50}{60} = 0.833$

Total No. of moles of the three substances

= 1.39 + 0.544 + 0.833 = 2.767

Mole fraction of water $= \frac{1.39}{2.797} = 0.502$

Mole fraction of ethanol $= \frac{0.544}{2.767} = 0.196$

Mole fraction of acetic acid $= \frac{0.833}{2.767} = 0.302$

Example 5. Find the molarity and molality of a 15% solution of H_2SO_4 (density of H_2SO_4 = 1.10 g/ml and molecular mass of H_2SO_4 = 98).

Solution. Molarity

100 ml of the solution contains = 15 g H_2SO_4

1000 ml of the solution contains $= \frac{15}{100} \times 1000$ g H_2SO_4 = 150 g H_2SO_4

or $= \frac{150}{98}$ moles or 1.53 moles

Hence molarity of the solution = 1.53 *M*

Molality

Mass of 100 ml of solution = 1000 × 1.10 (volume × density) = 1100 g

Mass of the acid – 150 g

Mass of water = 1100 – 150 = 950 g

950 g water contain = 1.53 moles of H_2SO_4

1000 g water contain $= \frac{1.53}{950} \times 1000 = 1.61$ mole

Hence molality of the solution $= 1.61$ m

Example 6. Calculate the mole fraction of water in a mixture of 12 g of water, 108 g of acetic acid and 92 g of ethyl alcohol.

Solution. Mass of water $= 12$ g

Mol. mass of water $= 18$ g

No. of moles of H_2O (n_A) $= \frac{12}{18} = 0.66$

Mass of acetic acid $= 108$ g

Mol. mass of acetic acid $= 60$ g

No. of moles of acetic acid (n_B) $= \frac{108}{60} = 1.8$

Mass of ethyl alcohol $= 92$ g

Mol. mass of ethyl alcohol $= 46$ g

No. of moles of ethyl alcohol (n_c) $= \frac{92}{46} = 2$

Total No. of moles, $n_A + n_B + n_C = 0.66 + 1.8 + 2$

Mole fraction of water $X_A = \frac{n_A}{n_A + n_B + n_C} = \frac{0.66}{0.66 + 1.8 + 2} = 0.148$

Example 7. 2.82 g of glucose (mol mass = 180) are dissolved in 30 g of water. Calculate the

(*i*) molality of the solution

(*ii*) mole fractions of (*a*) glucose (*b*) water.

Solution. (*i*) 30 g H_2O contains $= 2.82$ g of glucose

1000 g H_2O contains $= \frac{2.82}{30} \times 1000$ g of glucose

$= \frac{2.82 \times 1000}{30 \times 180}$ moles of glucose $= 0.52$ moles

Hence molality of the solution $= 0.52$ m

(*ii*) No. of moles of glucose $n_A = \frac{2.82}{180}$

No. of moles of water $n_B = \frac{30}{18}$

Mole fraction of glucose $(X_A) = \frac{2.82/180}{2.82/180 + 30/18} = 0.0093$

As $X_A - X_B = 1$

$\therefore$ $X_B = 1 - X_A$

Hence mole fraction of water $X_B = 1 - 0.0093 = 0.9907$

5.3. CHEMICAL POTENTIAL OR PARTIAL MOLAL FREE ENERGY

Let us suppose a system consists of n constituents with individual concentration n_1, n_2, n_3 ....etc (in moles). We want to study an extensive property of the system, like free energy, entropy etc. Let the property be represented as F.

Property F is a function not only of temperature and pressure but of the amounts of the different constituents as well, so that we can write.

$$F = f(T, P, n_1, n_2, n_3 ....) \quad ... (i)$$

If there is a small change in the temperature, pressure and the amounts of the constituents, then the change in the property F will be given by

$$dF = \left(\frac{\partial F}{\partial T}\right)_{P,n_1,n_2...} dT + \left(\frac{\partial F}{\partial P}\right)_{T,n_1,n_2...} dP$$

$$+\left(\frac{\partial F}{\partial n_1}\right)_{T,P,n_2,n_3...} dn_1 + \left(\frac{\partial F}{\partial n_2}\right)_{T,P,n_1,n_3...} dn_2 + ... \quad (ii)$$

On the right hand side, the first term gives the change in the value of F with temperature when pressure and composition are kept constant. The second term gives the change in the value of F with pressure when temperature and composition are kept constant. The other terms give the change in the value of F with a change in the amount of a constituent, when temperature, pressure and the amounts of the other constituents are kept constant.

If the temperature and pressure of the system are kept constant, then

$$dT = 0 \quad \text{and} \quad dP = 0$$

Thus eq. (*ii*) becomes

$$(dF)_{T,P} = \left(\frac{\partial F}{\partial n_1}\right)_{T,P,n_2,n_3...} dn_1 + \left(\frac{\partial F}{\partial n_2}\right)_{T,P,n_1,n_3...} dn_2 + ... \quad (iii)$$

Each derivative on the right hand side is called partial molar property and is represented by putting a bar over symbol of that particular property *i.e.*, $\overline{F}_1, \overline{F}_2$ for the Ist, 2nd component etc. respectively.

$$\text{Thus} \quad \left(\frac{\partial F}{\partial n_1}\right)_{T,P,n_2,n_3...} = \overline{F}_1; \quad \left(\frac{\partial F}{\partial n_2}\right)_{T,P,n_1,n_3...} = \overline{F}_2 \text{ etc.}$$

In general for any component i

$$\left(\frac{\partial F}{\partial n_i}\right)_{T,P,n_1,n_2...} = \overline{F}_i \quad (iv)$$

If we are studying free energy (G), we can replace F with G in eq. (*iv*)

$$\text{Thus,} \quad \left(\frac{\partial G}{\partial n_i}\right)_{T,P,n_1,n_2...} = \overline{G}_i \quad (v)$$

This quantity is called partial molar free energy or chemical potential and is usually represented by the symbol μ. Thus

$$\mu = \bar{G}_i = \left(\frac{\partial G}{\partial n_i}\right)_{T,P,n_1,n_2 \dots} \qquad (vi)$$

If $dn_i = 1$ mole, $\mu = (dG)_{T,P,n_1,n_2 \dots}$

Chemical potential may thus be defined as under:

The chemical potential of a constituent in a mixture is the increase in the free energy which takes place at constant temperature and pressure when 1 mole of that constituent is added to the system, keeping the amounts of all other constituents constant *i.e.*, when 1 mole of the constituent is added to such a large quantity of the system that its composition remains almost unchanged.

Equation (*iii*) with free energy as the extensive property can be written as

$$(dG)_{T,P} = \left(\frac{\partial G}{\partial n_1}\right)_{T,P,n_2,n_3 \dots} dn_1 + \left(\frac{\partial G}{\partial n_2}\right)_{T,P,n_1,n_3 \dots} dn_2 + \dots$$

or

$$(dG)_{T,P} = \mu_1\, dn_1 + \mu_2\, dn_2 + \dots \qquad (vii)$$

Equation (*vii*) on integration gives

$$G_{T,PN} = n_1\,\mu_1 + n_2\,\mu_2 + \dots$$

Where the subscript N stand for constant composition (with the number of moles of different constituents as n_1, n_2, n_3, etc.)

On the right hand side, the first term gives contribution of the first constituent to the total free energy of the system, the second term gives the contribution of the second constituent and so on. μ_1, μ_2 etc. give the *contribution per mole* to the total free energy. Hence chemical potential may also be defined as:

The chemical potential of a constituent in a mixture is its contribution per mole to the total free energy of the system of a constant composition at constant temperature and pressure.

It may noted that free energy is an extensive property but the chemical potential is an intensive property because it refers to the property per mole of the substance.

5.4. ACTIVITY AND ACTIVITY COEFFICIENTS

The chemical potential (μ) of an ideal gas is related to its pressure (p) by the expression

$$\mu = \mu_0 + RT \ln p \qquad \dots (i)$$

where μ_0 is the chemical potential in the standard state (i.e., when $p = 1$ atm).

In a mixture of ideal gases, the chemical potential of any constituent i in the mixture is related to its partial p_i as

$$\mu = \mu^\circ_i + RT \ln p_i \qquad \dots (ii)$$

where μ°_i is the chemical potential of that constituent in the standard state (i.e when $p_i = 1$ atm).

Similarly, in an *ideal solution*, the chemical potential of any component i is related to the concentration c_i of that component by the equation

$$\mu_i = \mu^\circ_i + RT \ln c_i \qquad \dots (iii)$$

However, we notice that eqs (*ii*) and (*iii*), are not obeyed by real (non-ideal) gases and real (non-ideal solutions)

To explain the behaviour of *real gases and real solutions,* G.N. Lewis (in 1901) introduced the term *fugacity* and *activity* as a substitute for pressure and concentration respectively.

Fugacity may be defined as a substitute for pressure to explain the behaviour of real gas and activity may be defined as the substitute for concentration to explain the behaviour of a non-ideal solution.

In a mixture of real gases, the chemical potential μ_i of any constituent i is given by

$$\mu_i = \mu^\circ_i + RT \ln f_i \qquad \text{... } (iv)$$

where f_i is the fugacity of the constituent i in the mixture

Similarly, in a non-ideal solution, the chemical potential of any component i is given by

$$\mu_i = \mu^\circ_i + RT \ln a_i \qquad \text{... } (v)$$

where a_i is the activity of component i in the solution.

To explain the behaviour of real gases, the pressure has to be corrected by multiplying with a suitable factor to get the fugacity *i.e.*,

$$f = \gamma \times p \qquad \text{... } (vi)$$

γ is called the activity coefficient of the gas.

From equation (*vi*), we have

$$\gamma = \frac{f}{p} \qquad \text{... } (vii)$$

Hence activity coefficient of a gas may be defined as the ratio of the fugacity of the gas in the given state to the pressure of the gas in the same state.

For the ideal gas, $\gamma = 1$,

Therefore $f = p$

Similarly, in a non-ideal solution, the concentration is corrected to give the activity

$$a = \gamma \times C \qquad \text{... } (viii)$$

γ represents the activity coefficient of that components in the solution whose concentration is C.

From equation (*viii*), we have

$$\gamma = \frac{a}{C} \qquad \text{... } (ix)$$

Hence activity *coefficient of any component in the solution may be defined as the activity of that component in the solution to the concentration of the same component in the solution.*

For an ideal solution (*i.e.*, for very dilute solution), $\gamma = 1$

Therefore $a = C$

If concentration is expressed in terms of mole/litre of mole fraction, we use the symbol f to denote activity coefficient.

$$f = \frac{a}{C} \quad \text{i.e., } a = C \times f \quad \text{or} \quad f = \frac{a}{x} \quad \text{i.e., } a = x \times f$$

where C represents molar concentration and x represents mole fraction.

On the other hand symbol γ is used for the activity coefficient if the concentration is expressed in terms of molality (m),

$$\gamma = \frac{a}{m} \quad \text{or} \quad a = m \times \gamma$$

5.5. RAOULT'S LAW

Raoult's law states :

In a solution the vapour pressure of a component at a given temperature is equal to the mole fraction of that component in the solution multiplied by the vapour pressure of that component in the pure state. Mathematically,

$$P_i = x_i \times P^o$$

Thus in a *binary solution*

Vapour pressure of the solvent in the solution

= Mole fraction of the solvent in solution × Vapour pressure of the pure solvent ... (*i*)

Now if the solute is non-volatile, it will not contribute to the total vapour pressure of the solution. Thus the vapour pressure of the solution will be the vapour pressure due to solvent in the solution only.

For such solutions,

Vapour pressure of the solution

= Mole fraction of the solvent in solution × Vapour pressure of the pure solvent

or in terms of symbols, we can write

$$p_s = x_1 \times p^o \qquad \text{... (ii)}$$

This can be rewritten in the form

$$\frac{p_s}{p^o} = x_1 \qquad \text{... (iii)}$$

If the solution contains n_2 moles of the solute dissolved in n_1 moles of the solvent, we have

Mole fraction of the solvent in solution (x_1)

$$= \frac{n_1}{n_1 + n_2}$$

Substituting this value in equation (*iii*), we get

$$\frac{p_s}{p^o} = \frac{n_1}{n_1 + n_2}$$

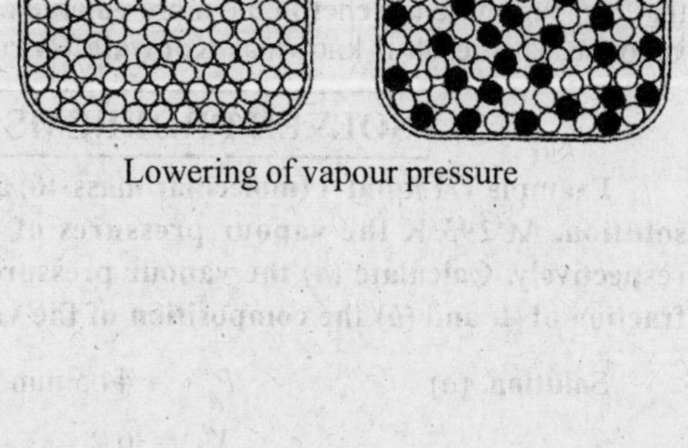

Lowering of vapour pressure

Subtracting each side from 1, we get

$$1 - \frac{p_s}{p^o} = 1 - \frac{n_1}{n_1 + n_2}$$

or $$\frac{p^o - p_s}{p^o} = \frac{n_2}{n_1 + n_2} \qquad \text{... (iv)}$$

In this expression, $p^o - p_s$ is the lowering of vapour pressure, $\frac{p^o - p_s}{p^o}$ is called relative lowering of vapour pressure, $\frac{n_2}{n_1 + n_2}$ represents the mole fraction of the solute in the solution. Hence the expression (*v*) may be expressed in words as follows :

The relative lowering of vapour pressure of a solution containing a non-volatile solute is equal to the mole fraction of the solute in the solution.

This is another definition of Raoult's law.

5.5.1. Relationship between the mole fractions of the components in the liquid Phase and in the vapour phase

Consider a binary solution of two components A and B. Let us suppose x_A and x_B are their mole fractions in the liquid phase and $P°_A$ and $P°_B$ are their vapour pressures in the pure state. The partial pressures of A and B will be given $p_A = x_A x°_A$ and $p_B = x_B p°_B$. Let us further suppose that y_A and y_B are the mole fractions of A and B in the vapour phase. y_A and y_B can be calculated as under:

$$y_A = \frac{p_A}{p_A + p_B} = \frac{x_A \overset{\circ}{p}_A}{x_A \overset{\circ}{p}_A + x_B \overset{\circ}{p}_B} \quad \text{... (i)}$$

$$= \frac{(1 - x_B)p^o{}_A}{(1 - x_B)p^o{}_A + x_B p^o{}_B} = \frac{(1 - x_B)p^o{}_A}{(p^o{}_B - p^o{}_A)\, x_B + p^o{}_A} \quad \text{... (ii)}$$

$$y_B = \frac{P_B}{P_A + P_B} = \frac{x_B P^o B}{x_A P^o A + x_B P^o B} \quad \text{... (iii)}$$

$$= \frac{x_B p^o B}{(p^o_B - P^o_A) x_B + p^o_A} \quad \text{... (iv)}$$

On dividing eqn. (*iii*) by eqn. (*i*). We get

$$\frac{y_B}{y_A} = \frac{x_B p^o_B}{x_A p^o_A} = \frac{x_B}{x_A} \times \frac{p^o_B}{p^o_A} \quad \text{... (v)}$$

Thus if B is more volatile than A, *i.e.* $p^o_B > p^o_A$ or $p^o_B / p^o_A > 1$, then $y_B/y_A > x_B/x_A$. This means that vapour phase is richer in the more volatile component B than the liquid phase from which it vaporise. This result is known as **Konwaloff's rule.**

SOLVED PROBLEMS ON RAOULT'S LAW

Example 1. Liquid A (molecular mass 46) and liquid *B* (molecular mass 18) form an ideal solution. At 293 K the vapour pressures of pure *A* and *B* are 44.5 and 17.5 mm of Hg respectively. Calculate (*a*) the vapour pressure of a solution of *A* in *B* containing 0.2 mole fraction of *A*, and (*b*) the composition of the vapour phase.

Solution. (*a*)

$$P^o_A = 44.5 \text{ mm}$$

$$X_A = 0.2$$

$$P_B{}^o = 17{\cdot}5 \text{ mm}$$

$$X_B = (1 - 0.2) = 0.8$$

According to Raoult's law

$$P_A = P^o_A . X_A$$

Partial Pressure of A, $P_A = P^o_A X_A = 44.5 \times 0.2 = 8.9$ mm.

Partial Pressure of B, $P_B = P_B^o X_B$

$= 17.5 \times 0.8$

$= 14.0$ mm.

Total pressure $P = P_A + P_B$

$= 8.9 + 14 = 22.9$ mm.

(*b*) In vapour phase

$$\text{Mole fraction of } A = \frac{\text{Partial Presure of A}}{\text{Total pressure}}$$

$$\text{or} \quad X_A = \frac{P_A}{P_A + P_B} = \frac{8.9}{8.9+14} = \frac{8.9}{22.9} = 0.39$$

$$X_A = 1 - 0.39 = 0.61.$$

Example 2. The vapour pressure of two pure liquids *A* and *B* are 15000 and 30000 Nm^{-2} at 298 K. Calculate the mole fraction of *A* and *B* in the vapour phase when an equimolar solution of the liquids is made.

Solution. $P_A^o = 15000 \text{ Nm}^{-2}$

$P_B^o = 30000 \text{ Nm}^{-2}$

In equimolar solution, mole fractions of A and B, *i.e.*, X_A and X_B are equal. Let

$X_A = X_B = 0.5$

$X_A = 0.5$

$X_B = 1 - 0.5 = 0.5$

Applying Raoult's law of ideal solution.

$p_A = P_A^o X_A$

$= 15000 \times 0.5 = 7500 \text{ Nm}^{-2}$

and $P_B = P_B^o X_B$

$= 30000 \times 0.5 = 15000 \text{ Nm}^{-2}$

Total pressure $P = P_A + P_B$

$= 7500 + 15000 = 22500 \text{ Nm}^{-2}$

In the vapour phase

$$\text{Mole fraction of } A = \frac{\text{Partial pressure A}}{\text{Total pressure}}$$

$$= \frac{7500}{22500} = 0.3333$$

Mole fraction of $B = 1 - 0.3333 = 0.6667$

Example 3. Benzene C_6H_6 (b.p. 353.1 K) and toluene C_7H_8 (b.p. 383.6 K) are two hydrocarbons that form a very nearly ideal solution. At 313 K, the vapour pressure of pure liquids are 160 mm Hg and 60 mm Hg respectively. Assuming an ideal solution behaviour, calculate the partial pressures of benzene and toluene and the total pressure over the following solutions:

(*i*) One made by combining equal number of toluene and benzene molecules.
(*ii*) One made by combining 4 mol of toluene and 1 mol of benzene.
(*iii*) One made by combining equal masses of toluene and benzene.

Solution. (*i*) When the number of molecules of toluene and benzene are equal, that means the number of moles of the two liquids are also equal :

Thus mole fraction of benzene $X_A = \frac{1}{1+1} = 0.5$

Mole fraction of toluene $X_B = 1 - 0.5 = 0.5$

According to Raoult's law

$$P = P^o . X$$

Partial pressure of benzene $P_A = P_A^o . X_A = 160 \times 0.5 = 80$ mm

Partial pressure of toluene $P_B = P_B^o X_B = 60 \times 0.5 = 30$ mm

Total vapour pressure $P_A + P_B = 80 + 30 = 110$ mm

(*ii*) Mole fraction of benzene $X_A = \frac{1}{1+4} = \frac{1}{5} = 0.2$

Mole fraction of toluene $X_B = 1 - 0.2 = 0.8$

Partial pressure of benzene $P_A = P^o{}_A X_A = 160 \times 0.2 = 32$ mm

Partial pressure of toluene $P_B = 60 \times 0.8 = 48$ mm

Total vapour pressure $P_A + P_B = 32 + 48 = 80$ mm

(*iii*) Here the masses of the two liquids are the same. Let the amount of each be *m* g

Then mole fraction of benzene $X_A = \frac{m/78}{m/78 + m/92} = 0.541$

Mole fraction of toluene $X_B = 1 - 0.541 = 0.459$

Partial pressure of benzene $P_A = P_A^o X_A = 160 \times 0.541$

$= 86.56$ mm of Hg

Partial pressure of toluene $P_B = P_B^o X_B$

$= 60 \times 0.459 = 27.54$ mm of Hg

Total vapour pressure $= 86.56 + 27.54 = 114.1$ mm

Example 4. The vapour pressures of pure components *A* and *B* are 120 mm and 96 mm Hg. What will be the partial pressures of the components and the total pressure when the solution contains 1 mole of component *A* and 4 mole of component *B* and the solution is ideal? What will be the composition in the vapour phase ?

Solution.

Here

$P_A^o = 120$ mm

$P_B^o = 96$ mm

Mole fraction of *A*, $X_A = \frac{\text{No. of moles of A}}{\text{No. of moles of A + no. of moles of B}}$

$= \frac{1}{1+4} = \frac{1}{5} = 0.2$

Mole fraction of *B*, $X_B = 1 - 0.2 = 0.8$

Partial pressure of component *A*, $P_A = P_A^o \times X_A$

$= 120 \times 0.2 = 24.0$ mm Hg

Partial pressure of B, $P_B = P_B^o \times X_B$

$= 96 \times 0.8 = 76.8$ mm

Total pressure $P_A + P_B = 24 + 76.8$ mm

$= 100.8$ mm

Mole fraction of A in the vapour phase $= \dfrac{P_A}{\text{Total pressure}} = \dfrac{24}{100.8} = 0.238$

Mole fraction of B in the vapour phase $= 1 - 0.238 = 0.762$

5.6. DERIVATION OF DUHEM-MARGULES EQUATION

Consider a liquid mixture of two components in equilibrium with their vapour. At *constant temperature and pressure,* the condition for an infinitesimal change of composition is given by Gibbs-Duhem equation in the form

$$n_1 d\mu_1 + n_2 d\mu_2 = 0 \quad \text{... (i)}$$

where n_1 and n_2 are the number of moles of the constituents 1 and 2 respectively in the solution and μ_1 and μ_2 are their chemical potentials.

Dividing eqn. (*i*) by $n_1 + n_2$, we get

$$\frac{n_1}{n_1+n_2} d\mu_1 + \frac{n_2}{n_1+n_2} d\mu_2 = 0$$

or

$$x_1 d\mu_1 + x_2 d\mu_2 = 0 \quad \text{... (ii)}$$

where x_1 and x_2 represent the mole fractions of the components 1 and 2 in the mixture (solution).

Now, the chemical potential of any component in the mixture depends upon temperature, pressure and composition of the mixture. Hence if the temperature and pressure are kept constant, the chemical potential depends only upon the composition. Then for an infinitesimal change of composition, it is possible to write

$$d\mu_1 = \left(\frac{\partial \mu_1}{\partial x_1}\right)_{T,P} dx_1 \quad \text{... (iii)}$$

and

$$d\mu_2 = \left(\frac{\partial \mu_2}{\partial x_2}\right)_{T,P} dx_2 \quad \text{... (iv)}$$

Substituting these values in eqn. (*iv*), we get

$$x_1\left(\frac{\partial \mu_1}{\partial x_1}\right)_{T,P} dx_1 + x_2\left(\frac{\partial \mu_2}{\partial x_2}\right)_{T,P} dx_2 = 0 \quad \text{... (v)}$$

Further we know that the sum of mole fractions is unity *i.e.*

$$x_1 + x_2 = 0$$

Hence

$$dx_1 + dx_2 = 0$$

or

$$dx_1 = -dx_2 \quad \text{... (vi)}$$

Substituting this value in eqn. (*v*) and simplifying, we get

$$x_1 \left(\frac{\partial \mu_1}{\partial x_1}\right) - x_2 \left(\frac{\partial \mu_2}{\partial x_2}\right)_{T,P} = 0$$

or
$$x_1 \left(\frac{\partial \mu_1}{\partial x_1}\right)_{T,P} = x_2 \left(\frac{\partial \mu_2}{\partial x_2}\right)_{T,P} \quad \text{... (vii)}$$

The chemical potential of any constituent of a liquid mixture is given by

$$\mu = \mu_o + RT \ln p \quad \text{... (viii)}$$

where p is the partial pressure of that constituent (assuming that the vapour behave like an ideal gas).

Differentiating eqn. (*viii*) *w.r.t.* x at constant temperature, we get

$$\frac{\partial \mu}{\partial x} = RT \frac{d \ln p}{dx}$$

Thus for the constituents 1 and 2, we can write

$$\frac{\partial \mu_1}{\partial x_1} = RT \frac{d \ln p_1}{dx_1} \quad \text{... (ix)}$$

and
$$\frac{\partial \mu_2}{\partial x_2} = RT \frac{d \ln p_2}{dx_2} \quad \text{... (x)}$$

Substituting these values in eqn. (*vii*) and simplifying, we get

$$x_1 \frac{d \ln p_1}{dx_1} = x_2 \frac{d \ln p_2}{dx_2} \quad \text{... (xi)}$$

or it can be written in the form

$$\frac{d \ln p_1}{d \ln x_1} = \frac{d \ln p_2}{d \ln x_2} \quad \text{... (xii)}$$

Equations (*xi*) and (*xii*) are two different forms of Duhem-Margules equation.

5.7. IDEAL AND NON-IDEAL SOLUTIONS

A solution of two components is said to be ideal of each components of the solution obeys Raoult's law at all temperature and concentrations. Thus, if two liquids A and B are mixed together, the solution formed by them will be ideal if

$$P_A = X_A \times P^\circ_A \quad \text{and} \quad P_A = X_B \times P^\circ_B$$

where P_A = vapour pressure of component A in the solution

X_A = mole fraction of the component A in the solution = $\frac{n_A}{n_A + n_B}$, where n_A and n_B are the number of moles of A and B respectively in the solution

p^o_A = vapour pressure of the liquid A in the pure state

p_B, x_B, p^o_B represent similar terms for the component B.

It is observed that the solution formed by two liquids A and B is ideal if the interactions among the molecules of A and B in the solution are just the same as among the molecules of pure A or B.

Also, it is observed that the *solution formed on mixing the two liquids is ideal if*

(*i*) there is no volume change on mixing i.e., $\Delta V_{mix} = 0$

(*ii*) there is no enthalpy change on mixing i.e., $\Delta H_{mix} = 0$.

For example, if 25 ml on *n*-hexane are mixed with 25 ml of *n*-heptane, the total volume of the solution is found to be 50 ml and the temperature of the solution is found to be same as that of pure constituents before mixing showing that no heat is evolved or absorbed on mixing. Such a solution is an ideal solution.

On the other hand, a solution formed by mixing two liquids is said to be non-ideal if it does not satisfy the above conditions i.e., it does not obey Raoult's law or the interaction of *A* and *B* molecules in the solution are not similar to those of pure *A* and pure *B* or $\Delta V_{\text{mix}} \neq 0$ and $\Delta H_{\text{mix}} \neq 0$.

Examples of non-ideal solution are water + sulphuric acid, water + nitric acid, ethanol + water etc.

5.8. THERMODYNAMIC PROPERTIES OF IDEAL SOLUTIONS

Suppose n_1 moles of pure component *A* are mixed with n_2 moles of pure component *B* to form a binary ideal solution. Then the change in free energy, volume, enthalpy and entropy is calculated as follows:

Free energy change on mixing.

Free energy change on mixing the pure components 1 and 2 is given by

$$\Delta G = (\text{Total free energy of the solution}) - (\text{Total free energy of the pure components})$$

$$= G_{\text{solution}} - G_{\text{pure components}} \quad (i)$$

Let G_1 be the free energy per mole of the component 1 in the solution and G_2 be the free energy per mole of the component 2 in the solution, then

$$G_{\text{solution}} = n_1 G_1 + n_2 G_2 \quad ... (ii)$$

Further of $\overline{G}_1$ is the free energy per mole of the pure component 1 and $\overline{G}_2$ is the free energy per mole of the pure component 2, then

$$G_{\text{pure component}} = n_1 \overline{G}_1 + n_2 \overline{G}_2 \quad ... (iii)$$

Substituting the values from equations (*ii*) and (*iii*) in equation (*i*), we get

$$\Delta G_{\text{mix}} = (n_1 G_1 + n_2 G_2) - (n_1 \overline{G}_1 + n_2 \overline{G}_2)$$

$$= n_1(\overline{G}_1 - G_1) + n_2(G_2 - \overline{G}_2) \quad ... (iv)$$

The free energy per mole of any component in the solution (non-ideal) is given by

$$G = G° + RT \ln a$$

where G° is the free energy per mole in the standard state and *a* is the activity of that component in the solution.

For component 1 in the solution, we have

$$G_1 = G_1^{\circ} + RT \ln a_1 \quad ... (v)$$

where G_1° is the free energy per mole of component 1 in the standard state.

As the standard state chosen for a liquid or a solid is the pure liquid or pure solid, therefore

$$G_1^{\circ} = \overline{G}_1. \qquad \{\because a = 1 \text{ for pure liquid or pure solid}\}$$

Putting this value in equation (*v*), we have

$$G_1 = \bar{G}_1 + RT \ln a_1 \quad \text{... (vi)}$$

Similarly for component 2 in the solution, we have

$$G_2 = \bar{G}_2 + RT \ln a_2 \quad \text{... (vii)}$$

Equations (*vi*) and (*vii*) can be rearranged as

$$G_1 - \bar{G}_1 = RT \ln a_1$$

$$G_2 - \bar{G}_2 = RT \ln a_2$$

Substituting these values in equation (*iv*), we get

$$\mathbf{\Delta G_{mix} = n_1 RT \ln a_1 + n_2 RT \ln a_2} \quad \text{... (viii)}$$

For ideal solution, the activity of any component in the solution (a) is equal to its mole fraction (x)

Thus, we have

$$a_1 = x_1 \text{ and } a_2 = x_2$$

Equation for ideal solution can be written as

$$\Delta G_{mix} = n_1 RT \ln x_1 + n_2 RT \ln x_2$$

For solutions of many components, we can write the following equations for real and ideal solutions.

$$\Delta G_{mix} = \Sigma n_i RT \ln a_i \text{ for real solution}$$

$$\Delta G_{mix} = \Sigma n_i RT \ln x_i \text{ for ideal solution}$$

Volume change on mixing

For ideal solution,

$$\Delta G_{mix} = n_1 RT \ln x_1 + n_2 RT \ln x_2.$$

Each term on the right hand side is independent of pressure. Differentiating with respect to pressure at constant temperature will give

$$\left(\frac{\partial(\Delta G_{mix})}{\partial P}\right)_T = 0 \quad \text{... (x)}$$

But

$$\left(\frac{\partial(\Delta G)}{\partial P}\right)_T = \Delta V$$

For the process of mixing, it can be written as

$$\left(\frac{\partial(\Delta G_{mix})}{\partial P}\right)_T = \Delta V_{mix} \quad \text{... (xi)}$$

Combining equations (*x*) and (*xi*), we get

$$\mathbf{\Delta V_{mix} = 0, \text{ for an ideal solution}}$$

i.e. when pure liquids are mixed to form an ideal solution, there is no change in volume on mixing.

Enthalpy change on mixing

For the ideal solution, again

$$\mathbf{\Delta G_{mix}} = n_1 RT \ln x_1 + n_2 RT \ln x_2.$$

Dividing throughout by T, we get

$$\frac{\Delta G_{mix}}{T} = n_1 R \ln x_1 + n_2 \text{R} \ln x_2$$

Now the terms on the right hand side are independent of temperature. Differentiating this equation with respect to temperature at constant pressure, we get

$$\left(\frac{\partial(\Delta G_{mix}/T)}{\partial T}\right)_P = 0 \qquad \text{... (xii)}$$

According to one of the forms of the Gibbs-Helmholtz equation (from Thermodynamics), we have

$$\left(\frac{\partial(\Delta G/T)}{\partial T}\right)_P = -\frac{\Delta H}{T^2}$$

or for the process of mixing, it can be written as

$$\left(\frac{\partial(\Delta G_{mix}/T)}{\partial T}\right)_P = -\frac{\Delta H_{mix}}{T^2} \qquad \text{... (xiii)}$$

Combining equations (*xii*) and (*xiii*), we get

$$\mathbf{\Delta H_{mix} = 0, \text{ for an ideal solution}}$$

i.e. no heat is evolved or absorbed when an ideal solution is formed.

Entropy change on mixing

$$\Delta G = \Delta H - T\,\Delta S$$

For the process of mixing, it can be written as

$$\Delta G_{mix} = \Delta H_{mix} - T\Delta S_{mix} \qquad \text{... (xiv)}$$

For an ideal solution, we have already derived that

$$\Delta G_{mix} = 0 \qquad \text{... (xv)}$$

Substituting this value in equation (*xiv*), we get

$$\Delta G_{mix} = -T\Delta S_{mix} \quad \text{or} \quad \Delta S_{mix} = -\frac{\Delta G_{mix}}{T} \qquad \text{... (xvi)}$$

Substituting the value of ΔG_{mix} from equation (*ix*) in equation (*xvi*), we get

$$\mathbf{\Delta S_{mix} = -n_1 R \ln x_1 - n_2 R \ln x_2} \qquad \text{... (xvii)}$$

It may be noted that since each of the mole fractions i.e x_1 and x_2 is less than 1, ln x_1 and ln x_2 will be negative quantities. Hence ΔS_{mix} will be positive i.e. the process of mixing is accompanied by increase of entropy. Since temperature is not involved in eq (*xvii*) the entropy change of mixing depends only upon the quantities of the components and is independent of the temperature.

To get ΔG_{mix} or ΔS_{mix} per mole of the mixture, divide both sides of eq. (*ix*) and (*xvii*) by n_1+n_2. We get

$$\Delta G_{mix} = x_1 RT \ln x_1 + x_2 RT \ln x_2$$

$$\Delta S_{mix} = -x_1 R \ln x_1 - x_2 R \ln x_2$$

All the above relation apply to mixture of perfect gases as well.

Example 1. Calculate the entropy of mixing when 15 g of benzene are mixed with 15 g of cyclohexane.

Solution. Entropy of mixing (ΔS_{mix}) is given by the equation

$$\Delta S_{mix} = -n_1 R \ln x_1 - n_2 R \ln x_2$$

$$= -2.303\, R\,(n_1 \log x_1 + n_2 \log x_2)$$

$$n_1 = \text{number of moles of component 1}$$

x_1 = mole fraction of component 1

n_2, x_2 are the corresponding values for component 2.

$$\text{No. of moles of benzene, } C_6H_6\,(n_1) = \frac{\text{Mass}}{\text{Mol.mass}} = \frac{15}{78} = 0.192$$

$$\text{No. of moles of benzene, } C_6H_6\,(n_1) = \frac{\text{Mass}}{\text{Mol.mass}} = \frac{15}{84} = 0.178$$

$$x_1 = \frac{n_1}{n_1 + n_2} = \frac{0.192}{0.192 + 0.178} = 0.5189$$

$$x_2 = 1 - x_1 = 0.4811$$

$$\Delta S_{mix} = -2.303 \times 8.314\,[(0.192 \log 0.5189 + 0.178 \log 0.4811)]$$

$$= -2.303 \times 8.314\,[0.192\,(-0.2849) + 0.178\,(-0.3178)]$$

$$= \mathbf{2.13\ JK^{-1}}$$

5.9. VAPOUR PRESSURE OF IDEAL SOLUTIONS

At any temperature, the molecules of a liquid have tendency to escape and form vapours and thus possess some vapour pressure. Similarly in case of solutions of two volatile liquids also, the molecules of both the liquids have the tendency to form vapours and hence possess some vapour pressure. The total vapour pressure of the solution (P_{sol}) is the sum of the partial pressure of the two components (p_A, p_B) i.e.,

$$P_{sol.} = p_A + p_B$$

In case of **ideal solutions**, the partial pressure of each component can be calculated by Raoult's law

$$p_A = x_A \times p_A^o \quad \text{and} \quad p_B = x_B \times p_B^o$$

where pA, p_A^o, p_B *and* p_B^o are all measured at same temperature.

Thus $$P_{sol.} = x_A\, p_A^o + x_B\, p_B^o$$

At any particular temperature, the value of $P_{sol.}$ can be determined experimentally for different mole fractions. The values p_A^o and p_B^o for the pure liquids being known at the same temperature, p_A and p_B can be calculated for any particular mole fraction. For an ideal solution, the straight line plots are obtained for vapour pressures *vs.* compositions as shown in Fig. 5.1 where dotted lines represent the plots of the partial pressures and the solid line that of the solution.

In these plots, for any mole fraction (say x), the total vapour pressure of the solution is the sum of the partial pressures of its components A and B.

Some examples of liquid pairs forming an ideal solution are n-hexane + n-heptane, ethyl bromide + ethyl iodide, n-butyl chloride + n-butyl bromide.

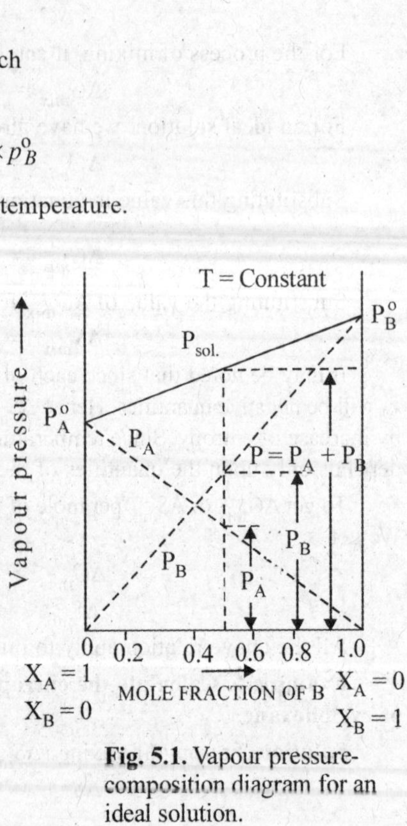

Fig. 5.1. Vapour pressure-composition diagram for an ideal solution.

5.10. VAPOUR PRESSURE VERSUS VAPOUR COMPOSITION CURVES FOR IDEAL SOLUTIONS

Fig . 51. gives total vapour pressure versus mole fraction of the component B in the solution. For any mole fraction of B in the solution (x_B), the mole fraction of B in the vapour phase (y_B) is usually not the same. The relationship between the two as derived in Art. 5.6 eq (*iv*) is

$$y_B = \frac{x_B\, p_B^o}{(p_B^o - p_A^o)\, x_B + p_A^o}$$

Thus for every definite value x_B, there will be definite value for y_B and the two will not be same except when $\overset{o}{p_A} = \overset{o}{p_B}$.

By calculating the values of y_B corresponding to the definite values of x_B, we can obtain vapour pressure – vapour composition curve for ideal solutions (Fig. 5.2). The straight line gives the plot of total vapour pressure versus mole fraction of component B in the solution.

While the curve below it gives the plot of total vapour pressure versus mole fraction of component B in the *vapour phase*. To obtain the composition of vapour phase corresponding to any mole fraction x_B of the solution, we move vertically upwards to the point P and then horizontally to the point Q and then vertically downwards to get y_B.

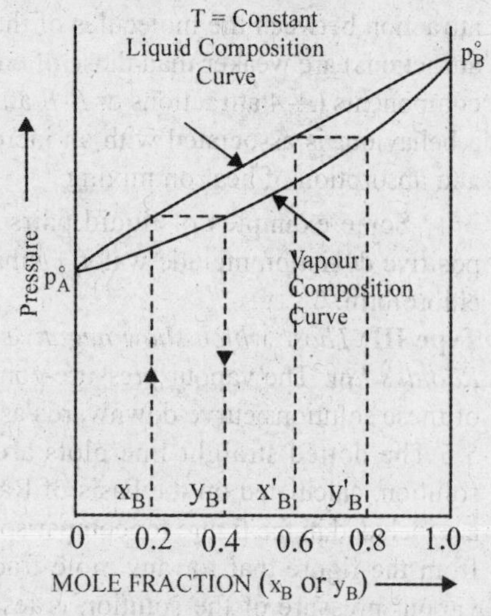

Fig 5.2 Liquid and vapour composition curves for an ideal solution.

From Fig. 5.2 we can readily observe that corresponding to any mole fraction in the solution, the vapour phase is richer in the more volatile component B. (For $x_B = 0.2$, $y_B = 0.4$, for $x'_B = 0.6, y'_B \simeq 0.8$).

5.11. VAPOUR PRESSURE-COMPOSITION CURVES FOR NON-IDEAL SOLUTIONS (DEVIATION FROM IDEAL BEHAVIOUR)

In case of non-ideal solutions, the plots of vapour pressure *vs.* composition (mole fraction) are curved lines instead of straight lines. Non-ideal solutions are divided into the following three types :

Type I: *Those which show small positive deviation from Raoult's law.* The vapour pressure-composition graphs of these solutions curve slightly upwards as shown in Fig. 5.3. The dotted straight line plots are for ideal solution, calculated by using Raoult's law.

Some examples of solutions of this type are benzene + toluene, water + methyl alcohol, carbon tetrachloride + cyclohexane etc.

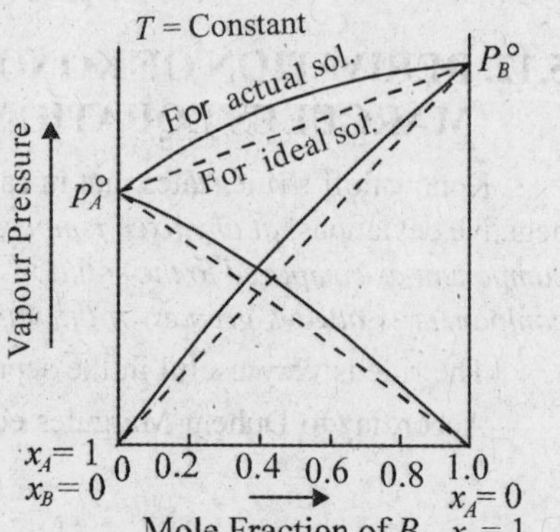

Fig. 5.3. Vapour pressure-composition plots for non-ideal solution showing small positive deviation.

Type II: *Those which show large positive deviation from Raoult's law.* The vapour pressure-composition plots of these solutions curve upwards considerably as shown in Fig. 5.4. The dotted straight line plots, included in the figure for comparison, are for ideal solution (calculated by means of Raoult's law).

It is obvious from Figs. 5.3 and 5.4 that for any mole fraction, the total vapour pressure of the solution is more than that for ideal solution. This means that the tendency for the molecules to escape from the solution is more than from the pure liquids. This indicates that in such solutions, the intermolecular forces of attraction between the molecules of the solution (*A-B* attractions) are weaker than those of either of the pure components (*A-A* attractions or *B-B* attractions). Such a behaviour is associated with an increase in volume and absorption of heat on mixing.

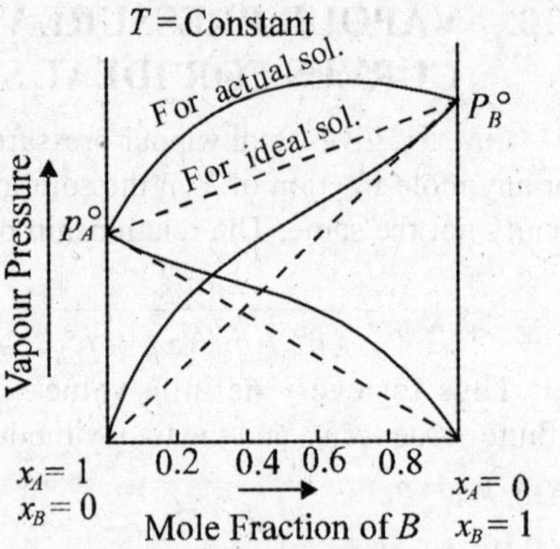

Fig. 5.4. Vapour pressure-composition plots for non-ideal solution showng large positive deviation

Some examples of liquid pairs showing large positive deviation include water + ethanol, ethanol + chloroform.

Type III: *Those which show negative deviation from Raoult's law.* The vapour pressure-composition plots of these solutions curve downwards as shown in Fig. 5.5 The dotted straight line plots are for the ideal solution, calculated on the basis of Raoult's law and are included in the figure for comparison. It is obvious from the figure that for any mole fraction, the total vapour pressure of the solution is less than that for ideal solution. This means that the tendency for the molecules to escape from the solution is less than that from the pure liquids. This indicates that in such solutions, the intermolecular forces of attraction between the molecules of the solution (*A-B* attractions) are stronger than those of either of the pure components (*A-A* attractions or *B-B* attractions). It is related to the contraction in volume and evolution of heat in making the solution.

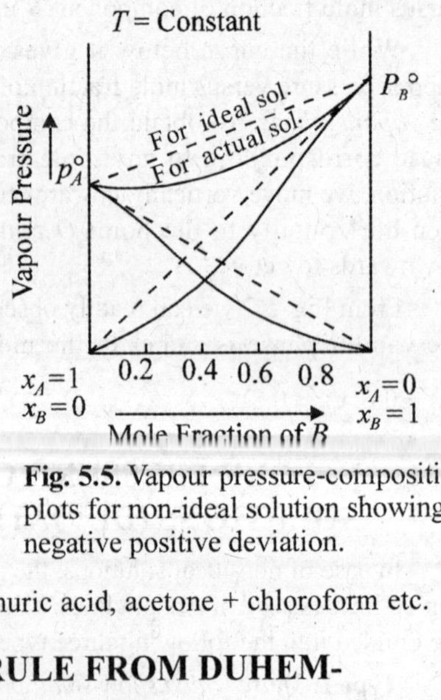

Fig. 5.5. Vapour pressure-composition plots for non-ideal solution showing negative positive deviation.

Some examples of the liquid pairs showing negative deviation are : water + nitric acid, water + sulphuric acid, acetone + chloroform etc.

5.12. DERIVATION OF KONOWALOFF'S RULE FROM DUHEM-MARGULES EQUATION

Konowaloff's rule states that in case of ideal solutions or solutions showing small positive or negative deviations, *at any fixed temperature, the vapour phase is always richer in the more volatile component as compared to the solution phase.* In other words, *the mole fraction of the more volatile component is always greater in the vapour phase than in the solution phase.*

The rule is very useful in the separation of binary liquid solutions by distillation.

According to Duhem-Margules equation,

$$x_1 = \frac{d \ln p_1}{dx_1} = x_2 \frac{d \ln p_2}{dx_2} \qquad \ldots (i)$$

It may be rewritten as $\dfrac{x_1}{dx_1}\dfrac{dp_1}{p_1} = \dfrac{x_2}{dx_2}\dfrac{dp_2}{p_2}$

or
$$\frac{x_1}{p_1}\frac{dp_1}{dx_1} = \frac{x_2}{p_2}\frac{dp_2}{dx_2} \quad ...(ii)$$

But $x_1 + x_2 = 1$ (sum of mole fractions = 1)

$\therefore \quad dx_1 + dx_2 = 0$ or $dx_1 = -dx_2$

$\therefore$ Equation (*ii*) becomes

$$\frac{x_1}{p_1}\frac{dp_1}{dx_1} = -\frac{x_2}{p_2}\frac{dp_2}{dx_1} \quad ...(iii)$$

or
$$\frac{dp_1}{dx_1} = -\frac{x_2 p_1}{x_1 p_2}\frac{dp_2}{dx_1} \quad ...(iv)$$

If P is the total vapour pressure, then

$$P = p_1 + p_2$$

$$\therefore \quad dP = dp_1 + dp_2$$

or
$$\frac{dP}{dx_1} = \frac{dp_1}{dx_1} + \frac{dp_2}{dx_1} \quad ...(v)$$

Substituting the value of dp_1/dx_1 from equation (*iv*), we get

$$\frac{dP}{dx_1} = -\frac{x_2 p_1}{x_1 p_2}\frac{dp_2}{dx_1} + \frac{dp_2}{dx_1}$$

$$= \frac{dp_2}{dx_1}\left(1 - \frac{x_2 p_1}{x_1 p_2}\right) \quad ...(vi)$$

Now as $\quad dx_1 = -dx_2$

$$\frac{dp_2}{dx_1} = -\frac{dp_2}{dx_2}$$

Thus $\frac{dp_2}{dx_1}$ is a negative quantity. Hence according to equation (*vi*), $\frac{dP}{dx_1}$ can be positive only if

$$\frac{x_2 p_1}{x_1 p_2} > 1$$

or
$$\frac{p_1}{p_2} > \frac{x_1}{x_2}$$

This means that the vapour is richer in component 1 (or *A*) than is the liquid from which it vaporises.

5.13. VAPOUR PRESSURE-LIQUID COMPOSITION AND VAPOUR PRESSURE-VAPOUR COMPOSITION CURVES FOR LIQUID MIXTURES

At any particular temperature, the composition of the vapour phase is not the same as that of the liquid phase. Hence we have different *vapour pressure-vapour composition curves for the two phases.* At any particular temperature, these can be obtained by measuring the vapour pressure as well as the composition of the vapour phase corresponding to each composition of the liquid mixture.

For the real solutions, there are following three types of vapour pressure-composition diagrams:

Type I: *Those in which total vapour pressure is intermediate between those of the pure components and do not show any maximum or minimum.*

Type II: *Those which show a maximum in the total vapour pressure curve.*

Type III: *Those which show a minimum in the total vapour pressure curve.*

The vapour pressure-liquid composition curves as well as the vapour pressure-vapour composition curves of all the three types are given in Fig. 5.6, in which liquid *B* is more volatile than liquid *A*

It may be observed that in all the three types, the vapour composition curve lies below the liquid composition curve. This is explained for each type as under :

(*i*) *In type I,* as *B* is more volatile than *A,* corresponding to composition *a* of the liquid phase, vapour phase will be richer in *B i.e.,* corresponding to composition *a′*.

(*ii*) *In type II,* upto the point *C,* behaviour is same as in Type I, at *C,* the liquid phase and the vapour phase have the same composition and after point *C*, the vapour phase is richer in *A* (or less rich in *B*) as seen from points *c* and *c′*.

(*iii*) *In type III,* upto the point *D,* the vapour phase is richer in *A* (or less rich in *B*) than the liquid phase, at the point *D* both have the same composition and after point *D,* the behaviour is similar to Type I. This can be seen from points *d, d′, e* and *e′*.

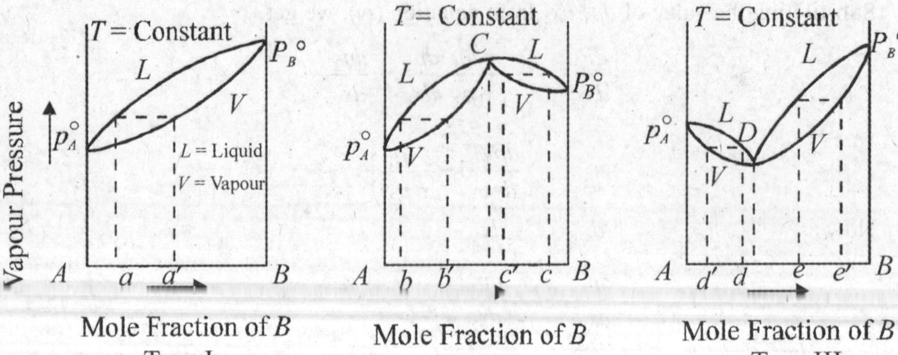

Fig. 5.6. Vapour pressure composition diagrams of binary miscible real solution at constant temperature.

5.14. BOILING POINT-COMPOSITION DIAGRAMS FOR NON-IDEAL SOLUTIONS

Boiling point is the temperature at which the total vapour pressure becomes equal to the atmospheric pressure. Thus a liquid or a solution having low vapour pressure will have a higher boiling point because it will have to be heated more to make the vapour pressure equal to the atmospheric pressure. Thus pure liquid *A* which has less vapour pressure will have a higher boiling point and the pure liquid *B* which has higher vapour pressure will have lower boiling point. In between these two extremes, a solution of any composition having lower vapour pressure will have higher boiling point. These facts help us to construct the boiling point-composition diagrams from the vapour pressure-composition diagrams as shown in Fig. 5.7.

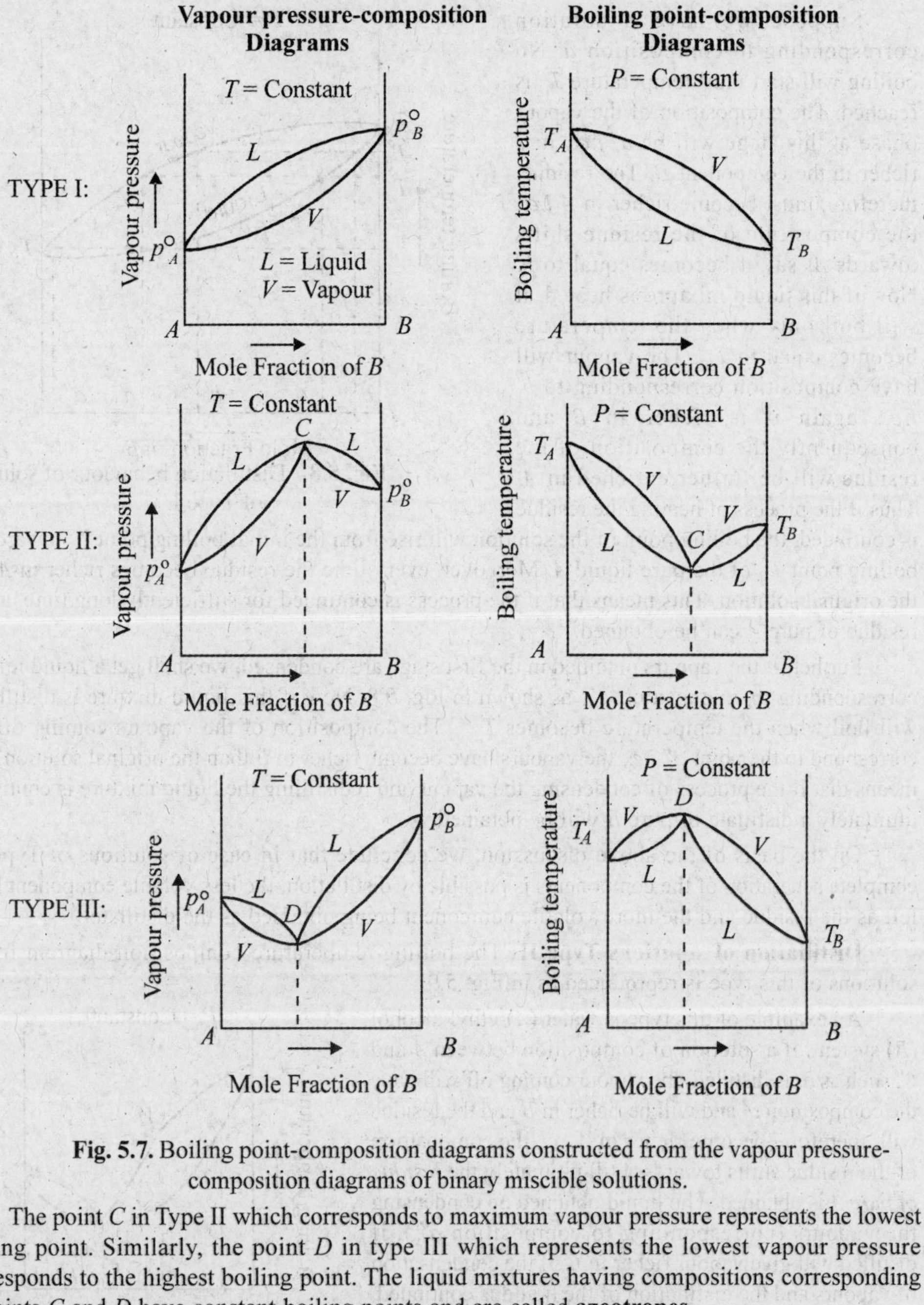

Fig. 5.7. Boiling point-composition diagrams constructed from the vapour pressure-composition diagrams of binary miscible solutions.

The point C in Type II which corresponds to maximum vapour pressure represents the lowest boiling point. Similarly, the point D in type III which represents the lowest vapour pressure corresponds to the highest boiling point. The liquid mixtures having compositions corresponding to points C and D have constant boiling points and are called **azeotropes**.

5.15. DISTILLATION OF BINARY MISCIBLE SOLUTIONS

Distillation of solutions of Type I–Fractional distillation: The boiling temperature-composition diagram for such a solution involving the liquids A and B in which B is more volatile than A is explained with the help of Fig. 5.8.

Suppose we have a solution corresponding to composition *a*. No boiling will start until temperature T_a is reached. The composition of the vapour phase at this stage will be a' *i.e.*, it is richer in the component *B*. The residue, therefore, must become richer in *A* *i.e.*, the composition of the residue shifts towards *A*, say it becomes equal to *b*. Now if this liquid mixture is heated, it will boil only when the temperature becomes equal to T_b. The vapour will have composition corresponding to b' *i.e.*, again it is richer in *B* and consequently the composition of the residue will be further enriched in *A*. Thus if the process of heating the residue is continued, the boiling point of the solution will rise from the initial boiling point T_a towards the boiling point T_A of the pure liquid *A*. Moreover, every time the residue becomes richer in *A* than the original solution. This means that if the process is continued for sufficiently long time, a final residue of pure *A* can be obtained.

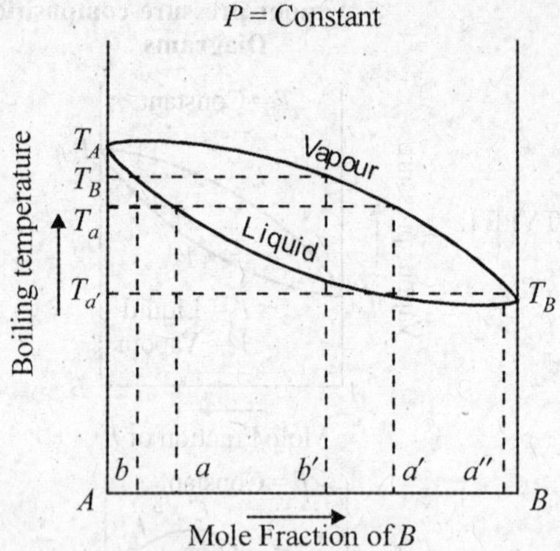

Fig. 5.8. Distillation behaviour of solutions of Type I.

Further, if the vapours obtained in the first stage are condensed, we shall get a liquid mixture corresponding to composition a', as shown in Fig. 5.8. Now if this liquid mixture is distilled, it will boil when the temperature becomes T_a'. The composition of the vapours coming off will correspond to the point a'' *i.e.*, the vapours have become richer in *B* than the original solution. This means that if the process of condensing the vapour and redistilling the liquid mixture is continued, ultimately a distillate of pure *B* will be obtained.

On the basis of the above discussion, we conclude that in case of solutions of Type I, a complete separation of the components is possible by distillation, the less volatile component being left as the residue and the more volatile component being obtained as the distillate.

Distillation of solutions Type II: The boiling temperature-composition diagram for the solutions of this type is reproduced as in Fig. 5.9.

An example of this type is water (*A*)-ethyl alcohol (*B*) system. If a solution of composition between *A* and *C*, such as *a*, is distilled, the vapour coming off will have the composition a' and will be richer in *B* and the residue will, therefore, become richer in *A* *i.e.*, the composition of the residue shifts towards *A* till ultimately the *residue of pure A* is obtained. The liquid obtained on condensing the vapours (corresponding to composition a'), if distilled will give vapour richer in *B*. If the condensation of vapours and the distillation of the liquid is continued, ultimately the vapours of composition *C* will be obtained. If these vapours are condensed and the solution distilled, the distillate obtained will have the same composition as the solution. Hence no further separation is possible by distillation.

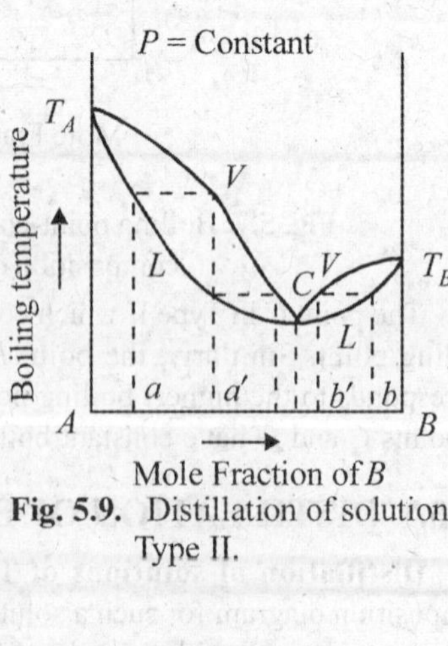

Fig. 5.9. Distillation of solutions of Type II.

Thus *"In case of solutions of Type II, a solution of composition between A and C on fractional distillation gives residue of pure A and a final distillate of composition C. No pure B can be recovered."*

Now consider the distillation of a solution having composition between C and B, say corresponding to the point b. The vapour coming off will be richer in A and so the residue will be richer in B. Hence on repeated distillations ultimately a residue of pure B and the final distillate of composition C will be obtained. Thus

For a solution of composition between C and B, on fractional distillation, a residue of pure B and a final distillate of composition C is obtained. No pure A can be recovered.

Distillation of solutions of Type III: The boiling temperature-composition diagram for solutions of this type, is explained with the help of Fig. 5.10.

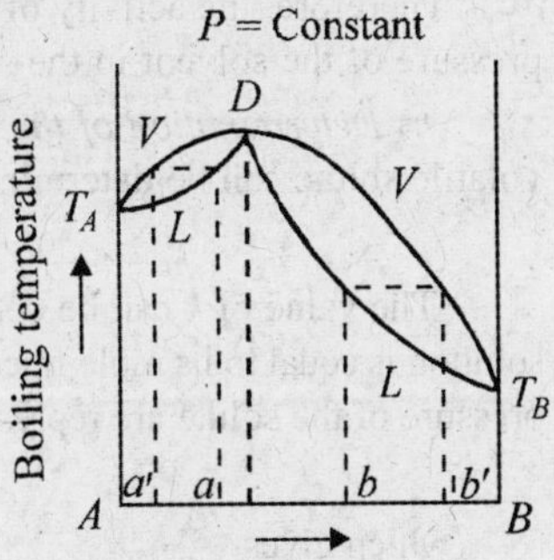

Fig. 5.10. Distillation of solutions of Type III.

The behaviour of solutions of Type III on distillation is analogous to that of solutions of Type II with the exception that the residues tend towards the constant boiling mixture corresponding to composition D whereas distillates tend towards the pure constituents. Further from the Fig. 5.10, it is clear that for a mixture having composition between A and D (say corresponding to a), the vapour coming off are richer in A and hence the residue is richer in B. In other words, the composition of the residue shifts towards D and ultimately becomes equal to that of D, the composition of the vapours shifts towards A and finally a distillate of pure A is obtained. Thus *for a mixture having composition between A and D, distillate of pure A is obtained and the residue has composition corresponding to point D* (*i.e.*, a constant boiling mixture with maximum boiling point). In a similar manner, it can be seen that *for a mixture having composition between D and B, ultimately distillate of pure B and residue of composition corresponding to D is obtained.*

5.16. AZEOTROPES

Mixtures of liquids which boil at constant temperature like a pure liquid such that the distillate has the same composition as that of the liquid mixture are called constant boiling mixtures or azeotropic mixtures or simply azeotropes. Evidently, the components of an azeotrope cannot be separated by fractional distillation. These azeotropes are of two types.

(*a*) **Minimum boiling (point) azeotropes:** These azeotropes are formed by those liquid pairs which show positive deviations from ideal behaviour. Such an azeotrope corresponds to an intermediate composition for which the total vapour pressure is the highest and hence the boiling point is the lowest. Such azeotropes have boiling points lower than either of the pure components. Ethyl alcohol-water is an example of this type of azeotrope.

(*b*) **Maximum boiling (point) azeotropes:** These azeotropes are formed by those liquid pairs which show negative deviations from ideal behaviour. Such an azeotrope corresponds to an intermediate composition for which the total vapour pressure is minimum and hence the boiling point is maximum. Such azeotropes have boiling points higher than either of the pure components. Water—HCl is an example of this type of azeotrope.

5.17. DETERMINATION OF ACTIVITIES AND ACTIVITY COEFFICIENTS

Determination of activity and activity coefficient by the **Vapour Pressure Method** is described below:

(*a*) *Determination of activity of the solvent in solution.* It is based on the principle that the activity of the solvent in the solution is directly proportional to its vapour pressure *i.e.*

$$a_1 = kp_1 \qquad \text{... } (i)$$

But the acivity of the pure solvent is unity i.e. when $a_1 = 1$, $p_1 = p_1^o$. Hence eqn. (*i*) gives

$$1 = kp_1^o \quad \text{or} \quad k = \frac{1}{p_1^o} \qquad \text{... } (ii)$$

Putting this value of k in eqn. (*i*), we get

$$a_1 = p_1/ p_1^o \qquad \text{... } (iii)$$

Therefore, the activity of the solvent in the solution is simply the ratio of the partial vapour pressure of the solvent in the solution to the vapour pressure of the solvent in the pure state.

(*b*) *Determination of the activity of the solute* (*volatile*) *in the solution.* The activity of the volatile solute can be determined by a method similar to the above. For the solute, we write

$$a_2 = kp_2$$

The value of k can be determined by making use of the fact that the activity of a very dilute solution is equal to its mole fraction. In such a dilute solution, if the activity, mole fraction and partial pressure of the solute are represented by $a_{2(0)}$, $x_{2(0)}$ and $p_{2(0)}$ respectively, then we have the relation

$$a_{2(0)} = x_{2(0)} = kp_{2(0)} \qquad \text{... } (v)$$

which gives

$$k = x_{2(0)}/p_{2(0)} \qquad \text{... } (vi)$$

Substituting this value in eqn. (*iv*) we get

$$a_2 = x_{2(0)}\, p_2/ p_{2(0)} \qquad \text{... } (vii)$$

Thus to know the activity a_2 of the solute at any given concentration in the solution, we should determine the partial pressure p_2 of the solute at this given concentration as well as the partial pressure $p_{2(0)}$ in a very dilute solution having a mole fraction $x_{2(0)}$. However, this method has the limitation that we cannot obtain accurate values of partial pressure $p_{2(0)}$ in a very dilute solution.

(*c*) *Determination of the activity of the solute in the solution when the solute may be volatile or non-volatile. According to Gibbs-Duhem equation, for a binary mixture*

$$n_1 d\mu_1 + n_1 d\mu_2 = 0 \qquad \text{... } (viii)$$

Chemical potential is given by the reaction

$$\mu = \mu_0 + RT \ln a$$

$$\therefore \quad d\mu = RT\, d \ln a$$

Hence eqn. (*viii*) can be written as

$n_1\, RT\, d \ln a_1 + n_2\, RT\, d \ln a_2 = 0$ ... (*ix*)

or $\quad n_1\, d \ln a_1 + n_2\, d \ln a_2 = 0$

where n_1 and n_2 are the number of moles of the solvent and solute respectively.

Eqn. (*ix*) can be rearranged as

$$d \ln a_2 = -\frac{n_1}{n_2} d \ln a_1$$

Intergrating between the limits represented by solutions of compositions n_1 and n_2 and n_1' and n_2' respectively, we get

$$\ln \frac{a_2}{a_2'} = \int_{n_1'}^{n_1} \ln \frac{n_2}{n_2} d \ln a_1 \qquad \text{... } (x)$$

By plotting n_1/n_2 against ln a_1, the area under the curve between the limit n_1 and n_1' gives the intergral on the right hand side. Hence the logarithm of the ratio and from there the ratio of the activities of the solute in the two solutions can be determined. Knowing the activity at one concentration, that at the other can be calculated.

5.18. COLLIGATIVE PROPERTIES

It has been observed that certain properties of dilute solutions are not dependent on the nature of the solute (non-volatile) presence in the solution. These properties depend upon the number of particle (or concentration) of the solute in the solution. Such properties are known as colligative properties and are listed below:

1. Lowering of vapour pressure of the solvent
2. Osmotic pressure of the solution
3. Elevation in boiling point of the solvent
4. Depression in freezing point of the solvent.

It has been observed that the validity of the above behaviour is limited to dilute solutions when they behave nearly as ideal solutions. As these properties are observed with non-volatile solutes, it can be visualised that the escaping tendency of the solvent is reduced in the presence of such solutes and consequently the vapour pressure of the solvent is lowered. With the lowering of vapour pressure of the solvent, we can explain the elevation in boiling point and depression in freezing point. An important application of the study of colligative properties is to determine the molecular masses of unknown substances and in the study of their molecular state in solution. In the following reactions, we shall take up a detailed study of each of the colligative properties.

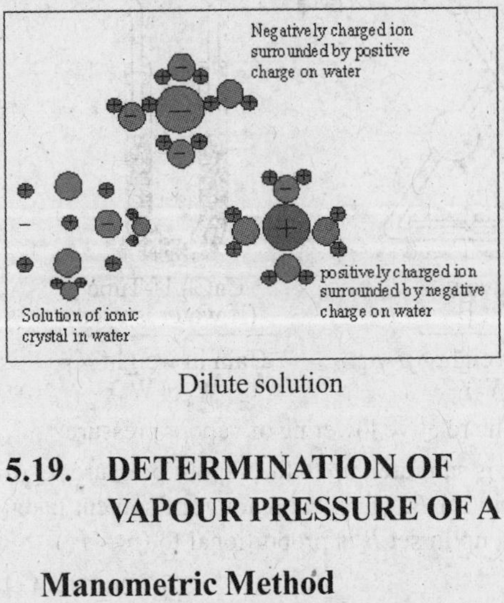

Dilute solution

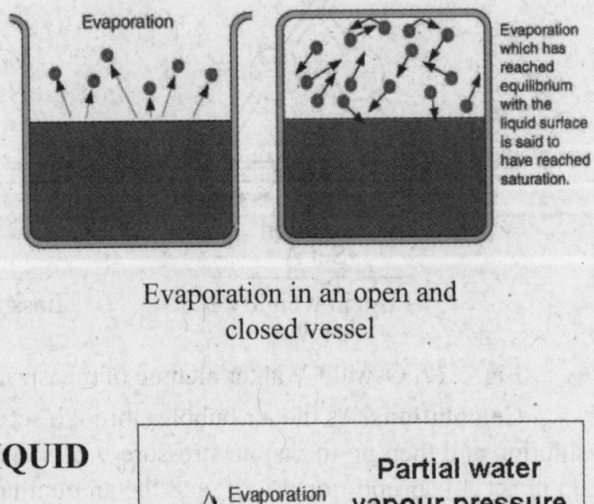

Evaporation in an open and closed vessel

5.19. DETERMINATION OF VAPOUR PRESSURE OF A LIQUID

Manometric Method

The vapour pressure of a liquid of solution can be measured with the help of a manometer (see Fig. 5.11). The bulb B is filled with the liquid or solution. The air in the connecting tube is then removed with a vacuum pump. When the stopcock is closed, the pressure inside is due only to the vapour evaporating from the solution

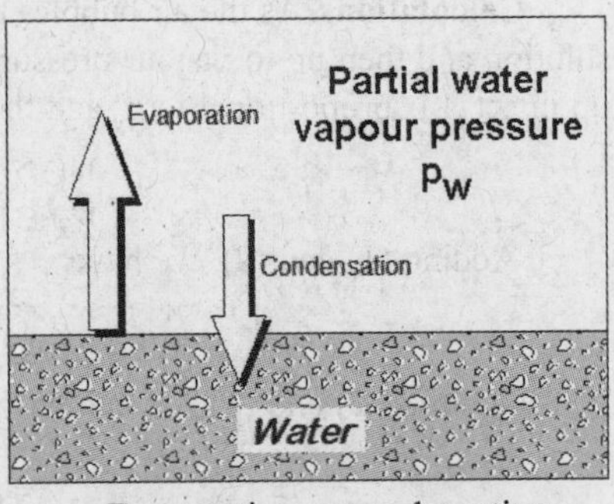

Evaporation vs condensation

or liquid. This method is generally used for aqueous solutions. The manometric liquid is usually mercury which has low volatility.

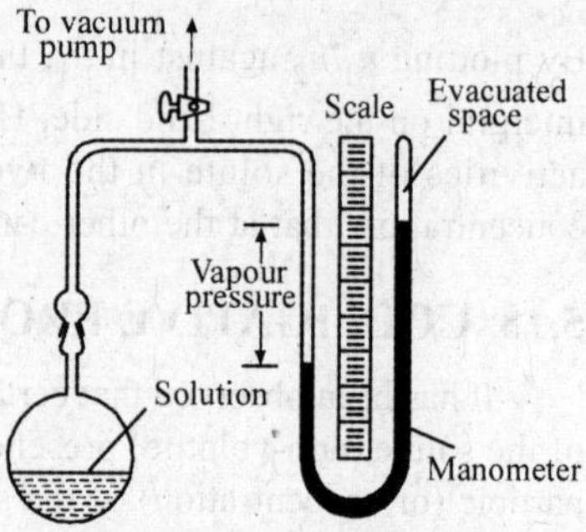

Fig.5. 11 Measurement of vapour of aqueous solution with a manometer

5.20. DETERMINATION OF LOWERING OF VAPOUR PRESSURE OF THE SOLVENT

Ostwald and Walker's Dynamic Method

(*Gas Saturation Method*)

In this method the relative lowering of vapour pressure can be determined straightaway. The measurement of the individual vapour pressures of a solution and solvent is thus eliminated.

Procedure. The apparatus used by Ostwald and Walker is shown in Fig. 5.12. It consists of two sets of bulbs:

(*a*) Set *A* containing the solution

(*b*) Set *B* containing the solvent

Each set is weighed separately. A slow stream of dry air is then drawn by suction pump through the two sets of bulbs. At the end of the operation, these sets are reweighed. From the loss of weight in each of the two sets, the lowering of vapour pressure is calculated. The temperature of the air the solution and the solvent must be kept constant throughout.

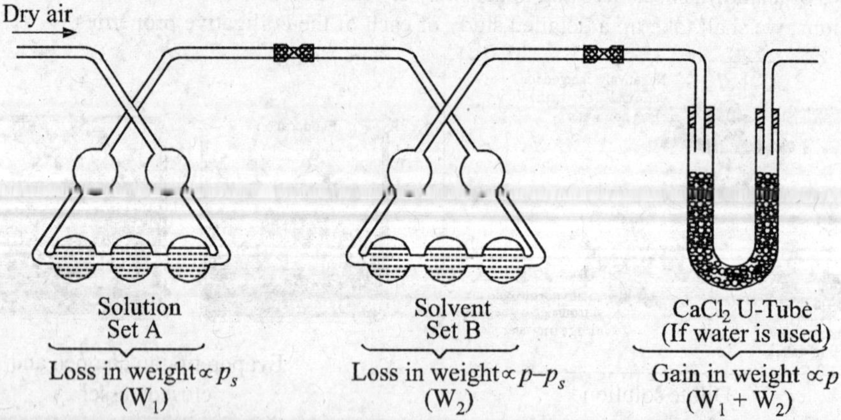

Fig. 5.12. Oswald-Walker method of measuring the relative lowering of vapour pressure

Calculations. As the air bubbles through set *A* it is saturated up to the vapour pressure p_s, of solution and then up to vapour pressure p of solvent in set *B*. Thus, the amount of solvent taken up in set *A* is proportional to p_s and the amount taken up in set *B* is proportional to $(p - p_s)$.

$$w_1 \propto p_s \qquad \text{... (1)}$$

$$w_2 \propto p - p_s \qquad \text{... (2)}$$

Adding (1) and (2), we have

$$w_1 + w_2 \propto p_s + p - p_s$$

$$\propto p \qquad \text{... (3)}$$

Dividing (2) and (3), we can write

$$\frac{p - p_s}{p} = \frac{w_2}{w_1 + w_2} \qquad \text{... (4)}$$

Knowing the loss of mass in set *B* (w_2) and the total loss of mass in the two sets ($w_1 + w_2$), we can find the relative lowering of vapour pressure from equation (4).

If water is the solvent used, a set of calcium chloride tubes is attached to the end of the apparatus to catch the escaping water vapour. Thus, the gain in mass of the $CaCl_2$- tubes is equal to ($w_1 + w_2$), the total loss of mass in sets *A* and *B*.

SOLVED EXAMPLE ON LOWERING OF VAPOUR PRESSURE

Example. A stream of dry air was passed through a bulb containing a solution of 7.50 g of an aromatic compound in 75.0 g of water and through another globe containing pure water. The loss in mass in the first globe was 2.810 g and in the second globe it was 0.054 g. Calculate the molecular mass of the aromatic compound (Mol mass of water = 18)

Solution. According to the theory of Ostwald-Walker method,

$$\frac{p - p_s}{p} = \frac{w_2}{w_1 + w_2}$$

In the present case,

w_1, loss of mass of solution = 2.810 g

w_2, loss of mass of solvent (water) = 0.054 g

Substituting values in the above equation

$$\frac{p - p_s}{p} = \frac{0.054}{2.810 + 0.054} = \frac{0.054}{2.864} = 0.0188$$

According to Raoult's law $\dfrac{p - p_s}{p} = \dfrac{w/m}{w/m + W/M}$

Substituting the values, we have

$$0.0188 = \frac{7.50/m}{7.50/m + 75.0/18} \quad \text{or} \quad m = 93.6$$

5.21. RELATION BETWEEN THE RELATIVE LOWERING OF VAPOUR PRESSURE AND THE MOLECULAR WEIGHT OF THE SOLUTE (RAOULT'S LAW).

Suppose vapour pressure of the pure solvent $A = P_A^{\circ}$

Let the solute be *B*.

Let the vapour pressure of the solvent in the solution = P_A

Since $P_A^{\circ} > P_A$

$\therefore$ Lowering of vapour pressure = $P_A^{\circ} - P_A$

Relative lowering of vapour pressure $= \dfrac{P_A^{\circ} - P_A}{P_A^{\circ}}$

According to Raoult's Law, relative lowering of vapour pressure is equal to the mole fraction of the solute in the solution.

Let mole fraction of the solute $B = X_B$.

$$\therefore \quad \frac{P_A^{\circ} - P_A}{P_A^{\circ}} = X_B \qquad \text{... } (i)$$

$$\text{Mole fraction of the solute} = \frac{n_2}{n_1 + n_2}$$

where n_2 = No. of moles of the solute

n_1 = No. of moles of the solvent

$$\therefore \quad X_B = \frac{n_2}{n_1 + n_2}$$

Thus relation (*i*) becomes

$$\frac{P_A^{\circ} - P_A}{P_A^{\circ}} = \frac{n_2}{n_1 + n_2} \qquad \text{... } (ii)$$

It is evident from (*ii*) that relative lowering of vapour pressure depends only upon mole fraction or molar concentration of the solute. Therefore, relative lowering of vapour pressure is a colligative property.

Molecular Mass of the Non-volatile Solute

Now, number of moles of the solute $n_2 = \dfrac{w}{m}$

where w = mass of the solute

m = molecular mass of the solute.

And number of moles of the solvent $n_1 = \dfrac{W}{M}$

where W = mass of the solvent.

M = molecular mass of the solvent.

The expression (*ii*) becomes

$$\frac{P_A^{\circ} - P_A}{P_A^{\circ}} = \frac{\frac{w}{m}}{\frac{w}{m} + \frac{W}{M}} \qquad \text{... } (iii)$$

In a dilute solution, n_2 is negligible as compared to n_1. Therefore neglecting n_2 (or w/m) in the denominator, we get from expression (*iii*)

$$\frac{P_A^{\circ} - P_A}{P_A^{\circ}} = \frac{\frac{w}{m}}{\frac{W}{M}} = \frac{w}{m} \times \frac{W}{M}$$

$$\text{or} \quad \frac{P_A^{\circ} - P_A}{P_A^{\circ}} = \frac{wM}{mW} \qquad \text{... } (iv)$$

Thus by measuring the lowering of vapour pressure of a solution, the molecular mass m of a solute in a given solution of a known concentration can be determined, if other quantities are known.

SOLVED PROBLEMS ON LOWERING OF VAPOUR PRESSURE

Example 1. A solution containing 6.0 gram of benzoic acid in 50 gram of ether ($C_2H_5.OC_2H_5$) has a vapour pressure equal to 5.466×10^4 Nm^{-2} at 300 K. Given that vapour pressure of ether at the same temperature is 5.893×10^4 NM^{-2}, calculate the molecular mass of benzoic acid.

Solution. Vapour pressure of ether (Solvent) = $P° = 5.893 \times 10^4$ Nm^{-2}

Vapour pressure of ether solution = $P = 5.466 \times 10^4$ Nm^{-2}

Molecular mass of solvent ($C_2H_5 - O - C_2H_5$) (M) = 74

Mass of solute (benzoic acid) w = 6 grams

Mass of solvent (ether) W = 50 grams

Let molecular mass of solute (benzoic acid) = m

Substituting the values in the relation

$$\frac{P_A^\circ - P}{P^\circ} = \frac{w}{m} \times \frac{M}{W}$$

$$\frac{5.893\times10^4 - 5.466\times10^4}{5.893\times10^4} = \frac{6\times74}{m\times50}$$

$$\frac{0.427\times10^4}{5.893\times10^4} = \frac{6\times74}{m\times50}$$

$$\frac{0.427}{5.893} = \frac{6\times74}{m\times50}$$

$$m = \frac{6\times74\times5.893}{50\times0.427}$$

$$m = 122.55$$

i.e., the molecular mass of solute (benzoic acid) = 122.55 amu.

Example 2. The vapour pressure of water at 293 K is 17.51 mm, lowering of vapour pressure of sugar solution is 0.0614 mm.

Calculate

(a) Relative lowering of vapour pressure.

(b) Vapour pressure of the solution.

(c) Mole fraction of water.

Solution. Vapour pressure of solvent (water) = 17.51

Let Vapour pressure of the solution= P (to be calculated)

$\therefore$ Lowering of Vapour pressure $= P° - P = 0.0614$ mm

(a) $\therefore$ Relative lowering of Vapour pressure $= \dfrac{p^\circ - p}{p^\circ} = \dfrac{0.0614}{17.51} = 0.00351$

(b) Vapour pressure of the solution $P = P° - (P° - P)$

$= 17.51 - (0.0614) = 17.4486$ mm

Now according to Raoult's Law

$$\frac{p^\circ - p}{p^\circ} = \text{mole fraction of the solute}$$

$$\frac{p^\circ - p}{p^\circ} = \frac{n_1}{n_1 + n_2} = x_2$$

$\therefore$ mole fraction of the solute $= \dfrac{p^\circ_{} - p}{p^\circ} = \dfrac{0.0614}{17.51} = 0.00351$

(*c*) Hence, mole fraction of the solvent = (1 – 0.00351) = 0.99649

Mole fraction of water = 0.99649

Example 3. The Vapour pressure of a 5% aqueous solution of non-volatile organic substances at 373 K is 745 mm. Calculate the molecular mass of the solute.

Solution. Weight of non-volatile organic solute, $w = 5$ g

Weight of solvent (water), $W = 95$ g

Molecular mass of solvent (water) $M = 18$

Molecular mass of non-volatile solute $m = ?$

P°, the Vapour pressure of the pure solvent (water) at 373 K = 760 mm

Vapour pressure of the solution $P = 745$ mm

Substituting the values in the relation,

$$\frac{p^\circ - p}{p^\circ} = \frac{w}{m} \times \frac{M}{W}$$

$$\frac{760 - 745}{760} = \frac{5 \times 18}{m \times 95}$$

or
$$m = \frac{5 \times 18 \times 760}{15 \times 95} = 48$$

Example 4. At 298 K, the vapour pressure of water is 23.75 mm of Hg. Calculate the vapour pressure at the same temperature over 5% aqueous solution of urea (NH_2 $CONH_2$).

Solution. This solution may be considered as a dilute solution and the approximate relation given below may be used

$$\frac{P_A^\circ - P_A}{P_A^\circ} = \frac{wM}{mW}$$

In the present case $P_A^\circ = 23.75$

$w = 5$ g Therefore $W = 100 - 5 = 95$ g

$M = 18$, $m = 60$ (mol. wt of urea)

Substituting these values in the equation above

$$\frac{23.75 - P_A}{23.75} = \frac{5 \times 18}{60 \times 95} \quad \text{or} \quad P_A = 23.375 \text{ mm.}$$

PROBLEMS FOR PRACTICE

1. The vapour pressure of an aqueous solution of cane sugar (mol. wt. 342) is 756 mm at 100° C. How many grams of sugar are present in 1000 g of water? [**Ans.** 100.4 g]

2. Dry air was passed through a solution containing 40 g of a solute in 90 g water and then through water. The loss in weight of water was 0.05 g. The wet air was then passed through sulphuric acid, whose weight increased by 2.0 g. What is the molecular weight of the dissolved substance? [**Ans.** 320]

3. The vapour pressure of water is 92 mm at 50°C. 18.1 g of urea are dissolved in 100 g of water. The vapour pressure is reduced by 5 mm. Calculate the molecular weight of urea.

[**Hint**. The solution is not dilute. Apply the relation $\frac{p^\circ - p_s}{p^\circ} = \frac{n_2}{n_1 + n_2}$]

4. The vapour pressure of 2.1% of an aqueous solution of a non-electrolyte at 100°C is 755 mm. Calculate the molecular weight of the solute. [**Ans.** 58.68]

5. The vapour pressure of water of 20°C is 17 mm. Calculate the vapour pressure of a solution containing 2 g of urea (mol. wt = 60) in 50 g of water. Assume that the solution is not dilute. [**Ans.** 16.799 mm]

6. A current of dry air was passed through a series of bulbs containing a solution of 3.458 g of a substance in 100 g of ethyl alcohol and then through pure ethyl alcohol. The loss in weight of former was 0.9675 g and in the later 0.055 g. Calculate the molecular weight of the solute. [**Ans.** 29.6]

7. Calculate the vapour pressure at 22°C of a 0.1 M solution of urea. The density of the solution may be taken as 1g/ml. The vapour pressure of pure water at 22° is 20 mm. [**Ans.** 19.96 mm]

[**Hint.** 0.1 M solution of urea means 0.1 mole *i.e.* 6.0 g of urea dissolved per liter of solution *i.e.* in 1000 g of solution ($\because$ density = 1 g/ml). Hence $w_2 = 60$ g, $w_1 = 1000 - 6.0 = 994.0$ g, $p^\circ = 20$ mm (Given), p_s = to be calculated]

8. A solution containing 6 g of benzoic acid in 50 g of ether ($C_2H_5OC_2H_5$) has a vapour pressure of 410 mm of mercury at 20°C. Given that the vapour pressure of ether at the same temperature is 442 mm of mercury, calculate the molecular weight of benzoic acid.[**Ans.** 122.56]

9. A current of dry air was passed through a bulb containing 26.66 g of an organic substance in 200 g of water, then through a bulb at the same temperature containing pure water and finally through a tube containing fused calcium chloride. The loss in weight of water bulb was 0.0870 g and the gain in the weight of $CaCl_2$ tube was 2.036 g. Calculate the molecular weight of the organic substance in the solution. [**Ans.** 53.8]

5.22. OSMOSIS PHENOMENON

Before coming to the phenomenon of osmosis, let us understand clearly about semipermeable membrane.

The membrane which allows the flow of solvent molecules through it but not the solute molecules, is called a semipermeable membrane. Examples are parchment, collodion, animal membranes etc. In nature the plant cell are protected by a semi permeable membrane. Chemically, we can get a semipermeable membrane of cupric ferrocyanide within the walls of a porous pot.

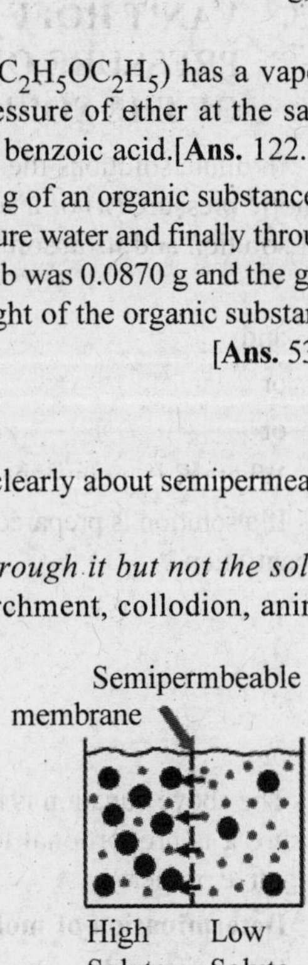

Osmosis

Osmosis and osmotic pressure: A solute tends to dissolve in a solvent as the most predominant randomness factor favours such a tendency. But when a pure solvent is separated from its solution by a semipermeable membrane, the molecules of the solvent diffuse through the semipermeable membrane into the solution. This is called **osmosis.** As a result of this transference of solvent to solution, the level of solvent decreases while that of solution increases. It may be demonstrated by taking solvent and solution in two compartments of a box separated by a semipermeable membrane as shown in Fig. 5.13.

As more and more of the solvent molecules pass through the membrane into the solution, the concentration of the solution falls gradually. After some time the hydrostatic pressure exerted by the solution column prevents the flow of more of solvent molecules and *osmosis* stops. i.e., there is no further rise in the level of solution in the column. Hence, osmotic pressure is defined as follows:

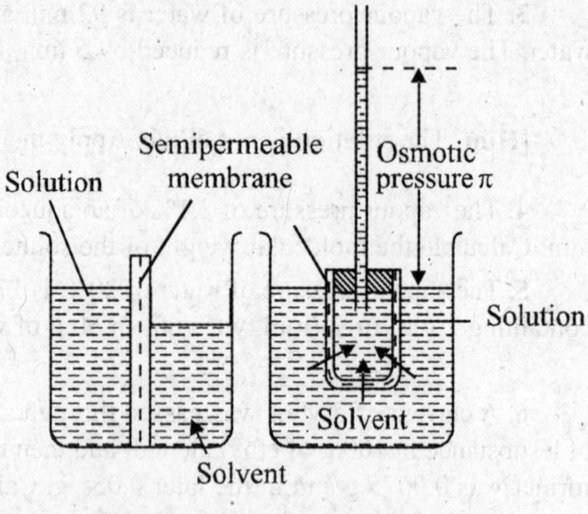

Fig. 5.13. Osmosis and Osmotic pressure.

***Osmotic pressure** is the equilibrium hydrostatic pressure exerted by the solution column which just prevents the flow of solvent molecules into the solution through a semipermeable membrane.*

***Osmosis** is a phenomenon by which the pure solvent molecules tend to diffuse through a semipermeable membrane into the solution.*

5.23. VAN'T HOFF RELATION BETWEEN THE OSMOTIC PRESSURE OF A SOLUTION AND MOLECULAR MASS OF THE SOLUTE

In dilute solutions the behaviour of solute molecules is similar to that of molecules of a gas. Osmotic pressure (π) of a solution is found to be directly proportional to the molar concentration C of solution and its absolute temperature T. Thus:

$$\pi \propto C$$

and $$\pi \propto T$$

or $$\pi \propto C.T$$

or $$\pi \propto R.C.T$$

Where R is a constant and its value is found to be same as that of gas constant.

If a solution is prepared by dissolving n moles of the solute in V litres of the solution, the molar concentration

$$C = \frac{n}{V} \text{ moles per liter}$$

$$\therefore \quad \pi = \frac{n}{V} RT$$

or $$\pi V = nRT$$

The above equation is known, as *Van't Hoff equation of dilute solution* and shows that osmotic pressure π is proportional to the molar concentration of the solute in the solution. Hence, it is a colligative property.

Determination of molecular mass from osmotic pressure

If w gram of the solute are dissolved in V liters of the solution and m is the molecular of the solute, then

$$n = w/m$$

Substituting this value in equation $\pi V = nRT$, we get

$$\pi V = \frac{w}{m} \times R.T$$

or
$$m = \frac{wR.T}{\pi V}$$

Then molecular mass m of the solute can be calculated from this equation if osmotic pressure π of the solution is known. The value of constant R is taken as 0.0821 litre-atmosphere per degree per mole when π is expressed in atmosphere and T in degree Kelvin.

5.23.1. Relation between the lowering of vapour pressure and osmotic pressure of a solution

The relation between the lowering of vapour pressure and the mole fraction of the solute is

$$\frac{p^\circ - p}{p^\circ} = \frac{w}{m} \times \frac{M}{W} \text{ (for a dilute solution)} \quad ... (i)$$

or
$$\frac{p^\circ - p}{p^\circ} = n \times \frac{M}{W} \quad ... (ii)$$

where n is the no. of moles of the solute.

The relation between the osmotic pressure and the no. of moles of the solute is given by

$$\pi = \frac{nRT}{V} \quad ... (iii)$$

From equation (*ii*)
$$n = \frac{p^\circ - p}{p^\circ} \times \frac{W}{M} \quad ... (iv)$$

From equation (*iii*)
$$n = \frac{\pi V}{RT} \quad ... (v)$$

Equating R.H.S of equations (*iv*) and (*v*)

$$\frac{p^\circ - p}{p^\circ} \frac{W}{M} = \frac{\pi V}{RT}$$

or
$$\frac{p^\circ - p}{p^\circ} = \frac{M}{W} \frac{\pi V}{RT}$$

or
$$\frac{p^\circ - p}{p^\circ} = \frac{M\pi}{(W/V).RT}$$

or
$$\frac{p^\circ - p}{p^\circ} = \frac{M\pi}{\rho.RT}$$

where M = Mol. mass of the solvent

ρ = Density of the solution

R = Solution (gas) constant

T = Temperature.

5.23.2. Interesting experiments to demonstrate the phenomenon of osmosis

Two interesting experiments to demonstrate the phenomenon of osmosis are being described here.

1. The egg experiment

The outer hard shell of two eggs of the same size is removed by dissolving in dilute hydrochloric acid. One of these is placed in distilled water and the other in saturated salt solution. After a few hours, it will be noticed that the egg placed in water swells and the one in salt solution shrinks. In the first case, water diffuses through the skin (a semipermeable membrane) into the egg which swells. In the second case, the concentration of the salt solution being higher than the material, the egg shrinks (Fig. 5.14).

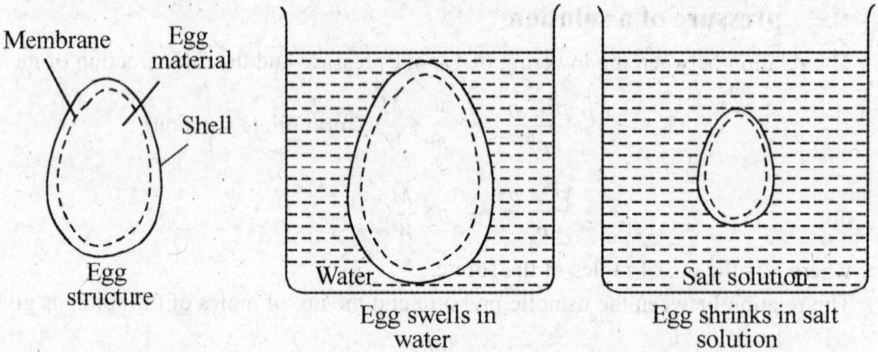

5.14. Demonstration of osmosis by Egg experiment.

2. Silica garden

Crystals of many salts *e.g.,* ferrous sulphate, nickel chloride, cobalt nitrate and ferric chloride are placed in a solution of water glass (sodium silicate). The layers of metallic silicates formed on the surface of crystals by double decomposition are semipermeable. The water from outside enters through these membranes which burst and form what we call a Silica Garden, It makes an interesting sight.

5.23. VAN'T HOFF THEORY OF DILUTE SOLUTION TAKING THE EXAMPLE OF OSMOSIS

Van't Hoff noted the striking resemblance between the behaviour of dilute solutions and gases. Dilute solutions obeyed laws analogous to the gas laws. To explain it van't Hoff visualised that gases consist of molecules moving in vacant space (or vacuum), while in the solutions the solute particles are moving in the solvent. The exact analogy between solutions and gases is illustrated with Fig. 5.15.

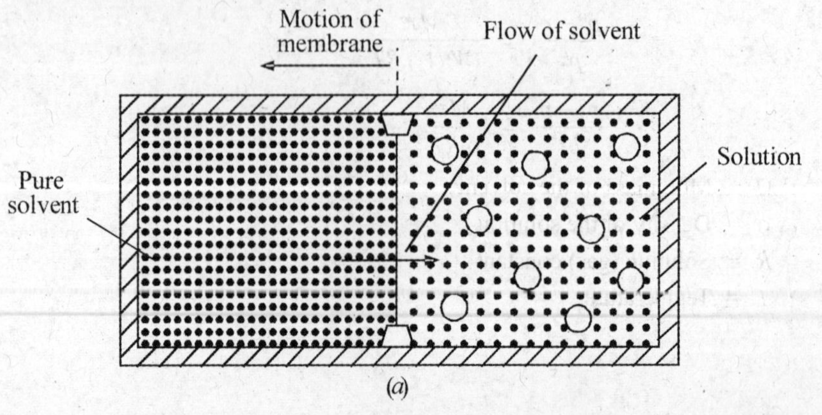

(*a*)

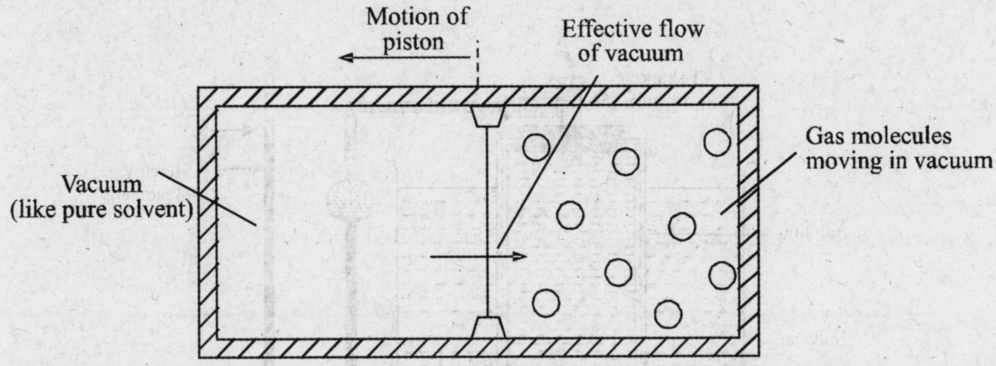

(*b*)

Fig. 5.15. The analogy between osmotic pressure and gas pressure

As shown in Fig. 5.15 (a), the pure solvent flows into the solution by osmosis across the semipermeable membrane. The solute molecules striking the membrane cause osmotic pressure and the sliding membrane is moved towards the solvent chamber. In case of a gas [Fig. 5.15 (*b*)], the gas molecules strike the piston and produce pressure that pushes it towards the empty chamber. Here it is the vacuum which moves into the gas. This demonstrates clearly that there is close similarity between a gas and a dilute solution.

Thinking on these lines, van't Hoff propounded his theory of dilute solution. The **van't Hoff Theory of Dilute Solutions** states that:

A substance in solution behaves exactly like a gas and the osmotic pressure of a dilute solution is equal to the pressure which the solute would exert if it were a gas at the same temperature occupying the same volume as the solution.

According to the van't Hoff theory of dilute solutions, all laws or relationships obeyed by gases would be applicable to dilute solutions.

From van't Hoff theory if follows that just as 1 mole of a gas occupying 22.4 liters at 0°C exert 1 atmosphere pressure so 1 mole of any solute dissolved in 22.4 litres would exert 1 atmosphere osmotic pressure.

5.24. DETERMINATION OF OSMOTIC PRESSURE

Prominent methods for the determination of osmotic pressure are described as under:

1. Pfeffer's method. The apparatus used by Peffor consists of semipermeable membrane of copper ferrocyanide, supported on the walls of a porous pot. A *T*-shaped wide tube is fitted into the porous pot as shown in Fig. 5.16. The side tube *S* of the *T*-tube is connected to a closed manometer containing mercury and nitrogen. A simple tube R is fitted at the top of the *T*-tube, through which the experimental solution is added into the porous pot till the porous pot and space above the mercury in the left limb of the manometer are completely filled. After filling, the tube *t* is sealed off. The pot is then placed in the pure solvent maintained at constant temperature. As a result of osmosis, the solvent passes into the solution. The highest pressure developed is recorded on the manometer.

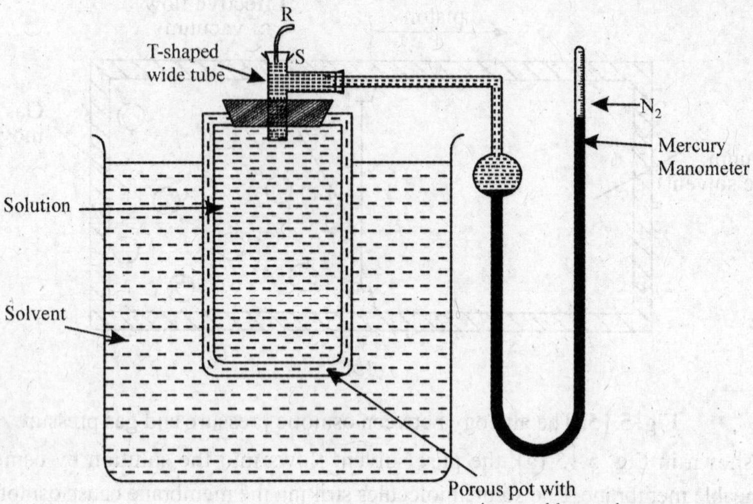

Fig. 5.16 Pfeffer's apparatus

2. Morse and Frazer's Method

In this method, the solvent was taken in the porous pot (having semipermeable membrane in its walls) and the solution was kept outside the porous pot in the vessel made of bronze and connected to a manometer at the top. A tube T open at both the ends was fixed to the porous pot, as shown in Fig. 5.17, to keep the porous pot filled with the solvent. Morse and Frazer prepared the semi-permeable membrane of better quality by electrolytic method to measure osmotic pressures upto 300 atmospheres.

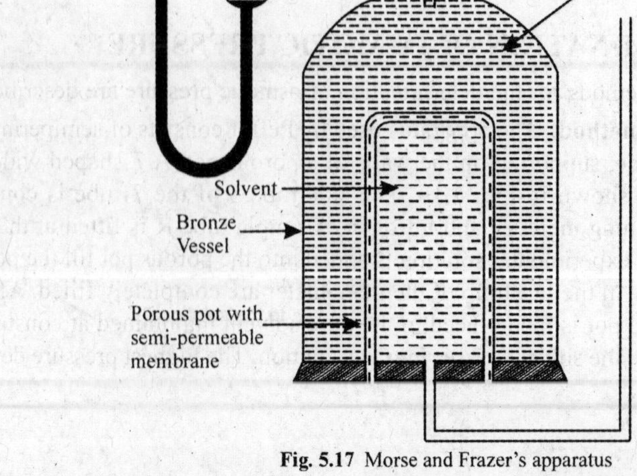

Fig. 5.17 Morse and Frazer's apparatus

3. Barkeley and Hartley method

The apparatus used for the determination of osmotic pressure by the above method is shown in Fig. 5.18.

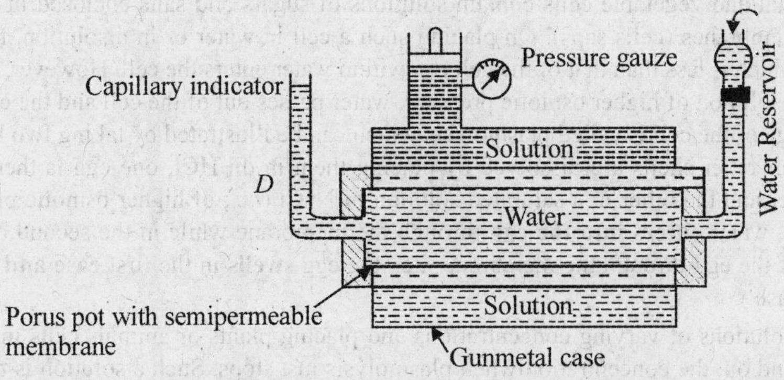

Fig. 25.18. Barkeley and Hartley methods for osmotic pressure determination.

It consists of a porous pot containing potassium ferrocyanide deposited on its walles and fitted into a metallic cylinder. A piston and a pressure gauge is attached to the metallic cylinder which is filled with the solution whose osmotic pressure is to be determined. The porous pot is fitted with a capillary tube and water reservoir on the opposite sides. Water from the porous pot moves towards the solution in the cylinder through the semipermeable membrane. As a result the level of water tends to fall down. External pressure is applied on the piston to such an extent that the water level in the capillary tube does not change. The magnitude of pressure applied can be read from the pressure gauge. This pressure is equal to the osmotic pressure.

5.24.1. Isotonic solutions

Solutions of equimolecular concentrations at the same absolute temperature have the same osmotic pressure. *Such solutions which have the same osmotic pressure at the same temperature are called isotonic solutions.*

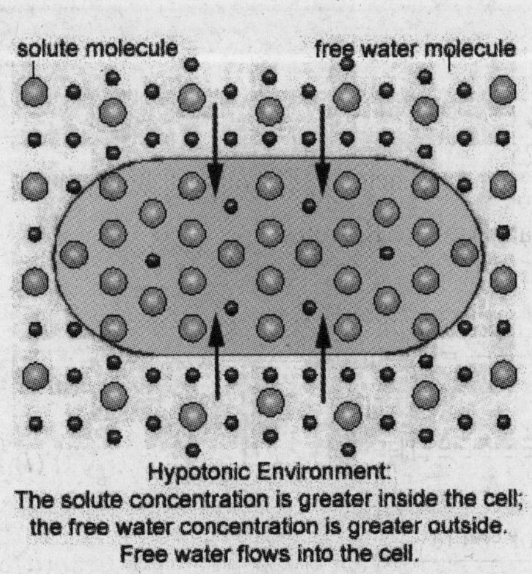

Hypotonic solution

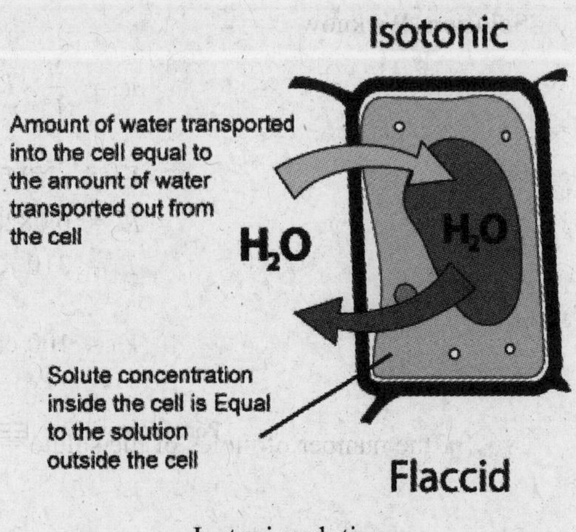

Isotonic solutions

5.24.2. Biological importance of osmosis

Osmosis plays a very important role in plants and animals i.e. living organisms, as described below;

1. Animal and vegetable cells contain solutions of sugars and salts enclosed in their semi-permeable membranes (cells saps). On placing such a cell in water or in a solution, the osmotic pressure of which is less than that of the cell sap within water enters the cell. However, on placing, the cells in a solution of higher osmotic pressure, water passes out of the cell and the cell shrinks. This shrinkage of the cell is called plasmolysis which can be illustrated by taking two hen eggs of the size. Their outer shells are dissolved by placing them in dil HCl, one egg is then placed in distilled water and the other in a saturated solution of NaCl (*i.e.*, of higher osmotic pressure). In the first case, water **enters into the egg** through the membrane while in the second case, **water comes out** of the egg through the membrane i.e., the egg **swells** in the first **case and shrinks** in the second case.

Using solutions of varying concentrations and placing plants or animals cells in them, it is possible to find out the concentration when plasmolysis just stops. Such a solution is then said to be **isotonic** (*i.e.*, having same osmotic pressure) with the cell solution. This process is thus helpful in measuring osmotic pressure of cell saps.

It is found that the red blood corpuscles and a 0.91% sol. of NaCl are isotonic.

2. Plant cell at their roots contain root hair, which are in contact with the soil, since the osmotic pressure inside the cell is higher, water from the soil flows into the cell by osmosis.

3. Damage done to the plant by use of excess of fertilizer is due to higher osmotic pressure of the fertilizer solution than that of the cell and makes the organs stretched and fully expanded.

4. Plant movements such as opening and closing of flowers, opening and closing of leaves etc. are also regulated by osmosis.

SOLVED PROBLEMS ON OSMOTIC PRESSURE

Example 1. Osmotic pressure of solution containing 7 grams of dissolved protein per 100 cm³ of a solution is 25 mm of Hg at body temperature (310 K). Calculate the molar mass of the protein (R = 0.08205 litre atm deg^{-1} mol^{-1}).

Solution. We know

$$\pi = \frac{n}{V} \times RT$$

Given

$$\pi = 25 \text{ mm} = \frac{25}{760} \text{ atmosphere } (\because 760 \text{ mm} = 1 \text{atmosphere})$$

$$R = 0.08205 \text{ L atmosphere K}^{-1} \text{ mol}^{-1}$$

$$T = 310\ K$$

$$V = 100 \text{ cm}^3 = \frac{100}{1000} = \frac{1}{10} \text{ litres}$$

$\therefore$ n, the number of moles of the solute $= \dfrac{\pi \times V}{RT}$ ... (*i*)

Also $n = \dfrac{\text{Wt. of solute}}{\text{Molecular mass of solute}}$ ... (*ii*)

From (*i*) and (*ii*)

$$\frac{\text{Wt. of solute}}{\text{Molecular mass of solute}} = \frac{\pi \times V}{RT}$$

Substituting the values, we have

$$\text{Molecular mass of solute} = \frac{\text{Wt. of solute} \times R \times T}{\pi \times V}$$

$$= \frac{7 \times 0.08205 \times 310}{\frac{25}{760} \times \frac{1}{10}}$$

$$\text{or molecular mass of the solute} = \frac{7 \times 0.08205 \times 310 \times 760 \times 10}{25} = 54126.7$$

$\therefore$ Molecular mass of the solute $=$ **54126.7**

Example 2. A solution of sucrose (Molecular mass 342) is prepared by dissolving 68.4 g of it per litre of the solution. What its osmatic pressure at 300 K? R = 0.082 litre atm K^{-1} mol^{-1}

Solution. We know, $\pi = \frac{n}{V} \times RT$

Given, molecular mass of the solute = 342

Wt. of the solute = 68.4 g

Volume V of the solution = 1 litre

T = 300 K

Osmotic pressure π = ?

R = 0.082 litre atm K^{-1} mol^{-1}

$$n\text{, the no. of moles of solute} = \frac{\text{Wt. of solute}}{\text{Molecular mass of solute}}$$

$$\text{or} \quad n = \frac{68.4}{342}$$

Substituting the values in the equation

$$\pi = \frac{n}{V} RT$$

$$\text{or} \quad \pi = \frac{68.4}{342} \times \frac{0.082 \times 300}{1} = \mathbf{4.22\ atm}$$

$\therefore$ Osmotic pressure = **4.22 atm**

Example 3. Calculate the osmotic pressure of a 5% solution of cane sugar at 288 K. R = 0.082 lit atm K^{-1} mol^{-1}

Solution. Molecular mass of the cane sugar ($C_{12}H_{22}O_{11}$) = 342

$\because$ The solution is 5%

$\therefore$ Wt . of sugar per litre = 50 g

$$\therefore \quad n\text{, the no. of moles} = \frac{\text{Wt. of solute}}{\text{Molecular mass of solute}} = \frac{50}{342}$$

$$V = 1 \text{ litre}$$
$$T = 288\ K$$

Substituting the values in the relation

$$\pi = \frac{n}{V} \cdot R.T$$

We have $$\pi = \frac{50}{342} \times \frac{0.082 \times 288}{1} = 3.45 \text{ atm}$$

$\therefore$ Osmotic pressure $= \mathbf{3.45\ atm}$

Example 4. At 298 K, 100 cm³ of a solution, containing 3.002 g of an unidentified solute, exhibits an osmotic pressure of 2.55 atmosphere. What would be the molecular mass of the solute?

Solution. $\pi V = nRT$ where n is the no. moles of the solute

Also $\pi V = \frac{w}{M} RT$ where w is the weight and M the molecular mass of the solute

or $$M = \frac{wRT}{\pi V}$$

Substituting the values

$$M = \frac{3.002 \times 0.0821 \times 298}{2.55 \times 0.1} \qquad (V = 100 \text{ cm}^3 = 0.1 \text{ litre})$$

$$M = 288 \text{ a.m.u.}$$

PROBLEMS FOR PRACTICE

1. Calculate the value of the constant R in litre atmosphere from the observation that solution containing 34.2 g of cane-sugar in one litre of water has an osmotic pressure of 2.405 atm at 20°C. **[Ans.** 0.0821]

2. Calculate the osmotic pressure at 25°C of a solution containing one gram of glucose ($C_6H_{12}O_6$) and one gram of sucrose ($C_{12}H_{22}O_{11}$) in 100 g of water. If it were not known that the solute was a mixture of glucose and sucrose, what would be the molecular weight of the solute corresponding to the calculated osmotic pressure. **[Ans.** 0.2074 atm, 235.8]

3. A solution of glucose containing 18g/litre had an osmotic pressure of 2.40 atm at 27°C. Calculate the molecular weight of glucose ($R = 0.082$ litre atm) **[Ans.** 183.5]

4. Calculate osmotic pressure of a solution containing 5 g of glucose in 100 ml of its solution at 17° C. (R = 0.0821 litre atm/mol/degree.)
[Ans. 6.61 atm]

5. A 6% solution of sucrose ($C_{12}H_{22}O_{11}$) is isotonic with a 3% solution of an unknown organic substance. Calculate the molecular weight of the unknown substance. **[Ans.** 171]

6. Calculate the osmotic pressure of solution obtained by mixing one litre of 7.5% solution of substance A (mol. wt = 75) and two litres of 3% solution of a substance B (mol. wt = 60) at 18°C. **[Ans.** 7.954 atm]

7. A 4% solution of cane-sugar gave an osmotic pressure of 208.0 cm of Hg at 15°C. Find its molecular weight. **[Ans.** 345.2]

8. Calculate the concentration of solution of glucose which is isotonic at the same temperature with a solution of urea containing 6.2 g/litre. **[Ans.** 18.6 g/litre]

SOLVED PROBLEM ON RELATIONSHIP BETWEEN OSMOTIC PRESSURE AND LOWERING OF VAPOUR PRESSURE

Example. *The vapour pressure of a solution containing 2.47 g of ethyl benzoate in 100 g of benzene (mol. wt 78 and density 0.8149 g/ml) was found to be 742.6 mm of Hg at 80° C while that of pure benzene at the same temperature, it is 751.86 mm of Hg. Calculate the osmotic pressure of the solution.*

Solution. It is given that

$$p_s = 742.6 \text{ mm}, p^\circ = 751.86 \text{ mm}, M = 78$$
$$\rho = 0.8149 \text{ g/ml} = 814.9/\text{g litre}$$
$$T = 80 + 273 = 353\ K$$

Taking $R = 0.0821$ litre atmosphere per degree per mol and substituting these values in the formula

$$\frac{p^\circ - p_s}{p^\circ} = \frac{M\pi}{\rho RT} \qquad \text{(Refer to section 5.23.1)}$$

$$\frac{751.86 - 742.60}{751.86} = \frac{78 \times \pi}{814.9 \times 0.0821 \times 353}$$

$$\text{Osmotic pressure, } \pi = \frac{9.26}{751.86} \times \frac{814.9 \times 0.0821 \times 353}{78} = \mathbf{3.73\ atm}$$

PROBLEM FOR PRACTICE

The vapour pressure of a solution of urea is 736.2mm at 100°C. What is the osmotic pressure of this solution at 15°C? **[Ans. 41. 1atm]**

5.25. THEORIES OF OSMOSIS

1. Molecular Sieves Theory

According to this theory, the membrane contains lots of fine pores and acts as a sort of molecular sieves. **Smaller solvent molecules can pass through the pores but the larger solute molecules cannot.** Solvent molecules flow from a region of higher solute concentration to one of lower concentration across such a membrane (Fig 5.19). But we observe that some membranes can act as sieves even when the solute molecules are smaller than the solvent molecules. This theory does not provide a satisfactory answer to this.

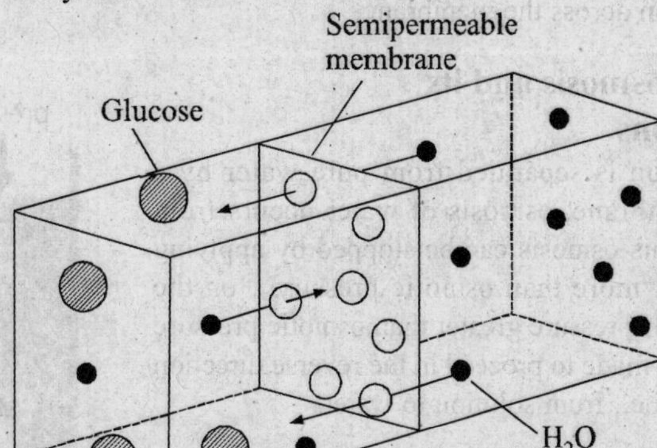

Fig. 5.19. A semipermeable membrane can separate particles on the basis of size. It allows the passage of small water molecules in both directions. But it prevents the passage of glucose molecules which are larger than water molecules.

Recently it has been shown that the pores or capillaries between the protein molecules constituting an animals membrane are lined with polar groups ($-COO^-$, $-NH_3^+$, $-S^{2-}$, etc.). Therefore, the membrane acts not simply as a sieve but also regulates the passage of solute molecules by electrostatic or 'chemical interactions'. In this way even solute molecules smaller than solvent molecules can be held back by the membrane.

2. Membrane solution theory

Membrane proteins bearing functional groups such as – COOH, – OH, – NH_2, etc., dissolve water molecules by hydrogen bonding or chemical intersection. Thus, membrane dissolves water from the pure water (solvent) forming what may be called '*membrane solution*'. The dissolved water flows into the solution across the membrane to equalise concentrations. In this way water molecules pass through the membrane while solute molecules being insoluble in the membrane do not.

3. Vapour Pressure theory

It suggests that a semipermeable membrane has many fine holes or capillaries. The walls of these capillaries are not wetted by water (solvent) or solution.

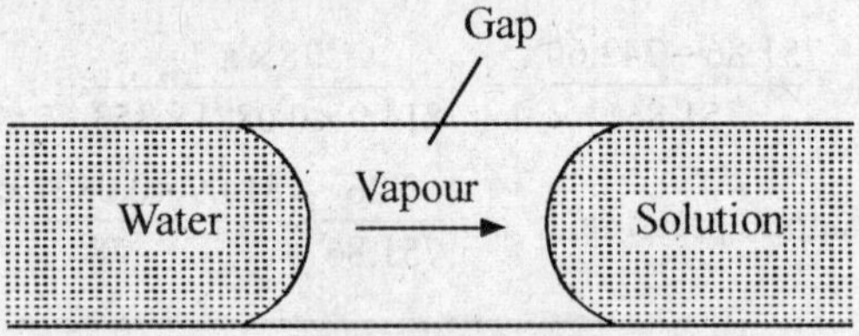

Fig. 5.20. Water vapours diffuse into solution across the gap in a capillary of the membrane.

Thus neither solution nor water can enter the capillaries. Therefore, each capillary will have in it solution at one end and water at the other, separated by a small gap (Fig. 5.20). Since the vapour pressure of a solution is lower than that of the pure solvent, the diffusion of vapour will occur across the gap from water side to solution side. This results in the transfer of water into the solution.

4. Membrane Bombardment theory

This theory suggest that osmosis result from an unequal bombardment pressure caused by solvent molecules on the two sides of the semipermeable membrane. On one side we have only solvent molecules while on the other side there are solute molecules occupying some of the surface area. Thus, there are fewer bombardments per unit area of surface on the solution side than on the solvent side. Hence the solvent molecules will diffuse more slowly through the membrane on the solution side than on the solvent side. The net result causes a flow of the solvent from the pure solvent to the solution across the membrane.

5.25.1. Reverse osmosis and its applications

When a solution is separated from pure water by a semipermeable membrane, osmosis of water occurs from water to solution. This osmosis can be stopped by applying pressure equal to or more than osmotic pressure , on the solution (Fig. 5.21). If pressure greater than osmotic pressure is applied, osmosis is made to proceed in the reverse direction to ordinary osmosis i.e., from solution to water.

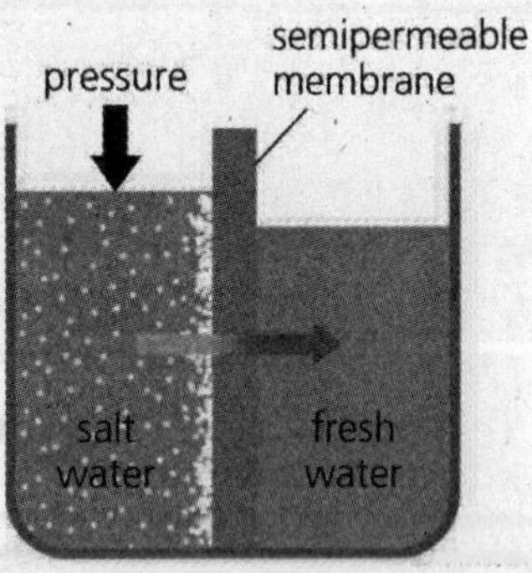

Principle of reverse osmosis

The osmosis taking place from solution to pure water by application of pressure greater than osmotic pressure, on the solution, is termed *Reverse Osmosis*.

This technology is used in the commercial production of water purifiers these days (R.O. Water purifiers)

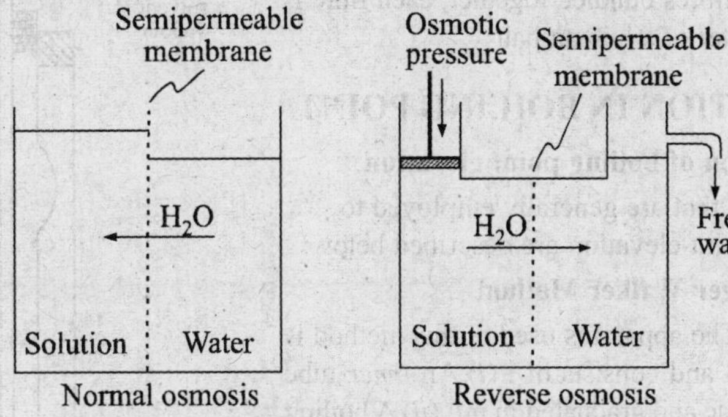

Fig. 5.21. Reverse osmosis versus ordinary osmosis

5.25.2. Desalination of sea water by hollow-fibre reverse osmosis

Reverse osmosis is used for the desalination of sea water for getting fresh drinking water. This is done with the help of hollow fibres (nylon or cellulose acetate) whose wall acts as semipermeable membrane. A hollow-fibre reverse osmosis unit is shown in Fig. 5.22.

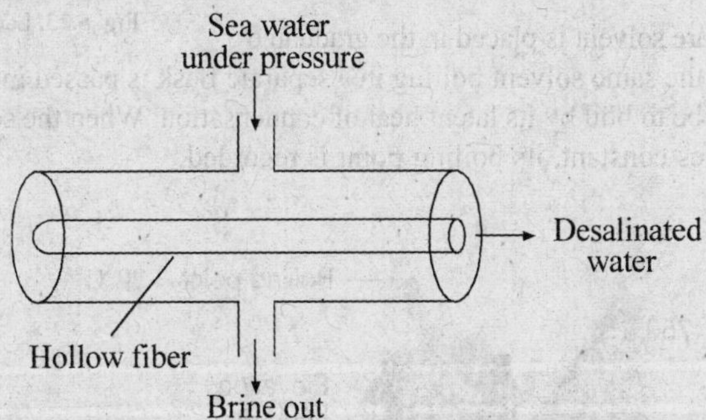

Fig. 5.22. Desalination of sea water by reverse osmosis in a hollow fibre unit

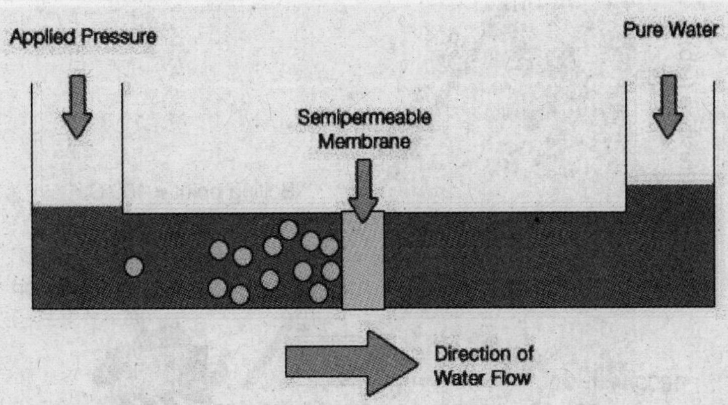

Design of a commercial R.O. Plant

Water is introduced under pressure around the hollow fibres. The fresh water is obtained from the inside of the fibre. In actual practice, each unit contains more than three million fibres bundled together, each fibre is of about the diameter of a human hair

5.26. ELEVATION IN BOILING POINT

Determination of boiling point elevation

Two methods that are generally employed to measure boiling point elevation are described below:

1. Landsberger-Walker Method

Apparatus. The apparatus used in this method is shown in Fig. 5.23 and consists of : (*i*) An *inner* tube with a hole in its side and graduated in ml; (*ii*) A boiling flask which sends solvent vapour into the graduated tube through a bulb with several holes (rose bulb), (*iii*) An *outer tube* which receives hot solvent vapour issuing from the side-hole of the inner tube; (*iv*) A *thermometer* reading to 0.1K, dipping in solvent or solution in the inner tube.

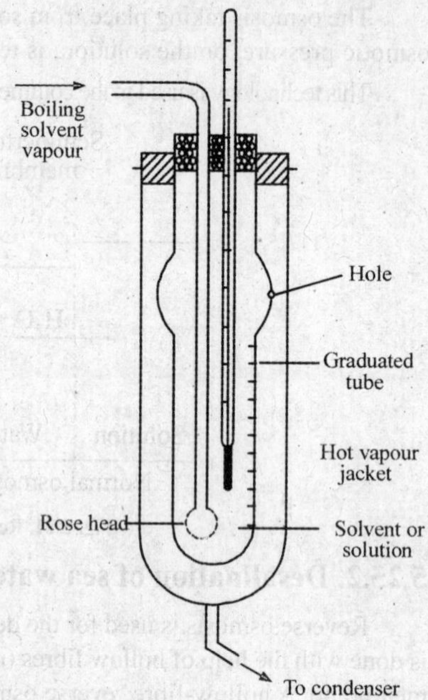

Fig. 5.23. Landsberger-Walker apparatus.

Procedure. Pure solvent is placed in the graduated tube and vapour of the same solvent boiling in a separate flask is passed into it. The vapour causes the solvent in the tube to boil by its latent heat of condensation. When the solvent starts boiling and temperature becomes constant, its boiling point is recorded.

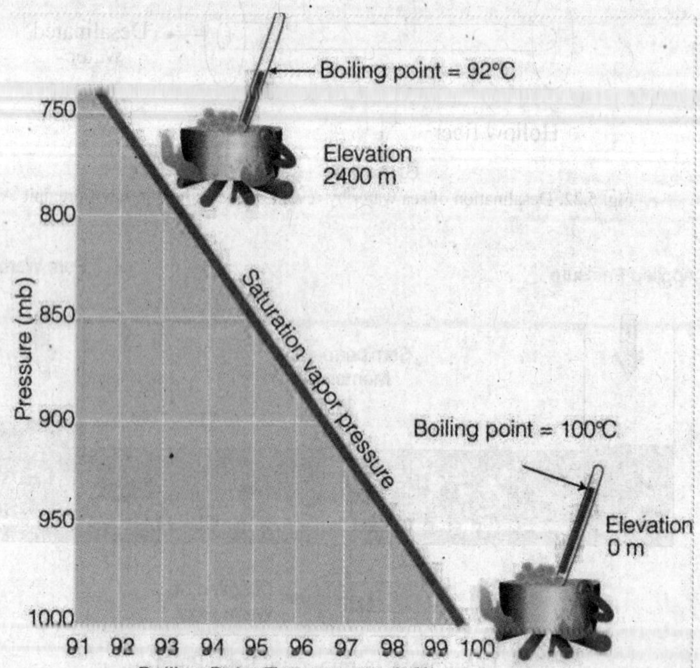

Variation in boiling point with atmospheric pressure

Now the supply of vapour is temporarily cut off and a weighed pellet of the solute is dropped into the solvent in the inner tube. The solvent vapour is again passed through until the boiling point of the solution is reached and this is recorded. The solvent vapour is then cut off, thermometer and rosehead raised out of the solution, and the volume of the solution read.

From a difference in the boiling points of solvent and solution, we can find the molecular weight of the solute by using the expression

$$m = \frac{1000 \times K_b \times w}{\Delta T_b \times W}$$

where w = weight of solute taken, W = weight of solvent which is given by the volume of solvent (or solution) measured in ml multiplied by the density of the solvent at its boiling point.

2. Cottrell's Method

Apparatus. It consists of : (*i*) a graduated *boiling tube* containing solvent or solution; (*ii*) a reflux condenser which returns the vapourised solvent to the boiling tube; (*iii*) a thermometer reading to 0.01 K, enclosed in a glass hood; (*iv*) A small inverted funnel with a narrow stem which branches into three jets projecting at the thermometer bulb. Fig. 5.24 (b) showing all the four components.

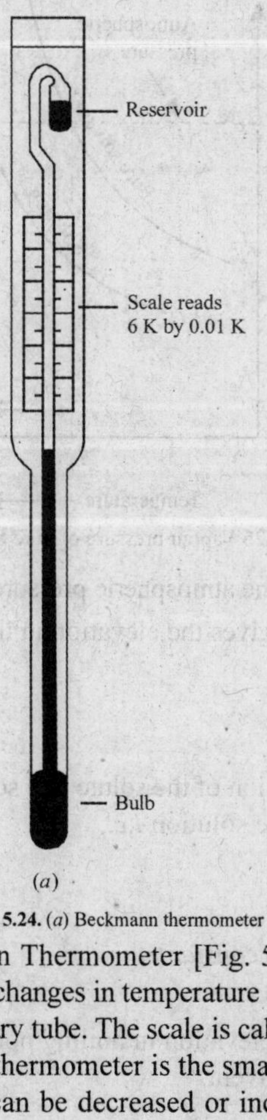

Fig. 5.24. (*a*) Beckmann thermometer reading to 0.01 K.

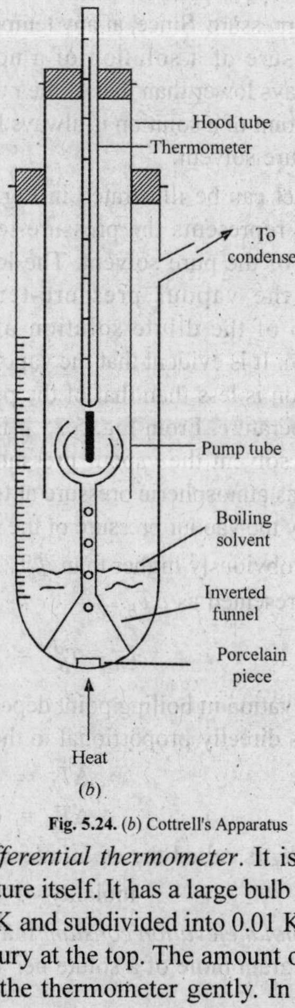

Fig. 5.24. (*b*) Cottrell's Apparatus

Beckmann Thermometer [Fig. 5.24 (*a*)]. It is a *differential thermometer*. It is designed to measure small changes in temperature and not the temperature itself. It has a large bulb at the bottom of a fine capillary tube. The scale is calibrated from 0 to 6 K and subdivided into 0.01 K. The unique feature of this thermometer is the small reservoir of mercury at the top. The amount of mercury in this reservoir can be decreased or increased by tapping the thermometer gently. In this way the thermometer is adjusted so that the level of mercury thread will show up at the middle of the scale when the instrument is placed in the boiling (or freezing) solvent.

Procedure. The apparatus is set up as shown in [Fig. 5.24 (*b*)]. Solvent is placed in the boiling tube with a porcelain piece lying in it. It is heated on a small flame. As the solution starts boiling, solvent vapour arising from the porcelain piece pump the boiling liquid into the narrow stem. Thus, a mixture of solvent vapour and boiling liquid is continuously sprayed around the thermometer bulb. The temperature soon becomes constant and the boiling point of the pure solvent is recorded.

Now a weighed amount of the solute is added to the solvent and the boiling point of the solution noted as the temperature becomes steady. Also, the volume of the solution in the boiling tube is noted. The difference of the boiling temperatures of the solvent and solute gives the elevation of boiling point. While calculating the molecular weight of the solute, the volume of solution is converted into mass by multiplying with density of solvent at its boiling point.

5.27. RELATION BETWEEN ELEVATION IN BOILING POINT OF THE SOLUTION AND THE MOLECULAR WEIGHT OF THE SOLUTE

The boiling point of a liquid is the temperature at which its vapour pressure becomes equal to the atmospheric pressure. Since, at any temperature, the vapour pressure of a solution of a non-volatile solute is always lower than that of the pure solvent, the boiling point of a solution is always higher than that of the pure solvent.

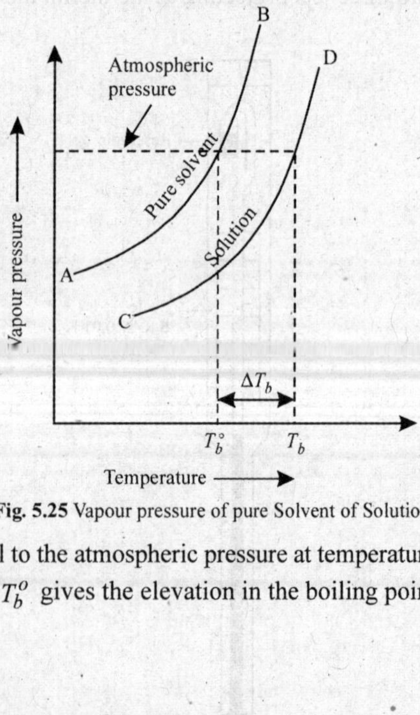

Fig. 5.25 Vapour pressure of pure Solvent of Solution.

This fact can be illustrated in Fig. 5.25. The upper curve represents the pressure-temperature relationship of the pure solvent. The lower curve represents the vapour pressure-temperature relationship of the dilute solution of a known concentration. It is evident that the vapour pressure of the solution is less than that of the pure solvent at each temperature. From Fig. 5.25 it is clear that for the pure solvent the vapour pressure becomes equal to the as atmospheric pressure at temperature T_b°. Similarly the vapour pressure of the solution is equal to the atmospheric pressure at temperature T_b, which is obviously higher than T_b^o. Therefore $T_b - T_b^o$ gives the elevation in the boiling point which is represented as ΔT_b.

$$\text{or} \qquad T_b - T_b^o = \Delta T_b$$

The elevation in boiling point depends upon concentration of the solute in a solution. In other words ΔT_b is directly proportional to the molality (m) of the solution *i.e.*,

$$\Delta T_b \propto \text{molality} \qquad \text{... }(i)$$

$$\text{or} \qquad \Delta T_b = K_b \times \text{molality}$$

where K_b is called the *molal elevation (ebullioscopic) constant.*

If $\quad \text{molality} = 1, \Delta T_b = K_b$

Thus, *molal elevation constant* may be defined as the elevation in boiling point of a solution containing l gram mole of a solute per 1000 gram of the solvent.

Molecular Mass of Non-volatile Solute

By definition,

$$\text{molality} = \frac{1000 \times w}{W \times m}$$

∴ eqn. (*i*) becomes

$$\Delta T_b = \frac{1000 \times K_b \times w}{W \times m}$$

$$m = \frac{1000 \times K_b \times w}{W \times \Delta T_b}$$

where ΔT_b = Elevation in boiling point.

K_b = molal elevation (Ebullioscopic) constant of the solvent.

m = molecular mass of the solute

w = weight of the solute in grams

W = weight of the solvent in grams.

5.27.1. Thermodynamic derivation of a relationship between elevation in boiling point and molecular weight of a non-volatile solute

Fig. 5.26. gives the vapour pressure-temperature curves for the solvent and solution. The horizontal dotted line drawn corresponding the external pressure cuts the solvent and the solution curves at the points E and F which correspond to temperatures T_0 and T_s respectively. Then by definition, T_0 is the boiling point of the pure solvent and T_s that of the solution. Obviously $T_s > T_0$. The increase $(T_s - T_0)$ is called the elevation in boiling point and is usually represented by ΔT_b. Thus

$$\Delta T_b = T_s - T_0 \qquad \text{... }(i)$$

The relationship between the elevation in boiling point and the concentration of the solution can be obtained by applying Clausius-Clapeyron equation and Raoult's law to the different conditions shown in Fig. 5.26.

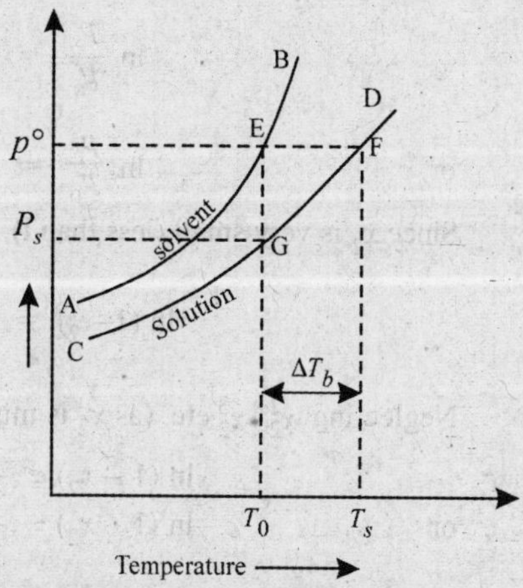

Fig. 5.26 Relationship between elevation in boiling point and lowering of vapour pressure

Clausius-Clapeyron equation which is applicable to a phase equilibrium, in the integrated form is given by

$$\ln \frac{p_2}{p_1} = \frac{\Delta H}{R}\left(\frac{1}{T_1} - \frac{1}{T_2}\right)$$

where p_1 is the vapour pressure at temperature T_1, p_2 is the vapour pressure at temperature T_2 and ΔH is the molar latent heat of transition.

Referring to Fig. 5.26 points G and F lie on the same solution curve. Suppose the vapour pressure of the pure solvent and the solution at temperature T_0 are p_0 and p_s respectively. Then from Fig. 5.26, we have

(*a*) Corresponding to point G, temperature = T_0, vapour pressure = p_s.

(*b*) Corresponding to point F, temperature = T_s, vapour pressure = p^o.

Hence for the equilibrium, solution $\rightleftharpoons$ vapour, according to Clausius-Clapeyron equation, we have (for the points G and F)

$$\ln \frac{p^o}{p_s} = \frac{\Delta H_v}{R}\left(\frac{1}{T_o} - \frac{1}{T_s}\right) \quad ... (ii)$$

where ΔH_v is the latent heat of vaporisation of one mole of the solvent from the solution. When the solution is dilute, ΔH_v is nearly equal to the latent heat of evaporation of the pure solvent.

Equation (*ii*) can be written as

$$\ln \frac{p^o}{p_s} = \frac{\Delta H_v}{R}\left(\frac{T_s - T_0}{T_0 T_s}\right)$$

or $$\ln \frac{p^o}{p_s} = \frac{\Delta H_v}{R}\frac{\Delta T_v}{T_b T_s} \qquad [\because T_s - T_0 = \Delta T_b] \quad ...(iii)$$

Further, when the solution is dilute T_s is nearly equal to T_0. Hence equation (*iii*) can written as

$$\ln \frac{p^o}{p_s} = \frac{\Delta H_v}{R}\frac{\Delta T_b}{T_0^2} \quad ...(iv)$$

According to Raoult's law, we have

$$\frac{p^o - p_s}{p^o} = x_2 \quad ...(v)$$

where x_2 is the mole fraction of the solute in the solution.

Equation (*v*) can be written as

$$1 - \frac{p_s}{p_o} = x_2 \quad \text{or} \quad \frac{p_s}{p^o} = 1 - x_2$$

$$\therefore \quad \ln \frac{p_s}{p_o} = \ln(1 - x_2)$$

or $$\ln \frac{p_s}{p_o} = -\ln(1 - x_2) \quad ...(vi)$$

Since x_2 is very small (less than l), the expansion of $\ln(1 - x_2)$ as an infinite series is given by

$$\ln(1 - x_2) = -x_2 - \frac{1}{2}x_2^2 - \frac{1}{3}x_2^3 ....$$

Neglecting x_2^2, x_2^3 etc. (as x_2 is much smaller than 1), we have

$$\ln(1 - x_2) \simeq -x_2 \quad ...(vii)$$

or $$-\ln(1 - x_2) \simeq -x_2$$

From eq. (*vi*) and (*vii*) $\ln \frac{p_s}{p_o} = x_2 \quad ...(viii)$

Substituting this value in equation (*iv*)

$$x_2 = \frac{\Delta H_v}{R}\frac{\Delta T_b}{T_0^2} \quad ...(ix)$$

or $$\boldsymbol{\Delta T_b = \frac{RT_0^2 x_2}{\Delta H_v}} \quad ...(x)$$

However the common practice in the studies on elevation in boiling point is not to express the concentrations in terms of mole fractions but in terms of *moles of the solute per 1000 g of the solvent i.e. in terms of molality, m*. Hence equation (*x*) is further modified as follows:

If n_2 moles of the solute are dissolved in n_1 moles of the solvent, the mole fraction of the solute (x_2) in the solution will be given by

$$x_2 = \frac{n_2}{n_1 + n_2}$$

As the solution is supposed to be dilute $n_2 << n_1$ so that in the denominator n_2 can be neglected in comparison to n_1. Hence for the dilute solution, above equation can be written as

$$x_2 = \frac{n_2}{n_1}$$

Further, if w_2 g of the solute is dissolved in w_1 g of the solvent and M_2 and M_1 are the molecular weights of the solute and solvent respectively, then

$$n_2 = \frac{w_2}{M_1} \text{ and } n_1 = \frac{w_1}{M_1}$$

Hence $$x_2 = \frac{n_2}{n_1} = \frac{w_2 / M_2}{w_1 / M_1} = \frac{w_2 \times M_1}{w_1 \times M_2}$$

Putting this value of x_2 in equation (*x*), we get

$$\Delta T_b = \frac{RT_0^2}{\Delta H_v} \frac{w_2 M_1}{w_1 M_2}$$

$$= \frac{RT_0^2}{\Delta H_v / M_1} \frac{w_2}{w_1 M_2} \quad \text{...(xi)}$$

Putting $\frac{\Delta H_v}{M_2} = l_v$, latent heat of vaporisation per gram of the solvent, equation (*xi*) becomes

$$\Delta T_b = \frac{RT_0^2}{l_v} \frac{w_2}{w_1 M_2} \quad \text{..(xii)}$$

Let us say molality of the solution is *m*. Thus *m* moles of the solute are dissolved in 1000 g of the solvent, we can write

$$\frac{w_2}{M_2} = m \text{ and } w_1 = 1000$$

Putting these values in equation (*xii*), we get

$$\Delta T_b = \frac{RT_0^2}{l_v} \frac{m}{1000} \quad \text{...(xiii)}$$

or $$\Delta T_b = \frac{RT_0^2}{1000 l_v} m \quad \text{...(xiv)}$$

For a given solvent, the quantity $\frac{RT_0^2}{1000 l_v}$ is a constant quantity because l_v, T_0 and R are constant. It is represented by K_b and is called **molal elevation constant** or **ebullioscopic constant**.

i.e.
$$K_b = \frac{RT_0^2}{1000\, l_v} \quad ...(xv)$$

Hence equation (*xiv*) can be written as

$$\Delta T_b = K_b \cdot m \quad ....(xvi)$$

If $m = 1, \Delta T_b = K_b$

Thus molal elevation constant may be defined as the elevation in boiling point when the molality of the solution is unity.

SOLVED PROBLEMS ON ELEVATION IN BOILING POINT

Example 1. A solution of 12.5 g of urea in 170 g of water gave boiling point elevation of 0.63 K. Calculate the molar mass of urea. K_b = 0.52 K kg mol^{-1}.

Solution. From the given data

Wt. of the solute, w = 12.5 g

Wt. of the solvent W = 170 g

Elevation of boiling point ΔT_b = 0.63 K

Elevation constant K_b = 0.52 K kg mol^{-1}

Let the molecular mass of solute (urea) = m

Calculation of molality

170 grams of water contain urea = 12.5 gram

$$\therefore \text{ 1000 grams of water contain urea} = \frac{1000 \times 12.5}{170} \text{ g}$$

$$= \frac{1000 \times 12.5}{170 \times m} \text{ mole}$$

$$\therefore \text{ molality (no. of mole in 1000 g of solvent)} = \frac{1000 \times 12.5}{170 \times m}$$

We know, $\Delta T_b = K_b \times \text{molality}$

Substituting the value in the above relation,

We have, $$0.63 = 0.52 \times \left(\frac{1000 \times 12.5}{170 \times m}\right)$$

$$m = \frac{0.52 \times 1000 \times 12.5}{170 \times 0.63} = 60.69 \text{ a.m.u.}$$

Example 2. A solution prepared from 0.3 g of an unknown non-volatile solute in 30.0g of CCl_4 boils at 350.392 K. Calculate the molecular mass of the solute. The boiling point of CCl_4 and its K_b values are 350.0 K and 5.03 respectively.

Solution. From the given data

Wt. of the solute, w = 0.3 g

Wt. of the solvent W = 30. 0 g

Elevation of boiling point = ΔT_b = 350.392 – 350.0 = 0.392 K

Elevation constant K_b = 5.03 K kg mol^{-1}

Calculation of molality

$$30 \text{ g of } CCl_4 \text{ contain} = 0.3 \text{ g of solute}$$

$$\therefore \quad 1000 \text{ g of } CCl_4 \text{ contain} = \frac{1000 \times 0.3}{30} \text{ g}$$

$$= \frac{1000 \times 0.3}{30 \times m} \text{ moles}$$

$$\therefore \quad \text{Molality (no. of moles in 1000 g of } CCl_4 = \frac{1000 \times 0.3}{30 \times m}$$

Substituting the values in the relation

$$\Delta T_b = K_b \times \text{molality}$$

$$0.92 = 5.03 \times \frac{1000 \times 0.3}{30 \times m}$$

$$m = \frac{5.03 \times 1000 \times 0.3}{30 \times 0.392} = 128.3$$

$$\therefore \quad \text{Molecular mass} = \mathbf{128.3}$$

Example 3. Find the b.p. of a solution containing 0.36 g of glucose ($C_6H_{12}O_6$) dissolved in 100 g of water (K_b= 0.52 K/m).

Solution.

$$\text{Mass of glucose (w)} = 0.36 \text{ g}$$

$$\text{Mass of water (W)} = 100 \text{ g}$$

$$\text{Mol. Mass of glucose (M)} = 180$$

$$\text{Molal elevation constant for water } (K_b) = 0.52$$

Substituting the values in the relation

$$\Delta T_b = \frac{1000 \times K_b \times w}{W \times m}$$

$$\text{or} \quad \Delta T_b = \frac{1000 \times 0.52 \times 0.36}{100 \times 180}$$

$$\text{Elevation in b.p.} = 0.0104$$

$$\text{B. P. of pure water} = 373 \text{ K}$$

$$\text{Hence b.p. of the solution} = 373 + 0.0104 = 373.0104 \text{ K}$$

Example 4. 10 g of a non-volatile solute when dissolved in 100g of benzene raises its b.p. by 1°. What is the molecular mass of the solute (K_b for benzene = 2.53 K mol^{-1})?

Solution. In this problem

$$\text{Mass of the solute (w)} = 10 \text{ g}$$

$$\text{Mass of the solvent (W)} = 100 \text{ g}$$

$$\text{Elevation in b.p. } (\Delta T_b) = 1°, K_b = 2.53$$

Substituting the values in the equation

$$m = \frac{1000 \times K_b \times w}{W \times \Delta T_b} = \frac{1000 \times 2.53 \times 10}{100 \times 1} = \mathbf{253}$$

Example 5. A solution containing 0.5126 g of naphthalene (mol mass 128) in 50 g of carbon tetrachloride yields a b.p. elevation of 0.402°C while a solution of 0.6216 g of an unknown solute

in the same weight of the solvent gives a b.p. elevation of 0.647°C. Find the molecular mass of the unknown solute.

Solution 1. Determination of K_b from the first data

$w = 0.5126$, $m = 128$, $W = 50$

$$K_b = \frac{m \times W \times \Delta T_b}{1000 \times w}$$

$$= \frac{128 \times 50 \times 0.402}{1000 \times 0.5126} = \mathbf{5.02}$$

2. Mol. mass of the unknown solute

$w = 0.6216$ g, W = 50 g, $\Delta T_b = 0.647$, $K_b = 5.02$

The value of K_b remains the same because the solvent is the same. Substituting the values in the equation

$$m = \frac{1000 \times K_b \times w}{W \times \Delta T_b}$$

$$= \frac{1000 \times 5.02 \times 0.6216}{50 \times 0.647} = \mathbf{96.46}$$

PROBLEMS FOR PRACTICE

1. What is the molecular weight of a non-volatile organic compound if the addition of 1.0g of it in 50·0 g of benzene raises the boiling point of benzene by 0·30°C ? K_b for benzene is 2·53°C per 1000 g of benzene. **[Ans.** 170]
2. A solution containing 36g of solute dissolved in one litre of water gave an osmotic pressure of 6·75 atmosphere at 27°C. The molal elevation constant of water is 0·52 K kg mol^{-1}. Calculate the boiling point of the solution. **[Ans.** 100·1425°C]
3. Calculate the molal boiling point constant for chloroform (M=119·4) from the fact than its boiling point is 61·2°C and its latent heat of vaporizatin is 59·0 cal/g. **[Ans.** 3·79°C]
4. A solution of 3·795 g sulphur in 100 g carbon disulphide (boiling point 46·30°C, ΔH_v= 6400 cal/mol) boils at 46·66°C. What is the formula of sulphur molecule in the solution? **[Ans.** S_8]

   [**Hint.** ΔH_v = 6400 cal/mol, $\therefore\ l_v = \frac{6400}{76}$ cal/g because mol wt of CS_2 = 76.

   First calculate K_b and then M_2.

   We get M_2 = 255·2. Hence atomicity of molecule = $\frac{255{\cdot}2}{32} = 8$]
5. When 1·80g of a non-volatile compound are dissolved in 25·0 g of acetone, the solution boils at 56·86°C while pure acetone boils at 56·38°C under the same atmospheric pressure. Calculate the molecular weight of the compound. The molal elevation constant for acetone is 1·72°K kg mol^{-1}. **[Ans.** 258]
6. What elevation in boiling point of alcohol is to be expected when 5 g of urea (mol. wt = 60) are dissolved in 75 g of it? The molal elevation constant for alcohol is 1·15°C per molality. **[Ans.** 1·28°C]

5.27.2. Relationship between elevation in boiling point and relative lowering of vapour pressure

We have the following relation between elevation in boiling point and molecular mass

$$\Delta T_b = \frac{1000 \times K_b\, w_2}{w_1 M_2} \qquad ...(i)$$

Also $$\frac{w_2}{M_2} = n_2 \quad ...(ii)$$

and $$\frac{w_1}{M_1} = n_1 \quad ...(iii)$$

Substituting the values of w_2 and w_1 for eq. (*ii*) and (*iii*) in (*i*), we get

we get $$\Delta T_b = \frac{1000 \times K_b n_2}{n_1 M_1} = \frac{1000 K_b}{M_1} \frac{n_2}{n_1} \quad ...(iv)$$

By Raoult's law, for dilute solutions, we have

$$\frac{\Delta p}{p^o} = \frac{n_2}{n_1}$$

where $\Delta p = p^\circ - p_s$ is the lowering of vapour pressure. From eqs (*iv*) and (*v*) we get

$$\Delta T_b = \frac{1000 \mathrm{K_b}}{\mathrm{M_1}} \frac{\Delta p}{p^o}$$

5.27.3. Relationship between elevation in boiling point and osmotic pressure

The equation for the osmotic pressure for n_2 moles of the solute dissolved in V litres of solution is

$$\pi V = n_2 RT$$

or $$n_2 = \frac{\pi V}{RT} \quad ...(i)$$

The equation for the elevation in boiling point is

$$\Delta T_b = \frac{1000 K_b w_2}{w_1 M_2} \quad ...(ii)$$

or $$\Delta T_b = \frac{1000 K_b w_2 / M_2}{w_1}$$

$$= \frac{1000 K_b \cdot n_2}{w_1} \quad ...(iii)$$

Substituting the value of n_2 from eq (*i*) in eq. (*iii*)

$$\Delta T_b = \frac{1000 K_b}{w_1} \times \frac{\pi V}{RT} = \frac{1000 K_b \cdot \pi}{(w_1 / V) RT} \quad ...(iv)$$

As the solution is dilute.

Volume of the solution (V) $\simeq$ Volume of the solvent.

$\therefore$ If d is the density of the solvent,

$$\frac{w_1}{V} = d$$

Substituting this value in equation, we get

$$\Delta T_b = \frac{1000 \mathrm{K_b} \pi}{dRT}$$

5.28. DEPRESSION IN FREEZING POINT

Methods for the determination of depression in freezing point

1. Beckmann's Method

Apparatus. It consists of : (*i*) *A freezing tube* with a side-arm to contain the solvent or solution, while the solute can be introduced through the side-arm ; (*ii*) An outer tube into which is fixed the freezing tube, the space in between providing an air jacket which ensures a slower and more uniform rate of cooling; (*iii*) A *large jar* containing a freezing mixture e.g., ice and salt with a stirrer (Fig. 5.27).

Procedure. About 20 g of the solvent is taken in the freezing point tube and the apparatus set up as shown in Fig. 5.27 so that the bulb of the thermometer is completely immersed in the solvent. Determine the freezing point of the solvent by directly cooling the freezing-point tube in the cooling bath.

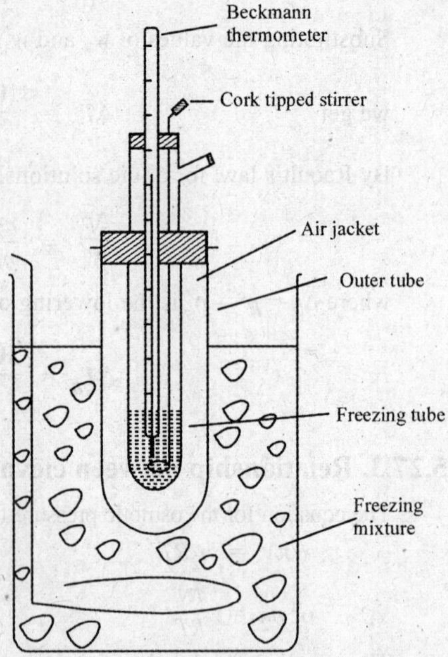

Fig. 5.27 Beckmann's Freezing-point apparatus

The freezing point of the solvent having been accurately determined, the solvent is remelted by removing the tube from the bath, and a weight amount (0.1–0.2 g) of the solute is introduced through the side tube. Now the freezing point of the solution is determined in the same way as that of the solvent. A further quantity of solute may then be added and another reading taken. Knowing the depression of the freezing point, the molecular weight of the solute can be determined by using the expression.

$$m = \frac{1000 \times K_f \times w}{\Delta T \times W}$$

2. Rast's Camphor Method (Cryoscopic method)

This method is used for determination of molecular weights of solutes which are soluble in molten camphor. The freezing point depressions are so large that an ordinary thermometer can also be used.

Pure camphor is powered and introduced into a capillary tube which is sealed at the upper end. This is tied along a thermometer and heated in a glycerol bath (see Fig. 5.28). The melting point of camphor is recorded. Then a weighed amount of solute and camphor (about 10 times as much) are melted in test-tube with the open end sealed. The solution of solute in camphor is cooled in air. After solidification, the mixture is powdered and introduced into a capillary tube which is sealed. Its melting point is recorded described before. The difference of the melting point of pure camphor and the mixture, gives the depression

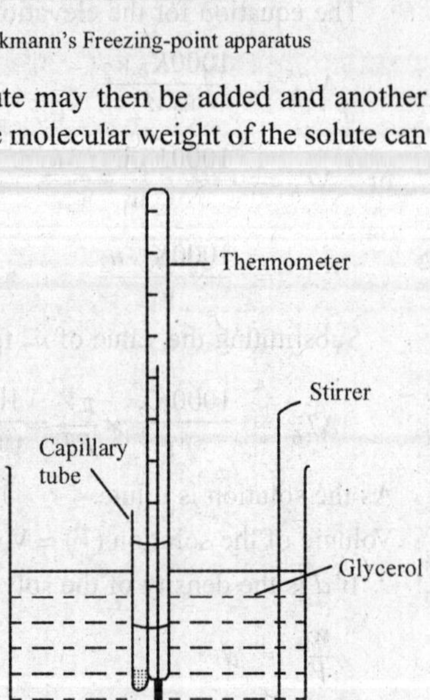

Fig. 5.28 Determination of depression of melting point by capillary method.

of freezing point. In modern practice, electrical heating apparatus (Fig. 5.29) is used for a quick determination of melting points of camphor as also the mixture.

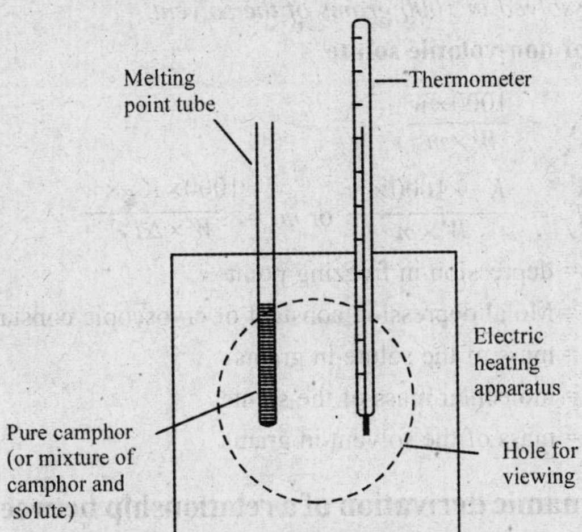

Fig. 5.29 Determination of depression of melting point by electrical apparatus

5.29. RELATION BETWEEN DEPRESSION IN FREEZING POINT AND MOLECULAR WEIGHT OF THE SOLUTE

Freezing point of a substance is the temperature at which solid and liquid states co-exists i.e., the two states have the same vapour pressure. The presence of a non-volatile solute lowers the vapour pressure of the solution. Thus liquid and solid states will have equal vapour pressure at a much lower temperature. Hence there is a depression in the freezing point. In other words, the freezing point of a solution is lower than that of the pure solvent as is clear from the Fig. 5.30.

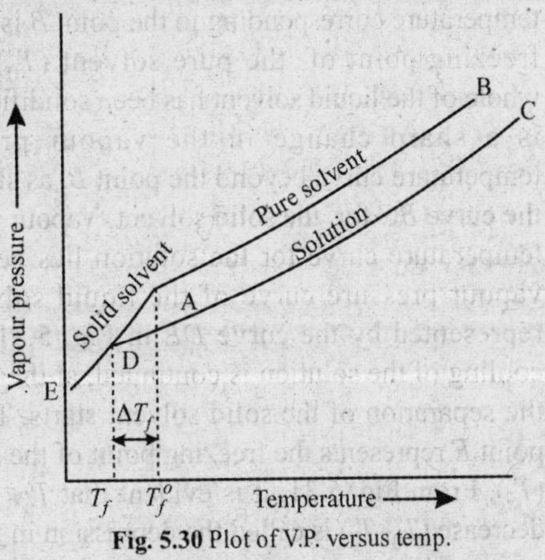

Fig. 5.30 Plot of V.P. versus temp.

The curve *AB* represents the vapour pressure curve of pure solvent. At point *A*, solvent co-exists with the liquid state. The temperature corresponding to point *A*, gives the freezing point of solvent. Similarly curve *CD* represents the vapour pressure curve of the solution in which point *D* corresponds to two states of the solvent existing simultaneously in the solution. Thus temperature corresponding to point *D* is the freezing point of solution which is clearly lower than the freezing point of the pure solvent. Hence there is a depression in the freezing point given by ΔT_f.

$$\Delta T_f = T_f^o - T_f$$

It has been found that depression in the freezing point is proportional to molal concentration of the solute in the solution i.e., molality as given below.

$$\Delta T_f = K_f \times \text{molality}$$

where K_f is constant known as *molal depression constant or cryoscopic constant of the solvent.*

If molality = 1 then $\Delta T_f = K_f$

Hence *molal depression constant may be defined as the depression in freezing point when one mole of the solute is dissolved in 1000 grams of the solvent.*

Molecular mass of non-volatile solute

$$\text{Since molality} = \frac{1000 \times w}{W \times m}$$

$$\therefore \quad \Delta T_f = \frac{K_f \times 1000 \times w}{W \times m} \quad \text{or } m = \frac{1000 \times K_f \times w}{W \times \Delta T_f}$$

Here ΔT_f = depression in freezing point

K_f = Molal depression constant or cryoscopic constant

w = mass of the solute in grams

m = molecular mass of the solute

W = mass of the solvent in grams

5.29.1. Thermodynamic derivation of a relationship between the freezing point and molecular weight of a non-volatile solute

The depression in freezing point of a solution can be explained on the basis of lowering of vapour pressure of the solution. If vapour pressure are plotted against temperature the curve *AB* is obtained for the liquid solvent as shown in Fig. 5.31. At the point *B*, the liquid solvent starts solidifying and hence the liquid solvent and the solid solvent are in equilibrium with each other. The temperature corresponding to the point *B* is thus the freezing point of the pure solvent (T_0). When whole of the liquid solvent has been solidified, there is a sharp change in the vapour pressure-temperature curve beyond the point *B*, as shown by the curve *BC* for the solid solvent. Vapour pressure temperature curve for the solution lies below the vapour pressure curve of the liquid solvent, as represented by the curve *DE* in Fig. 5.31 As the cooling of the solution is continued, at the point *E*, the separation of the solid solvent starts. Thus the point *E* represents the freezing point of the solution (T_s). From Fig. 5.31, it is evident that $T_s < T_0$. The decrease ($T_0 - T_s$) is called the depression in freezing point and is usually represented by ΔT_f. Thus

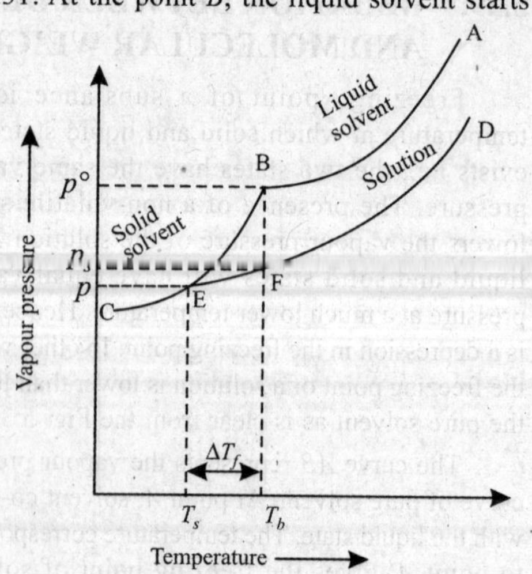

Fig. 5.31. Relationship between depression in freezing point and lowering of vapour pressure

$$\Delta T_f = T_0 - T_s \quad \text{...}(i)$$

The relationship between the depression in freezing point and the concentration of the solution can be derived by applying Clausius-Clapeyron equation and Raoult's law to the curves shown in Fig. 5.31.

Suppose the vapour pressures corresponding to the point *B*, *E* and *F* are $p°$, p and p_s respectively.

The equilibrium existing along the curve *BC* is

$$\text{Solid} \rightleftharpoons \text{Vapour}$$

The curve *BC* is, therefore, the sublimation curve of the solid solvent. For the points *B* and *E* lying on this curve, we have

(*a*) Corresponding to point *B*, temperature = T_0, vapour pressure = $p°$

(*b*) Corresponding to point *E*, temperature = T_s, vapour pressure = *p*.

Applying Clausius-Clapeyron equation corresponding to points *B* and *E*, we have

$$\ln \frac{p°}{p} = \frac{\Delta H_s}{R} \cdot \left(\frac{1}{T_s} - \frac{1}{T_0} \right)$$

where ΔH_s is the molar latent heat of sublimation of the solid solvent.

Rudolf Clausius

Equation (*ii*) can be written as

$$\ln \frac{p°}{p} = \frac{\Delta H_s}{R} \frac{T_0 - T_s}{T_s T_0}$$

$$= \frac{\Delta H_s}{R} \cdot \frac{\Delta T_f}{T_0 T_s} \quad ...(iii)$$

As the solution is dilute $T_s \simeq T_0$ and therefore equation (*iii*) becomes

$$\ln \frac{p°}{p} = \frac{\Delta H_s}{R} \cdot \frac{\Delta T_f}{T_0^2}$$

...(*iv*)

The equilibrium existing along the curve *DE* is

$$\text{Solution} \rightleftharpoons \text{Vapour}$$

The curve *DE* is thus the vaporisation curve of the solution. For the points *E* and *F* lying on this curve, we have

(*a*) Corresponding to point *E*, temperature = T_s, vapour pressure = $p°$

(*b*) Corresponding to point *F*, temperature = T_0, vapour pressure = p_s

Applying Clausius-Clapeyron equation to these conditions, we have

$$\ln \frac{p_s}{p} = \frac{\Delta H_v}{R} \left(\frac{1}{T_s} - \frac{1}{T_0} \right) = \frac{\Delta H_v}{R} \cdot \frac{T_0 - T_s}{T_s \cdot T_0}$$

$$= \frac{\Delta H_v}{R} \frac{\Delta T_f}{T_0^2} \quad \text{(Taking } T_s \simeq T_0\text{)} \quad ...(v)$$

where ΔH_v is the latent heat of vaporisation of one mole of the solvent from the solution and is nearly equal to the molar latent of vaporisation of the pure solvent for a dilute solution.

In order to get a relationship for the equilibrium

$$\text{Solid} \rightleftharpoons \text{Liquid}$$

subtract equation (*v*), from equation (*iv*), we get

$$\ln \frac{p°}{p} - \ln \frac{p_s}{p} = \frac{\Delta H_s}{R} \frac{\Delta T_f}{T_0^2} - \frac{\Delta H_v}{R} \frac{\Delta T_f}{T_0^2}$$

$$\text{or} \quad (\ln p^\circ - \ln p) - (\ln p_s - \ln p^\circ) = \frac{\Delta T_f}{RT_0^2}(\Delta H_s - \Delta H_v)$$

$$\ln p^\circ - \ln p_s = \frac{\Delta T_f}{RT_0^2}\Delta H_f$$

$$\ln \frac{p^\circ}{p_s} = \frac{\Delta T_f}{RT_0^2}\Delta H_f \qquad ...(vi)$$

where $\Delta H_s - \Delta H_v = \Delta H_f$ is the molar latent heat of fusion of the solid solvent.

The left hand term of equation (*vi*) can be expressed in terms of mole fraction by applying Raoult's law,

$$\frac{p^\circ - p_s}{p^\circ} = x_2 \qquad ...(vii)$$

where x_2 is the mole fraction of the solute in the solution.

Equation (*vii*) can be written as

$$1 - \frac{p_s}{p^\circ} = x_2$$

$$\text{or} \quad \frac{p_s}{p^\circ} = 1 - x_2$$

$$\text{or} \quad \ln \frac{p_s}{p^\circ} = \ln (1 - x_2)$$

$$\text{or} \quad \ln \frac{p_s}{p^\circ} = -\ln (1 - x_2) \qquad ...(viii)$$

The expansion of $\ln (1 - x_2)$ as an infinite series is given by

$$\ln (1 - x_2) = -x_2 - \frac{1}{2}x_2^2 - \frac{1}{3}x_2^3....$$

Neglecting x_2^2, x_2^3 etc. (because x_2 is very small, higher powers will be still smaller), we have

$$\ln (1 - x_2) = -x_2$$

Hence equation (*viii*) becomes

$$\ln \frac{p^\circ}{p_s} = x_2$$

Substituting this value in equation (*vi*), we get

$$x_2 = \frac{\Delta T_f}{RT_0^2} \cdot \Delta H_f$$

$$\text{or} \quad \Delta T_f = \frac{RT_0^2}{\Delta H_f} \cdot x_2 \qquad ...(ix)$$

This equation gives the relationship between depression in freezing point and the mole fraction x_2 of the solute in the solution.

To obtain the equation in terms of molality instead of mole fraction x_2, we have

$$x_2 = \frac{n_2}{n_1 + n_2}$$

where n_2 = number of moles of the solute in the solution

and n_1 = number of moles of the solvent in the solution.

For dilute solution $n_2 << n_1$ so that on neglecting n_2 in comparison to n_1, we have

$$x_2 = \frac{n_2}{n_1} = \frac{w_2 / M_2}{w_1 / M_1} = \frac{w_2 M_1}{w_1 M_2} \quad ...(x)$$

where w_2 = weight of the solute dissolved

M_2 = molecular weight of the solute

w_1 = weight of the solvent

M_1 = molecular weight of the solvent.

Putting the value of x_2 from equation (*x*) in equation (*ix*), we get

$$\Delta T_f = \frac{RT_0^2}{\Delta H_f} \frac{w_2 M_1}{w_1 M_2} \quad ...(xi)$$

$$= \frac{RT_0^2}{\Delta H_f / M_1} \cdot \frac{w_2}{w_1 M_2} \quad ...(xii)$$

$$= \frac{RT_0^2}{l_f} \cdot \frac{w_2}{w_1 M_2}$$

where $l_f = \frac{\Delta H_f}{M_1}$ is the latent heat of fusion *per gram* for the solid solvent.

If m is the molality of the solution *i.e.* m moles of the solute are dissolved in 1000 g of the solvent, then

$$\frac{w_2}{M_2} = m \text{ and } w_1 = 1000 \text{ g}$$

Putting these values in equation (*xii*), we get

$$\Delta T_f = \frac{RT_0^2}{l_f} \frac{m}{1000}$$

$$= \frac{RT_0^2}{l_f} \cdot m \quad ...(xiii)$$

Since T_0 and l_f for a given solvent are constant, the quantity $\frac{RT_0^2}{1000 l_f}$ in equation (*xiii*) must be a constant quantity. It is usually represented by K_f and is called **molal depression constant or cryoscopic constant.** Thus

$$K_f = \frac{RT_0^2}{1000 l_f} \quad ...(xiv)$$

Substituting this value in equation (*xiii*), we get

$$\boldsymbol{\Delta T_f = K_f m} \quad ...(xv)$$

If $m = 1$, $\Delta T_f = K_f$

Thus *molal depression constant may be defined as the depression in freezing point which takes place when the molality of the solution is unity.*

SOLVED PROBLEMS ON DEPRESSION IN FREEZING POINT

Example 1. Calculate the freezing point of a solution containing 0.520 g of glucose ($C_6H_{12}O_6$) in 80·2 gram of water. For water K_f = 1·86 K kg mol^{-1}.

Solution. Given

Wt. of the solute (glucose), w = 0·520 g

Wt. of solvent water, W = 80·2 g

Molar depression constant K_f = 1·86 K kg mol^{-1}

Molecular wt. of the solute $C_6H_{12}O_6$, m = 180

Calculation of molality

80·2 g of water contain glucose = 0·520 g

$$\therefore \quad 1000 \text{ g of water, contain glucose} = \frac{1000 \times 0.520}{80 \cdot 2}$$

$$\therefore \quad \text{Molality} = \frac{1000 \times 0.520}{80 \cdot 2 \times 180}$$

Applying the relation

$$\Delta T_f = K_f \times \text{molality}$$

$$= 1 \cdot 86 \times \frac{1000 \times 0.520}{80 \cdot 2 \times 180}$$

$\therefore$ Freezing point = 273 – 0·067 = 272·933K ($\because$ Freezing point of water = 273K)

Example 2. What is molal depression constant? How is it related to latent heat of fusion?

Solution. For definition of molal depression constant. Molal depression constant K_f is also given by the relation $K_f = \dfrac{RT_f^2}{1000 l_f}$ where

T_f = Freezing point of solvent

L_f = Molar latent heat of fusion

R = Gas constant

For water, T_f = 273 K, L_f = 336 J g^{-1}

Example 3. A solution of 1.25 g of a certain non-electrolyte in 20 g of water freezes at 271·95 K. Calculate the molecular mass of the solute. (K_f = 1·86 K kg mol^{-1})

Solution. Here we are given that:

Wt. of the solute w = 1·25 g

Wt. of solvent W = 20 g

Molar depression constant K_f = 1·86 K kg mol^{-1}

Depression in freezing point ΔT_f = 273 – 271·95 = 1·05 K

Calculation of molality

20g of water contain solute = 1·25 g

$$\therefore \quad 1000 \text{ g of water contain solute} = \frac{1000 \times 1 \cdot 25}{20}$$

Let the molecular mass of the solute = m

$\therefore$ Molality $= \dfrac{1000 \times 1{\cdot}25}{20 \times m}$

Substituting the values in the relation

$$\Delta T_f = K_f \times m$$

$$1.05 = 1{\cdot}86 \times \frac{1000 \times 1.25}{20 \times m}$$

$$\therefore \quad m = \frac{1{\cdot}86 \times 1000 \times 1{\cdot}25}{20 \times 1{\cdot}05} = 110{\cdot}71$$

Molecular mass of the solute (m) = **110·71**

Example 4. Find (*i*) the boiling point and (*ii*) the freezing point of a solution containing 0·52 g glucose ($C_6H_{12}O_6$) dissolved in 30·2g of water. For water K_b= 0·52 K kg mol^{-1}; K_f= 1·86 K kg mol^{-1}.

Solution. Given

Molecular mass of glucose m = 180

Wt. of solute w = 0·52 g

Wt. of solvent W = 80·2 g

Calculation of molality

80·2 g of the solvent contain solute = 0.52 g

$\therefore$ 1000 g of the solvent contain solute $= \dfrac{1000 \times 0{\cdot}52}{80{\cdot}2}$ g

$\therefore$ Molality $= \dfrac{1000 \times 0{\cdot}52}{80{\cdot}2 \times 180}$

We know, $\Delta T_b = K_b \times$ molality

$= 0{\cdot}52 \times \dfrac{1000 \times 1{\cdot}52}{80 \times 180}$ g $= 0{\cdot}018$K

$\therefore$ Boiling point $= 373 + 0{\cdot}018 = 373{\cdot}018$ K

$\because$ Boiling point of water = 373 K)

$\therefore$ Boiling point of solution = 373·018 K

For freezing point, $\Delta T_f = K_f \times$ molality

$= 1{\cdot}86 \times \dfrac{100 \times 0{\cdot}52}{80{\cdot}2 \times 180} = 0{\cdot}067$

Freezing point of H_2O = 273 K

$\therefore$ Freezing point of solution = 273 – 0·067 = **272·933 K**

Example 5. The normal freezing point of nitrobenzene $C_6H_5NO_2$ is 278·82 K. A 0·25 molal solution of a certain solute in nitrobenzene causes a freezing point depression of 2 degree. Calculate the value of K_f for nitrobenzene.

Solution. $\Delta T_f = K_f \cdot m$

$\Delta T_f = 2$

$m = 0.25$

Substituting the values in the first equation

$$2 = K_f \times 0.25$$

or $$K_f = \frac{2}{0.25} = 8 \text{ K kg mol}^{-1}$$

Example 6. A sample of camphor used in the Rast method of determination molecular masses had a melting point of 176.5°C. The melting point of a solution containing 0.522 g camphor and 0.0386 g of an unknown substance was 158.5°C. Find the molecular mass of the substance. K_f of camphor per kg is 37.7.

Solution. Applying the expression

$$m = \frac{1000 \times K_f \times w}{\Delta T \times W}$$

to the present case, we have

$$\Delta T = 176.5 - 158.5 = 18$$

or $$K_f = 37.7$$

$$w = 0.0386 \text{ g}$$

$$W = 0.522 \text{ g}$$

Substituting these values

$$m = \frac{1000 \times 37.7 \times 0.0386}{18 \times 0.522} = 154.8$$

Example 7. A solution of sucrose (molar mass = 342 g/mole) is prepared by dissolving 68.4 g in 1000 g of water. What is the

(*i*) Vapour pressure of solution at 293 K

(*ii*) Osmotic pressure at 293 K

(*iii*) Boiling point of the solution

(*iv*) Freezing point of the solution

The vapour pressure of water at 293 K is 17·5 mm, K_b = 0·52, K_f = 1·86.

Solution. (*i*) Mole fraction of sucrose $X_B = \frac{68.4/342}{68{\cdot}4/342 + 1000/18} = \frac{9}{2509}$

$$\frac{p_A^o - p_A}{p_A} = X_B$$

Vapour pressure of water $$p_A^o = 17{\cdot}5 \text{mm}$$

Substituting the values $$\frac{17{\cdot}5 - p_A}{17{\cdot}5} = \frac{9}{2509}$$

or $$17{\cdot}5 \times 2509 - 2509\, p_A = 9 \times 17{\cdot}5$$

or $$2509\, p_A = 43750$$

or $$p_A = 17{\cdot}43 \text{ mm}$$

(*ii*) $\pi V = nRT$, $V = 1$ litre, $n = \frac{68{\cdot}4}{342}$

$$\pi \times 1 = \frac{68{\cdot}4}{342} \times 0{\cdot}0821 \times 293$$

or $\pi = 4.8$ atmospheres

(*iii*) $\Delta T_b = K_b \times m$ where m is the molality

Molality $(m) = \dfrac{68\cdot4}{342}$, $K_b = 0\cdot52$

Substituting the values

$$\Delta T_b = 0\cdot52 \times \frac{68\cdot4}{342} = 0\cdot104$$

Boiling point of the solution = 373 + 0.104 = 373·104

(*iv*) $\Delta T_f = K_f \times m$

$m = \dfrac{68\cdot4}{342}$, $K_f = 1\cdot86$

$$\therefore \Delta T_f = 1\cdot86 \times \frac{68\cdot4}{342} = 0\cdot372$$

$\therefore$ Freezing point of the solution = 273 – 0·372 = **272.628K**

PROBLEMS FOR PRACTICE

1. Naphthalene (F.P. 80·1°C) has a molal freezing point constant of 6·82°C m^{-1}. A solution of 3·2 g in 100 g of naphthalene freezes at a temperature of 0·863°C less than pure naphthalene. What is the molecular formula of sulphur in naphthalene? [**Ans.** S_8]
2. The latent heat of fusion of ice is 79·7 cal/g. Calculate the molal depression constant of water. [**Ans.** 1·86 K kg mol^{-1}]
3. In winter season, ethylene glycol is added to water so that water may not freeze. Assuming ethylene glycol to be non-volatile, calculate the minimum amount of ethylene glycol that must be added to 6·0 kg of water to prevent it from freezing at –0·30°C. The molal depression constant of water is 1·86 kg mol^{-1}. [**Ans.** 60 g]
4. The freezing point of pure benzene is 5·40°C. A solution containing l g of a solute dissolved in 50 g of benzene freezes at 4·40°C. Calculate the molecular weight of the solute. The molecular depression constant of benzene with reference to 100 g is 50°C. [**Ans.** 100]
5. An aqueous solution contains 5% by weight of urea and 10% by weight of glucose. What will be its freezing point? Molal depression constant of water is 1·86°C/m. [**Ans.** –3.039°C]
6. An aqueous solution freezes at –0.2°C. What is the molality of the solution? Determine also
(*i*) elevation in the boiling point
(*ii*) lowering of vapour pressure at 25°C given that K_f = 1.86 kg mol^{-1} and K_b = 0.512° kg mol^{-1} and the vapour pressure of water at 25°C is 23.756 mm.
[**Ans.** Molality = 0.1075, ΔT_b = 0.055°C,
Lowering of vapour pressure = 0.046 m]

5.30. CAUSE OF ABNORMAL MOLECULAR MASSES OF SOLUTES IN SOLUTIONS

In certain situations, we find that the value of colligative property measured is greater or smaller than the value expected. Also since molecular mass is inversely related to the observed colligative property, the molecular mass calculated from the observed value of colligative property by applying the relevant relation, comes out to be different from the calculated molecular mass. We express it by saying that the solute is showing abnormal molecular mass.

Abnormal molecular mass of solute in solution is due to *association* or *dissociation* of molecules in the solution. These are explained as below.

(*a*) **Association:** If the molecules of a solute undergo association in solution, there will be decrease in the number of species. As a result there will be proportionate decrease in the value of each colligative property. The experimental value of molecular mass of the solute will be higher in such a case. This is because molecular mass is inversely proportional to the value of any colligative property.

For example acetic acid and benzoic acid both associate in benzene. Similarly, chloroacetic acid associates in naphthalene. The molecular mass of solute in such cases is higher than the molecular mass obtained from their molecular formulae. Thus the molecular mass of acetic acid (CH_3COOH) in benzene, as determined from freezing point depression is 118 instead of 60.

(*b*) **Dissociation:** Inorganic acids, bases and salts dissociate or ionise in solution. As a result of this, number of effective particles increases and therefore, the value of colligative property is also increased. Therefore, the experimental value of molecular mass of the solute will be lower in such a case. This is because molecular mass is inversely proportional to the value of any colligative property as already discussed.

Evidently the observed molecular mass will be higher in case of association and lower in case of dissociation.

5.31. VAN'T HOFF FACTOR

To account for abnormal cases, van't Hoff introduced a factor *i* **known as the van't Hoff factor**

$$i = \frac{\text{Observed colligative property}}{\text{Calculated (normal) colligative property}}$$

Van't Hoff

Since colligative properties vary inversely as the molecular mass of the solute, it follows that

$$i = \frac{\text{Calculated (normal) molecular mass}}{\text{Observed molecular mass}}$$

Relation between degree of association and van't Hoff factor

Let us consider the association of *n* molecules of a solute *A* to give one molecule of A_n

$$nA = A_n$$

Let the degree of association be α. If we start with 1 mole of *A*, the number of moles that associate is α and the number of moles that remain unchanged is $1 - \alpha$.

α moles on association will give α/n moles

$$\text{Total number of moles at equilibrium} = \frac{\alpha}{n} + (1 - \alpha)$$

van't Hoff factor is the ratio of observed colligative property to calculated colligative property. And colligative property is proportional to the number of moles of the solute. Hence

$$\text{van't Hoff factor } (i) = \frac{1-\alpha+\dfrac{\alpha}{n}}{1}$$

Relation between degree of dissociation and van't Hoff factor

Let us consider the dissociation of molecule to give *n* molecules or ions in solution

$$A \longrightarrow nB$$

Let us start with 1 mole of A and let α be the degree of dissociation.

At equilibrium

No. of moles of undissociated substance = $1 - \alpha$

No of moles of dissociated substance = $n\alpha$

(one molecule dissociates to give n molecules)

Total number of moles of solute at equilibrium = $1 - \alpha + n\alpha$.

Hence van't Hoff factor (i) = $\dfrac{1-\alpha+n\alpha}{1}$

SOLVED PROBLEMS ON ABNORMAL MOLECULAR MASSES AND VAN'T HOFF FACTOR

Example 1. 0.1 M solution of KNO_3 has an osmotic pressure of 4.5 atmosphere at 300 K. Calculate the apparent degree of dissociation of the salt.

Solution. π_{obs} = 4·5 atm C = 0.1 moles/litre

n = 2 because one KNO_3 molecule dissociates to give two ions, K^+ and NO_3^-

T = 300

Substituting the values in the equation

$$\pi = C.R.T.$$

$$\pi = 0.1 \times 0.082 \times 300$$

$$\pi = 2.6$$

$$i = \frac{\text{Observed osmotic pressure}}{\text{Calculated osmotic pressure}}$$

$$= \frac{4.5}{2.5} = 1.83 \qquad ...(1)$$

Also $\quad i = \dfrac{1-\alpha+n\alpha}{1}$

or $\quad i = 1 + \alpha(n-1)$ or $\alpha(n-1) = i - 1$

or $\quad \alpha = \dfrac{i-1}{n-1} \qquad ...(2)$

Substitute the value of i from (1) in (2)

$$\alpha = \frac{1.83-1}{2-1} \qquad \text{or } \alpha = 0.83$$

or $\quad = 83\%$

Example 2. Calculate the osmotic pressure of 20% anhydrous calcium chloride solution at 273 K, assuming that the solution is completely dissociated. (R = 0·082 lit. atm K^{-1} mol^{-1}).

Solution. Molecular mass of $CaCl_2$ = 40 + 71 = 111

Since solution is 20%

∴ Weight of $CaCl_2$ per litre = 200 g

$$V = 1 \text{ litre}$$

$$T = 273 \text{ K}$$

$$n = \frac{\text{wt. of solute}}{\text{molecular mass of solute}} = \frac{200}{111}$$

Substituting the values in the relation

$$\pi = \frac{n}{V} \times R.T.$$

we have $$\pi = \frac{200}{111} \times \frac{0.082 \times 273}{1} = 40.335 \text{ atm}$$

Since $CaCl_2$ dissolves to give 3 particles for each molecule on complete dissociation ($CaCl_2 \rightarrow Ca^{++} + 2Cl^-$)

$\therefore$ Observed osmotic pressure = Normal osmotic pressure × 3
= 40·335 × 3 = 121·00 atm

$\therefore$ Observed osmotic pressure = **121·00 atm.**

Example 3. 2.0 g of benzoic acid dissolved in 25·0g of benzene shows a depression in freezing point equal to 1·62 K. Molal depression constant (K_f) of benzene is 4·9 K kg mol^{-1}. What is the percentage association of the acid?

Solution. Mass of the solute w_2 = 2·0 g
Mass of solvent w_1 = 25·0 g
Observed ΔT_f = 1.62 K
K_f = kg mol^{-1}.

$\therefore$ Observed molecular mass of benzoic acid (solute)

$$M_2 = \frac{1000 \times K_f \times w_2}{\Delta T_f \times w_1} = \frac{1000 \times 4{\cdot}9 \times 2}{1{\cdot}62 \times 25{\cdot}0} = 242$$

Calculated molecular mass of benzoic acid (C_6H_5COOH) (By adding atomic masses)
= 72 + 5 + 12 + 32 + 1 = 122

$$\text{van't Hoff factor, } i = \frac{\text{Calculated mol. mass}}{\text{Observed mol. mass}} = \frac{122}{242} = 0{\cdot}504$$

If α is the degree of association of benzoic acid, then

$$2C_6H_5COOH \rightleftharpoons (C_6H_5COOH)_2$$

	$2C_6H_5COOH$	$(C_6H_5COOH)_2$
Initial moles	1	0
After association	$1-\alpha$	$\frac{\alpha}{2}$

$\therefore$ Total no. of moles after association $= 1 - \alpha + \frac{\alpha}{2} = 1 - \frac{\alpha}{2}$

$\therefore$ $$i = \frac{1 - \frac{\alpha}{2}}{1} = 0.504$$

or $$1 - \frac{\alpha}{2} = 0.504$$

or $$\alpha = (1 - 0{\cdot}504) \times 2 = 0{\cdot}496 \times 2 = \mathbf{0{\cdot}992}$$

or Per cent association = 0.992 × 100 = 99.2

PROBLEMS FOR PRACTICE

1. 0·5g KCl (mol. wt. 74·5) was dissolved in 100 g of water and the solution originally at 20°C froze at – 0·24°C. Calculate the percentage ionization of the salt. K_f for 1000 g water = 1·86°C. **[Ans.** 92%]

2. Phenol associates in benzene to a certain extent to form a dimer. A solution containing 2×10^{-2} kg of phenol in 1·0g of benzene has its freezing point depressed by 0·60 K. Calculate the fraction of phenol that has dimerised (K_f for benzene is 5·12 K kg mol^{-1}).

   [Ans. 0·733 or 73·3%]

3. A solution containing 2·2965 g of benzoic acid, C_6H_5COOH, in 20·27 g of benzene froze at a temperature 0·317°C below the freezing point of solvent. The freezing point of pure solvent is 5·5°C and its latent heat of fusion is 30·1 cal/g. Calculate
   (*a*) the apparent molecular weight of benzoic acid, and
   (*b*) its degree of association assuming that in this solution it forms double molecules.

   [Ans. 184·2, 67·4%]

4. A certain number of grams of a given substance in 100 g of benzene lower the freezing point by 1·28°C. The same weight of solute in 100 g of water gives a freezing point of – 1·395°. If the substance has normal molecular weight in benzene and is completely dissociated in water, into how many ions does a molecule of this substance dissociate when placed in water?

   [Ans. 3]

5. 0·3015 g of silver nitrate when dissolved in 28·40 g of water depressed the freezing point by 0·212°. To what extent is silver nitrate dissociated? (K_f for water = 1·85°C mol^{-1}). **[Ans.** 83·5%]

SOLVED CONCEPTUAL PROBLEMS

Problem 1. Out of one molar and one molal aqueous solutions, which one is more concentrated and why?

Solution. 1 M aqueous solution is more concentrated than 1 *m* aqueous solution. 1 M solution contains 1 moles of the solute in 1000 cc of the solution. This means that the solvent (water) present is less than 1000 cc because 1 mole of the solute is also present in the same volume. Hence 1000 g of the solvent contains more than one mole of the solute.

Problem 2. How are the mole fractions of the components in the liquid phase and in the vapour phase related to each other? What result follows from it and what is it called?

Solution. $$\frac{y_B}{y_A} = \frac{x_B}{x_A} \times \frac{p_B^\circ}{p_A^\circ}$$

Thus if B is more volatile than A, $p_B^o > p_A^o$ or $\frac{p_B^\circ}{p_A^\circ} > 1$ then $y_B/y_A > x_B/x_A$. This means that the vapour phase is richer in B than the liquid phase from which it vaporises. This is known as **Konowaloff's rule.**

Problem 3. Taking a suitable example, explain why some solutions show negative deviations.

Solution. Negative deviations are observed on mixing A and B if $A - B$ attractions are stronger than $A–A$ and $B–B$ attractions. For example, when chloroform and acetone are mixed, new forces of attraction due to hydrogen bonding, start operating between them, as shown below

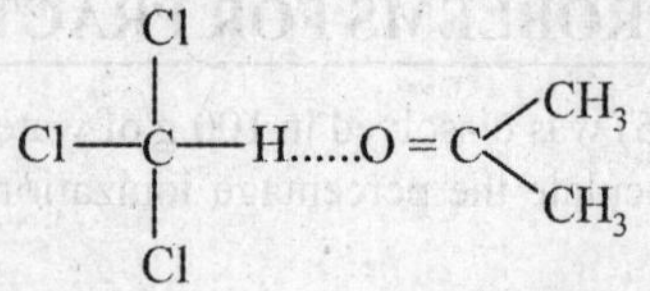

Chloroform Acetone

Thus, the molecules have a smaller tendency to escape into the vapour phase.

Problem 4. Out of the various methods of expressing concentration of a solution, which one are preferred over the other and why?

Solution. Molality, mole fraction and mass fraction are preferred over molarity, normality etc. because the former involves masses of the solutes and solvent which do not change with temperature. Molarity and normality involves volumes of the solution which change with temperature.

Problem 5. Taking a suitable example, explain why some solutions show positive deviations.

Solution. Solutions show positive deviations if on mixing the two components *A* and *B*, the *A* – *B* attractions are weaker than *A* – *A* and *B*–*B* attractions. For example, when chloroform is added to ethanol, the solution shows positive deviations. This is because there is hydrogen bonding in ethanol molecules which can be represented as

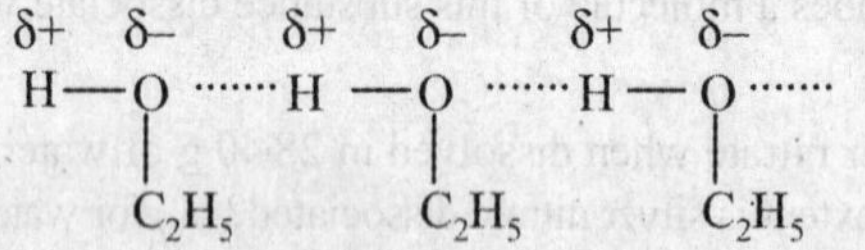

The chloroform molecules get in between the molecules of ethanol, thereby reducing the attractions between them.

Problem 6. How can you justify that osmotic pressure is a colligative property?

Solution. Osmotic pressure, $p = \frac{n}{V}RT$. Thus it depends only on the number of moles of the solute dissolved in a definite volume of the solution and there is no factor involving the nature of the solute. Hence it is a colligative property.

Problem 7. Of the following properties, which are colligative?

(i) Refractive index (ii) Vapour pressure (iii) Boiling point (iv) Depression in freezing point (v) Relative lowering of vapour pressure.

Solution. Depression in freezing point and relative lowering of vapour pressure.

Problem 8. What is meant by the statement "the osmotic pressure of a solution is 5·0 atmospheres?

Solution. This means that when the solution is separated from the solvent by a semipermeable membrane, there is a net flow of the solvent molecules from solvent to the solution through the semi-permeable membrane and that the pressure of the hydrostatic column set up is 5·0 atmosphere.

Problem 9. How is relative lowering of vapour pressure related to osmotic pressure?

Solution. $\frac{p^\circ - p_s}{p^\circ} = \frac{MP}{dRT}$. Hence relative lowering of V.P. ∝ Osmotic pressure.

Problem 10. How is elevation in boiling point related to (*i*) relative lowering of vapour pressure (*ii*) osmotic pressure.

Solution. The two relations are given as under

(*i*) $\Delta T_b = \frac{1000\, K_b\, \Delta p}{M_1 \quad p^\circ}$ *i.e.* $\Delta T_b \propto \frac{\Delta p}{p^\circ}$

(*ii*) $\Delta T_b = \frac{1000\, K_b}{dRT} \pi$ *i.e.* $\Delta T_b \propto \pi$.

Problem 11. Molecular weight of benzoic acid was determined by freezing point measurements using (i) its solution in water (ii) its solution in benzene. Will the results differ or not? Give reasons for your answer.

Solution. The results will differ. In water, benzoic acid dissociates. Thus, number of particles increases. As a result observed value of ΔT_f will be more than expected value. Hence, observed molecular mass will be less than the actual value. In benzene, benzoic acid will associate. The situation is just the reverse. No. of particles decreases. Hence observed ΔT_f will be less or molecular mass will be more.

Problem 12. Justify the statement that "dilute solutions behave in the same manner as gases" or "there exists an exact analogy between the dissolved state of the substance and its gaseous state."

Solution. Dilute solutions obey the same equation as ideal gas equation *i.e.* $PV = nRT$. Thus osmotic pressure of a dilute solution is equal to pressure which the solute would exert if it were a gas at the same temperature and occupied the same volume as the solution.

Problem 13. What are isotonic solutions? How are their molar concentrations related to each other?

Solution. Isotonic solutions are the solutions which have the same osmotic pressure. As $\pi = CRT$, at constant T, osmotic pressures will be equal when they have same molar concentration.

Problem 14. Explain why equimolar solutions of NaCl and cane sugar do not have the same osmotic pressure.

Solution. NaCl is an electrolyte which dissociates to give Na^+ and Cl^- ions. As a result, number of particles increases and so observed osmotic pressure is greater than expected. Canesugar (sucrose) is a non-electrolyte. It does not undergo any association or dissociation. Hence the expected osmotic pressure is observed.

Problem 15. Which colligative property is used for finding the molecular masses of polymers and why?

Solution. Polymers give very small values of colligative properties due to their large molecular masses. Δp, ΔT_b, ΔT_f etc. are too small to be measured while osmotic pressure can be measured to acceptable level of accuracy.

Problem 16. A solution of 4M HCl is expected to have osmotic pressure 197 atmosphere. How is that bottles containing this solution in the laboratory do not break?

Solution. Osmotic pressure has a meaning only when a solution is separated from the solvent by a semipermeable membrane. Thus, bottles containing this solution in the laboratory do not break.

Problem 17. Arrange the following in order of ascending values of their osmotic pressure

(i) 0·1 M Na_3PO_4 solution **(ii) 0·1 M sugar solution**

(iii) 0·1 M $BaCl_2$ solution **(iv) 0·1 M KCl solution**

Solution. 0·1 M Na_3PO_4 solution = 0·3 moles of Na^+ and 0·1 moles of PO_4^{3-} = Total 0·4 mole

0·1 M Sugar solution = 0·1 M Sucrose molecules = Total 0·1 mole

0·1 M $BaCl_2$ solution = 0·1 mole of Ba^{2+} and 0·2 mole of Cl^- = Total 0·3 moles

0·1 M KCl solution = 0·1 mole of K^+ + 0·1 mole of Cl^- = Total 0·2 mole. All these values are per litre of the solution.

Greater the concentration of particles in the solution, higher the osmotic pressure. Hence the values in the ascending order are 0·1 M sugar solution < 0·1 M KCl sol < 0·1 M $BaCl_2$ sol < 0·1M Na_3PO_4 solution.

Problem 18. Account for the following:

(*i*) Camphor is used as a solvent in the Rast method.

(*ii*) Benzoic acid in benzene shows less osmotic pressure than expected.

(*iii*) Boiling point of 0·1 m NaCl is greater than 0·1 m glucose solution.

Solution. (*i*) Camphor is used because it has high molal depression constant. We get a higher value of ΔT_b which helps in obtaining accurate results for molecular mass.

(*ii*) Benzoic acid undergoes association in benzene.

(*iii*) 0·1 m NaCl dissociates to form 0·2 m concentration of particles (Na^+ and Cl^- ions) whereas glucose remains as such.

Problem 19. Tell whether the osmotic pressure of M/10 solution of glucose to be same as that of M/10 solution of sodium chloride.

Solution. No. colligative property of a solution depends upon the number of moles of the compound or ions present per litre of the solution. M/10 solution of sodium chloride will contain twice the number of particles, as compared to M/10 glucose solution (one sodium chloride molecule gets dissociated into two ions). Consequently, M/10 sodium chloride will show nearly twice the osmotic pressure.

Problem 20. How and why does vapour pressure of a liquid depend upon the temperature?

Solution. Vapour pressure of a liquid increases with the increase of temperature. With the increase of temperature, the kinetic energy of the molecules on the surface increases (K.E = $\frac{3}{2}$RT). More and more molecules leave the surface of the liquid and are converted into vapours, thereby raising the vapour pressure.

EXERCISES
(Based on Question Papers of Different Universities)

Multiple Choice Questions (Choose the correct option)

1. A liquid boils when its vapour pressure becomes equal to

(*a*) one atmospheric pressure (*b*) zero

(*c*) very high (*d*) very low

2. The elevation in boiling point is given by the formula

$$\Delta T = K_b \times \frac{w}{m} \times \frac{1}{W}$$

where K_b is called

(*a*) boiling point constant (*b*) ebullioscopic constant

(*c*) molal elevation constant (*d*) all of these

3. The depression in freezing point is measured by using the formula $\Delta T = K_f \times \frac{w}{m} \times \frac{1}{W}$

where K_f is called

(*a*) molal depression constant (*b*) freezing point depression constant

(*c*) cryoscopic constant (*d*) all of these

4. The ratio of the colligative effect produced by an electrolyte solution to the corresponding effect for the same concentration of a non-electrolyte solution is known as

(*a*) degree of dissociation (*b*) degree of association

(*c*) activity coefficient (*d*) van't Hoff factor

5. Molal elevation constant is the boiling point elevation when _____of the solute is dissolved in one kg of the solvent.

(*a*) one gram (*b*) one kg

(*c*) one mole (*d*) none of these

6. Freezing point depression is measured by

(*a*) Beckmann's method (*b*) Rast's camphor method

(*c*) both (*d*) none of these

7. The weight percent of a solute in a solution is given by

(*a*) $\frac{\text{wt of the solvent}}{\text{wt of the solute}} \times 100$ (*b*) $\frac{\text{wt of the solute}}{\text{wt of the solvent}} \times 100$

(*c*) $\frac{\text{wt of the solute}}{\text{wt of the solution}} \times 100$ (*d*) $\frac{\text{wt of the solution}}{\text{wt of the solute}} \times 100$

8. In one molal solution that contains 0.5 mole of a solute there is

(*a*) 1000 g of solvent (*b*) 1000 ml of solvent

(*c*) 500 ml of solvent (*d*) 500 g of solvent

9. The law of the relative lowering or vapour pressure was given by

(*a*) van't Hoff (*b*) Ostwald

(*c*) Raoult (*d*) Henry

10. Which of the following is a colligative property?

(*a*) molar refractivity (*b*) optical rotation

(*c*) depression in freezing point (*d*) viscosity

11. The number of moles of a solute per kilogram of the solvent is called

(*a*) formality (*b*) normality

(*c*) molarity (*d*) molality

12. Which of the following does not depend upon the temperature?

(*a*) molarity (*b*) molality

(*c*) formality (*d*) normality

13. The study of depression in freezing point of a solution is called

(*a*) osmotic pressure (*b*) ebullioscopy

(*c*) cryoscopy (*d*) none of these

14. 36 g of glucose (molecular mass 180) is present in 500g of water, the molality of the solution is
(*a*) 0.2 (*b*) 0.4
(*c*) 0.8 (*d*) 1.0

15. The liquid mixtures which distil with a change in composition are called
(*a*) azeotropic mixtures (*b*) equilibrium mixtures
(*c*) zeotropic mixtures (*d*) non-schorating mixtures

ANSWERS

1. (*a*)	**2.** (*d*)	**3.** (*d*)	**4.** (*d*)	**5.** (*c*)	**6.** (*c*)
7. (*c*)	**8.** (*d*)	**9.** (*c*)	**10.** (*c*)	**11.** (*d*)	**12.** (*b*)
13. (*c*)	**14.** (*b*)	**15.** (*c*)			

Short Answer Questions

1. State and explain Raoult's law (*a*) for volatile solute (*b*) for non-volatile solutes.

2. Derive the relationship between mole fractions of the components in the liquid phase and those in the vapour phase. What result do you draw from it and what is this result called?

3. Define ideal and non-ideal solutions in as many ways as you can. Give at least two examples of each of them.

4. For an ideal solution, derive relationship between the mole fraction of a component in the liquid phase to that in the vapour phase. What do you conclude if out of the two components *A* and *B*, *B* is more volatile?

5. What are colligative properties? How can you say that 'Relative lowering of vapour pressure' is a colligative property?

6. Write the relationship between Relative lowering of vapour pressure and Osmotic pressure. Deduce Raoult's law from it.

7. Define osmotic pressure. How is it determined by Berkeley and Hartley's method?

8. Draw vapour pressure-composition diagrams and corresponding boiling point-composition diagrams for different types of non-ideal solutions.

9. Derive expression for molar free energy change of an ideal solution.

10. Describe Ostwald and Walker's method (transpiration method) for the determination of lowering of vapour pressure.

11. Derive the relationship between relative lowering of vapour pressure and osmotic pressure.

12. What do you understand by 'activity' of a component in a solution? How is it determined?

13. Prove that for an ideal solution, $\Delta V_{mixing} = 0$ and $\Delta H_{mixing} = 0$.

14. Briefly explain how real or non-ideal solutions show deviations from ideal behaviour.

15. What do you mean by molal elevation constant of a solvent? How is it related to the latent heat of vaporisation of the solvent? What are its units?

16. What is van't Hoff factor? How are the different expressions for colligative properties modified when the solute undergoes association or dissociation in the solvent?

17. Briefly describe Rast's method for the determination of molecular weight of a solute.

18. Derive the relationship between elevation in boiling point and relative lowering of vapour pressure.

19. What is the cause of elevation in boiling point? Explain clearly with the help of vapour-pressure temperature curve.

20. Explain with the help of vapour pressure-temperature diagram the cause for the depression in freezing point.
21. Using colligative properties, why are abnormal molecular weights observed in certain cases?
22. What is molal depression constant or ebullioscopic constant? How is it related to be enthalpy of fusion? What are its units?
23. Derive the relation between elevation in boiling point and osmotic pressure.
24. Derive the relation between depression in freezing point and lowering of vapour pressure.

General Questions

1. Briefly explain the terms–Chemical potential, Fugacity, Activity and Activity coefficient.
2. What are ideal and non-ideal solutions? Derive expressions for free energy change, entropy change, enthalpy change and volume change when n_1 moles of pure component 1 are mixed with n_2 moles of pure component 2 to form an ideal solution.
3. (*a*) Explain how non-ideal solutions show deviations from ideal behaviour or Raoult's law. Draw vapour pressure-composition diagrams and boiling point-composition diagrams for each of them.

   (*b*) What are Azeotropes? Briefly explain the types of azeotropes.
4. Briefly explain at least five different methods of expressing the concentration of a solution. Which out of these are preferred and why?
5. State and explain Raoult's law for volatile solutes as well as for non-volatile solutes. Derive the following:

   (*i*) Expression for total vapour pressure in terms of mole fractions of the components in the vapour phase.

   (*ii*) Relationship between the mole fractions of the components in the liquid phase and those in the vapour phase.
6. Briefly explain the following curves for ideal solutions.

   (*i*) Vapour pressure versus mole fraction of the component in the solution.

   (*ii*) Vapour pressure versus mole fraction of the component in the vapour phase.
7. What are 'colligative properties'? Briefly explain how lowering of vapour pressure is used in the calculation of molecular masses of solutes? Explain one method for the experimental determination of lowering of vapour pressure.
8. Why do we observe abnormal molecular masses of the solutes in certain cases when determined by studying colligative properties? What is van't Hoff factor? How is it used in the determination of degree of association and degree of dissociation of a solute?
9. Why there is a depression in freezing point when a non-volatile solute is dissolved in a solvent? Derive thermodynamically the relation between depression in freezing point and molecular mass of the solute.
10. Derive the following relationship between osmotic pressure and lowering of vapour pressure. Deduce Raoult's law from it.
11. Define osmotic pressure. How is it measured experimentally by Berkeley and Hartley's method? What are the advantages of this method over the other methods?
12. Briefly describe the following methods for the determination of depression in freezing point.

    (*i*) Beckmann's method (*ii*) Rast's method
13. Derive thermodynamically the expression for the Relative lowering of vapour pressure or Derive Raoult's law thermodynamically.

14. Define activity and activity coefficient. Describe vapour pressure method for their determination.

15. Derive the relationship between

(*i*) Depression in freezing point and lowering of vapour pressure.

(*ii*) Depression in freezing point and osmotic pressure.

16. Derive the relationship between

(*i*) Elevation in boiling point and relative lowering of vapour pressure.

(*ii*) Elevation in boiling point and osmotic pressure.

17. Briefly explain van't Hoff theory of dilute solutions. How does it help in the determination of molecular masses of solutes?

18. Why there is an elevation in boiling point when a non-volatile solute is dissolved in a solvent? Derive thermodynamically the relation between elevation in boiling point and molecular mass of the solute. Describe one method commonly used for the determination of elevation in boiling point.

19. Derive thermodynamically the expression for osmotic pressure, $\pi = CRT$.

6

Statistical Mechanics*

6.1. INTRODUCTION

Thermodynamics deals with the properties like internal energy, enthalpy, free energy, etc of the matter in bulk. It does not consider individual particles. On the other hand, quantum mechanics deals with individual particles (atoms or molecules) by assigning definite values of wave functions to these particles at any instant of time. It was thought that it should be possible to calculate the thermodynamics properties of the matter in bulk using the properties of individual particles by a suitable **averaging** method. This is how statistical mechanics, as a branch of science, was developed.

The method of averaging the behaviour of a large number of individual particles (atom or molecules) is called **statistical method**. As the laws of both classical mechanics and quantum mechanics can be applied to the particles, we call this branch of science as **statistical mechanics**. *Thus statistical mechanics is that branch of science which helps us to determine the macroscopic properties of a system (like internal energy, enthalpy, entropy, free energy, etc.) from its macroscopic properties by statistical methods.*

There are two types of statistical mechanics:

Classical statistical mechanics. If the laws of classical mechanics are applied, assuming definite positions and momenta for individual particles, it is known as classical statistical mechanics.

Quantum statistical mechanics. If the laws of quantum mechanics are applied, assuming quantised values of energy for individual particle it is known as quantum statistical mechanics.

We have Maxwell-Boltzmann statistics to study classical statistical mechanics. But, there are two approaches viz. Bose-Einstein statistics and Fermi-Dirac statistics to deal with quantum statistical mechanics. The distinguishing features of Maxwell-Boltzmann (M–B) statistics, Bose-Einstein (B–E) statistics and Fermi-Dirac (F–D) statistics are given in the following Table.

Table 6.1 Distinguishing features of the three statistics

M — B Statistics	*B—E Statistics*	*F—D Statistics*
1. The laws of *classical mechanics* are applied according to which individual atoms or molecules have definite positions and momenta.	1. The laws of *quantum mechanics* are applied according to which individual atoms or molecules have only quantized values of energy.	1. In this statistics also, laws of *quantum mechanics* are applied.
2. It treats the different particles as *distinguishable.*	2. It treats the different particles *indistinguishable.*	2. Like B—E statistics, it also treats the particles as *indistinguishable.*
3. According to this, *any number* of particles may occupy the same *energy level.*	3. Here also, any number of particles may occupy the same energy level.	3. In this case, *only one particle* is supposed to occupy a *particular energy level.*

* Now in U.G.C. Syllabus

M — B Statistics	*B — E Statistics*	*F — D Statistics*
4. The particles obeying M—B statistics are called *maxwellslet or boltzmannous.*	4. It is applied to those particles which have functions like photons or atoms and molecules having *even* no. of particles in the nucleus. Such particles are called *bosons.*	4. It is applied to those particles which have a symmetric wave functions like electrons, protons or atoms and molecules having *odd* no. of particles in the nucleus. Such particles are called *fermions.*

6.2. THE STATISTICAL METHOD

The method of averaging the behaviour of a large number of individuals is called a statistical method. In calculating the statistical average, it is not necessary to know the behaviour of any one individual. This is illustrated below with the help of following examples:

(*i*) The price index is calculated on the basis of the statistical average of rise in prices of a large number of essential commodities without laying emphasis on one particular commodity.

(*ii*) The death rate and the birth rate of any country are a statistical average of a large number of individuals without going into the exact details of the persons born or died in different places of a country.

(*iii*) Kinetic energy of a gas at a particular temperature is a statistical average of the kinetic energies of the individual molecules. Different molecules in a gas hit the walls of the vessel with different energies, but still the gas shows one particular pressure.

6.3. STATE OF THE SYSTEM

In **thermodynamics,** the state of a system is completely defined by specifying the various macroscopic properties of the system such as temperature, pressure, volume and composition. If any of these properties is changed, the system is said to be in a different state. These macroscopic properties are, therefore, called *state variables.*

If these properties are fixed, all other physical properties of the system get automatically fixed. Further, if the system is homogeneous and consists of a single substance, the composition is fixed (because it is 100%), and hence the state of the system depends only upon pressure, volume and temperature. It is not necessary to specify even these three thermodynamic properties because experiments have shown that these three properties of a simple homogeneous system of definite mass are related to one another. Thus out of P, V and T, if any two are fixed, the third is automatically fixed. Hence we conclude that the *state of a simple homogeneous system may be completely defined by specifying only two of the properties out of P, V and T.*

In **statistical mechanics,** the state of a system is completely described by specifying the state of the individual atoms or molecules. In classical statistics, the state of an individual particle is described by specifying its coordinates (x, y, z) and the components of its momentum (p_x, p_y, p_z) *along the three axes.* In quantum statistics, the state of the individual particles is described by specifying the particular quantum state (energy level) to which the particle belongs.

6.4. STIRLING FORMULA FOR A LARGE NUMBER N OF MOLECULES

In the study of statistical mechanics, we usually consider a system consisting of very large number of particles. If this number is represented by N, the calculation of $\ln N!$ becomes very laborious. Hence a simple approximation is used for the value of $\ln N!$ which is called **Stirling approximation**. This is achieved as follows:

We know that

$$N! = N \times (N-1)(N-2).......3 \times 2 \times 1$$

or
$$N! = 1 \times 2 \times ..... \quad (N-2) \times (N-1) \times N \quad ...(i)$$

$$\therefore \quad \ln N! = \ln 1 + \ln 2 + ...+ \ln (N-2) + \ln (N-1) + \ln N$$

$$= \sum_{m=1}^{N} \ln m \quad ...(ii)$$

In this summation, except for the first few terms whose values are small, as m increases and attains large values, the increase in the value of ln m with increase in the value of m by unity is very small. Hence in the above summation, ln m can be treated as continuous so that it is given by the area under the curve from $m = 1$ to $m = N$ obtained by plotting ln m *vs* m. This in turn is equal to the integration of ln $x\,dx$ between the limits $x = 1$ to $x = N$. Hence eqn. (*ii*) can be written as

$$\ln N! = \int_{x=1}^{x=N} \ln x \, dx$$

The integral on the R.H.S. can be solved by integration by parts. Thus

$$\ln N! = [x \ln x]_1^N - \int_1^N x.\frac{1}{x} dx$$

$$= N \ln N - N$$

Hence for large values of N

$$\boxed{\ln N! = N \ln N - N}$$

This is called Stirling's formula for large values of N.

6.5. ENSEMBLES

The word ensemble literally means collection. This term was used by Gibbs in 1900 to carry out averaging process conveniently. Averaging is a vital component of statistical mechanics.

A collection of a large number of systems which are identical with the system under consideration in a number of aspects such as total volume, total number of molecules, etc. is called an ensemble of system. They may, however, differ in some thermodynamic property such as energy etc.

Depending upon the thermodynamic variables kept constant, the ensembles are of different kinds. The most common are as follows:

(*i*) **Microcanonical ensemble.** If the original system is an isolated one having volume V, total number of molecules N and energy E, then we can have an ensemble of n such systems, each member (system) having the same value of N, V and E. This ensemble is shown schematically in Fig. 6.1. It can be set up by imagining rigid impermeable walls separating the system so that neither energy nor material particles can flow from one system to the other. Thus each system in this ensemble is like the isolated system in the thermodynamic sense. *Such an ensemble of systems in which each member has the same values of N, V and E is called a microcanonical ensemble.* In the microcanonical ensemble, shown in Fig. 6.1, there are 36 members ($n = 36$). However, in statistical mechanics, we consider a very large value for n and in a number of calculations, we take the limit $n \rightarrow \infty$.

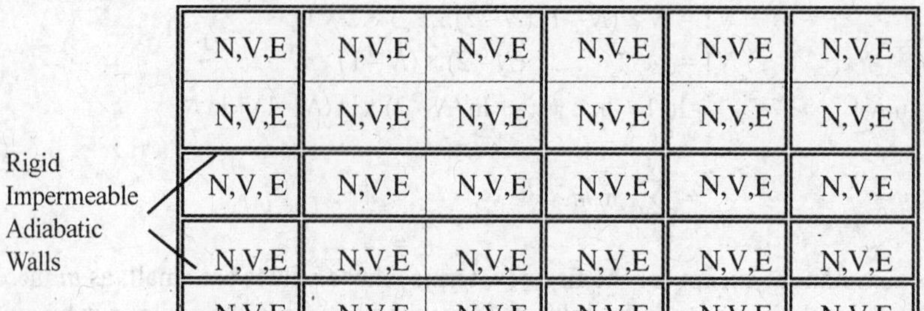

Fig. 6.1. Schematic representation of a microcanonical ensemble of n systems each with the same value of N, V and E.

(*ii*) **Canonical ensemble.** *If all the members of an ensemble have the same value for N, V and T, then it is called a canonical ensemble.* It can be set up by imagining rigid but conducting walls separating the different systems through which energy can pass but not the material particles. Thus each system in this ensemble is like a closed system in the thermodynamic sense. Such an ensemble is shown schematically in Fig. 6.2 (*a*). Due to the conducting walls, when the equilibrium is attained, each member of the ensemble has the same temperature T but may not have the same energy E. Hence diagrammatically, a canonical ensemble may also be represented as shown in Fig. 6.2 (*b*).

Canonical ensemble finds maximum use in statistical mechanics.

(*iii*) **Grand canonical ensemble.** In this ensemble, each system is an open system. Hence the matter can flow between the systems and the composition of each one may fluctuate. Consequently, the number of molecules in different systems is not same. However V, T and μ (chemical potential) for each component is same for each member of the ensemble.

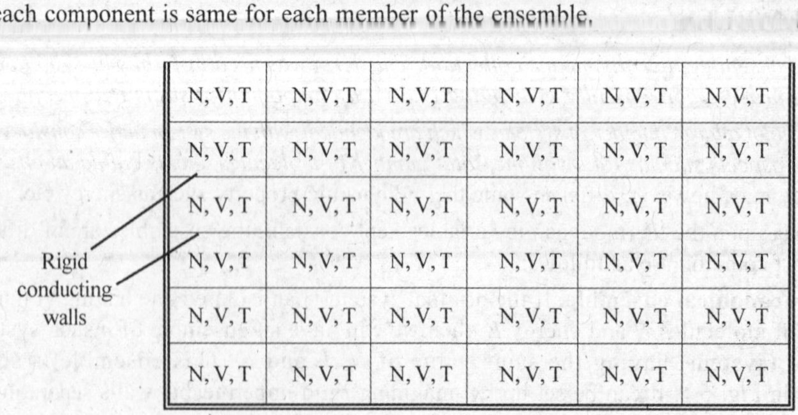

(*a*)

Fig. 6.2. Two different schematic representations of a canonical ensemble

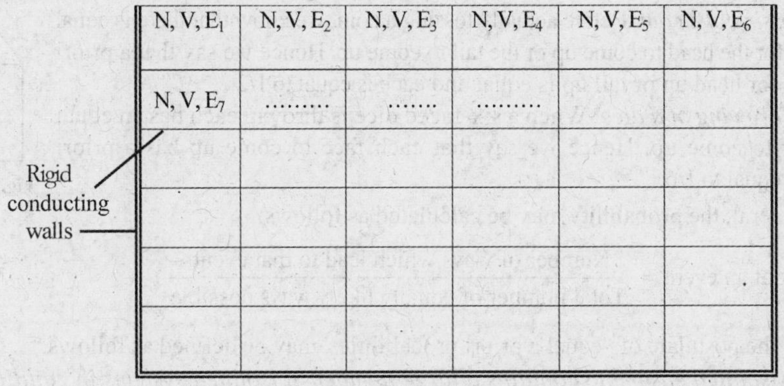

(*b*)

Fig. 6.2. Two different schematic representations of a canonical ensemble.

The features of the three types of ensembles may be grouped as under :

Types of ensemble	Common properties for each system
Microcanonical	N, V, E
Canonical	N, V, T
Grand canonical	μ, V, T

6.6. POSTULATES OF STATISTICAL MECHANICS

Statistical mechanics deals with both, the systems in equilibrium as well as the systems not in equilibrium.

For systems in equilibrium. To understand the postulate of statistical mechanics on which the study of systems in equilibrium is based, consider the canonical ensemble as represented in Fig. 6.2 (*a*) and (*b*).

After the equilibrium has been attained so that each system in the ensemble has the same values of *N, V* and *T,* suppose each system is insulated and hence isolated. Now different systems possess different energies (or some of them may possess the same energy, called the degenerate systems). These energies represented by E_1, E_2, E_3, ... etc. correspond to the eigen values of the energy for the molecules in the *N*-molecule system. In other words, they correspond to certain specific quantum states. After the equilibrium has been attained, suppose

n_1 systems possess energy E_1,

n_2 systems possess energy E_2,

n_3 systems possess energy E_3,

and so on.

This is called *'distribution'* of systems over different possible quantum states.

Now, as we are considering an isolated ensemble having total energy *E* and total number of systems equal to *n,* the above distribution must satisfy the following two conditions:

$$\Sigma n_i = n \qquad ...(i)$$

and

$$\Sigma n_i E_i = E \qquad ...(ii)$$

Many distributions N_1, N_2, N_3 ..... etc are possible which satisfy equations (*i*) and (*ii*). Each of these distributions has an equal probability. This postulate (or principle) is called the **"postulate of equal a priori probabilities"**. The word "a priori" means something which exists in our mind prior to its verification by actual observations. This concept may be illustrated with the following examples:

(*i*) *Tossing of a coin.* Before actually tossing a coin, we know that there is equal probability for the head to come up or the tail to come up. Hence we say that a priori probability for head up or tail up is equal and each is equal to 1/2.

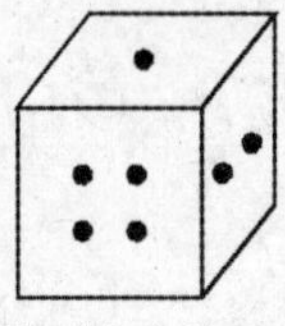

Fig. 6.3. A dice

(*ii*) *Throwing of a dice.* When a six-faced dice is thrown, each has an equal probability to come up. Hence we say that each face to come up has a priori probability equal to 1/6.

In general, the probability may be calculated as follows:

$$\text{Probability of an event} = \frac{\text{Number of ways which lead to that event}}{\text{Total number of equally likely ways possible}}$$

Thus the postulate of 'equal a priori probabilities' may be defined as follows:

If an isolated ensemble contains a large number of isolated systems in equilibrium then each system has an equal probability to exist in any one of the possible quantum states.

For systems not in equilibrium. The postulate of statistical mechanics for systems not in equilibrium is as follows:

If an isolated system in an ensemble is not in equilibrium it is not found with equal probability in the different possible quantum states; it then tends to change with time until it ultimately attains the equilibrium where it has an equal possibility to exist in any one of the possible quantum states.

There is no proof of these postulates. However, the values of the thermodynamics properties calculated on the basis of these postulates agree well with the experimental measurements. It gives enough proof of the validity of these postulates. And nothing has been reported against these postulates.

6.7. MAXWELL-BOLTZMANN DISTRIBUTION LAW

Maxwell-Boltzmann distribution law can be derived as under:

Consider a *system* of N *identical but distinguishable* particles such as gas molecules at temperature T. Suppose the total volume of the system is V and total energy is E. Then, all the molecules do not have the same energy. Suppose n_1 molecules are present in the level with energy ε_1, n_2 in the level with energy ε_2, n_3 in the level with energy ε_3 and so on. In each case, the following two conditions must be satisfied:

$$\Sigma n_i = N \text{ (total number of molecules)} \qquad \text{... (}i\text{)}$$

and

$$\Sigma n_i \varepsilon_i = E \text{ (total energy of the system)} \qquad \text{...(}ii\text{)}$$

where n_i is the number of molecules in the ith level with energy ε_i.

Many distributions are possible which satisfy the above two conditions. Now let us first see what is the total number of such possible distributions. The situation is similar to putting N numbered balls into n boxes such that n_1 are in the first box, n_2 in the second box and so on. Mathematically, the number of possible arrangements is given by the relation.

$$G = \frac{N!}{n_1!n_2!....n_n!}$$

When applied to the system of N molecules, the number of possible distributions is usually represented by W and is given by

$$W = \frac{N!}{n_1!n_2!....n_n!} \qquad \text{... (}iii\text{)}$$

W is called the **thermodynamic probability** of the system, because it is found that most of the thermodynamic properties are intimately related to the value of W.

According to principles of statistical mechanics, the most probable distribution is one that makes W the maximum. This condition for maximum probability may be written as

$$d \ln W = 0 \qquad ...(iv)$$

$d \ln W$ stands for differential of $\ln W$ which is equal to $\frac{1}{W}$ and for maximum value of W, $\frac{1}{W} \to 0$.

Taking logarithm of both sides of equation (*iii*), we get

$$\ln W = \ln N! - [\ln n_1! + \ln n_2! + ... + \ln n_n!]$$

$$= \ln N! - \Sigma \ln n_i! \qquad ... (v)$$

Applying Stirling's formula,

$$\ln N! = N \ln N - N$$

and $$\ln n_i! = n_i \ln n_i - n_i$$

eqn. (v) takes the form

$$\begin{aligned} \ln W &= N \ln N - N - \Sigma(n_i \ln n_i - n_i) \\ &= N \ln N - N - \Sigma n_i \ln n_i + \Sigma n_i \\ &= N \ln N - N - \Sigma n_i \ln n_i + N \quad [\because \Sigma n_i = N, eq. (i)] \\ &= N \ln N - \Sigma n_i \ln n_i \qquad ... (vi) \end{aligned}$$

Differentiating both sides of this equation, we get

$$d \ln W = - \Sigma\, d\,(n_i \ln n_i) \qquad [d\,(N \ln N) = 0, \text{ because } N \text{ is constant}]$$

$$= -\Sigma\left[n_i \times \frac{1}{n_i} + (\ln n_i \times 1) \right] dn_i$$

($n_i \ln n_i$ has been differentiated as a product of two functions)

$$= -\Sigma\,(\ln n_i + 1)\, dn_i \qquad ... (vii)$$

But $d \ln W = 0$ (according to eqn. (*iv*))

$\therefore$ Eqn. (*vii*) gives

$$\Sigma\,(\ln n_i + 1)\, dn_i = 0 \qquad ...(viii)$$

For the system under consideration, we have the following relations:

$$\Sigma\, n_i = N = \text{constant}$$

and $$\Sigma\, n_i \varepsilon_i = E = \text{constant}$$

Differentiation of these equations gives

$$\Sigma\, dn_i = 0 \qquad ...(ix)$$

and $$\Sigma\, \varepsilon_i\, dn_i = 0 \qquad ...(x)$$

Equations (*viii*), (*ix*) and (*x*) represent the conditions which must be satisfied simultaneously for the most probable distribution.

Multiplying equations (*ix*) and (*x*) by the undetermined constants α' and β respectively and adding to eqn. (*viii*), we get

$$\Sigma\,(\ln n_i + 1 + \alpha' + \beta\varepsilon_i)\, dn_i = 0 \qquad ...(xi)$$

This is called method of *Langrange undetermined multipliers* according to which the constraints (restrictions/conditions) are multiplied with undetermined multipliers and added to differential of the function that we want to maximise.

Putting $\alpha' + 1 = \alpha$, another constant, eqn. (*xi*) takes the form

$$\Sigma\,(\ln n_i + \alpha + \beta\varepsilon_i)\, dn_i = 0 \qquad ... (xii)$$

Now within the restrictions imposed on the system for constant N and E values, the variations in n_i are independent of each other and need not be zero, *i.e.,* $dn_i \neq 0$. Consequently, equation (*xii*) will be satisfied only if

$$\ln n_i + \alpha + \beta\varepsilon_i = 0$$

or
$$\ln n_i = -(\alpha + \beta\varepsilon_i) \qquad ...(xii\ (a))$$

or
$$n_i = e^{-(\alpha+\beta\varepsilon_i)} \qquad ...(xii\ (b))$$

or
$$\boldsymbol{n_i} = \frac{\mathbf{1}}{e^{\alpha+\beta\varepsilon_i}} \qquad ...(xiii)$$

James Clerk Maxwell

This expression is called **Maxwell-Boltzmann distribution law**. It gives the most probable distribution of molecules among the various possible energy states at equilibrium for a system of constant total energy.

Ludwig Boltzmann

Eqn. (*xiii*) can be written in the form.

$$n_i = e^{-\alpha} \, . \, e^{-\beta\varepsilon_i} \qquad ...(xiv)$$

The value of $e^{-\alpha}$ can be found from eqn. (*i*) as :

$$N = \Sigma n_i = \Sigma e^{-\alpha} . e^{-\beta\varepsilon_i}$$
$$= e^{-\alpha} . \Sigma e^{-\beta\varepsilon_i}$$

$$\therefore \quad e^{-\alpha} = \frac{N}{\Sigma e^{-\beta\varepsilon_i}} \qquad ...(xv)$$

Substituting this value in eqn. (*xiv*), we get

$$n_i = \frac{Ne^{-\beta\varepsilon_i}}{\Sigma e^{-\beta\varepsilon_i}} \qquad ...(xvi)$$

By applying eqn. (*ii*) to the energy of an ideal gas, we can see that

$$\beta = \frac{1}{kT} \qquad ...(xvii)$$

where k is Boltzmann constant.

Substituting this value of β in eqn. (*xvi*), we get

$$n_i = \frac{Ne^{-\varepsilon_i / kT}}{\Sigma e^{-\varepsilon_i / kT}}$$

which can be written as

$$\frac{\boldsymbol{n_i}}{\boldsymbol{N}} = \frac{\boldsymbol{e^{-\varepsilon_i/kT}}}{\sum \boldsymbol{e^{-\varepsilon_i/kT}}} \qquad ...(xviii)$$

This expression is called **Boltzmann distribution law.** This form of equation gives the most probable distribution *i.e.*, at temperature T it gives the fraction of the total number of molecules which in the most probable state will possess the energy ε_i.

In the derivation of the above expression, it has been assumed that each energy level is non-degenerate *i.e.,* consists of a single level. However, in many cases it may not be so. Corresponding to each eigen value of the energy there may be a number of eigen functions (called the eigen states). *The number of eigen states corresponding to a particular energy* ε_i *is called* **degeneracy** or **statistical weight**. It is usually represented by g_i, (for the level with energy ε_i). Thus the eqn. (*xviii*) can be written in a more general form as follows:

$$\frac{n_i}{N} = \frac{g_i e^{-\varepsilon_i/kT}}{\Sigma g_i e^{-\varepsilon_i/kT}} \quad \text{.... } (xix)$$

Maxwell-Boltzmann distribution law in the form $n_i = g_i \exp(-a - be_i)$

If degeneracy is taken into consideration then the number of possible distributions (thermodynamic probability W) of N **distinguishable** particles in the different energy levels such that the level with energy ε_i has g_i states (any of which may be occupied by a particle with energy ε_i) and n_i is the number of particles which have energy ε_i, is given by

$$W = N!\frac{g_0^{n_0}\, g_1^{n_1} \ldots g_n^{n_n}}{n_0 \,!\, n_1 \ldots n_n\,!} = N!\prod \frac{g_i^{n_i}}{n_i\,!} \quad (\Pi \text{ stands for product of the terms}) \quad \text{... } (i)$$

From eqn. (*i*), $\ln W = \ln N\,! + \Sigma(\ln g_i^{n_i} - \ln n_i\,!)$

$$= N \ln N - N + \Sigma\,(n_i \ln g_i - n_i \ln n_i + n_i)$$

(applying Stirling's formula)

$$= N \ln N + \Sigma n_i \ln \frac{g_i}{n_i} \quad (\because\ \Sigma n_i = N) \text{ ...}(ii)$$

Derivation. Differentiating both sides of eqn. (*ii*), we get (keep in mind that N is constant)

$$\begin{aligned} d \text{ in } W &= \Sigma d\left(n_i \ln \frac{g_i}{n_i}\right) \\ &= \Sigma\, d\,(n_i \ln g_i - n_i \ln n_i) \\ &= \Sigma \ln g_i\, dn_i - \Sigma\, n_i\, d \ln n_i - \Sigma \ln n_i\, dn_i \\ &= \Sigma \ln g_i\, dn_i - \Sigma\, dn_i - \Sigma \ln n_i\, dn_i \end{aligned} \quad \text{...}(iii)$$

But $\quad d \ln W = 0$

$\therefore$ Eqn. (*iii*) gives

$$\Sigma \ln g_i\, dn_i - \Sigma dn_i - \Sigma \ln n_i\, dn_i = 0 \quad \text{...}(iv)$$

For the system under consideration

$$\Sigma\, n_i = N = \text{constant}$$

$$\Sigma\, n_i\, \varepsilon_i = E = \text{constant}$$

therefore, differentiation of these equations give

$$\Sigma\, dn_i = 0 \quad \text{....}(v)$$

$$\Sigma \varepsilon_i dn_i = 0 \quad \text{... } (vi)$$

Substituting the value from eqn. (*v*) into eqn. (*iv*), we get

$$\Sigma \ln g_i \, dn_i - \Sigma \ln n_i \, dn_i = 0 \quad \text{... (vii)}$$

Applying the method of Langrange undetermined multipliers *i.e.*, multiplying the constraints (*v*) and (*vi*) by undetermined constants α and β and subtracting from eqn. (*vii*), we have

$$\Sigma (\ln g_i - \ln n_i - \alpha - \beta\varepsilon_i) \, dn_i = 0 \quad \text{...(viii)}$$

or

$$\Sigma \left(\ln \frac{g_i}{n_i} - \alpha - \beta\varepsilon_i \right) dn_i = 0$$

Now within the restrictions imposed on the system for constant N and E value, $dn_i \neq 0$. Hence

$$\ln \frac{g_i}{n_i} - \alpha - \beta\varepsilon_i = 0 \quad \text{or} \quad \ln \frac{g_i}{n_i} = \alpha + \beta\varepsilon_i$$

$$\ln \frac{g_i}{n_i} = -(\alpha + \beta\varepsilon_i) \quad \text{or} \quad \frac{n_i}{g_i} = e^{-(\alpha + \beta\varepsilon_i)}$$

$$n_i = g_i \, e^{-(\alpha+\beta\varepsilon_i)} \quad \text{or} \quad n_i = \frac{g_i}{e^{\alpha+\beta\varepsilon_i}}$$

6.8. STATISTICAL THERMODYNAMICS

The branch of science dealing with the calculation of thermodynamic properties of the systems using the methods of statistical mechanics is called statistical thermodynamics. In other words, statistical thermodynamics deals with the calculation of the macroscopic properties of the systems such as pressure, entropy, internal energy, Gibbs free energy *etc.* from the microscopic properties *i.e.*, properties of the individual molecules such as their position and momenta (in classical statistics) and their quantum states (in quantum statistics).

6.9. RELATIONSHIP BETWEEN ENTROPY AND THERMODYNAMIC PROBABILITY ($S = k \ln W$)

Thermodynamic probability is the number of microstates corresponding to the given macroscopic state.

From thermodynamics, we know that when a system with constant volume and energy is in equilibrium, the entropy is maximum. According to statistical mechanics, for such a system in equilibrium, the probability is also maximum. Hence it was suggested by Boltzmann that a relationship between entropy and probability must exist.

The relationship between entropy and probability can be obtained by considering the mixing of two systems and remembering that the total entropy (S) is the sum of entropies (S_1 and S_2) of the individual systems and the total probability (W) is the product of the probabilities (W_1 and W_2) of the individual systems.

Thus, we have the following relationships:

$$S = f(W) \quad \text{... (i)}$$

$$S = S_1 + S_2 \quad \text{... (ii)}$$

$$W = W_1 \times W_2 \quad \text{... (iii)}$$

Combining these equations, we get

$$f(W) = S_1 + S_2$$

or

$$f(W_1 \times W_2) = f(W_1) + f(W_2) \quad \text{... (iv)}$$

This relationship is true only if the function f stands for the logarithm, ln. (like log AB = log A + log B) Hence the relationship between entropy (S) and the thermodynamic probability (W) must be logarithmic. Hence we may write

$$S = k_1 \ln W + k_2 \qquad ...(v)$$

where k_1 and k_2 are constants.

The constant k_2 is found to be zero as discussed below:

Applying eqn. (v) to a perfect crystal at 0 K, $S = 0$ (third law of thermodynamics). Further for a perfect crystal at 0K, there is only one arrangement possible. Hence $W = 1$. Substituting these values in equation (v), we get $k_2 = 0$. So equation (v) becomes

$$\boldsymbol{S = k_1 \ln W} \qquad ...(vi)$$

The constant k_1 is found to be equal to the Boltzmann constant, k (*viz.* gas constant per molecule, R/N). Hence equation (vi) becomes

$$\boldsymbol{S = k \ln W} \qquad ...(vii)$$

6.10. ENTROPY AND DISORDER

Consider a system existing in a macrostate having only one microstate *i.e.*, $W = 1$. This means that the particles of the system can be arranged in only one way and in no other way. Thus in this state we have *full and definite* information about the particles of the system. In such a case, the system is said to be in a *perfect order*. For example a system at 0 K should be in a state of perfect order.

Let us now consider a macrostate of the system for which $W = 2$. In this case, the system can exist in any of the two equally probable microstates. We cannot say with certainty in which of the two states the system exists at any instant of time. Thus there is a certain amount of loss of information about the system. In this case, we can only say that the system can exist in either of the two microstates with a probability = 1/2. *This statement of uncertainty or the loss of information about the state of a system is known as disorder.*

Thus increase in the value of W means increase in the amount of disorder or in other words, decrease in the amount of information as to which microstate the system occupies. Hence *W can be taken as measure of the disorder in a system and is sometimes called the disorder number of a given macrostate. From the relation $S = k \ln W$, we infer that entropy is connected with the disorder of the system.*

6.11. PARTITION FUNCTION

According to Boltzmann distribution law, the fraction of molecules which in the most probable state at temperature T possesses the energy ε_i is given by

$$\frac{n_i}{N} = \frac{g_i e^{-\varepsilon_i/kT}}{\Sigma g_i e^{-\varepsilon_i/kT}} \qquad ...(i)$$

where n_i is the number of molecules having energy ε_i at temperature T, N is the total number of molecules and g_i is the degeneracy (or statistical weight) of the energy level ε_i.

The denominator of the above equation which gives the sum of the terms $g_i e^{-\varepsilon_i/kT}$ for all the energy levels is called the partition function. It is usually represented by Q. Thus

$$\textbf{Partition function} \quad Q = \Sigma g_i e^{-\varepsilon_i/kT} \qquad ...(ii)$$

The partition function is so called because it indicates how the particles, are distributed among the various energy levels.

6.12. EXPRESSION FOR THERMODYNAMIC FUNCTIONS IN TERMS OF PARTITION FUNCTION

The expressions for the thermodynamic functions are generally obtained in terms of the partition function Q and its derivatives

$$(\delta \ln Q / \delta T)_V \text{ and } (\delta^2 \ln Q/\delta T^2)_V$$

The expressions for these derivatives are found to be as follows:

$$\left(\frac{\partial \ln Q}{\partial T}\right)_V = \frac{1}{T}\left(\frac{Q'}{Q}\right) \quad \text{... (i)}$$

and
$$\left(\frac{\partial^2 \ln Q}{\partial T^2}\right)_V = \frac{1}{T^2}\left[\left(\frac{Q''}{Q}\right) - \left(\frac{Q'}{Q}\right)^2 - 2\left(\frac{Q'}{Q}\right)\right] \quad \text{... (ii)}$$

where
$$Q' = \Sigma g_i \left(\frac{\varepsilon_i}{kT}\right) e^{-\varepsilon_i / kT} \quad \text{... (iii)}$$

and
$$Q'' = \Sigma g_i \left(\frac{\varepsilon_i}{kT}\right)^2 e^{-\varepsilon_i / kT} \quad \text{... (iv)}$$

(1) **Expression for Energy.** If the molecules are non-interacting *i.e.,* the gas is ideal, the total energy E of the system is given by

$$E = \Sigma\, n_i\, \varepsilon_i \quad \text{...(v)}$$

where n_i is the number of particles present in the level of energy ε_i relative to the ground state.

From equations (*i*) and (*ii*), sec. 6.11 we have

$$\frac{n_i}{N} = \frac{g_i e^{\varepsilon_i / kT}}{Q}$$

i.e.,
$$n_i = \frac{N}{Q} g_i e^{-\varepsilon_i / kT} \quad \text{... (vi)}$$

Substituting this value in equation (*v*), we get

$$E = \frac{N}{Q} \Sigma \frac{\varepsilon_i}{kT} \varepsilon_i g_i e^{-\varepsilon_i / kT} \quad \text{... (vii)}$$

which can be rearranged as

$$E = \frac{NkT}{Q} \Sigma \frac{\varepsilon_i}{kT} g_i e^{-\varepsilon_i / kT} \quad \text{... (viii)}$$

If one mole of the gas is considered, the total number of molecules N will be equal to Avogadro's number so that $Nk = R$, the molar gas constant. The eqn. (*viii*) becomes

$$E = \frac{RT}{Q} \Sigma g_i \left(\frac{\varepsilon_i}{kT}\right) e^{-\varepsilon_i / kT} \quad \text{... (ix)}$$

But
$$\Sigma g_i \left(\frac{\varepsilon_i}{kT}\right) e^{-\varepsilon_i / kT} = Q'$$
[according to eqn. (*iii*) above]

∴ Eqn. (*ix*) can be written as

$$\boldsymbol{E = RT\left(\frac{Q'}{Q}\right)} \quad \text{... (x)}$$

$$E_e = RT^2 \left(\frac{\partial \ln Q_e}{\partial T} \right)_V \quad \text{... (vi)}$$

where Q_t, Q_r, Q_v and Q_e represent the partition functions for translational, rotational, vibrational and electronic energies.

Substituting these values in equation (*i*), we get

$$E = RT^2 \left[\left(\frac{\partial \ln Q_t}{\partial T} \right)_V + \left(\frac{\partial \ln Q_r}{\partial T} \right)_V + \left(\frac{\partial \ln Q_v}{\partial T} \right)_V + \left(\frac{\partial \ln Q_e}{\partial T} \right)_V \right]$$

$$= RT^2 \left[\frac{\partial \ln(Q_t Q_r Q_v Q_e)}{\partial T} \right] \quad \text{...(vii)}$$

Comparing this equation with equation (*ii*), we find that

$$Q = Q_t Q_r Q_v Q_e \quad \text{...(viii)}$$

6.14. EXPRESSION FOR TRANSLATIONAL PARTITION FUNCTION

Partition function of a system is given by

$$Q = \Sigma g_i e^{-\varepsilon_i / kT} \quad \text{... (i)}$$

For translational energy, $g_i = 1$. Representing the translational energy of a molecule by ε_t, the translational partition function, Q_t will be given by

$$Q_t = \Sigma e^{-\varepsilon_i / kT} \quad \text{...(ii)}$$

Further we know that the translational energy of a molecule moving in a rectangular box of dimensions *a, b,* and *c* is given by

$$\varepsilon_t = \frac{h^2}{8\,m} \left(\frac{n_x^2}{a^2} + \frac{n_y^2}{b^2} + \frac{n_z^2}{c^2} \right)$$

Substituting this value in eqn. (*ii*), we get

$$Q_t = \Sigma \exp\left[-\frac{h^2}{8\,mkT} \left(\frac{n_x^2}{a^2} \right) \right] \exp\left[-\frac{h^2}{8\,mkT} \left(\frac{n_y^2}{b^2} \right) \right] \exp\left[-\frac{h^2}{8\,mkT} \left(\frac{n_z^2}{c^2} \right) \right] \quad \text{... (iii)}$$

Let us consider the motion in *x*-direction only. The partition Q_x for this motion will be

$$Q_x = \sum_{n_x=0}^{n_x=\infty} \exp\left[-\frac{h^2}{8\,mkT} \left(\frac{n_x^2}{a^2} \right) \right] \quad \text{...(iv)}$$

As the energy levels are very closely spaced, the summation in the equation may be replaced by integration *i.e.*, we have

$$Q_x = \int_0^\infty \exp\left[-\frac{h^2}{8\,mkT} \left(\frac{n_x^2}{a^2} \right) \right] dn_x = \frac{(2\pi mkT)^{1/2}}{h} a \quad \text{... (v)}$$

Similarly, for the motion in the *y* and *z*-directions, we will have

$$Q_y = \frac{(2\pi mkT)^{1/2}}{h} b \quad \text{and} \quad Q_z = \frac{(2\pi mkT)^{1/2}}{h} c$$

Hence from equation (*iii*)

$$Q_t = Q_x . Q_y . Q_z = \frac{(2\pi mkT)^{3/2}}{h^3} (abc)$$

or
$$Q_t = \frac{(2\pi mkT)^{3/2}}{h^3} V \qquad ...(vi)$$
where V is the volume of the container occupied by the particle.

SOLVED PROBLEMS ON TRANSLATIONAL PARTITION FUNCTION

Example 1. Calculate translational partition function for helium molecule confined in a vessel 10^2 cm^3 volume at 300 K. Given that $k = 1.38 \times 10^{-23}$ J K^{-1}, $h = 6.625 \times 10^{-34}$ J s. Mass of He atom $m = 4$ amu, $N = 6.023 \times 10^{23}$.

Solution.
$$m = \frac{4}{6.023\times10^{23}} \times \frac{1}{1000}\text{ kg} = \frac{4}{6.023\times10^{26}}\text{ kg}$$
$$k = 1.38\times10^{-23}\text{ JK}^{-1} = 1.38\times10^{-23}\text{ kg m}^2\, s^{-2}\text{ K}^{-1}$$
$$h = 6.625\times10^{-34}\text{ Js} = 6.625\times10^{-34}\text{ kg m}^2\, s^{-1}$$
$$V = 10^2\text{ cm}^3 = \frac{10^2}{100\times100\times100}\, m^2 = 10^{-4}\, m^3$$

The translational partition function for He molecule is given by $Qt\text{ (He)} = \left[\frac{2\pi mkT}{h^2}\right]^{3/2}$

Putting various values in the above expression

$$Qt\text{(He)} = \left[\frac{2\pi \times \left(\frac{4}{6.023\times10^{26}}\text{ kg}\right)(1.38\times10^{-22}\text{ gm}^2\text{ s}^{-2}\text{ K}^{-1})\times300\text{K}}{(6.625\times10^{-34}\text{ kgm}^2\text{ s}^{-1})^2}\right]^{3/2} \times 10^{-4}\text{ m}^3$$

$\therefore \quad Qt\text{ (He)} = \mathbf{7.808\times10^{26}}$

Example 2. Calculate the translational partition function of NO molecule at 300 K in a volume 1000 m^3. Assuming the gas to behave ideally.

Solution. Here V = 1000 cm^3 = 10^3 cm^3, T = 300 K, m = Mass of NO molecule $= \frac{30}{6.023\times10^{23}}$ g

$$h = 6.625\times10^{-27}\text{ erg s} = 6.625\times10^{-27}\text{ g cm}^2\text{ cm}^2\text{ s}^{-1}$$
$$k = 1.38\times10^{-16}\text{ erg K}^{-1} = 1.38\times10^{-16}\text{ g cm}^2\text{ s}^{-2}\text{ K}^{-1}$$
$$[\because 1\text{ erg} = 1\text{ g cm}^2\text{ s}^{-2}]$$

The translational partition function of NO molecule is given by the expression
$$Q_t = \left[\frac{2\pi\, mkT}{h^2}\right]^{3/2} .\text{V} \qquad ...(i)$$

Substituting various values in equation (i), we get

$$Q_t = \left[\frac{2\times\pi\times\left(\dfrac{30g}{6.023\times10^{23}}\right)(1.38\times10^{-16}\ \text{gcm}^2\ \text{s}^{-2}\text{K}^{-1})\times300\text{K}}{(6.625\times10^{-27}\ \text{Kgm}^2\ \text{s}^{-1})^2}\right]\times10^3\ \text{cm}^3$$

$$= \mathbf{1.6038 \times 10^{29}}$$

Example 3. Calculate the ratio of Q_t of helium molecule (He) to that of hydrogen (H_2) at the same temperature and volume.

Solution. For helium molecule (He), m = 4 a.m.u.

$$Q_t\ (\text{He}) = \left(\frac{2\pi\times4\times kT}{h^2}\right)^{3/2}.V \quad ...(i)$$

For hydrogen molecule *molecule* (H_2), m = 2 a.m.u.

$$Q_t\ (H_2) = \left(\frac{2\pi\times2\times kT}{h^2}\right)^{3/2}.V \quad ...(ii)$$

Dividing (*i*) by (*ii*)

$$\frac{Q_t\ (\text{He})}{Q_t\ (H_2)} = \frac{\left(\dfrac{2\pi\times4\times kT}{h^2}\right)^{3/2}.V}{\left(\dfrac{2\pi\times2\times kT}{h^2}\right)^{3/2}.V} = 2^{3/2} = 2\sqrt{2} = 2(1.414)$$

or
$$\frac{Q_t\ (\text{He})}{Q_t\ (H_2)} = \mathbf{2.828}$$

In other words, Q_t (He) is 2.828 times larger than Q_t (H_2).

6.15. EXPRESSION FOR ROTATIONAL PARTITION FUNCTION (Q_r)

The rotational energy of a molecule, ε_r depends upon the rotational quantum number, J. Further, for rotation

$$g_i = (2J + 1)$$

Thus the general partition function for rotation is given by

$$Q_r = \Sigma(2J+1)e^{-\varepsilon_r/kT} \quad ...(i)$$

The simplest model for a diatomic molecule is the rigid rotator, the energy for which is given by the relation

$$\varepsilon_r = J(J+1)\frac{h^2}{8\pi^2 I} \quad ...(ii)$$

where J is the rotational quantum number and I is the moment of inertia of the rotator.

Substituting this value of ε_r in equation (*i*), we obtain

$$Q_r = \Sigma(2J+1)e^{-aJ(J+1)} \quad ...(iii)$$

where
$$a = \frac{h^2}{8\pi^2 IkT} \quad ...(iv)$$

Assuming that the energy levels are very closely spaced, the summation in equation (*iii*) may be replaced by integration.

$$Q_r = \int_0^\infty (2J+1)e^{-aJ(J+1)} \quad \text{... (v)}$$

Solution of this equation gives

$$Q_r = \frac{1}{a}$$

Substituting the value of *a* from equation (*iv*), we get

$$\boldsymbol{Q_r = \frac{8\pi^2 IkT}{h^2}} \quad \text{... (vi)}$$

Considering the symmetry effect of the molecule, the above expression is modified to

$$\boldsymbol{Q_r = \frac{8\pi^2 IkT}{\sigma h^2}} \quad \text{...(vii)}$$

where σ is called *symmetry number.* Its value is 1 for unsymmetrical molecules like HCl, CO etc. and it is 2 for symmetrical molecules like O_2, Cl_2 etc.

6.16. EXPRESSION FOR VIBRATIONAL PARTITION FUNCTION (Q_v)

Considering the diatomic molecule as a simple harmonic oscillator, its vibrational energy is given by

$$e_v' = (v + 1/2)\, h\nu_0 \quad \text{... (i)}$$

where *v* is the vibrational quantum number and ν_0 is the fundamental vibrational frequency.

Equation (*i*) can be rewritten as

$$e_v' = \frac{1}{2}h\nu_0 + vh\nu_0$$

$$= \varepsilon_0 + \varepsilon_v$$

where $\varepsilon_0 = \frac{1}{2}h\nu_0$ is the zero point energy of the oscillator and

$\varepsilon_v = vh\nu_0$ is the energy of the oscillator relative to the ground state.

Further for vibration $g_i = 1$. Substituting these values of g_i and ε_v the vibrational partition function becomes

$$Q_v = \sum_{v=0}^{\infty} e^{-vh\nu_0/kT} \quad \text{... (ii)}$$

Putting $h\nu_0/kT = x$, this equation becomes

$$Q_v = \sum_{v=0}^{\infty} e^{-vx} \quad \text{...(iii)}$$

$$= 1 + e^{-x} + e^{-2x} + ...$$

$$= (1 - e^{-x})^{-1}$$

$$= \frac{1}{1 - e^{-x}}$$

or
$$Q_v = \frac{1}{1 - e^{-h\nu_0/kT}} \qquad ...(iv)$$

6.17. EEXPRESSION FOR ELECTRONIC PARTITION FUNCTION(Q_e)

The electronic partition function of an atom or a molecule is given by

$$Q_e = \Sigma g_i e^{-\varepsilon_e/kT} \qquad ...(i)$$

where ε_e is the electronic energy relative to the ground state.

The above summation is carried over the energy levels instead of the states. Hence we write

$$Q_e = \sum_i g_i e^{-\varepsilon_i/kT} \qquad ...(ii)$$

If the degeneracy of the ground state level is g_0, (for which $\varepsilon_i = 0$) and that of the first excited state is g_1 (for which we can write $\varepsilon_i = \varepsilon_1$) and so on, the above expression can be written as

$$Q_e = g_0 + g_i e^{-\varepsilon_1/kT} + ... \qquad ...(iii)$$

6.18. STATISTICAL CALCULATION OF THERMODYNAMIC FUNCTIONS

The total value of any thermodynamic property is the sum of translational, rotational, vibrational and electronic contributions.

Let us consider a systems of perfect gases.

(*i*) **Translational energy.** Translational energy is given by

$$E_t = RT^2 \left(\frac{\partial \ln Q_t}{\partial T} \right)_V \qquad ...(i)$$

Let us now find the value of $(\partial \ln Q_t/\partial T)_V$

The expression for translational partition function is

$$Q_t = \frac{(2\pi mkT)^{3/2}}{h^3} V$$

Taking logarithm of both sides, we get

$$\ln Q_t = \ln \frac{(2\pi mkT)^{3/2}}{h^3} V$$

$$= \ln \frac{(2\pi mk)^{3/2} V}{h^3} + \ln T^{3/2}$$

$$= \frac{(2\pi mk)^{3/2} V}{h^3} + \frac{3}{2} \ln T \qquad ...(ii)$$

Differentiating w.r.t T at constant V, we get

$$\left(\frac{\partial \ln Q_t}{\partial T} \right)_v = \frac{3}{2} \left(\frac{1}{T} \right) \qquad ...(iii)$$

Substituting this value in equation (*i*), we get

$$E_t = RT^2 \times \frac{3}{2}\left(\frac{1}{T}\right)$$

or

$$\boldsymbol{E_t = \frac{3}{2}RT} \qquad ...(iv)$$

(*ii*) **Translational enthalpy.** Enthalpy is given by

$$H = E + PV$$

For 1 mole of an ideal gas, $PV = RT$ so that

$$H = E + RT$$

For translational enthalpy, we can write

$$H_t = E_t + RT$$

Substituting the value of E_t from equation (*iv*), we get

$$H_t = \frac{3}{2}RT + RT$$

i.e.,

$$\boldsymbol{H_t = \frac{5}{2}RT} \qquad ...(v)$$

(*iii*) **Translational Heat Capacity at Constant Volume.**

By definition $$C_{v(t)} = \left(\frac{\partial E_t}{\partial T}\right)_v = \left[\frac{\partial\left(\frac{3}{2}RT\right)}{\partial T}\right]_V$$

i.e.,

$$\boldsymbol{C_{v(t)} = \frac{3}{2}R} \qquad ...(vi)$$

(*iv*) **Translational Heat Capacity at Constant Pressure.**

By definition $$C_{p(t)} = \left(\frac{\partial H_t}{\partial T}\right)_p = \left[\frac{\partial\left(\frac{5}{2}RT\right)}{\partial T}\right]_p$$

i.e.,

$$\boldsymbol{C_{p(t)} = \frac{5}{2}R} \qquad ...(vii)$$

(*v*) **Translational entropy.** The expression for entropy is

$$S - S_0 = RT\left(\frac{\partial \ln Q}{\partial T}\right)_v + R\ln Q - R\ln Q_0 \qquad \text{(Sec 6.12)} \qquad ...(viii)$$

where S_0 is the entropy and Q_0 is the value of Q at $T = 0$.

For translational motion,

$$S_0 = R\ln Q_0 - R\ln N + R$$

Substituting this value in equation (*viii*), we get

$$S_t = \text{RT}\left[\frac{\partial \ln Q_t}{\partial T}\right]_v + \text{R}\ln\frac{Q_t}{N} + \text{R} \qquad ...(ix)$$

$$= R\left[T\left(\frac{\partial \ln Q_t}{\partial T}\right)_v + \ln Q_t - \ln N + 1\right]$$

Substituting the values of $\ln Q_t$ and $\left(\frac{\partial \ln Q_t}{\partial T}\right)_v$ from equations (*ii*) and (*iii*), we get

$$S_t = R\left[T\cdot\frac{3}{2}\left(\frac{1}{T}\right) + \ln\frac{(2\pi mk)^{3/2}}{h^3}V + \frac{3}{2}\ln T - \ln N + 1\right]$$

$$= R\left[\frac{3}{2} + \ln\left(\frac{(2\pi mk)}{h^2}\right)^{3/2} V + \frac{3}{2}\ln T - \ln N + 1\right]$$

$$= R\left[\frac{5}{2} + \ln\left(\frac{(2\pi mk)}{h^2}\right)^{3/2} V + \frac{3}{2}\ln T - \ln N\right]$$

$$= R\left[\frac{5}{2} + \ln\left(\frac{(2\pi mkT)}{h^2}\right)^{3/2} + \ln\frac{V}{N}\right]$$

or

$$\mathbf{S_t = R\left[\frac{5}{2} + \ln\left\{\left(\frac{(2\pi mkT)}{h^2}\right)^{3/2}\frac{V}{N}\right\}\right]} \qquad ...(x)$$

This equation can be expressed in another form by putting $V = RT/P$ (for one mole of an ideal gas) so that

$$\frac{V}{N} = \frac{RT}{NP} = \frac{kT}{P}$$

where $R/N = k$ is Boltzmann's constant. Hence we have

or

$$\mathbf{S_t = R\left[\frac{5}{2} + \ln\left\{\left(\frac{(2\pi mkT)}{h^2}\right)^{3/2}\frac{kT}{P}\right\}\right]} \qquad ...(xi)$$

Equations (*x*) and (*xi*), are known as **Sackur-Tetrode** equations.

The above equations are sometimes written in a number of convenient forms which may be obtained from the above equations as explained below:

Putting mass of the molecule, $m = M/N$ (*i.e.*, Molecular mass of the gas/Avogadro's number and expanding, eqn. (*x*) can be written as

$$S_t = \left[\frac{5}{2} + \frac{3}{2}(\ln 2\pi + \ln M - \ln N + \ln k + \ln T - \ln h^2 + \ln V - \ln N)\right]$$

$$= R\left(\frac{5}{2} + \frac{3}{2}\ln 2\pi - \frac{5}{2}\ln N + \frac{3}{2}\ln k - 3\ln h\right) + \left(\frac{3}{2}\ln M + \frac{3}{2}\ln T + \ln V\right)$$

or

$$\mathbf{S_t = C_1 + R\left(\frac{3}{2}\ln M + \frac{3}{2}\ln T + \ln V\right)} \qquad ...(xii)$$

where

$$C_1 = R\left(\frac{5}{2} + \frac{3}{2}\ln 2\pi - \frac{5}{2}\ln N + \frac{3}{2}\ln k - 3\ln h\right) = \text{constant}$$

$$= -11.073 \text{ cal degree}^{-1} = -46.329 \text{ JK}^{-1}\text{ mol}^{-1}$$

Similarly, eqn. (*xi*) on expanding takes up the form

$$S_t = C_2 + R\left(\frac{3}{2}\ln M + \frac{5}{2}\ln T - \ln P\right) \quad \text{...(xiii)}$$

where $C_2 = C_1 + R \ln R$ = constant (Remember $R = Nk$)

$= -2.315$ cal degree^{-1} = -9.686 JK^{-1} mol^{-1}.

If standard state conditions are taken, the value obtained is for standard entropy ($S°$). Under these conditions, putting $P = 101.32$ kPa and $T = 298.15$ K, the above equation is simplified to the form

$$S_t^0 = 195 + \frac{3}{2}R\ln M \ (\text{JK}^{-1}\text{mol}^{-1}) \quad \text{...(xiv)}$$

(*vi*) **Translational contribution to Helmholtz free energy.** By definition

$$A_t = E_t - TS_t$$

Substituting the values of E_t and S_t from equations (*iv*) and (*xi*), we get

$$A_t = \frac{3}{2}RT - RT\left[\frac{5}{2} + \ln\left\{\left(\frac{(2\pi mkT)}{h^2}\right)^{3/2}\frac{kT}{P}\right\}\right]$$

or

$$A_t = -RT\left[1 + \ln\left\{\left(\frac{(2\pi mkT)}{h^2}\right)^{3/2}\frac{kT}{P}\right\}\right] \quad \text{...(xv)}$$

(*vii*) **Translational contribution to Gibb's free energy.** By definition

$$G_t = H_t - TS_t$$

Substituting the values of H_t and S_t from equation (*v*) and (*xi*), we get

$$G_t = \frac{5}{2}RT - RT\left[\frac{5}{2} + \ln\left\{\left(\frac{(2\pi mkT)}{h^2}\right)^{3/2}\frac{kT}{P}\right\}\right]$$

or

$$G_t = -RT\ln\left\{\left(\frac{(2\pi mkT)}{h^2}\right)^{3/2}\frac{kT}{P}\right\} \quad \text{... (xvi)}$$

6.19. EQUATION OF STATE OF AN IDEAL GAS FROM PARTITION FUNCTION

The relation between pressure 1 and molecular partition function Q for 1 mole of an ideal gas is

$$P = RT\left(\frac{\partial \ln Q}{\partial V}\right)_T \quad \text{...(i)}$$

Only Q_t is volume dependent partition function. The other partition such as rotation, vibrational and electronic are independent of volume.

∴ Equation (*i*) can be written as

$$P = RT\left(\frac{\partial \ln Q_t}{\partial V}\right)_T \quad \text{...(ii)}$$

But

$$Q_t = \left(\frac{2\pi\, m\, kT}{h^3}\right)^{3/2} - V$$

Putting in (*ii*) the value of Q_t

$$P = RT\left[\frac{\partial}{\partial V}\left[\ln\frac{(2\pi m\,kT)^{3/2}.V}{h^3}\right]\right]_T$$

$$= RT\left[\frac{\partial}{\partial V}\left[\ln\frac{(2\pi m\,kT)^{3/2}}{h^3} + \ln V\right]\right]_T$$

$$P = RT\left[0+\frac{1}{V}\right] = \frac{RT}{V}$$

or

$$\boxed{PV = RT} \qquad \ldots(iii)$$

which is an ideal gas equation for 1 more of an ideal

For n moles of an ideal gas, the equation of state become PV = n RT

Example 1. Calculate the translational contribution to $H°$, $S°$ and $G°$ for oxygen gas at 298 K. Boltzmann constant $k = 1.381\times10^{-23}$ and $h = 6.626 \times 10^{-34}$.

Solution. Mass of the oxygen molecule is

$$m = \frac{\text{Mol. mass of } O_2}{\text{Avogadro's No.}} = \frac{32}{6.022\times10^{23}}\text{g} = 5.314 \times 10^{-23}\text{g}$$

$$= 5.314 \times10^{-26}\,\text{kg}$$

$$T = 298\text{ K}$$

$$P = 10^5\text{ Nm}^{-2}.$$

(*i*) $H_t^0 = \frac{5}{2}RT = \frac{5}{2}(8.314\text{ JK}^{-1}\text{ mol}^{-1})\,(298\text{ K})$

$= \mathbf{6.194\ kJ\ mol^{-1}}.$

$$(ii)\ S_t^0 = R\left[\frac{5}{2}+\ln\left\{\left(\frac{(2\pi mkT)}{h^2}\right)^{3/2}\frac{kT}{P}\right\}\right]$$

$$= R\left[\frac{5}{2}+\ln\left(\frac{(2\pi mkT)}{h^2}\right)^{3/2}+\ln\frac{kT}{P}\right]$$

$$= R\left[\frac{5}{2}+\frac{3}{2}\ln\left(\frac{(2\pi mkT)}{h^2}\right)+\ln\frac{kT}{P}\right]$$

$$= R\left[\frac{5}{2}+\frac{3}{2}\times2.303\log\frac{2\pi mkT}{h^2}+2.303\log\frac{kT}{P}\right]$$

$$= 8.314\left[\frac{5}{2}+\frac{3\times2.303}{2}\log\frac{2\times3.143\times5.314\times10^{-26}\times1.381\times10^{-23}\times298}{(6.626\times10^{-34})^2} + 2.303\log\frac{1.381\times10^{-23}\times298}{10^5}\right]$$

$$= 8.314\left[\frac{5}{2}+3.4545\log(3.131\times10^{21})+2.303\log(4.115\times10^{-26})\right]$$

$$= 8.314\left[\frac{5}{2}+3.4545\times(21.4956)+2.303\times(-25.3857)\right]$$

$$= 8.314\,[2.5 + 74.2565 - 58.4633]$$
$$= 8.314 \times 18.2932$$
$$= \mathbf{152.09\ JK^{-1}\ mol^{-1}}.$$

(iii) $G_t^0 = H_t^0 - TS_t^0$

$$= 6.194 - 298 \times 0.15209$$
$$= \mathbf{-39.129\ kJ\ mol^{-1}}$$

6.20. ROTATIONAL CONTRIBUTION

(i) **Rotational energy.** The rotational energy is given by the equation (Refer to eq. (xi) sec. 6.12)

$$E_r = RT^2\left(\frac{\partial \ln Q_r}{\partial T}\right)_V \qquad ...(i)$$

The rotational partition function (Q_r) is given by the equation (Refer to eq. (vii), sec. 6.15)

$$Q_r = \frac{8\pi^2 IkT}{\sigma h^2}$$

Taking logarithm of both sides

$$\ln Q_r = \ln\left(\frac{8\pi^2 Ik}{\sigma h^2}\right) + \ln T \qquad ...(ii)$$

$$\therefore \qquad \left(\frac{\partial \ln Q_r}{\partial T}\right)_V = \frac{1}{T} \qquad ...(iii)$$

Substituting this value in equation (i), we get

$$E_r = RT^2 \times \frac{1}{T}$$

or
$$E_r = RT \qquad ...(iv)$$

(ii) **Rotational enthalpy.** Equation for enthalpy is

$$H_r = RT^2\left(\frac{\partial \ln Q_r}{\partial T}\right)_P$$

But as derived above

$$\left(\frac{\partial \ln Q_r}{\partial T}\right)_P = \frac{1}{T}$$

$$H_r = RT \qquad ...(v)$$

(iii) **Rotational Heat Capacity at constant volume.** By definition

$$C_{v(r)} = \left(\frac{\partial E_r}{\partial T}\right)_V = \left(\frac{\partial (RT)}{\partial T}\right)_V$$

i.e.,
$$C_{v(r)} = R \qquad ...(vi)$$

(iv) **Rotational Heat Capacity at constant pressure.** By definition

$$C_{P(r)} = \left(\frac{\partial H_r}{\partial T}\right)_P = \left(\frac{\partial (RT)}{\partial T}\right)_P$$

i.e., $C_{P(r)} = R$ ...(*vii*)

(*v*) **Rotational entropy.** Expression for entropy is (Refer to eq (*xviii*), sec 6.12)

$$S - S_0 = RT\left(\frac{\partial \ln Q}{\partial T}\right)_V + R\ln Q - R\ln Q_0 \qquad ...(viii)$$

where S_0 is the entropy and Q_0 is the value of Q at $T = 0$.

For rotational (as well as vibrational) motion, it is found that

$$S_0 = R\ln Q_0$$

Hence eqn. (*viii*) becomes

$$S_r = RT\left(\frac{\partial \ln Q_r}{\partial T}\right)_V + R\ln Q_r \qquad ...(ix)$$

But
$$Q_r = \frac{8\pi^2 IkT}{\sigma h^2}$$

$\therefore$
$$\ln Q_r = \ln\frac{8\pi^2 IkT}{\sigma h^2} = \ln\frac{8\pi^2 Ik}{\sigma h^2} + \ln T$$

and
$$\left(\frac{\partial \ln Q_r}{\partial T}\right)_V = \frac{1}{T}$$

Substituting these values in equation (*ix*), we get

$$S_r = RT\cdot\frac{1}{T} + R\ln\frac{8\pi^2 IkT}{\sigma h^2}$$

or
$$S_r = R\left[1 + \ln\frac{8\pi^2 IkT}{\sigma h^2}\right] \qquad ...(x)$$

or
$$S_r = R\ln\frac{eT}{\sigma}\frac{8\pi^2 Ik}{h^2} \qquad (\because 1 = \ln e)$$

or
$$S_r = R\ln\frac{eT}{\sigma\alpha} \qquad ...(xi)$$

where $\alpha = h^2/8\pi^2 Ik$

(*vi*) **Rotational contribution to Helmoltz free energy** . By definition.

$$A_r = E_r - TS_r \qquad ...(xii)$$

Substituting the values of E_r and S_r, A_r can be calculated.

(*vii*) **Rotational contribution to Gibb's free energy.** By definition

$$G_r = H_r - TS_r$$

Substituting the values of H_r and S_r, we get

$$G = RT - RT\ln\frac{eT}{\sigma\alpha}$$

$$= RT\left(1 - \ln\frac{eT}{\sigma\alpha}\right)$$

$$= -RT\left(\ln\frac{eT}{\sigma\alpha} - 1\right)$$

$$= -RT\left(\ln\frac{eT}{\sigma\alpha} - \ln e\right)$$

$$G_r = -RT\ln\frac{eT}{\sigma\alpha}$$

Example 1. Find the rotational partition function for Cl_2 molecule at 25°C and 1 atmospheric pressure.

Given (*i*) Internuclear distance Cl — Cl = 1.988Å.

(*ii*) Reduced mass μ_{Cl_2} = 17.4894 a.m.u.

Solution. Step I. [*To calculate moment of inertia I*]

$$\mu_{Cl_2} = 17.4894 \text{ a.m.u.} = \frac{17.4894}{6.023\times10^{23}} g$$

$$r_{Cl_2} = 1.988 \text{ Å} = 1.988 \times 10^{-8} \text{ cm}$$

$$I = \mu r^2 = \left(\frac{17.4894}{6.023\times10^{23}} g\right) \times (1.988 \times 10^{-8} \text{ cm})^2$$

$$I = 1.1476\times10^{-38} \text{ g cm}^2$$

Step II. [*To calculate Q_r*]

$$Q_r = \frac{8\pi^2 I\, kT}{\sigma h^2}$$

$$\sigma = 2, T = 298, k = 1.38 \times 10^{-16} \text{ g cm}^2 \text{ s}^{-2} \text{ K}^{-1}$$

$$h = 6.625\times10^{-27} \text{ g cm}^2 \text{ s}^{-1}$$

$$Q_r = \frac{8\pi^2 (1.1476\times10^{-38} g \text{ cm}^2)(1.38\times10^{-16} g \text{ cm}^2 \text{ s}^{-2} \text{ K}^{-1})\, 298 \text{ K}}{2\times(6.625\times10^{-27} g \text{ cm}^2\text{s}^{-1})^2}$$

$$Q_r = \mathbf{424.84}$$

Example2. The rotational constant (B) of gaseous HCl, determined from microwave spectroscopy is 10.59 cm^{-1}, calculate the rotational partition function of HCl at (*i*) 100 K and (*ii*) 500 K.

Solution. Rotational constant B = 10.59 cm^{-1}

To convert this value to Joules, we multiply by *hc*

$\therefore$ $$B = (10.59 \text{ cm}^{-1})(6.625\times10^{-34} \text{ Js})(3\times10^{10} \text{ cm s}^{-1})$$

$$\sigma = 1$$

(*i*) $$\therefore Q_r = \frac{kT}{\sigma B} = \frac{1.38\times10^{-23} \text{ JK}^{-1}\times 100\text{K}}{1\times10.59 \text{ cm}^{-1}\times6.625\times10^{-34} \text{ Js}\times3\times10^{10} \text{ cm s}^{-1}}$$

$$Q_r = \mathbf{6.56} \text{ at } 100 \text{ K}$$

(*ii*) $$Q_r = \frac{1.38\times10^{-23} \text{ JK}^{-1}\times 500\text{K}}{1\times10.59 \text{ cm}^{-1}\times 6.625\times10^{-34} \text{ Js}\times3\times10^{10} \text{ cm s}^{-1}}$$

$$= \mathbf{32.8} \text{ at } 500 \text{ K}$$

Example 3. Calculate the rotational contributions to H^o, S^o and G^o for oxygen gas at 298 K. The moment of inertia for O_2 (g), $I = 1.937 \times 10^{-46}$ kg m^2.

Solution. (*i*) $H_r^0 = RT = (8.314 \text{ JK}^{-1} \text{ mol}^{-1}) \times (298\text{K})$

$= \mathbf{2.477 \text{ kJ mol}^{-1}}$

(*ii*)
$$S_r^0 = R \ln \frac{eT}{\sigma\alpha}$$

$$\alpha = \frac{h^2}{8\pi^2 Ik}$$

$$= \frac{(6.626\times10^{-34}\,\text{Js})^2}{8\times(3.143)^2\times(1.937\times10^{-46}\,\text{kg m}^2)\times(1.384\times10^{-23}\,\text{JK}^{-1})} = 2.507\text{K}$$

$e = 2.7183$

$\sigma = 2$ for O_2

$$\therefore \quad S_r^0 = 8.314\times2.303\log\frac{2.718\times298}{2\times2.507}$$

$= 8.314 \times 2.303 \log 161.5$

$= \mathbf{42.28 \text{ kJ mol}^{-1}}$.

(*iii*)
$$G_r^0 = H_r^0 - TS_r^0$$

$= 2.477 - 298 \times 0.04228$

$= \mathbf{-10.12 \text{ kJ mol}^{-1}}$.

Example 4. Calculate S_r for O_2 in a mole of O_2 at 298 K. Given I for O_2 is 1.937×10^{-46} kg m^2.

Solution. Rotational entropy is given by the equation

$$S_r = R\left(1 + \ln\frac{8\pi^2 I kT}{\sigma h^2}\right) \quad ...(i)$$

$\sigma = 2$ (for O_2); $I = 1.937 \times 10^{-46}$ kg m^2

$T = 298$ K; $k = 1.38 \times 10^{-23}$ kg m^2 s^{-2} K^{-1}

$h = 6.625 \times 10^{-34}$ kg m^2 s^{-1}

Substituting various values in (*i*) we have

$$S_r = R\left[1 + \ln\left(\frac{8\times\pi^2\times1.937\times10^{-46}\times1.38\times10^{-23}\times298}{2\times(6.625\times10^{-34})^2}\right)\right]$$

$= 8.314\,[1 + 4.2726] = 8.314 \times 5.2726 \text{ JK}^{-1} \text{ mol}^{-1}$ **[1*EU* = *1 JK*$^{-1}$]**

$\therefore \quad S_r = 43.84$ EU mol^{-1}

6.21. VIBRATIONAL CONTRIBUTION

(*i*) Vibrational energy. The vibrational energy is given by the following equation

$$E_v = RT^2\left(\frac{\partial \ln Q_v}{\partial T}\right)_v \quad ...(i)$$

The vibrational partition function is given by

$$Q_v = \frac{1}{1-e^{-h\nu_0/kT}}$$

where ν_0, is the fundamental frequency of vibration. ...(*ii*)

Putting $\frac{h\nu_0}{k} = \beta$, eqn. (*ii*) becomes

$$Q_v = \frac{1}{1-e^{-\beta/T}} \qquad ...(iii)$$

Substituting this value in equation (*i*) and solving, we get

$$\boxed{E_v = RT\frac{\beta/T}{e^{\beta/T}-1}} \qquad ...(iv)$$

where $\beta = \frac{h\nu_0}{k} = \frac{hc\bar{\nu}}{k}$ in which $\bar{\nu}$ is the frequency in wave number *i.e.* cm^{-1}.

(*ii*) **Vibrational Enthalpy.** It is found to be same *viz.*

$$\boxed{H_v = RT\frac{\beta/T}{e^{\beta/T}-1}} \qquad ...(v)$$

(*iii*) **Vibrational Heat Capacity at constant volume.** By definition

$$C_{v(v)} = \left(\frac{\partial E_v}{\partial T}\right)_V$$

Using the value of E_v derived in equation (*iv*), we get

$$C_{v(v)} = R\left(\frac{\beta}{T}\right)^2 \frac{e^{\beta/T}}{(e^{\beta/T}-1)^2} \qquad ...(vi)$$

(*iv*) **Vibrational Heat Capacity at constant pressure.** It is found to be same as $C_{v(v)}$ *viz.*

$$\boxed{C_{p(v)} = R\left(\frac{\beta}{T}\right)^2 \frac{e^{\beta/T}}{(e^{\beta/T}-1)^2}} \qquad ...(vii)$$

(*v*) **Vibrational entropy.** Vibrational entropy is given by

$$S_v = RT\left(\frac{\partial \ln Q_v}{\partial T}\right)_v + R\ln Q_v \qquad ...(viii)$$

Substituting $Q_v = (1-e^{-\beta/T})^{-1}$ and solving, we get

$$\boxed{S_v = R\left[\frac{\beta/T}{(e^{\beta/T}-1)} - \ln(1-e^{-\beta/T})\right]} \qquad ...(ix)$$

(*vi*) **Vibrational contribution to Helmholtz free energy.** By definition

$$A_v = E_v - T S_v$$

Substituting the values of E_v and S_v, A_v can be calculated.

(*vii*) **Vibrational contribution to Gibb's free energy.** By definition

$$G_v = H_v - T S_v$$

Substituting the values of H_v and S_v and simplifying, we get

$$G_v = RT \ln (1 - e^{-\beta/T})$$

Example 1. Fundamental vibrational frequency of O_2 at 27°C is s^{-1}, calculate Q_v for O_2. Assuming that O_2 molecule behaves as a simple harmonic oscillator.

Solution. The vibrational partition function is given by

$$Q_v = \frac{1}{1 - e^{-h\nu/kT}}$$

Here $\nu = 4.74 \times 10^{-13}\ s^{-1}$, $h = 6.625 \times 10^{-27}$ ergs

$k = 1.38 \times 10^{-16}$ erg K^{-1}, T = 300 K

$$\therefore \quad \frac{h\nu}{kT} = \frac{(6.625 \times 10^{-27}) \times (4.74 \times 10^{-10})}{(1.38 \times 10^{-16}) \times (300)}$$

or $$\frac{h\nu}{kT} = 7.585$$

Now $$Q_v = \frac{1}{1 - e^{-7.585}} = \frac{1}{1 - 0.000508}$$

$$= \frac{1}{0.99492} = 1.0005$$

Example 2. For CO_2 the characteristic vibration temperature θ_v is 3084 K at 300 K. Calculate θ_v for CO_2.

Solution. Q_v in terms of θ_v is given by

$$Q_v = \frac{1}{1 - e^{-\theta_v/T}}$$

$$\theta_v = 3084\ k\ \&\ T = 300\ K$$

$$\therefore \quad Q_v = \frac{1}{1 - e^{-3084/300}} = \frac{1}{1 - 0.0000343} = \frac{1}{0.9999657}$$

$$Q_v = \mathbf{1.0000343}$$

Example 3. Calculate the vibrational contribution to *H*°, *S*° and *G*° for oxygen gas at 298 K. The vibrational frequency is 1580.25 cm^{-1}.

Solution. (*i*) $$H_v^0 = RT \frac{\beta/T}{e^{\beta/T} - 1}$$

$$\beta = \frac{hc\bar{\nu}}{k} = \frac{(6.626 ´ 10^{-34}\ Js)\ (3.0 ´ 10^8\ ms^{-1})\ (1.5802 ´ 10^5\ m^{-1})}{1.381 ´ 10^{-23}\ JK^{-1}}$$

$= 2274.5 \text{ K}$

$$\therefore \quad H_v^0 = (8.314 \text{JK}^{-1}\text{mol}^{-1})(298\text{K})\left(-\frac{2274.5/298}{e^{2274.5/298}-1}\right)$$

$$= 8.314 \times 298 \times \frac{7.63}{e^{7.63}-1}$$

$$= 8.314 \times 298 \times \frac{7.63}{2041}$$

[To solve $e^{7.63}$, Put $x = e^{7.6}$, $\ln x = 7.63$]

or $2.303 \log x = 7.63$ or $\log x = \frac{7.63}{2.303} = 3.31$ $\therefore x = 2.042 \times 10^3 = 2042$]

$= \mathbf{9.2 J\ mol^{-1}}$

(ii) $$S_v^0 = R\left[\frac{\beta/T}{(e^{\beta/T}-1)} - \ln(1-e^{-\beta/T})\right] = 8.314\left[\left(\frac{7.63}{e^{7.63}-1}\right) - 2.303\log(1-e^{-7.63})\right]$$

$$= 8.314\left[\frac{7.63}{2041} - 2.303\log(1)\right] = \mathbf{0.031 JK^{-1}mol^{-1}}$$

(iii) $G_v^0 = H_v^0 - TS_v^0 = 9.2 - 298 \times (0.031)$

$= \mathbf{-0.038\ Jmol^{-1}}$

Example 4. Calculate standard vibrational entropy of a mole of HCl gas, given $x = 4.1$ cm K.

Solution. We know that

$$S_v = \frac{Rx}{e^x - 1} - R\ln(1-e^{-x})$$

$$x = 4.1$$

$$e^x = e^{4.1} = 60.34$$

$$e^{-x} = e^{-4.1} = 0.01657, \text{ R} = 1.987 \text{ cal deg}^{-1} \text{ mole}^{-1}$$

$$S_v = \frac{1.987 \times 4.1}{60.34 - 1} - 1.987 \ln(1 - 0.01657)$$

$$= \frac{8.1465}{59.34} - 1.987 \ln 0.9834$$

$$= 0.1374 - 1.987(-0.0167)$$

$$= 0.1373 + 0.03326$$

$$\therefore \quad S_v = \mathbf{0.17055\ eu\ mol^{-1}}$$

6.22. ELECTRONIC CONTRIBUTION TO ENTROPY

The electronic partition function is given by the equation

$$Q_e = g_0 + g_1 e^{-\varepsilon_1/kT} + ... \quad ...(i)$$

The electronic contribution to entropy is then given by

$$S_e^0 = R \ln Q_e$$

$$= R \ln(g_0 + g_1 e^{-\varepsilon_1/kT} + ...)$$

If the energies of the excited states are all large with respect to kT, only the degeneracy g_0 of the ground state has to be taken into account so that we have

$$\boxed{S_e^0 = R \ln g_0}$$

Example 1. Calculate the electronic contribution to the entropy $S°$ for O_2 (g) at 298 K. Given that the electronic ground state is triplet and the energy of the first excited electronic state is so high that it has not to be considered.

Solution. Here $g_0 = 3$

$$\therefore \quad S_e^0 = 8.314 \ln 3$$
$$= 8.314 \times 2.303 \log 3$$
$$= \mathbf{9.135\ JK^{-1}\ mol^{-1}}.$$

Example2. The energy of the first three energy levels of F-atom determined spectroscopically are:

Term Symbol	Frequency in ω (cm^{-1})
$^2P_{3/2}$	0
$^2P_{1/2}$	404.0
$^2P_{5/2}$	102406.5

Calculate the electronic partition function of F-atom at 1500 K.

Solution. The degeneracies of the above electronic levels are calculated as follows :

For $\quad {}^2P_{3/2}.J = \frac{3}{2}$

$$= (2J+1) = \left(2 \times \frac{3}{2} + 1\right) = 4$$

For $\quad {}^2P_{1/2}.J = \frac{1}{2}$

$$g_e(1) = \left(2 \times \frac{5}{2} + 1\right) = 2$$

For $\quad {}^2P_{5/2}.J = \frac{5}{2}$

$$g_e(2) = \left(2 \times \frac{5}{2} + 1\right) = 6$$

The electronic partition function of F-atom is given by

$$Q_e = g_e(0) + g_e(1)e^{-hc\omega_1/kT} + g_e(2)e^{-hc\omega_2/kT} + \ldots\ldots\ldots$$

$$\omega_1 = 404 \text{ cm}^{-1}$$

$$\omega_2 = 102,406.5 \text{ cm}^{-1}$$

$$\frac{hc\omega_1}{kT} = \frac{1.439 \times 404}{1500} = 0.3876$$

$$\frac{hc\omega_2}{kT} = \frac{1.439 \times 102,406.5}{1500} = 98.2419$$

Now $\quad Q_e = g_e(0) + g_e(1)e^{-0.386} + g_e(2)\, e^{-98.2419}$

$$= 4 + 2 \times 0.67868 + 6 \times 2.15816 \times 10^{-43}$$
$$= 4 + 1.35736 + 0$$
$$Q_e = 5.35736$$

Example 3. For Tellurium vapours, there is low lying electronic state 0.94 eV above the ground state. Calculate Q_e at 10,000 K. If degeneracies of the ground is 2 and of the first excited state is 4.

Solution. $\varepsilon_e(0) = 0 ; \varepsilon_e(1) = 0.94\ eV$

As $1\ eV = 1.602 \times 10^{-19}$ J

$\therefore$ $0.94\ eV = 1.602 \times 10^{-19} \times 0.94 = 1.50588 \times 10^{-19}$ J

$$\varepsilon_e(1) = 1.50588 \times 10^{-10} \text{ J}$$
$$g_e(0) = 2 ; g_e(1) = 4$$
$$\frac{\varepsilon_e(1)}{kT} = \frac{1.50588 \times 10^{-19} \text{ J}}{1.38 \times 10^{-23} \text{ JK}^{-1} \times 10{,}000 \text{ K}} = 1.0912$$

The electronic partition function Q_e is given by

$$Q_e = g_{e(0)} + g_{e(1)} e^{-\varepsilon_e(1)/kT}$$
$$= 2 + 4 \times e^{-1.0912}$$
$$= 2 + 4\,(0.3858)$$
$$Q_e = 2 + 1.5432 = 3.5432$$

$\therefore$ $Q_e = 3.5432$

Example 4. In Si-atom at 25°C, the ground and two excited states have frequencies (ω) in terms cm^{-1} are respectively 10, 77.15 and 223.31 cm^{-1}. The corresponding degeneracies are 1,3 and 5. Determine Q_e for Si-atom.

Solution. $g_e(0) = 1, g_e(1) = 3 ; g_e(2) = 5$

$$\varepsilon_e(0) = 10 ; \omega_1 = 77.15 ; \omega = 223.31$$
$$\frac{\varepsilon_e(1)}{kT} = \frac{hc\omega_1}{kT} = \frac{1.439 \times 77.15}{298} = 0.36725$$
$$\frac{\varepsilon_e(2)}{kT} = \frac{hc\omega_2}{kT} = \frac{1.439 \times 223.31}{298} = 1.07883$$

The electronic partition function Q_e is given by

$$Q_e = g_e(0) + g_e(1)\, e^{-0.3725} + g_e(2)^{-1.07883}$$
$$= 1 + 3\, e^{-0.3725} + 5e^{-1.07883}$$
$$Q_e = 1 + 2.067 + 1.69995$$

$\therefore$ $Q_e = 4.76695$

SOLVED CONCEPTUAL PROBLEMS

Problem 1. Write expressions for Bose-Einstein and Fermi-Dirac distribution laws. How does Maxwell-Boltzmann's law follow from these?

Solution. According to Bose-Einstein distribution law, the most probable distribution of N particles among the various levels is given by

$$n_i = \frac{g_i}{e^{\alpha+\beta\varepsilon_i} - 1} \qquad ...(i)$$

Here n_i is the number of molecules in the ith level with energy ε_i, g_i represents the number of eigen functions corresponding to the energy ε_i and is called the degeneracy or statistical weight. α and β are undetermined constants.

According to **Fermi-Dirac** distribution law, the most probable distribution is given by

$$n_i = \frac{g_i}{e^{\alpha+\beta\varepsilon_i} + 1} \qquad ...(ii)$$

It may be noted that in the eqns. (*i*) and (*ii*) if n_i's are small which can be so when the denominator is large *i.e.*, $e^{\alpha+\beta\varepsilon_i}$ is large as compared to 1, then 1 can be neglected in comparison to the exponential term, then these equations reduce to

$$n_i = \frac{g_i}{e^{\alpha + \beta\varepsilon_i}}$$

which is the same as Maxwell-Boltzmann's law.

Example 2. Prove that the constant k_1 in the expression $S = k_1 \ln W$ is Boltzmann's constant, K.

Solution. Thermodynamic probability, W is given by the relation

$$W = \frac{N!}{n_1!n_2!....n_n!}$$

This gives the total number of arrangements (distributions) possible for N particles among n energy levels (taking the particles to the distinguishable).

However, as each energy level may correspond to a number of eigen states, it is essential to introduce the degeneracy or statistical weight corresponding to each energy level. Also as identical particles are indistinguishable, this factor should also be taken into consideration. With these two modifications, we have

$$W = \frac{g_0^{n_0}\, g_1^{n_1}\, g_2^{n_2} ....g_n^{n_n}}{n_0!n_1!n_2!n_n!} = \Pi\, \frac{g_i^{n_i}}{n_i!}$$

where n_0 is the number of particles in ground state energy level ε_0 with degeneracy g_0.

n_1 is the number of particles in the energy level ε_1 with degeneracy g_1 and so on.

$$\ln W = (n_0 \ln g_0 + n_1 \ln g_1 + ... n_n \ln g_0) - (\ln n_0! + \ln n_1! + ... \ln n_n!)$$
$$= \Sigma\, n_i \ln g_i - [(n_0 \ln n_0 - n_0) + (n_1 \ln n_1 - n_1) + ... (n_n \ln n_n - n_n)]$$

(by Stirling formula *viz.*, $\ln n_i = n_i \ln n_i - n_i$)

or
$$\ln W = \Sigma n_i \ln g_i - \Sigma\, n_i \ln n_i + (n_0 + n_1 + \quad ..... n_n)$$
$$= \Sigma n_i \ln g_i - \Sigma\, n_i \ln n_i + N \qquad ...(i)$$

But according to Boltzmann's distribution law

$$\frac{n_i}{N} = \frac{g_i e^{-\varepsilon_i / kT}}{\Sigma g_i e^{-\varepsilon_i / kT}}$$

Applying to it translational energy, we can write

$$\frac{n_i}{N} = \frac{g_i e^{-\varepsilon_i / kT}}{Q_t} \quad \text{(where } Q_t \text{= translational partition function)}$$

or

$$n_i = \frac{N g_i e^{-\varepsilon_i / kT}}{Q_t}$$

Substituting this value in the second term of eqn. (*i*), we get

$$\ln W = \Sigma n_i \ln g_i - \Sigma n_i \ln \left[\left(\frac{N}{Q_t}\right) g_i e^{-\varepsilon_i / kT}\right] + N$$

$$= \Sigma n_i \ln g_i - \Sigma n_i \ln \left(\frac{N}{Q_t}\right) - \Sigma n_i \ln g_i + \Sigma \frac{n_i \varepsilon_i}{kT} + N$$

$$= -N \ln \left(\frac{N}{Q_t}\right) + \frac{1}{kT} \Sigma n_i \varepsilon_i + N$$

$$= N \ln \left(\frac{Q_t}{N}\right) + \frac{1}{kT} \Sigma n_i \varepsilon_i + N$$

putting $\Sigma n_i \varepsilon_i = E_t = RT^2 \left(\frac{\partial \ln Q}{\partial T}\right)_V$, we get

$$\ln W = N \ln \left(\frac{Q_t}{N}\right) + \frac{RT}{k} \left(\frac{\partial \ln Q}{\partial T}\right)_V + N$$

Multiplying both sides by Boltzmann's constant, *k*, we get

$$k \ln W = Nk \ln \left(\frac{Q_t}{N}\right) + RT \left(\frac{\partial \ln Q}{\partial T}\right)_V + Nk$$

$$= R \ln \left(\frac{Q_t}{N}\right) + RT \left(\frac{\partial \ln Q}{\partial T}\right)_V + R \qquad ...(ii)$$

But for the translation entropy,

$$S_t = R \ln \left(\frac{Q_t}{N}\right) + RT \left(\frac{\partial \ln Q}{\partial T}\right)_V + R \qquad ...(iii)$$

Comparing equations (*ii*) and (*iii*), we have

$S_t = k \ln W$ (*i.e.*, $k_1 = k$)

or in general $S = k \ln W$

Example 3. Derive the value of β used in statistical mechanics.

Solution. From Maxwell-Boltzmann distribution law,

$$n_i = \frac{g_i}{e^{\alpha + \beta \varepsilon_i}} = g_i e^{-\alpha - \beta \varepsilon_i} = g_i e^{-\alpha} . e^{-\beta \varepsilon_i} \qquad ...(i)$$

$$\therefore \qquad \Sigma n_i = \Sigma g_i e^{-\alpha} e^{-\beta\varepsilon_i}$$

$$\text{or} \qquad N = e^{-\alpha} \Sigma g_i e^{-\beta\varepsilon_i} \qquad (\because \Sigma n_i = N) \qquad ...(ii)$$

$$\therefore \qquad e^{-\alpha} = \frac{N}{\Sigma g_i e^{-\beta\varepsilon_i}} = \frac{N}{Q} \qquad ...(iii)$$

where $Q = \Sigma g_i e^{-\beta\varepsilon_i}$ is the molecular partition function.

Substituting the value of $e^{-\alpha}$ from eqn. (*iii*) into eqn. (*i*), we get

$$n_i = \frac{N g_i e^{-\beta\varepsilon_i}}{Q} \qquad ...(iv)$$

We know that the number of possible distributions (thermodynamic probability, W) of N distinguishable particles is given by

$$W = N! \, \Pi \frac{g_i^{n_i}}{n_i!} \qquad ...(v)$$

Taking logarithm of both the sides, we get

$$\begin{aligned} \ln W &= \ln N! + \Sigma (n_i \ln g_i - \ln n_i!) \\ &= N \ln N - N + \Sigma (n_i \ln g_i - n_i \ln n_i + n_i) \text{ (by applying Stirling's formula)} \\ &= N \ln N + \Sigma n_i \ln g_i - \Sigma n_i \ln n_i \qquad ...(vi) \end{aligned}$$

Taking logarithm of both sides of eqn. (*iv*), we get

$$\ln n_i = \ln N + \ln g_i - \ln Q - \beta\varepsilon_i \qquad ...(vii)$$

Substituting this value in eqn. (*vi*), we get

$$\begin{aligned} \ln W &= N \ln N + \Sigma n_i \ln g_i - \Sigma n_i (\ln N + \ln g_i - \ln Q - \beta\varepsilon_i) \\ &= N \ln N + \Sigma n_i \ln g_i - \Sigma n_i \ln N - \Sigma n_i \ln g_i + \Sigma n_i \ln Q + \beta \Sigma n_i \varepsilon_i \\ &= N \ln N + \Sigma n_i \ln g_i - N \ln N - \Sigma n_i \ln g_i + N \ln Q + \beta E \quad (\text{where } E = \Sigma n_i \varepsilon_i) \\ &= N \ln Q + \beta E \qquad ...(viii) \end{aligned}$$

Substituting this value in the Boltzmann's equation viz.,

$S = k \ln W$, we get

$$S = Nk \ln Q + k\beta E \qquad ...(ix)$$

From first law of thermodynamics

$$q = dE + PdV \qquad ...(x)$$

From second law of thermodynamics

$$dS = \frac{q}{T} \text{ i.e., } q = TdS \qquad ...(xi)$$

Combining equations (*x*) and (*xi*), we get

$$dE = TdS - PdV \qquad ...(xii)$$

At constant volume, $dV = 0$ so that

$$(dE)_v = (TdS)_V$$

$$\text{or} \qquad \left(\frac{\partial S}{\partial E}\right)_V = \frac{1}{T} \qquad ...(xiii)$$

Differentiating (*ix*) with respect to E at constant volume, we get

$$\left(\frac{\partial S}{\partial E}\right)_V = \frac{Nk}{Q}\left(\frac{\partial Q}{\partial E}\right)_V + k\beta + kE\left(\frac{\partial \beta}{\partial E}\right)_V$$

$$= \frac{Nk}{Q}\left(\frac{\partial Q}{\partial \beta}\right)\left(\frac{\partial \beta}{\partial E}\right)_V + k\beta + kE\left(\frac{\partial \beta}{\partial E}\right)_V \quad ...(xiv)$$

Now, as already mentioned, the molecular partition function is

$$Q = \Sigma g_i e^{-\beta\varepsilon_i}$$

$$\therefore \quad = g_1 e^{-\beta\varepsilon_1} + g_2 e^{-\beta\varepsilon_2} + ...$$

Differentiating w.r.t. β, we get

$$\frac{dQ}{d\beta} = g_1 e^{-\beta\varepsilon_1}(-\varepsilon_1) + g_2 e^{-\beta\varepsilon_2}(-\varepsilon_2) + ...$$

Taking $\varepsilon_1 = \varepsilon_2 .... = \bar{\varepsilon} = E/N$, we get

$$\frac{dQ}{d\beta} = -\frac{E}{N}(g_1 e^{-\beta\varepsilon_1} + g_2 e^{-\beta\varepsilon_2} + ...)$$

$$= -\frac{E}{N}\sum g_i e^{-\beta\varepsilon_i}$$

$$= -\frac{E}{N}Q \quad ...(xv)$$

Substituting this value in (*xiv*), the first and the last term cancel out. Hence we get

$$\left(\frac{\partial S}{\partial E}\right)_V = k\beta \quad ...(xvi)$$

Comparing with equation (*xiii*), we get

$$k\beta = \frac{1}{T}$$

or

$$\beta = \frac{1}{kT} \quad ...(xvii)$$

Example 4. Why standard third law entropies are less than statistical entropies for CO and N_2O, whereas it is reverse for $(CH_3)_2$ Cd and CH_3 CCl_3 ?

Solution. Statistical and third laws standard entropies of the four substances are given as under in Table 6.2

Table 6.2. Comparison of standard entropies calculated from statistical mechanics and third law

Substance	$S°(JK^{-1}\ mol^{-1})$	
	Statistical	*Third law*
CO	197.9	193.4
N_2O	219.9	215.2
$CH_2 - Cd - CH_3$	290.5	302.9
$CH_3 - CCl_3$	290.5	302.9

CO and N_2O have third law entropies lower than the statistical values.

The explanation for this is that at $T = 0$, the crystals of these substances are not perfect. Hence the assumption that $S^0 = 0$ at 0 K is not correct. The explanation for higher third law entropies in case of $(CH_3)_2Cd$ and CH_3CCl_3 is that the rotational contribution towards entropy has been taken for the molecule as a whole whereas actually in such molecules, groups of atoms can rotate as a unit about other group to which they are joined by single bonds. *e.g.*, CH_3 around the bond linking it to Cd or CCl_3 around the bond linking it to CH_3.

Example 5. What is residual entropy? How is it calculated?

Solution. *The entropy which the crystal of a substance has at T = 0, is called the residual entropy.* It can be calculated by applying Boltzmann's formula *viz.*, $S = k \ln W$ as follows :

Suppose a sample consists of N molecules and each molecule can have two orientations that are equally probable. Then the same energy can be achieved in 2^N different ways *i.e.*, $W = 2^N$. Hence

$$S = k \ln 2^N = kN \ln 2$$

If 1 mol of the sample is taken, $kN = R$. Then

$$S = R \ln 2 = 2.303\, R \log 2 = 5.85 \text{ JK}^{-1} \text{ mol}^{-1}$$

Thus residual molar entropy = 5.85 $\text{JK}^{-1} \text{ mol}^{-1}$

In general if a molecule can have p possible orientations with about the same energy then residual molar entropy will be

$$S_R = R \ln p$$

For example, $FClO_3$ molecule can have four possible orientations with approximately same energy, hence its residual entropy will be $= R \ln 4 = 11.5 \text{ JK}^{-1} \text{ mol}^{-1}$.

Example 6. In terms of wave function, to which type is B–E statistics applicable and to which type F–D statistics is applicable?

Solution. *B–E* statistics is applicable to those systems which have symmetric wave functions while *F – D* statistics is applicable to those systems which have asymmetric wave functions.

Example 7. Write expression for Boltzmann distribution law taking degeneracy of states into consideration. What do different symbols signify?

Solution. Boltzmann distribution law is

$$\frac{n_i}{N} = \frac{g_i e^{-\varepsilon_i / kT}}{\sum g_i e^{-\varepsilon_i / kT}}$$

where n_i = no. of systems which in the most probable state have energy ε_i

N = total no. of molecules

g_i = statistical weight, *i.e.*, degeneracy of the energy level ε_i

k = Boltzmann constant

T = temperature in kelvin

Example 8. What is the main difference between microcanonical and canonical ensemble in statistical mechanics?

Solution. In a microcanonical ensemble, each member of the ensemble has the same value of *N*, *V* and *E* and each system is like an isolated system in the thermodynamic sense. In a canonical ensemble, all the members have the same value of *N*, *V* and *T* and each system behaves like a closed system.

Example 9. Write expression for Maxwell-Boltzmann distribution law taking degeneracy of states into consideration. What do different symbols signify?

Solution. Maxwell-Boltzmann distribution law is

$$n_i = \frac{g_i}{e^{\alpha+\beta\varepsilon_i}}$$

where g_i = degeneracy of the energy level ε_i

n_i = no. of molecules having energy ε_i

α and β = undetermined constants

Example 10. Definite postulate of equal a priori probability for systems in equilibrium.

Solution. It states as follows:

If an isolated ensemble contains a large number of isolated systems in equilibrium, then each system has an equal probability to exist in any one of the possible quantum states.

Example 11. Name the formula used to write approximate value of ln N! when N is large number. Write formula.

Solution. Stirling's formula is used. The formula is

$$\ln N! = N \ln N - N$$

Example 12. What is 'partition function' in statistical mechanics? Why is it so called?

Solution. In the expression for Boltzmann distribution law, *viz*

$$\frac{n_i}{N} = \frac{g_i e^{-\varepsilon_i/kT}}{\sum g_i e^{-\varepsilon_i/kT}}$$

the summation term of the denominator is called partition function (represented by Q or Z). It is so called because it indicates how the particles are distributed among the various energy states.

Example 13. Write expressions for Bose-Einstein and Fermi-Dirac statistics. Under what condition these expressions reduce to Maxwall-Boltzmann law?

Solution. $B - E$ distribution law is

$$n_i = \frac{g_i}{e^{\alpha+\beta\varepsilon_i} - 1}$$

$F - D$ distribution law is

$$n_i = \frac{g_i}{e^{\alpha+\beta\varepsilon_i} + 1}$$

If n_i's are small which is so when denominator is large, *i.e.*, $e^{\alpha + \beta\varepsilon i}$ is large as compared to 1, then 1 can be neglected compared to the exponential term. The above equations then reduce to the form

$$n_i = \frac{g_i}{e^{\alpha+\beta\varepsilon_i}}$$

This expression is a same as Maxwell-Boltzmann law.

Example 14. What is the expression for translational partition function? What do different symbols represent?

Solution.
$$Q_t = \frac{(2\pi mkT)^{3/2}}{h^3} V$$

m = mass of the molecule

k = Boltzmann constant

h = Planck's constant

T = temperature in K

V = volume of the container occupied by the particle.

Example 15. Write Sackur-Tetrode equation for the translational entropy.

Solution. $$S_t = R\left[\frac{5}{2}+\ln\left(\frac{2\pi mkT}{h^2}\right)^{3/2}\frac{V}{N}\right]$$

or $$S_t = R\left[\frac{5}{2}+\ln\left(\frac{2\pi mk}{h^2}\right)^{3/2}\frac{kT}{P}\right]$$

Example 16. What is Residual entropy? How is it related to the number of possible orientations of the molecule with about the same energy.

Solution. The entropy which the crystal of a substance possesses at $T = 0$ is called the residual entropy. It is related to the number of possible orientations (p) of the molecule as

$$S_R = R \ln p.$$

Example 17. Why are third law entropies less than the statistical entropies of CO and N_2O whereas it is reverse for $(CH_3)_2$ Cd and CH_3 CCl_3?

Solution. For CO and N_2O, their crystals are not perfect even at $T = 0$. They have some entropy even at $T = 0$. $(CH_3)_2$ Cd and CH_3 CCl_3 molecules not only rotate as a whole but groups of atoms can rotate as a unit about other group to which they are joined by single bonds. Thus, some additional rotational entropy is present in such molecules.

EXERCISES
(Based on Question Papers of Different Universities)

Multiple Choice Questions (Choose the correct option)

1. The stirling formula used to write the approximate value of ln N! is

(*a*) ln N! = N – N ln N (*b*) ln N ! = N ln N – N

(*c*) ln N! = N – ln N (*d*) ln N ! = N ln N

2. Statistics which is applicable to those systems which have asymmetric wave functions is called

(*a*) F – D statistics (*b*) B–E statistics

(*c*) M – B statistics (*d*) None

3. In a microcanonical ensemble, each member has the same value of

(*a*) N, V, T (*b*) N, V, P

(*c*) N, V, E (*d*) P, V, E

4. Residual entropy is the entropy which the crystal of a substance possesses at

(*a*) T = 100 (*b*) t = 0

(*c*) t = 100 (*d*) T = 0

5. Third law entropy is more than the statistical entropy in the case of

(*a*) $(CH_3)_2$ Cd (*b*) CO

(*c*) NO (*d*) N_2O

6. If g_o is the degeneracy of the ground state and the energies of the excited states are very high, what is the expression for electronic contribution to the entropy?

(a) $S_e^o = \frac{1}{R} \ln g_o$ (b) $S_e^o = R \ln g_o$

(c) $S_e^o = R - R \ln g_o$ (d) $S_e^o = \frac{R}{\ln g_o}$

7. Symmetry number for unsymmetrical molecules like HC is

(a) 4 (b) 3

(c) 2 (d) 1

ANSWERS

1. (b) **2.** (a) **3.** (c) **4.** (d) **5.** (a) **6.** (b)
7. (d)

Short Answer Questions

1. What do you understand by 'statistical method'? Briefly explain with suitable examples.
2. Define the postulate of equal a priori probabilities
   (i) for a system in equilibrium, and (ii) for a system not in equilibrium
3. Derive the following expression for the heat capacity (C_υ) in terms of partition function, Q

$$C_v = \frac{R}{T^2}\left\{\frac{\partial \ln Q}{\partial(1-T)^2}\right\}_v$$

4. Using statistical mechanics, prove that the internal energy of an ideal monatomic gas is $\frac{3}{2}RT$.
5. Write the expressions for translational, rotational and vibrational partition functions. What do the different symbols signify?
6. What is Residual entropy? Starting from the relation between entropy and probability, derive expression for residual entropy.
7. What is 'Statistical mechanics'? What are the two main points of difference between classical statistical mechanics and 'quantum statistical mechanics'?
8. How is the state of a system described in (i) thermodynamics (ii) statistical mechanics?
9. What is 'Statistical thermodynamics'? Give the basic relation which links thermodynamics with statistical mechanics.
10. Applying statistical mechanics, derive expression for contribution of entropy for ideal monatomic gases (Sackur-Tetrode equation).
11. Explain the difference between microcanonical ensemble, canonical ensemble and grand canonical ensemble.
12. Derive the following expression for the energy of the system in terms of partition function, Q

$$E = RT^2\left(\frac{\partial \ln Q}{\partial T}\right)_\nu$$

General Questions

1. What is 'ensemble' as used in statistical mechanics? Briefly explain the following:
(*i*) Microcanonical ensemble (*ii*) Canonical ensemble (*iii*) Grand canonical ensemble
2. Derive Boltzmann distribution law in the form $\frac{n_i}{N} = \frac{g_i e^{-\varepsilon_i/kT}}{\Sigma g_i e^{-\varepsilon_i/kT}}$
3. What is partition function? Derive expressions for the following in terms of partition function:
(*i*) Energy (*ii*) Enthalpy (*iii*) Heat capacity C_υ and C_P
(*iv*) Entropy (*v*) Helmholtz free energy (*vi*) Gibbs free energy
4. Derive the expression for the rotational partition function *viz*, $Q_r = \frac{8\pi^2 IkT}{\sigma h^2}$
5. Derive expression for the rotational contributions to the following thermodynamic functions:
(*i*) Energy (*ii*) Enthalpy (*iii*) Heat capacity C_υ or C_P (*iv*) Entropy
6. Write the general expressions for:
(*i*) Vibrational entropy in terms of partition function.
(*ii*) Partition function Q_υ
Hence derive the expression for S_υ not involving Q_υ.
7. Compare the important features of Maxwell-Boltzmann, Bose-Einstein in and Fermi-Dirac statistics.
8. Derive Maxwell-Boltzmann distribution law in the form $n_i = g_i \exp(-\alpha - \beta\,\varepsilon_i)$
9. Derive expression for vibrational contribution to the energy of a system.
10. Derive the relationship between entropy and probability. How can you justify that probability is a measure of disorder of the system?
11. Derive expression for the translational contribution to the following thermodynamic properties.
(*i*) Energy (*ii*) Enthalpy (*iii*) Entropy
12. Derive the expression for translational partition function *viz*., $Q_t = \frac{(2\pi mkT)^{3/2}}{h^3}$
13. Prove that the complete partition function for a system is the product of translational, rotational, vibrational and electronic partition functions.
14. Derive the expression for the vibrational partition function *viz*., $Q_v = \frac{1}{1-e^{-h\nu_0/kT}}$
15. Derive Maxwell-Boltzmann distribution law.
16. Define 'postulate of equal a priori probabilities' for system in equilibrium. Explain how has this postulate been arrived at?
17. Derive Stirling's formula for a large number (*N*) of particles.
18. Derive the value of β used in statistical mechanics.
19. Prove that the rate constant k_1 in the expension $S = k_1$ law is boltzmann's constant K.

APPENDIX 1

GREEK APLPHABET TABLE

Capital	*Small*	*Pronunication*	*English equivalent*
Α	α	alpha	a
Β	β	beta	b
G	γ	gamma	g, n
Δ	δ	delta	d
Ε	ε	eosilon	e
Ζ	ζ	zeta	z
Η	η	eta	$\bar{e}$
Θ	θ	theta	th
Ι	ι	iota	i
Κ	κ	kappa	k
Λ	λ	lamba	l
Μ	μ	mu	m
Ν	ν	nu	n
Ξ	ξ	xi	x
Ο	ο	omricon	o
Π	π	pi	p
Ρ	ρ	rho	r, rh
Σ	σ	sigma	s
Τ	τ	tau	t
Υ	υ	upsilon	y, u
Φ	φ	phi	ph
Χ	χ	chi	ch
Ψ	ψ	psi	ps
Ω	ω	omega	$\bar{o}$

APPENDIX 2

Values of some common physical and chemical constants

Quantity	*Symbol*	*Values of CGS units*	*Values in SI units*
Velocity of ligth	C	2.997925×10^{10} cm / sec	2.997925×10^{8} m sec^{-1}
(in vaccum)		$\simeq 3.0 \times 10^{10}$ cm / sec	$\simeq 3.0 \times 10^{8}$ m sec
Planck's constant	h	6.6262×10^{-27} erg sec	6.6262×10^{-34} J sec
		$\simeq 6.62 \times 10^{-27}$ erg sec	$\simeq 6.62 \times 10^{-34}$ J sec
Avogadro number	N_0 or N	6.022169×10^{23} mole^{-1}	6.022169×10^{23} mol^{-1}
		$\simeq 6.022 \times 10^{23}$ mole^{-1}	$\simeq 6.022 \times 10^{23}$ mol^{-1}
Gas constant	R	0.082053 lit atm K^{-1} mole^{-1}	8.3144 JK^{-1} mol^{-1}
		$\simeq$ 0.0821 lit atm K^{-1} mole^{-1}	$\simeq$ 8.314 JK^{-1} mol^{-1}
		or	
		8.3144×10^{7} ergs K^{-1} mole^{-1}	
		$\simeq$ 0.0821 lit atm K^{-} mole^{-1}	
Molar volume at	V	22.414 litres	0.022414 m^{3}
NTP		$\simeq$ 22.4 litres	$\simeq$ 0.0224 m^{3}
Faraday's constant	F	96487 coulombs / equivalent	96487 C mol^{-1}
		$\simeq$ 96500 coulombs / equivalent	$\simeq$ 96500 C mol^{-1}
Electronic charge	e	1.6022×10^{-19} coulombs	1.6022×10^{-19} C
Electron rest mass	m_e	9.109558×10^{-28} g	9.109558×10^{-31} kg
		$\simeq 9.11 \times 10^{-28}$ g	$\simeq 9.11 \times 10^{-31}$ kg
Proton rest mass	m_p	1.672614×10^{-24} g	1.672614×10^{-27} kg
		$\simeq 1.67 \times 10^{-24}$ g	$\simeq 1.67 \times 10^{-27}$ kg
Neutron rest mass	m_n	1.67492×10^{-24} g	1.67492×10^{-27} kg
		$\simeq 1.67 \times 10^{-24}$ g	$\simeq 1.67 \times 10^{-27}$ kg